T0210735

Advances in Global Change Research

Volume 74

This book series has been accepted for inclusion in SCOPUS.

Advances in Global Change Research

Aims and Scopes

This book series is aimed at addressing a range of environmental issues through state-of-the-art and/or interdisciplinary approaches. The books in the series can either be monographs or edited volumes based, for example, on the outcomes of conferences and workshops, or by invitation of experts. The topics that the series can consider publishing include, but are not limited to:

Physical and biological elements of earth system science, in particular

- Climate change
- Biodiversity
- Sea-level rise
- Paleo-climates and paleo-environments

Social aspects of global change

- Environmentally-triggered migrations
- Environmental change and health
- Food security and water availability
- Access to essential resources in a changing world

Economic and policy aspects of global change

- Economic impacts
- Cost-benefit analyses
- Environmental governance
- Energy transition

Methodologies for addressing environmental issues

- Planetary data analysis
- Earth observations from space
- Proxy data analyses
- Numerical modeling
- Statistical analyses

Solutions to global environmental problems

- Sustainability
- Ecosystem services
- Climate services
- Technological/engineering solutions

Books in the series should be at least 200 pages in length, and include a table of contents.

Images supplied in colour will be reproduced in both the print and electronic versions of the book at no cost to the author/editor.

Manuscripts should be provided as a Word document and will be professionally typeset at no cost to the authors/editors.

All contributions to an edited volume should undergo standard peer review to ensure high scientific quality, while monographs should also be reviewed by at least two experts in the field.

Manuscripts that have undergone successful review should then be prepared according to the Publisher's guidelines manuscripts: https://www.springer.com/gp/authors-editors/book-authors-editors/book-manuscript-guidelines

Miguel Montoro Girona · Hubert Morin ·
Sylvie Gauthier · Yves Bergeron
Editors

Boreal Forests in the Face of Climate Change

Sustainable Management

Editors
Miguel Montoro Girona
Université du Québec en
Abitibi-Témiscamingue
Amos, QC, Canada

Hubert Morin
Département des Sciences fondamentales
Université du Québec à Chicoutimi
Chicoutimi, QC, Canada

Sylvie Gauthier
Laurentian Forestry Centre
Natural Resources Canada
Québec, QC, Canada

Yves Bergeron
Université du Québec à Montréal
Montréal, QC, Canada

ISSN 1574-0919   ISSN 2215-1621 (electronic)
Advances in Global Change Research
ISBN 978-3-031-15990-9   ISBN 978-3-031-15988-6 (eBook)
https://doi.org/10.1007/978-3-031-15988-6

This Springer imprint is published by the registered company Springer Nature Switzerland AG
The registered company address is: Gewerbestrasse 11, 6330 Cham, Switzerland

For our students and the next generation of scientists

Humans are resilient.
With science, solidarity, and creativity, we can adapt.
We can be the change!

Miguel Montoro Girona
Sylvie Gauthier
Hubert Morin
Yves Bergeron

Book Presentation

Why This Project?

We are living in a critical moment. Ecosystems are changing more quickly than anticipated, and climate change has become the greatest challenge facing humanity. Forest ecosystems provide essential resources and services for the development and subsistence of societies around the world. The boreal forest covers a worldwide belt of 14 million km^2 and represents approximately 25% of the world's forest area. Two-thirds of this surface is managed for wood production, and this biome supplies 37% of the world's timber. Boreal forests have a key role in the climate system and its modification through processes such as carbon sequestration, nutrient cycling, hydrology, and albedo changes.

Ecosystem-based management has been the main approach for attaining sustainable forest management in the boreal biome; however, these practices have yet to be fully adapted to climate change and the associated impacts. A new conceptual framework integrating climate change is therefore needed. Adaptation and mitigation strategies are critical for sustaining boreal forests under future conditions. Moreover, the results of sustainable management in the boreal forest of the last 20 years must be evaluated and, if required, alternative practices be introduced. This book presents new reflections, strategies, and recommendations for academics, students, and forest managers to guide future forest management, identify the challenges facing the second-largest terrestrial biome on Earth, and define new research avenues required to face these challenges.

Editors-in-Chief

Miguel Montoro Girona, Groupe de recherche en écologie de la MRC-Abitibi (GREMA), Forest Research Institute, Université du Québec en Abitibi-Témiscamingue; Restoration Ecology Group, Department of Wildlife, Fish and Environmental Studies, Swedish University of Agricultural Sciences (SLU), Umeå, Sweden; and the Centre for Forest Research, Université du Québec à Montréal, miguel.montoro@uqat.ca

Sylvie Gauthier, Natural Resources Canada, Canadian Forest Service, Laurentian Forestry Centre, Sylvie.gauthier@nrcan-rncan.gc.ca

Hubert Morin, Département des Sciences fondamentales, Université du Québec à Chicoutimi and the Centre for Forest Research, Université du Québec à Montréal, hubert_morin@uqac.ca

Yves Bergeron, Forest Research Institute, Université du Québec en Abitibi-Témiscamingue and Département des sciences biologiques, Université du Québec à Montréal, yves.bergeron@uqat.ca

Miguel M. Girona

Being born in the second-largest protected area in Europe within the largest forested regions in Andalucía, Spain, and growing up in direct contact with nature have conditioned my life, my personal values, my worldview, and my career. I am from the *Parque Natural de las Sierras de Cazorla, Segura y Las Villas* (Jaén, Spain). I lived in a small house in the forest with my grandparents, my brother, my dogs, and my telescope. I owe 99% of who I am, my sensitivity, my animal instinct, my madness, and an endless number of traits to those mountains that city dwellers may struggle to understand because when you are immersed in an environment such as this corner of Andalucía, your relationship with nature becomes your way of understanding the world: a life of breathing the mountain air, walking on stones, feeling the wind, bathing in rivers, lying on the grass, looking at the sky, watching the Griffon vultures… There is an invisible chain that keeps me tied to that place and those cliffs because this place is responsible for my passion for nature, my curiosity about ecological research, my sensibility to environmental problems, and my living of life as a permanent adventure where dreaming with open eyes and without fear is essential!

For this reason, when I told my family that I was going to study my bachelor's degree in environmental sciences, it was not a surprise. During those years, I explored my interest in botany and forest birds, by conducting floral inventories in Sierra Nevada National Park and undertaking migration bird surveys in Tarifa and Donana. My first research experience was in the Department of Wildlife Toxicology at the Veterinarian Faculty of the University of Murcia, where I worked four years as an intern evaluating the impact of heavy metals and organochlorines on birds of prey and Mediterranean cetaceans (Life Project EC). I became fascinated with research and wanted to earn a Ph.D. to become a researcher; however, in the summer before I finished my bachelor studies, a massive wildfire burned the forest close to my house. It was a strong inflection point in my young career, and I experienced a dilemma between choosing *intervention* or *research*. Nature is experiencing change that needs an immediate answer, one that cannot wait for tomorrow; science, however, requires time to study and understand the problems before proposing innovative solutions. I solved this personal dilemma by studying a second bachelor's degree in forest engineering at the University of Huelva and a master's degree in land-use planning and geographical information systems at the University of Sevilla, where I learned the complexity of natural disturbances and the challenges of forest management.

Before finishing my master's studies, I began working at the Government of Andalusía as a forest engineer in wildfire management and evaluated projects related to biodiversity, national parks, and forest management that sought to obtain environmental grants from the European Commission. I also guided companies in adapting to new environmental policies. However, after four years in government, I began to feel that I was not solving many problems, and my life was becoming routine where I could not express my originality. I then had another inflection point. As a kid, I had a poster of Banff National Park in my room, and I had always wished to visit and study the Canadian forest, the largest and wildest of forest ecosystems. I came across an offer from a university in Québec to work on a great Ph.D. project, and I decided to leave my country, my family, my work, and my house to start a new professional adventure in Canada and follow the dreams of my inner kid.

Many people I have met ask: How does a Spaniard end up working in Canadian boreal forest management? My answer is that Europeans have a great deal of experience in forest ecosystem degradation because of our long history of forest exploitation. We harvested the forest to build cities, cook our meals, heat our homes, and build warships, the Sevilla Cathedral, Notre Dame de Paris. Thus, we know very well the many interests to consider when balancing forest management: forest companies, tourism, biodiversity conservation, hunting, fishing, natural and anthropic disturbances, and local needs. I decided to move to Canada, as this country has an opportunity to not repeat the same mistakes we made in Europe in regard to our natural resources.

During my Ph.D., I evaluated the potential of partial cutting as a silvicultural tool for achieving sustainable management in the boreal forest. The most exciting part of research is to see your results for the first time and feel useful as someone searching for solutions to improve the world. Over the course of my doctoral studies, everything was an exciting challenge: developing my project, dealing with research dogmas, mastering new skills (dendrochronology, statistics), and publishing, nevermind integrating myself into a new country, learning French, and being 6000 km from my family... All of these factors were crucial for becoming a resilient and positive-minded researcher (and also a little bit crazy and funny, of course), able to take on numerous tasks with few resources. However, the best aspect of my Ph.D. was to discover and work in the Canadian boreal forest, even more wild and larger than I had imagined, and begin the journey of better understanding boreal forest functioning to establish sustainable management practices.

When I finished my Ph.D., many people told me that with a Ph.D. in forest ecology I would only find work in the fast-food industry! Nonetheless, I applied for a postdoctoral position with the Swedish University of Agricultural Sciences (SLU) as part of the restoration ecology group at Umeå. I was to model future management scenarios involving silvicultural management, wood production, and moose browsing. At the same time, I combined this Swedish research with a second postdoctoral position in Canada to study new paleoecological tools, apply a dendroecological approach to the reconstruction of insect outbreaks, and evaluate the impact of insect outbreaks on regeneration. Being a postdoc at SLU was an amazing research and human experience to discover European boreal forests. As an international postdoc, I matured as a scientist and had the opportunity to create my first projects, supervise my first students, and establish my first international collaborations (with colleagues from Canada, Sweden, Finland, Spain, USA, Brazil, Italy, and France). I realized then that being a researcher and professor would be my ideal profession!

My experience as a postdoc was relatively short because 16 months after my Ph.D. defense, I saw an offer for an assistant professorship in forest sciences at UQAT. This was my first application for a tenure-track position, and I could barely believe it when I received the news that I had been selected for this position! I also realized that it was the beginning of many things: securing funding, building my lab, creating my team … and six months later, COVID-19 arrived! However, if you work hard and give the best of you, sometimes magic does occur, everything becomes easier, and the impossible gets done… I got some funding, created my lab with an amazing group of passionate interns, Ph.D., and M.Sc. students, and developed new and exciting research projects. I also founded a research group in ecology (GREMA) that focuses on finding solutions to adapt regional forest management to climate change.

The ecology of natural and anthropic disturbances is my primary research subject, and these events are the major drivers controlling the structure and function of forest ecosystems. These drivers also interact. If we are aiming to adapt forest management to climate change, it is crucial to understand these disturbances and their impact on forests at multiple scales. My conception of science holds that it is a key tool to answer fundamental and applied problems. Being a researcher makes me feel useful each day because serious problems that have faced humanity have been solved through science by anonymous superheroes working in the shadows to find solutions to serious challenges, often without social or economic recognition. For this reason, I find contributing to the creation of a new generation of researchers most exciting. However, I do not wish to only be a professor or a researcher. I hope to be a reference for the new generation of international students, for LGTBQ people, and for early-career researchers to tell them that everything is possible when you are curious and have an imagination for research, passion as a fuel, and the ability to dream with eyes wide open. For me, research is cooperation, teamwork, excellence, quality, solidarity, creativity, innovation, originality, and PASSION. It is an art… The new generation must know that to face the challenges of our planet, science will need more help. We need motivated people. We need talent and innovation to change the world. BE THE CHANGE. So I hope to continue sharing this adventure with you and change the traditional model of researcher!

Sylvie Gauthier

As a child, I was very interested in the TV program *Atomes et Galaxies*, a science show that aired on Radio-Canada. In particular, the researchers who studied animal behavior fascinated me. Then in high school, I was lucky enough to have science teachers in chemistry, physics, and ecology who continued to develop my interest in the nature of things. I went to Cégep to study pure sciences and then hesitated between physics and ecology. I enrolled at UQAM in ecology, discovered the world of plants and became aware of the many emerging environmental problems. In my last year of undergraduate studies, just before starting the fall semester, I received a job offer from an ecological center in the Montérégie region, southwest of Montréal, Québec. My main task was to describe the forest vegetation of the area surrounding the center. I contacted Daniel Gagnon, a recently hired biology professor at UQAM, to ask if I could pursue an honors thesis on the topic. What a pleasure to spend all autumn walking in the forest; carrying a map, compass, and notebook; and describing the vegetation of these forests!

Having the opportunity to spend part of my time in the forest and then return to the office to analyze the collected data and write a report suited me perfectly. I was passionate enough about this that I accepted Daniel's offer to undertake a master's degree under his direction to study the forest vegetation of the Laurentian foothills.

I had the chance to be a teaching assistant for a plant ecology course at the research station. Here, I discussed my interest in evolution with Yves Bergeron, and together we designed a Ph.D. project on jack pine population genetics, which I would complete at the University of Montréal under the supervision of Jean-Pierre Simon. Yves was my co-director and ever-inspiring mentor. The work aimed to assess whether jack pine populations on islands that had experienced less severe fires than those on the mainland had adapted to these conditions. The work involved dendrochronology, lab work in genetics, and a fair amount of field and lab measurements. Three more or less long field seasons near Lake Duparquet, Québec: long days of work on the lake or along the Chemin de la Mine, followed by long evenings of stargazing with colleagues after sharing dinner in an old-fashioned, multibedroom house in which we had set up binoculars and microscopes to measure the collected tree cores. The friendships that emerged from that time are irreplaceable.

At the time of my doctorate, Yves Bergeron, Daniel Coderre, and Daniel Gagnon had created the GREF (*Groupe de recherche en écologie forestière*), which aimed to translate ecological knowledge of forests for forest management purposes. Under the impetus of this group, which would later become the *Centre d'étude de la forêt/Centre for Forest Research* (CEF/CFR)—the largest forest research group in Canada—I had the opportunity, very early in my career, to participate in this new dynamic setting where ecologists, biologists, and forest engineers began to work closer together to influence the management of this ecosystem.

During my postdoctoral studies, I participated in the first disturbance dynamics meeting, which took place in Sweden, where I spent several days with a small group of researchers, including Yves Bergeron, Hubert Morin, Réjean Gagnon, Ed Johnson, Pierre Richard, Luc Sirois, and several students and postdocs, including Ola Engelmark, Annika Hoffgaard, David Paré, Louis De Grandpré, Jacques Tardif, and Sherry Gutsell. I have since collaborated with several of them. After this postdoctorate directed by Francine Tremblay and Sylvie Laliberté, I was fortunate to be recruited by the Canadian Forest Service (CFS) within the fire ecology and behavior group. I participated in prescribed burn experiments over several years, which taught me much about fire behavior. I got to know Mike Flannigan and Mike Wotton with whom I still enjoy working.

I also continued to consolidate my collaboration with Yves Bergeron. We both participated in the emergence of the Sustainable Forest Management Network and developed links with our Fennoscandian colleagues, including Timo Kuuluvainen, whom we met in 1996 during a tour of Sweden and Finland. With Yves, we were perhaps among the last researchers to be able to describe with our many students the natural forest fire regimes in Québec and eastern Ontario while attempting to translate this information into ecosystem-based forest management strategies. Under the impetus of the network and the collaborative project that I was leading, the idea was born that we would then produce a book. This book, *Aménagement écosystémique en forêt boréale/Ecosystem management in the boreal forest*, which involved more than 60 authors of various skills, ages, and backgrounds, remains one of the accomplishments for which I am most proud.

Very often, I was one of the few women working in the team. For a long time, I was the only female researcher as part of the CFS community working on forest fires. As recently as on the Northern Limit Committee, I was the sole woman on a committee of 17. At times, I felt quite alone with few female role models to inspire me and to talk to. Fortunately, my male colleagues have always supported and encouraged me to take on exciting challenges. And as times are changing, I have more and more female colleagues having differing expertises and experiences with whom I work and share. A diversity of role models now exists to encourage everyone to be themselves.

As a theater lover, I have always liked a passage from Bertolt Brecht's *Galileo's Life* in which Galileo says that science only makes sense if it improves the condition of Man. As a civil servant, I tried to make the knowledge we gained useful in forest and fire management, among other things, by often publishing popular-science texts or presenting the results to diverse audiences. At the CFS, I slowly became the only female senior researcher through my work on ecosystem-based forest management, fire regimes and management, climate change impacts and adaptation, and the sustainability of forest practices. I have contributed to the assessment of the potential effects of climate change on the Canadian forest and forestry sector. In one of the richest periods of my career, I participated in the committee studying the northern limit of forest allocation to management and flew by helicopter over the vast expanses of water and forest between the 49th and 53rd parallels. In the lab at that time, I worked happily with many male and female professionals, colleagues, students, and post-docs from multiple backgrounds who inspired me with their passion, dedication, and dynamism on various projects related to boreal forest dynamics and management. I was invited to write about the future of the circumboreal forest in the prestigious journal *Science*, which allowed me to reflect on the fate of this biome in the face of future climate change. It became clear to me that both forest management and conservation are parts of the tools we have for maintaining healthy boreal ecosystems in the future. This is particularly true if the practices we develop are rooted in a good understanding of the ecosystem and if management is bound within the productive capacity of these ecosystems to face future disturbances.

I sincerely enjoyed my job, taking advantage of the opportunities it brought me to work often in nature, for more or less long periods. I enjoyed living with colleagues, often in fairly rugged accommodations, where we had the pleasure of sharing not only thoughts on forestry past and present, but also many meals, drinks, and songs. What I enjoyed most was participating in team projects, where together we achieved more than any of us could have done alone. And that, in the end, is what counts.

As human relationships allow us to believe that we can change things, I hope that the collaboration that led to this book will enable us to make progress in the management and conservation of the boreal forest that is so dear to me.

Hubert Morin

Short Biography of a Pseudoecologist

The first time I heard about ecology was in the early 1970s when I was 16 years old. A group of young people, who today would be called hippies, had obtained a "youth perspective" project. The government of the time proposed to have unemployed youth work during the summer on community-based projects. This particular project consisted of developing an "ecological" trail in the beautiful backcountry of the lower Laurentians. It was mainly to get young people to work on a trail in the forest. Until then, there had not been much ecology to this work. These young people were camping on land adjacent to where I lived, and they decided to educate us about ecology. Although no one knew exactly what it meant, the term was interpreted at the time as something cool that would save the planet, no less. I was about to enter college and, like most people, I had no idea what I was going to do.

Being much of a dreamer and often lost in thought, the field of ecology appealed to me, as I was obviously going to save the planet—the more it changes, the more it stays the same. So I headed to the Cégep du Vieux Montréal, participating in strikes, outdoor activities, and, oh, a bit of studying. Then, as a natural extension, I headed to UQAM's biology–ecology option, the only program in Québec to offer a complete session (five courses) in the field. No research station existed at the time, so our courses were given in a hunting and fishing outfitter's lodge. Through strikes and protests, I developed my artistic and carpentry talents, and I spent much time pursuing outdoor activities, participating in several expeditions to the Great North and northern Quebec, Baffin Island, the Torngat Mountains... I was not a model student focused on a career goal ... not at all! I was a pretty average student, and it wasn't until the last year of my bachelor's degree that I realized I might be able to get a scholarship to do a master's degree. Far be it from me to do a Ph.D. If I got a scholarship, I would do a master's degree that was very pragmatic and not very scientific...

So, there I was looking for a project. Because I didn't have a mentor who had lit a flame earlier during my studies, I had very broad interests ranging from ethology to brain development. Far be it from me to do plant ecology. My courses in that area had been very ordinary, and my botany course, riddled with strikes, was a real disaster. At the time, I would have preferred to work on large mammals...very spectacular... However, no one had a grant at the time to take on students. A friend told me about a researcher at Université Laval who was interested in northern and mountain ecology. Mountains. Now that interested me. I met him and applied for a grant to work on caribou feeding in a new park in the Gaspésie region of Québec! Alas, the project changed, but I received a new grant to work on the vegetation of Mont Jacques-Cartier.

So. Vegetation belts. Now that's very vegetal... and I was terrible in vegetation! Serge Payette, my master's supervisor, was an excellent botanist, a soil scientist by training, and an terrific geomorphologist. Because I had many other occupations and didn't attend classes very often, I was quite a bad master's student. I quickly realized that I had the choice of dropping out or getting serious about improving my knowledge of plant ecology to reach the very high standards that were expected of me. I may be a dreamer, but I am also very curious and quite hardheaded! It was only in the second year of my master's degree (Serge was very patient) that I really got involved in research. I met fantastic colleagues in the lab who were more serious than I was, and I was being supervised by Serge Payette, an incredible ecologist and very inspiring person. I learned to work better, to be innovative, to have a critical mind and imagination to develop hypotheses that may initially seem far-fetched, and to open my mind to research. It was at this point that I caught the research bug.

I finished my master's degree and began my doctorate under the direction of Serge Payette, studying questions in northern Quebec that combined dendrochronology, geomorphology, and plant dynamics. At that time (the early 1980s), his team was already interested in climate change and its impacts on vegetation in sensitive environments at treeline and the migration of tree species during the Holocene. This field, however, was a revelation to me.

The doctorate is a step that calls upon all of one's resources, and the team around you and the human side of things are a massive part of this endeavor. These years were very formative on both a scientific and human level. I hadn't finished my Ph.D. when I got a call from Réjean Gagnon in Chicoutimi, Québec. He had just got a job at the university there, and he was working on the dynamics of the boreal forest. He was alone in his field and so offered me a six-month postdoctorate position. Réjean was filled with ideas, and as I had to wait for my thesis corrections and my defense before taking advantage of a postdoctoral fellowship in Sweden that I had obtained, I moved to Chicoutimi for six months… and I never left! Of course, I did go to Sweden several times, just not to do a postdoc.

It was an exciting time. I was in a small team where your expertise was recognized and where you felt important. Everything had to be built: the lab, the projects, the research center, the funding opportunities… I owe much to this wonderful team, who, among others, included Réjean Gagnon and Daniel Lord (our trio known as the three bearded men), and over the years we established an enviable reputation in regard to understanding boreal forest dynamics, particularly for black spruce. Because Réjean was doing much work on this species, I decided to work on the dynamics of balsam fir stands and the growth of black spruce. I had no idea at the time that I was going to open a novel research axis focused on the impacts of budworm on the dynamics of boreal forest communities that would be important for forestry stakeholders and the region and have a worldwide impact. Boreal spruce and fir forests are synonymous with spruce budworm, a more important disturbance than fire in the northeastern North America. Understanding the dynamics of recurrent epidemics and their relationship with old-growth forests requires reconstructing budworm outbreak history both in the recent past, through dendrochronology, and in the more distant past over the Holocene. This historical perspective is essential for better appreciating the impact of outbreaks on the landscape, dissecting the fire–climate–outbreak relationship, and carrying out sustainable ecosystem management. It was therefore necessary to innovate approaches in both dendrochronology and insect paleoecology. Thus, we have established major innovative techniques, such as the identification of Holocene epidemic periods using macroremains, budworm feces, and microremains, such as Lepidoptera scales found in lake sediments. The development of these techniques has opened up previously inaccessible research niches, such as the relationship between fire frequency, spruce budworm outbreaks, vegetation composition, and climate change. These novel techniques have led to international collaborations, as scale-covered Lepidoptera are active insects found in all ecosystems around the world.

Our disappointment at not being able to adequately explain the relationships between boreal tree growth and environmental variables through dendrochronology alone led us to examine the fine-scale relationships between intra-annual ring growth, xylogenesis, and environmental variables. To cover the entire boreal forest, we installed a north–south transect, stretching from 48° to 54°N, of black spruce sites (22 years of data) and four other plots in balsam fir stands (25 years of data). A weather tower and electronic dendrometers collect data continuously at each plot, and tree-ring microsamples are collected weekly during the growing season. With the help of national and provincial funding agencies (CFI, NSERC-DRC, NSERC-Industrial Chair, FRQNT…), the Consortium de recherche sur la forêt boréale commerciale (now the Centre de recherche sur la boréalie), and the UQAC Foundation, we have been able to maintain these stations and build up one of the most complete growth databases in the world. In addition to the numerous theses that have contributed to and relied on this database, these data have allowed us to collaborate with researchers around the world, including from Russia, China, France, Spain, Italy, Germany, Sweden, Finland, and Norway.

These achievements could never have existed had I not been able to count on a team of students, professionals, and technicians, and, of course, my family, who support me and our particular family conditions. I currently hold an NSERC-Industrial Research Chair on the growth of black spruce and the influence of spruce budworm on the landscape. My atypical background is probably why I try to give all students a chance, whether they are super-achievers with scholarships or not.

Do I define myself as a good ecologist? Maybe. Maybe not. Perhaps more like an artist in ecology with original and often surprising ideas. I do think, however, I have contributed to the understanding of boreal forest dynamics in the context of climate change.

Yves Bergeron

I have often wondered what led me to make ecosystem-based management the focus of my scientific career. As I had lived in the city, my connection to the forest was not very strong. However, I remember that I loved nature and collected plants and butterflies behind the Boulevard shopping center in Montréal. It was much later, during my bachelor studies at the Université de Montréal, that I had the chance to have courses at the city's botanical garden and got the bug for botany and forestry. As a young student in biology, I had the great fortune to take a course entitled *Plant Ecology and Forest Management*. This course was taught by Jean-Pierre Simon, who was doing advanced research on the ecophysiology and ecogenetics of plants, and André Bouchard, then curator of the Montreal Botanical Garden, who was interested in the ecology of forest communities and, above all, in conservation and land management. Both aspects excited me: the scientific curiosity to understand plant ecology and the need to apply this knowledge to concrete situations. I had finally found my way. From the beginning, I was immersed in this antagonism between fundamental research (curiosity-driven research) and applied research, a divide that still separates our scientific community. From that moment, I knew that I would navigate between both worlds at the risk of being dreamy for some and too down-to-earth for others.

Under the direction of André Bouchard, I began advanced studies in ecological classification. At the time, I was greatly influenced by the ecological studies carried out by the *Capital Nature* team in the context of the hydroelectric development project in James Bay, northern Québec. I started my doctoral thesis, which focused on the ecological classification of forests in an area in the Abitibi region, with the aim of understanding boreal forest development to possibly guide how to manage these forests. The Abitibi region would also become my main research territory for the rest of my career. The trigger for ecosystem-based forest management was made at that time. While I was inventorying virgin forests in Abitibi, I was confronted with the first mechanized salvage cuttings that were taking place following a severe spruce budworm epidemic. I could not ignore that forestry was now the dominant disturbance; not studying its effects on the ecosystem would no longer be possible.

My thesis allowed me to make two important observations. First, I had started with a rather static idea of what factors explained the presence of a particular forest on a site, and I had to accept that I only had part of the solution. Forests are, in fact, very dynamic and react strongly to natural disturbances such as fire or insect outbreaks. Second, the forests around my thesis study sites were clear-cut. It was explained to me that these forests were overgrown and too susceptible to mortality from the budworm epidemic occurring at the time. I thus learned two critical lessons: the forests were dynamic and the action of humans could, to a certain extent, resemble the actions of natural disturbances.

As a young professor at UQAM, I decided to return to the Abitibi but this time accompanied by students. Over the years, we—I say *we* because this process involved dozens of master and doctoral students, several collaborators, and many postdoctoral fellows—reconstructed how natural disturbances had established the present-day forests and, knowing this, how we could develop closer-to-nature forestry.

It must be said that at the time the ice rink was almost empty, as foresters had almost completely abandoned ecology to emphasize the economic vocation of forests. It was also the development of the Université du Québec network that made it possible to bring into the discussion the knowledge developed in the regions where forest exploitation occurred. The regional industrial, governmental, and citizen actors did not see the forest only as a resource but also as a living environment, and they were eager for research to be undertaken in their forests.

My students were the ones who initially developed knowledge on the dynamics of natural disturbances and then compared disturbances with the effects of forest management. A hundred students later, we had accumulated enough knowledge on natural ecosystems to put it to use in forest management. We first demonstrated this at the Lac Duparquet teaching and research forest, an 80 km^2 area managed by UQAT and UQAM. We were also able to convince our industrial partners, who needed to certify their forest products as sustainable.

Our main discovery was to highlight that there were certain similarities between forest cutting and forest fires. Indeed, large fires in the boreal forest, much like logging, leave areas where a good proportion of the trees die. However, our work also demonstrated that the expected 70-to-100-year harvesting rotation was much shorter than the preindustrial return period for fires. Logging therefore rejuvenated the landscape to the point of eliminating a significant proportion of old-growth forest. As many organisms depend on old-growth forests, the consequences of mainly younger forests for biodiversity in the boreal forest are enormous. We had therefore identified a major problem; but we did not want to stop there. We wanted to propose viable solutions. To do this, we worked with our partners to develop silvicultural approaches, such as partial cutting, to maintain the structure of older forests while letting forest exploitation to continue.

We were also interested in the processes operating during disturbances. Fire burns the organic layer and makes nutrients accessible. Cutting, on the other hand, especially on frozen soils in winter, does not disturb the organic layer very much. In some cases, the new forest cannot grow postcutting as it would postfire. Thus, regrowth requires a mixing of the organic matter or even a prescribed burning. Nature is complex, and one solution does not solve all problems; hence, either partial cutting or full cutting with soil mixing must be undertaken at the right place and time.

Our common determination contributed significantly in convincing forest managers that natural disturbance regimes could inspire forest management strategies. Thus, ecosystem-based management is now part of the forestry regime in Québec as well as elsewhere in Canada and the world. I am also proud, as co-chair of the Northern Limit Committee, that Québec is also one of the few jurisdictions in the world that has set a northern limit to forest management for ecological reasons based on scientific knowledge.

The game is not over, however, as production forestry regularly resurfaces because of economic pressures. Climate change and carbon sequestration by the forest are becoming important issues that are often used as a reason to return to intensively managed, carbon-fixing forests. In this context, it is becoming increasingly important to argue that ecosystem-based management remains quite likely the best solution to accompany ecosystems toward future trajectories. This book will make a significant contribution to that effort.

I remain confident about the future. My greatest pride is to have trained and perhaps inspired several students, some of whom are currently researchers or forest managers. Many contribute to this book and others will use it. This is a good demonstration that we have succeeded collectively to put more science into forestry practices to the great benefit of the forest and society.

The Birth of the Idea

The idea for a multiauthor, multidisciplinary book began one night in the summer of 2017 during the final stages of writing my Ph.D. thesis. I had been searching for existing literature to place my project within the context of contemporary and future climate change. It dawned on me that there lacked a conceptual framework for adapting boreal forest management to climate change. This realization was surprising, and even somewhat worrisome, because such a framework should be a scientific priority given the major climate-related consequences expected for the future boreal forest. The next morning, I met Hubert Morin, my Ph.D. supervisor, to discuss this concern. I proposed a collective project at the biome scale to include this new context and fill this gap in boreal forest science. Together, we discussed the main elements required for this large-scale collaboration:

(1) Needing a new paradigm in boreal forest management The current state of ecosystem-based management must be reassessed to determine how best to incorporate climate change as a key driver within boreal ecosystems. This work would therefore discuss the lessons learned so far and consider new issues, paradigms, and previously neglected challenges to identify future directions.

(2) Expanding on the success of *Ecosystem-Based Management (EBM) in the Boreal Forest (2009)* This book represents a major contribution to boreal forest science, in both theoretical and applied sciences, and has served as a practical guide for achieving sustainable forest management in the boreal forest. Moreover, this book provides an example of cooperation and scientific collaboration among researchers at the leading edge of boreal forest ecology. Nonetheless, this book focused mainly on eastern Canada, and many topics were addressed only marginally, e.g., social aspects, restoration, and climate change. This new project would fill these missing gaps in boreal forest–related research and cover the boreal biome at a global scale. The subjects would include

- Climate change
- Complex adaptive systems
- Social aspects

- Ecophysiology
- Ecological restoration
- Invasive species
- Landscape-scale modeling
- Terrestrial–aquatic interactions

(3) Gaining insight from end-of-career researchers and pursuing new ideas from early-career scientists The authors and collaborators of the 2009 EBM book have had long and productive professional careers. In the next few years, most of these researchers will retire. It is therefore critical to compile this vast accumulation of knowledge and insight. These end-of-career researchers offer new ideas to be pursued and provide suggestions on how to face the challenges of managing the boreal forest under climate change. The next generation of boreal forest researchers has emerged under this wise counsel, and they offer promising avenues of thought and application. This book aims to harness this rich pool of experience and novel thinking.

After my meeting with Hubert, I wrote the outline of this project and the book proposal. I sent an e-mail to Sylvie Gauthier, Yves Bergeron, Christian Messier, Louis de Grandpré, Jean-Claude Ruel, Timo Kuuluvainen, Joakim Hjältén, Marie-Josée Fortin, Anouchka Hof, Rupert Seidl, Nicole Fenton, Elise Filotas, Pierre Drapeau, Yan Boulanger, Nelson Thiffault, Tuomas Aakala, and Sergio Rossi to validate this idea. My pleasant surprise was that all my colleagues answered this e-mail enthusiastically, justifying the need for such a book, and offered their participation and implication as authors and associate editors in this project. Thus, the project was born, and research, motivation, work, collaboration, and engagement came together to make the book a reality.

Leading the compilation of this book has been rewarding both as a great scientific experience in conceptualizing a new framework and as a personal experience in working with colleagues from around the boreal biome and coordinating the ideas of 148 authors.

Miguel M. Girona

The Philosophy and Spirit of This Book

This book not only offers new directions for boreal forest management but also serves as an example of research collaboration and intergenerational knowledge of the boreal biome, where our values of diversity, equity, and inclusion were at the forefront when creating the team of editors and authors and during the overall process.

International The need to carry out scientific studies at a more global scale requires us to structure this new book to address issues at the biome level because we face similar challenges and problems in North America and Eurasia. Thus, 148 authors representing 94 research groups and institutions from 20 countries became involved in this project.

Intergenerational In this book, we analyze the past and present but also look to the future. We have thus created an intergenerational book (very experienced and early-career researchers) incorporating a rich knowledge of accumulated past work and novel ideas driving boreal science. This approach also helps young researchers to be involved in collaborating as part of a large-scale project.

Women in forestry We applied a gender perspective and ensured the participation of women in the project as editors, authors, and collaborators; women represent 40%–50% of the authors and associate editors, thus making this work one of the first gender-equal forestry books.

Why Read This Book?

- **Innovation** We provide a new definition of ecosystem-based management, a new framework that integrates additional topics in forest management, such as social aspects, ecophysiology, restoration, aquatic systems, and biodiversity. Most existing forest ecology books focus primarily on silviculture and natural disturbances.
- **Scale** We provide a novel biome-scale perspective and synthesis, as most books on the boreal forest focus solely on North America or specific northern countries.

We include research and contributions from around the boreal biome, covering Russia, Scandinavia, and North America.

- **Structure** To cover the main topics in sustainable forest management, we structure the book as specialized parts having multiple chapters that deal with the subject at hand, e.g., parts on natural disturbances, biodiversity, new trends and technologies, silviculture, and social issues.
- **High-quality experts** The editors and authors of the book are leading researchers in their respective fields involving the boreal forest.
- **Approach and utility** We combine a fundamental and applied perspective throughout the book. We also provide original syntheses and data compilations from around the boreal biome. Moreover, this book aims to be a practical guide for stakeholders to apply sustainable forest management practices in a changing world and provide students with a state-of-the-science portrait of trends, challenges, and novel research avenues in boreal forest science.

Audience

- Scholars and students (bachelor, master, and doctorate students) in the fields of boreal forest ecology, restoration, biodiversity conservation, natural resources, and engineering
- Practitioners and stakeholders involved in the planning and management of forests and forest resources, natural disturbances, silviculture, and the use of forest products
- Forest agencies, research institutes, and government ministries able to provide public services and decision-support tools for forest management
- International and interdisciplinary researchers involved in modeling, climate change, economics, and social sciences

Acknowledgments

This book project was funded through a FODAR inter-reseaux (UQAT-UQAM-UQAR-UQO-UQAC-TÉLUQ) initiative of the Université du Québec with a strategic and interuniversity collaboration grant obtained by Miguel M. Girona, with Christian Messier (UQO), Dominique Arsenault (UQAR), Élise Filotas (TELUQ), Hubert Morin (UQAC), Sergio Rossi (UQAC), Guillaume Groshois (UQAT), Yves Bergeron (UQAT-UQAM), and Pierre Drapeau (UQAM) as collaborators. The writing of the book benefited from the activities of the International Laboratory on Cold Forests, https://forets-froides.org.

Chapter 1 We wish to acknowledge the contribution of Elise Imbeau and Dominique Boucher for their help with the figures. Special thanks to Murray Hay for his careful editing the text. Fruitful discussions with many colleagues from universities, forest management teams in industry, and government are also acknowledged.

Chapter 2 The authors thank the funding received to develop this chapter. Miguel M. Girona obtained funding from the Natural Sciences and Engineering Research Council of Canada (NSERC)-Alliance to understand the dynamics of spruce budworm (ALLRP 558267-20), and an NSERC Discovery grant to reconstruct the regime of natural disturbances (RGPIN-2022-05423). Normunds Stivrins received funding through the University of Latvia project "Studies of the fire impact on the bog environment and recovery" in partnership with JSC "Latvia's State Forests" and an Estonian Research Council grant PRG323. Niina Kuosmanen was funded by project 323065 from the Academy of Finland. Fabio Gennaretti received funding through an NSERC Discovery grant (RGPIN-2021-03553).

Chapter 5 The authors thank the reviewers Therese Löfroth and Louis Imbeau for their constructive comments on an earlier version of the manuscript. We thank Rein Drenkhan for background information on pathogens in relation to climate change, Marjorie Wilson for providing the photo of the spruce grouse, and Asko Lõhmus, Maarja Kõrkjas, and Liina Remm for sharing the graph from their unpublished study involving microhabitat development.

Chapter 7 This study was undertaken as part of the research program of the Forest Research Institute of the Université du Québec en Abitibi-Témiscamingue, the Industrial Research Chair–NSERC for black spruce growth and the influence of the spruce budworm on landscape variability in the boreal zone at the Université du Québec à Chicoutimi, and the State Research Program of Forest Research Institute of Karelian Research Centre, Russian Academy of Sciences. The authors also thank the two reviewers for their helpful comments that greatly improved the quality of the manuscript. We thank Valentina Buttò for the watercolor image used in Fig. 7.4a.

Chapter 8 A number of colleagues affiliated with universities and the ministère des Forêts, de la Faune et des Parcs du Québec commented on preliminary versions of this chapter: Yves Bergeron (UQAT), Michel Campagna (MFFP), Guillaume Cyr (MFFP), Sophie Dallaire (MFFP), Marie-Claude Lambert (MFFP, Fig. 8.4), Sonia Légaré (MFFP), Maxence Martin (UQAC), Catherine Périé (MFFP, Fig. 8.6), Marie-Andrée Vaillancourt (MFFP), Ana Verhulst-Casanova (UQAT), and Stephen Yamasaki (Bureau du Forestier en chef, Québec). Del Meidinger (British Columbia) and Tuomas Aakala (Finland) provided perspectives from outside Québec and helped structure the presented information. Lisa Bajolle and Thomas Suranyi created the pollen diagram and the other variables in Fig. 8.5. Karen Grislis greatly improved the English of an earlier draft. Denise Tousignant supervised the administrative process of the English revision. Our sincere thanks to each of these individuals.

Chapter 9 This study was funded by an NSERC Discovery grant to Annie Deslauriers, the Programme de Financement de la Recherche et Développement en Aménagement Forestier (MFFP, Québec, Canada). Lorena Balducci was funded by a MITACS Accelerate Fellowship in collaboration with SOPFIM.

Chapter 10 We are grateful to the many researchers who measured and published the data used here to illustrate site-level carbon balances in the circumboreal region. This work used eddy covariance data acquired and shared by the FLUXNET community, including the networks AmeriFlux, AfriFlux, AsiaFlux, CarboAfrica, CarboEuropeIP, CarboItaly, CarboMont, ChinaFlux, Fluxnet-Canada, GreenGrass, ICOS, KoFlux, LBA, NECC, OzFlux-TERN, TCOS-Siberia, and USCCC. The ERA-Interim reanalysis data were provided by ECMWF and processed by LSCE. The processing and harmonization of the FLUXNET eddy covariance data were carried out by the European Fluxes Database Cluster, AmeriFlux Management Project, and Fluxdata project of FLUXNET, with the support of CDIAC, the ICOS Ecosystem Thematic Center, and the OzFlux, ChinaFlux, and AsiaFlux offices.

Chapter 11 This work was supported by the Russian Ministry of Science and Higher Education (projects #FSRZ-2020-0010 and #FSRZ-2020-0014). Vladimir V. Shishov and Margarita I. Popkova appreciate the support of the Russian Science Foundation (Project #18-14-00072; preliminary data analysis and model development).

Chapter 12 The authors wish to thank the Natural Sciences and Engineering Research Council of Canada, Fonds québécois de la recherche sur la nature et les technologies, Syndicat des producteurs des bois du Saguenay-Lac-Saint-Jean, Forêt d'enseignement et de recherche Simoncouche, and the Fondation de l'Université du Québec à Chicoutimi (Sergio Rossi); the scholarship fund of the Forest Research Institute (Marcin Klisz); the Academy of Finland (grants 337549, 347782) and the Ministry of Agriculture and Forestry of Finland (grant VN/28414/2021) (Anna

Lintunen); the TRP 122 Translational-Research-Program Softwood for the Future (Austrian Science Fund); and MOBJD588: How dry is too dry? Quantifying the adverse effects of droughts for European forests across the last two decades (Estonian Research Agency) (Jan-Peter George).

Chapter 13 The authors thank the Department of Interior Northeast Climate Adaptation Science Center, the USDA National Institute of Food and Agriculture McIntire-Stennis Cooperative Forestry Research Program, the Rubenstein School of Environment and Natural Resources, the University of Vermont, the Department of Forest Resources, the University of Minnesota, the USDA Forest Service Northern Research Station, and the Northern Institute of Applied Climate Science.

Chapter 15 The authors thank the Edmund Hayes Professorship in Silviculture Alternatives.

Chapter 16 The funding for this project was obtained by Miguel M. Girona from an NSERC-Alliance–Silviculture grant UQAT-UQAC ALLRP 557166–20 and a MITACS scholarship obtained by Louiza Moussaoui. We thank the Fonds de Recherche du Québec—Nature et Technologies (FQRNT), the Programme de mise en valeur des ressources forestières (MFFPQ), the MRC d'Abitibi, and Natural Resources Canada for the funding to set up the experimental areas. We also acknowledge the contributions from Boisaco, GreenFirst, Résolu Saguenay, and the MRC-Abitibi and input from R. Gagnon, E. Dussault-Chouinard, G. Grosbois, M.J. Tremblay, D. Laprise, M. Cusson, E. Pamerleau-Couture, C. Gosselin, F. Marchand, S. Rossi, C. Krause, and P. Meek.

Chapter 17 We thank Philip J. Burton for his valuable comments on an earlier version of the manuscript.

Chapter 18 We are grateful to Jon Andersson, Petri Martikainen, and David Bell for allowing the use of their photos and figures.

Chapter 19 This research was funded by the Swedish Environmental Protection Agency grant NV-03728-17.

Chapter 21 This research was sponsored by the Swedish Research Council FORMAS (grant 2017-00826), the Finnish Academy project "Confronting sustainability: governing forests and fisheries in the Arctic" grant 333231, the Social Sciences and Humanities Research Council (grant RNH01176), and the Kone Foundation "Diversities of the Environmental Movement in Russia," (grant 2020 5986).

Chapter 25 Funding was provided by FORMAS through a research grant to Anouschka R. Hof (grant 2016-01072). We thank Nathan De Jager and an anonymous reviewer for helpful suggestions to an earlier draft.

Chapter 26 We acknowledge support from the NASA Arctic Boreal Vulnerability Experiment (ABoVE) grants NNX17AE44G and 80NSSC19M0112 to Scott J. Goetz and a DoD Strategic Environmental Research Initiative Program (SERDP) contract RC18-1183 to Scott J. Goetz.

Chapter 28 We acknowledge funding from the USDA McIntire-Stennis Forest Research Program (William S. Keeton, P.I.) and the Natural Sciences and Engineering Research Council of Canada Discovery grant RGPIN 2018-06156 (Élise Filotas).

Chapter 29 The authors thank Élise Imbeau (illustrator), Jan Karlsson (Umeå University), Hélène Masclaux (Université de Bourgogne), Éric Capo (Institut de Ciències del Mar), MRC-Abitibi, Natural Sciences and Engineering Research Council of Canada (NSERC)-Alliance, Ministère des Relations internationales et de la Francophonie du Québec, Arctic Council of Ministers, the Swedish Council for Environment, Agricultural Sciences and Spatial Planning (FORMAS), Centre d'étude de la forêt (CEF), and the Groupe de recherche interuniversitaire en limnologie (GRIL).

Chapter 30 We thank Alain Leduc and Pierre Drapeau for insightful comments that greatly improved this manuscript. Loïc D'Orangeville acknowledges funding from an NSERC Discovery grant RGPIN-2019-04353 and the New Brunswick Innovation Foundation (NBIF) RIF 2019-029.

Chapter 31 This work was supported by the Natural Sciences and Engineering Research Council of Canada (NSERC)-Alliance (ALLRP 557166-20; ALLRP 556815-20; ALLRP 558267-20), NSERC Discovery grant (RGPIN-2022-05423), the Social Sciences and Humanities Research Council of Canada (SSHRC), the Fonds de recherche du Québec-Nature et Technologies (FRQNT), the Ouranos Consortium on Regional Climatology and Adaptation to Climate Change, and the Northern Scientific Training Program (NSTP), SmartForests FCI, FORMAS: the Swedish Research Council for Sustainable Development, FODAR inter-reseaux UQAT-UQAM-UQAR-UQO-UQAC-TÉLUQ, the Ella and Georg Ehrnrooth Foundation, and the Excellence scholarship programs. The authors also thank Akib Hasan and Anoj Subedi (GREMA-UQAT), Milla Rautio (UQAC), and Marc-André Gemme (UQAT). We also thank the companies and institutions of Materiaux Blanchet, Scerie Landrienne, Boisaco, GreenFirst, Resolu Produit Forestiers, FPInnovations, MRC-Abitibi, Forêt d'enseignement et de recherche du lac Duparquet, Centre d'étude de la forêt (CEF), and the Groupe de recherche interuniversitaire en limnologie (GRIL).

The editors wish to thank Elise Imbeau for the drawings separating the book parts and multiple figures in Chaps. 1, 2, 16, 29 and 31. They also thank Louis de Grandpré, who was involved as an editor until his retirement, Frederic Doyon, Marc-André Gemme for the cover design, and finally Murray Hay (Maxafeau Editing Services, www.maxafeau.com) for the copy editing and compilation of the chapters.

Associate Editors, Contributors, and Reviewers

Associate Editors

Part I: Introduction

S. Ellen Macdonald
Timo Kuuluvainen

Part II: Natural Disturbances

Tuomas Aakala
Ekaterina Shorohova
Daniel Kneeshaw
Jean Claude Ruel
Kaysandra Waldron

Part III: Biodiversity

Nicole J. Fenton
Therese Löfroth
Pierre Drapeau

Part IV: Response of Functional Traits

Sergio Rossi
Annie Deslauriers
Christoforos Pappas

Part V: Silviculture as a Tool to Promote Forest Resilience

Nelson Thiffault

Patricia Raymond
Magnus Löf
Klaus Puettmann

Part VI: Ecological Restoration

Joakim Hjältén
Anne Tolvanen

Part VII: Forest Management and Society

Sara Teitelbaum
Hugo Asselin

Part VIII: New Tools for Monitoring Climate Change Effects

Marie-Josée Fortin
Rupert Seidl
Anouschka R. Hof
Udayalakshmi Vepakomma

Part IX: Trends and Challenges

Élise Filotas
Christian Messier
Loïc D'Orangeville
Guillaume Grosbois

Contributors

Tuomas Aakala School of Forest Sciences, University of Eastern Finland, Finland, e-mail: tuomas.aakala@uef.fi

Adam A. Ali Institut des Sciences de l'Évolution, Montpellier, France, e-mail: ahmed-adam.ali@umontpellier.fr

Jon Andersson Borcasts, Sweden, e-mail: jon.pm.andersson@outlook.com

Núria Aquilué Centre for Forest Research, Université du Québec à Montréal, Canada, e-mail: nuria.aquilue@ctfc.cat

Dominique Arseneault Département de biologie, chimie et géographie, Université du Québec à Rimouski, Canada, e-mail: Dominique_Arseneault@uqar.ca
    Centre for Forest Research, Université du Québec à Montréal, Canada

Alberto Arzac Siberian Federal University, Russia, e-mail: arzac@gmail.com

Hugo Asselin École d'études autochtones/School of Indigenous Studies, Université du Québec, Canada en Abitibi-Témiscamingue, e-mail: hugo.asselin@uqat.ca

Flurin Babst School of Natural Resources and the Environment & Laboratory of Tree-Ring Research, University of Arizona, USA, e-mail: babst@arizona.edu

Lorena Balducci Département des Sciences fondamentales, Université du Québec à Chicoutimi, Canada, e-mail: Lorena1_Balducci@uqac.ca

Guillaume Bastille-Rousseau School of Biological Sciences, Southern Illinois University, USA, e-mail: gbr@siu.edu

Annie-Claude Bélisle Forest Research Institute, Université du Québec en Abitibi-Témiscamingue, Canada, e-mail: annieclaude.belisle@uqat.ca
Centre for Forest Research, Université du Québec à Montréal, Canada

Yves Bergeron Forest Research Institute, Université du Québec en Abitibi-Témiscamingue and the Département des sciences biologiques, Université du Québec à Montréal Canada, e-mail: Yves.Bergeron@uqat.ca

Martin Berggren Department of Physical Geography and Ecosystem Science, Lund University, Sweden, e-mail: martin.berggren@nateko.lu.se

Logan T. Berner School of Informatics, Computing, and Cyber Systems, Northern Arizona, University, USA, e-mail: logan.berner@nau.edu

Tone Birkemoe Faculty of Environmental Sciences and Natural Resource Management, Norwegian University of Life Sciences, Norway, e-mail: tone.birkemoe@nmbu.no

Jean-François Bissonnette Département de géographie, Faculté de foresterie, de géographie et de géomatique, Université Laval, Canada, e-mail: jean-francois.bissonnette@ggr.ulaval.ca

Denis Blouin Faculté de foresterie, géographie et géomatique, Université Laval, Canada, e-mail: deblo27@ulaval.ca

Laura Boisvert-Marsh Great Lakes Forestry Centre, Canadian Forest Service, Natural Resources Canada, Canada, e-mail: laura.boisvert-marsh@nrcan-rncan.gc.ca

Arun Bosé Swiss Federal Institute for Forest, Snow and Landscape Research WSL, Switzerland, e-mail: arun.bose@wsl.ch
Forestry and Wood Technology Discipline, Khulna University, Bangladesh

Mathieu Bouchard Département des sciences du bois et de la forêt, Université Laval, Canada, e-mail: mathieu.bouchard@sbf.ulaval.ca

Étienne Boucher Département de géographie, GEOTOP-UQAM, Université du Québec à Montréal, Canada, e-mail: boucher.etienne@uqam.ca

Yan Boulanger Natural Resources Canada, Canadian Forest Service, Laurentian Forestry Centre, Canada, e-mail: yan.boulanger@NRCan-RNCan.gc.ca

Marie-Hélène Brice Institut de recherche en biologie végétale, Université de Montréal et Jardin botanique de Montréal, Canada, e-mail: marie-helene.brice@umontreal.ca

Chris Brimacombe Department of Ecology & Evolutionary Biology, University of Toronto, Canada, e-mail: chris.brimacombe@mail.utoronto.ca

Jakub W. Bubnicki Mammal Research Institute, Polish Academy of Sciences, Poland, e-mail: kbubnicki@ibs.bialowieza.pl

Philip J. Burton Ecosystem Science & Management, University of Northern British Columbia, Canada, e-mail: Phil.Burton@unbc.ca

Debojyoti Chakraborty Department of Forest Growth, Silviculture and Genetics, Austrian Research Centre for Forests (BFW), Austria, e-mail: debojyoti.chakraborty@bfw.gv.at

Aurélie Chalumeau Direction de la recherche forestière, ministère des Forêts, de la Faune et des Parcs du Québec, Canada, e-mail: Aurelie.chalumeau@mffp.gouv.qc.ca

Emeline Chaste Université de Lorraine, AgroParisTech, France, e-mail: emelinechaste6@hotmail.com

Phil Comeau Department of Renewable Resources, University of Alberta, Canada, e-mail: phil.comeau@ualberta.ca

Denis Cormier FPInnovations, Canada, e-mail: Denis.Cormier@fpinnovations.ca

Pierre-Luc Couillard Direction de la recherche forestière, ministère des Forêts, de la Faune et des Parcs du Québec, Canada, e-mail: Pierre-Luc.Couillard@mffp.gouv.qc.ca

Branislav Cvjetkovic Faculty of Forestry, University of Banja Luka, Bosnia and Herzegovina, e-mail: branislav.cvjetkovic@sf.unibl.org

Anthony W. D'Amato Rubenstein School of Environment and Natural Resources, University of Vermont, USA, e-mail: awdamato@uvm.edu

Victor Danneyrolles Groupe de recherche en écologie de la MRC-Abitibi (GREMA), Forest Research Institute, Université du Québec en Abitibi-Témiscamingue, Canada, e-mail: victor.danneyrolles@uqat.ca
     Centre for Forest Research, Université du Québec à Montréal, Canada

Paul del Giorgio Department of Biological Sciences, Université du Québec à Montréal, Canada, e-mail: del_giorgio.paul@uqam.ca

Annie Deslauriers Département des Sciences fondamentales, Université du Québec à Chicoutimi, Canada, e-mail: annie_deslauriers@uqac.ca

Olalla Díaz-Yáñez ETH Zürich, Department of Environmental Systems Science, Forest Ecology, Switzerland, e-mail: olalla.diaz@gmail.com

Loïc D'Orangeville Faculty of Forestry and Environmental Management, University of New Brunswick, Canada, e-mail: loic.dorangeville@unb.ca

Pierre Drapeau Centre for Forest Research and UQAT-UQAM Chair in Sustainable Forest Management, Université du Québec à Montréal, Canada, e-mail: drapeau.pierre@uqam.ca

Matthew J. Duveneck Harvard Forest, Harvard University, USA, e-mail: mduveneck@gmail.com

Mats Dynesius Department of Wildlife, Fish, and Environmental Sciences, Swedish University of Agricultural Sciences, Sweden, e-mail: mats.dynesius@slu.se

Marine Elbakidze Faculty of Forest Sciences, Swedish University of Agricultural Sciences, Sweden, e-mail: marine.elbakidze@slu.se
Faculty of Geography, Ivan Franko National University of Lviv, Ukraine

Simone Fatichi Department of Civil and Environmental Engineering, College of Design and Engineering National University of Singapore, Singapore, e-mail: ceesimo@nus.edu.sg

Nicole J. Fenton Forest Research Institute, Université du Québec en Abitibi-Témiscamingue, Canada, e-mail: nicole.fenton@uqat.ca
Centre for Forest Research, Université du Québec à Montréal, Canada

Angelo Fierravanti Département des Sciences fondamentales, Université du Québec à Chicoutimi, Canada, e-mail: angelo.f3050@gmail.com

Élise Filotas Centre for Forest Research, Université du Québec à Montréal, Canada, e-mail: efilotas@teluq.ca
Département Science et Technologie, Université TÉLUQ, Canada

Marie-Josée Fortin Department of Ecology & Evolutionary Biology, University of Toronto, Canada, e-mail: mariejosee.fortin@utoronto.ca

Adrianna C. Foster Climate and Global Dynamics Laboratory, National Center for Atmospheric Research, USA, e-mail: afoster@ucar.edu

Sylvie Gauthier Natural Resources Canada, Canadian Forest Service, Laurentian Forestry Centre, Canada, e-mail: Sylvie.gauthier@nrcan-rncan.gc.ca

Fabio Gennaretti Groupe de recherche en écologie de la MRC-Abitibi (GREMA), Forest Research Institute, Université du Québec en Abitibi-Témiscamingue, Canada, e-mail: fabio.gennaretti@uqat.ca

Jan-Peter George Tartu Observatory, Faculty of Science and Technology, University of Tartu, Estonia, e-mail: jan.peter.george@ut.ee

Martin P. Girardin Natural Resources Canada, Canadian Forest Service, Laurentian Forestry Centre, Canada, e-mail: martin.girardin@canada.ca

Willem Goedkoop Department of Aquatic Sciences and Assessment, Swedish University of Agricultural Sciences, Sweden, e-mail: Willem.Goedkoop@slu.se

Scott J. Goetz School of Informatics, Computing, and Cyber Systems, Northern Arizona University, USA, e-mail: scott.goetz@nau.edu

Michael Grabner University of Natural Resources and Life Sciences—BOKU, Austria, e-mail: michael.grabner@boku.ac.at

Pierre Grondin Direction de la recherche forestière, ministère des Forêts, de la Faune et des Parcs du Québec, Canada, e-mail: Pierre.Grondin@mffp.gouv.qc.ca

Guillaume Grosbois Groupe de recherche en écologie de la MRC-Abitibi (GREMA), Forest Research Institute, Université du Québec en Abitibi-Témiscamingue, Canada, e-mail: guillaume.grosbois@uqat.ca
    Department of Aquatic Sciences and Assessment, Swedish University of Agricultural Sciences, Sweden

Aino Hämäläinen Department of Ecology, Swedish University of Agricultural Sciences, Sweden, e-mail: aino.hamalainen@slu.se

Linnea Hansson Skogforsk, The Forestry Research Institute of Sweden, Sweden, e-mail: linnea.hansson@skogforsk.se

Minhui He Tianyuanju, China, e-mail: hmh0503lb@163.com

Alison Hester James-Hutton Institute, UK, e-mail: alison.hester@hutton.ac.uk

Dmitry Himelbrant Department of Botany, Faculty of Biology, Saint Petersburg State University, Russia, e-mail: d.himelbrant@spbu.ru
    Laboratory of Lichenology and Bryology, Komarov Botanical Institute of Russian Academy of Sciences, Russia

Joakim Hjältén Department of Wildlife, Fish, and Environmental Studies, Swedish University of Agricultural Sciences, Sweden, e-mail: Joakim.Hjalten@slu.se

Karin Hjelm Southern Swedish Forest Research Centre, Swedish University of Agricultural Sciences, Sweden, e-mail: karin.hjelm@slu.se

Anouschka R. Hof Wildlife Ecology and Conservation Group, Wageningen University, The Netherlands, e-mail: Anouschka.Hof@wur.nl
Department of Wildlife, Fish, and Environmental Studies, Swedish University of Agricultural Sciences, Sweden

Juha Honkaniemi Natural Resources Institute Finland (Luke), Finland, e-mail: juha.honkaniemi@luke.fi

Malcolm Itter Department of Environmental Conservation, University of Massachusetts, USA Amherst, e-mail: mitter@umass.edu

Bengt-Gunnar Jonsson Department of Natural Sciences, Mid Sweden University, Sweden, e-mail: bengt-gunnar.jonsson@miun.se
Department of Wildlife, Fish, and Environmental Studies, Swedish University of Agricultural Sciences, Sweden

William S. Keeton Rubenstein School of Environment and Natural Resources and Gund Institute for Environment, University of Vermont, USA, e-mail: william.keeton@uvm.edu

Sanghyun Kim Groupe de recherche en écologie de la MRC-Abitibi (GREMA), Forest Research Institute, Université du Québec en Abitibi-Témiscamingue, Canada
Centre for Forest Research, Université du Québec à Montréal, Canada

Stefan Klesse Forest Resources and Management, Swiss Federal Research Institute for Forest, Snow and Landscape Research WSL, Switzerland, e-mail: stefan.klesse@wsl.ch

Marcin Klisz Dendrolab IBL, Department of Silviculture and Genetics of Forest Trees, Forest Research Institute, Poland, e-mail: m.klisz@ibles.waw.pl

Daniel Kneeshaw Centre for Forest Research, Université du Québec à Montréal, Canada, e-mail: kneeshaw.daniel@uqam.ca

Matti Koivula Natural Resources Institute Finland (Luke), Finland, e-mail: matti.koivula@luke.fi

Jari Kouki School of Forest Sciences, University of Eastern Finland, Finland, e-mail: jari.kouki@uef.fi

Niko Kulha Natural Resources Institute Finland (Luke), Finland, e-mail: niko.kulha@luke.fi
Zoological Museum, Biodiversity Unit, University of Turku, Finland

Niina Kuosmanen Department of Geosciences and Geography, University of Helsinki, Finland, e-mail: niina.kuosmanen@helsinki.fi

Timo Kuuluvainen Department of Forest Sciences, University of Helsinki, Finland, e-mail: timo.kuuluvainen@helsinki.fi

Danny Chun Pong Lau Department of Aquatic Sciences and Assessment, Swedish University of Agricultural Sciences, Sweden, e-mail: danny.lau@slu.se
  Department of Ecology and Environmental Science, Umeå University, Sweden

Alain Leduc Département des sciences biologiques, Université du Québec à Montréal, Canada, e-mail: leduc.alain@uqam.ca

Patrick R. N. Lenz Natural Resources Canada, Canadian Forest Service, Laurentian Forestry Centre, Canada, e-mail: patrick.lenz@canada.ca

Gun Lidestav Swedish University for Agricultural Sciences, Sweden, e-mail: gun.lidestav@slu.se

Anna Lintunen Institute for Atmospheric and Earth System Research, e-mail: anna.lintunen@helsinki.fi

Magnus Löf Southern Swedish Forest Research Centre, Swedish University of Agricultural Sciences, Sweden, e-mail: Magnus.Lof@slu.se

Therese Löfroth Department of Wildlife, Fish, and Environmental Sciences, Swedish University of Agricultural Sciences, Sweden, e-mail: therese.lofroth@slu.se

Piret Lõhmus Institute of Ecology and Earth Sciences, University of Tartu, Estonia, e-mail: piret.lohmus@ut.ee

Melissa Lucash Department of Geography, University of Oregon, USA, e-mail: mlucash@uoregon.edu

Johanna Lundström Department of Forest Resource Management, Swedish University of Agricultural Sciences, Sweden, e-mail: johanna.lundstrom@slu.se

Jean-Martin Lussier Natural Resources Canada, Canadian Forest Service, Canadian Wood Fibre Centre, Canada, e-mail: jean-martin.lussier@canada.ca

S. Ellen Macdonald Department of Renewable Resources, University of Alberta, Canada, e-mail: ellen.macdonald@ualberta.ca

Maxence Martin Forest Research Institute, Université du Québec en Abitibi-Témiscamingue, Canada, e-mail: maxence.martin@uqat.ca
  Centre for Forest Research, Université du Québec à Montréal, Canada
  Département des Sciences Fondamentales, Université du Québec à Chicoutimi, Canada

Richard Massey School of Informatics, Computing, and Cyber Systems, Northern Arizona University, USA, e-mail: rm885@nau.edu

Konrad Mayer University of Natural Resources and Life Sciences—BOKU, Austria, e-mail: konrad.mayer@gmail.com

Constance L. McDermott Environmental Change Institute, School of Geography and the Environment, University of Oxford, UK, e-mail: constance.mcdermott@ouce.ox.ac.uk

Christian Messier Département des sciences naturelles, Institut des sciences de la forêt tempérée (ISFORT), University of Québec en Outaouais, Canada, e-mail: christian.messier@uqo.ca
    Centre for Forest Research, Université du Québec à Montréal, Canada

Grzegorz Mikusiński School for Forest Management, Swedish University of Agricultural Sciences, Sweden, e-mail: grzegorz.mikusinski@slu.se

Miguel Montoro Girona Groupe de recherche en écologie de la MRC-Abitibi (GREMA), Forest Research Institute, Université du Québec en Abitibi-Témiscamingue, Canada, e-mail: miguel.montoro@uqat.ca
    Restoration Ecology Group, Department of Wildlife, Fish, and Environmental Studies, Swedish University of Agricultural Sciences, Sweden
    Centre for Forest Research, Université du Québec à Montréal, Canada

Hubert Morin Département des sciences fondamentales, Université du Québec à Chicoutimi, Canada, e-mail: hubert_morin@uqac.ca

Claude Morneau Direction de la recherche forestière, ministère des Forêts, de la Faune et des Parcs du Québec, Canada, e-mail: Claude.Morneau@mffp.gouv.qc.ca

Louiza Moussaoui Groupe de recherche en écologie de la MRC-Abitibi (GREMA), Forest Research Institute, Université du Québec en Abitibi-Témiscamingue, Canada, e-mail: louiza.moussaoui@uqat.ca

Lionel Navarro Département des sciences fondamentales, Université du Québec à Chicoutimi, Canada, e-mail: lionel.navarro@uqac.ca

Petri Nummi Department of Forest Sciences, University of Helsinki, Finland, e-mail: petri.nummi@helsinki.fi

Brian J. Palik USDA Forest Service Northern Research Station, USA, e-mail: brian.palik@usda.gov

Christoforos Pappas Department of Civil Engineering, University of Patras, Greece, e-mail: cpappas@upatras.gr
    Centre for Forest Research, Université du Québec à Montréal, Canada
    Département Science et Technologie, Téluq, Université du Québec, Canada

John R. Parkins Department of Resource Economics and Environmental Sociology, University of Alberta, Canada, e-mail: jparkins@ualberta.ca

Athanasios Paschalis Department of Civil and Environmental Engineering, Imperial College London, UK, e-mail: a.paschalis@imperial.ac.uk

Richard L. Peters Laboratory of Plant Ecology, Department of Plants and Crops, Faculty of Bioscience Engineering, Ghent University, Belgium, e-mail: richard.peters@unibas.ch

    Forest Is Life, TERRA Teaching and Research, Gembloux Agro Bio-Tech, University of Liège, Belgium

    Physiological Plant Ecology, Department of Environmental Sciences, University of Basel, Switzerland

Véronique Poirier Direction de la recherche forestière, ministère des Forêts, de la Faune et des Parcs du Québec, Canada, e-mail: Veronique.Poirier@mffp.gouv.qc.ca

Margarita I. Popkova Siberian Federal University, Russia, e-mail: popkova.marg@gmail.com

Jeanne Portier Swiss Federal Institute for Forest, Snow and Landscape Research WSL, Switzerland, e-mail: jeanne.portier@wsl.ch

Klaus J. Puettmann Department of Forest Ecosystems and Society, Oregon State University, USA, e-mail: Klaus.Puettmann@oregonstate.edu

Patricia Raymond Direction de la recherche forestière, ministère des Forêts, de la Faune et des Parcs du Québec, Canada, e-mail: patricia.raymond@mffp.gouv.qc.ca

Maureen G. Reed UNESCO Co-Chair in Biocultural Diversity, Sustainability, Reconciliation and Renewal, School of Environment and Sustainability, University of Saskatchewan, Canada, e-mail: mgr774@mail.usask.ca

Cécile C. Remy Augsburg University, Institute of Geographie, Germany, e-mail: cecile.remy@geo.uni-augsburg.de

Pierre J. H. Richard Département de géographie, Université de Montréal, Canada, e-mail: pierrejhrichard@sympatico.ca

Sergio Rossi Département des Sciences fondamentales, Université du Québec à Chicoutimi, Canada, e-mail: sergio.rossi@uqac.ca

Jean-Claude Ruel Faculté de foresterie, géographie et géomatique, Université Laval, Canada, e-mail: jean-claude.ruel@sbf.ulaval.ca

Kadri Runnel Institute of Ecology and Earth Sciences, University of Tartu, e-mail: kadri.runnel@ut.ee

    Department of Ecology, Swedish University of Agricultural Sciences, Institute for Ecology, Sweden

Lars Rytter Forestry Research Institute of Sweden, Sweden, e-mail: Lars.P.Rytter@gmail.com

Rupert Seidl School of Life Sciences, Technical University of Munich, Germany, e-mail: rupert.seidl@tum.de

Heikki Seppä Department of Geosciences and Geography, University of Helsinki, Finland, e-mail: heikki.seppa@helsinki.fi

Vladimir V. Shishov Siberian Federal University, Russia, e-mail: vlad.shishov@gmail.com

Ekaterina Shorohova Forest Research Institute of the Karelian Research Center, Russian Academy of Science, Russia, e-mail: shorohova@cs13334.spb.cdu
    Saint Petersburg State Forest Technical University, Russia
    Natural Resources Institute Finland (Luke), Finland

Anatoly Shvidenko IIASA, Austria, e-mail: shvidenk@iiasa.ac.at
    Center of Ecology and Productivity of Forests, Russia

A. John Sinclair Natural Resources Institute, University of Manitoba, Canada, e-mail: John.Sinclair@umanitoba.ca

Jörgen Sjögren Department of Wildlife, Fish, and Environmental Sciences, Swedish University of Agricultural Sciences, Sweden, e-mail: jorgen.sjogren@slu.se

Martin-Hugues St-Laurent Département de biologie, chimie et geographie, Université du Québec à Rimouski, Canada, e-mail: Martin-hugues_St-laurent@uqar.ca
    Centre for Forest Research, Université du Québec à Montréal, Canada

Normunds Stivrins Faculty of Geography, University of Latvia, Latvia, e-mail: normunds.stivrins@gmail.com
    Department of Geology, Tallinn University of Technology, Estonia

Johan Svensson Department of Wildlife, Fish, and Environmental Sciences, Swedish University of Agricultural Sciences, Sweden, e-mail: johan.svensson@slu.se

Bruce Talbot Department of Forest and Wood Science, Faculty of AgriSciences, Stellenbosch University, South Africa, e-mail: bruce@sun.ac.za
    Division of Forest and Forest Resources, Norwegian Institute of Bioeconomy Research, Norway

Sara Teitelbaum Département de sociologie, Université de Montréal, Canada, e-mail: sara.teitelbaum@umontreal.ca

Nelson Thiffault Natural Resources Canada, Canadian Forest Service, Canadian Wood Fibre Centre, Canada, e-mail: nelson.thiffault@canada.ca
    Centre for Forest Research, Université du Québec à Montréal, Canada
    IUFRO Task Force on Resilient Planted Forests Serving Society and Bioeconomy, Austria

Groupe de recherche en écologie de la MRC-Abitibi (GREMA), Forest Research Institute, Université du Québec en Abitibi-Témiscamingue, Canada
Natural Resources Institute Finland (Luke), Finland

Anne Tolvanen Natural Resources Institute Finland (Luke), Finland, e-mail: anne.tolvanen@luke.fi

Jean-Pierre Tremblay Département de biologie, Centre d'étude de la forêt et centre d'études nordiques, Université Laval, Canada, e-mail: jean-pierre.tremblay@bio.ulaval.ca

Junior A. Tremblay Wildlife Research Division, Environment and Climate Change Canada, Canada, e-mail: Junior.Tremblay@ec.gc.ca
Centre for Forest Research, Université du Québec à Montréal, Canada
Faculté de Foresterie, Géographie et Géomatique, Université Laval, Canada

Maria Tysiachniouk University of Eastern Finland, Finland, e-mail: Maria.Tysiachniouk@uef.fi
Nelson Institute, University of Wisconsin-Madison, USA

Nina Ulanova Faculty of Biology, Moscow State University, Russia, e-mail: nulanova@mail.ru

Eugene A. Vaganov Siberian Federal University, Russia, e-mail: eavaganov@hotmail.com

Marie-Andrée Vaillancourt Direction de l'expertise sur la faune terrestre, l'herpétofaune et l'avifaune, ministère des Forêts, de la Faune et des Parcs du Québec, Canada, e-mail: marie-andree.vaillancourt@mffp.gouv.qc.ca

Udayalakshmi Vepakomma FPInnovations, Canada, e-mail: Udayalakshmi.vepakomma@fpinnovations.ca

Martijn Versluijs The Helsinki Lab of Ornithology (HelLO), Finnish Museum of Natural History, University of Helsinki, Finland, e-mail: martijnversluijs@hotmail.com

Kaysandra Waldron Natural Resources Canada, Canadian Forest Service, Laurentian Forestry Centre, Canada, e-mail: Kaysandra.waldron@NRCan-RNCan.gc.ca

Märtha Wallgren Department of Wildlife, Fish, and Environmental Sciences, Swedish University of Agricultural Sciences, Sweden, e-mail: martha.wallgren@slu.se

Jiejie Wang Faculty of Forestry and Environmental Management, University of New Brunswick, Canada, e-mail: jiejie.wang@unb.ca
Natural Resources Canada, Canadian Forest Service—Atlantic Forestry Centre, Canada

Beat Wermelinger Swiss Federal Institute for Forest, Snow and Landscape Research WSL, Switzerland, e-mail: beat.wermelinger@wsl.ch

Isabelle Witté Centre for Forest Research, Université du Québec à Montréal, Canada, e-mail: isabelle.witte@mnhn.fr
    PatriNat, OFB-CNRS-MNHN, France

Bao Yang Key Laboratory of Desert and Desertification, Northwest Institute of Eco- Environment and Resources, Chinese Academy of Sciences, China, e-mail: yangbao@lzb.ac.cn

Xianliang Zhang College of Forestry, Hebei Agricultural University, China, e-mail: zhxianliang@126.com

Reviewers

Kaysandra Waldron, Jeane Portier, Niko Kulha, Louis Imbeau, Therese Löfroth, Ellen Macdonald, Aino Hämäläinen, Martijn Versluijs, Alessio Giovannelli, Christophoros Pappas, Roberto Pilli, Annie Deslauriers, Fabrizio Carteni, Fabio Gennaretti, Samuel Royer Tardif, Patricia Raymond, Philippe Nolet, Nathalie Isabel, Dominik Thom, Maxence Martin, Vladan Ivetic, Louiza Moussaoui, Phil Burton, Jörg Müller, Pierre Grondin, Núria Aquilué, Émiline Chaste, Tim Horstoke, Samuel Roturier, Solagne Nadeau, Nicole Fenton, Mark Stoddart, Rob Kozak, Jean- François Dissonnette, Annie Booth, Olalla Díaz-Yáñez, Anouschka Hof, Nathan De Jager, Philippe Marchand, Joanne White, Osvaldo Valeria, Johan Svensson, Hélène Masleaux, Eric Rosa, Alain Leduc, and Pierre Drapeau.

Contents

Part III   Biodiversity

Part IV   Response of Functional Traits

Part I
Introduction

Chapter 1
Ecosystem Management of the Boreal Forest in the Era of Global Change

Sylvie Gauthier, Timo Kuuluvainen, S. Ellen Macdonald,
Ekaterina Shorohova, Anatoly Shvidenko, Annie-Claude Bélisle,
Marie-Andrée Vaillancourt, Alain Leduc, Guillaume Grosbois,
Yves Bergeron, Hubert Morin, and Miguel Montoro Girona

S. Gauthier (✉)
Natural Resources Canada, Canadian Forest Service, Laurentian Forestry Centre, 1055 rue du
PEPS, P.O. Box 10380, Stn. Sainte-Foy, Québec, QC G1V 4C7, Canada
e-mail: Sylvie.gauthier@nrcan-rncan.gc.ca

T. Kuuluvainen
Department of Forest Sciences, University of Helsinki, P.O. Box 27, 00014 Helsinki, Finland
e-mail: timo.kuuluvainen@helsinki.fi

S. E. Macdonald
Department of Renewable Resources, University of Alberta, Edmonton, AB T6G 2H1, Canada
e-mail: emacdona@ualberta.ca

E. Shorohova
Forest Research Institute of the Karelian Research Center, Russian Academy of Science,
Pushkinskaya str. 11, Petrozavodsk 185910, Russia
e-mail: shorohova@es13334.spb.edu; ekaterina.shorokhova@luke.fi

Saint Petersburg State Forest Technical University, Institutsky str. 5, Saint Petersburg 194021,
Russia

Natural Resources Institute Finland (Luke), Latokartanonkaari 9, FI-00790 Helsinki, Finland

A. Shvidenko
IIASA, Schlossplatz 1, A-2361 Laxenburg, Austria
e-mail: shvidenk@iiasa.ac.at

Center of Ecology and Productivity of Forests, Profsoyuznaya st. 84/32 bldg. 14,
Moscow 117997, Russia

A.-C. Bélisle · Y. Bergeron
Forest Research Institute, Université du Québec en Abitibi-Témiscamingue, 445 boul. de
l'Université, Rouyn-Noranda, QC J9X 5E4, Canada
e-mail: annieclaude.belisle@uqat.ca

Y. Bergeron
e-mail: yves.bergeron@uqat.ca

© The Author(s) 2023   3
M. M. Girona et al. (eds.), *Boreal Forests in the Face of Climate Change*,
Advances in Global Change Research 74,
https://doi.org/10.1007/978-3-031-15988-6_1

1.1 Introduction

The boreal forest is a vast biome encompassing approximately one-third (30%) of the world's forest area. It harbors about half of the world's remaining natural and near-natural forests and provides important ecological, economic, social, and cultural services and values that benefit human communities (Burton et al., 2010; Gauthier et al., 2015a). Although the diversity of tree species in boreal forests is low relative to that of other biomes, the forests' structural and compositional variability and the diversity of ecological interaction networks are high (Burton, 2013; Isaev, 2012, 2013; Kuuluvainen & Siitonen, 2013). The genetic diversity of tree species is generally high with most species being wind pollinated and characterized by large population sizes; this genetic diversity provides a foundation for an adaptive capacity in the face of fluctuating environmental conditions and ongoing climate change (Aitken et al., 2008).

Landscape diversity in the boreal biome reflects the influence of site variation, the effect of natural disturbances of varying type, severity, and extent, and the resulting

A.-C. Bélisle · Y. Bergeron · M. M. Girona
Centre for Forest Research, Université du Québec à Montréal, P.O. Box 8888, Stn. Centre-Ville, Montréal, QC H3C 3P8, Canada

M.-A. Vaillancourt
Direction de l'expertise sur la Faune Terrestre, l'herpétofaune et l'avifaune, Ministère des Forêts, de la Faune et des Parcs, 880, Chemin Sainte-Foy, Québec, QC G1S 4X4, Canada
e-mail: marie-andree.vaillancourt@mffp.gouv.qc.ca

A. Leduc · Y. Bergeron
Département des Sciences Biologiques, Université du Québec à Montréal, P.O. Box 8888, Stn. Centre-Ville, Montréal, QC H3C 3P8, Canada
e-mail: leduc.alain@uqam.ca

G. Grosbois · M. M. Girona
Groupe de Recherche en Écologie de la MRC-Abitibi, Forest Research Institute, Université du Québec en Abitibi-Témiscamingue, Amos Campus, 341, rue Principale Nord, Amos, QC J9T 2L8, Canada
e-mail: guillaume.grosbois@uqat.ca

M. M. Girona
e-mail: miguel.montoro@uqat.ca

G. Grosbois
Department of Aquatic Sciences and Assessment, Swedish University of Agricultural Sciences (SLU), P.O. Box 7050, SE-75007 Uppsala, Sweden

H. Morin
Département des Sciences Fondamentales, Université du Québec à Chicoutimi, 555 boul. de l'université, Chicoutimi, QC G7H 2B1, Canada
e-mail: hubert_morin@uqac.ca

M. M. Girona
Department of Wildlife, Fish, and Environmental Studies, Swedish University of Agricultural Sciences (SLU), SE-901 83 Umeå, Sweden

dynamic processes of ecosystem succession (Fig. 1.1; Chap. 3; Kneeshaw et al., 2018; Shorohova et al., 2011). Fire, insects, wind, beaver, and severe drought events are among the most important natural disturbances in the boreal forest (Chap. 24; Girardin et al., 2006; Johnson, 1992; Labrecque-Foy et al., 2020; Lavoie et al., 2021). Because the boreal biome is located at northern latitudes, it is subject to more rapid and severe effects from climate change than more southern forests. The boreal forest is already affected by changing climate as evidenced by drought as well as fires and insect outbreaks being more frequent and severe (e.g., Hanes et al., 2019; Navarro et al., 2018b; Safranyik et al., 2010; Seidl et al., 2017; Chap. 9). High-latitude regions are associated with cold climates and short growing seasons; thus, tree growth and decomposition processes are relatively slow (Chap. 11). This slow decay of organic matter results in a large stock of deadwood and carbon in the soil. Therefore, the boreal zone can have substantial disturbance-related feedback effects on CO_2 emissions (Chap. 10; Ameray et al., 2021; Bradshaw & Warkentin, 2015; Pan et al., 2011).

Although human population density in the boreal forest is low, two-thirds (2/3) of forested boreal regions are under some form of management, mainly for wood production. These forests account for 33% of lumber and 25% of paper products within the global export market (Burton et al., 2010). In the latter decades of the twentieth century, increased concerns about the effect of forest management on ecosystem functioning, the loss of biodiversity and a change in social and cultural values toward forests drove a paradigm shift toward an ecosystem approach (EA) to forest management (Franklin, 1997). Forest ecosystem management (FEM) principles have since been adopted in many jurisdictions in the boreal forest (Gauthier et al., 2009; Perera et al., 2004; Shvidenko et al., 2017).

Today, however, we are challenged with implementing FEM approaches in the context of global climate change, which affects tree growth and regeneration, causes dieback due to drought, and favors more frequent and severe natural disturbances (Gauthier et al., 2015a). Forests are also increasingly affected by the cumulative impacts of previous management practices, disturbance by other industries, and the consequences of other stresses (e.g., pollution). Hence, there is an urgent need to revisit and adapt the FEM concept to address these new and often synergetic challenges.

In this introductory chapter, we provide the background and context for understanding the emergence and evolution of forest management paradigms. We define the FEM concept and describe the approaches undertaken in its implementation to manage/restore boreal forests within different regions. We then set the stage for discussing the potential effects of global change and the suggested paradigm shifts to FEM.

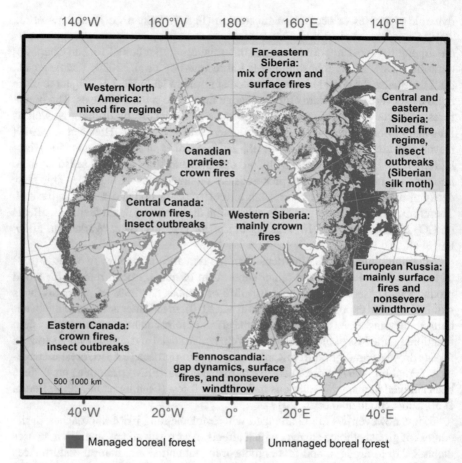

Fig. 1.1 The main disturbance dynamics within the boreal forest regions. Understanding natural disturbance dynamics and their ecological roles is indispensable for forest ecosystem management. Modified from Gauthier et al. (2015b; Reprinted with permission from AAAS) and Shorohova et al. (2011; CC BY-SA 4.0 licence)

Box 1.1 Forest Management Approaches Referred to in This Book

Sustained yield management (SY) Sustained yield management focuses on ensuring a continuous supply of resources (typically timber) that can be exploited over the long term. In the boreal forest, it often entails applying even -aged management and regulating forest age structure to ensure a constant, even flow of timber. Although sustainable forest management (SFM) aims to maintain more ecosystem services than SY, the SY paradigm remains part of forest ecosystem management (FEM) in many parts of the managed boreal biome (Luckert & Williamson, 2005).

Forest ecosystem management (FEM) Both SFM and EA approaches have been crucial in shifting the dominant paradigm of FEM at the beginning of the twenty-first century. Across regions of the boreal forest, FEM has evolved differently because of differences in historical and management contexts.

Natural disturbance–based management (NDBM)/Natural range of variability (NRV) The NDBM/NRV approach developed in North America aims to maintain resilient ecosystems by establishing management approaches on a solid understanding of natural disturbance regimes (NDBM). The presumption is that, despite human management and use, forests will maintain their key intrinsic structures, species communities, and ecological processes. In turn, this approach supports maintaining a continuous flow of the desired ecological, social, and economic values. The approach is based on the idea that current forest ecosystems have evolved under specific disturbance regimes (fire, insects, wind, etc.) that have driven forest dynamics, species composition, and overall biodiversity at the genetic to landscape scale.

Attention has been given to regimes prevailing before European colonization to identify the reference conditions for implementing NDBM and during analogous climates experienced at various periods in the Holocene (Chap. 2; De Grandpré et al., 2018; Gauthier et al., 2009; Landres et al., 1999; Montoro Girona et al., 2018b; Morin et al., 2009; Navarro et al., 2018a, b; Swetnam et al., 1999). These efforts aimed to define a baseline upon which the current state of regional forest landscapes can be compared while considering the inherent variability induced by these regimes. The framework is based on characterizing the NRV of several elements of the disturbance regime, such as disturbance type, frequency, size, spatial pattern, severity, and specificity, and then using this knowledge as a guide to implementing management strategies that will maintain the health of the ecosystem (Keane et al., 2009; Landres et al., 1999; Montoro Girona 2017). This approach permits comparing managed and natural ecosystems and landscapes (Grondin et al., 2018) and helps establish management or restoration targets (Fig. 1.2).

FEM also often involves using a combination of *coarse-* and *fine-filter* approaches. Coarse-filter strategies are implemented on a large spatial and temporal frame of reference, i.e., larger than the stand level, with the understanding that the time/space continuity is essential for some attributes. The coarse-filter approach aims to maintain the various forest habitats representative of natural forest landscapes and some of their key characteristics. Such an approach seeks to conserve most of the biological diversity. Fine-filter strategies are implemented through stand-level management or conservation to protect rare species or those having particular and known habitat requirements. The hierarchy of coarse- and fine-filter approaches explicitly acknowledges that stand-level actions affect the landscape over time, altering characteristics such as forest composition and age structure. Finally, to ensure that objectives set under a FEM system are achieved, monitoring is crucial for assessing the implemented management system's success or failure and measuring the responses of target organisms to management (Drapeau et al., 2009). The results from this monitoring should then feed into future refinements, and new scientific knowledge should be incorporated through an adaptive management framework.

Retention forestry This forest management approach is based on retaining structures and organisms, such as living and dead trees and small areas of intact forest, both for harvesting and the longer term. Retention forestry aims to achieve temporal and spatial continuity in forest structure, composition, and the processes that promote biodiversity and sustain ecological functions at different spatial scales (Gustafsson et al., 2012). Retention is applied at various levels, from very low levels (Kuuluvainen et al., 2019) to up to 40% of the standing stock (Beese et al., 2019; Montoro Girona et al., 2019; Scott et al., 2019).

Continuous-cover forestry Continuous-cover management involves managing a forest without the use of clear-cutting. Harvesting is typically based on a single tree or group selection, and a significant portion of canopy trees is retained (Felton et al., 2016; Kim et al., 2021; Sharma et al., 2016; Sténs et al., 2019). This produces forests having an uneven-aged structure.

Restorative management Restorative management prioritizes ecological restoration while simultaneously harvesting for profit. This management approach represents the first step toward forest ecosystem management in regions where intensive forest management has decreased or has degraded biodiversity and ecosystem functions (Vanha-Majamaa et al., 2007).

Zoning approach (TRIAD) The TRIAD zoning approach, proposed by Seymour and Hunter (1992), has forested landscapes divided into three zones, each subjected to different management objectives (Burton, 1995; Nitschke & Innes, 2005). The reserve portion is devoted to conservation purposes, whereas the intensive management portion focuses on timber production and can potentially compensate for the lower timber yields because of the presence of conservation areas. Between these two endmembers of the production/conservation spectrum lies a multiple-use zone where extensive management is conducted. Management in this area does not focus solely on timber production but also includes maintaining some important elements for biodiversity (Montigny & MacLean, 2006). The overall objective is to sustain the forest to support the needs of society (Seymour & Hunter, 1999). The actual size of the respective zones is specific to each landscape (Burton, 1995; Harvey et al., 2009). For instance, if maintaining old growth is not possible or too expensive, more area can be preserved.

Intensive forest management (IFM) IFM aims to increase or maximize the value, volume, or both of desired forest components, often timber. Attaining this goal involves such practices as density regulation, regeneration control, silvicultural intervention, and genetic improvement (Bell et al., 2000). Silviculture applied to reach these goals focuses on practices designed to accelerate stand development and improve stand value and yield: site preparation, the planting of species matched to site conditions, and vegetation management timed to maximize early growth. IFM can include natural regeneration but with density regulation. It often requires a series of actions during the rotation to achieve growth and yield objectives (Bell et al., 2000). Sweden and Finland have implemented this approach successfully for almost all their managed forests. Although increasing productivity is the main goal of intensive forest management, it can also be done in the context of maintaining or restoring diversity.

Extensive forest management (EFM) EFM is a management approach that does not rely on a series of interventions to attain growth and yield objectives. Instead, it focuses on protecting the forest from the primary natural disturbances, such as fire and insects, and relies partly on natural regeneration to provide the next forest. Silvicultural interventions focus mainly on attaining a minimum density with desired species composition and maintaining a given age-class distribution (Bell et al., 2000). This form of forest management is used in large areas of Canada and Russia.

Conservation area (adapted from the IUCN glossary) These are areas of various sizes (from the stand to the landscape scale) dedicated to protecting, caring, managing, and maintaining ecosystems, habitats, wildlife species, and populations. The creation of these spaces aims to safeguard natural conditions for their long-term preservation by conserving ecosystems and natural habitats and maintaining viable populations of species.

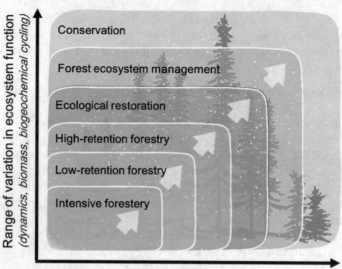

Fig. 1.2 The natural range of variability (NRV) is a means of framing or implementing sustainable forest management (SFM). Management approaches can be schematized in a conceptual hierarchy, in terms of species composition and ecosystem structure, in relation to their degree of overlap with NRV. Overlap is lowest for intensive plantation-type management but increases with retention forestry and ecological restoration; the latter is required in cases where the forest has been degraded by long-term intensive management or other uses (e.g., mining). Different management types can be combined within the same forest management unit. For instance, the TRIAD zoning approach (Messier et al., 2009), in which intensive management can increase the yield per hectare in some portions of the landscape, can be applied to decrease timber production pressure on other portions of the forest where extensive forest management is applied. Under the TRIAD approach, intensive and extensive management zones—along with conservation areas—are all included in the landscape in varying proportions, with each contributing to meet the goals of FEM

1.2 A Brief History of Boreal Forest Management Paradigms

1.2.1 The Early Era of Forest Management

Despite the extensive geographic spread of the boreal forest across the Northern Hemisphere, numerous commonalities exist among the ecological and management challenges for boreal countries. The main harvesting methods (clear-cutting) and silvicultural practices (single-cohort management, site preparation, planting, stand tending) are similar throughout the circumboreal forest. Common issues related to this management approach include landscape fragmentation, the loss of mature and old-growth forests, the homogenization of forest structure and tree species composition, and forest susceptibility to the impacts of climate change. However, boreal countries also differ in their forest exploitation histories and their forest management cultures, policies, and priorities. When evaluating the current situation and challenges, it is essential to consider the respective forest management histories (See Chap. 31). Here, we briefly describe the historical background and development of forest management in Canada, Sweden and Finland, and Russia.

1.2.2 Canada

In Canada, boreal landscapes were and remain inhabited by First Nations. Traditional Indigenous livelihood relies on forest resources for hunting, trapping, gathering, and various provisioning and cultural services (Chap. 20). Traditional land management is based on deep ecological knowledge and aims to maintain the capacity of the land to sustain life (Feit, 2001). For instance, fire was used in some regions of boreal Canada until the 1950s to maintain blueberry patches, attract wildlife within strategic areas, and prepare the soil for planting (e.g., Berkes & Davidson-Hunt, 2006; Lewis & Ferguson, 1988). The transition to commercial forestry has, however, restricted the forest management role of First Nations.

Large-scale commercial harvesting of forests began in the early nineteenth century, focusing on conifer species used for construction, firewood, and shipbuilding (Drushka, 2003; Gaudreau, 1998). During the nineteenth century, Canadian forestry entered its administrative period, responding to the need for a regulatory approach to better preserve timber supplies and safeguard the stability of the forest industry. By the end of the century, most provincial jurisdictions had adopted forestry policies, thereby establishing the first forest management regimes, which now form the basis of current policies. The Canadian Forest Service, a federal research agency, was established in 1899, and the University of Toronto inaugurated the first forestry school in Canada in 1907. Moreover, between 1871 and 1921, 11 treaties were signed between the Crown and First Nations to open the land for settlement in the south and secure access to natural resources in the north (Crown–Indigenous Relations and Northern Affairs Canada, 2020).

An impending decline in timber capital in Canada first became apparent at the onset of the twentieth century; this precipitated a transition to the era of *sustained yield forest management* (see Box 1.1; Bouthillier, 1998; Canadian Forest Service, 1998; Drushka, 2003). This management approach, also called *fully regulated forest*, involves compartmentally managing for an even forest age-class distribution, which theoretically ensures a regular and constant supply of similar wood volume over time. In the boreal forest, sustained yield forestry developed under an even-aged management system, using primarily clear-cutting and controlling forest age structure via management units. Under this system, forests are scheduled for harvest when volume increase levels off (maximum mean annual increment); this corresponds to a stand age of 50 to 150 years, depending on the forest type and location (Duchesne, 1994; Stadt et al., 2014).

This stand-wise even-aged management approach emphasized *normalizing* the boreal forest stand age distribution by the targeted harvesting of *overmature* stands, considered less productive. This approach also aimed for the long-term sustainability of timber supply by ensuring that annual harvests did not exceed what the forest produced. Thus, sustained yield management aimed to harvest a regular amount of timber and ensure the preservation of the forest capital. Nonetheless, in Canada, with its vast expanses of unmanaged forest, forestry has been mostly extensive since the Second World War.

Forest management is more intensive in Sweden and Finland and has a longer history relative to the Canadian context. Although clear-cutting and planting are common in both regions, Canadian forest management places greater reliance on natural regeneration and less use of intensive management approaches, such as early stand tending, fertilization, and thinning. In many regions of Canada, the forest industry continues to rely exclusively on primary forests, which have not been previously subjected to organized forest management.

1.2.3   Sweden and Finland

In this vast geographic area, and more recently in its northern parts, the Indigenous Sami people were among the first forest dwellers and users. Although their population size was relatively small, their mobile reindeer herding culture impacted forests (Josefsson et al., 2009). Since the Middle Ages, the regional human population has increased, and the boreal forest has been used for diverse purposes. Major influences include charcoal production for the large-scale mining industry (especially in Sweden), shipbuilding, tar production, and slash-and-burn agriculture (especially in Finland). Other extensive and important uses of the forest included domestic-use cuttings for firewood and building material as well as cattle herding in forests surrounding settlements.

Multiple impacts due to selective cutting, the careless use of fire, and cattle herding in forests prevented forest regeneration, leading to the regional scarcity or even depletion of timber by the nineteenth century. This development sparked fears of a permanent loss of these forests (Keto-Tokoi & Kuuluvainen, 2014; Östlund et al., 1997). At the same time, the *timber frontier* moved north along rivers in search of pristine forests and timber that could be floated to sawmills on the coast (Östlund & Norstedt, 2021).

The local and regional depletion of forest resources, combined with increased demands for wood as the forest industry expanded after the mid-nineteenth century, culminated in the need to organize forestry more effectively in terms of regulations, administration, and the education of forest managers. Sweden established a forestry institute in Stockholm in 1828 to train forestry professionals (Puettmann et al., 2009). In Finland, the Evo forestry school began to educate professional foresters in 1858. Legislation was also established to halt the careless use of forests. In Finland, for example, the 1928 Law on Private Forests by and large outlawed clear-cutting, allowing this practice only for special reasons. In the late-1940s, however, the interpretation of the law took a 180° shift; selective logging was outlawed, and only clear-cutting coupled with subsequent regeneration was allowed.

This development was linked to the establishment of the pulp and paper industry after the Second World War when smaller and lower quality timber also became merchantable. This change, coupled with low-cost fossil fuels and advances in harvesting technologies, led to a large-scale transition from selective harvesting to clear-cutting and even-aged forest management (Östlund et al., 1997; Siiskonen, 2007). At the same time, government-directed public funds into forestry infrastructure, such as building road networks and improving forest regeneration techniques and silvicultural practices.

In Finland, the large-scale ditching of forested peatlands was initiated to increase forest growth and raw material supply for the forest industry in the future (Keto-Tokoi & Kuuluvainen, 2014). The post–Second World War economic and construction boom led to the large-scale clear-cutting of natural or near-natural forest in both Sweden and Finland. As part of the terms of the peace treaty, Finland had to pay reparations to the Soviet Union and forest industry products formed part of this compensation, further increasing the extensive clear-cutting of natural and near-natural forest, especially in northern Finland, in the late 1940s and 1950s. Strict national laws and forest policies drove the development of forestry practices. Still, there was strong opposition among private forest owners, who had selectively harvested their forests for decades.

In both Sweden and Finland, the most significant change in forest utilization and management occurred when the formerly dominant selective cutting practices were rejected and even-aged management driven by clear-cut harvesting and regeneration by planting or seeding became the dominant method. This management model was favorable for the influential and economically important pulp and paper industry and hence formed a key part of the national forest policies, where increasing timber yield was the primary goal.

1.2.4 Russia

Historically, human-forest interactions in boreal Russia were minimal owing to the lack of roads and the sparse human population scattered across the vast expanses of forest. Northwestern Russia was an exception to this pattern, as forests were closer to settlements. Since the fifteenth century, human activities in the boreal forest of this region have included slash-and-burn cultivation, the use of wood for buildings and heating, and the production of tar, potash, salt, and charcoal for industry. The first legislation related to forest harvesting dates from the early eighteenth century when large-diameter trees along rivers were required to supply Peter the Great's shipbuilding program (Fedorchuk et al., 2005; Redko, 1981; Sokolov, 2006). The first forestry university in Russia was established in 1803 in St. Petersburg.

Over the last centuries, forest management in Russia has been closely linked to the country's dramatic political and economic changes. Whereas traditional forestry in Russia had obvious German roots, the second half of the nineteenth century and the beginning of the twentieth century were periods of rapid development of national forest science and increased study of natural forest ecosystems, forest management, and silviculture (Morozov, 1924; Orlov, 1927, 1928a, b). At the onset of the First World War, however, only 5% of Russian forests had been inventoried and had developed forest management plans; another 13% had been surveyed for different goals (Kozlovsky, 1959). The 1923 Forest Code acknowledged various functions of forests (protection, conservation, cultural and commercial uses) and formed the basis for further classification of forests into major functional forest management categories.

Around 1930, extensive management began to restore and industrialize the economy, normalizing the harvest of the most productive and accessible stands, the preferential selection of the most valuable tree size and quality, and the use of natural and assisted regeneration (Fedorchuk et al., 2005). Typical forestry involved large-scale "concentrated" clear-felling with 50–100 ha harvesting areas and, in many cases, substantially larger surfaces (Aksenov et al., 1999; Fedorchuk et al., 2005; Kozubov & Taskaev, 2000) until the second half of the 1960s, whereas other features of extensive forest management remain in application (Sokolov, 2006). These concentrated harvesting areas were not conventional clear-cuts, as foresters left behind large uncut patches of various sizes and individual trees of unused species or individual trees having bad stem quality (Baranov, 1954; Solntsev, 1950). Moreover, in the *incomplete clear fellings*, 61–90% of the stand growing stock was harvested (Melekhov, 1966), representing a retention level of up to 40%. This model, however, decreased the growing stock or altered stand composition over large areas in the managed parts of the boreal zone. These changes, combined with large fires in post-harvesting areas, encouraged the logging of new previously uncut regions in Russia.

1.3   New Forest Paradigm After Sustained Yield Management

Intensive even-aged forest management and the sustained yield approach have provided a sustained supply of wood fiber for industry, as reflected by the success of Sweden and Finland in increasing forest yield. Toward the end of the twentieth century, throughout the boreal biome, the cumulative adverse ecological effects of even-aged management with clear-cut harvesting began to draw attention (Franklin, 1989). These negative consequences include the simplification of forest structures, the disappearance of old, large trees, and the decline in the amount of dead-wood (Chap. 5). Sustained yield management based on the "fully regulated forest" paradigm began to be questioned for its inability to maintain forest values and resources other than timber.

Short harvest rotations with clear-cutting were shown to fundamentally alter ecosystem structure compared with conditions produced through natural distur-bances; the latter are more variable in terms of frequency, severity, and extent than traditional harvesting approaches. Particular concern involved managed forest land-scapes becoming fragmented because of the loss of older and more structurally heterogeneous forests, which dominate landscapes under longer, or less severe, natural disturbance regimes (Cyr et al., 2009; Franklin, 1997; Kuuluvainen, 2009; Östlund et al., 1997). Most managed boreal forest stands suffered declines in dead-wood, reduced structural heterogeneity, and, in some cases, tree species diversity (Chap. 6; Shorohova et al., 2019; Siitonen, 2001). In many regions, young, struc-turally homogeneous stands with early successional species began to dominate managed forest landscapes. This change was accompanied by a reduction in the area hosting older, structurally complex stands dominated by later successional species and large living and dead trees (Cyr et al., 2009; Kuuluvainen & Gauthier, 2018; Shvidenko & Nilsson, 1996).

These concerns were accompanied by a growing scientific knowledge related to (1) the relationships between forest structure, stand age, and biodiversity; (2) the importance of biological legacies in forest regeneration and succession; (3) the critical role of deadwood in forest ecosystem functioning and biodiversity; (4) the importance of natural disturbances as key ecological drivers within forest landscapes; and (5) the relationship between biodiversity and forest productivity, resistance, and resilience (Angelstam, 1998; Bergeron & Fenton, 2012; Bergeron et al., 2017a, b; Burton, 2013; D'Amato et al., 2017; Franklin, 1997; Gauthier et al., 2009; Gustafsson & Perhans, 2010; Lavoie et al., 2019; Montoro Girona et al., 2016).

Together with increased public and market awareness of the importance of sustaining the economic, ecological, and social/cultural values of forests, these concerns led to the emergence of a new forest management paradigm. The term *sustainable forest management* (SFM) was coined in the "Forest Principles" arising from the United Nations Conference on the Environment and Development (UNCED; i.e., the Rio Earth Summit) in 1992. In the subsequent years, countries collabo-rated to define SFM criteria and indicators (Wilkie et al., 2003). At the Conference of the Parties of the Convention on Biological Diversity (CBD) held in Jakarta in 1995, participants identified the *ecosystem approach* (EA)—an integrated strategy for conserving and sustaining land, water, and biological resources—as the primary framework for actions under CBD (Box 1.2). Both approaches have been very influential in developing forest ecosystem management in the boreal biome.

The fundamental difference between FEM and traditional forest management lies in the former's focus on managing the forest as an integrated, holistic, ecological entity existing at multiple spatial and temporal scales. FEM explicitly incorporates planning for what is to be extracted and for the full range of economic, ecological, and social/cultural values to be maintained within the landscape. Thus, this approach considers not only forest structure and composition but also ecological processes such as biogeochemical cycling, forest regeneration, species migration patterns, carbon sequestration, and ecosystem resistance and resilience (Gauthier et al., 2009; Palik et al., 2020). Although the definition of the concepts and practical applications vary from one jurisdiction to another, common principles, characteristics, and goals are shared among most national frameworks (see Box 1.2; Christensen et al., 1996; Galindo-Leal & Bunnell, 1995; Gauthier et al., 2009; Grumbine, 1994; Kimmins, 2004).

Box 1.2 Origins of the Sustainable Forest Management/Ecosystem Approach

The "Forest Principles" arising from the United Nations Conference on the Environment and Development (UNCED, i.e., the Rio Earth Summit) in 1992 helped define the concept of sustainable forest management (SFM), which was subsequently adhered to and developed by many countries. Conceptually, SFM aims to balance the ecological, economic, and sociocultural pillars of forest management. The goal of SFM is to provide integrated benefits to all, including safeguarding local livelihoods, protecting biodiversity and other ecological services provided by forests, reducing rural poverty, and mitigating some of the effects of climate change. Despite variations in definitions among countries, several criteria serve as common targets for SFM. These include: (1) the maintenance of the extent of forest resources; (2) the conservation of biological diversity (genetic, species, landscapes); (3) the conservation/enhancement of forest health and vitality; (4) the maintenance of forest productivity; (5) the maintenance of the ecological functions of forests, such as water cycling, carbon cycling, and interactions with climate; (6) the maintenance of socioeconomic benefits from forest resources.

At the Conference of the Parties of the Convention on Biological Diversity (CBD, 1995), the ecosystem approach (EA) was proposed as a framework for conserving biodiversity and ensuring the sustainable use of ecosystem resources. Its development continued until 2000 with the framing of an integrated strategy for conserving and sustaining land, water, and biological resources (Wilkie et al., 2003). The CBD (2000) defines EA as:

> a strategy for the integrated management of land, water and living resources that promotes conservation and sustainable use in an equitable way. [EA is] based on the application of appropriate scientific methodologies focused on levels of biological organization, which encompass the essential structure, processes, functions and interactions among organisms and their environment. It recognizes that humans, with their cultural diversity, are an integral part of many ecosystems.

Several principles of EA are similar to those proposed in SFM, whereas other principles focus more on ecosystem complexity and functioning. Important elements are that EA should: (1) consider management effects on adjacent ecosystems; (2) prioritize the maintenance of ecosystem structure and function; (3) manage the ecosystem at appropriate temporal and spatial scales relevant to long-term management objectives; (4) establish a balance between conservation and the use of biodiversity; and (5) consider all forms of information be it scientific, traditional ecological knowledge (TEK), etc. "…overall, SFM and EA express similar goals and ambitions for forest management focussing on environmental, social and economic sustainability, and on generating and maintaining benefits for both present and future generations." (Wilkie et al., 2003). In Canada, the Canadian Council of Forest Ministers adopted the SFM principles in 1995 (CCFM, 1995).

1.4 Implementing Sustainable Forest Management Within Boreal Regions: Approaches, Successes, and Shortfalls

Over the past three to four decades, boreal jurisdictions have agreed to the SFM principles and have more or less succeeded in implementing FEM within forestry policies, regulations, and planning.

1.4.1 Canada

In North America, both SFM and FEM emerged out of the ideas of ecological forestry of the Harvard Forest developed in the 1940s (D'Amato et al., 2017). These ideas were modified further and became known as *new forestry* or NDBM (Franklin, 1989; Gauthier et al., 2009; Hunter, 1993). These concepts have since been implemented partly (the late 1990s) by forest managers by fitting these approaches into traditional planning schemes of forest management (Box 1.1; Harvey et al., 2003).

Since the 1990s, the implementation of FEM in the boreal forest of Canada has been deeply rooted in an understanding of past disturbance regimes (NDBM) and the natural range of variability (NRV; Box 1.1) of these events (Gauthier et al., 2009; Ontario Ministry of Natural Resources, 2001; Perera et al., 2004). This was considered a precautionary coarse-filter approach, as without a proper understanding of ecological mechanisms, maintaining natural forest conditions within the NRV was perceived as a suitable means of preserving the ecological structure, function, and resilience in forested landscapes (Cissel et al., 1999; Hunter, 1993). The NRV concept aims to maintain the characteristics of managed stands and landscapes within the historical natural range of variability (Cissel et al., 1999; Landres et al., 1999). Although the implementation of FEM has differed among Canadian jurisdictions, commonalities have emerged. These similar ideas are notably because of the existence of the NSERC (Natural Sciences and Engineering Research Council) *Sustainable Forest Management network* (SFMn), a large research–industry partnership, which existed between 1995 and 2010 (https://sfmn.ualberta.ca/about-us/ consulted 26 April 2021).

One of the FEM framework elements aimed to facilitate "the formulation of environmental issues and the development of targets that have to be sustained or achieved within the implemented management system" (Gauthier et al., 2009). With the transition toward FEM, several attributes and processes manipulated by forest management were identified as vulnerable because of past management approaches. It was also recognized that long-term planning over large areas was needed to ensure the maintenance or restoration of these attributes (Table 1.1). These identified attributes included (1) the proportion of different forest age classes (old-growth versus young forest) and their spatial distribution across the landscape; (2) the landscape pattern of forest composition at the stand and landscape levels and associated dynamics; (3) variable internal stand structure; the retention of biological legacies such as deadwood or the pit and mound aspects of soils; (4) soil fertility and site productivity (Gauthier et al., 2009). The fire regime was the main disturbance regime on which the FEM was based in Canada (Bergeron et al., 1999, 2002; Ontario Ministry of Natural Resources, 2001; Vaillancourt et al., 2009). More recently, low and moderate severity disturbances (wind, insect, and low severity fire) have been recognized as contributing to NRV and have been slowly incorporated into FEM (Chap. 4; Bergeron et al., 2017a, b; De Grandpré et al., 2018; Lavoie et al., 2021; Stockdale et al., 2016). For instance, it is now recognized that although both fire and insect outbreaks over the Holocene have co-occurred at a regional level, outbreaks were more frequent when fire frequency was low (Chap. 2; Navarro et al., 2018b). These disturbances also strongly influence forest dynamics, impacting the amount, composition, and structure of old forests (Martin et al., 2019, 2020). In short, the characterization of the range of variability in past disturbance return intervals, severity, and extent over the last few centuries serves to set targets for maintaining or recovering particular forest characteristics, e.g., successional stages (old forest), forest composition (shade-tolerant species), and forest structure (Table 1.1; Chap.7).

Several experimental studies examining the effects of partial harvesting and variable retention have been established in various regions of Canada (Chap. 16; Box 1.1; Brais et al., 2004; Fenton et al., 2013; Montoro Girona et al., 2016; Ruel et al., 2007; Spence et al., 1999), and the knowledge gained from these research projects has slowly been implemented into operational practice. Assessment of the impacts of these treatments on biodiversity, forest regeneration and dynamics, deadwood dynamics, soils, and carbon storage (for up to approximately 15 years post-harvest) has provided considerable insight into the ecological structure, functioning, and dynamics of these forests. Retention or partial harvesting has been shown as a means of meeting FEM objectives (e.g., Bartels et al., 2018; Fenton et al., 2013; Franklin et al., 2018; Montoro Girona et al., 2016, 2017, 2018a, 2019; Moussaoui et al., 2020; Pinzon et al., 2016; Thorpe & Thomas, 2007; Work et al., 2010). The results are

Table 1.1 Natural disturbance–based management (NDBM)/natural range of variation (NRV) targets for addressing sustainable forest management (SFM) issues identified at the end of the 1990s (Gauthier et al., 2009) for which a FEM framework would help achieve. Current implementation approaches for different regions of the circumboreal forest are also presented

NDBM/NRV targets to address the issues	SFM issues and potential effects	Application in management in different regions of the boreal forest			
		Eastern Canada (Québec)	Western Canada (Alberta)	Sweden–Finland	Russia
Maintain the age-class structure within the observed and past NRV	Decrease in mature and old forest	Age-class structure designed to maintain some (minimal) quantities of old forest (on the basis of NRV) and to avoid exceeding a certain level of young forest	Rotation length designed to match natural disturbance rates	No NRV-based targets; fully regulated forest age-class structure on the basis of clear-cut rotation	Maintain the age-class structure (even-aged vs. uneven-aged forest; all age classes) to approach a quasi-normal age-class distribution
	Increase in young forest		Targets for areas in different age classes within regions	Some continuous-cover management	
			Requirement to maintain some areas without human intervention		

(continued)

Table 1.1 (continued)

NDBM/NRV targets to address the issues	SFM issues and potential effects	Application in management in different regions of the boreal forest			
		Eastern Canada (Québec)	Western Canada (Alberta)	Sweden–Finland	Russia
Maintain forest composition within the past NRV (sometimes based on knowledge of fire cycles)	Increase of shade-intolerant species	Experimental use of mixed silviculture (SAFE, etc.)	Requirement to regenerate with species harvested	No NRV-based targets; deciduous tree species mixture is promoted in forest regeneration in all site types	Natural regeneration after clear-felling; the share of pioneer shade-intolerant species remains high
	Decrease of old-growth species	Some operational mixed-species plantations	Some efforts to regenerate to mixedwood stands and regulations to facilitate regeneration	Some continuous-cover management	To protect old-growth species, there are no forestry activities in reserves and clear-cutting is forbidden in protected forests
	Homogenization of stand composition	Adjustment of silviculture scenarios (education)	Alternative silviculture approaches to maintain mixedwoods (e.g., understory retention, underplanting)	Old-forest protection; preserving key biotopes and leaving retention trees on clear-cuts	Selection felling systems
		Extended rotations or partial cuts where long-lived species are present; silvicultural actions to maintain or ensure regeneration			Preserving key biotopes in all categories of forest
					Smaller sizes of harvested areas
					Mixed silvicultural systems; Assisting natural regeneration with planting under the tree canopy and/or after harvesting

(continued)

Table 1.1 (continued)

NDBM/NRV targets to address the issues	SFM issues and potential effects	Application in management in different regions of the boreal forest			
		Eastern Canada (Québec)	Western Canada (Alberta)	Sweden–Finland	Russia
Emulation of fire-size distribution (young forest) and distances among large forest tracts	Change in connectivity within and between open and closed stands	Some emulation of the size and shape of young forest and residual patches resulting from fire	Some emulation of the size and shape of patches resulting from wildfires	No NRV-based emulation of fire-size distribution	Spatial landscape planning Forest monitoring, improvement of disturbance management strategies (e.g., fire and large-scale insect outbreaks)
	Fragmentation of forest landscapes and loss of old-growth stands	Minimum amount of large forest tracts to be maintained (massifs)	Requirement to maintain areas of a certain size without human disturbance to support interior- and old-growth-dependent species (e.g., woodland caribou)	Landscape ecological planning in state and company lands; forest protection based on spatial optimization	Preserving key biotopes and ecotones (e.g., between forest and peatlands)
	Rate of cumulative disturbances exceeding the capacity of the stand/landscape	Some consideration of fire and insect disturbances in annual allowable cut computation	Some consideration of fire and insect disturbances in annual allowable cut computation	Cuttings mostly in line with sustainable yield limits; concerns about cumulative impacts on ecological sustainability (biodiversity)	Conservation of intact forest landscapes
			Consideration of the total disturbance footprint in an area (e.g., oil & gas and forestry) through the biodiversity management framework		Introduction of special forest management regimes in landscapes containing stands of old-growth forest

(continued)

Table 1.1 (continued)

NDBM/NRV targets to address the issues	SFM issues and potential effects	Application in management in different regions of the boreal forest			
		Eastern Canada (Québec)	Western Canada (Alberta)	Sweden–Finland	Russia
Emulation of stand structure on the basis of forest dynamics and disturbance regime severity	Decreases in uneven-aged forest and complex stand structure	Experimental partial cuttings	Requirement to leave some areas uncut (e.g., riparian buffers, key wildlife habitat) and leave a minimum amount of green-tree retention within harvest blocks	Leaving retention trees, preserving key biotopes, continuous-cover management	Retaining unharvested patches within harvested areas and on their edges
	Decreases in deadwood quantity and quality	Extended rotations	Retention harvesting; avoidance of harvesting in key wildlife habitat	Leaving small amounts of retention trees, protecting deadwood in harvesting operations, protecting key biotopes	Seed trees are, in most cases, left permanently
	Loss of ecological attributes specific to naturally disturbed stands due to suppression or salvage logging	Green-tree retention of various forms and amounts in some of clear-cuts	Unburned patches and some burned forest left during salvage logging; some insect-disturbed areas left unsalvaged	Some prescribed and restoration burnings, retention trees, protection of key biotopes	Almost all deadwood is left on harvested areas
		Guidelines to leave disturbed areas (1/3) of various severity untouched after any fire event			

(continued)

Table 1.1 (continued)

NDBM/NRV targets to address the issues	SFM issues and potential effects	Application in management in different regions of the boreal forest			
		Eastern Canada (Québec)	Western Canada (Alberta)	Sweden–Finland	Russia
Maintain site regenerative capacity and productivity	Poor regeneration or growth leading to low density, unproductive stands	Soil scarification on paludified sites	Replanting of conifers	Soil scarification, protection of undergrowth in harvesting	Preserving undergrowth during harvesting
		Some plantations or enrichment plantings	Regeneration surveys to quantify the growth of regenerated stands		Tending natural regeneration and complemented by planting, if not successful
			Standards for minimum density and spacing in regenerated stands		
	Decline in soil fertility	Obligation to leave branches and leaves on poor sites	Fire slash often left on site	Soil scarification, prescribed burning, maintenance of deciduous mixtures	All logging slash is left

slowly being applied to operational harvesting, forest management planning, and government policy (Jetté et al., 2013; Ontario Ministry of Natural Resources, 2001).

Despite the push for implementing a FEM framework, several elements of this paradigm remain unaddressed, and not all elements of the framework have been implemented (Table 1.1; Van Damme et al., 2014). In some Canadian jurisdictions, targets exist for maintaining a minimal proportion of forest older than a certain age, and some constraints have been produced related to the acceptable amount of young forest within various land units (Table 1.1; Alberta Sustainable Resource Development, 2006; Bergeron et al. 2017a, b; Bouchard et al., 2015; Jetté et al., 2013). Elsewhere, harvesting rotation cycles are designed to be aligned with the mean average fire return interval of the regional forest (DeLong, 2007).

Some Canadian jurisdictions have developed requirements to regenerate stands having the same composition as the original harvested forest. These requirements include efforts to regenerate mixedwood stands (Alberta; see Table 1.1). Retention harvesting (Box 1.1) is adopted increasingly to maintain stand structural heterogeneity, deadwood amounts, and key habitat features such as old, large trees. Maintaining forest productivity is approached through strict requirements for regenerating to sufficient density and monitoring to ensure early stand growth (Québec, Alberta; see Table 1.1). In some areas, there are considerations to maintain mixed stands, although true mixedwood management is uncommon (Chap. 15). In terms of spatial configuration, the shape and size of cutblocks have been modified in many instances to emulate the patterns created by natural fires (Ontario Ministry of Natural Resources, 2001). The conservation of key species is approached by conserving key habitats and maintaining some larger areas lacking human disturbance. Efforts are also undertaken to maintain the within-stand structure through partial cutting and tree retention (Table 1.1).

Although some FEM elements based on the NDBM/NRV approach have been applied, FEM has yet to be fully implemented. For example, despite both the importance of preserving old forests or forests with recognized old-growth attributes and the recorded increase in green-tree retention harvesting, forest management continues to operate predominantly under a single cohort, even-aged management system with low-retention clear-cut harvesting and short rotation cycles. This system tends to reduce the proportion of older forest stands while homogenizing the forest structure (Bergeron et al., 2006; Bouchard & Garet, 2014; Dhital et al., 2015). Stand-level considerations remain largely the focus of planning and management processes, and the focus continues to lie mostly on structures to a much greater degree than on processes. Moreover, although there is recognition of the importance of monitoring the effects of silviculture and management practices to determine whether the objectives for biodiversity and forest productivity have been achieved, this has only been partially fulfilled in operational landscapes (Chap. 14).

The consideration of First Nations values and rights in forest management is developing through various mechanisms in Canada. Co-management initiatives were launched through Canada's Model Forest program (1992–2007) (Bullock et al.,

2017). The program aimed to define and implement sustainable forest management at the local and operational scales through a collaborative exercise (Bullock et al., 2017). The program generated an important research effort in both the natural and social sciences (Bonnell, 2012) and led to some lasting partnerships; for example, the Prince Albert Model Forest, inaugurated in 1992, is co-managed by a group of stakeholders, including First Nations, federal and provincial agencies, research agencies, and industry (Bouman et al., 1996). Its success is attributed to the implication of First Nations at all levels of governance.

The signing of modern treaties and agreements between First Nations and levels of government provides another mechanism. The *James Bay and Northern Québec Agreement* was the first modern treaty in Canada (1975). The treaty led to *La Paix des Braves Agreement*, negotiated between the Grand Council of the Cree (*Eeyou Istchee*) and the Québec Government in 2002. The forestry chapter's spirit enhanced the importance of the Cree traditional lifestyle, sustainable development, and the consultation process within Eeyou Istchee, the land of the Cree. This treaty initiated the monitoring and regulation of timber harvesting at the trapline scale, per local land use and management. It also officialized the roles and responsibilities of the tallyman, often a family elder, as the trapline manager (Whiteman, 2004). Despite some successes, many challenges remain for considering First Nations values and rights in forest management. They include the conciliation of values and knowledge (Asselin, 2015), the consideration of Indigenous land use in forest planning and monitoring (Bélisle & Asselin, 2021; Bélisle et al., 2021; Saint-Arnaud et al., 2009), and the adaptation of governance structures for First Nations to be involved at all decision-making steps.

1.4.2   Sweden and Finland

In Sweden and Finland, the pathways toward FEM have differed from those of Canada. These differences between the chosen FEM approaches of both regions partly reflect conditions and restrictions determined by differences in forest-use histories and ownership structures. In Canada, boreal forests are primarily state-owned, and harvesting has, until now, involved mainly primary forests rented to forestry companies as long-term concessions; this organization facilitated the development of landscape-level coarse-filter management approaches. In Sweden and Finland, on the other hand, implementation was mainly fine-grained, reflecting the long history of intensive forest use, where pristine forests have largely disappeared, and most harvesting occurs within secondary or human-influenced—to varying degrees—forest. Moreover, the distribution of forest ownership among numerous private forest owners hampers the development of larger-scale approaches.

The first marked initiative was the introduction of the ASIO-model based on fire occurrense (Absent, Seldom, Infrequent, Often; Angelstam, 1998). This approach was based on the assumptions of natural fire regime effects on forest structure and dynamics (Angelstam, 1998; Kuuluvainen & Grenfell, 2012). Although influential as a pedagogical tool, the model's implementation in the field was only vaguely based on reference conditions. One problem was the lack of a proper understanding of natural fire ecology (Berglund & Kuuluvainen, 2021). Thus, instead of coarse-filter approaches, the focus mainly fell on biodiversity conservation by protecting ecologically valuable but relatively small-scale features, such as *woodland key habitats* (Timonen et al., 2010). Although the definition varies somewhat between countries, these are typically small—moist, fertile sites hosting high biodiversity and that are seldom naturally disturbed. Because they are small and sparsely located across the landscape, the ability of species to move between patches can be restricted; thus, the capacity of these patches to protect species populations from a metapopulation perspective has been questioned (Hanski, 2000).

Another approach to compensate for the adverse ecological impacts of clear-cut timber harvesting involves leaving retention trees during harvesting operations (Box 1.1; Gustafsson et al., 2012; Simonsson et al., 2015). However, the applied tree retention is typically low; for example, in Sweden–Finland, it is common to leave only a small number of trees (5–10 per ha) (Kuuluvainen et al., 2019). As the retention level strongly influences species responses, the low retention levels do not provide the habitat quality and continuity needed for declining and red-listed forest species, notably as many are dependent on old living trees and coarse woody debris. The accumulated research evidence suggests that current retention levels are too low to provide credible positive effects on biodiversity (Kim et al., 2021; Kuuluvainen et al., 2019).

Together, tree retention practices, protection of woodland key habitats, and conservation areas have been called the *hierarchical multiscale approach* to biodiversity conservation (Gustafsson & Perhans, 2010). However from the 1990s onward, the practices have been mainly fine-filter or *precision-conservation* approaches, which aim to protect valuable small-scale habitats and the associated biodiversity. In contrast, forest management has focused less on the large-scale ecosystem components, forest structures, and processes, i.e., the coarse-filter approach. Thus, actions related to biodiversity conservation are generally not part of any integrated ecosystem-based management framework but instead are implemented as separate measures on top of the intensive, business-as-usual even-aged management system (Kuuluvainen et al., 2019).

Research efforts to develop coarse-filter-inspired management based on natural disturbances have been put forward. An example is the DISTDYN project. This project involves an experimental setting specifically designed to emulate natural disturbance patterns in harvesting (Koivula et al., 2014). The focus is on large-scale (150–200 ha) blocks or "landscapes," each characterized by a different scale of harvesting units (from single tree selective cuts to clear-cutting) and retention level, derived from recent research on natural disturbance dynamics (Kuuluvainen & Aakala, 2011).

Despite the ongoing implementation of SFM strategies and practices, the managed forest landscapes in Sweden and Finland face considerable challenges. Biodiversity loss remains a serious concern, and habitat loss and fragmentation continue to drive the ecological degradation in boreal forests. In Sweden and Finland, the long history of intensive forest management for timber production has reduced habitat quality and connectivity. In Finland, for example, there are currently 816 endangered forest species (Hyvärinen et al., 2019), and the extinction debt of forest species because of forest management is estimated at around 1,000 species (Hyvärinen et al., 2019). This loss of biodiversity is likely to adversely affect the functioning of forest ecosystems (i.e., decomposition of organic matter, nutrient cycling, and carbon sequestration) and the capacity of forests to provide ecosystem services (Duffy, 2009). The main drivers of biodiversity decline are the loss of natural forest habitats, including those lost through wildfire (Bergeron & Fenton, 2012; Koivula & Vanha-Majamaa, 2020; Nordén et al., 2013). Growing concerns about biodiversity loss in Swedish–Finnish forests (Granström, 2001; Kouki et al., 2001; Hyvärinen et al. 2019) have heightened the importance of maintaining and even restoring biodiversity (Kuuluvainen, 2009). Although the last 20 years have been witness to several retention and restoration experiments (Halme et al., 2013; Koivula & Vanha-Majamaa, 2020; Vanha-Majamaa et al., 2007), the knowledge produced from these studies has yet to be implemented at a larger scale (Koivula & Vanha-Majamaa, 2020; Kuuluvainen et al., 2019).

1.4.3 Russia

Russia took a different path in implementing SFM because of the significant sociopolitical changes of the past 50 years. The Soviet period of forest management left a diverse legacy. On the one hand, the Soviet system produced a well-developed forest science and professional education structure. It established sound systems of forest inventory and management, forest regeneration, and protection against disturbances. Forests also had a relatively high political profile for some periods, such as during Stalin's plan of nature transformation (1948–1953) (Koldanov, 1992), and the Soviet system improved our understanding of the role of forests in a changing world. On the other hand, the Soviet political and economic system was incapable of generating a forest strategy able to address the challenges of a rapidly changing world. Political pressure, inappropriate forest statistics, misleading information about the availability of forest resources, and ignored regional natural and sociocultural variation in forest structure and functions hampered the development of state forest policy.

The dramatic political, social, and economic changes in Russia after the 1990s worsened the situation with the reforms introduced by the Forest Code published in 2006. Currently, forests in Russia are owned by the state and are leased to private forest companies. Forest management is regulated by the Forest Code of 2006—although many subsequent corrections have been made—and numerous federal and regional laws and regulations. The practice of forest leases does not, however, correspond to sustainable forest management principles. As a result, the governance and protection of forests have deteriorated significantly. Areas in which major silvicultural treatments have been implemented have decreased two to four times relative to areas in the 1990s (FAO, 2012; Petrov, 2013; Shvidenko et al., 2017; Shutov, 2006). In some jurisdictions, the amount of available timber resources has become depleted. There are currently intense debates on these issues within Russian industry, government, and academia.

Russia is a member of both the Montréal and pan-European processes on criteria and indicators for sustainable forest management. Most boreal forests used for wood production are certified according to national Forest Stewardship Council (FSC) standards (Elbakidze et al., 2011). Although some appropriate decisions have been made, none of the top-level decisions during the last three decades have been fulfilled completely.

All Russian forests are divided into protective, commercial (exploitable), and reserve forests. Protective forests are divided into four categories, each having different management regimes—from the complete prohibition of any harvest to varying levels of restriction—and aim to protect natural areas as well as water supply and quality through providing protective belts of forest along transport ways or in cities, forest parks, urban forests, and other valued forests, e.g., anti-erosion forests, forests growing in steppe, forest–tundra, and high mountains. Most of the forest estate lies within the commercial category. The forest inventory data estimates this area at approximately 40% of the total boreal forest area within the country. Diverse categories of protective forests comprise 26% of the total forest area. Reserve forests are practically unmanaged territories (around 210 million ha in 2010), as they are not planned to be harvested within at least the next 20 years.

Since 1978, in addition to the particular state-level protected areas, key biotopes (forests of 0.1–1,000 ha), which can occur in protective, commercial, and reserve forests, remain partly or entirely unmanaged; for example, habitats of rare species or old-growth forests are completely unmanaged. Clear-felling is forbidden in all types of critical biotopes. The key biotopes and elements preserved in NW Russia are similar to woodland critical habitats in NSF and the Baltic (Latvia, Estonia, Lithuania) countries (Timonen et al., 2010). The main types of key biotopes include (1) forest patches around peatlands, small lakes, and springs; uneven-aged forest patches; (2) gaps after windthrows; (3) regionally rare tree species; (4) old trees; (5) trees with bird nests and hollows; (6) snags; and (7) high stumps and large downed deadwood of different decay classes. Since 2001, biodiversity conservation has been actively incorporated into forest management per forest certification criteria (Chap. 21). In addition to the mandatory forest management restrictions within key biotopes, some nonmandatory protected key biotopes and key elements (retention forest patches and individual structures) with possible buffer zones around these protected areas are also left unharvested (Shorohova et al., 2019 and citations therein). Evidence related to the quantity of key biotopes and elements is scarce. One case study of ten FSC-certified forest companies demonstrated that the area of key biotopes inside clear-cut areas (data from 2005 to 2014) varied from 1 to 13% with a mean of 6%; therefore, most key biotopes are protected outside the areas planned for harvesting (Ilina & Rodionov, 2017).

The practice of leaving retention tree patches and critical elements in harvesting areas began with model forests in 2000 (Elbakidze et al., 2010; Romanyuk et al., 2001) and later became common in NW Russia. Since the 1990s, selective logging has become more common. After 2000, the share of selective harvest in NW Russia varied among regions, ranging from 2 to 58% with a mean of 22% (Federal Forestry Agency, 2013).

The growing decline in forest resources in European Russia and southern Siberia has brought into question the sustainability of harvests at the regional scale (Shvidenko & Nilsson, 1996). The annual allowable cut (AAC) assessment is based on the sustained yield model derived from the German classical school (Antanaitis et al., 1985; Sukhikh, 2006). The inconsistency of this approach has been demonstrated (Sheingauz, 2007), with one of the main critiques being the lack of integration of several important issues, such as the impact of natural disturbances, the uneven-aged nature of forest stands (Shvidenko & Nilsson, 1996), and regional variation in timber demand. There exists a means of accounting for these issues within AAC calculations (Sheingauz, 2007); however, this calculation has not been implemented in practice.

Multiple studies have shown that the officially established AAC (about 650–700 million $m^3 \cdot year^{-1}$ for all of Russia during the last decade) is about twice as high as the potential sustainable harvesting level should be, according to the SFM principles (Sokolov, 1997; Sukhikh, 2006). Therefore, the official information on the significant underutilization of AAC in Russia in recent decades must be cited with caution. Significant hidden overharvesting was typical for individual forest enterprises in northern European Russia, south-central Siberia, and the Russian Far East between 1950 and 1990 (Koldanov, 1992; Sheingauz, 2007).

Increasing wood production and a shift to intensive forest management (Karjalainen et al., 2009; Karvinen et al., 2011) have been much discussed over the last 30 years. Alternatively, adaptive management for maximizing resilience and the sustainability of forests under climate change has been recommended (Chap. 13; Chapin et al., 2007; Karpachevsky, 2007; Naumov et al., 2017; Nordberg et al., 2013). The concept promotes selective felling practices and preserving key biotopes and elements in parallel with research and monitoring of the results of their practical implementation. Its implementation, however, is affected by discrepancies between existing forestry regulations and sustainability (Karpachevsky, 2007; Kulikova et al., 2017; Sinkevich et al., 2018; Yanitskaya & Shmatkov, 2009). The diverse natural and socioeconomic conditions across the country and the variable legacies from past forestry activities should be considered in forest management planning (Lukina et al., 2015; Naumov et al., 2017; Shvarts, 2003; Shvidenko & Schepaschenko, 2011; Sinkevich et al., 2018).

1.5   Role and Need for a Restoration Framework

If the forest is heavily used and degraded, sustainable ecosystem management for multiple ecosystem values and services is not directly possible (see the definition of FEM, Box 1.1). This is the case in some southern boreal regions, especially in Fennoscandia, where forest use has been most intensive and long lasting (Berglund & Kuuluvainen, 2021; Kuuluvainen, 2009). In these cases, a lengthy restoration period may be required before FEM is possible (Fig. 1.2; Halme et al., 2013; Seymour,

2005). This long period occurs because forest landscapes show considerable inertia to changes in management, and there can be significant time delays in attaining favorable management status goals, depending on the level of restoration activities and the past use of the forest.

Finland and Sweden provide examples of a situation where restoration is needed before FEM becomes possible (Fig. 1.2; Chap. 18). Boreal forest management has been intensive in these regions and based on even-aged forest management and clear-cut harvesting. This practice, combined with short cutting rotations relative to natural disturbance cycles, has produced landscapes of young, structurally simplified forests that fall outside the NRV of the regional natural heterogeneous landscapes, which are characterized by old uneven-aged forests, big trees, abundant deadwood, and a relatively high structural variability (Kuuluvainen, 2009). Here, restoration using natural disturbance–based management is needed before FEM can be applied (Berglund & Kuuluvainen, 2021).

At present, restoration has been carried out in protected areas for habitat management purposes (Similä & Junninen, 2012). The first controlled burning for restoration purposes in Finland, and possibly anywhere in Europe, was conducted on a small, wooded island surrounded by peatland in Patvinsuo National Park in 1989. Twenty years later, the burned site is a hotspot for polypore fungi, hosting many red-listed species (Similä & Junninen, 2012). Experiences from such experiments can also be used for restoring managed forests (Vanha-Majamaa et al., 2007).

Although heavily exploited for a long time in its southern parts, the boreal zone still encompasses half of the world's unexploited forests (Burton et al., 2010). These large areas of relatively unmanaged boreal forest are found in Canada and Russia. Over the last 50 years, however, harvest operations have increased significantly in Canada, reaching the highest ratio of cutting globally by the end of the 1990s (Perrow & Davy, 2002). Consequently, Canadian restoration goals focus on protecting natural forests (passive restoration), restoring degraded areas related to mining, and applying sustainable forest management practices. Recently, some experiments to restore the natural forest structure have used commercial thinning operations to convert plantations from even-aged to irregular or uneven-aged stands (Schneider et al., 2021). Similarly, Thibeault et al. (submitted) also demonstrate that planting conifers to replace fallow lands not only maintains carbon sequestration capacity but also contributes to counteracting the decrease in native conifers observed since colonization in northern Québec (Marchais et al., 2020).

In Russia, there have been only a few studies on ecological restoration, with research focused on broadleaf forests (Korotkov, 2017), peatlands (Minayeva et al., 2017), and individual species (Baerselman, 2002). Green desertification, a form of degradation, has been observed in the northern bioclimatic zones of boreal Asian Russia (Yefremov & Shvidenko, 2004). Ongoing climate change has increased the area burned as well as fire frequency and severity (Shvidenko & Schepaschenko, 2013), which has led to the marked transformation of forest ecotopes. In harsh environmental conditions, e.g., on permafrost, in mountains, and within zonal ecotones, such burned areas cannot restore their productive potential and forest cover for decades or even centuries without human assistance. Similar regeneration failures have also been reported in Canada (Whitman et al., 2019) and are expected to increase in the future (Splawinski et al., 2019).

We are therefore in urgent need of effective methods for restoring forests impacted by intensive management or other human disturbances. Nonetheless, ecological restoration is far from a straightforward template-based model, especially considering the uncertainties caused by ongoing global change. These changes are likely to affect (directly and indirectly) terrestrial ecosystems, but restoration planners rarely account for such future impacts. Restoration ecology requires novel approaches and more interdisciplinary scientific collaboration to address these new challenges. Global change occurs at multiple scales, as do degradation and restoration; thus, it is necessary to consider species, processes, and interactions from the microhabitat to landscape scale to ensure efficacy and success in future management approaches. In the light of global change, the priority lies not only on conserving but also on restoring forest ecosystems, taking their resilience to global change into account (Chap. 17). Even if restoration represents a major challenge in boreal forests, the research effort in this field is limited relative to that in other biomes, e.g., tropical forests. We therefore need to apply ecosystem-based management strategies and implement effective practices to restore degraded forest systems if we want to safeguard forest biodiversity and ecosystem services (Chap. 25; Aronson & Alexander, 2013; Hof & Hjältén, 2018; Moen et al., 2014).

1.6   A New Context Challenging the FEM Paradigm

1.6.1   Climate Change in the Boreal Forest

Boreal forests are experiencing rapid climate change and increased pressure from resource extraction and land use. As the boreal biome is located at higher latitudes, it is particularly affected by the changing climate (Bush & Lemmen, 2019; IPCC, 2014; Price et al., 2013); for example, modified climate patterns are already affecting regional disturbance regimes (Hanes et al., 2019; Safranyik et al., 2010; Seidl et al., 2017). By the end of the twenty-first century, under the business-as-usual IPCC climate scenario (RCP8.5), the average temperature of the boreal biome is predicted to rise from −4.3 to 4.2 °C, with some regions attaining average increases of 10 °C (based on the data of Thrasher et al. (2012) with the CanESM2). In Russia, for example, under the RCP8.5 scenario, the average annual temperature is expected to increase from 6 to 9 °C by 2100 over much of the country (even higher in some regions), and uncertain, yet likely small, increases of the precipitation are predicted in continental Russia. Similarly, only a slight increase in total precipitation is projected during this period in other extensive areas of the boreal zones.

These changes are likely to be accompanied by changing disturbance regimes having a diversity of potential outcomes. In most regions where fire is an important disturbance agent, the number of fires and the annual area burned are expected to increase (Boulanger et al., 2014; IPCC, 2014). In Russia, for instance, recent evidence points to a new fire regime of greater area burned and an increased fire frequency and severity (Bartalev & Stytsenko, 2021; Bartalev et al., 2015, 2020), which has led to the destruction of forest resources of dozens of forest enterprises. Disturbances such as fire are already limiting commercial forestry in many boreal forest areas (Gauthier et al., 2015b), and forestry activities are expected to be even more limited as climate change–related disturbances increase (Boucher et al., 2018; Hof et al., 2021). Moreover, direct impacts of heat waves (e.g., central Russia in 2010, western Siberia in 2012, northern central Siberia in 2013) may substantially decrease forest productivity in Russian boreal forests because of higher temperatures and greater water stress (Bastos et al., 2014). Drought frequency is expected to rise, and the overall regional climate is projected to become dryer, resulting in potential effects on forest productivity (Girardin et al., 2016; Shvidenko et al., 2017; Tchebakova et al., 2009).

Although future climate change may be more conducive to insect outbreaks (e.g., Navarro et al., 2018b; Régnière et al., 2012; Safranyik et al., 2010)—allowing the insects to migrate north or east of their current range—it may also favor a lack of synchroneity with their hosts' phenologies (Pureswaran et al., 2015), thereby reducing their potential effect. However, recent work suggests that insects can evolve rapidly to synchronize with hosts (Bellemin-Noël et al., 2021; Pureswaran et al., 2019). Thus, invasive insects could produce outbreaks in regions where a cold climate previously prevented their colonization (Kharuk et al., 2019; Safranyik et al., 2010).

Moreover, although current human population densities in most boreal regions remain relatively low, land use and excessive natural resource exploitation add further stresses to the boreal biome (Gauthier et al., 2015a). Development-related air pollution represents another potential stressor (Bytnerowicz et al., 2007). Landscape fragmentation is increased through the cumulative effects of land-use activities, including forest harvesting, urbanization, transportation infrastructure, energy and mineral development (e.g., Chap. 19; Schneider et al., 2003). Market forces and global events also reduce or heighten the pressure on forest resources—the 2008 economic recession provided an example when global economic forces lowered harvesting levels in Canada. Such socioeconomic hazards and random elements may compound the climate change–related impacts by reducing the forest's adaptive capacity (Millar et al., 2007). These events also render the entire socioecological forest system even more unpredictable (Nocentini et al., 2017). All these effects have consequences on our ability to manage forests sustainably in the future.

1.6.2   Challenging the FEM Paradigm

As the extent of potential impacts of climate change on forests became increasingly evident by the early 2000s, the scientific community began to present some criticisms of FEM and propose alternative management approaches (Messier et al., 2019; Millar et al., 2007). A prominent critique of FEM relates to the relevance of using the past NRV as a management reference. The main questions centered on whether establishing baseline conditions from past conditions could create ecosystems ill adapted to rapidly evolving, non-analog future conditions (Millar et al., 2007).

Millar et al. (2007) identified three types of adaptive strategies to help forest ecosystems face future climate conditions: resistance, resilience, and transition. First, heightening forest resistance requires management strategies and practices that focus on maintaining or restoring forest conditions that are of high value to society. Such an example would be maintaining specific forest conditions to help preserve an endangered species or a high-value plantation. Second, bolstering forest resilience demands actions that ensure forests preserve their ability to return to the desired state. The return to the closed forest state after disturbance in areas where successive disturbances can cause regeneration failure is one crucial resilience aspect to focus on (Blatzer et al. 2021; Kuuluvainen & Gauthier, 2018; Splawinski et al., 2019). The third strategy involves helping ecosystems adapt to projected future conditions. One common example of such a strategy is related to assisted migration, where seedlings from populations adapted to future climatic conditions for the region are used in plantations or as seed sources (Chap. 30; Pedlar et al., 2012; Ste-Marie et al., 2011). Several frameworks, tools, and field guides have since been developed to help forest managers analyze the vulnerability of particular forest ecosystems to future change, and to prepare management plans and silviculture practices to address upcoming changes (Chap. 12; Edwards et al., 2015; Gauthier et al., 2014; Handler et al., 2020; Nagel et al., 2017; Swanston et al., 2016).

Aquatic environments are another neglected aspect of FEM. These water bodies contribute to the high complexity of boreal forests and are essential to forest functioning (Chap. 29). Aquatic environments provide essential resources for terrestrial species, such as irreplaceable habitats for the larval stages of multiple species and the export of essential fatty acids and nutrients toward terrestrial fauna and flora (Fritz et al., 2019; Martin-Creuzburg et al., 2017). Water-covered lands represent about 30% of the world's boreal forest area, ranking the boreal biome as one of the world's major sources of freshwater (Benoy et al., 2007). Terrestrial and aquatic environments are in constant interaction in the boreal landscape. Whereas most organic matter and energy fluxes are sourced from the forest and then transported to aquatic habitats by precipitation, freshet, and wind (Solomon et al., 2015; Tanentzap et al., 2017), freshwater to land fluxes are greater in terms of energy and nutritional quality (Gladyshev et al., 2019). Terrestrial organic matter traveling from land to aquatic environments is processed by aquatic food webs (Grosbois et al., 2020; McMeans et al., 2015) and returned to terrestrial environments via respiration (Lapierre et al., 2013) or animal movements, e.g., the emergence of aquatic insects, as *boomerang fluxes* (Scharnweber et al., 2014). Aquatic environments are therefore an integral part of boreal forest functioning at the landscape scale and contribute to the complexity of the boreal forest; thus, they are components that must be considered within any future FEM framework.

The recognition of forest ecosystems as complex adaptive systems has also become part of the conceptual sphere of forest management. This shift in thinking arose from the understanding that many feedback loops characterize forest ecosystems, each strongly influenced by their initial conditions, for which the outcomes have a relatively low level of predictability (Nocentini et al., 2017). This approach acknowledges the diversity of stand responses; therefore, silviculture implemented under this concept should not aim to homogenize forest stands but rather adapt to the stands themselves (Nocentini et al., 2017).

These approaches question the command-and-control idea used in traditional forestry, a practice that has simplified forest structure to render the system more fragile and vulnerable in the face of stressors such as pollution, climate change, and fragmentation (Messier et al., 2019; Millar et al., 2007; Nocentini et al., 2017). Moreover, the complex adaptive system framework stresses that the future is highly uncertain, and the entire system outcomes have low predictability (Chap. 28; Messier et al., 2019; Millar et al., 2007; Nocentini et al., 2017). Thus, a portfolio approach is required (Gauthier et al., 2014; Millar et al., 2007), i.e., the use of a diversity of solutions to address one particular challenge. An example of this approach would be using a mixture of provenances when replanting a post-disturbance area to ensure some trees will be successful under future conditions. This approach contrasts markedly with more deterministic and optimization strategies, which work best under a set of known conditions. Permanent outcome monitoring is considered a vital tool for selecting, controlling, and correcting forest management decisions. At first glance, these novel approaches proposed to adapt forests to future climate change may seem quite different in their respective philosophies from the original FEM concepts. Nonetheless, many of the principles of the FEM approaches remain essential and can be complemented by these novel approaches (Messier et al., 2019). Management based on the past natural range of variability will remain adequate in certain regions or for selected periods. For instance, in the boreal forest in northwestern Québec, projected burn rates remain within the natural range of variability of the past 8,000 years (Fig. 1.3). They thus can serve as a basis for management into the century. However, new situations could emerge that profoundly change natural ecosystems, notably in regions dominated by fire-adapted species (Baltzer et al., 2021).

This book examines the concepts of FEM in the context of global change. The chapters in this book also identify potential conceptual improvements and adjustments required to address the challenge of future global change and associated uncertainties. Therefore, this book aims to revise the principles of FEM to ensure managed forests remain resilient in the face of future changes. To achieve this goal, we build a new framework in collaboration with forest researchers studying all regions of the boreal biome and highlight new issues, challenges, and trends in forest management in a changing world. We also provide novel paradigms for the future of boreal forest management, including the need to consider social concerns (Chaps. 21 and 22), the interactions between forest and aquatic ecosystems (Chap. 29), the role of ecological restoration (Chaps. 17 and 18), the potential of new tools facing climate change (Chaps. 26 and 27), the complexity of forest ecosystems (Chap. 28), and the challenges and trends facing the future (Chap. 31).

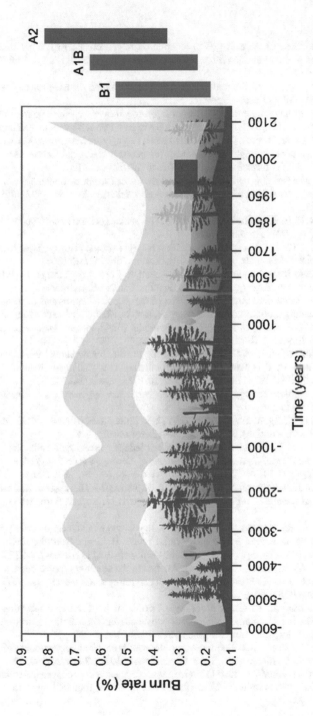

Fig. 1.3 Long-term variability of the burn rate for the boreal forest of northwestern Québec. The projected future burn rate. The *red box* on the figure represents the current burn rate. The projected future burn rates (*B1*, *A1B*, *A2*) are within the natural range of variability although at the upper margin, suggesting that FEM principles can apply in the region for the next century. Projections for 2100 of annual area burned were made using simulated monthly drought-code data collected from an ensemble of 19 global climate models and forcing experiments run against three IPCC scenarios (*B1*, *A1B*, *A2*). Modified with permission of CSIRO Publishing from Bergeron et al. (2010); permission conveyed through Copyright Clearance Center, Inc.

References

Aitken, S. N., Yeaman, S., Holliday, J. A., et al. (2008). Adaptation, migration or extirpation: Climate change outcomes for tree populations. *Evolutionary Applications, 1*(1), 95–111. https://doi.org/10.1111/j.1752-4571.2007.00013.x.

Aksenov, D., Karpachevskiy, M., Lloyd, S., et al. (1999). *The last of the last: the old-growth forests of boreal Europe.* Helsinki: Taiga Rescue Network.

Alberta Sustainable Resource Development. (2006). *Alberta forest management planning standard: version 4.1* (p. 114). Edmonton: Public Lands and Forests Division—Forest Management Branch.

Ameray, A., Bergeron, Y., Valeria, O., et al. (2021). Forest carbon management: A review of silvicultural practices and management strategies across boreal, temperate and tropical forests. *Current Forestry Reports, 7*(4), 245–266. https://doi.org/10.1007/s40725-021-00151-w.

Angelstam, P. K. (1998). Maintaining and restoring biodiversity in European boreal forests by developing natural disturbance regimes. *Journal of Vegetation Science, 9*(4), 593–602. https://doi.org/10.2307/3237275.

Antanaitis, V. V., Djaltuvas, R. P., & Mazheika, Y. F. (1985). *Forest management at soil-typological background* [in Russian] (p. 200). Agropromizdat.

Aronson, J., & Alexander, S. (2013). Ecosystem restoration is now a global priority: Time to roll up our sleeves. *Restoration Ecology, 21*, 293–296. https://doi.org/10.1111/rec.12011.

Asselin, H. (2015). Indigenous forest knowledge. In K. H.-S. Peh, R. Corlett, & Y. Bergeron (Eds.), *Routledge handbook of forest ecology* (pp. 586–596). London: Earthscan, Routledge.

Baerselman, F. (2002). The large herbivore initiative: An Eurasian conservation and restoration programme for a key species group in ecosystems (Europe, Russia, Central Asia and Mongolia). In B. Redecker, W. Härdtle, P. Finck, U. Riecken, & E. Schröder (Eds.), *Pasture landscapes and nature conservation* (pp. 303–312). Springer.

Baltzer, J. L., Day, N. J., Walker, X. J., et al. (2021). Increasing fire and the decline of fire adapted black spruce in the boreal forest. *Proceedings of the National Academy of Sciences of the United States of America, 118*(45), e2024872118. https://doi.org/10.1073/pnas.2024872118.

Baranov, N. I. (1954). *Incomplete clear felllings and their economic importance* [in Russian]. Central Forest Research Institute.

Bartalev, S. A., Stytsenko, F. V., Egorov, V. A., et al. (2015). Satellite-based estimate of the death of Russian forest from fire [in Russian]. *Forestry Science (Lesovedenie), 2*, 83–94.

Bartalev, S. A., & Stytsenko, F. V. (2021). Fire-induced forest lethality assessment using data on seasonal distribution of burnt area. *Contemporary Problems of Ecology, 14*(7), 711–716.

Bartalev, S., Shvidenko, A., & Held, A. (2020). Natural forest disturbances. In P. Leskinen, M. Lindner, P. J. Verkerk, G. J. Nabuurs, J. Van Brusselen, E. Kulikova, M. Hassegawa, B. Lerink (Eds.), *Russian forests and climate change. What science can tell us 11.* Sarjanr: European Forest Institute.

Bartels, S. F., Macdonald, S. E., Johnson, D., et al. (2018). Bryophyte abundance, diversity, and composition after retention harvest in boreal mixedwood forest: Evidence from the EMEND experiment. *Journal of Applied Ecology, 55*, 947–957. https://doi.org/10.1111/1365-2664.12999.

Bastos, A., Gouveia, C.M., Trigo, R.M., et al. (2014). Analysing the spatio-temporal impacts of the 2003 and 2010 extreme heatwaves on plant productivity in Europe. *Biogeosciences, 11*(13), 3421–3435. https://doi.org/10.5194/bg-11-3421-2014.

Beese, W. J., Deal, J., Dunsworth, B. G., et al. (2019). Two decades of variable retention in British Columbia: A review of its implementation and effectiveness for biodiversity conservation. *Ecological Processes, 8*, 33. https://doi.org/10.1186/s13717-019-0181-9.

Bélisle, A. C., & Asselin, H. (2021). A collaborative typology of boreal Indigenous landscapes. *Canadian Journal of Forest Research, 51*(8), 1–10. https://doi.org/10.1139/cjfr-2020-0369.

Bélisle, A. C., Wapachee, A., & Asselin, H. (2021). From landscape practices to ecosystem services: Landscape valuation in Indigenous contexts. *Ecological Economics, 179*, 106858. https://doi.org/10.1016/j.ecolecon.2020.106858.

Bell, F. W., Pitt, D. G., Irvine, M., et al. (2000). *Intensive forest management in Ontario: Intensive forest management science workshop summary* (p. 45). Sault Ste. Marie: Ontario Forest Research Institute.

Bellemin-Noël, B., Bourassa, S., Despland, E., et al. (2021). Improved performance of the eastern spruce budworm on black spruce as warming temperatures disrupt phenological defenses. *Global Change Biology, 27*, 3358–3366. https://doi.org/10.1111/gcb.15643.

Benoy, G., Cash, K., McCauley, E., et al. (2007). Carbon dynamics in lakes of the boreal forest under a changing climate. *Environmental Reviews, 15*, 175–189. https://doi.org/10.1139/a07-006.

Bergeron, Y., & Fenton, N. J. (2012). Boreal forests of eastern Canada revisited: Old growth, nonfire disturbances, forest succession, and biodiversity. *Botany, 90*(6), 509–523. https://doi.org/10.1139/b2012-034.

Bergeron, Y., Harvey, B., Leduc, A., et al. (1999). Forest management guidelines based on natural disturbance dynamics: Stand-and forest-level considerations. *The Forestry Chronicle, 75*(1), 49–54. https://doi.org/10.5558/tfc75049-1.

Bergeron, Y., Cyr, D., Drever, C. R., et al. (2006). Past, current and future fire frequency in Quebec's commercial forests: Implications for the cumulative effects of harvesting and fire on age-class structure and natural disturbance-based management. *Canadian Journal of Forest Research, 36*, 2737–2744. https://doi.org/10.1139/x06-177.

Bergeron, Y., Cyr, D., Girardin, M. P., et al. (2010). Will climate change drive 21st century rates in Canadian boreal forest outside of its natural variability: Collating global climate model experiments with sedimentary charcoal data. *International Journal of Wildland Fire, 19*, 1127–1139. https://doi.org/10.1071/WF09092.

Bergeron, J. A. C., Pinzon, J., Odsen, S., et al. (2017a). Ecosystem memory of wildfires affects multi-taxa biodiversity resilience in boreal mixedwood forest after retention harvest. *Oikos, 12*, 1738–1747. https://doi.org/10.1111/oik.04208.

Bergeron, Y., Vijayakumar, D. B. I. P., Ouzennou, H., et al. (2017b). Projections of future forest age-class structure under the influence of fire and harvesting: Implications for forest management in the boreal forest of eastern Canada. *Forestry, 90*(4), 485–495. https://doi.org/10.1093/forestry/cpx022.

Bergeron, Y., Leduc, A., Harvey, B., et al. (2002). Natural fire regime: A guide for sustainable management of the Canadian boreal forest. *Silva Fennica, 6*(1), 81–95. https://doi.org/10.14214/sf.553.

Berglund, H., & Kuuluvainen, T. (2021). Representative boreal forest habitats in northern Europe, and a revised model for ecosystem management and biodiversity conservation. *Ambio, 50*, 1003–1017. https://doi.org/10.1007/s13280-020-01444-3.

Berkes, F., & Davidson-Hunt, I. J. (2006). Biodiversity, traditional management systems, and cultural landscapes: Examples from the boreal forest of Canada. *International Social Science Journal, 58*(187), 35–47. https://doi.org/10.1111/j.1468-2451.2006.00605.x.

Bonnell, B. (2012). Trends in research and collaboration in the Canadian model forest network, 1993–2010. *The Forestry Chronicle, 88*(3), 274–282. https://doi.org/10.5558/tfc2012-054.

Bouchard, M., Boucher, Y., Belleau, A., et al. (2015). *Modélisation de la variabilité naturelle de la structure d'âge des forêts du Québec*. In Ministère des Forêts, de la Faune, et des Parcs, Direction de la recherche forestière (ed) Mémoire de recherche forestière. Québec: Gouvernement du Québec.

Bouchard, M., & Garet, J. (2014). A framework to optimize the restoration and retention of large mature forest tracts in managed boreal landscapes. *Ecological Applications, 24*(7), 1689–1704. https://doi.org/10.1890/13-1893.1.

Boucher, D., Boulanger, Y., Aubin, I., et al. (2018). Current and projected cumulative impacts of fire, drought, and insects on timber volumes across Canada. *Ecological Applications, 28*(5), 1245–1259. https://doi.org/10.1002/eap.1724.

Boulanger, Y., Gauthier, S., & Burton, P. J. (2014). A refinement of models projecting future Canadian fire regimes using homogeneous fire regime zones. *Canadian Journal of Forest Research, 44*(4), 365–376. https://doi.org/10.1139/cjfr-2013-0372.

Bouman, O. T., Langen, G., & Bouman, C. E. (1996). Sustainable use of the boreal Prince Albert model forest in Saskatchewan. *The Forestry Chronicle, 72*(1), 63–72. https://doi.org/10.5558/tfc 72063-1.

Bouthillier, L. (1998). *Brève histoire du régime forestier québécois*. Faculté de foresterie et de géomatique, Université Laval, Québec.

Bradshaw, C. J. A., & Warkentin, I. G. (2015). Global estimates of boreal forest carbon stocks and flux. *Global Planet Change, 128*, 24–30. https://doi.org/10.1016/j.gloplacha.2015.02.004.

Brais, S., Harvey, B. D., Bergeron, Y., et al. (2004). Testing forest ecosystem management in boreal mixedwoods of northwestern Quebec: Initial response of aspen stands to different levels of harvesting. *Canadian Journal of Forest Research, 34*(2), 431–446. https://doi.org/10.1139/ x03-144.

Bullock, R., Jastremski, K., & Reed, M. G. (2017). Canada's model forests 20 years on: Towards forest and community sustainability? *Natural Resources Forum, 41*(3), 156–166. https://doi.org/ 10.1111/1477-8947.12129.

Burton, P. J. (1995). The Mendelian compromise: A vision for equitable land use allocation. *Land Use Policy, 12*(1), 63–68.

Burton, P. J. (2013). Exploring complexity in boreal forests. In C. Messier, K. J. Puettmann, & K. D. Coates (Eds.), *Managing forests as complex adaptive systems: Building resilience to the challenge of global change* (pp. 79–109). Routledge.

Burton P. J., Bergeron Y., Bogdanski, B. E. C., et al. (2010). Sustainability of boreal forests and forestry in a changing environment. In G. Mery, P. Katila, G. Galloway, R. I. Alfaro, M. Kanninen, M. Lobovikov, J. Varjo (Eds.), *Forests and society—responding to global drivers of change* (pp. 249–282). International Union of Forest Research Organizations, IUFRO.

Bush, E., & Lemmen, D. S. (Eds.). (2019). *Canada's changing climate report* (p. 444). Ottawa: Government of Canada.

Bytnerowicz, A., Omasa, K., & Paoletti, E. (2007). Integrated effects of air pollution and climate change on forests: A northern hemisphere perspective. *Environmental Pollution, 147*(3), 438–445. https://doi.org/10.1016/j.envpol.2006.08.028

Canadian Council of Forest Ministers (CCFM). (1995). *Criteria and indicators of sustainable forest management in Canada* (p. 162). Ottawa: Canadian Forest Service.

Canadian Forest Service. (1998). *The state of Canada's forests 1997–1998: The people's forests.* Ottawa: Canadian Forest Service Headquarters, Policy, Planning and International Affairs Branch (Ed.) Natural Resources Canada, p. 108.

Chapin, F. S., Danell, K., Elmqvist, T., et al. (2007). Managing climate change impacts to enhance the resilience and sustainability of Fennoscandian forests. *Ambio, 36*(7), 528–533. https://doi. org/10.1579/0044-7447(2007)36[528:MCCITE]2.0.CO;2.

Christensen, N. L., Bartuska, A. M., Brown, J. H., et al. (1996). The report of the Ecological Society of America committee on the scientific basis for ecosystem management. *Ecological Applications, 6*, 665–691. https://doi.org/10.2307/2269460.

Cissel, J. H., Swanson, F. J., & Weisberg, P. J. (1999). Landscape management using historical fire regimes: Blue River, Oregon. *Ecological Applications, 9*(4), 1217–1231. https://doi.org/10.1890/ 1051-0761(1999)009[1217:LMUHFR]2.0.CO;2.

Convention on Biological Diversity (CBD). (1995). *Report on the second meeting of the conference of the parties to the convention on biological diversity*. United Nations Environment Programme, UNEP/CBD/COP/2/19. https://www.cbd.int/doc/meetings/cop/cop-02/official/cop-02-19-en.pdf

Convention on Biological Diversity (CBD). (2000). *Report on the fifth meeting of the conference of the parties to the convention on biological diversity*. United Nations Environment Programme, UNEP/CBD/COP/5/23. https://www.cbd.int/doc/meetings/cop/cop-05/official/cop-05-23-en.pdf

Crown-Indigenous Relations and Northern Affairs Canada. (2020). *The Numbered Treaties (1871–1921)*. Ottawa: Government of Canada.

Cyr, D., Gauthier, S., Bergeron, Y., et al. (2009). Forest management is driving the eastern North American boreal forest outside its natural range of variability. *Frontiers in Ecology Environment, 7*(10), 519–524. https://doi.org/10.1890/080088.

D'Amato, A. W., Palik, B. J., Franklin, J. F., et al. (2017). Exploring the origins of ecological forestry in North America. *Journal of Forestry, 115*(2), 126–127. https://doi.org/10.5849/jof.16-013.

De Grandpré, L., Waldron, K., Bouchard, M., et al. (2018). Incorporating insect and wind disturbances in a natural disturbance-based management framework for the boreal forest. *Forests, 9*, 471. https://doi.org/10.3390/f9080471.

DeLong, S. C. (2007). Implementation of natural disturbance-based management in northern British Columbia. *The Forestry Chronicle, 83*(3), 338–346. https://doi.org/10.5558/tfc83338-3.

Dhital, N., Raulier, F., Bernier, P. Y., et al. (2015). Adaptation potential of ecosystem-based management to climate change in the eastern Canadian boreal forest. *Journal of Environmental Planning Management, 58*(12), 2228–2249. https://doi.org/10.1080/09640568.2014.978079.

Drapeau, P., Leduc, A., Kneeshaw, D. D., et al. (2009). An adaptive framework for monitoring ecosystem management in the boreal black spruce forest. In S. Gauthier, M. A. Vaillancourt, A. Leduc, L. De Grandpre, D. D. Kneeshaw, H. Morin, P. Drapeau, & Y. Bergeron (Eds.), *Ecosystem management in the boreal forest* (pp. 343–372). Presses de l'Université du Québec.

Drushka, K. (2003). *Canada's forest: A history.* McGill-Queen's University Press.

Duchesne, L. C. (1994). Defining Canada's old-growth forests—problems and solutions. *The Forestry Chronicle, 70*(6), 739–744. https://doi.org/10.5558/tfc70739-6.

Duffy, J. E. (2009). Why biodiversity is important to the functioning of real-world ecosystems. *Frontiers in Ecology and the Environment, 7*(8), 437–444. https://doi.org/10.1890/070195.

Edwards, J. E., Pearce, C., Ogden, O. E., et al. (2015). *Climate change and sustainable forest management in Canada : A guidebook for assessing vulnerability and mainstreaming adaptation into decision making* (p. 161). Canadian Council of Forest Ministers, Climate Change Task Force.

Elbakidze, M., Angelstam, P. K., Sandström, C., et al. (2010). Multi-stakeholder collaboration in Russian and Swedish model forest initiatives: Adaptive governance toward sustainable forest management? *Ecology and Society, 15*(2), 14. https://doi.org/10.5751/ES-03334-150214.

Elbakidze, M., Angelstam, P., Andersson, K., et al. (2011). How does forest certification contribute to boreal biodiversity conservation? Standards and outcomes in Sweden and NW Russia. *Forest Ecology and Management, 262*(11), 1983–1995. https://doi.org/10.1016/j.foreco.2011.08.040.

Federal Forestry Agency. (2013). *Annual report on the status and use of forest resources in the Russian Federation for 2012* (p. 123). Moscow: Russian Federal Forestry Agency.

Fedorchuk, V., Neshataev, V., & Kuznetsova, M. (2005). *Forest ecosystems of the north-western regions of Russia: Typology, dynamics, management features* [in Russian] (p. 382). Saint Petersburg Polytechnical University Press.

Feit, H. A. (2001). Hunting, nature and metaphor: Political and discursive strategies in James Bay Cree resistance and autonomy. In J. A. Grim (Ed.), *Indigenous traditions and ecology: The interbeing of cosmology and community* (pp. 411–452). Harvard University Press.

Felton, A., Gustafsson, L., Roberge, J. M., et al. (2016). How climate change adaptation and mitigation strategies can threaten or enhance the biodiversity of production forests: Insights from Sweden. *Biological Conservation, 194*, 11–20. https://doi.org/10.1016/j.biocon.2015.11.030.

Fenton, N. J., Imbeau, L., Work, T., et al. (2013). Lessons learned from 12 years of ecological research on partial cuts in black spruce forests of northwestern Québec. *The Forestry Chronicle, 89*(3), 350–359. https://doi.org/10.5558/tfc2013-065.

Food and Agriculture Organization of the United Nations (FAO). (2012). *The Russian Federation forest sector: Outlook study to 2030.* Rome: Food and Agriculture Organization of the United Nations, p. 84.

Franklin, J. F. (1997). Ecosystem management: An overview. In M. S. Boyce & A. W. Haney (Eds.), *Ecosystem management: Applications for sustainable forest and wildlife resources* (pp. 21–53). Yale University Press.

Franklin, C. M. A., Macdonald, S. E., & Nielsen, S. E. (2018). Combining aggregated and dispersed tree retention harvesting for conservation of vascular plant communities. *Ecological Applications, 28*(7), 1830–1840. https://doi.org/10.1002/eap.1774.

Franklin, J. (1989). Toward a New Forestry. *American Forests, 95*, 1–8.

Fritz, K. A., Whiles, M. R., & Trushenski, J. T. (2019). Subsidies of long-chain polyunsaturated fatty acids from aquatic to terrestrial environments via amphibian emergence. *Freshwater Biology, 64*(5), 832–842. https://doi.org/10.1111/fwb.13266.

Galindo-Leal, C., & Bunnell, F. L. (1995). Ecosystem management: Implications and opportunities of a new paradigm. *The Forestry Chronicle, 71*, 601–606. https://doi.org/10.5558/tfc71601-5.

Gaudreau, G. (1998). *Les récoltes des forêts publiques au Québec et en Ontario: 1840–1900.* McGill-Queen's University Press.

Gauthier, S., Vaillancourt, M. A., Leduc, A., et al. (Eds.). (2009). *Ecosystem management in the boreal forest* (p. 572). Presses de l'Universite du Quebec.

Gauthier, S., Bernier, P., Burton, P. J., et al. (2014). Climate change vulnerability and adaptation in the managed Canadian boreal forest. *Environmental Reviews, 22*(3), 256–285. https://doi.org/10.1139/er-2013-0064.

Gauthier, S., Bernier, P. Y., Boulanger, Y., et al. (2015a). Vulnerability of timber supply to projected changes in fire regime in Canada's managed forests. *Canadian Journal of Forest Research, 45*, 1439–1447. https://doi.org/10.1139/cjfr-2015-0079.

Gauthier, S., Bernier, P., Kuuluvainen, T., et al. (2015b). Boreal forest health and global change. *Science, 349*, 819–822. https://doi.org/10.1126/science.aaa9092.

Girardin, M. P., Bouriaud, O., Hogg, E. H., et al. (2016). No growth stimulation of Canada's boreal forest under half-century of combined warming and CO_2 fertilization. *Proceedings of the National Academy of Sciences of the United States of America, 113*(52), E8406–E8414. https://doi.org/10.1073/pnas.1610156113.

Girardin, M.-P., Tardif, J. C., Flannigan, M. D., et al. (2006). Synoptic-scale atmospheric circulation and boreal Canada summer drought variability of the past three centuries. *Journal of Climate, 19*(10), 1922–1947. https://doi.org/10.1175/jcli3716.1.

Gladyshev, M. I., Gladysheva, E. E., & Sushchik, N. N. (2019). Preliminary estimation of the export of omega-3 polyunsaturated fatty acids from aquatic to terrestrial ecosystems in biomes via emergent insects. *Ecological Complexity, 38*, 140–145. https://doi.org/10.1016/j.ecocom.2019.03.007.

Granström, A. (2001). Fire management for biodiversity in the European boreal forest. *Scandinavian Journal of Forest Research, 16*, 62–69. https://doi.org/10.1080/028275801300090627.

Grondin, P., Gauthier, S., Poirier, V., et al. (2018). Have some landscapes in the eastern Canadian boreal forest moved beyond their natural range of variability? *Forest Ecosystems, 5*, 30. https://doi.org/10.1186/s40663-018-0148-9.

Grosbois, G., Vachon, D., Del Giorgio, P. A., et al. (2020). Efficiency of crustacean zooplankton in transferring allochthonous carbon in a boreal lake. *Ecology, 101*(6), e03013. https://doi.org/10.1002/ecy.3013.

Grumbine, R. E. (1994). What is ecosystem management? *Conservation Biology, 8*, 27–38. https://doi.org/10.1046/j.1523-1739.1994.08010027.x.

Gustafsson, L., & Perhans, K. (2010). Biodiversity conservation in Swedish forests: Ways forward for a 30-year-old multi-scaled approach. *Ambio, 39*(8), 546–554. https://doi.org/10.1007/s13280-010-0071-y.

Gustafsson, L., Baker, S. C., Bauhus, J., et al. (2012). Retention forestry to maintain multifunctional forests: A world perspective. *BioScience, 62*(7), 633–645. https://doi.org/10.1525/bio.2012.62.7.6.

Halme, P., Allen, K. A., Auniņš, A., et al. (2013). Challenges of ecological restoration: Lessons from forests in northern Europe. *Biological Conservation, 167*, 248–256. https://doi.org/10.1016/j.biocon.2013.08.029.

Handler, S., Marcinkowski, K., Janowiak, M., et al. (2020). *Climate change field guide for northern Wisconsin forests: Site-level considerations and adaptation* (p. 98). Houghton: US Department of Agriculture, Northern Forests Climate Hub Technical Report #3–2.

Hanes, C. C., Wang, X., Jain, P., et al. (2019). Fire-regime changes in Canada over the last half century. *Canadian Journal of Forest Research, 49*(3), 256–269. https://doi.org/10.1139/cjfr-2018-0293.

Hanski, I. (2000). Extinction debt and species credit in boreal forests: Modelling the consequences of different approaches to biodiversity conservation. *Annales Zoologici Fennici, 37,* 271–280.

Harvey, B. D., Nguyen-Xuan, T., Bergeron, Y., et al. (2003). Forest management planning based on natural disturbance and forest dynamics. In P. J. Burton, C. Messier, D. W. Smith, & W. L. Adamowicz (Eds.), *Towards sustainable management of the boreal forest* (pp. 395–432). NRC Research Press.

Harvey, B. D., Bergeron, Y., Leduc, A., et al. (2009). Forest ecosystem management in the boreal mixedwood forest of western Québec: An example from the Lake Duparquet forest. In S. Gauthier, M. A. Vaillancourt, A. Leduc, L. De Grandpré, D. D. Kneeshaw, H. Morin, P. Drapeau, & Y. Bergeron (Eds.), *Ecosystem management in the boreal forest* (pp. 449–478). Presses de l'Université du Québec.

Hof, A. R., Montoro Girona, M., Fortin, M. -J., et al. (2021). Editorial: Using landscape simulation models to help balance conflicting goals in changing forests. *Frontiers in Ecology and Evolution* 9. https://doi.org/10.3389/fevo.2021.795736.

Hof, A. R., & Hjältén, J. (2018). Are we restoring enough? Simulating impacts of restoration efforts on the suitability of forest landscapes for a locally critically endangered umbrella species. *Restoration Ecology, 26*(4), 740–750. https://doi.org/10.1111/rec.12628.

Hunter, M. (1993). Natural fire regimes as spatial models for managing boreal forests. *Biological Conservation, 65,* 115–120. https://doi.org/10.1016/0006-3207(93)90440-C.

Hyvärinen, E., Juslén, A., Kemppainen, E., et al. (2019). *Suomen lajien uhanalaisuus— Punainen kirja 2019/The 2019 Red List of Finnish Species.* Ympäristöministeriö and Suomen ympäristökeskus/Ministry of the Environment and Finnish Environment Institute.

Ilina, O., & Rodionov, A. O. (2017). The ways to preserve forest environment and mosaics of forest landscapes during timber harvesting [in Russian]. *LesPromInform, 128*(6).

Intergovernmental Panel on Climate Change (IPCC) (ed). (2014). *Climate Change 2014: Synthesis report. Contribution of Working Groups I, II and III to the fifth assessment report of the Intergovernmental Panel on Climate Change* (p. 151). Geneva: IPCC.

Isaev, A. S. (Ed.). (2012). *Diversity and dynamics of forest ecosystems in Russia,* [in Russian] (Vol. 1, p. 460). KMK Scientific Publishing.

Isaev, A. S. (Ed.). (2013). *Diversity and dynamics of forest ecosystems in Russia,* [in Russian] (Vol. 2, p. 460). KMK Scientific Publishing.

Jetté, J. P., Leblanc, M., Bouchard, M., et al. (2013). *Intégration des enjeux écologiques dans les plans d'aménagement forestier intégré, Partie II—Élaboration de solutions aux enjeux, Québec* (p. 159). Gouvernement du Québec, ministère des Ressources naturelles, Direction de l'aménagement et de l'environnement forestiers, Québec.

Johnson, E. A. (1992). *Fire and vegetation dynamics: Studies from the North American boreal forest.* Cambridge University Press.

Josefsson, T., Hörnberg, G., & Östlund, L. (2009). Long-term human impact and vegetation changes in a boreal forest reserve: Implications for the use of protected areas as ecological references. *Ecosystems, 12*(6), 1017–1036. https://doi.org/10.1007/s10021-009-9276-y.

Karjalainen, T., Leinonen, T., Gerasimov, Y., et al. (2009). Intensification of forest management and improvement of wood harvesting in Northwest Russia. *Working Papers of the Finnish Forest Research Institute* 110:151.

Karpachevsky, M. (2007). Legislative tools for biodiversity conservation during forest fellings. *Sustainable Forest Use, 13*(1), 18–23.

Karvinen, S., Välkky, E., Torniainen, T., et al. (2011). *Northwest Russian forest sector in a nutshell* (p. 138). Sastamala: Finnish Forest Research Institute.

Keane, R. E., Hessburg, P. F., Landres, P. B., et al. (2009). The use of historical range and variability (HRV) in landscape management. *Forest Ecology and Management, 258*, 1025–1037. https://doi.org/10.1016/j.foreco.2009.05.035.

Keto-Tokoi, P., & Kuuluvainen, T. (2014). *Primeval forests of Finland, cultural history, ecology and conservation* (p. 302). Helsinki: Maahenki.

Kharuk, V. I., Shushpanov, A. S., Petrov, I. A., et al. (2019). Fir (*Abies sibirica* Ledeb.) mortality in mountain forests of the Eastern Sayan Ridge, Siberia. *Contemporary Problems of Ecology, 12*(4), 299–309. https://doi.org/10.1134/S199542551904005X.

Kim, S., Axelsson, E. P., Girona, M. M., et al. (2021). Continuous-cover forestry maintains soil fungal communities in Norway spruce dominated boreal forests. *Forest Ecology and Management, 480*, 118659. https://doi.org/10.1016/j.foreco.2020.118659.

Kimmins, J. P. (2004). Emulating natural forest disturbances: What does this mean? In A. H. Perera, L. J. Buse, & M. G. Weber (Eds.), *Emulating natural forest landscape disturbances: Concepts and applications* (pp. 8–28). Columbia University Press.

Kneeshaw, D. D., Burton, P. J., De Grandpré, L., et al. (2018). Is management or conservation of old growth possible in North American boreal forests? In A. M. Barton & W. S. Keeton (Eds.), *Ecology and recovery of eastern old-growth forests* (pp. 139–157). Island Press.

Koivula, M., & Vanha-Majamaa, I. (2020). Experimental evidence on biodiversity impacts of variable retention forestry, prescribed burning, and deadwood manipulation in Fennoscandia. *Ecological Processes, 9*, 11. https://doi.org/10.1186/s13717-019-0209-1.

Koivula, M., Kuuluvainen, T., Hallman, E., et al. (2014). Forest management inspired by natural disturbance dynamics (DISTDYN)—a long-term research and development project in Finland. *Scandinavian Journal of Forest Research, 29*, 579–592. https://doi.org/10.1080/02827581.2014.938110.

Koldanov, V. Y. (1992). *Essays on the history of Soviet forest management* [in Russian] (p. 256). Ecology Publications.

Korotkov, V. N. (2017). Basic concepts and methods of restoration of natural forests in Eastern Europe. *Russian Journal of Ecosystem Ecology, 2*(1), 1–18.

Kouki, J., Löfman, S., Martikainen, P., et al. (2001). Forest fragmentation in Fennoscandia: Linking habitat requirements of wood-associated threatened species to landscape and habitat changes. *Scandinavian Journal of Forest Research, 16*, 27–37. https://doi.org/10.1080/028275801300090564.

Kozlovsky, B. A. (Eds.). (1959). Forest management during the years of Soviet power [in Russian]. In *Forest management during the years of Soviet power* (pp 3–48). Moscow: Lesproject, Ministry of Agriculture of the USSR.

Kozubov, G. M., & Taskaev, A. I. (Eds.). (2000). *Forestry and forest resources of the Komi Republic [in Russian].* Institute of Biology, Komi Science Centre, Ural Division (p. 512). Russian Academy of Sciences. Design Information.

Kulikova, E., Ivannikova, T., & Shmatkov, N. (2017). The conference "Sustainable forest use: Regulations, management, problems and solutions" [in Russian]. *Sustainable Forest Use, 49*(1), 2–14.

Kuuluvainen, T. (2009). Forest management and biodiversity conservation based on natural ecosystem dynamics in northern Europe: The complexity challenge. *Ambio, 38*, 309–315. https://doi.org/10.1579/08-A-490.1.

Kuuluvainen, T., & Gauthier, S. (2018). Young and old forest in the boreal: Critical stages of ecosystem dynamics and management under global change. *Forest Ecosystems, 5*(1), 26. https://doi.org/10.1186/s40663-018-0142-2.

Kuuluvainen, T., & Grenfell, R. (2012). Natural disturbance emulation in boreal forest ecosystem management: Theories, strategies and a comparison with conventional even-aged management. *Canadian Journal of Forest Research, 42*, 1185–1203. https://doi.org/10.1139/x2012-064.

Kuuluvainen, T., Lindberg, H., Vanha-Majamaa, I., et al. (2019). Low-level retention forestry, certification, and biodiversity: Case Finland. *Ecological Processes, 8*, 47. https://doi.org/10.1186/s13717-019-0198-0.

Kuuluvainen, T., & Aakala, T. (2011). Natural forest dynamics in boreal Fennoscandia: A review and classification. *Silva Fennica, 45*(5), 823–841. https://doi.org/10.14214/sf.73.

Kuuluvainen, T., & Siitonen, J. (2013). Fennoscandian boreal forests as complex adaptive systems. Properties, management challenges and opportunities. In C. Messier, K. J. Puettman, & K. D. Coates (Eds.), *Managing forests as complex adaptive systems. Building resilience to the challenge of global change* (pp. 244–268). London: Routledge, The Earthscan forest library.

Labrecque-Foy, J.-P., Morin, H., & Girona, M. M. (2020). Dynamics of territorial occupation by North American beavers in canadian boreal forests: A novel dendroecological approach. *Forests, 11*(2), 221. https://doi.org/10.3390/f11020221.

Landres, P. B., Morgan, P., & Swanson, F. J. (1999). Overview of the use of natural variability concepts in managing ecological systems. *Ecological Applications, 9*(4), 1179–1188.

Lapierre, J.-F., Guillemette, F., Berggren, M., et al. (2013). Increases in terrestrially derived carbon stimulate organic carbon processing and CO_2 emissions in boreal aquatic ecosystems. *Nature Communications, 4*(1), 2972. https://doi.org/10.1038/ncomms3972.

Lavoie, J., Montoro Girona, M., & Morin, H. (2019). Vulnerability of conifer regeneration to spruce budworm outbreaks in the eastern Canadian boreal forest. *Forests, 10*(10), 850. https://doi.org/10.3390/f10100850.

Lavoie, J., Montoro Girona, M., Grosbois, G., et al. (2021). Does the type of silvicultural practice influence spruce budworm defoliation of seedlings? *Ecosphere, 12*(4), 17. https://doi.org/10.1002/ecs2.3506.

Lewis, H. T., & Ferguson, T. A. (1988). Yards, corridors, and mosaics: How to burn a boreal forest. *Human Ecology Interdisciplinary Journal, 16*(1), 57–77. https://doi.org/10.1007/BF01262026.

Luckert, M. K., & Williamson, T. (2005). Should sustained yield be part of sustainable forest management? *Canadian Journal of Forest Research, 35*(2), 356–364. https://doi.org/10.1139/x04-172.

Lukina, N. V., Isaev, A. S., Kryshen, A. M., et al. (2015). Priorities in the development of forest science as a basis for sustainable forest management [in Russian]. *Russian Forest Science, 4*, 243–254.

Marchais, M., Arseneault, D., & Bergeron, Y. (2020). Composition changes in the boreal mixedwood forest of western Quebec since Euro-Canadian settlement. *Frontiers in Ecology and Evolution, 8*, 126. https://doi.org/10.3389/fevo.2020.00126.

Martin, M., Morin, H., & Fenton, N. J. (2019). Secondary disturbances of low and moderate severity drive the dynamics of eastern Canadian boreal old-growth forests. *Annals of Forest Science, 76*, 108. https://doi.org/10.1007/s13595-019-0891-2.

Martin, M., Montoro Girona, M., & Morin, H. (2020). Driving factors of conifer regeneration dynamics in eastern Canadian boreal old-growth forests. *PLoS ONE, 15*, e0230221. https://doi.org/10.1371/journal.pone.0230221.

Martin-Creuzburg, D., Kowarik, C., & Straile, D. (2017). Cross-ecosystem fluxes: Export of polyunsaturated fatty acids from aquatic to terrestrial ecosystems via emerging insects. *Science of the Total Environment, 577*, 174–182. https://doi.org/10.1016/j.scitotenv.2016.10.156.

McMeans, B. C., Koussoroplis, A.-M., Arts, M. T., et al. (2015). Terrestrial dissolved organic matter supports growth and reproduction of *Daphnia magna* when algae are limiting. *Journal of Plankton Research, 37*(6), 1201–1209. https://doi.org/10.1093/plankt/fbv083.

Melekhov, I. (1966). *Final fellings* [in Russian] (p. 374). Moscow: Forest Industry Publication.

Messier, C., Tittler, R., Kneeshaw, D. D., et al. (2009). TRIAD zoning in Quebec: Experiences and results after 5 years. *The Forestry Chronicle, 85*(6), 885–896. https://doi.org/10.5558/tfc85885-6.

Messier, C., Bauhus, J., Doyon, F., et al. (2019). The functional complex network approach to foster forest resilience to global changes. *Forest Ecosystems, 6*(1), 21. https://doi.org/10.1186/s40663-019-0166-2.

Millar, C. I., Stephenson, N. L., & Stephens, S. L. (2007). Climate change and forests of the future: Managing in the face of uncertainty. *Ecological Applications, 17*(8), 2145–2151. https://doi.org/10.1890/06-1715.1.

Minayeva, T. Y., Bragg, O. M., & Sirin, A. A. (2017). Towards ecosystem-based restoration of peatland biodiversity. *Mires and Peat, 19*(1), 1–36. https://doi.org/10.19189/MaP.2013.OMB.150.

Moen, J., Rist, L., Bishop, K., et al. (2014). Eye on the taiga: Removing global policy impediments to safeguard the boreal forest. *Conservation Letters, 7*(4), 408–418. https://doi.org/10.1111/conl.12098.

Montigny, M. K., & MacLean, D. A. (2006). Triad forest management: Scenario analysis of forest zoning effects on timber and non-timber values in New Brunswick, Canada. *The Forestry Chronicle, 82*, 496–511. https://doi.org/10.5558/tfc82496-4.

Montoro Girona, M., Morin, H., Lussier, J. M., et al. (2016). Radial growth response of black spruce stands ten years after experimental shelterwoods and seed-tree cuttings in boreal forest. *Forests, 7*, 240. https://doi.org/10.3390/f7100240.

Montoro Girona, M., Rossi, S., Lussier, J. M., et al. (2017). Understanding tree growth responses after partial cuttings: A new approach. *PLoS ONE, 12*(2), e0172653. https://doi.org/10.1371/journal.pone.0172653.

Montoro Girona, M., Lussier, J. M., Morin, H., et al. (2018a). Conifer regeneration after experimental shelterwood and seed-tree treatments in boreal forests: Finding silvicultural alternatives. *Frontiers in Plant Science, 9*, 1145. https://doi.org/10.3389/fpls.2018.01145.

Montoro Girona, M., Navarro, L., & Morin, H. (2018b). A secret hidden in the sediments: Lepidoptera scales. *Frontiers in Ecology and Evolution, 6*, 2. https://doi.org/10.3389/fevo.2018.00002.

Montoro Girona, M., Morin, H., Lussier, J.-M., et al. (2019). Post-cutting mortality following experimental silvicultural treatments in unmanaged boreal forest stands. *Frontiers in Forests and Global Change, 2*, 4. https://doi.org/10.3389/ffgc.2019.00004.

Montoro Girona, M. (2017). *À la recherche de l'aménagement durable en forêt boréale: croissance, mortalité et régénération des pessières noires soumises à différents systèmes sylvicoles.* Ph.D. thesis, Université du Québec à Chicoutimi, Chicoutimi.

Morin, H., Laprise, D., Simon, A. A., et al. (2009). Spruce budworm outbreak regimes in in eastern North America. In S. Gauthier, M. A. Vaillancourt, A. Leduc, L. De Grandpré, D. D. Kneeshaw, H. Morin, P. Drapeau, & Y. Bergeron (Eds.), *Ecosystem management in the boreal forest* (pp. 156–182). Les Presses de l'Université du Québec.

Morozov, G. F. (1924). *Forest doctrine* [in Russian] (p. 406). Gosizdat, Moscow.

Moussaoui, L., Leduc, A., Montoro Girona, M., et al. (2020). Success factors for experimental partial harvesting in unmanaged boreal forest: 10-year stand yield results. *Forests, 11*, 1199. https://doi.org/10.3390/f11111199.

Nagel, L. M., Palik, B. J., Battaglia, M. A., et al. (2017). Adaptive silviculture for climate change: A national experiment in manager-scientist partnerships to apply an adaptation framework. *Journal of Forestry, 115*(3), 167–178. https://doi.org/10.5849/jof.16-039.

Naumov, V., Angelstam, P., & Elbakidze, M. (2017). Satisfying rival forestry objectives in the Komi Republic: Effects of Russian zoning policy change on wood production and riparian forest conservation. *Canadian Journal of Forest Research, 47*, 1339–1349. https://doi.org/10.1139/cjfr-2016-0516.

Navarro, L., Harvey, A. É., Ali, A., et al. (2018a). A Holocene landscape dynamic multiproxy reconstruction: How do interactions between fire and insect outbreaks shape an ecosystem over long time scales? *PLoS ONE, 13*(10), e0204316. https://doi.org/10.1371/journal.pone.0204316.

Navarro, L., Morin, H., Bergeron, Y., et al. (2018b). Changes in spatiotemporal patterns of 20th century spruce budworm outbreaks in eastern Canadian boreal forests. *Frontiers in Plant Science, 9*, 1905. https://doi.org/10.3389/fpls.2018.01905.

Nitschke, C. R., Innes, J. L. (2005). The application of forest zoning as an alternative to multiple use forestry. In J. L. Innes, G. M. Hickey & H. F. Hoen (Eds.), *Forestry and environmental change: Socioeconomic and political dimensions*. Oxford: CABI.

Nocentini, S., Buttoud, G., Ciancio, O., et al. (2017). Managing forests in a changing world: The need for a systemic approach. A review. *Forest Systems, 26*(1), eR01. https://doi.org/10.5424/fs/2017261-09443.

Nordberg, M., Angelstam, P., Elbakidze, M., et al. (2013). From logging frontier towards sustainable forest management: Experiences from boreal regions of NorthWest Russia and North Sweden. *Scandinavian Journal of Forest Research, 28*(8), 797–810. https://doi.org/10.1080/02827581.2013.838993.

Nordén, J., Penttilä, R., Siitonen, J., et al. (2013). Specialist species of wood-inhabiting fungi struggle while generalists thrive in fragmented boreal forests. *Journal of Ecology, 101*(3), 701–712. https://doi.org/10.1111/1365-2745.12085.

Ontario Ministry of Natural Resources. (2001). *Forest management guide for natural disturbance pattern emulation, Version 3.1*. In Ontario Ministry of Natural Resources (ed) Toronto: Queen's Printer for Ontario, p. 40.

Orlov, M. M. (1927). Elements of forest practice. *Forest regulation, vol. 1* [in Russian]. Leningrad: Forestry, Forest Industry and Fuel, p. 428.

Orlov, M. M. (1928a). Elements of forest practice. *Forest regulation, vol. 2* [in Russian]. Leningrad: Forestry, Forest Industry and Fuel, p. 326.

Orlov, M. M. (1928b). Elements of forest practice. *Forest regulation, vol. 3* [in Russian]. Leningrad: Forestry, Forest Industry and Fuel, p. 348.

Östlund, L., & Norstedt, G. (2021). Preservation of the cultural legacy of the indigenous Sami in northern forest reserves—Present shortcomings and future possibilities. *Forest Ecology and Management, 502*, 119726. https://doi.org/10.1016/j.foreco.2021.119726.

Östlund, L., Zackrisson, O., & Axelsson, A. L. (1997). The history and transformation of a Scandinavian boreal forest landscape since the 19th century. *Canadian Journal of Forest Research, 27*(8), 1198–1206. https://doi.org/10.1139/x97-070.

Palik, B. J., D'Amato, A. W., Franklin, J. F., et al. (2020). *Ecological silviculture: Foundations and applications*. Waveland Press.

Pan, Y., Birdsey, R. A., Fang, J., et al. (2011). A large and persistent carbon sink in the world's forests. *Science, 333*(6045), 988–993. https://doi.org/10.1126/science.1201609.

Pedlar, J. H., McKenney, D. W., Aubin, I., et al. (2012). Placing forestry in the assisted migration debate. *BioScience, 62*(9), 835–842. https://doi.org/10.1525/bio.2012.62.9.10.

Perera, A. H., Buse, L. J., & Weber, M. G. (Eds.). (2004). *Emulating natural forest landscape disturbances: Concepts and applications* (p. 352). Columbia University Press.

Perrow, M. R., & Davy, A. J. (Eds.). (2002). *Handbook of ecological restoration* (p. 444). Cambridge University Press.

Petrov, A. P. (2013). Forest policy: Branch and regional priorities in the development of the forest sector. *Lesnoe Khozyaĭstvo [Forest Management], 2*, 7–10.

Pinzon, J., Spence, J. R., Langor, D. W., et al. (2016). Ten-year responses of ground-dwelling spiders to retention harvest in the boreal forest. *Ecological Applications, 26*, 2579–2597. https://doi.org/10.1002/eap.1387.

Price, D. T., Alfaro, R. I., Brown, K. J., et al. (2013). Anticipating the consequences of climate change for Canada's boreal forest ecosystems. *Environmental Reviews, 21*(4), 322–365. https://doi.org/10.1139/er-2013-0042.

Puettmann, K. J., Coates, K. D., & Messier, C. C. (2009). *A critique of silviculture: Managing for complexity*. Island Press.

Pureswaran, D. S., De Grandpré, L., Paré, D., et al. (2015). Climate-induced changes in host tree-insect phenology may drive ecological state-shift in boreal forests. *Ecology, 96*, 1480–1491. https://doi.org/10.1890/13-2366.1.

Pureswaran, D. S., Neau, M., Marchand, M., et al. (2019). Phenological synchrony between eastern spruce budworm and its host trees increases with warmer temperatures in the boreal forest. *Ecology and Evolution, 9*(1), 576–586. https://doi.org/10.1002/ece3.4779.

Redko, G. I. (1981). *The history of forestry in Russia* [in Russian]. Moskow State Forest University Publication.

Régnière, J., Powell, J., Bentz, B., et al. (2012). Effects of temperature on development, survival and reproduction of insects: Experimental design, data analysis and modeling. *Journal of Insect Physiology, 58*(5), 634–647. https://doi.org/10.1016/j.jinsphys.2012.01.010.

Romanyuk, B., Zagidullina, A., & Knize, A. (2001) *Planning forestry on a nature conservation basis* [in Russian]. World Wildlife Fund, Pskov Model Forest.

Ruel, J. C., Roy, V., Lussier, J. M., et al. (2007). Mise au point d'une sylviculture adaptée à la forêt boréale irrégulière. *The Forestry Chronicle, 83*(3), 367–374. https://doi.org/10.5558/tfc83367-3.

Safranyik, L. A. L., Carroll, A. L., Régnière, J., et al. (2010). Potential for range expansion of mountain pine beetle into the boreal forest of North America. *The Canadian Entomologist, 142*(5), 415–442. https://doi.org/10.4039/n08-CPA01.

Saint-Arnaud, M., Asselin, H., Dubé, C., et al. (2009). Developing criteria and indicators for Aboriginal forestry: Mutual learning through collaborative research. In M. G. Stevenson & D. C. Natcher (Eds.), *Changing the culture of forestry in Canada: Building effective institutions for Aboriginal engagement in sustainable forest management* (pp. 85–105). Canadian Circumpolar Institute Press.

Scharnweber, K., Vanni, M. J., Hilt, S., et al. (2014). Boomerang ecosystem fluxes: Organic carbon inputs from land to lakes are returned to terrestrial food webs via aquatic insects. *Oikos, 123*(12), 1439–1448. https://doi.org/10.1111/oik.01524

Schneider, R. R., Stelfox, J. B., Boutin, S., et al. (2003). Managing the cumulative impacts of land uses in the Western Canadian Sedimentary Basin: A modeling approach. *Conservation Ecology, 7*(1), 8. https://doi.org/10.5751/ES-00486-070108.

Schneider, R., Franceschini, T., Duchateau, E., et al. (2021). Influencing plantation stand structure through close-to-nature silviculture. *European Journal of Forest Research, 140*(3), 567–587. https://doi.org/10.1007/s10342-020-01349-6.

Scott, R. E., Neyland, M. G., & Baker, S. C. (2019). Variable retention in Tasmania, Australia: Trends over 16 years of monitoring and adaptive management. *Ecological Processes, 8*(1), 23. https://doi.org/10.1186/s13717-019-0174-8.

Seidl, R., Thom, D., Kautz, M., et al. (2017). Forest disturbances under climate change. *Nature Climate Change, 7*(6), 395–402. https://doi.org/10.1038/nclimate3303.

Seymour, R. S., & Hunter, M. L. (1992). *New forestry in eastern spruce-fir forests: Principles and applications to Maine.* Orono: University of Maine.

Seymour, R. S., & Hunter, M. L. (1999). Principles of ecological forestry. In M. L. Hunter (Ed.), *Maintaining Biodiversity in Forest Ecosystems* (pp. 22–62). Cambridge University Press.

Seymour, R. S. (2005). Integrating natural disturbance parameters into conventional silvicultural systems: Experience from the Acadian forest on northeastern North America. In C. E. Peterson & D. A. Maguire (Eds.), *Balancing ecosystem values: Innovative experiments for sustainable forestry* (pp. 41–48), General Technical Report 635. Portland: US Department of Agriculture, Forest Service Pacific Northwest Research Station.

Sharma, A., Bohn, K., Jose, S., et al. (2016). Even-aged vs. uneven-aged silviculture: Implications for multifunctional management of southern pine ecosystems. *Forestry, 7*:86. https://doi.org/10.3390/f7040086.

Sheingauz, A. S. (2007). Forest use—continues and even, or economically stipulated? [in Russian]. *Forest Inventory and Planning, 1*(37), 157–167.

Shorohova, E., Sinkevich, S., Kryshen, A., et al. (2019). Variable retention forestry in European boreal forests in Russia. *Ecological Processes, 8*, 34. https://doi.org/10.1186/s13717-019-0183-7.

Shorohova, E., Kneeshaw, D., Kuuluvainen, T., et al. (2011). Variability and dynamics of old-growth forests in the circumboreal zone: Implications for conservation, restoration and management. *Silva Fennica, 45*(5), 785–806. https://doi.org/10.14214/sf.72.

Shutov, I. V. (2006). *Degradation of forest management in Russia* (p. 97). Saint Petersburg: Saint Petersburg Forest Research Institute.

Shvarts, E. A. (2003). Forestry, economic development and biodiversity: Rejecting myths of the past [in Russian]. *Sustainable Forest Use, 2*, 2–7.

Shvidenko, A. Z., Schepaschenko, D. G., Kraxner, F., et al. (2017). Transition to sustainable forest management in Russia: Theoretical and methodological backgrounds [in Russian]. *Siberian Journal of Forest Science, 6*, 3–25. https://doi.org/10.15372/SJFS20170601.

Shvidenko, A., & Nilsson, S. (1996). Are Russian forests disappearing? *Unasilva, 1*(48), 57–64.

Shvidenko, A., & Schepaschenko, D. (2011). What do we know about Russian forests today? [in Russian]. *Forest Inventory and forest Planning, 1–2*(45–46), 153–172.

Shvidenko, A., & Schepaschenko, D. (2013). Climate change and wildfires in Russia. *Contemporary Problems of Ecology, 6*(7), 683–692. https://doi.org/10.1134/S199542551307010X.

Siiskonen, H. (2007). The conflict between traditional and scientific forest management in the 20th century Finland. *Forest Ecology and Management, 249*, 125–133. https://doi.org/10.1016/j.for eco.2007.03.018.

Siitonen, J. (2001). Forest management, coarse woody debris and saproxylic organisms: Fennoscandian boreal forests as an example. *Ecological Bulletins, 49*, 11–41.

Similä, M., & Junninen, K. (Eds.). (2012). *Ecological restoration and management—best practices from Finland* (p. 50). Metsähallitus Natural Heritage Services.

Simonsson, P., Gustafsson, L., & Östlund, L. (2015). Retention forestry in Sweden: Driving forces, debate and implementation 1968–2003. *Scandinavian Journal of Forest Research, 30*, 154–173. https://doi.org/10.1080/02827581.2014.968201.

Sinkevich, S. M., Sokolov, A. I., Ananycv, V. A., et al. (2018). On the regulatory framework for intensification of forestry [in Russian]. *Siberian Journal of Forest Science, 4*, 66–75.

Sokolov, V. A. (1997). *Basics of forest management in Siberia* [in Russian] (p. 308). Krasnoyarsk: Russian Academy of Sciences, Siberian Branch Publishing House.

Sokolov, A. I. (2006). *Forest regeneration of harvesting areas in northwestern Russia* [in Russian] (p. 215). Petrozavodsk: Siberian Branch of the Russian Academy of Sciences.

Solntsev, Z. Y. (1950). Cuttings and regeneration in the forests of the III category in the northern and northwestern regions of the European part of the USSR [in Russian]. In *Proceedings of the scientific conference on forestry in the Karelian Finnish Republic* (pp. 56–71).

Solomon, C. T., Jones, S. E., Weidel, B. C., et al. (2015). Ecosystem consequences of changing inputs of terrestrial dissolved organic matter to lakes: Current knowledge and future challenges. *Ecosystems, 18*, 376–389. https://doi.org/10.1007/s10021-015-9848-y.

Spence, J. R., Volney, W. J. A., Lieffers, V. J., et al. (1999). The Alberta EMEND project: recipe and cooks' argument In T. S. Veeman, D. W. Smith, B. G. Purdy, F. J. Salkie & G. A. Larkin (Eds.), *Science and practice: sustaining the boreal forest. Proceedings of the 1999 Sustainable Forest Management Network Conference, Sustainable Forest Management Network* (pp. 583–590). Edmonton: University of Alberta.

Splawinski, T. B., Cyr, D., Gauthier, S., et al. (2019). Analyzing risk of regeneration failure in the managed boreal forest of North-western Quebec. *Canadian Journal of Forest Research, 49*, 680–691. https://doi.org/10.1139/cjfr-2018-0278.

Stadt, K. J., Nunifu, T., & Aitkin, D. (2014). *Mean annual increment standards for crow forest management units*. Edmonton: Government of Alberta, Environment and Sustainable Resource Development, p.38.

Ste-Marie, C. A., Nelson, E. A., Dabros, A., et al. (2011). Assisted migration: Introduction to a multifaceted concept. *The Forestry Chronicle, 87*(6), 724–730. https://doi.org/10.5558/tfc201 1-089.

Sténs, A., Roberge, J. M., Löfmarck, E., et al. (2019). From ecological knowledge to conservation policy: A case study on green tree retention and continuous-cover forestry in Sweden. *Biodiversity and Conservation, 28*, 3547–3574. https://doi.org/10.1007/s10531-019-01836-2.

Stockdale, C., Flannigan, M., & Macdonald, S. E. (2016). Is the END (emulation of natural disturbance) a new beginning? A critical analysis of the use of fire regimes as the basis of forest ecosystem management with examples from the Canadian western Cordillera. *Environmental Reviews, 24*(3), 233–243. https://doi.org/10.1139/er-2016-0002.

Sukhikh, V. I. (2006). On improving the methodology of estimating the size of main felling in forests [in Russian]. *Lesnoe Khozyaĭstvo [Forest Management], 6*, 30–35.

Swanston, C. W., Janowiak, M. K., Brandt, L. A., et al. (2016). *Forest adaptation resources: climate change tools and approaches for land managers* (General Technical Report. NRS-GTR-87–2). Newtown Square: US Department of Agriculture, Forest Service, Northern Research Station, p. 161.

Swetnam, T. W., Allen, C. D., & Betancourt, J. L. (1999). Applied historical ecology: Using the past to manage for the future. *Ecological Applications, 9*(4), 1189–1206. https://doi.org/10.1890/1051-0761(1999)009[1189:AHEUTP]2.0.CO;2.

Tanentzap, A. J., Kielstra, B. W., Wilkinson, G. M., et al. (2017). Terrestrial support of lake food webs: Synthesis reveals controls over cross-ecosystem resource use. *Science Advances, 3*(3), e1601765. https://doi.org/10.1126/sciadv.1601765.

Tchebakova, N. M., Parfenova, E. I., & Soja, A. J. (2009). The effects of climate, permafrost and fire on vegetation change in Siberia in a changing climate. *Environmental Research Letters, 4*(4), 045013. https://doi.org/10.1088/1748-9326/4/4/045013.

Thorpe, H. C., & Thomas, S. C. (2007). Partial harvesting in the Canadian boreal: Success will depend on stand dynamic responses. *The Forestry Chronicle, 83*, 319–325. https://doi.org/10.5558/tfc83319-3.

Thrasher, B., Maurer, E. P., McKellar, C., et al. (2012). Technical Note: Bias correcting climate model simulated daily temperature extremes with quantile mapping. *Hydrology and Earth System Sciences, 16*(9), 3309–3314. https://doi.org/10.5194/hess-16-3309-2012.

Timonen, J., Siitonen, J., Gustafsson, L., et al. (2010). Woodland key habitats in northern Europe: Concepts, inventory and protection. *Scandinavian Journal of Forest Research, 25*, 309–324. https://doi.org/10.1080/02827581.2010.497160.

Vaillancourt, M. A., De Grandpré, L., Gauthier, S., et al. (2009). How can natural disturbances be a guide for forest ecosystem management? In S. Gauthier, M. A. Vaillancourt, A. Leduc, L. De Grandpré, D. D. Kneeshaw, H. Morin, P. Drapeau, & Y. Bergeron (Eds.), *Ecosystem management in the boreal forest* (pp. 39–56). Presses de l'Université du Québec.

Van Damme, L., Burkhardt, R., Plante, L., et al. (2014). *Status report on ecosystem-based management (EBM): Policy barriers and opportunities for EBM in Canada*. Prepared for the Canadian Boreal Forest Agreement. KBM Resources Group.

Vanha-Majamaa, I., Lilja, S., Ryömä, R., et al. (2007). Rehabilitating boreal forest structure and species composition in Finland through logging, dead wood creation and fire: The EVO experiment. *Forest Ecology and Management, 250*(1–2), 77–88. https://doi.org/10.1016/j.foreco.2007.03.012.

Whiteman, G. (2004). The impact of economic development in James Bay, Canada: The Cree tallymen speak out. *Organization & Environment, 17*(4), 425–448. https://doi.org/10.1177/1086026604270636.

Whitman, E., Parisien, M. A., Thompson, D. K., et al. (2019). Short-interval wildfire and drought overwhelm boreal forest resilience. *Science and Reports, 9*(1), 18796. https://doi.org/10.1038/s41598-019-55036-7.

Wilkie, M. L., Holmgren, P., Castañeda, F. (2003). *Sustainable forest management and the ecosystem approach: Two concepts, one goal*. In Forest Management Working Papers (Working Paper FM 25). Rome: Forest Resources Development Service, Forest Resources Division, Food and Agriculture Organization of the United Nations.

Work, T. T., Jacobs, J. M., Spence, J. R., et al. (2010). High levels of green-tree retention are required to preserve ground beetle biodiversity in boreal mixedwood forests. *Ecological Applications, 20*, 741–751. https://doi.org/10.1890/08-1463.1.

Yanitskaya, T., & Shmatkov, N. (2009). Joint opinion of public environmental organisations and Russian forest business on the improvement of law related to sustainable forest management. *Sustainable Forest Use, 3*(22), 42–44.

Yefremov, D., & Shvidenko, A. (2004). Long-term impacts of catastrophic forest fires in Russia's Far East and their contribution to global processes. *International Forest Fire News, 32*, 43–49.

Part II
Natural Disturbances

Chapter 2
Millennial-Scale Disturbance History of the Boreal Zone

Tuomas Aakala, Cécile C. Remy, Dominique Arseneault, Hubert Morin, Martin P. Girardin, Fabio Gennaretti, Lionel Navarro, Niina Kuosmanen, Adam A. Ali, Étienne Boucher, Normunds Stivrins, Heikki Seppä, Yves Bergeron, and Miguel Montoro Girona

T. Aakala
School of Forest Sciences, University of Eastern Finland, P.O. Box 111, FI-80101 Joensuu, Finland
e-mail: tuomas.aakala@uef.fi

F. Gennaretti · M. M. Girona (✉)
Groupe de Recherche en Écologie de la MRC-Abitibi, Forest Research Institute, Université du Québec en Abitibi-Témiscamingue, Amos Campus, 341, rue Principale Nord, Amos, QC J9T 2L8, Canada
e-mail: miguel.montoro@uqat.ca

F. Gennaretti
e-mail: fabio.gennaretti@uqat.ca

C. C. Remy
Institute of Geography, Augsburg University, Alter Postweg 118, 86159 Augsburg, Germany
e-mail: cecile.remy@geo.uni-augsburg.de

D. Arseneault
Centre d'études nordiques, Centre d'étude de la forêt, Département de biologie, chimie et géographie, Université du Québec à Rimouski, 300, allée des Ursulines, Rimouski, QC G5L 3A1, Canada
e-mail: Dominique_Arseneault@uqar.ca

H. Morin · L. Navarro
Département des Sciences Fondamentales, Université du Québec à Chicoutimi, 555 boul. de l'université, Chicoutimi, QC G7H 2B1, Canada
e-mail: hubert_morin@uqac.ca

L. Navarro
e-mail: lionel.navarro@uqac.ca

M. P. Girardin
Natural Resources Canada, Canadian Forest Service, Laurentian Forestry Centre, 1055 rue du PEPS, P.O. Box 10380, Stn. Sainte-Foy, Québec, QC G1V 4C7, Canada
e-mail: martin.girardin@canada.ca

D. Arseneault · Y. Bergeron · M. M. Girona
Centre for Forest Research, Université du Québec à Montréal, P.O. Box 8888, Stn. Centre-Ville, Montréal, QC H3C 3P8, Canada

© The Author(s) 2023
M. M. Girona et al. (eds.), *Boreal Forests in the Face of Climate Change*,
Advances in Global Change Research 74,
https://doi.org/10.1007/978-3-031-15988-6_2

Abstract Long-term disturbance histories, reconstructed using diverse paleoecological tools, provide high-quality information about pre-observational periods. These data offer a portrait of past environmental variability for understanding the long-term patterns in climate and disturbance regimes and the forest ecosystem response to these changes. Paleoenvironmental records also provide a longer-term context against which current anthropogenic-related environmental changes can be evaluated. Records of the long-term interactions between disturbances, vegetation, and climate help guide forest management practices that aim to mirror "natural" disturbance regimes. In this chapter, we outline how paleoecologists obtain these long-term data sets and extract paleoenvironmental information from a range of sources. We demonstrate how the reconstruction of key disturbances in the boreal forest, such as fire and insect outbreaks, provides critical long-term views of disturbance-climate-vegetation interactions. Recent developments of novel proxies are highlighted to illustrate advances in reconstructing millennial-scale disturbance-related dynamics and how this new information benefits the sustainable management of boreal forests in a rapidly changing climate.

N. Kuosmanen · H. Seppä
Department of Geosciences and Geography, University of Helsinki, P.O. Box 64 (Gustaf Hällströmin katu 2), 00014 Helsinki, Finland
e-mail: niina.kuosmanen@helsinki.fi

H. Seppä
e-mail: heikki.seppa@helsinki.fi

A. A. Ali
Institut des Sciences de l'Évolution Montpellier, UMR 5554 CNRS-IRD-Université Montpellier-EPHE, Montpellier, France
e-mail: ahmed-adam.ali@umontpellier.fr

É. Boucher
Département de géographie, GEOTOP-UQAM, Université du Québec à Montréal, P.O. Box 8888, Stn. Centre-Ville, Montréal, QC H3C 3P8, Canada
e-mail: boucher.etienne@uqam.ca

N. Stivrins
Faculty of Geography, University of Latvia, Rīga 1004, Latvia
e-mail: normunds.stivrins@gmail.com

Department of Geology, Tallinn University of Technology, Ehitajate tee 5, 19086 Tallinn, Estonia

Y. Bergeron
Forest Research Institute, Université du Québec en Abitibi-Témiscamingue, Rouyn-Noranda, QC J9X 5E4, Canada
e-mail: yves.bergeron@uqat.ca

Département des Sciences Biologiques, Université du Québec à Montréal, P.O. Box 8888, Stn. Centre-Ville, Montréal, QC H3C 3P8, Canada

M. M. Girona
Department of Wildlife, Fish, and Environmental Studies, Swedish University of Agricultural Sciences (SLU), 901 83 Umeå, Sweden

2.1 Introduction

Understanding the complex interactions between abiotic and biotic factors and the impact of these factors on the structure of forest communities across space and time is crucial for emulating natural disturbance regimes in sustainable forest management strategies. Disentangling past relationships between biotic and abiotic factors has historically been challenging. Paleoecological and dendroecological approaches serve as the primary means of reconstructing past dynamics, disturbance regimes, and the biotic and abiotic interactions within boreal ecosystems. Tree rings and the preserved accumulations of peat and lake sediments are the main archives that record past environmental conditions within the boreal region. Tree-ring properties and the preserved accumulations of fossil pollen, charcoal, lepidopteran scales, and spores within peat and sediment records serve as proxies of past environmental conditions. Careful interpretation of these proxy tools and their interactions then provides insight into long-term, i.e., the Holocene, patterns of climate, vegetation, and disturbance regimes. All paleoecological approaches and their proxy tools hold intrinsic advantages and disadvantages; combined, however, they offer a powerful tool for building our understanding of boreal ecosystem functioning.

This long term perspective holds two major advantages. First, rare disturbances or those having a long return interval—relative to the human lifespan and the existing observational record—require a longer reference period to record their occurrence and importance. Second, we are living in a critical, "non-analog" moment in terms of ecology and climate change; therefore, longer time frames offer the possibility of indirectly observing a wider range of climatic conditions and the related response of vegetation and disturbance regimes. Paleoenvironmental data can guide projections of how changing environmental conditions will affect future forest ecology and disturbance regimes.

From the perspective of sustainable forest management, silvicultural interventions can be placed within the same framework as disturbances (see Chap. 1). The consequences of silviculture on forest structures at various spatial scales depend on the characteristics of the given intervention (see Chaps. 13, 16). If we consider that species have adapted to these natural conditions, understanding how past forest structure and composition have responded to specific disturbances can provide insight into how forest management could be improved to maintain those structural and compositional characteristics necessary for preserving biodiversity. In the boreal forest, fire and insect outbreaks, because of their frequency and potential severity, are the major determinants of boreal forest dynamics. Paleoecological methods able to reconstruct this pair of disturbances are now well established and continue to be refined. In this chapter, we provide an overview of the paleoecological approaches able to decipher past records of fire and insect disturbance. This chapter synthesizes the current state of knowledge related to the long-term records of insect and fire disturbances in the boreal forest. We illustrate the potential of new proxies and demonstrate the importance of millennial-scale reconstructions of disturbances for improving our understanding of current and future forest dynamics. Finally, we explain how this knowledge has implications for forest management in the context of future climate change.

2.2   Fire History Reconstruction

Fire is a major disturbance agent in the boreal forest, and future climate warming is projected to increase its frequency and severity in many parts of this biome. Interactions between climate, fire, vegetation, and, in particular, the forest response to changing fire regimes are difficult to predict because of the long timescales associated with these changes. Thus, reconstructing a regional fire—through documentary, observational, and remote-sensing data—becomes essential for extending time series. Fire histories involve the analysis of fire regime characteristics, i.e., fire occurrence, frequency, areal extent, and severity, over the long term. These fire histories also provide a context within which we can evaluate current fire observations. Given the complex interactions between climate, fire, vegetation, and humans, there is increasing recognition by ecologists, restoration planners, and forest managers of the value of the long-term perspectives provided by paleofire records. Understanding the causes and consequences of fire provides a more solid foundation for developing appropriate management guidelines, mitigating the loss of forest ecosystem services, and improving predictions of future fire activity in a changing climate (Waito et al., 2018).

Climate conditions and vegetation characteristics control fires in boreal forests (Girardin & Terrier, 2015; Krawchuk & Cumming, 2011). In the boreal forest, for example, vegetation flammability and fire propagation rates are higher in needleleaf forest stands than in broadleaf forest stands. Needleleaf forest species produce flammable resins and have a lower leaf moisture content. Human-ignited fires have also strongly influenced vegetation dynamics in these forests over thousands of years; this human influence has shaped the current vegetation and fire activity in the boreal zone (Waito et al., 2018). Moreover, active fire suppression policies in populated regions of boreal Canada during the mid to late twentieth century decreased fire activity, leading to accumulations of forest fuel and a higher risk of future catastrophic fires (Parisien et al., 2020).

2.2.1   Studying Fire Histories at Millennial Time Scales

Fire histories are reconstructed using proxies from two main archives: (1) tree ring−based methods, which rely on fire-induced damage in trees and the age of new (even-aged) postfire forest stands; and (2) fire-related charcoal particles deposited onto−and subsequently buried within−soil, peat, or lake sediments.

2.2.1.1   Tree Rings

In general, forest fire reconstructions using tree rings rely on two primary approaches (Niklasson & Granström, 2000), namely using tree rings to date fire scars and examining the age structure of forest stands. Dating fire scars assesses low-intensity fires, which damage the tree cambium without killing the tree. This damage to the cambium leaves a distinct scar; the timing of the related fire event is then determined from the location of the fire scar in the sequence of annual tree rings (Fig. 2.1). When a fire occurs during the growing season, the scar's location within the annual ring can even be used to date the event at a subannual temporal resolution, distinguishing, for instance, early−, late−, and dormant-season fires. An individual tree can hold numerous fire scars and thus record the geographic location and timing of multiple fires. Samples used for dating fire scars are commonly (and preferably) the cross-section of tree stems; however, where possibilities for sampling are limited, such as in strictly protected forests, increment cores extracted from the stem can be used.

Stand initiation dates based on tree rings provide another means of dating forest fires. This approach relies on the premise that a fire event leads to a pulse of regenerating trees. These pulses can often be observed after surface fires in those stands recording fire scars; however, they are particularly useful for dating high-intensity fires in which no trees survive to preserve fire scars. Aging the cohort of postfire regeneration then gives the approximate year of the most recent high-intensity fire.

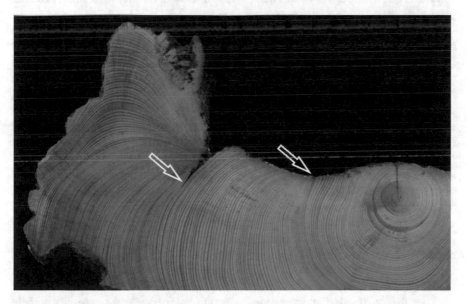

Fig. 2.1 A partial cross-section extracted from a fire-scarred Scots pine (*Pinus sylvestris* L.) collected from northeastern Finland. The *arrows* indicate the locations of fire scars dated dendrochronologically at 1296 and 1227 CE. *Photo credit* Tuomas Aakala

2.2.1.2 Charcoal in Forest Soils

The temporal extent of fire records using stand initiation dates is limited to the most recent fire. This is particularly limiting in locations where the regional fire regime often involves stand-replacing fires. In these conditions, information related to past recurring fires at a given locality can be gained from studying charcoal deposited in forest soils. For this approach, samples of organic matter and mineral soil are collected (Payette et al., 2012). Charcoal fragments greater or equal to 2 mm in diameter are assumed to have been produced in situ; they thus represent local fires (Asselin & Payette, 2005). The fire year is then determined by the radiocarbon dating of a selected number of randomly selected charcoal pieces. Although the temporal resolution of this soil charcoal–based method is rather coarse, it may greatly extend the temporal scale of fire histories initially developed using tree ring–based reconstructions.

2.2.1.3 Charcoal in Lake Sediments and Peat

Lake sediments are natural "hard drives" that record the environmental conditions and events affecting the surrounding landscape over time (Dodd & Stanton, 1990). The stored information in this ecological hard drive must be interpreted using proxy indicators found within the sediment record (Bigler & Hall, 2002; Mauquoy & Van Geel, 2007). An effective paleoindicator must be abundant, easy to identify, and well preserved over sufficiently long periods (see Sect. 2.3.1).

Charcoal originating from forest fires can be transported by wind and water to a lake or peat deposit. These pieces then sink and settle onto the bottom of lakes or fall onto the surface of peat. They eventually become buried and preserved in lake sediments and peat accumulations. These sediments archive past fires and can be recovered by extracting a sediment or peat core. For longer lake sediment records, cores are typically extracted from the deepest portion of a lake (Fig. 2.2). More recent sediments, found higher in the sedimentary record, are closer to the water–sediment interface. These less dense sediments have a higher water content and, as they are more easily disturbed, must be collected separately using a free-falling gravity corer, such as the Kajak-Brinkhurst or Willner-type corer. These separate cores are then correlated against one another to produce a composite record using, for instance, ^{210}Pb or the sedimentary properties recorded within each core. The sampling of lakes is conducted in winter using the frozen lake surface as a platform or using a raft during ice-free months. Peat can be sampled from bogs, mires, forested peatlands (Magnan et al., 2018), or small forest hollows (Fig. 2.3), the latter being paludified depressions inside forest stands (Bradshaw, 1988). Peat sequences are usually extracted using a Russian corer or a Wardenaar sampler.

After their extraction and transport to the laboratory, sediment cores are usually sliced into continuous subsamples that are at least 1 cm^3 in volume. In practice, this typically means subsampling at 0.5 or 1 cm intervals along the core. These subsamples are then processed to recover charcoal particles and, often, related proxy tools, e.g., pollen, diatoms, macrofossils, and sediment samples for loss-on-ignition

Fig. 2.2 (*left*) Winter sampling of lake sediments of Lake Huard, Ontario, Canada. Raynald Julien, Adam A. Ali, and Hans Asnong are present in the photo. (*right*) Sediment (gyttja) recovered from Lake Araisu, Latvia, showing varves (annual laminations). *Photo credits* Adam A. Ali (*left*), Normunds Stivrins (*right*)

(LOI) and grain-size analysis (Birks & Birks, 2006). In fire history reconstructions, charcoal pieces are usually categorized according to their size. Charcoal size reflects the distance traveled by a particle from its origin to the sediment archive. In lake sediments, charcoal fragments larger than 160 μm indicate local fires, whereas pieces smaller than 160 μm are sourced from fire events having occurred 0 to 40 km around the sampled lake (Higuera et al., 2010; Oris et al., 2014). A similar particle size−distance interpretation is applied to charcoal recovered from mires and bogs. Charcoal records from small forest hollows, however, usually originate from local fires (<100 s m distant) and are preferably used to reconstruct local or stand-level fire histories (Bradshaw, 1988).

Fire occurrence is typically based on the position of the charcoal layer within the sediment core. Chronological control of the sediment record, and thus the dating of fire events or periods, is commonly through radiocarbon dating of wood charcoal, plant macroremains, or bulk gyttja recovered from the core. The age-depth model derived from the obtained dates then provides an estimated age for each subsample. The temporal resolution of the collected subsample therefore depends on the thickness of the subsample and the sediment deposition rate, i.e., the number of years represented by a 0.5 or 1 cm thick subsample. Sites having a higher sedimentation rate permit a higher resolution of analysis, i.e., fewer years combined within a given sample. The time series of charcoal particle abundance is then typically analyzed

Fig. 2.3 Recovering a peat core from a small forest hollow. (*left*) Richard Bradshaw, Heikki Seppä, and Oleg Kuznetsov working with a Russian corer in the Russian Karelia region. (*right*) Peat core collected with a Russian corer. Charcoal bands are the darker strips observed near the end of the core (toward the bottom of the image). *Photo credits* Niina Kuosmanen

with statistical approaches that aim to distinguish past fire events from background levels of charcoal deposition (Higuera et al., 2010).

Box 2.1 Varved Lake Sediments

Annually laminated lake sediments, also known as varves (Fig. 2.2), are a special, albeit rare, type of lake sediment record. Here, an annual record of sediment deposition is distinguishable, making it possible to date deposited material at an annual and even seasonal resolution, similar to the resolution of tree rings, although varved records can often extend much further back in time. The seasonality within varves is produced by intra-annual changes in the materials deposited from the water column or transported from within the lake catchment area. In addition to a clear seasonality in deposition, other prerequisite conditions include sufficient incoming organic-inorganic material, no disturbance of the deposited material (e.g., through bioturbation), and anoxic conditions at the lake bottom. Once a laminated sequence is determined to represent annual layers (varves), multiple environmental proxies can then be applied, combining the advantages of centennial–millennial length lake records typical to organic sediments with the annual resolution and dating accuracy of tree rings. For fire histories, reconstructions from varved lake sediments have demonstrated the influence of humans and the environment on fire activity over long timescales in boreal Europe (Pitkänen & Huttunen, 1999) and Alaska (Gaglioti et al., 2016). This technique has also been used to validate the use of charcoal in the sediment record in general by comparing the deposition of charcoal in the varves with the fire scar record found in the vicinity of the sampled lake (Clark, 1988).

2.2.2 Limitations and Potential of Fire Reconstruction Methods

Each archive and proxy has its particular advantages and shortcomings; the nature of these depends on the required information or specific question being asked by the researcher (Remy et al., 2018; Waito et al., 2015). Tree-ring analyses remain the most accurate method for reconstructing local- and landscape-scale fire histories; in the boreal forest, however, these reconstructions are limited to the recent past (i.e., <1,000 years; Oris et al., 2014; Wallenius et al., 2010). Tree-ring analyses are also limited in the types of fires that can be dated. Fire scars require that trees survive the fires, and fire scars are also rarely formed in trees that are maladapted to frequent fires. In European boreal forests, for example, Scots pine (*Pinus sylvestris* L.) and Siberian larch (*Larix sibirica* Ledeb.) are useful for dating fires from scars, whereas Norway spruce (*Picea abies* (L.) H. Karst.) and deciduous trees are usually not. In high-intensity and tree-killing fires, stand initiation dates provide valuable information, but the data are limited to the most recent fire (however, see Sect. 2.4 discussing subfossil trees). Finally, although tree-ring records are ubiquitous in boreal forests, forest management based on clear-cutting tends to remove these biological archives in managed areas of boreal forests. Hence, the spatial extent of these reconstructions in such locations is more limited, and the study material is often less available than in unmanaged forest areas.

When fire history is investigated at longer millennial timescales or in regions where tree-ring proxies are unavailable, the selection of archives and proxies depends on the study objectives and the targeted spatiotemporal scale. Charcoal from lake sediments and large peatlands allows the reconstruction of long-term fire histories at a larger spatial scale. Nonetheless, several sites must be analyzed to reliably uncover regional trends in the reconstructions. Furthermore, taphonomic biases specific to each proxy, e.g., effects related to transportation, charcoal mixing, and the quality of charcoal preservation over time, must be minimized. This includes, for instance, avoiding sites showing visible signs of disturbance at the top of the peat (when sampling peatlands) or lakes having a substantial sediment influx. Excluding lakes that contain varved sediment records (Box 2.1), lake sediment−based fire reconstructions identify low-frequency trends rather than individual fire events. This lower-resolution state relates to the dating uncertainty of the age−depth models. This low resolution also occurs as a given charcoal peak within a charcoal series can encompass more than a single fire. Moreover, the low−density nature of the uppermost (i.e., the most recent) lake sediments leads to fewer charcoal fragments being recovered at these shallow depths in the sediment record, leading to a possible underestimation of the number of detected fires in the recent past (Lehman, 1975).

Charcoal records from soil and peat deposits in small forest hollows provide information on past fires at the local scale, and charcoal layers in peat sediments offer a reliable record of in situ fires within a single forest stand. However, these peat records of fire events suffer from the same limitations in temporal resolution as lake sediments. Furthermore, fire events can also consume/destroy the uppermost peat layers during exceptional droughts.

Paleofire studies continue to pursue novel methodological advances to refine current proxy tools and to develop new avenues. Recent studies have used charcoal morphology (morphotypes) to identify the fuel type−herbs, grass, wood, leaves (broadleaf versus coniferous)−and determine the material burned in a given fire, thereby providing a more complete portrait of the reconstructed fire regime. Stivrins et al. (2019) recently used *Neurospora* fungal spores to complement the charcoal-based fire record. *Neurospora* spp. produce spores after forest fires, and these spores can be identified within the sediment sequence.

Fire reconstructions are also being improved by integrating a wider set of data derived from various proxies of past fire and environmental conditions. A fire history, combined with detailed descriptions of past vegetation changes inferred from pollen and macrofossil records from the same sediment cores (Colombaroli et al., 2009), provides an ecosystem-level assessment of the effects of fire. Moreover, combining this paleofire and paleoecological information with multiproxy, high-resolution centennial- to millennial-scale climate reconstructions−including both temperature and precipitation−and modern observational data can offer details regarding the long-term trajectories in fire activity and identify the associated drivers (Girardin et al., 2013b, 2019).

2.2.3 Fire in the North American Boreal Forest

The fire history in the boreal region of eastern North America has been particularly well documented (Fig. 2.4). The regional Holocene fire history can be divided into four periods. The earliest period (ca. 10,000−8,000 ± 500 years BP) corresponds to the *afforestation phase* during which fire activity began to increase owing to the progressive regeneration of vegetation following the retreat of the Laurentide ice sheet (Dyke, 2004; Liu, 1990). Between ca. 7,500 and 3,500 years BP, the

Holocene thermal maximum (also called Holocene climatic optimum) was characterized by hotter and drier conditions, which favored an increase in fire activity (Viau & Gajewski, 2009). A colder and moister climatic phase, the *neoglacial period*, then followed, lasting until the last two centuries, during which fire activity was relatively reduced in boreal forests (Cayer & Bhiry, 2014; Viau & Gajewski, 2009). The most recent *industrial period* (starting ca. AD 1850), marked by anthropogenic warming, has generally witnessed an increase in fire activity (De Groot et al., 2013; Krawchuk et al., 2009). Nonetheless, a decreased fire frequency observed in some regions (Drobyshev et al., 2014; Larsen, 1996) underlines the spatial heterogeneity of fire activity across North American boreal forests related to regional and local abiotic and biotic conditions (Remy et al., 2017b).

Stand composition has also altered the Holocene fire regimes in eastern boreal North America (Fig. 2.5). The early Holocene *afforestation phase* was characterized by more frequent and larger fires in the temperate deciduous forest than those within the boreal coniferous forest owing to the earlier retreat of the ice sheet in southern latitudes (Blarquez et al., 2015). During the *Holocene thermal maximum*, fire frequency and, to a lesser extent, the amount of biomass burned were greater in the coniferous forest than in the deciduous forest because of the higher abundance of fire-prone species in the former (Gaboriau et al., 2020; Girardin et al., 2013a). This relatively higher fire activity in coniferous forests decreased slightly during the neoglacial period to reach levels similar to those within the deciduous forest. This neoglacial shift is best explained by a shorter fire season in the coniferous forest related to the cooler conditions and the larger amount of precipitation falling as snow during this period (Ali et al., 2012; Remy et al., 2017a; Turetsky et al., 2011). In deciduous forests, the amount of biomass burned increased slightly during the neoglacial period, and fire frequency reached its Holocene maximum for deciduous forests at this time. A higher abundance of fire-prone coniferous forest species colonizing from higher latitudes—resulting from colder and moister conditions—can explain this increased fire activity (Blarquez et al., 2015; Girardin et al., 2013a; Remy et al., 2019). An absence of a large increase in fire activity over the last centuries in eastern boreal North America in both coniferous and deciduous forests results from a combination of a less favorable climate for fire and anthropogenic fire suppression (Bergeron & Archambault, 1993; Bergeron et al., 2001; Blarquez et al., 2015).

Interregional comparisons of fire reconstructions improve our understanding of the mechanisms leading to long-term changes in fire activity, especially when the sites vary in their environmental characteristics. Nonetheless, interactions between climate changes and vegetation dynamics behind the extreme fire events experienced over the past two decades and their consequences on forest regeneration remain poorly understood. Thus, a new challenge in paleoecology is detecting and focusing on past extreme fire events to understand their causes and consequences to improve predictions and mitigate future risks. Several studies have begun to address this issue and have highlighted the *Medieval Warm Period*, a period characterized by particularly warm temperatures during which unusual peaks of fire activity occurred within various regions of the boreal forest (Girardin et al., 2019). Further studies

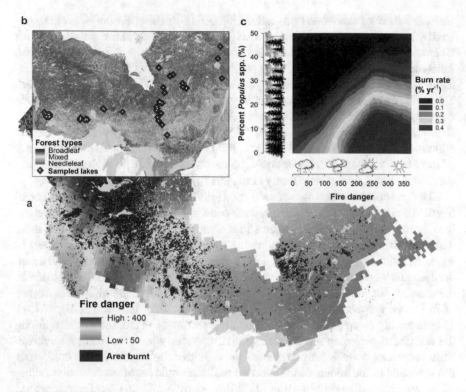

Fig. 2.4 Simplified representation of interactions between fire, vegetation, and climate within Canadian forests. **a** Mean seasonal fire danger across Canada for the 1900–2017 period and areas burned 1981–2017 (*dark red*). Fire danger includes the additive effects of seasonal drought severity and the duration of the snow-free period (equivalent to the fire season length), with higher values reflecting a greater seasonal fire danger. **b** The dominance of needleleaf trees in northern Ontario and northwestern Québec, Canada. Data was obtained from Beaudoin et al. (2014) at 250 m resolution land cover classes. The dimensionless scale covers needleleaf-dominated (*dark green*) to broadleaf-dominated (*dark brown*) areas, with lakes sampled for fire history reconstruction using charcoal records (*red diamonds*). **c** An empirical model of the burn rate, i.e., percentage of burned area per year for a given region, as a function of fire danger (in **a**) and percentage cover of broadleaf *Populus* species **b**. Fire-prone conditions exist when the fire danger is high and the percentage of *Populus* spp. in the regional landscapes is less than 30%. Adapted by permission from Springer Nature from Girardin and Terrier (2015)

focusing on this warm period at multiple locations in the boreal forest could improve our understanding of the environmental processes involved in extreme fires.

Enhanced fire activity is projected for the twenty-first century as temperatures rise (Flannigan et al., 2009; Jolly et al., 2015). Anticipated consequences from the increased fire activity include changes to wildlife habitat, increased carbon emissions, heightened threats to human safety and infrastructure (e.g., injury, death, property loss, reduced clean air and water supplies), and greater economic losses for the forestry sector, losses that may include fewer commercial products and timber supplies (Brecka et al., 2018; Gauthier et al., 2015; Walker et al., 2018).

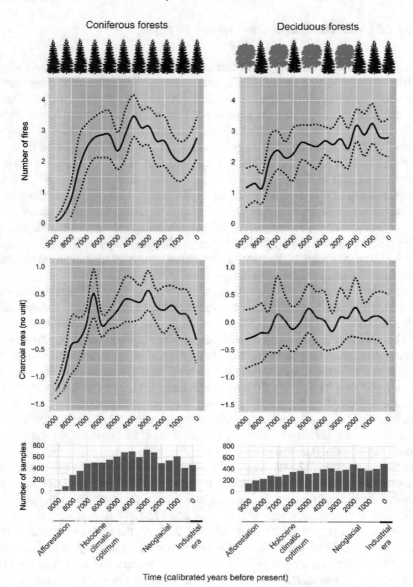

Fig. 2.5 The reconstructed Holocene fire regime history of the boreal deciduous and coniferous forests of eastern Canada as derived from the analysis of lacustrine charcoal deposits. Plots of the number of fires and the charcoal areas indicate the fire occurrence and the area burned, respectively. General temperature patterns over the Holocene include cooler (*light blue*) and warmer (*light red*) periods. The trees illustrate the most abundant species over the Holocene in terms of deciduous (*gray*) or coniferous (*black*) trees

2.2.4   Fire in the European Boreal Forest

Studies of past fires in European boreal forests have revealed a complex, mixed-severity fire regime that varies in both time and space and is influenced by climate, vegetation, landscape structure, and human activities. Fire histories are particularly well studied in regions within the western portions of European boreal forests, particularly on the Fennoscandian Shield. Here, fire shows several broad-scale patterns during the Holocene. The analysis of 69 individual fire records recovered from lake sediments spread across Fennoscandia revealed that early Holocene fire frequencies peaked 8,500 to 6,000 years BP. Fire frequency then declined until starting a rising trend ca. 4,000 years BP (Molinari et al., 2020). This early Holocene pattern reflects the well-resolved climate variability over similar time frames showing that the warmest part of the Holocene, the Holocene thermal maximum, and the changes in fire activity coincide very well. The trend of more frequent fires in the region over the last 4,000 years is driven by an increased human influence related to greater human population densities and changes in forest use.

In addition to the climate-driven patterns in fire occurrence, millennial fire history reconstructions illustrate a long-term interaction between vegetation and fire. After the Holocene thermal maximum, the most conspicuous change in forest composition in the boreal forest in Europe involved the expansion of spruce, which began in eastern Fennoscandia ca. 6,500 years BP and has continued in western Fennoscandia over the last two millennia. Paleoecological records of charcoal in organic sediments from remote sites having limited human influence demonstrated that the expansion of spruce coincided with a marked decrease in fire occurrence (Ohlson et al., 2011). Nonetheless, it remains unclear whether the expansion of spruce represented the cause or the consequence—a changing microclimate or fuel type and distribution—of reduced fire activity (Ohlson et al., 2011).

In the more southern hemiboreal and boreonemoral zones, the emerging picture of the Holocene fire trends similarly differs from the expected pattern of a climate-only forcing; the observed pattern confirms the importance of interactions with vegetation. A detailed lake sediment record from this zone showed that fire frequency was relatively high 9,500 to 8,000 years BP (Fig. 2.6). As the climate became warmer and drier around 8,000 years BP, fire frequency decreased notably. This observation contrasts with the expected causal link between the climate and fire frequency in the boreal zone; however, it may be explained by a change in the vegetation and fuel type (Feurdean et al., 2017). During the warm and dry period, 8,000 to 5,000 years BP, the populations of temperate deciduous broadleaf tree species (e.g., hazel, oak, lime, and elm) in the southern part of the boreal zone increased and replaced the boreal tree species. This major shift in forest composition reduced the regional fire frequency because these deciduous species are less flammable than conifers. It is also possible that greater shade in the dense deciduous forest and the higher moisture content of the leaves favored a reduced fire frequency (Feurdean et al., 2017), similar to the effects of spruce expansion in more northern regions (Ohlson et al., 2011).

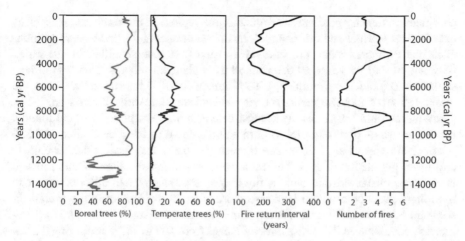

Fig. 2.6 Vegetation and fire frequency in the southernmost edges of the European boreal forest; fire return intervals increased, and the number of fires decreased as vegetation shifted toward a greater presence of temperate trees during the Holocene thermal maximum, despite the climate being warmer and drier. Figure redrawn from Feurdean et al. (2017) with permission from Elsevier

Approaching the modern period, the human influence on forests and the fire regime becomes increasingly evident within the sediment record (Molinari et al., 2020); this pattern is also observed in tree ring–based records where the longest reconstructions extend 700 to 1,000 years BP (Niklasson & Granström, 2000; Rolstad et al., 2017; Wallenius et al., 2010). The timing of this increased human influence varies between regions, very much related to human settlement and lifestyle changes (Wallenius, 2011). In particular, the increase and eventual decline in slash-and burn agriculture, widely practiced over much of Finland, has been identified as a driver of the onset and cessation of high fire activity; similar patterns have been identified across the region, from southeastern Norway in the west (Rolstad et al., 2017) to the Komi Republic in the east (Drobyshev et al., 2004). This anthropogenic influence is reflected by increased fire frequencies, smaller fire sizes, and a greater number of early−season fires (Niklasson & Granström, 2000; Rolstad et al., 2017). Climate continues to be a driver, in particular during exceptionally dry years and periods when Fennoscandian forests experience a greater area of forest burned (Aakala et al., 2018; Drobyshev et al., 2016). In most of Fennoscandia, this period of human-induced high fire activity has receded over the past 100 to 250 years because of changes in forest use, land tenure, and, more recently, the greater development of infrastructure and fire suppression (Rolstad et al., 2017; Wallenius, 2011), giving way to the modern fire regime (see Chap. 3).

In addition to this temporal variability, long-term fire reconstructions have demon-strated a latitudinal gradient of more frequent fires in the south to less frequent fires in the north (Drobyshev et al., 2014). This gradient has a climatic origin (Larjavaara et al., 2005); however, except for the mountainous areas, the fire gradient also follows a population density gradient over much of the region. In the southern areas, characterized by warmer and drier conditions in the summer and a greater human influence, the estimated mean fire interval over the past several millennia has varied between 70 and 95 years, as determined from charcoal records from varved lake sediments (Pitkänen & Huttunen, 1999; Tolonen, 1978). In northern Sweden, a fire interval about of about 80 years has been obtained from tree rings (Zackrisson, 1977), and a millennium-long tree-ring reconstruction in northern boreal Finland found a mean fire cycle—time required to burn an area equal to the area studied—of 350 years (Wallenius et al., 2010). The Finnish site has a less fire-conducive climate and lower population density than more southern, fire-prone sites. Within forested landscapes, characteristics such as fire breaks, topography, and differences in soil hydrology produce a within-landscape variability in the Holocene fire record. In southern Fennoscandian and western Russian boreal forests, for example, the sediment charcoal–based fire return interval ranges from 109 to 237 years during the last 11,000 years (Stivrins et al., 2019), whereas nearby sites are without any evidence of fires (Kuosmanen et al., 2014). A similar type of spatial variability in fire history is recorded in the eastern parts of the European boreal forests in the Ural Mountains (Barhoumi et al., 2020). Tree ring–based reconstructions tell a similar story with substantially different fire return intervals in various parts of a landscape, depending on soil hydrology (Aakala, 2018).

2.3   Millennial Insect Outbreak History

The detailed patterns of fire history described in the preceding sections reflect the predominance of wildfire as the most commonly studied disturbance in boreal paleoenvironmental research (Bergeron et al., 2010; Flannigan et al., 2001). Our understanding of millennial-scale natural disturbances has traditionally revolved around the role of fire in influencing forest dynamics, despite an understanding that disturbances interact and operate at multiple scales and that in many locations, insect outbreaks, rather than wildfires, are the major drivers of forest landscapes. Over the short term, insect outbreaks and plant diseases can damage extensive areas of forest and produce significant economic losses. Insect outbreaks are one of the most influential factors shaping modern boreal forest diversity (McCullough et al., 1998). As with fire, insects contribute to the regeneration of the forest mosaic. In contrast to fire, however, insects affect stands selectively by, for example, targeting old and vulnerable trees.

Various insect defoliators, composed mainly of lepidopterans, affect boreal stands. These defoliators include the forest tent caterpillar, *Malacosoma disstria* (Hübner), the hemlock looper, *Lambdina fiscellaria* (Guénée), and the spruce budworm (SBW), *Choristoneura fumiferana* (Clemens). The latter has the greatest influence within the boreal region owing to its very extensive distribution and marked effect on North American boreal forests. SBW is a defoliating lepidopteran native to coniferous forests in Canada and the northeastern United States. This species is responsible for the largest area of damage in the North American boreal forest for insect defoliators. Its primary hosts are balsam fir (*Abies balsamea* (L.) Mill.), white spruce (*Picea glauca* (Moench) Voss), and, to a lesser extent, red spruce (*Picea rubens* Sarg.) and black spruce (*Picea mariana* (Mill.) BSP). The univoltine cycle of this moth consists of an egg stage, diapause, six larval instars, pupation, and an adult stage (moth). This last stage is relatively short (two weeks), during which the insect spends all its time looking for a mate. If successful in finding a mate, the females then lay their eggs. Balsam fir may die after three or four consecutive years of severe defoliation (Bergeron et al., 1995; MacLean, 1984), whereas secondary hosts suffer crown and branch mortality and growth reduction of up to 75% (MacLean, 1984; Nealis & Régnière, 2004). In the province of Québec (Canada), the forest surface affected by this species of Lepidoptera over the last century is twice the size of the state of California (Navarro et al., 2018c). SBW outbreaks have major ecological effects and result in important economic consequences through the loss of forest productivity (Shorohova et al., 2011).

Despite the scale and significance of this natural disturbance agent, we remain limited in our knowledge regarding the frequency and severity of SBW outbreaks at a multimillennial scale and understanding how these outbreaks relate to climate and other disturbances, such as fire. Given that variations in temperature and precipitation affect an organism's survival, reproduction cycle, and spatial dispersion (Dale et al., 2001), it is critical to understand the links between SBW outbreaks and climate to better understand the potential of SBW outbreaks under future climate change scenarios (Klapwijk et al., 2013; Volney & Fleming, 2000). Paleoenvironmental records of these insect outbreaks can therefore offer a long-term perspective of SBW outbreaks and shed light on the periodicity, synchronicity, and consequences of past insect outbreaks improve our understanding of the spatiotemporal patterns of SBW in relation to climate (Berguet et al., 2021; Jardon et al., 2003; Navarro et al., 2018a). Until recently, however, the lack of effective proxies and methods for reconstructing insect-related disturbances led to a severely neglected and oversimplified understanding of the frequency, intensity, and impacts of past insect outbreaks on the forest landscape. In the following section, we summarize recent advances in the paleoenvironmental reconstruction of past insect outbreaks, specifically those of SBW, in the boreal forest.

2.3.1 Insect Outbreak Reconstruction

2.3.1.1 Dendrochronology

Dendroecological approaches have been applied to the reconstruction of past insect outbreaks. Tree rings provide indirect measurements of insect activity; years of unusually narrow or otherwise anatomically abnormal tree rings can be related to insect outbreaks. These tree ring–based approaches have helped reconstruct outbreaks of numerous insects, including outbreaks of the forest tent caterpillar

(Cooke & Roland, 2000; Sutton & Tardif, 2007), the larch sawfly (*Pristiphora erich-sonii* (Htg.); Jardon et al., 1994; Girardin et al., 2001, 2002; Nehemy & Laroque, 2018), the larch budmoth (*Zeiraphera diniana* Gn.; Weber, 1997; Rolland et al., 2001), the western and eastern spruce budworm (Boulanger et al., 2012; De Grandpré et al., 2019; Flower et al., 2014; Krause, 1997; Morin & Laprise, 1990; Navarro et al., 2018c; Swetnam & Lynch, 1993), and the jack pine budworm (*Choristoneura pinus* Free; Volney, 1988), as well as outbreaks of the geometrid moths *Epirrita autum-nata* Borkh (Babst et al., 2010) and *Operophtera brumata* L. (Hoogesteger, 2006; Tikkanen & Roininen, 2001; Young et al., 2014).

The reconstruction of insect outbreak regimes at the landscape scale is a major challenge, as aerial surveys of defoliation have been available only since the 1960s—covering only one major outbreak in the last century—and are concentrated mainly in the balsam fir area; thus, the use of dendrochronological approaches becomes essential. Similar to tree ring–based studies of fire history, a major limitation in dendrochronological reconstructions of insect outbreaks is the maximum age of host trees. This is particularly true for trees affected by eastern spruce budworm, as this insect often kills its host. Cross-dating has been helpful when using dead trees, found either in the field or as lumber in old buildings, to extend tree-ring chronologies (Boulanger & Arseneault, 2004; Boulanger et al., 2012; Krause, 1997); in North America, however, there are few historical buildings, which limits the longest chronologies to the last 400 years. The available tree-ring series for extensive areas of the eastern Canadian boreal forest extend only to the early twentieth century (Navarro et al., 2018c). Subfossil trees, buried stems recovered from peatlands or lakes, can extend dendrochronological records further back in time. Simard et al. (2002), for example, studied a small peat bog surrounded by host trees of spruce budworm and found evidence of outbreaks between 4,170 and 4,740 years BP. A more extensive use of subfossil trees from lakes appears promising, as highlighted by a recently published 800-year chronology of SBW outbreaks relying on subfossil stems (Morin et al., 2020). Nonetheless, long-term local and regional chronologies remain unavailable for extensive areas.

2.3.1.2 Macrofossils

Macrofossils are plants and animal parts preserved in the sediment record and are visible without using a microscope; they include cones, leaves, seeds, stems, exoskeletons, teeth, and bones. These indicators confirm the nearby presence of these organisms and are powerful tools for reconstructing insect outbreaks. Head capsules, pupae, and other insect remains preserved in the sedimentary record can serve as proxies of past SBW (and other species) outbreaks (Bhiry & Filion, 1996; Davis & Anderson, 1980). Most body parts of the caterpillar or butterfly stage are nonetheless fragile and often recycled very rapidly within the soil humus layer (Potelle, 1995). SBW feces (frass pellets), however, are well-preserved macrofossils (Fig. 2.7). During heavy budworm infestations, fecal pellets can rain down continuously from infested trees to the ground. The feces can be identified to the species level, and parts of balsam fir needles within the fecal matrixes remain identifiable (Potelle, 1995).

At present, the longest budworm macrofossil profile covers more than 8,200 calibrated (cal.) years BP (Simard et al., 2006). Spruce budworm feces began accumulating at the study site around 8,240 cal. years BP and were observed throughout the profile. Budworm feces peaks occur at ca. 6,775 cal. years BP and 6,550 cal. years BP. Three other sampled bogs from the same region also demonstrate two or three periods of higher feces abundance during the Holocene (Simard et al., 2011); however, these periods of higher insect macrofossil abundance are not synchronized, indicating that episodes of high spruce budworm abundance varied between locations. This initial evidence also suggested that peaks of high spruce budworm abundance were rare over the course of the Holocene. Although Simard et al. (2006, 2011) found only a few peaks of spruce budworm feces during the Holocene, these were the first studies to identify budworm outbreaks over the Holocene.

Macrofossils as indicators of insect outbreaks have some significant limitations. Similar to the lake- or peat-based paleoecological methods presented above, macrofossils collected from sedimentary records do not provide high-resolution reconstructions; identified periods of high budworm populations can encompass several outbreaks. Furthermore, questions have been raised regarding feces preservation over time owing to greater decomposition with age, biasing against older outbreaks. Moreover, insect macrofossils represent only a local signal, and study sites are limited to

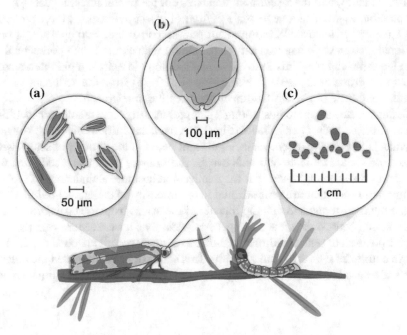

Fig. 2.7 Proxies from lepidopterans used to reconstruct insect outbreaks; **a** wing scales, **b** cephalic capsules, and **c** feces

the few, sporadically scattered locations where balsam fir, the preferred host of the spruce budworm, grow near the sampled peat bogs and lakes. Finally, this approach requires a sizable amount of sample material, and the extraction of macrofossils is a very laborious manual task.

2.3.1.3   Microfossils: Lepidopteran Scales as a Novel Paleoindicator

The lack of robust, abundant, and nondecomposing proxies has limited previous paleoecological reconstructions of SBW outbreaks. During the current SBW infestation in the Saguenay-Lac-Saint-Jean region (Québec), however, large quantities of adult moth scales were observed in the water column of regional lakes. These lepidopteran scales are released as an individual moth dies (around 150,000 scales per insect). These scales, transported by the wind and water, land on the lake surface and eventually settle onto the lake bottom to become part of the sediment record. The chitinous composition of these scales favors their preservation in the sediment, and their abundance in the sediment indicates the relative timing and intensity of the outbreaks (Montoro Girona et al., 2018b).

There are several advantages to this new proxy. The identification of spruce budworm scales is less problematic than that of spruce budworm feces, as the scales are chitinous, and their long-term preservation in lake sediments is excellent. Numerous lakes, and their sediment archives, dot the boreal forest landscape; thus, it is possible to produce a large-scale portrait of insect outbreaks. Moreover, only a small amount of material is required for sample preparation and analysis (1 cm^3), and lepidopteran scale analysis can be combined with charcoal and pollen analyses from the same sample. This innovative methodology to extract lepidopteran scales (Montoro Girona et al., 2018b; Navarro et al., 2018b) from the sediment samples circumvents some of the limitations of the feces-based approach.

Distinct scale morphologies among lepidopteran taxa permit taxonomic identification of the scale to the species level (Fig. 2.8). Given that billions of spruce budworm individuals live during an outbreak, significant peaks in the number of scales within a lake core should indicate outbreak events. Preliminary work using sediment traps and short cores demonstrated that the relative and absolute abundances of scales in the traps and sediment are proportional to the intensity of the annual defoliation of the surrounding forest and that the transfer of the scales from the lake surface to the lake bottom occurs over a few days, generally less than a week. Moreover, the stratigraphic position of scales within a well-dated sediment record matched the timing of known outbreaks (Navarro et al., 2018b). This series of tests confirmed the potential for a scale-based reconstruction of lepidopteran outbreaks from the sediment record.

Fig. 2.8 Potential of lepidopteran scales as a paleoindicator of insect outbreaks. Scales are composed by chitin and are thus difficult to degrade. **a** Comparison of four well-preserved scales extracted from a lake sediment core with spruce budworm (SBW) morphotypes generated through shape measurements of thousands of SBW specimen scales. **b** Wing scales organized like roof tiles. **c** Diversity of wing scale morphotypes. *Photo credits* **a** Montoro Girona et al. (2018b; CC BY 1.0); **b**, **c** Emy Tremblay and Miguel Montoro Girona

2.3.2   Holocene History of Insect Outbreaks and Consequences for Understanding Outbreak-Fire-Climate-Vegetation Interactions

Navarro et al. (2018c) identified 87 significant peaks in scale abundance over the last 8,000 years. These results contrasted markedly with those of the SBW feces–based record, which recorded few events over the Holocene. The lepidopteran scale record indicates a pattern of highly variable but consistently present budworm populations over the Holocene. Pairing the scale record with microcharcoal and pollen records revealed that the frequency of outbreak events was inversely correlated with the frequency of fire events (Fig. 2.9). When the periods of high budworm populations in the four feces diagrams produced by Simard et al. (2011) are combined, the frequency of outbreaks produces an inverse relationship with published fire frequency events. Therefore, the spruce budworm feces record recovered from peat deposits did not contain all outbreaks that occurred at the sampling site; this absence from the peat record likely relates to the easily degradable nature of the SBW feces in the peat archive.

The use of lepidopteran scales has heightened our ability to understand outbreak dynamics during the Holocene and their relationship to fire, climate, and forest structure across the landscape. We are currently sampling several lakes in the mixed forest—the current center of SBW distribution—and the black spruce forest—the modern northern distribution of the insect—to better understand the links between SBW outbreaks and forest structure. Balsam fir abundance fluctuates in relation to other species, and these fluctuations relate mainly to climate and fire as indicated by the fluctuations of fire-adapted species, such as jack pine, and charcoal abundance over the Holocene (Bergeron & Leduc, 1998). Our initial results support the earlier finding of an inverse relationship between outbreak frequency and fire (Fig. 2.9). A drier climate appears to induce a higher fire frequency. This shift favors the installation of fire-adapted species (e.g., jack pine) that are not hosts of budworm. The frequency of detectable outbreaks therefore decreases. In contrast, a more humid climate—most likely a warmer humid climate—leads to a lower fire occurrence, thereby favoring the maturing of forests where SBW host trees, such as balsam fir

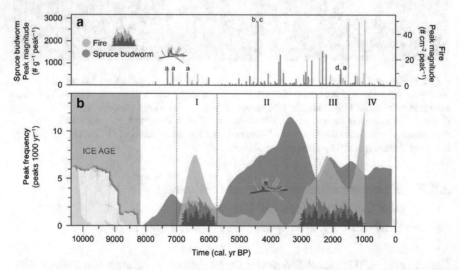

Fig. 2.9 Interactions between fire and insect outbreaks over the Holocene in the province of Québec, Canada. **a** Magnitude of fire and lepidopteran scale peaks (#·cm^{-2}·peak^{-1}). Letters correspond to peaks identified from other proxy records; *a* Simard et al. (2006), *b* Anderson et al. (1986), *c* Bhiry and Filion (1996), and *d* Jasinski and Payette (2007). **b** The frequency of fire and SBW outbreaks (peaks·1,000·yr^{-1}). To extract fire events from the charcoal stratigraphy, we defined a background component (C$_{back}$) using a Lowess smoothing that was robust to outliers and with a 500-year smoothing window. C$_{back}$ was extracted from the interpolated series of raw data (C$_{int}$) to define a peak series (C$_{peak}$) as a residual of C$_{int}$ − C$_{back}$. Each peak exceeds the 99th percentile threshold of the residual of C$_{int}$ − C$_{back}$. Roman numbers correspond to the disturbance interaction steps during the Holocene: *I* and *IV* represent periods where fire was the dominant disturbance; *II* (6,200–2,500 cal. yr BP) had insect outbreaks as the main disturbance, and period *III* experienced fire and outbreaks at a similar frequency

and white spruce, proliferate. Thus, this specific forest composition leads to periods of higher SBW abundance.

The recent dynamics of SBW outbreaks and the observed changes to fire regimes reflect millennia of interactions between the insects and their environment. Thus, understanding these complex dynamics requires that we use approaches such as dendrochronology and paleoecology to improve our understanding of the frequency and severity of epidemic periods over the longest possible period. Some studies have demonstrated that SBW was already present and abundant in stands lacking balsam fir, suggesting a high SBW activity on this insect's secondary host, namely black spruce (Simard et al., 2006). Moreover, this high abundance was observed during a relatively warm period of the Holocene (7,000–6,000 cal. yr BP), suggesting a phenological synchronization between insect and host (Fig. 2.9).

Using periods of growth suppression in dendroecological series, Navarro et al. (2018a) identified three insect outbreaks in eastern Canadian forests over the last century; these outbreaks differed in their respective spatiotemporal pattern, duration, and severity. The first outbreak (AD 1905–1930) affected up to 40% of the studied trees, initially synchronizing from local infestations and then migrating to more northern stands. The second outbreak (AD 1935–1965) was the longest lasting, although the least severe, with only up to 30% of trees affected by SBW activity. The third event (AD 1968–1988) was the shortest; however, it was also the most severe and extensive, affecting nearly 50% of trees and 70% of the study area. This most recent event was identified for the first time at the limit of the commercial forest, illustrating a northward shift of the SBW distribution area during the twentieth century. This observation provided the first documented evidence of how climate change influences the current spatiotemporal patterns of SBW outbreaks (Navarro et al., 2018c).

However, dendroecological reconstructions of past outbreaks have assumed that only defoliation is responsible for the sustained growth suppression in the host trees. Recent work illustrates that periods of climate-related growth suppressions can precede or co-occur with insect disturbances (De Grandpré et al., 2019). It is therefore possible that some of these reconstructed outbreak periods are, in fact, confounding effects of climatic periods unfavorable for growth (Gennaretti et al., 2018; Girardin et al., 2014, 2019). More research must be carried out to differentiate the effect of defoliation and climate-related growth suppressions in dendroecological series to improve reconstructions of the spatial and temporal dynamics of past insect outbreaks.

Outbreak reconstructions provide strong support for the hypothesis that SBW has been present and influencing forest dynamics in the boreal forest Québec throughout the Holocene (Simard et al., 2006). SBW abundance and outbreaks are strongly correlated with the presence of its primary hosts; this presence is itself influenced by climatic variations and fire regimes. This information will be essential for building predictive models of SBW outbreaks in the face of climate change. The early and late Holocene were characterized by a relatively high fire frequency; the greater number of fires may have restricted the development of severe epidemics by reducing the number of mature hosts in the landscape. These results also suggest that epidemics would have been much more frequent and possibly more severe in the mid-Holocene when fires had less (or more local) influence on the landscape, highlighting the importance of intermediate severity disturbances in the forest landscape and providing insight for ecosystem-based management to adapt the silvicultural practices to this type of disturbance, e.g., applying partial harvest in locations having longer fire intervals (Bose et al., 2014; Martin et al., 2020; Montoro Girona et al., 2016; Moussaoui et al., 2020).

2.4   An Archive of Boreal Forest Dynamics: Subfossil Trees

Subfossil trees, where available, offer another means of reconstructing disturbance histories at an annual resolution. The potential information preserved in the tree-ring records of subfossil trees includes the temporal patterns of tree recruitment and mortality, the occurrence and timing of forest disturbances, and the interannual variations in forest productivity and climate over timescales ranging from centuries to a few millennia (Gennaretti et al., 2014b). Subfossil trees provide information mainly in terms of local stand-scale forest dynamics; however, their tree-ring patterns may also be imprinted by regional- and hemispheric-scale climate signals, thereby allowing reconstructions of past climate variability and the influence of main climate forcing agents, e.g., solar, orbital, and volcanic influences; see Gennaretti et al. (2014a). The varying ages of the preserved subfossil trees can also extend the regional tree-ring records beyond the period covered by living trees.

The preservation of these paleoenvironmental archives requires exceptional depositional conditions for the trees to experience minimal decay (Fig. 2.10 a, b). These settings include anoxic sediments (peat, lake, and river sediments) or in sites where arid or cold conditions limit insect and microbial activity on the dead tree trunks (Eronen et al., 2002; Gennaretti et al., 2014b; Hantemirov & Shiyatov, 2002; Spurk et al., 2002). Subfossil trees can sometimes be dated from their depositional context with variable precision, although the main interest in their use stems from the analysis of their tree rings to determine the exact calendar years of ring formation through the dendrochronological cross-dating of ring-width patterns against a "master chronology" (see Box 2.2 on master chronologies).

The systematic or exhaustive sampling of subfossil stems at a single site can reveal several tree generations of stand-scale forest dynamics acting in response to local disturbances (Fig. 2.10). Subfossil tree records collected from peatlands, lakes, and rivers in the eastern Canadian boreal forest highlight the long-lasting consequences of individual fire events (Arseneault & Payette, 1997; Arseneault & Sirois, 2004; Arseneault et al., 2007; Gennaretti et al., 2014c). These effects include shifts to treeless environments, changes in stem density, and the exclusion of fire-sensitive tree species. The sampling of subfossil logs from several sites across a relatively broad region allows large-scale patterns and processes to be documented, including latitudinal or altitudinal shifts of tree line (Helama et al., 2005; Kullman, 1995) and

Fig. 2.10 Typical examples of subfossil trees in lakes of the eastern Canadian boreal forest. **a** A dominant living tree prone to be recruited as a dead tree in the littoral zone of a small lake surrounded by an old-growth black spruce forest. This tree fell into the lake two years after this picture was taken **b** A dense accumulation of subfossil trees in the littoral zone of the same lake as **a**. The subfossil stems in the photo have been accumulating continuously over the last two millennia. **c** A charred lateral branch of a cross-dated tree indicates that the corresponding tree died during a fire several centuries ago. **d** Impact of a stand-killing fire (*vertical dashed line*) on the recruitment of individual subfossil stems (*horizontal bars*). More than a century is generally needed for the postfire recovery of the lakeshore forest and subsequent inputs of new tree trunks into the littoral zone. Subfossils recruited before the fire event must be cross dated using subfossil trees from another shore segment or nearby lake. Such fire-induced recruitment gaps often limit the development of millennial master chronologies for the North American boreal forest. *Photo credits* **a–c** Dominique Arsenault **d** Modified permission from John Wiley & Sons, Inc. (Journal of Ecology © 2013 British Ecological Society) from Gennaretti et al. (2014b)

long-term reconstructions of defoliating-insect outbreaks (Esper et al., 2007). At a more local scale, sunken cut logs, deposited in river sediments during the timber-driving era in eastern Canada, provide evidence of the nineteenth and twentieth-century logging history in the corresponding watershed. Information obtained from these sunken logs testifies to the progressive changes in logging activities from the preferential cutting of large pine and spruce stems in the nineteenth century to a more generalized exploitation of all conifer species following the development of the pulp and paper industry at the turn of the twentieth century (Boucher et al., 2009).

Box 2.2 Master Chronologies

Developing a long "master chronology" from subfossil trees is a long, difficult, and expensive task, requiring the analysis of several hundred to thousands of trees and their respective tree-ring records. Master chronologies longer than 7,000 years have been developed from subfossil stems recovered from lakes in northern Fennoscandia by sampling about 1,000 Scots pine (Eronen et al., 2002; Grudd et al., 2002), and more than 2,000 black spruce were needed to develop a 1,300-year master chronology for the eastern Canadian boreal forest (Gennaretti et al., 2014b). The general idea is to use these hundreds of overlapping tree-ring records to produce an average characteristic tree-ring sequence for a region. Undated tree-ring records of an individual tree or a group of trees ("floating" tree-ring series) can then be matched to the patterns of the master chronology to obtain a precise dating (cross-dating process) of the individual records. Successive dating of older trees permits the temporal extension of the master chronology. The challenge of developing a master chronology in the boreal forest stems from the high frequency of stand-killing forest fires, which limits the temporal continuity of tree-ring chronologies in this fire-prone region (Fig. 2.10c; Arseneault et al., 2013; Gennaretti et al., 2014c). Thus, many trees and sites are required to build a long, boreal master chronology that extends through periods where severe stand-killing fires burned specific stands, but not all sites.

2.5 Looking Toward the Future

Anthropogenic environmental changes are pushing global forest ecosystems toward *non-analog* states, including disturbance regimes not previously encountered in the period of recorded human history. The use of the various environmental signals stored in biological archives, such as tree rings and lake and peat sediments, can provide critical information on past changes in environmental conditions, the associated changes in disturbances, and how forest ecosystems have responded to these shifts. This multiproxy paleoenvironmental approach is particularly important for slowly occurring processes, which require centennial- to millennial-scale measurements to be noted and assessed. Paleoecological information provides a long-term context for the observed changes and a means of testing models and simulations that fall beyond environmental conditions observed in recorded history. Paleoecology can also explain how the current ecosystem structures have developed, e.g., the long-term patterns of fire occurrence related to human activities. Combining paleoenvironmental approaches at sites within the boreal forest has improved our understanding of the interactions between various disturbances over the Holocene, the role of insect outbreaks on landscape dynamics, and the interactions between climate, vegetation, fire, and insects over the short term, i.e., last 200 years, and plurimillennial scales.

The development of alternative paleoenvironmental proxy indicators remains active, including the use of fungal spores for reconstructing the occurrence of pathogenic fungi, the recovery of insect remains other than the abovementioned lepidopteran scales (Schafstall et al., 2020), and the application of ancient DNA and molecular biomarker analyses to the sedimentary record (Crump, 2021; Dubois & Jacob, 2016). Specific proxy records may serve complementary and related purposes. Living and subfossil tree-ring chronologies hold information on long-term forest disturbances at an annual resolution and may also be used to study climate variability over the chronological coverage. Subfossil samples could also potentially improve our understanding of climate-related changes in forest productivity in commercial forests by providing information related to past tree growth, an element needed to improve the forecasts of future forest productivity under climate warming. Finally, the long-term patterns of forest response to natural disturbances can help develop more sustainable forest management strategies. Central to this framework is that harvest methods should emulate patterns of natural disturbance to thereby minimize the differences between managed and natural forests (Kuuluvainen, 2002; Montoro Girona et al., 2018a). The development of these management methods requires a thorough understanding of patterns, consequences, and long-term variability of fire, insect outbreaks, and other natural disturbances to adapt silvicultural practices to future shifts in climate.

References

Aakala, T. (2018). Forest fire histories and tree age structures in varrio and maltio strict nature reserves, northern finland. *Boreal Environment Research, 23*, 209–219.

Aakala, T., Pasanen, L., Helama, S., et al. (2018). Multiscale variation in drought controlled historical forest fire activity in the boreal forests of eastern Fennoscandia. *Ecological Monographs, 88*, 74–91. https://doi.org/10.1002/ecm.1276.

Ali, A. A., Blarquez, O., Girardin, M. P., et al. (2012). Control of the multimillennial wildfire size in boreal North America by spring climatic conditions. *Proceedings of the National Academy of Sciences of the United States of America, 109*, 20966–20970. https://doi.org/10.1073/pnas.120 3467109.

Anderson, R. S., Davis, R. B., Miller, N. G., et al. (1986). History of late- and post-glacial vegetation and disturbance around Upper South Branch Pond, northern Maine (USA). *Canadian Journal of Botany, 64*, 1977–1986. https://doi.org/10.1139/b86-262.

Arseneault, D., & Payette, S. (1997). Reconstruction of millennial forest dynamics from tree remains in a subarctic tree line peatland. *Ecology, 78*, 1873–1883. https://doi.org/10.1890/0012-9658.

Arseneault, D., & Sirois, L. (2004). The millennial dynamics of a boreal forest stand from buried trees. *Journal of Ecology, 92*, 490–504. https://doi.org/10.1111/j.0022-0477.2004.00887.x.

Arseneault, D., Boucher, É., & Bouchon, É. (2007). Asynchronous forest-stream coupling in a fire-prone boreal landscape: Insights from woody debris. *Journal of Ecology, 95*, 789–801. https://doi.org/10.1111/j.1365-2745.2007.01251.x.

Arseneault, D., Dy, B., Gennaretti, F., et al. (2013). Developing millennial tree ring chronologies in the fire-prone North American boreal forest. *Journal of Quaternary Science, 28*, 283–292. https://doi.org/10.1002/jqs.2612.

Asselin, H., & Payette, S. (2005). Detecting local-scale fire episodes on pollen slides. *Review of Palaeobotany and Palynology, 137*, 31–40. https://doi.org/10.1016/j.revpalbo.2005.08.002.

Babst, F., Esper, J., & Parlow, E. (2010). Landsat TM/ETM+ and tree-ring based assessment of spatiotemporal patterns of the autumnal moth (*Epirrita autumnata*) in northernmost Fennoscandia. *Remote Sensing of Environment, 114*, 637–646. https://doi.org/10.1016/j.rse.2009.11.005.

Barhoumi, C., Ali, A. A., Peyron, O., et al. (2020). Did long-term fire control the coniferous boreal forest composition of the northern Ural region (Komi Republic, Russia)? *Journal of Biogeography, 47*, 2426–2441. https://doi.org/10.1111/jbi.13922.

Beaudoin, A., Bernier, P.Y., Guindon, L., et al. (2014). Mapping attributes of Canada's forests at moderate resolution through kNN and MODIS imagery. *Canadian Journal of Forest Research 44*, 521–532. https://doi.org/10.1139/cjfr-2013-0401.

Bergeron, Y., & Archambault, S. (1993). Decreasing frequency of forest fires in the southern boreal zone of Quebec and its relation to global warming since the end of the "Little Ice Age." *The Holocene, 3*, 255–259. https://doi.org/10.1177/095968369300300307.

Bergeron, Y., & Leduc, A. (1998). Relationships between change in fire frequency and mortality due to spruce budworm outbreak in the southeastern Canadian boreal forest. *Journal of Vegetation Science, 9*, 492–500. https://doi.org/10.2307/3237264.

Bergeron, Y., Leduc, A., Joyal, C., et al. (1995). Balsam fir mortality following the last spruce budworm outbreak in northwestern Quebec. *Canadian Journal of Forest Research, 25*, 1375–1384. https://doi.org/10.1139/x95-150.

Bergeron, Y., Gauthier, S., Kafka, V., et al. (2001). Natural fire frequency for the eastern Canadian boreal forest: Consequences for sustainable forestry. *Canadian Journal of Forest Research, 31*, 384–391. https://doi.org/10.1139/x00-178.

Bergeron, Y., Cyr, D., Girardin, M. P., et al. (2010). Will climate change drive 21st century burn rates in Canadian boreal forest outside of its natural variability: Collating global climate model experiments with sedimentary charcoal data. *International Journal of Wildland Fire, 19*, 1127. https://doi.org/10.1071/WF09092.

Berguet, C., Martin, M., Arseneault, D., et al. (2021). Spatiotemporal dynamics of 20th-century spruce budworm outbreaks in eastern Canada: Three distinct patterns of outbreak severity. *Frontiers in Ecology and Evolution, 8*, 544088. https://doi.org/10.3389/fevo.2020.544088.

Bhiry, N., & Filion, L. (1996). Mid-Holocene hemlock decline in eastern North America linked with phytophagous insect activity. *Quaternary Research, 45*, 312–320. https://doi.org/10.1006/qres.1996.0032.

Bigler, C., & Hall, R. I. (2002). Diatoms as indicators of climatic and limnological change in Swedish lapland: A 100-lake calibration set and its validation for paleoecological reconstructions. *Journal of Paleolimnology, 27*, 97–115. https://doi.org/10.1023/A:1013562325326.

Birks, H. H., & Birks, H. J. B. (2006). Multi-proxy studies in palaeolimnology. *Vegetation History and Archaeobotany, 15*(4), 235–251. https://doi.org/10.1007/s00334-006-0066-6.

Blarquez, O., Ali, A. A., Girardin, M. P., et al. (2015). Regional paleofire regimes affected by non-uniform climate, vegetation and human drivers. *Scientific Reports, 5*, 13356. https://doi.org/10.1038/srep13356.

Bose, A. K., Harvey, B. D., Brais, S., et al. (2014). Constraints to partial cutting in the boreal forest of Canada in the context of natural disturbance-based management: A review. *Forestry, 87*(1), 11–28. https://doi.org/10.1093/forestry/cpt047.

Boucher, Y., Arseneault, D., & Sirois, L. (2009). Logging history (1820–2000) of a heavily exploited southern boreal forest landscape: Insights from sunken logs and forestry maps. *Forest Ecology and Management, 258*, 1359–1368. https://doi.org/10.1016/j.foreco.2009.06.037.

Boulanger, Y., & Arseneault, D. (2004). Spruce budworm outbreaks in eastern Quebec over the last 450 years. *Canadian Journal of Forest Research, 34*, 1035–1043. https://doi.org/10.1139/x03-269.

Boulanger, Y., Arseneault, D., Morin, H., et al. (2012). Dendrochronological reconstruction of spruce budworm (*Choristoneura fumiferana*) outbreaks in southern Quebec for the last 400 years. *Canadian Journal of Forest Research, 42*, 1264–1276. https://doi.org/10.1139/X2012-069.

Bradshaw, R. H. W. (1988). Spatially-precise studies of forest dynamics. In B. Huntley & T. Webb (Eds.), *Vegetation history* (pp. 725–751). Springer.

Brecka, A. F. J., Shahi, C., & Chen, H. Y. H. (2018). Climate change impacts on boreal forest timber supply. *Forest Policy and Economics, 92*, 11–21. https://doi.org/10.1016/j.forpol.2018.03.010.

Cayer, D., & Bhiry, N. (2014). Holocene climate and environmental changes in western subarctic Quebec as inferred from the sedimentology and the geomorphology of a lake watershed. *Arctic, Antarctic, and Alpine Research, 46*(1), 55–65. https://doi.org/10.1657/1938-4246.46.1.55.

Clark, J. S. (1988). Particle motion and the theory of charcoal analysis: Source area, transport, deposition, and sampling. *Quaternary Research, 30*(1), 67–80. https://doi.org/10.1016/0033-5894(88)90088-9.

Colombaroli, D., Tinner, W., Van Leeuwen, J., et al. (2009). Response of broadleaved evergreen mediterranean forest vegetation to fire disturbance during the holocene: Insights from the peri-Adriatic region. *Journal of Biogeography, 36*(2), 314–326. https://doi.org/10.1111/j.1365-2699.2008.01987.x.

Cooke, B. J., & Roland, J. (2000). Spatial analysis of large–scale patterns of forest tent caterpillar outbreaks. *Ecoscience, 7*, 410–422. https://doi.org/10.1080/11956860.2000.11682611.

Crump, S. E. (2021). Sedimentary ancient DNA as a tool in paleoecology. *Nature Reviews Earth & Environment, 2*, 229. https://doi.org/10.1038/s43017-021-00158-8.

Dale, V. H., Joyce, L. A., McNulty, S., et al. (2001). Climate change and forest disturbances: Climate change can affect forests by altering the frequency, intensity, duration, and timing of fire, drought, introduced species, insect and pathogen outbreaks, hurricanes, windstorms, ice storms, or landslides. *BioScience, 51*(9), 723–734. https://doi.org/10.1641/0006-3568(2001)051[0723:Ccafd]2.0.Co;2.

Davis, R. B., & Anderson, R. S. (1980). A new parameter for paleoecological reconstruction: Head capsules of forest tree defoliator microlepidopterans in lake sediment. In Institute of Quaternary Studies (Ed.) *Abstracts and Program of the 6th Biennial Meeting of the American Quaternary Association* (p. 62). Orono: University of Maine.

De Grandpré, L., Kneeshaw, D. D., Perigon, S., et al. (2019). Adverse climatic periods precede and amplify defoliator-induced tree mortality in eastern boreal North America. *Journal of Ecology, 107*, 452–467. https://doi.org/10.1111/1365-2745.13012.

De Groot, W. J., Flannigan, M. D., & Cantin, A. S. (2013). Climate change impacts on future boreal fire regimes. *Forest Ecology and Management, 294*, 35–44. https://doi.org/10.1016/j.foreco.2012.09.027.

Dodd, J. R., & Stanton, R. J. (1990). *Paleoecology: Concepts and applications* (p. 528). John Wiley and Sons.

Drobyshev, I., Niklasson, M., Angelstam, P., et al. (2004). Testing for anthropogenic influence on fire regime for a 600-year period in the Jaksha area, Komi Republic, East European Russia. *Canadian Journal of Forest Research, 34*(10), 2027–2036. https://doi.org/10.1139/x04-081.

Drobyshev, I., Granström, A., Linderholm, H. W., et al. (2014). Multi-century reconstruction of fire activity in northern European boreal forest suggests differences in regional fire regimes and their sensitivity to climate. *Journal of Ecology, 102*(3), 738–748. https://doi.org/10.1111/1365-2745.12235.

Drobyshev, I., Bergeron, Y., de Vernal, A., et al. (2016). Atlantic SSTs control regime shifts in forest fire activity of Northern Scandinavia. *Scientific Reports, 6*, 22532. https://doi.org/10.1038/srep22532.

Dubois, N., & Jacob, J. (2016). Molecular biomarkers of anthropic impacts in natural archives: A review. Frontiers in Ecology and Evolution, 4(92). https://doi.org/10.3389/fevo.2016.00092

Dyke, A. S. (2004) An outline of North American deglaciation with emphasis on central and northern Canada. In J. Ehlers, & P. L. Gibbard, (Eds.) *Quaternary glaciations—extent and chronology: Part II: North America.* Developments in Quaternary science 2 (pp. 373–424). Amsterdam: Elsevier.

Eronen, M., Zetterberg, P., Briffa, K. R., et al. (2002). The supra-long scots pine tree-ring record for Finnish Lapland: Part 1, chronology construction and initial inferences. *The Holocene, 12,* 673–680. https://doi.org/10.1191/0959683602hl580rp.

Esper, J., Büntgen, U., Frank, D. C., et al. (2007). 1200 years of regular outbreaks in alpine insects. *Proceedings of the Royal Society B: Biological Sciences, 274*(1610), 671–679. https://doi.org/10.1098/rspb.2006.0191.

Feurdean, A., Veski, S., Florescu, G., et al. (2017). Broadleaf deciduous forest counterbalanced the direct effect of climate on Holocene fire regime in hemiboreal/boreal region (NE Europe). *Quaternary Science Reviews, 169,* 378–390. https://doi.org/10.1016/j.quascirev.2017.05.024.

Flannigan, M., Campbell, I., Wotton, M., et al. (2001). Future fire in Canada's boreal forest: Paleoecology results and general circulation model—regional climate model simulations. *Canadian Journal of Forest Research, 31,* 854–864. https://doi.org/10.1139/cjfr-31-5-854.

Flannigan, M., Stocks, B., Turetsky, M., et al. (2009). Impacts of climate change on fire activity and fire management in the circumboreal forest. *Global Change Biology, 15*(3), 549–560. https://doi.org/10.1111/j.1365-2486.2008.01660.x.

Flower, A., Gavin, D. G., Heyerdahl, E. K., et al. (2014). Drought-triggered western spruce budworm outbreaks in the interior Pacific Northwest: A multi-century dendrochronological record. *Forest Ecology and Management, 324,* 16–27. https://doi.org/10.1016/j.foreco.2014.03.042.

Gaboriau, D. M., Remy, C. C., Girardin, M. P., et al. (2020). Temperature and fuel availability control fire size/severity in the boreal forest of central Northwest Territories, Canada. *Quaternary Science Reviews, 250,* 106697. https://doi.org/10.1016/j.quascirev.2020.106697.

Gaglioti, B. V., Mann, D. H., Jones, B. M., et al. (2016). High-resolution records detect human-caused changes to the boreal forest wildfire regime in interior Alaska. *The Holocene, 26*(7), 1064–1074. https://doi.org/10.1177/0959683616632893.

Gauthier, S., Bernier, P., Kuuluvainen, T., et al. (2015). Boreal forest health and global change. *Science, 349*(6250), 819–822. https://doi.org/10.1126/science.aaa9092.

Gennaretti, F., Arseneault, D., & Bégin, Y. (2014a). Millennial stocks and fluxes of large woody debris in lakes of the North American taiga. *Journal of Ecology, 102,* 367–380. https://doi.org/10.1111/1365-2745.12198.

Gennaretti, F., Arseneault, D., & Bégin, Y. (2014b). Millennial stocks and fluxes of large woody debris in lakes of the North American taiga. *Journal of Ecology, 102,* 367–380. https://doi.org/10.1111/1365-2745.12198.

Gennaretti, F., Arseneault, D., Nicault, A., et al. (2014c). Volcano-induced regime shifts in millennial tree-ring chronologies from northeastern North America. *Proceedings of the National Academy of Sciences of the United States of America, 111,* 10077–10082. https://doi.org/10.1073/pnas.132 4220111.

Gennaretti, F., Boucher, E., Nicault, A., et al. (2018). Underestimation of the Tambora effects in North American taiga ecosystems. *Environmental Research Letters, 13*(3), 034017. https://doi.org/10.1088/1748-9326/aaac0c.

Girardin, M. P., & Terrier, A. (2015). Mitigating risks of future wildfires by management of the forest composition: An analysis of the offsetting potential through boreal Canada. *Climatic Change, 130*(4), 587–601. https://doi.org/10.1007/s10584-015-1373-7.

Girardin, M. P., Tardif, J., & Bergeron, Y. (2001). Radial growth analysis of Larix laricina from the Lake Deparquet area, Quebec, in relation to climate and larch sawfly outbreaks. *Ecoscience, 8,* 127–138. https://doi.org/10.1080/11956860.2001.11682638.

Girardin, M. P., Tardif, J., & Bergeron, Y. (2002). Dynamics of eastern larch stands and its relationships with larch sawfly outbreaks in the northern Clay Belt of Quebec. *Canadian Journal of Forest Research, 32,* 206–216. https://doi.org/10.1139/x01-185.

Girardin, M. P., Ali, A. A., Carcaillet, C., et al. (2013a). Vegetation limits the impact of a warm climate on boreal wildfires. *New Phytologist, 199*(4), 1001–1011. https://doi.org/10.1111/nph.12322.

Girardin, M. P., Ali, A. A., Carcaillet, C., et al. (2013b). Fire in managed forests of eastern Canada: Risks and options. *Forest Ecology and Management, 294*, 238–249. https://doi.org/10.1016/j.foreco.2012.07.005.

Girardin, M. P., Guo, X. J., De Jong, R., et al. (2014). Unusual forest growth decline in boreal North America covaries with the retreat of Arctic sea ice. *Global Change Biology, 20*(3), 851–866. https://doi.org/10.1111/gcb.12400.

Girardin, M. P., Portier, J., Remy, C. C., et al. (2019). Coherent signature of warming-induced extreme sub-continental boreal wildfire activity 4800 and 1100 Years BP. *Environmental Research Letters, 14*(12), 124042. https://doi.org/10.1088/1748-9326/ab59c9.

Grudd, H., Briffa, K. R., Karlén, W., et al. (2002). A 7400-year tree-ring chronology in northern Swedish Lapland: Natural climatic variability expressed on annual to millennial timescales. *The Holocene, 12*, 657 665. https://doi.org/10.1191/0959683602hl578rp.

Hantemirov, R. M., & Shiyatov, S. G. (2002). A continuous-multimillennial ring-width chronology in Yamal, northwestern Siberia. *The Holocene, 12*, 717–726. https://doi.org/10.1191/095968360 2hl585rp.

Helama, S., Lindholm, M., Timonen, M., et al. (2005). Mid- and late Holocene tree population density changes in northern Fennoscandia derived by a new method using megafossil pines and their tree-ring series. *Journal of Quaternary Science, 20*, 567–575. https://doi.org/10.1002/jqs.929.

Higuera, P. E., Gavin, D. G., Bartlein, P. J., et al. (2010). Peak detection in sediment-charcoal records: Impacts of alternative data analysis methods on fire history interpretations. *International Journal of Wildland Fire, 19*(8), 996–1014. https://doi.org/10.1071/WF09134.

Hoogesteger, J. (2006). *Tree ring dynamics in mountain birch.* Licentiate thesis, Swedish University of Agricultural Sciences.

Jardon, Y., Filion, L., & Cloutier, C. (1994). Tree ring evidence for endemicity of the larch sawfly in North America. *Canadian Journal of Forest Research, 24*, 742–747. https://doi.org/10.1139/x94-098.

Jardon, Y., Morin, H., & Dutilleul, P. (2003). Periodicity and synchronism of spruce budworm outbreaks in Quebec. *Canadian Journal of Research, 33*(10), 1947–1961. https://doi.org/10.1139/x03-108.

Jasinski, J. P. P., & Payette, S. (2007). Holocene occurrence of Lophodermium piceae, a black spruce needle endophyte and possible paleoindicator of boreal forest health. *Quaternary Research, 67*(1), 50–56. https://doi.org/10.1016/j.yqres.2006.07.008.

Jolly, W. M., Cochrane, M. A., Freeborn, P. H., et al. (2015). Climate-induced variations in global wildfire danger from 1979 to 2013. *Nature Communications, 6*, 7537. https://doi.org/10.1038/ncomms8537.

Klapwijk, M. J., Csóka, G., Hirka, A., et al. (2013). Forest insects and climate change: Long-term trends in herbivore damage. *Ecology and Evolution, 3*, 4183–4196. https://doi.org/10.1002/ece3.717.

Krause, C. (1997). The use of dendrochronological material from buildings to get information about past spruce budworm outbreaks. *Canadian Journal of Forest Research, 27*, 69–75. https://doi.org/10.1139/x96-168.

Krawchuk, M. A., & Cumming, S. G. (2011). Effects of biotic feedback and harvest management on boreal forest fire activity under climate change. *Ecological Applications, 21*(1), 122–136. https://doi.org/10.1890/09-2004.1.

Krawchuk, M. A., Cumming, S. G., & Flannigan, M. D. (2009). Predicted changes in fire weather suggest increases in lightning fire initiation and future area burned in the mixedwood boreal forest. *Climatic Change, 92*, 83–97. https://doi.org/10.1007/s10584-008-9460-7.

Kullman, L. (1995). Holocene tree-limit and climate history from the Scandes Mountains, Sweden. *Ecology, 76*, 2490–2502. https://doi.org/10.2307/2265823.

Kuosmanen, N., Fang, K., Bradshaw, R. H. W., et al. (2014). Role of forest fires in Holocene stand-scale dynamics in the unmanaged taiga forest of northwestern Russia. *The Holocene, 24,* 1503–1514. https://doi.org/10.1177/0959683614544065.

Kuuluvainen, T. (2002). Natural variability of forests as a reference for restoring and managing biological diversity in boreal Fennoscandia. *Silva Fennica, 36*(1), 552. https://doi.org/10.14214/sf.552.

Larjavaara, M., Pennanen, J., & Tuomi, T. J. (2005). Lightning that ignites forest fires in Finland. *Agricultural and Forest Meteorology, 132*(3–4), 171–180. https://doi.org/10.1016/j.agrformet.2005.07.005.

Larsen, C. P. S. (1996). Fire and climate dynamics in the boreal forest of northern Alberta, Canada, from AD 1850 to 1989. *The Holocene, 6*(4), 449–456. https://doi.org/10.1177/095968369600600407.

Lehman, J. T. (1975). Reconstructing the rate of accumulation of lake sediment: The effect of sediment focusing. *Quaternary Research, 5,* 541–550. https://doi.org/10.1016/0033-5894(75)90015-0.

Liu, K. B. (1990) Holocene paleoecology of the boreal forest and Great Lakes-St. Lawrence forest in Northern Ontario. *Ecological Monographs, 60*(2), 179–212. https://doi.org/10.2307/1943044.

MacLean, D. A. (1984). Effects of spruce budworm outbreaks on the productivity and stability of balsam fir forests. *The Forestry Chronicle, 60,* 273–279. https://doi.org/10.5558/tfc60273-5.

Magnan, G., Le Stum-Boivin, E., Garneau, M., et al. (2018). Holocene vegetation dynamics and hydrological variability in forested peatlands of the Clay belt, eastern Canada, reconstructed using a palaeoecological approach. *Boreas, 48*(1), 131–146. https://doi.org/10.1111/bor.12345.

Martin, M., Montoro Girona, M., & Morin, H. (2020). Driving factors of conifer regeneration dynamics in eastern Canadian boreal old-growth forests. *PLoS ONE, 15,* e0230221. https://doi.org/10.1371/journal.pone.0230221.

Mauquoy, D., & Van Geel, B. (2007). Plant macrofossil methods and studies: Mire and peat macros. In S. A. Elias (Ed.), *Encyclopedia of Quaternary science* (pp. 2315–2336). Elsevier Science.

McCullough, D. G., Werner, R. A., & Neumann, D. (1998). Fire and insects in northern and boreal forest ecosystems of North America. *Annual Review of Entomology, 43,* 107–127. https://doi.org/10.1146/annurev.ento.43.1.107.

Molinari, C., Carcaillet, C., Bradshaw, R. H. W., et al. (2020). Fire-vegetation interactions during the last 11,000 years in boreal and cold temperate forests of Fennoscandia. *Quaternary Science Reviews, 241,* 106408. https://doi.org/10.1016/j.quascirev.2020.106408.

Montoro Girona, M., Morin, H., Lussier, J. M., et al. (2016). Radial growth response of black spruce stands ten years after experimental shelterwoods and seed-tree cuttings in boreal forest. *Forests, 7,* 240. https://doi.org/10.3390/f7100240.

Montoro Girona, M., Lussier, J. M., Morin, H., et al. (2018a). Conifer regeneration after experimental shelterwood and seed-tree treatments in boreal forests: Finding silvicultural alternatives. *Frontiers in Plant Science, 9,* 1145. https://doi.org/10.3389/fpls.2018.01145.

Montoro Girona, M., Navarro, L., & Morin, H. (2018b). A secret hidden in the sediments: Lepidoptera scales. *Frontiers in Ecology and Evolution, 6,* 2. https://doi.org/10.3389/fevo.2018.00002.

Morin, H., Gagnon, R., Lemay, A., et al. (2020). Revisiting the relationship between spruce budworm outbreaks and forest dynamics over the Holocene in Eastern North America based on novel proxies. In E. A. Johnson & K. Miyanishi (eds.), *Plant disturbance ecology: The process and the response* 2nd edition (pp. 463–488). San Diego: Academic Press.

Morin, H., & Laprise, D. (1990). Histoire récente des épidémies de la Tordeuse des bourgeons de l'épinette au nord du lac Saint-Jean (Québec): Une analyse dendrochronologique. *Canadian Journal of Forest Research, 20,* 1–8. https://doi.org/10.1139/x90-001.

Moussaoui, L., Leduc, A., Montoro Girona, M., et al. (2020). Success factors for experimental partial harvesting in unmanaged boreal forest: 10-year stand yield results. *Forests, 11,* 1199. https://doi.org/10.3390/f11111199.

Navarro, L., Harvey, A. É., Ali, A., et al. (2018a). A Holocene landscape dynamic multiproxy reconstruction: How do interactions between fire and insect outbreaks shape an ecosystem over long time scales? *PLoS ONE, 13*, e0204316. https://doi.org/10.1371/journal.pone.0204316.

Navarro, L., Harvey, A. É., & Morin, H. (2018b). Lepidoptera wing scales: A new paleoecological indicator for reconstructing spruce budworm abundance. *Canadian Journal of Forest Research, 48*, 302–308. https://doi.org/10.1139/cjfr-2017-0009.

Navarro, L., Morin, H., Bergeron, Y., et al. (2018c). Changes in spatiotemporal patterns of 20th century spruce budworm outbreaks in eastern Canadian boreal forests. *Frontiers in Plant Science, 9*, 1905. https://doi.org/10.3389/fpls.2018.01905.

Nealis, V. G., & Régnière, J. (2004). Insect-host relationships influencing disturbance by the spruce budworm in a boreal mixedwood forest. *Canadian Journal of Forest Research, 34*, 1870–1882. https://doi.org/10.1139/X04-061.

Nehemy, M. F., & Laroque, C. P. (2018). Tree-ring analysis of larch sawfly (*Pristiphora erichsonii* (Hartig)) defoliation events and hydrological growth suppression in a peatland. *Dendrochronologia, 51*, 1–9. https://doi.org/10.1016/j.dendro.2018.06.006.

Niklasson, M., & Granström, A. (2000). Numbers and sizes of fires: Long-term spatially explicit fire history in a Swedish boreal landscape. *Ecology, 81*, 1484–1499. https://doi.org/10.1890/0012-9658(2000)081[1484:NASOFL]2.0.CO;2.

Ohlson, M., Brown, K. J., Birks, H. J. B., et al. (2011). Invasion of Norway spruce diversifies the fire regime in boreal European forests. *Journal of Ecology, 99*(2), 395–403. https://doi.org/10.1111/j.1365-2745.2010.01780.x.

Oris, F., Ali, A. A., Asselin, H., et al. (2014). Charcoal dispersion and deposition in boreal lakes from 3 years of monitoring: Differences between local and regional fires. *Geophysical Research Letters, 41*, 6743–6752. https://doi.org/10.1002/2014GL060984.

Parisien, M. A., Barber, Q. E., Hirsch, K. G., et al. (2020). Fire deficit increases wildfire risk for many communities in the Canadian boreal forest. *Nature Communications, 11*(1), 2121. https://doi.org/10.1038/s41467-020-15961-y.

Payette, S., Delwaide, A., Schaffhauser, A., et al. (2012). Calculating long-term fire frequency at the stand scale from charcoal data. *Ecosphere, 3*(7), 1–16. https://doi.org/10.1890/ES12-00026.1.

Pitkänen, A., & Huttunen, P. (1999). A 1300-year forest-fire history at a site in eastern Finland based on charcoal and pollen records in laminated lake sediment. *The Holocene, 9*(3), 311–320. https://doi.org/10.1191/095968399667329540.

Potelle, B. (1995). *Potentiel de l'analyse des macrorestes pour détecter les épidémies de la tordeuse des bourgeons de l'épinette dans des sols de sapinières boréales.* M.Sc. thesis, Université du Québec à Chicoutimi.

Remy, C. C., Hély, C., Blarquez, O., et al. (2017a). Different regional climatic drivers of Holocene large wildfires in boreal forests of northeastern America. *Environmental Research Letters, 12*(3), 035005. https://doi.org/10.1088/1748-9326/aa5aff.

Remy, C. C., Lavoie, M., Girardin, M. P., et al. (2017b). Wildfire size alters long-term vegetation trajectories in boreal forests of eastern North America. *Journal of Biogeography, 44*(6), 1268–1279. https://doi.org/10.1111/jbi.12921.

Remy, C. C., Fouquemberg, C., Asselin, H., et al. (2018). Guidelines for the use and interpretation of palaeofire reconstructions based on various archives and proxies. *Quaternary Science Reviews, 193*, 312–322. https://doi.org/10.1016/j.quascirev.2018.06.010.

Remy, C. C., Senici, D., Chen, H. Y. H., et al. (2019). Coniferization of the mixed-wood boreal forests under warm climate. *Journal of Quaternary Science, 34*(7), 509–518. https://doi.org/10.1002/jqs.3136.

Rolland, C., Baltensweiler, W., & Petitcolas, V. (2001). The potential for using Larix decidua ring widths in reconstructions of larch budmoth (*Zeiraphera diniana*) outbreak history: Dendrochronological estimates compared with insect surveys. *Trees, 15*, 414–424. https://doi.org/10.1007/s004680100116.

Rolstad, J., Blanck, Y. L., & Storaunet, K. O. (2017). Fire history in a western Fennoscandian boreal forest as influenced by human land use and climate. *Ecological Monographs, 87*(2), 219–245. https://doi.org/10.1002/ecm.1244.

Schafstall, N., Whitehouse, N., Kuosmanen, N., et al. (2020). Changes in species composition and diversity of a montane beetle community over the last millennium in the High Tatras, Slovakia: Implications for forest conservation and management. *Palaeogeography, Palaeoclimatology, Palaeoecology, 555*, 109834. https://doi.org/10.1016/j.palaeo.2020.109834.

Shorohova, E., Kneeshaw, D. D., Kuuluvainen, T., et al. (2011) Variability and dynamics of old-growth forests in the circumboreal zone: Implications for conservation, restoration and management. *Silva Fennica, 45*(5), 785–806. https://doi.org/10.14214/sf.72.

Simard, I., Morin, H., & Potelle, B. (2002). A new paleoecological approach to reconstruct long-term history of spruce budworm outbreaks. *Canadian Journal of Forest Research, 32*, 428–438. https://doi.org/10.1139/x01-215.

Simard, I., Morin, H., & Lavoie, C. (2006). A millennial-scale reconstruction of spruce budworm abundance in Saguenay, Quebec, Canada. *The Holocene, 16*, 31–37. https://doi.org/10.1191/095 9683606hl904rp.

Simard, S., Morin, H., & Krause, C. (2011). Long-term spruce budworm outbreak dynamics reconstructed from subfossil trees. *Journal of Quaternary Science, 26*, 734–738. https://doi.org/10.1002/jqs.1492.

Spurk, M., Leuschner, H. H., Baillie, M. G. L., et al. (2002). Depositional frequency of German subfossil oaks: Climatically and non-climatically induced fluctuations in the Holocene. *The Holocene, 12*, 707–715. https://doi.org/10.1191/0959683602hl583rp.

Stivrins, N., Aakala, T., Ilvonen, L., et al. (2019). Integrating fire-scar, charcoal and fungal spore data to study fire events in the boreal forest of northern Europe. *The Holocene, 29*, 1480–1490. https://doi.org/10.1177/0959683619854524.

Sutton, A., & Tardif, J. C. (2007). Dendrochronological reconstruction of forest tent caterpillar outbreaks in time and space, western Manitoba, Canada. *Canadian Journal of Forest Research, 37*, 1643–1657. https://doi.org/10.1139/X07-021.

Swetnam, T. W., & Lynch, A. M. (1993). Multicentury, regional-scale patterns of western spruce budworm outbreaks. *Ecological Monographs, 63*, 399–424. https://doi.org/10.2307/2937153.

Tikkanen, O. P., & Roininen, H. (2001). Spatial pattern of outbreaks of *Operophtera brumata* in eastern Fennoscandia and their effects on radial growth of trees. *Forest Ecology and Management, 146*, 45–54. https://doi.org/10.1016/S0378-1127(00)00451-5.

Tolonen, M. (1978). Palaeoecology of annually laminated sediments in Lake Ahvenainen, S. Finland. I. Pollen and charcoal analyses and their relation to human impact. *Annales Botanici Fennici, 15*(3), 177–208.

Turetsky, M. R., Kane, E. S., Harden, J. W., et al. (2011). Recent acceleration of biomass burning and carbon losses in Alaskan forests and peatlands. *Nature Geoscience, 4*(1), 27–31. https://doi.org/10.1038/ngeo1027.

Viau, A. E., & Gajewski, K. (2009). Reconstructing millennial-scale, regional paleoclimates of boreal Canada during the Holocene. *Journal of Climate, 22*(2), 316–330. https://doi.org/10.1175/2008JCLI2342.1.

Volney, W. J. A. (1988). Analysis of historic jack pine budworm outbreaks in the Prairie provinces of Canada. *Canadian Journal of Forest Research, 18*, 1152–1158. https://doi.org/10.1139/x88-177.

Volney, W. J. A., & Fleming, R. A. (2000). Climate change and impacts of boreal forest insects. *Agriculture, Ecosystems & Environment, 82*, 283–294. https://doi.org/10.1016/S0167-8809(00)00232-2.

Waito, J., Girardin, M. P., Tardif, J. C., et al. (2015). Fire and climate: Using the past to predict the future. In K.S.-H. Peh. R. T. Corlett & Y. Bergeron (Eds.), *Routledge handbook of forest ecology* (pp. 473–487). Routledge.

Waito, J., Girardin, M. P., Tardif, J. C., et al. (2018). Recent fire activity in the boreal eastern interior of North America is below that of the past 2000 yr. *Ecosphere, 9*(6), e02287. https://doi.org/10.1002/ecs2.2287.

Walker, X. J., Rogers, B. M., Baltzer, J. L., et al. (2018). Cross-scale controls on carbon emissions from boreal forest megafires. *Global Change Biology, 24*(9), 4251–4265. https://doi.org/10.1111/gcb.14287.

Wallenius, T. (2011) Major decline in fires in coniferous forests-reconstructing the phenomenon and seeking for the cause. *Silva Fennica, 45*, 139–155. https://doi.org/10.14214/sf.36.

Wallenius, T. H., Kauhanen, H., Herva, H., et al. (2010). Long fire cycle in northern boreal Pinus forests in Finnish Lapland. *Canadian Journal of Forest Research, 40*, 2027–2035. https://doi.org/10.1139/X10-144.

Weber, U. M. (1997). Dendroecological reconstruction and interpretation of larch budmoth (*Zeiraphera diniana*) outbreaks in two central alpine valleys of Switzerland from 1470–1990. *Trees, 11*, 277–290. https://doi.org/10.1007/PL00009674.

Young, A. B., Cairns, D. M., Lafon, C. W., et al. (2014). Geometrid moth outbreaks and their climatic relations in northern Sweden. *Arctic, Antarctic, and Alpine Research, 46*, 659–668. https://doi.org/10.1657/1938-4246-46.3.659.

Zackrisson, O. (1977). Influence of forest fires on the North Swedish boreal forest. *Oikos, 29*, 22–32. https://doi.org/10.2307/3543289.

Chapter 3
Natural Disturbances from the Perspective of Forest Ecosystem-Based Management

Ekaterina Shorohova, Tuomas Aakala, Sylvie Gauthier, Daniel Kneeshaw, Matti Koivula, Jean-Claude Ruel, and Nina Ulanova

Abstract Natural disturbances drive forest dynamics and biodiversity at different spatial and temporal scales. Forests in the boreal biome are shaped by several types of disturbance, including fire, windthrow, and insect outbreaks, that vary in frequency, extent, severity, and specificity. In managed forests, disturbances also affect the amount and quality of available timber. Ecosystem management uses information on disturbance regimes as a guide to finding a balance between ecological, economic, and social viewpoints. In this chapter, we review current knowledge on disturbance regimes in boreal forests and discuss some implications for managing the impact and risk of disturbances in the context of forest ecosystem management and restoration.

E. Shorohova (✉)
Forest Research Institute of the Karelian Research Center, Russian Academy of Science, Pushkinskaya str. 11, Petrozavodsk 185910, Russia
e-mail: shorohova@es13334.spb.edu; ekaterina.shorokhova@luke.fi

Saint Petersburg State Forest Technical University, Institutsky str. 5, Saint Petersburg 194021, Russia

E. Shorohova · M. Koivula
Natural Resources Institute Finland (Luke), Latokartanonkaari 9, FI-00790 Helsinki, Finland
e-mail: matti.koivula@luke.fi

T. Aakala
School of Forest Sciences, University of Eastern Finland, P.O. Box 111, FI-80101 Joensuu, Finland
e-mail: tuomas.aakala@uef.fi

S. Gauthier
Natural Resources Canada, Canadian Forest Service, Laurentian Forestry Centre, 1055 rue du PEPS, P.O. Box 10380, Stn. Sainte-Foy, Québec, QC G1V 4C7, Canada
e-mail: Sylvie.gauthier@nrcan-rncan.gc.ca

D. Kneeshaw
Centre for Forest Research, Université du Québec à Montréal, P.O. Box 8888, Stn. Centre-Ville, Montréal, QC H3C 3P8, Canada
e-mail: kneeshaw.daniel@uqam.ca

© The Author(s) 2023
M. M. Girona et al. (eds.), *Boreal Forests in the Face of Climate Change*,
Advances in Global Change Research 74,
https://doi.org/10.1007/978-3-031-15988-6_3

Box. 3.1 Definitions of Terms Used in the Chapter

A *disturbance* is defined as a relatively discrete event that affects the structure of an ecosystem, community, or population and that modifies resources, substrate availability, or the physical environment (Pickett & White, 1985).

A *disturbance regime* consists of a combination of all characteristics generated by one or several disturbance agents acting within a given land area. Some principal descriptors related to natural disturbance regimes are listed below.

Intensity: the physical force of the event per area per unit of time (e.g., heat, wind speed)

Severity: the impact of the disturbance on an organism, community, or ecosystem (e.g., tree mortality)

Duration: the time (minutes to years) from the beginning to the end of a single disturbance event

Frequency: the proportion of area affected annually. Return interval = 1/frequency

Specificity: the selective nature of a disturbance agent toward one or several types of habitat or species

3.1 Fire

Fire is a dominant disturbance in circumboreal forests (Gauthier et al., 2015b), and it has been the basis for many emulation and restoration strategies. Circumboreal fire regimes are, however, highly variable (Buryak et al., 2003; Furyaev, 1996; Gromtsev, 2002; Rogers et al., 2015; Sofronov & Volokitina, 1990). In North American boreal forests, crown fires dominate (Rogers et al., 2015; Wooster & Zhang, 2004), although fire severity, i.e., the magnitude of the impact of fire on living plants and the soil organic layer, varies within and between events as well as within a fire season (April to October; Guindon et al., 2021). Eurasian boreal forests are shaped by mixed-severity fire regimes, where variation in fire severity is driven by climate and weather, vegetation, and characteristics of the soil and bedrock (Gromtsev, 2008; Sofronov & Volokitina, 1990; Valendik & Ivanova, 2001). Flammability is similarly dependent on the above factors and is often inversely related to severity, i.e., easily ignited areas

J.-C. Ruel
Faculté de Foresterie, Géographie et Géomatique, Université Laval, Pavillon Abitibi-Price, 2405 rue de la terrasse, Québec, QC G1V 0A6, Canada
e-mail: jean-claude.ruel@sbf.ulaval.ca

N. Ulanova
Faculty of Biology, Moscow State University, Leninskie Gory 1-12, Moscow 119991, Russia
e-mail: nulanova@mail.ru

are often subject to low-intensity surface fires. Flammability and fire severity can also be influenced by the occurrence of other disturbances, which affect the quality and quantity of fuel; for example, when poorly flammable forests are disturbed by an insect outbreak, they may become more flammable and, consequently, the resulting fire may display a crown-fire behavior.

As the characteristics of current fire regimes vary widely between forest regions in Canada and Eurasia, we detail below the regimes for Canadian and Eurasian boreal forests separately. We further distinguish the Fennoscandian boreal forests of Finland, Sweden, and Norway, as their respective fire regimes and fire-related management challenges differ greatly from the rest of the boreal zone.

3.1.1 Current Fire Regimes

3.1.1.1 Canada

The annual area burned varies markedly in Canadian boreal forests (Boulanger et al., 2014; Hanes et al., 2019). On average, 8,000 fires burn around 2 million ha of forest across the country each year (Gauthier et al., 2015a; Hanes et al., 2019). For fires larger than 200 ha (data covering 1959–2015), 85% were ignited by lightning (Hanes et al., 2019), whereas smaller fires may include a greater share of human-caused fires (Cardil et al., 2019). The regional annual burn rates (i.e., the fraction of the region that burns on average every year, compiled for 1959–1999) can vary from approximately 0.05% to 0.1% per year in northern and eastern regions to 1.5% per year in western and central Canada; this corresponds to return intervals of 2000 and 67 years, respectively (Boulanger et al., 2014). This relative interregional difference is expected to persist with climate change, whereas the total area burned is predicted to increase (Boulanger et al., 2014). Most fires are small, whereas a few large lightning-ignited fires are responsible for most of the area burned (Hanes et al., 2019).

In Canada, debates continue in regard to fire frequency and the influence of stand age, fuel types, and site conditions versus that of climate and weather (Bessie & Johnson, 1995; Cumming, 2001; Erni et al., 2018; Héon et al., 2014; Lefort et al., 2003). Under a given regional fire regime, deciduous forests are less likely to burn than coniferous ones (Bernier et al., 2016), and young and low-biomass forests are less likely to burn than older and high-biomass ones. In regions having the highest burn rate (1.5% per year), the return interval—the inverse of burn rate—is 66 years. Young (<30 years) deciduous forests and old (>90 years) coniferous forests have burn rates of 0.14% and 2.82% per year, respectively, whereas in regions experiencing the lowest regional burn rate (0.05% per year and a 2000-year return interval), the respective burn rate would be between 0.005% and 0.09% per year. As most future projections of fire burn rate are based only on future climatic conditions, accounting for this variation in fire selectivity can markedly change the outcome of projections, notably in areas where the burn rate is projected to be above 1% (Boulanger et al., 2017).

The occurrence of successive fires at short intervals may cause the regeneration failure of many tree species, contributing to a shift from a closed forest cover to an open woodland (Payette & Delwaide, 2018). This scenario occurs, for example, when young forests burn before a propagule bank—which can ensure post-disturbance forest recovery—has been constituted. With the projected increase in fire frequencies across Canada, regeneration failure may become more common in some stand types (Baltzer et al., 2021; Splawinski et al., 2019).

3.1.1.2 Russia

The annual area burned in Russia is considerably greater than that in North American boreal forests. In 2020, for example, 35,134 forest fires burned 16.44 million ha. However, there is a strong geographic gradient with over 90% of burned areas situated east of the Ural Mountains, i.e., in the Asian portion of Russia (Sofronov & Volokitina, 1990). In Siberia, 83% of fires occur in eastern Siberia and the Far East, whereas 17% occur in western Siberia. Western Siberia is characterized by a low-frequency fire regime. Variations in climate and vegetation drive these differences; frequent surface fires characterize the eastern part with easily flammable *light conif- erous* forests dominated by Scots pine (*Pinus sylvestris* L.) and Siberian larch (*Larix gmelinii* Rupr. and *L. sibirica* Ledeb.) (Buryak et al., 2003; Korovin, 1996), whereas the western region burns less intensively, consisting mostly of *dark coniferous* forests dominated by Siberian spruce (*Picea obovata* Ledeb.), Siberian fir (*Abies sibirica* Ledeb.), and Siberian pine (*Pinus sibirica* Du Tour.). Climate also imposes a latitu- dinal gradient within the Siberian region; for example, in Siberian larch forests, the mean fire return interval increases with latitude, from 80 years at 64°N to about 200 years near the Arctic Circle and about 300 years near the northern range limit of larch forests (71°N) (Kharuk et al., 2016a). Among vegetation types, recently harvested southern boreal forests are considered as the most flammable of all Siberian forests, mainly because logging slash burns easily (Valendik et al., 2013). These differences in fire frequency are inversely related to fire severity; forest types that burn often are mostly subject to surface fires, characterized by low fuel loads and tree species adapted to survive frequent fires. However, fire severity varies greatly even within a given landscape type (Fig. 3.1a). Whereas surface fires are generally more common than crown fires, patchy crown fires can represent 50% of the total area burned during severe fire seasons (Belov, 1976; Valendik & Ivanova, 2001).

West of the Ural Mountains, in the European boreal forests of Russia, fire return intervals can vary among landscapes from 40 to more than 200 years, depending on site conditions (Melekhov, 1971; Zyabchenko, 1984), dominant tree species, land-forms, and bedrock (Gromtsev, 2008). The variation in natural fire regimes is driven by differences in superficial deposits and topography that create a landscape mosaic with varying flammability and fuels. Similar to the Siberian part of the boreal forest, pine-dominated forests burn with a higher frequency, typically as surface fires but occasionally as crown fires. The most fire-prone pine forests tend to burn at least twice per century as surface fires and three to four times per millennium as crown

Fig. 3.1 Varying fire severity in boreal forests. **a** Six years after a mixed-severity fire of ca. 4,000 ha in the dark coniferous forests of the Eastern Sayan Mountains, Siberia, Russia; **b** crown fire in the coniferous boreal forest of eastern Canada; **c** patchily burned area six years after a surface fire in the north-eastern boreal primeval rocky Scots pine forest, Karelia, Russia; **d** burning for ecological restoration in a southern boreal Scots pine forest, Finland. *Photo credits* **a** Ilkka Vanha-Majamaa, **b** Société de Protection des forêts contre les feux (SOPFEU), **c** Daria Glazunova, **d** Erkki Oksanen/LUKE archive

fires, whereas the less fire-prone pine forests burn with higher severity as crown fires, one to three times every 300 years (Gromtsev, 2008). Forests dominated by Norway spruce (*Picea abies* (L.) H. Karst.) burn as either intense crown fires triggered by severe droughts at a mean return period of once or twice per millennium or as more frequent but lower-severity ground fires (Gromtsev, 2008).

Fires in Russian boreal forests are ignited by lightning strikes or by humans. In Siberia, the occurrence probability of lightning-ignited fires varies with the type of terrain (Shishikin et al., 2012). In European Russia, where there is a higher human population density and easier accessibility to the forest than in Asian Siberia, humans are responsible for igniting more than 65% of fires (Conard & Ivanova, 1997; Shishikin et al., 2012); however, regional variation in the causes of ignition is great. For example, in northern larch stands, about 90% of wildfires are of natural origin (Ivanova & Ivanov, 2004), whereas in southern boreal forests, notably in the Khakasia region, 80% of fires are caused by campfires, the burning of logging slash on harvested areas, and the agricultural burning of grasses (Shishikin et al., 2012).

3.1.1.3   Norway, Sweden, and Finland

The boreal part of Fennoscandian forests outside Russia (i.e., Norway, Sweden, and Finland; NSF) has a natural fire regime resembling that of the adjacent Russian Fennoscandia, where diverse landscape conditions result in variable fire regimes in terms of fire frequency, size, and severity (Engelmark, 1987; Gromtsev, 2008). However, the current fire regime differs significantly from that of the boreal forest of European Russia and from the more active fire regime of the past because of human influence (Pinto et al., 2020; Rolstad et al., 2017). Forest fires were previously common in all three countries but have declined considerably in frequency from historical levels 150–250 years ago (Chap. 2; Rolstad et al., 2017; Wallenius, 2011). These changes were not associated with climatic shifts (Aakala et al., 2018; Rolstad et al., 2017) but rather with changes in cultural practices and land tenure. The mechanization of firefighting and the development of a dense forest road network have also influenced the efficacy of the active suppression of surface fires (Wallenius, 2011).

The number of fires and the area burned is currently small, having declined dramatically during the twentieth century. For example, from 2007 to 2016, the average annually burned area was 496 ha in Finland, 842 ha in Norway, and 2,876 ha in Sweden, corresponding roughly to 0.002, 0.007, and 0.01% of the forested area of each country (San-Miguel-Ayanz et al., 2018). The area burned annually, however, varies considerably. A notable example is the two peak fire years in Sweden in 2014 and 2018, during which 12,600 and 22,400 ha burned, respectively (MSB, 2020). Although human influence on fire ignition has declined over the last century, most fires are still ignited by humans.

3.1.2   Fire and Forest Management

Fires and forest management are linked in various ways, including fuel management and the use of fires to guide forest ecosystem management. Fire suppression strategies and forest fire policies differ markedly around the circumboreal region, as does the role of fires in managing forest ecosystems.

3.1.2.1   Fire Suppression Policies and Practices

In Canada, fire management agencies have been established in every province and territory. In regions where forest management licenses are active, these agencies aim to minimize the number of large fires and their adverse effects on people, property, and timber (Stocks, 2013; Stocks & Martell, 2016). With early fire detection systems to locate small fires, e.g., infrared satellite and aerial flyover monitoring, and the use of initial attack strategies to contain fires, the aim is to extinguish fires at a small final size (2–4 ha; Martell & Sun, 2008). Despite such fire management systems in place, slightly more than 3% of fires become larger than 200 ha and are responsible for almost 97% of the area burned (Hanes et al., 2019). Cardil et al. (2019) showed that fire suppression success is greater in regions of mixed boreal forest because of the presence of deciduous species that are less flammable than resinous ones, with 82% to 92% of fires extinguished before they reach 3 ha compared with 53% to 77% in regions of coniferous boreal forest. Large fires occur on extreme fire-weather days when fuels are dry and winds favor fire spread. Since 1959, the area burned by larger fires has increased on average by about 350 ha per year (Hanes et al., 2019). By 2100, the annual area burned in Canada is projected to increase by two to four times (Boulanger et al., 2014; Coogan et al., 2019). Such situations may overwhelm the capacity of fire management agencies (Wotton et al., 2010) and result in a substantial increase in fire management expenditures (Hope et al., 2016).

In Russia, forest fire monitoring is based chiefly on satellite-derived information compiled and analyzed by the Federal Forest Service (since 2001) and on reports from the Federal State Agency "Central Base for aerial forest protection Avialesookhrana." It should be noted that up to 100 million ha of unused agricultural lands are now overgrown by forests in Russia. Such forests represent approximately 10% of all forests in Russia, although they are not officially referred to as forests (Shmatkov & Yaroshenko, 2018). Because these patches do not have an official forest status, fires within these forests are not classified as forest fires and are not officially monitored, with a notable exception being volunteer monitoring organized by Greenpeace.

Regional forest fire centers and aerial forest protection offices are responsible for forest fire protection in Russia; however, these organizations lack resources, both in terms of labor and equipment. During severe fire situations, local forest organizations—including forest companies—are obliged to participate in extinguishing fires. Forest fire protection in protected areas, urban forests, and military forests is organized respectively by the staff of protected areas, regional authorities, and the Russian Ministry of Emergency Situations. Responsibility for fire protection on former agricultural lands has not been allocated to any entity, except in cases deemed as "high emergency."

Currently, fires in about 45% of Russian boreal forests are not extinguished because of their remoteness and often low accessibility. Regional authorities define these *control zones* without explicit, law-based principles. The proportion of fires left to burn without intervention by firefighters in the spring–summer of 2020 varied regionally and monthly, ranging between 50 and 98% of the total number of forest fires.

In NSF, the detailed implementation of fire detection and suppression differs among the countries; however, the overall aim is to actively suppress all fires. Satellite detection and reconnaissance flights are used for the early detection of fires. Given that most forest areas are easily accessible because of the dense network of forest roads, fire suppression is generally efficient, which is reflected by the limited area that burns annually (see above).

3.1.2.2   Fire as a Driver for Forest Management

At a global level, ecosystem management and habitat restoration increasingly seek inspiration from natural disturbances, particularly fire, to support both an economically viable forest industry and biodiversity in managed forests. Four main descriptors of fire regime (annual burn rate, fire size, severity, and, more recently, specificity) form the basis of ecosystem management and restoration in boreal forests (Gauthier et al., 2009; Koivula & Vanha-Majamaa, 2020; Lindberg et al., 2020; Shishikin et al. 2012). In Canada, the annual burn rate helps define management targets in terms of even-aged vs. uneven-aged forests within landscapes. The variation in annual burn rates also strongly influences the amount of old-growth forest present in a given landscape (Bergeron et al., 2001; Weir et al., 2000), which in turn influences forest composition at the landscape level. Therefore, the amount of old-growth forest to be maintained in different boreal regions is defined on the basis of past fire regimes (Bouchard et al., 2015; DeLong, 2007). Fire-size distribution provides insights into the spatial configuration of different forest types and ages across the landscape (Gauthier et al., 2004; Perron et al., 2009), whereas variation in fire severity influences the retention strategies applied in harvested areas.

In Russia, the mosaic of forest patches and landscapes stemming from variable fire return intervals and burn severities (Kharuk et al., 2016a) leads to differences in the economic and nature conservation value of forests. Whereas emphasis has been traditionally placed on fire suppression, there are now calls for region-specific forest fire management policies. Fire policies should recognize the beneficial functions of fire for pyrophilous and deadwood-dependent species as well as its role in forest successional processes (Furyaev, 1996). This shift would imply replacing the current fire exclusion policy with a policy that allows for natural low-intensity fires and prescribed burning to reduce fire hazards and promote biodiversity (Davidenko et al., 2003; Goldammer, 2013). From a biodiversity perspective, implementing fire management strategies in protected areas, where actions may vary from fire prevention and suppression to doing nothing, is of particular importance (Kuleshova, 2002; Shishikin et al., 2012). Regional policy guidelines for fire should be based on the scientific knowledge of the (1) landscape-specific fire regimes; (2) regional- and landscape-specific effects of fire and postfire succession on biodiversity; and (3) socioeconomic conditions, including human population density, road networks, economic factors, agricultural use of fire, and forestry activities. Depending on the region or landscape, different strategies can be prescribed: (1) fire prevention, e.g., through establishing fire breaks and education; (2) suppression, including the control, monitoring, and fighting of fires whenever possible; (3) localization of ignited fires; (4) controlled or prescribed burning; and (5) regulation of postfire successional processes by applying different restoration measures. Multilevel educational actions designed for target groups from preschool children to university students and local people also play an important role in fire management (Kuleshova, 2002; Shishikin et al., 2012).

In NSF, there have been a few instances in which fire ecology has been used as a guide for developing sustainable forest management strategies; these are similar to the

Canadian ecosystem management approach. Perhaps the most well-known approach in the European context has been the ASIO model (Angelstam, 1998), which divides landscapes into four categories according to how frequently the forest burns under natural conditions: rarely (Aldrig), seldom (Sällan), infrequently (Ibland), or often (Ofta). In the mixed-severity fire regime, this frequency is often inversely related to fire intensity. The idea is that the forest management strategy applied in a given area is tailored according to the category in which the area is classified, emulating the stand age structure that would naturally occur in the area. This model has been applied in forest management planning by some large forest owners in Sweden (Angelstam, 1998). In Finland, ASIO has been used in conjunction with the landscape ecological planning of public lands. However, its role has been small and limited primarily to identifying parts of the landscape that almost never burn (Karvonen et al., 2001). Recently, Berglund and Kuuluvainen (2021) outlined a refined version of the ASIO model relying on an improved understanding of how forest fires shape the boreal forests of NSF.

In practical forest management in NSF, understanding fire ecology has been more commonly used as a silvicultural tool and for managing biodiversity rather than as a management template or guideline. The use of fire in silviculture was common after the 1950s when prescribed burning following clear-cutting was widely used as a regeneration tool. However, controlled prescribed burning is expensive; therefore, its popularity has declined. For example, the area of annual prescribed burns in Finland was around 35,000 ha in the 1950s, whereas it is currently only a few hundred hectares (Lindberg et al., 2020). As a consequence, habitats and structures previously maintained by frequent fires—mostly early successional habitats with abundant legacies such as burnt wood—have greatly declined (Kontula & Raunio, 2019). These types of fire-dependent habitats are currently being created by the prescribed burning of single or groups of retention trees in clear-cut areas—used to promote biodiversity and soil preparation—and through restoration burning (Lindberg et al., 2020). However, despite the benefits of fire for biodiversity, the areas burned annually in NSF remain small relative to the past natural fire regime.

Over the years, salvage logging of burned areas has globally gained importance to compensate for the impact of fire on timber availability (Nappi et al., 2004; Thorn et al., 2018). This practice negatively impacts the diversity of species occupying postfire forests, adding to the negative impacts of fire on diversity (Cobb et al., 2011). Habitat conditions, e.g., shadiness, and associated species communities appear altered less by insects than by fire or windthrow, whereas subsequent salvage logging renders these environments similar. Moreover, salvage logging reduces forest-species richness more in insect-disturbed than in fire- or windthrow-disturbed forests—reductions to about 57% versus 70–75% from the post-disturbance level (Thorn et al., 2020). Guidelines for retention within salvage logging areas have been proposed for reducing the negative impacts of such practices (Nappi et al., 2011; Thorn et al., 2020).

3.2   Wind

3.2.1   Susceptibility to Wind Damage

Windthrow occurs when wind speed is strong enough to override tree-root resistance to uprooting or stem resistance to trunk breakage. Wind is a common disturbance in a variety of biomes, from boreal (Ulanova, 2000) to temperate (Canham et al., 2001; Fischer et al. 2013) to tropical forests (Putz et al., 1983). Wind-induced disturbances vary in frequency, size, and severity both between and within biomes (Everham & Brokaw, 1996). In different parts of the boreal forest, windthrow return intervals vary from decades to a few hundred or thousand years (De Grandpré et al., 2018; Smolonogov, 1995; Waldron et al., 2013).

Wind damage can recur regularly at small scales and low severity or occur less often but at a large scale and high severity (Miller, 1985). Severe damage is associated with infrequent major storms and has led to significant efforts to document damaged areas and timber losses (Grayson, 1989; Ruel & Benoit, 1999; Valinger et al., 2014). Nonetheless, although less spectacular and often poorly documented, small-scale windthrow events can have significant consequences for forest management (Rollinson, 1987). In a compilation covering 29 European countries, Seidl et al. (2014) estimated that wind damaged 32.3 million $m^3 \cdot yr^{-1}$ of timber during the first decade of the twenty-first century.

Windthrow severity is influenced by interactions between wind speed, topographic and edaphic conditions, disturbance history, and the current characteristics of forest stands and landscapes (Everham & Brokaw, 1996; Ruel et al., 1998; Saad et al., 2017). Shallow and poorly drained soils restrict rooting depth, which leads to lower tree resistance to uprooting. However, soil properties interact with tree species and stand attributes; for example, jack pine (*Pinus banksiana* Lamb.) is more resistant than black spruce (*Picea mariana* (Mill.) BSP) on relatively deep soils but not on shallow soils, which prevent the development of deep roots. On the other hand, black spruce is inherently shallow-rooted and is thus better adapted to shallow soils (Élie & Ruel, 2005). Old-growth Norway spruce–dominated stands on rich soils consist of large trees with flagged crowns and a shallow root system. Consequently, they are more vulnerable to windthrow than, for instance, Scots pine or birch forests that are more deeply rooted (Karpachevsky et al., 1999; Skvortsova et al., 1983; Ulanova, 2000).

The most important characteristics influencing stand vulnerability to windthrow are tree species composition, size, and age structure. Tree-pulling studies allow a quantitative comparison of species resistance to windthrow (Achim et al., 2005; Nicoll et al., 2006; Peltola et al., 2000). Wood properties and the presence of decay (notably because of *Heterobasidion* fungi) strongly influence the resistance of trees to stem breakage (Rich et al., 2007). In eastern Canada, for instance, balsam fir (*Abies balsamea* (L.) Mill) has been consistently ranked as the most windthrow-prone tree species in large part owing to a high level of decay (Ruel, 2000). Among European boreal tree species, Norway spruce is the most sensitive to uprooting, whereas aspen

(*Populus tremula*) is damaged mainly by stem breakage (Skvortsova et al., 1983). In primeval European boreal forests, susceptibility to windthrow decreases with tree age structure from even-aged to all-aged stands, increases with the proportion of deciduous species, and decreases with site fertility (Fedorchuk et al., 2012; Karpachevsky et al., 1999; Shorohova et al., 2008).

3.2.1.1 Windthrow Impacts

Immediate and long-term windthrow impacts include (1) an abrupt or continuous change of forest structure with an increased share of broken and/or uprooted trees and deadwood; (2) pedoturbation (soil-mixing) with the creation of pit-and-mound systems (Fig. 3.2); (3) a change in microclimate; and (4) a change in stand vulnerability to subsequent disturbances (Chap. 4; Fischer et al., 2013; Šamonil et al., 2010; Schaetzl et al., 1989; Skvortsova et al., 1983; Ulanova, 2000). In old-growth forests, pit-and-mound systems may cover an area of up to 90% and remain visible for up to 200–500 years. In high-severity windthrow, environmental conditions, notably light availability and soil moisture, are strongly modified, and water balance can even change across the entire landscape (Karpachevsky et al., 1999). At the landscape scale, low- or moderate-severity windthrow results in a scattered pattern tree mortality of various modes (uprooting, stem breakage, or the formation of snags) (Fig. 3.3), a complex fine-scale mosaic of living and dead trees, and windthrow gaps that vary from 0.05 ha to a few hectares in size and have a variable pit-and-mound topography (Fedorchuk et al., 2012; Schaetzl et al., 1989; Shorohova et al. 2008; Skvortsova et al., 1983).

Spatial patterns of wind-induced tree mortality lead to multiple post-windthrow successional pathways in forest ecosystems, depending on the interplay between windthrow severity and stand attributes, including tree age structure, tree species composition, and site productivity (Meigs et al., 2017). Biotic and abiotic factors influence the succession of post-windthrow regeneration (Fischer et al., 2013; Girard et al., 2014; Ulanova, 2000). Coniferous tree species successfully regenerate where less than 60% of trees die in a stand (Petukhov & Nemchinova, 2015) and the surface area of windthrow pits covers less than 15% (Ulanova & Cherednichenko, 2012).

3.2.2 Wind and Forest Ecosystem Management

Windthrow generates timber loss due to falls and wounds on trees, and windthrow often results in subsequent biotic disturbances, such as bark beetle outbreaks or fungal infestations. In mountain regions, windthrow may increase the risk of avalanche and rockfall and consequently threaten human settlement and infrastructure (Schönenberger et al., 2005). Although extended rotation and partial cutting are important ecosystem management strategies (Bélisle et al., 2011; Montoro Girona et al., 2016), increasing rotation length can lead to more windthrow. Thus, wind damage

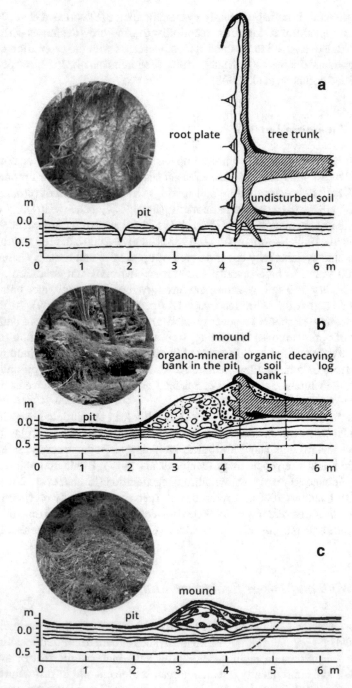

Fig. 3.2 Pit-and-mound complexes at **a** 20 to 30 years, **b** 50 to 60 years, **c** more than 100 years post-windthrow in southern boreal forests. Modified from (Ulanova, 2000) with permission from Elsevier. *Photo credits* **a** Aleksandr Gladyshev, **b** Ilkka Vanha-Majamaa, **c** Anna Ruokolainen

Fig. 3.3 Diverse coarse woody debris after a patchy windthrow in a primeval forest in the Vepssky Reserve, Russia. *Photo credit* Alexandr Korepin

tends to increase with tree age because of the increased tree height and incidence of decay (Ruel, 1995), posing a constraint when applying extended rotations to short-lived species. Increasing intertree spacing through partial cutting heightens the wind load on trees (Gardiner et al., 1997). There are many examples of increased wind damage after partial cutting (Chap. 16; Becquey & Riou-Nivert, 1987; Cremer et al., 1982; Montoro Girona et al., 2019; Ruel & Gardiner, 2019). Windthrow losses can, however, be minimized through windthrow hazard evaluation tools, including decision keys, empirical models, or hybrid/mechanistic models (Gardiner et al., 2008). In recent decades, progress has been made in developing such tools, especially in the modeling of complex stands (Gardiner et al., 2008; Hale et al., 2012).

In post-disturbance situations, forest management strategies include (1) allowing natural successional development; (2) salvage logging followed by natural regeneration; and (3) salvage logging with subsequent soil preparation and tree planting (Brang et al., 2004; Fischer & Fischer, 2012; Fischer et al., 2002; Lässig & Močalov, 2000; Močalov & Lässig, 2002; Schönenberger, 2002; Soukhovolsky et al., 2012). A study from a Bavarian national park in Germany, comparing successional dynamics after windfall on permanent plots, demonstrated that salvage logging triggers natural secondary succession through intermediate phases having a dominance of birch or aspen (Fischer et al., 2002). The costliest silvicultural treatments allow for the regeneration of mixed conifer–deciduous forests, although without predisturbance natural mosaics (Fischer & Fischer, 2012; Lässig & Močalov, 2000). In mountain forests, "doing nothing" may ensure natural protection against snow avalanches and rockfall (Schönenberger et al., 2005).

The natural landscape–specific regime of wind disturbance can be considered as a basis, or a reference, for ecosystem-based forest management and ecological restoration. In landscapes dominated by small- and medium-scale windthrow, gap felling or variable retention felling can be recommended (Koivula et al., 2014). If salvage logging must be used for some economic or public safety reasons, post-windthrow attributes of known ecological importance, such as deadwood, living trees, and micro-topography, should be retained within salvaged cutblocks, with some proportion of windthrow exempted from logging operations (Thorn et al., 2020; Waldron et al., 2013). Mimicking partial windthrows in wind-prone forests by conducting partial cuts can increase the likelihood of subsequent wind damage. However, a widely shared opinion among foresters in Finland is that damage risks are generally higher in even-aged than in uneven-aged management regimes, with the notable exception of root-rot infestations in Norway spruce forests (Nevalainen, 2017).

An additional challenge in incorporating wind disturbance into forest management is related to alterations of future disturbance regimes. Storms characterized by high wind speeds are more common in autumn and early winter in northern Europe and eastern Canada, periods where the frozen topsoil "anchors" trees in the ground, thereby decreasing the chances of treefall. Because of climate warming, however, periods of unfrozen soil are predicted to lengthen, resulting in a poorer anchoring of trees in a season of severe winds. Moreover, the frequency of autumn or early-winter windstorms may increase; thus, windstorm-caused timber damage could become more common and widespread (Gregow et al., 2011; Saad et al., 2017). Indeed, in Europe, the level of damage by wind, reported by Seidl et al. (2014) for the first decade of the twenty-first century, increased 140% compared with wind damage between 1971 and 1981. Between 1950 and 2000, more than 50% of natural tree mortality in Europe was due to windthrow, whereas biotic factors were responsible for 16% (Schelhaas et al., 2003). Although biotic factors appear relatively minor from this perspective, they can be locally devastating (Hlásny et al., 2019). These percentages are likely to change in the near future, however, as windthrow, drought, and insect outbreaks are predicted to increase, particularly for the boreal region (Seidl et al., 2020).

3.3 Insects

3.3.1 Insect Outbreaks and Their Characteristics

Forest insect outbreaks occur in all major forest ecosystems throughout the world but cause the most damage in high-latitude forests. Unlike fire and wind disturbances, insects are often specific in nature, such that only a limited number of host-tree species—usually a single genus or family—are affected (Bentz et al., 2020). This specificity also implies that certain attributes (frequency, size, severity) used to characterize fire regimes do not apply directly to insects (De Grandpré et al., 2018). For example, although insect outbreaks may affect a larger total area than fire or wind-storm, being specific to certain host-tree species, qualities, and sizes, these events lead to partial mortality except in pure host-species stands (Raffa et al., 2015). Thus, the losses of timber volume may be less than after windthrow or fire (Kneeshaw et al., 2015).

Contrary to wildfire, the return interval (the inverse of frequency) for insect outbreaks is usually calculated on the basis of insect population dynamics rather than the time required to affect a given area. Spruce budworm (*Choristoneura fumiferana*) outbreaks in eastern Canada occur every 30 to 40 years (Jardon et al., 2003; Morin et al., 2009; Navarro et al., 2018), a return interval similar to that of the mountain pine beetle (*Dendroctonus ponderosae*) (Alfaro et al., 2010). For insect species usually affecting only small areas, population return intervals are rarely calculated. Examples of such species include the European spruce bark beetle (*Ips typographus*), the gypsy moth (*Lymantria dispar*), and the oak processionary moth (*Thaumetopoea processionea*) (Bentz et al., 2020).

The severity of an insect outbreak can be expressed as the number or proportion of infected trees. Aerial surveys of areas affected by the spruce budworm give stand-level severity estimates based on annual defoliation. In Québec, these classes are 0–33% (light), 34–66% (moderate), and 67–100% (severe) (MFFP, 2019). If defoliation is less than 33%, tree growth is minimally affected (Chen et al., 2017). As the spruce budworm only eats current year (new) foliage and trees carry five to seven years of foliage, multiple subsequent years of infestation are required for the spruce budworm to kill a tree (Lavoie et al., 2021). For example, removing all foliage on a tree requires five years of 100% defoliation of new foliage, although a tree may die before the cumulative defoliation reaches 500%. This rule of thumb is useful for translating defoliation into mortality. Severity has also been measured through dendrochronological records by inspecting reductions in tree growth rings (Robert et al., 2018; Thomas et al., 2002).

Tree mortality is another useful indicator of outbreak severity. In mild outbreaks, only growth reduction may occur, whereas severe outbreaks result in detectable tree mortality. There is no accepted standard of the level of mortality required for an outbreak to be considered severe. The mountain pine beetle, for example, feeds on the phloem of living trees but can only successfully reproduce if it kills the tree and eliminates its defenses (Safranyik et al., 2010). In contrast, many other insect species can reach high population numbers (and thus outbreak conditions) while primarily affecting only tree growth. The forest tent caterpillar (*Malacosoma disstria*), the jack pine budworm (*Choristoneura pinus*), and the oak processionary moth, for example, rarely directly kill their host trees (Man & Rice, 2010; Sands, 2017). Thus, outbreaks causing any mortality may be considered severe for these species (Cooke et al., 2012).

Given the host specificity of herbivorous insects and their feeding preferences (defoliation of some or all leaves versus feeding on phloem or xylem), insects cause various forms of damage to trees. Hence, forest management based on the emulation of tree structure and microclimatic conditions resulting from insect disturbances must focus on parameters other than the impacted area or return interval. For the spruce budworm, Baskerville (1975) suggested that the insect acts as a *super silvi-culturist* in releasing advance, i.e., pre-established, regeneration. Bouchard et al. (2006), Kneeshaw and Bergeron (1998), Reinikainen et al. (2012), and Burton et al. (2015) showed that outbreaks of defoliators are essential for maintaining the structural diversity of forests. Other authors have also evaluated the influence of insects on tree regeneration and, therefore, the future composition of forests within various site types. The mountain pine beetle, for instance, can act as an agent that removes and kills large older lodgepole pine (*Pinus contorta* Douglas ex Loudon) and, in turn, releases space and resources for the smaller stems of lodgepole pine or favor the recruitment of other tree species (Kayes & Tinker, 2012).

3.3.2 Forest Ecosystem Management and Insect Outbreaks

Lessons from insect outbreaks suggest that if forest management aims to emulate tree structures resulting from these outbreaks, forest managers should avoid monocultures and even-tree-size stands and favor tree diversity. These features would benefit wildlife diversity and decrease the likelihood of future outbreaks, as suitable host trees for these specialists would be less abundant. Koivula et al. (2014) suggest that partial cutting could emulate insect disturbances as most insect disturbances cause only partial mortality. Currently, forest managers preferentially harvest the most valuable companion tree species at maturity (Blais, 1983; Kneeshaw et al., 2021; Sonntag, 2016). Recent work suggests that insects tend to attack large contiguous blocks of host-tree species with greater synchrony and severity; therefore, breaking up such large blocks may be an effective pest management strategy at the landscape scale (Robert et al., 2012, 2018, 2020). As the ranges of many insects are currently expanding, managers should be aware that large blocks of monocultures should be eliminated or reduced to avoid increasing forest vulnerability to outbreaks (Kneeshaw et al., 2021).

Climate change may affect the population dynamics of different insect species, alter outbreak frequencies, and facilitate range shifts to more northern latitudes and higher elevations. Range expansions of forest insect pests may lead to widespread mortality of trees within the insect's new range. However, they may also be associated with contractions in other parts of the range (Régnière et al., 2012). Insect population density is regulated by density-dependent and density-independent factors, such as weather conditions and forest ecosystem characteristics (Isaev et al., 2017). Increases in temperature, especially in winter months, and drier conditions may contribute to increases in bark beetle populations and the ability of these beetles to overcome the defense mechanisms of trees (Raffa et al., 2015; Romashkin et al., 2020). Droughts have also been implicated by stressing trees and rendering them more vulnerable to bark beetle attacks, as has been observed for European spruce bark beetle outbreaks (Maslov, 2010). However, drought effects on defoliators remain equivocal (Itter et al., 2019; Kolb et al., 2016). Recent reviews have attempted to predict the effects of climate change on future insect outbreaks (Jactel et al., 2012; Kolb et al., 2016;

Pureswaran et al., 2018). These studies indicate that, despite expectations of greater outbreaks, responses are complex, and positive and negative feedback will probably occur (Haynes et al., 2014). In other words, some outbreaking insects may cause more damage whereas others will cause less, and this—combined with range contractions and expansions—adds much uncertainty to projections of future insect influence on forests.

3.4 Pathogens

Many pathogens influence trees by reducing tree growth and vitality (Hicke et al., 2012) by acting as predisposing agents to a number of other disturbances. Several pathogen species also kill trees directly. Because of its harsh climate, the boreal zone has previously been beyond the distribution of many pathogens. Consequently, their role in the disturbance regimes of natural forests has been overlooked. Certain species of fungi may play a significant role in the dynamics of old-growth forests in northern Fennoscandia (Lännenpää et al., 2008) in causing the small-scale mortality of individual trees or small groups of trees. Hence, at the landscape scale, pathogens occur frequently, but their impacts are of low severity and spatially scattered.

Many pathogen species are strictly host specific (Zhou & Hyde, 2001). Partly because of this host specificity, their role in intensively managed, monospecific, and structurally homogeneous forests appears greater than in natural forests (Storozhenko, 2001). However, trees in continuous-cover forest management appear to suffer from *Heterobasidion* infestations to a greater degree than those growing in standard even-aged management because of logging-caused damage to retained trees (Piri & Valkonen, 2013) and difficult root and stump removal. Fungi of the genera *Heterobasidion* and *Armillaria* are considered particularly problematic for forestry in the boreal zone (Garbelotto & Gonthier, 2013); as they spread through roots, trees in the next generation are easily infected.

The most aggressive fungal pathogen causing root rot in naturally regenerated coniferous boreal forests is *Armillaria borealis* Marx. & Korh. (Pavlov, 2015). Soil conditions determine the activity of and disturbance severity caused by *Armillaria* and *Heterobasidion* spp. (Fig. 3.4; Pavlov, 2015).

In European Russia, the bacterial dropsy diseases on birch (*Betula* spp.) and coniferous tree species, caused by *Erwinia multivora* Scz.-Parf., have increased during the last decades (Voronin, 2018). These bacterial diseases are triggered by drought and anomalous thaw events, causing fungal outbreaks in Siberian fir and pine forests (Voronin, 2018).

Climate is an important driver of disease outbreaks, influencing the disturbance agent directly or indirectly through host susceptibility (Sturrock et al., 2011). Changing climate may generate conditions favorable to pathogens by extending periods of growth and reproduction or causing phenological changes that may result in a greater overlap of host susceptibility and pathogen aggressiveness. *Heterobasidion* and *Phytophtora* species are expected to benefit from a warming climate

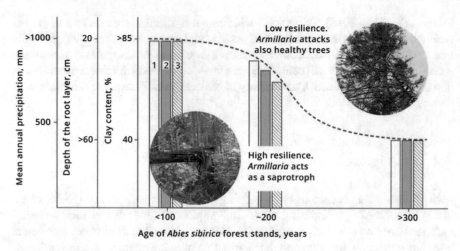

Fig. 3.4 Effects of drought and soil conditions on the resilience of Siberian fir trees against the pathogenic fungus *Armillaria mellea* s.l. Redrawn by permission from Springer Nature from Pavlov (2015). *Photo credit* Ilkka Vanha-Majamaa

(Pavlov, 2015). Similarly, the widespread *Armillaria* has the potential to increase in significance in boreal forests (Dempster, 2017). Like insects, pathogens are also candidates for invasive spread through human influence (Dukes et al., 2009).

3.5 Drought-induced Forest Decline

Whereas past disturbance studies have rarely considered drought, it is now recognized as a potent disturbance agent that can reduce tree growth (Itter et al., 2019), increase the vulnerability of trees to defoliation (Cooke & Roland, 2007), and drive tree mortality (De Grandpré et al., 2019). Mechanisms of drought-induced mortality include hydraulic failure, xylem embolism, and increased vulnerability to biotic disturbance agents, such as insects, fungi, and bacteria (Anderegg et al., 2013; Kharuk et al., 2016b; Voronin, 2018). Repeated drought events can weaken trees and decrease their resilience to subsequent drought events and secondary disturbance agents (DeSoto et al., 2020; Haynes et al., 2014; Pavlov, 2015).

Cases of drought-induced disturbances of varying severity have been reported across the boreal region (Chaps. 11 and 30; Michaelian et al., 2011; Pavlov, 2015; Zamolodchikov, 2012). In boreal forests, patchy drought-induced mortality is typical, especially in spruce-dominated primeval forests (Aakala & Kuuluvainen, 2011; Khakimulina et al., 2016) (Fig. 3.5). Similar patterns of decline and mortality of the "dark conifers" *Abies sibirica* and *Pinus sibirica* have been recorded in the southern Siberian Mountains and Baikal Mountains (Kharuk et al., 2013a). Birch mortality, caused by prolonged drought, has been documented within the Trans-Baikal forest–steppe (Kharuk et al., 2013b). Notably, all reported cases of mortality of "dark conifers" in Russia have coincided with drought episodes, often accompanied by insect outbreaks (Kharuk et al., 2016b). However, interactions between drought and insect outbreaks are complex, especially for defoliating insects. Haynes et al. (2014) showed that outbreaks of only one of five forest insect pests in Germany were influenced by drought over the past centuries. Similarly, in North America, Itter et al. (2019) could not find an interaction between the growth reductions caused by two different defoliators and drought. On the other hand, De Grandpré et al. (2019) suggested that drought preceded spruce budworm–caused mortality. Another study indicates that bark beetles respond directly to climate change, whereas the evidence for defoliators is equivocal (Kolb et al., 2016). In addition to stressing host trees, drought can impact the insect itself. Thus, the effect of climate change on future insect outbreaks is difficult to predict. Outbreaks could become more severe if the trees are more negatively affected than the insect pests, or outbreaks could decrease in amplitude and severity if insects are more negatively affected than trees (Pureswaran et al., 2018).

From a forest management perspective, species response to drought is a key issue. In European boreal forests, the shallow-rooted overstory Norway spruce, which suffers from drought events over large spatial scales, is predicted to be strongly affected in the future (Kharuk et al., 2016b). In North America, jack pine is considered one of the least vulnerable tree species to drought (Peng et al., 2011). An inclusion of drought-resistant trees in forest management has been proposed as a strategy to mitigate the impacts of drought on forest productivity. However, before advocating large-scale switches from drought-sensitive to drought-tolerant species, it is

Fig. 3.5 Patchy drought-induced mortality of Norway spruce trees in a primeval forest of the Vepssky Forest Reserve, Russia. *Photo credit* Aleksandr Korepin

necessary to point out the complexity of these relationships. Aubin et al. (2018) used traits to identify drought-resistant trees in western Canada and classified trembling aspen (*Populus tremuloides* Michx.) as being highly sensitive to drought, whereas in eastern Canada, trembling aspen is considered one of the most drought-tolerant species (D'Orangeville et al., 2018; Héon-Grenier, 2020). Moreover, D'Orangeville et al. (2018) showed that although species differ in drought sensitivity, the marked intraspecific variability in this respect underlines the overriding effect of site. The severity of drought and other disturbances is also related to elevation, terrain topography, slope steepness, and aspect (Kharuk et al., 2013a). Even the least vulnerable trees will experience high rates of growth loss and mortality following drought if they are growing on shallow soils.

Deep snowpacks in boreal forests ensure that soil water is recharged annually; thus, cumulative soil moisture deficits may be limited and, consequently, minimize the effects of drought on boreal trees (Oogathoo et al., 2020; Léger-Beaulieu et al. In-Review). The timing of dry conditions during a season is also an essential factor to consider. D'Orangeville et al. (2016) have shown that spring droughts can benefit boreal soils subject to cold and wet conditions, whereas summer droughts can have more negative effects. Sánchez-Pinillos et al. (2022) also show that subsequent low-severity droughts can cause greater mortality than severe droughts.

Drought has been an increasingly common phenomenon over recent decades and is projected to be even more frequent and severe in the future. However, its effects are complex, especially its interactions with other disturbances. As tempting as it is to identify and favor drought-resistant species, site factors should be the primary consideration when predicting future impacts. Thus, from a management perspective, foresters should learn from forest vulnerability to drought. In particular, species and site conditions should be considered in silvicultural decisions, as certain sites may be at high risk of drought and should not be managed for timber production.

3.6 Snow and Ice

Snow and ice are often included in the list of typical disturbance agents of the boreal forest, but their effects have rarely been quantified. Ice storms are a major meteorological hazard in midlatitude regions (Cheng et al., 2007). They occur when freezing rain accumulates on trees, and the weight of the accumulated ice breaks the branches and stems. This can cause widespread damage in temperate forests, but these events are less common in boreal forests. Nonetheless, Markham et al. (2019) documented such an event in jack pine forests in Manitoba, where over 2,000 km^2 were damaged by ice in 2010. Similar events have been recorded in Manitoba in 1930 and 1958, showing that ice storms are also a potentially important disturbance agent in parts of the boreal zone.

The impacts of snow and ice on trees and forests resemble those of windstorms (Peltola et al., 1999; see also Sect. 3.2) in that they mechanically cause tree boles and branches to break. The breaking can also occur in interaction with high winds that exert further forces on the stem. At the tree level, tree architecture and wood properties play a role. Ice storms and snow do not usually kill all trees in a stand (Markham et al., 2019), but they may change species composition, size structure, and stand spatial structure (Jalkanen & Konocpka, 2007; Nykänen et al., 1997). The accumulation of snow and ice and the resulting damage on trees across spatial scales from stands to regions depend on weather, e.g., cold and warm fronts, precipitation, air temperature, wind speed and direction, and locality, e.g., continentality, topography, altitude, and water table height (Barry & Chorley, 2010).

With changing climate, snow damage patterns are predicted to change (Kilpeläinen et al., 2010). As climate change brings about more extreme weather events and warmer conditions in the early winter and spring, the occurrence of ice storms is also likely to increase in the North American boreal forest (Cheng et al., 2007).

3.7 Concluding Remarks

Temporal and spatial descriptors of all disturbance types in the boreal forest vary in time and space and are thus difficult to emulate, predict, and control in an ecosystem management framework. The relative importance of different disturbance agents and the variability of current and future disturbance regimes within the boreal region require developing programs for ecosystem management and ecological restoration at a regional level. Decades of research have shown that the landscape scale should be better considered in ecosystem management (Patry et al., 2017). For instance, the current level of harvesting may, at least locally, be close to (or even beyond) the capacity of the system to cope with the combined effects of fire and harvesting, let alone climate alterations. Future climatic conditions are projected to become more conducive to several disturbance types, including fire, windthrow, insect outbreaks, and drought. Hence, maintaining the current level of harvesting in the future may be challenging (Boucher et al., 2018; Gauthier et al., 2015a). Assessments of the implementation of ecosystem management approaches are crucial in mitigating the future impacts of increasing disturbance frequency on forest ecosystems.

Partial harvesting, especially with the retention of deadwood and *habitat trees* (exceptionally large, usually scattered, individual trees in a stand), can maintain structural forest features similar to stands affected by insects or windstorms, and these features are crucial for hundreds of threatened forest species (Gustafsson et al., 2020; Kneeshaw et al., 2011; Koivula & Vanha-Majamaa, 2020). Descriptors of severity and specificity may provide a template for developing policies for maintaining biological legacies in post-harvest and salvage-logged forests (De Grandpré et al., 2018; Nappi et al., 2011).

Episodic disturbances may foster ecosystem adaptations to the effects of ongoing and future climatic change by increasing structural diversity with cascading positive effects on biodiversity, edaphic conditions, biogeochemical cycles, and increased heterogeneity across various spatial scales. Allowing some forests to be shaped by natural processes may be congruent with multiple goals of forest management, even in densely settled and developed countries (Kulakowski et al., 2017). Emulating natural disturbances and successional dynamics at landscape and regional scales should be used to maintain the natural variability in old-growth attributes over time (Shorohova et al., 2011).

Addressing all forest ecosystem services calls for developing regional strategies to integrate disturbances into ecosystem management, with actions varying from prevention, control, and post-disturbance management to passive "rewilding" to the active emulation of disturbances. These actions have the combined goal of restoring ecosystem resilience by maintaining tree stand composition, age-class distribution, and natural-like structures.

References

Aakala, T., & Kuuluvainen, T. (2011). Summer droughts depress radial growth of *Picea abies* in pristine taiga of the Arkhangelsk province, northwestern Russia. *Dendrochronologia, 29*(2), 67–75. https://doi.org/10.1016/j.dendro.2010.07.001.

Aakala, T., Pasanen, L., Helama, S., et al. (2018). Multiscale variation in drought controlled historical forest fire activity in the boreal forests of eastern Fennoscandia. *Ecological Monographs, 88*, 74–91. https://doi.org/10.1002/ecm.1276.

Achim, A., Ruel, J. C., Gardiner, B. A., et al. (2005). Modeling the vulnerability of balsam fir forests to wind damage. *Forest Ecology and Management, 204*, 35–50. https://doi.org/10.1016/j.foreco.2004.07.072.

Alfaro. R. I., Campbell, E., & Hawkes. B. C. (2010). *Historical frequency, intensity and extent of mountain pine beetle disturbance in British Columbia.* Mountain Pine Beetle Working Paper 2009–30 (p. 52). Victoria: Pacific Forestry Centre.

Anderegg, L. D. L., Anderegg, W. R. L., & Berry, J. A. (2013). Not all droughts are created equal: Translating meteorological drought into woody plant mortality. *Tree Physiology, 33*(7), 701–712. https://doi.org/10.1093/treephys/tpt044.

Angelstam, P. K. (1998). Maintaining and restoring biodiversity in European boreal forests by developing natural disturbance regimes. *Journal of Vegetation Science, 9*, 593–602. https://doi.org/10.2307/3237275.

Aubin, I., Boisvert-Marsh, L., Kebli, H., et al. (2018). Tree vulnerability to climate change: Improving exposure-based assessments using traits as indicators of sensitivity. *Ecosphere, 9*(2), e02108. https://doi.org/10.1002/ecs2.2108.

Baltzer, J. L., Day, N. J., Walker, X. J., et al. (2021). Increasing fire and the decline of fire adapted black spruce in the boreal forest. *Proceedings of the National Academy of Sciences of the United States of America, 118*(45), e2024872118. https://doi.org/10.1073/pnas.2024872118.

Barry, R. G., & Chorley, R. J. (2010). *Atmosphere, weather, and climate.* London: Routledge, Taylor & Francis.

Baskerville, G. L. (1975). Spruce budworm: Super silviculturist. *The Forestry Chronicle, 51*(4), 138–140. https://doi.org/10.5558/tfc51138-4.

Becquey, J., & Riou-Nivert, P. (1987). L'existence de zones de stabilite des peuplements. Consequences sur la gestion. *Revue forestière française, 39*, 323–334. https://doi.org/10.4267/2042/25804.

Bélisle, A. C., Gauthier, S., Cyr, D., et al. (2011). Fire regime and old-growth boreal forests in central Quebec, Canada: An ecosystem management perspective. *Silva Fennica, 45*, 889–908. https://doi.org/10.14214/sf.77.

Belov, S. V. (1976). *Forest pyrology* [in Russian]. St. Petersburg: Leningrad Forest Technical Academy.

Bentz, B., Pierluigi, P., Delb, H., et al. (2020). Advances in understanding and managing insect pests of forest trees. In J. A. Stanturf (Ed.), *Achieving sustainable management of boreal and temperate forests* (pp. 515–585). Cambridge: Burleigh Dodds Science Publishing Ltd.

Bergeron, Y., Gauthier, S., Kafka, V., et al. (2001). Natural fire frequency for the eastern Canadian boreal forest: Consequences for sustainable forestry. *Canadian Journal of Forest Research, 31,* 384–391. https://doi.org/10.1139/x00-178.

Berglund, H., & Kuuluvainen, T. (2021). Representative boreal forest habitats in northern Europe, and a revised model for ecosystem management and biodiversity conservation. *Ambio, 50,* 1003–1017. https://doi.org/10.1007/s13280-020-01444-3.

Bernier, P. Y., Gauthier, S., Jean, P. O., et al. (2016). Mapping local effects of forest properties on fire risk across Canada. *Forests, 7,* 157. https://doi.org/10.3390/f7080157.

Bessie, W. C., & Johnson, E. A. (1995). The relative importance of fuels and weather on fire behavior in subalpine forests. *Ecology, 76*(3), 747–762. https://doi.org/10.2307/1939341.

Blais, J. R. (1983). Trends in the frequency, extent, and severity of spruce budworm outbreaks in eastern Canada. *Canadian Journal of Forest Research, 13*(4), 539–547. https://doi.org/10.1139/x83-079.

Bouchard, M., Kneeshaw, D., & Bergeron, Y. (2006). Forest dynamics after successive spruce budworm outbreaks in mixedwood forests. *Ecology, 87*(9), 2319–2329. https://doi.org/10.1890/0012-9658(2006)87[2319:FDASSB]2.0.CO;2.

Bouchard, M., Boucher, Y., Belleau, A., et al. (2015). *Modélisation de la variabilité naturelle de la structure d'âge des forêts du Québec* (p. 175). Québec: Mémoire de recherche forestière, Direction de la recherche forestière ministère de la Forêt, de la Faune et des Parcs, Gouvernement du Québec.

Boucher, D., Boulanger, Y., Aubin, I., et al. (2018). Current and projected cumulative impacts of fire, drought and insects on timber volumes across Canada. *Ecological Applications, 28*(5), 1245–1259.

Boulanger, Y., Gauthier, S., & Burton, P. J. (2014). A refinement of models projecting future Canadian fire regimes using homogeneous fire regime zones. *Canadian Journal of Forest Research, 44*(4), 365–376. https://doi.org/10.1139/cjfr-2013-0372.

Boulanger, Y., Girardin, M. P., Bernier, Y., et al. (2017). Changes in mean forest age in Canada's forests could limit future increases in area burned but compromise potential harvestable conifer volume. *Canadian Journal of Forest Research, 47*(6), 755–764. https://doi.org/10.1139/cjfr-2016-0445.

Brang, P., Schönenberger, V., & Fischer, A. (2004). Reforestation in Central Europe: Lessons from multi-disciplinary field experiments. *Forest, Snow and Landscape Research, 78*(1/2), 53–69.

Burton, P. J., Svoboda, M., Kneeshaw, D., et al. (2015). Options for promoting the recovery and rehabilitation of forests affected by severe insect outbreaks. In J. A. Stanturf (Ed.), *Restoration of boreal and temperate forests* (pp. 495–517). Boca Raton: CRC Press.

Buryak, L., Luzganov, A., Matveev, P., et al. (2003). *Impact of surface fires on the formation of light-coniferous forests of southern central Siberia* [in Russian]. Krasnoyarsk: Siberian State Technological University.

Canham, C. D., Papaik, M. J., & Latty, E. F. (2001). Interspecific variation in susceptibility to windthrow as a function of tree size and storm severity for northern temperate tree species. *Canadian Journal of Forest Research, 31,* 1–10. https://doi.org/10.1139/x00-124.

Cardil, A., Lorente, M., Boucher, D., et al. (2019). Factors influencing fire suppression success in the province of Quebec (Canada). *Canadian Journal of Forest Research, 49,* 531–542. https://doi.org/10.1139/cjfr-2018-0272.

Chen, C., Weiskittel, A., Bataineh, M., et al. (2017). Evaluating the influence of varying levels of spruce budworm defoliation on annualized individual tree growth and mortality in Maine, USA and New Brunswick, Canada. *Forest Ecology and Management, 396,* 184–194. https://doi.org/10.1016/j.foreco.2017.03.026.

Cheng, C. S., Auld, H., Li, G., et al. (2007). Possible impacts of climate change on freezing rain in south-central Canada using downscaled future climate scenarios. *Natural Hazards and Earth Systems Sciences, 7*, 71–87. https://doi.org/10.5194/nhess-7-71-2007.

Cobb, T. P., Morissette, J. L., Jacobs, J. M., et al. (2011). Effects of postfire salvage logging on deadwood-associated beetles. *Conservation Biology, 25*, 94–104. https://doi.org/10.1111/j.1523-1739.2010.01566.x.

Conard, S. G., & Ivanova, G. A. (1997). Wildfire in Russian boreal forests–potential impacts of fire regime characteristics on emissions and global carbon balance estimates. *Environmental Pollution, 98*(3), 305–313. https://doi.org/10.1016/S0269-7491(97)00140-1.

Coogan, S. C., Robinne, F. N., Jain, P., et al. (2019). Scientists' warning on wildfire–a Canadian perspective. *Canadian Journal of Forest Research, 49*(9), 1015–1023. https://doi.org/10.1139/cjfr-2019-0094.

Cooke, B. J., & Roland, J. (2007). Trembling aspen responses to drought and defoliation by forest tent caterpillar and reconstruction of recent outbreaks in Ontario. *Canadian Journal of Forest Research, 37*(9), 1586–1598. https://doi.org/10.1139/X07-015.

Cooke, B. J., MacQuarrie, C. J., & Lorenzetti, F. (2012). The dynamics of forest tent caterpillar outbreaks across east-central Canada. *Ecography, 35*(5), 422–435. https://doi.org/10.1111/j.1600-0587.2011.07083.x.

Cremer, K. W., Borough, C. J., McKinnel, F. H., et al. (1982). Effects of stocking and thinning on wind damage in plantations. *New Zealand Journal of Forest Science, 12*, 245–268.

Cumming, S. G. (2001). Forest type and wildfire in the Alberta boreal mixedwood: What do fires burn? *Ecological Applications, 11*(1), 97–110. https://doi.org/10.1890/1051-0761(2001)011[0097:FTAWIT]2.0.CO;2.

D'Orangeville, L., Duchesne, L., Houle, D., et al. (2016). Northeastern North America as a potential refugium for boreal forests in a warming climate. *Science, 352*, 1452–1455. https://doi.org/10.1126/science.aaf4951.

D'Orangeville, L., Maxwell, J., Kneeshaw, D., et al. (2018). Drought timing and local climate determine the sensitivity of eastern temperate forests to drought. *Global Change Biology, 24*(6), 2339–2351. https://doi.org/10.1111/gcb.14096.

Davidenko, E. P., Furyaev, V. V., Sukhinin, A. I., et al. (2003). Fire management needs in Russia's boreal forest zone. *3rd International Wildland Fire Congress*, Sydney.

De Grandpré, L., Waldron, K., Bouchard, M., et al. (2018). Incorporating insect and wind disturbances in a natural disturbance-based management framework for the boreal forest. *Forests, 9*, 471. https://doi.org/10.3390/f9080471.

De Grandpré, L., Kneeshaw, D. D., Perigon, S., et al. (2019). Adverse climatic periods precede and amplify defoliator-induced tree mortality in eastern boreal North America. *Journal of Ecology, 107*, 452–467. https://doi.org/10.1111/1365-2745.13012.

DeLong, S. C. (2007). Implementation of natural disturbance-based management in northern British Columbia. *The Forestry Chronicle, 83*(3), 338–346. https://doi.org/10.5558/tfc83338-3.

Dempster, W. R. (2017). Impact of climate on juvenile mortality and *Armillaria* root disease in lodgepole pine. *The Forestry Chronicle, 93*, 148–160. https://doi.org/10.5558/tfc2017-021.

DeSoto, L., Cailleret, M., Sterck, F., et al. (2020). Low growth resilience to drought is related to future mortality risk in trees. *Nature Communications, 11*(1), 545. https://doi.org/10.1038/s41467-020-14300-5.

Dukes, J. S., Pontius, J., Orwig, D., et al. (2009). Responses of insect pests, pathogens, and invasive plant species to climate change in the forests of northeastern North America: What can we predict? *Canadian Journal of Forest Research, 39*, 231–248. https://doi.org/10.1139/X08-171.

Élie, J. G., & Ruel, J. C. (2005). Windthrow hazard modelling in boreal forests of black spruce and jack pine. *Canadian Journal of Forest Research, 35*, 2655–2663. https://doi.org/10.1139/x05-189.

Engelmark, O. (1987). Fire history correlations to forest type and topography in northern Sweden. *Annales Botanici Fennici, 24*, 317–324.

Erni, S., Arseneault, D., & Parisien, M. A. (2018). Stand age influence on potential wildfire ignition and spread in the boreal forest of northeastern Canada. *Ecosystems, 21*(7), 1471–1486. https://doi.org/10.1007/s10021-018-0235-3.

Everham, E. M., & Brokaw, N. V. L. (1996). Forest damage and recovery from catastrophic wind. *Botanical Review, 62*, 113–185. https://doi.org/10.1007/BF02857920.

Fedorchuk, V. N., Shorohov, A. A., Shorohova, E. V., et al. (2012). *Primeval spruce dominated forest landscapes: Structure, dynamics, and resilience* [in Russian]. Saint-Petersburg: Saint-Petersburg Polytechnical University Press.

Fischer, A., & Fischer, H. S. (2012). Individual-based analysis of tree establishment and forest stand development within 25 years after wind throw. *European Journal of Forest Research, 131*, 493–501. https://doi.org/10.1007/s10342-011-0524-2.

Fischer, A., Lindner, M., Abs, C., et al. (2002). Vegetation dynamics in Central European forest ecosystems (near-natural as well as managed) after storm events. *Folia Geobotanica, 37*, 17–32. https://doi.org/10.1007/BF02803188.

Fischer, A., Marshall, P., & Camp, A. (2013). Disturbances in deciduous temperate forest ecosystems of the northern hemisphere: Their effects on both recent and future forest development. *Biodiversity and Conservation, 22*, 1863–1893. https://doi.org/10.1007/s10531-013-0525-1.

Furyaev, V. V. (1996). *Rol'pozharov v protsesse lesoobrazovaniya* (The role of fires in the forest-forming process) (p. 253). Novosibirsk: Nauka.

Garbelotto, M., & Gonthier, P. (2013). Biology, epidemiology, and control of *Heterobasidion* species worldwide. *Annual Review of Phytopathology, 51*, 39–59. https://doi.org/10.1146/annurev-phyto-082712-102225.

Gardiner, B. A., Stacey, G. R., Belcher, R. E., et al. (1997). Field and wind tunnel assessments of the implications of respacing and thinning for tree stability. *Forestry, 70*, 233–252. https://doi.org/10.1093/forestry/70.3.233.

Gardiner, B., Byrne, K., Hale, S., et al. (2008). A review of mechanistic modelling of wind damage risk to forests. *Forestry, 81*, 447–463. https://doi.org/10.1093/forestry/cpn022.

Gauthier, S., Nguyen, T., Bergeron, Y., et al. (2004). Developing forest management strategies based on fire regimes in northwestern Quebec. In A. H. Perera, L. J. Buse, & M. G. Weber (Eds.), *Emulating natural forest landscape disturbances: Concepts and applications* (pp. 219–229). New York: Columbia University Press.

Gauthier, S., Vaillancourt, M. A., Leduc, A., et al. (2009). *Ecosystem management in the boreal forest* (p. 392). Québec: Laurentian Forestry Centre, Canadian forest service, Natural Resources Canada.

Gauthier, S., Bernier, P. Y., Boulanger, Y., et al. (2015a). Vulnerability of timber supply to projected changes in fire regime in Canada's managed forests. *Canadian Journal of Forest Research, 45*, 1439–1447. https://doi.org/10.1139/cjfr-2015-0079.

Gauthier, S., Bernier, P., Kuuluvainen, T., et al. (2015b). Boreal forest health and global change. *Science, 349*, 819–822. https://doi.org/10.1126/science.aaa9092.

Girard, F., De Grandpré, L., & Ruel, J. C. (2014). Partial windthrow as a driving process of forest dynamics in old-growth boreal forests. *Canadian Journal of Forest Research, 44*, 1165–1176. https://doi.org/10.1139/cjfr-2013-0224.

Goldammer, J. G. (Ed.). (2013). *Prescribed burning in Russia and neighboring temperate-boreal Eurasia: A publication of the global fire monitoring center (GFMC)* (p. 325). Remagen-Oberwinter: Kessel Publishing House.

Grayson, A. J. (1989). The 1987 storm: Impacts and responses. *Forestry Commission Bulletin* (Vol. 87, p. 42). London: His Majesty's Stationery Office.

Gregow, H., Peltola, H., Laapas, M., et al. (2011). Combined occurrence of wind, snow loading and soil frost with implications for risks to forestry in Finland under the current and changing climatic conditions. *Silva Fennica, 45*, 35–54. https://doi.org/10.14214/sf.30.

Gromtsev, A. (2002). Natural disturbance dynamics in the boreal forests of European Russia: A review. *Silva Fennica, 36*, 41–55. https://doi.org/10.14214/sf.549.

Gromtsev, A. (2008). *Osnovy landshaftnoj ekologii evropejskikh tayezhnykh lesov Rossii* (Basics of landscape ecology of Russia's European boreal forests). Petrozavodsk: Karelian Centre of Russian Academy of Science.

Guindon, L., Gauthier, S., Manka, F., et al. (2021). Trends in wildfire burn severity across Canada, 1985 to 2015. *Canadian Journal of Forest Research, 51*(9), 1230–1244. https://doi.org/10.1139/cjfr-2020-0353.

Gustafsson, L., Hannerz, M., Koivula, M., et al. (2020). Research on retention forestry in Northern Europe. *Ecological Processes, 9*, 3. https://doi.org/10.1186/s13717-019-0208-2.

Hale, S. E., Gardiner, B. A., Wellpott, A., et al. (2012). Wind loading of trees: Influence of tree size and competition. *European Journal of Forest Research, 131*, 203–217. https://doi.org/10.1007/s10342-010-0448-2.

Hanes, C. C., Wang, X., Jain, P., et al. (2019). Fire-regime changes in Canada over the last half century. *Canadian Journal of Forest Research, 49*(3), 256–269. https://doi.org/10.1139/cjfr-2018-0293.

Haynes, K. J., Allstadt, A. J., & Klimetzek, D. (2014). Forest defoliator outbreaks under climate change: Effects on the frequency and severity of outbreaks of five pine insect pests. *Global Change Biology, 20*(6), 2004–2018. https://doi.org/10.1111/gcb.12506.

Héon-Grenier, D. (2020). *Analyse des patrons géographique de la mortalité des arbres au Québec.* M.Sc. thesis, Université du Québec à Montréal.

Héon, J., Arseneault, D., & Parisien, M. A. (2014). Resistance of the boreal forest to high burn rates. *Proceedings of the National Academy of Sciences of the United States of America, 111*(38), 13888–13893. https://doi.org/10.1073/pnas.1409316111.

Hicke, J. A., Allen, C. D., Desai, A. R., et al. (2012). Effects of biotic disturbances on forest carbon cycling in the United States and Canada. *Global Change Biology, 18*, 7–34. https://doi.org/10.1111/j.1365-2486.2011.02543.x.

Hlásny, T., Krokene, P., Liebhold, A., et al. (2019). Living with bark beetles: Impacts, outlook and management options. *From science to policy* (Vol. 8, p. 52). European Forest Institute.

Hope, E. S., McKenney, D. W., Pedlar, J. H., et al. (2016). Wildfire suppression costs for Canada under a changing climate. *PLoS ONE, 11*(8), e0157425. https://doi.org/10.1371/journal.pone.0157425.

Isaev, A. S., Soukhovolsky, V. G., Tarasova, O. V., et al. (2017). *Forest insect population dynamics, outbreaks and global warming effects.* Hoboken: Wiley.

Itter, M. S., D'Orangeville, L., Dawson, A., et al. (2019). Boreal tree growth exhibits decadal-scale ecological memory to drought and insect defoliation, but no negative response to their interaction. *Journal of Ecology, 107*, 1288–1301. https://doi.org/10.1111/1365-2745.13087.

Ivanova, G. A., & Ivanov, V. A. (2004). The fire regime in the forests of central Siberia [in Russian]. In Furyaev, V. V. (Ed.), *Forest fire management at regional level* (pp. 147–150). Moscow: Alex.

Jactel, H., Petit, J., Desprez-Loustau, M. L., et al. (2012). Drought effects on damage by forest insects and pathogens: A meta-analysis. *Global Change Biology, 18*(1), 267–276. https://doi.org/10.1111/j.1365-2486.2011.02512.x.

Jalkanen, R., & Konocpka, B. (2007). Snow-packing as a potential harmful factor on *Picea abies, Pinus sylvestris* and *Betula pubescens* at high altitude in northern Finland. *Forest Pathology, 28*, 373–382. https://doi.org/10.1111/j.1439-0329.1998.tb01191.x.

Jardon, Y., Morin, H., & Dutilleul, P. (2003). Periodicite et synchronisme des epidemies de la tordeuse des bourgeons au Quebec. *Canadian Journal of Forest Research, 33*(10), 1947–1961. https://doi.org/10.1139/x03-108.

Karpachevsky, L. O., Kuraeva, E. N., Minaeva, T. Y., et al. (1999). Regeneration processes after severe windthrows in spruce forests [in Russian]. In O. V. Smirnova & E. S. Shaposhnikov (Eds.), *Successional processes in Russian Reserves and problems of biodiversity conservation* (pp. 380–387). St. Petersburg: Russian Botanical Society.

Karvonen, L., Eisto, K., Korhonen, K.-M., et al. (2001). *Alue-ekologinen suunnittelu Metsähallituksessa-Yhteenvetoraportti vuosilta 1996–2000* [in Finnish]. Vantaa: Metsähallitus.

Kayes, L. J., & Tinker, D. B. (2012). Forest structure and regeneration following a mountain pine beetle epidemic in southeastern Wyoming. *Forest Ecology and Management, 263*, 57–66. https://doi.org/10.1016/j.foreco.2011.09.035.

Khakimulina, T., Fraver, S., & Drobyshev, I. (2016). Mixed-severity natural disturbance regime dominates in an old-growth Norway spruce forest of northwest Russia. *Journal of Vegetation Science, 27*, 400–413. https://doi.org/10.1111/jvs.12351.

Kharuk, V. I., Im, S. T., Oskorbin, P. A., et al. (2013a). Siberian pine decline and mortality in southern Siberian mountains. *Forest Ecology and Management, 310*, 312–320. https://doi.org/10.1016/j.foreco.2013.08.042.

Kharuk, V. I., Ranson, K. J., Oskorbin, P. A., et al. (2013b). Climate induced birch mortality in the trans-Baikal lake region, Siberia. *Forest Ecology and Management, 289*, 385–392. https://doi.org/10.1016/j.foreco.2012.10.024.

Kharuk, V. I., Dvinskaya, M. L., Petrov, I. A., et al. (2016a). Larch forests of Middle Siberia: Long-term trends in fire return intervals. *Regional Environmental Change, 16*, 2389–2397. https://doi.org/10.1007/s10113-016-0964-9.

Kharuk, V. I., Im, S. T., Petrov, I. A., et al. (2016b). Decline of dark coniferous stands in Baikal region. *Contemporary Problems of Ecology, 9*, 617–625. https://doi.org/10.1134/S1995425516050073.

Kilpeläinen, A., Gregow, H., Strandman, H., et al. (2010). Impacts of climate change on the risk of snow-induced forest damage in Finland. *Climatic Change, 99*, 193–209. https://doi.org/10.1007/s10584-009-9655-6.

Kneeshaw, D. D., & Bergeron, Y. (1998). Canopy gap characteristics and tree replacement in the southeastern boreal forest. *Ecology, 79*(3), 783–794. https://doi.org/10.1890/0012-9658(1998)079[0783:CGCATR]2.0.CO;2.

Kneeshaw, D. D., Harvey, B. D., Reyes, G. P., et al. (2011). Spruce budworm, windthrow and partial cutting: Do different partial disturbances produce different forest structures? *Forest Ecology and Management, 262*, 482–490. https://doi.org/10.1016/j.foreco.2011.04.014.

Kneeshaw, D., Sturtevant, B. R., Cooke, B., et al. (2015). Insect disturbances in forest ecosystems. In K. S.-H. Peh, R. T. Corlett, & Y. Bergeron (Eds.), *Routledge handbook of forest ecology* (pp. 109–129). Abington: Routledge Handbooks Online.

Kneeshaw, D. D., Sturtevant, B. R., De Grandpré, L., et al. (2021). The vision of managing for pest-resistant landscapes: Realistic or utopic? *Current Forestry Reports, 7*(2), 97–113. https://doi.org/10.1007/s40725-021-00140-z.

Koivula, M., & Vanha-Majamaa, I. (2020). Experimental evidence on biodiversity impacts of variable retention forestry, prescribed burning, and deadwood manipulation in Fennoscandia. *Ecological Processes, 9*, 11. https://doi.org/10.1186/s13717-019-0209-1.

Koivula, M., Kuuluvainen, T., Hallman, E., et al. (2014). Forest management inspired by natural disturbance dynamics (DISTDYN)-a long-term research and development project in Finland. *Scandinavian Journal of Forest Research, 29*, 579–592. https://doi.org/10.1080/02827581.2014.938110.

Kolb, T. E., Fettig, C. J., Ayres, M. P., et al. (2016). Observed and anticipated impacts of drought on forest insects and diseases in the United States. *Forest Ecology and Management, 380*, 321–334. https://doi.org/10.1016/j.foreco.2016.04.051.

Kontula, T., & Raunio, A. (Eds.). (2019). *Threatened habitat types in Finland 2018-Red list of habitats results and basis for assessment* (p. 258). Helsinki: Finnish Environment Institute and Ministry of the Environment.

Korovin, G. N. (1996). Analysis of the distribution of forest fires in Russia. In J. G. Goldammer & V. V. Furyaev (Eds.), *Fire in ecosystems of boreal Eurasia* (pp. 112–128). Netherlands, Dordrecht: Springer.

Kulakowski, D., Seidl, R., Holeksa, J., et al. (2017). A walk on the wild side: Disturbance dynamics and the conservation and management of European mountain forest ecosystems. *Forest Ecology and Management, 388*, 120–131. https://doi.org/10.1016/j.foreco.2016.07.037.

Kuleshova, L. V. (Ed). (2002). *Monitoring soobshchestv na garyakh i upravlenie pozharami v zapovednikakh* (Monitoring of communities on the fire-sites and control of fires in nature reserves). Moscow: Vseross. Nauchno-Issled. Inst. Prirody.

Lännenpää, A., Aakala, T., Kauhanen, H., et al. (2008). Tree mortality agents in pristine Norway spruce forests in northern Fennoscandia. *Silva Fennica, 42*, 151–163. https://doi.org/10.14214/sf.468.

Lässig, R., & Močalov, S. A. (2000). Frequency and characteristics of severe storms in the Urals and their influence on the development, structure and management of the boreal forests. *Forest Ecology and Management, 135*, 179–194. https://doi.org/10.1016/S0378-1127(00)00309-1.

Lavoie, J., Montoro Girona, M., Grosbois, G., et al. (2021). Does the type of silvicultural practice influence spruce budworm defoliation of seedlings? *Ecosphere, 12*(4), 17. https://doi.org/10.1002/ecs2.3506.

Lefort, P., Gauthier, S., & Bergeron, Y. (2003). The influence of fire weather and land use on the fire activity of the Lake Abitibi area, eastern Canada. *Forestry Sciences, 49*(4), 509–521.

Léger-Beaulieu, C., D'Orangeville, L., Houle, D., et al. (*in-review*) Experimental warming and drying reveals high stress resistance in jack pine versus reduced carbon uptake and growth in black and white spruce. *Tree Physiology*.

Lindberg, H., Punttila, P., & Vanha-Majamaa, I. (2020). The challenge of combining variable retention and prescribed burning in Finland. *Ecological Processes, 9*, 4. https://doi.org/10.1186/s13717-019-0207-3.

Man, R., & Rice, J. A. (2010). Response of aspen stands to forest tent caterpillar defoliation and subsequent overstory mortality in northeastern Ontario, Canada. *Forest Ecology and Management, 260*(10), 1853–1860. https://doi.org/10.1016/j.foreco.2010.08.032.

Markham, J., Adorno, B. V., & Weisser, N. (2019). Determinants of mortality in *Pinus banksiana* (Pinaceae) stands during an ice storm and its effect on stand spatial structure. *The Journal of the Torrey Botanical Society, 146*, 111–118. https://doi.org/10.3159/TORREY-D-18-0002.1.

Martell, D. L., & Sun, H. (2008). The impact of fire suppression, vegetation, and weather on the area burned by lightning-caused forest fires in Ontario. *Canadian Journal of Forest Research, 38*(20046), 1547–1563. https://doi.org/10.1139/X07-210.

Maslov, A. D. (2010). *European spruce bark beetle and the decline of spruce forests* (p. 138). Moscow: All-Russian Research Institute of Silviculture and Mechanization of Forestry.

Meigs, G. W., Morrissey, R. C., Bače, R., et al. (2017). More ways than one: Mixed-severity disturbance regimes foster structural complexity via multiple developmental pathways. *Forest Ecology and Management, 406*, 410–426. https://doi.org/10.1016/j.foreco.2017.07.051.

Melekhov, I. S. (1971). *On the patterns and periodicity of forest burnability* [in Russian]. Collected papers on forestry and forest chemistry, Archangelsk.

Michaelian, M., Hogg, E. H., Hall, R. J., et al. (2011). Massive mortality of aspen following severe drought along the southern edge of the Canadian boreal forest. *Global Change Biology, 17*(6), 2084–2094. https://doi.org/10.1111/j.1365-2486.2010.02357.x.

Miller, K. F. (1985). Windthrow hazard classification. *U.K. forestry commission* (No. 85). London.

Ministère des forêts de la faune et des parcs (MFFP). (2019). *Aires infestées par la tordeuse des bourgeons de l'épinette au Québec en 2019* (p. 32). Direction de la protection des forêts, Gouvernement du Québec.

Močalov, S. A., & Lässig, R. (2002). Development of two boreal forests after large-scale windthrow in the Central Urals. *Forest, Snow and Landscape Research, 77*, 171–186.

Montoro Girona, M., Morin, H., Lussier, J. M., et al. (2016). Radial growth response of black spruce stands ten years after experimental shelterwoods and seed-tree cuttings in boreal forest. *Forests, 7*, 240. https://doi.org/10.3390/f7100240.

Montoro Girona, M., Morin, H., Lussier, J.-M., et al. (2019). Post-cutting mortality following experimental silvicultural treatments in unmanaged boreal forest stands. *Frontiers in Forests and Global Change, 2*, 4. https://doi.org/10.3389/ffgc.2019.00004.

Morin, H., Laprise, D., Simard, A. A., et al. (2009). Spruce budworm outbreak regimes in eastern North America. In S. Gauthier, M. A. Vaillancourt, A. Leduc, L. De Grandpre, D. D. Kneeshaw,

H. Morin, P. Drapeau, & Y. Bergeron (Eds.), *Ecosystem management of the boreal forest* (pp. 155–182). Québec: Les Presses de l'Université du Quebec.

Myndigheten för samhällsskydd och beredskap (MSB). (2020). *MSB: Statistik-och analysverktyg IDA* (MSB's statistics and analysis tool IDA). Swedish Civil Contingencies Agency.

Nappi, A., Drapeau, P., & Savard, J. P. (2004). Salvage logging after wildfire in the boreal forest: Is it becoming a hot issue for wildlife? *The Forestry Chronicle, 80*(1), 67–74. https://doi.org/10.5558/tfc80067-1.

Nappi, A., Dery, S., Bujold, F., et al. (2011). *Harvesting in burned forest–issues and orientations for Ecosystem-based management* (p. 47). Québec: Direction de l'environnement et de la protection des forêts, ministère des Ressources naturelles et de la Faune, Gouvernement du Québec.

Navarro, L., Morin, H., Bergeron, Y., et al. (2018). Changes in spatiotemporal patterns of 20th century spruce budworm outbreaks in eastern Canadian boreal forests. *Frontiers in Plant Science, 9*, 1905. https://doi.org/10.3389/fpls.2018.01905.

Nevalainen, S. (2017). Comparison of damage risks in even- and uneven-aged forestry in Finland. *Silva Fennica, 51*, 1741. https://doi.org/10.14214/sf.1741.

Nicoll, B. C., Gardiner, B. A., Rayner, B., et al. (2006). Anchorage of coniferous trees in relation to species, soil type and rooting depth. *Canadian Journal of Forest Research, 36*, 1871–1883. https://doi.org/10.1139/x06-072.

Nykänen, M.-L., Broadgate, M., Kellomäki, S., et al. (1997). Factors affecting snow damage of trees with particular reference to European conditions. *Silva Fennica, 31*(2), 5618. https://doi.org/10.14214/sf.a8519.

Oogathoo, S., Houle, D., Duchesne, L., et al. (2020). Vapour pressure deficit and solar radiation are the major drivers of transpiration of balsam fir and black spruce tree species in humid boreal regions, even during a short-term drought. *Agricultural and Forest Meteorology, 291*, 108063. https://doi.org/10.1016/j.agrformet.2020.108063.

Patry, C., Kneeshaw, D., Aubin, I., et al. (2017). Intensive forestry filters understory plant traits over time and space in boreal forests. *Forestry, 90*(3), 436–444.

Pavlov, I. N. (2015). Biotic and abiotic factors as causes of coniferous forests dieback in Siberia and Far East. *Contemporary Problems of Ecology, 8*(4), 440–456. https://doi.org/10.1134/S1995425515040125.

Payette, S., & Delwaide, A. (2018). Tamm review: The North-American lichen woodland. *Forest Ecology and Management, 417*, 167–183. https://doi.org/10.1016/j.foreco.2018.02.043.

Peltola, H., Kellomäki, S., Väisänen, H., et al. (1999). A mechanistic model for assessing the risk of wind and snow damage to single trees and stands of Scots pine, Norway spruce, and birch. *Canadian Journal of Forest Research, 29*, 647–661. https://doi.org/10.1139/x99-029.

Peltola, H., Kellomäki, S., Hassinen, H., et al. (2000). Mechanical stability of Scots pine, Norway spruce and birch: An analysis of tree-pulling experiments in Finland. *Forest Ecology and Management, 135*, 143–153. https://doi.org/10.1016/S0378-1127(00)00306-6.

Peng, C., Ma, Z., Lei, X., et al. (2011). A drought-induced pervasive increase in tree mortality across Canada's boreal forests. *Nature Climate Change, 1*, 467–471. https://doi.org/10.1038/nclimate1293.

Perron, N., Bélanger, L., & Vaillancourt, M. A. (2009). Spatial structure of forest stands and remnants under fire and timber harvesting regimes. In S. Gauthier, M. A. Vaillancourt, A. Leduc, L. De Grandpré, D. D. Kneeshaw, H. Morin, P. Drapeau, & Y. Bergeron (Eds.), *Ecosystem management in the boreal forest* (pp. 129–154). Québec: Presses de l'Université du Québec.

Petukhov, I. N., & Nemchinova, A. V. (2015). Windthrows in forests of Kostroma Oblast and the neighboring lands in 1984–2011. *Contemporary Problems of Ecology, 8*, 901–908. https://doi.org/10.1134/S1995425515070094.

Pickett, S. T. A., & White, P. S. (1985). *The ecology of natural disturbances and patch dynamics* (p. 472). San Diego: Academic Press.

Pinto, G. A. S. J., Rousseu, F., Niklasson, M., et al. (2020). Effects of human-related and biotic landscape features on the occurrence and size of modern forest fires in Sweden. *Agricultural and Forest Meteorology, 291*, 108084. https://doi.org/10.1016/j.agrformet.2020.108084.

Piri, T., & Valkonen, S. (2013). Incidence and spread of *Heterobasidion* root rot in uneven-aged Norway spruce stands. *Canadian Journal of Forest Research, 43*(9), 872–877. https://doi.org/10.1139/cjfr-2013-0052.

Pureswaran, D. S., Roques, A., & Battisti, A. (2018). Forest insects and climate change. *Current Forestry Reports, 4*(2), 35–50. https://doi.org/10.1007/s40725-018-0075-6.

Putz, F. E., Coley, P. D., Lu, K., et al. (1983). Uprooting and snapping of trees: Structural determinants and ecological consequences. *Canadian Journal of Forest Research, 13*(5), 1011–1020. https://doi.org/10.1139/x83-133.

Raffa, K. F., Grégoire, J. C., & Staffan Lindgren, B. (2015). Natural history and ecology of bark beetles. In F. E. Vega, & R. W. Hofstetter (Eds.), *Bark beetles. Biology and ecology of native and invasive species* (pp. 1–40). San Diego: Elsevier.

Régnière, J., St-Amant, R., & Duval, P. (2012). Predicting insect distributions under climate change from physiological responses: Spruce budworm as an example. *Biological Invasions, 14*(8), 1571–1586. https://doi.org/10.1007/s10530-010-9918-1.

Reinikainen, M., D'Amato, A. W., & Fraver, S. (2012). Repeated insect outbreaks promote multi-cohort aspen mixedwood forests in northern Minnesota, USA. *Forest Ecology and Management, 266*, 148–159. https://doi.org/10.1016/j.foreco.2011.11.023.

Rich, R. L., Frelich, L. E., & Reich, P. B. (2007). Wind-throw mortality in the southern boreal forest: Effects of species, diameter and stand age. *Journal of Ecology, 95*, 1261–1273. https://doi.org/10.1111/j.1365-2745.2007.01301.x.

Robert, L. E., Kneeshaw, D., & Sturtevant, B. R. (2012). Effects of forest management legacies on spruce budworm (*Choristoneura fumiferana*) outbreaks. *Canadian Journal of Forest Research, 42*(3), 463–475. https://doi.org/10.1139/x2012-005.

Robert, L. E., Sturtevant, B. R., Cooke, B. J., et al. (2018). Landscape host abundance and configuration regulate periodic outbreak behavior in spruce budworm (*Choristoneura fumiferana* Clem.). *Ecography, 41*(9), 1556–1571. https://doi.org/10.1111/ecog.03553.

Robert, L. E., Sturtevant, B. R., Kneeshaw, D., et al. (2020). Forest landscape structure influences the cyclic-eruptive spatial dynamics of forest tent caterpillar outbreaks. *Ecosphere, 11*(8), e03096. https://doi.org/10.1002/ecs2.3096.

Rogers, B. M., Soja, A. J., Goulden, M. L., et al. (2015). Influence of tree species on continental differences in boreal fires and climate feedbacks. *Nature Geoscience, 8*(3), 228–234. https://doi.org/10.1038/ngeo2352.

Rollinson, T. J. D. (1987). Thinning control of conifer plantation in Great Britain. *Annales Des Sciences Forestières, 44*, 25–34. https://doi.org/10.1051/forest:19870103.

Rolstad, J., Blanck, Y. L., & Storaunet, K. O. (2017). Fire history in a western Fennoscandian boreal forest as influenced by human land use and climate. *Ecological Monographs, 87*, 219–245. https://doi.org/10.1002/ecm.1244.

Romashkin, I., Neuvonen, S., & Tikkanen, O. P. (2020). Northward shift in temperature sum isoclines may favour *Ips typographus* outbreaks in European Russia. *Agricultural and Forest Entomology, 22*, 238–249. https://doi.org/10.1111/afe.12377.

Ruel, J. C. (1995). Understanding windthrow: Silvicultural implications. *The Forestry Chronicle, 71*, 434–445. https://doi.org/10.5558/tfc71434-4.

Ruel, J. C. (2000). Factors influencing windthrow in balsam fir forests: From landscape studies to individual tree studies. *Forest Ecology and Management, 135*, 169–178. https://doi.org/10.1016/S0378-1127(00)00308-X.

Ruel, J. C., & Benoit, R. (1999). Analyse du chablis du 7 novembre 1994 dans les regions de Charlevoix et de la Gaspesie, Quebec, Canada. *The Forestry Chronicle, 75*, 293–301. https://doi.org/10.5558/tfc75293-2.

Ruel, J. C., & Gardiner, B. (2019). Mortality patterns after different levels of harvesting of old-growth boreal forests. *Forest Ecology and Management, 448*, 346–354. https://doi.org/10.1016/j.foreco.2019.06.029.

Ruel, J. C., Pin, D., & Cooper, K. (1998). Effect of topography on wind behaviour in a complex terrain. *Forestry, 71*, 261–265. https://doi.org/10.1093/forestry/71.3.261.

Saad, C., Boulanger, Y., Beaudet, M., et al. (2017). Potential impact of climate change on the risk of windthrow in eastern Canada's forests. *Climatic Change, 143*, 487–501. https://doi.org/10.1007/s10584-017-1995-z.

Safranyik, L., Carroll, A. L., Régnière, J., et al. (2010). Potential for range expansion of mountain pine beetle into the boreal forest of North America. *The Canadian Entomologist, 142*, 415–442. https://doi.org/10.4039/n08-CPA01.

Šamonil, P., Kra'l, K., & Hort, L. (2010). The role of tree uprooting in soil formation: A critical literature review. *Geoderma, 157*, 65–79. https://doi.org/10.1016/j.geoderma.2010.03.018.

San-Miguel-Ayanz, J., Durrant, T., Boca, R., et al. (2018). *Forest fires in Europe, Middle East and North Africa 2017* (p. 139). Ispra: European Commission, Joint Research Centre.

Sánchez-Pinillos, M., D'Orangeville, L., Boulanger, Y., Comeau, P., Wang, J., Taylor, A. R., & Kneeshaw, D. (2022). Sequential droughts: A silent trigger of boreal forest mortality. *Global Change Biology, 28*, 542–556. https://doi.org/10.1111/gcb.15913.

Sands. R. J. (2017). *The population ecology of oak processionary moth.* Ph.D. thesis, University of Southampton.

Schaetzl, R. J., Johnson, D. L., Burns, S. F., et al. (1989). Tree uprooting: Review of terminology process and environmental implications. *Canadian Journal of Forest Research, 19*, 1–11. https://doi.org/10.1139/x89-001.

Schelhaas, M. J., Nabuurs, G. J., & Schuck, A. (2003). Natural disturbances in the European forests in the 19th and 20th centuries. *Global Change Biology, 9*, 1620–1633. https://doi.org/10.1046/j.1365-2486.2003.00684.x.

Schönenberger, W. (2002). Post windthrow stand regeneration in Swiss mountain forests: The first ten years after the 1990 storm Vivian 2002. *Forest, Snow and Landscape Research, 77*(1), 61–80.

Schönenberger, W., Noack, A., & Thee, P. (2005). Effect of timber removal from windthrow slopes on the risk of snow avalanches and rockfall. *Forest Ecology and Management, 213*, 197–208. https://doi.org/10.1016/j.foreco.2005.03.062.

Seidl, R., Schelhaas, M. J., Rammer, W., et al. (2014). Increasing forest disturbances in Europe and their impact on carbon storage. *Nature Climate Change, 4*, 806–810. https://doi.org/10.1038/nclimate2318.

Seidl, R., Honkaniemi, J., Aakala, T., et al. (2020). Globally consistent climate sensitivity of natural disturbances across boreal and temperate forest ecosystems. *Ecography, 43*(7), 967–978. https://doi.org/10.1111/ecog.04995.

Shishikin, A. S., Ivanov, V. A., Ponomarev, V. I., et al. (2012). *Fire danger mitigation: A strategy for protected areas of the Altai-Sayan ecoregion* (p. 61). Krasnoyarsk: United Nations development programme.

Shmatkov, N., & Yaroshenko, A. (2018). *Call for a clear legal status for forests on abandoned agricultural landscapes.* Moscow: Joint statement, WWF-Russia and Greenpeace.

Shorohova, E., Fedorchuk, V. N., Kuznetsova, M. L., et al. (2008). Wind induced successional changes in pristine boreal *Picea abies* forest stands: Evidence from long-term permanent plot records. *Forestry, 81*, 335–359. https://doi.org/10.1093/forestry/cpn030.

Shorohova, E., Kneeshaw, D., Kuuluvainen, T., et al. (2011). Variability and dynamics of old-growth forests in the circumboreal zone: Implications for conservation, restoration and management. *Silva Fennica, 45*, 785–806. https://doi.org/10.14214/sf.72.

Skvortsova, E. B., Ulanova, N. G., & Basevich, V. F. (1983). *Ecological role of windthrows* [in Russian] (p. 190). Moscow: Lesnaya promyshlennost.

Smolonogov, E. P. (1995). Forest forming process and windthrows [in Russian]. In Y. M. Alesenkov (Ed.), *Posledstviya katastroficheskogo vetrovala dlya lesnykh ecosystem (Consequences of a catastrophic windthrow for forest ecosystems)* (pp. 12–17). Ekaterinburg: URO RAS.

Sofronov, M. A., & Volokitina, A. V. (1990) *Division of taiga zone into pyrological districts* [in Russian] (p. 205). Novosibirsk: Nauka.

Sonntag, P. (2016). *Attack of the budworms: The current infestation threatens Canadian forests.* Toronto: The Walrus.

Soukhovolsky, V., Mochalov, S., Zoteeva, E., et al. (2012). Early stages of forest restoration after windthrow in Ural (Russia): Observations and mathematical models. *Tree and Forestry Science and Biotechnology, 6*(1), 69–74.

Splawinski, T. B., Cyr, D., Gauthier, S., et al. (2019). Analyzing risk of regeneration failure in the managed boreal forest of northwestern Quebec. *Canadian Journal of Forest Research, 49*(6), 680–691. https://doi.org/10.1139/cjfr-2018-0278.

Stocks, B. J. (2013). *Evaluating past, current and future wildland fire load trends in Canada* (p. 51). Sault Ste Marie: Stocks Wildfire Investigations Ltd.

Stocks, B., & Martell, D. (2016). Forest fire management expenditures in Canada 1970–2013. *The Forestry Chronicle, 92*(3), 298–306. https://doi.org/10.5558/tfc2016-056.

Storozhenko, V. G. (2001). Structure of biotrophic fungal communities in forest ecosystems [in Russian]. In V. Storozhenko, V. Krutov, & N. Selochnik (Eds.), *Fungal communities in forest ecosystems* (pp. 224–251). Moscow, Petrozavodsk: Karelian Research Centre of RAS.

Sturrock, R. N., Frankel, S. J., Brown, A. V., et al. (2011). Climate change and forest diseases. *Plant Pathology, 60*, 133–149. https://doi.org/10.1111/j.1365-3059.2010.02406.x.

Thomas, F. M., Blank, R., & Hartmann, G. (2002). Abiotic and biotic factors and their interactions as causes of oak decline in Central Europe. *Forest Pathology, 32*(4–5), 277–307. https://doi.org/10.1046/j.1439-0329.2002.00291.x.

Thorn, S., Bässler, C., Brandl, R., et al. (2018). Impacts of salvage logging on biodiversity: A meta-analysis. *Journal of Applied Ecology, 55*(1), 279–289. https://doi.org/10.1111/1365-2664.12945.

Thorn, S., Chao, A., Georgiev, K. B., et al. (2020). Estimating retention benchmarks for salvage logging to protect biodiversity. *Nature Communications, 11*(1), 4762. https://doi.org/10.1038/s41467-020-18612-4.

Ulanova, N. G. (2000). The effects of windthrow on forests at different spatial scales: A review. *Forest Ecology and Management, 135*, 155–168. https://doi.org/10.1016/S0378-1127(00)00307-8.

Ulanova, N. G., & Cherednichenko, O. V. (2012). Patch dynamics of herb layer vegetation after catastrophic windthrow in a mixed spruce forest [in Russian]. *Izvestiya of the Samara Center of the Russian Academy of Sciences, 14*(1–5), 1399–1402.

Valendik, E. N., & Ivanova, G. A. (2001). Fire regimes in the forests of Siberia and Far East [in Russian]. *Russian Forest Sciences, 4*, 69–76.

Valendik, E. N., Goldammer, J. G., Kisilyakov, Y. K., et al. (2013). Prescribed burning in Russia. In J. G. Goldammer (Ed.), *Prescribed burning in Russia and neighboring temperate-boreal Eurasia* (pp. 13–148). Remagen-Oberwinter: Kessel Publishing House.

Valinger, E., Kempe, G., & Fridman, J. (2014). Forest management and forest state in southern Sweden before and after the impact of storm Gudrun in the winter of 2005. *Scandinavian Journal of Forest Research, 29*, 466–472. https://doi.org/10.1080/02827581.2014.927528.

Voronin, V. I. (2018). Bacterial infections of the coniferous in the Baikal forests: Causes and risks of epiphythetics [in Russian]. In *Proceedings of the All-Russian Scientific Conference with International Participation and Schools of Young Scientists, "Mechanisms of Resistance of Plants and Microorganisms to Unfavorable Environments" (Parts I, II)* (pp. 9–12). Annual Meeting, Society of Plant Physiologists of Russia, Irkutsk.

Waldron, K., Ruel, J. C., & Gauthier, S. (2013). The effects of site characteristics on the landscape-level windthrow regime in the North Shore region of Quebec, Canada. *Forestry, 86*, 159–171. https://doi.org/10.1093/forestry/cps061.

Wallenius, T. (2011). Major decline in fires in coniferous forests-reconstructing the phenomenon and seeking for the cause. *Silva Fennica, 45*, 139–155. https://doi.org/10.14214/sf.36.

Weir, J. M. H., Johnson, E. A., & Miyanishi, K. (2000). Fire frequency and the spatial age mosaic of the mixed-wood boreal forest in western Canada. *Ecological Applications, 10*(4), 1162–1177. https://doi.org/10.1890/1051-0761(2000)010[1162:FFATSA]2.0.CO;2.

Wooster, M. J., & Zhang, Y. H. (2004). Boreal forest fires burn less intensely in Russia than in North America. *Geophysical Research Letters* 31. https://doi.org/10.1029/2004GL020805.

Wotton, B. M., Nock, C. A., & Flannigan, M. D. (2010). Forest fire occurrence and climate change in Canada. *International Journal of Wildland Fire, 19*(3), 253–271. https://doi.org/10.1071/WF0 9002.

Zamolodchikov, D. G. (2012). An estimate of climate related changes in tree species diversity based on the results of forest fund inventory. *Biology Bulletin Reviews, 2*(2), 154–163. https://doi.org/ 10.1134/S2079086412020119.

Zhou, D., & Hyde, K. D. (2001). Specificity, host-exclusivity, and host-recurrence in saprobic fungi. *Mycological Research, 105,* 1449–1457. https://doi.org/10.1017/S0953756201004713.

Zyabchenko, S. S. (1984). *Pine forests of the European North* [in Russian] (p. 248). Leningrad: Nauka.

Chapter 4
Selected Examples of Interactions Between Natural Disturbances

Jean-Claude Ruel, Beat Wermelinger, Sylvie Gauthier, Philip J. Burton, Kaysandra Waldron, and Ekaterina Shorohova

Abstract Understanding natural disturbance regimes and their impacts is crucial in designing ecosystem management strategies. However, disturbances do not always occur in isolation; the occurrence of one disturbance influences the likelihood or the effect of another. In this chapter, we illustrate the importance of disturbance interactions by focusing on a subset of interactions present in different parts of the boreal forest. The selected interactions include insects and wind, insects and fire, and wind and fire. The potential consequences of climate change on these interactions are also discussed.

J.-C. Ruel (✉)
Faculté de Foresterie, Géographie et Géomatique, Université Laval, Pavillon Abitibi-Price, 2405 rue de la terrasse, Québec, QC G1V 0A6, Canada
e-mail: Jean-Claude.Ruel@sbf.ulaval.ca

B. Wermelinger
Swiss Federal Institute for Forest, Snow and Landscape Research WSL, Zürcherstrasse 111, 8903 Birmensdorf, Switzerland
e-mail: Beat.Wermelinger@wsl.ch

S. Gauthier · K. Waldron
Natural Resources Canada, Canadian Forest Service, Laurentian Forestry Centre, 1055 rue du PEPS, P.O. Box 10380, Stn. Sainte-Foy, Québec, QC G1V 4C7, Canada
e-mail: Sylvie.gauthier@nrcan-rncan.gc.ca

K. Waldron
e-mail: Kaysandra.waldron@NRCan-RNCan.gc.ca

P. J. Burton
Ecosystem Science and Management, University of Northern British Columbia, 4837 Keith Avenue, Terrace, BC V8G 1K7, Canada
e-mail: Phil.Burton@unbc.ca

M. M. Girona et al. (eds.), *Boreal Forests in the Face of Climate Change*,
Advances in Global Change Research 74,
https://doi.org/10.1007/978-3-031-15988-6_4

4.1 Introduction

A forest ecosystem management approach that mimics natural forest dynamics requires a solid understanding of natural disturbance regimes. The previous chapters (Chaps. 2 and 3) have provided information on disturbance regimes and how climate change influences them. However, disturbances do not always act in isolation but often interact (Buma, 2015; De Grandpré et al., 2018). Some disturbances such as insect outbreaks and windstorms increase the raw material (food or fuel) upon which other disturbances can build and consequently augment their importance. Conversely, certain events such as fires and landslides remove or reduce the available biotic material on which subsequent disturbances can act, thus decreasing the occurrence of future disturbances. The marked heterogeneity of disturbances and general patterns generated by interacting disturbances can lead to complex disturbance regimes and landscapes (Cannon et al., 2019; Sturtevant & Fortin, 2021).

Interactions can take two major forms: (1) the occurrence of one disturbance influences the likelihood and impact of a second event, and (2) a disturbance influences the capacity of the ecosystem to recover from a previous event (Buma, 2015). Both forms may occur simultaneously. Infrequent, large disturbances would normally produce minimal long-term change, so long as they remain within the natural range of variability for disturbance frequency and severity (Kulha et al., 2020). Compound disturbances that occur within the period where the ecosystem is recovering from the initial disturbance may lead, however, to the long-term alteration of communities (Jasinski & Payette, 2005; Paine et al., 1998; Splawinski et al., 2019). Ecosystem recovery can also be compromised when a disturbance occurs in a community already affected by a chronic stress, e.g., drought, a situation that may become more common in the context of climate change (Jactel et al., 2012). However, there are also cases where a disturbance may reduce the probability, intensity, or severity of subsequent disturbances (Cannon et al., 2019). The amplifying or buffering nature of these interactions can even vary with the particular response variable (Cannon et al., 2019).

To assess disturbance interactions, we must discuss both the implicated mechanisms and their respective impacts on the ecosystem. Different forms of disturbance can affect various ecosystem components, and we require a means of describing these effects. Buma (2015) has suggested focusing on the legacies from each disturbance and the mechanisms involved. Traditionally, the amount of canopy removed has been used to describe disturbance severity in forests; however, Roberts (2007) suggested

E. Shorohova
Forest Research Institute of the Karelian Research Center, Russian Academy of Science, Pushkinskaya str. 11, Petrozavodsk 185910, Russia
e-mail: shorohova@es13334.spb.edu; ekaterina.shorokhova@luke.fi

Saint Petersburg State Forest Technical University, Institutsky str. 5, Saint Petersburg 194021, Russia

Natural Resources Institute Finland (Luke), Latokartanonkaari 9, 00790 Helsinki, Finland

that this measure is insufficient by itself to fully describe the impact of forest disturbances. This is particularly relevant when considering disturbance interactions. As an alternative, Roberts (2007) suggested describing disturbance severity along three axes: (1) percentage of canopy removed, (2) percentage of understory removed, and (3) percentage of forest floor and soil removed or disrupted.

In this chapter, we illustrate specific regional interactions between the main natural disturbances of the boreal forest and discuss the potential effects of global change on these interactions. We recognize that we do not touch upon all possible interactions and regions; however, we believe that focusing on these selected cases can help design future ecosystem management strategies. Finally, we highlight some knowledge gaps and their associated research needs.

4.2 Windthrow and Insects

Interactions between windthrow and insects are common in the boreal forest biome. The implicated tree and insect species vary geographically, as does the nature of the interactions. Windthrow and insect disturbances can interact in two different manners. Insect damage can open the stand, exposing trees to higher wind speeds (Gardiner et al., 1997) and making them more susceptible to windthrow. Examples of this type of interaction include infestations of the spruce budworm (*Choristoneura fumiferana* Clemens) in northeastern North America and the mountain pine beetle (*Dendroctonus ponderosae* Hopkins) in western North America. Nonetheless, windthrow and insects can also interact in a reverse manner. Windthrow can generate an ample supply of breeding material, supporting a population increase of some bark beetle species, which can then switch to attack living trees (e.g., Havašová et al., 2017).

4.2.1 Windthrow and Defoliators

In the absence of fire, windthrow and outbreaks of spruce budworm represent the main disturbances in the boreal forest of eastern Canada. Three major spruce budworm outbreaks occurred during the twentieth century (Navarro et al., 2018), mostly affecting forests dominated by balsam fir (*Abies balsamea* (L.) Mill.). Vulnerability to spruce budworm–related defoliation differs among tree species, balsam fir being the most vulnerable, followed by white spruce (*Picea glauca* (Moench) Voss), red spruce (*Picea rubens* Sarg.), and black spruce (*Picea mariana* (Mill.) BSP). Pines and hardwoods are unaffected.

Windthrow is a common feature of the Canadian boreal forest. Although the return period of total windthrow may exceed 4,000 years (Bouchard et al., 2009; Waldron et al., 2013), partial windthrow can be more frequent (Waldron et al., 2013).

Windthrow occurrence varies with wind exposure as well as soil and stand characteristics (Ruel, 1995, 2000). Vulnerability also depends on species, stem taper, and rooting depth. Balsam fir, which is the most vulnerable tree species to the spruce budworm, is also one of the species most vulnerable to windthrow in eastern Canada.

Defoliating insect outbreaks of intermediate severity or outbreaks occurring in mixed-species stands typically cause partial canopy mortality. Because of the increased tree spacing resulting from this partial mortality, wind load on residual trees is increased, potentially leading to windthrow (Girard et al., 2014; Morin, 1990). Spruce budworm–related defoliation can also lead to a reduction of the fine root biomass of surviving trees (Morin, 1990). This reduced size of the root system affects a tree's resistance to overturning. Taylor and MacLean (2009) documented an increase in wind-driven mortality 11 to 25 years after a spruce budworm–related defoliation. In mixedwood stands, hardwoods surviving a spruce budworm outbreak are also more wind resistant, thereby limiting the potential for disturbance interactions. In their study of Newfoundland forests that had previously been attacked by hemlock looper (*Lambdina fiscellaria* Guenée), Arsenault et al. (2016) reported a greater incidence of mappable windthrow patches. In both cases, the increased exposure of surviving trees after widespread insect-caused defoliation provoked elevated levels of windthrow.

Spruce budworm has a major but variable effect on the main forest canopy and generally a minor effect on the understory, although advance regeneration may be somewhat affected (Nie et al., 2018). The impact on the forest floor is, however, generally negligible. Because balsam fir is more vulnerable than other species, this perturbation can reduce the proportion of this species in the canopy over the short term; however, as a shade-tolerant species having relatively few seedbed requirements, balsam fir generally dominates the advance regeneration in mixed coniferous stands. Hence, the impact of the spruce budworm on the longer-term tree species composition tends to be minor (Girard et al., 2014).

When spruce budworm damage does lead to windthrow, the additional consequences on the understory layer tend to be limited (Girard et al., 2014); however, substantial changes occur on the forest floor. The creation of a pit and mound microtopography by windthrow disrupts the herbaceous layer and exposes mineral soil or mixtures of mineral soil and organic material. This microtopography contributes to an increased post-disturbance seedbed heterogeneity, which can improve seedling establishment and increase plant species richness (Ulanova, 2000). Given the aggressiveness of balsam fir regeneration, however, balsam fir typically remains the main tree species and may even increase its relative abundance (Fig. 4.1; Girard et al., 2014; Morin, 1990). The effect may differ within stands that have yet to reach the understory reinitiation stage (sensu Oliver, 1980). In these latter stands, the seedling bank is not yet developed, and the insect can reduce the production of seeds, thereby compromising new seedling establishment (Côté & Bélanger, 1991).

Climate change may modify the phenology of both the tree host and the insect. This modification could lead to the expansion of the insect's range and increase damage severity (Pureswaran et al., 2015). Climate warming is expected to reduce the period when soils are frozen in most regions of eastern Canada. Although there

Fig. 4.1 Stand originating from the combined action of spruce budworm and windthrow in eastern Québec, Canada. *Photo credit* François Girard

is as of yet no clear evidence of an increased occurrence of strong wind events in the boreal forest, budworm-impacted stands may become more exposed to the strong winds that typically occur in late fall, without benefiting from the increased resistance to overturning provided by a frozen soil and snowpack (Saad et al., 2017). This interaction could become more important in the future because of the possible extension of the area vulnerable to outbreaks. This frozen soil–windthrow–insect interaction could therefore become a significant issue in the parts of eastern Canada, where a low occurrence of fires and an associated abundance of uneven-aged stands would see an increased use of partial cuttings, further heightening the vulnerability of these stands to windthrow (Anyomi & Ruel, 2015).

4.2.2 Windthrow and Bark Beetles

Mass outbreaks of bark beetles are natural events, particularly in the long-term dynamics of coniferous forests. Bentz et al. (2010) identified 14 species of bark beetles that have the potential to cause landscape-level mortality of trees making up western North American forests. In European forests, 8% of all forest damage is caused by bark beetles (Schelhaas et al., 2003). The most destructive species in Europe is the spruce bark beetle (*Ips typographus* (L.)) (Wermelinger, 2004). This beetle almost exclusively colonizes Norway spruce trees (*Picea abies* (L.) Karst.). In central Europe, generally two generations of spruce bark beetle develop per year, whereas in Fennoscandia and at elevations above approximately 1,500 m asl, only one generation per year develops.

During its latency phase at normal population levels, the spruce bark beetle develops at low densities under the bark of dead trunks or stumps. Because of poor phloem quality, interspecific competition with other bark dwellers, and mortality imposed by natural enemies, bark beetle populations remain low (Raffa et al., 2008). Under these conditions, healthy conifers are generally not colonized because the trees can physically and chemically defend themselves against these attacking insects by releasing resins containing toxic terpenoid compounds (Krokene, 2015).

However, windthrow in a spruce-dominated forest changes the situation for the spruce bark beetle. The fallen trees offer an ample supply of fresh, poorly defended bark, easily colonized by adult beetles (Eriksson et al., 2005). The still-soft and nutrient-rich phloem of the windthrown trees provides a high-quality substrate for the development of the bark beetle offspring. The beetles quickly propagate in the windthrown timber, and their population levels increase. However, at higher latitudes and under endemic conditions, small windthrow patches may produce too few bark beetles to allow the subsequent attack of adjacent living trees (Eriksson et al., 2007). Depending on local conditions, the phloem of windthrown trees becomes desiccated after one to three years and thereafter unsuitable for further colonization (Dodds et al., 2019; Wermelinger, 2004).

When the spruce bark beetle attains very high population levels, it attacks living trees. The beetles initially target nearby trees, particularly those within 250 m of the windthrown stems (Fig. 4.2; Havašová et al., 2017; Seidl & Blennow, 2012). These trees may have root damage caused by the storm, and the previously shaded stems become exposed to detrimental irradiation from the sun, i.e., sunburn. During this time, further infestation spots caused by single overthrown trees emerge in the stand interior. The bark beetles increase their population levels further and become sufficiently abundant to overcome the defenses of even healthy trees through the mass attack throughout the stand. Only at this point—extremely high populations of adult beetles—can bark beetles successfully colonize vigorous trees. The insects profit from an almost infinite supply of living trees, containing high-quality phloem and without competing phloem feeders except for their conspecifics. With the positive feedback of a high-reproductive output and successful colonization through mass attack, the populations reach a self-sustaining dynamic that may last for several years. A compilation of the most significant outbreaks of the last few decades in central Europe and Scandinavia revealed that the propagations last between 5 and 12 years (Wermelinger & Jakoby, 2019). Often, these outbreak dynamics are sustained by dry and hot summers, recurring smaller disturbances (e.g., heavy snowfall events), and heavy seed masts, all of which deplete the energy reserves of the trees (Nüsslein et al., 2000). Post-windthrow spruce bark beetle outbreaks lasting two to six years have been reported for different boreal forest regions in Russia (Maslov, 2010). In Sweden, peak infestations were attained in the third summer after the windthrow event (Kärvemo et al., 2014; Schroeder & Lindelöw, 2002), a timing that also holds for higher elevations in the Alps (Wermelinger, 2004).

Fig. 4.2 Damage from the bark beetle (*Ips typographus*) in a subalpine spruce forest in Switzerland. A windthrow event gave rise to a subsequent infestation of the adjacent spruce trees. *Photo credit* Beat Wermelinger

Outbreaks cease for various reasons. These include consistently high host-tree resistance stemming from sufficient precipitation and relatively cooler weather, increased bark beetle mortality because of intraspecific competition, natural enemies, human control measures (Stadelmann et al., 2013), a decreasing supply of host trees, or a combination of these factors (Marini et al., 2017). Bark beetle populations eventually fall below the critical threshold required for successfully attacking live trees, and the mass infestation thus ends.

The transition from colonizing low-defense substrates in latency to infesting high-defense trees in the eruptive phase is most commonly triggered by an abiotic disturbance. In Europe, windthrow is the main trigger of bark beetle outbreaks (cf. Table 1 in Wermelinger & Jakoby, 2019). More recently, pronounced dry spells have also led to large-scale infestations by temporarily compromising the defense capacities of living spruce trees. Moreover, higher temperatures have allowed the production of a third generation of bark beetles in central Europe (Jakoby et al., 2019) and a second in Scandinavia (Jönsson et al., 2009; Neuvonen & Viiri, 2017). The interactions between tree resistance and bark beetle population size are crucial for the dynamics of an infestation (Fig. 4.3). The number of simultaneously attacking beetles required to successfully colonize a tree is positively related to the health and vigor of the tree (Mulock & Christiansen, 1986; Nelson & Lewis, 2008). With climate change, the projected increased frequency of hot and dry summers (and possibly more storm events) will favor an increased spruce tree mortality and a distinct decline of this tree species in central Europe (Jakoby et al., 2019; Jönsson et al., 2007).

In North American forests, the mountain pine beetle is the most devastating bark beetle. This coleopteran has a similar biology to that of the European spruce bark

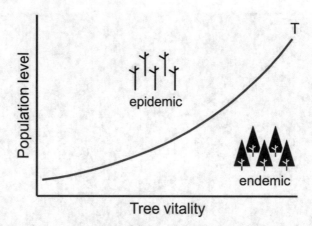

Fig. 4.3 Relationship between tree vitality and population levels of bark beetles required to cross the threshold (T) separating endemic and epidemic stages. Modified from Wermelinger and Jakoby (2019) with permission from Haupt Verlag

beetle; however, it is restricted to living pine tree hosts, and outbreaks are often provoked by drought or a series of mild winters. During an unprecedented mass propagation in the decades at the turn of the twenty-first century, the mountain pine beetle expanded its natural outbreak range toward the northeast and to higher elevations. During this process, the beetle also colonized—in addition to its preferred host of lodgepole pine (*Pinus contorta var. latifolia* Dougl. ex. Loud.)—whitebark pine (*P. albicaulis* Engelm.) and jack pine (*P. banksiana* Lamb.) (Raffa et al., 2013). Several years of higher temperatures, which reduced levels of winter mortality for this beetle, and the large-scale availability of old, even-aged, and drought-stressed pine forests (Logan & Powell, 2001; Taylor & Carroll, 2004) allowed the growth and spread of these extensive and long-lasting outbreaks (Stahl et al., 2006).

Contrary to the previous situations in which windthrow or climate anomalies triggered bark beetle outbreaks, surprisingly little attention has been given to the reverse interaction, namely the effect of bark beetle attack on subsequent windthrow events. As in other cases of insect attack in mixed-species stands, the death of pine trees exposes nonhost trees to higher wind loads, making them more vulnerable to wind. In addition, trees killed by the mountain pine beetle will gradually lose functional integrity in their stem and roots, resulting in the eventual deterioration and collapse of dead trees over time. Furthermore, the rate of bole collapse may be accelerated by windsway, and high winds are often directly responsible for the collapse of beetle-killed trees.

The annual monitoring of tree health and condition during and after an outbreak of mountain pine beetle in a forest stand dominated by lodgepole pine in central British Columbia provides an example of windthrow following a bark beetle outbreak. An eddy flux tower documented a remarkable recovery of the stand's function as a net carbon sink within three years of the outbreak despite 90% of the tree layer having died (Brown et al., 2012). The status of these dead trees over 15 years (Fig. 4.4) showed a clear uptick in the rate of bole collapse between 2013 and 2014. This collapse—31% of the total number of fallen trees—was related to a wind event in December 2013 that had sustained average wind speeds of 36 km/h. Unfortunately, there were no nearby lodgepole pine stands unaffected by the mountain pine beetle for comparison purposes. Nonetheless, the observed treefall rates were much higher than background rates in stands dominated by living trees. It is interesting to note that stronger winds had been recorded several times in 2009 but without a noticeable increase in windthrow. Wind drag is reduced soon after a tree dies as needles gradually fall off; however, the stem and root resistance remain unaffected during the initial three years after a beetle attack. This example illustrates the importance of case-specific and dynamic lag effects in detecting and understanding disturbance interactions (Burton et al., 2020). Furthermore, in accelerating post-beetle tree collapse, wind contributes to the accumulation of boles resting on the ground and having contact with the forest floor. This accumulation further accelerates fungal attack and decomposition and elevates the rate of CO_2 release due to tree decay (Kaytor, 2016). Although inadequately documented, greater concentrations of fallen beetle-killed trees could plausibly result in more intense wildfires (Jenkins et al., 2012), especially if fallen trees are *jack-strawed*, i.e., collapsed at multiple intersecting angles with many boles elevated above the ground and staying well dried, leading to a three-way interaction between insects, wind, and fire.

Bark beetle attacks in living stands markedly reduce canopy cover, especially in pure stands of host species; however, the understory and the forest floor are normally minimally affected. When dead trees are subsequently damaged by wind, the level

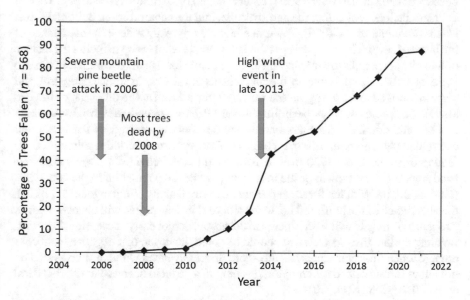

Fig. 4.4 Progression of treefall after a mountain pine beetle attack of a lodgepole pine stand in central British Columbia, Canada, accelerated by extreme winds in late 2013. Unpublished data from Dale Seip and Vanessa Foord, Government of British Columbia

of soil disturbance is typically less than the soil disturbance experienced with the windthrow of living trees, as the root systems of the dead trees will have degraded to some extent. The amount of advance regeneration then likely influences stand regeneration.

When a bark beetle outbreak develops from wind-killed trees, the on-site effects will mostly reflect the wind's impact: a reduced canopy cover, a minor effect on the understory, and the local perturbation of the forest floor. However, the consequences of the beetle outbreak will extend to neighboring stands, where it will mostly affect canopy cover. After the collapse of the infestation, the development of a new stand in managed forests depends largely on silvicultural measures such as planting and fostering preferred tree species. In unmanaged or extensively managed forests, natural regeneration and future stand development depend on multiple factors, including elevation, climate conditions, the spatial magnitude of the infestation, the density of advance regeneration, and the proximity of seed trees. In high-elevation natural spruce forests, even-aged spruce stands are likely to develop.

4.3   Fire Interaction with Other Disturbances

Large fires are frequent in the boreal forest of North America. In Canada, for instance, almost 2 million ha of forests burn every year on average, with some years experiencing more than 8 million ha burned (Hanes et al., 2019). Broad-scale fires have also been increasing in size since 1959 (Hanes et al., 2019). These disturbances are one of the main factors controlling the age structure, and thus composition, of boreal forest stands (Gauthier et al., 2009). Fire can eliminate or greatly reduce the abundance of the hosts of several insect species at the landscape level, thereby reducing the chance of these insect populations exploding to an epidemic level in the region. This scenario is the case for balsam fir, one of the major hosts of the eastern spruce budworm, and a species not adapted to regenerate quickly after a fire. On the other hand, fire can also favor species such as trembling aspen (*Populus tremuloides* Michx.), which can become dominant over large areas and be possibly affected by the forest tent caterpillar (*Malacosoma disstria* Hübner). The large areas of lodgepole pine originating from the 1880–1920 fires in British Columbia provided a large continuous food source for the growth of the mountain pine beetle population in the late 1990s (Burton, 2010). With the forecasted increase in fire activity (Boulanger et al., 2014), the abundance of the preferred hosts for different insect species will change. Shorter fire-return intervals will likely increase the proportion of early successional species, favoring certain insect species. It should be noted, however, that negative feedbacks emerge when fire-return intervals are short. Successive fires within a short period do not allow forest fuels to accumulate, resulting in a de facto decrease in fire risk (Erni et al., 2018; Heon et al., 2014).

Short intervals between successive fires can be responsible for a change in ecosystem state in which closed forests are replaced by open woodlands owing to a lack of seeds for forest regeneration (Jasinski & Payette, 2005; Splawinski et al.,

2019). Short intervals between insect disturbances and fire can also produce such a state change (Simard & Payette, 2001).

Surface fires are frequent in the European boreal forest (Shorohova et al., 2011), and surviving trees are subjected to subsequent windthrow and bark beetle attacks (Fig. 4.5; Ananyev et al., 2016). A similar situation has been described for surface fires and insect outbreaks in Siberian *light coniferous* forests dominated by Siberian larch (*Larix sibirica* Ledeb.); within two years of a fire, 18–30% of surviving trees were attacked by insects (Isaev, 1962). In *dark coniferous* forests, dominated by Siberian pine and fir (*Pinus sibirica* Du Tour and *Abies sibirica* Ledeb., respectively), forest fires typically induce fungal diseases (Pavlov, 2015) and insect outbreaks (Kharuk et al., 2016, 2017).

Fire ignition, spread, behavior, and burned area can also be affected by the fuel inputs from tree and shrub mortality stemming from other disturbances, such as drought, insect outbreaks, and windthrow. The pulse created by this influx of dead and dry fuel can vary in duration depending on the tree species, the rate of the mortality process, and the regional weather/climate conditions. Recent drought has caused significant episodes of mortality, increasing the fuel load available to burn when the weather becomes conducive for fire, thereby increasing fire intensity (Ruthrof et al., 2016). Windthrow can increase the probability, intensity, and/or severity of subsequent fires by increasing the fuel load (Fig. 4.6). In Siberia, outbreaks of the Siberian silkmoth (*Dendrolimus sibiricus* Tschetv.) increase the risk of fires. Forest stands affected by an outbreak burn seven times more frequently than unaffected

Fig. 4.5 Bark beetle–induced decline and wind breakage six years after a surface fire in a mixed primeval boreal forest, Vodlozersky National Park, Russia. *Photo credit* Ilkka Vanha-Majamaa

stands, and the burned area in a silkmoth-affected forest is 20 times larger than in unaffected stands (Kharuk & Antamoshkina, 2017). On the other hand, fire kills the natural enemies of silkmoth and thus can trigger an outbreak (Baksheeva et al., 2019).

Interactions between fire and other disturbances can also affect the postfire recovery potential by removing propagule sources (Cannon et al., 2019). However, these interactions are complex and influenced by disturbance intensity and severity, which are often highly heterogeneous, and the interval between disturbances. The degree to which plant community recovery reflects the compound effects of a bark beetle outbreak and fire disturbance depends strongly on fire severity (Edwards et al., 2015). In some situations, such as low-intensity fire conditions, buffering effects can also be observed, i.e., by reducing fuels that otherwise might support more severe

Fig. 4.6 Fuel load in a black spruce stand after severe windthrow in eastern Québec, Canada. *Photo credit* Kaysandra Waldron

fires, thereby conferring a degree of forest resistance to subsequent disturbances for some time (Cannon et al., 2017, 2019).

The action of fire strongly influences the legacies of interactions between fire and other disturbances. Windthrow involves a reduction of forest cover, an increase in soil disturbance, and some modification of the understory layer (Waldron et al., 2013). Insect defoliation has a similar effect on the forest cover but with less effect on the understory layer and forest floor. However, when one of these disturbances precedes a fire event, its effects will become relatively minor relative to that of fire, which has a dominant impact on all constituents. When surface fires lead to windthrow or bark beetle attacks, the second disturbance will add to the canopy reduction, although the understory and the forest floor will remain dominated by the impact of fire.

4.4 Interactions Between Natural Disturbances and Forest Management Practices

In addition to interactions between natural disturbances, forest management practices can also interact with disturbances. For instance, partial cutting is a central element of many forest ecosystem management strategies (Bergeron & Harvey, 1997). By opening the forest canopy, standing trees become exposed to higher wind speeds, and damage due to increased wind has often been observed (Coates et al., 2018; Hanell & Ottoson-Lofvenius, 1994; Hautala & Vanha-Majamaa, 2006; Montoro Girona et al., 2019; Ruel & Gardiner, 2019). Salvage logging often follows high-severity disturbances and greatly modifies the legacies from natural disturbances, including the removal of residual living trees, the reduction of snags and downed woody debris, and the added disturbance to the understory and forest floor. These added effects can damage advance regeneration, increase fire risk (Donato et al., 2006), and decrease biodiversity (Thorn et al., 2018). Salvage logging in black spruce–dominated stands defoliated by the spruce budworm could increase the defoliation of black spruce regeneration. In turn, this loss of regeneration would influence stand development trajectories and eventually increase future stand vulnerability to the insect (Cotton-Gagnon et al., 2018). Partial cutting in defoliated stands would also lead to increases in regeneration defoliation although to a lesser extent than in clear-cut stands (Lavoie et al., 2021).

4.5 Conclusion

This chapter has focused on disturbance interactions where the occurrence of one disturbance influences the likelihood of another. We have shown that the interactions may significantly impact ecosystem processes and attributes. There are knowledge gaps in our understanding of these effects, and further research is required. For example, other instances of compound disturbances can occur when two consecutive disturbances occur in sequence, without a causal relationship between them. These can also lead to significant effects on ecosystems, but their occurrence and impacts are less predictable without a direct relationship. The lack of suitable controls typically constrains evaluations of the interactive or compound effects of natural disturbances; that is, the effects of disturbance A without disturbance B and those of disturbance B without disturbance A are difficult to study under similar conditions as experienced for the interacting disturbances. As climate and weather influence many of these interactions, further research should target the possible effects of climate change on these interactions. Research should also expand the temporal scale being analyzed because studies of disturbance interactions are often conducted opportunistically and over the short term, covering only the initial years post-disturbance. The age structure of forest landscapes further influences the vulnerability of certain stands to disturbances and their possible interactions, an effect that must be better documented and understood.

References

Ananyev, V. A., Moshnikov, S. A., Timofeeva, V. V., et al. (2016). Monitoring of primeval forests after fires in the "Vodlozersky" National Park [in Russian]. (pp. 74–77). *The role of science in solving regional developmental problems Proceedings of the scientific conference.*

Anyomi, K., & Ruel, J. C. (2015). A multi-scale analysis of the effects of alternative silvicultural treatments on windthrow within balsam fir dominated stands. *Canadian Journal of Forest Research, 45,* 1739–1747. https://doi.org/10.1139/cjfr-2015-0221.

Arsenault, A., LeBlanc, R., Earle, E., et al. (2016). Unravelling the past to manage Newfoundland's forests for the future. *The Forestry Chronicle, 92,* 487–502. https://doi.org/10.5558/tfc2016-085.

Baksheeva, E. O., Morozov, A. S., Mikhailov, P. V., et al. (2019). The analysis of fire activities in Eniseysk forest district during the Siberian moth outbreak [in Russian]. (pp. 24–29). *Forest and chemical complexes: problems and solutions Proceedings of the scientific conference.*

Bentz, B. J., Regniere, J., Fettig, C. J., et al. (2010). Climate change and bark beetles of the western United States and Canada: Direct and indirect effects. *BioScience, 60,* 602–613. https://doi.org/10.1525/bio.2010.60.8.6.

Bergeron, Y., & Harvey, B. (1997). Basing silviculture on natural ecosystem dynamics: An approach applied to the southern boreal mixedwood forest of Quebec. *Forest Ecology and Management, 92,* 235–242. https://doi.org/10.1016/S0378-1127(96)03924-2.

Bouchard, M., Pothier, D., & Ruel, J. C. (2009). Stand-replacing windthrow in the boreal forests of eastern Quebec. *Canadian Journal of Forest Research, 39,* 481–487. https://doi.org/10.1139/X08-174.

Boulanger, Y., Gauthier, S., & Burton, P. J. (2014). A refinement of models projecting future Canadian fire regimes using homogeneous fire regime zones. *Canadian Journal of Forest Research, 44,* 365–376. https://doi.org/10.1139/cjfr-2013-0372.

Brown, M. G., Black, T. A., Nesic, Z., et al. (2012). The carbon balance of two lodgepole pine stands recovering from mountain pine beetle attack in British Columbia. *Agricultural and Forest Meteorology, 153,* 82–93. https://doi.org/10.1016/j.agrformet.2011.07.010.

Buma, B. (2015). Disturbance interactions: Characterization, prediction, and the potential for cascading effects. *Ecosphere, 6,* 70. https://doi.org/10.1890/ES15-00058.1.

Burton, P. J. (2010). Striving for sustainability and resilience in the face of unprecedented change: The case of the mountain pine beetle outbreak in British Columbia. *Sustainability, 2*, 2403–2423. https://doi.org/10.3390/su2082403.

Burton, P. J., Jentsch, A., & Walker, L. R. (2020). The ecology of disturbance interactions. *BioScience, 70*(10), 854–870. https://doi.org/10.1093/biosci/biaa088.

Cannon, J. B., Peterson, C. J., O'Brien, J. J., et al. (2017). A review and classification of interactions between forest disturbance from wind and fire. *Forest Ecology and Management, 406*, 381–390. https://doi.org/10.1016/j.foreco.2017.07.035.

Cannon, J. B., Henderson, S. K., Bailey, M. H., et al. (2019). Interactions between wind and fire disturbance in forests: Competing amplifying and buffering effects. *Forest Ecology and Management, 436*, 117–128. https://doi.org/10.1016/j.foreco.2019.01.015.

Coates, K. D., Hall, E. C., & Canham, C. D. (2018). Susceptibility of trees to windthrow storm damage in partially harvested complex-structured multi-species forests. *Forests, 9*, 199. https://doi.org/10.3390/f9040199.

Côté, S., & Bélanger, L. (1991). Variation de la régénération preetable dans les sapinières boréales en fonction de leurs caractéristiques écologiques. *Canadian Journal of Forest Research, 21*, 1779–1795. https://doi.org/10.1139/x91-246.

Cotton-Gagnon, A., Simard, M., De Grandpre, L., et al. (2018). Salvage logging during spruce budworm outbreaks increases defoliation of black spruce regeneration. *Forest Ecology and Management, 430*, 421–430. https://doi.org/10.1016/j.foreco.2018.08.011.

De Grandpré, L., Waldron, K., Bouchard, M., et al. (2018). Incorporating insect and wind disturbances in a natural disturbance-based management framework for the boreal forest. *Forests, 9*(8), 471. https://doi.org/10.3390/f9080471.

Dodds, K. J., DiGirolomo, M. F., & Fraver, S. (2019). Response of bark beetles and woodborers to tornado damage and subsequent salvage logging in northern coniferous forests of Maine, USA. *Forest Ecology and Management, 450*, 117489. https://doi.org/10.1016/j.foreco.2019.117489.

Donato, D. C., Fontaine, J. B., Campbell, J. L., et al. (2006). Post-wildfire logging hinders regeneration and increases fire risk. *Science, 311*, 352. https://doi.org/10.1126/science.1122855.

Edwards, M., Krawchuk, M. A., & Burton, P. J. (2015). Short-interval disturbance in lodgepole pine forests, British Columbia, Canada: Understory and overstory response to mountain pine beetle and fire. *Forest Ecology and Management, 338*, 163–175. https://doi.org/10.1016/j.foreco.2014.11.011.

Eriksson, M., Pouttu, A., & Roininen, H. (2005). The influence of windthrow area and timber characteristics on colonization of wind-felled spruces by *Ips typographus* (L.). *Forest Ecology and Management, 216*, 105–116. https://doi.org/10.1016/j.foreco.2005.05.044.

Eriksson, M., Neuvonen, S., & Roininen, H. (2007). Retention of wind-felled trees and the risk of consequential tree mortality by the European spruce bark beetle *Ips typographus* in Finland. *Scandinavian Journal of Forest Research, 22*, 516–523. https://doi.org/10.1080/02827580701800466.

Erni, S., Arseneault, D., & Parisien, M. A. (2018). Stand age influence on potential wildfire ignition and spread in the boreal forest of northeastern Canada. *Ecosystems, 21*, 1471–1486. https://doi.org/10.1007/s10021-018-0235-3.

Gardiner, B. A., Stacey, G. R., Belcher, R. E., et al. (1997). Field and wind tunnel assessments of the implications of respacing and thinning for tree stability. *Forestry, 70*, 233–252. https://doi.org/10.1093/forestry/70.3.233.

Gauthier, S., Leduc, A., Bergeron, Y., et al. (2009). Fire frequency and forest management based on natural disturbances. In S. Gauthier, M. A. Vaillancourt, A. Leduc, L. De Grandpre, D. D. Kneeshaw, H. Morin, P. Drapeau, & Y. Bergeron (Eds.), *Ecosystem management in the boreal forest* (pp. 39–56). Québec: Presses de l'Université du Québec.

Girard, F., De Grandpre, L., & Ruel, J. C. (2014). Partial windthrow as a driving process of forest dynamics in old-growth boreal forests. *Canadian Journal of Forest Research, 44*, 1165–1176. https://doi.org/10.1139/cjfr-2013-0224.

Hanell, B., & Ottoson-Lofvenius, M. (1994). Windthrow after shelterwood cutting in *Picea abies* peatland forests. *Scandinavian Journal of Forest Research, 9*, 261–269. https://doi.org/10.1080/02827589409382839.

Hanes, C. C., Wang, X., Jain, P., et al. (2019). Fire-regime changes in Canada over the last half century. *Canadian Journal of Forest Research, 49*(3), 256–269. https://doi.org/10.1139/cjfr-2018-0293.

Hautala, H., & Vanha-Majamaa, I. (2006). Immediate tree uprooting after retention-felling in a coniferous boreal forest in Fennoscandia. *Canadian Journal of Forest Research, 36*, 3167–3172. https://doi.org/10.1139/x06-193.

Havašová, M., Ferenčík, J., & Jakuš, R. (2017). Interactions between windthrow, bark beetles and forest management in the Tatra national parks. *Forest Ecology and Management, 391*, 349–361. https://doi.org/10.1016/j.foreco.2017.01.009.

Heon, J., Arseneault, D., & Parisien, M. A. (2014). Resistance of the boreal forest to high burn rates. *Proceedings of the National Academy of Sciences of the United States of America 111*, 13888–13893. https://doi.org/10.1073/pnas.1409316111.

Isaev, A. S. (1962). The role of stem borers in the decline of larch forests after fires [in Russian]. *Special issue of the Working Paper Series of Siberian Technological Institute, 29*, 70–78.

Jactel, H., Petit, J., Desprez-Loustau, M. L., et al. (2012). Drought effects on damage by forest insects and pathogens: A meta-analysis. *Global Change Biology, 18*, 267–276. https://doi.org/10.1111/j.1365-2486.2011.02512.x.

Jakoby, O., Lischke, H., & Wermelinger, B. (2019). Climate change alters elevational phenology patterns of the European spruce bark beetle (*Ips typographus*). *Global Change Biology, 25*, 4048–4063. https://doi.org/10.1111/gcb.14766.

Jasinski, J. P., & Payette, S. (2005). The creation of alternative stable states in the southern boreal forest, Quebec, Canada. *Ecological Monographs, 75*, 561–583. https://doi.org/10.1890/04-1621.

Jenkins, M. J., Page, W. G., Hebertson, E. G., et al. (2012). Fuels and fire behavior dynamics in bark beetle-attacked forests in western North America and implications for fire management. *Forest Ecology and Management, 275*, 23–34. https://doi.org/10.1016/j.foreco.2012.02.036.

Jönsson, A. M., Harding, S., Bärring, L., et al. (2007). Impact of climate change on the population dynamics of *Ips typographus* in southern Sweden. *Agricultural and Forest Meteorology, 146*, 70–81. https://doi.org/10.1016/j.agrformet.2007.05.006.

Jönsson, A. M., Appelberg, G., Harding, S., et al. (2009). Spatio-temporal impact of climate change on the activity and voltinism of the spruce bark beetle, *Ips typographus*. *Global Change Biology, 15*, 486–499. https://doi.org/10.1111/j.1365-2486.2008.01742.x.

Kärvemo, S., Rogell, B., & Schroeder, M. (2014). Dynamics of spruce bark beetle infestation spots: Importance of local population size and landscape characteristics after a storm disturbance. *Forest Ecology and Management, 334*, 232–240. https://doi.org/10.1016/j.foreco.2014.09.011.

Kaytor, B. (2016). Decomposition and carbon loss in lodgepole pine (*Pinus contorta* var. *latifolia*) wood following attack by mountain pine beetle (*Dendroctonus ponderosae*). M.Sc. thesis, University of Northern British Columbia.

Kharuk, V. I., & Antamoshkina, O. A. (2017). Impact of silkmoth outbreak on taiga wildfires. *Contemporary Problems of Ecology, 10*(5), 556–562. https://doi.org/10.1134/S199542551705 0055.

Kharuk, V. I., Demidko, D. A., Fedotova, E. V., et al. (2016). Spatial and temporal dynamics of Siberian silk moth large-scale outbreak in dark-needle coniferous tree stands in Altai. *Contemporary Problems of Ecology, 9*(6), 711–720. https://doi.org/10.1134/S199542551606 6007X.

Kharuk, V. I., Im, S. T., Petrov, I. A., et al. (2017). Climate-induced mortality of Siberian pine and fir in the Lake Baikal watershed, Siberia. *Forest Ecology and Management, 384*, 191–199. https://doi.org/10.1016/j.foreco.2016.10.050.

Krokene, P. (2015). Conifer defense and resistance to bark beetles. In F. E. Vega & R. W. Hofstetter (Eds.), *Bark beetles: Biology and ecology of native and invasive species* (pp. 177–207). San Diego: Academic Press.

Kulha, N., Pasanen, L., Holmström, L., et al. (2020). The structure of boreal old-growth forests changes at multiple spatial scales over decades. *Landscape Ecology, 35*, 843–858. https://doi.org/10.1007/s10980-020-00979-w.

Lavoie, J., Montoro Girona, M., Grosbois, G., et al. (2021). Does the type of silvicultural practice influence spruce budworm defoliation of seedlings? *Ecosphere, 12*(4), 17. https://doi.org/10.1002/ecs2.3506.

Logan, J. A., & Powell, J. A. (2001). Ghost forests, global warming, and the mountain pine beetle (Coleoptera: Scolytidae). *American Entomologist, 47*, 160–172. https://doi.org/10.1093/ae/47.3.160.

Marini, L., Okland, B., Jonsson, A. M., et al. (2017). Climate drivers of bark beetle outbreak dynamics in Norway spruce forests. *Ecography, 40*, 1426–1435. https://doi.org/10.1111/ecog.02769.

Maslov, A. D. (2010). *European spruce bark beetle and the decline of spruce forests* (p. 138). Moscow: All-Russian Research Institute of Silviculture and Mechanization of Forestry.

Montoro Girona, M., Morin, H., Lussier, J.-M., et al. (2019). Post-cutting mortality following experimental silvicultural treatments in unmanaged boreal forest stands. *Frontiers in Forests and Global Change, 2*, 4. https://doi.org/10.3389/ffgc.2019.00004.

Morin, H. (1990). Analyse dendro-écologique d'une sapinière issue d'un chablis dans la zone boréale (Québec). *Canadian Journal of Forest Research, 20*, 1753–1758. https://doi.org/10.1139/x90-233.

Mulock, P., & Christiansen, E. (1986). The threshold of successful attack by *Ips typographus* on *Picea abies*: A field experiment. *Forest Ecology and Management, 14*, 125–132. https://doi.org/10.1016/0378-1127(86)90097-6.

Navarro, L., Morin, H., Bergeron, Y., et al. (2018). Changes in spatiotemporal patterns of 20th century spruce budworm outbreaks in eastern Canadian boreal forests. *Frontiers in Plant Science, 9*, 1905. https://doi.org/10.3389/fpls.2018.01905.

Nelson, W. A., & Lewis, M. A. (2008). Connecting host physiology to host resistance in the conifer-bark beetle system. *Theoretical and Applied Genetics, 1*. 160–177 https://doi.org/10.1007/s12080-008-0017-1.

Neuvonen, S., & Viiri, H. (2017). Changing climate and outbreaks of forest pest insects in a cold northern country, Finland. In K. Latola & H. Savela (Eds.), *The Interconnected Arctic-UArctic Congress 2016* (pp. 49–59). Cham: Springer International Publishing.

Nie, Z., MacLean, D. A., & Taylor, A. R. (2018). Forest overstory composition and seedling height influence defoliation of understory regeneration by spruce budworm. *Forest Ecology and Management, 409*, 353–360. https://doi.org/10.1016/j.foreco.2017.11.033.

Nüsslein, S., Faisst, G., Weissbacher, A., et al. (2000). Zur Waldentwicklung im Nationalpark Bayerischer Wald 1999. Buchdrucker-Massenvermehrung und Totholzflächen im Rachel-Lusen-Gebiet. (On forest development in the Bavarian Forest National Park 1999. Mass proliferation of bookworms and deadwood areas in the Rachel-Lusen area) [in German]. *Berichte des Bayerischen Landesanstalt für Wald und Forstwirtschaft*, LWF 25

Oliver, C. D. (1980). Forest development in North America following major disturbances. *Forest Ecology and Management, 3*, 153–168. https://doi.org/10.1016/0378-1127(80)90013-4.

Paine, R. T., Tegner, M. J., & Johnson, E. A. (1998). Compounded perturbations yield ecological surprises. *Ecosystems, 1*, 535–545. https://doi.org/10.1007/s100219900049.

Pavlov, I. N. (2015). Biotic and abiotic factors as causes of coniferous forests dieback in Siberia and Far East. *Contemporary Problems of Ecology, 8*(4), 440–456. https://doi.org/10.1134/S1995425515040125.

Pureswaran, D. S., De Grandpré, L., Paré, D., et al. (2015). Climate-induced changes in host tree–insect phenology may drive ecological state-shift in boreal forests. *Ecology, 96*, 1480–1491. https://doi.org/10.1890/13-2366.1.

Raffa, K. F., Aukema, B. H., Bentz, B. J., et al. (2008). Cross-scale drivers of natural disturbances prone to anthropogenic amplification: The dynamics of bark beetle eruptions. *BioScience, 58*, 501–517. https://doi.org/10.1641/B580607.

Raffa, K. F., Powell, E. N., & Townsend, P. A. (2013). Temperature-driven range expansion of an irruptive insect heightened by weakly coevolved plant defenses. *Proceedings of the National Academy of Sciences of the United States of America, 110,* 2193–2198. https://doi.org/10.1073/pnas.1216666110.

Roberts, M. R. (2007). A conceptual model to characterize disturbance severity in forest harvests. *Forest Ecology and Management, 242,* 58–64. https://doi.org/10.1016/j.foreco.2007.01.043.

Ruel, J.-C. (1995). Understanding windthrow: Silvicultural implications. *The Forestry Chronicle, 71,* 434–445. https://doi.org/10.5558/tfc71434-4.

Ruel, J.-C. (2000). Factors influencing windthrow in balsam fir forests: From landscape studies to individual tree studies. *Forest Ecology and Management, 135,* 169–178. https://doi.org/10.1016/S0378-1127(00)00308-X.

Ruel, J.-C., & Gardiner, B. (2019). Mortality patterns after different levels of harvesting of old-growth boreal forests. *Forest Ecology and Management, 448,* 346–354. https://doi.org/10.1016/j.foreco.2019.06.029.

Ruthrof, K. X., Fontaine, J. B., Matusick, G., et al. (2016). How drought-induced forest die-off alters microclimate and increases fuel loadings and fire potentials. *International Journal of Wildland Fire, 25,* 819–830. https://doi.org/10.1071/WF15028.

Saad, C., Boulanger, Y., Beaudet, M., et al. (2017). Potential impact of climate change on the risk of windthrow in eastern Canada's forests. *Climatic Change, 143,* 487–501. https://doi.org/10.1007/s10584-017-1995-z.

Schelhaas, M. J., Nabuurs, G. J., & Schuck, A. (2003). Natural disturbances in the European forests in the 19th and 20th centuries. *Global Change Biology, 9,* 1620–1633. https://doi.org/10.1046/j.1365-2486.2003.00684.x.

Schroeder, L. M., & Lindelöw, A. (2002). Attacks on living spruce trees by the bark beetle *Ips typographus* (Col. Scolytidae) following a storm-felling: A comparison between stands with and without removal of wind-felled trees. *Agricultural and Forest Entomology, 4,* 47–56. https://doi.org/10.1046/j.1461-9563.2002.00122.x.

Seidl, R., & Blennow, K. (2012). Pervasive growth reduction in Norway spruce forests following wind disturbance. *PLoS ONE, 7,* e33301. https://doi.org/10.1371/journal.pone.0033301.

Shorohova, E., Kneeshaw, D., Kuuluvainen, T., et al. (2011). Variability and dynamics of old-growth forests in the circumboreal zone: Implications for conservation, restoration and management. *Silva Fennica, 45,* 785–806. https://doi.org/10.14214/sf.72.

Simard, M., & Payette, S. (2001). Black spruce decline triggered by spruce budworm at the southern limit of lichen woodland in eastern Canada. *Canadian Journal of Forest Research, 31*(12), 2160–2172. https://doi.org/10.1139/x01-160.

Splawinski, T. B., Cyr, D., Gauthier, S., et al. (2019). Analyzing risk of regeneration failure in the managed boreal forest of northwestern Quebec. *Canadian Journal of Forest Research, 49,* 680–691. https://doi.org/10.1139/cjfr-2018-0278.

Stadelmann, G., Bugmann, H., Meier, F., et al. (2013). Effects of salvage logging and sanitation felling on bark beetle (*Ips typographus* L.) infestations. *Forest Ecology and Management, 305,* 273–281. https://doi.org/10.1016/j.foreco.2013.06.003.

Stahl, K., Moore, R. D., & McKendry, I. G. (2006). Climatology of winter cold spells in relation to mountain pine beetle mortality in British Columbia, Canada. *Climate Research, 32,* 13–23. https://doi.org/10.3354/cr032013.

Sturtevant, B. R., & Fortin, M.-J. (2021). Understanding and modeling forest disturbance interactions at the landscape level. *Frontiers in Ecology and Evolution, 9.* https://doi.org/10.3389/fevo.2021.653647.

Taylor, S. L., & MacLean, D. A. (2009). Legacy of insect defoliators: Increased wind-related mortality two decades after a spruce budworm outbreak. *Forestry Sciences, 55,* 256–267.

Taylor, S. W., & Carroll, A. L. (2004). Disturbance, forest age, and mountain pine beetle outbreak dynamics in BC: A historical perspective. In T. L. Shore, J. E. Brooks, & J. E. Stone (Eds.), *Mountain pine beetle symposium: Challenges and solutions. Natural Resources Canada* (pp. 41–51). Victoria: Canadian Forest Service, Pacific Forest Centre.

Thorn, S., Bässler, C., Brandl, R., et al. (2018). Impacts of salvage logging on biodiversity: A meta-analysis. *Journal of Applied Ecology, 55*(1), 279–289. https://doi.org/10.1111/1365-2664.12945.

Ulanova, N. G. (2000). The effects of windthrow on forests at different spatial scales: A review. *Forest Ecology and Management, 135*, 155–167. https://doi.org/10.1016/S0378-1127(00)00307-8.

Waldron, K., Ruel, J.-C., & Gauthier, S. (2013). The effects of site characteristics on the landscape-level windthrow regime in the North Shore region of Quebec, Canada. *Forestry, 86*, 159–171. https://doi.org/10.1093/forestry/cps061.

Wermelinger, B. (2004). Ecology and management of the spruce bark beetle *Ips typographus*-a review of recent research. *Forest Ecology and Management, 202*, 67–82. https://doi.org/10.1016/j.foreco.2004.07.018.

Wermelinger, B., & Jakoby, O. (2019). Borkenkafer. In T. Wohlgemuth, A. Jentsch, & R. Seidl (Eds.), *Storungsokologie* (pp. 236–255). Bern: Haupt.

Chapter 5
Living Trees and Biodiversity

Aino Hämäläinen, Kadri Runnel, Grzegorz Mikusiński, Dmitry Himelbrant, Nicole J. Fenton, and Piret Lõhmus

Abstract Living trees are fundamental for boreal forest biodiversity. They contribute to stand structural diversity, which determines the range of habitat niches available for forest-dwelling species. Specific characteristics of living trees, such as species, age, and presence of microhabitats, determine how species utilize trees for food, as nesting places, or as growing substrates. This chapter explores the associations between living trees and aboveground biodiversity, reviews the factors such as soil productivity, hydrological regime, stand successional stage, and forestry activities that influence the characteristics of living trees and stand structural diversity, and presents the consequences of current and future climate change on boreal biodiversity.

A. Hämäläinen (✉) · K. Runnel
Department of Ecology, Swedish University of Agricultural Sciences, Institute for Ecology, Box 7044, 750 07 Uppsala, Sweden
e-mail: aino.hamalainen@slu.se

K. Runnel
e-mail: kadri.runnel@ut.ee

K. Runnel · P. Lõhmus
Institute of Ecology and Earth Sciences, University of Tartu, J. Liivi tn 2, 50409 Tartu linn, Tartumaa, Estonia
e-mail: piret.lohmus@ut.ee

G. Mikusiński
School for Forest Management, Swedish University of Agricultural Sciences, 739 21 Skinnskatteberg, Sweden
e-mail: grzegorz.mikusinski@slu.se

D. Himelbrant
Faculty of Biology, Department of Botany, Saint Petersburg State University, Universitetskaya emb. 7-9, 199034 Saint Petersburg, Russia
e-mail: d.himelbrant@spbu.ru; himelbrantde@binran.ru

Laboratory of Lichenology and Bryology, Komarov Botanical Institute of Russian Academy of Sciences, Professor Popov str. 2, 197376 Saint Petersburg, Russia

© The Author(s) 2023
M. M. Girona et al. (eds.), *Boreal Forests in the Face of Climate Change*,
Advances in Global Change Research 74,
https://doi.org/10.1007/978-3-031-15988-6_5

5.1 Introduction

Living trees play a crucial role in supporting biodiversity and providing ecosystem functions in boreal forests. Numerous forest-dwelling species are directly dependent on living trees; for example, trees provide a substrate for epiphytic and saprotrophic species, which in boreal forests consist mainly of bryophytes, lichens, and fungi. Trees also provide shelter and nesting places for invertebrates, mammals, and birds. Trees are equally a fundamental part of forest food webs as their foliage, flowers, seeds, and bark are food sources for various species. Finally, living trees are crucial as they form the structure of forest stands, which in turn influences the number of different habitat niches available and, thus, biodiversity (Chase & Leibold, 2003).

This chapter describes the linkages between living trees and their role in developing stand structure in boreal forests and generating aboveground biodiversity. Most existing studies on these topics come from Europe (especially Fennoscandia) and North America, whereas less research has been conducted in Asian boreal forests. Although we aim to cover the entire boreal region, our main focus in this chapter will be on the regions from which a greater amount of knowledge is available. The discussion of stand structural diversity is limited to the features directly connected with living trees, such as tree species diversity, canopy structure, and tree-related microhabitats. We then explain the main natural factors that influence the structural diversity both at the tree and stand scales, such as climate, primary productivity, and stand succession. In addition, we describe anthropogenic disturbances and related changes, most importantly commercial rotation forestry, that shape the structural diversity of living trees and the associated biodiversity in the boreal region. Finally, we discuss the potential effects of future climate change.

5.2 Structural Diversity of Living Trees

The structural diversity of living trees (hereafter *structural diversity*) in a forest, as well as how it changes during forest succession, is an important driver of biodiversity. Structural diversity can be described as tree species richness, variability in tree age and size distribution, the occurrence of several canopy layers *vertical diversity*, and the presence of canopy gaps or denser patches of trees *horizontal diversity*, as well as smaller-scale variations, e.g., foliage density (Franklin & Van Pelt, 2004). Traditional successional models suggested that structural diversity was higher in old boreal forests, whereas young stands that originated naturally from a stand-replacing

N. J. Fenton
Forest Research Institute, Université du Québec en Abitibi-Témiscamingue, 445 boul. de l'Université, Rouyn-Noranda, Québec, QC J9X 5E4, Canada
e-mail: nicole.fenton@uqat.ca

Centre for Forest Research, Université du Québec à Montréal, P.O. Box 8888, Stn. Centre-Ville, Montréal, QC H3C 3P8, Canada

disturbance were more uniform in structure (e.g., Brassard & Chen, 2006). It is now clear, however, that in a variety of forest types, high diversity can be present at any age because of variable seedbed quality and the presence of legacy trees post-disturbance (Kuuluvainen, 2002; Lecomte et al., 2006; Martin et al., 2019). Furthermore, even when a classical succession sequence exists, the age at which structural diversification begins depends on growing conditions and differs among stand types. For example, in Canadian boreal forests, white spruce (*Picea glauca*) stands develop a high structural diversity after 160 years of age, whereas structural enrichment begins after only 80 years in balsam poplar (*Populus balsamifera*) stands (Timoney & Robinson, 1996).

In boreal forests, tree species richness is low compared with temperate or tropical regions (Esseen et al., 1997). Mid- and late-successional boreal forest stands are typically dominated by a few coniferous species; in western Europe, these dominant taxa are Norway spruce (*Picea abies*) and Scots pine (*Pinus sylvestris*), and in eastern Europe and Asia, they are Siberian and Yezo spruce (*Pinus obovata* and *P. jezoensis*, respectively), Siberian fir (*Abies sibirica*), Scots and Siberian pine (*Pinus sibirica*), and Siberian and Dahurian larch (*Larix sibirica* and *L. gmelinii*, respectively) (Shorohova et al., 2011). In North America, the dominant conifers are white and black spruce (*Picea glauca* and *P. mariana*, respectively) and tamarack/larch (*Larix laricina*). In central and eastern North America, balsam fir (*Abies balsamea*) and jack pine (*Pinus banksiana*) are also present, and in western North America, alpine fir (*Abies lasiocarpa*) and lodgepole pine (*Pinus contorta*) are significant components of the forest (Larsen, 1980). The most important broadleaf tree species throughout the boreal region are birches (*Betula* spp.) and aspens (*Populus* spp.). Broadleaf trees occur mainly in young stands or as an admixture in the older, conifer-dominated stands (e.g., Bergeron, 2000), although there are some exceptions, such as Erman's birch (*Betula ermanii*), which also forms stands in older successional stages in the Russian Far East.

Tree species richness is generally correlated with the other aspects of structural diversity (Juchheim et al., 2020). There is a functional feedback loop, where a horizontally diverse stand structure with canopy gaps and openings can promote the establishment of light-demanding tree species in mature stands and thus promote tree species richness (Brassard & Chen, 2006; Kuuluvainen, 1994). In turn, species-rich stands are usually structurally diverse, as tree species vary in size and physical construction. Moreover, other habitat qualities that influence forest-dwelling species, such as soil conditions and water availability, are also affected by tree species and may therefore be more heterogeneous in mixed stands than in monospecific ones (Barbier et al., 2008).

Furthermore, tree species richness is important for the diversity of forest-dwelling species, as many are associated with a particular tree species or tree species group. Reasons for these associations are various; for example, tree species–specific bark characteristics are important for epiphytes (Ellis, 2012) and bark-dwelling invertebrates (Nicolai, 1986), and the chemistry and nutritional qualities of wood, foliage, and seeds affect the species that use these resources as a food source. The strength of the associations varies from a preference to strict specialization, but in general, species that consume foliage or other soft tissues of trees tend to be stricter in their specialization than species utilizing bark or wood (Sundberg et al., 2019). For most boreal forest trees, the total number of associated species is unknown. An exception is a recent report by Sundberg et al. (2019), which listed the number of associated species for all Swedish tree species. According to the report, the most common indigenous tree species, Norway spruce, was host to the highest number (1,100) of associated species, but note that this number includes species on both living trees and deadwood. However, less common tree species can also be important for biodiversity if they provide specific, valuable habitats for forest-dwelling species. For example, many species throughout the boreal region, particularly epiphytes and cavity-nesting birds, are associated with aspens, which provide these taxa more favorable habitats than the more common coniferous trees (e.g., Boudreault et al., 2000; Cadieux & Drapeau, 2017; Kivinen et al., 2020).

In addition to the specific tree species, the habitat value of a single tree is affected by its age and size; old and large trees are particularly important for biodiversity. Although tree age and size are often correlated, there are exceptions; on low-productivity sites, in particular, old trees can remain small but have a high biodiversity value (Cecile et al., 2013). With aging, trees develop specific characteristics and microhabitats, such as different bark structures or holes and cavities, that are important for various species (Michel & Winter, 2009); for example, rough bark typical of old trees hosts more arthropod species than smooth bark associated with younger trees (Nicolai, 1986), and certain epiphytic lichens are specifically associated with thick bark or deep bark crevices (Ranius et al., 2008). Table 5.1 presents further examples of important tree-scale microhabitats. In addition, old trees are valuable because they have been available for colonization for a longer time than young trees, which increases the chances for the establishment of dispersal-limited sessile biota (Ellis, 2012). Tree size, in turn, can be important because larger trees provide more habitat space and, in some cases, improved habitat quality; for example, many cavity-excavating birds prefer larger-diameter trees that are more stable and, because of thicker cavity walls, offer safer nesting places (Remm et al., 2006).

5.3 Factors Influencing the Structural Diversity of Living Trees

Natural key factors determining the structural diversity of living trees in unmanaged boreal forests are successional stage, disturbance history, and site productivity (Liira & Kohv, 2010; Moussaoui et al., 2016). The natural disturbance dynamics and stand successional sequence in boreal forests are explained elsewhere in this book (Chap. 3); therefore, we focus here more on the effects of productivity and its interaction with succession. For forestry purposes, productivity is usually defined as the ability to produce wood biomass per unit area over a given time (Bontemps &

Table 5.1 Important microhabitats occurring on living trees in boreal forests, with examples of associated species

Microhabitat	Occurrence and examples of species utilizing the microhabitat
Tree cavities and rot holes	Cavities are created by excavating birds (woodpeckers, in the boreal region) or, more rarely, by wood-decaying fungi. Woodpecker-excavated cavities are more prevalent on large broadleaf trees, and decay-induced cavities are more common on old trees (Andersson et al., 2018; Parsons et al., 2003). Cavities provide habitats for many insects and other invertebrates and nesting and hiding places for various birds and some mammals, such as bats (Esseen et al., 1997). Humid and shaded hollows are also colonized by some lichens, especially calicioids (Tibell, 1999)
Broken or irregular treetops	Broken or irregular tops of large trees provide nesting places for large birds of prey, e.g., Golden Eagle (*Aquila chrysaetos*) and Osprey (*Pandion haliaetus*) (Kuuluvainen, 2002)
Dead branches and treetops	Dead branches of living trees are an abundant microhabitat: they can constitute half of the total deadwood surface area in managed conifer-dominated boreal forests (Svensson et al., 2014). They serve as a substrate for saproxylic insects, fungi, and lichens (Larrieu et al., 2018), including rare species, e.g., the lichen *Erioderma pedicellatum* on dead branches of old spruces in the Russian Far East, Alaska, and eastern North America (Fig. 5.3c; Lauriault & Wiersma, 2020; Tagirdzhanova et al., 2019)
Bark pockets	Partially loose pieces of bark form a pocket between the bark and the tree trunk. These pockets are essential for many invertebrates and, if sufficiently large, can also be used as nesting places by some birds (e.g., treecreepers, Certhidae) or as day roosts by bats (Winter & Möller, 2008)
Cracks, scars, and bark loss	Cracks on the trunk provide nesting and hiding places for invertebrates, e.g., spiders and flat bugs; larger ones can also serve for birds and bats (Michel & Winter, 2009). Cracks and scars also host some crustose lichens and various fungi and microorganisms (Roll-Hansen and Roll-Hansen, 1980). Exposed wood is used by saproxylic invertebrates, fungi, and lichens (Larrieu et al., 2018)
Fire scars	Charred, exposed wood resulting from earlier forest fires occurs on fire-resistant conifers (e.g., *Pinus sylvestris*) and usually on larger trees, which are more likely to survive a fire. The fire scars serve as a substrate for species specialized on charred wood, e.g., the lichens *Carbonicola anthracophila*, *C. myrmecina*, and *Hertelidea botryosa* (Fig. 5.3d; Andersson et al., 2009; Lõhmus & Kruustük, 2010)
Resin and sap flows	Resinoses can host specialized fungi, such as *Chaenothecopsis* spp. (Titov, 2006) or the discomycetes *Sarea resinae* and *S. difformis* (Beimforde et al., 2020). Sap flows provide a food source, e.g., for several beetle species

(continued)

Table 5.1 (continued)

Microhabitat	Occurrence and examples of species utilizing the microhabitat
Cankers and burls	Cankers and burls provide substrates for epiphytic bryophytes and lichens. They are also used by certain Lepidoptera (Larrieu et al., 2018)
Witch brooms	Brooms are dense branch growths caused by pathogenic fungi, e.g., *Chrysomyxa arctostaphyli* on white and black spruce (Paragi, 2010). They provide nest sites and food sources for arthropods, birds, and small mammals, such as red squirrels (*Tamiasciurus hudsonicus*) (Tinnin et al., 1982)
Fungal fruiting bodies	Fungal fruiting bodies are used as habitat or food sources by various insects, such as beetles (Jonsell et al., 2001). They also host other fungi, e.g., calicioid fungi, such as *Phaeocalicium polyporaeum* (Titov, 2006), and lichens, e.g., *Bacidina* spp. and *Chaenotheca* spp.
Epiphytes	Epiphytic lichens and bryophytes provide habitats for various invertebrates, such as spiders, mites, and tardigrades (Ellis, 2012) and host lichenicolous fungi and epibryophytic lichens, e.g., *Mycobilimbia* spp.

Bouriaud, 2014) and combines the effects of soil conditions—nutrient availability and hydrology—and regional temperature (Fig. 5.1). The latter is an important factor in the boreal forest because in high-latitude or high-altitude regions, low temperatures can restrict the rate of cell division in trees (Rossi et al., 2007) and lead to slower tree growth. Another significant factor is paludification, i.e., the accumulation of soil organic matter, which can reduce productivity, especially in old boreal stands. In eastern Canada, for example, paludification can decrease black spruce productivity by 50% to 80%, particularly during the first centuries after a fire (Simard et al., 2007).

In general, the structural diversity of living trees increases with soil productivity in natural conditions (e.g., Boucher et al., 2006; Liira & Kohv, 2010). At higher site productivity, total stand volume is larger, and greater numbers of tree species and larger ranges of diameters and heights can co-occur. For example, in Estonian hemiboreal forests, high-productivity spruce–deciduous mixed stands have twice the stand volume, a significantly higher tree species richness, and a greater number of tree diameter classes than low-productivity Scots pine stands (Liira & Kohv, 2010; Lõhmus, 2004). Furthermore, the speed of structural development depends on productivity; trees grow faster on fertile soils, which accelerates the development toward stand complexity (Boucher et al., 2006; Larson et al., 2008). In boreal Canada, Boucher et al. (2006) found the tree size diversity of >200-year-old low-productivity black spruce stands to be low, whereas productive black spruce stands had an uneven-sized structure at a younger age. Productivity also affects the recruitment rate of new tree species, particularly for shade-tolerant coniferous species. The recruitment of these species occurs faster in more productive sites and may not happen at all at low-productivity sites (Boucher et al., 2006; Larson et al., 2008).

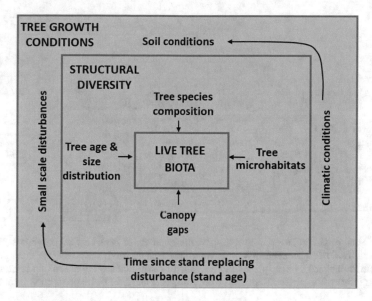

Fig. 5.1 Conceptual scheme of stand-scale predictors for the diversity of tree-dwelling biota. Some growth conditions are related to each other; all growth conditions influence the various elements of structural diversity, which in turn determine the live-tree biota

The occurrence of large or old trees is primarily determined by stand age, although large trees are also found more frequently in more productive sites. Because many tree-scale microhabitats develop with increasing age and decreasing tree vitality, their occurrence is generally correlated with that of old trees; however, various processes influence the dynamics of microhabitat development and loss (Fig. 5.2).

In addition to the processes listed above, the structural diversity of living trees and the associated biodiversity are subject to human-induced changes, particularly forest management. Management effects are often greatest in the most productive and economically valuable stand types (Martin et al., 2020). The effects vary depending on the applied practices, but typically, management simplifies the structural diversity of living trees and decreases habitat diversity.

In particular, intensely managed stands, i.e., clear-cut, reinitiated through planting, and thinned several times before final harvest, have a uniform structure lacking the multiple canopy layers, gaps, and other small-scale variations typical of natural stands (Cyr et al., 2009; Esseen et al., 1997). Because clear-cutting targets primarily old stands, a rejuvenation of the forest landscape has occurred throughout the boreal region (Bergeron et al., 2006). This rejuvenation is due to the short rotation periods used in even-aged forestry (roughly 70–120 years for conifer stands). Therefore, the number of large and old trees, and consequently the number of associated species, is low in production forests (Linder & Östlund, 1998). Even shorter rotation periods may be adopted in the future as the risk of pest and storm damage in older stands

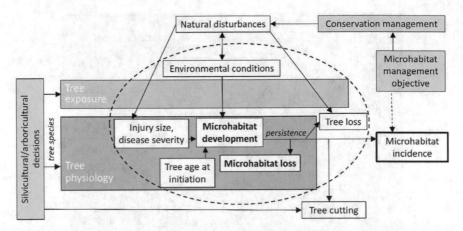

Fig. 5.2 Factors and processes contributing to the dynamics of tree-scale microhabitats. The rates of microhabitat development and loss determine the frequency of their occurrence; these processes are, in turn, influenced by tree species and physiology, environmental conditions, and disturbances, as well as forest management. Reprinted from Kõrkjas et al. (2021) with permission from Elsevier

is expected to increase because of climate change (Felton et al., 2016; Gauthier et al., 2014). Furthermore, the frequency and severity of natural disturbances such as fires, windthrow, and insect outbreaks are predicted to increase with climate change (e.g., Seidl et al., 2020). These enhanced disturbances may further decrease the proportion of old forests, especially if the frequency and intensity of salvage logging also increase (Lindenmayer et al., 2008).

Management also shapes tree species diversity because species of higher economic value are promoted by planting and thinning. In Fennoscandia, this has led to a decrease in broadleaf and mixed stands and increased conifer stands, with negative consequences for the associated biodiversity (Östlund et al., 1997). The planting of non-native or hybrid trees, such as lodgepole pine in Fennoscandia (Elfving et al., 2001), can have even more drastic effects on biodiversity because stands of non-native trees often host different communities of associated species than native stands (Bäcklund et al., 2016; Roberge & Stenbacka, 2014). The planting of non-native trees is relatively rare in the boreal region but may become more common in the future if non-native trees are deemed more profitable under a changing climate or more resilient against new pests and pathogens that will colonize the boreal region as the climate warms (Felton et al., 2016).

Climate change–related risks are also likely to affect the management of native trees. Broadleaf trees, for example, may be promoted as they are expected to be less susceptible to many climate-related risks. This scenario could have positive effects on biodiversity in regions where the number of broadleaf trees has been reduced (Felton et al., 2016). A warmer climate can also result in hemiboreal and temperate tree species migrating northward, affecting the distribution of native boreal trees (Gauthier et al., 2015).

5.4 Living Trees and the Diversity of Forest-Dwelling Species

In the previous sections, we provided an overview of how the structural diversity of living trees can affect boreal forest–dwelling species. The remainder of the chapter will illustrate these processes in more detail by presenting three groups of forest-dwelling organisms having different relationships with living trees: (1) epiphytic lichens and bryophytes, which depend on living trees as a substrate; (2) forest-dwelling birds, which utilize trees for foraging and nesting places; and (3) understory plants and epigeic lichens, which are influenced indirectly by living trees through stand microclimate, for example.

5.4.1 Epiphytic Lichens and Bryophytes

In most parts of the boreal region, epiphyte communities are formed solely by lichens and bryophytes; ferns as epiphytes are rare and occur only locally, e.g., in Norwegian boreal and boreonemoral rainforests (DellaSala et al., 2011). Lichens tend to grow on the whole tree, whereas bryophytes mostly form wefts, mats, tufts, or pendants on the lower parts of tree trunks and branches (Marmor et al., 2013; Tarasova et al., 2017). Although epiphyte biomass in boreal forests is relatively modest compared with tree foliar biomass (Botting et al., 2008), epiphytes contribute to nutrient and mineral cycling (e.g., Botting et al., 2008; Knops et al., 1991), participate in forest wood webs (e.g., Pettersson et al., 1995), and provide valuable nesting material for birds and mammals (Hayward & Rosentreter, 1994; Wesołowski & Wierzcholska, 2018).

In general, the epiphyte diversity in boreal Europe and eastern Canada is relatively well known and rich (>500 species), whereas fewer studies have examined other Canadian regions and northern Asia. Both lichens and bryophytes include obligate and facultative epiphyte species. Among macrolichens, pendulous taxa of the genera *Alectoria*, *Bryoria*, *Evernia*, *Ramalina*, and *Usnea* dominate on conifer branches (Fig. 5.3a; Esseen et al., 1997). Diverse assemblages of crustose lichens, which are often host-tree specific, inhabit both smooth and coarse bark of tree trunks (e.g., Androsova et al., 2018; Hyvärinen et al., 1992; McMullin et al., 2008). Examples of obligate epiphytic bryophytes include the mosses *Orthotrichum obtusifolium* and *Pylaisiella polyantha* and the liverworts *Frullania* spp., which all grow on aspens in the Canadian boreal forest. Tree bases and exposed roots of living trees are inhabited mainly by facultative epiphytes, e.g., *Ptilidium pulcherrimum* and *Pleurozium schreberi*, which often form a special "bryophyte sock" at the base of the tree. The bryophyte sock can house multiple species, and although most of these species can also grow on other substrates, tree bases are often the most abundant and, therefore, the most critical microhabitat.

Fig. 5.3 **a** Pendulous lichens, e.g., witch's hair lichen *Alectoria sarmentosa*, dominate on conifer branches throughout the boreal region. Broadleaf trees host many specialized lichen species, e.g., **b** lungwort lichen *Lobaria pulmonaria*, here on *Salix caprea*. Certain lichen species require specific microhabitats; **c** the globally threatened boreal felt lichen *Erioderma pedicellatum* occurs on dead spruce branches, whereas **d** the small clam lichen *Carbonicola anthracophila* colonizes only burned bark or wood, here accompanied by a generalist tube lichen *Hypogymnia physodes*. *Photo credits* **a b** Aino Hämäläinen, **c** Dmitry Himelbrant, and **d** Piret Lõhmus

The quality of living trees as epiphyte habitat is determined by tree characteristics that affect the availability of light, water, and nutrients. Bark structure and chemistry are particularly important in this respect (Brodo, 1973; Gustafsson & Eriksson, 1995). Rough bark with furrows and crevices can provide a large variety of microhabitats and favors specific assemblages of epiphytes such as calicioid lichens (*pin lichens*) (Holien, 1996). Furthermore, rough bark generally has better water holding capacity, which increases the water supply for epiphytes and favors their establishment (Snäll et al., 2004). Another critical factor is bark stability; unstable, easily exfoliating bark is an unfavorable substrate for most epiphytes. In addition to the structural factors, bark pH has a strong effect on epiphytes, as most species have a specific pH range that they can tolerate (Bates, 1992; Brodo, 1973). Bark pH can influence, for example, the availability of nutrients (Bates, 1992), the germination of bryophyte spores (Wiklund & Rydin, 2004), or the competition for space between micro- and macrolichens (Hyvärinen et al., 1992).

Bark characteristics differ among tree species, which leads to distinct epiphyte assemblages (e.g., Kuusinen, 1996). Most epiphytic species show at least some degree of preference for certain host tree species (Ellis, 2012; Kuusinen, 1996), although the association with tree species is often rather loose. Tree species differ in the total number of epiphytic species and the number of specialized species they host. In European boreal forests, for example, *Populus tremula* and *Salix caprea* are considered particularly important for epiphyte diversity, as they have a high number of specialist epiphytes (Kuusinen, 1996). Similarly, in North America, *Populus tremuloides* hosts a high epiphyte richness (Bartels & Chen, 2015). Bark characteristics are further affected by tree age; the bark of older trees is usually thicker and rougher than that of young trees. Concurrently, epiphytic communities change as the trees age. Although both young and old trees have associated epiphyte species, species richness is usually highest on old trees (e.g., Lie et al., 2009). In addition to the specific bark qualities, a longer colonization time can contribute to the high diversity on old trees (Lie et al., 2009).

Rotation forestry generally negatively affects epiphyte diversity, as it decreases the overall diversity of live-tree habitats and the amount of important substrates, such as old deciduous trees. Climate change is also expected to affect boreal forest epiphytes. Changes in temperature and precipitation can directly affect epiphyte growth and survival (e.g., Smith et al., 2018). However, the effects vary among species (Löbel et al., 2018) and forest types (Barbé et al., 2020). Furthermore, indirect effects through changes in tree species composition, disturbance regimes, or forestry practices are also likely. For example, climate change may lead to an intensification of forestry through shorter rotations or increased demand for biofuels (Felton et al., 2016), which can have negative impacts, particularly for epiphytes dependent on old-growth forest characteristics. Together, these effects and interactions can change epiphyte assemblages in complex and unexpected ways (Smith et al., 2018).

5.4.2 Forest Birds

Birds have developed many ecological adaptations to forest environments that allow them to utilize a very broad spectrum of habitats (Villard & Foppen, 2018). Moreover, because of their high mobility, birds as a group are capable of rapid responses to habitat changes in forest environments (Wesołowski & Fuller, 2012). Their ability to fly allows them to fully explore the three-dimensionality of forests created by trees, leading to an exceptionally high diversity of bird species assemblages in forests (Flade, 1994; James & Wamer, 1982; MacArthur & MacArthur, 1961).

Living trees provide birds with food, nesting sites, and shelter but are also used as a physical structure for several other activities, including perching, singing, and courting. Some boreal birds consume different vegetative parts of trees. For example, the Eurasian Western Capercaillie (*Tetrao urogallus*) and Spruce Grouse (*Falcipennis canadensis*, Fig. 5.4) in North America are highly dependent on the needles of coniferous trees during winter. The winter diet of Hazel Grouse in Eurasia consists almost exclusively of catkins, buds, and twigs of alder, birch, and other broadleaf trees. The Great Spotted Woodpecker (*Dendrocopos major*), Eurasian Siskin (*Spinus spinus*), and several crossbill species (*Loxia* spp.) with a circumboreal distribution are specialized in eating seeds from mostly coniferous trees, and waxwings (*Bombycilla* spp.) are dependent on berries provided by trees. Furthermore, invertebrates on living trees are a key food source for many bird species. The rapidly increasing biomass of herbivorous and predatory invertebrates in the spring secures nourishment for most arboreal passerines, both resident and migratory, and provides necessary proteins for the development of their chicks. In addition, saproxylic invertebrates occurring in living trees are an important food of forest birds, especially woodpeckers (Picidae), nuthatches (Sittidae), treecreepers (Certhidae), and tits (Paridae).

Many forest birds use trees as nesting sites. This behavior encompasses common passerines like the Common Chaffinch (*Fringilla coelebs*) in Eurasia or the Least Flycatcher (*Empidonax minimus*) in North America that usually build open nests in tree canopies; however, it also includes large birds of prey such as Osprey (*Pandion haliaetus*), which often build their nests on the tops of living trees. Living trees and snags are particularly important for cavity-nesting birds, including primary excavators, i.e., woodpeckers, and species using existing cavities created by woodpeckers and decay. In boreal forests, decay-formed cavities are rare, and therefore most cavities are made by woodpeckers (Andersson et al., 2018; Cockle et al., 2011). Thus, since several dozen bird species in boreal forests are obligate secondary cavity nesters, woodpeckers can be considered keystone species. For most forest birds, living trees and shrubs also provide shelter for both day- and nighttime roosting on branches, underneath the low branches or in cavities. For example, in a Norwegian study, Finne et al. (2000) found that 90% of daytime roosting sites of capercaillie were located underneath the low branches of Norway spruce trees.

Fig. 5.4 Male Spruce Grouse (*Falcipennis canadensis*) on jack pine (*Pinus banksiana*); jack pine needles are a key food item for this species. *Photo credit* Marjorie Wilson

The intersection of the species-habitat dependencies discussed above, interspecific interactions, the regional species pool, and landscape context leads to a highly variable bird assemblage for a particular boreal forest stand. The post-disturbance successional gradient will generate a range of bird assemblages from one of open or early-succession species (often generalist and migratory species) to specialized residents dependent on old-growth stands, generally characterized by a high structural diversity (Drapeau et al., 2000; Helle & Mönkkönen, 1986; Imbeau et al., 1999; Jansson & Andrén, 2003). Structural differences, including the dominance of particular tree species, are greater and have a more marked effect on differentiating bird assemblages in the Nearctic than in the western Palearctic; this is possibly explained by the lower level of specialization in the latter (Mönkkönen, 1994).

A greater number of native tree species and a longer time since a stand-replacing disturbance generate a higher diversity of bird assemblages. In particular, the presence of broadleaf trees heightens bird species diversity, especially in boreal Europe (Jansson & Andrén, 2003). The occurrence of large and old trees provides nesting and foraging opportunities for many species, including specialized cavity nesters and large birds of prey. In addition, multilayered stands have richer bird assemblages than stands with even-aged woody vegetation (e.g., Klein et al., 2020). Thus, from a conservation and management perspective, promoting these features of structural diversity can support a higher diversity of forest bird assemblages. For example, green tree retention applied during the logging of boreal forests may positively affect several species, but it can only partially counteract the loss of mature forest (Basile et al., 2019; Price et al., 2020; Söderström, 2009; Venier et al., 2015). There is a need for more research on how different natural processes interact with human actions across landscapes and in the context of climate change (Mikusiński et al., 2018). A recent study in the boreal forests of Alberta (Cadieux et al., 2020) predicted declines in bird species associated with older coniferous forests because of climate change, and this process is expected to be accelerated by forestry. Similarly, in Finnish boreal forests, habitat alteration due to forestry compounds the negative impact of climate change on bird assemblages (Virkkala, 2016). To counteract such adverse developments, Stralberg et al. (2019) proposed large-scale recovery plans and adaptive forest management, the designation of critical habitat, and land protection. These measures are based mainly on the appropriate management of living trees at stand and landscape levels.

5.4.3   Plants and Lichens in Forest Understories

Despite not directly growing on living trees, vascular plants, bryophytes, and lichens in boreal forest understories are nevertheless influenced by the structural diversity that governs the microclimate and nutrient availability on the forest floor. In boreal forests, particularly in older conifer-dominated stands, epigeic bryophytes and lichens prevail in the forest understory rather than vascular plants (Bergeron & Fenton, 2012; Esseen et al., 1997). Bryophytes dominate the forest floor in mesic sites, whereas lichens are more common in drier sites and in the most northern parts of the boreal region (Esseen et al., 1997). Vascular plants are more abundant in younger stands and nutrient-rich broadleaf stands.

Canopy tree species composition is an important determinant of understory diversity, and stands of different tree species typically host distinct understory assemblages. Canopy closure and, concurrently, light availability can differ notably among tree species, depending on the size and arrangement of leaves (Barbier et al., 2008). In eastern Canadian boreal forests, for example, the highest light levels are observed in *Betula* stands; the light levels then decrease in the order of *Populus*, *Pinus*, *Picea*, *Abies*, and *Thuja* stands (Messier et al., 1998). Furthermore, tree species composition affects water availability in the understory, as the amount of throughfall and water absorbed by tree roots varies among tree species. For example, throughfall is generally higher in broadleaf stands than coniferous ones and in stands of early-successional rather than late-successional tree species (Barbier et al., 2008). In addition, nutrient availability is generally greater in broadleaf and mixed stands than in coniferous stands because of the presence of nutrient-rich leaf litter (Hart & Chen, 2006). Therefore, vascular plants that benefit from higher levels of light and nutrients are usually more diverse and abundant in broadleaf and mixed stands. Bryophytes and lichens, in turn, suffer from increased competition from vascular plants and are affected by the leaf litter of broadleaf trees, which can inhibit their establishment and growth (Bartels & Chen, 2013). The cover and diversity of bryophytes and lichens are therefore generally higher in conifer-dominated stands.

In addition to tree species composition, microclimate on the forest floor is influenced by variations in canopy openness. Large canopy gaps and other openings that lead to increased light levels positively affect vascular plants and some species of lichen, e.g., *Cladonia* spp. (Boudreault et al., 2013). Bryophytes, which are generally shade tolerant, do not benefit from increased light; instead, their cover decreases in large canopy gaps because of intensified competition with vascular plants. However, since bryophytes require high moisture levels, they can benefit from small canopy gaps, which do not have notably higher light levels but rather greater water availability owing to higher throughfall and the lower transpiration by live trees (Hart & Chen, 2006; Muscolo et al., 2014). In addition, small gaps formed by tree uprooting provide an important microhabitat for bryophytes (e.g., *Schistostega pennata*) and lichens (e.g., *Chaenotheca furfuracea*) in the form of root plates and bare mineral soil (Jonsson & Esseen, 1990; Lõhmus et al., 2010).

The relationships described above make epigeic species vulnerable to changes in canopy closure and tree species composition because of, for example, forest management. In Sweden, managed forests have become denser, which has led to an increased proportion of shade-adapted plants in the understory (Hedwall et al., 2019). In North America, however, most studies have shown that the vascular plant flora of boreal forests (MacDonald et al., 2015), and to some degree epigeic lichens (Lafleur et al., 2016), are resilient in the face of most forest management regimes. Bryophytes, however, are more vulnerable (MacDonald et al., 2015; Paquette et al., 2016). Climate change is expected to affect understory species directly through altered temperature and precipitation and decreased snow cover. The latter, accompanied by frost damage, is predicted to be one of the most significant climate change–related factors influencing boreal forest understories and could lead to altered species composition and notable decreases in the abundance of dominant species (e.g., Kreyling et al., 2012).

5.5 Conclusions

Live trees are essential for the biodiversity of boreal forests. Although forest-dwelling species groups have different relationships to live trees and therefore require different tree habitats or characteristics, the diversity of all groups generally increases with a greater structural diversity of live trees. Rotation forestry simplifies this diversity and negatively impacts the various species that depend on live trees. In the past decades, however, different restoration and management methods, such as retention harvest, have been developed to mitigate these negative impacts. Promoting habitat diversity at both the tree and stand scales (e.g., retaining trees of various species, ages, and containing different microhabitats) by creating a variable canopy structure or applying continuous cover forestry can help maintain the diversity of forest-dwelling species in managed stands. In turn, higher biodiversity can increase forest ecosystem resilience to climate change (Drever et al., 2006; Loreau, 2000).

Climate change will affect living trees and the associated biodiversity through various direct (e.g., higher disturbance severity and frequency) and indirect (e.g., changes in forest management) pathways. Although the responses of boreal tree species to climate change have been examined in various studies (e.g., Boulanger et al., 2017; Peng et al., 2011), the reaction of other forest-dwelling aboveground organism groups is less well known. With various complex effects and interactions, the overall changes in communities (e.g., species abundance, diversity, and distribution) are difficult to predict (but see Villén-Peréz et al., 2020). Finally, if forest management practices are modified because of climate change adaptation or mitigation (e.g., an increased proportion of broadleaf trees), the effects on forest-dwelling species should be assessed and compensated where necessary.

References

Andersson, J., Domingo Gomez, E., Michon, S., et al. (2018). Tree cavity densities and characteristics in managed and unmanaged Swedish boreal forest. *Scandinavian Journal of Forest Research, 33*, 233–244. https://doi.org/10.1080/02827581.2017.1360389.

Andersson, L., Alexeeva, N., & Kuznetsova, E. (2009). *Survey of biologically valuable forests in northwestern European Russia. Identification manual of species to be used during surveys at the stand level* [in Russian] (Vol. 2, p. 38). Saint-Petersburg: Pobeda Publishing.

Androsova, V. I., Tarasova, V. N., & Gorshkov, V. V. (2018). Diversity of lichens and allied fungi on Norway spruce (*Picea abies*) in the middle boreal forests of Republic of Karelia (Russia). *Folia Cryptogamica Estonica, 55*, 133–149. https://doi.org/10.12697/fce.2018.55.14.

Bäcklund, S., Jönsson, M., Strengbom, J., et al. (2016). A pine is a pine and a spruce is a spruce–the effect of tree species and stand age on epiphytic lichen communities. *PLoS ONE, 11*, e0147004. https://doi.org/10.1371/journal.pone.0147004.

Barbé, M., Bouchard, M., & Fenton, N. J. (2020). Examining boreal forest resilience to temperature variability using bryophytes: Forest type matters. *Ecosphere, 11*, e03232. https://doi.org/10.1002/ecs2.3232.

Barbier, S., Gosselin, F., & Balandier, P. (2008). Influence of tree species on understory vegetation diversity and mechanisms involved–a critical review for temperate and boreal forests. *Forest Ecology and Management, 254*, 1–15. https://doi.org/10.1016/j.foreco.2007.09.038.

Bartels, S. F., & Chen, H. Y. H. (2013). Interactions between overstorey and understorey vegetation along an overstorey compositional gradient. *Journal of Vegetation Science, 24*, 543–552. https://doi.org/10.1111/j.1654-1103.2012.01479.x.

Bartels, S. F., & Chen, H. Y. H. (2015). Species dynamics of epiphytic macrolichens in relation to time since fire and host tree species in boreal forest. *Journal of Vegetation Science, 26*, 1124–1133. https://doi.org/10.1111/jvs.12315.

Basile, M., Mikusiński, G., & Storch, I. (2019). Bird guilds show different responses to tree retention levels: A meta-analysis. *Global Ecology and Conservation, 18*, e00615. https://doi.org/10.1016/j.gecco.2019.e00615.

Bates, J. W. (1992). Mineral nutrient acquisition and retention by bryophytes. *Journal of Bryology, 17*, 223–240. https://doi.org/10.1179/jbr.1992.17.2.223.

Beimforde, C., Schmidt, A. R., Rikkinen, J., et al. (2020). *Sareomycetes cl. nov.*: A new proposal for placement of the resinicolous genus *Sarea* (*Ascomycota, Pezizomycotina*). *Fungal Systematics and Evolution, 6*, 25–37. https://doi.org/10.3114/fuse.2020.06.02.

Bergeron, Y. (2000). Species and stand dynamics in the mixed woods of Quebec's southern boreal forest. *Ecology, 81*, 1500–1516. https://doi.org/10.1890/0012-9658(2000)081[1500:SASDIT]2.0.CO;2.

Bergeron, Y., & Fenton, N. J. (2012). Boreal forests of eastern Canada revisited: Old growth, nonfire disturbances, forest succession, and biodiversity. *Botany, 90*(6), 509–523. https://doi.org/10.1139/b2012-034.

Bergeron, Y., Cyr, D., Drever, C. R., et al. (2006). Past, current, and future fire frequencies in Quebec's commercial forests: Implications for the cumulative effects of harvesting and fire on age-class structure and natural disturbance-based management. *Canadian Journal of Forest Research, 36*, 2737–2744. https://doi.org/10.1139/x06-177.

Bontemps, J. D., & Bouriaud, O. (2014). Predictive approaches to forest site productivity: Recent trends, challenges and future perspectives. *Forestry, 87*, 109–128. https://doi.org/10.1093/forestry/cpt034.

Botting, R. S., Campbell, J., & Fredeen, A. L. (2008). Contrasting arboreal and terrestrial macrolichen and bryophyte communities in old-growth sub-boreal spruce forests of central British Columbia. *The Bryologist, 111*, 607–619. https://doi.org/10.1639/0007-2745-111.4.607.

Boucher, D., Gauthier, S., & De Grandpré, L. (2006). Structural changes in coniferous stands along a chronosequence and a productivity gradient in the northeastern boreal forest of Québec. *Ecoscience, 13*, 172–180. https://doi.org/10.2980/i1195-6860-13-2-172.1.

Boudreault, C., Gauthier, S., & Bergeron, Y. (2000). Epiphytic lichens and bryophytes on *Populus tremuloides* along a chronosequence in the southwestern boreal forest of Québec, Canada. *The Bryologist, 103*, 725–738. https://doi.org/10.1639/0007-2745(2000)103[0725:ELABOP]2.0.CO;2.

Boudreault, C., Zouaoui, S., Drapeau, P., et al. (2013). Canopy openings created by partial cutting increase growth rates and maintain the cover of three *Cladonia* species in the Canadian boreal forest. *Forest Ecology and Management, 304*, 473–481. https://doi.org/10.1016/j.foreco.2013.05.043.

Boulanger, Y., Taylor, A. R., Price, D. T., et al. (2017). Climate change impacts on forest landscapes along the Canadian southern boreal forest transition zone. *Landscape Ecology, 32*, 1415–1431. https://doi.org/10.1007/s10980-016-0421-7.

Brassard, B. W., & Chen, H. Y. H. (2006). Stand structural dynamics of North American boreal forests. *Critical Reviews in Plant Sciences, 25*(2), 115–137. https://doi.org/10.1080/07352680500348857.

Brodo, I. M. (1973). Substrate ecology. In V. Ahmadjian & M. E. Hale (Eds.), *The lichens* (pp. 401–441). New York: Academic Press.

Cadieux, P., & Drapeau, P. (2017). Are old boreal forests a safe bet for the conservation of the avifauna associated with decayed wood in eastern Canada? *Forest Ecology and Management, 385*, 127–139. https://doi.org/10.1016/j.foreco.2016.11.024.

Cadieux, P., Boulanger, Y., Cyr, D., et al. (2020). Projected effects of climate change on boreal bird community accentuated by anthropogenic disturbances in western boreal forest, Canada. *Diversity and Distributions, 26*(6), 668–682. https://doi.org/10.1111/ddi.13057.

Cecile, J., Silva, L. R., & Anand, M. (2013). Old trees: Large and small. *Science, 339*, 904–905. https://doi.org/10.1126/science.339.6122.904-c.

Chase, J., & Leibold, M. (2003). *Ecological niches* (p. 212). Chicago: University of Chicago Press.

Cockle, K. L., Martin, K., & Wesołowski, T. (2011). Woodpeckers, decay, and the future of cavity-nesting vertebrate communities worldwide. *Frontiers in Ecology and the Environment, 9*, 377–382. https://doi.org/10.1890/110013.

Cyr, D., Gauthier, S., Bergeron, Y., et al. (2009). Forest management is driving the eastern North American boreal forest outside its natural range of variability. *Frontiers in Ecology and the Environment, 7*(10), 519–524. https://doi.org/10.1890/080088.

DellaSala, D. A., Alaback, P., Drescher, A., et al. (2011). Temperate and boreal rainforest relicts of Europe. In D. A. DellaSala (Ed.), *Temperate and boreal rainforests of the world: Ecology and conservation* (pp. 154–180). Washington: Island Press.

Drapeau, P., Leduc, A., Giroux, J. F., et al. (2000). Landscape-scale disturbances and changes in bird communities of boreal mixed-wood forests. *Ecological Monographs, 70*, 423–444. https://doi.org/10.1890/0012-9615(2000)070[0423:LSDACI]2.0.CO;2.

Drever, C. R., Peterson, G., Messier, C., et al. (2006). Can forest management based on natural disturbances maintain ecological resilience? *Canadian Journal of Forest Research, 36*, 2285–2299. https://doi.org/10.1139/x06-132.

Elfving, B., Ericsson, T., & Rosvall, O. (2001). The introduction of lodgepole pine for wood production in Sweden–a review. *Forest Ecology and Management, 141*, 15–29. https://doi.org/10.1016/S0378-1127(00)00485-0.

Ellis, C. J. (2012). Lichen epiphyte diversity: A species, community and trait-based review. *Perspectives in Plant Ecology, Evolution and Systematics, 14*, 131–152. https://doi.org/10.1016/j.ppees.2011.10.001.

Esseen, P. A., Ehnström, B., Ericson, L., et al. (1997). Boreal forests. *Ecological Bulletins, 46*, 16–47.

Felton, A., Gustafsson, L., Roberge, J. M., et al. (2016). How climate change adaptation and mitigation strategies can threaten or enhance the biodiversity of production forests: Insights from Sweden. *Biological Conservation, 194*, 11–20. https://doi.org/10.1016/j.biocon.2015.11.030.

Finne, M. H., Wegge, P., Eliassen, S., et al. (2000). Daytime roosting and habitat preference of capercaillie *Tetrao urogallus* males in spring - the importance of forest structure in relation to anti-predator behavior. *Wildlife Biology, 6*, 241–249. https://doi.org/10.2981/wlb.2000.022.

Flade, M. (1994). *Die Brutvogelgemeinschaften Mittel- und Norddeutschlands: Grundlagen für den Gebrauch vogelkundlicher Daten in der Landschaftsplanung.* IHW-Verlag

Franklin, J. F., & Van Pelt, R. (2004). Spatial aspects of structural complexity in old-growth forests. *Journal of Forestry, 102*, 22–28.

Gauthier, S., Bernier, P., Burton, P. J., et al. (2014). Climate change vulnerability and adaptation in the managed Canadian boreal forest. *Environmental Reviews, 22*, 256–285. https://doi.org/10.1139/er-2013-0064.

Gauthier, S., Bernier, P., Kuuluvainen, T., et al. (2015). Boreal forest health and global change. *Science, 349*(6250), 819–822. https://doi.org/10.1126/science.aaa9092.

Gustafsson, L., & Eriksson, I. (1995). Factors of importance for the epiphytic vegetation of aspen *Populus tremula* with special emphasis on bark chemistry and soil chemistry. *Journal of Applied Ecology, 32*, 412–424. https://doi.org/10.2307/2405107.

Hart, S. A., & Chen, H. Y. H. (2006). Understory vegetation dynamics of North American boreal forests. *Critical Reviews in Plant Sciences, 25*(4), 381–397. https://doi.org/10.1080/07352680600819286.

Hayward, G. D., & Rosentreter, R. (1994). Lichens as nesting material for northern flying squirrels in the northern Rocky Mountains. *Journal of Mammalogy, 75*, 663–673. https://doi.org/10.2307/1382514.

Hedwall, P. O., Gustafsson, L., Brunet, J., et al. (2019). Half a century of multiple anthropogenic stressors has altered northern forest understory plant communities. *Ecological Applications, 29*, e01874. https://doi.org/10.1002/eap.1874.

Helle, P., & Mönkkönen, M. (1986). Annual fluctuations of land bird communities in different successional stages of boreal forest. *Annales Zoologici Fennici, 23*, 269–280.

Holien, H. (1996). Influence of site and stand factors on the distribution of crustose lichens of the caliciales in a suboceanic spruce forest area in central Norway. *The Lichenologist, 28*, 315–330. https://doi.org/10.1006/lich.1996.0029.

Hyvärinen, M., Halonen, P., & Kauppi, M. (1992). Influence of stand age and structure on the epiphytic lichen vegetation in the middle-boreal forests of Finland. *The Lichenologist, 24*, 165–180. https://doi.org/10.1017/S0024282900500073.

Imbeau, L., Savard, J. P. L., & Gagnon, R. (1999). Comparing bird assemblages in successional black spruce stands originating from fire and logging. *Canadian Journal of Zoology, 77*, 1850–1860. https://doi.org/10.1139/z99-172.

James, F. C., & Wamer, N. O. (1982). Relationships between temperate forest bird communities and vegetation structure. *Ecology, 63*, 159–171. https://doi.org/10.2307/1937041.

Jansson, G., & Andrén, H. (2003). Habitat composition and bird diversity in managed boreal forests. *Scandinavian Journal of Forest Research, 18*, 225–236. https://doi.org/10.1080/02827581.2003.9728293.

Jonsell, M., Nordlander, G., & Ehnström, B. (2001). Substrate associations of insects breeding in fruiting bodies of wood-decaying fungi. *Ecological Bulletins, 49*, 173–194.

Jonsson, B. G., & Esseen, P. A. (1990). Treefall disturbance maintains high bryophyte diversity in a boreal spruce forest. *Journal of Ecology, 78*, 924–936. https://doi.org/10.2307/2260943.

Juchheim, J., Ehbrecht, M., Schall, P., et al. (2020). Effect of tree species mixing on stand structural complexity. *Forestry, 93*, 75–83. https://doi.org/10.1093/forestry/cpz046.

Kivinen, S., Koivisto, E., Keski-Saari, S., et al. (2020). A keystone species, European aspen (*Populus tremula* L.), in boreal forests: Ecological role, knowledge needs and mapping using remote sensing. *Forest Ecology and Management, 462*, 118008. https://doi.org/10.1016/j.foreco.2020.118008.

Klein, J., Thor, G., Low, M., et al. (2020). What is good for birds is not always good for lichens: Interactions between forest structure and species richness in managed boreal forests. *Forest Ecology and Management, 473*, 118327. https://doi.org/10.1016/j.foreco.2020.118327.

Knops, J. M. H., Nash, T. H., III., Boucher, V. L., et al. (1991). Mineral cycling and epiphytic lichens: Implications at the ecosystem level. *The Lichenologist, 23*, 309–321. https://doi.org/10.1017/S0024282991000452.

Kõrkjas, M., Remm, L., & Lõhmus, A. (2021). Development rates and persistence of the microhabitats initiated by disease and injuries in live trees: A review. *Forest Ecology and Management, 482*, 118833. https://doi.org/10.1016/j.foreco.2020.118833.

Kreyling, J., Haei, M., & Laudon, H. (2012). Absence of snow cover reduces understory plant cover and alters plant community composition in boreal forests. *Oecologia, 168*, 577–587. https://doi.org/10.1007/s00442-011-2092-z.

Kuuluvainen, T. (1994). Gap disturbance, ground microtopography, and the regeneration dynamics of boreal coniferous forests in Finland: A review. *Annales Zoologici Fennici, 31*, 35–51.

Kuuluvainen, T. (2002). Natural variability of forests as a reference for restoring and managing biological diversity in boreal Fennoscandia. *Silva Fennica, 36*(1), 552. https://doi.org/10.14214/sf.552.

Kuusinen, M. (1996). Epiphyte flora and diversity on basal trunks of six old-growth forest tree species in southern and middle boreal Finland. *The Lichenologist, 28*, 443–463. https://doi.org/10.1006/lich.1996.0043.

Lafleur, B., Zouaoui, S., Fenton, N. J., et al. (2016). Short-term response of *Cladonia* lichen communities to logging and fire in boreal forests. *Forest Ecology and Management, 372*, 44–52. https://doi.org/10.1016/j.foreco.2016.04.007.

Larrieu, L., Paillet, Y., Winter, S., et al. (2018). Tree related microhabitats in temperate and Mediterranean European forests: A hierarchical typology for inventory standardization. *Ecological Indicators, 84*, 194–207. https://doi.org/10.1016/j.ecolind.2017.08.051.

Larsen, J. A. (1980). Boreal communities and ecosystems: The broad view. In T. T. Kozlowski (Ed.), *The boreal ecosystem. Physiological ecology: A series of monographs, texts, and treatises* (pp. 128–236). New York: Academic Press.

Larson, A. J., Lutz, J. A., Gersonde, R. F., et al. (2008). Potential site productivity influences the rate of forest structural development. *Ecological Applications, 18*, 899–910. https://doi.org/10.1890/07-1191.1.

Lauriault, P., & Wiersma, Y. F. (2020). Identifying important characteristics for critical habitat of boreal felt lichen (*Erioderma pedicellatum*) in Newfoundland, Canada. *The Bryologist, 123*, 412–420. https://doi.org/10.1639/0007-2745-123.3.412.

Lecomte, N., Simard, M., Fenton, N., et al. (2006). Fire severity and long-term ecosystem biomass dynamics in coniferous boreal forests of eastern Canada. *Ecosystems, 9*, 1215–1230. https://doi.org/10.1007/s10021-004-0168-x.

Lie, M., Arup, U., Grytnes, J. A., et al. (2009). The importance of host tree age, size and growth rate as determinants of epiphytic lichen diversity in boreal spruce forests. *Biodiversity and Conservation, 18*, 3579–3596. https://doi.org/10.1007/s10531-009-9661-z.

Liira, J., & Kohv, K. (2010). Stand characterisation and biodiversity indicators along the productivity gradient in boreal forests: Defining a critical set of indicators for the monitoring of habitat nature quality. *Plant Biosystems, 144*, 211–220. https://doi.org/10.1080/11263500903560868.

Lindenmayer, D. B., Burton, P. J., & Franklin, J. F. (2008). *Salvage logging and its ecological consequences* (p. 246). Washington: Island Press.

Linder, P., & Östlund, L. (1998). Structural changes in three mid-boreal Swedish forest landscapes, 1885–1996. *Biological Conservation, 85*, 9–19. https://doi.org/10.1016/S0006-3207(97)00168-7.

Löbel, S., Mair, L., Lönnell, N., et al. (2018). Biological traits explain bryophyte species distributions and responses to forest fragmentation and climatic variation. *Journal of Ecology, 106*, 1700–1713. https://doi.org/10.1111/1365-2745.12930.

Lõhmus, E. (2004). *Eesti metsakasvukohatüübid*. Tartu: Eesti Loodusfoto.

Lõhmus, P., & Kruustük, K. (2010). Lichens on burnt wood in Estonia: A preliminary assessment. *Folia Cryptogamica Estonica, 47*, 37–41.

Lõhmus, P., Turja, K., & Lõhmus, A. (2010). Lichen communities on treefall mounds depend more on root-plate than stand characteristics. *Forest Ecology and Management, 260*, 1754–1761. https://doi.org/10.1016/j.foreco.2010.07.056.

Loreau, M. (2000). Biodiversity and ecosystem functioning: Recent theoretical advances. *Oikos, 91*, 3–17. https://doi.org/10.1034/j.1600-0706.2000.910101.x.

MacArthur, R. H., & MacArthur, J. W. (1961). On bird species diversity. *Ecology, 42*, 594–598. https://doi.org/10.2307/1932254.

MacDonald, R. L., Chen, H. Y., Bartels, S. F., et al. (2015). Compositional stability of boreal understorey vegetation after overstorey harvesting across a riparian ecotone. *Journal of Vegetation Science, 26*, 733–741. https://doi.org/10.1111/jvs.12272.

Marmor, L., Tõrra, T., Saag, L., et al. (2013). Lichens on *Picea abies* and *Pinus sylvestris*–from tree bottom to the top. *The Lichenologist, 45*, 51–63. https://doi.org/10.1017/S0024282912000564.

Martin, M., Morin, H., & Fenton, N. J. (2019). Secondary disturbances of low and moderate severity drive the dynamics of eastern Canadian boreal old-growth forests. *Annals of Forest Science, 76*(4), 108. https://doi.org/10.1007/s13595-019-0891-2.

Martin, M., Boucher, Y., Fenton, N. J., et al. (2020). Forest management has reduced the structural diversity of residual boreal old-growth forest landscapes in eastern Canada. *Forest Ecology and Management, 458*, 117765. https://doi.org/10.1016/j.foreco.2019.117765.

McMullin, R. T., Duinker, P. N., Cameron, R. P., et al. (2008). Lichens of coniferous old-growth forests of southwestern Nova Scotia, Canada: Diversity and present status. *The Bryologist, 111*, 620–637. https://doi.org/10.1639/0007-2745-111.4.620.

Messier, C., Parent, S., & Bergeron, Y. (1998). Effects of overstory and understory vegetation on the understory light environment in mixed boreal forests. *Journal of Vegetation Science, 9*, 511–520. https://doi.org/10.2307/3237266.

Michel, A. K., & Winter, S. (2009). Tree microhabitat structures as indicators of biodiversity in Douglas-fir forests of different stand ages and management histories in the Pacific Northwest, U.S.A. *Forest Ecology and Management, 257*, 1453–1464. https://doi.org/10.1016/j.foreco.2008.11.027.

Mikusiński, G., Fuller, R. J., & Roberge, J.-M. (2018). Future forests, avian implications and research priorities. In G. Mikusinski, R. J. Fuller, & J.-M. Roberge (Eds.), *Ecology and conservation of forest birds. Ecology, Biodiversity and Conservation Series* (pp. 508–536). Cambridge University Press.

Mönkkönen, M. (1994). Diversity patterns in palaearctic and nearctic forest bird assemblages. *Journal of Biogeography, 21*, 183–195. https://doi.org/10.2307/2845471.

Moussaoui, L., Fenton, N. J., Leduc, A., et al. (2016). Deadwood abundance in post-harvest and post-fire residual patches: An evaluation of patch temporal dynamics in black spruce boreal forest. *Forest Ecology and Management, 368*, 17–27. https://doi.org/10.1016/j.foreco.2016.03.012.

Muscolo, A., Bagnato, S., Sidari, M., et al. (2014). A review of the roles of forest canopy gaps. *Journal of Forest Research, 25*, 725–736. https://doi.org/10.1007/s11676-014-0521-7.

Nicolai, V. (1986). The bark of trees: Thermal properties, microclimate and fauna. *Oecologia, 69*, 148–160. https://doi.org/10.1007/BF00399052.

Östlund, L., Zackrisson, O., & Axelsson, A. L. (1997). The history and transformation of a Scandinavian boreal forest landscape since the 19th century. *Canadian Journal of Forest Research, 27*(8), 1198–1206. https://doi.org/10.1139/x97-070.

Paquette, M., Boudreault, C., Fenton, N., et al. (2016). Bryophyte species assemblages in fire and clear-cut origin boreal forests. *Forest Ecology and Management, 359*, 99–108. https://doi.org/10.1016/j.foreco.2015.09.031.

Paragi, T. F. (2010). Density and size of snags, tree cavities, and spruce rust brooms in Alaska boreal forest. *Western Journal of Applied Forestry, 25*, 88–95. https://doi.org/10.1093/wjaf/25.2.88.

Parsons, S., Lewis, K. J., & Psyllakis, J. M. (2003). Relationships between roosting habitat of bats and decay of aspen in the sub-boreal forests of British Columbia. *Forest Ecology and Management, 177*, 559–570. https://doi.org/10.1016/S0378-1127(02)00448-6.

Peng, C., Ma, Z., Lei, X., et al. (2011). A drought-induced pervasive increase in tree mortality across Canada's boreal forests. *Nature Climate Change, 1*, 467–471. https://doi.org/10.1038/nclimate1293.

Pettersson, R. B., Ball, J. P., Renhorn, K. E., et al. (1995). Invertebrate communities in boreal forest canopies as influenced by forestry and lichens with implications for passerine birds. *Biological Conservation, 74*, 57–63. https://doi.org/10.1016/0006-3207(95)00015-V.

Price, K., Daust, K., Lilles, E., et al. (2020). Long-term response of forest bird communities to retention forestry in northern temperate coniferous forests. *Forest Ecology and Management, 462*, 117982. https://doi.org/10.1016/j.foreco.2020.117982.

Ranius, T., Johansson, P., Berg, N., et al. (2008). The influence of tree age and microhabitat quality on the occurrence of crustose lichens associated with old oaks. *Journal of Vegetation Science, 19*, 653–662. https://doi.org/10.3170/2008-8-18433.

Remm, J., Lõhmus, A., & Remm, K. (2006). Tree cavities in riverine forests: What determines their occurrence and use by hole-nesting passerines? *Forest Ecology and Management, 221*, 267–277. https://doi.org/10.1016/j.foreco.2005.10.015.

Roberge, J. M., & Stenbacka, F. (2014). Assemblages of epigaeic beetles and understory vegetation differ between stands of an introduced pine and its native congener in boreal forest. *Forest Ecology and Management, 318*, 239–249. https://doi.org/10.1016/j.foreco.2014.01.026.

Roll-Hansen, F., & Roll-Hansen, H. (1980). Microorganisms which invade *Picea abies* in seasonal stem wounds. *Forest Pathology, 10*, 396–410. https://doi.org/10.1111/j.1439-0329.1980.tb00057.x.

Rossi, S., Deslauriers, A., Anfodillo, T., et al. (2007). Evidence of threshold temperatures for xylogenesis in conifers at high altitudes. *Oecologia, 152*, 1–12. https://doi.org/10.1007/s00442-006-0625-7.

Seidl, R., Honkaniemi, J., Aakala, T., et al. (2020). Globally consistent climate sensitivity of natural disturbances across boreal and temperate forest ecosystems. *Ecography, 43*(7), 967–978. https://doi.org/10.1111/ecog.04995.

Shorohova, E., Kneeshaw, D., Kuuluvainen, T., et al. (2011). Variability and dynamics of old-growth forests in the circumboreal zone: Implications for conservation, restoration and management. *Silva Fennica, 45*, 785–806. https://doi.org/10.14214/sf.72.

Simard, M., Lecomte, N., Bergeron, Y., et al. (2007). Forest productivity decline caused by successional paludification of boreal soils. *Ecological Applications, 17*, 1619–1637. https://doi.org/10.1890/06-1795.1.

Smith, R. J., Nelson, P. R., Jovan, S., et al. (2018). Novel climates reverse carbon uptake of atmospherically dependent epiphytes: Climatic constraints on the iconic boreal forest lichen *Evernia mesomorpha. American Journal of Botany, 105*, 266–274. https://doi.org/10.1002/ajb2.1022.

Snäll, T., Hagström, A., Rudolphi, J., et al. (2004). Distribution pattern of the epiphyte *Neckera pennata* on three spatial scales–importance of past landscape structure, connectivity and local conditions. *Ecography, 27*, 757–766. https://doi.org/10.1111/j.0906-7590.2004.04026.x.

Söderström, B. (2009). Effects of different levels of green and dead tree retention on hemi-boreal forest bird communities in Sweden. *Forest Ecology and Management, 257*, 215–222. https://doi.org/10.1016/j.foreco.2008.08.030.

Stralberg, D., Berteaux, D., Drever, C. R., et al. (2019). Conservation planning for boreal birds in a changing climate: A framework for action. *Avian Conservation and Ecology, 14*(1), 13. https://doi.org/10.5751/ACE-01363-140113.

Sundberg, S., Carlberg, T., Sandström, J., et al. (Eds.). (2019). *Värdväxters betydelse för andra organismer-med fokus på vedartade värdväxter* (p. 52). Uppsala: ArtDatabanken, Sveriges lantbruksuniversitet.

Svensson, M., Dahlberg, A., Ranius, T., et al. (2014). Dead branches on living trees constitute a large part of the dead wood in managed boreal forests, but are not important for wood-dependent lichens. *Journal of Vegetation Science, 25*, 819–828. https://doi.org/10.1111/jvs.12131.

Tagirdzhanova, G., Stepanchikova, I. S., Himelbrant, D. E., et al. (2019). Distribution and assessment of the conservation status of *Erioderma pedicellatum* in Asia. *The Lichenologist, 51*, 575–585. https://doi.org/10.1017/S0024282919000380.

Tarasova, V. N., Obabko, R. P., Himelbrant, D. E., et al. (2017). Diversity and distribution of epiphytic lichens and bryophytes on aspen (*Populus tremula*) in the middle boreal forests of Republic of Karelia (Russia). *Folia Cryptogamica Estonica, 54*, 125–141. https://doi.org/10.12697/fce.2017.54.16.

Tibell, L. (1999). Caliciales. In T. Athi, P. M. Jørgensen, H. Kristinsson, R. Moberg, U. Søchting, & G. Thor (Eds.), *Nordic lichen flora. Vol. 1: Introduction.* Calicioid lichens and fungi 1 (pp. 20–71). Naturcentrum AB.

Timoney, K. P., & Robinson, A. L. (1996). Old-growth white spruce and balsam poplar forests of the Peace River Lowlands, Wood Buffalo National Park, Canada: Development, structure, and diversity. *Forest Ecology and Management, 81*, 179–196. https://doi.org/10.1016/0378-1127(95)03645-8.

Tinnin, R. O., Hawksworth, F. G., & Knutson, D. M. (1982). Witches' broom formation in conifers infected by *Arceuthobium* spp.: An example of parasitic impact upon community dynamics. *American Midland Naturalist, 107*, 351–359. https://doi.org/10.2307/2425385.

Titov, A. N. (2006). *Mycocalicioid fungi (the order Mycocaliciales) of Holarctic.* Moscow: KMK.

Venier, L. A., Dalley, K., Goulet, P., et al. (2015). Benefits of aggregate green tree retention to boreal forest birds. *Forest Ecology and Management, 343*, 80–87. https://doi.org/10.1016/j.foreco.2015.01.024.

Villard, M. A., & Foppen, R. (2018). Ecological adaptations of birds to forest environments. In G. Mikusiński, J.-M. Roberge, & R. J. Fuller (Eds.), *Ecology and conservation of forest birds.* Ecology, biodiversity and conservation series. Cambridge University Press.

Villén-Peréz, S., Heikkinen, J., Salemaa, M., et al. (2020). Global warming will affect the maximum potential abundance of boreal plant species. *Ecography, 43*, 801–811. https://doi.org/10.1111/ecog.04720.

Virkkala, R. (2016). Long-term decline of southern boreal forest birds: Consequence of habitat alteration or climate change? *Biodiversity and Conservation, 25*, 151–167. https://doi.org/10.1007/s10531-015-1043-0.

Wesołowski, T., & Fuller, R. (2012). Spatial variation and temporal shifts in habitat use by birds at the European scale. In R. Fuller (Ed.), *Birds and habitat: Relationships in changing landscapes* (pp. 63–92). Cambridge: Cambridge University Press.

Wesołowski, T., & Wierzcholska, S. (2018). Tits as bryologists: Patterns of bryophyte use in nests of three species cohabiting a primeval forest. *Journal of Ornithology, 159*, 733–745. https://doi.org/10.1007/s10336-018-1535-2.

Wiklund, K., & Rydin, H. (2004). Ecophysiological constraints on spore establishment in bryophytes. *Functional Ecology, 18*, 907–913. https://doi.org/10.1111/j.0269-8463.2004.00906.x.

Winter, S., & Möller, G. C. (2008). Microhabitats in lowland beech forests as monitoring tool for nature conservation. *Forest Ecology and Management, 255*, 1251–1261. https://doi.org/10.1016/j.foreco.2007.10.029.

Chapter 6
Deadwood Biodiversity

Therese Löfroth, Tone Birkemoe, Ekaterina Shorohova, Mats Dynesius,
Nicole J. Fenton, Pierre Drapeau, and Junior A. Tremblay

Abstract Deadwood is a key component for biodiversity and ecosystem services in
boreal forests; however, the abundance of this critical element is declining worldwide.
In natural forests, deadwood is produced by tree death due to physical disturbances,
senescence, or pathogens. Timber harvesting, fire suppression, and salvage logging
reduce deadwood abundance and diversity, and climate change is expected to bring
further modifications. Although the effects of these changes are not yet fully understood, restoring a continuous supply of deadwood in boreal forest ecosystems is
vital to reverse the negative trends in species richness and distribution. Increasing
the availability of deadwood offers a path to building resilient forest ecosystems for
the future.

T. Löfroth (✉) · M. Dynesius
Department of Wildlife, Fish, and Environmental Studies, Swedish University of Agricultural
Sciences, 901 83 Umeå, Sweden
e-mail: therese.lofroth@slu.se

M. Dynesius
e-mail: mats.dynesius@slu.se

T. Birkemoe
Faculty of Environmental Sciences and Natural Resource Management, Norwegian University of
Life Sciences, Ås, Norway
e-mail: tone.birkemoe@nmbu.no

E. Shorohova
Forest Research Institute of the Karelian Research Center, Russian Academy of Science,
Pushkinskaya str. 11, Petrozavodsk 185910, Russia
e-mail: shorohova@es13334.spb.edu; ekaterina.shorokhova@luke.fi

Saint Petersburg State Forest Technical University, Institutsky str. 5, Saint Petersburg 194021,
Russia

Natural Resources Institute Finland (Luke), Latokartanonkaari 9, 00790 Helsinki, Finland

N. J. Fenton
Forest Research Institute, Université du Québec en Abitibi-Témiscamingue, 445 boul. de
l'Université, Rouyn-Noranda, Québec, QC J9X 5E4, Canada
e-mail: nicole.fenton@uqat.ca

© The Author(s) 2023
M. M. Girona et al. (eds.), *Boreal Forests in the Face of Climate Change*,
Advances in Global Change Research 74,
https://doi.org/10.1007/978-3-031-15988-6_6

6.1 Introduction

Deadwood, which includes standing dead trees, stumps, and downed logs in both terrestrial and aquatic habitats, is an important driver of biodiversity in boreal forests (Fig. 6.1; Thorn et al., 2020b). Deadwood abundance and its composition—characterized by deadwood diameter, decay class, tree species, and position—influence the diversity and abundance of a variety of organisms, including bryophytes, lichens, fungi, beetles, birds, and mammals (Fig. 6.2; Stokland et al., 2012). Saproxylic species live in and/or feed on deadwood for at least some part of their life cycle. They use deadwood as a direct or indirect food source (e.g., herbivores, detrivores, fungivores, predators, parasitoids) and/or as a nesting site or shelter. Epixylic species, such as bryophytes and lichens, live on the deadwood surface, and tree seedlings often establish on decomposing downed logs (Stokland et al., 2012). The ecosystem services that deadwood-associated organisms provide, e.g., decomposition, nutrient turnover, and pollination, make them an integral component of the boreal food web (Harmon, 2021; Müller et al., 2020). The extensive variability in deadwood-related habitats favors a high diversity of specialized species and intricate species interactions. Consequently, any anthropogenic disturbance that changes the abundance and diversity of deadwood alters this biodiversity and ecosystem functioning. The main factor currently influencing deadwood abundance in boreal forests is large-scale intensive forestry, including biofuel harvesting (Hof et al., 2018), although the influence of climate change is also growing (Cadieux et al., 2020; Tremblay et al., 2018). In this chapter, we first review deadwood characteristics and dynamics across the boreal biome. We then provide a brief overview of the various groups of organisms associated with the specific forms of deadwood in terrestrial and aquatic habitats. Finally, we examine the anthropogenic factors, including forestry and climate change, that alter deadwood forms and abundance in boreal forests.

N. J. Fenton · J. A. Tremblay
Centre for Forest Research, Université du Québec à Montréal, P.O. Box 8888, Stn. Centre-Ville, Montréal, QC H3C 3P8, Canada

P. Drapeau
Centre for Forest Research and UQAT-UQAM Chair in Sustainable Forest Management, Université du Québec à Montréal, P.O. Box 8888, Stn. Centre-Ville, Montréal, QC H3C 3P8, Canada
e-mail: drapeau.pierre@uqam.ca

J. A. Tremblay
Wildlife Research Division, Environment and Climate Change, 801-1550, avenue d'Estimauville, Québec, QC G1J 0C3, Canada
e-mail: junior.tremblay@ec.gc.ca

Faculté de Foresterie, Géographie et Géomatique, Université Laval, Pavillon Abitibi-Price, 2405 rue de la terrasse, Québec, QC G1V 0A6, Canada

Fig. 6.1 Examples of deadwood types in boreal forests; **a** kelo tree, **b** fallen tree, **c** dead top caused by the fungus *Cronartium flaccidum* or *Peridermium pini*. *Photo credits* **a**, **c** Ekaterina Shorohova, **b** Therese Löfroth

6.2 Deadwood Composition and Dynamics in Natural Forests

The volume and diversity of deadwood vary greatly with site productivity, tree species composition, forest age, and disturbance history (Table 6.1; Martin et al., 2018; Shorohova & Kapitsa, 2015). High productivity sites, producing more and larger trees, also produce more abundant and larger deadwood (Shorohova & Kapitsa, 2015). Tree species vary in size, wood quality, and dominant mortality mode (i.e., uprooting, decline, or stem breakage), eventually resulting in different types of deadwood (Müller et al., 2020). Special deadwood qualities are formed from injured and slow-growing trees that form dense and resin-rich wood (Fig. 6.1). Deadwood dynamics include the generation and loss (e.g., through combustion, decomposition, and overgrowth by vegetation) of deadwood. The cumulative effects of these processes, adding and removing deadwood, are reflected in the deadwood volume and composition in a given area (Fig. 6.3).

Late-seral and post-disturbance forests are the two most deadwood-rich habitats in natural boreal forest landscapes (Siitonen, 2001; Stokland et al., 2012). These stands are shaped by small-scale mortality processes that provide a relatively constant recruitment of recently dead trees (Aakala et al., 2008; Boulanger & Sirois, 2006),

Fig. 6.2 Many species depend on deadwood for larval development, foraging, or nesting. Examples of the deadwood-dependent species include **a** the buprestid beetle (*Chalcophora mariana*) that depends on large pine trunks in sun-exposed habitats, **b** the Black-backed Woodpecker (*Picoides arcticus*), nesting in a dead tree, and **c** the Pileated Woodpecker (*Dryocopus pileatus*) nesting in a trembling aspen (*Populus tremuloides*). *Photo credits* **a** Kristina Viklund, **b** David Tremblay, **c** Réjean Deschênes

which have been identified as critical ecological attributes to many support specialist forest-related species (Martin et al., 2020, 2021). Larger-scale disturbances concentrated in time and space, such as fire, storm felling, and insect outbreaks, produce large pulses of deadwood (Bergeron et al., 2004; Taylor & MacLean, 2007). Following these pulses, deadwood volume decreases over the next 50–100 years before gradually increasing, although to a much lower level, when either these stands reach maturity (Harmon, 2021) or another disturbance occurs (Fig. 6.4). After intense disturbances in late-seral boreal forests, deadwood volumes might exceed hundreds of $m^3 \cdot ha^{-1}$ and average 210 $m^3 \cdot ha^{-1}$ (Table 6.1; Shorohova & Kapitsa, 2015). Less severe disturbances also affect deadwood quality by injuring trees; these trees later produce tar-rich deadwood that, in turn, provides critical resources for specialized fungi, wood-boring arthropods, and avian and arthropod predators (Nappi et al., 2010).

Standing dead trees (*snags*) often constitute a significant part of the basal area of trees in natural boreal forests (Nilsson et al., 2002). Tree species differ in standing times after death, leading to variable snag dynamics across boreal forests (Aakala et al., 2008; Taylor & MacLean, 2007). In eastern Canada, Angers et al. (2011) found that the standing time of snags ranged from 15 years for trembling aspen (*Populus tremuloides*) to close to 30 years for jack pine (*Pinus banksiana*). In Europe, tree species such as Scots pine (*Pinus sylvestris*) can form barkless tar-rich snags having a characteristic hard and silvery-gray surface; these snags are known as *kelo trees* (Fig. 6.1). The formation and decay of kelo trees require centuries, and these snags

Table 6.1 Examples of estimated deadwood volumes across the boreal biome for various mature unmanaged forest types experiencing different disturbance regimes

Region	Forest type	Disturbance regime	Mean (± SE) total deadwood volume (m^3·ha^{-1})	Range (m^3·ha^{-1})	References
Northwestern Russia	Spruce-dominated	Wind/gap dynamics	147.7 ± 10.8	–	Shorohova and Kapitsa (2015)
Northwestern Russia	Pine-dominated	Fire	74.4 ± 13.1	–	Shorohova and Kapitsa (2015)
Western Canada	Mixedwood	Fire	76.1 ± 41.5 (SD)	3–93	Work et al. (2004)
Western Canada	Coniferous	Fire	93.9 ± 18.9 (SD)	–	Work et al. (2004)
Eastern Canada	Black spruce (*Picea mariana*)	Fire		3–155	Martin et al. (2018)
Eastern Canada	Mixedwood	Fire/Postfire succession		0–708	Hély et al. (2000)
Finland/western Russia	Scots pine (*Pinus sylvestris*)–dominated	Fire		117	Rouvinen and Kouki (2002)
Fennoscandia	Coniferous	Fire/gap dynamics		20–120	Siitonen (2001)
Finland and northwestern Russia	Norway spruce (*Picea abies*)	Gap dynamics	60	41–170	Aakala (2010)
Eastern Fennoscandia	Scots pine (*Pinus sylvestris*)	–	69.5	22.2–158.7	Karjalainen and Kuuluvainen (2002)
Eastern Canada	Black spruce (*Picea mariana*)	–	71.3 ± 11.2 (>90 y) 49.5 ± 14.2 (<90 y)	–	Tremblay et al. (2009)

form a distinct habitat for specialized wood fungi and lichens (Niemelä et al., 2002; Santaniello et al., 2017). The extremely slow recruitment of kelo trees and their suitability as firewood have made them a rarity in modern landscapes, and long-term conservation and restoration strategies are needed for these unique habitats (Kuuluvainen et al., 2017).

In natural forests, deadwood is lost through consumption by fire, decomposition by fungi, bacteria, and animals, and overgrowth by ground vegetation. Although forest fires create deadwood, they also consume existing pre-fire deadwood (Hyde et al.,

Fig. 6.3 Several natural processes impact deadwood dynamics, including wildfires, which add (by damaging living trees) and consume deadwood; **a** hollow trunk after fire, **b** overgrowth is an important process that incorporates deadwood into the soil. *Photo credits* Ekaterina Shorohova

2011). The nonfire decomposition rate of deadwood varies with climate, site conditions, tree species, deadwood size (Shorohova & Kapitsa, 2016), and the composition of the decomposer community (Bani et al., 2018). In addition, many wood attributes, including annual ring width, wood density, and chemical composition (e.g., resin content), affect the decomposition rate (Edman et al., 2006; Venäläinen et al., 2003). Finally, the burial of downed deadwood within the soil organic layer affects, for example, accessibility to deadwood for colonization by saproxylic insects and its utility as habitat for epixylic bryophytes. Burial is faster in sites with a soft organic layer, such as peat, and with fast-covering ground vegetation, such as vascular plants and *Sphagnum* mosses (Fig. 6.3; Dynesius et al., 2010). More than a quarter of the carbon originating from deadwood in boreal forests is estimated to be stored in buried, downed deadwood. Therefore, although this wood is no longer important for aboveground biodiversity, it continues to perform important ecosystem functions, such as nutrient cycling and carbon storage (Stokland et al., 2016). Deadwood burial is affected by several factors that can be altered by climate change, such as microclimate and the depth of the organic layer (Dynesius et al., 2010; Stokland et al., 2016).

Fig. 6.4 Deadwood dynamics for forests under even-aged management, after a stand-replacing fire in natural forest, and in a late-seral natural forest

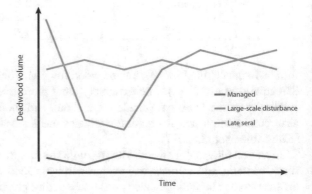

Burial also interacts with decomposition, as buried deadwood typically decomposes at a much slower rate than aboveground deadwood (Stokland et al., 2016).

6.3 Deadwood Substrates and Their Associated Biodiversity

Deadwood provides numerous ecological niches, and a multitude of species interact within the deadwood food web (Fig. 6.5). Species assemblage composition and richness in deadwood are affected by tree species, sun exposure, decay stage, wood density and diameter, type of rot, cause of death, and whether the stem is standing, lying, charred, or in contact with the ground (Hägglund & Hjältén, 2018; Johansson et al., 2017; Kushnevskaya & Shorohova, 2018).

A well-known example of niche differentiation linked to deadwood diameter is bark beetles selecting wood on the basis of bark thickness. For example, the emerald ash borer (*Agrilus planipennis*) attacks ash trees that have a stem diameter within a limited range (Timms et al., 2006), and the six-toothed beetle (*Ips sexdentatus*) is restricted to the base of old large-diameter pines (Gilbert et al., 2005). Some saproxylic organisms function as keystone species, e.g., the primary spruce bark beetle (*Ips typographus*), which affects more than 100 associated species (Weslien, 1992). Several invertebrate species use deadwood primarily as nesting sites. Solitary bees and wasps dig their nest tunnels into soft decaying wood or use tunnels made by other insects. Social wasps and honeybees build their nests in hollow trees, and ants

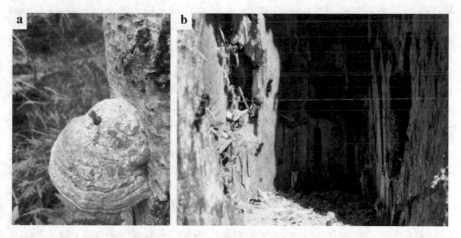

Fig. 6.5 Species interactions are an essential part of deadwood dynamics. **a** The lepidopteran *Scardia boletella* (Tineidae) seen here recently hatched from the saproxylic tinder fungus (*Fomes fomentarius*). *Scardia boletella* is vulnerable to forest management and is listed as endangered in Norway. **b** Carpenter ants (*Camponotus* sp.) in a dead balsam fir (*Abies balsamea*), here excavated by the Pileated Woodpecker (*Dryocopus pileatus*). *Photo credits* **a** Tone Birkemoe, **b** Pierre Drapeau

(e.g., *Camponotus* spp.) excavate their nests into living or dead trees (King et al., 2018; Westerfelt et al., 2015). Many dipterans use emergence holes and galleries from other insects as nest sites (Dennis et al., 2018).

Below we describe the biodiversity associated with three types of fully dead trees: snags, logs, and underwater deadwood.

6.3.1   Dead Standing Trees (Snags)

Lichens thrive on snags and often on decorticated stems, such as kelo trees. In Fennoscandia and the Pacific Northwest of North America, more than 40% of the analyzed lichen species occurred on deadwood, and 10% were found solely on dead-wood (Spribille et al., 2008). The pin lichens (calicioid lichens) are the most special-ized of the wood-living lichens, and for many species in this group, kelo trees consti-tute the main substrate (Santaniello et al., 2017). The specific lichen flora varies with snag decay, and there is a distinct shift in assemblage makeup, from species that colonized before tree death to species colonizing the barkless surface after bark loss. After bark loss, the tree species becomes less important as a factor influencing the lichen flora (Lõhmus & Lõhmus, 2001).

The decay processes of standing deadwood are intrinsically related to saproxylic insect activity (Siitonen, 2001), particularly for wood-feeding species that colonize dying and dead trees. Saint-Germain et al. (2007) found that wood-feeding insects were most abundant in black spruce at the beginning of the decay sequence on fresh snags, whereas they observed opposite wood-feeding insect patterns in aspen, as insects reached large numbers in snags at the middle- to late-decay stages. These results highlight the importance of considering the entire range of decay classes of standing deadwood for conservation planning in managed forest landscapes. In a study of the substrate requirements of red-listed saproxylic invertebrates in Sweden, Jonsell et al. (1998) found that a high proportion of these species require sun-exposed deadwood, of which snags are a significant proportion. Typically, snags are inhabited by species that thrive in drier and more exposed habitats (Hjältén et al., 2012); an example is the beetle *Peltis grossa* that thrives in snags within clear-cuts (Weslien et al., 2011). Hence, leaving standing, sun-exposed deadwood in clear-cuts could be an effective means of increasing the breeding substrates for saproxylic invertebrates; however, such retention strategies may not be adequate for species living in late-seral forests that depend on shaded sites.

Insects colonizing snags and dying trees are critical food resources for wood-peckers (Hammond & Theimer, 2020). Forest stands after natural disturbances often represent significant foraging opportunities for these bird species; this includes burned areas (Nappi et al., 2010; Versluijs et al., 2020) and forest stands affected by insect outbreaks (Rota et al., 2015). However, landscapes characterized by high amounts of late-seral forest are also important and provide snags continuously over time (Martin et al., 2021; Nappi et al., 2015). Individual woodpecker species specialize in specific decay stages of dying and dead trees (Hammond & Theimer,

2020), again underlining the importance of a continuous input of standing deadwood to ensure a steady supply of suitable foraging trees for the woodpecker community (Nappi et al., 2015).

Dying trees and snags provide cavities for nesting, roosting, and denning for 10–40% of species of birds (including the families Picidae and Sittidae and the order Stringiformes) and mammals (from the order Microchiroptera and the families Mustelidae and Sciuridae) in forest ecosystems (Kotowska et al., 2020; Parsons et al., 2003). In boreal forests, most cavities are produced by avian excavators, mostly woodpeckers, whereas few cavities originate through natural tree-decay processes (Wesołowski & Martin, 2018). At least 1878 species worldwide (18.1% of all bird species in the world) nest in tree cavities, and at least 338 of these species use cavities created by woodpeckers (Picidae) (van der Hoek et al., 2017). Cavity-using communities form interspecific hierarchical networks called nest webs (Martin & Eadie, 1999), where cavity-bearing tree species, cavity-producing agents (excavators and decay processes), and nonexcavating cavity users interact. Cavities are created every year, reused over time, change as they age (Edworthy et al., 2018), and are formed both in living trees showing signs of decay and in dead trees (Drapeau et al., 2009; Edworthy et al., 2018). Aspens (*Populus* spp.) are particularly important cavity-bearing trees—either natural or excavated cavities—in both North American and European boreal forests (Andersson et al., 2018; Parsons et al., 2003).

6.3.2 Downed Deadwood (Logs)

The characteristics of lying and standing deadwood differ. Downed deadwood is generally moister because it has more extensive ground contact and is shaded; sun-exposed logs also occur, particularly after a severe disturbance. Logs contribute to the structural diversity of the forest floor and provide nest sites, food, and cover for both mammals and amphibians (Fauteux et al., 2012). Because downed deadwood in boreal areas is hidden in winter under snow, its importance as foraging sites for woodpeckers and other birds is more limited than for snags; however, downed deadwood is still used, and its use is likely underestimated (Tremblay et al., 2010).

Large-diameter logs host a higher number of species (and specialized species) than small-diameter logs (Juutilainen et al., 2011). The larger, longer-lasting, and more varied deadwood habitat of larger logs partially explains this difference, often offering larger proportions of heartwood, an important habitat for some species. For example, the hairy pine borer beetle (*Tragosoma depsarium*) inhabits pine logs larger than 25 cm in diameter having large proportions of heartwood and a slow decay rate (Wikars, 2004). The polypore fungus *Fomitopsis rosea* also occurs more frequently in larger logs and is favored by the higher wood density of slow-growing trees (Edman et al., 2006). However, in old-growth forests having a high availability of deadwood, the total log volume per hectare rather than log size may be more critical for species diversity and composition, as demonstrated for mosses, liverworts, and lichens growing on logs (Kushnevskaya & Shorohova, 2018).

The saproxylic insects of logs include several feeding guilds that shift in dominance as the log decomposes. First, the cambium consumers and their associated predators and parasitoids colonize the freshly fallen tree, followed after a few years by wood borers, which feed on the wood, and fungivores (Gibb et al., 2013). Species feeding on the short-lived but nutritious cambium have short generation times, whereas larger wood-boring taxa have life cycles lasting 3–15 years, e.g., *Tragosoma depsarium*, *Pytho kolwensis*, and *Chalcophora mariana* (Fig. 6.2; Siitonen & Saaristo, 2000; Wikars, 2004).

Logs are essential hibernation sites for many epigeic and litter-dwelling arthropods, e.g., carabid beetles and mollusks. For wood ants and small mammals, logs can also serve as pathways when foraging (Boucher et al., 2015; Westerfelt et al., 2015).

Logs harbor a rich flora of wood-decaying fungi. There is a turnover of fungal species during log decomposition, as wood density and C:N ratios decrease and moisture and lignin contents increase (Rajala et al., 2011). In unmanaged boreal forests, Ascomycetes colonize recently fallen spruce logs, whereas Basidiomycete fungi—responsible for brown rot—peak during the intermediate decay stages. White rot fungi constitute approximately one-fifth of all fungal species in decomposing logs, except at the latest decay stages when ectomycorrhizal fungi become dominant (Rajala et al., 2012). Bark attached to logs also hosts diverse fungal communities that vary during decomposition (Kazartsev et al., 2018).

Many boreal bryophytes (i.e., mosses and liverworts) grow on fallen logs, and logs are often bryophyte biodiversity hot spots. Several ecological groups of deadwood-associated bryophytes can be distinguished (Kushnevskaya et al., 2007). *Facultative epiphytes* grow on the lowest parts of living tree trunks and also colonize other parts of fallen logs until the midstages of decay (e.g., *Ptilidium pulcherrimum*). *Epixylic specialists* grow mainly on logs and stumps. Some species (e.g., *Lophozia ciliata*) colonize the bark, whereas others colonize softened barkless logs (e.g., *Crossocalyx hellerianus*, *Riccardia palmata*, and *Lophocolea heterophylla*). *Opportunistic generalists* colonize at any stage of decay, and *epigeic species* normally cover the forest floor but overgrow the logs as they decay (Dynesius et al., 2010; Kushnevskaya & Shorohova, 2018).

6.3.3 Deadwood in Water

Deadwood has several vital ecological functions in aquatic environments. Deadwood alters river flow and serves as habitat for fish. In lakes, woody debris in the littoral zone has proven important for prey fish abundance and predatory fish growth potential (Ahrenstorff et al., 2009). Driftwood, i.e., stranded deadwood, harbors a rich fungal flora and contributes significantly to the deadwood biodiversity in coastal regions (Blanchette et al., 2016). In forest ecosystems, deadwood can link terrestrial and aquatic habitats. A key species here is the beaver, which creates substantial amounts of deadwood both directly by felling trees and indirectly by flooding forests. The deadwood in water includes rare deadwood types (e.g., aspen in Fennoscandia) in areas otherwise rarely affected by stand-replacing disturbances (Thompson et al., 2016).

6.4   Species Interactions in Deadwood

Insects and fungi are the most species-rich taxa in deadwood. Coexisting at least since the late Silurian (Misof et al., 2014; Sherwood-Pike & Gray, 1985), this tight coupling has resulted in reciprocal adaptations and intricate interactions with profound impacts on deadwood dynamics; however, relatively little is known about these interactions (Birkemoe et al., 2018), possibly because of the cryptic nature of fungi. One important and direct interaction is through feeding (Fig. 6.5a), which is likely to have a significant functional importance for biodiversity and deadwood decomposition. Fungi live on insects as parasites, pathogens, and mutualists; insects feed on various forms of fungi, including yeast cells, mycelia, and fruiting bodies. Many insect species feed directly on fungi, and adding fungi to the diet might be essential for beetles that feed primarily on wood, which has low nutritional value. For instance, it has been calculated that the longhorn beetle *Stictoleptura rubra* would require 40–85 years to reach adulthood if all its nutrients were obtained from wood, or 13–28 times longer than its maximum recorded life-cycle length (Filipiak & Weiner, 2014).

Bark and ambrosia beetles bring their mutualistic fungi to the colonized trees; however, recent studies indicate that insects also disperse nonmutualistic wood-decaying fungi (Jacobsen et al., 2018; Seibold et al., 2019). A study identifying wood-living fungi from beetles landing on recently cut wood showed that the networks between beetles and fungi were comparable in strength to seed dispersal networks (Jacobsen et al., 2018) and thus of potential importance for deadwood biodiversity. However, the study must be replicated in other systems to determine whether these findings can be generalized.

Insects also farm fungi, as observed in termites and several ambrosia beetles. Conversely, some fungi protect insects by reinforcing nest-wall structures (Schlick-Steiner et al., 2008), fighting microbial pathogens (Flórez et al., 2015), or degrading tree defenses that would otherwise be detrimental to insects. Indirect interactions, where fungi or insects modify the deadwood habitat, could also significantly affect the insect/fungal communities and their functions.

In addition to the interactions between insects and fungi, multiple interactions exist within these two highly diverse groups. Among insects, predator/prey interactions are important for regulating populations. For instance, the bark beetle *Tomicus piniperda* produces fewer offspring when predators are present in high numbers (Schroeder & Weslien, 1994). Fungi in deadwood live in constant chemical warfare with other fungi (Hiscox et al., 2018), and the war zones can be observed as dark lines in deadwood. Many fungi also feed on other fungi (Maurice et al., 2021). Insects also compete, facilitate colonization, and produce priority effects, i.e., when initially colonizing species determine what species can colonize later. Priority effects have been observed for insects and fungi in the succession and colonization of deadwood (Weslien et al., 2011).

Various other taxa also interact within deadwood. Bacteria is an often overlooked but vital part of the deadwood community. Bacteria interact with wood-decaying fungi (Johnston et al., 2016), affecting the wood decay rate and fungal species composition. Recently, nematode parasites of insects were also found to modify the wood decay rate (Davis & Prouty, 2019). Larger organisms, such as shrews, mice, and woodpeckers, forage for invertebrates in deadwood.

6.5 Forest Management Impact on the Deadwood Profile

Forest management significantly impacts the abundance and diversity of deadwood in boreal forests. Even-aged management, converting deadwood-rich, uneven-aged mature, and old forests into even-aged stands, remains the most common forestry harvesting approach. Furthermore, forest management promotes fast-growing healthy trees and reduces the abundance of slow-growing, injured, and unhealthy trees. Consequently, forest management has caused a decline of many boreal deadwood-associated species (Siitonen, 2001). Deadwood species are generally lower in managed forests than in natural ones, and harvesting intensity and time amplify these differences (Junninen et al., 2006). Consequently, regions with a long forest management history have smaller populations of specialized deadwood-associated species found in fewer sites (Müller et al., 2013; Nordén et al., 2013). For example, the amount of deadwood in intensively managed forests in Fennoscandia is considerably lower (4–10 $m^3 \cdot ha^{-1}$) than that found in natural stands (Fig. 6.4) (for volumes in natural stands, see Table 6.1). Timber harvesting by thinning and final felling has caused this decline, exacerbated by the more recent practice of extracting woody debris left after harvesting for biofuels (Hof et al., 2018). In contrast, less intensive management, in which stands are allowed to self-thin during development and smaller pieces of deadwood generated during harvest are left on-site, helps reduce declines in deadwood abundance. Thus, significant volumes of deadwood have been documented in Russian managed forests, with an average of 28.0 $m^3 \cdot ha^{-1}$ (Malysheva et al., 2019) and ranging from 1 to more than 100 $m^3 \cdot ha^{-1}$ in the Novgorod region (Shorohova & Tetioukhin, 2003). Forestry also changes the diameter and decay stage distributions of deadwood and disrupts the recruitment of new large deadwood items (Martin et al., 2021). In unmanaged forests, large stems often constitute the majority of deadwood volume. For example, in spruce-dominated boreal old-growth forest, large-diameter (>30 cm) dead trees can comprise 42–54% of the volume, whereas smaller-diameter stems (<10 cm) represent only 1.7–2.7% (Nilsson et al., 2002; Siitonen, 2001); this pattern is reversed in younger to middle-aged managed forests (Stenbacka et al., 2010).

6.6 Effects of Climate Change on Deadwood and Associated Biodiversity

Timber harvesting has disfavored deadwood-associated species for decades if not centuries, particularly in Europe (Stokland et al., 2012). More recently, climate change has put additional stress on forest ecosystems and their biodiversity, as a changing climate alters nutrient cycles and disturbance regimes (Tremblay et al., 2018; Venäläinen et al., 2020). Climate change mitigation may also impact forestry practices; for example, logging and biofuel harvesting may increase to substitute for fossil-based products.

Changes in nutrient cycles and disturbance regimes will influence deadwood abundance and diversity in various ways. In boreal areas, forests are expected to grow either faster or slower depending on site-specific conditions (Marchand et al., 2019; Miquelajauregui et al., 2019), altering the input of deadwood and deadwood quality, e.g., fast growth versus slow growth. The decomposition rate is also expected to increase in locations where temperature is currently a limiting factor (Davidson & Janssens, 2006); however, the effect on the decomposition rate will largely depend on local habitat factors (Bradford et al., 2014).

Natural disturbances (fire, windthrow, and insect outbreak) are generally expected to accelerate in the future, albeit showing large geographic variability (Chap. 3); this may limit the long-term development of deadwood in some areas, as trees may not have sufficient time to grow large before another severe disturbance strikes (Kuulu-vainen & Gauthier, 2018; Seidl et al., 2020). Moreover, the projected reduction of large deadwood because of climate change will be exacerbated by forest management policies having caused a skewed age-class distribution with a low proportion of old forest (Berglund & Kuuluvainen, 2021; Lamarre & Tremblay, 2021). Shorter distur bance return intervals may accelerate the development toward higher proportions of young stands and small-diameter deadwood.

These changes in deadwood abundance, type, and diversity affect deadwood-associated species in multiple manners. First, the quantity of insects that emerge from a burned tree is proportional to tree diameter (Saint-Germain et al., 2008), resulting in cascading effects on predators that depend on this resource, e.g., the Black-backed Woodpecker (*Picoides arcticus*) (Nappi & Drapeau, 2009; Nappi et al., 2010; Tremblay et al., 2020). These effects contribute to a projected decline of up to 92% in the potential productivity of the Black-backed Woodpecker under the worst-case climate forcing scenario (RCP8.5) (Tremblay et al., 2018). Second, a lower abundance of large trees related to an increased fire frequency would reduce the area of forests housing large standing or downed deadwood stems, which are typically associated with a high species diversity of bryophytes and lichens (Dittrich et al., 2014).

Climate change will also alter the distribution and phenology of wood-inhabiting species, affecting interactions and food chains. Higher temperatures and a longer growing season will affect insect phenology. Depending on the length of the growing season, the European spruce bark beetle (*Ips typographus*) can produce one to six generations per year, and the area prone to outbreaks of this species is expanding northward (Romashkin et al., 2020). Likewise, warmer winters in western North America have contributed to continuous outbreaks of the mountain pine beetle (*Dendroctonus ponderosae*) on lodgepole pine (*Pinus contorta*) and other pine species, turning the forests from carbon sinks into carbon sources (Kurz et al., 2008). Indirect consequences of climate change also include potential effects on insect-fungus interactions. For example, phenological mismatches (i.e., relevant species life stages no longer co-occur) may emerge between insects and fungi. Insect visits of fungal fruiting bodies, and thus spore dispersal, may be disrupted given that sporulation is generally determined by environmental cues other than insect emergence and flight. Similarly, controls on populations may be disrupted if predators and their prey no longer co-occur (Ekholm et al., 2020). Such phenological mismatches will add uncertainty to the current difficulty in predicting insect outbreaks, including significant uncertainties in terms of outbreak duration, intensity, and spatial variation (Biedermann et al., 2019; Boulanger et al., 2016).

6.7 Conclusions and Future Perspectives

Deadwood profiles, the frequencies and severity of natural disturbances, and management history vary across the boreal zone. Deadwood abundance and quality have decreased dramatically in many managed areas, especially in northern Europe, and deadwood-associated biodiversity has declined accordingly. Moreover, climate change will likely affect the formation and dynamics of deadwood to produce concomitant effects on deadwood-associated organisms and the intricate interactions and networks associated with this habitat.

A means of adapting to this massive challenge is implementing new silviculture approaches that mimic natural disturbance patterns and their effects on standing and fallen deadwood. The retention and maintenance of these biological legacies is a tenet of ecosystem-based management, which proposes a diversification of forestry practices (Gauthier et al., 2009). These approaches include longer rotations, retention forestry, continuous-cover silviculture, and enhanced patch retention of living and dead trees in clear-cuts (Felton et al., 2020). When included in forest landscape planning, such approaches are likely to attenuate habitat alteration and biodiversity loss associated with conventional forest management (Fig. 6.6; Berglund & Kuuluvainen, 2021; Drapeau et al., 2016). For deadwood management, this will necessarily require incorporating baseline data on deadwood dynamics and recruitment and the biodiversity it supports (Tremblay et al., 2015). Given that deadwood dynamics—including recruitment and decay processes—are tree species–specific and support taxonomic and functional diversity, live and deadwood retention strategies will have to be flexible in regard to the dominant tree species or forest cover types under management (Angers et al., 2011). Careful and rigorous planning in managed boreal forest landscapes is thus vital to account for a wide range of tree ages and sizes, species, and deadwood decay stages to ensure a steady supply of suitable substrates for biodiversity (Drapeau et al., 2009; Edworthy et al., 2018).

In severely impoverished areas, ecological restoration is needed to maintain deadwood-associated biodiversity. Prescribed burning, tree felling, pushing over, and girdling, together with green tree retention, have already proven successful (Hägglund & Hjältén, 2018; Hägglund et al., 2020), and future research must further evaluate the effects of such efforts on biodiversity (Fig. 6.6). It is possible to restore some deadwood qualities such as intermediate size classes and early decay stages in the short term. Recent reviews (Koivula & Vanha-Majamaa, 2020; Sandström et al., 2019) show that the artificial addition of deadwood supports a wide range of saproxylic species; however, the species composition on artificially created wood differs from communities in trees that died naturally. Some species require the active creation and conservation of their specific habitats, such as thick-diameter deadwood, slow-grown wood, resin-rich wood, wood from injured trees, and wood in forests with continuous canopy cover. For substrates and qualities that require a long time to regenerate (e.g., large logs of late-decay stage, wood from old trees), it is essential to conserve what is left, but also to implement artificial aging, for example by partial bark removal.

Conflicts between deadwood restoration and pest and fire management can occur because deadwood is often regarded as a source of pest species and wildfire fuel. Salvage logging may lead to ecological traps, i.e., species are attracted to a habitat that is too degraded for their survival (Hale & Swearer, 2016), for saproxylic organisms on burned areas, and pest control can reduce the recruitment of deadwood (Thorn et al., 2020a, 2020b). More research is needed to better balance these conflicting goals in an era of climate change.

Management practices that ensure the continuous availability of deadwood in managed boreal forests require monitoring and modeling. Modeling the potential of different forest types to produce and maintain deadwood could be a means forward,

Fig. 6.6 In heavily managed landscapes, the restoration of deadwood may be necessary. Restoration examples include **a** created high stumps and retention trees at a clear-felling site and **b** the scouring of tree stems to reduce tree vitality. *Photo credits* **a** Therese Löfroth, **b** Joakim Hjältén

and accurate models are vital for reliable estimates of deadwood volume and production in both natural and managed stands (e.g., Mikkonen et al., 2020). To conclude, both forestry and climate change are interactive challenges for conserving the biodiversity of deadwood-associated species (Tremblay et al., 2018). Conservation and restoration efforts must be designed appropriately to provide a continuous supply of highly variable forms of deadwood.

References

Aakala, T. (2010). Coarse woody debris in late-successional *Picea abies* forests in northern Europe: Variability in quantities and models of decay class dynamics. *Forest Ecology and Management*, *260*, 770–779. https://doi.org/10.1016/j.foreco.2010.05.035.

Aakala, T., Kuuluvainen, T., Gauthier, S., et al. (2008). Standing dead tree and their decay-class dynamics in the northeastern boreal old-growth forests of Quebec. *Forest Ecology and Management*, *255*, 410–420. https://doi.org/10.1016/j.foreco.2007.09.008.

Ahrenstorff, T. D., Sass, G. G., Helmus, M. R. (2009). The influence of littoral zone coarse woody habitat on home range size, spatial distribution, and feeding ecology of Largemouth Bass (*Micropterus salmoides*). *Hydrobiologia*, *623*, 223–233. https://doi.org/10.1007/s10750-008-9660-1.

Andersson, J., Domingo Gomez, E., Michon, S., et al. (2018). Tree cavity densities and characteristics in managed and unmanaged Swedish boreal forest. *Scandinavian Journal of Forest Research*, *33*, 233–244. https://doi.org/10.1080/02827581.2017.1360389.

Angers, V. A., Gauthier, S., Drapeau, P., et al. (2011). Tree mortality and snag dynamics in North American boreal tree species after a wildfire: A long-term study. *International Journal of Wildland Fire*, *20*, 751–763. https://doi.org/10.1071/WF10010.

Bani, A., Pioli, S., Ventura, M., et al. (2018) The role of microbial community in the decomposition of leaf litter and deadwood. *Applied Soil Ecology*, *126*, 75–84. https://doi.org/10.1016/j.apsoil.2018.02.017.

Bergeron, Y., Flannigan, M., Gauthier, S., et al. (2004). Past, current and future fire frequency in the Canadian boreal forest: Implications for sustainable forest management. *Ambio*, *33*(6), 356–360. https://doi.org/10.1579/0044-7447-33.6.356.

Berglund, H., & Kuuluvainen, T. (2021). Representative boreal forest habitats in northern Europe, and a revised model for ecosystem management and biodiversity conservation. *Ambio, 50*, 1003–1017. https://doi.org/10.1007/s13280-020-01444-3.

Biedermann, P. H. W., Müller, J., Grégoire, J.-C., et al. (2019). Bark beetle population dynamics in the Anthropocene: Challenges and solutions. *Trends in Ecology & Evolution*, *34*(10), 914–924. https://doi.org/10.1016/j.tree.2019.06.002.

Birkemoe, T., Jacobsen, R. M., Sverdrup-Thygeson, A., et al. (2018). Insect-fungus interactions in dead wood systems. In M. D. Ulyshen (Ed.), *Saproxylic insects: Diversity, ecology and conservation* (pp. 377–427). Cham: Springer International Publishing.

Blanchette, R. A., Held, B. W., Hellmann, L., et al. (2016). Arctic driftwood reveals unexpectedly rich fungal diversity. *Fungal Ecology*, *23*, 58–65. https://doi.org/10.1016/j.funeco.2016.06.001.

Boucher, P., Hébert, C., Francoeur, A., et al. (2015). Postfire succession of ants (Hymenoptera: Formicidae) nesting in dead wood of northern boreal forest. *Environmental Entomology*, *44*, 1316–1327. https://doi.org/10.1093/ee/nvv109.

Boulanger, Y., & Sirois, L, (2006). Postfire dynamics of black spruce coarse woody debris in northern boreal forest of Quebec. *Canadian Journal of Forest Research*, *36*, 1770–1780. https://doi.org/10.1139/x06-070.

Boulanger. Y., Gray, D. R., Cooke, B. J., et al. (2016). Model-specification uncertainty in future forest pest outbreak. *Global Change Biology, 22,* 1595–1607. https://doi.org/10.1111/gcb.13142.

Bradford, M. A., Warren, II R. J., Baldrian, P., et al. (2014). Climate fails to predict wood decomposition at regional scales. *Nature Climate Change, 4,* 625–630. https://doi.org/10.1038/nclima te2251.

Cadieux, P., Boulanger, Y., Cyr, D., et al. (2020). Projected effects of climate change on boreal bird community accentuated by anthropogenic disturbances in western boreal forest, Canada. *Diversity and Distributions, 26,* 668–682. https://doi.org/10.1111/ddi.13057.

Davidson, E. A., & Janssens, I. A. (2006). Temperature sensitivity of soil carbon decomposition and feedbacks to climate change. *Nature, 440,* 165–173. https://doi.org/10.1038/nature04514.

Davis, A. K., & Prouty, C. (2019) The sicker the better: Nematode-infected passalus beetles provide enhanced ecosystem services. *Biology Letters, 15,* 20180842. https://doi.org/10.1098/rsbl.2018. 0842.

Dennis, R. W. J., Malcolm, J. R., Smith, S. M., et al. (2018). Response of saproxylic insect communities to logging history, tree species, stage of decay, and wood posture in the central Nearctic boreal forest. *Journal of Forestry Research,* 29:1365–1377. https://doi.org/10.1007/s11676-017-0543-z.

Dittrich, S., Jacob, M., Bade, C., et al. (2014). The significance of deadwood for total bryophyte, lichen, and vascular plant diversity in an old-growth spruce forest. *Plant Ecology, 215,* 1123–1137. https://doi.org/10.1007/s11258-014-0371-6.

Drapeau, P., Nappi, A., Imbeau, L., et al. (2009). Standing deadwood for keystone bird species in the eastern boreal forest: Managing for snag dynamics. *The Forestry Chronicle, 85,* 227–234. https://doi.org/10.5558/tfc85227-2.

Drapeau, P., Villard, M.-A., Leduc, A., et al. (2016). Natural disturbance regimes as templates for the response of bird species assemblages to contemporary forest management. *Diversity and Distributions, 22*(4), 385–399. https://doi.org/10.1111/ddi.12407.

Dynesius, M., Gibb, H., Hjältén, J. (2010). Surface covering of downed logs: Drivers of a neglected process in dead wood ecology. *PLoS ONE, 5,* e13237. https://doi.org/10.1371/journal.pone.001 3237.

Edman, M., Moller, R., & Ericson, L. (2006). Effects of enhanced tree growth rate on the decay capacities of three saprotrophic wood-fungi. *Forest Ecology and Management, 232,* 12–18. https://doi.org/10.1016/j.foreco.2006.05.001.

Edworthy, A. B., Trzcinski, M. K., Cockle, K. L., et al. (2018). Tree cavity occupancy by nesting vertebrates across cavity age. *The Journal of Wildlife Management, 82,* 639–648. https://doi.org/10.1002/jwmg.21398.

Ekholm, A., Tack, A. J. M., Pulkkinen, P., et al. (2020). Host plant phenology, insect outbreaks and herbivore communities-the importance of timing. *Journal of Animal Ecology, 89,* 829–841. https://doi.org/10.1111/1365-2656.13151.

Fauteux, D., Imbeau, L., Drapeau, P., et al. (2012). Small mammal responses to coarse woody debris distribution at different spatial scales in managed and unmanaged boreal forests. *Forest Ecology and Management, 266,* 194–205. https://doi.org/10.1016/j.foreco.2011.11.020.

Felton, A., Löfroth, T., Angelstam, P., et al. (2020). Keeping pace with forestry: Multi-scale conservation in a changing production forest matrix. *Ambio, 49,* 1050–1064. https://doi.org/10.1007/s13280-019-01248-0.

Filipiak, M., & Weiner, J. (2014). How to make a beetle out of wood: Multi-elemental stoichiometry of wood decay, xylophagy and fungivory. *PLoS ONE, 9,* e115104. https://doi.org/10.1371/journal. pone.0115104.

Flórez, L. V., Biedermann, P. H., Engl, T., et al. (2015). Defensive symbioses of animals with prokaryotic and eukaryotic microorganisms. *Natural Product Reports, 32,* 904–936. https://doi. org/10.1039/C5NP00010F.

Gauthier, S., Vaillancourt, M. A., Leduc, A., et al. (Eds.). (2009). *Ecosystem management in the boreal forest* (p. 572). Québec: Presses de l'Universite du Quebec.

Gibb, H., Johansson, T., Stenbacka, F., et al. (2013). Functional roles affect diversity-succession relationships for boreal beetles. *PLoS ONE, 8*, e72764. https://doi.org/10.1371/journal.pone.007 2764.

Gilbert, M., Nageleisen, L. M., Franklin, A., et al. (2005). Post-storm surveys reveal large-scale spatial patterns and influences of site factors, forest structure and diversity in endemic bark-beetle populations. *Landscape Ecology, 20*, 35–49. https://doi.org/10.1007/s10980-004-0465-y.

Hägglund, R., & Hjältén, J. (2018). Substrate specific restoration promotes saproxylic beetle diversity in boreal forest set-asides. *Forest Ecology and Management, 425*, 45–58. https://doi.org/10. 1016/j.foreco.2018.05.019.

Hägglund, R., Dynesius, M., Löfroth, T., et al. (2020). Restoration measures emulating natural disturbances alter beetle assemblages in boreal forest. *Forest Ecology and Management, 462*, 117934. https://doi.org/10.1016/j.foreco.2020.117934.

Hale, R., & Swearer, S. E. (2016). Ecological traps: Current evidence and future directions. *Proceedings of the Royal Society B: Biological Sciences, 283*(1824), 20152647. https://doi.org/10.1098/ rspb.2015.2647.

Hammond, R. L., & Theimer, T. C. (2020). A review of tree-scale foraging ecology of insectivorous bark-foraging woodpeckers in North America. *Forest Ecology and Management, 478*, 118516. https://doi.org/10.1016/j.foreco.2020.118516.

Harmon, M. E. (2021). The role of woody detritus in biogeochemical cycles: Past, present, and future. *Biogeochemistry, 154*, 349–369. https://doi.org/10.1007/s10533-020-00751-x.

Hély, C., Bergeron, Y., & Flannigan, M. D. (2000). Effects of stand composition on fire hazard in mixed-wood Canadian boreal forest. *Journal of Vegetation Science, 11*, 813–824. https://doi.org/ 10.2307/3236551.

Hiscox, J., O'Leary, J., & Boddy, L. (2018). Fungus wars: Basidiomycete battles in wood decay. *Studies in Mycology, 89*, 117–124. https://doi.org/10.1016/j.simyco.2018.02.003.

Hjältén, J., Stenbacka, F., Pettersson, R. B., et al. (2012). Micro and macro-habitat associations in saproxylic beetles: Implications for biodiversity management. *PLoS ONE, 7*(7), e41100. https:// doi.org/10.1371/journal.pone.0041100.

Hof, A. R., Löfroth, T., Rudolphi, J., et al. (2018). Simulating long-term effects of bioenergy extraction on dead wood availability at a landscape scale in Sweden. *Forests, 9*(8), 457. https:// doi.org/10.3390/f9080457.

Hyde, J. C., Smith, A. M. S., Ottmar, R. D., et al. (2011). The combustion of sound and rotten coarse woody debris: A review. *International Journal of Wildland Fire, 20*, 163–174. https://doi.org/10. 1071/WF09113.

Jacobsen, R. M., Sverdrup-Thygeson, A., Kauserud, H., et al. (2018). Exclusion of invertebrates influences saprotrophic fungal community and wood decay rate in an experimental field study. *Functional Ecology, 32*, 2571–2582. https://doi.org/10.1111/1365-2435.13196.

Johansson, T., Gibb, H., Hjältén, J., et al. (2017). Soil humidity, potential solar radiation and altitude affects boreal beetle assemblages in dead wood. *Biological Conservation, 209*, 107–118. https:// doi.org/10.1016/j.biocon.2017.02.004.

Johnston, S. R., Boddy, L., & Weightman, A. J. (2016). Bacteria in decomposing wood and their interactions with wood-decay fungi. *FEMS Microbiology Ecology, 92*, 92. https://doi.org/10. 1093/femsec/fiw179.

Jonsell, M., Weslien, J., & Ehnstrom, B. (1998). Substrate requirements of red-listed saproxylic invertebrates in Sweden. *Biodiversity and Conservation, 7*, 749–764. https://doi.org/10.1023/A: 1008888319031.

Junninen, K., Simila, M., Kouki, J., et al. (2006). Assemblages of wood-inhabiting fungi along the gradients of succession and naturalness in boreal pine-dominated forests in Fennoscandia. *Ecography, 29*, 75–83. https://doi.org/10.1111/j.2005.0906-7590.04358.x.

Juutilainen, K., Halme, P., Kotiranta, H., et al. (2011). Size matters in studies of dead wood and wood-inhabiting fungi. *Fungal Ecology, 4*, 342–349. https://doi.org/10.1016/j.funeco.2011.05.004.

Karjalainen, L., & Kuuluvainen, T. (2002). Amount and diversity of coarse woody debris within a boreal forest landscape dominated by *Pinus sylvestris* in Vienansalo wilderness, eastern Fennoscandia. *Silva Fennica, 36*(1), 147–167. https://doi.org/10.14214/sf.555.

Kazartsev, I., Shorohova, E., Kapitsa, E., et al. (2018). Decaying *Picea abies* log bark hosts diverse fungal communities. *Fungal Ecology, 33*, 1–12. https://doi.org/10.1016/j.funeco.2017.12.005.

King, J., Warren, R. J. I., Maynard, D., et al. (2018). Ants: Ecology and impacts in dead wood. In M. D. Ulyshen (Ed.), *Saproxylic insects-diversity. Ecology and conservation. Zoological monographs* (pp. 237–262). Berlin: Springer.

Koivula, M., & Vanha-Majamaa, I. (2020). Experimental evidence on biodiversity impacts of variable retention forestry, prescribed burning, and deadwood manipulation in Fennoscandia. *Ecological Processes, 9*, 11. https://doi.org/10.1186/s13717-019-0209-1.

Kotowska, D., Zegarek, M., Osojca, G., et al. (2020). Spatial patterns of bat diversity overlap with woodpecker abundance. *PeerJ, 8*, e9385. https://doi.org/10.7717/peerj.9385.

Kurz, W. A., Dymond, C. C., Stinson, G., et al. (2008). Mountain pine beetle and forest carbon feedback to climate change. *Nature, 452*, 987–990. https://doi.org/10.1038/nature06777.

Kushnevskaya, E., & Shorohova, E. (2018). Presence of bark influences the succession of cryptogamic wood-inhabiting communities on conifer fallen logs. *Folia Geobotanica, 53*, 175–190. https://doi.org/10.1007/s12224-018-9310-y.

Kushnevskaya, H., Mirin, D., & Shorohova, E. (2007). Patterns of epixylic vegetation on spruce logs in late-successional boreal forest. *Forest Ecology and Management, 250*, 25–33. https://doi.org/10.1016/j.foreco.2007.03.006.

Kuuluvainen, T., & Gauthier, S. (2018). Young and old forest in the boreal: Critical stages of ecosystem dynamics and management under global change. *Forest Ecosystems, 5*, 26. https://doi.org/10.1186/s40663-018-0142-2.

Kuuluvainen, T., Aakala, T., & Varkonyi, G. (2017). Dead standing pine trees in a boreal forest landscape in the Kalevala National Park, northern Fennoscandia: Amount, population characteristics and spatial pattern. *Forest Ecosystems, 4*, 12. https://doi.org/10.1186/s40663-017-0098-7.

Lamarre, V., & Tremblay, J. A. (2021). Occupancy of the American Three-toed Woodpecker in a heavily-managed boreal forest of eastern Canada. *Diversity, 13*, 35. https://doi.org/10.3390/d13010035.

Lõhmus, P., & Lõhmus, A. (2001). Snags, and their lichen flora in old Estonian peatland forests. *Annales Botanici Fennici, 38*, 265–280.

Malysheva, N. A., Filipchuk, A. N., Zolina, T. A., et al. (2019). Quantitative assessment of coarse woody debris in the forests of the Russian Federation according to the SFI data [in Russian]. In *Forestry Information Electronic Weblog [Lesokhozyaistvennaya informatsia Elektronnyy Setevoy Zhurnal-in Russian]* (Vol. 1, pp. 101–128).

Marchand, W., Girardin, M. P., Hartmann, H., et al. (2019). Taxonomy, together with ontogeny and growing conditions, drives needleleaf species' sensitivity to climate in boreal North America. *Global Change Biology, 25*, 2793–2809. https://doi.org/10.1111/gcb.14665.

Martin, K., & Eadie, J. M. (1999). Nest webs: A community-wide approach to the management and conservation of cavity-nesting forest birds. *Forest Ecology and Management, 115*, 243–257. https://doi.org/10.1016/S0378-1127(98)00403-4.

Martin, M., Fenton, N., & Morin, H. (2018). Structural diversity and dynamics of boreal old-growth forests case study in Eastern Canada. *Forest Ecology and Management, 422*, 125–136. https://doi.org/10.1016/j.foreco.2018.04.007.

Martin, M., Girona, M. M., & Morin, H. (2020). Driving factors of conifer regeneration dynamics in eastern Canadian boreal old-growth forests. *PLoS ONE, 15*(7), e0230221. https://doi.org/10.1371/journal.pone.0230221.

Martin, M., Tremblay, J. A., Ibarzabal, J., et al. (2021). An indicator species highlights continuous deadwood supply is a key ecological attribute of boreal old-growth forests. *Ecosphere, 12*, e03507. https://doi.org/10.1002/ecs2.3507.

Maurice, S., Arnault, G., Nordén, J., et al. (2021). Fungal sporocarps house diverse and host-specific communities of fungicolous fungi. *The ISME Journal, 15*, 1445–1457. https://doi.org/10.1038/s41396-020-00862-1.

Mikkonen, N., Leikola, N., Halme, P., et al. (2020). Modeling of dead wood potential based on tree stand data. *Forests, 11*, 913. https://doi.org/10.3390/f11090913.

Miquelajauregui, Y., Cumming, S. G., & Gauthier, S. (2019). Short-term responses of boreal carbon stocks to climate change: A simulation study of black spruce forests. *Ecological Modelling, 409*, 108754. https://doi.org/10.1016/j.ecolmodel.2019.108754.

Misof, B., Liu, S., Meusemann, K., et al. (2014). Phylogenomics resolves the timing and pattern of insect evolution. *Science, 346*(6210), 763–767. https://doi.org/10.1126/science.1257570.

Müller, J., Jarzabek-Müller, A., & Bussler, H. (2013). Some of the rarest European saproxylic beetles are common in the wilderness of Northern Mongolia. *Journal of Insect Conservation, 17*, 989–1001. https://doi.org/10.1007/s10841-013-9581-9.

Müller, J., Ulyshen, M., Seibold, S., et al. (2020). Primary determinants of communities in deadwood vary among taxa but are regionally consistent. *Oikos, 129*(10), 1579–1588. https://doi.org/10.1111/oik.07335.

Nappi, A., & Drapeau, P. (2009). Reproductive success of the black-backed woodpecker (*Picoides arcticus*) in burned boreal forests: Are burns source habitats? *Biological Conservation, 142*, 1381–1391. https://doi.org/10.1016/j.biocon.2009.01.022.

Nappi, A., Drapeau, P., Saint-Germain, M., et al. (2010). Effect of fire severity on long-term occupancy of burned boreal conifer forests by saproxylic insects and wood-foraging birds. *International Journal of Wildland Fire, 19*, 500–511. https://doi.org/10.1071/WF08109.

Nappi, A., Drapeau, P., & Leduc, A. (2015). How important is dead wood for woodpeckers foraging in eastern North American boreal forests? *Forest Ecology and Management, 346*, 10–21. https://doi.org/10.1016/j.foreco.2015.02.028.

Niemelä, T., Wallenius, T., & Kotiranta, H. (2002). The kelo tree, a vanishing substrate of specified wood-inhabiting fungi. *Polish Botanical Journal, 47*, 91–101.

Nilsson, S. G., Niklasson, M., Hedin, J., et al. (2002). Densities of large living and dead trees in old-growth temperate and boreal forests. *Forest Ecology and Management, 161*, 189–204. https://doi.org/10.1016/S0378-1127(01)00480-7.

Nordén, J., Penttilä, R., Siitonen, J., et al. (2013). Specialist species of wood-inhabiting fungi struggle while generalists thrive in fragmented boreal forests. *Journal of Ecology, 101*, 701–712. https://doi.org/10.1111/1365-2745.12085.

Parsons, S., Lewis, K. J., & Psyllakis, J. M. (2003). Relationships between roosting habitat of bats and decay of aspen in the sub-boreal forests of British Columbia. *Forest Ecology and Management, 177*, 559–570. https://doi.org/10.1016/S0378-1127(02)00448-6.

Rajala, T., Peltoniemi, M., Hantula, J., et al. (2011). RNA reveals a succession of active fungi during the decay of Norway spruce logs. *Fungal Ecology, 4*, 437–448. https://doi.org/10.1016/j.funeco.2011.05.005.

Rajala, T., Peltoniemi, M., Pennanen, T., et al. (2012). Fungal community dynamics in relation to substrate quality of decaying Norway spruce (*Picea abies* [L.] Karst.) logs in boreal forests. *FEMS Microbiology Ecology, 81*, 494–505. https://doi.org/10.1111/j.1574-6941.2012.01376.x.

Romashkin, I., Neuvonen, S., & Tikkanen, O. P. (2020). Northward shift in temperature sum isoclines may favour *Ips typographus* outbreaks in European Russia. *Agricultural and Forest Entomology, 22*, 238–249. https://doi.org/10.1111/afe.12377.

Rota, C. T., Rumble, M. A., Lehman, C. P., et al. (2015). Apparent foraging success reflects habitat quality in an irruptive species, the Black-backed Woodpecker. *The Condor, 117*(2), 178–191. https://doi.org/10.1650/condor-14-112.1.

Rouvinen, S., & Kouki, J. (2002). Spatiotemporal availability of dead wood in protected old-growth forests: A case study from boreal forests in eastern Finland. *Scandinavian Journal of Forest Research, 17*(4), 317–329. https://doi.org/10.1080/02827580260138071.

Saint-Germain, M., Drapeau, P., & Buddle, C. M. (2007). Host-use patterns of saproxylic phloeophagous and xylophagous Coleoptera adults and larvae along the decay gradient in standing dead black spruce and aspen. *Ecography*, *30*, 737–748. https://doi.org/10.1111/j.2007.0906-7590. 05080.x.

Saint-Germain, M., Drapeau, P., & Buddle, C. M. (2008). Persistence of pyrophilous insects in fire-driven boreal forests: Population dynamics in burned and unburned habitats. *Diversity and Distributions*, *14*, 713–720. https://doi.org/10.1111/j.1472-4642.2007.00452.x.

Sandström, J., Bernes, C., Junninen, K., et al. (2019). Impacts of dead wood manipulation on the biodiversity of temperate and boreal forests. A systematic review. *Journal of Applied Ecology*, *56*, 1770–1781. https://doi.org/10.1111/1365-2664.13395.

Santaniello, F., Djupstrom, L. B., Ranius, T., et al. (2017). Large proportion of wood dependent lichens in boreal pine forest are confined to old hard wood. *Biodiversity and Conservation*, *26*, 1295–1310. https://doi.org/10.1007/s10531-017-1301-4.

Schlick-Steiner, B. C., Steiner, F. M., Konrad, H., et al. (2008). Specificity and transmission mosaic of ant nest-wall fungi. *Proceedings of the National Academy of Sciences of the United States of America*, *105*, 940–943. https://doi.org/10.1073/pnas.0708320105.

Schroeder, L. M., & Weslien, J. (1994). Reduced offspring production in bark beetle *Tomicus piniperda* in pine bolts baited with ethanol and α-pinene, which attract antagonistic insects. *Journal of Chemical Ecology*, *20*, 1429–1444. https://doi.org/10.1007/BF02059871.

Seibold, S., Müller, J., Baldrian, P., et al. (2019). Fungi associated with beetles dispersing from dead wood–Let's take the beetle bus! *Fungal Ecology*, *39*, 100–108. https://doi.org/10.1016/j.fun eco.2018.11.016.

Seidl, R., Honkaniemi, J., Aakala, T., et al. (2020). Globally consistent climate sensitivity of natural disturbances across boreal and temperate forest ecosystems. *Ecography*, *43*(7), 967–978. https:// doi.org/10.1111/ecog.04995.

Sherwood-Pike, M. A., & Gray, J. (1985). Silurian fungal remains: Probable records of the Class Ascomycetes. *Lethaia*, *18*, 1–20. https://doi.org/10.1111/j.1502-3931.1985.tb00680.x.

Shorohova, E., & Kapitsa, E. (2015). Stand and landscape scale variability in the amount and diversity of coarse woody debris in primeval European boreal forests. *Forest Ecology and Management*, *356*, 273–284. https://doi.org/10.1016/j.foreco.2015.07.005.

Shorohova, E., & Kapitsa, E. (2016). The decomposition rate of non-stem components of coarse woody debris (CWD) in European boreal forests mainly depends on site moisture and tree species. *European Journal of Forest Research*, *135*, 593–606. https://doi.org/10.1007/s10342-016-0957-8.

Shorohova, E., & Tetioukhin, S. (2003). Natural disturbances and the amount of large trees, deciduous trees and coarse woody debris in the forests of Novgorod Region, Russia. *Ecological Bulletins*, *51*, 137–147.

Siitonen, J. (2001). Forest management, coarse woody debris and saproxylic organisms: Fennoscandian boreal forest as an example. *Ecological Bulletins*, *49*, 11–41.

Siitonen, J., & Saaristo, L. (2000). Habitat requirements and conservation of *Pytho kolwensis*, a beetle species of old-growth boreal forest. *Biological Conservation*, *94*, 211–220. https://doi.org/ 10.1016/S0006-3207(99)00174-3.

Spribille, T., Thor, G., Bunnell, F. L., et al. (2008). Lichens on dead wood: Species-substrate relationships in the epiphytic lichen floras of the Pacific Northwest and Fennoscandia. *Ecography*, *31*, 741–750. https://doi.org/10.1111/j.1600-0587.2008.05503.x.

Stenbacka, F., Hjältén, J., Hilszczanski, J., et al. (2010). Saproxylic and non-saproxylic beetle assemblages in boreal spruce forests of different age and forestry intensity. *Ecological Applications*, *20*, 2310–2321. https://doi.org/10.1890/09-0815.1.

Stokland, J. N., Siitonen, J., & Jonsson, B. G. (2012). *Biodiversity in dead wood*. Cambridge: Cambridge University Press.

Stokland, J. N., Woodall, C. W., Fridman, J., et al. (2016). Burial of downed deadwood is strongly affected by log attributes, forest ground vegetation, edaphic conditions, and climate zones. *Canadian Journal of Forest Research*, *46*, 1451–1457. https://doi.org/10.1139/cjfr-2015-0461.

Taylor, S. L., & MacLean, D. A. (2007). Dead wood dynamics in declining balsam fir and spruce stands in New Brunswick, Canada. *Canadian Journal of Forest Research*, *37*, 750–762. https:// doi.org/10.1139/X06-272.

Thompson, S., Vehkaoja, M., & Nummi, P. (2016). Beaver-created deadwood dynamics in the boreal forest. *Forest Ecology and Management, 360*, 1–8. https://doi.org/10.1016/j.foreco.2015.10.019.

Thorn, S., Chao, A., Georgiev, K. B., et al. (2020a). Estimating retention benchmarks for salvage logging to protect biodiversity. *Nature Communications, 11*(1), 4762. https://doi.org/10.1038/s41 467-020-18612-4.

Thorn, S., Seibold, S., Leverkus, A. B., et al. (2020b). The living dead: Acknowledging life after tree death to stop forest degradation. *Frontiers in Ecology and the Environment, 18*, 505–512. https://doi.org/10.1002/fee.2252.

Timms, L. L., Smith, S. M., & De Groot, P. (2006). Patterns in the within-tree distribution of the emerald ash borer *Agrilus planipennis* (Fairmaire) in young, green-ash plantations of southwestern Ontario, Canada. *Agricultural and Forest Entomology, 8*, 313–321. https://doi.org/10. 1111/j.1461-9563.2006.00311.x.

Tremblay, J. A., Ibarzabal, J., Dussault, C., et al. (2009). Habitat requirements of breeding Black-backed Woodpeckers (*Picoides arcticus*) in managed, unburned boreal forest. *Avian Conservation and Ecology, 4*(1), 2.

Tremblay, J. A., Ibarzabal, J., & Savard, J.-P. L. (2010). Foraging ecology of black-backed woodpeckers (*Picoides arcticus*) in unburned eastern boreal forest stands. *Canadian Journal of Forest Research, 40*, 991–999. https://doi.org/10.1139/X10-044.

Tremblay, J. A., Savard, J. P. L., & Ibarzabal, J. (2015). Structural retention requirements for a key ecosystem engineer in conifer-dominated stands of a boreal managed landscape in eastern Canada. *Forest Ecology and Management, 357*, 220–227. https://doi.org/10.1016/j.foreco.2015. 08.024.

Tremblay, J. A., Boulanger, Y., Cyr, D., et al. (2018). Harvesting interacts with climate change to affect future habitat quality of a focal species in eastern Canada's boreal forest. *PLoS ONE, 13*, e0191645. https://doi.org/10.1371/journal.pone.0191645.

Tremblay, J. A., Dixon, R. D., Saab, V. A., et al. (2020). Black-backed Woodpecker (*Picoides arcticus*), version 1.0. In P. G. Rodewald (Ed.), *Birds of the world*. Ithaca: Cornell Lab of Ornithology.

van der Hoek, Y., Gaona, G. V., & Martin, K. (2017). The diversity, distribution and conservation status of the tree-cavity-nesting birds of the world. *Diversity and Distributions, 23*, 1120–1131. https://doi.org/10.1111/ddi.12601.

Venäläinen, M., Harju, A. M., Kainulainen, P., et al. (2003). Variation in the decay resistance and its relationship with other wood characteristics in old Scots pines. *Annals of Forest Science, 60*, 409–417. https://doi.org/10.1051/forest:2003033.

Venäläinen, A., Lehtonen, I., Laapas, M., et al. (2020). Climate change induces multiple risks to boreal forests and forestry in Finland: A literature review. *Global Change Biology, 26*, 4178–4196. https://doi.org/10.1111/gcb.15183.

Versluijs, M., Eggers, S., Mikusinski, G., et al. (2020). Foraging behavior of the Eurasian Three-toed Woodpecker (*Picoides tridactylus*) and its implications for ecological restoration and sustainable boreal forest management. *Avian Conservation and Ecology, 15*, 6. https://doi.org/10.5751/ACE-01477-150106.

Weslien, J. (1992). The arthropod complex associated with *Ips typographus* (L.) (Coleoptera, Scolytidae): Species composition, phenology and impact on bark beetle productivity. *Entomologica Fennica, 3*, 205–213. https://doi.org/10.33338/ef.83730.

Weslien, J., Djupström, L. B., Schroeder, M., et al. (2011). Long-term priority effects among insects and fungi colonizing decaying wood. *Journal of Animal Ecology, 80*, 1155–1162. https://doi.org/ 10.1111/j.1365-2656.2011.01860.x.

Wesołowski, T., & Martin, K. (2018). Tree holes and hole-nesting birds in European and North American forests. In G. Mikusinski, J. M. Roberge, & R. J. Fuller (Eds.), *Ecology and conservation of forest birds* (pp. 79–134). Cambridge: Cambridge University Press.

Westerfelt, P., Widenfalk, O., Lindelow, A., et al. (2015). Nesting of solitary wasps and bees in natural and artificial holes in dead wood in young boreal forest stands. *Insect Conservation and Diversity, 8*, 493–504. https://doi.org/10.1111/icad.12128.

Wikars, L. O. (2004). Habitat requirements of the pine wood-living beetle *Tragosoma depsarium* (Coleoptera: Cerambycidae) at log, stand and landscape scale. *Ecological Bulletins, 51*, 287–294.

Work, T. T., Shorthouse, D. P., Spence, J. R., et al. (2004). Stand composition and structure of the boreal mixedwood and epigaeic arthropods of the Ecosystem Management Emulating Natural Disturbance (EMEND) landbase in northwestern Alberta. *Canadian Journal of Forest Research, 34*, 417–430. https://doi.org/10.1139/x03-238.

Chapter 7
Embracing the Complexity and the Richness of Boreal Old-Growth Forests: A Further Step Toward Their Ecosystem Management

Maxence Martin, Ekaterina Shorohova, and Nicole J. Fenton

Abstract Boreal old-growth forests are specific and often undervalued ecosystems, as they present few of the structural attributes that usually define old forests in the collective culture. Yet, these ecosystems are characterized by exceptional naturalness, integrity, complexity, resilience, as well as structural and functional diversity. They therefore serve as biodiversity hot spots and provide crucial ecosystem services. However, these forests are under significant threat from human activities, causing a rapid and large-scale reduction in their surface area and integrity. The multiple values associated with boreal old-growth forests should be therefore better acknowledged and understood to ensure the sustainable management of boreal landscapes.

M. Martin (✉) · N. J. Fenton
Forest Research Institute, Université du Québec en Abitibi-Témiscamingue, 445 boul. de l'Université, Rouyn-Noranda, QC J9X 5E4, Canada
e-mail: maxence.martin@uqat.ca

N. J. Fenton
e-mail: nicole.fenton@uqat.ca

M. Martin
Département des Sciences Fondamentales, Université du Québec à Chicoutimi, 555 boul. de l'Université, Chicoutimi, QC G7H 2B1, Canada

M. Martin · N. J. Fenton
Centre for Forest Research, Université du Québec à Montréal, P.O. Box 8888, Stn. Centre-Ville, Montreal, QC H3C 3P8, Canada

E. Shorohova
Forest Research Institute of the Karelian Research Center, Russian Academy of Science, Pushkinskaya str. 11, Petrozavodsk 185910, Russia
e-mail: shorohova@es13334.spb.edu; ekaterina.shorokhova@luke.fi

Saint Petersburg State Forest Technical University, Institutsky str. 5, Saint Petersburg 194021, Russia

Natural Resources Institute Finland (Luke), Latokartanonkaari 9, FI-00790 Helsinki, Finland

© The Author(s) 2023
M. M. Girona et al. (eds.), *Boreal Forests in the Face of Climate Change*,
Advances in Global Change Research 74,
https://doi.org/10.1007/978-3-031-15988-6_7

191

7.1   The Old-Growth Forest Concept: A General Overview

Forests considered as "natural" have an important place in our collective conscious-
ness for cultural, ethical, spiritual, artistic, and aesthetic reasons (Frelich & Reich,
2003; Kimmins, 2003; Pesklevits et al., 2011; Satterfield, 2002). Interest in these
forests has also grown continuously during the twentieth century as a source of
inspiration for establishing sustainable management strategies (Puettmann et al.,
2009). Theories of how to maintain forest biodiversity and ecosystem services include
forest management based on natural disturbance dynamics (Gauthier et al., 2009) and
managed stands containing forest-specific structural elements that are considered as
natural references (Bauhus et al., 2009; Halme et al., 2013).

Many terms considered synonymous with *natural forest* have long been used,
e.g., primary, primeval, pristine, old-growth, virgin, mature, natural, overmature,
original, or intact forest (Wirth et al., 2009); however, each of these terms represents
a different ecological concept, and further clarification of the terminology has gradually taken place over the twentieth century (Frelich & Reich, 2003; Wirth et al.,
2009). The concept of *old-growth forest* was one of the most important concepts to
attract the attention of the scientific community, managers, and the general public,
as it relates to many important current issues related to forests: (1) intrinsic value
(e.g., academic, cultural, spiritual), (2) exigencies of "closer to nature" management,
conservation, and restoration strategies, and (3) their role in addressing the challenges of climate change and biodiversity loss (Frelich & Reich, 2003; Kuuluvainen
et al., 2017; Pesklevits et al., 2011; Wirth et al., 2009).

Defining what an old-growth forest is, and by extension what is not, is nevertheless
particularly complex. Many definitions have been proposed over time, some of which
are now debated. These definitions can be grouped into seven main classes (Frelich &
Reich, 2003; Issekutz, 2020; Kimmins, 2003; Kneeshaw & Gauthier, 2003; Wirth
et al., 2009):

Structural a stand that has reached a certain age; the presence of many old trees with
large diameters; a high volume of deadwood of all decay classes; a high vertical and
horizontal complexity; the presence of trees of all ages.

Dynamic: a stand under gap dynamics; the stand has reached the final stage of succession; a stand age greater than the return interval of primary disturbances, i.e., disturbances of high severity that reinitiate forest succession; the age of trees exceeding
their average life expectancy.

Scale: a continuous forest having a small human footprint over a sufficiently large
area (forest track, forest massif).

Functional or *biogeochemical*: the net primary productivity is equal to or less than
zero; the *climax* concept; a trophic network reaching a given threshold of complexity;
the presence of all stages of deadwood degradation.

Economic: a forest that has exceeded the optimum age for harvesting; the volume of
commercial timber has reached a peak and is now stable or declining.

Aesthetic: an impressive forest; invites humility and spirituality.

Other definitions: undisturbed by humans; covers a minimum area.

Each of these definitions has its specific limitations and therefore represents a different view of what can be considered *old growth*. The main criticisms generally relate to their arbitrary nature, the difficulty of integrating some of these thresholds into daily management, and the existence of counterexamples that limit their universality (Kimmins, 2003; Pesklevits et al., 2011; Wirth et al., 2009). Other definitions, such as the concept of climax defined by Odum (1969), are now generally considered too reductionist and consequently ecologically irrelevant (Wirth & Lichstein, 2009). Similarly, the degree of human footprint in a forest is more a question of naturalness rather than *old-growthness*, even if the two concepts are often linked (Frelich & Reich, 2003). Old-growth forests are not necessarily *primary*, i.e., a forest of high naturalness almost undisturbed by anthropogenic activities, and, conversely, not all primary forests are old growth. For example, a primary forest that recently burned due to a wildfire caused by lightning can still be considered primary after the disturbance, as this disturbance does not influence its naturalness. Conversely, a previously managed stand that has returned to an old-growth state is not a primary forest because of its history, although its abandonment progressively increased its naturalness. From a more philosophical perspective, the very concept of old-growth forest is arbitrary, artificially classifying forest ecosystems (Pesklevits et al., 2011). Therefore, it is now accepted that a universal definition is neither possible nor necessarily desirable. On the contrary, definitions of old-growth forests need to be adapted to the ecological context of the region under study (Frelich & Reich, 2003; Pesklevits et al., 2011). There may therefore be a diversity of definitions restricted to a local scale. Hunter and White (1997) offer a less precise but more general definition that is commonly used: an old-growth forest is relatively old and minimally disturbed by natural and anthropogenic disturbances.

Moreover, the term *old growth* actually describes a wide diversity of forests in terms of structure, tree species composition, and disturbance history, even within a restricted area (Martin et al., 2018; Meigs et al., 2017; Shorohova & Kapitsa, 2015). Combining all these attributes influences habitat characteristics markedly at the local scale (Kozák et al., 2021). It also underscores the importance of the spatial extent and continuity of old-growth forests, as small and insulated old-growth stands are not a surrogate for large old-growth areas (Moussaoui et al., 2016; Schmiegelow & Mönkkönen, 2002). For these reasons, it is crucial to consider that the forests designated as old growth often contain a diversity of structures and composition. Oliver and Larson (1996) thus proposed to distinguish *true* old-growth forests—all the trees of the first cohort have died and have been replaced by new shade-tolerant trees—from *transition* old-growth forests in which some individuals of the first cohort are still present. This approach still has its limitations, notably its relatively arbitrary nature; however, it distinguishes between different types of old-growth forest. Nevertheless, the concept of transition or true old-growth forests can group forests with very different structures and compositions (Martin et al., 2018). We therefore propose a hierarchical definition of old-growth forests, highlighting the complexity and limits of this concept while making it adaptable to different operational contexts (Table 7.1). In this chapter, we will use the general definition of old-growth forests as "relatively old and little disturbed by natural and anthropogenic disturbances" because of the great diversity of contexts covered.

The importance of old-growth forests as biodiversity hot spots is widely recognized. They contain many structural features that are absent or rare in younger and managed stands, such as deadwood of various sizes and decay stages, large trees, and high structural complexity (Franklin et al., 2002; Wirth et al., 2009). The diversity of old-growth forest attributes and structures within the same forest tract is also an essential factor in explaining their importance for biodiversity, as they provide a wide range of habitats (De Grandpré et al., 2018; Schowalter, 2017). The high degree of forest continuity (i.e., the length of time an area has been continuously wooded) that defines old-growth forests is also vital for low-dispersal and disturbance-sensitive species that can require decades or centuries to recolonize a stand after a severe

Table 7.1 Proposition for a hierarchical definition of old-growth forests, depending on the spatial scale and the research question addressed. A definition integrating several levels adapted to different contexts allows for recognizing the complexity of this ecological concept while offering the possibility of adjusting to possible particular cases

Scale	Question/Motivation	Definition
General definition (broad concept, international scale)	What do we generally consider as "old growth"?	Hunter and White (1997) definition: "Relatively old and little disturbed by natural and anthropogenic disturbances"
Intermediate-scale definition (country, continent, or biome scale)	Which forests can be considered as *old growth* in a given area?	Definition based on the most relevant ecological criteria at the level of the concerned territory, with attention to possible specific cases
Coarse small-scale definition (local scale)	How can we roughly distinguish between different old-growth forest types within the same landscape?	Distinction between *transition* and *true* old-growth forests proposed by Oliver and Larson (1996), with the possibility of adjusting the thresholds depending on the context (see, for example, Kneeshaw and Gauthier (2003) and Martin et al. (2018))
Fine small-scale definition (local scale)	How can we finely distinguish between different old-growth forest types within the same landscape?	Consideration of the range of structural characteristics and tree species composition that old-growth forests can take depending on the successional process, the action of natural disturbances, and the influence of abiotic conditions

disturbance (McMullin & Wiersma, 2019). However, many forest species are dependent on younger forests (Drapeau et al., 2003; Fenton & Bergeron, 2011), while certain attributes often associated with old-growth forests may also be abundant in young forests that have been recently disturbed, e.g., deadwood (Donato et al., 2012). Overall, primary forests generally contain stands of all ages, the proportions of which depend on the natural disturbance regime (Kneeshaw et al., 2018). Thus, old-growth forests do not necessarily host maximum species diversity.

The importance of old-growth forests is not limited to their role as habitats for biodiversity. In the context of climate change, the importance of old-growth forests for the long-term sequestration of atmospheric carbon is, for example, a concrete ecosystem service, acting for the benefit of all (Lafleur et al., 2018; Vedrova et al., 2018). Watson et al. (2018) also listed many services provided by intact forests and, by extension, old-growth forests, such as regulating local and regional weather regimes, buffering against the transmission of new diseases, and providing a source of yet unexplored scientific knowledge. The cultural value attributed to old-growth forests, whether in terms of aesthetics, intrinsic value, or spirituality, should also be considered (Kimmins, 2003; Satterfield, 2002). The tensions and conflicts regularly observed for issues related to the management and protection of old-growth forests, e.g., between economic and environmental actors, can be partly explained by the strong cultural and social values attributed to these forests (Kimmins, 2003; Pesklevits et al., 2011; Satterfield, 2002). Although this chapter focuses mainly on old-growth forests through the perspective of forest ecology and management, we also invite the reader to explore the insights from the social sciences and humanities on this subject.

7.2 Can the Distinctive Characteristics of Boreal Forests Help Us Rethink Old-Growth Forests?

Boreal old-growth forests are one of the counterexamples limiting the relevance of broad-scale old-growth definitions. Forests in this biome are generally characterized by a relatively low diversity of tree species, and many of these species are also found at the beginning or end of forest succession (Angelstam & Kuuluvainen, 2004; Harvey et al., 2002; Shorohova et al., 2011). This particularity challenges standard forest succession models, where the replacement of pioneer shade-intolerant species by shade-tolerant species is one of the conditions defining the old-growth stage (Oliver & Larson, 1996; Wirth et al., 2009). Harsh climatic conditions and low site fertility also limit tree growth and size, resulting in stands defined by a relatively simple vertical structure compared with what is commonly expected from old-growth forests (Bergeron & Harper, 2009). Martin et al. (2020b) highlighted the numerous similarities among the vertical structures of even-aged and old-growth black spruce (*Picea mariana*)–dominated stands. However, this type of structure may also partially result from a particular regeneration dynamic, where the black spruce understory remains limited as long as the canopy is not disturbed (Martin et al., 2020d). A similar pattern has also been observed in Norway spruce (*Picea abies*) forests in Finland and Russia (Shorohova et al., 2008, 2009). Moreover, the process of paludification, i.e., the gradual thickening of the organic horizon under poor drainage conditions (Fenton et al., 2005), can markedly reduce stand productivity (Bergeron and Fenton 2012). This process eventually creates forests composed of trees of very small diameter and height despite their old age. The productivity decline caused by paludification in boreal old-growth forests is nevertheless generally restricted to specific environmental conditions; old-growth forests situated on sufficiently drained sites retain their structure over the centuries (Pollock and Payette 2010; Shorohova et al., 2008).

Tree diameter and stand volume are also relatively low compared with forests of other biomes (Fig. 7.1). The presence of very large living or dead trees—generally defined by a diameter at breast height between 70 and 100 cm; (Gosselin & Larrieu, 2020; Spies & Franklin, 1991)—is often considered as one of the key attributes of old-growth forests, for either ecological or cultural reasons, e.g., because very large trees give a sense of greatness and oldness (Kimmins, 2003; Paillet et al., 2017; Wirth et al., 2009). Some types of boreal old-growth forest can contain trees of notable size (e.g., diameter at breast height >40 cm), such as Scots pine (*Pinus sylvestris*), Norway spruce, or balsam fir (*Abies balsamea*)–white birch (*Betula papyrifera*) stands (Desponts et al., 2004; Lilja & Kuuluvainen, 2005; Shorohova et al., 2009). Yet, very large trees can be rare if not completely absent from many boreal old-growth forests, as observed in eastern Canada (Bergeron & Harper, 2009). These examples illustrate that tree size is unreliable for defining old-growth forests, as this attribute can vary enormously from one boreal old-growth forest type to another (Fig. 7.1). Moreover, large trees in old boreal forests are mainly softwood species, as hardwood species are generally pioneer taxa. The value of large trees for ecological, economic, and aesthetic reasons has been often emphasized (Lindenmayer et al., 2014; Lutz et al., 2018; Paillet et al., 2019), explaining their importance in the discussions related to old-growth forests. Poplar species (e.g., *Populus tremula* in Eurasia and *Populus tremuloides* in Canada) are often the larger hardwood species than can be found in boreal landscapes, even though these fast-growing species are generally restricted to the youngest successional stages and specific abiotic conditions (Hardenbol et al., 2020; Harvey et al., 2002). It should be noted, however, that some counterexamples of multicohort *Populus tremuloides* do exist (Cumming et al., 2000). Similarly, hardwood species can sometimes be found mixed in small proportions with conifers in some old-growth boreal forests because of natural disturbances (Bergeron & Harper, 2009; Vehmas et al., 2009), thus increasing habitat diversity in these forests.

The small tree size in boreal forests thereby limits deadwood volume and large log density in boreal old-growth stands. This scarcity of deadwood can be reinforced by the rapid burial of fallen dead trees in soils dominated by moss species (Stokland et al. 2016). Boreal old-growth forests can therefore contain a large volume of almost-intact deadwood within the soil organic layer, although not immediately apparent at the surface. Hence, several studies in eastern Canada found no significant changes in visible deadwood volume in old-growth forests of different ages or differing in

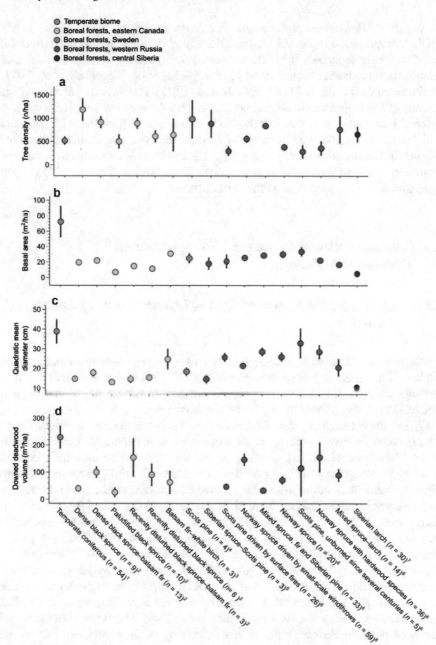

Fig. 7.1 Mean (*circles*) and 95% confidence intervals (*vertical black bars*) for **a** tree density, **b** basal area, **c** quadratic mean diameter, and **d** downed deadwood volume among temperate and coniferous old-growth forests from the literature review of Burrascano et al. (2013) and those of boreal old-growth forests differing in terms of location, tree species composition, and disturbance history. 1 Burrascano et al. (2013), 2 Martin et al. (2018), 3 Desponts et al. (2004), 4 Lundqvist et al. (2019), 5 Stavrova et al. (2020), 6 Shorohova and Kapitsa (2015), 7 Bondarev (1997). Survey methodologies, e.g., the minimum size of sampled trees, may vary between sources

the degree of first-cohort replacement (Bergeron & Harper, 2009; Martin et al., 2018). Marked variability in abiotic conditions (e.g., drainage and surficial deposits) and disturbance dynamics (e.g., disturbance agent, severity, and recurrence) also characterizes the boreal biome at the local to global scale (Kneeshaw et al., 2011; Kuuluvainen & Aakala, 2011; Shorohova et al., 2011). Deadwood dynamics, either in terms of input, decomposition, or burial, can thereby vary markedly between two different locations or periods (Aakala, 2011; Shorohova & Kapitsa, 2015; Stokland et al. 2016). Although a small volume of visible deadwood may define some boreal old-growth forests, other nearby old-growth stands can hold a substantial volume of deadwood (>150 m³/ha) (Martin et al., 2018; Shorohova & Kapitsa, 2015), again highlighting the diverse nature of these ecosystems.

7.3 The Exceptional Ecological Value of Boreal Old-Growth Forests

7.3.1 An Important Aspect of the Last Great Tracts of Intact Forest

Primary forests continue to decline rapidly and often represent small and isolated patches within degraded areas (Potapov et al., 2017; Sabatini et al., 2018). Most of the large tracts of remnant primary forests are now in remote boreal and tropical regions (Fig. 7.2) (Achard et al., 2009; Kuuluvainen et al., 2017; Potapov et al., 2017). For the boreal zone, most of these forests are found in Canada and Russia, as forestry activities have drastically modified most northern Fennoscandian forest landscapes (Sabatini et al., 2018). Because natural disturbances are essentially defined by a low to moderate severity in northern Fennoscandia, old-growth forests were initially abundant in preindustrial landscapes (Shorohova et al., 2011). Nevertheless, forestry activities caused a rapid decline and almost complete loss of their surface area (Östlund et al., 1997; Sabatini et al., 2018). In northern Fennoscandia, the remaining old-growth forests are thus often isolated in small areas and poorly accessible territories (Sabatini et al., 2018; Svensson et al., 2020). Consequently, populations of many forest species have declined sharply through the loss and fragmentation of their habitats (Esseen et al., 1992). Restoration strategies have since been successfully implemented, but conservation remains far more effective than having to restore altered ecosystems (Halme et al., 2013). The example of the boreal forests of northern Fennoscandia is thus a warning of the ecological risks of the disappearance of old-growth forests and must therefore be considered in areas where these forests are still present.

Not all primary boreal forests are, however, old-growth forests, and their abundance may significantly vary from one region to another, depending on the characteristics of the natural disturbance regime (Shorohova et al., 2011). Because of the generally random distribution of wildfires, the main primary disturbance in boreal

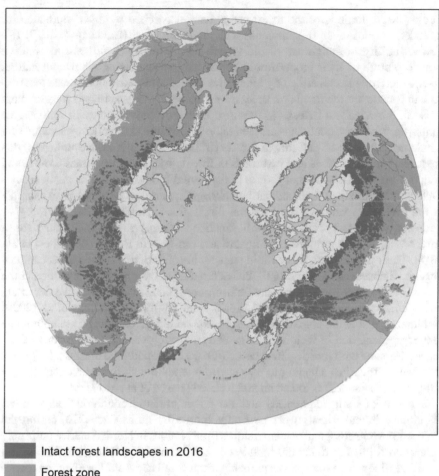

■ Intact forest landscapes in 2016

■ Forest zone

□ National boundaries

□ Oceans and seas

Fig. 7.2 Intact forest landscapes (*dark green*) in the forested regions (*light green*) in the Northern Hemisphere as determined by Potapov et al. (2017). Intact forest landscapes are defined as a "seamless mosaic of forests and associated natural treeless ecosystems that exhibit no remotely detected signs of human activity or habitat fragmentation and are large enough to maintain all native biological diversity, including viable populations of wide-ranging species" (Potapov et al., 2008). Most remaining intact forests are situated in the northern areas of the forested zone, hence in the boreal biome

forests, large portions of the forest can remain untouched by this disturbance for extended periods, even in landscapes with recurring fires (Kneeshaw et al., 2018). The variability in definitions of old-growth forests makes it difficult to obtain a general picture of their proportion in boreal landscapes. Using the overall natural disturbance regimes identified by Shorohova et al. (2011), we can estimate that old-growth forests are the dominant successional stage in the eastern and western parts of North America and Eurasia, as the fire cycles are relatively long. In contrast, the central parts of these continents are generally defined by a shorter fire cycle, implying a reduced presence of old-growth forests (Belleau et al., 2007). Nevertheless, this abundance of continuous and vast tracts of forests with high naturalness containing both old-growth and younger forests in the boreal biome is vital for many species (Venier et al., 2018). Woodland caribou (*Rangifer tarandus caribou*) is an example of the biological value of large natural forest areas. This subspecies of *Rangifer tarandus*, which is specific to North America, requires vast (>1,000 km^2) forest areas having a high proportion of mature and old-growth forests (Kneeshaw et al., 2018). Their populations are declining rapidly because of direct and indirect anthropogenic disturbances modifying the characteristics of their habitats, in particular the loss of old-growth forest areas and unfavorable predation dynamics (Venier et al., 2014). Similarly, Schmiegelow and Mönkkönen (2002) and Cadieux et al. (2020) highlighted that avian communities dependent on old boreal forests are vulnerable to the fragmentation of their habitats caused by the rejuvenation of the boreal landscapes. Species with low dispersal capacity, e.g., arthropods and nonvascular plants, may have difficulties adapting to new environmental conditions where the residual old-growth forest area is too small and isolated (Barbé et al., 2017).

Because of their remoteness and the harsh climatic conditions, large tracts of primary boreal forests were generally spared by human activities during the Holocene. Archaeological evidence shows First Nations in North America purposely influenced forest fire dynamics long before European colonization, but this anthropogenic influence on forests was probably limited and had a similar effect on forests as wildfires (Munoz & Gajewski, 2010). In contrast, forests considered primary or highly natural in areas characterized by a continuous human presence over centuries, such as Europe or northwestern Russia, may show ancient traces of forest management, deforestation, and agricultural activity (Jaroszewicz et al., 2019; Shorohova et al., 2019a). Although not immediately visible, they can durably modify certain environmental conditions and thus the associated biodiversity (Dambrine et al., 2007).

An essential part of boreal old-growth forests therefore belongs to vast, continuous massifs of highly natural forests. This temporal and spatial continuity is critical for biodiversity (McMullin & Wiersma, 2019; Venier et al., 2018). Such forests, however, are becoming increasingly rare because of the impact of human activities, old-growth forests generally being the first stands to disappear (Aksenov et al., 1999; Cyr et al., 2009; Martin et al., 2020a). Maintaining or restoring large areas of intact forest containing a high proportion of old-growth forest in boreal landscapes must be prioritized for maintaining associated habitats (Sabatini et al., 2020; Venier et al., 2018).

7.3.2 High Habitat Diversity Characterizes Boreal Old-Growth Forests

Although boreal old-growth forests generally contain few large trees and few tree species, these ecosystems are characterized by a high diversity of structures (Fig. 7.3). For example, Martin et al. (2018) identified 11 different old-growth forests types within a 2,200 km² area, defined by specific structures, e.g., canopy cover, basal area, the volume of downed wood debris, and composition (varying proportions of black spruce and balsam fir). This diversity observed within a relatively restricted landscape resulted from different environmental conditions, e.g., surficial deposits and drainage, and disturbance history, e.g., time since the last high-severity and moderate-severity secondary disturbances (Martin et al., 2018, 2020c). These results are in line with other studies that highlighted the heterogeneity in stem diameter distribution, aboveground biomass, and tree species composition observed in the boreal old-growth forests of eastern Canada (McCarthy & Weetman, 2007; Moussaoui et al., 2019; Portier et al., 2018). A high internal diversity for these ecosystems has also been demonstrated in northern Fennoscandia and Russia, where tree species composition, disturbance regime, and abiotic composition can greatly vary among landscapes (Shorohova et al., 2009, 2011). Primary forests dominated by Norway spruce in the alpine regions of eastern Europe (Kozák et al., 2021; Meigs et al., 2017; Trotsiuk et al., 2014) also provide examples of old-growth dynamics, where such reference forests are now almost entirely absent. Boreal old-growth forests are therefore dynamic and diverse ecosystems, from the circumboreal to the local scale. The scarcity of obvious attributes, e.g., very large and tall trees, may nevertheless challenge the recognition of these ecosystems, particularly within large and remote areas where forest surveys are based mainly on remote sensing. For example, Martin et al. (2020b) highlighted that the vast majority of boreal old-growth forests in Québec, Canada, were not identified as such in provincial surveys, probably because the applied size and structure criteria, defined for use in temperate forests, are unsuitable for boreal forests.

The secondary disturbance regime of boreal forests (i.e., a disturbance of low to moderate severity that does not reinitiate forest succession) is defined by a high diversity in its nature, periodicity, spatiality, and severity from the local to the circumboreal scale (Chap. 3; De Grandpré et al., 2018; Kuuluvainen & Aakala, 2011; Shorohova et al., 2011). The characteristics of the disturbance history at the local scale characterize part of forest structural attributes, hence the habitats it contains (Martin et al., 2020c; Meigs et al., 2017) (Fig. 7.4). For example, spruce budworm (*Choristoneura fumiferana*) outbreaks in eastern Canada are top-down disturbances, first killing tall balsam fir trees and progressing toward the understory and spruce species as the outbreak increases in severity (Morin et al., 2009). Windthrow will also kill the tallest trees first. Relative to spruce budworm outbreaks, this latter disturbance is less species-specific and creates fewer snags, produces more fallen logs, and generates more tips and mounds, the latter providing habitats for many forest species (De Grandpré et al., 2018). In contrast, surface fires will generally kill the

◀**Fig. 7.3** Old-growth forests can represent a wide diversity of structures, composition, and dynamics at the local and circumboreal scales. **a** Balsam fir (*Abies balsamifera*) forest in eastern Canada that was severely disturbed 40 years ago by a spruce budworm outbreak. This disturbance produced a stand having a relatively simple diameter structure despite a multicohort age structure and a large deadwood volume; **b** a large white spruce (*Picea glauca*) surrounded by smaller balsam fir trees in eastern Canada; **c** a mixed Siberian larch (*Larix sibirica*), Scots pine (*Pinus sylvestris*), and Norway spruce (*Picea abies*) forest in western Russia; **d** a primeval northern boreal Scots pine forest driven by periodic surface fires; **e** a Norway spruce forest in eastern Russia recently disturbed by moderate-severity windthrows, creating a diversity of soil microhabitats; **f** a dense black spruce (*Picea mariana*)–balsam fir forest in eastern Canada driven by low-severity disturbances; **g** a paludified black spruce forest in eastern Canada. Trees are generally small, but their age can often exceed 250 years; **h** buried deadwood pieces at various stages of decay in a black spruce forest in eastern Canada. A very large portion of deadwood in boreal old-growth forests can be hidden in the soil organic layer; **i** a large Siberian pine (*Pinus sibirica*) log in eastern Russia. These stems may require more than 1,000 years to decompose, sequestering carbon and nutrients and providing a habitat for many wood-inhabiting species; **j** a white spruce with a large wound exposing sapwood in eastern Canada. The disturbance dynamics and the presence of trees from all ages and sizes in old-growth forests favor the development of tree-related microhabitats, necessary for many species. *Photo credits* **a b f–h j** Maxence Martin, **c–e i** Ekaterina Shorohova

understory but preserve the overstory, particularly in pine forests (Mosseler et al., 2003; Shorohova et al., 2009). Secondary disturbances, however, also greatly vary in severity, even for a same disturbance event within a restricted area (Khakimulina et al., 2016; Martin et al., 2019). This variability creates complex matrixes of old-growth forest structures (Kulha et al., 2020; Kuuluvainen et al., 2014). This complexity is also reinforced by variable local abiotic factors that may increase stand sensitivity to a certain disturbance type, e.g., windthrow and hilly topography, or favor tree species that are more susceptible to specific disturbance agents, e.g., balsam fir and spruce budworm outbreaks (De Grandpré et al., 2018). Current knowledge about the complexity of secondary disturbances is, however, still limited. Kuuluvainen and Aakala (2011) classified secondary disturbance severity into three classes of forest dynamics: gap, patch, cohort. They stressed that patch dynamics, i.e., canopy openings between 200 and 10,000 m², have been little studied in Scandinavian forests. Hart and Kleinman (2018) expressed a similar concern, highlighting that moderate-severity disturbances, also called intermediate-severity disturbances, have been generally overlooked in favor of low-severity (e.g., gap dynamics) or high-severity (e.g., crown fire) disturbances. The cumulative impact of disturbances over the centuries, such as recurrent insect epidemics of moderate severity, on the structure and dynamics of old-growth forests also remains to be determined (Martin et al., 2019). Similarly, the dichotomy between young/simple forests following high-severity disturbances and old/complex forests boosted by low-severity disturbances is questioned, and more elaborate successional models are now proposed (Donato et al., 2012; Meigs et al., 2017). Therefore, although there is growing interest in the complex effects of secondary disturbances in boreal landscapes, our knowledge of these dynamics is incomplete.

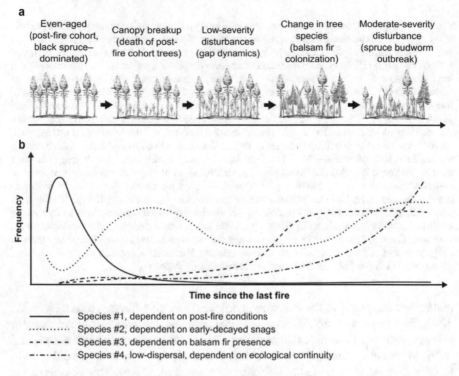

Fig. 7.4 a An example of the changes in structure and composition that an old-growth forest in eastern Canada may follow over time; **b** possible changes in the frequency of species defined by different ecologies in this old-growth forest as a function of its change

Overall, the diversity in stand structural attributes and forest history creates a wide variety of habitats that can differ markedly from one stand to another. A greater vertical and structural complexity favors more diversity in forest species such as invertebrates and epixylics (Desponts et al., 2004; Rheault et al., 2009; Schowalter, 2017). Deadwood-related species often depend on specific substrate characteristics, e.g., tree species, decay stage, contact with the ground, cause of death, and size (Janssen et al., 2011; McMullin et al., 2010; Stokland et al., 2012), and the high abundance and diversity of deadwood at the stand scale often results in greater deadwood-related species richness (Lassauce et al., 2011; Wagner et al., 2014). Kozák et al. (2021), for example, underscored that the characteristics of the secondary disturbance regime, e.g., severity, frequency, and time since the last disturbance, strongly influence saproxylic beetle diversity in alpine forests dominated by Norway spruce. Similarly, trees that survived a disturbance can also act as refugia for low-dispersal species, such as lichens, facilitating their recolonization of a disturbed area (Zemanová et al., 2017). However, the link between biodiversity and old-growth forest attributes can be unclear. Forest age, i.e., time since the last stand-replacing disturbance, for example, is a commonly applied indicator. Some species, e.g., birds and nonvascular plants, may be highly dependent on this variable, whereas other taxa may not (Drapeau et al., 2003; Fenton & Bergeron, 2011). This variability can be partly explained by species-specific requirements in terms of habitat, ecological continuity, and landscape. Moreover, two old-growth forests of the same age can be defined by very different structures because of local abiotic conditions and disturbance history (Martin et al., 2018), implying markedly different habitats (Martin et al., 2021a). Similarly, landscape structure around a given old-growth forest can also strongly influence its use by some forest species, e.g., birds and mammals (Faille et al., 2010; Schmiegelow & Mönkkönen, 2002; Tremblay et al., 2015). For these reasons, although forest age is a relevant indicator, it is not sufficient on its own.

Many abiotic, historical, and spatial factors can hence influence the characteristics of old-forest habitats and their attractiveness to boreal species. The complexity of the interactions between these factors can make it difficult to identify clear links between the structural/functional biodiversity and specific structural/ecological attributes (Burrascano et al. 2018; Kozák et al., 2021; Larrieu et al., 2018). Experiments in close-to-nature silvicultural practices can be effective in gaining a direct understanding of how disturbance dynamics can influence biodiversity (Fenton et al., 2013; Franklin et al., 2019; Koivula and Vanha-Majamaa 2020). Given that these are anthropogenic disturbances, e.g., use of heavy machinery, wood removal, and soil disturbance, it can be difficult to compare them with natural disturbances. Yet, certain types of old-growth forests, especially those more productive stands—and therefore of greater economic value—are more threatened by human activities than others (Martin et al., 2021b). This pattern is consistent with a general trend worldwide where the remaining intact forests are the least attractive for human use (Joppa & Pfaff, 2009). It implies that the risk of losing specific habitats, e.g., habitats observed almost exclusively in the most productive forests, could go unnoticed if boreal old-growth forests continue to be considered homogeneous ecosystems. Although the example of the boreal forests of northern Fennoscandia gives us an idea of the general consequences of old-growth forest rarefaction, it may not be sufficient to accurately understand what can be potentially lost. Therefore, a better understanding of the factors influencing the structural and habitat diversity of old forests, both at the stand and landscape scales, is necessary to maintain the associated biodiversity.

7.4 Ecosystem-Based Management of Boreal Old-Growth Forests: Where to Start?

7.4.1 Accurately Identifying Old-Growth Forests

One of the main challenges for sustainable management of old-growth forests is the need for operational definitions. Defining those stands that can be considered (or not) as *old growth* has been a recurrent issue for forest ecologists and managers

(Hendrickson, 2003; Pesklevits et al., 2011; Wirth et al., 2009). This difficulty is reinforced for boreal forests, where the scarcity of obvious old-growth attributes further complexifies this task (Bergeron & Harper, 2009; Martin et al., 2020b). Issekutz (2020), for instance, underscored that only six of the Canadian provinces (Québec excluded) have an operational definition and that there was little consistency between these definitions (e.g., variable age thresholds between and within provinces, use of different indicators of stand or landscape structure). In Québec, forests are classified as old growth if their age exceeds 80 years in the balsam fir–white birch bioclimatic domain and 100 years in the black spruce–feathermoss bioclimatic domain (MFFP 2016). This definition can be considered close to those of other provinces for boreal forests (e.g., Ontario, Saskatchewan), although some differences remain (Issekutz, 2020). Nonetheless, some stands can attain the old-growth stage well before (Cumming et al., 2000) or after (Kneeshaw & Gauthier, 2003; Martin et al., 2018) the age thresholds used by Québec. This issue mainly concerns boreal territories where primary forests are still abundant. In the case of northern Fennoscandia, where most forests are managed, detailed knowledge of these landscapes' history facilitates the identification of remnant old-growth forests.

Accurately identifying these ecosystems at a large scale is also challenging. Because of the limited longevity of some boreal species, taking core samples to determine tree age provides only limited information (Garet et al., 2012; Kneeshaw & Gauthier, 2003). The typical suppression period for trees in old-growth forests is sometimes challenging to account for, causing an underestimation of forest age (Krause and Morin 2005; Marchand & DesRochers, 2016). The remoteness and vastness of boreal landscapes nevertheless prevent exhaustive field surveys in many regions. Aerial photographic surveys can cover large areas and have often been used; however, their accuracy in identifying old-growth forests has been questioned (Martin et al., 2020b). Therefore, it remains necessary to assess whether this misclassification is an inherent limitation of this inventory method or whether it stems from the use of unsuitable criteria for boreal forests. Other techniques have been explored over the last decades for a more straightforward and more accurate classification of forest ecosystems, in particular the use of LiDAR (light detection and ranging). This technology has provided promising results in identifying old-growth structures in temperate forests (Kane et al., 2010; Torresan et al., 2016). However, related studies involving boreal forests are currently lacking, as current predictive models of forest age generally end at a relatively early age, e.g., 100 or 160 years (Maltamo et al., 2020; Wylie et al., 2019). Multispectral airborne imagery can also complement LiDAR for identifying and discriminating old-growth forests within boreal landscapes (Zhang et al., 2017). Although promising, these methods still require further studies to evaluate their ability to accurately evaluate the *old-growthness* of boreal stands and identify the structural diversity of these stands at a fine scale. Developing new innovative and effective forest survey tools able to discern the complexity of boreal landscapes at a fine scale is essential for ensuring the sustainable management of old-growth forests.

7.4.2 Reducing Anthropogenic Pressure on Old-Growth Forests

Anthropogenic activities severely degrade and fragment boreal old-growth forests, in particular by applying short-rotation, i.e., having a shorter return period than the natural primary disturbance regime, and clear-cut-based forestry (Aksenov et al., 1999; Cyr et al., 2009; Kuuluvainen & Gauthier, 2018). Other disturbances, such as mining and oil and gas extraction, can also severely damage natural boreal landscapes (Venier et al., 2014). Industrial-scale forestry is nonetheless a particularly specific disturbance, as it generally targets old-growth forests first (Bouchard and Pothier 2011; Martin et al., 2020a; Östlund et al., 1997). The scarcity of remnant old-growth forests in northern Fennoscandia provides a striking example of the possible consequences of forest management that excessively exploits old-growth forests. Without explicit constraints favoring old-growth forest protection, simulations for eastern Canadian forests show that these ecosystems will disappear in the coming decades (Bergeron et al., 2017; Didion et al., 2007). Therefore, it is urgent to decrease the logging pressure on old-growth forests in landscapes where they are still present, as it is easier to protect than to restore the systems (Halme et al., 2013).

In addition to the reduction of old-growth areas, clear-cut-based forestry also leads to changes in tree species composition (Bouchard and Pothier 2011; Boucher & Grondin, 2012; Kuuluvainen et al., 2017), landscape homogenization and fragmentation (Faille et al., 2010; Schmiegelow & Mönkkönen, 2002), and a decrease in dead wood richness (Jonsson & Siitonen, 2012; Moussaoui et al., 2016). In the context of ecosystem-based management, new management strategies are necessary to maintain the habitats and services related to boreal forests. A combination of clear-cuts having a longer rotation, careful salvage logging, active forest restoration, retention forestry, and continuous-cover forestry, coupled with an investment in disturbance suppression, are the leading proposals for achieving a balance between sustainable wood provision and environmental objectives (Bauhus et al., 2009; Eyvindson et al., 2021; Halme et al., 2013; Kuuluvainen, 2009; Leduc et al., 2015; Shorohova et al., 2019b). For example, numerous recent experiments underscore the efficacy of low-intensity continuous-cover forestry to maintain old-growth attributes and the associated biodiversity (Fenton et al., 2013; Franklin et al., 2019; Koivula and Vanha-Majamaa 2020).

However, the economic feasibility of these alternative strategies is still debated, particularly for remote boreal regions, and will undoubtedly be an important social question in the coming years (Kneeshaw et al., 2018). In regions already heavily modified by forestry practices, it has been demonstrated that economic benefits outweigh the costs (Eyvindson et al., 2021; Ruel et al., 2013). The benefits of ecosystem-based management alternatives may be significantly reinforced by targeting high-quality wood products for harvest (Rijal et al., 2018) or by including ecosystem services in the financial balance (Anielski & Wilson, 2005). Nevertheless, a certain precaution is required in developing alternative management strategies; for example, by removing deadwood, salvage logging can negatively affect species that depend on this habitat (Nappi et al., 2004; Thorn et al., 2018; Waldron et al., 2013). Continuous-cover forestry can lose its benefits if the harvest rate is too high (Fenton et al., 2013; Franklin et al., 2019). Similarly, the extension of road networks through primary forest landscapes to apply alternative methods can also accentuate problems of fragmentation and the modification of trophic networks (Venier et al., 2018). Finally, strategies for conserving intact boreal old-growth forest tracts mainly on the basis of area thresholds are insufficient, as some old-growth forest types—often those with the highest market value—can be under much greater pressure than others (Martin et al., 2020a, 2021b; Fig. 7.5).

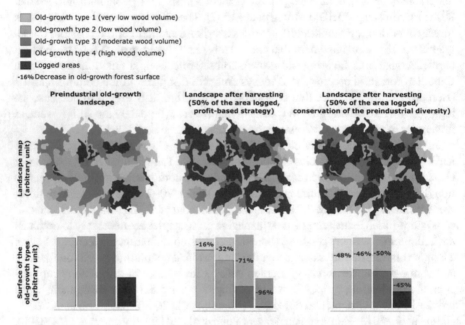

Fig. 7.5 Conceptual scheme of the impact of logging on old-growth forests if only their surface area and not their structural diversity are considered. In this example, old-growth forests can be divided into four types (e.g., different structures and/or composition) distributed along a gradient of merchantable wood volume and, therefore, a gradient of economic interest. A conservation strategy that aims to maintain 50% old-growth forest could then have a very different impact on the residual landscape depending on the criteria used to select the harvested forests: profit-based strategy (objective of maximum profitability from logging, and the more economically valuable old-growth forests are logged first) or the conservation of preindustrial diversity (the proportions of the various old-growth forest types are the same before and after logging)

7.4.3 What is the Place for Boreal Old-Growth Forests in an Uncertain Future?

Climate change will markedly modify boreal landscapes in terms of, for example, disturbance dynamics and tree species composition and hence the characteristics of the oldest forests (Bouchard et al., 2019; Gauthier et al., 2015; Seidl et al., 2017). Nonetheless, evaluating the impacts of climate change on boreal old-growth forests remains challenging, and these impacts can strongly vary between territories and periods (see Chap. 31). Overall, an increase in natural disturbance recurrence and severity is expected, which may markedly decrease the abundance and functionality of boreal old-growth forests (Kuuluvainen & Gauthier, 2018). Boreal tree species, in particular softwoods, may also become less competitive than boreal and sub-boreal hardwood species, implying significant changes in tree species composition (Bouchard et al., 2019). Yet, forest management may remain the leading cause of the loss of boreal old-growth forest surfaces in such areas as eastern Canada, for example (Bergeron et al., 2017). Moreover, it has been emphasized that these forests provide and will continue to provide precious ecosystem services, in particular carbon sequestration, which can help mitigate the impacts of climate change (Lafleur et al., 2018; Thom et al., 2019; Vedrova et al., 2018). Kalliokoski et al. (2020) also recently highlighted that maintaining a continuous forest cover was more effective for carbon sequestration and cooling than increased harvesting rates. However, some boreal regions may become a carbon source because of lower stand productivity and increased fire activity (Miquelajauregui et al., 2019; Walker et al., 2020). This scenario underscores the need to adapt management strategies to local climatic characteristics. Nevertheless, it is generally assumed that maintaining a high diversity of structures and composition in the forest landscape will increase forest resilience and facilitate the adaptation of management practices to new environmental conditions (Augustynczik et al., 2019; Gauthier et al., 2009; Seidl et al., 2017). Old-growth forests will therefore play a key role in adapting ecosystems and human societies to a changing environment (Halpin & Lorimer, 2016; Kuuluvainen & Gauthier, 2018; Leduc et al., 2015). For these reasons, it is essential to protect the remaining old-growth forests and ensure that their functionality, e.g., connectivity, productivity, and diversity, is maintained (Halme et al., 2013; Smith, 2020; Chap. 31). In areas where these ecosystems are now absent, it will be necessary to apply active policies to restore elements and dynamics related to old-growth forests (Bauhus et al., 2009; Halme et al., 2013; Kuuluvainen & Gauthier, 2018; Sabatini et al., 2020; Smith, 2020).

7.5 Conclusions

Old-growth boreal forests are complex ecosystems of high value for biodiversity and human societies. However, these ecosystems are often still described in a reductionist way that does not consider their richness. Old-growth forests are under serious threat, and many measures are urgently needed to protect these forests and their associated habitats and services. A fully effective ecosystem-based management must emphasize the size, connectivity, diversity, and functionality of old-growth forests. The operational tools to achieve this objective remain nevertheless lacking and must be developed rapidly. Similarly, the pressure exerted by human activities on these ecosystems must be significantly and urgently reduced, including in areas where old-growth forests remain abundant. Numerous alternative management practices to short-rotation clear-cuts hold promise and could fulfill this objective. Improved management and protection of old-growth forests go beyond the sphere of forest managers and ecologists alone and involves much broader socioeconomic considerations.

References

Aakala, T. (2011). Temporal variability of deadwood volume and quality in boreal old-growth forests. *Silva Fennica, 45*(5), 969–981.

Achard, F., Eva, H., Mollicone, D., et al. (2009). Detecting intact forests from space: hot spots of loss, deforestation and the UNFCCC. In C. Wirth, G. Gleixner, & M. Heimann (Eds.), *Old-growth forests: Function, fate and value.* Ecological Studies 207 (pp. 411–428). Springer.

Aksenov, D., Karpachevskiy, M., Lloyd, S., et al. (1999). The last of the last: the old-growth forests of boreal Europe (p. 67). Taiga Rescue Network.

Angelstam, P., & Kuuluvainen, T. (2004). Boreal forest disturbance regimes, successional dynamics and landscape structures–a European perspective. *Ecological Bulletins, 51,* 117–136.

Anielski, M., & Wilson, S. (2005). *Counting Canada's natural capital: Assessing the real value of Canada's ecosystem services* (p. 76). Canadian Boreal Initiative.

Augustynczik, A. L. D., Asbeck, T., Basile, M., et al. (2019). Diversification of forest management regimes secures tree microhabitats and bird abundance under climate change. *Science of the Total Environment, 650,* 2717–2730. https://doi.org/10.1016/j.scitotenv.2018.09.366.

Barbé, M., Fenton, N. J., & Bergeron, Y. (2017). Are post-fire residual forest patches refugia for boreal bryophyte species? Implications for ecosystem based management and conservation. *Biodiversity and Conservation, 26*(4), 943–965. https://doi.org/10.1007/s10531-016-1281-9.

Bauhus, J., Puettmann, K., & Messier, C. (2009). Silviculture for old-growth attributes. *Forest Ecology and Management, 258*(4), 525–537. https://doi.org/10.1016/j.foreco.2009.01.053.

Belleau, A., Bergeron, Y., Leduc, A., et al. (2007). Using spatially explicit simulations to explore size distribution and spacing of regenerating areas produced by wildfires: recommendations for designing harvest agglomerations for the Canadian boreal forest. *The Forestry Chronicle, 83*(1), 72–83. https://doi.org/10.5558/tfc83072-1.

Bergeron, Y., & Fenton, N. J. (2012). Boreal forests of eastern Canada revisited: old growth, nonfire disturbances, forest succession, and biodiversity. *Botany, 90*(6), 509–523. https://doi.org/10.1139/b2012-034.

Bergeron, Y., & Harper, K. A. (2009). Old-growth forests in the Canadian boreal: the exception rather than the rule? In Wirth, C., Gleixner, G., Heimann, M. (Eds.) *Old-growth forests: Function, fate and value.* Ecological Studies 207 (pp. 285–300). Berlin, Heidelberg: Springer.

Bergeron, Y., Vijayakumar, D. B. I. P., Ouzennou, H., et al. (2017). Projections of future forest age class structure under the influence of fire and harvesting: Implications for forest management in the boreal forest of eastern Canada. *Forestry, 90*(4), 485–495. https://doi.org/10.1093/forestry/cpx022.

Bondarev, A. (1997). Age distribution patterns in open boreal Dahurican larch forests of Central Siberia. *Forest Ecology and Management, 93*(3), 205–214. https://doi.org/10.1016/S0378-1127(96)03952-7.

Bouchard, M., & Pothier, D. (2011). Long-term influence of fire and harvesting on boreal forest age structure and forest composition in eastern Quebec. *Forest Ecology and Management, 261*(4), 811–820. https://doi.org/10.1016/j.foreco.2010.11.020.

Bouchard, M., Aquilue, N., Perié, C., et al. (2019). Tree species persistence under warming conditions: A key driver of forest response to climate change. *Forest Ecology and Management, 442*, 96–104. https://doi.org/10.1016/j.foreco.2019.03.040.

Boucher, Y., & Grondin, P. (2012). Impact of logging and natural stand-replacing disturbances on high-elevation boreal landscape dynamics (1950–2005) in eastern Canada. *Forest Ecology and Management, 263*, 229–239. https://doi.org/10.1016/j.foreco.2011.09.012.

Burrascano, S., Keeton, W. S., Sabatini, F. M., et al. (2013). Commonality and variability in the structural attributes of moist temperate old-growth forests: A global review. *Forest Ecology and Management, 291*, 458–479. https://doi.org/10.1016/j.foreco.2012.11.020.

Burrascano, S., de Andrade, R. B., Paillet, Y., et al. (2018). Congruence across taxa and spatial scales: Are we asking too much of species data? *Global Ecology and Biogeography, 27*(8), 980–990. https://doi.org/10.1111/geb.12766.

Cadieux, P., Boulanger, Y., Cyr, D., et al. (2020). Projected effects of climate change on boreal bird community accentuated by anthropogenic disturbances in western boreal forest, Canada. *Diversity and Distributions, 26*(6), 668–682. https://doi.org/10.1111/ddi.13057.

Cumming, S. G., Schmiegelow, F. K. A., & Burton, P. J. (2000). Gap dynamics in boreal aspen stands: Is the forest older than we think? *Ecological Applications, 10*(3), 744–759.

Cyr, D., Gauthier, S., Bergeron, Y., et al. (2009). Forest management is driving the eastern North American boreal forest outside its natural range of variability. *Frontiers in Ecology and the Environment, 7*(10), 519–524. https://doi.org/10.1890/080088.

Dambrine, E., Dupouey, J. L., Laut, L., et al. (2007). Present forest biodiversity patterns in France related to former Roman agriculture. *Ecology, 88*(6), 1430–1439. https://doi.org/10.1890/05-1314.

De Grandpré, L., Waldron, K., Bouchard, M., et al. (2018). Incorporating insect and wind disturbances in a natural disturbance-based management framework for the boreal forest. *Forests, 9*(8), 471. https://doi.org/10.3390/f9080471.

Desponts, M., Brunet, G., Bélanger, L., et al. (2004). The eastern boreal old-growth balsam fir forest: a distinct ecosystem. *Canadian Journal of Botany, 82*, 830–849. https://doi.org/10.1139/B04-063.

Didion, M., Fortin, M. J., & Fall, A. (2007). Forest age structure as indicator of boreal forest sustainability under alternative management and fire regimes: A landscape level sensitivity analysis. *Ecological Modelling, 200*(1–2), 45–58. https://doi.org/10.1016/j.ecolmodel.2006.07.011.

Donato, D. C., Campbell, J. L., Franklin, J. F. (2012). Multiple successional pathways and precocity in forest development: Can some forests be born complex? *Journal of Vegetation Science, 23*(3), 576–584. https://doi.org/10.1111/j.1654-1103.2011.01362.x.

Drapeau, P., Leduc, A., Savard, J. P., et al. (2003). Les communautés d'oiseaux des vieilles forêts de la pessière à mousses de la ceinture d'argile : Problèmes et solutions face à l'aménagement forestier. *The Forestry Chronicle, 79*(3), 531–540. https://doi.org/10.5558/tfc79531-3.

Esseen, P., Ehnström, B., Ericson, L., et al. (1992). Boreal forests---the focal habitats of Fennoscandia. In L. Hansson (Ed.), *Ecological principles of nature conservation*. Conservation ecology series: Principles, practices and management (pp. 252–325). Springer.

Eyvindson, K., Du, R., Triviño, M., et al. (2021). High boreal forest multifunctionality requires continuous cover forestry as a dominant management. *Land Use Policy, 100*, 104918. https://doi. org/10.1016/j.landusepol.2020.104918.

Faille, G., Dussault, C., Ouellet, J.-P., et al. (2010) Range fidelity: The missing link between caribou decline and habitat alteration? *Biological Conservation, 143*(11), 2840–2850. https://doi.org/10. 1016/j.biocon.2010.08.001.

Fenton, N. J., & Bergeron, Y. (2011). Dynamic old-growth forests? A case study of boreal black spruce forest bryophytes. *Silva Fennica, 45*, 983–994.

Fenton, N. J., Lecomte, N., Légaré, S., et al. (2005). Paludification in black spruce (*Picea mariana*) forests of eastern Canada: Potential factors and management implications. *Forest Ecology and Management, 213*(1–3), 151–159. https://doi.org/10.1016/j.foreco.2005.03.017.

Fenton, N. J., Imbeau, L., Work, T., et al. (2013). Lessons learned from 12 years of ecological research on partial cuts in black spruce forests of northwestern Quebec. *The Forestry Chronicle, 89*(3), 350–359. https://doi.org/10.5558/tfc2013-065.

Franklin, C. M. A., Macdonald, S. E., Nielsen, S.E. (2019). Can retention harvests help conserve wildlife? Evidence for vertebrates in the boreal forest. *Ecosphere, 10*(3), e02632. https://doi.org/ 10.1002/ecs2.2632.

Franklin, J. F., Spies, T. A., Van Pelt, R., et al. (2002). Disturbances and structural development of natural forest ecosystems with silvicultural implications, using Douglas-fir forests as an example. *Forest Ecology and Management, 155*, 399–423. https://doi.org/10.1016/S0378-112 7(01)00575-8.

Frelich, L. E., Reich, P. B. (2003). Perspectives on development of definitions and values related to old-growth forests. *Environmental Reviews, 11*, S9–S22. https://doi.org/10.1139/a03-011.

Garet, J., Raulier, F., Pothier, D., et al. (2012). Forest age class structures as indicators of sustainability in boreal forest: Are we measuring them correctly? *Ecological Indicators, 23*, 202–210. https://doi.org/10.1016/j.ecolind.2012.03.032.

Gauthier, S., Vaillancourt, M. A., Leduc, A., et al. (Eds.). (2009). *Ecosystem management in the boreal forest* (p. 572). Presses de l'Universite du Quebec.

Gauthier, S., Bernier, P., Kuuluvainen, T., et al. (2015). Boreal forest health and global change. *Science, 349*(6250), 819–822. https://doi.org/10.1126/science.aaa9092.

Gosselin, F., & Larrieu, L. (2020). Developing and using statistical tools to estimate observer effect for ordered class data: The case of the IBP (Index of Biodiversity Potential). *Ecological Indicators, 110*, 105884. https://doi.org/10.1016/j.ecolind.2019.105884.

Halme, P., Allen, K. A., Auniņš, A., et al. (2013). Challenges of ecological restoration: Lessons from forests in northern Europe. *Biological Conservation, 167*, 248–256. https://doi.org/10.1016/ j.biocon.2013.08.029.

Halpin, C. R., & Lorimer, C. G. (2016). Trajectories and resilience of stand structure in response to variable disturbance severities in northern hardwoods. *Forest Ecology and Management, 365*, 69–82. https://doi.org/10.1016/j.foreco.2016.01.016.

Hardenbol, A. A., Junninen, K., & Kouki, J. (2020). A key tree species for forest biodiversity, European aspen (*Populus tremula*), is rapidly declining in boreal old-growth forest reserves. *Forest Ecology and Management, 462*, 118009. https://doi.org/10.1016/j.foreco.2020.118009.

Hart, J. L., & Kleinman, J. S. (2018). What are intermediate-severity forest disturbances and why are they important? *Forests, 9*(9), 579. https://doi.org/10.3390/f9090579.

Harvey BD, Leduc A, Gauthier S, et al (2002) Stand-landscape integration in natural disturbance-based management of the southern boreal forest. *Forest Ecology and Management, 155*(1–3), 369–385. https://doi.org/10.1016/S0378-1127(01)00573-4.

Hendrickson, O. (2003). Old-growth forests: Data gaps and challenges. *The Forestry Chronicle, 79*(3), 645–651. https://doi.org/10.5558/tfc79645-3.

Hunter, M. L., & White, A. S. (1997). Ecological thresholds and the definition of old-growth stands. *Natural Areas Journal, 4*(17), 292–296.

Issekutz, P. B. (2020). *A critical evaluation of old-growth forest definitions in Canada.* BSc honours thesis, Dalhousie University.

Janssen, P., Hébert, C., & Fortin, D. (2011). Biodiversity conservation in old-growth boreal forest: black spruce and balsam fir snags harbour distinct assemblages of saproxylic beetles. *Biodiversity and Conservation, 20*(13), 2917–2932. https://doi.org/10.1007/s10531-011-0127-8.

Jaroszewicz, B., Cholewińska, O., Gutowski, J. M., et al. (2019). Białowieża forest—a relic of the high naturalness of European forests. *Forests, 10*(10), 849. https://doi.org/10.3390/f10100849.

Jonsson, B. G., & Siitonen, J. (2012). Dead wood and sustainable forest management. In J. N. Stokland, J. Siitonen, & B. G. Jonsson (Eds.), *Biodiversity in dead wood* (pp. 303–337). Cambridge University Press.

Joppa, L. N., & Pfaff, A. (2009). High and far: Biases in the location of protected areas. *PLoS ONE, 4*(12), e8273. https://doi.org/10.1371/journal.pone.0008273.

Kalliokoski, T., Bäck, J., Boy, M., et al. (2020). Mitigation impact of different harvest scenarios of Finnish forests that account for albedo, aerosols, and trade-offs of carbon sequestration and avoided emissions. *Frontiers in Forests and Global Change, 3*, 112. https://doi.org/10.3389/ffgc.2020.562044

Kane, V. R., Bakker, J. D., McGaughey, R. J., et al. (2010). Examining conifer canopy structural complexity across forest ages and elevations with LiDAR data. *Canadian Journal of Forest Research, 40*(4), 774–787. https://doi.org/10.1139/X10-064.

Khakimulina, T., Fraver, S., & Drobyshev, I. (2016). Mixed-severity natural disturbance regime dominates in an old-growth Norway spruce forest of northwest Russia. *The Journal of Vegetation Science, 27*(2), 400–413. https://doi.org/10.1111/jvs.12351.

Kimmins, J. P. (2003). Old-growth forest: An ancient and stable sylvan equilibrium, or a relatively transitory ecosystem condition that offers people a visual and emotional feast? Answer—it depends. *The Forestry Chronicle, 79*(3), 429–440. https://doi.org/10.5558/tfc79429-3.

Kneeshaw, D. D., & Gauthier, S. (2003). Old growth in the boreal forest: A dynamic perspective at the stand and landscape level. *Environmental Reviews, 11*, S99–S114. https://doi.org/10.1139/a03-010.

Kneeshaw, D. D., Bergeron, Y., & Kuuluvainen, T. (2011). Forest ecosystem structure and disturbance dynamics across the circumboreal forest. In A. C. Millington, M. Blumler, & U. Schickhoff (Eds.), *The SAGE handbook of biogeography* (pp. 263–280). SAGE.

Kneeshaw, D. D., Burton, P. J., De Grandpré, L., et al. (2018). Is management or conservation of old growth possible in North American boreal forests? In A. M. Barton & W. S. Keeton (Eds.), *Ecology and recovery of eastern old-growth forests* (pp. 139–157). Island Press.

Koivula, M., & Vanha-Majamaa, I. (2020). Experimental evidence on biodiversity impacts of variable retention forestry, prescribed burning, and deadwood manipulation in Fennoscandia. *Ecological Processes, 9*, 11. https://doi.org/10.1186/s13717-019-0209-1.

Kozák, D., Svitok, M., Wiezik, M., et al. (2021). Historical disturbances determine current taxonomic, functional and phylogenetic diversity of saproxylic beetle communities in temperate primary forests. *Ecosystems, 24*(1), 37–55. https://doi.org/10.1007/s10021-020-00502-x

Krause, C., & Morin, H. (2005). Adventive-root development in mature black spruce and balsam fir in the boreal forests of Quebec, Canada. *Canadian Journal of Forest Research, 35*(11), 2642–2654. https://doi.org/10.1139/x05-171.

Kulha, N., Pasanen, L., Holmström, L., et al. (2020). The structure of boreal old-growth forests changes at multiple spatial scales over decades. *Landscape Ecology, 35*(4), 843–858. https://doi.org/10.1007/s10980-020-00979-w.

Kuuluvainen, T. (2009). Forest management and biodiversity conservation based on natural ecosystem dynamics in Northern Europe: The complexity challenge. *AMBIO: A Journal of the Human Environment, 38*(6), 309–315. https://doi.org/10.1579/08-A-490.1.

Kuuluvainen, T., & Aakala, T. (2011). Natural forest dynamics in boreal Fennoscandia: a review and classification. *Silva Fennica, 45*(5), 823841. https://doi.org/10.14214/sf.73.

Kuuluvainen, T., & Gauthier, S. (2018). Young and old forest in the boreal: critical stages of ecosystem dynamics and management under global change. *Forest Ecosystems, 5*, 26. https://doi.org/10.1186/s40663-018-0142-2.

Kuuluvainen, T., Wallenius, T. H., Kauhanen, H., et al. (2014). Episodic, patchy disturbances characterize an old-growth *Picea abies* dominated forest landscape in northeastern Europe. *Forest Ecology and Management, 320*, 96–103. https://doi.org/10.1016/j.foreco.2014.02.024.

Kuuluvainen, T., Hofgaard, A., Aakala, T., et al. (2017). North Fennoscandian mountain forests: History, composition, disturbance dynamics and the unpredictable future. *Forest Ecology and Management, 385*, 140–149. https://doi.org/10.1016/j.foreco.2016.11.031.

Lafleur, B., Fenton, N. J., Simard, M., et al. (2018). Ecosystem management in paludified boreal forests: enhancing wood production, biodiversity, and carbon sequestration at the landscape level. *Forest Ecosystems, 5*(1), 27. https://doi.org/10.1186/s40663-018-0145-z.

Larrieu, L., Gosselin, F., Archaux, F., et al. (2018). Cost-efficiency of cross-taxon surrogates in temperate forests. *Ecological Indicators, 87*, 56–65. https://doi.org/10.1016/j.ecolind.2017.12.044.

Lassauce, A., Paillet, Y., Jactel, H., et al. (2011). Deadwood as a surrogate for forest biodiversity: Meta-analysis of correlations between deadwood volume and species richness of saproxylic organisms. *Ecological Indicators, 11*(5), 1027–1039. https://doi.org/10.1016/j.ecolind.2011.02.004.

Leduc, A., Bernier, P. Y., Mansuy, N., et al. (2015). Using salvage logging and tolerance to risk to reduce the impact of forest fires on timber supply calculations. *Canadian Journal of Forest Research, 45*(4), 480–486. https://doi.org/10.1139/cjfr-2014-0434.

Lilja, S., & Kuuluvainen, T. (2005). Structure of old *Pinus sylvestris* dominated forest stands along a geographic and human impact gradient in mid-boreal Fennoscandia. *Silva Fennica, 39*(3), 407–428. https://doi.org/10.14214/sf.377.

Lindenmayer, D. B., Laurance, W. F., Franklin, J. F., et al. (2014). New policies for old trees: Averting a global crisis in a keystone ecological structure. *Conservation Letters, 7*(1), 61–69. https://doi.org/10.1111/conl.12013.

Lundqvist, L., Ahlström, M. A., Petter Axelsson, E., et al. (2019). Multi-layered Scots pine forests in boreal Sweden result from mass regeneration and size stratification. *Forest Ecology and Management, 441*, 176–181. https://doi.org/10.1016/j.foreco.2019.03.044.

Lutz, J. A., Furniss, T. J., Johnson, D. J., et al. (2018). Global importance of large-diameter trees. *Global Ecology and Biogeography, 27*(7), 849–864. https://doi.org/10.1111/geb.12747.

Maltamo, M., Kinnunen, H., Kangas, A., et al. (2020). Predicting stand age in managed forests using National Forest Inventory field data and airborne laser scanning. *Forest Ecosystems, 7*(1), 44. https://doi.org/10.1186/s40663-020-00254-z.

Marchand, W., & DesRochers, A. (2016). Temporal variability of aging error and its potential effects on black spruce site productivity estimations. *Forest Ecology and Management, 369*, 47–58. https://doi.org/10.1016/j.foreco.2016.02.034.

Martin, M., Fenton, N. J., Morin, H. (2018). Structural diversity and dynamics of boreal old-growth forests case study in Eastern Canada. *Forest Ecology and Management, 422*, 125–136. https://doi.org/10.1016/j.foreco.2018.04.007.

Martin, M., Morin, H., Fenton, N. J. (2019). Secondary disturbances of low and moderate severity drive the dynamics of eastern Canadian boreal old-growth forests. *Annals of Forest Science, 76*(108), 108. https://doi.org/10.1007/s13595-019-0891-2.

Martin, M., Boucher, Y., Fenton, N. J., et al. (2020a). Forest management has reduced the structural diversity of residual boreal old-growth forest landscapes in Eastern Canada. *Forest Ecology and Management, 458*, 1–10. https://doi.org/10.1016/j.foreco.2019.117765.

Martin, M., Fenton, N. J., & Morin, H. (2020b). Boreal old-growth forest structural diversity challenges aerial photographic survey accuracy. *Canadian Journal of Forest Research, 50*, 155–169. https://doi.org/10.1139/cjfr-2019-0177.

Martin, M., Krause, C., Fenton, N. J., et al. (2020c). Unveiling the diversity of tree growth patterns in boreal old-growth forests reveals the richness of their dynamics. *Forests, 11*, 252. https://doi. org/10.3390/f11030252.

Martin, M., Montoro Girona, M., & Morin, H. (2020d). Driving factors of conifer regeneration dynamics in eastern Canadian boreal old-growth forests. *PLoS ONE, 15*(7), e0230221. https:// doi.org/10.1371/journal.pone.0230221.

Martin, M., Fenton, N. J., & Morin, H. (2021a). Tree-related microhabitats and deadwood dynamics form a diverse and constantly changing mosaic of habitats in boreal old-growth forests. *Ecological Indicators, 128*, 107813. https://doi.org/10.1016/j.ecolind.2021.107813.

Martin, M., Grondin, P., Lambert, M.-C., et al. (2021b). Compared to wildfire, management practices reduced old-growth forest diversity and functionality in primary boreal landscapes of eastern Canada. *Frontiers in Forests and Global Change, 4*. https://doi.org/10.3389/ffgc.2021.639397.

McCarthy, J. W., & Weetman, G. (2007). Stand structure and development of an insect-mediated boreal forest landscape. *Forest Ecology and Management, 241*(1–3), 101–114. https://doi.org/10. 1016/j.foreco.2006.12.030.

McMullin, R. T., & Wiersma, Y. F. (2019). Out with OLD growth, in with ecological continNEWity: New perspectives on forest conservation. *Frontiers in Ecology and the Environment, 17*(3),176– 181. https://doi.org/10.1002/fee.2016.

McMullin, R. T., Duinker, P. N., Richardson, D. H. S., et al. (2010). Relationships between the structural complexity and lichen community in coniferous forests of southwestern Nova Scotia. *Forest Ecology and Management, 260*(5), 744–749. https://doi.org/10.1016/j.foreco.2010.05.032.

Meigs, G. W., Morrissey, R. C., Bače, R., et al. (2017). More ways than one: Mixed-severity disturbance regimes foster structural complexity via multiple developmental pathways. *Forest Ecology and Management, 406*, 410–426. https://doi.org/10.1016/j.foreco.2017.07.051.

Ministère des Forêts de la Faune et des Parcs (MFFP). (2016). *Intégration des enjeux écologiques dans les plans d'aménagement forestier intégré de 2018–2023, Cahier 2.1–Enjeux liés à la structure d'âge des forêts.* (p. 67). Quebec: Direction de l'aménagement et de l'environnement forestiers (ed) Gouvernement du Québec.

Miquelajauregui, Y., Cumming, S. G., & Gauthier, S. (2019). Sensitivity of boreal carbon stocks to fire return interval, fire severity and fire season: a simulation study of black spruce forests. *Ecosystems, 22*(3), 544–562. https://doi.org/10.1007/s10021-018-0287-4.

Morin, H., Laprise, D., Simon, A. A., et al. (2009). Spruce budworm outbreak regimes in in eastern North America. In S. Gauthier, M. A. Vaillancourt, A. Leduc, L. De Grandpré, D. D. Kneeshaw, H. Morin, P. Drapeau, & Y. Bergeron (Eds.), *Ecosystem management in the boreal forest* (pp. 156–182). Les Presses de l'Universite du Quebec.

Mosseler, A., Thompson, I., & Pendrel, B. A. (2003). Overview of old-growth forests in Canada from a science perspective. *Environmental Reviews, 11*, S1–S7. https://doi.org/10.1139/a03-018.

Moussaoui, L., Fenton, N. J., Leduc, A., et al. (2016). Deadwood abundance in post-harvest and post-fire residual patches: An evaluation of patch temporal dynamics in black spruce boreal forest. *Forest Ecology and Management, 368*, 17–27. https://doi.org/10.1016/j.foreco.2016.03.012.

Moussaoui, L., Leduc, A., Fenton, N. J., et al. (2019). Changes in forest structure along a chronosequence in the black spruce boreal forest: Identifying structures to be reproduced through silvicultural practices. *Ecological Indicators, 97*, 89–99. https://doi.org/10.1016/j.ecolind.2018. 09.059.

Munoz, S. E., & Gajewski, K. (2010). Distinguishing prehistoric human influence on late-Holocene forests in southern Ontario, Canada. *The Holocene, 20*(6), 967–981. https://doi.org/10.1177/095 9683610362815.

Nappi, A., Drapeau, P., & Savard, J. P. L. (2004). Salvage logging after wildfire in the boreal forest: Is it becoming a hot issue for wildlife? *The Forestry Chronicle, 80*(1), 67–74. https://doi.org/10. 5558/tfc80067-1.

Odum, E. P. (1969). The strategy of ecosystem development. *Science, 164*(3877), 262–270. https:// doi.org/10.1126/science.164.3877.262.

Oliver, C. D., & Larson, B. C. (1996). *Forest stand dynamics.* John Wiley & Sons Inc.

Östlund, L., Zackrisson, O., & Axelsson, A. L. (1997). The history and transformation of a Scandinavian boreal forest landscape since the 19th century. *Canadian Journal of Forest Research, 27*(8), 1198–1206. https://doi.org/10.1139/x97-070.

Paillet, Y., Archaux, F., Boulanger, V., et al. (2017). Snags and large trees drive higher tree microhabitat densities in strict forest reserves. *Forest Ecology and Management, 389*, 176–186. https://doi.org/10.1016/j.foreco.2016.12.014.

Paillet, Y., Debaive, N., Archaux, F., et al. (2019). Nothing else matters? Tree diameter and living status have more effects than biogeoclimatic context on microhabitat number and occurrence: An analysis in French forest reserves. *PLoS ONE, 14*(5), e0216500. https://doi.org/10.1371/journal.pone.0216500.

Pesklevits, A., Duinker, P. N., & Bush, P. G. (2011) Old-growth forests: Anatomy of a wicked problem. *Forests, 2*(1), 343–356. https://doi.org/10.3390/f2010343.

Pollock, S. L., Payette, S. (2010). Stability in the patterns of long-term development and growth of the Canadian spruce-moss forest. *Journal of Biogeography, 37*(9), 1684–1697. https://doi.org/10.1111/j.1365-2699.2010.02332.x.

Portier, J., Gauthier, S., Cyr, G., et al. (2018). Does time since fire drive live aboveground biomass and stand structure in low fire activity boreal forests? Impacts on their management. *Journal of Environmental Management, 225*(April), 346–355. https://doi.org/10.1016/j.jenvman.2018.07.100.

Potapov, P., Hansen, M. C., Laestadius, L., et al. (2017). The last frontiers of wilderness: Tracking loss of intact forest landscapes from 2000 to 2013. *Science Advances, 3*(1), e1600821. https://doi.org/10.1126/sciadv.1600821.

Potapov, P., Yaroshenko, A., Turubanova, S., et al. (2008). Mapping the world's intact forest landscapes by remote sensing. *Ecology & Society, 13*(2), 51. https://doi.org/10.5751/ES-02670-130251.

Puettmann, K. J., Coates, K. D., & Messier, C. (2009). *A critique of silviculture: Managing for complexity* (p. 190). Island Press.

Rheault, H., Bélanger, L., Grondin, P., et al. (2009) Stand composition and structure as indicators of epixylic diversity in old-growth boreal forests. *Ecoscience, 16*(2), 183–196. https://doi.org/10.2980/16-2-3216.

Rijal, B., LeBel, L., Martell, D. L., et al. (2018). Value-added forest management planning: A new perspective on old-growth forest conservation in the fire-prone boreal landscape of Canada. *Forest Ecology and Management, 429*, 44–56. https://doi.org/10.1016/j.foreco.2018.06.045.

Ruel, J.-C., Fortin, D., & Pothier, D. (2013). Partial cutting in old-growth boreal stands: An integrated experiment. *The Forestry Chronicle, 89*(03), 360–369. https://doi.org/10.5558/tfc2013-066.

Sabatini, F. M., Burrascano, S., Keeton, W. S., et al. (2018). Where are Europe's last primary forests? *Diversity and Distributions, 24*(10), 1426–1439. https://doi.org/10.1111/ddi.12778.

Sabatini, F. M., Keeton, W. S., Lindner, M., et al. (2020). Protection gaps and restoration opportunities for primary forests in Europe. *Diversity and Distributions, 26*(12), 1646–1662. https://doi.org/10.1111/ddi.13158.

Satterfield, T. (2002). *Anatomy of a conflict: Identity, knowledge, and emotion in old-growth forests*. UBC Press.

Schmiegelow, F. K. A., & Mönkkönen, M. (2002). Habitat loss and fragmentation in dynamic landscapes: Avian perspectives from the boreal forest. *Ecological Applications, 12*(2), 375–389.

Schowalter, T. (2017). Arthropod diversity and functional importance in old-growth forests of North America. *Forests, 8*(4), 97. https://doi.org/10.3390/f8040097.

Seidl, R., Thom, D., Kautz, M., et al. (2017). Forest disturbances under climate change. *Nature Climate Change, 7*(6), 395–402. https://doi.org/10.1038/nclimate3303.

Shorohova, E., Fedorchuk, V., Kuznetsova, M., et al. (2008). Wind-induced successional changes in pristine boreal *Picea abies* forest stands: Evidence from long-term permanent plot records. *Forestry, 81*(3), 335–359. https://doi.org/10.1093/forestry/cpn030.

Shorohova, E., & Kapitsa, E. (2015). Stand and landscape scale variability in the amount and diversity of coarse woody debris in primeval European boreal forests. *Forest Ecology and Management, 356*, 273–284. https://doi.org/10.1016/j.foreco.2015.07.005.

Shorohova, E., Kapitsa, E., Ruokolainen, A., et al. (2019a). Types and rates of decomposition of *Larix sibirica* trees and logs in a mixed European boreal old-growth forest. *Forest Ecology and Management, 439*, 173–180. https://doi.org/10.1016/j.foreco.2019.03.007.

Shorohova, E., Kneeshaw, D. D., Kuuluvainen, T., et al. (2011). Variability and dynamics of old-growth forests in the circumboreal zone: Implications for conservation, restoration and management. *Silva Fennica, 45*(5), 785–806. https://doi.org/10.14214/sf.72.

Shorohova, E., Kuuluvainen, T., Kangur, A., et al. (2009). Natural stand structures, disturbance regimes and successional dynamics in the Eurasian boreal forests: a review with special reference to Russian studies. *Annals of Forest Science, 66*(2), 201. https://doi.org/10.1051/forest/2008083.

Shorohova, E., Sinkevich, S., Kryshen, A., et al. (2019b). Variable retention forestry in European boreal forests in Russia. *Ecological Processes, 8*, 34. https://doi.org/10.1186/s13717-019-0183-7.

Smith, R. B. (2020). *Enhancing Canada's climate change ambitions with natural climate solutions.* Vedalia Biological Inc.

Spies, T. A., & Franklin, J. F. (1991). The structure of natural young, mature, and old-growth Douglas-fir forests in Oregon and Washington. In L. F. Luggiero, K. B. Aubry, A. B. Carey, & M. H. Huff (Eds.), *Wildlife and vegetation of unmanaged Douglas-fir forests* (pp. 91–109). Portland: U.S. Department of Agriculture, Forest Service, Pacific Northwest Research Station.

Stavrova, N. I., Gorshkov, V. V., Katjutin, P. N., et al. (2020). The structure of northern Siberian spruce-Scots pine forests at different stages of post-fire succession. *Forests, 11*, 558. https://doi.org/10.3390/f11050558.

Stokland, J. N., Siitonen, J., & Jonsson, B. G. (Eds.). (2012). *Biodiversity in dead wood* (p. 509). Cambridge University Press.

Stokland, J. N., Woodall, C. W., Fridman, J., et al. (2016). Burial of downed deadwood is strongly affected by log attributes, forest ground vegetation, edaphic conditions, and climate zones. *Canadian Journal of Forest Research, 46*(12), 1451–1457. https://doi.org/10.1139/cjfr-2015-0461.

Svensson, J., Bubnicki, J. W., Jonsson, B. G., et al. (2020). Conservation significance of intact forest landscapes in the Scandinavian Mountains Green Belt. *Landscape Ecology, 35*(9), 2113–2131. https://doi.org/10.1007/s10980-020-01088-4.

Thom, D., Golivets, M., Edling, L., et al. (2019). The climate sensitivity of carbon, timber, and species richness covaries with forest age in boreal-temperate North America. *Global Change Biology, 25*, 2446–2458. https://doi.org/10.1111/gcb.14656.

Thorn, S., Bässler, C., Brandl, R., et al. (2018). Impacts of salvage logging on biodiversity: A meta-analysis. *Journal of Applied Ecology, 55*(1), 279–289. https://doi.org/10.1111/1365-2664.12945.

Torresan, C., Corona, P., Scrinzi, G., et al. (2016). Using classification trees to predict forest structure types from LiDAR data. *Annals of Forest Research, 59*(2), 281–298. https://doi.org/10.15287/afr.2016.423.

Tremblay, J. A., Savard, J. P. L., & Ibarzabal, J. (2015). Structural retention requirements for a key ecosystem engineer in conifer-dominated stands of a boreal managed landscape in eastern Canada. *Forest Ecology and Management, 357*, 220–227. https://doi.org/10.1016/j.foreco.2015.08.024.

Trotsiuk, V., Svoboda, M., Janda, P., et al. (2014). A mixed severity disturbance regime in the primary *Picea abies* (L.) Karst. forests of the Ukrainian Carpathians. *Forest Ecology and Management, 334*, 144–153. https://doi.org/10.1016/j.foreco.2014.09.005.

Vedrova, E. F., Mukhortova, L. V., & Trefilova, O. V. (2018). Contribution of old growth forests to the carbon budget of the boreal zone in central Siberia. *Bulletin of the Russian Academy of Sciences, 45*(3), 288–297. https://doi.org/10.1134/S1062359018030111.

Vehmas, M., Kouki, J., & Eerikäinen, K. (2009). Long-term spatio-temporal dynamics and historical continuity of European aspen (*Populus tremula* L.) stands in the Koli National Park, eastern Finland. *Forestry, 82*(2), 135–148. https://doi.org/10.1093/forestry/cpn044.

Venier, L. A., Thompson, I. D., Fleming, R., et al. (2014). Effects of natural resource development on the terrestrial biodiversity of Canadian boreal forests. *Environmental Reviews, 22*(4), 457–490. https://doi.org/10.1139/er-2013-0075.

Venier, L. A., Walton, R., Thompson, I. D., et al. (2018). A review of the intact forest landscape concept in the Canadian boreal forest: its history, value, and measurement. *Environmental Reviews, 26*(4), 369–377. https://doi.org/10.1139/er-2018-0041.

Wagner, C., Schram, R. T., McMullin, R. T., et al. (2014). Lichen communities in two old-growth pine (*Pinus*) forests. *Lichenologist, 46*(5), 697–709. https://doi.org/10.1017/S00242829 1400022X.

Waldron, K., Ruel, J.-C., & Gauthier, S. (2013). Forest structural attributes after windthrow and consequences of salvage logging. *Forest Ecology and Management, 289*, 28–37. https://doi.org/10.1016/j.foreco.2012.10.006.

Walker, X. J., Baltzer, J. L., Bourgeau-Chavez, L., et al. (2020). Patterns of ecosystem structure and wildfire carbon combustion across six ecoregions of the North American boreal forest. *Frontiers in Forests and Global Change, 3*, 87. https://doi.org/10.3389/ffgc.2020.00087.

Watson, J. E. M., Evans, T., Venter, O., et al. (2018). The exceptional value of intact forest ecosystems. *Nature Ecology and Evolution, 2*(4), 599–610. https://doi.org/10.1038/s41559-018-0490-x.

Wirth, C., & Lichstein, J. W. (2009). The imprint of species turnover on old-growth forest carbon balances-Insights from a trait-based model of forest dynamics. In C. Wirth, G. Gleixner, & M. Heimann (Eds.), *Old-growth forests: Function, fate and value*. Ecological Studies 207 (pp. 81–113). Springer

Wirth, C., Messier, C., Bergeron, Y., et al. (2009). Old-growth forest definitions: A pragmatic view. In C. Wirth, G. Gleixner, & M. Helmann (Eds.), *Old-growth forests: Function, fate and value*. Ecological Studies 207 (pp. 11–33). Springer.

Wylie, R. R. M., Woods, M. E., & Dech, J. P. (2019). Estimating stand age from airborne laser scanning data to improve models of black spruce wood density in the boreal forest of Ontario. *Remote Sensing, 11*(17), 2022. https://doi.org/10.3390/rs11172022.

Zemanová, L., Trotsiuk, V., Morrissey, R. C., et al. (2017). Old trees as a key source of epiphytic lichen persistence and spatial distribution in mountain Norway spruce forests. *Biodiversity and Conservation, 26*(8), 1943–1958. https://doi.org/10.1007/s10531-017-1338-4.

Zhang, W., Hu, B., Woods, M., et al. (2017) Characterizing forest succession stages for wildlife habitat assessment using multispectral airborne imagery. *Forests, 8*(7), 234. https://doi.org/10.3390/f8070234.

Chapter 8
Ecological Classification in Forest Ecosystem Management: Links Between Current Practices and Future Climate Change in a Québec Case Study

Pierre Grondin, Marie-Hélène Brice, Yan Boulanger, Claude Morneau, Pierre-Luc Couillard, Pierre J. H. Richard, Aurélie Chalumeau, and Véronique Poirier

Abstract Climate change is expected to profoundly impact boreal forests, ranging from changes in forest composition and productivity to modifications in disturbance regimes. These climate-induced changes represent a major challenge for forest ecosystem management, as information based on ecological classification may no longer provide a straightforward guide for attaining management goals in the future. In this chapter, we examine how climate change could influence the use of ecological classification and by what means this approach can continue to be relevant for guiding the ongoing development of management practices. We address these questions by first describing ecological classification, using the example of Québec's classification system, and then showing its importance in forest ecosystem management. Using a forest landscape in Québec as a case study, we then look at how climate change could affect boreal forest ecosystems by presenting a detailed, multi-step analysis that considers climate analogs, habitat suitability, and changes in forest composition. We show that at the end of the century, the vegetation of the *Abies-Betula* western subdomain will not change sufficiently to resemble that of its climate analog, currently located ~500 km to the south. Changes in fire frequency and severity could

P. Grondin (✉) · A. Chalumeau · V. Poirier
Direction de la recherche forestière, ministère des Forêts, de la Faune et des Parcs du Québec, 2700 rue Einstein, Québec, QC G1P 3W8, Canada
e-mail: pierre.grondin@mffp.gouv.qc.ca

A. Chalumeau
e-mail: Aurelie.chalumeau@mffp.gouv.qc.ca

V. Poirier
e-mail: veronique.poirier@mffp.gouv.qc.ca

M.-H. Brice
Jardin botanique de Montréal, 4101 Sherbrooke Est, Montréal, QC H1X 2B2, Canada
e-mail: marie-helene.brice@umontreal.ca

Département de sciences biologiques, Institut de recherche en biologie végétale, Université de Montréal, 4101 Sherbrooke Est, Montréal, QC H1X 2B2, Canada

© The Author(s) 2023
M. M. Girona et al. (eds.), *Boreal Forests in the Face of Climate Change*,
Advances in Global Change Research 74,
https://doi.org/10.1007/978-3-031-15988-6_8

significantly modify forest dynamics and composition. Consequently, the potential vegetation and the successional pathways defined under the current climate could change and follow new successional trajectories. This possible reality forces us to question some fundamental aspects of ecological classification. However, we argue that ecological classification can still provide a valuable framework for future forest management, particularly in continuing to recognize the various types of ecosystems present along toposequences. Given the changes expected in forest vegetation composition and dynamics, future variability and uncertainty must be integrated into the current stable classification units and predictable successional trajectories of ecological classification.

8.1 Introduction

Ecological classification provides complete descriptions of biophysical forest features and their historical variability at various spatial scales (global, regional, local). This ecological information is necessary to inform forest ecosystem management, which consists of implementing silvicultural strategies that aim to ensure the long-term maintenance of ecosystem functions and, consequently, the social and economic benefits that forest ecosystems provide to society (Chap. 1; Gauthier et al., 2009). However, climate change is expected to affect growth rates, disturbance regimes, species distributions, and, ultimately, biodiversity (Boulanger et al., 2014, 2016; Périé & de Blois, 2016; Périé et al., 2014; Price et al., 2013). These changes raise serious concerns for forest ecosystem management if knowledge from ecological classification can no longer provide a reliable guide for meeting future management objectives (Millar et al., 2007; Nagel et al., 2017; Puettmann, 2011).

In this chapter, we first introduce ecological classification and the concept of potential vegetation (Sect. 8.2.1) and then characterize the principles of the Québec hierarchical classification system (Sect. 8.2.2). We then consider historical variability

Y. Boulanger
Natural Resources Canada, Canadian Forest Service, Laurentian Forestry Centre, 1055 rue du PEPS, P.O. Box 10380, Stn. Sainte-Foy, Québec, QC G1V 4C7, Canada
e-mail: yan.boulanger@NRCan-RNCan.gc.ca

C. Morneau · P.-L. Couillard
Direction des inventaires forestiers, ministère des Forêts, de la Faune et des Parcs du Québec, 5700 4e Avenue Ouest, Québec, QC G1H 6R1, Canada
e-mail: Claude.Morneau@mffp.gouv.qc.ca

P.-L. Couillard
e-mail: Pierre-Luc.Couillard@mffp.gouv.qc.ca

P. J. H. Richard
Département de géographie, Université de Montréal, Complexe des sciences, P.O. Box 6128, Stn. Centre-Ville, Montréal, QC H3C 3J7, Canada
e-mail: pierrejhrichard@sympatico.ca

at regional and local scales (Sect. 8.2.3) and the importance of ecological classification in forest ecosystem management (Sect. 8.2.4). Next, using a three-step analysis, we indicate how climate change could modify boreal forest ecosystems and ecological classification (Sect. 8.3.1). In the final section (Sect. 8.3.2), we discuss the issues and challenges raised by climate change and illustrate how the ecological classification framework could integrate future variability in forest dynamics in the context of risk management. Throughout the chapter, we focus on the boreal biome, using the case study of the *Abies balsamea–Betula papyrifera* western subdomain (84,567 km^2) in Québec, Canada (hereafter the *Abies-Betula* w. subdomain).

8.2 Ecological Classification of Ecosystems

We define ecological classification as "an analytical process that consists of delineating and defining ecosystems, at different spatial scales, on the basis of the abiotic and biotic factors that govern their development (climate, disturbances, physical environment, plant succession) along ecological gradients for the purposes of land resource management" (Bailey, 2009; Barnes et al., 1982; Rowe & Sheard, 1981; Sims et al., 1996; Whittaker, 1962). The ecological classification of ecosystems has developed progressively in parts of the world, including in Canada, e.g., the provinces of Québec (Fig. 8.1; MFFP 2021; Saucier et al., 2009, 2010), British Columbia (Banner et al., 1993; MacKenzie & Meidinger, 2018; Pojar et al., 1987) and others (Baldwin et al., 2020), as well as in the United States (Bailey, 2009), Scandinavia (Engelmark & Hytteborn, 1999; Sjörs, 1999), and northeastern Asia (Krestov et al., 2009).

8.2.1 Ecological Classification: Potential Vegetation

Ecological classification is based on the science of phytosociology, which originated more than a century ago when Flahault and Schröter (1910) defined a plant association or community as "a vegetal grouping of determined floristic composition, presenting a uniform physiognomy that grows in uniform site conditions." From the school of phytosociology emerged the concept of potential natural vegetation (hereafter potential vegetation). This concept refers to *a stable state* or *late-successional vegetation stage* at a particular site under the existing climatic and edaphic conditions (Baldwin et al., 2020; Blouin & Grandtner, 1971; Braun-Blanquet, 1972; Clements, 1936; Daubenmire, 1968; Küchler, 1964; Tuxen 1956 in Härdtle, 1995). Gradually, although mainly in the early 1970s, interest focused on understanding natural disturbances, particularly fire frequency and severity, and the impacts of disturbance on vegetation dynamics (Damman, 1964; Heinselman, 1973; Rowe & Scotter, 1973; White, 1979). For example, in some cases, the late-successional stage is never reached because of recurrent disturbances, such as fire, and only the early- or mid-successional

a Continental scale: circumboreal zonation

b Continental scale: vegetation zones (Arctic, boreal, temperate)

c Continental scale: vegetation subzones of the boreal zone (forest tundra, open boreal forest, closed boreal forest)

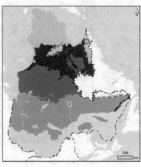

d Supraregional scale: *Abies balsamea–Betula papyrifera* bioclimatic domain and its western and eastern subdomains; Fig 8.3 shows all the domains

e Regional scale: ecological regions of the *Abies balsamea–Betula papyrifera* western subdomain

f Landscape scale: regional landscapes and ecological districts for ecological region 5C

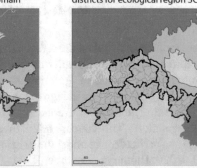

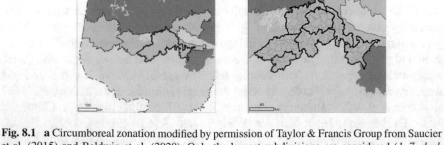

Fig. 8.1 **a** Circumboreal zonation modified by permission of Taylor & Francis Group from Saucier et al. (2015) and Baldwin et al. (2020). Only the largest subdivisions are considered (*1–7*, *dark green*). *1* North European boreal, *2* Western Siberian boreal, *3* Central Siberian boreal, *4* Northeastern Siberian boreal, *5* Alaska–Yukon boreal, *6* West-central North American boreal, *7* Eastern North American boreal. **b–f** Ecological classifications of the ministère des Forêts, de la Faune et des Parcs du Québec (MFFP 2021 [CC-BY 4.0]; Saucier et al., 2009, 2010, with permission from Multimondes and AgroParisTech, respectively)

stage is attained after disturbances (Bergeron et al., 2014; Cogbill, 1985; Couillard et al. 2016; Dansereau, 1957; Frégeau et al., 2015; Frelich & Reich, 1995; Grondin & Gosselin, 2013; Grandtner, 1966; White, 1979). Anthropogenic disturbances are also considered an important factor that can modify the dynamics of potential vegetation (Danneyrolles et al., 2020; Härdtle, 1995; Laflamme et al., 2016; Seastedt et al., 2008). Québec's ecological classification system defines potential vegetation as "a classification unit that includes the complete set of communities associated to the late-successional stage on a given site," thus integrating vegetation dynamics within the concept (Saucier et al., 2009). Potential vegetation is defined on the basis of a particular set of tree species (early- and late-successional species) and understory indicator species that grow together on similar site conditions. Potential vegetation is considered permanent if soil and climatic conditions and the disturbance regime do not change (Saucier et al., 2009). It is mainly on the basis of this predictable successional trajectory that the concept has been challenged (Loidi & Fernández-González, 2012).

The concept of potential vegetation is widely used worldwide as the basic unit of vegetation classification systems (Härdtle, 1995; Loidi & Fernández-González, 2012; Sims et al., 1996). The distribution of potential vegetation (Fig. 8.2) can be illustrated along a toposequence, which divides the landscape into relatively homogeneous sections with respect to topography (slope, elevation), microclimate, drainage, surficial deposits, nutrient regime, soil type, disturbance regime, and forest dynamics (Banner et al., 1993; Barnes et al., 1982; Blouin et al., 2008; Rey, 1960; Rowe & Sheard, 1981; White 1973). This combined use of the physical environment and forest dynamics is termed a toposequence–chronosequence approach (Damman, 1964). Potential vegetation typical of mesic sites—zonal soils sensu Baldwin et al. (2020) and Pojar et al. (1987)—reflects the regional climate (temperature and precipitation). Other potential vegetation characterizing the toposequence is associated with regional climate and reflects the influence of specific local environmental characteristics (e.g., *Picea-Sphagnum* potential vegetation, which grows on poorly drained sites). Toposequences present a synthetic view of the ecological information about a landscape that can be used to produce an ecological map of potential vegetation. Such maps present a global view of the distribution of vegetation over a specific territory (Blouin & Grandtner, 1971; Küchler, 1964).

8.2.2 Ecological Classification: A Hierarchical Approach

Various hierarchical classifications have been developed for describing and managing ecosystems at different spatial scales, and each has its own nomenclature and levels (Bailey, 2009; Baldwin et al., 2020; Banner et al., 1993; Klijn & Udo de Haes, 1994; MacKenzie and Meidinger 2017; Powell, 2000; Sims et al., 1996). The ecological classification system developed in Québec (MFFP 2021; Saucier et al., 2009, 2010) was influenced by European phytosociological classification (Grandtner, 1966) and

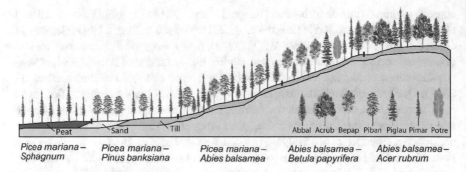

Picea mariana –   Picea mariana –   Picea mariana –   Abies balsamea –   Abies balsamea –
Sphagnum   Pinus banksiana   Abies balsamea   Betula papyrifera   Acer rubrum

Fig. 8.2 Zonation of potential vegetation illustrated along a toposequence characterizing the southern part of the *Abies balsamea–Betula papyrifera* western subdomain, Québec. The species codes are defined in Fig. 8.7

Australian landscape classification (Jurdant et al., 1977). Québec's ecological classification system consists of multiple hierarchy levels corresponding to various spatial scales. At the *continental scale* (~1,000,000 km²), vegetation zones and subzones correspond to very large territories characterized by a vegetation physiognomy and floristic composition that reflect macroclimatic conditions controlled by latitude and continentality. The boreal zone in Québec extends from the northern temperate to Arctic zones and is part of the eastern North American boreal floristic subdivision (Fig. 8.1a and b). It is subdivided from south to north into three subzones based on vegetation structure: closed boreal forest, open boreal forest, and forest–tundra (Fig. 8.1c). This latitudinal zonation is common in the boreal biome, which forms a forest belt around the Northern Hemisphere (Fig. 8.1a). At the *supraregional scale* (~100,000 km²), the bioclimatic domains and subdomains of Québec encompass large areas characterized by relatively homogeneous mesoclimatic conditions that are associated with the dominant potential vegetation of mesic sites and a natural disturbance regime (Fig. 8.1d). At the *regional scale* (~10,000 km²), each ecological region is characterized by the relative abundance of a potential vegetation and their respective physical environments (ecological types); for example, in the *Abies balsamea–Betula papyrifera* domain, ecological region 5B differs from region 5C by having a less rugged relief and much larger areas of organic deposits. This physical environment leads to a greater abundance of *Picea-Sphagnum* potential vegetation on wet organic soils and a lower abundance of *Abies-Betula* potential vegetation, which is normally typical of mesic sites in this bioclimatic domain. At the *landscape scale* (~100–1,000 km²), the regional landscape unit is a relatively homogeneous portion of land in terms of relief, altitude, surficial deposits, hydrography, and potential vegetation (Fig. 8.1f). The ecological district, delimited according to the same criteria as the regional landscapes but at a finer scale, is the basic mapping unit at all higher levels described above. The *local scale* (0.1–1 km²) is characterized by the potential vegetation, ecological type, and forest type. The ecological type combines a potential vegetation type and a physical environment type and is considered a stable ecological unit. The forest type describes the observed vegetation using physiognomy, overstory tree species composition, and understory indicator species of site conditions.

In Québec, the boreal zone is subdivided into four bioclimatic domains distributed north to south along a temperature gradient (Fig. 8.3; Grondin et al., 2007, 2014; MFFP 2021; Richard, 1987; Saucier et al., 2009). The southernmost *Abies balsamea–Betula papyrifera* domain comprises mainly mixed boreal forests. The *Picea mariana*–moss domain, located to the north of this southernmost domain, is characterized by closed coniferous boreal forests. Further north, the *Picea mariana–lichen* domain is characterized by open forests, whereas the northernmost forest–tundra domain consists of a mosaic of forest and treeless communities where, with increasing latitude, forests are increasingly scattered and confined to sheltered locations up to the Arctic tree line (MFFP 2021; Payette, 1992; Rowe & Scotter, 1973). Most bioclimatic domains in Québec are subdivided further into two subdomains (west and east) because of differences in precipitation, physical features, and the dominant natural disturbance regimes, which produce differences in the dominant vegetation (Couillard et al., 2019; Grondin et al., 2007). The climate of the western subdomains is more continental and drier, therefore more susceptible to fire than the eastern subdomains, which have a maritime influence, as illustrated by ombrothermal areas (Fig. 8.4a; Rey, 1960; Richard, 1978).

At the local scale, the domain, subdomain, and ecological region are all characterized by a range of potential vegetation distributed along toposequences. In the toposequence of the southern part of the *Abies-Betula* w. subdomain (Fig. 8.2), the warmer hilltops host a mixed boreal vegetation (*Betula papyrifera*, *Abies balsamea*, *Picea glauca*, *P. mariana*) with the presence of *Acer rubrum*, which forms the *Abies balsamea–Acer rubrum* potential vegetation (hereafter *Abies-Acer*). The mesic midslopes (zonal soils) are occupied by the potential vegetation that best reflects the climatic conditions of the *Abies balsamea–Betula papyrifera* bioclimatic domain, which is the *Abies-Betula* potential vegetation. The lower subhydric, and often stoniest, slopes are occupied by the *Picea mariana–Abies balsamea* potential vegetation (*Picea-Abies*), and *Pinus banksiana* may also be present. Finally, the flatter areas host the *Picea mariana–Pinus banksiana* potential vegetation (*Picea-Pinus*) on till or sand, whereas the wet organic soils support the *Picea-Sphagnum* potential vegetation.

8.2.3 Ecological Classification: A Historical Perspective

A historical perspective (millennial scale) that considers the range of natural variability of vegetation can enhance the description of ecological classification units and forest dynamics. The concept of natural variability was introduced in the 1990s to incorporate an understanding of past spatial and temporal variability into ecosystem management (Cissel et al., 1999; Morgan et al., 1994; Powell, 2000; Swanson et al., 1994). Keane et al. (2009) defined the natural range of variability as "the variation of

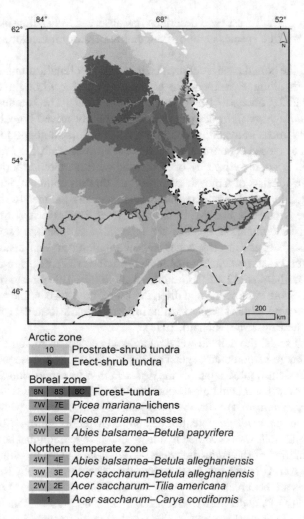

Arctic zone
- 10   Prostrate-shrub tundra
- 9   Erect-shrub tundra

Boreal zone
- 8N | 8S | 8C   Forest–tundra
- 7W | 7E   *Picea mariana*–lichens
- 6W | 6E   *Picea mariana*–mosses
- 5W | 5E   *Abies balsamea*–*Betula papyrifera*

Northern temperate zone
- 4W | 4E   *Abies balsamea*–*Betula alleghaniensis*
- 3W | 3E   *Acer saccharum*–*Betula alleghaniensis*
- 2W | 2E   *Acer saccharum*–*Tilia americana*
- 1   *Acer saccharum*–*Carya cordiformis*

Fig. 8.3 Vegetation zonation in Québec according to the level of zone, domain (*1–10*) and subdomain (*W*, west; *E*, east; *N*, north; *S*, south; *C*, coastal) (MFFP 2021 [CC-BY 4.0]; Saucier et al., 2009, with permission from Multimondes). The codes (*1–10*) are used in Fig. 8.4 to define the ombrothermal area of the bioclimatic subdomains. The *red line* is the northern limit of commercially productive forest (MRN 2013a)

historical ecosystem characteristics and processes (vegetation, disturbance regimes, climate) over time and space scales that are appropriate for management application." An approach that considers the natural range of variability assumes (1) that ecosystems are complex and have a range of conditions that are self-sustaining. Beyond that range they move into disequilibrium; and (2) historical conditions can indicate ecosystem health (Keane et al., 2009; Kuuluvainen, 2002; Morgan et al., 1994).

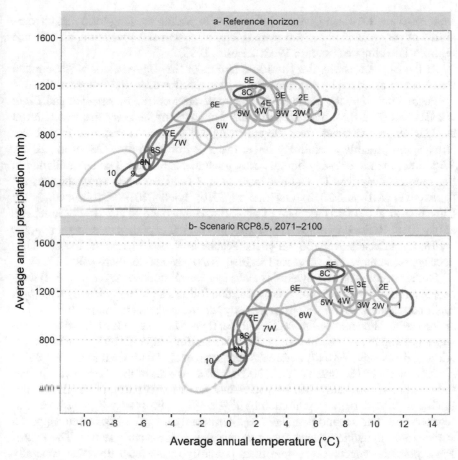

Fig. 8.4 Ombrothermal area of the bioclimatic subdomains of Québec's boreal forest (Fig. 8.3) for **a** the reference horizon (1969–1990) and **b** the scenario under RCP8.5 for 2100. The subdomain codes are composed of the latitudinal gradient (*1–10*) and the longitudinal gradient (*E:east* or *W:west*). *Ellipses* trace the 95% confidence level with a multivariate *t*-distribution of the data generated from global climate simulations (https://www.climateinteractive.org/tools/world-climate-sim ulation/). Some subdomains exhibit a wide climatic range because of their large area (*6–10*) or the highly variable habitat conditions (*5E*). The climate of the coastal forest–tundra subdomain (*8C*) overlaps with the southern portion of the *Picea mariana*–mosses subdomain because its vegetation is strongly influenced by rocky substrate and strong winds

Spatial variability is generally based on ecological units of the ecological classification determined at regional (Shorohova et al., 2011) to local scales (MacLean et al., 2021). Temporal scales define the short-, medium-, and long-term variability of forest composition, natural disturbances (mainly fire), and climate. This variability is reconstructed using dendrochronology and paleoecology (Chap. 2). Such information of historical conditions helps justify and validate management strategies elaborated in the context of climate change (Gillson & Marchant, 2014; Grondin et al., 2020;

Marcisz et al., 2018). In Québec, information regarding the long-term natural variability of ecosystems is derived from studies of postglacial vegetation, disturbance regimes, and climate (Payette, 1993; Richard, 1993).

At the *regional scale*, the Holocene history of the *Abies-Betula* w. subdomain was first characterized by an afforestation period, followed by the Holocene climatic optimum, marked by the expansion of the temperate species *Pinus strobus* and *Thuja occidentalis* (Fig. 8.5). Subsequently, the abundance of these species was reduced as their range contracted over the following 3,000 years in response to a cooler climate and possibly a relatively lower frequency of large fires (Ali et al., 2012). With time, the modern-day boreal forest became established. The last millennium has been characterized by a slight increase in *P. banksiana* and an opening of the forest cover (Ali et al., 2012; Asselin et al., 2016; Bajolle, 2019; Bajolle et al., 2018; Carcaillet et al., 2001; Fréchette et al., 2018; Hennebelle et al., 2018; Larochelle et al., 2018; Richard et al., 2020). Today, this region experiences recurrent spruce budworm outbreaks every 30–40 years (Saucier et al., 2009), and the fire cycle has been approximately 275 years for the 1890–2020 period (Couillard et al., 2022).

Studies of recent forest dynamics at the *local scale* (toposequence, Fig. 8.2) indicate that the *Abies-Betula* potential vegetation follows a pathway characterized by a high abundance of *B. papyrifera* in the early postfire successional stages followed by an increasing abundance of *A. balsamea* over time. Mature stands of *A. balsamea* are subject to gap dynamics, and some stands are severely affected by spruce budworm (*Choristoneura fumiferana*) outbreaks (Bergeron et al., 2014; Couillard et al., 2012; Navarro et al., 2018; Saucier et al., 2009). The dynamics of the *Picea-Abies* potential vegetation can follow one of two pathways. The first pathway has fires maintaining a stand dynamic dominated by *P. mariana,* with scattered *A. balsamea* in regeneration. The second pathway has a sufficiently long fire interval to support a successional dynamic toward *Picea mariana–Abies balsamea* stands. The *Picea-Pinus* potential vegetation is controlled primarily by frequent fires that generally favor the cycling of the same plant communities (recurrence dynamics), i.e., fires enable the post-disturbance recovery process to produce stands resembling those of the predisturbance state (Fig. 8.2; Couillard et al. 2016; Frégeau et al., 2015; Martin et al., 2018). This potential vegetation can be subject to a regeneration failure that shifts closed-canopy stands toward lichen woodlands. Various causal factors can explain this transformation; these factors include fire intensity (severe or light), successive fires within a short period, and the cumulative effect of an insect outbreak or logging immediately followed by a fire. For these areas, the return of closed-canopy stands is not possible without direct intervention, such as planting (Couillard et al., 2021; Girard et al., 2008; Pinno et al., 2013; Schab et al., 2021; Splawinski et al., 2019).

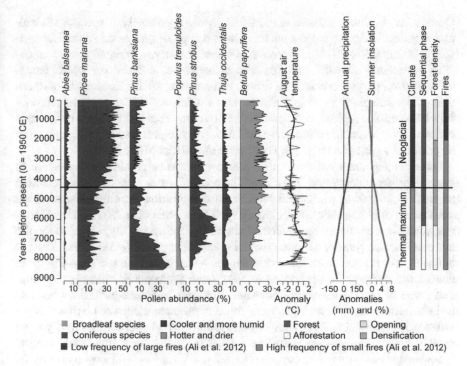

Fig. 8.5 Postglacial vegetation and environmental data of the *Abies balsamea–Betula papyrifera* western subdomain, Québec. Pollen curves for selected coniferous (*5*) and broadleaf (*2*) tree taxa from Lake Lili (Bajolle, 2019) are coupled with a chironomid-based reconstruction of August air temperature at Lake Aurélie (Bajolle et al. 2018), a millennial reconstruction of annual precipitation and summer sunshine for the entire subdomain (Fréchette et al., 2018), and a regional summary of past climate, sequential cover type, forest density, and fires according to Bajolle (2019). Note that some taxa of the diagram are greatly underrepresented by the pollen they shed compared to their abundance in the field (e.g., *A. balsamea*, *P. tremuloides*, *T. occidentalis*), whereas others are highly overrepresented (e.g., *P. strobus*, *B. papyrifera*). The maximum values of the *P. banksiana* pollen curve at the base of the sequence (7,000–8,500 cal yr BP) reflect mainly a long-distance, northerly transport of extraregional pollen during the afforestation stage

8.2.4 Ecological Classification: A Foundation for Forest Ecosystem Management

Ecological classification is a framework that provides users with a straightforward structure to practice ecosystem-based management, facilitate modeling of the spatial distribution of ecosystems, and evaluate the impacts of changing climate and environmental conditions on vegetation (MacKenzie & Meidinger, 2018; Sims et al., 1996). For example, MacLean et al. (2021) present a new approach for establishing management guidelines defined for each potential vegetation in Nova Scotia. The forest management system developed in Québec also relies on potential vegetation

(Grondin et al., 2003; Saucier et al., 2010). More specifically, a specific silvicultural scenario is developed for each ecological type or groups of ecological type (MFFP 2020; MRN 2013b). Moreover, Québec forestry is oriented toward calculating the annual allowable cut for each management unit and for which potential vegetation plays a crucial role (Fortin & Langevin, 2010). Recently, a northern limit of commercially productive forest was established on the basis of relationships between fire regime, regeneration, and the time required for trees and forests to reach a merchantable volume, as calculated by ecological district and considering the potential vegetation (Fig. 8.3; Gauthier et al., 2015b; MRN 2013a).

Some management activities also consider reference conditions that serve to ensure long-term ecosystem health. In Québec, reference conditions for defining targets for old-growth proportions and forest composition were established at the spatial scale of bioclimatic subdomains (Fig. 8.3; Boucher et al., 2011). These reference conditions are based on studies of natural variability, mainly related to the fire regime of the last century or sometimes an even longer period (~200 years). The fire regime is documented from stand origin maps—which date the last fire—and from which a fire cycle and a proportion of old-growth forest are calculated. The long-term range of natural variability does not determine reference conditions because the links between long-term history and forest management must be clarified, as per Hennebelle et al. (2018). They showed that a relatively high proportion of old-growth forest existed throughout the Holocene in Québec. Such information would suggest that estimates based on the recent past are similar to those based on long-term history.

By using ecological classification information to manage Québec's forest ecosystems, forest managers (1) describe the natural forest, (2) compare it to the managed forest to highlight the differences induced by forest management, (3) translate this information into ecological issues to be resolved and (4) integrate these issues into management plans to reduce differences between natural and managed forests (Harvey et al., 2009; MFFP 2017). In the *Abies-Betula* w. subdomain, for example, the reference conditions indicate that even-aged mature stands (older than 80 years) represent 30% of the subdomain compared with 31% for uneven-aged old-growth forests (Boucher et al., 2011). These thresholds serve as guidelines to define targets for ecosystem management plans (MFFP 2015). However, Martin et al. (2021) showed that in the *Picea mariana*–mosses domain, when local features of potential vegetations are not sufficiently considered in forest landscapes, old-growth forests are preserved essentially on the least productive sites, such as forested wetlands (*Picea-Sphagnum* potential vegetation); the most productive sites are harvested. The system of protected areas in Québec, in which a variety of forest stands are protected, and relatively small, protected areas (i.e., biological shelters), which protect in particular old-growth forest, partially avoid the pattern of old-growth forests being found only on the least productive sites. It is clear that local conditions must be better integrated within targets set at the landscape scale, particularly with respect to complex forest structures and their attributes (e.g., deadwood). In the context of ecosystem management, adequate protection and adapted silviculture (partial cutting) of old-growth forests specific to each potential vegetation are key to ensuring the resilience and adaptive capacity of boreal forest ecosystems under global change (Martin et al., 2018; Shorohova et al., 2011).

Overall, ecological classification is a valuable tool for summarizing spatiotemporal information to help guide effective conservation and forest management practices (Keane et al., 2009; Loidi & Fernández-González, 2012; MacLean et al., 2021). However, ecological classification was developed assuming a relatively stable climate and that potential vegetation recovers its initial structure and composition after a disturbance (Loidi & Fernández-González, 2012; Powell, 2000; Saucier et al., 2009). Future climate change challenges the validity of predictable successional pathways and stable potential vegetation dynamics. This assumed predictability requires a reanalysis of ecological classification in light of the uncertainties associated with climate change by considering different successional trajectories for a given potential vegetation. Some mechanisms already exist in Québec's ecological classification system, particularly in forest mapping activities, to account for observed changes in potential vegetation, such as shifts from closed-canopy stands (*Picea-Pinus* potential vegetation) to lichen woodlands (*Picea mariana*–lichens) caused by regeneration failure. Over the past 15 years, several studies have attempted to improve our understanding of natural variability at regional and local scales (Sect. 8.2.3; e.g., Couillard et al. 2016; Fréchette et al., 2018). These studies have greatly contributed to our knowledge of forest dynamics and the concepts of stability and predictability associated with Québec's ecological classification. However, vegetation changes that can be anticipated because of ongoing climate change have yet to be considered. The following section reflects on strategies for bringing ecological classification into the future.

8.3 Ecological Classification and Climate Change

Future environmental conditions are expected to differ markedly from present-day conditions. This shift will occur faster and at a greater magnitude than at any time since the end of the last glaciation (IPCC 2021). Certain projections have ecosystems falling beyond their natural range of variability (Seastedt et al., 2008). In North America, Keane et al. (2020) projected significant changes for potential vegetation in Montana, United States, along an altitudinal gradient. Wang et al. (2012) showed that suitable habitats in British Columbia, Canada, for grasslands, dry forests, and moist continental cedar–hemlock forests would expand and habitats for boreal, subalpine, and alpine ecosystems would decrease. Moreover, in the adjacent province of Alberta, wildfire activity is projected to accelerate the conversion of about half the province's upland mixed and coniferous forests to more climate-adapted deciduous woodlands and grasslands by 2100 (Stralberg et al., 2018). In Siberia, the expansion of dark coniferous forest—currently found in the southern boreal forest—into light coniferous forest (*Larix laricina* northern forest, often underlain by permafrost) has also been documented (Gauthier et al., 2015a; Kharuk et al., 2007).

The consequences of climate change for Canada's boreal forests will be numerous; these impacts include changes in forest composition and growth rates, shifts in disturbance regimes, and, ultimately, altered biodiversity levels (Price et al., 2013). These climate-induced changes in forest ecosystems thus represent a major challenge for forest management. Climate projections must be translated into concrete information and guidelines to prepare and assist managers in adapting forest management plans to these changes (Forestier en chef, 2020; Gauthier et al., 2014, 2015a; Millar et al., 2007; Nagel et al., 2017; Puettmann, 2011; Thiffault et al., 2021). In this context, climate-induced changes could be forecast from the information and models available at the different levels of the ecological classification system already in use in forest ecosystem management. The following questions are thus appropriate to consider in the discussion below:

- Question 1. Will climate change modify vegetation along the various ecological classification scales, i.e., from bioclimatic subdomains to potential vegetation?
- Question 2. Can knowledge derived from ecological classification help respond to the challenges inherent to climate change?

8.3.1   A Three-Step Analysis to Characterize Changes in Climate, Species Sustainability, and Forest Dynamics

In the following section, we present an analytical approach that will address these two questions using information generated at different levels of ecological classification. This approach proceeds through several steps that we consider to be prerequisites for determining silvicultural strategies in the context of climate change. Our method can be adapted according to specific regional and local contexts. The aim is to illustrate how climate change may impact forests at the relevant scale by focusing on the *Abies balsamea–Betula papyrifera* w. subdomain in the province of Québec and comparing current vegetation patterns with those developed under the RCP8.5 scenario from now to 2150. As explained by the IPCC (2021),

> Representative Concentration Pathways (RCP) are scenarios that include time series of emissions and concentrations of the full suite of greenhouse gases (GHGs) and aerosols and chemically active gases, as well as land use/land cover … RCPs usually refer to the portion of the concentration pathway extending up to 2100, for which Integrated Assessment Models produced corresponding emission scenarios.

The RCP8.5 scenario is generally considered the most severe anthropogenic climate forcing scenario. Indeed, under this pathway, the forcing reaches 8.5 W·m^{-2} in 2100 and continues to increase for some time afterward. Under this scenario, the temperature in the studied bioclimatic subdomain will increase 7–8 °C and precipitation 7% by 2100 (Boulanger & Pascual Puigdevall, 2021). The *Abies-Betula* w. subdomain will thus likely experience a longer fire season and more fire-prone weather conditions, which could increase the annual area burned by 2 to 4 times by the end of the century, as projected for various regions in Canada (Boulanger et al., 2014). Ecological processes including tree growth, regeneration, interspecific competition, and forest productivity could also be affected by these climate changes; these altered processes could ultimately modify the forest assemblage currently defining each potential vegetation (Boulanger & Pascual Puigdevall, 2021). Our approach involves acquiring information about future climate analogs (Step 1), defining potential impacts on tree species–habitat suitability at the subdomain scale (Step 2), and finally considering local information at the potential vegetation scale to assess changes in forest dynamics (Step 3).

8.3.1.1 Step 1. Identifying Contemporary Climate Analogs

A climate analog is a simple tool for visualizing the magnitude of climate change at a relatively large spatial scale. This tool "identifies locations for which historical climate is similar to the anticipated future climate at a reference location" (Grenier et al., 2013). Here, the approach shows how the temperature and precipitation of the *Abies-Betula* w. subdomain (Fig. 8.3) will be modified under RCP8.5 during the 2071–2100 period (code 5 W in Fig. 8.4b). Under this scenario, this subdomain will experience a climate analogous to that currently experienced by the *Acer saccharum–Carya cordiformis* bioclimatic subdomain (code 1 in Fig. 8.4b), located ~500 km to the south. Northern temperate hardwoods characterize this latter area, with several thermophilous species (e.g., *Quercus*, *Carya*, *Tilia*) reaching the northernmost limit of their range and small-gap dynamics being the most common natural disturbance. Therefore, this climatically analogous region includes very different tree species assemblages and experiences markedly different disturbance dynamics than those currently observed in the *Abies-Betula* w. subdomain.

8.3.1.2 Step 2. Assessing Future Tree Species Habitat Suitability at the Regional Scale

Because the climate analog provides only partial insight into the impacts of climate change on forest ecosystems, it is important to identify to what extent the projected climate in the *Abies-Betula* w. subdomain will be suitable for species currently observed in the area and those taxa that could potentially migrate into the subdomain. Niche models (Boisvert-Marsh et al., 2014; Périé & de Blois, 2016; Périé et al., 2014) are used to assess the climatic vulnerability of the dominant tree species of the subdomain (Fig. 8.6). Under the RCP8.5 climate scenario for 2071–2100, niche models project that much of this subdomain will become a less favorable habitat for most conifers and *B. papyrifera*. *Populus tremuloides* habitat will remain present at a similar abundance (status quo), whereas the situation appears more critical for *P. banksiana*, as the model indicates a less favorable habitat over much of the subdomain. *Acer rubrum*, a thermophilous species that favors the warmer hilltops (Fig. 8.2) of the southern part of the western subdomain, has a distinct profile from the other species as new suitable habitats will become available. These changes in habitat

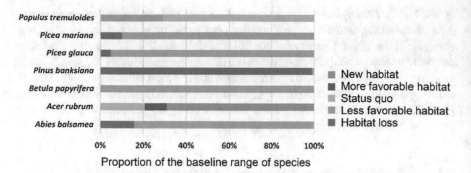

Fig. 8.6 The impact of climate change on tree habitat suitability under scenario RCP8.5 (2071–2100) as defined by niche models. The baseline (1969–1990) range of a species is the total area (km^2) of all cells where the baseline average model predicted a suitable habitat for that species in the *Abies balsamea–Betula papyrifera* western subdomain and within the *Abies balsamea–Betula papyrifera* potential vegetation

suitability suggest that species that have traditionally defined entire regional vegetation assemblages could become less adapted to their particular regions, resulting in significant impacts on ecosystems.

8.3.1.3   Step 3. Assessing Local Change in Forest Composition and Dynamics

How will the climate-induced vulnerability of these species affect the dynamics of specific potential vegetation? Evaluating such impacts requires a more thorough assessment of climate-induced changes on processes at the stand- (e.g., interspecific competition, soil characteristics, productivity) and landscape- (e.g., seed dispersal, natural disturbances) scales, which govern finer-scale forest dynamics. Differences in soil characteristics, topographical position, and current tree species assemblages may thus impact how specific potential vegetation will respond to climate change. Climate-induced changes in natural disturbance regimes are another key component to consider. Simulating changes in the disturbance regime enables assessing possible future alterations in the natural range of variability (Fig. 8.5), which is not explicitly accounted for within niche models. Using a forest landscape model (LANDIS-II; Scheller et al., 2007), we projected the future long-term compositional change of two contrasting potential vegetation types characteristic of the *Abies-Betula* w. subdomain, i.e., *Abies-Betula* and *Picea-Pinus* potential vegetation. We simulated 130 years, starting in 2020, under baseline and RCP8.5 climate scenarios. To assess the impact of changes in the natural disturbance regime, we simulated future forest landscapes under RCP8.5, considering the current fire regime (status quo) and a future fire regime under climate change projections.

We found that under the RCP8.5 projected climate conditions and regardless of the fire regime, i.e., current or projected fires (Fig. 8.7), there is a significant decrease in total biomass after 2070 for both potential vegetation types, primarily caused by a large decrease in the biomass of *A. balsamea* and/or *P. mariana*, the two dominant boreal species. These decreases in total biomass are likely to be even greater when considering the concomitant climate-induced change in fire regimes (Boulanger et al., 2014, 2016). Our analysis distinguishes different alterations in forest dynamics specific to each potential vegetation. For instance, *A. rubrum* and *B. papyrifera* are projected to sharply increase in the *Abies-Betula* potential vegetation, whereas *P. tremuloides* biomass will remain stable or increase only slightly. The increase of *A. rubrum* is greater in the current fire regime than in the projected fire regime. In the *Picea-Pinus* potential vegetation, however, *P. tremuloides* sharply increases, whereas *A. rubrum* remains a minor component of the forest landscape. Still here, *A. rubrum* is greater in the current fire regime and after 2100. Therefore, by 2100, the *Abies-Betula* potential vegetation will be dominated mainly by conifers accompanied by *B. papyrifera*, *P. tremuloides*, and *A. rubrum*. Beyond 2100, and given the projected fires, broadleaf species (*P. tremuloides*, *B. papyrifera*, *A. rubrum*) will dominate. In the *Picea-Pinus* potential vegetation, *P. tremuloides* and *P. banksiana* will be the most common species. The projected dynamics for both potential vegetation types are consistent with contemporary trends found in other studies (Boisvert-Marsh et al., 2014; Brice et al., 2019; Fisichelli et al., 2014; Iverson et al., 2008; Jain et al., 2021).

8.3.2 Ecological Classification and Climate Change: The Main Issues

Several conclusions can be drawn from this multistep analysis to answer both questions posed above. First, we are interested in specifying how climate change will modify the current vegetation (Question 1). At the regional scale, the future vegetation of the *Abies-Betula* w. subdomain will not correspond to that identified by its climate analog by the end of the century (Fig. 8.4 and Step 1). Given dispersal limitations (Iverson et al., 2011) and forest inertia (Brice et al., 2020), it is unlikely that the thermophilous species associated with the *Acer saccharum–Carya cordiformis* subdomain will keep pace with the >500 km northward migration of the ombrothermal area (Boulanger & Pascual Puigdevall, 2021; Taylor et al., 2017). With climate niches of most boreal species deteriorating within the *Abies-Betula* w. subdomain, this scenario will likely introduce a *climate debt* (Taylor et al., 2017), with current species assemblages being strongly maladapted to future climate conditions. Under such conditions, it is improbable that regional vegetation will be at a climate equilibrium, at least for the next several decades.

At the local scale (Fig. 8.7), our model (LANDIS-II) also suggests that increased disturbance rates will cause a decline in the biomass of coniferous species and provide pioneer deciduous tree species a competitive advantage because they can reproduce vegetatively after a disturbance (Boulanger et al., 2019; Landhäusser et al., 2010). If stands of *Abies-Betula* potential vegetation become dominated by *P. tremuloides*

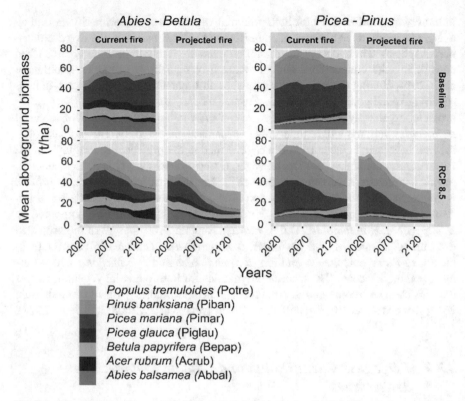

Fig. 8.7 Mean stand aboveground biomass of forest species in the *Abies balsamea–Betula papyrifera* western subdomain according to the LANDIS-II model and projections using the RCP8.5 scenario. The panels represent the *Abies-Betula* potential vegetation in midslope (*left panel*) and the *Picea-Pinus* potential vegetation (*right panel*) in flat areas (Fig. 8.2). In the *bottom left chart* of each panel, the current fire regime remains relatively unchanged in the future. In the *bottom right chart* of each panel, the fire regime is altered in relation to projected climate change

and *A. rubrum*, will this community be considered (1) a mid-successional stage of the *Abies-Betula* potential vegetation with a greater abundance of *A. rubrum* than expected under current climate conditions, (2) a transition stage toward an *Abies-Acer* potential vegetation, or (3) a new potential vegetation? Knowledge of forest dynamics under climate change is not sufficiently advanced to project which of the three possibilities will prevail in the future.

Furthermore, we project that warmer conditions should favor the growth and reproduction of thermophilous species (e.g., *Acer* spp.), which will then be more likely to outcompete boreal species in the potential vegetation where they are currently thriving. Such conditions could favor the migration of *Abies-Acer* potential vegetation to mid-hillside, a topographic position where conditions are not currently suitable to support such potential vegetation in the *Abies-Betula* w. bioclimatic subdomain (Fig. 8.2).

For *Picea-Pinus* potential vegetation, dynamic modeling (LANDIS-II) suggests a significant amount of *P. banksiana* will be maintained (Fig. 8.7), even when niche models predict a less favorable habitat for this species. This divergence relates to the niche models relying only on climate, whereas LANDIS-II considers several other elements, including the impacts of the fire regime.

Climate change may therefore affect vegetation as we know it today at different scales of ecological classification, from the bioclimatic subdomain to the potential vegetation level. At the same time, a great deal of uncertainty remains regarding future climate and vegetation changes. The high uncertainty means that anticipated changes in vegetation under climate change will likely be intermediate, falling between those projected under baseline conditions (minimum change) and those of RCP8.5 (maximum change). The variability of climate scenario projections remains great given the highly unpredictable global, political, and social conditions that will exert a strong influence (IPCC 2021). Moreover, changes in climate conditions are likely to evolve over several decades; therefore, it could take decades or even centuries for highly *inertial* forest ecosystems, i.e., those resistant to change, to modify significantly under future climate conditions. Finally, climate and forest models each have their predictive limitations and hence intrinsic uncertainty in addition to the uncertainty of climate scenarios.

How can knowledge of natural variability (Sect. 8.2.3; Fig. 8.5) help us better interpret future vegetation changes? Past patterns in vegetation illustrate previous dynamics that can recur in the future and, therefore, help researchers formulate hypotheses about the limits of variability that species have experienced in the past and that may be exceeded in the future (Gillson & Marchant, 2014; Grondin et al., 2020; Marcisz et al., 2018).

1. In the *Abies-Betula* w. subdomain, both types of modeling, corresponding to the niche models (Step 2) and the LANDIS-II model (Step 3), suggest that by 2100 there will be fewer favorable habitats for conifers and a gradual decline in their biomass. If the abundance of conifers becomes relatively low, as predicted for RCP8.5, then there will be fewer conifers in the landscape than over the past several thousand years (Fig. 8.5). It is difficult to compare the Holocene abundance of coniferous species with that expected in the future; thus, we cannot determine whether future landscapes will move beyond the range of their past natural variability.

2. *Populus tremuloides* and *Pinus banksiana* are predicted to be abundant in the future (Fig. 8.7), albeit within a different context compared with the early Holocene when fires were frequent, and vegetation was in the afforestation stage (Fig. 8.5). After the afforestation period, and mainly from 5,000 to 2,000 yr BP, broadleaf species (*B. papyrifera*) remained abundant; the future *Abies-Betula* w. subdomain may present a similar vegetation pattern (Fréchette et al., 2018).

3. Long-term regional natural variability indicates that *P. banksiana* has increased over the last millennium in western Québec (Fig. 8.5). This trend should continue under climate change with an increased fire incidence given the great ability of *P. banksiana* to renew itself in a recurrence dynamic (Couillard et al. 2016;

Frégeau et al., 2015; Payette et al., 2012). LANDIS-II modeling also supports this hypothesis, although there is a significant decrease in biomass, indicating that the stands will be younger. It is difficult to comment on the spatial variation of *Pinus banksiana* stands; however, one hypothesis holds that this species will increase in geographic extent (Schab et al., 2021; Splawinski et al., 2019). All these dynamics depend on the initial species composition at each site (greater or lower abundance of *P. banksiana*), the fire cycle, and the species' response to a potentially shortened cycle.

4. Finally, the Holocene history of the *Abies-Betula* w. subdomain shows that *P. strobus* and *T. occidentalis* were historically well represented in the landscape during warmer climatic periods (Fig. 8.5). Both species are not considered in contemporary modeling because of their low current abundance. On the other hand, habitat models suggest that habitats will be suitable for the expansion of these species (Périé et al., 2009). This migration will nonetheless depend on multiple factors, e.g., the inertia of the vegetation.

Overall, comparing past and future scenarios allows us to better situate the potential magnitude of future vegetation change (e.g., the decreased abundance of coniferous species), base future projections on specific patterns observed during the Holocene, e.g., broadleaf species such as *Betula papyrifera*, and propose hypotheses regarding future vegetation change, e.g., *P. banksiana*, *P. strobus*, *T. occidentalis*.

Can knowledge generated by ecological classification (Sect. 8.1) be used to address the challenges inherent to climate change (Sect. 8.2)? (Question 2). Despite projected changes in forest ecosystem composition and dynamics at various spatial scales, we argue that current ecological classification can still provide a useful framework to inform future forest management (Loidi & Fernández-González, 2012). The characteristics that currently define potential vegetation at the local scale (soil, surficial deposits, position, and slope) will continue to develop independently in future ecological domains, e.g., *Abies-Betula* versus *Picea-Pinus*. The assemblages within each potential vegetation will evolve according to the competitive abilities of its various species. The future toposequence will continue to show a gradation of potential vegetation along a physiographic gradient. Each potential vegetation will have a specific future successional pathway, albeit different from the current pathway (Fig. 8.7). Moreover, each potential vegetation distributed along the toposequence will continue to be distinct from other assemblages, and the links with the current vegetation will likely remain for a long duration. Studies on how ecosystem inertia could limit future vegetation dynamics must be undertaken. Overall, the hierarchy of ecological classification (regional vs. local) and the specificity of each potential vegetation will continue to be paramount for strategic forest planning under climate change.

8.4 Conclusions

It is evident from our first question (Question 1) that climate change will modify the principles of ecological classification. Given the expected modifications in forest vegetation composition and dynamics, we must integrate the expected variability and uncertainty of forest vegetation composition and dynamics into the current stable and predictable potential vegetation dynamics. The results presented in this chapter, through a three-step analysis, provide a preliminary view of the links between current ecological classification and forest management under climate change. This information is limited in scope and should be interpreted with caution. New knowledge, more complete than that currently available, will be added rapidly, especially through projects that address potential vegetation dynamics under climate change. In regard to the possible use of ecological classification in the context of climate change (Question 2), we documented that each potential vegetation could change and be characterized by its own successional trajectory. Thus, we must maintain and manage this diversity. Ecological classification can and must assist forest managers in estimating potential future changes despite inherent variability and uncertainty. The challenge is to select target species and silvicultural scenarios at the regional and local levels on the basis of the different trajectories that current forest stands are projected to follow under increased temperatures and precipitation amounts and changing disturbance regimes. "Decision makers within any institution, therefore, have to find their own way through sometimes conflicting information and face the prospect of planning with and for uncertainty" (Gauthier et al., 2014)

References

Ali, A. A., Blarquez, O., Girardin, M. P., et al. (2012). Control of the multimillennial wildfire size in boreal North America by spring climatic conditions. *Proceedings of the National Academy of Sciences of the United States of America, 109*, 20966–20970. https://doi.org/10.1073/pnas.120 3467109.

Asselin, M., Grondin, P., Lavoie, M., et al. (2016). Fires of the last millenium led to landscapes dominated by early successional species in Québec's clay belt boreal forest, Canada. *Forests, 7*, 205. https://doi.org/10.3390/f7090205.

Bailey, R. G. (2009). *Ecosystem geography: From ecoregions to sites* (p. 251). Springer.

Bajolle, L. (2019). *Reconstitution des paléotempératures holocènes de la forêt boréale coniférienne de l'ouest du Québec basée sur une approche multi-indicateurs.* Ph.D. thesis, Université du Québec en Abitibi-Témiscamingue.

Bajolle, L., Larocque-Tobler, I., Ali, A. A., et al. (2018). A chironomid-inferred Holocene temperature record from a shallow Canadian boreal lake: Potentials and pitfalls. *Journal of Paleolimnology, 61*, 69–84. https://doi.org/10.1007/s10933-018-0045-9.

Baldwin, K., Allen, L., Basquill, S., et al. (2020). *Vegetation zones of Canada: A biogeoclimatic perspective* Information Report, GLC-X-25 (p. 163). Sault Ste. Marie: Natural Resources Canada, Canadian Forest Service, Great Lakes Forestry Centre.

Banner, A., MacKenzie, W., Haeussler, S., et al. (1993). *A field guide to site identification and interpretation for the Prince Rupert forest region* (p. 26). Land Management Handbook. Victoria: British Columbia Ministry of Forest Research Branch.

Barnes, B. V., Pregitzer, K. S., Spies, T. A., et al. (1982). Ecological forest site classification. *Journal of Forestry, 80*(8), 493–498. https://doi.org/10.1093/jof/80.8.493.

Bergeron, Y., Chen, H. Y. H., Kenkel, N. C., et al. (2014). Boreal mixedwood stand dynamics: Ecological processes underlying multiple pathways. *The Forestry Chronicle, 90*, 202–213. https://doi.org/10.5558/tfc2014-039.

Blouin, J.-L., & Grandtner, M. M. (1971). *Étude écologique et cartographie de la végétation du comté de Rivière-du-Loup.* Mémoire de recherche forestière 6 (p. 371). Québec: Gouvernement du Québec, ministère des Terres et Forêts, Service de la recherche.

Blouin, J., Berger, J.-P., Landry, Y., et al. (2008). *Guide de reconnaissance des types écologiques des régions écologiques 5b-Coteaux du réservoir Gouin, 5c-Collines du haut Saint-Maurice et 5d-Collines ceinturant le lac Saint-Jean.* Québec: Gouvernement du Québec, ministère des Ressources naturelles et de la Faune, Forêt-Québec, Direction des inventaires forestiers.

Boisvert-Marsh, L., Périé, C., & de Blois, S. (2014). Shifting with climate? Evidence for recent changes in tree species distribution at high latitudes. *Ecosphere, 5*, art83. https://doi.org/10.1890/ES14-00111.1.

Boucher, Y., Bouchard, M., Grondin, P., et al. (2011). *Le registre des états de référence: intégration des connaissances sur la structure, la composition et la dynamique des paysages forestiers naturels du Québec méridional.* Mémoire de recherche forestière 161 (p. 21). Québec: Gouvernement du Québec, ministère des Forêts, de la Faune et des Parcs, Direction de la recherche forestière.

Boulanger, Y., & Pascual Puigdevall, J. (2021). Boreal forests will be more severely affected by projected anthropogenic climate forcing than mixedwood and northern hardwood forests in eastern Canada. *Landscape Ecology, 36*, 1725–1740. https://doi.org/10.1007/s10980-021-01241-7.

Boulanger, Y., Gauthier, S., & Burton, P. J. (2014). A refinement of models projecting future Canadian fire regimes using homogeneous fire regime zones. *Canadian Journal of Forest Research, 44*, 365–376. https://doi.org/10.1139/cjfr-2013-0372.

Boulanger, Y., Taylor, A. R., Price, D. T., et al. (2016). Climate change impacts on forest landscapes along the Canadian southern boreal forest transition zone. *Landscape Ecology, 32*, 1415–1431. https://doi.org/10.1007/s10980-016-0421-7.

Boulanger, Y., Arseneault, D., Boucher, Y., et al. (2019). Climate change will affect the ability of forest management to reduce gaps between current and presettlement forest composition in southeastern Canada. *Landscape Ecology, 34*, 159–174. https://doi.org/10.1007/s10980-018-0761-6.

Braun-Blanquet, J. (1972). *Plant sociology.* Hafner Publishing Company.

Brice, M.-H., Cazelles, K., Legendre, P., et al. (2019). Disturbances amplify tree community responses to climate change in the temperate–boreal ecotone. *Global Ecology and Biogeography, 28*(11), 1668–1681. https://doi.org/10.1111/geb.12971.

Brice, M.-H., Vissault, S., Vieira, W., et al. (2020). Moderate disturbances accelerate forest transition dynamics under climate change in the temperate-boreal ecotone of eastern North America. *Global Change Biology, 26*, 4418–4435. https://doi.org/10.1111/gcb.15143.

Carcaillet, C., Bergeron, Y., Richard, P. J. H., et al. (2001). Change of fire frequency in the eastern Canadian boreal forests during the Holocene: Does vegetation composition or climate trigger the fire regime? *Journal of Ecology, 89*, 930–946. https://doi.org/10.1111/j.1365-2745.2001.00614.x.

Cissel, J. H., Swanson, F. J., & Weisberg, P. J. (1999). Landscape management using historical fire regimes: Blue River, Oregon. *Ecological Applications, 9*, 1217–1231. https://doi.org/10.1890/1051-0761(1999)009[1217:LMUHFR]2.0.CO;2.

Clements, F. E. (1936). Nature and structure of the climax. *Journal of Ecology, 24*(1), 252–284. https://doi.org/10.2307/2256278.

Cogbill, C. V. (1985). Dynamics of the boreal forests of the Laurentian Highlands, Canada. *Canadian Journal of Forest Research, 15*, 252–261. https://doi.org/10.1139/x85-043.

Couillard, P.-L., Payette, S., & Grondin, P. (2012). Recent impact of fire on high-altitude balsam fir forests in south-central Quebec. *Canadian Journal of Forest Research, 42*, 1289–1305. https://doi.org/10.1139/x2012-081.

Couillard, P.-L., Frégeau, M., Payette, S., et al. (2016). *Dynamique et variabilité naturelle de la pessière à mousses au nord de la région du Lac-Saint-Jean* (p. 35). Québec: Gouvernement du Québec, ministère des Forêts, de la Faune et des Parcs, Secteur des forêts, Direction des inventaires forestiers.

Couillard, P.-L., Payette, S., Lavoie, M., et al. (2019). La forêt boréale du Québec: Influence du gradient longitudinal. *Le Naturaliste Canadien, 143*(2), 18–32. https://doi.org/10.7202/1060052ar.

Couillard, P.-L., Payette, S., Lavoie, M., et al. (2021). Precarious resilience of the boreal forest of eastern North America during the Holocene. *Forest Ecology and Management, 485*, 118954. https://doi.org/10.1016/j.foreco.2021.118954.

Couillard, P. L., Bouchard, M., Laflamme, J., et al. (2022). *Zonage des régimes de feux du Québec méridional.* Mémoire de recherche forestière 189 (p. 48). Québec: Gouvernement du Québec, ministère des Forêts, de la Faune et des Parcs, Direction de la recherche forestière.

Damman, A. W. (1964). *Some forest types of central Newfoundland and their relation to environmental factors.* Society of American Foresters, U.S. Monograph 8.

Danneyrolles, V., Dupuis, S., Boucher, Y., et al. (2020). *Utilisation couplée des archives d'arpentage et de la classification écologique pour affiner les cibles de composition dans l'aménagement écosystémique des forêts tempérées du Québec.* Mémoire de recherche forestière 183 (p. 36). Québec: Gouvernement du Québec, ministère des Forêts, de la Faune et des Parcs, Direction de la recherche forestière.

Dansereau, P. (1957). *Biogeography, an ecological perspective.* The Ronald Press Company.

Daubenmire, R. (1968). *Plant communities, a textbook of plant synecology.* Harper & Row.

Engelmark, O., & Hytteborn, H. (1999). Coniferous forest. In H. Rydin, P. Snoeijs, M. Diekmann (Eds.), *Swedish plant geography, Acta Phytogeographica Suecica 84*, 55–74.

Fisichelli, N. A., Frelich, L. E., & Reich, P. B. (2014). Temperate tree expansion into adjacent boreal forest patches facilitated by warmer temperatures. *Ecography, 37*, 152–161. https://doi.org/10.1111/j.1600-0587.2013.00197.x.

Flahault, C., & Schröter, C. (1910). Rapport sur la nomenclature phytogéographique. In E. D. Wilderman (Ed.), *Actes du IIIème Congrès International de Botanique* (pp. 131–142). Bruxelles: De Boeck.

Forestier en chef. (2020). *Intégration des changements climatiques et développement de la capacité d'adaptation dans la détermination des niveaux de récolte au Québec* (p. 60). Roberval: Gouvernement du Québec.

Fortin, M., Langevin, L. (2010). *ARTÉMIS-2009: Un modèle de croissance basé sur une approche par tiges individuelle pour les forêts du Québec.* Mémoire de recherche forestière 156 (p. 48). Québec: Gouvernement du Québec, ministère des Ressources naturelles et de la Faune, Direction de la recherche forestière.

Fréchette, B., Richard, P. J. H., Grondin, P., et al. (2018). *Histoire postglaciaire de la végétation et du climat des pessières et des sapinières de l'ouest du Québec.* Mémoire de recherche forestière 179 (p. 165). Québec: Gouvernement du Québec, ministère des Ressources naturelles et de la Faune, Direction de la recherche forestière.

Frégeau, M., Payette, S., & Grondin, P. (2015). Fire history of the central boreal forest in eastern North America reveals stability since the mid-Holocene. *The Holocene, 25*(12), 1912–1922. https://doi.org/10.1177/0959683615591361.

Frelich, L. E., & Reich, P. B. (1995). Spatial patterns and succession in a Minnesota southernboreal forest. *Ecological Monographs, 65*, 325–346. https://doi.org/10.2307/2937063.

Gauthier, S., Vaillancourt, M. A., Leduc, A., et al. (Eds.). (2009). *Ecosystem management in the boreal forest* (p. 572). Presses de l'Universite du Quebec.

Gauthier, S., Bernier, P., Burton, P. J., et al. (2014). Climate change vulnerability and adaptation in the managed Canadian boreal forest. *Environmental Reviews, 22*, 256–285. https://doi.org/10.1139/er-2013-0064.

Gauthier, S., Bernier, P., Kuuluvainen, T., et al. (2015a). Boreal forest health and global change. *Science, 349*, 819–822. https://doi.org/10.1126/science.aaa9092.

Gauthier, S., Raulier, F., Ouzennou, H., et al. (2015b). Strategic analysis of forest vulnerability to risk related to fire: An example from the coniferous boreal forest of Quebec. *Canadian Journal of Forest Research, 45*, 553–565. https://doi.org/10.1139/cjfr-2014-0125.

Gillson, L., & Marchant, R. (2014). From myopia to clarity: Sharpening the focus of ecosystem management through the lens of palaeoecology. *Trends in Ecology & Evolution, 29*, 317–325. https://doi.org/10.1016/j.tree.2014.03.010.

Girard, F., Payette, S., & Gagnon, R. (2008). Rapid expansion of lichen woodlands within the closed-crown boreal forest zone over the last 50 years caused by stand disturbances in eastern Canada. *Journal of Biogeography, 35*, 529–537. https://doi.org/10.1111/j.1365-2699.2007.018 16.x.

Grandtner, M. M. (1966). *La végétation forestière du Québec méridional* (p. 216). Québec: Les Presses de l'Université Laval.

Grenier, P., Parent, A. C., Huard, D., et al. (2013). An assessment of six dissimilarity metrics for climate analogs. *Journal of Applied Meteorology and Climatology, 52*, 733–752. https://doi.org/10.1175/JAMC-D-12-0170.1.

Grondin, P., & Gosselin, J. (2013). La dynamique des peuplements et les végétations potentielles. In ministère des Ressources naturelles (MRN) (Ed.) *Le guide sylvicole du Québec, Tome 1, Les fondements biologiques de la sylviculture* (pp. 297–393). Québec: Les Publications du Québec.

Grondin, P., Saucier, J.-P., Blouin, J., et al. (2003). *Information écologique et planification forestière au Québec*. Note de recherche forestière 118 (p. 9). Québec: Direction de la recherche forestière, ministère des Ressources naturelles, de la Faune et des Parcs.

Grondin, P., Noël, J., & Hotte, D. (2007). *L'intégration de la végétation et de ses variables explicatives à des fins de classification et de cartographie d'unités homogènes du Québec méridional*. Mémoire de recherche forestière 150 (p. 62). Québec: Gouvernement du Québec, ministère des Ressources naturelles et de la Faune du Québec, Direction de la recherche forestière.

Grondin, P., Gauthier, S., Borcard, D., et al. (2014). A new approach to ecological land classification for the Canadian boreal forest that integrates disturbances. *Landscape Ecology, 29*, 1–16. https://doi.org/10.1007/s10980-013-9961-2.

Grondin, P., Fréchette, B., Lavoie, M., et al. (2020). L'histoire postglaciaire de la végétation boréale: son utilité en aménagement écosystémique. Avis de recherche forestière 149. Québec: Gouvernement du Québec, ministère des Forêts, de la Faune et des Parcs, Direction de la recherche forestière.

Härdtle, W. (1995). On the theoretical concept of the potential natural vegetation and proposals for an up-to-date modification. *Folia Geobotanica, 30*, 263–276. https://doi.org/10.1007/BF0280 3708.

Harvey, B. D., Bergeron, Y., Leduc, A., et al. (2009). Forest ecosystem management in the boreal mixedwood forest of western Québec: An example from the Lake Duparquet forest. In S. Gauthier, M. A. Vaillancourt, A. Leduc, L. De Grandpré, D. D. Kneeshaw, H. Morin, P. Drapeau, & Y. Bergeron (Eds.), *Ecosystem management in the boreal forest* (pp. 449–478). Presses de l'Université du Québec.

Heinselman, M. L. (1973). Fire in the virgin forests of the Boundary Waters Canoe Area, Minnesota. *Quaternary Research, 3*, 329–382. https://doi.org/10.1016/0033-5894(73)90003-3.

Hennebelle, A., Grondin, P., Aleman, J. C., et al. (2018). Using paleoecology to improve reference conditions for ecosystem-based management in western spruce-moss subdomain of Québec. *Forest Ecology and Management, 430*, 157–165. https://doi.org/10.1016/j.foreco.2018.08.007.

Intergovernmental Panel on Climate Change (IPCC). (2021). *Climate Change 2021: The physical science basis. Contribution of Working Group I to the Sixth Assessment Report of the Intergovernmental Panel on Climate Change*. Cambridge: Cambridge University Press.

Iverson, L. R., Prasad, A. M., Matthews, S. N., et al. (2008). Estimating potential habitat for 134 eastern US tree species under six climate scenarios. *Forest Ecology and Management, 254*, 390–406. https://doi.org/10.1016/j.foreco.2007.07.023.

Iverson, L. R., Prasad, A. M., Matthews, S. N., et al. (2011). Lessons learned while integrating habitat, dispersal, disturbance, and life-history traits into species habitat models under climate change. *Ecosystems, 14*, 1005–1020. https://doi.org/10.1007/s10021-011-9456-4.

Jain, P., Khare, S., Sylvain, J.-D., et al. (2021). Predicting the location of maple habitat under warming scenarios in two regions at the northern range in Canada. *Forest Science, 67*(4), 446–456. https://doi.org/10.1093/forsci/fxab012

Jurdant, M., Bélair, J. L., Gerardin, V., et al. (1977). *L'inventaire du capital-nature: méthode de classification et de cartographie écologique du territoire (3ème approximation)* (p. 202). Québec: Service des études écologiques régionales, Direction générale des terres, Pêches et Environnement Canada.

Keane, R. E., Hessburg, P. F., Landres, P. B., et al. (2009). The use of historical range and variability (HRV) in landscape management. *Forest Ecology and Management, 258*, 1025–1037. https://doi.org/10.1016/j.foreco.2009.05.035.

Keane, R. E., Holsinger, L., & Loehman, R. (2020). Bioclimatic modeling of potential vegetation types as an alternative to species distribution models for projecting plant species shifts under changing climates. *Forest Ecology and Management, 477*, 118498. https://doi.org/10.1016/j.foreco.2020.118498.

Kharuk, V., Ranson, K., & Dvinskaya, M. (2007). Evidence of evergreen conifer invasion into Larch dominated forests during recent decades in central Siberia. *Eurasian Journal of Forest Research, 10*, 163–171.

Klijn, F., & Udo de Haes, H. A. (1994). A hierarchical approach to ecosystems and its implications for ecological land classification. *Landscape Ecology, 9*, 89–104. https://doi.org/10.1007/BF00124376.

Krestov, P., Ermakov, N. B., Osipov, S. V., et al. (2009). Classification and phytogeography of Larch forest in Northeast Asia. *Folia Geobotanica, 44*, 323–363. https://doi.org/10.1007/s12224-009-9049-6.

Küchler, A. W. (1964). *Potential natural vegetation of the conterminous United States* (p. 116). New York: American Geographical Society Special Publication No. 36.

Kuuluvainen, T. (2002). Natural variability of forests as a reference for restoring and managing biological diversity in boreal Fennoscandia. *Silva Fennica, 36*, 97–125. https://doi.org/10.14214/sf.552.

Laflamme, J., Munson, A. D., Grondin, P., et al. (2016). Anthropogenic disturbances create a new vegetation toposequence in the Gatineau river valley, Quebec. *Forests, 7*, 254. https://doi.org/10.3390/f7110254.

Landhäusser, S. M., Deshaies, D., & Lieffers, V. J. (2010). Disturbance facilitates rapid range expansion of aspen into higher elevations of the Rocky Mountains under a warming climate. *Journal of Biogeography, 37*(1), 68–76. https://doi.org/10.1111/j.1365-2699.2009.02182.x.

Larochelle, É., Lavoie, M., Grondin, P., et al. (2018). Vegetation and climate history of Quebec's mixed boreal forest suggests greater abundance of temperate species during the early- and mid-Holocene. *Botany, 96*, 437–448. https://doi.org/10.1139/cjb-2017-0182.

Loidi, J., & Fernández-González, F. (2012). Potential natural vegetation: Reburying or reboring. *Journal of Vegetation Science, 23*, 596–604. https://doi.org/10.1111/j.1654-1103.2012.01387.x.

MacKenzie, W. H., & Meidinger, D. V. (2018). The biogeoclimatic ecosystem classification approach: An ecological framework for vegetation classification. *Phytocoenologia, 48*, 203–213. https://doi.org/10.1127/phyto/2017/0160.

MacLean, D. A., Taylor, A. R., Neily, P. D., et al. (2021). Natural disturbances regimes for implementation of ecological forestry: A review and case study from Nova Scotia, Canada. *Environmental Reviews, 30*(1), 128–158. https://doi.org/10.1139/er-2021-0042.

Marcisz, K., Vannière, B., & Blarquez, O. (2018). Taking fire science and practice to the next level: Report from the PAGES Global Paleofire Working Group Workshop 2017 in Montreal, Canada-Paleofire knowledge for current and future ecosystem management. *Open Quaternary, 4*, 1–7. https://doi.org/10.5334/oq.44.

Martin, M., Fenton, N. J., & Morin, H. (2018). Structural diversity and dynamics of boreal old-growth forests case study in Eastern Canada. *Forest Ecology and Management, 422*, 125–136. https://doi.org/10.1016/j.foreco.2018.04.007.

Martin, M., Grondin, P., Lambert, M., et al. (2021). Compared to wildfire, management practices reduced old-growth forest diversity and functionality in primary boreal landscapes of Eastern Canada. *Frontiers in Forests and Global Change, 4*, 1–16. https://doi.org/10.3389/ffgc.2021.639397.

Millar, C. I., Stephenson, N. L., & Stephens, S. L. (2007). Climate change and forests of the future: Managing in the face of uncertainty. *Ecological Applications, 17*, 2145–2151. https://doi.org/10.1890/06-1715.1.

Ministère des Forêts de la Faune et des Parcs (MFFP). (2015). *Stratégie d'aménagement durable des forêts* (p. 56). Québec: ministère des Forêts, de la Faune et des Parcs, Québec.

Ministère des Forêts, de la Faune et des Parcs (MFFP). (2017). *Intégration des enjeux écologiques dans les plans d'aménagement forestier intégré de 2018–2023, Cahier 1–Concepts généraux liés à l'aménagement écosystémique des forêts* (p. 30). Québec: Direction de l'aménagement et de l'environnement forestiers.

Ministère des Forêts, de la Faune et des Parcs (MFFP). (2020). *Plan d'aménagement forestier intégré tactique 2018–2023 révisé en 2020. Région Nord-du-Québec* (p. 198). Québec: Gouvernement du Québec, ministère des Forêts, de la Faune et des Parcs, Direction générale du secteur nord-ouest.

Ministère des Forêts, de la Faune et des Parcs (MFFP). (2021). *Classification écologique du territoire québécois* (p. 11). Québec: Gouvernement du Québec, ministère des Forêts, de la Faune et des Parcs, Direction des inventaires forestiers.

Ministère des Ressources naturelles (MRN). (2013a). *Rapport du Comité scientifique chargé d'examiner la limite nordique des forêts attribuables* (p. 148). Québec: Gouvernment du Québec, Secteur des Forêts.

Ministère des Ressources naturelles (MRN) (Ed.). (2013b). *Les concepts et l'application de la sylviculture* (p. 744). Les Publications du Québec.

Morgan, P., Aplet, G. H., Haufler, J. B., et al. (1994). Natural range of variability, a useful tool for evaluating ecosystem change. *Journal of Sustainable Forestry, 2*, 87–111. https://doi.org/10.1300/J091v02n01_04.

Nagel, L. M., Palik, B. J., Battaglia, M. A., et al. (2017). Adaptive silviculture for climate change: A national experiment in manager-scientist partnerships to apply an adaptation framework. *Journal of Forestry, 115*, 167–178. https://doi.org/10.5849/jof.16-039.

Navarro, L., Morin, H., Bergeron, Y., et al. (2018). Changes in spatiotemporal patterns of 20th century spruce budworm outbreaks in eastern Canadian boreal forests. *Frontiers in Plant Science, 9*, 1905. https://doi.org/10.3389/fpls.2018.01905.

Payette, S. (1992). Fire as a controlling process in the North American Boreal forest. In H. H. Shugart, R. Leemans, & G. B. Bonan (Eds.), *A systems analysis of the global boreal forest* (pp. 145–169). Cambridge University Press.

Payette, S. (1993). The range limit of boreal tree species in Quebec-Labrador: An ecological and palaeoecological interpretation. *Review of Palaeobotany and Palynology, 79*, 7–30. https://doi.org/10.1016/0034-6667(93)90036-T.

Payette, S., Delwaide, A., Schaffhauser, A., et al. (2012). Calculating long-term fire frequency at the stand scale from charcoal data. *Ecosphere, 3*, 59. https://doi.org/10.1890/ES12-00026.1.

Périé, C., & de Blois, S. (2016). Dominant forest tree species are potentially vulnerable to climate change over large portions of their range even at high latitudes. *PeerJ, 4*, e2218. https://doi.org/10.7717/peerj.2218.

Périé, C., de Blois, S., & Lambert, M.-C. (2009). Atlas interactif: Changements climatiques et habitats des arbres [base de données] Québec: Gouvernement du Québec, ministère des Forêts de la Faune et des Parcs, Direction de la recherche forestière.

Périé, C., de Blois, S., Lambert, M.-C., et al. (2014). *Effets anticipés des changements climatiques sur l'habitat des espèces arborescentes au Québec.* Mémoire de recherche forestière 173 (p. 46). Québec: Gouvernement du Québec, ministère des Ressources naturelles, Direction de la recherche forestière.

Pinno, B. D., Errington, R. C., & Thompson, D. K. (2013). Young jack pine and high severity fire combine to create potentially expansive areas of understocked forest. *Forest Ecology and Management, 310,* 517–522. https://doi.org/10.1016/j.foreco.2013.08.055.

Pojar, J., Klinka, K., & Meidinger, D. V. (1987). Biogeoclimatic ecosystem classification in British Columbia. *Forest Ecology and Management, 22,* 119–154. https://doi.org/10.1016/0378-112 7(87)90100-9.

Powell, D. C. (2000). *Potential vegetation, disturbance, plant succession, and other aspects of forest ecology.* Technical Publication F14-SO-TP-09–00US (p. 88). Portland: U.S. Department of Agriculture, Forest Service, Pacific Northwest Region.

Price, D. T., Alfaro, R. I., Brown, K. J., et al. (2013). Anticipating the consequences of climate change for Canada's boreal forest ecosystems. *Environmental Reviews, 21,* 322–365. https://doi.org/10.1139/er-2013-0042.

Puettmann, K. J. (2011). Silvicultural challenges and options in the context of global change: "simple" fixes and opportunities for new management approaches. *Journal of Forestry, 109*(6), 321–331. https://doi.org/10.1093/jof/109.6.321.

Rey, P. (1960). *Essai de phytocinétique biogéographique* (p. 399). Centre national de la Recherche scientifique.

Richard, P. (1978). Aires ombrothermiques des principales unités de végétation du Québec. *Naturaliste Canadien, 105,* 195–207.

Richard, P. J. H. (1987). Le couvert végétal au Québec-Labrador et son histoire postglaciaire. Notes et Documents 87-01. Département de géographie, Université de Montréal.

Richard, P. J. H. (1993). Origine et dynamique postglaciaire de la forêt mixte au Québec. *Review of Palaeobotany and Palynology, 79,* 31–68. https://doi.org/10.1016/0034-6667(93)90037-U.

Richard, P. J. H., Fréchette, B., Grondin, P., et al. (2020). Histoire postglaciaire de la végétation de la forêt boréale du Québec et du Labrador. *Naturaliste Canadien, 144,* 63–76. https://doi.org/10.7202/1070086ar.

Rowe, J. S., & Scotter, G. W. (1973). Fire in the boreal forest. *Quaternary Research, 3,* 444–464. https://doi.org/10.1016/0033-5894(73)90008-2.

Rowe, J. S., & Sheard, J. W. (1981). Ecological land classification: A survey approach. *Environmental Management, 5,* 451–464. https://doi.org/10.1007/BF01866822.

Saucier, J.-P., Grondin, P., Robitaille, A., et al. (2009). Écologie forestière. Carte bioclimatique de référence du Québec. In Ordre des ingénieurs forestiers du Québec (Ed.), *Manuel de foresterie,* 2nd edition (pp. 165–316; pp. SC-10). Montréal: Éditions MultiMondes.

Saucier, J.-P., Gosselin, J., Morneau, C., et al. (2010). Utilisation de la classification de la végétation dans l'aménagement forestier au Québec. *Revue Forestrière Française LXI, I*(3–4), 428–438.

Saucier, J.-P., Baldwin, K., Krestov, P., et al. (2015). Boreal Forests. In K. S.-H. Peh, R. T. Corlett, & Y. Bergeron (Eds.), *Routledge handbook of forest ecology* (pp. 7–29). Routledge.

Schab, A., Gauthier, S., Pascual, J., et al. (2021). Modeling paludification and fire impacts on the forest productivity of a managed landscape using valuable indicators: The example of the Clay Belt. *Canadian Journal of Forest Research, 51,* 1347–1356. https://doi.org/10.1139/cjfr-2020-0386.

Scheller, R. M., Domingo, J. B., Sturtevant, B. R., et al. (2007). Design, development, and application of LANDIS-II, a spatial landscape simulation model with flexible temporal and spatial resolution. *Ecological Modelling, 201,* 409–419. https://doi.org/10.1016/j.ecolmodel.2006.10.009.

Seastedt, T. R., Hobbs, R. J., & Suding, K. N. (2008). Management of novel ecosystems: Are novel approaches required? *Frontiers in Ecology and the Environment, 6*, 547–553. https://doi.org/10.1890/070046.

Shorohova, E., Kneeshaw, D., Kuuluvainen, T., et al. (2011). Variability and dynamics of old-growth forests in the circumboreal zone: Implications for conservation, restoration and management. *Silva Fennica, 45*, 785–806. https://doi.org/10.14214/sf.72.

Sims, R. A., Corns, I. G. W., & Klinka, K. (1996). Introduction-global to local: Ecological land classification. *Environmental Monitoring and Assessment, 39*(1), 1–10. https://doi.org/10.1007/BF00396130.

Sjörs, H. (1999). Swedish plant geography: 1. The background: Geology, climate and zonation. In H. Rydin, P. Snoeijs, M. Diekmann (Eds.). *Swedish plant geography, Acta Phytogeographica Suecica 84*, 5–14.

Splawinski, T. B., Cyr, D., Gauthier, S., et al. (2019). Analyzing risk of regeneration failure in the managed boreal forest of northwestern Quebec. *Canadian Journal of Forest Research, 49*(6), 680–691. https://doi.org/10.1139/cjfr-2018-0278.

Stralberg, D., Wang, X., Parisien, M.-A., et al. (2018). Wildfire-mediated vegetation change in boreal forests of Alberta, Canada. *Ecosphere, 9*(3), e02156. https://doi.org/10.1002/ecs2.2156.

Swanson, F. J., Jones, J. A., Wallin, D. O., et al. (1994). Natural variability-implications for ecosystem management. In M. E. Jensen, P. S. Bourgeron (Eds.), *Volume II: Ecosystem management: principles and applications* (pp. 80–94). General Technical Report PNW-GTR-318. Portland: U.S. Department of Agriculture, Forest Service, Pacific Northwest Research Station.

Taylor, A. R., Boulanger, Y., Price, D. T., et al. (2017). Rapid 21st century climate change projected to shift composition and growth of Canada's Acadian Forest Region. *Forest Ecology and Management, 405*, 284–294. https://doi.org/10.1016/j.foreco.2017.07.033.

Thiffault, N., Raymond, P., Lussier, J. M., et al. (2021). Adaptative sylviculture for climate change: From concepts to reality. Report on a symposium held at Carrefour Forêts 2019. *The Forestry Chronicle, 97*, 13–27. https://doi.org/10.5558/tfc2021-004.

Wang, T., Campbell, E., O'Neil, G. A., et al. (2012). Projecting future distribution of ecosystem climate niches: Uncertainties and management applications. *Forest Ecology and Management, 279*, 128–140. https://doi.org/10.1016/j.foreco.2012.05.034.

White, P. S. (1979). Pattern, process, and natural disturbance in vegetation. *Botanical Review, 45*, 229–299. https://doi.org/10.1007/BF02860857.

Whittaker, R. H. (1962). Classification of natural communities. *Botanical Review, 28*, 1–239. https://doi.org/10.1007/BF02860872.

Part IV
Response of Functional Traits

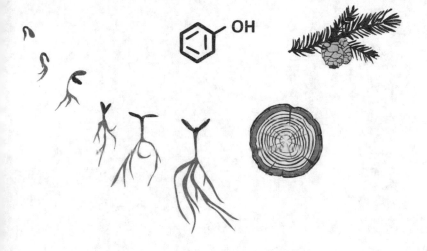

Chapter 9
Changes in Water Status and Carbon Allocation in Conifers Subjected to Spruce Budworm Defoliation and Consequences for Tree Mortality and Forest Management

Annie Deslauriers, Lorena Balducci, Angelo Fierravanti, and Mathieu Bouchard

Abstract The ability of forests to provide ecosystem services and renewable goods faces several challenges related to insect defoliation. Spruce budworm outbreaks represent one of the major natural disturbances in the boreal forest of eastern North America. In this chapter, we will focus on the effects of defoliation by eastern spruce budworm in balsam fir and black spruce trees. We first describe tree water status depending on the duration of defoliation. We then present the response of springtime starch reserves and radial growth at different levels of defoliation. We summarize four mechanisms to explain mortality under defoliation and the consequences for forest management.

A. Deslauriers (✉) · L. Balducci · A. Fierravanti
Département des Sciences fondamentales, Université du Québec à Chicoutimi, 555 boul. de l'université, Chicoutimi, QC G7H 2B1, Canada
e-mail: Annie_Deslauriers@uqac.ca

L. Balducci
e-mail: Lorena1_Balducci@uqac.ca

A. Fierravanti
e-mail: angelo.f3050@gmail.com

M. Bouchard
Département des sciences du bois et de la forêt, Université Laval, 2405 Rue de la Terrasse, Québec, QC G1V 0A6, Canada
e-mail: mathieu.bouchard@sbf.ulaval.ca

© The Author(s) 2023
M. M. Girona et al. (eds.), *Boreal Forests in the Face of Climate Change*,
Advances in Global Change Research 74,
https://doi.org/10.1007/978-3-031-15988-6_9

9.1   Introduction

Insect outbreaks, fire, and drought are the major disturbances in forest ecosystems (Seidl et al., 2017). Worldwide, about 345 million ha of forests are affected annually by such disturbances (van Lierop et al., 2015). Because of their impact on growth and survival, insect outbreaks have serious economic and ecological implications for the boreal ecosystem (Sturtevant et al., 2015). Outbreaks of eastern spruce budworm (*Choristoneura fumiferana* (Clemens), hereafter SBW) occur periodically in eastern North America (Morin et al., 2010; Zhang et al., 2014). Cyclical SBW outbreaks during the twentieth century caused more than a 50% annual productivity loss and widespread mortality in spruce and fir species (Morin, 1994; Pureswaran et al., 2016). Since 2006, there has been a new and ongoing SBW outbreak in Québec, Canada, affecting more than 8 million ha of forest in 2018 (Bouchard et al., 2018; MFFP 2018).

Climate factors, such as temperature and precipitation, determine the distribution of SBW (Pureswaran et al., 2015). Ongoing climate change thus alters the interactions between plants and insects (Fleming & Volney, 1995; Haynes et al., 2014; Singer & Parmesan, 2010), directly influencing the frequency of outbreaks and their spatial distribution (Despland, 2018; Foster et al., 2013). Moreover, under global warming scenarios, future outbreaks will last six years longer than at present and produce 15% greater defoliation (Gray, 2008), thereby influencing the soil–plant-atmosphere continuum and the associated forest carbon, water, and energy dynamics (Balducci et al., 2020; Fierravanti et al., 2019; Liu et al., 2018).

Given the ongoing global change, a 20 to 40% reduction in soil water content is expected in eastern North America (Houle et al., 2015). During the twentieth century, SBW outbreaks increased in duration and severity in eastern Canada (Navarro et al., 2018). However, insect defoliation may offset the negative impact of water deficit on tree growth, thereby reducing mortality (Bouzidi et al., 2019; Itter et al., 2019). Nonetheless, the interactive effect of water availability and defoliation on tree water status remains puzzling (Quentin et al., 2012), especially when considering the short-term chewing effect of SBW (Bouzidi et al., 2019) and the long duration (ca. 10 years) of their outbreaks (Candau et al., 1998; Gray et al., 2000), which affect long-term water uptake and tree growth (Balducci et al., 2020).

Outbreaks result in a tree mortality of approximately 50% (Bergeron et al., 1995) and volume losses of 32–48% (Ostaff & MacLean, 1995) Thus, outbreaks play a significant role in the carbon (C) flux of forests, with losses estimated at 2.87 t C·ha^{-1} over the course of an outbreak. Insect outbreaks represent a major disturbance that affects the entire physiology of a tree. The changes in carbon allocation patterns under defoliation have important physiological consequences, including the modification of bud burst (Deslauriers et al., 2019), a drastic reduction in radial growth rate, i.e., wood formation (Bouzidi et al., 2019; Deslauriers et al., 2015), and increased tree mortality (Fierravanti et al., 2019). As the mortality of defoliated trees depends on many factors, including water status and carbon allocation, disentangling these effects can help predict tree mortality under defoliation.

This chapter aims to describe the effect of defoliation by SBW on the water status and carbon allocation of balsam fir (*Abies balsamea* L. Mill.) and black spruce (*Picea mariana* (Mill.) BSP). We first describe the short- and long-term effects of budworm defoliation on the tree and soil water status. We then describe how growth reduction following defoliation is associated with a reduction in carbon reserves. Last, we describe how changes in tree water status and carbon allocation affect tree mortality and discuss the related implications for forest management.

9.2 Change in the Water Status of Trees and Soils During Budworm Defoliation

Depending on the specific plant–insect interaction, the direct effects of defoliation on plant water status range from positive to negative (Aldea et al., 2005; Nardini et al., 2012; Pittermann et al., 2014). An improvement in tree water potential, i.e., an increase in leaf water potential, has been observed at the beginning of a defoliation event in partially defoliated eucalyptus trees (Eyles et al., 2013; Quentin et al., 2011, 2012). In contrast, insect attacks on broadleaf species reduce leaf size and, therefore, midday water potential, leading to morphological and physiological changes that are similar to drought responses (Nabity et al., 2009; Peschiutta et al., 2016). Similarly, lower water content has been observed on defoliated twigs in balsam fir (Deslauriers et al., 2015), black spruce (Bouzidi et al., 2019), and Scots pine (*Pinus sylvestris* L.) (Salmon et al., 2015), indicating a higher evaporative demand in defoliated trees and a consequent decrease in tree water status during defoliation. The responses of defoliated conifers to water must therefore consider the cumulative effect of defoliation over both shorter and longer periods to better understand the impact on tree mortality.

9.2.1 Short-Term Effects

The plant water status of black spruce saplings defoliated by eastern SBW shows time-dependent effects (Bouzidi et al., 2019). During defoliation, i.e., when the budworm is feeding vigorously on the needles, a higher evaporative demand and lower midday leaf water potential (Ψ_{md}) were observed, whereas there was an opposite pattern after the period of defoliation (Fig. 9.1) (Bouzidi et al., 2019). This was closely related to the timing of measurements linked with the period of larvae feeding. Although plant water status changes rapidly during and after defoliation, no effect of defoliation has been observed on soil moisture (Bouzidi et al., 2019).

In the short term, Ψ_{md} is often lower in defoliated plants than nondefoliated controls, indicating a higher evaporative demand during or immediately following defoliation (Salmon et al., 2015). Eyles et al. (2013) also report a decrease in Ψ_{md}

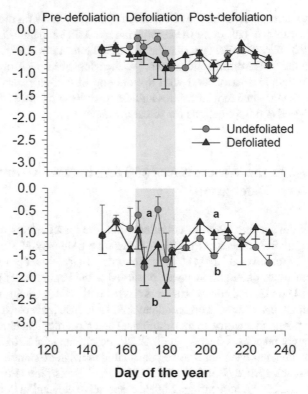

Pre-defoliation Defoliation Post-defoliation

Fig. 9.1 Predawn (Ψpd, MPa) and midday (Ψmd, MPa) leaf water potential of black spruce saplings subjected to defoliation. The shaded gray areas indicate defoliation periods. Different letters indicate significant differences ($P < 0.05$) in the defoliated treatments per sampling period. Modified by permission of Springer Nature from Bouzidi et al. (2019)

in saplings subjected to different levels of defoliation—50% or 100% apical bud damage. The decrease of Ψ_{md} occurs over a short period (approximately two weeks) and is caused by the mechanical chewing action of larvae during active defoliation (Bouzidi et al., 2019). In contrast, predawn leaf water potential (Ψ_{md}) is not affected during or after defoliation by SBW (Fig. 9.1). Budworm feeding habits damage many growing needles, resulting in a loss of turgor or even localized cavitation because of the entry of air into the water conduits of damaged needles; this likely decreases leaf water potential. However, an opposite pattern is observed afterward during the post-defoliation period when leaf Ψmd is higher in nondefoliated saplings (Fig. 9.1); this observation agrees with similar patterns for defoliated *Larix decidua* Mill., *Pinus strobus* L., and *Quercus velutina* Lam. (Vanderklein & Reich, 2000; Wiley et al., 2013). When direct defoliation ceases, the reduced leaf area leads to lower transpiring woody tissues (Schmid et al., 2017; Wiley et al., 2013), thereby decreasing water transport. In the short term, therefore, defoliation affects the plant–water relationship both negatively (during active defoliation by growing instars) and positively (after defoliation).

9.2.2   Long-Term Effects

An event of several years of needle loss reveals a different response than that observed for short-term defoliation, i.e., one year. We define long-term effects as mean various defoliation intensities lasting between 5 and 15 years. To study the long-term effects of recurrent defoliation on the water status of trees and soil, we measured soil volumetric water content (VWC), shoot relative water content (SWC), and midday water potential (Ψ_{md}) along a defoliation gradient in black spruce and balsam fir during a two-year period (2014–2015) at several sites in Québec, Canada (Balducci et al., 2020). The decrease in SWC and Ψ_{md} with increased total tree defoliation reveal that the plant water status reflects the quantity of foliage loss in mature trees (Fig. 9.2). In the long term, the water potential of both fir and spruce (Ψ_{md}) decreases with greater defoliation; values range from −1.01 MPa at 0% defoliation to a maximum of −1.84 MPa in completely defoliated plants. Plant water status lowers as defoliation increases (Fig. 9.2). At low defoliation levels between 5 and 15%, Ψ_{md} increases from −0.96 to −0.87 MPa but becomes more negative at higher defoliation levels (>20%), suggesting a threshold effect. Therefore, at the beginning of defoliation (<20%), the reduced leaf area shows a similar effect to the post-defoliation period (Fig. 9.1), also leading to lower leaf transpiration (Schmid et al., 2017; Wiley et al., 2013).

In defoliated trees, Ψ_{md} is also more negative (reaching values of between −0.96 and −1.36 MPa) with an increasing vapor pressure deficit (VPD) (ranging from 0.1 to 0.8 kPa, Fig. 9.2). High VPD decreases the midday water potential, i.e., an increase in water tension in the xylem. Trees having a more negative water potential because of high VPD are also more likely to lose hydraulic conductivity, with a consequent increase in the risk of xylem embolism (Tyree & Sperry, 1989) and eventual death by hydraulic failure (Adams et al., 2017; Anderegg et al., 2013). In two consecutive years of defoliation, an increase in defoliation also resulted in a significant decrease

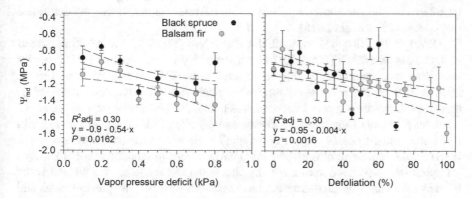

Fig. 9.2 Midday water potential (Ψ_{md}) expressed as a function of vapor pressure deficit (VPD) and defoliation (%) in both balsam fir and black spruce. The *solid lines* represent the fitted linear regression, the *dashed lines* refer to the 95% confidence interval for the data points, and the *vertical bars* represent the standard deviation of the mean. Modified with permission from Elsevier from Balducci et al. (2020)

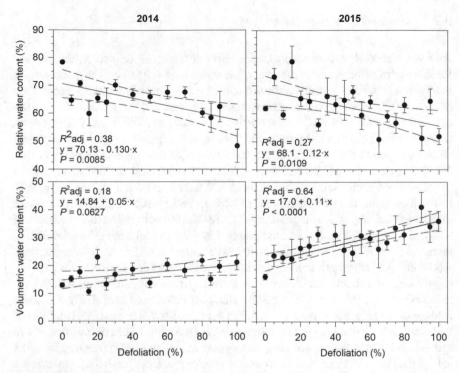

Fig. 9.3 Soil volumetric water content (VWC, %) and shoot relative water content (SWC, %) expressed as a function of defoliation for sampled trees in 2014 and 2015. Data includes the fitted linear regression (*solid black line*), the 95% confidence intervals (*dashed lines*), and the standard error of the individual data points (*vertical bars*)

in the SWC (Fig. 9.3). Similar results have been found in the twigs of mature balsam fir, where the SWC decreased by 8% in defoliated trees compared with nondefoliated trees (Deslauriers et al., 2015).

Although Ψ_{md} dropped as defoliation increased (Fig. 9.2), soil VWC increased in both 2014 and 2015 (Fig. 9.3). In both years, VWC was close to 14% in lightly defoliated trees, whereas, in fully defoliated trees, VWC increased from 17% in 2014 to 25% in 2015. These results for the plant and soil water content of defoliated trees therefore suggest a reduced water transfer throughout the soil–plant–air continuum. As also observed by Dietze and Matthes (2014), a slight increase in soil water content occurs following defoliation. Such an increase can be explained by (i) a loss of foliage at a greater rate than fine root production and (ii) reduced transpiration. Defoliated trees markedly alter water storage in the soil and the water balance of a stand by changing the ratio between water input (precipitation) and output (evapotranspiration) (Dietze & Matthes, 2014; Hata et al., 2016).

When defoliation increases canopy transparency, such as via canopy openness (Hata et al., 2016), less precipitation is intercepted by the canopy, and, consequently, the soil receives more water (Sun et al., 2015). As vegetation cover reflects water movement in the soil–plant sphere, a higher percentage of cover (a larger leaf area index, LAI) increases evaporation from interception, thereby compensating for the reduced evaporation from the soil (Fatichi & Pappas, 2017). Defoliation thus changes the water storage, i.e., the difference between precipitation and evapotranspiration (Viglizzo et al., 2016) through a reduced interception and canopy evaporation. An ecophysiological framework applied to ponderosa pine (*Pinus ponderosa* Dougl. ex C. Lawson) in western North America showed that canopy transparency at higher defoliation levels increases soil moisture in response to reduced canopy transpiration and dryness of the soil surface because of increased evaporation (Dietze & Matthes, 2014). In the boreal forest of eastern Québec, however, temperatures are generally low and precipitation abundant, i.e., greater than soil evaporation (Gauthier et al., 2015); this is particularly true in boreal forest peatlands where the contribution of transpiration to evapotranspiration is limited to 1% (Warren et al., 2018).

9.3 Carbon Allocation in Conifers Defoliated by Budworm

The intra-annual carbon allocation is severely modified during defoliation (Figs. 9.4 and 9.5) because recurrent defoliation represents a special case of carbon source limitation (Körner, 2015). The changes will be reflected in carbon sink priority, which is ordered according to organ priorities with seed production having the highest priority and reserves the lowest (Minchin & Lacointe, 2005). In conifers, carbon reserves in the form of starch increase before the resumption of shoot and stem growth (Desalme et al., 2017; Hoch et al., 2003; Little, 1970; Martinez-Vilalta et al., 2016), and the highest amounts of starch are found in the needles (Deslauriers et al., 2015, 2019). Regardless of host species (balsam fir, white spruce, or black spruce), recurrent defoliation prevents the accumulation of starch during the spring (Deslauriers et al., 2015, 2019), thus reflecting sink priority. The reduction in the buildup of carbon reserves is followed by an earlier bud burst, promoting needle growth at the expense of cambium activity; thus, ring width decreases. Multiple studies on conifers have illustrated the changes occurring in the carbon reserves of nondefoliated and defoliated trees, the latter having a lower starch content (Webb, 1980, 1981; Webb & Karchesy, 1977).

In conifers (except for larch), the older needles assimilate and store carbon, especially during the period preceding bud burst and growth. Under defoliation conditions, any change in carbon balance thus occurs initially in the starch reserves of older needles (Deslauriers et al., 2019; Vanderklein & Reich, 1999) because of reduced starch accumulation during the spring (Wiley & Helliker, 2012) before defoliation (Fig. 9.4). Therefore, the defoliation level closely corresponds to differences in springtime starch reserves (Fig. 9.5) and radial growth (Fig. 9.6); both variables decrease at higher defoliation levels. Within a wider annual context, the metabolism of starch and soluble sugars during the year (Figs. 9.4 and 9.5) becomes biologically meaningful compared with growth activity and bud dormancy. After bud break, starch levels slowly decrease during the summer to sustain primary and secondary growth

(Deslauriers et al., 2019; Webb & Karchesy, 1977); however, when the starch reserves fail to increase under defoliation, radial growth decreases proportionally.

The absence of starch buildup during the spring reveals the changes occurring in carbon allocation under defoliation. Rather than allocating carbon to starch reserves during the spring (Figs. 9.4 and 9.5), this carbon is allocated to shoot growth, which has a higher sink strength, at the expense of radial stem growth and reserve accumulation; this change represents a shift in the trees' priorities for physiological resources to ensure survival (Deslauriers et al., 2019). In deciduous conifers such as larch (*Larix decidua* Mill.) affected by budmoth (*Zeiraphera griseana* Hübner), however, replenishment of carbon storage for refoliation is prioritized (Peters et al., 2020).

In conifers, newly assimilated carbon by older needles is mainly allocated to the canopy during primary growth (85% of the allocated ^{13}C), with only a minor fraction (1.6%) translocated to the lower stem (Heinrich et al., 2015). During primary growth, the substantial loss of needles under defoliation produces a carbon limitation in other parts of the tree, mainly in the stem wood where starch starvation can occur (Fig. 9.5, *right*). Carbon allocation to the roots is also severely impaired after defoliation (Li et al., 2002; Reich et al., 1993). As bud growth competes with other sinks, such as wood formation (Antonucci et al., 2015; Deslauriers et al., 2016; Huang et al., 2014; Traversari et al., 2018) and root growth (Reich et al., 1993), prioritizing carbon allocation to the buds has a positive impact on bud opening and successive shoot

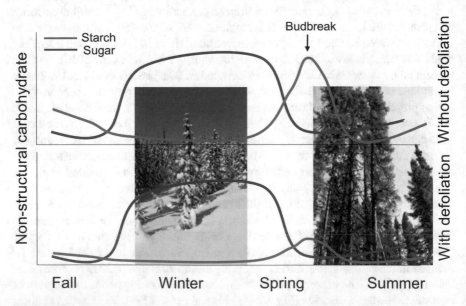

Fig. 9.4 Variations of starch (*orange*) and sugar (*blue*) concentrations over a year in the shoots of boreal conifers. The cases of no defoliation (*top*) and after several years of defoliation (*bottom*) are presented. Bud break generally occurs around the maximum of starch accumulation during spring. Under recurrent years of defoliation, needle loss impairs starch accumulation during the spring and can reduce soluble sugar concentrations in several tree organs (e.g., needles, twigs, stem, and roots). *Photo credits* Annie Deslauriers

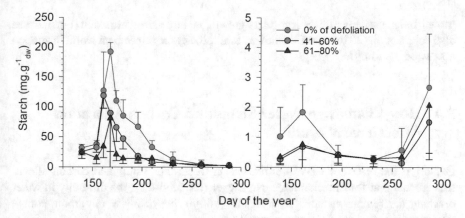

Fig. 9.5 Variation in starch ($mg·g^{-1}_{dw}$) measured in leaf (*right*) and stem wood (*left*) of balsam fir during the growing season for the different defoliation classes (nondefoliated, moderately defoliated, and heavily defoliated). *Vertical bars* represent the standard deviation among six trees. Note the different scales of the vertical axes. Modified from Deslauriers et al. (2015), CC BY 4.0 license

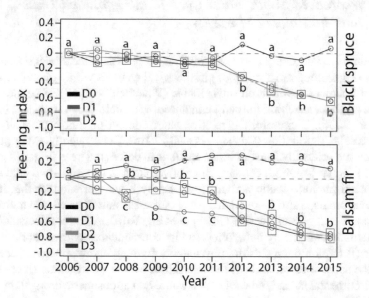

Fig. 9.6 Tree-ring widths in trees belonging to the four defoliation classes in black spruce and balsam fir, from measurements obtained between 2006 and 2015. *D0* represents nondefoliated trees, and *D1*, *D2*, and *D3* represent 1%–33%, 33%–66%, and 66%–100% defoliation, respectively. Different letters indicate significant differences between defoliation classes. Rectangles indicate situations where data points sharing the same letter are overlapping on the plots. Reproduced from Fierravanti et al. (2019), CC BY 4.0 license

growth but negatively affects secondary growth, i.e., tree-ring formation (Fig. 9.6; see also Peters et al. (2020) and Deslauriers et al. (2015) for impacts on wood formation and wood anatomy).

9.4 How Changes in Water Status and Carbon Allocation Affect Tree Mortality

Exploring the effect of SBW defoliation on tree water status and carbon allocation reveals that four mechanisms can trigger tree mortality: the difficulty in water reaching the remaining needles, hydraulic failure, the decrease in carbon storage during spring, and the decrease in radial growth.

9.4.1 Decreased Shoot Water Content in Defoliated Trees and the Related Mortality

Needle damage and death due to defoliation are visible effects of insect feeding, and these signs integrate a series of physiological changes within plants. First, they represent a direct constraint on transpiration (Pincebourde et al., 2006), linked to a decrease in water reaching the remaining needles as defoliation increases (Sack & Holbrook, 2006). Therefore, the relative water content of shoots gradually drops with increasing defoliation (Deslauriers et al., 2015). Consequently, water potential decreases and gradually impairs water relations in defoliated trees (Fig. 9.7; also see Sect. 9.1). In the twentieth century, cumulative climatic stresses—two cold springs that reduced photosynthetic activity followed by a warmer summer that induced higher evapotranspiration—preceded SBW outbreaks and induced tree mortality in black spruce and balsam fir (De Grandpré et al., 2019). A better understanding of water status imbalance by defoliation and its consequences on tree mortality must consider (i) tree ontogeny, as the physiology of young and mature trees can differ within the same species, and (ii) the interaction between defoliation (biotic stress) and environmental stress (reduced soil moisture and atmospheric drought, including a high VPD.

Under controlled conditions, four-year-old saplings of fir, white spruce, and black spruce affected by severe defoliation after a full year of SBW outbreak show mainly bud phenological shifts. This alteration modifies the carbon allocation within the trees to primary rather than secondary growth (Deslauriers et al., 2019) and thereby ensures tree survival. Similarly, in young black spruce, a mild defoliation intensity (~40%) observed in current-year defoliation does not compromise tree survival, although water status is negatively affected during active insect feeding (Bouzidi et al., 2019). Under concomitant defoliation and drought stresses, however, the effect of drought prevails over defoliation-related stress. Drought alone increases the sapling mortality

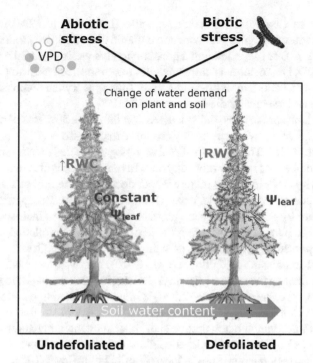

Fig. 9.7 Summary of the effect of environmental (*VPD*, vapor pressure deficit) and biotic stresses (spruce budworm) on the water status of trees (*RWC*, root water content, Ψ_{leaf}, leaf water potential) and soil in defoliated and nondefoliated trees. Symbols (−) and (+) represent the limiting and optimal soil water availability, respectively. Modified with permission from Elsevier from Balducci et al. (2020)

of black spruce immediately after the stress period. Nondefoliated drought-related saplings show higher mortality rates during the first two weeks after resuming irrigation (0.3 dead saplings·day^{-1}) compared with defoliated drought-related saplings (0 dead saplings·day^{-1}) (Bouzidi et al., 2019). The detrimental effects observed in needles during active defoliation become positive for tree water balance in the first few weeks. After two weeks, the mortality rate in defoliated saplings remains similar (0.05–0.1 dead saplings·day^{-1}) to defoliated drought-related spruce. This difference in mortality rates occurs mainly because of the water storage effect in defoliated plants, as the reduced foliar biomass temporarily decreases water loss through transpiration.

As the hydraulic pathway from roots to canopy can be jeopardized by defoliation, trees maintain leaf water potential within functional limits to avoid critical levels of hydraulic failure and death (Benito Garzón et al., 2018). The hydraulic safety margin (HSM), which is the difference between the minimum leaf water potential measured in the plants (Ψ_{min}) and that inducing 12 and 50% embolism, acts as an internal hydraulic buffer and provides information about the thresholds of leaf water potential inducing hydraulic failure. These physiological mechanisms involved in the resistance to cavitation are species-specific (Delzon et al., 2010); plants can differ in their safety margins. Atmospheric demand also negatively influences water potential (VPD). Trees experiencing a greater negative water potential because of an increase in VPD, i.e., increased summer temperatures and decreased air humidity, are also more likely to lose hydraulic conductivity following xylem embolism (Tyree & Sperry, 1989) and eventually die (Adams et al., 2017).

In young black spruce trees, the xylem tension induces a 50% loss of conductivity ($P50$) to -4.26 MPa, whereas the xylem air entry pressure ($P12$) is -2.9 MPa (Balducci et al., 2015). Therefore, the HSM of young defoliated trees is large (Bouzidi et al., 2019), explaining the low mortality rate. Juvenile wood is also more resistant to xylem failure by embolism than mature wood, decreasing the former's vulnerability to cavitation relative to that of older trees (Domec et al., 2009). This greater resistance to xylem failure by embolism in younger trees is mainly due to their smaller xylem cell diameter. In mature stands during the period of active defoliation, shoot water potentials surpassed Ψ_{min} thresholds, inducing 12% xylem embolism in mature black spruce but not in balsam fir (Balducci et al., 2020). The Ψ_{min} reached -2.95 MPa, which corresponds to the onset of xylem embolism in balsam fir—average xylem air entry pressure, $P12$, being around -2.8 MPa in nondefoliated trees. Mature balsam fir might therefore have experienced greater xylem embolism during defoliation compared with defoliated black spruce and, thus, showing a narrower HSM under defoliation and a higher risk of hydraulic failure than the latter. Trees show a narrow HSM; therefore, they operate close to the hydraulic failure level (Choat et al., 2012). But populations that experience a narrower HSM are strongly associated with higher mortality (Benito Garzón et al., 2018). Conifers in North America, Europe, and Australia exhibit a large interspecific variability of cavitation resistance (Delzon et al., 2010). Although the phenotypic variability is considered negligible in the point of critical loss of xylem conductivity, intraspecific variability in $P50$ has recently been reported in some species (Anderegg, 2015; Benito Garzón et al., 2018). In this context, the HSM for young black spruce is greater than for mature black spruce, which is greater than that for mature balsam fir. Therefore, the mortality rate under defoliation follows an opposite pattern, suggesting that in future scenarios of more frequent and intense drought events and SBW attacks, balsam fir could be more vulnerable than black spruce.

The integration of knowledge regarding water status and the hydraulic conductivity of trees can help us understand mortality dynamics in natural conditions (Martinez-Vilalta et al., 2019). In mature trees, SBW outbreaks of longer duration and higher intensity limit tree survival (Chen et al., 2017; MacLean, 1980). Several studies agree that mature firs are more sensitive, i.e., higher tree mortality, after five years of severe defoliation by SBW than other host species, e.g., white and black spruce (MacLean, 1980, 2016; Virgin & MacLean, 2017). However, species susceptibility to SBW depends on the stand, with higher mortality rates in balsam fir–dominated stands than in mixed boreal stands, although mortality remains high for the host species (Bouchard et al., 2005). A greater abundance of hardwood in stands significantly reduces the dispersion of the second instars (L2) of SBW, thereby helping limit balsam fir defoliation (MacLean, 2016; Zhang et al., 2020). A similar consensus exists for the survival of balsam fir regeneration, where the type of forest overstory (as hardwood, softwood, and mixed) reduces or increases seedling mortality rate to 17% and 24% to 26%, respectively (Nie et al., 2018). Furthermore, seedling height is likely associated with a larval density increase, as taller seedlings of balsam fir and black spruce are more defoliated than shorter seedlings (10%) (Cotton-Gagnon et al., 2018; Lavoie et al., 2019; Nie et al., 2018). Balsam fir at 1.2 m height has a 50% probability of sustaining severe defoliation, whereas for black spruce the same probability of severe defoliation is attained only when seedlings are taller than 3.5 m. Thus, balsam fir is 15% more defoliated than black spruce (Cotton-Gagnon et al., 2018; Lavoie et al., 2021; Nie et al., 2018). Nonetheless, taller seedlings grown in open conditions should show greater survival than seedlings under shaded conditions, even if the defoliation level reaches 75% (Nie et al., 2018).

9.4.2 Decreased Carbon Storage and Growth in Defoliated Trees Leads to Greater Mortality

Boreal stands of spruce and fir can tolerate extended periods of spruce budworm defoliation because trees allocate most of their carbon resources to the production of new shoots and needles (Deslauriers et al., 2019; Piene, 1989a, b) rather than storage (Figs. 9.4 and 9.5) or stem radial growth (Fig. 9.6). This allocation strategy allows trees to endure several years of defoliation before eventually succumbing when very few needles remain. Compensatory mechanisms are also used to maximize carbon gain under defoliation, such as longer needle retention (Doran et al., 2017) and greater epicormic bud production (Piene, 1989a).

Under defoliation, drastic growth reductions in dying balsam fir trees occur simultaneously with declines in starch reserves, especially in May and June. A nominal logistic regression can calculate the probability of tree mortality on the basis of starch concentrations (Fig. 9.8). The starch concentration in needles in May explained the increase in tree mortality, with the minimum values indicating mortality during spring (Fig. 9.8). The starch concentration at which the probability of plant mortality exceeds 50% is significant in May, having a value of 28 $mg \cdot g^{-1}_{dw}$ of starch in needles (Fig. 9.8). This threshold is reached only in defoliation classes higher than 66%, although the probability of mortality begins to increase between 33 and 66% defoliation (Fierravanti et al., 2019).

By using canonical correlation, Fierravanti et al. (2019) showed how defoliation intensity is inversely correlated to changes in both growth (Fig. 9.6) and carbon allocation (Fig. 9.8). In balsam fir, mortality is related to both reduced radial growth (i.e., as measured by tree-ring width) and starch reserves; however, a direct causal link requires further testing, including carbon allocation at the whole tree level in relation to the tree water status (see Sect. 9.1). The physiological mechanisms that lead to mortality and the associated change in growth rates remain poorly known, although mortality is preceded by growth reductions in 84% of cases (Cailleret et al.,

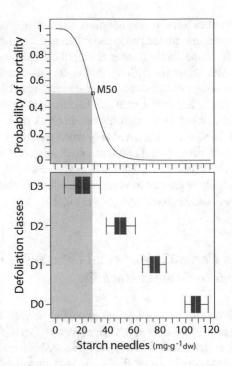

Fig. 9.8 Logistic function linking starch concentrations in May with the probability of balsam fir mortality. The box plot represents starch concentrations across the different defoliation classes. The gray rectangles represent the threshold of tree mortality at a probability of 50%. *D0* represents nondefoliated trees, and *D1*, *D2*, and *D3* represent 1%–33%, 33%–66%, and 66%–100% defoliation, respectively. Reproduced from Fierravanti et al. (2019), CC BY 4.0 license

2017), as observed in balsam fir (Fig. 9.6). Moreover, for shade-tolerant species such as balsam fir, the probability of tree mortality is a function of the growth rate with increasing size (Das et al., 2016; De Grandpré et al., 2019; Kneeshaw et al., 2006). In the absence of environmental changes, the common assumption is that the tree growth rate is inversely associated with the probability of mortality and the slowest-growing trees having a higher probability of dying (Das et al., 2016; De Grandpré et al., 2019; Kneeshaw et al., 2006). Co-occurring past disturbances—climatic and biotic stress events—result in a reduced tree growth (Das et al., 2016; De Grandpré et al., 2019; Kneeshaw et al., 2006); thus, a lower growth rate appears to be an early signal of tree vigor in boreal conifers.

9.5 Implications for Forest Management

Improved knowledge of the changes in water status and carbon allocation of fir and spruce in relation to water deficits, meteorological conditions, and defoliation will help refine current forest management practices to reduce the impacts of SBW outbreaks. A better understanding of the factors driving mortality is particularly critical in the case of SBW-caused mortality in eastern Canada, as trees that die following SBW defoliation generally decay too quickly to be salvaged and used in conventional transformation processes such as pulp and paper or sawmills.

We highlight that over several years of defoliation, a tree's capacity to absorb and conduct water decreases as defoliation increases. Over the long term, i.e., 5 to 15 years (as described above), defoliated trees are therefore more susceptible to drought, potentially generating synchronous mortality in defoliated trees when severe drought conditions occur. Moreover, larger trees, which tend to have a higher portion of their energy budget devoted to supporting their existing biomass, are probably even more exposed to such events. This reinforces the idea that the most effective means of preventing defoliation-caused mortality, other than spraying insecticides, is probably to harvest susceptible trees relatively early during outbreak development, e.g., less than five years after defoliation is first observed at the stand level. The sudden bursts of mortality likely to occur during eventual droughts can therefore be anticipated and prevented, at least in part.

Although the main harvesting method in a SBW outbreak context is clear-cutting (followed or not by plantation), partial harvesting can be used under particular conditions. We suggest that when partial cuts are used, black spruce—less likely to be severely defoliated and more likely to maintain hydraulic conductivity—and smaller or younger trees should be left unfelled, as they are less likely to suffer mortality. To further refine guidelines for partial harvesting in a SBW outbreak context, future studies should look at differences in susceptibility to defoliation and water deficit in trees that belong to a range of size classes (from seedlings to large trees) and canopy positions, i.e., acclimated to different shade levels.

References

Adams, H. D., Barron-Gafford, G. A., Minor, R. L., et al. (2017). Temperature response surfaces for mortality risk of tree species with future drought. *Environmental Research Letters, 12*, 115014. https://doi.org/10.1088/1748-9326/aa93be.

Aldea, M., Hamilton, J. G., Resti, J. P., et al. (2005). Indirect effects of insect herbivory on leaf gas exchange in soybean. *Plant, Cell and Environment, 28*, 402–411. https://doi.org/10.1111/j.1365-3040.2005.01279.x.

Anderegg, W. R. L. (2015). Spatial and temporal variation in plant hydraulic traits and their relevance for climate change impacts on vegetation. *New Phytologist, 205*, 1008–1014. https://doi.org/10.1111/nph.12907.

Anderegg, W. R. L., Plavcová, L., Anderegg, L. D. L., et al. (2013). Drought's legacy: Multiyear hydraulic deterioration underlies widespread aspen forest die-off and portends increased future risk. *Global Change Biology, 19*(4), 1188–1196. https://doi.org/10.1111/gcb.12100.

Antonucci, S., Rossi, S., Deslauriers, A., et al. (2015). Synchronisms and correlations of spring phenology between apical and lateral meristems in two boreal conifers. *Tree Physiology, 35*, 1086–1094. https://doi.org/10.1093/treephys/tpv077.

Balducci, L., Deslauriers, A., Giovannelli, A., et al. (2015). How do drought and warming influence survival and wood traits of *Picea mariana* saplings? *Journal of Experimental Botany, 66*, 377–389. https://doi.org/10.1093/jxb/eru431.

Balducci, L., Fierravanti, A., Rossi, S., et al. (2020). The paradox of defoliation: Declining tree water status with increasing soil water content. *Agricultural and Forest Meteorology, 290*, 108025. https://doi.org/10.1016/j.agrformet.2020.108025.

Benito Garzón, M., González Muñoz, N., Wigneron, J.-P., et al. (2018). The legacy of water deficit on populations having experienced negative hydraulic safety margin. *Global Ecology and Biogeography, 27*(3), 346–356. https://doi.org/10.1111/geb.12701.

Bergeron, Y., Leduc, A., Joyal, C., et al. (1995). Balsam fir mortality following the last spruce budworm outbreak in northwestern Quebec. *Canadian Journal of Forest Research, 25*, 1375–1384. https://doi.org/10.1139/x95-150.

Bouchard, M., Kneeshaw, D., & Bergeron, Y. (2005). Mortality and stand renewal patterns following the last spruce budworm outbreak in mixed forests of western Quebec. *Forest Ecology and Management, 204*, 297–313. https://doi.org/10.1016/j.foreco.2004.09.017.

Bouchard, M., Martel, V., Régnière, J., et al. (2018). Do natural enemies explain fluctuations in low-density spruce budworm populations? *Ecology, 99*, 2047–2057. https://doi.org/10.1002/ecy.2417.

Bouzidi, H. A., Balducci, L., Mackay, J., et al. (2019). Interactive effects of defoliation and water deficit on growth, water status, and mortality of black spruce (*Picea mariana* (Mill.) B.S.P.). *Annals of Forest Science, 76*, 21. https://doi.org/10.1007/s13595-019-0809-z.

Cailleret, M., Jansen, S., Robert, E. M. R., et al. (2017). A synthesis of radial growth patterns preceding tree mortality. *Global Change Biology, 23*(4), 1675–1690. https://doi.org/10.1111/gcb.13535.

Candau, J., Fleming, R. A., & Hopkin, A. (1998). Spatiotemporal patterns of large-scale defoliation caused by the spruce budworm in Ontario since 1941. *Canadian Journal of Forest Research, 28*, 1733–1741. https://doi.org/10.1139/x98-164.

Chen, C., Weiskittel, A., Bataineh, M., et al. (2017). Evaluating the influence of varying levels of spruce budworm defoliation on annualized individual tree growth and mortality in Maine, USA and New Brunswick, Canada. *Forest Ecology and Management, 396*, 184–194. https://doi.org/10.1016/j.foreco.2017.03.026.

Choat, B., Jansen, S., Brodribb, T. J., et al. (2012). Global convergence in the vulnerability of forests to drought. *Nature, 491*, 752–755. https://doi.org/10.1038/nature11688.

Cotton-Gagnon, A., Simard, M., De Grandpre, L., et al. (2018). Salvage logging during spruce budworm outbreaks increases defoliation of black spruce regeneration. *Forest Ecology and Management, 430*, 421–430. https://doi.org/10.1016/j.foreco.2018.08.011.

Das, A. J., Stephenson, N. L., & Davis, K. P. (2016). Why do trees die? Characterizing the drivers of background tree mortality. *Ecology, 97*, 2616–2627. https://doi.org/10.1002/ecy.1497.

De Grandpré, L., Kneeshaw, D. D., Perigon, S., et al. (2019). Adverse climatic periods precede and amplify defoliator-induced tree mortality in eastern boreal North America. *Journal of Ecology, 107*, 452–467. https://doi.org/10.1111/1365-2745.13012.

Delzon, S., Douthe, C., Sala, A., et al. (2010). Mechanism of water-stress induced cavitation in conifers: Bordered pit structure and function support the hypothesis of seal capillary-seeding. *Plant, Cell and Environment, 33*, 2101–2111. https://doi.org/10.1111/j.1365-3040.2010.02208.x.

Desalme, D., Priault, P., Gerant, D., et al. (2017). Seasonal variations drive short-term dynamics and partitioning of recently assimilated carbon in the foliage of adult beech and pine. *New Phytologist, 213*, 140–153. https://doi.org/10.1111/nph.14124.

Deslauriers, A., Caron, L., & Rossi, S. (2015). Carbon allocation during defoliation: Testing a defense-growth trade-off in balsam fir. *Frontiers in Plant Science, 6*, 338. https://doi.org/10.3389/fpls.2015.00338.

Deslauriers, A., Huang, J. G., Balducci, L., et al. (2016). The contribution of carbon and water in modulating wood formation in black spruce saplings. *Plant Physiology, 170*, 2072–2084. https://doi.org/10.1104/pp.15.01525.

Deslauriers, A., Fournier, M. P., Carteni, F., et al. (2019). Phenological shifts in conifer species stressed by spruce budworm defoliation. *Tree Physiology, 39*, 590–605. https://doi.org/10.1093/treephys/tpy135.

Despland, E. (2018). Effects of phenological synchronization on caterpillar early-instar survival under a changing climate. *Canadian Journal of Forest Research, 48*, 247–254. https://doi.org/10.1139/cjfr-2016-0537.

Dietze, M. C., & Matthes, J. H. (2014). A general ecophysiological framework for modelling the impact of pests and pathogens on forest ecosystems. *Ecology Letters, 17*, 1418–1426. https://doi.org/10.1111/ele.12345.

Domec, J. C., Warren, J. M., Meinzer, F. C., et al. (2009). Safety factors for xylem failure by implosion and air-seeding within roots, trunks and branches of young and old conifer trees. *IAWA Journal, 30*, 101–120. https://doi.org/10.1163/22941932-90000201.

Doran, O., MacLean, D. A., & Kershaw, J. A. (2017). Needle longevity of balsam fir is increased by defoliation by spruce budworm. *Trees, 31*, 1933–1944. https://doi.org/10.1007/s00468-017-1597-4.

Eyles, A., Pinkard, E. A., Davies, N. W., et al. (2013). Whole-plant versus leaf-level regulation of photosynthetic responses after partial defoliation in *Eucalyptus globulus* saplings. *Journal of Experimental Botany, 64*, 1625–1636. https://doi.org/10.1093/jxb/ert017.

Fatichi, S., & Pappas, C. (2017). Constrained variability of modeled T:ET ratio across biomes. *Geophysical Research Letters, 44*, 6795–6803. https://doi.org/10.1002/2017GL074041.

Fierravanti, A., Rossi, S., Kneeshaw, D., et al. (2019). Low non-structural carbon accumulation in spring reduces growth and increases mortality in conifers defoliated by spruce budworm. *Frontiers for Global Change, 2*, 15. https://doi.org/10.3389/ffgc.2019.00015.

Fleming, R. A., & Volney, W. J. A. (1995). Effects of climate change on insect defoliator population processes in Canada's boreal forest: Some plausible scenarios. *Water, Air, and Soil Pollution, 82*, 445–454. https://doi.org/10.1007/BF01182854.

Foster, J. R., Townsend, P. A., & Mladenoff, D. J. (2013). Mapping asynchrony between gypsy moth egg-hatch and forest leaf-out: Putting the phenological window hypothesis in a spatial context. *Forest Ecology and Management, 287*, 67–76. https://doi.org/10.1016/j.foreco.2012.09.006.

Gauthier, S., Bernier, P., Kuuluvainen, T., et al. (2015). Boreal forest health and global change. *Science, 349*(6250), 819–822. https://doi.org/10.1126/science.aaa9092.

Gray, D. R. (2008). The relationship between climate and outbreak characteristics of the spruce budworm in eastern Canada. *Climatic Change, 87*, 361–383. https://doi.org/10.1007/s10584-007-9317-5.

Gray, D. R., Régnière, J., & Boulet, B. (2000). Analysis and use of historical patterns of spruce budworm defoliation to forecast outbreak patterns in Quebec. *Forest Ecology and Management, 127*(1), 217–231. https://doi.org/10.1016/S0378-1127(99)00134-6.

Hata, K., Kawakami, K., & Kachi, N. (2016). Increases in soil water content after the mortality of non-native trees in oceanic island forest ecosystems are due to reduced water loss during dry periods. *Science of the Total Environment, 545–546*, 372–380. https://doi.org/10.1016/j.scitotenv.2015.12.007.

Haynes, K. J., Allstadt, A. J., & Klimetzek, D. (2014). Forest defoliator outbreaks under climate change: Effects on the frequency and severity of outbreaks of five pine insect pests. *Global Change Biology, 20*, 2004–2018. https://doi.org/10.1111/gcb.12506.

Heinrich, S., Dippold, M. A., Werner, C., et al. (2015). Allocation of freshly assimilated carbon into primary and secondary metabolites after in situ ^{13}C pulse labelling of Norway spruce (*Picea abies*). *Tree Physiology, 35*, 1176–1191. https://doi.org/10.1093/treephys/tpv083.

Hoch, G., Richter, A., & Körner, C. (2003). Non-structural carbon compounds in temperate forest trees. *Plant, Cell and Environment, 26*, 1067–1081. https://doi.org/10.1046/j.0016-8025.2003.01032.x.

Houle, D., Paquette, A., Côté, B., et al. (2015). Impacts of climate change on the timing of the production season of maple syrup in eastern Canada. *PLoS ONE, 10*, e0144844. https://doi.org/10.1371/journal.pone.0144844.

Huang, J. G., Deslauriers, A., & Rossi, S. (2014). Xylem formation can be modeled statistically as a function of primary growth and cambium activity. *New Phytologist, 203*, 831–841. https://doi.org/10.1111/nph.12859.

Itter, M. S., D'Orangeville, L., Dawson, A., et al. (2019). Boreal tree growth exhibits decadal-scale ecological memory to drought and insect defoliation, but no negative response to their interaction. *Journal of Ecology, 107*(3), 1288–1301. https://doi.org/10.1111/1365-2745.13087.

Kneeshaw, D. D., Kobe, R. K., Coates, K. D., et al. (2006). Sapling size influences shade tolerance ranking among southern boreal tree species. *Journal of Ecology, 94*, 471–480. https://doi.org/10.1111/j.1365-2745.2005.01070.x.

Körner, C. (2015). Paradigm shift in plant growth control. *Current Opinion in Plant Biology, 25*, 107–114. https://doi.org/10.1016/j.pbi.2015.05.003.

Lavoie, J., Montoro Girona, M., & Morin, H. (2019). Vulnerability of conifer regeneration to spruce budworm outbreaks in the eastern Canadian boreal forest. *Forests, 10*(10), 850. https://doi.org/10.3390/f10100850.

Lavoie, J., Montoro Girona, M., Grosbois, G., et al. (2021). Does the type of silvicultural practice influence spruce budworm defoliation of seedlings? *Ecosphere, 12*(4), 17. https://doi.org/10.1002/ecs2.3506.

Li, M., Hoch, G., & Korner, C. (2002). Source/sink removal affects mobile carbohydrates in *Pinus cembra* at the Swiss treeline. *Trees, 16*, 331–337. https://doi.org/10.1007/s00468-002-0172-8.

Little, C. H. A. (1970). Derivation of the springtime starch increase in balsam fir (*Abies balsamea*). *Canadian Journal of Botany, 48*, 1995–1999. https://doi.org/10.1139/b70-291.

Liu, Z., Peng, C., De Grandpré, L., et al. (2018). Development of a new TRIPLEX-insect model for simulating the effect of spruce budworm on forest carbon dynamics. *Forests, 9*, 513. https://doi.org/10.3390/f9090513.

MacLean, D. A. (2016). Impacts of insect outbreaks on tree mortality, productivity, and stand development. *The Canadian Entomologist, 148*, S138–S159. https://doi.org/10.4039/tce.2015.24.

MacLean, D. A. (1980). Vulnerability of fir-spruce stands during uncontrolled spruce budworm outbreaks: A review and discussion. *The Forestry Chronicle, 56*(5), 13–221. 5558/tfc56213-5.

Martinez-Vilalta, J., Sala, A., Asensio, D., et al. (2016). Dynamics of non-structural carbohydrates in terrestrial plants: A global synthesis. *Ecological Monographs, 86*, 495–516. https://doi.org/10.1002/ecm.1231.

Martinez-Vilalta, J., Anderegg, W. R. L., Sapes, G., et al. (2019). Greater focus on water pools may improve our ability to understand and anticipate drought-induced mortality in plants. *New Phytologist, 223*, 22–32. https://doi.org/10.1111/nph.15644.

Minchin, P. E., & Lacointe, A. (2005). New understanding on phloem physiology and possible consequences for modelling long-distance carbon transport. *New Phytologist, 166*, 771–779. https://doi.org/10.1111/j.1469-8137.2005.01323.x.

Ministère des Forêts de la Faune et des Parcs (MFFP). (2018). Aires infestées par la tordeuse des bourgeons de l'épinette au Québec en 2018-Version 1.0. Québec: Direction de la protection des forêts, Gouvernement du Québec.

Morin, H., Jardon, Y., & Simard, S. (2010). Détection et reconstitution des épidemies de la tordeuse des bourgeons de l'épinette (*Choristoneura fumiferana* Clem.) a l'aide de la dendrochronologie. In S. Payette, L. Filion (Eds.), *La dendroécologie: Principes, methodes et applications* (pp. 415–436). Québec: Presses de l'Universite Laval.

Morin, H. (1994). Dynamics of balsam fir forests in relation to spruce budworm outbreaks in the Boreal Zone of Quebec. *Canadian Journal of Forest Research, 24*(4), 730–741. 1139/x94-097.

Nabity, P. D., Zavala, J. A., & DeLucia, E. H. (2009). Indirect suppression of photosynthesis on individual leaves by arthropod herbivory. *Annals of Botany, 103*, 655–663. https://doi.org/10.1093/aob/mcn127.

Nardini, A., Pedà, G., & Rocca, N. (2012). Trade-offs between leaf hydraulic capacity and drought vulnerability: Morpho-anatomical bases, carbon costs and ecological consequences. *New Phytologist, 196*, 788–798. https://doi.org/10.1111/j.1469-8137.2012.04294.x.

Navarro, L., Morin, H., Bergeron, Y., et al. (2018). Changes in spatiotemporal patterns of 20th century spruce budworm outbreaks in eastern Canadian boreal forests. *Frontiers in Plant Science, 9*, 1905. https://doi.org/10.3389/fpls.2018.01905.

Nie, Z., MacLean, D. A., & Taylor, A. R. (2018). Forest overstory composition and seedling height influence defoliation of understory regeneration by spruce budworm. *Forest Ecology and Management, 409*, 353–360. https://doi.org/10.1016/j.foreco.2017.11.033.

Ostaff, D. P., & MacLean, D. A. (1995). Patterns of balsam fir foliar production and growth in relation to defoliation by spruce budworm. *Canadian Journal of Forest Research, 25*, 1128–1136. https://doi.org/10.1139/x95-125.

Peng, C., Ma, Z., Lei, X., et al. (2011). A drought-induced pervasive increase in tree mortality across Canada's boreal forests Nat. *Nature Climate Change, 1*, 467–471. https://doi.org/10.1038/nclimate1293.

Peschiutta, M. L., Bucci, S. J., Scholz, F. G., et al. (2016). Compensatory responses in plant-herbivore interactions: Impacts of insects on leaf water relations. *Acta Oecologica, 73*, 71–79. https://doi.org/10.1016/j.actao.2016.03.005.

Peters, R. L., Miranda, J. C., Schonbeck, L., et al. (2020). Tree physiological monitoring of the 2018 larch budmoth outbreak: Preference for leaf recovery and carbon storage over stem wood formation in *Larix decidua*. *Tree Physiology, 40*, 1697–1711. https://doi.org/10.1093/treephys/tpaa087.

Piene, H. (1989a). Spruce budworm defoliation and growth loss in young balsam fir: Defoliation in spaced and unspaced stands and individual tree survival. *Canadian Journal of Forest Research, 19*, 1211–1217. https://doi.org/10.1139/x89-185.

Piene, H. (1989b). Spruce budworm defoliation and growth loss in young balsam fir: Recovery of growth in spaced stands. *Canadian Journal of Forest Research, 19*, 1616–1624. https://doi.org/10.1139/x89-244.

Pincebourde, S., Frak, E., Sinoquet, H., et al. (2006). Herbivory mitigation through increased water-use efficiency in a leaf-mining moth apple tree relationship. *Plant, Cell and Environment, 29*, 2238–2247. https://doi.org/10.1111/j.1365-3040.2006.01598.x.

Pittermann, J., Lance, J., Poster, L., et al. (2014). Heavy browsing affects the hydraulic capacity of *Ceanothus rigidus* (Rhamnaceae). *Oecologia, 175*, 801–810. https://doi.org/10.1007/s00442-014-2947-1.

Pureswaran, D. S., De Grandpré, L., Paré, D., et al. (2015). Climate-induced changes in host tree–insect phenology may drive ecological state-shift in boreal forests. *Ecology, 96*, 1480–1491. https://doi.org/10.1890/13-2366.1.

Pureswaran, D. S., Johns, R., Heard, S. B., et al. (2016). Paradigms in eastern spruce budworm (Lepidoptera: Tortricidae) population ecology: A century of debate. *Environmental Entomology, 45*, 1333–1342. https://doi.org/10.1093/ee/nvw103.

Quentin, A. G., O'Grady, A. P., Beadle, C. L., et al. (2011). Responses of transpiration and canopy conductance to partial defoliation of *Eucalyptus globulus* trees. *Agricultural and Forest Meteorology, 151*, 356–364. https://doi.org/10.1016/j.agrformet.2010.11.008.

Quentin, A. G., O'Grady, A. P., Beadle, C. L., et al. (2012). Interactive effects of water supply and defoliation on photosynthesis, plant water status and growth of *Eucalyptus globulus* Labill. *Tree Physiology, 32*, 958–967. https://doi.org/10.1093/treephys/tps066.

Reich, P. B., Walters, M. B., Krause, S. C., et al. (1993). Growth, nutrition and gas exchange of *Pinus resinosa* following artificial defoliation. *Trees, 7*, 67–77. https://doi.org/10.1007/BF00225472.

Sack, L., & Holbrook, N. M. (2006). Leaf hydraulics. *Annual Review of Plant Biology, 57*, 361–381. https://doi.org/10.1146/annurev.arplant.56.032604.144141.

Salmon, Y., Torres-Ruiz, J. M., Poyatos, R., et al. (2015). Balancing the risks of hydraulic failure and carbon starvation: A twig scale analysis in declining Scots pine. *Plant, Cell and Environment, 38*, 2575–2588. https://doi.org/10.1111/pce.12572.

Schmid, S., Palacio, S., & Hoch, G. (2017). Growth reduction after defoliation is independent of CO_2 supply in deciduous and evergreen young oaks. *New Phytologist, 214*, 1479–1490. https://doi.org/10.1111/nph.14484.

Seidl, R., Thom, D., Kautz, M., et al. (2017). Forest disturbances under climate change. *Nature Climate Change, 7*(6), 395–402. https://doi.org/10.1038/nclimate3303.

Singer, M. C., & Parmesan, C. (2010). Phenological asynchrony between herbivorous insects and their hosts: Signal of climate change or pre-existing adaptive strategy? *Philosophical Transactions of the Royal Society of London. Series B, Biological Sciences, 365*, 3161–3176. https://doi.org/10.1098/rstb.2010.0144.

Sturtevant, B. R., Cooke, B. J., Kneeshaw, D. D., et al. (2015). Modeling insect disturbance across forested landscapes: Insights from the spruce budworm. In A. H. Perera, B. R. Sturtevant, & L. J. Buse (Eds.), *Simulation modeling of forest landscape disturbances* (pp. 93–134). Springer International Publishing.

Sun, G., Caldwell, P. V., & McNulty, S. G. (2015). Modelling the potential role of forest thinning in maintaining water supplies under a changing climate across the conterminous United States. *Hydrological Processes, 29*, 5016–5030. https://doi.org/10.1002/hyp.10469.

Traversari, S., Emiliani, G., Traversi, M. L., et al. (2018). Pattern of carbohydrate changes in maturing xylem and phloem during growth to dormancy transition phase in *Picea abies* (L.) Karst. *Dendrobiology, 80*, 12–23. https://doi.org/10.12657/denbio.080.002.

Tyree, M. T., & Sperry, J. S. (1989). Vulnerability of xylem to cavitation and embolism. *Annual Review of Plant Physiology and Plant Molecular Biology, 40*, 19–36. https://doi.org/10.1146/annurev.pp.40.060189.000315.

van Lierop, P., Lindquist, E., Sathyapala, S., et al. (2015). Global forest area disturbance from fire, insect pests, diseases and severe weather events. *Forest Ecology and Management, 352*, 78–88. https://doi.org/10.1016/j.foreco.2015.06.010.

Vanderklein, D. W., & Reich, P. B. (1999). The effect of defoliation intensity and history on photosynthesis, growth and carbon reserves of two conifers with contrasting leaf lifespans and growth habits. *New Phytologist, 144*, 121–132. https://doi.org/10.1046/j.1469-8137.1999.00496.x.

Vanderklein, D. W., & Reich, P. B. (2000). European larch and eastern white pine respond similarly during three years of partial defoliation. *Tree Physiology, 20*, 283–287. https://doi.org/10.1093/treephys/20.4.283.

Viglizzo, E. F., Jobbágy, E. G., Ricard, M. F., et al. (2016). Partition of some key regulating services in terrestrial ecosystems: Meta-analysis and review. *Science of the Total Environment, 562*, 47–60. https://doi.org/10.1016/j.scitotenv.2016.03.201.

Virgin, G. V. J., & MacLean, D. A. (2017). Five decades of balsam fir stand development after spruce budworm-related mortality. *Forest Ecology and Management, 400*, 129–138. https://doi.org/10.1016/j.foreco.2017.05.057.

Warren, R. K., Pappas, C., Helbig, M., et al. (2018). Minor contribution of overstorey transpiration to landscape evapotranspiration in boreal permafrost peatlands. *Ecohydrology, 11*, e1975. https://doi.org/10.1002/eco.1975.

Webb, W. L. (1980). Starch content of conifers defoliated by the Douglas-fir tussock moth. *Canadian Journal of Forest Research, 10*, 535–540. https://doi.org/10.1139/x80-087.

Webb, W. L. (1981). Relation of starch content to conifer mortality and growth loss after defoliation by the Douglas-fir tussock moth. *Forest Sciences, 27*, 224–232.

Webb, W. L., & Karchesy, J. J. (1977). Starch content of Douglas-fir defoliated by the tussock moth. *Canadian Journal of Forest Research, 7*, 186–188. https://doi.org/10.1139/x77-026.

Wiley, E., & Helliker, B. (2012). A re-evaluation of carbon storage in trees lends greater support for carbon limitation to growth. *New Phytologist, 195*, 285–289. https://doi.org/10.1111/j.1469-8137.2012.04180.x.

Wiley, E., Huepenbecker, S., Casper, B. B., et al. (2013). The effects of defoliation on carbon allocation: Can carbon limitation reduce growth in favour of storage? *Tree Physiology, 33,* 1216–1228. https://doi.org/10.1093/treephys/tpt093.

Zhang, X., Lei, Y., Ma, Z., et al. (2014). Insect-induced tree mortality of boreal forests in under a changing climate. *Ecology and Evolution, 4,* 2384–2394. https://doi.org/10.1002/ece3.988.

Zhang, B., MacLean, D. A., Johns, R. C., et al. (2020). Hardwood-softwood composition influences early-instar larval dispersal mortality during a spruce budworm outbreak. *Forest Ecology and Management, 463,* 118035. https://doi.org/10.1016/j.foreco.2020.118035.

Chapter 10
A Circumpolar Perspective on the Contribution of Trees to the Boreal Forest Carbon Balance

Christoforos Pappas, Flurin Babst, Simone Fatichi, Stefan Klesse, Athanasios Paschalis, and Richard L. Peters

Abstract Partitioned estimates of the boreal forest carbon (C) sink components are crucial for understanding processes and developing science-driven adaptation and mitigation strategies under climate change. Here, we provide a concise tree-centered overview of the boreal forest C balance and offer a circumpolar perspective on the contribution of trees to boreal forest C dynamics. We combine an *ant's-eye* view, based on quantitative in situ observations of C balance, with a *bird's-eye* perspective on C dynamics across the circumboreal region using large-scale data sets. We conclude with an outlook addressing the trajectories of the circumboreal C dynamics in response to projected environmental changes.

C. Pappas (✉)
Department of Civil Engineering, University of Patras, 26504 Rio Patras, Greece
e-mail: cpappas@upatras.gr

Centre for Forest Research, Université du Québec à Montréal, P.O. Box 8888, Stn. Centre-Ville, Montréal, QC 3C 3P8, Canada

Département Science et Technologie, Université TÉLUQ, 5800 rue Saint-Denis, Montréal, QC 2S 3L5, Canada

F. Babst
School of Natural Resources and the Environment, Laboratory of Tree-Ring Research, The University of Arizona, 1046 E Lowell St, Tucson, AZ 85721, USA
e-mail: babst@arizona.edu

S. Fatichi
Department of Civil and Environmental Engineering, College of Design and Engineering, National University of Singapore, 1 Engineering Drive 2, Block E1A #05-12, Singapore 117576, Singapore
e-mail: ceesimo@nus.edu.sg

S. Klesse
Forest Dynamics, Swiss Federal Research Institute for Forest, Snow and Landscape Research WSL, Zürcherstrasse 111, 8903 Birmensdorf, Switzerland
e-mail: stefan.klesse@wsl.ch

A. Paschalis
Department of Civil and Environmental Engineering, Imperial College London, South Kensington Campus London, Skempton Building, SW7 2AZ London, UK
e-mail: a.paschalis@imperial.ac.uk

© The Author(s) 2023
M. M. Girona et al. (eds.), *Boreal Forests in the Face of Climate Change*,
Advances in Global Change Research 74,
https://doi.org/10.1007/978-3-031-15988-6_10

271

10.1 Introduction

One of the many services that forest ecosystems offer to humanity is the absorption of atmospheric CO_2 and its conversion to chemical energy stored as biomass. Plant biomass, which has about 50% carbon (C) content, represents the largest biomass pool of the biosphere (Bar-On et al., 2018). In the Anthropocene, characterized by continuously increasing anthropogenic CO_2 emissions, this forest C sink has been a particularly important yet vulnerable component of the terrestrial C balance and a fundamental ecosystem service. Forests have a pronounced effect on the Earth's biogeochemical cycles and climate system (e.g., Bonan, 2008; Ciais et al., 2019; Friedlingstein, 2015). The latest estimates report that the land C sink amounts to about three Pg $C \cdot yr^{-1}$ (where 1 Pg equals 10^{15} g), which corresponds to an average of 25 g $C \cdot yr^{-1}$ stored per square meter of vegetated land (Fatichi et al., 2019; Le Quéré et al., 2018). Attributing the terrestrial C sink to specific components and their inherent processes is important and topical, not only for a better understanding and modeling of the biogeochemical cycles under climate change but also for applying science-driven adaptation and mitigation strategies (Duncanson et al., 2019; Friedlingstein et al., 2020; Grassi et al., 2018; Kurz et al., 2013).

Natural management solutions, like tree planting and forest regrowth or densification, have been advocated as viable approaches for removing a fraction of the continuously increasing anthropogenic CO_2 emissions (Cook-Patton et al., 2020). However, concerns have been raised regarding the efficacy of such natural solutions, as biophysical (e.g., energy, water, and nutrient availability) and ecological (e.g., self-thinning) theory provides well-known constraints on a tree's capacity to grow, thrive, and sequester C (Baldocchi & Peñuelas, 2019; Popkin, 2019). Moreover, the long-term strength of this C sink is also shaped by disturbance dynamics and the turnover rates of the different C pools, e.g., leaves and roots, which span a large range of temporal scales, i.e., C residence times, rather than tree growth per se (Bugmann & Bigler, 2011; Carvalhais et al., 2014; Clemmensen et al., 2013; Friend et al., 2014; Harmon, 2001; Körner, 2017; Yu et al., 2019). Thus, disentangling the C balance pools could facilitate a robust and quantitative description of the C sink strength and its fate under global change (Anderegg et al., 2020; Brienen et al., 2020; Büntgen et al., 2019; Duchesne et al., 2019; Pugh et al., 2019b).

The circumboreal region represents a massive amount of land (ca. 33% of the Earth's forested area); thus, understanding and quantifying its C balance is extremely

R. L. Peters
Laboratory of Plant Ecology, Department of Plants and Crops, Faculty of Bioscience Engineering, Ghent University, Coupure links 653, 9000 Ghent, Belgium
e-mail: richard.peters@unibas.ch

Forest Is Life, TERRA Teaching and Research, Gembloux Agro Bio-Tech, University of Liège, Avenue de la Faculté d'Agronomie, B-5030 Gembloux, Belgium

Physiological Plant Ecology, Department of Environmental Sciences, University of Basel, Schönbeinstrasse 6, CH-4056 Basel, Switzerland

critical for both science and policymaking. Boreal forests constitute the largest C storage pool in the extratropics, with a total biomass stock of 270 Pg C, the largest fraction of which is stored belowground—218 Pg C, as soil organic matter, litter, fine and coarse roots (Pan et al., 2013). About 0.74 trillion trees grow in the boreal region (Crowther et al., 2015), with a total living aboveground biomass (overstory vegetation) of approximately 54 Pg C, having an average biomass density of 5 kg C·m^{-2} (Pan et al., 2013). Ongoing environmental changes, including a higher frequency and intensity of transient disturbances (e.g., droughts, heat waves, fires, and insect outbreaks) and persistent changes (e.g., permafrost thaw and the warming and associated shifts in spring and autumn phenology) affect boreal forest C sink strength (Amiro et al., 2010; Bradshaw & Warkentin, 2015; Pugh et al., 2019a; Seidl et al., 2017). Such changes in the C cycle also impact the intrinsically coupled water and energy cycles and could trigger numerous feedbacks to the climate system at local, regional, and global scales.

Numerical models, supported by novel monitoring techniques, allow us to quantify the forest C balance and its components. However, observations and mechanistic models, e.g., the biochemical model of photosynthesis (Farquhar et al., 1980), are much more refined for aboveground processes than belowground ones. This reflects a bias imposed by the ease of measuring certain variables, e.g., aboveground vegetation monitoring with direct measurements, forest surveys, and remote sensing, compared with the variables and dynamics of belowground biogeochemical processes, which are more challenging to observe (e.g., Körner, 2018). From a C balance perspective, whereas tree stature, leaf area, and the biochemical properties of foliage define the C supply (source), it is the interplay with the belowground resources, e.g., water and nutrient availability, and the environment, e.g., temperature, that shape tree growth (sink) and ecosystem C partitioning at intra- and interannual time scales (e.g., Fatichi et al., 2019).

Here, we provide a concise overview of the magnitude and variability of boreal forest C fluxes and offer a circumpolar perspective on the contribution of trees to the boreal forest C balance. We deliberately provide a tree-centered perspective and focus mainly on vegetation physiological processes (e.g., Babst et al., 2021). Yet, forest management, natural disturbances (e.g., fires, insect outbreaks, and permafrost thaw), and forest demography, all of which are not discussed here, are fundamental coregulators of boreal forest structure and functioning and have pronounced implications for the boreal forest C sources and sinks (Anderegg et al., 2020; Brassard & Chen, 2006; McDowell et al., 2020). We combine an *ant's-eye* view, based on quantitative in situ observations of the component C fluxes and pools within three boreal forests—spanning from eastern Siberia to Finland to central Canada—with a *bird's-eye* perspective on C dynamics across the circumboreal region using spatially explicit state-of-the-art remote-sensing and machine-learning data sets. The compiled literature illustrates the importance of tree physiological processes in shaping the tree-level C balance and the interplay with belowground processes in regulating the C balance at the landscape and ecosystem levels. We conclude with an outlook on the potential trajectories of C dynamics in the circumboreal region in response to the projected increase in air temperature, vapor pressure deficit, and atmospheric CO_2 concentration, on the basis of a tree-level physiological understanding and large-scale observational evidence.

10.2   Forest Carbon Balance and Its Components

The terrestrial C sink is often quantified by net ecosystem production (NEP), which is the difference between two large fluxes: the C assimilated through photosynthesis (C input), which at the ecosystem level is referred to as gross ecosystem production (GEP), and the C that is released to the atmosphere as ecosystem respiration (R_{eco}; C output), i.e., the sum of autotrophic (R_a) and heterotrophic (R_h) respiration components (Fig. 10.1; Chapin et al., 2006; Fatichi et al., 2019; Keenan & Williams, 2018; Manzoni et al., 2018). If the balance of these two fluxes is positive, i.e., NEP = GEP – R_{eco} > 0, then the ecosystem is a C sink. The net primary production (NPP = GEP – R_a) corresponds to the assimilated C that is invested into structural biomass growth, root exudates, nonstructural carbohydrates, and biogenic volatile organic compounds. When NPP is expressed as a fraction of GEP, it provides an integrated metric of ecosystem carbon use efficiency (CUE), i.e., CUE = NPP/GEP = 1 – (R_a/GEP) (Fatichi et al., 2019; Manzoni et al., 2018). Several methods exist for measuring and inferring the components of the C balance (Fig. 10.1). They include (1) the eddy covariance technique (Baldocchi, 2020) used to measure the forest-stand level net exchange of CO_2 between the land surface and the atmosphere, i.e., net ecosystem exchange (NEE), where NEE = –NEP in the absence of lateral C fluxes and major disturbances, and to infer the component C fluxes, i.e., GEP and R_{eco} (Reichstein et al., 2005); (2) repeated inventories of permanent forest monitoring plots to derive area-based C fluxes and pools (Anderson-Teixeira et al., 2015; Clark et al., 2001); (3) gas exchange chambers, e.g., enclosing specific soil patches or tree organs or entire plants or trees (Drake et al., 2019); and (4) biometric methods for reconstructing aboveground tree growth at the forest-stand level from tree-ring widths, allometric relationships, and information on forest-stand demography (e.g., Campioli et al., 2016). In addition, recent technological advancements in airborne and satellite remote sensing, e.g., NASA's GEDI mission; https://gedi.umd.edu/, and multivariate data synthesis initiatives, e.g., the Forest Observation System initiative, http://forest-observation-system.net/, offer spatially explicit products of C fluxes and pools worldwide (Rodríguez-Veiga et al., 2019; Schepaschenko et al., 2019; Thurner et al., 2014).

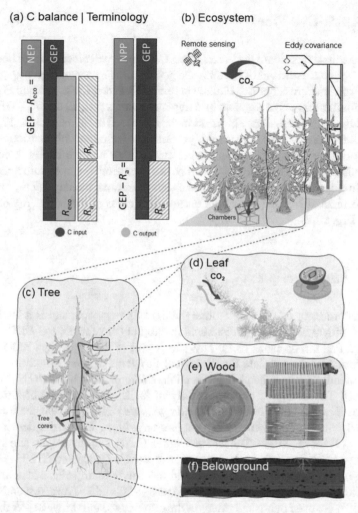

Fig. 10.1   a An overview of the terrestrial carbon (C) balance and common measuring techniques at the **b** ecosystem, **c** tree, and organ level, including **d** leaf, **e** wood, and **f** roots. **a** Ecosystem-level C input (gross ecosystem production; GEP) and output (ecosystem respiration, R_{eco}, and its components, namely autotrophic, R_a, and heterotrophic, R_h, respiration) and the resulting C fluxes (net primary production (NEP) and net primary productivity (NPP)); **b** ecosystem-level C fluxes and pools can be estimated through remote sensing and the eddy covariance technique, and, when combined with tree-level measurements and chamber techniques, a partition of C fluxes and pools can be derived; **c** tree rings, when combined with auxiliary biometric measurements, can reconstruct aboveground tree biomass increments, thereby assessing how **d** leaf-level C assimilation is translated into **e** stem biomass or **f** translocated to belowground tree C pools

10.3   Ant's-Eye View

To illustrate the dynamics of the boreal forest C balance, we selected three sites having well-documented estimates of C fluxes and pools that cover an extensive longitudinal gradient from eastern Siberia to Finland to central Canada (Table 10.1 and Fig. 10.2). We compiled estimates of R_{eco} and its component fluxes (i.e., R_a and R_h), GEP, NEP, NPP, and the aboveground tree biomass increments (AGBi) at these sites (Fig. 10.1). These values were derived with the eddy covariance technique, biometric/inventory-based approaches, gas exchange chamber methods, or a combination thereof. The selected sites provide not only a circumboreal perspective across three continents but also include distinct boreal tree species compositions, namely larch-, pine-, and spruce-dominated landscapes, which occupy vast areas of the boreal region (Table 10.1 and Fig. 10.2).

10.3.1   Site Description

At the eastern limit of the longitudinal gradient, the example site is a larch (*Larix cajanderi* Mayr) dominated forest stand in Siberia (fluxnet ID: RU-SKP; Yakutsk Spasskaya Pad larch; Table 10.1). This site is part of the Spasskaya Pad Scientific Forest Station of the Institute for Biological Problems of the Cryolithozone and is situated 20 km north of Yakutsk in the Republic of Sakha, Russia (62.26°N, 129.17°E; 246 m asl). The overstory consists mainly of larch, with sporadic occurrences of willow (*Salix* spp.) and silver birch (*Betula pendula* Roth), whereas the forest floor is covered with dense cowberry (*Vaccinium vitis-idaea* L.). Stand density is 840 stem·ha^{-1} (Tei et al., 2019). The leaf area index (LAI), canopy height, and stand age are equal to 1.5 m^2·m^{-2}, 20 m, and 190 yr, respectively (Kotani et al., 2014). The site lies in the continuous permafrost zone and has sandy loam soils. Mean annual air temperature and total precipitation at the site are –9 °C and 237 mm, respectively (Tei et al., 2019). We used published values of the growing season (May–September) GEP from Tei et al. (2019), collected between 2004 and 2014 with the eddy covariance technique. A detailed description of the micrometeorological and eddy covariance instrumentation at the site and the eddy covariance processing methods used for deriving GEP are provided in Ohta et al. (2008, 2014), Kotani et al. (2014), and Tei et al. (2019). Ecosystem respiration at the site was estimated as 75% of GEP, and NEP amounts to 25% of GEP (Kotani et al., 2014). Furthermore, R_h was estimated as 52% of GEP (Sawamoto et al., 2003), and R_a was then calculated as $R_{eco} - R_h$. This allowed us to derive partitioned C fluxes at the site for the 2004–2014 period. We complemented these variables with tree-ring-derived AGBi estimates for the 1990–2013 period (S. Klesse; unpublished results, https://www.ncei.noaa.gov/acc ess/paleo-search/study/34312). Specifically, AGBi time series were derived with a biometric approach (Babst et al., 2014a, b) by combining ring widths from tree cores sampled in 2014, species-specific allometry, and stand density information

Table 10.1 An overview of the three examined boreal forest stands, including their location, climatic conditions (mean annual temperature, MAT, and total precipitation, MAP), vegetation characteristics (leaf area index, LAI), and carbon fluxes and pools, namely autotrophic (R_a) and heterotrophic respiration (R_h), ecosystem respiration (R_{eco}), gross ecosystem production (GEP), net primary production (NPP), net ecosystem production (NEP), and aboveground biomass increment (AGBi). The evaluated period is in parentheses

Location			
Site name	Yakutsk Spasskaya Pad Larch	Hyytiälä	Southern Old Black Spruce
Site ID	RU-SKP	FI-HYY	CA-OBS
Longitude (°)	129.17	24.50	−105.12
Latitude (°)	62.26	61.85	53.98
Elevation (m a.s.l.)	256	181	629
DOI	https://doi.org/10.18140/FLX/1440243	https://doi.org/10.18140/FLX/1440158	https://doi.org/10.18140/FLX/1440044
Climatic conditions			
MAT (°C)	−9[a]	4.4[b]	1.4
MAP (mm)	237[a]	604[a]	428
Vegetation characteristics			
Overstory vegetation	*Larix cajanderi*	*Pinus sylvestris*	*Picea mariana*
Density (stems·ha⁻¹)	840[a]	1,373 (2008)[e]	5,921 (2016)[j]
Basal area (m²·ha⁻¹)	24[b]	23 (2008)[e]	33 (2016)[j]
Stand age (yr)	190[b]	47 (2009)[f]	140[k]
Canopy height (m)	20[b]	15 (2008)[e]	16 (2005)[l]
LAI (m²·m⁻²)	1.5[b]	2.5 1995[g]	3.8 (2005)[l]
Carbon fluxes and pools			
GEP (g C·m⁻²·yr⁻¹)	635 ± 73 (2004–2014)[a]	1,109 ± 152 (1996–2014)[h]	809 ± 50 (1999–2015)[j,m]

(continued)

Table 10.1 (continued)

Location			
NEP (g C·m⁻²·yr⁻¹)	159 ± 18 (2004–2014)[a,b]	246 ± 148 (1996–2014)[h]	40 ± 23 (1999–2015)[j,m]
NPP (g C·m⁻²·yr⁻¹)	489 ± 56 (2004–014) based on GEP and R_a	643 ± 84 (1996–2014)[i]	405 ± 27 (1999–2015)[j,m]
R_{eco} (g C·m⁻²·yr⁻¹)	477 ± 55 (2004–2014)[a,b]	927 ± 83 (1996–2014)[h]	769 ± 56 (1999–2015)[j,m]
R_h (g C·m⁻²·yr⁻¹)	330 ± 38 (2004 –2014)[c]	401 ± 54 (1996–2014)[i]	365 ± 14 (1999–2015)[j,m]
R_a (g C·m⁻²·yr⁻¹)	146 ± 17 (2004–2014)[f]; using R_{eco} and R_h from above	526 ± 69 (1996–2014)[i]	404 ± 45 (1999–2015)[j,m]
AGBi (g C·m⁻²·yr⁻¹)	26 ± 9 (1990–2013)[d]	97 ± 12 (1996–2009)[f]	71 ± 7 (1999 − 2015)[j,m]
AGBi/GEP (%)	4.1	8.3	8.8

[a]Tei et al. (2019), [b]Kotani et al. (2014), [c]Sawamoto et al. (2003), [d]this study, [e]Ilvesniemi et al. (2009), [f]Babst et al. (2014a), [g]Mencuccini and Bonosi (2001), [h]FLUXNET2015, [i]Pumpanen et al. (2015), [j]Pappas et al. (2020), [k]Krishnan et al. (2008), [l]Chen et al. (2006), [m]Liu et al. (2019)

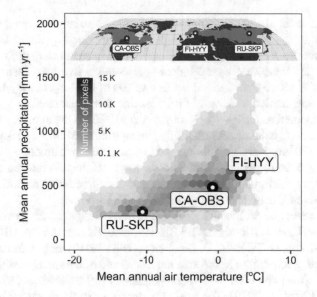

Fig. 10.2 Climate envelope of the boreal forest ecozone and the location of the three example boreal forest sites in the two-dimensional climate space of mean annual air temperature and total precipitation; *hexagons*, color-coded with the density of 5′ resolution pixels, define the climate space of mean annual air temperature and total precipitation, as quantified using the WorldClim version 2.1 data set (long-term mean for 1970–2000; Fick & Hijmans, 2017). The circumpolar boreal forest extent (*dark green* in the upper panel) follows the definition of the World Wildlife Fund terrestrial ecoregions map and was downloaded from http://glad.geog.umd.edu/projects/gtm/boreal/data.html. *Note* that a lower cutoff of 100 pixels was applied to enhance the figure's clarity

(Sawamoto et al., 2003). Mass values were converted into C units using a 47% C concentration proposed in the literature for larch stemwood (Alexander et al., 2012).

In Europe, we selected a Scots pine (*Pinus sylvestris* L.)–dominated forest in southern Finland, located 200 km north of Helsinki (Hyytiälä; 61.85°N, 24.30°E; 181 m asl; Table 10.1). This is a long-term experimental forest site operated by the University of Helsinki (Station for Measuring Forest Ecosystem-Atmosphere Relations, SMEAR II; fluxnet ID: FI-HYY). Long-term (1996–2014) mean annual air temperature and total precipitation at the site are 4.4 °C and 604 mm, respectively. Stand density is 1,373 stems ha^{-1}, and average canopy height, LAI, and stand age are 15 m, 2.5 m^2 m^{-2} and 47 years, respectively (Babst et al., 2014a; Ilvesniemi et al., 2009; Mencuccini & Bonosi, 2001). Overstory vegetation consists of Scots pine (>95%) with the sporadic occurrence of Norway spruce (*Picea abies* (L.) H. Karst) and silver birch. The understory vegetation is considered negligible because of local forest management practices. The ground cover consists mainly of blueberry (*Vaccinium myrtillus* L.), cowberry, small-reed (*Calamagrostis epigejos* (L.) Roth), heather (*Calluna vulgaris* (L.) Hull), and feather mosses (*Pleurozium schreberi* (Brid.) Mitt., *Dicranum polysetum* Sw.). The soils are mineral on glacial till over a homogeneous bedrock; sporadic peat soils are found in the depressions (Kolari et al.,

2009; Pumpanen et al., 2015). Here, we used eddy covariance–derived estimates of NEP, GEP, and R_{eco} for the 1996–2014 period, as provided by the FLUXNET2015 data set (Pastorello et al., 2020). Combining these eddy covariance–derived estimates with the detailed site-level forest C balance presented in Ilvesniemi et al. (2009) and Pumpanen et al. (2015), we derived average values of NPP, R_a, and R_h. More specifically, R_a was estimated as 45% of GEP; R_h was then calculated as $R_{eco} - R_a$, and NPP was computed as GEP – R_a (Pumpanen et al., 2015). Moreover, tree contribution to the forest C balance was quantified using the AGBi time series presented in Babst et al., (2014a, 2014b). This allowed us to derive partitioned C fluxes at the site for the 1996–2014 period. Detailed descriptions of the micrometeorological and eddy covariance instrumentation at the site and data processing documentation can be found in Rannik et al. (2002) and Vesala et al. (2005).

The Canadian site is located near the southern edge of the boreal forest in Saskatchewan, Canada (Southern Old Black Spruce, fluxnet ID: CA-OBS; 53.98°N, 105.12°W; 629 m asl; Table 10.1). Long-term (1981–2010) mean annual air temperature and total precipitation in the area are 1.4 °C and 428 mm, respectively (Pappas et al., 2020). Stand age is ca. 140 years—initiated after a fire in ca. 1879 (Krishnan et al., 2008; Liu et al., 2019). The overstory consists mainly of mature black spruce (*Picea mariana* (Mill.) BSP) mixed with sporadic mature eastern larch (*Larix laricina* (Du Roi) K. Koch). The average LAI at the site is 3.8 $m^2 \cdot m^{-2}$ (Chen et al., 2006). Wild rose (*Rosa woodsii* Lindl.) and Labrador tea (*Rhododendron groenlandicum* (Oeder) Kron & Judd) are the dominant woody shrubs, and the ground cover consists mainly of mixed feather mosses (*Hylocomium splendens* (Hedw.) Schimp., *Pleurozium schreberi*, and *Ptilium crista-castrensis* (Hedw.) De Not.) with some peat moss (*Sphagnum* spp.) and lichen (*Cladina* spp.) (Gaumont-Guay et al., 2014). The soil is moderately to poorly drained with a 20 to 30 cm thick peat layer overlying waterlogged sand (Barr et al., 2012; Gower et al., 1997; Griffis et al., 2003). The recent study of Liu et al. (2019) provides a detailed overview of the partitioned C fluxes at the site during the last decades (1999–2017), including GEP, NEP, R_{eco}, R_a, and R_h, by combining long-term eddy covariance measurements with automatic chamber measurements. Furthermore, Pappas et al. (2020) present reconstructed estimates of AGBi at the site using a biometric approach, i.e., tree cores collected with a biomass upscaling design and species-specific allometry, similar to the one followed in the AGBi estimates at the other two sites, detailed above. Here, we used published values of the partitioned C fluxes and pools from Pappas et al. (2020), focusing on 1999 to 2015.

10.3.2   *Ecosystem Carbon Balance and the Impact of Environmental Conditions*

The selected forest sites offer a perspective on the magnitude and variability of boreal forest C fluxes and pools and illustrate the numerous interactions between biotic, e.g., species composition, and abiotic factors, e.g., local climate and stand demography, that can shape site-level C dynamics (Fig. 10.3). The long-term measurements revealed that the study sites were C sinks during the last decade, i.e., NEP > 0, yet the sink strength at each site varied from 40 ± 23 g C·m^{-2}·yr^{-1} at CA-OBS to 246 ± 148 g C·m^{-2}·yr^{-1} at FI-HYY (hereafter, the presented values correspond to mean $\pm$ one standard deviation; Table 10.1).

The long-term mean annual GEP across the three locations followed a climatological gradient (Figs. 10.2 and 10.3); the highest value of GEP corresponded to the site having the most favorable climatological conditions, i.e., highest mean annual air temperature and precipitation at FI-HYY; GEP $= 1,169 \pm 152$ g C·m^{-2}·yr^{-1}, whereas the lowest values of GEP coincided with the Siberian site where mean annual air temperature and precipitation were lowest (RU-SKP; GEP $- 635 \pm 73$ g C·m^{-2}·yr^{-1}; Figs. 10.2 and 10.3). Similarly, AGBi across FI-HYY, CA-OBS, and RU-SKP ranged

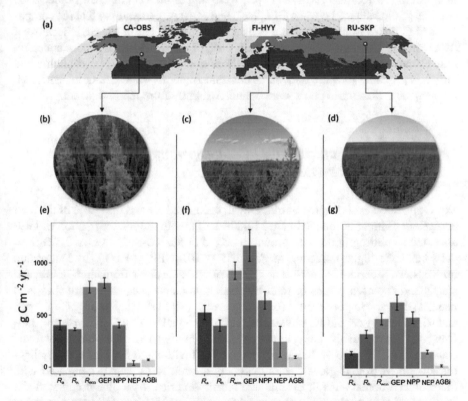

Fig. 10.3 An overview of the carbon balance components at the three reference boreal forest sites. **a** Circumboreal forest extent with the three sites indicated in *red*; **b**, **c**, and **d** the forest canopy as seen from the top of the eddy covariance tower at the sites; **e**, **f**, and **g** carbon balance components as compiled from the literature, namely autotrophic (R_a) and heterotrophic respiration (R_h), ecosystem respiration (R_{eco}), gross ecosystem production (GEP), net primary production (NPP), net ecosystem production (NEP), and aboveground biomass increment (AGBi)

from 97 ± 12 g C·m^{-2}·yr^{-1} to 71 ± 7 g C·m^{-2}·yr^{-1} to 26 ± 9 g C·m^{-2}·yr^{-1}, respectively, following the decrease in mean annual precipitation and temperature across the sites (Fig. 10.2). However, apart from the prevailing climatological conditions, both ecosystem-level production (GEP) and tree-level stem growth (AGBi) can be affected by numerous other factors explaining the reported differences across the examined forest stands. Such factors include site-specific biophysical constraints, e.g., stand density and LAI, as well as differences in demography and species composition, tree ontogeny (e.g., stand age ranging from 190 yr at RU-SKP to 47 yr at FI-HYY), and phenology (e.g., deciduous larch at RU-SKP versus evergreen spruce and pine at CA-OBS and FI-HYY, respectively), and soil biochemistry (Table 10.1). The impact of these biotic and abiotic factors manifests in the cross-site variability of the long-term component C fluxes. Importantly, while R_{eco}, R_a, and R_h decreased across the three sites with a pattern comparable to that of GEP, NEP and NPP did not follow this pattern. At RU-SKP, NEP and NPP were higher than at CA-OBS (Table 10.1) because of the balance between GEP, R_{eco}, and R_a. It is worth noting that the components of C fluxes were derived as a constant fraction of GEP at RU-SKP because of data scarcity. Thus, the estimates of NEP, NPP, R_{eco}, R_a, and R_h are more uncertain than estimates at both FI-HYY—where NEP and GEP values were available and only NPP, R_a, and R_h were derived as constant fractions—and CA-OBS, where all component C fluxes and pools were obtained directly from measurements.

10.3.3 Limited Contribution of Aboveground Tree Growth to the Ecosystem Carbon Sink

The C allocated to aboveground tree growth accounted for less than 10% of the total ecosystem C input at the study sites (Table 10.1), with the lowest fraction of GEP allocated to aboveground stem growth reported at the RU-SKP site (AGBi/GEP = 4.1%) and the highest fraction at CA-OBS (AGBi/GEP = 8.8%). In central and northeastern Siberia, Kajimoto et al. (2010) provided insights from larch-dominated stands and reported values of total biomass increments (i.e., stem, branches, and needles) and aboveground NPP for young stands (<30 yr) equal to 54 g C·m^{-2}·yr^{-1} and 95 g C·m^{-2}·yr^{-1} and for old stands (>100 yr) equal to 15 g C·m^{-2}·yr^{-1} and 34 g C·m^{-2}·yr^{-1}, respectively. The estimates of AGBi reported here for the old intact forest stand at RU-SKP (190 yr; Table 10.1) agree well with these findings and highlight the low aboveground growth rates of Siberian larch forest stands compared with the southern boreal sites in Europe and North America. At a pine-dominated site in Siberia, Lloyd et al. (2002) reported AGBi/GEP $\approx$ 16%, which is comparatively larger than our estimate at RU-SKP but within the expected range of AGBi/GEP variability in the boreal region (Pappas et al., 2020). Permafrost conditions and waterlogging in eastern Siberia have been identified as key factors affecting ecosystem-level C fluxes and forest structure and health (Kotani et al., 2014; Ohta et al., 2014). The estimates of AGBi and the resulting fractions of AGBi/GEP are also well constrained at CA-OBS and agree well with previously published studies and auxiliary observations, e.g., forest inventories and remote-sensing products, from the same site, e.g., AGBi $\approx$ 110 g C·m^{-2}·yr^{-1}, therefore 11% of GEP (Malhi et al., 1999), and other boreal forest sites in North America (see Pappas et al. (2020) and references therein). Some discrepancy exists, however, in the AGBi estimates at FI-HYY. Ilvesniemi et al. (2009) reported a larger contribution of C accumulation into the aboveground tree biomass at FI-HYY, i.e., AGBi $\approx$ 240 g C·m^{-2}·yr^{-1} for the period 1995 to 2008, when compared with the estimates from Babst et al., (2014a, 2014b) used here. This discrepancy could be associated with methodological differences between the two studies, especially in regard to stand density and species-specific biomass allometry. Furthermore, Lagergren et al. (2019) combined eddy covariance observations with annual biomass increments in a mixed pine-spruce forest in the southern part of the boreal region of Sweden and found that out of the 957 g C·m^{-2}·yr^{-1} of mean annual GEP, stem growth accounted for ~14%, i.e., AGBi $\approx$ 137 g C·m^{-2}·yr^{-1}. The low fraction of GEP allocated to aboveground tree growth in the boreal region underlines the relatively high rates of R_a, the important belowground tree C investments, and also the role of understory vegetation for the boreal forest C balance (Bradshaw & Warkentin, 2015; Clemmensen et al., 2013; Hart & Chen, 2006). Additionally, boreal forest soils are often nutrient poor, potentially explaining the priority for trees to invest in belowground components, including fine roots, root exudates, and mycorrhizae (Chapin, 1980; Clemmensen et al., 2013; Vicca et al., 2012).

10.4   Bird's-Eye View

We complemented the in situ measurements of the component C fluxes and pools with long-term estimates of GEP and NPP from two recently developed global data sets. We used mean (2000–2019) annual (a) GEP derived by upscaling eddy covariance data from the FLUXNET2015 data set (Pastorello et al., 2020) with random-forest machine-learning algorithms (Zeng et al., 2020) and (b) NPP from MODIS remote-sensing algorithms (https://lpdaac.usgs.gov/products/mod17a3hgfv006/; Running & Zhao, 2019). Both products were resampled to a 0.1° spatial resolution and extracted over the circumboreal region (http://glad.geog.umd.edu/projects/gfm/boreal/data. html). Although such global products lack the direct constraint and accuracy of in situ observations, as presented in the previous section, they can illustrate relevant spatial patterns and provide a "bird's eye" circumboreal perspective on forest C dynamics.

Across the circumboreal region, GEP and NPP are 692 ± 299 g C·m^{-2}·yr^{-1} and 319 ± 163 g C·m^{-2}·yr^{-1}, respectively. The circumboreal distribution of GEP and NPP frames an envelope for the in situ partitioned C fluxes presented here and in the recent meta-analysis of Pappas et al. (2020). More specifically, across the 21 boreal forest sites synthesized in Pappas et al. (2020), GEP and NPP were 898 ± 251 g C·m^{-2}·yr^{-1} and 373 ± 153 g C·m^{-2}·yr^{-1}, respectively, with the estimates at the three sites examined here falling within this range (Fig. 10.4). Yet, it is worth mentioning that the site-specific GEP and NPP observations correspond mainly to southern boreal forest sites; this explains their higher values compared with those estimated across the entire circumboreal region (Fig. 10.4).

The inferred remotely sensed CUE across the circumboreal region varied around 0.47 ± 0.15 (Fig. 10.4), with the site-level estimates compiled in Pappas et al. (2020) pointing toward a lower mean value (i.e., 0.40 ± 0.12). At the three sites examined here, CUE was equal to 0.77, 0.55, and 0.50 for RU-SKP, FI-HYY, and CA-OBS,

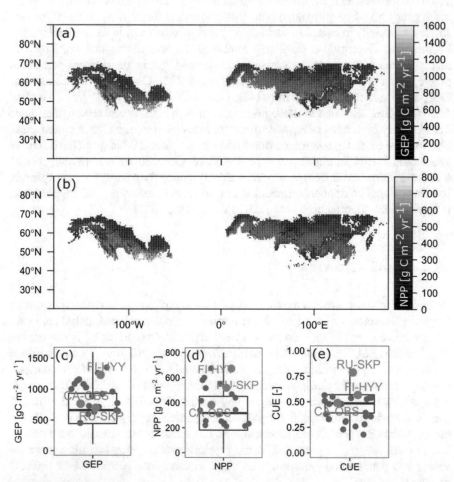

Fig. 10.4 Circumboreal distribution of mean annual (2000–2019) **a** gross ecosystem production (GEP), estimated using machine-learning and eddy covariance data (Zeng et al., 2020) and **b** net primary production (NPP) obtained from remote sensing (MODIS/TerraMOD17A3HGFv006; https://lpdaac.usgs.gov/products/mod17a3hgfv006/; Running & Zhao, 2019). The spatial resolution of the presented maps is 0.1°. Box plots of **c** GEP, **d** NPP, and **e** carbon use efficiency (CUE), CUE = NPP/GEP, across the circumboreal region using data presented in subplots **a** and **b**. Points within the box plots denote the estimates at the three example locations (*orange*) and a synthesis of boreal forest sites (*gray*) from the meta-analysis of Pappas et al. (2020)

respectively (Fig. 10.4). The high value of CUE at RU-SKP and the circumboreal estimates of CUE from the global products of NPP and GPP falling outside the interquartile range are most likely explained by observational uncertainties. More specifically, partitioned respiration fluxes (R_h and R_a) at RU-SKP were not estimated with independent methods but rather using fixed fractions of GEP, e.g., for R_h. Although such fixed fractions are supported by the available site-specific literature and are also expected because of biophysical constraints—R_h and R_a should be large and constrained fractions of productivity (Baldocchi & Peñuelas, 2019)—the resulting CUE at RU-SKP seems unrealistically high. Moreover, in situ GEP estimates in the circumboreal region are generally scarce, thereby resulting in poorly constrained predictions using machine-learning techniques across these regions, potentially explaining the reported discrepancy.

Theoretical, observational, and modeling insights point toward a well-constrained variability of CUE worldwide, with a mean value around 0.46 (Collalti & Prentice, 2019; De Lucia et al., 2007; Dewar et al., 1998; Fatichi et al., 2019; Landsberg et al., 2020; Litton et al., 2007; Van Oijen et al., 2010; Waring et al., 1998). Boreal forests are expected to manifest lower CUE values relative to those of other biomes because of lower overall productivity and higher rates of C loss during the dormant season (De Lucia et al., 2007). As recently summarized by Landsberg et al. (2020), CUE estimates outside the 0.4 to 0.6 range are unlikely and can be associated with an inaccurate assessment of NPP, especially in regard to belowground components such as fine roots, root exudates, and mycorrhizae (Chapin, 1980; Clemmensen et al., 2013; Vicca et al., 2012). However, the variability of ecosystem-level CUE in the 0.4 to 0.6 range can be explained by site-level differences in soil nutrition, stand demography, and stand ontogeny. More specifically, with increasing stand age and canopy closure, CUE is expected to decrease because of a saturation in GEP and higher rates of R_a (Brassard & Chen, 2006; De Lucia et al., 2007; Odum, 1969). Thus, site-level stand demography, disturbance history, and differences in soil fertility could explain some of the spatial patterns illustrated in Fig. 10.4.

10.5 Synthesis and Outlook

The circumboreal region provides habitat for flora and fauna and crucial ecological, social, economic, and spiritual services (e.g., Gauthier et al., 2015; Trumbore et al., 2015). In addition to these benefits, boreal forests provide a substantial C sink and regulate numerous land–atmosphere interactions and feedbacks to the climate system, e.g., sensible/latent heat, albedo feedback, and runoff/evapotranspiration partitioning; Bonan, 2008; Reichstein et al., 2013. Thus, a predictive understanding of boreal forests' growth trajectories could provide quantitative insights for comprehensive climate policy recommendations (Bonan, 2016; Kurz et al., 2013). The scientific community has put forward interdisciplinary methodological toolboxes for understanding the components of the boreal forest C balance, including in situ detailed monitoring, remote-sensing observations, and numerical modeling. Some of these multivariate sources of information were summarized here to illustrate the magnitude and variability of the C balance components in the circumboreal region. The estimates provide only a "static" perspective based on the long-term average boreal forest C balance from the last decades. Attributing how individual aspects of climate change have affected the temporal dynamics of the C balance in the circumboreal region remains uncertain across different levels of spatial organization, spanning from the tissue and tree levels to integrated responses at the forest-stand, landscape, and regional levels.

Climate change manifests itself in the circumboreal region primarily through warmer air temperatures (ACIA, 2005; IPCC 2013; Serreze & Francis, 2006). Such changes directly affect boreal forest structure and function (Gauthier et al., 2015; Soja et al., 2007). Multivariate observations point toward a significant increase in heterotrophic respiration because of increased air temperatures during the last decades (Bond-Lamberty et al., 2018). In addition, warmer temperatures have been associated with a lengthening of the growing season, including shifts in spring and autumn phenology, which results in the *greening* of the circumboreal region and an increase in GEP, and, potentially, NPP (Forkel et al., 2016; Kauppi et al., 2014; Piao et al., 2020; Tagesson et al., 2020; Zhu et al., 2016). However, such patterns have not always been confirmed at local scales, when, for example, tree-ring records are analyzed (Duchesne et al., 2019; Giguère-Croteau et al., 2019; Girardin et al., 2016; Marchand et al., 2018). Furthermore, warmer temperatures in the circumpolar region lead to a northward advance of the boreal forest–tundra transition zone and shrub expansion in tundra ecosystems (Elmendorf et al., 2012; Gareth Rees et al., 2020; Myers-Smith et al., 2011). The beneficial effect of air temperature increases on boreal tree function might be transitory, however, because of additional factors limiting growth and productivity in the long term (Babst et al., 2019; D'Orangeville et al., 2018).

Increasing atmospheric CO_2 concentrations and nitrogen deposition have been associated with a possible stimulation of tree growth, i.e., the fertilization effect and the greening of the Northern Hemisphere. This CO_2 fertilization effect implies that other factors affecting forest productivity were partially or not at all limiting. This viewpoint has been challenged recently (Peñuelas et al., 2017; Wang et al., 2020) because additional factors are likely limiting tree growth beyond C availability, including nutrient and water limitations (Peters et al., 2021; Yuan et al., 2019). The projected changes in major climatic variables are expected to shift tree growth climate sensitivity in boreal forests, e.g., reducing the importance of temperature/energy limitations and increasing the importance of water limitations (Babst et al., 2019; D'Orangeville et al., 2018), potentially altering C turnover rates and sink strength. The land C sink results from the interplay between C assimilation and respiration; thus, the net result of changes in climatic variables in the boreal forest C balance and sink strength remains uncertain (Naidu & Bagchi, 2021; Schurgers et al., 2018; Zhu et al., 2019).

Forest productivity (and tree growth) is a key aspect that contributes to the forest C balance and the C sink strength, but it is not the only one. An increase in productivity does not necessarily translate into an increase in forest C sink strength. The latter depends largely on the C residence times, i.e., at the tree level, to which tissues/compartments the assimilated C will be allocated (e.g., long-lived tissues like stemwood or short-lived tissues such as leaves and fine roots) or at the forest-stand or larger spatial scales, to the frequency and intensity of natural disturbances (e.g., fires, insects, tree mortality) and forest management strategies. The interplay of residence times, e.g., turnover rates, and productivity, e.g., tree growth, ultimately controls the long-term C source-sink strength (Harmon, 2001; Körner, 2017; Peñuelas et al., 2017). The two processes often present a positive covariation at the tree level, i.e., fast-growing strategies are also associated with faster turnover rates (lower longevity; Brienen et al., 2020; Reich, 2014). Yet, at the ecosystem level, stand demography and community interactions shape the net effect, making extrapolations of the forest C sink strength challenging when done solely on the basis of tree physiology (Anderegg et al., 2020; Harmon, 2001; McDowell et al., 2020). Although the boreal forest landscapes have always been dynamic, reports now cite changes in the frequency and intensity of disturbances such as wildfires, insect outbreaks, and droughts (Bond-Lamberty et al., 2007; Brandt et al., 2013; Goetz et al., 2012; Peng et al., 2011) or chronic changes in forest structure and function because of permafrost thaw (e.g., Helbig et al., 2016). When combined with management practices (harvesting; Ameray et al., 2021; Ceccherini et al., 2020), such changes can substantially affect the C balance of the circumboreal region and C sink strength (Seidl et al., 2020).

Although aboveground tree growth only represents a minor fraction of the forest C balance in the boreal region, because of low productivity and a short growing season, aboveground physiological processes trigger numerous feedbacks of belowground components, e.g., energy partitioning/albedo feedback and thus regulation of soil temperature. Furthermore, tree growth represents a major C-storing compartment for climate-meaningful time scales and contributes to the main inflow of C from the atmosphere to belowground C pools and to ecosystem-level C dynamics. A holistic monitoring and modeling approach is required to better constrain future trajectories of C in boreal ecosystems and delineate climate-sustainable forest management strategies. This can be better achieved with partitioned estimates of the C balance components as summarized in this study.

References

ACIA. (2005). *Impacts of a warming arctic: Arctic climate impact assessment*. Cambridge University Press.

Alexander, H. D., Mack, M. C., Goetz, S., et al. (2012). Carbon accumulation patterns during post-fire succession in cajander larch (*Larix cajanderi*) forests of siberia. *Ecosystems, 15*(7), 1065–1082. https://doi.org/10.1007/s10021-012-9567-6.

Ameray, A., Bergeron, Y., Valeria, O., et al. (2021). Forest carbon management: A review of silvicultural practices and management strategies across boreal, temperate and tropical forests. *Current Forestry Reports , 7*(4), 245–266. https://doi.org/10.1007/s40725-021-00151-w.

Amiro, B.D., Barr, A. G., Barr, J. G. et al. (2010). Ecosystem carbon dioxide fluxes after disturbance in forests of North America. *Journal of Geophysical Research: Biogeosciences, 115*(G4), G00K02 https://doi.org/10.1029/2010JG001390.

Anderegg, W. R. L., Trugman, A. T., Badgley, G., et al. (2020). Climate-driven risks to the climate mitigation potential of forests. *Science, 368*(6497), eaaz7005 https://doi.org/10.1126/science.aaz 7005.

Anderson-Teixeira, K. J., Davies, S. J., Bennett, A. C., et al. (2015). CTFS-ForestGEO: A worldwide network monitoring forests in an era of global change. *Global Change Biology, 21*(2), 528–549. https://doi.org/10.1111/gcb.12712.

Babst, F., Bouriaud, O., Alexander, R., et al. (2014a). Toward consistent measurements of carbon accumulation: A multi-site assessment of biomass and basal area increment across Europe. *Dendrochronologia, 32*(2), 153–161. https://doi.org/10.1016/j.dendro.2014.01.002.

Babst, F., Bouriaud, O., Papale, D., et al. (2014b). Above-ground woody carbon sequestration measured from tree rings is coherent with net ecosystem productivity at five eddy-covariance sites. *New Phytologist, 201*(4), 1289–1303. https://doi.org/10.1111/nph.12589.

Babst, F., Friend, A. D., Karamihalaki, M., et al. (2021). Modeling ambitions outpace observations of forest carbon allocation. *Trends in Plant Science, 26*(3), 210–219. https://doi.org/10.1016/j.tpl ants.2020.10.002.

Babst, F., Bouriaud, O., Poulter, B. et al. (2019). Twentieth century redistribution in climatic drivers of global tree growth. *Science Advances, 5*(1), eaat4313. https://doi.org/10.1126/sciadv.aat4313.

Baldocchi, D. D. (2020). How eddy covariance flux measurements have contributed to our understanding of *Global Change Biology. Global Change Biology, 26*(1), 242–260. https://doi.org/10.1111/gcb.14807.

Baldocchi, D., & Peñuelas, J. (2019). The physics and ecology of mining carbon dioxide from the atmosphere by ecosystems. *Global Change Biology, 25*(4), 1191–1197. https://doi.org/10.1111/gcb.14559.

Bar-On, Y. M., Phillips, R., & Milo, R. (2018). The biomass distribution on Earth. *Proceedings of the National Academy of Sciences of the United States of America, 115*(25), 6506–6511. https://doi.org/10.1073/pnas.1711842115.

Barr, A. G., van der Kamp, G., Black, T. A., et al. (2012). Energy balance closure at the BERMS flux towers in relation to the water balance of the white gull creek watershed 1999–2009. *Agricultural and Forest Meteorology, 153*, 3–13. https://doi.org/10.1016/j.agrformet.2011.05.017.

Bonan, G. B. (2008). Forests and climate change: Forcings, feedbacks, and the climate benefits of forests. *Science, 320*(5882), 1444–1449. https://doi.org/10.1126/science.1155121.

Bonan, G. B. (2016). Forests, climate, and public policy: A 500-year interdisciplinary odyssey. *Annual Review of Ecology, Evolution, and Systematics, 47*, 97–121. https://doi.org/10.1146/ann urev-ecolsys-121415-032359.

Bond-Lamberty, B., Peckham, S. D., Ahl, D. E., et al. (2007). Fire as the dominant driver of central Canadian boreal forest carbon balance. *Nature, 450*(7166), 89–92. https://doi.org/10.1038/nature 06272.

Bond-Lamberty, B., Bailey, V. L., Chen, M., et al. (2018). Globally rising soil heterotrophic respiration over recent decades. *Nature, 560*(7716), 80–83. https://doi.org/10.1038/s41586-018-0358-x.

Bradshaw, C. J. A., & Warkentin, I. G. (2015). Global estimates of boreal forest carbon stocks and flux. *Global and Planetary Change, 128*, 24–30. https://doi.org/10.1016/j.gloplacha.2015.02.004.

Brandt, J. P., Flannigan, M. D., Maynard, D. G., et al. (2013). An introduction to Canada's boreal zone: Ecosystem processes, health, sustainability, and environmental issues. *Environmental Reviews, 21*(4), 207–226. https://doi.org/10.1139/er-2013-0040.

Brassard, B. W., & Chen, H. Y. H. (2006). Stand structural dynamics of North American boreal forests. *Critical Reviews in Plant Sciences, 25*(2), 115–137. https://doi.org/10.1080/073526805 00348857.

Brienen, R. J. W., Caldwell, L., Duchesne, L., et al. (2020). Forest carbon sink neutralized by pervasive growth-lifespan trade-offs. *Nature Communications, 11*(1), 4241. https://doi.org/10.1038/s41467-020-17966-z.

Bugmann, H., & Bigler, C. (2011). Will the CO_2 fertilization effect in forests be offset by reduced tree longevity? *Oecologia, 165*(2), 533–544. https://doi.org/10.1007/s00442-010-1837-4.

Büntgen, U., Krusic, P. J., Piermattei, A., et al. (2019). Limited capacity of tree growth to mitigate the global greenhouse effect under predicted warming. *Nature Communications, 10*(1), 2171. https://doi.org/10.1038/s41467-019-10174-4.

Campioli, M., Malhi, Y., Vicca, S., et al. (2016). Evaluating the convergence between eddy-covariance and biometric methods for assessing carbon budgets of forests. *Nature Communications, 7*(5), 13717. https://doi.org/10.1038/ncomms13717.

Carvalhais, N., Forkel, M., Khomik, M., et al. (2014). Global covariation of carbon turnover times with climate in terrestrial ecosystems. *Nature, 514*(7521), 213–217. https://doi.org/10.1038/nature13731.

Ceccherini, G., Duveiller, G., Grassi, G., et al. (2020). Abrupt increase in harvested forest area over Europe after 2015. *Nature, 583*(7814), 72–77. https://doi.org/10.1038/s41586-020-2438-y.

Chapin, F. S., III. (1980). The mineral nutrition of wild plants. *Annual Review of Ecology and Systematics, 11*, 233–260. https://doi.org/10.1146/annurev.es.11.110180.001313.

Chapin, F. S., III., Woodwell, G. M., Randerson, J. T., et al. (2006). Reconciling carbon-cycle concepts, terminology, and methods. *Ecosystems, 9*(7), 1041–1050. https://doi.org/10.1007/s10021-005-0105-7.

Chen, J. M., Govind, A., Sonnentag, O., et al. (2006). Leaf area index measurements at Fluxnet-Canada forest sites. *Agricultural and Forest Meteorology, 140*(1–4), 257–268. https://doi.org/10.1016/j.agrformet.2006.08.005.

Ciais, P., Tan, J., Wang, X., et al. (2019). Five decades of northern land carbon uptake revealed by the interhemispheric CO_2 gradient. *Nature, 568*(7751), 221–225. https://doi.org/10.1038/s41586-019-1078-6.

Clark, D. A., Brown, S,, Kicklighter, D W, et al (2001). Measuring net primary production in forests: Concepts and field methods. *Ecological Applications, 11*(2), 356–370. https://doi.org/10.1890/1051-0761(2001)011[0356:MNPPIF]2.0.CO;2.

Clemmensen, K. E., Bahr, A., Ovaskainen, O., et al. (2013). Roots and associated fungi drive long-term carbon sequestration in boreal forest. *Science* 339(6127), 1615–1618 https://doi.org/10.1016/b978-0-408-01434-2.50020-6; https://doi.org/10.1126/science.1231923.

Collalti, A., & Prentice, I. C. (2019). Is NPP proportional to GPP? Waring's hypothesis 20 years on. *Tree Physiology, 39*(8), 1473–1483. https://doi.org/10.1093/treephys/tpz034.

Cook-Patton, S. C., Leavitt, S. M., Gibbs, D., et al. (2020). Mapping carbon accumulation potential from global natural forest regrowth. *Nature, 585*(7826), 545–550. https://doi.org/10.1038/s41586-020-2686-x.

Crowther, T. W., Glick, H. B., Covey, K. R., et al. (2015). Mapping tree density at a global scale. *Nature, 525*(7568), 201–205. https://doi.org/10.1038/nature14967.

De Lucia, E. H., Drake, J. E., Thomas, R. B., et al. (2007). Forest carbon use efficiency: Is respiration a constant fraction of gross primary production? *Global Change Biology, 13*(6), 1157–1167. https://doi.org/10.1111/j.1365-2486.2007.01365.x.

Dewar, R. C., Medlyn, B. E., & McMurtrie, R. E. (1998). A mechanistic analysis of light and carbon use efficiencies. *Plant, Cell and Environment, 21*(6), 573–588. https://doi.org/10.1046/j.1365-3040.1998.00311.x.

D'Orangeville, L., Houle, D., Duchesne, L., et al. (2018). Beneficial effects of climate warming on boreal tree growth may be transitory. *Nature Communications, 9*(1), 3213. https://doi.org/10.1038/s41467-018-05705-4.

Drake, J. E., Tjoelker, M. G., Aspinwall, M. J., et al. (2019). The partitioning of gross primary production for young *Eucalyptus tereticornis* trees under experimental warming and altered water availability. *New Phytologist, 222*(3), 1298–1312. https://doi.org/10.1111/nph.15629.

Duchesne, L., Houle, D., Ouimet, R., et al. (2019). Large apparent growth increases in boreal forests inferred from tree-rings are an artefact of sampling biases. *Scientific Reports, 9*(1), 6832. https://doi.org/10.1038/s41598-019-43243-1.

Duncanson, L., Armston, J., Disney, M., et al. (2019). The importance of consistent global forest aboveground biomass product validation. *Surveys in Geophysics, 40*(4), 979–999. https://doi.org/10.1007/s10712-019-09538-8.

Elmendorf, S. C., Henry, G. H. R., Hollister, R. D., et al. (2012). Plot-scale evidence of tundra vegetation change and links to recent summer warming. *Nature Climate Change, 2*(6), 453–457. https://doi.org/10.1038/nclimate1465.

Farquhar, G. D., von Caemmerer, S., & Berry, J. A. (1980). A biochemical model of photosynthetic CO_2 assimilation in leaves of C3 species. *Planta, 149*(1), 78–90. https://doi.org/10.1007/BF00386231.

Fatichi, S., Pappas, C., Zscheischler, J., et al. (2019). Modelling carbon sources and sinks in terrestrial vegetation. *New Phytologist, 221*(2), 652–668. https://doi.org/10.1111/nph.15451.

Fick, S. E., & Hijmans, R. J. (2017). WorldClim 2: New 1-km spatial resolution climate surfaces for global land areas. *International Journal of Climatology, 37*, 4302–4315. https://doi.org/10.1002/joc.5086.

Forkel, M., Carvalhais, N., Rödenbeck, C., et al. (2016). Enhanced seasonal CO_2 exchange caused by amplified plant productivity in northern ecosystems. *Science, 351*(6274), 696–699. https://doi.org/10.1126/science.aac4971.

Friedlingstein, P., O'Sullivan, M., Jones, M. W., et al. (2020). Global carbon budget 2020. *Earth System Science Data, 12*(4), 3269–3340. https://doi.org/10.5194/essd-12-3269-2020.

Friedlingstein, P. (2015). Carbon cycle feedbacks and future climate change. *Philosophical Transactions of the Royal Society A: Mathematical, Physical and Engineering Sciences, 373*, 1–22 https://doi.org/10.1098/not.

Friend, A. D., Lucht, W., Rademacher, T. T., et al. (2014). Carbon residence time dominates uncertainty in terrestrial vegetation responses to future climate and atmospheric CO_2. *Proceedings of the National Academy of Sciences of the United States of America, 111*(9), 3280. https://doi.org/10.1073/pnas.1222477110.

Gaumont-Guay, D., Black, T. A., Barr, A. G., et al. (2014). Eight years of forest-floor CO_2 exchange in a boreal black spruce forest: Spatial integration and long-term temporal trends. *Agricultural and Forest Meteorology, 184*, 25–35. https://doi.org/10.1016/j.agrformet.2013.08.010.

Gauthier, S., Bernier, P., Kuuluvainen, T., et al. (2015). Boreal forest health and global change. *Science, 349*, 819–822. https://doi.org/10.1126/science.aaa9092.

Giguère-Croteau, C., Boucher, É., Bergeron, Y., et al. (2019). North America's oldest boreal trees are more efficient water users due to increased [CO_2], but do not grow faster. *Proceedings of the National Academy of Sciences of the United States of America, 116*(7), 2749–2754. https://doi.org/10.1073/pnas.1816686116.

Girardin, M. P., Bouriaud, O., Hogg, E. H., et al. (2016). No growth stimulation of Canada's boreal forest under half-century of combined warming and CO_2 fertilization. *Proceedings of the National Academy of Sciences of the United States of America, 113*(52), E8406–E8414. https://doi.org/10.1073/pnas.1610156113.

Goetz, S. J., Bond-Lamberty, B., Law, B. E., et al. (2012). Observations and assessment of forest carbon dynamics following disturbance in North America. *Journal of Geophysical Research, 117*(G2), G02022. https://doi.org/10.1029/2011JG001733.

Gower, S. T., Vogel, J. G., Norman, M. et al. (1997). Carbon distribution and aboveground net primary production in aspen, jack pine, and black spruce stands in Saskatchewan and Manitoba, Canada. *Journal of Geophysical Research, 102*(D24), 29029–29041. https://doi.org/10.1029/97JD02317.

Grassi, G., House, J., Kurz, W. A., et al. (2018). Reconciling global-model estimates and country reporting of anthropogenic forest CO_2 sinks. *Nature Climate Change, 8*(10), 914–920. https://doi.org/10.1038/s41558-018-0283-x.

Griffis, T. J., Black, T. A., Morgenstern, K., et al. (2003). Ecophysiological controls on the carbon balances of three southern boreal forests. *Agricultural and Forest Meteorology, 117*(1–2), 53–71. https://doi.org/10.1016/S0168-1923(03)00023-6.

Harmon, M. E. (2001). Carbon sequestration in forest; addressing the scale question. *Journal of Forestry, 99*, 24–29. https://doi.org/10.1093/jof/99.4.24.

Hart, S. A., & Chen, H. Y. H. (2006). Understory vegetation dynamics of North American boreal forests. *Critical Reviews in Plant Sciences, 25*(4), 381–397. https://doi.org/10.1080/07352680600819286.

Helbig, M., Pappas, C., & Sonnentag, O. (2016). Permafrost thaw and wildfire: Equally important drivers of boreal tree cover changes in the Taiga Plains, Canada. *Geophysical Research Letters, 43*(4), 1598–1606. https://doi.org/10.1002/2015GL067193.

Ilvesniemi, H., Levula, J., Ojansuu, R., et al. (2009). Long-term measurements of the carbon balance of a boreal Scots pine dominated forest ecosystem. *Boreal Environment Research, 14*(4), 731–753.

Intergovernmental Panel on Climate Change (IPCC). (2013). *Climate change 2013: The physical science basis. Contribution of Working Group 1 to the fifth assessment report of the Intergovernmental Panel on Climate Change.* Cambridge: Cambridge University Press.

Kajimoto, T., Osawa, A., Usoltsev, V. A., et al. (2010). Biomass and productivity of Siberian larch forest ecosystems. In A. Osawa, O. A. Zyryanova, Y. Matsuura, T. Kajimoto, R. W. Wein (Eds.), *Permafrost ecosystems: Siberian larch forests.* Ecological Studies 209 (pp. 99–122). Dordrecht: Springer.

Kauppi, P. E., Posch, M., & Pirinen, P. (2014). Large impacts of climatic warming on growth of boreal forests since 1960. *PLoS ONE, 9*(11), e111340. https://doi.org/10.1371/journal.pone.0111340.

Keenan, T. F., & Williams, C. A. (2018). The terrestrial carbon sink. *Annual Review of Environment and Resources, 43*(1), 219–243. https://doi.org/10.1146/annurev-environ-102017-030204.

Kolari, P., Kulmala, L., Pumpanen, J., et al. (2009). CO_2 exchange and component CO_2 fluxes of a boreal Scots pine forest. *Boreal Environment Research, 14*(4), 761–783.

Körner, C. (2017). A matter of tree longevity. *Science, 355*(6321), 130–131. https://doi.org/10.1126/science.aal2449.

Körner, C. (2018). Concepts in empirical plant ecology. *Plant Ecology Diversity, 11*(4), 405–428. https://doi.org/10.1080/17550874.2018.1540021.

Kotani, A., Kononov, A. V., Ohta, T., et al. (2014). Temporal variations in the linkage between the net ecosystem exchange of water vapour and CO_2 over boreal forests in eastern Siberia. *Ecohydrology, 7*(2), 209–225. https://doi.org/10.1002/eco.1449.

Krishnan, P., Black, T. A., Barr, A. G., et al. (2008). Factors controlling the interannual variability in the carbon balance of a southern boreal black spruce forest. *Journal of Geophysical Research, 113*(D9), D09109. https://doi.org/10.1029/2007JD008965.

Kurz, W. A., Shaw, C. H., Boisvenue, C., et al. (2013). Carbon in Canada's boreal forest–A synthesis. *Environmental Reviews, 21*(4), 260–292. https://doi.org/10.1139/er-2013-0041.

Lagergren, F., Jönsson, A. M., Linderson, H., et al. (2019). Time shift between net and gross CO_2 uptake and growth derived from tree rings in pine and spruce. *Trees, 33*(3), 765–776. https://doi.org/10.1007/s00468-019-01814-9.

Landsberg, J. J., Waring, R. H., & Williams, M. (2020). The assessment of NPP/GPP ratio. *Tree Physiology, 40*(6), 695–699. https://doi.org/10.1093/treephys/tpaa016.

Le Quéré, C., Andrew, R. M., Friedlingstein, P., et al. (2018). Global carbon budget 2017. *Earth System Science Data, 10*(1), 405–448. https://doi.org/10.5194/essd-10-405-2018.

Litton, C. M., Raich, J. W., & Ryan, M. G. (2007). Carbon allocation in forest ecosystems. *Global Change Biology, 13*(10), 2089–2109. https://doi.org/10.1111/j.1365-2486.2007.01420.x.

Liu, P., Black, T. A., Jassal, R. S., et al. (2019). Divergent long-term trends and interannual variation in ecosystem resource use efficiencies of a southern boreal old black spruce forest 1999–2017. *Global Change Biology, 25*(9), 3056–3069. https://doi.org/10.1111/gcb.14674.

Lloyd, J., Shibistova, O., Zolotoukhine, D., et al. (2002). Seasonal and annual variations in the photosynthetic productivity and carbon balance of a central Siberian pine forest. *Tellus B Chemical and Physical Meteorology, 54*(5), 590–610. https://doi.org/10.3402/tellusb.v54i5.16689.

Malhi, Y., Baldocchi, D. D., & Jarvis, P. G. (1999). The carbon balance of tropical, temperate and boreal forests. *Plant, Cell and Environment, 22*(6), 715–740. https://doi.org/10.1046/j.1365-3040.1999.00453.x.

Manzoni, S., Čapek, P., Porada, P., et al. (2018). Reviews and syntheses: Carbon use efficiency from organisms to ecosystems–definitions, theories, and empirical evidence. *Biogeosciences, 15*(19), 5929–5949. https://doi.org/10.5194/bg-15-5929-2018.

Marchand, W., Girardin, M. P., Gauthier, S., et al. (2018). Untangling methodological and scale considerations in growth and productivity trend estimates of Canada's forests. *Environmental Research Letters, 13*(9), 093001. https://doi.org/10.1088/1748-9326/aad82a.

McDowell, N. G., Allen, C. D., Anderson-Teixeira, K., et al. (2020). Pervasive shifts in forest dynamics in a changing world. *Science, 368*(6494), eaaz9463. https://doi.org/10.1126/science.aaz9463.

Mencuccini, M., & Bonosi, L. (2001). Leaf/sapwood area ratios in Scots pine show acclimation across Europe. *Canadian Journal of Forest Research, 31*(3), 442–456. https://doi.org/10.1139/x00-173.

Myers-Smith, I. H., Forbes, B. C., Wilmking, M., et al. (2011). Shrub expansion in tundra ecosystems: Dynamics, impacts and research priorities. *Environmental Research Letters, 6*(4), 045509. https://doi.org/10.1088/1748-9326/6/4/045509.

Naidu, D. G. T., & Bagchi, S. (2021). Greening of the earth does not compensate for rising soil heterotrophic respiration under climate change. *Global Change Biology, 27*(10), 2029–2038. https://doi.org/10.1111/gcb.15531.

Odum, E. P. (1969). The strategy of ecosystem development. *Science, 164*(3877), 262–270. https://doi.org/10.1126/science.164.3877.262.

Ohta, T., Maximov, T. C., Dolman, A. J., et al. (2008). Interannual variation of water balance and summer evapotranspiration in an eastern Siberian larch forest over a 7-year period (1998–2006). *Agricultural and Forest Meteorology, 148*(12), 1941–1953. https://doi.org/10.1016/j.agrformet.2008.04.012.

Ohta, T., Kotani, A., Iijima, Y., et al. (2014). Effects of waterlogging on water and carbon dioxide fluxes and environmental variables in a Siberian larch forest, 1998–2011. *Agricultural and Forest Meteorology, 188*, 64–75. https://doi.org/10.1016/j.agrformet.2013.12.012.

Pan, Y., Birdsey, R., Phillips, O. L., et al. (2013). The structure, distribution, and biomass of the world's forests. *Annual Review of Ecology, Evolution, and Systematics, 44*(1), 593–622. https://doi.org/10.1146/annurev-ecolsys-110512-135914.

Pappas, C., Maillet, J., Rakowski, S., et al. (2020). Aboveground tree growth is a minor and decoupled fraction of boreal forest carbon input. *Agricultural and Forest Meteorology, 290*, 108030. https://doi.org/10.1016/j.agrformet.2020.108030.

Pastorello, G., Trotta, C., Canfora, E., et al. (2020). The FLUXNET2015 dataset and the ONEFlux processing pipeline for eddy covariance data. *Scientific Data, 7*(1), 225. https://doi.org/10.1038/s41597-020-0534-3.

Peng, C., Ma, Z., Lei, X., et al. (2011). A drought-induced pervasive increase in tree mortality across Canada's boreal forests. *Nature Climate Change, 1*, 467–471. https://doi.org/10.1038/nclimate1293.

Peñuelas, J., Ciais, P., Canadell, J. G., et al. (2017). Shifting from a fertilization-dominated to a warming-dominated period. *Nature Ecology and Evolution, 1*(10), 1438–1445. https://doi.org/10.1038/s41559-017-0274-8.

Peters, R. L., Steppe, K., Cuny, H. E., et al. (2021). Turgor-a limiting factor for radial growth in mature conifers along an elevational gradient. *New Phytologist, 229*(1), 213–229. https://doi.org/10.1111/nph.16872.

Piao, S., Wang, X., Park, T., et al. (2020). Characteristics, drivers and feedbacks of global greening. *Nature Reviews Earth and Environment, 1*(1), 14–27. https://doi.org/10.1038/s43017-019-0001-x.

Popkin, G. (2019). How much can forests fight climate change? *Nature, 565*, 280–282. https://doi.org/10.1038/d41586-019-00122-z.

Pugh, T. A. M., Arneth, A., Kautz, M., et al. (2019a). Important role of forest disturbances in the global biomass turnover and carbon sinks. *Nature Geoscience, 12*(9), 730–735. https://doi.org/10.1038/s41561-019-0427-2.

Pugh, T. A. M., Lindeskog, M., Smith, B., et al. (2019b). Role of forest regrowth in global carbon sink dynamics. *Proceedings of the National Academy of Sciences of the United States of America, 116*(10), 4382–4387. https://doi.org/10.1073/pnas.1810512116.

Pumpanen, J., Kulmala, L., Lindén, A., et al. (2015). Seasonal dynamics of autotrophic respiration in boreal forest soil estimated by continuous chamber measurements. *Boreal Environment Research, 20*(5), 637–650.

Rannik, Ü., Altimir, N., Raittila, J., et al. (2002). Fluxes of carbon dioxide and water vapour over Scots pine forest and clearing. *Agricultural and Forest Meteorology, 111*(3), 187–202. https://doi.org/10.1016/S0168-1923(02)00022-9.

Rees, W. G., Hofgaard, A., Boudreau, S., et al. (2020). Is subarctic forest advance able to keep pace with climate change? *Global Change Biology, 26*(7), 3965–3977. https://doi.org/10.1111/gcb.15113.

Reich, P. (2014). The world-wide "fast-slow" plant economics spectrum: A traits manifesto. *Journal of Ecology, 102*, 275–301. https://doi.org/10.1111/1365-2745.12211.

Reichstein, M., Falge, E., Baldocchi, D., et al. (2005). On the separation of net ecosystem exchange into assimilation and ecosystem respiration: Review and improved algorithm. *Global Change Biology, 11*(9), 1424–1439. https://doi.org/10.1111/j.1365-2486.2005.001002.x.

Reichstein, M., Bahn, M., Ciais, P., et al. (2013). Climate extremes and the carbon cycle. *Nature, 500*(7462), 287–295. https://doi.org/10.1038/nature12350.

Rodríguez Veiga, P., Quegan, S., Carreiras, J., et al. (2019). Forest biomass retrieval approaches from earth observation in different biomes. *International Journal of Applied Earth Observation and Geoinformation, 77*, 53–68. https://doi.org/10.1016/j.jag.2018.12.008.

Running, S., & Zhao, M. (2019). MOD17A3HGF MODIS/Terra net primary production gap-filled yearly L4 Global 500 m SIN Grid V006. NASA EOSDIS Land Processes DAAC.

Sawamoto, T., Hatano, R., Shibuya, M., et al. (2003). Changes in net ecosystem production associated with forest fire in taiga ecosystems, near Yakutsk, Russia. *Soil Science and Plant Nutrition, 49*(4), 493–501. https://doi.org/10.1080/00380768.2003.10410038.

Schepaschenko, D., Chave, J., Phillips, O. L., et al. (2019). The forest observation system, building a global reference dataset for remote sensing of forest biomass. *Scientific Data, 6*(1), 198. https://doi.org/10.1038/s41597-019-0196-1.

Schurgers, G., Ahlstrom, A., Arneth, A., et al. (2018). Climate sensitivity controls uncertainty in future terrestrial carbon sink. *Geophysical Research Letters, 45*(9), 4329–4336. https://doi.org/10.1029/2018GL077528.

Seidl, R., Thom, D., Kautz, M., et al. (2017). Forest disturbances under climate change. *Nature Climate Change, 7*(6), 395–402. https://doi.org/10.1038/nclimate3303.

Seidl, R., Honkaniemi, J., Aakala, T., et al. (2020). Globally consistent climate sensitivity of natural disturbances across boreal and temperate forest ecosystems. *Ecography, 43*(7), 967–978. https://doi.org/10.1111/ecog.04995.

Serreze, M. C., & Francis, J. A. (2006). The Arctic amplification debate. *Climatic Change, 76*(3–4), 241–264. https://doi.org/10.1007/s10584-005-9017-y.

Soja, A. J., Tchebakova, N. M., French, N. H. F., et al. (2007). Climate-induced boreal forest change: Predictions versus current observations. *Global and Planetary Change, 56*, 274–296. https://doi.org/10.1016/j.gloplacha.2006.07.028.

Tagesson, T., Schurgers, G., Horion, S., et al. (2020). Recent divergence in the contributions of tropical and boreal forests to the terrestrial carbon sink. *Nature Ecology and Evolution, 4*(2), 202–209. https://doi.org/10.1038/s41559-019-1090-0.

Tei, S., Sugimoto, A., Kotani, A., et al. (2019). Strong and stable relationships between tree-ring parameters and forest-level carbon fluxes in a Siberian larch forest. *Polar Science, 21*(1), 146–157. https://doi.org/10.1016/j.polar.2019.02.001.

Thurner, M., Beer, C., Santoro, M., et al. (2014). Carbon stock and density of northern boreal and temperate forests. *Global Ecology and Biogeography, 23*(3), 297–310. https://doi.org/10.1111/geb.12125.

Trumbore, S., Brando, P., & Hartmann, H. (2015). Forest health and global change. *Science, 349*(6250), 814–818. https://doi.org/10.1126/science.aac6759.

Van Oijen, M., Schapendonk, A., & Höglind, M. (2010). On the relative magnitudes of photosynthesis, respiration, growth and carbon storage in vegetation. *Annals of Botany, 105*(5), 793–797. https://doi.org/10.1093/aob/mcq039.

Vesala, T., Suni, T., Rannik, Ü., et al. (2005). Effect of thinning on surface fluxes in a boreal forest. *Global Biogeochemical Cycles, 19*(2), GB2001. https://doi.org/10.1029/2004GB002316.

Vicca, S., Luyssaert, S., Peñuelas, J., et al. (2012). Fertile forests produce biomass more efficiently. *Ecology Letters, 15*(6), 520–526. https://doi.org/10.1111/j.1461-0248.2012.01775.x.

Wang, S., Zhang, Y., Ju, W., et al. (2020). Recent global decline of CO_2 fertilization effects on vegetation photosynthesis. *Science, 370*(6522), 1295–1300. https://doi.org/10.1126/science.abb7772.

Waring, R. H., Landsberg, J. J., & Williams, M. (1998). Net primary production of forests: A constant fraction of gross primary production? *Tree Physiology, 18*(2), 129–134. https://doi.org/10.1093/treephys/18.2.129.

Yu, K., Smith, W. K., Trugman, A. T., et al. (2019). Pervasive decreases in living vegetation carbon turnover time across forest climate zones. *Proceedings of the National Academy of Sciences of the United States of America, 116*(49), 24662–24667. https://doi.org/10.1073/pnas.1821387116.

Yuan, W., Zheng, Y., Piao, S., et al. (2019). Increased atmospheric vapor pressure deficit reduces global vegetation growth. *Science Advances, 5*(8), eaax1396. https://doi.org/10.1126/sciadv.aax1396.

Zeng, J., Matsunaga, T., Tan, Z. H., et al. (2020). Global terrestrial carbon fluxes of 1999–2019 estimated by upscaling eddy covariance data with a random forest. *Scientific Data, 7*(1), 313. https://doi.org/10.1038/s41597-020-00653-5.

Zhu, Z., Piao, S., Myneni, R. B., et al. (2016). Greening of the earth and its drivers. *Nature Climate Change, 6*(8), 791–795. https://doi.org/10.1038/nclimate3004.

Zhu, P., Zhuang, Q., Welp, L., et al. (2019). Recent warming has resulted in smaller gains in net carbon uptake in northern high latitudes. *Journal of Climate, 32*(18), 5849–5863. https://doi.org/10.1175/jcli-d-18-0653.1.

Chapter 11
Experimental and Theoretical Analysis of Tree-Ring Growth in Cold Climates

Vladimir V. Shishov, Alberto Arzac, Margarita I. Popkova, Bao Yang, Minhui He, and Eugene A. Vaganov

Abstract The medium- and long-term projections of global climate models show the effects of global warming will be most pronounced in cold climate areas, especially in the high latitudes of the Northern Hemisphere. The consequences could involve a higher probability of global natural disasters and a higher uncertainty as to plant response to climate risk. In this chapter, we describe life under a cold climate, particularly in relation to forest ecosystems, species distribution, and local conditions in the Northern Hemisphere. We analyze recent climate trends and how the ongoing and future climate changes can affect the sensitivity of conifer species, the most common tree form in the boreal regions. We combine experimental data and theoretical process-based simulations involving tree-ring width, tree-ring density, and wood anatomy. This combined approach permits assessing a longer tree-ring record that overlaps with direct instrumental climate observations. The latter are currently experiencing the divergence problem in which tree-ring growth has diverged from the trends of the main climatic drivers. Given that most process-based models are multidimensional, the parameterization described in this chapter is key for obtaining reliable tree growth simulations connected with a site-specific climate, tree species, and the individual trajectory of tree development. Our approach combining experimental and theoretical approaches in xylogenesis is of interest to forest ecologists, physiologists, and wood anatomists.

V. V. Shishov (✉) · A. Arzac · M. I. Popkova · E. A. Vaganov
Siberian Federal University, Svobodny pr., 79, Krasnoyarsk 660041, Russia
e-mail: vlad.shishov@gmail.com

A. Arzac
e-mail: aarzac@gmail.com

M. I. Popkova
e-mail: popkova.marg@gmail.com

E. A. Vaganov
e-mail: eavaganov@hotmail.com

© The Author(s) 2023
M. M. Girona et al. (eds.), *Boreal Forests in the Face of Climate Change*,
Advances in Global Change Research 74,
https://doi.org/10.1007/978-3-031-15988-6_11

11.1 Cold Climates and Terrestrial Ecosystems: Definitions and Examples

The Earth's climate can be classified on the basis of various criteria, and many classification systems have been proposed, including the aridity index, the Holdridge life zone classification, and the respective climate classifications of Alisov, Berg, Köppen, and Lauer (Critchfield, 1966). Because climate is a major controlling factor of biological ecosystems, climate classification is closely correlated with biome distributions. Of the climate classifications, the Köppen system is one of the most widely applied (Beck et al., 2018). This approach, based on the thresholds and seasonality of monthly surface air temperatures and precipitation, divides the Earth's climates into 5 main classes and 30 subgroups (i.e., subclusters). The extensive use of Köppen's system relates to climate being long recognized as a major driver of global vegetation distributions (Beck et al., 2018; Vaganov et al., 2006; Woodward & Williams, 1987; Yang et al., 2017). Therefore, from the Köppen system, we can identify four main groups of boreal/subboreal climates in the Northern Hemisphere: subarctic (boreal), wet continental, hemiboreal, and cold semiarid climate (Fig. 11.1).

In the Northern Hemisphere, boreal forests represent 65% of the land area covered by vegetation, and a significant portion of that distribution is dominated by conifer species (about 70% of the forests). Cold climate conditions characterize these ecosystems, and the associated main conifer species are very sensitive to climatic conditions. Therefore, coniferous trees are of particular interest to the scientific community as the study of their climate sensitivity offers insight into plant physiology and ecology, particularly concerning tree-ring growth response to climatic forcing (Briffa et al., 1998; D'Arrigo et al., 2006, 2008; Esper & Frank, 2009; Kirdyanov et al., 2020; Rossi et al., 2013, 2016; Tumajer et al., 2021a; Vaganov et al., 1999, 2011).

11.2 Recent Trends in Climate and Their Influence on the Seasonal Growth of Trees

The interest in climate change is driven by the extraordinary contemporary changes in the Earth's climate system. These changes are manifested by the globally increasing surface air temperature, albeit with regional differences (IPCC, 2007, 2014). Increasing concentrations of atmospheric greenhouse gases, in particular CO_2,

B. Yang
Key Laboratory of Desert and Desertification, Northwest Institute of Eco-Environment and Resources, Chinese Academy of Sciences (CAS), 320 Donggang West Road, Lanzhou, Gansu 730000, China
e-mail: yangbao@lzb.ac.cn

M. He
Shenguo Road, Shanxi Province 712100, Yangling, China
e-mail: hmh0503lb@163.com

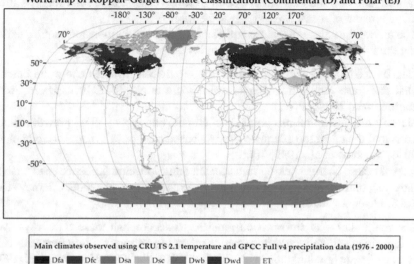

Fig. 11.1 Main classes of cold climates and the representative Köppen climate classification subtypes: subarctic or boreal (*Dfc, Dwc, Dsc, Dfd, Dwd, and Dsd*), wet continental (*Dsa, Dsb, Dwa, Dwb, Dfa, Dfb*), hemiboreal (*Dfb, Dsb*), and cold semiarid climate (*ET*). Modified from Beck et al. (2018), CC BY license

are driving this warming. Two periods in global CO_2 concentrations are evident in the twentieth-century record: increasing concentrations at a relatively low rate (pre-1960s) and a heightened annual increase in CO_2 concentrations since (IPCC, 2007, 2014; Jones et al., 2001; Thorne et al., 2003). This increase is also reflected by rising mean temperatures, which are most striking in the Northern Hemisphere (IPCC, 2007, 2014; Jones, 2002; Jones et al., 2001; Jones & Briffa, 1992). This recent temperature uptick, relative to the earlier, more stable temperature pattern (known as the *hockey stick* temperature record), can be attributed to the anthropogenic-related greenhouse gas and land-use changes (IPCC, 2007; Jones et al., 2001).

Medium- and long-term global climate projections show that warming will be most pronounced at higher latitudes, especially in the Northern Hemisphere (IPCC, 2007). In scenario B1 (one of the lower projections of increased greenhouse gases), global temperatures are projected to increase by 0.8 °C in 2020–2029, whereas this increase will be 2 °C in 2090–2099 IPCC, 2007. In colder climate regions, the projected increase for 2020–2029 and 2090–2099 is 1–2.5 °C and 3–4.5 °C, respectively. The most conservative scenario A2 projects a 4–6.5 °C increase at high latitudes by 2090–2099.

These projections of warming lead to numerous questions. How will the current and projected climate change affect forest ecosystems, particularly those in cold climates? What are the mechanisms of such impacts? Have similar changes occurred in the past, and how did they affect tree-ring growth? How might such changes affect the development of forest ecosystems in the future?

The answers to these questions are complex and ambiguous and require undertaking the following dendroecological and dendrophysiological tasks:

- identify meaningful statistical relationships between climate factors and interannual and intra-annual tree-ring growth (Briffa et al., 2002; Fritts, 1976; Rossi et al., 2008a, 2008b, 2013);
- identify the functional mechanisms of climate and woody plant growth interactions and then develop adequate process-based tree growth models (Guiot et al., 2014; Vaganov et al., 1999, 2006);
- forecast these changes taking into account anthropogenic influences and analyze the effect of these changes on forest ecosystems (Briffa et al., 2008; Charney et al., 2016; He et al., 2018a).

The main source of information required to accomplish these tasks is derived from tree rings, which record information related to various environmental factors, including climate (Anchukaitis et al., 2012; Briffa et al., 2002; Vaganov et al., 1999, 2006). Most tree species in cold regions are extremely sensitive to climate (Kirdyanov et al., 2003; Rossi et al., 2016; Shishov et al., 2016), and climatic factors account for 40–70% of the variability observed in the anatomical traits of tree rings (Vaganov et al., 2006). The successful resolution of these tasks requires using different (preferably independent) tree-ring characteristics, each recording specific (and different) information about tree growth patterns and the environmental factors affecting these patterns (Arzac et al., 2019; Gennaretti et al., 2017a; Puchi et al., 2019). This research commonly analyzes tree growth based on tree-ring width (Briffa et al., 2002; Cook & Kairiukstis, 1990; Fritts, 1976). Complementary tree-ring traits are also highly useful when analyzing the seasonal growth patterns of boreal trees. Vaganov et al., (1999, 2006, 2011) found a strong positive correlation between the temperature at different periods of the growing season—in particular at the start of the growing season—and cell size, cell wall thickness, and maximum density in the rings of larch collected from northern Eurasia. The relationship between coniferous tree-ring structural parameters and climate has also been analyzed for various regions of the globe. The use of complementary independent tree-ring traits—derived from those already being used via mathematical and statistical transformations—leads to novel information about the tree environment (Briffa et al., 2008).

Dendroclimatic analyses, using temperature-sensitive tree-ring chronologies obtained from a network of dendroclimatic monitoring stations distributed across the high latitudes of northern Eurasia and North America, indicate a heterogeneous response of woody plant growth to temperature increases (Briffa et al., 1998; D'Arrigo et al., 2008). Spatiotemporal analyses of long tree-ring chronologies from the mid to high latitudes of the Northern Hemisphere have revealed a distortion in the relationship between positive temperature trends and observed tree-ring growth after the 1960s in some regions where temperature is the principal growth-limiting factor (Briffa et al., 1998; Büntgen et al., 2021; D'Arrigo et al., 2006, 2008; Driscoll et al., 2005; Esper & Frank, 2009; Kirdyanov et al., 2020). This distortion, known as the *divergence problem*, reflects a change in tree growth with warmer summer temperatures.

Understanding and resolving this problem is critical for developing adequate statistical models for reconstructing climatic variables from dendrochronological data (Briffa et al., 1998).

A possible cause of this divergence may be the temperature-dependent drought stress of woody plants, which is particularly pronounced for fast-growing trees (Barber et al., 2000). This conclusion derives from a comparative analysis of dendrochronological data with a climate index representing a linear combination of temperature and precipitation (Barber et al., 2000). Another hypothesis for the divergence problem is a decrease in tree-ring growth when the temperature reaches a physiological threshold, thereby limiting growth (Hoch & Körner, 2003; Wilmking et al., 2004). Moreover, the current warming in the Northern Hemisphere is unprecedented over the last 2000 years (IPCC, 2007, 2014). Combining data from measurements of tree-ring width and maximum tree-ring density with simulations using the process-based tree-ring growth Vaganov-Shashkin model (VS-model) (Vaganov et al., 2006) has demonstrated a relationship between the decline in the sensitivity of trees and a positive trend in winter precipitation in subarctic Siberia between 1960 and 1995 (Vaganov et al., 1999). This observation led to the hypothesis of a shift in the start of the tree growth season to later dates because of a delayed melting of snow cover and, consequently, a decreased sensitivity of the trees to temperature change (Vaganov et al., 1999). A shift in temperature as a factor limiting tree-ring growth has also been observed in Alaska (Lloyd & Fastie, 2002). Finally, a most recent hypothesis proposes that the significant discrepancy between the annual growth of temperature-sensitive woody plants and summer temperature can be explained by the nonlinear dynamics of a low-frequency component of incoming solar radiation, which is closely correlated with the concentration of aerosol elements in the atmosphere at high latitudes in the Northern Hemisphere (Fig. 11.2; Büntgen et al., 2021; Kirdyanov et al., 2020).

Most of the abovementioned methods for estimating trends of a nonclimatic nature share a distortion (over- or underestimation) of the *true* index values of the dendrochronological series under selected conditions (Melvin, 2004). This distortion is pronounced at the ends of time series (Melvin, 2004) and is characteristic of a power law, polynomial approximations of degree P ($P \leq 5$) (Cook & Kairiukstis, 1990), as well as the commonly used low-frequency cubic spline (Melvin, 2004). Such inconsistencies can significantly alter both the statistical response function of woody plants to principal climatic factors (Melvin, 2004) and likely the tree-ring simulations based on various process-based tree-ring models (Guiot et al., 2014). Another reason for the divergence may relate to a *sampling bias* (Brienen et al., 2012; Duchesne et al., 2019).

However, the causes may be much more complex, combining the interaction of limiting and accelerating growth factors. The large number of noncontradictory

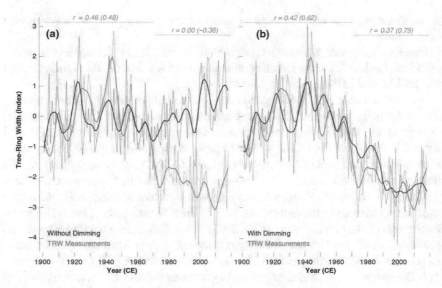

Fig. 11.2 Measured (*green*) and simulated boreal tree-ring width (*TRW*) indices **a** without (*red line with smoothing*) and **b** with (*black line with smoothing*) dimming that includes the low-frequency component of incoming solar radiation in the process-based forward model (VS-lite). Reproduced with permission from John Wiley and Sons from Kirdyanov et al. (2020)

hypotheses[1] relating to the same problem indicates that all information contained in the characteristics of the tree ring cannot be extracted correctly on the basis of mathematical and statistical approaches used in dendrochronology, dendrophysiology, and wood anatomy. Therefore, there is a need to develop new theoretical tools (models), tested with direct high-quality measurements and experimental analyses, that can adequately assess the influence of external factors on the growth of woody plants.

11.3 Experimental Analysis of Tree-Ring Growth in Cold Climates

Tree growth and survival are severely affected by climate, having significant consequences on their contribution to forest dynamics and carbon fluxes (Frank et al., 2015). Tree-ring structure depends on a complex cell formation process following successive phases of development (Rathgeber et al., 2016), controlled by external and internal factors occurring during the growing season (Dengler, 2001). Thus, in conifers, different cell development phases lead to intraseasonal changes in the anatomical characteristics of tracheids during the growing season, from wide and thin-walled earlywood cells to narrow and thick-walled latewood cells (Fig. 11.3).

---

[1] There is a well-known theorem in mathematical logic that holds that it is not possible to prove the contradictory nature of a hypothesis within the framework of the theory in which it is put forward.

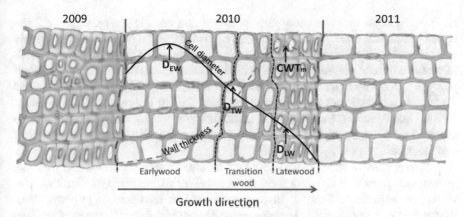

Fig. 11.3 Diagram of a *Pinus sylvestris* xylem cross section indicating parameters including the cell diameter in the earlywood (D_{EW}), the transition zone (D_{TW}), latewood (D_{LW}), maximum cell wall thickness (CWT_m), and tracheidograms of the cell diameter (*solid black line*) and cell wall thickness (*dashed black line*). The limits of the 2010 tree ring are identified (*between the vertical dashed blue lines*) as are the earlywood, transition wood, and latewood zones (*vertical dashed black lines*). Reproduced from Arzac et al. (2018a) with permission from Elsevier

Moreover, each type of cell has different functions. Thus, earlywood cells ensure efficient sap transport, whereas latewood cells favor mechanical stability (Björklund et al., 2017). The ratio between earlywood and latewood is critical for maintaining the balance between the structure and function of the xylem and plays a crucial role in tree water–carbon interactions (Domec & Gartner, 2002).

The intraseasonal change in xylem morphology might be driven by the need for structural reinforcement by latewood (Sperry et al., 2006), photoperiod (Gyllenstrand et al., 2007), or changes in water availability (Olano et al., 2012) throughout the growing season. In addition, the low availability of carbohydrates in the cambium at the beginning of the growing season allows a longer period for cell enlargement and limited wall deposition, whereas at the end of the growing season, a higher availability results in cells having a smaller lumen and wider walls (Cartenì et al., 2018). In temperature-sensitive regions, temperature can control cell production and radial cell expansion (Vaganov et al., 1999). Moreover, secondary wall thickening and the formation of latewood cells are also driven by temperature, as reflected by the strong correlation between maximum latewood density and summer temperature (Vaganov et al., 1999).

Because different cells are formed during distinct periods over the growing season, they encode the environmental information during their formation on a weekly to a seasonal basis (Bryukhanova et al., 2013; Kirdyanov et al., 2003; Vaganov et al., 1999). Thus, tree-ring structure depends highly on the timing and magnitude of the climatic events and conditions occurring during the cell developmental phases (Castagneri et al., 2017; Rathgeber, 2017), thereby being relevant when extracting the environmental information encoded in the tree rings. A full tree ring encodes information at an annual resolution (showing interannual variability from ring to ring). Earlywood and latewood parts of a tree ring encode the seasonal or intra-annual variability, and xylem cell traits encode environmental information at a weekly resolution. Moreover, xylem cell traits, such as cell size, wall thickness, and the wall-to-lumen ratio, can present different detailed seasonal information depending on their position within the ring and be used to identify the principal factors controlling tree growth or to reconstruct past climatic conditions (Fonti & Jansen, 2012; Vaganov, 1990). Understanding the specific climatic factors affecting tree-ring formation is essential to assess the impact of changing climatic seasonality on tree-ring structure and functioning.

Ongoing climate change will likely have diverse impacts on the various tree-ring sectors and functions depending on the seasonality of the changes. For example, when trees experience drought conditions, they reduce transpiration to protect their tissues from extensive water loss and avoid hydraulic failure (Irvine et al., 1998). However, these physiological responses affect the capacity of the tree to photoassimilate atmospheric carbon and maintain the turgor pressure of the growing cell, which modifies the amount, size, structure (Fonti et al., 2010; Steppe et al., 2015), and functioning of a forming tree ring; this response represents an important legacy for future tree performance, e.g., biomass production and resilience capacity (Anderegg et al., 2015). Moreover, changes in climate conditions may also influence the phenological patterns of tree growth via a lengthening or shortening of the growing season, with important consequences for forest productivity (Arzac et al., 2021a).

In cold regions, the tree growing season spans late spring to late summer, whereas trees remain dormant in autumn and winter. The onset of tree growth requires a minimum temperature threshold, which generally occurs somewhere between April and June, depending on the forest's location and sufficient soil moisture to maintain the process over the growing season (Kramer, 1964). Therefore, the temperature before xylogenesis (early spring) would be expected to promote an earlier onset of growth and larger growth rings. Low temperatures are also linked to the production of smaller cells (Zhirnova et al., 2020) and a limited carbon assimilation, resulting in important ecological consequences globally. Warming trends affect the stability and diversity of global forest ecosystems at various spatiotemporal scales. However, temperature is not the only factor controlling tree growth in cold environments. Although late spring and early summer temperatures are considered as the main drivers of boreal forest growth, soil moisture availability is likely to become a critical factor even in the coldest environments (Arzac et al., 2018b, 2019; Tabakova et al., 2020).

Reduced summer precipitation and increased temperatures will favor increased transpiration (Babushkina et al., 2015); this scenario will reduce turgor pressure, which eventually induces the formation of smaller and thicker latewood-like cells. Climatic factors constraining tree growth in temperate and boreal environments shift along a gradient, passing from water shortages, limiting growth at lower altitudes/latitudes, to colder temperatures limiting growth at higher latitudes (Babst et al., 2013; Hellmann et al., 2016). Therefore, temperature is a main limiting factor controlling tree growth in cold environments (Vaganov et al., 2006). Studies carried out on contrasting environmental conditions, e.g., high-elevation sites in the Mediter-ranean region and permafrost sites in northern Siberia, have shown the relevant effect of water availability later in the growing season (Arzac et al., 2016, 2019).

Various techniques can be applied (von Arx et al., 2016) to evaluate the effect of climate on tree growth and tree-ring structure. These approaches can investigate at the cellular to tree-ring level and include the in situ monitoring of intra-annual dynamics of wood formation using dendrometers. Measuring ring width and early-wood and latewood width require the xylem structure in the wood sample to be clearly visible. The widths of the tree ring, earlywood, and latewood are then usually measured by direct observation of the wood samples by using specialized tree-ring measurement systems, e.g., LinTab and Velmex, or by measuring digitized wood samples, e.g., CooRecorder/CDendro, and WinDENDRO. However, if the study aims to obtain more detailed information on xylem cell traits, quantitative wood anatomical (QWA) methods are applied (von Arx et al., 2016). QWA involves the production and analysis of thin histological preparations of wooden material, thereby allowing the study of many parameters in the xylem and also cambial activity. QWA methods follow a series of successive steps, including (1) microsection preparation (sectioning, staining, fixation); (2) the digitizing of anatomical sections, e.g., using a slide scanner or a camera mounted on an optical microscope; and (3) the measure-ments of the cell structures by specialized software, e.g., ROXAS, WinCELL, and AutoCellRow (Dyachuk et al., 2020; von Arx et al., 2016). Finally, in the case of in situ monitoring of the intra-annual dynamics of wood formation, dendrometers provide automatic measurements of changes in stem diameter at various temporal scales.

Typically, dendrochronological studies assess the effect of climate conditions on tree growth by correlating tree-ring width indices (or other parameters) with meteoro-logical data, e.g., temperature, precipitation, cloudiness, soil moisture, and/or wind, at a daily or monthly resolution. Such analyses determine the main environmental factors—and the role of the magnitude and timing of these factors—controlling tree growth and structure. Although the first steps to unveil climate effects are rela-tively simple, e.g., response functions, the subsequent statistical analyses are now quite sophisticated and can include a large number of diverse parameters to obtain a more comprehensive understanding of the mechanisms involved in tree-ring forma-tion. For example, general additive mixed models identify differences between the various tree-ring parameters as a function of several variables, including site loca-tion, tree age, and target year (Zuur et al., 2009). Because this type of statistical model provides a broader view of the parameters controlling tree growth, they are increasingly applied to dendrochronological studies. In addition, modeling tree-ring growth as a theoretical approach is commonly used to simulate cell growth rates, determine ring structure and cell phenology, evaluate the effect of climatic limiting factors, and forecast tree growth under future climate scenarios.

Significant warming trends affect the phenology and physiology of trees and the geographical distribution of different types of boreal forests (Barber et al., 2000; Cuny et al., 2015; Menzel et al., 2006). Warmer temperatures also likely trigger an earlier onset of growth because of more favorable conditions in usually cold climate regions

(He et al., 2018a; Menzel et al., 2006; Yang et al., 2017). These changes are reflected in tree rings and xylem structure; for example, increased tree radial growth of *Pinus sylvestris* L. has been observed in stands from western to eastern Siberia (Tabakova et al., 2020). Arzac et al. (2019) obtained similar results in central Yakutia, in which the ring width and latewood width of *Larix cajanderi* Mayr and *P. sylvestris* have increased over the recent decades in response to warmer temperatures. Nevertheless, the climate sensitivity of *P. sylvestris* could decrease because of changes in climate seasonality; thus, in southern Siberia, current climate seasonality changes positively impact both the hydraulic efficiency (by increasing the diameter of the earlywood cells) and the latewood width of wood produced (Arzac et al., 2018a).

In terms of xylem structure, seasonal variations in climate have clearly affected xylem cell differentiation, and therefore, total ring structure (Cuny & Rathgeber, 2016). Thus, favorable temperatures at the beginning of the growing season may contribute to the extension of cambial activity during the formation of earlywood (Rossi et al., 2013), whereas low temperatures at the end of the growing season constrain cell wall deposition during the formation of latewood (Cuny & Rathgeber, 2016; Zhirnova et al., 2020). Beyond the critical role of temperature for tree growth, precipitation signals are very strong at critical tree growth stages for both earlywood and latewood (Babushkina et al., 2018).

11.4 Tree-Ring Process-Based Models as Tools for Analyzing Climate Influence on Long-Term Tree-Ring Growth

One of the main objectives of dendrochronology and wood anatomy is the study of the year-to-year variability in the qualitative and quantitative characteristics of tree-ring growth and the identification of environmental factors that determine this variability over the long term, i.e., up to several decades, throughout the cold climate boreal zone (Vaganov et al., 2006, 2011). Seasonal direct observations of the xylogenesis of conifers and the appropriate statistical analysis are unique sources of information for understanding the processes occurring during the formation of tree rings (Vaganov et al., 2006). The direct and experimental observations of tree-ring formation contribute significantly to a deeper understanding of tree growth response to environmental conditions (Rossi et al., 2008a, 2008b, 2013). However, this kind of analysis often requires weekly monitoring, sampling, and measuring; this requirement is extremely labor intensive and generally unfeasible over vast territories. Without belittling the experimental and theoretical significance of direct observations of the xylogenesis of conifers, unfortunately such data in most cases can cover only a few seasons (2–4 years), with rare exceptions. Even 15-year xylogenesis observations (Buttò et al., 2020) do not ensure that, during the analysis, a long-term phenomenon, e.g., the divergence problem, does not occur (Kirdyanov et al., 2020). Moreover, the network of xylogenesis observations is significantly inferior to the spatial network of dendrochronological data, which covers the main forest biomes of the boreal zone.

Thus, estimating the differentiation time of cambium and xylem by process-based modeling is a possible tool that adequately extrapolates local xylogenesis analyses over widespread territories (Guiot et al., 2014). Given the significant increase in the quantity of tree-ring (including anatomical wood) data, there is an ongoing need to develop methods and software able to automatically identify and process all forms of biological information obtained from tree rings. Performing an adequate simulation of tree-ring cell structure makes it possible to separate the climate-driven component from other external (e.g., forest fires, insect outbreaks, snow avalanches) and internal (e.g., seasonal hormone variability, age-dependent trends) factors in tree-ring growth and understand the principal processes during the formation of tree rings over the long term (Shishov et al., 2021; Vaganov et al., 2006).

Process-based models describe tree growth on the basis of climate forcing and local nonclimatic environments, such as tree competition, insect outbreaks, and fires (Guiot et al., 2014). These multidimensional models can describe nonlinear interactions between tree growth and environments. In most processed-based tree growth models, climate variables are considered the primary global drivers of spatiotemporal growth variability (Guiot et al., 2014; Misson, 2004; Ogée et al., 2009; Peters et al., 2021; Vaganov et al., 2006). These models are useful for understanding the growth processes under investigation and finding new patterns reflecting the interaction of environmental factors with biological processes occurring within woody plants.

For example, the model MAIDEN (Modeling and Analysis In DENdroecology) and its modification MAIDENiso simulate annual tree-ring increments, carbon and oxygen isotope compositions based on daily CO_2 atmospheric concentrations, precipitation, and minimum and maximum air temperatures. These models evaluate carbon assimilation and allocation within various global forest stands, including those in cold climate regions (Gennaretti et al., 2017a; Lavergne et al., 2017; Rezsöhazy et al., 2020). Two modifications exist for MAIDEN: one developed for Mediterranean forests (Gea-Izquierdo et al., 2015) and one for boreal tree species (Gennaretti et al., 2017a, 2017b). This model can estimate daily photosynthesis and allocate the daily available carbon and stored nonstructural carbohydrates to different pools, i.e., leaves, roots, stem.

The stem growth and wood formation model of Drew and Downes (2015) is a potential candidate for use in cold climate conditions. The model uses CABALA-estimated daily variables—daily minimum and maximum leaf water potential, carbohydrate allocated to stem, stand density, tree height, and crown length (Battaglia et al., 2004)—as inputs to predict tracheid size, cell wall thickness, and microfiber angles in a cell on the basis of cambial activity and carbohydrate balance (Drew & Downes, 2015). Growth is limited by the daily osmotic potential of cell growth; this parameter links cell wall turgor, water, and carbohydrate balance. From these cell simulations, the model can estimate wood density. The model distinguishes between radial and longitudinal cell expansion.

One of the most modern and comprehensive models is the turgor-driven growth model of Peters et al. (2021), which estimates the growth dynamics of most tree tissues. The model has already been tested successfully along an elevation gradient involving different climate conditions. In this model, tissue development is limited by water balance and depends on cell growth. In turn, cell growth is limited by the water and temperature balance and depends on the turgor of cell walls. The variables required as inputs to the model are tree-specific allometric characteristics, hourly tree physiological measurements, and micrometeorological data and parameters. Incorporating these variables disentangles reversible, i.e., daily shrinkage and swelling as a result of water transport, from irreversible diameter growth (Peters et al., 2021).

The Vaganov-Shashkin tree-ring simulation model (VS-model) is one of the most applied model. This model is often used because of the minimal requirements for inputs and has already been used in various environments from warm semiarid and temperate conditions (Anchukaitis et al., 2006; Evans et al., 2006; Jevšenak et al., 2021; Touchan et al., 2012; Tumajer et al., 2021a, 2021b) to cold climates (Belousova et al., 2021; Buttò et al., 2020; He et al., 2017, 2018b; Popkova et al., 2018; Shishov et al., 2016; Tychkov et al., 2019; Vaganov & Shashkin, 2000; Vaganov et al., 2006, 2011; Yang et al., 2017). As a significant simplification of the VS-model, the VS-lite version accepts monthly temperature and precipitation data and offers the best choice for estimating a nonlinear tree-ring response to changing climate on a global scale (Tolwinski-Ward et al., 2011, 2013). However, the VS-lite is not formally a process-based model; it can be considered as a mathematical operator having some biological basis that effectively estimates a monthly-scale nonlinear relationship between climatic and dendrochronological data sets (Guiot et al., 2014).

Below we consider some of the issues concerning the tree-ring growth simulations in a cold climate that are common to most process-based models, using the VS-model, which we use as an example.

The VS-model is based on several assumptions (Vaganov et al., 2006):

The main target of external (climatic) influence is the cambial zone, the zone of actively dividing cells. The external influence affects the linear growth rate of cambial cells (and the cell cycle).

The main external factors affecting the growth rate of cambial cells are daily average temperature, day length (closely correlated with solar irradiation), and soil moisture. Day length is determined by latitude, solar declination, and day of the year.

The tree-ring growth rate is positively correlated with the number of new cells in the enlargement zone and their sizes. Therefore, the growth rate variations predetermine mainly the anatomical characteristics of the tracheids being formed, i.e., radial diameter. The principle of limiting factors—Liebig's principle; Ebelhar et al. (2008)—is used to estimate the growth rate.

The model simulates only climatically induced tree-ring width and structural variations. Therefore, the model can insulate climatic forcing from other local environmental effects, i.e., fires, insect outbreaks.

The basic algorithm of the VS-model (Fig. 11.4) involves the input of daily climatic data (temperature, precipitation, and solar irradiation), the calculation of integral tree-ring growth rate G from the climatic data, and estimation of cell production (number of cells formed during the growing season) and their radial sizes by the integral (environmental) tree-ring growth rate G.

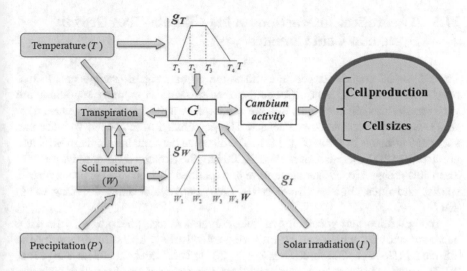

Fig. 11.4 Flowchart of the Vaganov-Shashkin model algorithm, where G, the integral growth rate, and $g_I(t)$, $g_T(t)$, $g_W(t)$, the partial growth rates, depend on solar irradiance (or day length) I, temperature T, and soil moisture W, respectively

The principal factors affecting the growth rate, i.e., air temperature, soil moisture, and day length, are used as inputs to the model. The cell production and their radial sizes are estimated on the basis of the integral (or environmental; see Anchukaitis et al. 2020) growth rate values at each time step (day of the year) t, which are calculated using the principle of limiting factors:

$$G(t) = g_I(t) * min\{g_T(t), g_W(t)\},$$

where $G(t)$ is the integral growth rate, $g_I(t)$, $g_T(t)$, $g_W(t)$ are the partial growth rates dependent on day length (solar irradiance) I, temperature T, and soil water content W, respectively.

The model estimates the daily water balance on the basis of the accumulated precipitation in the soil (with or without snowmelt), transpiration (temperature-dependent), and runoff (Thornthwaite & Mather, 1955). Day length (or incoming sunlight) is determined by the model according to the latitude at which the meteorological station or dendrochronological site is located (Vaganov et al., 2006). The number of cells formed per growing season and their sizes are then calculated on the basis of the integral (environmental) growth rate (Anchukaitis et al., 2020; Belousova et al., 2021; Shishov et al., 2021).

11.5   Theoretical Interactions in the Climate–Tree Growth System in Cold Climates

The VS-model was developed to estimate the climate signal (component) in tree rings (Vaganov et al., 2006). Generally, trees growing in extreme conditions are very sensitive to climate; thus, climate can explain up to 60–65% of annual tree-ring variability measured over decades. The percentage of explained variance can vary depending on location (Fig. 11.5). The distance between the southern MIN and northern PlatPO Siberian sites is about 2,200 km; the explained variance varies from 36 to 50%, respectively. The percentage of explained variance also depends on tree species and microclimates (Buttò et al., 2020; Popkova et al., 2020; Yang et al., 2017).

Another important specificity of cold climates is tree phenology, particularly cambium activity, which can be estimated effectively over the long-term by the VS-model (He et al., 2018b; Jevšenak et al., 2021; Tumajer et al., 2021a; Yang et al., 2017). Relative to more temperate conditions, the period of ring formation is shorter in cold climates; the cambium of conifer species is active between 180 and 240 days per year in semiarid Mediterranean Tunisia (Touchan et al., 2012) or Spain (Tumajer et al., 2021b) and just during 50–65 days per year in the extreme cold forest–tundra region of northern Yakutia, Russia (unpublished data). In both cases, ring cells pass through all stages of xylogenesis.

Moreover, the VS-model simulation can also reveal long-term trends in cambium phenology. For example, the period of cambium activity has become longer because of climate warming in the cold semiarid part of the Tibetan Plateau (Yang et al., 2017) and, as a result, significant negative (positive) shifts have been observed at the onset (end) of the growing season (Fig. 11.6).

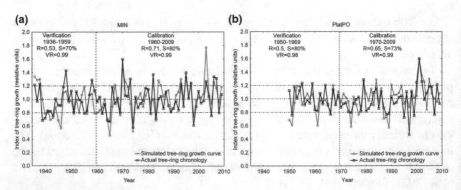

Fig. 11.5 The observed tree-ring chronology (*solid black line*) and simulated chronology (*solid gray line*) **a** for 1936–2009 at the MIN site (southern forest–steppe of central Siberia) and **b** for 1950–2009 at the PlatPO site (taiga of central Siberia). Average index for tree-ring growth and standard deviation are included (*dashed horizontal lines*). Reprinted by permission from Springer Nature from Tychkov et al. (2019)

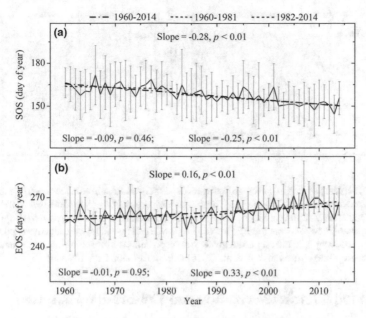

Fig. 11.6 Cambium phenology in a cold semiarid area of the Tibetan Plateau, northwestern China; **a** the average onset of the growing season (*SoS*) and **b** end of the growing season (*EoS*) for 1960–2014, as obtained using the VS-model. *Dashed lines* indicate linear trends for 1960–2014 (*black line*), 1982–2014 (*red line*), and 1960–1981 (*blue line*). Error bars indicate the standard deviation among the 20 composite sites. Significant ($P < 0.01$) advancing (delaying) trends in SoS (EoS) were detected for the periods 1960–2014 and 1982–2014. During 1960–1981, however, a nonsignificant ($P > 0.05$) trend was identified. Reproduced with permission from Yang et al. (2017)

Although most tree species are very sensitive to seasonal temperature variations in cold climates (Cook & Kairiukstis, 1990), this sensitivity can change over the growing season, i.e., two conifer species located in the same cold habitat can vary in their respective response to a similar temperature and moisture regime (Fig. 11.7). Spruce growth is always limited by temperature (Fig. 11.7b); however, soil moisture becomes a critical component controlling tree-ring growth in the middle of the growing season for larch, even in permafrost conditions (Fig. 11.7a).

Finally, because of the high percentage of explained variance in tree rings in cold climates (Briffa, 2000; Briffa et al., 1998; Vaganov et al., 1999) and the climate-oriented outputs of the VS-model (Anchukaitis et al., 2020; Vaganov et al., 2006), it is possible to effectively reconstruct the long-term seasonal cambial kinetics and their timing (Popkova et al., 2018; Shishov et al., 2021). This is also possible for temperate habitats, where trees are less sensitive to thermal conditions (Tumajer et al., 2021a). Generally, cold climate trees show less seasonal cell production than trees in temperate climate environments (Vaganov et al., 2006).

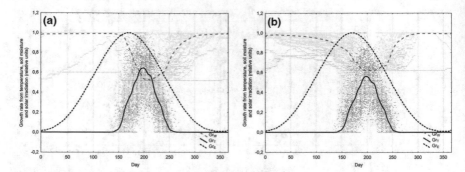

Fig. 11.7 Typical patterns of partial growth rates dependent on solar irradiance $Gr_E(t)$ (*black dotted line*), soil moisture $Gr_W(t)$ (*dashed gray line*), and temperature $Gr_T(t)$ (*solid black line*) between 1950 and 2009 for two northern Siberia taiga sites with **a** Siberian larch (*Larix gmelinii* (Rupr.)) and **b** Siberian spruce (*Picea obovata* Ledeb.) trees. The growth rates are fitted by a negative exponentially weighted smoothing. The lowest partial growth rate represents the most limiting factor for each day of the year. Reprinted from Shishov et al. (2016) with permission from Elsevier

11.6   Model Parameterization and Calibration Features

The main problem for most process-based models is the large number of model parameters that must be reasonably re-estimated for each habitat. Therefore, an adequate parameterization of the models is needed using appropriate experimental (ecobiological) and theoretical (mathematical) approaches. First, the obtained values of the model parameters should be reasonably interpreted by the nature of processes involved in the model (Shishov et al., 2016; Tychkov et al., 2019) and direct field observations (Buttò et al., 2020; Jevšenak et al., 2021; Tumajer et al., 2021a, 2021b). Second, even the most sophisticated mathematical optimizations, i.e., Bayesian approaches (Anchukaitis et al., 2020), or differential evolution (Kirdyanov et al., 2020) in a multidimensional parameter space cannot ensure that mathematically optimal parameters providing the best fit between the observed and simulated tree-ring growth are not artificial. To resolve these issues, we suggest using a two-step parameterization procedure: (1) visual (manual or semiautomatic) parameterization (Shishov et al., 2016) to obtain reasonable initial values of the model parameters that are ecobiologically interpreted; and (2) mathematical multidimensional optimization limited by the neighborhood of the obtained parameter values.

The visual parameterization approach described here could be applied to most process-based models. The new version of the model is VS-oscilloscope-online (http://vs-web.sfu-kras.ru:8080/), analogous in nature to a physical oscilloscope (Fig. 11.8). The VS-oscilloscope models (visualizes) the nonlinear tree-ring growth response to climate variability while assessing the contribution of each climatic variable (temperature or soil moisture) to the daily variability of seasonal dynamics of tree-ring formation (Tumajer et al., 2021a).

VS-oscilloscope-online is a web-graphical interface software based on Lazarus code (Shishov et al., 2016; Tychkov et al., 2019) with the potential use of the

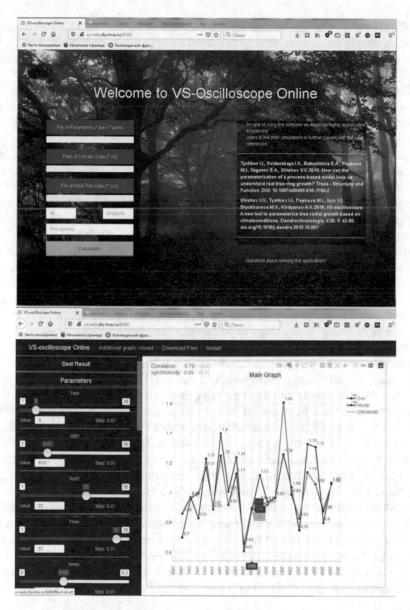

Fig. 11.8 Presentation of the VS-oscilloscope; (*upper panel*) the *Open Data* browser tab and (*lower panel*) the *Parameterization* browser tab

MATLAB version of the VS-model (Anchukaitis et al., 2020). The web-designed tree-ring simulation system can be considered as a cross-platform application; therefore, the VS-model can be used on computers regardless of the installed operating system (e.g., Windows, Linux, Mac iOS).

The basic idea of the visual parameterization (VS-oscilloscope) is to select the optimal parameter values to achieve a maximum correlation and synchronicity from a visual assessment of the synchrony between the simulated growth curve and the actual tree-ring chronology (Shishov et al., 2016).

The values of all parameters are fixed (or held constant), and only one parameter is modified, e.g., T_{min}, the minimum temperature for the onset of tree-ring formation or first cambial cell division (Vaganov et al., 2006). Changing the position of the bottom T_{min} to the right (left) increases (decreases) the value of this parameter in degrees (°C), which is then used to estimate the growth rate. Changing the position of the bottom for the analyzed parameter leads to a recalculation of the simulated growth curve, accounting for the parameter's new value (Fig. 11.8b, *right panel*).

The VS-oscilloscope's virtual display (Fig. 11.8b, *right panel*) shows the actual tree-ring chronology downloaded from a *.crn file (*red line*), the new simulated growth curve (*blue*), and the previous modeled curve (*green*). Iterations with new parameter values are performed until the maximum correlation between the growth curve and tree-ring chronology is obtained. The procedure is repeated for all parameters of the VS-model (Shishov et al., 2016).

The VS-oscilloscope control unit interface contains two tabs for inputting raw data (Fig. 11.8a) and adjusting growth parameter values (Fig. 11.8b). Raw data containing daily values of temperature and cumulative precipitation and values of initial tree-ring chronology, latitude, longitude, and tree species are loaded using a special ASCII format. The VS-oscilloscope-online operates with 19 parameters associated with the local conditions of the woody plant growth. According to the VS-model algorithm (Fig. 11.4), the parameters can be divided into two groups: those essential for calculating the integral growth rate of tree rings (19 parameters) and those necessary to determine cambial activity (17 parameters). The values of 19 parameters can be changed manually in the *Model Parameterization* panel (Fig. 11.8b, *left panel*).

If the user has questions while using the VS-oscilloscope-online, they can use the web link *Click Me* to access a ZIP archive of an input data example and user guide (Fig. 11.8a). When the application has finished, the best model obtained by the optimal parameter settings is saved by the system and downloaded in *.csv, *.dat, or *.xlsx format as a ZIP archive (Fig. 11.8b, use the *Best Result* and *Download Files* links).

When the calculations are completed, several files are created: (1) a chronology file (e.g., crns.*), which contains the simulated growth index (Model), actual tree-ring chronologies for the study area (Crn), Z-scores of simulated (NMOD) and actual (NCRN) chronologies, the number of days when the minimum temperature for growth start is reached (T_{min}), the day of the year when a growing season starts (SoS) and ends (EoS) (2) a growth rate file (e.g., rates.*), having a table containing dates of the year (Date), days of the year (Day), an integral growth rate (Gr, relative units), a partial growth rate (Gr$_W$, relative units) dependent on soil moisture, a partial growth rate (Gr_T, relative units) dependent on temperature, a partial growth rate (Gr_W, relative units) dependent on soil moisture, and a partial growth rate (Gr_E, relative units) dependent on solar irradiance, daily temperature (Tem, °C), daily precipitation (Prec, mm), estimated soil moisture Sm (v/v), and daily transpiration (Tr, mm).

These output data can be used to analyze long-term trends in cambium phenology (Yang et al., 2017), seasonal tree-ring growth patterns dependent on climatic variability (Buttò et al., 2020; He et al., 2017; Jevšenak et al., 2021; Tychkov et al., 2019), and seasonal cambium kinetics due to the cambium block of the VS-model (Anchukaitis et al., 2020; Belousova et al., 2021; Shishov et al., 2021; Tumajer et al., 2021a).

This parameterization procedure allows us to assess the impact of each individual climatic factor (temperature, soil moisture, and solar irradiance) on tree-ring growth and interactively control the intervals of valid model parameter values on the basis of available direct observations of the physical processes described by the growth simulation model (Vaganov et al., 2006). Such approaches provide a better understanding of the growth process of tree rings and greatly facilitate working with the multidimensional VS-model.

The main goal of any parameterization is to obtain the best fit of the simulated values to the observed direct measurements by selecting certain parameter values of the model. At the same time, in the context of tree-ring modeling, the selected parameters should not conflict with the biological principles of growth and field parameters obtained for the different environmental conditions of the studied forest stands (Buttò et al., 2020; Tumajer et al., 2021b; Tychkov et al., 2019). Solving this problem by the direct mathematical optimization of multidimensional parameter space is problematic, given the high probability of attaining a local optimum that generates artificial outcomes (Evans et al., 2006; Tolwinski-Ward et al., 2013).

Along with the Bayesian approach (Anchukaitis et al., 2020; Tolwinski-Ward et al., 2013), the differential evolution (DE) approach is a good candidate and is one of the fastest optimization methods (Price et al., 2005). The basic concept of DE is to obtain optimal values for multidimensional, real-valued functions of parameters on the basis of *genetic mutations* of a specially generated parameter set (see Storn & Price, 1997 for more details). Moreover, DE does not use the gradient of the problem being optimized; thus, DE does not require the optimization problem to be differentiable, as required by classic optimization methods such as gradient descent and quasi-Newton methods. Thus, DE can be applied to a wide suite of process-based tree-ring models. This optimization is already incorporated into the VS-lite and VS-models (Kirdyanov et al., 2020; Tumajer et al., 2021b).

Following the above-described two-step parameterization procedure will significantly limit the risk of obtaining inadequate model parameters; therefore, theoretical (simulated) tree-ring growth patterns under climatic forcing will be more reliable.

11.7   Conclusions

In the last decades, the Earth's climate, particularly in the higher latitudes of the Northern Hemisphere, has changed markedly as increases in average annual temperatures are melting glaciers and raising global sea levels (IPCC, 2007, 2014). In addition to warming, there is also an imbalance in natural systems, leading to changes in rainfall patterns, temperature anomalies, and an increased frequency of extreme events, such as hurricanes, floods, and droughts (Anderegg et al., 2015; IPCC, 2007, 2014). According to the medium- and long-term projections of global climate models, the effects of global warming will be most pronounced in areas characterized by cold climates, especially in the high latitudes of the Northern Hemisphere (IPCC, 2007, 2014). The consequences of this warming include a higher probability of global natural disasters, greater uncertainty in plant response to climate (IPCC, 2007, 2014), and decreased diversity in many forest ecosystems (Anderegg et al., 2015). The experimental and theoretical study of tree rings and their formation is a key source of information for revealing nonlinear relationships between climatic factors and seasonal tree-ring growth. These studies contribute to a better understanding of forest ecosystem processes and how they are affected by global climate change (Vaganov et al., 2006, 2011). Moreover, a greater understanding of the tree-ring growth processes is essential when developing adequate models for reconstructing climatic variables from dendrochronological data (Esper & Frank, 2009). Conifers are particularly sensitive to climatic conditions and are of special interest for studying the nonlinear response of tree growth to climatic influences in cold climates (Rossi et al., 2013, 2016; Vaganov et al., 1999, 2006).

The most effective research combines experimental approaches, including wood anatomy data of tree rings (Fonti et al., 2010) and the respective advanced statistical analyses (Babst et al., 2013; Cuny et al., 2013), with theoretical process-based simulations (Anchukaitis et al., 2020; Cuny et al., 2015; Gennaretti et al., 2017a; Peters et al., 2021; Vaganov et al., 2006) capable of extending experimentation over the long-term to ensure an overlap with direct instrumental climate observations (Briffa et al., 1998). Given that most process-based models are multidimensional and involve many parameters (Guiot et al., 2014), parameterization is a crucial issue for obtaining reliable tree growth simulations connected with local climate and species and the individual trajectory of tree development (Tychkov et al., 2019). Accordingly, we have detailed the example of the two-step parameterization procedure of the VS-model, which can be used in most models. The potential of tree-ring growth simulations can be applied to various research fields, including tree phenology, forest management, and carbon cycle analysis through the annual estimation of the absolute volume of woody biomass of forest ecosystems (Vaganov et al., 2006).

References

Anchukaitis, K. J., Evans, M. N., Kaplan, A., et al. (2006). Forward modeling of regional scale tree-ring patterns in the southeastern United States and the recent influence of summer drought. *Geophysical Research Letters, 33*(4), L04705. https://doi.org/10.1029/2005GL025050.

Anchukaitis, K. J., Breitenmoser, P., Briffa, K. R., et al. (2012). Tree rings and volcanic cooling. *Nature Geoscience, 5*(12), 836–837. https://doi.org/10.1038/ngeo1645.

Anchukaitis, K. J., Evans, M. N., Hughes, M. K., et al. (2020). An interpreted language implementation of the Vaganov-Shashkin tree-ring proxy system model. *Dendrochronologia, 60*, 125677. https://doi.org/10.1016/j.dendro.2020.125677.

Anderegg, W. R. L., Hicke, J. A., Fisher, R. A., et al. (2015). Tree mortality from drought, insects, and their interactions in a changing climate. *New Phytologist, 208*(3), 674–683. https://doi.org/10.1111/nph.13477.

Arzac, A., Garcia-Cervigon, A. I., Vicente-Serrano, S. M., et al. (2016). Phenological shifts in climatic response of secondary growth allow *Juniperus sabina* L. to cope with altitudinal and temporal climate variability. *Agricultural and Forest Meteorology, 217*, 35–45. https://doi.org/10.1016/j.agrformet.2015.11.011.

Arzac, A., Babushkina, E., Fonti, P., et al. (2018a). Evidences of wider latewood in *Pinus sylvestris* from a forest-steppe of Southern Siberia. *Dendrochronologia, 49*, 1–8. https://doi.org/10.1016/j.dendro.2018.02.007.

Arzac, A., Rozas, V., Rozenberg, P., et al. (2018b). Water availability controls *Pinus pinaster* xylem growth and density: A multi-proxy approach along its environmental range. *Agricultural and Forest Meteorology, 250–251*, 171–180. https://doi.org/10.1016/j.agrformet.2017.12.257.

Arzac, A., Popkova, M., Anarbekova, A., et al. (2019). Increasing radial and latewood growth rates of *Larix cajanderi* Mayr. and *Pinus sylvestris* L. in the continuous permafrost zone in Central Yakutia (Russia). *Annals of Forest Science, 76*(4), 96. https://doi.org/10.1007/s13595-019-0881-4.

Arzac, A., Tabakova, M. A., Khotcinskaia, K., et al. (2021a). Linking tree growth and intra-annual density fluctuations to climate in suppressed and dominant *Pinus sylvestris* L. trees in the forest-steppe of Southern Siberia. *Dendrochronologia, 67*, 125842. https://doi.org/10.1016/j.dendro.2021.125842.

Arzac, A., Tychkov, I., Rubtsov, A., et al. (2021b). Phenological shifts compensate warming-induced drought stress in southern Siberian Scots pines. *European Journal of Forest Research, 140*(6), 1487–1498. https://doi.org/10.1007/s10342-021-01412-w.

Babst, F., Poulter, B., Trouet, V., et al. (2013). Site- and species-specific responses of forest growth to climate across the European continent. *Global Ecology and Biogeography, 22*(6), 706–717. https://doi.org/10.1111/geb.12023.

Babushkina, E. A., Vaganov, E. A., Belokopytova, L. V., et al. (2015). Competitive strength effect in the climate response of Scots pine radial growth in south-central Siberia forest-steppe. *Tree-Ring Research, 71*(2), 106–117. https://doi.org/10.3959/1536-1098-71.2.106.

Babushkina, E. A., Belokopytova, L. V., Kostyakova, T. V., et al. (2018). Earlywood and latewood features of *Pinus sylvestris* in semiarid natural zones of South Siberia. *Russian Journal of Ecology, 49*, 209–217. https://doi.org/10.1134/S1067413618030013.

Barber, V. A., Juday, G. P., & Finney, B. P. (2000). Reduced growth of Alaskan white spruce in the twentieth century from temperature-induced drought stress. *Nature, 405*, 668–673. https://doi.org/10.1038/35015049.

Battaglia, M., Sands, P., White, D., et al. (2004). CABALA: A linked carbon, water and nitrogen model of forest growth for silvicultural decision support. *Forest Ecology and Management, 193*, 251–282. https://doi.org/10.1016/j.foreco.2004.01.033.

Beck, H. E., Zimmermann, N. E., McVicar, T. R., et al. (2018). Present and future Koppen-Geiger climate classification maps at 1-km resolution. *Science Data, 5*, 180214. https://doi.org/10.1038/sdata.2018.214.

Belousova, D. A., Shishov, V. V., Babushkina, E. A., et al. (2021). VS-cambium-developer: A new approach to modeling the functioning of the cambial zone of conifers under the influence of environmental factors. *Russian Journal of Ecology, 52*(5), 358–367. https://doi.org/10.1134/S10 67413621050040.

Björklund, J., Seftigen, K., Schweingruber, F., et al. (2017). Cell size and wall dimensions drive distinct variability of earlywood and latewood density in Northern Hemisphere conifers. *New Phytologist, 216*, 728–740. https://doi.org/10.1111/nph.14639.

Brienen, R. J. W., Gloor, E., & Zuidema, P. A. (2012). Detecting evidence for CO_2 fertilization from tree ring studies: The potential role of sampling biases. *Global Biogeochemical Cycles, 26*(1), GB1025. https://doi.org/10.1029/2011GB004143.

Briffa, K. R. (2000). Annual climate variability in the Holocene: Interpreting the message of ancient trees. *Quaternary Science Reviews, 19*, 87–105. https://doi.org/10.1016/S0277-3791(99)00056-6.

Briffa, K. R., Schweingruber, F. H., Jones, P. D., et al. (1998). Reduced sensitivity of recent tree-growth to temperature at high northern latitudes. *Nature, 391*, 678–682. https://doi.org/10.1038/35596.

Briffa, K. R., Osborn, T. J., Schweingruber, F. H., et al. (2002). Tree-ring width and density data around the Northern Hemisphere: Part 1, local and regional climate signals. *The Holocene, 12*, 737–757. https://doi.org/10.1191/0959683602hl587rp.

Briffa, K. R., Shishov, V. V., Melvin, T. M., et al. (2008). Trends in recent temperature and radial tree growth spanning 2000 years across northwest Eurasia. *Philosophical Transactions of the Royal Society of London. Series B: Biological Sciences, 363*(1501), 2269–2282. https://doi.org/10.1098/rstb.2007.2199.

Bryukhanova, M. V., Kirdyanov, A. V., Prokushkin, A. S., et al. (2013). Specific features of xylogenesis in Dahurian larch, *Larix gmelinii* (Rupr.) Rupr., growing on permafrost soils in Middle Siberia. *Russian Journal of Ecology, 44*, 361–366. https://doi.org/10.1134/S1067413613050044.

Büntgen, U., Kirdyanov, A. V., Krusic, P. J., et al. (2021). Arctic aerosols and the 'divergence problem' in dendroclimatology. *Dendrochronologia, 67*, 125837. https://doi.org/10.1016/j.dendro.2021.125837.

Buttò, V., Shishov, V., Tychkov, I., et al. (2020). Comparing the cell dynamics of tree-ring formation observed in microcores and as predicted by the Vaganov-Shashkin model. *Frontiers in Plant Science, 11*, 1268. https://doi.org/10.3389/fpls.2020.01268.

Cartenì, F., Deslauriers, A., Rossi, S., et al. (2018). The physiological mechanisms behind the earlywood-to-latewood transition: A process-based modeling approach. *Frontiers in Plant Science, 9*(July), 1053. https://doi.org/10.3389/fpls.2018.01053.

Castagneri, D., Fonti, P., von Arx, G., et al. (2017). How does climate influence xylem morphogenesis over the growing season? Insights from long-term intra-ring anatomy in *Picea abies*. *Annals of Botany, 119*(6), 1011–1020. https://doi.org/10.1093/aob/mcw274.

Charney, N. D., Babst, F., Poulter, B., et al. (2016). Observed forest sensitivity to climate implies large changes in 21st century North American forest growth. *Ecology Letters, 19*(9), 1119–1128. https://doi.org/10.1111/ele.12650.

Cook, E. R., & Kairiukstis, L. A. (1990). *Methods of dendrochronology: Applications in the environmental sciences* (p. 394). Dordrecht, Netherlands: Springer.

Critchfield, H. (1966). *General climatology*. Prentice Hall Inc.

Cuny, H. E., & Rathgeber, C. B. K. (2016). Xylogenesis: Coniferous trees of temperate forests are listening to the climate tale during the growing season but only remember the last words! *Plant Physiology, 171*(1), 306–317. https://doi.org/10.1104/pp.16.00037.

Cuny, H. E., Rathgeber, C. B. K., Kiessé, T. S., et al. (2013). Generalized additive models reveal the intrinsic complexity of wood formation dynamics. *Journal of Experimental Botany, 64*(7), 1983–1994. https://doi.org/10.1093/jxb/ert057.

Cuny, H. E., Rathgeber, C. B. K., Frank, D., et al. (2015). Woody biomass production lags stem-girth increase by over one month in coniferous forests. *Nature Plants, 1*(11), 15160. https://doi.org/10.1038/nplants.2015.160.

D'Arrigo, R., Wilson, R., & Jacoby, G. (2006). On the long-term context for late twentieth century warming. *Journal of Geophysical Research, 111*, D03103. https://doi.org/10.1029/200 5JD006352.

D'Arrigo, R., Wilson, R., Liepert, B., et al. (2008). On the "divergence problem" in northern forests: A review of the tree-ring evidence and possible causes. *Global and Planetary Change, 60*, 289–305. https://doi.org/10.1016/j.gloplacha.2007.03.004.

Dengler, N. G. (2001). Regulation of vascular development. *Journal of Plant Growth Regulation, 20*(1), 1–13. https://doi.org/10.1007/s003440010008.

Domec, J. C., & Gartner, B. L. (2002). How do water transport and water storage differ in coniferous earlywood and latewood? *Journal of Experimental Botany, 53*, 2369–2379. https://doi.org/10.1093/jxb/erf100.

Drew, D. M., & Downes, G. (2015). A model of stem growth and wood formation in *Pinus radiata*. *Trees, 29*, 1395–1413. https://doi.org/10.1007/s00468-015-1216-1.

Driscoll, W. W., Wiles, G. C., D'Arrigo, R. D., et al. (2005). Divergent tree growth response to recent climatic warming, Lake Clark National Park and Preserve, Alaska. *Geophysical Research Letters, 32*, L20703. https://doi.org/10.1029/2005GL024258.

Duchesne, L., Houle, D., Ouimet, R., et al. (2019). Large apparent growth increases in boreal forests inferred from tree-rings are an artefact of sampling biases. *Scientific Reports, 9*(1), 6832. https://doi.org/10.1038/s41598-019-43243-1.

Dyachuk, P., Arzac, A., Peresunko, P., et al. (2020). AutoCellRow (ACR)—A new tool for the automatic quantification of cell radial files in conifer images. *Dendrochronologia, 60*, 125687. https://doi.org/10.1016/j.dendro.2020.125687.

Ebelhar, S. A., Chesworth, W., & Paris, Q. (2008). Law of the minimum. In W. Chesworth (Ed.), *Encyclopedia of soil science* (pp. 431–437). Springer.

Esper, J., & Frank, D. (2009). Divergence pitfalls in tree-ring research. *Climatic Change, 94*(3), 261. https://doi.org/10.1007/s10584-009-9594 2.

Evans, M. N., Reichert, B. K., Kaplan, A., et al. (2006). A forward modeling approach to paleoclimatic interpretation of tree-ring data. *Journal of Geophysical Research, 111*(G3), G03008. https://doi.org/10.1029/2006JG000166.

Fonti, P., & Jansen, S. (2012). Xylem plasticity in response to climate. *New Phytologist, 195*, 734–736. https://doi.org/10.1111/j.1469-8137.2012.04252.x.

Fonti, P., von Arx, G., Garcia-Gonzalez, I., et al. (2010). Studying global change through investigation of the plastic responses of xylem anatomy in tree rings. *New Phytologist, 185*(1), 42–53. https://doi.org/10.1111/j.1469-8137.2009.03030.x.

Frank, D., Reichstein, M., Bahn, M., et al. (2015). Effects of climate extremes on the terrestrial carbon cycle: Concepts, processes and potential future impacts. *Global Change Biology, 21*(8), 2861–2880. https://doi.org/10.1111/gcb.12916.

Fritts, H. C. (1976). *Tree rings and climate*. London: Academic Press.

Gea-Izquierdo, G., Guibal, F., Joffre, R., et al. (2015). Modelling the climatic drivers determining photosynthesis and carbon allocation in evergreen Mediterranean forests using multiproxy long time series. *Biogeosciences, 12*(12), 3695–3712. https://doi.org/10.5194/bg 12-3695-2015.

Gennaretti, F., Gea-Izquierdo, G., Boucher, E., et al. (2017a). Ecophysiological modeling of photosynthesis and carbon allocation to the tree stem in the boreal forest. *Biogeosciences, 14*, 4851–4866. https://doi.org/10.5194/bg-14-4851-2017.

Gennaretti, H., Naulier, S., & Bégin, A. (2017b). Bayesian multiproxy temperature reconstruction with black spruce ring widths and stable isotopes from the northern Quebec taiga. *Climate Dynamics, 49*(11–12), 4107–4119. https://doi.org/10.1007/s00382-017-3565-5.

Guiot, J., Boucher, E., & Gea-Izquierdo, G. (2014). Process models and model-data fusion in dendroecology. *Frontiers in Ecology and Evolution, 2*, 1–12 https://doi.org/10.3389/fevo.2014.00052.

Gyllenstrand, N., Clapham, D., Kallman, T., et al. (2007). A Norway spruce FLOWERING LOCUS T homolog is implicated in control of growth rhythm in conifers. *Plant Physiology, 144*, 248–257. https://doi.org/10.1104/pp.107.095802.

He, M., Shishov, V., Kaparova, N., et al. (2017). Process-based modeling of tree-ring formation and its relationships with climate on the Tibetan Plateau. *Dendrochronologia, 42*, 31–41. https://doi.org/10.1016/j.dendro.2017.01.002.

He, M., Yang, B., Shishov, V., et al. (2018a). Relationships between wood formation and cambium phenology on the Tibetan plateau during 1960–2014. *Forests, 9*(2), 1–13. https://doi.org/10.3390/f9020086.

He, M., Yang, B., Shishov, V., et al. (2018b). Projections for the changes in growing season length of tree-ring formation on the Tibetan Plateau based on CMIP5 model simulations. *International Journal of Biometeorology, 62*(4), 631–641. https://doi.org/10.1007/s00484-017-1472-4.

Hellmann, L., Agafonov, L., Ljungqvist, F. C., et al. (2016). Diverse growth trends and climate responses across Eurasia's boreal forest. *Environmental Research Letters, 11*(7), 074021. https://doi.org/10.1088/1748-9326/11/7/074021.

Hoch, G., & Körner, C. (2003). The carbon charging of pines at the climatic treeline: A global comparison. *Oecologia, 135*, 10–21. https://doi.org/10.1007/s00442-002-1154-7.

Intergovernmental Panel on Climate Change (IPCC). (Ed.). (2007). *Climate Change 2007: Synthesis report. Contribution of Working Groups I, II and III to the fourth assessment report of the Intergovernmental Panel on Climate Change* (p. 104). Geneva: IPCC.

Intergovernmental Panel on Climate Change (IPCC) (Ed.). (2014). *Climate Change 2014: Synthesis report. Contribution of Working Groups I, II and III to the fifth assessment report of the Intergovernmental Panel on Climate Change* (p. 151). Geneva: IPCC.

Irvine, J., Perks, M. P., Magnani, F., et al. (1998). The response of *Pinus sylvestris* to drought: Stomatal control of transpiration and hydraulic conductance. *Tree Physiology, 18*(6), 393–402. https://doi.org/10.1093/treephys/18.6.393.

Jevšenak, J., Tychkov, I., Gričar, J., et al. (2021). Growth-limiting factors and climate response variability in Norway spruce (*Picea abies* L.) along an elevation and precipitation gradients in Slovenia. *International Journal of Biometeorology, 65*(2), 311–324. https://doi.org/10.1007/s00484-020-02033-5.

Jones, P. D. (2002). Changes in climate and variability over the last 1000 years. *International Geophysics, 83*, 133–142. https://doi.org/10.1016/S0074-6142(02)80162-0.

Jones, P. D., & Briffa, K. R. (1992). Global surface air temperature variations during the twentieth century: Part 1, Spatial, temporal and seasonal details. *The Holocene, 2*, 165–179. https://doi.org/10.1177/095968369200200208.

Jones, P. D., Osborn, T. J., & Briffa, K. R. (2001). The evolution of climate over the last millennium. *Science, 292*(5517), 662–667. https://doi.org/10.1126/science.1059126.

Kirdyanov, A., Hughes, M., Vaganov, E., et al. (2003). The importance of early summer temperature and date of snow melt for tree growth in the Siberian Subarctic. *Trees, 17*(1), 61–69. https://doi.org/10.1007/s00468-002-0209-z.

Kirdyanov, A. V., Krusic, P. J., Shishov, V. V., et al. (2020). Ecological and conceptual consequences of Arctic pollution. *Ecology Letters, 23*(12), 1827–1837. https://doi.org/10.1111/ele.13611.

Kramer, P. (1964). The role of water in wood formation. In M. Zimmermann (Ed.), *The formation of wood in forest trees* (pp. 519–532). New York: Academic Press.

Lavergne, A., Gennaretti, F., Risi, C., et al. (2017). Modelling tree ring cellulose $\delta^{18}O$ variations in two temperature-sensitive tree species from North and South America. *Climate of the Past, 13*, 1515–1526. https://doi.org/10.5194/cp-13-1515-2017.

Lloyd, A. H., & Fastie, C. L. (2002). Spatial and temporal variability in the growth and climate response of treeline trees in Alaska. *Climatic Change, 52*, 481–509. https://doi.org/10.1023/A:1014278819094.

Melvin, T. (2004). *Historical growth rates and changing climatic sensitivity of boreal conifers*. Ph.D. thesis, University of East Anglia.

Menzel, A., Sparks, T. H., Estrella, N., et al. (2006). European phenological response to climate change matches the warming pattern. *Global Change Biology, 12*(10), 1969–1976. https://doi.org/10.1111/j.1365-2486.2006.01193.x.

Misson, L. (2004). MAIDEN: A model for analyzing ecosystem processes in dendroecology. *Canadian Journal of Forest Research, 34*(4), 874–887. https://doi.org/10.1139/x03-252.

Ogée, J., Barbour, M. M., Wingate, L., et al. (2009). A single-substrate model to interpret intra-annual stable isotope signals in tree-ring cellulose. *Plant, Cell and Environment, 32*, 1071–1090. https://doi.org/10.1111/j.1365-3040.2009.01989.x.

Olano, J., Eugenio, M., Garcia-Cervigon, A. I., et al. (2012). Quantitative tracheid anatomy reveals a complex environmental control of wood structure in continental Mediterranean climate. *International Journal of Plant Sciences, 173*(2), 137–149. https://doi.org/10.1086/663165.

Peters, R. L., Steppe, K., Cuny, H. E., et al. (2021). Turgor—A limiting factor for radial growth in mature conifers along an elevational gradient. *New Phytologist, 229*, 213–229. https://doi.org/10.1111/nph.16872.

Popkova, M. I., Vaganov, E. A., Shishov, V. V., et al. (2018). Modeled tracheidograms disclose drought influence on *Pinus sylvestris* tree-rings structure from Siberian forest-steppe. *Frontiers in Plant Science, 9*(August), 1144. https://doi.org/10.3389/fpls.2018.01144.

Popkova, M. I., Shishov, V. V., Vaganov, E. A., et al. (2020). Contribution of xylem anatomy to tree-ring width of two larch species in permafrost and non-permafrost zones of Siberia. *Forests, 11*, 1343. https://doi.org/10.3390/f11121343.

Price, K., Storn, R. M., & Lampinen, J. A. (2005). *Differential evolution.* Berlin, Heidelberg: Springer.

Puchi, P. F., Castagneri, D., Rossi, S., et al. (2019). Wood anatomical traits in black spruce reveal latent water constraints on the boreal forest. *Global Change Biology, 26*(3), 1767–1777. https://doi.org/10.1111/gcb.14906.

Rathgeber, C. B. K. (2017). Conifer tree-ring density inter-annual variability—anatomical, physiological and environmental determinants. *New Phytologist, 216*(3), 621–625. https://doi.org/10.1111/nph.14763.

Rathgeber, C. B. K., Cuny, H. E., & Fonti, P. (2016). Biological basis of tree-ring formation: A crash course. *Frontiers in Plant Science, 7*(734). https://doi.org/10.3389/fpls.2016.00734.

Rezsohazy, J., Goosse, H., Guiot, J., et al (2020). Application and evaluation of the dendroclimatic process-based model MAIDEN during the last century in Canada and Europe. *Climate of the Past, 16*, 1043–1059. https://doi.org/10.5194/cp-16-1043-2020.

Rossi, S., Deslauriers, A., Anfodillo, T., et al. (2008a). Age-dependent xylogenesis in timberline conifers. *New Phytologist, 177*(1), 199–208. https://doi.org/10.1111/j.1469-8137.2007.02235.x.

Rossi, S., Deslauriers, A., Gricar, J., et al. (2008b). Critical temperatures for xylogenesis in conifers of cold climates. *Global Ecology and Biogeography, 17*(6), 696–707. https://doi.org/10.1111/j.1466-8238.2008.00417.x.

Rossi, S., Anfodillo, T., Cufar, K., et al. (2013). A meta-analysis of cambium phenology and growth: Linear and non-linear patterns in conifers of the northern hemisphere. *Annals of Botany, 112*(9), 1911–1920. https://doi.org/10.1093/aob/mct243.

Rossi, S., Anfodillo, T., Čufar, K., et al. (2016). Pattern of xylem phenology in conifers of cold ecosystems at the Northern Hemisphere. *Global Change Biology, 22*(11), 3804–3813. https://doi.org/10.1111/gcb.13317.

Shishov, V. V., Tychkov, I. I., Popkova, M. I., et al. (2016). VS-oscilloscope: A new tool to parameterize tree radial growth based on climate conditions. *Dendrochronologia, 39*, 42–50. https://doi.org/10.1016/j.dendro.2015.10.001.

Shishov, V. V., Tychkov, I. I., Anchukaitis, K. J., et al. (2021). A band model of cambium development: Opportunities and prospects. *Forests, 12*(10), 1361.

Sperry, J. S., Hacke, U. G., & Pittermann, J. (2006). Size and function in conifer tracheids and angiosperm vessels. *American Journal of Botany, 93*(10), 1490–1500. https://doi.org/10.3732/ajb.93.10.1490.

Steppe, K., Sterck, F., & Deslauriers, A. (2015). Diel growth dynamics in tree stems: Linking anatomy and ecophysiology. *Trends in Plant Science, 20*(6), 335–343. https://doi.org/10.1016/j.tplants.2015.03.015.

Storn, R., & Price, K. (1997). Differential evolution—a simple and efficient heuristic for global optimization over continuous spaces. *Journal of Global Optimization, 11*, 341–359. https://doi.org/10.1023/A:1008202821328.

Tabakova, M., Arzac, A., Martinez, E., et al. (2020). Climatic factors controlling *Pinus sylvestris* radial growth along a transect of increasing continentality in southern Siberia. *Dendrochronologia, 62*, 125709. https://doi.org/10.1016/j.dendro.2020.125709.

Thorne, P. W., Jones, P. D., Tett, S. F. B., et al. (2003). Probable causes of late twentieth century tropospheric temperature trends. *Climate Dynamics, 21*, 573–591. https://doi.org/10.1007/s00382-003-0353-1.

Thornthwaite, C. W., & Mather, J. R. (1955). *The water balance* (p. 104). Centerton: Laboratory of Climatology, Drexel Institute of Technology.

Tolwinski-Ward, S. E., Evans, M. N., Hughes, M. K., et al. (2011). Erratum to: An efficient forward model of the climate controls on interannual variation in tree-ring width. *Climate Dynamics, 36*(11), 2441–2445. https://doi.org/10.1007/s00382-011-1062-9.

Tolwinski-Ward, S. E., Anchukaitis, K. J., & Evans, M. N. (2013). Bayesian parameter estimation and interpretation for an intermediate model of tree-ring width. *Climate of the Past, 9*(4), 1481–1493. https://doi.org/10.5194/cp-9-1481-2013.

Touchan, R., Shishov, V. V., Meko, D. M., et al. (2012). Process based model sheds light on climate sensitivity of Mediterranean tree-ring width. *Biogeosciences, 9*(3), 965–972. https://doi.org/10.5194/bg-9-965-2012.

Tumajer, J., Kašpar, J., Kuželová, H., et al. (2021a). Forward modeling reveals multidecadal trends in cambial kinetics and phenology at treeline. *Frontiers in Plant Science, 12*(32), 613643. https://doi.org/10.3389/fpls.2021.613643.

Tumajer, J., Shishov, V. V., Ilyin, V. A., et al. (2021b). Intra-annual growth dynamics of Mediterranean pines and junipers determines their climatic adaptability. *Agricultural and Forest Meteorology, 311*, 108685. https://doi.org/10.1016/j.agrformet.2021.108685.

Tychkov, I. I., Sviderskaya, I. V., Babushkina, E. A., et al. (2019). How can the parameterization of a process-based model help us understand real tree-ring growth? *Trees, 33*(2), 345–357. https://doi.org/10.1007/s00468-018-1780-2.

Vaganov, E. A. (1990). The tracheidogram method in tree-ring analysis and its application. In E. Cook & L. Kairiukstis (Eds.), *Methods of dendrochronology* (pp. 63–76). Dordrecht: Springer.

Vaganov, E. A., & Shashkin, A. V. (2000). *The growth and structure of annual rings* [in Russian]. Novosibirsk: Nauka Publishing House.

Vaganov, E. A., Hughes, M. K., Kirdyanov, A. V., et al. (1999). Influence of snowfall and melt timing on tree growth in subarctic Eurasia. *Nature, 400*, 149–151. https://doi.org/10.1038/22087.

Vaganov, E. A., Hughes, M. K., & Shashkin, A. V. (2006). *Growth dynamics of conifer tree rings: Images of past and future environments* (p. 357). Berlin, Heidelberg: Springer.

Vaganov, E. A., Anchukaitis, K. J., & Evans, M. N. (2011). How well understood are the processes that create dendroclimatic records? A mechanistic model of the climatic control on conifer tree-ring growth dynamics. In M. K. Hughes, T. W. Swetnam, & H. F. Diaz (Eds.), *Dendroclimatology: Progress and prospects* (pp. 37–75). Dordrecht: Springer.

von Arx, G., Crivellaro, A., Prendin, A. L., et al. (2016). Quantitative wood anatomy–Practical guidelines. *Frontiers in Plant Science, 7*, 781. https://doi.org/10.3389/fpls.2016.00781.

Wilmking, M., Juday, G. P., Barber, V. A., et al. (2004). Recent climate warming forces contrasting growth responses of white spruce at treeline in Alaska through temperature thresholds. *Global Change Biology, 10*, 1724–1736. https://doi.org/10.1111/j.1365-2486.2004.00826.x.

Woodward, F. I., & Williams, B. G. (1987). Climate and plant distribution at global and local scales. *Vegetatio, 69*, 189–197. https://doi.org/10.1007/BF00038700.

Yang, B., He, M., Shishov, V., et al. (2017). New perspective on spring vegetation phenology and global climate change based on Tibetan Plateau tree-ring data. *Proceedings of the National Academy of Sciences of the United States of America, 114*(27), 6966–6971. https://doi.org/10.1073/pnas.1616608114.

Zhirnova, D. F., Babushkina, E. A., Belokopytova, L. V., et al. (2020). To which side are the scales swinging? Growth stability of Siberian larch under permanent moisture deficit with periodic droughts. *Forest Ecology and Management, 459*, 117841. https://doi.org/10.1016/j.foreco.2019.117841.

Zuur, A., Ieno, E., Walker, N., et al. (2009). *Mixed effects models and extensions in ecology with R.* New York: Springer.

Chapter 12
Functional Traits of Boreal Species and Adaptation to Local Conditions

Marcin Klisz, Debojyoti Chakraborty, Branislav Cvjetković,
Michael Grabner, Anna Lintunen, Konrad Mayer, Jan-Peter George,
and Sergio Rossi

Abstract Species continuity under the harsh climatic conditions of the boreal forest requires trees to ensure the functioning of two main life processes, namely growth and reproduction. However, species survival becomes a challenge when environmental conditions become unstable and reach the taxa's ecological tolerance limit. Survival in an unstable environment is possible through the concurring processes of phenotypic plasticity and local adaptation; each process has its advantages and shortcomings. Local adaptation allows attaining the best possible fitness under conditions of limited gene flow and strong directional selection, leading to specific adaptations to the local environment; however, there is a risk of maladaptation when conditions suddenly change. In turn, phenotypic plasticity provides trees an advantage when weather events change rapidly and enables a response expressed by the production of different phenotypes by the same genotype. However, this process is expensive in terms of costs in maintenance and causes developmental instability within the individual. Boreal trees utilize both processes as reflected in variations in their functional traits within the same species. In this chapter, we address the main life processes, presenting the variability of functional traits of flowering and seed production, xylem conductivity, bud and cambium phenology, as well as transpiration and photosynthesis, as a consequence of the interaction of genotype and environment. We describe the practical consequences of a variation in functional traits, as expressed in chemical and mechanical wood properties. Finally, we outline applications and perspectives for managing boreal forests in a context of heterogeneous and changing environmental conditions.

M. Klisz (✉)
Dendrolab IBL, Department of Silviculture and Genetics of Forest Trees, Forest Research Institute, ul. Braci Leśnej nr 3, 05-090 Raszyn, Mazovia, Poland
e-mail: m.klisz@ibles.waw.pl

D. Chakraborty
Department of Forest Growth, Silviculture and Genetics, Austrian Research Centre for Forests (BFW), Seckendorff-Gudent-Weg 8, 1131 Vienna, Austria
e-mail: debojyoti.chakraborty@bfw.gv.at

© The Author(s) 2023
M. M. Girona et al. (eds.), *Boreal Forests in the Face of Climate Change*,
Advances in Global Change Research 74,
https://doi.org/10.1007/978-3-031-15988-6_12

12.1 Introduction

12.1.1 Populations and Local Adaptations

To understand how functional traits vary across ecotypes, we first define an ecotype, explain how the concept has evolved, and identify the main drivers of its variation. The terms *ecotype* and *provenance* are often used in parallel, especially in forest sciences, because the phenotypic variation within tree species has been used in forestry for centuries. The Swedish researcher Olof Langlet, who focused primarily on forestry, was one of the early pioneers in provenance research, although there had been considerable work conducted since the mid-eighteenth century in Europe (Langlet, 1971a). Since most of this early work was undertaken in France, it is not surprising that the word provenance is the French term meaning "origin." We can better understand the term origin (and provenance) from a simple experiment by the French Navy in the middle of the eighteenth century. As *Inspecteur Général de la Marine* (Inspector General of the Navy), Henri Louis Duhamel du Monceau (b. 1700–d. 1782) became responsible for improving the quality of ship masts for the French Navy. To find more suitable pylon material, he collected seeds from pines growing in various locations (Scotland, Latvia, and central Europe) and planted them in the same site in France, the precursor of a common garden. At that time, it was thought that the seeds belonged to individuals from different species because of their miscellaneous appearances. However, a nephew of Duhamel du Monceau continued the work of

B. Cvjetković
Faculty of Forestry, University of Banja Luka, Vojvode Stepe Stepanovića Blvd 75A, 78000 Banja Luka, Bosnia and Herzegovina
e-mail: branislav.cvjetkovic@sf.unibl.org

M. Grabner · K. Mayer
University of Natural Resources and Life Sciences, Vienna, Konrad Lorenz Strasse 24, 3430 Tulln, Austria
e-mail: michael.grabner@boku.ac.at

K. Mayer
e-mail: konrad.mayer@gmail.com

A. Lintunen
Institute for Atmospheric and Earth System Research, University of Helsinki, P.O. Box 64, 00014 Helsinki, Finland
e-mail: anna.lintunen@helsinki.fi

J.-P. George
Tartu Observatory, Faculty of Science and Technology, University of Tartu, Observatooriumi 1, Nõo parish, 61602 Tõravere, Estonia
e-mail: jan.peter.george@ut.ee

S. Rossi
Département des Sciences fondamentales, Université du Québec à Chicoutimi, 555 boul. de l'université, Chicoutimi, QC G7H 2B1, Canada
e-mail: sergio.rossi@uqac.ca

his uncle and reported, 40 years later, that they were rather varieties of one species, namely Scots pine (*Pinus sylvestris* L.) (Langlet, 1971b). This early and ground-breaking experiment served as the starting point for many other investigations across the globe that led to significant improvements in our understanding of how genetics and the environment shape the geographic variation of plant phenotypes.

There are two major interacting forces, genetic and environmental, involved in creating the geographic variation in tree traits. Genetics define what is intrinsic to the tree and create heritable variations that can be passed to successive generations. The unit of observation is usually defined as the gene, which is the carrier of all essential information at the DNA level, often resulting in an expressed amino acid and subsequent complex molecular organic compounds (Pearson, 2006). The environment represents all abiotic and biotic agents conditioning a tree during its lifetime. Genecology, the interaction between genetics and environment, is the research of provenance variation or, more precisely, the discipline analyzing the environmental effects on the spatiotemporal variation of genotypes and phenotypes. Although our definition differs slightly from that of Langlet (1971b), we aim to provide a handy and intuitive description of this term for applied ecologists rather than provide an exhaustive and incontestable theoretical definition.

Going back to the abovementioned French experiment from 1745, we raise the question as to what factor caused the Scots pine trees to differ greatly in form despite growing in the same environment in Duhamel's French garden. The answer is simple: the gene expression under specific environmental conditions. The genetic background of the transplanted seeds still matched their original locations and represented an assembly of well-adapted genes developed under specific environmental pressures. Thus, the trees grew as per the environmental conditions at their site of origin. Thus, variation at the trait level for a given species, i.e., growth or frost resistance, usually coincides with the variation in allele frequencies at the DNA level among provenances. The relationship between changes in mean trait values such as bud flush or height growth across environmental gradients, e.g., per degree of mean annual temperature or degree of latitude, is named a cline. The steepness and general shape of these provenance clines can vary considerably among species, traits, and geographical scales (Alberto et al., 2013). In an extensive literature review, Alberto et al. (2013) analyzed genecological clines for several tree species, including those of the boreal forest. They described evident clines between the timing of bud burst and origin, i.e., the latitude of seed sources, in Norway spruce (*Picea abies* Karst.) and sessile oak; however, they found weak clines among white spruce (*Picea glauca* (Moench) Voss) provenances. This demonstrates that tree species can be differentiated unequally, a factor that must be accounted for when intraspecific variation is used in seed transfer. Important genecological features associated with boreal tree populations are frost hardiness, timings of bud set, and early survival of seedlings, as growing conditions are characterized by harsh winters and short growing seasons. In Scots pine, for example, these traits show a strong adaptive divergence even at moderate geographical scales. Thus, the traits of Scots pine populations from northern Finland are easily distinguished from populations originating from southern Finland (Savolainen et al., 2004).

The field of genecology and provenance research has recently started to attract scientists because of its putative usability in adaptive forest management (Aitken et al., 2008). Consequently, provenance research combined with improvements in molecular genetics and climate modeling has already resulted in revised seed source zonations and novel recommendations for forest genetic resources when climate is considered.

12.1.2 Inter- and Intrapopulation Variability

We can define historical gene flow among populations as being related to the exchange of genetic information—through sexual reproduction—during the postglacial recolonization of newly accessible and suitable habitats. The traits of populations originating from glacial refugia, e.g., the Iberian Peninsula and the Carpathian Mountains, often differ from those of populations at the leading edge of the northern boreal zone when analyzed using neutral genetic markers. For instance, Scots pine (*Pinus sylvestris* L.) populations from southern Spain and Norway spruce (*Picea abies* (L.) H. Karst.) populations from the Carpathian Mountains differ in their genetic structure from Scandinavian and Baltic populations of both taxa (Sinclair et al., 1999; Tollefsrud et al., 2008). To explain this diversity pattern, we should consider the tree populations as islands that frequently exchange a certain number of migrants with each other. The rate of exchange, i.e., the number of exchanged migrants or gene copies per generation, consequently determines the similarity of populations. This can be expressed by using common F-statistics, such as F_{ST} (Wright, 1943). The maximum value of F_{ST} is 1 and indicates no exchange of genetic information over a sufficiently long period among populations. An F_{ST} of 0 raises questions about whether we are observing different populations because they exchange so much genetic information that divergence in space or time would be extremely unlikely. For Scots pine and Norway spruce, F_{ST} reaches 0.8 and 0.6, respectively (Sinclair et al., 1999; Tollefsrud et al., 2008), suggesting that most neutral variability is indeed found among, rather than within, populations.

Another common phenomenon leading to variability in space is the stochastic loss of gene variants through founder events during colonization. This random loss of alleles in space and time, also known as genetic drift (Lande, 1976), is generally portrayed as a bottle filled with differently colored beads representing alleles or genotypes. Each time a subsample of these beads is removed through the bottleneck, i.e., the founder event, the relative proportions of colors can change unpredictably. Colors that initially appeared at lower probabilities, i.e., rare alleles, are more likely to disappear after such a bottleneck. Hence, it is thought that boreal tree populations are often genetically less diverse than southern populations because they have experienced more removals before their arrival to the boreal regions—the *southern richness but northern purity* paradigm of Hewitt (2000). A closer look at Norway spruce and Scots pine reveals, however, that this pattern is only partly true, as both species appear to have maintained their intrapopulation variability in the boreal portion of their distribution despite the extensive distance from glacial refugia (Savolainen et al., 2011; Tollefsrud et al., 2008).

Historical gene flow and genetic drift are neutral processes, and their contribution to the cold-climate adaptations of Norway spruce and Scots pine is probably marginal. In Scots pine, adaptation to cold environments in the boreal north has likely been happening by directional selection from standing genetic variation within populations (Savolainen et al., 2011), which is a necessary process to consider in the context of adaptation to climate change. Nevertheless, neutral processes, such as gene flow and drift, harbor critical implications for adaptive forest management, e.g., defining conservation goals for rear-edge tree populations, because the adaptation to novel environmental conditions, such as climate change, requires certain thresholds of minimum genetic diversity to allow the populations to adapt (Fady et al., 2016). This awareness has also led to legal frameworks, e.g., national forest policies, that recommend a minimum number of mother trees for seed harvesting or establishing clonal seed orchards. Rather than the result of adaptive processes, gene flow and drift have partly shaped the current diversity patterns, providing the raw material for evolution and selection.

The environment remains an important factor affecting variation within and among populations. Contrary to gene flow and drift, selection is targeted, shifting mean phenotypic values toward specific directions, the phenotypic optimum. One of the fundamentals of natural selection is based on the idea of Darwin (2003) that individuals exhibiting a phenotype conferring greater survival and reproduction in a particular environment contribute to the next generation with more offspring (and hence more gene copies). We can identify several adaptive phenotypes in boreal species, such as greater cold hardiness or specific bud set, as observed for Scots pine (Savolainen et al., 2004) and Norway spruce (Oleksyn et al., 1998) populations. The term adaptive indicates that a certain amount of the phenotypic variation is heritable and provides a fitness advantage to the individual. A commonly used measure for the degree of heritable variation in a trait is the narrow-sense heritability (h^2). Only traits showing significant heritability within populations can improve breeding to obtain more resilient genotypes (Louzada & Fonseca, 2002). As for the abovementioned F_{ST}, we define a measure partitioning quantitative or adaptive variability within and among populations by including heritability. This measure is named Q_{ST} (Q represents quantitative and implies that the variability in the studied trait is adaptive). It relates the total heritable variation to the heritable variation found among populations (Spitze, 1993). Where an F_{ST} of 0.9 suggests a limited exchange of genetic information between populations because of limited gene flow, e.g., a mountain cascade, a similar Q_{ST} suggests that populations experienced spatially varying selection pressures and hence differ markedly in their average values for a certain trait, e.g., cold hardiness, between southern and northern provenances. Whereas heritability can be used as a broad surrogate for breeding success, Q_{ST} may inform on the success of applying assisted gene flow in adaptive forest management. Only populations with a significant Q_{ST} for a specific trait can assure adaptation success by transferring putatively preadapted genotypes. Traits related to adapting to cold environments usually show strong quantitative trait differentiation among populations, underlining their evolutionary importance in species with boreal distributions (Savolainen, 1996).

Inter- and intrapopulation variability can also exist for the ability of phenotypes to adjust rapidly to a changing environment, referred to as *variation in phenotypic plasticity* or *genotype-by-environment interaction*. Variation in plasticity is also the result of a targeted process, although not necessarily an adaptive one (Merilä & Hendry, 2014). Variation in plasticity is much more challenging to assess, particularly for trees, and therefore its contribution to climatic adaptation remains largely uncertain.

In this section, we have seen that both inter- and intrapopulation variability in boreal tree species are important features for the current diversity from neutral or adaptive points of view. However, neutral and adaptive processes are not mutually exclusive but simultaneously affect variability, sometimes making it difficult to disentangle the two forces to estimate the evolutionary outcome of a species. Figure 12.1 provides a simple framework to understand the two forces and their implications for the adaptive management of boreal tree populations.

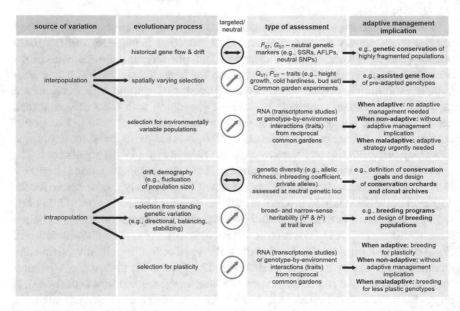

Fig. 12.1 Overview of inter- and intrapopulation diversity measures (*pink column*), the underlying evolutionary processes (*green column*), whether the evolutionary process in neutral (*horizontal black arrow*) or targeted (*angled green arrow*) types of assessment (*yellow column*), and some implications for adaptive forest management (*blue column*)

12.2 Functional Traits

12.2.1 Ecophysiological Responses to Short Growing Seasons and Harsh Winters

In the previous section, we discussed the interaction between genetics and the environment. In this section, we focus on how boreal trees respond to the environment during their lifespans and how these functional and structural responses are coordinated. This area of research is called tree ecophysiology. In more detail, we discuss how boreal trees respond to an environment characterized by short growing seasons and harsh winters in terms of photosynthetic production, growth, wood anatomy, water transport, and frost tolerance. Understanding these responses is important because boreal forests are typically temperature limited and are thus considered especially sensitive to climate warming.

Compared with trees in most temperate and tropical ecosystems, boreal trees are small relative to their age and have a relatively low net primary production of about 270–540 g C.m^{-2}.yr^{-1} (Kolari et al., 2009; Luyssaert et al., 2007). In boreal forests, 80%–95% of gross primary production is used for ecosystem respiration, and soil processes are responsible for a large share of total ecosystem respiration (Kolari et al., 2009; Luyssaert et al., 2007). The annual solar radiation and temperature cycle regulate the photosynthetic processes and timing of tree growth, i.e., the time between the spring thaw and the autumn freeze determines the amount of annual tree growth (Jarvis & Linder, 2000).

The light-saturated rate of photosynthesis is affected by air temperature because enzymatic processes involved in photosynthesis are temperature driven (Farquhar et al., 1980). In addition, the physiological state of the photosynthetic machinery follows changes in temperature with a time lag of a few days (Kolari et al., 2014). There are, however, other indirect responses of photosynthesis to temperature. When plants take in CO_2 for photosynthesis through stomatal openings in leaves, they lose water to the atmosphere. Xylem transport and water uptake by roots must provide sufficient water supply for transpiration because otherwise, stomata will close to prevent excessive dehydration of the plant. Hölttä et al. (2017) developed a whole tree–level theoretical framework to explain stomatal behavior and presented a model linking carbon source (leaf gas exchange), carbon sink (sugar utilization), and soil water uptake relations through xylem and phloem transport. The model simulations showed that when sink strength decreases with lower ambient temperatures—as per the well-known temperature dependence of plant respiration—this leads to a higher leaf sugar concentration and further limits photosynthesis (Hölttä et al., 2017). In addition to air temperature, soil temperature is also critical for photosynthesis. Stomatal conductance and photosynthesis decrease sharply when the soil temperature is less than approximately 8 °C in boreal conifers (e.g., Mellander et al., 2004). Under these conditions, insufficient water is available for trees because the cold soil limits the capacity of trees to extract and transport water from the soil to the canopy, thereby reducing canopy conductance and photosynthesis (Lintunen et al., 2020).

Although the boreal climate is typically rather moist, soil moisture could become a critical factor for boreal trees in the future. Reich et al. (2018), in a three-year open-air warming experiment with 11 temperate and boreal tree species, showed that an increase of 3.4 °C increased light-saturated net photosynthesis in moist soils only. Therefore, low soil moisture reduces or even reverses the potential benefits of climate warming for photosynthesis in boreal environments during drought and regularly occurring modestly dry periods.

How then are growth and photosynthesis linked to wood anatomy? Trees scale leaf and xylem areas to couple transpiration and photosynthesis with xylem water transport, and some species are known to acclimate their leaf to xylem area ratio in response to climatic conditions. Petit et al. (2018) analyzed climate effects on the scaling of leaf and xylem areas in branches of Scots pine, Norway spruce, silver birch (*Betula pendula* Roth), and common aspen (*Populus tremula* L.) sampled across a continental transect in Europe. They found that the scaling of cumulative leaf and xylem areas axially from the branch apex down along the main branch axis is constant irrespective of species across Europe. Trees in the cold boreal region and dry southern Europe keep their functional balance between water transport and transpiration by maintaining their biomass allocation to leaf and xylem areas according to their growth rate. Specifically, allocation to leaf area is relatively higher for reduced growth rates because older growth rings are kept functional to maintain xylem conductance.

Cold temperatures in the boreal region limit photosynthesis and tree growth. Moreover, coping with cold winter periods is crucial for trees to survive at these latitudes, as these trees experience frequent freeze–thaw events during the winter. The freezing of xylem sap has several consequences for trees. The first is freeze–thaw-induced embolism. Gases dissolved in the xylem sap form bubbles during freezing, and these bubbles are at risk of expanding and creating embolisms upon thawing under tension (Sperry & Sullivan, 1992). Embolism prevents water transport and decreases xylem conductivity and is thus harmful to trees, although some tree species can restore the hydraulic system in spring (Mayr et al., 2020). The second important consequence of freezing for trees is frost-induced cellular damage, i.e., extreme winter dehydration or cell membrane rupture caused by ice crystal formation within the living cells (Thomashow, 1998). Living cells either avoid freezing by deep supercooling or tolerate subzero conditions by extracellular freezing. Extracellular freezing is visible, e.g., as shrinkage of woody tissues (Lintunen et al., 2017), and is typical in boreal species that experience temperatures colder than -40 °C (Fujikawa & Kuroda, 2000). In extracellular freezing, water is withdrawn from the cells into the apoplast, increasing the intracellular solute concentration and protecting the cell sap from freezing.

Winter acclimation is essential for frost tolerance. Winter acclimation is how plants prepare themselves for winter conditions and become hardy. During the winter acclimation stage, protective downregulation of photosynthetic light reactions in boreal Scots pine and Norway spruce are stronger in spring than in autumn (Linkosalo et al., 2014). Relative to spruce, pine down-regulates photosynthetic light reactions earlier in autumn and reactivates them later in spring. This pattern suggests that spruce benefits more than pine from warm spring temperatures by increasing photosynthetic production during warm spells; however, spruce is more vulnerable to frost damage if the temperature cools markedly after a warm spell (Linkosalo et al., 2014).

12.2.2 Flowering and Seed Production

The reproduction process of tree species is conditioned by cyclical timings of flowering and seed production, i.e., masting. The published literature provides many definitions of masting, including the synchronous production of seeds over long intervals by a plant population and the episodic synchronous production of large seed crops by plant populations (Janzen, 1976; Kelly, 1994). In general, masting occurs at the population level when abundant seed production is synchronized among individuals (Kelly, 1994). Masting is variable between years and synchronous among populations (Kelly, 1994; Poncet et al., 2009). The main evolutionary advantage of masting for wind-pollinated species, e.g., boreal taxa, is an improved pollination efficacy (Moreira et al., 2014; Nilsson & Wastljung, 1987). There are limited data regarding the masting frequency of boreal species (Ascoli et al., 2017). The masting pattern may vary from one to two years for eastern hemlock (*Tsuga canadensis* (L.) Carrière) (Ruth, 1974), three years for jack pine (*Pinus banksiana* Lamb.) (OECD 2010), two to six years for black spruce (*Picea mariana* (Mill.) BSP) (Viereck & Johnston, 1990) and seven to eight years for Douglas fir (*Pseudotsuga menziesii* (Mirb.) Franco.) (Stein & Owston, 2002). Norway spruce and Sitka spruce (*Picea sitchensis* (Bong.) Carrière) produce cones sporadically, with four years between mast crops. During this nonmasting period, cone density is very low or cone production is absent (Broome et al., 2007). In Sitka spruce populations in Norway, mast years occur at three- to five-year intervals in western Norway, whereas in the northern regions, mast years occur on a far more irregular basis (Nygaard & Øyen, 2017). Multiple hypotheses have been proposed to explain masting patterns (Pearse et al., 2016). In the following section, we detail the two most common hypotheses explaining masting for boreal species—weather conditions and resource budgets.

Environmental conditions are the drivers of pollen and seed production. Geburek et al. (2012) investigated pollen production of some boreal species (*Picea, Larix, Abies, Betula, Populus,* and *Alnus*) and distinguished masting and nonmasting pollen producers. Trees of the first type produce high amounts of pollen only before a masting event; they only mast when the pollination period is synchronized with favorable weather conditions. Nonmasting species produce pollen every year. Some boreal North American and European species show intraspecific spatiotemporal synchroneity in masting over vast areas (Koenig & Knops, 1998), with synchronous masting for the same boreal species occurring over distances of 500 to 1,000 km between sites (Gallego Zamorano et al., 2016; Koenig & Knops, 1998).

A relationship between masting and ecological parameters permits predicting cone production, and multiple authors have investigated potential correlations between specific environmental conditions and masting years. Gallego Zamorano et al. (2016) studied four boreal species in Finland—silver birch, downy birch (*Betula pubescens*

Ehrh.), Norway spruce, and rowanberry (*Sorbus aucuparia* L.). Flowering was affected positively by May temperatures but negatively by previous-year temperatures. The spatial synchrony covered up to 1,000 km owing to synchronous weather conditions. The influence of larger-scaled events caused by global climate changes can also affect masting; for example, the El Niño-Southern Oscillation (ENSO) and Atlantic Multidecadal Oscillation (AMO) influence masting in species such as white spruce (Ascoli et al., 2019).

There is evidence that masting reduces the radial growth of trees. Selås et al. (2002) observed a negative correlation between tree-ring width and seed masting for Norway spruce. Hacket-Pain et al. (2019) found the same results but only for *super producers*, i.e., trees having exceptional masting, and Ascoli et al. (2019) observed a positive correlation between drought events and masting in white spruce.

Wildfire strongly influences masting patterns (Charron & Greene, 2002). The environmental prediction hypothesis holds that some wood boreal species produce abundant seed after the first year of wildfires, which ensures a high survival rate of young trees (Peters et al., 2005; Wirth et al., 2008). This hypothesis applies primarily to plant species in fire-prone regions, whereby woody plants produce serotinous fruits, which release their seeds after wildfires (Kelly, 1994; Pearse et al., 2016). The projected increased occurrence, amplitude, and severity of fires across the boreal ecosystems will likely increase seed production in the boreal zone (Shvidenko & Schepaschenko, 2013).

Other hypotheses of masting behavior are based on the physiological aspects of plant life. The resource budget model illustrates that masting requires more resources than plants can secure over a single year, hence the resulting periodicity in fructification. There is little empirical support for the commonly stated hypothesis that plants store carbohydrates over several years to expend in a high-seed year. Plants can allocate carbohydrates from growth in high-seed years, and seed crops are more probably limited by nitrogen or phosphorus (Pearse et al., 2016). The source depletion hypothesis describes the occurrence of masting years through the accumulation and storage of resources. Masting occurs once a specific threshold is reached. This hypothesis agrees with some of the abovementioned results, which also support spatial synchronism through environmental factors caused by large-scale weather conditions. According to this hypothesis, mast seeding leads to resource depletion, and the threshold must be reached again through resource accumulation to produce the next masting.

12.2.3   Wood Properties

The marked variability in wood properties is the complex result of genetics, stand-level conditions, and short-term weather events. Wood properties, such as tracheid or fiber length, strength, stiffness, and density, are not only important for the tree as a living organism but are also essential for the use of wood as a material.

Tracheid, or fiber, length varies within a single tree ring, between juvenile and adult wood, and among individual trees and species. It is under strong genetic control but can also be influenced by silvicultural practices. Tracheid length increases from pith to bark (Bannan, 1967), and changes in growth rates (and the available time for growth) cause intra-annual differences that result in longer tracheid lengths toward latewood. Tracheid length influences various hydraulic parameters of living trees (Choat et al., 2008); it also has technological implications, e.g., paper quality (Wimmer et al., 2002). More important, however, is the microfibrillar angle, which is associated closely with tracheid length (Donaldson, 2008). Low microfibrillar angles are associated with high wood strength and stiffness. Environmental factors influence the produced angles, although the angles do show significant heritability (Donaldson, 2008).

Adequate mechanical wood characteristics are needed to support the tree architecture and the use of wood products. Whereas there are many different measures, such as hardness and tensile, bending, compression, and impact strength, these parameters are all strongly tied to wood density. In addition to ring width, wood density is the most commonly studied parameter of wood. Wood density measures the total amount of cell wall material per volume (Wimmer, 1995). A higher density in softwood results mainly from a greater presence of latewood within the tree ring (Kort, 1993), which can vary in Scots pine, for example, from 19 to 50% (Wimmer, 1995). Ring width, which affects the latewood percentage, strongly affects wood density, with the higher latewood proportions observed in smaller tree rings (Rathgeber, 2017). There are, however, exceptions to this general assumption. Rossi et al. (2015) found a higher density, associated with a higher latewood percentage and higher values for mechanical traits, in the wider rings at a lower latitude site of boreal black spruce. Under the shorter vegetation periods at more northern latitudes, cell wall deposition and latewood formation—two processes related to temperature—are reduced (Cuny & Rathgeber, 2016). Therefore, lower temperatures at higher latitudes may reduce carbon allocation in latewood, leading to a lower wood density, even when small tree rings are produced (Rossi et al., 2015).

Usually, ring width decreases with age, whereas the latewood percentage and wood density of softwood increases. The effect of climate on earlywood density is of particular interest (Grabner et al., 2010) owing to its close link with plant hydraulics. A potential consequence of low earlywood density is that negative water pressure exceeds the fracture limits of the wood, and radial cracks or stem cracks can therefore occur (Grabner & Wimmer, 2006). Unfortunately, information regarding earlywood density, using X-ray densitometry, is rare in dendrochronological studies. Maximum density at high-altitude or high-latitude sites is mainly affected by summer temperature, the period when the wood is formed.

In addition to the environmental influences, wood formation and, therefore, wood quality are under strong genetic control (Cuny & Rathgeber, 2016). The analysis of different species of *Abies* (and particularly *Abies alba*) provenances growing in eastern Austria demonstrated the influence of species and provenance on wood-quality traits (George et al., 2015). The observed variation of wood traits and their responses to drought revealed that the genus (*Abies*) explains between 10 and 20% of the variance in wood density but only a negligible amount of the ring-width variance. In contrast, provenances (of *A. alba*) are responsible for 10–15% of the variation in ring width. Nonetheless, provenance explains only a nonsignificant proportion of the variance of ring density (George et al., 2015).

Wood quality is also influenced by ecological parameters and forest management decisions (Jaakkola et al., 2006). Pamerleau-Couture et al. (2019) found differences in wood properties dependent on stand structure and forest practices, e.g., partial cutting, in the boreal forest (Montoro Girona et al., 2016).

12.2.4 *Anomalies in Xylem Structure*

In boreal climates, abnormally low temperatures may damage the tissues surrounding the cambium, disrupting cell division and differentiation; this damage can manifest itself as deformed and collapsed tracheid or traumatic parenchyma cells. When frost events occur in the middle or at the end of the growing season, before the ending of lignification, incompletely lignified earlywood or latewood tracheids are formed to produce what is termed light rings or frost rings (Gindl et al., 2000). Anomalies in xylem structure also appear when trees are subjected to high spring and early summer temperatures when the latewood-like cells in earlywood are formed. Moreover, rain-fall following drought periods may stimulate the formation of earlywood-like cells in the latewood ring zone. These xylem anomalies are classified as intra-annual density fluctuations (IADF) and can be formed in both temperate and boreal species (George et al., 2019; Klisz et al., 2019).

Although the main drivers of anomalies in xylem structure are extreme climatic events, there is evidence that the predisposition of trees to the formation of IADF, frost rings, and light rings is, to some extent, genetically controlled and can lead to the adaptation or maladaptation of genotypes to drought and frost (Battipaglia et al., 2016; Birgas & Colombo, 2001). Moreover, intraspecific variation in growth reaction to climatic anomalies has been demonstrated for several gymnosperms growing in boreal and temperate climates, e.g., lodgepole pine (*Pinus contorta* Dougl. ex Loud.), Douglas fir, European larch (*Larix decidua* Mill.), silver fir, and Norway spruce (George et al., 2019; Klisz et al., 2016; Montwé et al., 2018). For example, boreal provenances of Douglas fir originating from cold-climate areas in Canada had lower IADF frequencies in latewood than temperate provenances when trees had been growing in a warm and dry common garden where drought occurred more frequently (George et al., 2019). This pattern for IADF strongly suggests that the formation of IADF has a genetic origin and results from a long-term adaptation of provenances to different climatic conditions. Assessment of IADF in regard to adaptive forest management will probably become more critical in the future, as boreal species are vulnerable to frequent drought periods (Isaac-Renton & Montwé 2018).

From the assumption that genetic and environmental components control xylogenesis, we can conclude that abnormal xylem structures, namely frost and light rings, result from gene expression under extreme climatic events. Confirmation of this hypothesis involved studies of the cold adaptation in 20 lodgepole pine provenances growing in a common garden located in southern interior British Columbia, Canada (Montwé et al., 2018). Provenances adapted to colder environments with larger temperature amplitudes and shorter growing seasons show less susceptibility to frost damage preceding meristem activity; however, they are more susceptible to frost damage at the beginning of the growing season (Fig. 12.2). In general, a strong geographic cline for frost damage and incompletely lignified tracheids is noticeable, suggesting the importance of considering cold adaptation when long-distance seed transfer is introduced to minimize risks in assisted migration (Bansal, 2015).

From a functional point of view, IADF is a consequence of an adaptive strategy to maximize water-use efficiency (earlywood-like cells) or avoid hydraulic failure (latewood-like cells) (Pacheco et al., 2016). Considering the evidence of genetic variations in the morphofunctional adaptation of xylem structure manifested in IADF (George et al., 2019; Klisz et al., 2016, 2019), there are solid premises for incorporating xylem functional traits into assisted migration strategies. Studies on tree species from cold (Norway spruce) or alpine climates (European larch) note a clear geographical trend in IADF formation, which may be even more noticeable in the northern regions under boreal climate conditions (Fig. 12.3a–d). A general latitudinal trend is evident for Norway spruce in the frequency of earlywood-like (IADF E) and latewood-like (IADF L) cell structures, with more frequent IADF L in the southern part of the transect than in the north and an opposite trend for IADF E (Fig. 12.3c, d). This observed gradient confirms the abovementioned findings for the non-native Douglas fir, where boreal provenances had a lower IADF L frequencies compared with warmer provenances from more southern regions (George et al., 2019; Fig. 12.3e, f). A similar pattern can be observed for European larch, although in this case, the incomplete representation of the species distribution prompts greater caution when drawing general conclusions (Fig. 12.3a, b).

Furthermore, a higher frequency of IADF types testifies to a more pronounced genetic determination of IADF frequency, which can be clearly seen in Norway spruce and European larch growing under marginal conditions (Klisz et al., 2016, 2019); however, this is almost imperceptible under conditions close to the optimal species requirements (George et al., 2019). Given that few observations of geographical clines in IADF formation are available, the variation between populations increases as a function of the south-to-north dimension of the species occurrence. Nevertheless, this hypothesis requires thorough testing under boreal climate conditions through multienvironmental provenance trials.

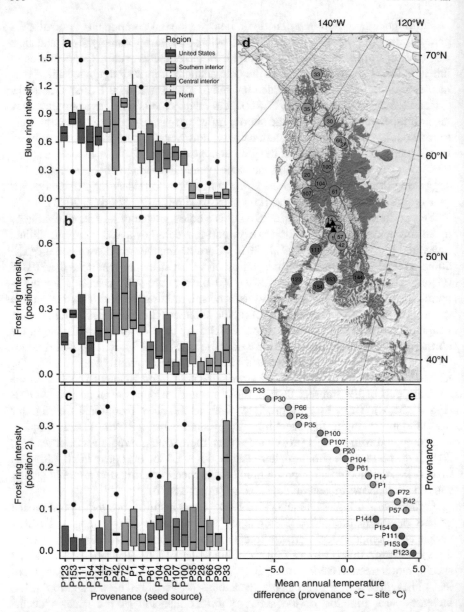

Fig. 12.2 Blue- and frost-ring intensities by provenance. Blue-rings **a** frost-rings position 1 (first cells of earlywood indicative of fall frost damage), **b** frost-rings position 2 (later in the earlywood indicative of spring frost damage), **c** the distribution and medians of light-ring and frost-ring intensities, where provenances are sorted by the mean annual temperature of their source climate (*warmest to the left*, *n* = 117). Provenances are colored according to their region (United States and regions of British Columbia, Canada) and labeled by identification number. (*Column right side*) **d** Location of provenances and regions as well as the range of lodgepole pine (*dark gray shading*) and the location of test sites sampled in this study (*black triangles*); **e** the difference of the provenance source climate to the average climate of the test sites indicates the degree and direction of climate transfers. Modified from Montwé et al. (2018), CC BY license

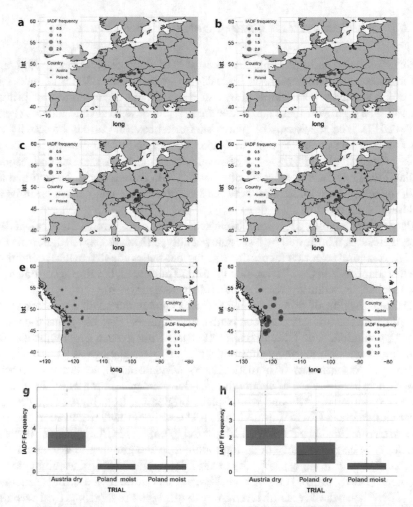

Fig. 12.3 Within-species variation in intra-annual density fluctuations (IADF) for three boreal conifers: **a**, **b** European larch (*Larix decidua*) **c**, **d** Norway spruce (*Picea abies*), and **e**, **f** Douglas fir (*Pseudotsuga. menziesii*). The *left* panels (**a**, **c**, **e**) display earlywood IADF, and the *right* panels (**b**, **d**, **f**) show latewood IADF. The *bottom* panels present overall IADF frequencies (earlywood + latewood) per trial site for **g** European larch and **h** Norway spruce. Douglas fir data were obtained from a provenance trial in eastern Austria (Traismauer); therefore, no trial site is shown in **e** and **f**. The illustrated data were compiled from two different data sets. Data set 1 is from George et al. (2019), Federal Research and Training Centre for Forests (BFW), Vienna, Austria. Data set 2 is from the Forest Research Institute (IBL), Poland. *Black symbols* show the trial sites (*triangles*: IBL; *squares*: BFW). *Red* and *blue circles* indicate the mean IADF frequency for each analyzed provenance. Data from the Austrian trial sites were reanalyzed to provide stabilized IADF frequencies for comparability, following Osborn et al. (1997). Boxplots show the combined IADF frequencies per trial site (earlywood + latewood) for each species. We kindly acknowledge the help of Michael Grabner, Konrad Mayr, and Filipe Campelo for IADF detection in the Austrian data. Modified from Klisz et al., (2016, 2019) CC-BY license and with permission from Elsevier, respectively, and George et al. (2019), CC-BY license

12.2.5   Phenology: The Case of Boreal Black Spruce

In boreal climates, most physiological processes of plants occur during a short lapse of time when the temperature is favorable to growth. The activity of meristems involves a sequence of multiple stages of development or maturation of primary and secondary growth; these stages last from a few days to several months (Perrin et al., 2017). The meristems of plants alternate between periods of activity and rest, following an annual cycle. The beginning and end of the growing season are the key times involving a trade-off between environmental constraints and resource availability. They mark the period of the year when resources can be acquired and used, reflecting an optimization between frost avoidance and carbon assimilation (Allevato et al., 2019).

Phenology, the study of seasonal biological cycles, results from climatically driven gene expression manifested by a specific phenotype (Man & Lu, 2010; Perrin et al., 2017). As natural selection favors those genotypes better adapted to local conditions, specific adaptive traits, which include the timing of growth, develop according to the gradual changes occurring with latitude (Morgenstern, 1969).

The reactivation of primary growth is well known and studied in black spruce both in natural stands and common garden experiments (Fig. 12.4); this trait varies among populations with latitude (Blum, 1988). Having a broad geographic distribution, boreal tree species show substantial phenotypic variations at a regional scale in response to variations in climatic factors (Andalo et al., 2005); this response arises from a combination of phenotypic plasticity and genetic variation. In common garden experiments, black and white spruce originating from higher latitudes have an earlier bud break (Rossi & Isabel, 2017). The earlier growth reactivation observed in northern provenances reflects an adaptation to colder environments. It allows the meristems to be active early in spring to lengthen the growing season as much as possible. Trees from higher latitudes or altitudes require less heat accumulation for bud break (Blum, 1988). Buds of boreal species reactivate in late spring (Antonucci et al., 2015) when the day length is relatively long, nights are short, and frost events are unlikely. At high latitudes, the trees are adapted to colder conditions, and bud development is more rapid than for southern provenances under similar thermal conditions (Körner, 2003). The genotypes originating from colder climates have developed a high metabolic activity and strict developmental regime (Körner, 2012). This evolutionary strategy maximizes safety. An early and quick growth ensures favorable conditions for photosynthesis, especially at higher latitudes where the photoperiod changes markedly between the equinoxes compared with the photoperiod at lower latitudes (Rossi et al., 2006).

For evergreen conifers of cold climates, the dates of growth reactivation of cambium and buds are closely correlated (Antonucci et al., 2015; Rossi et al., 2009). Perrin et al. (2017) observed that black spruce provenances with early bud flush have an early reactivation of cambial activity and xylem cell differentiation. Similar trends

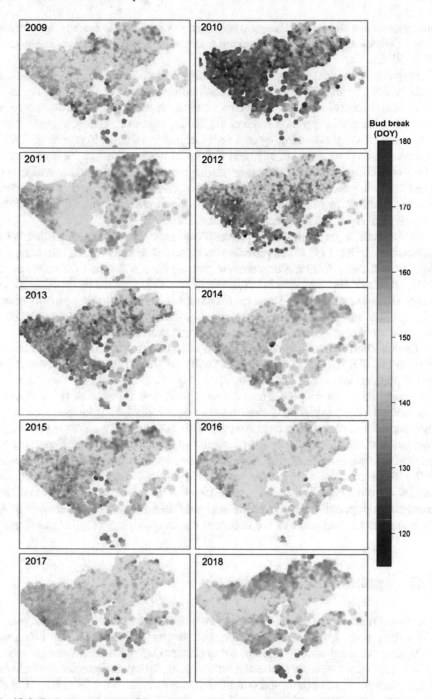

Fig. 12.4 Spring reactivation of the primary meristem represented by bud break dates (day of the year, DOY) across boreal black spruce stands in Québec, Canada. Reproduced from Khare et al. (2019), CC BY license

are reported for the end of the growing season. These results diverge for Norway spruce, where xylem phenology among provenances does not differ (Kalliokoski et al., 2012).

Both environmental and genetic factors affect cambium phenology (Fukatsu & Nakada, 2018), although with different contributions (Perrin et al., 2017). The environment is more critical for spring events, whereas autumnal events and dormancy are controlled mainly by genes (Cooke et al., 2012). Moreover, genetics most control the timing of the final enlarging and wall-thickening of cells. Perrin et al. (2017) observed that provenance, family, and individual tree explained a high proportion of the variability in cambial phenology during the summer and autumn, which indicated that endogenous factors are strongly involved in growth cessation. In conifers, growth cessation is genetically predetermined, and environmental effects on this phenological event are marginal (Cooke et al., 2012).

Differences in bud set among provenances mirror the dynamics of bud break (Johnsen & Seiler, 1996). The provenances classified as early spring also complete their growth early, resulting in a similar growing season duration (Silvestro et al., 2019). Xylem cell production in black spruce is mainly affected by growing season length (Rossi et al., 2014). Thus, no difference in annual tree-ring growth occurs among black spruce provenances growing in a common garden. All trees produce the same amount of xylem, regardless of their respective timing of flushing.

The variation in bud phenology within black spruce provenances is also relevant (Perrin et al., 2017; Rossi & Bousquet, 2014; Silvestro et al., 2019). A wide variability within populations is confirmed in other boreal species, such as silver birch (Rousi & Heinonen, 2007), Scots pine (Hurme et al., 1997), and Douglas fir (Li & Adams, 1993). Each local population is constituted by genetically heterogeneous individuals showing both earlier and later phenology. This heterogeneity allows a part of the population to endure unexpected, unfavorable climatic events, thereby ensuring the survival of some individuals under changing environmental conditions (Hurme et al., 1997). Such a reservoir of genetic variation within populations is highly valuable for the adaptive capacity of local populations. The richness in genetic variation ensures that populations evolve in the next generations according to changes in the environmental conditions that might exceed the limits of physiological plasticity.

12.3   Applications and Perspectives

Functional traits act as templates through which the basic components of a plant's life history (e.g., growth, reproduction, and survival) operate. Hence, functional traits have attracted much attention for understanding the mechanisms governing plant community assembly (HilleRisLambers et al., 2012), the interactions between trait variation and environmental conditions (Matías et al., 2018; Wainwright et al., 2019), and how inter- and intraspecific variation in functional traits connect with those mechanisms maintaining biodiversity (Chesson, 2000, 2012). Such an understanding is crucial for formulating strategies for adapting forests and associated plant communities to climate change (Pérez-Ramos et al., 2012).

12.3.1   Climate Adaptation

Decades of scientific research have provided convincing evidence of observed and likely impacts of human-induced climate change on forest systems. In Europe, the effects of climate warming on forests include changes in forest productivity (Reyer et al., 2014), tree species distribution, and the economic value of forests (Dyderski et al., 2018; Hanewinkel et al., 2013), as well as intensified disturbance regimes (Seidl et al., 2017) and droughts.

Over the millennia, the boreal forest has adapted to short growing seasons, low summer temperatures, and a limited nitrogen supply (Kellomäki & Väisänen, 1997; Mäkipää et al., 1999). Warmer temperatures and increases in nitrogen supply could potentially lead to greater forest growth and productivity in the northern boreal regions, whereas the higher requirements for water in the southern boreal regions may limit such a productivity increase (Briceño-Elizondo et al., 2006; Peng et al., 2011). Moreover, a warmer climate and longer growing period combined with water stress should promote the propagation of insects and other parasites (Wermelinger, 2004). These factors will likely produce novel conditions to which the tree species and populations are poorly adapted. Current forest management must incorporate adaptive strategies that aim to reduce forest vulnerability and enhance forest resistance and resilience (Bolte et al., 2009).

Adaptation is an adjustment of natural or human systems as a response to actual or expected climate changes or their effects. Adaptive capacity is the ability of the system to adjust to novel conditions, take advantage of opportunities, or respond to consequences. Adaptation can be classified as either *autonomous* or *planned* (Schoene & Bernier, 2012). Autonomous adaptations are usually reactive and rely on existing knowledge and technology to respond to changing climate conditions, whereas planned adaptations are anticipatory responses aiming to alter the adaptive capacity of forests (Schoene & Bernier, 2012).

A portfolio of different adaptive management strategies has been discussed at the stand level. These strategies include (1) the conservation of forest structures, (2) active adaptation, and (3) passive adaptation (Bolte et al., 2009; Schoene & Bernier, 2012). The conservation of forest structures aims to maintain the current structural and compositional status of forests. This strategy is believed to increase the vulnerability of the forests to catastrophic disturbance events (Harris et al., 2006; Jandl et al., 2019) but may enable the manager to attain the original management targets. Active adaptation refers to the use of silvicultural measures to alter stand structures and composition to increase the adaptability of the forests to climate change. Such measures may include adjusting the rotating period, changing species composition, and using adapted provenances (Kellomäki et al., 2008). Passive adaptation uses spontaneous adaptation processes in natural succession and species migration (Aitken & Bemmels, 2015). Other adaptive measures have received much attention in recent years and are currently subjects of intense debate; they include assisted migration (Marris, 2008) and assisted gene flow (Aitken & Whitlock, 2013). Whereas assisted migration aims to facilitate the colonization of forest tree species within new habitats having a suitable climate, assisted gene flow aims to manage the translocation of individuals within the current species range to facilitate a rapid adaptation to climate change and improve the long-term prospects of trees and related communities.

For the boreal forest, there exists a variety of adaptive management strategies, such as promoting species mixtures, reducing the rotation length of current stands, planting alternative genotypes or new species in anticipation of future climate, minimizing the fragmentation of habitat, and maintaining connectivity (Gauthier et al., 2014). In particular, the boreal forest is currently experiencing new disturbance regimes, including stand-replacing fires and the outbreak of herbivorous insects, e.g., the mountain pine beetle (*Dendroctonus ponderosae* Hopkins) and the spruce bark beetle (*Dendroctonus rufipennis* Kirby) (Bernier et al., 2016). The conversion of vulnerable stands from coniferous to mixed stands (with broadleaf species) has been suggested as a major adaptive management option to respond to such disturbance agents (Astrup et al., 2018). Forest management solutions focused only on wood volume and productivity risk failure in a rapidly changing climate because they ignore the trade-offs between productivity and traits such as cold tolerance and drought tolerance. Therefore, forest managers should resort to diverse strategies of adaptive management accounting for diverse functional traits and their trade-offs (Park et al., 2014).

It is increasingly understood that climate change adaptation is intricately related to forest sustainability and the principles found within the Montreal Process (Ogden & Innes, 2007) and now the global sustainability goals (Hazarika & Jandl, 2019). Effective adaptation of the forest management system will revolve around including risk management in planning processes, selecting robust and diversified adaptation actions, and adopting an adaptive management framework. Monitoring is always regarded as an action that is central to implementing adaptive forest management (Gauthier et al., 2014).

12.3.2 Assisted Migration

Forest tree populations are likely to respond to climate change in three possible ways: (1) they migrate to track their ecological niche; (2) they adapt to the new conditions in their current locations; and (3) they go locally extinct (Aitken et al., 2008). The evidence suggests that tree species have undergone range shifts, migration, and extinction during past glacial and postglacial periods. Such changes continue today with the warming-related poleward and altitudinal migration of tree species and populations (Dyderski et al., 2018).

Natural migration over long distances is a slow process and often requires several generations and centuries for long-lived trees. Migration rates during the postglacial age are estimated at 100–500 m · y^{-1} (Williams & Dumroese, 2013). Ongoing climate change is expected to occur at a much faster rate, requiring tree populations to migrate faster than tree migration during the glacial period to remain within the species' environmental envelope. According to Tchebakova et al. (2006), some boreal populations of Scots pine would need to move 700–1,500 km north to track the climate projections for 2100. Geographic barriers and habitat fragmentation pose additional challenges to the intrinsic ability of trees to migrate, making some species vulnerable to extinction (Sáenz-Romero et al., 2012).

Forest trees have evolved at the species and population levels to adapt to the local environment in which they grow (Aitken & Whitlock, 2013; Kreyling et al., 2014). Such local adaptations lead to genetically diverse populations, with traits that enable these taxa to adapt to their local environment's biotic and abiotic factors (e.g., growing season and outbreaks of fire and insects). Examples of such adaptive traits include the timing and rate of growth, resistance to frost damage or drought stress, masting patterns, and dispersal distances and timings. Climate change will disrupt the link between climate and the local adaptation of forest tree populations, thereby creating physiological stresses that can lead to mismatches between the population and climate, known as the adaptation lag (Aitken & Whitlock, 2013). Thus, climate change will produce novel conditions to which the tree populations may not adapt. Phenotypic plasticity or the ability of the plant to respond to environmental change may mitigate the impacts of this decoupling to a certain extent, although it may be ineffective in cases of stress induced by extreme environmental events (Mátyás et al., 2010; Neuner et al., 2015).

Because of the limited migration and slow adaptation rate of trees, human-facilitated realignment will be required to match populations to the environment to which the trees are adapted (Aitken et al., 2008; Pedlar et al., 2012; Williams & Dumroese, 2013). Such facilitated movement is commonly known as assisted migration, assisted colonization, assisted relocation, or facilitated migration (Fig. 12.5). In particular, assisted population migration (assisted genetic migration or assisted gene flow) refers to moving seed sources or populations to new locations within the historical species range. Assisted range expansion refers to moving seed sources or populations from their current range to suitable areas beyond the historical species range, facilitating or simulating a natural dispersal. Assisted species migration (or assisted long-distance migration) moves seed sources or populations to a location far beyond the historical species range, beyond the natural ability of dispersion of the species for crossing natural geographical barriers.

Researchers and foresters have revisited historical provenance trials of forest tree species to understand intraspecific variations in climate adaptation and to plan assisted migration worldwide. Such experiments involve planting different species populations in a common environment, acting as a space-for-time substitute to study climate change (Kapeller et al., 2013; Leites et al., 2012). Several studies based on such provenance experiments conclude that tree populations often grow in suboptimal conditions; thus, a facilitated movement of such populations may be desirable to

Fig. 12.5 Representations of the general concepts of assisted migration. Three main strategies are shown: assisted population migration, assisted range expansion, and assisted species migration

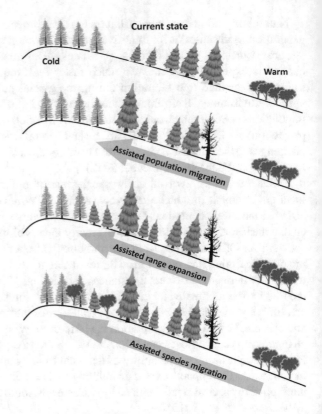

ensure their fitness under a changing climate (Chakraborty et al., 2016; Isaac-Renton et al., 2014; Rehfeldt et al., 2014; Wang et al., 2006). For example, whitebark pine (*Pinus albicaulis*) experiences better growth and germination when moved 800 km to the north of its current range, where seed sources from Oregon and Washington states performed well in locations in northwestern British Columbia (McLane & Aitken, 2012). Assisted migration is already incorporated into forest management policy in some countries, including Canada (Marris, 2009). A system similar to assisted migration, known as predictive provenance, is being used in the United Kingdom. This approach aims to match the current seed sources of native and non-native tree species to predicted climate in the future (Whittet et al., 2016).

Assisted migration does not necessarily need to be implemented widely. It should depend on a variety of criteria, such as the vulnerability of the native tree population, the provided ecosystem services, the current risks to stands, and the overall management goal of the landowner. A clear management strategy should be the first step for evaluating the need for assisted migration (Aitken & Bemmels, 2015). Assisted migration raises legal concerns about the trade and utilization of forest reproductive material. Seed and reproductive material for reforestation, traditionally sourced locally, may need to be adapted under climate change. However, legislation at the local and national levels may pose obstacles to such a transfer of plant material. In some cases, the concept of assisted migration may lead to conflicts with conservation principles because of the likelihood of the increased use of non-native species and ecotypes, which may be potentially invasive (Aitken et al., 2008). The uncertainty in climate predictions over the century is another issue for assisted migration. Assisted migration may be more expensive than traditional regeneration practices, especially under conditions where natural regeneration is commonly practiced. Despite the uncertainties and challenges, it is often pertinent to evaluate the outcomes of inaction as a management option. There is no concrete evidence to believe that inaction would reduce the vulnerability of current tree populations to climate change. Although novel climates also can reveal the adaptive potential of populations, such adaptive variations will require time. However, if climate change causes the current forest populations to die or become too weak to produce healthy seeds, it might be impossible or unaffordable to assist their migration in the future.

12.4   Conclusions: The Importance of Ecotypes Under a Changing Environment

Climatic influences can modify many anatomical and chemical wood properties. Extreme events, such as droughts, heat waves, storms, late frosts, and flooding, substantially affect the metabolism of trees and lead to irreversible responses in wood formation in both its morphological and chemical structure (Bräuning et al., 2016; Grabner & Wimmer, 2006). It is difficult to determine the threshold for the triggering of certain modifications in wood anatomy or wood structure in relation to the occurrence of an extreme climatic event. Many stress factors induce complex responses, meaning that the measured effect has an unknown relationship with the stimulus; for example, storms may cause mechanical damage or bending in trees, inducing the formation of reaction wood or traumatic resin ducts. However, the susceptibility of trees to such events depends not only on the magnitude and frequency of these events but also on tree size, tree mechanical strength, and tree position within a stand.

The forest industry must become aware of the climatic influences on forest development and the related consequences on wood quality. The frequency and intensity of extreme climatic events are expected to increase (Salinger, 2005; Schär et al., 2004). In the worst-case scenario, the decline in wood quality because of extreme events may be so severe that it constrains the wood industry to find new provenances for their resources, replace wood by nonwood or nonsustainable raw materials, or develop new wood processing technologies.

Over the twenty-first century, the boreal forest will experience the greatest increases in temperature among all forest biomes worldwide. Projected temperature increases range between 4 and 11 °C, accompanied by a much less pronounced increase in precipitation; the combination represents a major threat to the health of this ecosystem (Gauthier et al., 2015). These changes will not only modify the disturbance regime but also increase drought stress and drought-related mortality (Gauthier et al., 2015; Peng et al., 2011).

A strong influence of climate (drought) on earlywood density (besides mean ring density) can be seen by studying the influence of pointer years, such as drought years (Grabner et al., 2010). The microdensity profiles of Norway spruce trees grown in a dry region of Austria showed a slightly increasing trend of earlywood density. During drought periods (1992–1995, 2000–2003), these trees experienced an increased earlywood density (Grabner et al., 2010), a dramatically reduced ring width, and an increased mean ring density because of higher latewood percentages.

Climate change alters forest productivity through CO_2 fertilization and the lengthening of the growing season. In the boreal environment, these changes lead to the northward migration of the tree line and an increased tree and shrub cover (IPCC, 2019). Likewise, a significant increase in the annual growth of boreal forests in Finland has been found (Kauppi et al., 2014). However, Girardin et al. (2016) found no consistent growth response over 60 years in boreal forests across Canada. In high-latitude regions, warming will also increase drought, wildfire, and insect outbreaks (IPCC, 2019). Because of these environmental changes, foresters will need to develop new management strategies, and the wood industry may struggle to secure a resource that meets process requirements and market demands. Genetic improvement of wood-quality traits by selecting suitable tree provenances represents one option to alleviate the possible decrease in wood quality.

References

Aitken, S. N., & Bemmels, J. B. (2015). Time to get moving: Assisted gene flow of forest trees. *Evolutionary Applications, 9*, 271–290. https://doi.org/10.1111/eva.12293.

Aitken, S. N., & Whitlock, M. C. (2013). Assisted gene flow to facilitate local adaptation to climate change. *Annual Review of Ecology, Evolution, and Systematics, 44*, 367–388. https://doi.org/10.1146/annurev-ecolsys-110512-135747.

Aitken, S. N., Yeaman, S., Holliday, J. A., et al. (2008). Adaptation, migration or extirpation: Climate change outcomes for tree populations. *Evolutionary Applications, 1*, 95–111. https://doi.org/10.1111/j.1752-4571.2007.00013.x.

Alberto, F. J., Aitken, S. N., Alia, R., et al. (2013). Potential for evolutionary responses to climate change—Evidence from tree populations. *Global Change Biology, 19*, 1645–1661. https://doi.org/10.1111/gcb.12181.

Allevato, E., Saulino, L., Cesarano, G., et al. (2019). Canopy damage by spring frost in European beech along the Apennines: Effect of latitude, altitude and aspect. *Remote Sensing of Environment, 225*, 431–440. https://doi.org/10.1016/j.rse.2019.03.023.

Andalo, C., Beaulieu, J., & Bousquet, J. (2005). The impact of climate change on growth of local white spruce populations in Quebec, Canada. *Forest Ecology and Management, 205*, 169–182. https://doi.org/10.1016/j.foreco.2004.10.045.

Antonucci, S., Rossi, S., Deslauriers, A., et al. (2015). Synchronisms and correlations of spring phenology between apical and lateral meristems in two boreal conifers. *Tree Physiology, 35*, 1086–1094. https://doi.org/10.1093/treephys/tpv077.

Ascoli, D., Maringer, J., Hacket-Pain, A., et al. (2017). Two centuries of masting data for European beech and Norway spruce across the European continent. *Ecology, 98*(5), 1473–1473. https://doi. org/10.1002/ecy.1785.

Ascoli, D., Hacket-Pain, A., LaMontagne, J. M., et al. (2019). Climate teleconnections synchronize *Picea glauca* masting and fire disturbance: Evidence for a fire-related form of environmental prediction. *Journal of Ecology, 108*, 1186–1198. https://doi.org/10.1111/1365-2745.13308.

Astrup, R., Bernier, P. Y., Genet, H., et al. (2018). A sensible climate solution for the boreal forest. *Nature Climate Change, 8*, 11–12. https://doi.org/10.1038/s41558-017-0043-3.

Bannan, M. W. (1967). Anticlinal divisions and cell length in conifer cambium. *Forest Products Journal, 17*, 63–69.

Bansal, S., St Clair, J. B., Harrington, C. A., et al. (2015). Impact of climate change on cold hardiness of Douglas-fir (*Pseudotsuga menziesii*): Environmental and genetic considerations. *Global Change Biology, 21*, 3814–3826. https://doi.org/10.1111/gcb.12958.

Battipaglia, G., Campelo, F., Vieira, J., et al. (2016). Structure and function of intra–annual density fluctuations: Mind the gaps. *Frontiers in Plant Science, 7*. https://doi.org/10.3389/fpls.2016. 00595.

Bernier, P., Gauthier, S., Jean, P. O., et al. (2016). Mapping local effects of forest properties on fire risk across Canada. *Forests, 7*(8), 157. https://doi.org/10.3390/f7080157.

Blum, B. M. (1988). Variation in the phenology of bud flushing in white and red spruce. *Canadian Journal of Forest Research, 18*, 315–319. https://doi.org/10.1139/x88-048.

Bolte, A., Ammer, C., Lof, M., et al. (2009). Adaptive forest management in central Europe: Climate change impacts, strategies and integrative concept. *Scandinavian Journal of Forest Research, 24*, 473–482. https://doi.org/10.1080/02827580903418224.

Bräuning, A., De Ridder, M., Zafirov, N., et al. (2016). Tree-ring features: Indicators of extreme event impacts. *IAWA Journal, 37*, 206–231. https://doi.org/10.1163/22941932-20160131.

Briceño-Elizondo, E., Garcia-Gonzalo, J., Peltola, H., et al. (2006). Sensitivity of growth of Scots pine, Norway spruce and silver birch to climate change and forest management in boreal conditions. *Forest Ecology and Management, 232*, 152–167. https://doi.org/10.1016/j.foreco.2006. 05.062.

Broome, A., Hendry, S., & Peace, A. (2007). Annual and spatial variation in coning shown by the Forest Condition Monitoring programme data for Norway spruce, Sitka spruce and Scots pine in Britain. *Forestry, 80*, 17–28. https://doi.org/10.1093/forestry/cpl046.

Chakraborty, D., Wang, T., Andre, K., et al. (2016). Adapting Douglas-fir forestry in Central Europe: Evaluation, application, and uncertainty analysis of a genetically based model. *European Journal of Forest Research, 135*, 919–936. https://doi.org/10.1007/s10342-016-0984-5.

Charron, I., & Greene, D. F. (2002). Post-wildfire seedbeds and tree establishment in the southern mixedwood boreal forest. *Canadian Journal of Forest Research, 32*, 1607–1615. https://doi.org/ 10.1139/x02-085.

Chesson, P. (2000). Mechanisms of maintenance of species diversity. *Annual Review Ecology Systematics, 31*(1), 343–366. https://doi.org/10.1146/annurev.ecolsys.31.1.343.

Chesson, P. (2012). Species competition and predation. In R. A. Meyers (Ed.), *Encyclopedia of sustainability science and technology* (pp. 10061–10085). Springer.

Choat, B., Cobb, A. R., & Jansen, S. (2008). Structure and function of bordered pits: New discoveries and impacts on whole-plant hydraulic function. *New Phytologist, 177*, 608–626. https://doi.org/ 10.1111/j.1469-8137.2007.02317.x.

Cooke, J. E., Eriksson, M. E., & Junttila, O. (2012). The dynamic nature of bud dormancy in trees: Environmental control and molecular mechanisms. *Plant, Cell and Environment, 35*, 1707–1728. https://doi.org/10.1111/j.1365-3040.2012.02552.x.

Cuny, H. E., & Rathgeber, C. B. K. (2016). Xylogenesis: Coniferous trees of temperate forests are listening to the climate tale during the growing season but only remember the last words! *Plant Physiology, 171*, 306–317. https://doi.org/10.1104/pp.16.00037.

Darwin, C. (2003). *On the origin of species, 1859*. London: Routledge.

Donaldson, L. (2008). Microfibril angle: Measurement, variation and relationships—A review. *IAWA Journal, 29*, 345–386. https://doi.org/10.1163/22941932-90000192.

Dyderski, M. K., Paź, S., Frelich, L. E., et al. (2018). How much does climate change threaten European forest tree species distributions? *Global Change Biology, 24*, 1150–1163. https://doi.org/10.1111/gcb.13925.

Fady, B., Cottrell, J., Ackzell, L., et al. (2016). Forests and global change: What can genetics contribute to the major forest management and policy challenges of the twenty-first century? *Regional Environmental Change, 16*(4), 927–939. https://doi.org/10.1007/s10113-015-0843-9.

Farquhar, G. D., von Caemmerer, S., & Berry, J. A. (1980). A biochemical model of photosynthetic CO_2 assimilation in leaves of C_3 species. *Planta, 149*(1), 78–90. https://doi.org/10.1007/BF0038 6231.

Fujikawa, S., & Kuroda, K. (2000). Cryo-scanning electron microscopic study on freezing behavior of xylem ray parenchyma cells in hardwood species. *Micron, 31*, 669–686. https://doi.org/10.1016/S0968-4328(99)00103-1.

Fukatsu, E., & Nakada, R. (2018). The timing of latewood formation determines the genetic variation of wood density in *Larix kaempferi*. *Trees, 32*, 1233–1245. https://doi.org/10.1007/s00468-018-1705-0.

Gallego Zamorano, J., Hokkanen, T., & Lehikoinen, A. (2016). Climate-driven synchrony in seed production of masting deciduous and conifer tree species. *Journal Plant Ecology, 11*(2), 180–188. https://doi.org/10.1093/jpe/rtw117.

Gauthier, S., Bernier, P., Burton, P. J., et al. (2014). Climate change vulnerability and adaptation in the managed Canadian boreal forest. *Environmental Reviews, 22*, 256–285. https://doi.org/10.1139/er-2013-0064.

Gauthier, S., Bernier, P., Kuuluvainen, T., et al. (2015). Boreal forest health and global change. *Science, 349*, 819–822. https://doi.org/10.1126/science.aaa9092.

Geburek, T., Hiess, K., Litschauer, R., et al. (2012). Temporal pollen pattern in temperate trees: Expedience or fate? *Oikos, 121*, 1603–1612. https://doi.org/10.1111/j.1600-0706.2011.20140.x.

George, J.-P., Schueler, S., Karanitsch-Ackerl, S., et al. (2015). Inter- and intra-specific variation in drought sensitivity in *Abies* spec. and its relation to wood density and growth traits. *Agricultural and Forest Meteorology, 214–215*, 430–443. https://doi.org/10.1016/j.agrformet.2015.08.268.

George, J.-P., Grabner, M., Campelo, F., et al. (2019). Intra-specific variation in growth and wood density traits under water-limited conditions: Long-term-, short-term-, and sudden responses of four conifer tree species. *Science of the Total Environment, 660*, 631–643. https://doi.org/10.1016/j.scitotenv.2018.12.478.

Gindl, W., Grabner, M., & Wimmer, R. (2000). The influence of temperature on latewood lignin content in treeline Norway spruce compared with maximum density and ring width. *Trees, 14*, 409–414. https://doi.org/10.1007/s004680000057.

Girardin, M. P., Bouriaud, O., Hogg, E. H., et al. (2016). No growth stimulation of Canada's boreal forest under half-century of combined warming and CO_2 fertilization. *Proceedings of the National Academy of Sciences of the United States of America, 113*, E8406–E8414. https://doi.org/10.1073/pnas.1610156113.

Grabner, M., & Wimmer, R. (2006). Variation of different tree-ring parameters in samples from each terminal shoot of a Norway spruce tree. *Dendrochronologia, 23*(3), 111–120. https://doi.org/10.1016/j.dendro.2005.11.001.

Grabner, M., & Karanitsch-Ackerl, S. (2010). Schüler S (2010) The influence of drought on density of Norway spruce. In J. Kudela & R. Lagana (Eds.), *6th International Symposium of Wood Structure and Properties* (pp. 27–32). Arbora Publishers.

Hacket-Pain, A., Ascoli, D., Berretti, R., et al. (2019). Temperature and masting control Norway spruce growth, but with high individual tree variability. *Forest Ecology and Management, 438*, 142–150. https://doi.org/10.1016/j.foreco.2019.02.014.

Hanewinkel, M., Cullmann, D. A., Schelhaas, M. J., et al. (2013). Climate change may cause severe loss in the economic value of European forest land. *Nature Climate Change, 3*, 203–207. https://doi.org/10.1038/nclimate1687.

Harris, J. A., Hobbs, R. J., Higgs, E., et al. (2006). Ecological restoration and global climate change. *Restoration Ecology, 14*, 170–176. https://doi.org/10.1111/j.1526-100x.2006.00136.x.

Hazarika, R., & Jandl, R. (2019). The nexus between the Austrian forestry sector and the sustainable development goals: A review of the interlinkages. *Forests, 10*, 205. https://doi.org/10.3390/f10030205.

Hewitt, G. (2000). The genetic legacy of the Quaternary ice ages. *Nature, 405*, 907–913. https://doi.org/10.1038/35016000.

HilleRisLambers, J., Adler, P. B., Harpole, W. S., et al. (2012). Rethinking community assembly through the lens of coexistence theory. *Annual Review of Ecology, Evolution, and Systematics, 43*(1), 227–248. https://doi.org/10.1146/annurev-ecolsys-110411-160411.

Hölttä, T., Lintunen, A., Chan, T., et al. (2017). A steady-state stomatal model of balanced leaf gas exchange, hydraulics and maximal source-sink flux. *Tree Physiology, 37*, 851–868. https://doi.org/10.1093/treephys/tpx011.

Hurme, P., Repo, T., Savolainen, O., et al. (1997). Climatic adaptation of bud set and frost hardiness in Scots pine (*Pinus sylvestris*). *Canadian Journal of Forest Research, 27*(5), 716–723. https://doi.org/10.1139/x97-052.

Intergovernmental Panel on Climate Change (IPCC). (2019). Summary for policymakers. In P. R. Shukla, J. Skea, E. C. Buendia, V. Masson-Delmotte, H.-O. Pörtner, D. C. Roberts, P. Zhai, R. Slade, S. Connors, R. V. Diemen, M. Ferrat, E. Haughey, S. Luz, S. Neogi, M. Pathak, J. Petzold, J. P. Pereira, P. Vyas, E. Huntley, K. Kissick, M. Belkacemi & J. Malley (Eds.), *Climate change and land: an IPCC special report on climate change, desertification, land degradation, sustainable land management, food security, and greenhouse gas fluxes in terrestrial ecosystems*. Geneva: Intergovernmental Panel on Climate Change.

Isaac-Renton, M. G., Roberts, D. R., Hamann, A., et al. (2014). Douglas-fir plantations in Europe: A retrospective test of assisted migration to address climate change. *Global Change Biology, 20*, 2607–2617. https://doi.org/10.1111/gcb.12604.

Isaac-Renton, M., Montwé, D., Hamann, A., et al. (2018). Northern forest tree populations are physiologically maladapted to drought. *Nature Communications, 9*, 5254. https://doi.org/10.1038/s41467-018-07701-0.

Jaakkola, T., Makinen, H., & Saranpaa, P. (2006). Wood density of Norway spruce: Responses to timing and intensity of first commercial thinning and fertilisation. *Forest Ecology and Management, 237*, 513 521. https://doi.org/10.1016/j.foreco.2006.09.083.

Jandl, R., Spathelf, P., Bolte, A., et al. (2019). Forest adaptation to climate change–Is non-management an option? *Annals of Forest Science, 76*, 48. https://doi.org/10.1007/s13595-019-0827-x.

Janzen, D. H. (1971). Seed predation by animals. *Annual Review of Ecology, Evolution, and Systematics, 2*, 465–492. https://doi.org/10.1146/annurev.es.02.110171.002341.

Janzen, D. H. (1976). Why bamboos wait so long to flower. *Annual Review of Ecology, Evolution, and Systematics, 7*, 347–391. https://doi.org/10.1146/annurev.es.07.110176.002023.

Jarvis, P., & Linder, S. (2000). Constraints to growth of boreal forests. *Nature, 405*, 904–905. https://doi.org/10.1038/35016154.

Johnsen, K. H., Seiler, J. R., & Major, J. E. (1996). Growth, shoot phenology and physiology of diverse seed sources of black spruce: II. 23-year-old field trees. *Tree Physiology, 16*, 375–380. https://doi.org/10.1093/treephys/16.3.375.

Kalliokoski, T., Reza, M., Jyske, T., et al. (2012). Intra-annual tracheid formation of Norway spruce provenances in southern Finland. *Trees, 26*(2), 543–555. https://doi.org/10.1007/s00468-011-0616-0.

Kapeller, S., Schuler, S., Huber, G., et al. (2013). Provenance trials in alpine range—Review and perspectives for applications in climate change. In G. A. Cerbu, M. Hanewinkel, G. Gerosa, & R. Jandl (Eds.), *Management strategies to adapt alpine space forests to climate change risks*. IntechOpen.

Kauppi, P. E., Posch, M., & Pirinen, P. (2014). Large impacts of climatic warming on growth of boreal forests since 1960. *PLoS ONE, 9*, e111340–e111340. https://doi.org/10.1371/journal.pone. 0111340.

Kellomäki, S., Peltola, H., Nuutinen, T., et al. (2008). Sensitivity of managed boreal forests in Finland to climate change, with implications for adaptive management. *Philosophical Transactions of the Royal Society of London. Series B, Biological Sciences, 363*, 2341–2351. https://doi. org/10.1098/rstb.2007.2204.

Kellomäki, S., & Väisänen, H. (1997). Modelling the dynamics of the forest ecosystem for climate change studies in the boreal conditions. *Ecological Modelling, 97*, 121–140. https://doi.org/10. 1016/s0304-3800(96)00081-6.

Kelly, D. (1994). The evolutionary ecology of mast seeding. *Trends in Ecology & Evolution, 9*, 465–470. https://doi.org/10.1016/0169-5347(94)90310-7.

Khare, S., Drolet, G., Sylvain, J. D., et al. (2019). Assessment of spatio-temporal patterns of black spruce bud phenology across Quebec based on MODIS-NDVI time series and field observations. *Remote Sensing, 11*, 2745. https://doi.org/10.3390/rs11232745.

Klisz, M., Koprowski, M., Ukalska, J., et al. (2016). Does the genotype have a significant effect on the formation of intra-annual density fluctuations? A case study using *Larix decidua* from northern Poland. *Frontiers in Plant Science, 7*, 691. https://doi.org/10.3389/fpls.2016.00691.

Klisz, M., Ukalska, J., Koprowski, M., et al. (2019). Effect of provenance and climate on intra-annual density fluctuations of Norway spruce *Picea abies* (L.) Karst. in Poland. *Agricultural and Forest Meteorology, 269–270*, 145–156. https://doi.org/10.1016/j.agrformet.2019.02.013.

Koenig, W. D., & Knops, J. M. H. (1998). Scale of mast-seeding and tree-ring growth. *Nature, 396*, 225–226. https://doi.org/10.1038/24293.

Kolari, P., Kulmala, L., Pumpanen, J., et al. (2009). CO_2 exchange and component CO_2 fluxes of a boreal Scots pine forest. *Boreal Environment Research, 14*, 761–783.

Kolari, P., Chan, T., Porcar-Castell, A., et al. (2014). Field and controlled environment measurements show strong seasonal acclimation in photosynthesis and respiration potential in boreal Scots pine. *Frontiers in Plant Science, 5*, 717. https://doi.org/10.3389/fpls.2014.00717.

Körner, C. (2003). *Alpine plant life: Functional plant ecology of high mountain ecosystems*. Berlin, Heidelberg: Springer.

Körner, C. (2012). *Alpine treelines—Functional ecology of the global high elevation tree limits*. Basel: Springer.

Kort, I. (1993). Wood production and latewood percentage of Douglas-fir from different stands and vitality classes. *Canadian Journal of Forest Research, 23*, 1480–1486. https://doi.org/10.1139/ x93-185.

Kreyling, J., Buhk, C., Backhaus, S., et al. (2014). Local adaptations to frost in marginal and central populations of the dominant forest tree *Fagus sylvatica* L. as affected by temperature and extreme drought in common garden experiments. *Ecology and Evolution, 4*, 594–605. https://doi.org/10. 1002/ece3.971.

Lande, R. (1976). Natural selection and random genetic drift in phenotypic evolution. *Evolution, 30*(2), 314–334 https://doi.org/10.2307/2407703.

Langlet, O. (1971a). Revising some terms of intra-specific differentiation. *Hereditas, 68*, 277–279. https://doi.org/10.1111/j.1601-5223.1971.tb02402.x.

Langlet, O. (1971b). Two hundred years genecology. *Taxon, 20*, 653–721. https://doi.org/10.2307/ 1218596.

Leites, L. P., Rehfeldt, G. E., Robinson, A. P., et al. (2012). Possibilities and limitations of using historic provenance tests to infer forest species growth responses to climate change. *Natural Resource Modeling, 25*, 409–433. https://doi.org/10.1111/j.1939-7445.2012.00129.x.

Li, P., & Adams, W. T. (1993). Genetic control of bud phenology in pole-size trees and seedlings of coastal Douglas-fir. *Canadian Journal of Forest Research, 23*, 1043–1051. https://doi.org/10. 1139/x93-133.

Linkosalo, T., Heikkinen, J., Pulkkinen, P., et al. (2014). Fluorescence measurements show stronger cold inhibition of photosynthetic light reactions in Scots pine compared to Norway spruce as well as during spring compared to autumn. *Frontiers in Plant Science, 5*, 264. https://doi.org/10.3389/fpls.2014.00264.

Lintunen, A., Lindfors, L., Nikinmaa, E., et al. (2017). Xylem diameter changes during osmotic stress, desiccation and freezing in *Pinus sylvestris* and *Populus tremula*. *Tree Physiology, 37*, 491–500. https://doi.org/10.1093/treephys/tpw114.

Lintunen, A., Paljakka, T., Salmon, Y., et al. (2020). The influence of soil temperature and water content on belowground hydraulic conductance and leaf gas exchange in mature trees of three boreal species. *Plant, Cell and Environment, 43*, 532–547. https://doi.org/10.1111/pce.13709.

Louzada, J. L. P. C., & Fonseca, F. M. A. (2002). The heritability of wood density components in *Pinus pinaster* Ait. and the implications for tree breeding. *Annals of Forest Science, 59*, 867–873. https://doi.org/10.1051/forest:2002085.

Luyssaert, S., Inglima, I., Jung, M., et al. (2007). CO_2 balance of boreal, temperate, and tropical forests derived from a global database. *Global Change Biology, 13*(12), 2509–2537. https://doi.org/10.1111/j.1365-2486.2007.01439.x.

Makipää, R., Karjalainen, T., Pussinen, A., et al. (1999). Effects of climate change and nitrogen deposition on the carbon sequestration of a forest ecosystem in the boreal zone. *Canadian Journal of Forest Research, 29*, 1490–1501. https://doi.org/10.1139/x99-123.

Man, R., & Lu, P. (2010). Effects of thermal model and base temperature on estimates of thermal time to bud break in white spruce seedlings. *Canadian Journal of Forest Research, 40*, 1815–1820. https://doi.org/10.1139/X10-129.

Marris, E. (2008). Moving on assisted migration. *Nature Climate Change, 1*, 112–113. https://doi.org/10.1038/climate.2008.86.

Marris, E. (2009). Forestry: Planting the forest of the future. *Nature, 459*, 906–908. https://doi.org/10.1038/459906a.

Matías, L., Godoy, O., Gómez-Aparicio, L., et al. (2018). An experimental extreme drought reduces the likelihood of species to coexist despite increasing intransitivity in competitive networks. *Journal of Ecology, 106*(3), 826–837. https://doi.org/10.1111/1365-2745.12962.

Mátyás, C., Berki, I., Czúcz, B., et al. (2010). Future of beech in southeast Europe from the perspective of evolutionary ecology. *Acta Silvatica et Lignaria Hungarica, 6*, 91–110.

Mayr, S., Schmid, P., Beikircher, B., et al. (2020). Die hard: Timberline conifers survive annual winter embolism. *New Phytologist, 226*(1), 13–20. https://doi.org/10.1111/nph.16304.

McLane, S. C., & Aitken, S. N. (2012). Whitebark pine (*Pinus albicaulis*) assisted migration potential: Testing establishment north of the species range. *Ecological Applications, 22*, 142–153. https://doi.org/10.1890/11-0329.1.

Mellander, P. E., Bishop, K., & Lundmark, T. (2004). The influence of soil temperature on transpiration: A plot scale manipulation in a young Scots pine stand. *Forest Ecology and Management, 195*, 15–28. https://doi.org/10.1016/j.foreco.2004.02.051.

Merilä, J., & Hendry, A. P. (2014). Climate change, adaptation, and phenotypic plasticity: The problem and the evidence. *Evolutionary Applications, 7*, 1–14. https://doi.org/10.1111/eva.12137.

Montoro Girona, M., Morin, H., Lussier, J. M., et al. (2016). Radial growth response of black spruce stands ten years after experimental shelterwoods and seed-tree cuttings in boreal forest. *Forests, 7*(10), 1–20. https://doi.org/10.3390/f7100240.

Montwé, D., Isaac-Renton, M., Hamann, A., et al. (2018). Cold adaptation recorded in tree rings highlights risks associated with climate change and assisted migration. *Nature Communications, 9*, 1574. https://doi.org/10.1038/s41467-018-04039-5.

Moreira, X., Abdala-Roberts, L., Linhart, Y. B., et al. (2014). Masting promotes individual- and population-level reproduction by increasing pollination efficiency. *Ecology, 95*, 801–807. https://doi.org/10.1890/13-1720.1.

Morgenstern, E. K. (1969). Genetic variation in seedlings of *Picea mariana* (Mill.) BSP. I Correlation with ecological factors. *Silvae Genetica, 18*, 151–161.

Neuner, S., Albrecht, A., Cullmann, D., et al. (2015). Survival of Norway spruce remains higher in mixed stands under a dryer and warmer climate. *Global Change Biology, 21*, 935–946. https://doi.org/10.1111/gcb.12751.

Nilsson, S. G., & Wastljung, U. (1987). Seed predation and cross-pollination in mast-seeding beech (*Fagus sylvatica*) patches. *Ecology, 68*, 260–265. https://doi.org/10.2307/1939256.

Nygaard, P. H., & Øyen, B.-H. (2017). Spread of the introduced sitka spruce (*Picea sitchensis*) in coastal Norway. *Forests, 8*(1), 24. https://doi.org/10.3390/f8010024.

Organisation for Economic Co-operation and Development (OECD). (2010). *Section 2—Jack pine (Pinus banksiana)* (p. 34). Paris: OECD Publishing.

Ogden, A. E., & Innes, J. (2007). Incorporating climate change adaptation considerations into forest management planning in the boreal forest. *International Forestry Review, 9*, 713–733. https://doi.org/10.1505/ifor.9.3.713.

Oleksyn, J., Tjoelker, M. G., & Reich, P. B. (1998). Adaptation to changing environment in Scots pine populations across a latitudinal gradient. *Silva Fennica, 32*, 129–140. https://doi.org/10.14214/sf.691.

Pacheco, A., Camarero, J. J., & Carrer, M. (2016). Linking wood anatomy and xylogenesis allows pinpointing of climate and drought influences on growth of coexisting conifers in continental Mediterranean climate. *Tree Physiology, 36*, 502–512. https://doi.org/10.1093/treephys/tpv125.

Pamerleau-Couture, E., Rossi, S., Pothier, D., et al. (2019). Wood properties of black spruce (*Picea mariana* (Mill.) BSP) in relation to ring width and tree height in even- and uneven-aged boreal stands. *Annals of Forest Science, 76*, 43 https://doi.org/10.1007/s13595-019-0828-9.

Park, A., Puettmann, K., Wilson, E., et al. (2014). Can boreal and temperate forest management be adapted to the uncertainties of 21st century climate change? *Critical Reviews in Plant Sciences, 33*, 251–285. https://doi.org/10.1080/07352689.2014.858956.

Pearse, I. S., Koenig, W. D., & Kelly, D. (2016). Mechanisms of mast seeding: Resources, weather, cues, and selection. *New Phytologist, 212*, 546–562. https://doi.org/10.1111/nph.14114.

Pearson, H. (2006). Genetics: What is a gene? *Nature, 441*, 398–401. https://doi.org/10.1038/441398a.

Pedlar, J. H., McKenney, D. W., Aubin, I., et al. (2012). Placing forestry in the assisted migration debate. *BioScience, 62*, 835–842. https://doi.org/10.1525/bio.2012.62.9.10.

Peng, C., Ma, Z., Lei, X., et al. (2011). A drought-induced pervasive increase in tree mortality across Canada's boreal forests. *Nature Climate Change, 1*, 467–471. https://doi.org/10.1038/nclimate1293.

Pérez-Ramos, I. M., Matías, L., Gómez-Aparicio, L., et al. (2019). Functional traits and phenotypic plasticity modulate species coexistence across contrasting climatic conditions. *Nature Communications, 10*(1), 2555. https://doi.org/10.1038/s41467-019-10453-0.

Perrin, M., Rossi, S., & Isabel, N. (2017). Synchronisms between bud and cambium phenology in black spruce: Early-flushing provenances exhibit early xylem formation. *Tree Physiology, 37*, 593–603. https://doi.org/10.1093/treephys/tpx019.

Peters, V. S., Macdonald, S. E., & Dale, M. R. T. (2005). The interaction between masting and fire is key to white spruce regeneration. *Ecology, 86*, 1744–1750. https://doi.org/10.1890/03-0656.

Petit, G., von Arx, G., Kiorapostolou, N., et al. (2018). Tree differences in primary and secondary growth drive convergent scaling in leaf area to sapwood area across Europe. *New Phytologist, 218*(4), 1383–1392. https://doi.org/10.1111/nph.15118.

Poncet, B. N., Garat, P., Manel, S., et al. (2009). The effect of climate on masting in the European larch and on its specific seed predators. *Oecologia, 159*, 527–537. https://doi.org/10.1007/s00442-008-1233-5.

Rathgeber, C. B. K. (2017). Conifer tree-ring density inter-annual variability—Anatomical, physiological and environmental determinants. *New Phytologist, 216*, 621–625. https://doi.org/10.1111/nph.14763.

Rehfeldt, G. E., Leites, L. P., Bradley St Clair, J., et al. (2014). Comparative genetic responses to climate in the varieties of *Pinus ponderosa* and *Pseudotsuga menziesii*: Clines in growth potential. *Forest Ecology and Management, 324*, 138–146. https://doi.org/10.1016/j.foreco.2014.02.041.

Reich, P. B., Sendall, K. M., Stefanski, A., et al. (2018). Effects of climate warming on photosynthesis in boreal tree species depend on soil moisture. *Nature, 562*, 263–267. https://doi.org/10.1038/s41586-018-0582-4.

Reyer, C., Lasch-Born, P., Suckow, F., et al. (2014). Projections of regional changes in forest net primary productivity for different tree species in Europe driven by climate change and carbon dioxide. *Annals of Forest Science, 71*, 211–225. https://doi.org/10.1007/s13595-013-0306-8.

Rossi, S., & Bousquet, J. (2014). The bud break process and its variation among local populations of boreal black spruce. *Frontiers in Plant Science, 5*, 574. https://doi.org/10.3389/fpls.2014.00574.

Rossi, S., & Isabel, N. (2017). Bud break responds more strongly to daytime than night-time temperature under asymmetric experimental warming. *Global Change Biology, 23*, 446–454. https://doi.org/10.1111/gcb.13360.

Rossi, S., Deslauriers, A., Anfodillo, T., et al. (2006). Conifers in cold environments synchronize maximum growth rate of tree-ring formation with day length. *New Phytologist, 170*, 301–310. https://doi.org/10.1111/j.1469-8137.2006.01660.x.

Rossi, S., Rathgeber, C. B. K., & Deslauriers, A. (2009). Comparing needle and shoot phenology with xylem development on three conifer species in Italy. *Annals of Forest Science, 66*, 206. https://doi.org/10.1051/forest/2008088.

Rossi, S., Girard, M. J., & Morin, H. (2014). Lengthening of the duration of xylogenesis engenders disproportionate increases in xylem production. *Global Change Biology, 20*, 2261–2271. https://doi.org/10.1111/gcb.12470.

Rossi, S., Cairo, E., Krause, C., et al. (2015). Growth and basic wood properties of black spruce along an alti-latitudinal gradient in Quebec, Canada. *Annals of Forest Science, 72*, 77–87. https://doi.org/10.1007/s13595-014-0399-8.

Rousi, M., & Heinonen, J. (2007). Temperature sum accumulation effects on within-population variation and long-term trends in date of bud burst of European white birch (*Betula pendula*). *Tree Physiology, 27*, 1019–1025. https://doi.org/10.1093/treephys/27.7.1019.

Ruth, R. H. (1974). *Tsuga* (Endl.) Carr. Hemlock. In C. S. Schopmeyer (Ed.), *Seeds of woody plants in the United States* (pp. 819–827). Agriculture Handbook No. 450. Washington: U.S. Department of Agriculture.

Sáenz-Romero, C., Rehfeldt, G. F., Duval, P., et al. (2012). *Abies religiosa* habitat prediction in climatic change scenarios and implications for monarch butterfly conservation in Mexico. *Forest Ecology and Management, 275*, 98–106. https://doi.org/10.1016/j.foreco.2012.03.004.

Salinger, M. J. (2005). Increasing climate variability and change: Reducing the vulnerability. *Climatic Change, 70*, 1–3. https://doi.org/10.1007/s10584-005-4243-x.

Savolainen, O. (1996). Pines beyond the polar circle: Adaptation to stress conditions. *Euphytica, 92*, 139–145. https://doi.org/10.1007/BF00022839.

Savolainen, O., Bokma, F., Garca-Gil, R., et al. (2004). Genetic variation in cessation of growth and frost hardiness and consequences for adaptation of *Pinus sylvestris* to climatic changes. *Forest Ecology and Management, 197*, 79–89. https://doi.org/10.1016/j.foreco.2004.05.006.

Savolainen, O., Kujala, S. T., Sokol, C., et al. (2011). Adaptive potential of northernmost tree populations to climate change, with emphasis on scots pine (*Pinus sylvestris* L.). *Journal of Heredity, 102*(5), 526–536. https://doi.org/10.1093/jhered/esr056.

Schär, C., Vidale, P. L., Lüthi, D., et al. (2004). The role of increasing temperature variability in European summer heatwaves. *Nature, 427*(6972), 332–336. https://doi.org/10.1038/nature02300.

Schoene, D. H. F., & Bernier, P. Y. (2012). Adapting forestry and forests to climate change: A challenge to change the paradigm. *Forest Policy and Economics, 24*, 12–19. https://doi.org/10.1016/j.forpol.2011.04.007.

Seidl, R., Thom, D., Kautz, M., et al. (2017). Forest disturbances under climate change. *Nature Climate Change, 7*, 395–402. https://doi.org/10.1038/nclimate3303.

Selås, V., Piovesan, G., Adams, J. M., et al. (2002). Climatic factors controlling reproduction and growth of Norway spruce in southern Norway. *Canadian Journal of Forest Research, 32*, 217–225. https://doi.org/10.1139/x01-192.

Shvidenko, A. Z., & Schepaschenko, D. G. (2013). Climate change and wildfires in Russia. *Contemporary Problems of Ecology, 6*, 683–692. https://doi.org/10.1134/s199542551307010x.

Silvestro, R., Rossi, S., Zhang, S., et al. (2019). From phenology to forest management: Ecotypes selection can avoid early or late frosts, but not both. *Forest Ecology and Management, 436*, 21–26. https://doi.org/10.1016/j.foreco.2019.01.005.

Sinclair, W. T., Morman, J. D., & Ennos, R. A. (1999). The postglacial history of Scots pine (*Pinus sylvestris* L.) in western Europe: Evidence from mitochondrial DNA variation. *Molecular Ecology, 8*, 83–88. https://doi.org/10.1046/j.1365-294X.1999.00527.x.

Sperry, J. S., & Sullivan, J. E. (1992). Xylem embolism in response to freeze-thaw cycles and water stress in ring-porous, diffuse-porous, and conifer species. *Plant Physiology, 100*, 605–613. https://doi.org/10.1104/pp.100.2.605.

Spitze, K. (1993). Population structure in *Daphnia obtusa*: Quantitative genetic and allozymic variation. *Genetics, 135*, 367–374. https://doi.org/10.1093/genetics/135.2.367.

Stein, W. I., & Owston, P. W. (2002). *Pseudotsuga* Carr., Douglas-fir. In F. T. Bonner & R. G. Nisley (Eds.), *Woody plant seed manual* (pp. 1–38). Washington: U.S. Department of Agriculture Forest Service.

Tchebakova, N. M., Rehfeldt, G. E., & Parfenova, E. I. (2006). Impacts of climate change on the distribution of *Larix* spp. and *Pinus sylvestris* and their climatypes in Siberia. *Mitigation and Adaptation Strategies for Global Change, 11*, 861–882. https://doi.org/10.1007/s11027-005-9019-0.

Thomashow, M. F. (1998). Role of cold-responsive genes in plant freezing tolerance. *Plant Physiology, 118*, 1–8. https://doi.org/10.1104/pp.118.1.1.

Tollefsrud, M. M., Kissling, R., Gugerli, F., et al. (2008). Genetic consequences of glacial survival and postglacial colonization in Norway spruce: Combined analysis of mitochondrial DNA and fossil pollen. *Molecular Ecology, 17*(18), 4134–4150. https://doi.org/10.1111/j.1365-294X.2008.03893.x.

Viereck, L. A., & Johnston, W. F. (1990). *Picea mariana* (Mill.) BSP black spruce. In R. M. Burns & B. H. Honkala (Eds.), *Silvics of North America* (pp. 227–237). Washington: U.S. Department of Agriculture Forest Service.

Wainwright, C. E., HilleRisLambers, J., Lai, H. R., et al. (2019). Distinct responses of niche and fitness differences to water availability underlie variable coexistence outcomes in semi-arid annual plant communities. *Journal of Ecology, 107*(1), 293–306. https://doi.org/10.1111/1365-2745.13056.

Wang, T., Hamann, A., Yanchuk, A., et al. (2006). Use of response functions in selecting lodgepole pine populations for future climates. *Global Change Biology, 12*, 2404–2416. https://doi.org/10.1111/j.1365-2486.2006.01271.x.

Wermelinger, B. (2004). Ecology and management of the spruce bark beetle Ips typographus—A review of recent research. *Forest Ecology and Management, 202*, 67–82. https://doi.org/10.1016/j.foreco.2004.07.018.

Whittet, R., Cavers, S., Cottrell, J., et al. (2016). Seed sourcing for woodland creation in an era of uncertainty: An analysis of the options for Great Britain. *Forestry, 90*(2), 163–173. https://doi.org/10.1093/forestry/cpw037.

Williams, M. I., & Dumroese, R. K. (2013). Preparing for climate change: Forestry and assisted migration. *Journal of Forestry, 111*, 287–297. https://doi.org/10.5849/jof.13-016.

Wimmer, R. (1995). Intra-annual cellular characteristics and their implications for modeling softwood density. *Wood and Fiber Science, 27*(4), 413–420.

Wimmer, R., Downes, G. M., Evans, R., et al. (2002). Direct effects of wood characteristics on pulp and handsheet properties of *Eucalyptus globulus*. *Holzforschung, 56*, 244–252. https://doi.org/10.1515/hf.2002.040.

Wirth, C., Lichstein, J. W., Dushoff, J., et al. (2008). White spruce meets black spruce: Dispersal, postfire establishment, and growth in a warming climate. *Ecological Monographs, 78*, 489–505. https://doi.org/10.1890/07-0074.1.

Wright, S. (1943). Isolation by distance. *Genetics, 28*, 114–138. https://doi.org/10.1093/genetics/28.2.114.

Part V
Silviculture as a Tool to Promote Forest Resilience

Chapter 13
Building a Framework for Adaptive Silviculture Under Global Change

Anthony W. D'Amato, Brian J. Palik, Patricia Raymond,
Klaus J. Puettmann, and Miguel Montoro Girona

Abstract Uncertainty surrounding global change impacts on future forest conditions has motivated the development of silviculture strategies and frameworks focused on enhancing potential adaptation to changing climate and disturbance regimes. This includes applying current silvicultural practices, such as thinning and mixed-species and multicohort systems, and novel experimental approaches, including the deployment of future-adapted species and genotypes, to make forests more resilient to future changes. In this chapter, we summarize the general paradigms and approaches associated with adaptation silviculture along a gradient of strategies ranging from resistance to transition. We describe how these concepts have been operationalized and present potential landscape-scale frameworks for allocating different adaptation intensities as part of functionally complex networks in the face of climate change.

A. W. D'Amato (✉)
Rubenstein School of Environment and Natural Resources, University of Vermont, 81 Carrigan Drive, Burlington, VT 05405, USA
e-mail: awdamato@uvm.edu

B. J. Palik
USDA Forest Service Northern Research Station, 1831 Hwy. 169 E, Grand Rapids, MN 55744, USA
e-mail: brian.palik@usda.gov

P. Raymond
Direction de la recherche forestière, ministère des Forêts, de la Faune et des Parcs du Québec, 2700 rue Einstein, Québec, QC G1P 3W8, Canada
e-mail: patricia.raymond@mffp.gouv.qc.ca

K. J. Puettmann
Department of Forest Ecosystems and Society, Oregon State University, 321 Richardson Hall, Corvallis, OR 97331, USA
e-mail: Klaus.Puettmann@oregonstate.edu

© The Author(s) 2023
M. M. Girona et al. (eds.), *Boreal Forests in the Face of Climate Change*,
Advances in Global Change Research 74,
https://doi.org/10.1007/978-3-031-15988-6_13

13.1 Introduction

Silvicultural systems have long been intended to represent a working hypothesis adapted over time to address unanticipated changes in treatment outcomes or the impacts of exogenous factors, including natural disturbances and changing market conditions and objectives (Smith, 1962). Nevertheless, silvicultural approaches have assumed a general level of predictability in outcomes, with risks avoided or minimized through a top-down control of ecosystem attributes and properties, such as dominant tree species and genotypes, stand densities, soil fertility, and age structures (Palik et al., 2020; Puettmann et al., 2009). The increasing departure of environmental conditions from those under which many of these silvicultural practices and systems were developed has led to an explicit need for adaptive silvicultural approaches that account for future uncertainty and novelty in forests around the globe (Millar et al., 2007; Puettmann, 2011).

This chapter summarizes the general frameworks and approaches for developing silvicultural strategies that confer adaptation to forest ecosystems in the face of novel dynamics, including changes in disturbance and climate regimes and the proliferation of nonnative species. Experience with adaptation silviculture is in its infancy compared with traditional applications. Therefore, our focus is primarily on early outcomes of operational-scale experiments and demonstrations and landscape- and regional-scale simulations of long-term dynamics under adaptive silvicultural approaches. Our goal is to introduce new conceptual frameworks for adaptive silviculture as context for the chapters in Sect. 13.5 of this book. Although the discussion is focused on temperate and boreal ecosystems in North America and Europe, the conceptual frameworks are appropriate for many different forest ecosystems around the globe. Subsequent chapters provide more detail on specific facets of managing for adaptation in boreal ecosystems, including the role of plantation silviculture and tree improvement (Chap. 14), management for mixed species and structurally complex conditions (Chap. 15), and large-scale experiments inspired by natural disturbance emulation (Chap. 16).

M. M. Girona
Groupe de Recherche en Écologie de la MRC-Abitibi, Forest Research Institute, Université du Québec en Abitibi-Témiscamingue, 341, rue Principale Nord, Amos Campus, Amos, QC J9T 2L8, Canada
e-mail: miguel.montoro@uqat.ca

Department of Wildlife, Fish, and Environmental Studies, Swedish University of Agricultural Sciences (SLU), SE-901 83 Umeå, Sweden

Centre for Forest Research, Université du Québec à Montréal, P.O. Box 8888, Stn. Centre-Ville, Montréal, QC H3C 3P8, Canada

13.1.1   Silvicultural Challenges in the Face of Climate Change

Historically, silvicultural approaches and practices have reflected changing economic and social conditions (Puettmann et al., 2009). In contrast, ecological conditions have been sufficiently constant that foresters did not see the need to alter silvicultural approaches to accommodate changing ecological conditions. As a result, climate change enhances existing challenges and adds novel complexities to silviculture, given the limited experience of managing forests in a rapidly changing environment (Table 13.1). From a forest management perspective, the overarching challenge for addressing global change is to deal with trends and the uncertainty of future climate and disturbance regimes and the associated ecosystem dynamics and societal demands (Puettmann, 2011). In this context, selected aspects of climate change are being predicted rather consistently, e.g., the general increase in temperature and growing season length in boreal forests. Other aspects of climate change provide additional challenges of high uncertainty, including the magnitude of temperature increases among regions. Even more challenging are predictions of, for example, contrasting and variable precipitation patterns (Alotaibi et al., 2018).

The degree of certainty of future conditions influences the ability to prepare and minimize negative impacts (Meyers & Bull, 2002; Puettmann & Messier, 2020). Silvicultural practices directly aimed at accommodating temperature increases, for example, can be implemented relatively easily (Chmura et al., 2011; Hemery, 2008; Park et al., 2014), for instance favoring species or genotypes adapted to the projected

Table 13.1 Categories of silvicultural challenges with examples, confidence in current predictions, and conceptual basis from the ecological literature for respective silvicultural practices

Challenge	Example	Confidence	Conceptual basis
Changing growing conditions	Increased temperature, longer growing season	High	Tree and stand vigor (Camarero et al., 2018), niche theory (Wiens et al., 2009)
Uncertainty of predicted trends	Changes in the seasonality of precipitation	Low	Insurance hypothesis (Yachi & Loreau, 1999)
Unpredicted events	Changing population dynamics of existing insect or fungal species	Low	Insurance hypothesis
Scale mismatch—long term	Time needed to change stand structure or species mixtures	High	Niche dynamics (Brokaw & Busing, 2000)
Scale mismatch—short term	Species or provenances selected to fit in future climates cannot grow under current climate	High	Niche dynamics

temperatures. In contrast, foresters have less confidence when selecting specific silvi-cultural practices to accommodate novel disturbance regimes or an altered seasonality of precipitation patterns. In such cases, multiple practices may be promising, but the specific selections can only be viewed as *bet hedging* (sensu Meyers & Bull, 2002), which is based conceptually on the insurance hypothesis (Yachi & Loreau, 1999).

Another challenge arises through a temporal-scale mismatch. Forests are slow to respond to many silvicultural manipulations, e.g., conversion from single to multiple canopy layers will likely take several decades. Thus, managing for changing condi-tions requires a certain lead time (Biggs et al., 2009), an unlikely scenario with the immediacy of future climate and other global changes. At the same time, managing for future climate conditions can result in short-term incompatibilities or mismatches that may generate near-term undesirable outcomes in regard to ecosystem produc-tivity and structure (Wilhelmi et al., 2017) and lead to failures (e.g., regeneration) that may be viewed as too risky in reforestation activities.

Natural disturbances are crucial elements to consider in any silvicultural planning because of their substantial economic and ecological implications and potentially significant impact on forest productivity, carbon sequestration, and timber supplies (Flint et al., 2009; Kurz et al., 2008; Seidl et al., 2014). Climate change predictions indicate that the effects on boreal ecosystems will be profound, and natural distur-bance cycles (e.g., fire, insect outbreaks, and windthrow) will generally increase in frequency and severity (Seidl et al., 2017). These projections introduce a potentially massive *new* challenge to silvicultural planning. For example, the first evidence of the northward movement of spruce budworm (*Choristoneura fumiferana*) outbreaks has recently been reported combined with an increase in the frequency and level of damage during the last century; these findings indicated climate change to be the main cause of the altered spatiotemporal patterns of spruce budworm outbreaks in eastern Canadian boreal forests (Navarro et al., 2018). Climate change is expected to expand the range of natural disturbance variability in forest ecosystems beyond those under which past strategies, including ecosystem-based management (Chris-tensen et al., 1996), have been developed. Thus, a better understanding of how forest landscapes will respond to alterations in natural disturbances is needed to mitigate negative effects and adapt boreal forest management strategies to projections of climate change.

Vulnerability assessments of ecosystem attributes that quantify sensitivity to projected climate changes and the adaptive capacity to respond to these and distur-bance impacts (Mumby et al., 2014) have become a common strategy for addressing uncertainty. These assessments are also used to guide where adaptive silviculture may have the greatest benefit for meeting long-term management goals (Gauthier et al., 2014). In practical terms, the vulnerability of a forest type is based on the degree of climate and disturbance impacts expected in a region and the ability of the forest to respond to those impacts without a major change in forest conditions in terms of structure and function (Janowiak et al., 2014). Just as with actual climate change, vulnerability can vary regionally stemming from differences in biophysical settings within the stand because of variable tree-level conditions (e.g., resource availability and tree species, size, and age) and temporally owing to ontogenetic shifts in tree-

and stand-level conditions (Daly et al., 2010; Frey et al., 2016; Nitschke & Innes, 2008). Therefore, adaptation strategies must be tailored to regional and within-stand vulnerabilities and be flexible to account for changing vulnerabilities over time. For example, adaptation strategies applied to ecosystems having a low vulnerability may resemble current management practices and be designed to maintain current forest conditions and refugia (Thorne et al., 2020). In contrast, strategies applied to highly vulnerable ecosystems, such as those in boreal regions where natural migration is expected to be outpaced by climate change and disturbance impacts (Aubin et al., 2018), may need to employ deliberate actions to increase adaptive capacity. In the latter case, silvicultural strategies may look very different from current practices. The following section outlines general adaptation strategies broadly recognized for addressing climate change. However, their appropriateness for any given situation must be informed by regional- and site-level vulnerability assessments and overall management goals. The remaining sections present outcomes of adaptation approaches specific to temperate and boreal forests.

13.2 General Adaptation Strategies

Generally, active adaptation practices are categorized into *resistance, resilience,* and *transition* (also referred to as *response*) strategies (Millar et al., 2007; Table 13.2; Fig. 13.1). Note that passive adaptation, while included in Table 13.2, is not discussed further in this chapter, given our focus on active management strategies. Nevertheless, passive approaches, including reserve designation and protection, remain important strategies in the portfolio of options for addressing climate change impacts. Although presented as discrete categories, adaptation strategies fall along a continuum, such that implementation of an adaptation approach may involve elements of two or three categories (Nagel et al., 2017). Moreover, aspects of the tactics and outcomes associated with adaptation strategies are often conceptualized within a complex adaptive-systems framework (Puettmann & Messier, 2020; Puettmann et al., 2009), with structural and functional outcomes and associated multiscale feedbacks created by resistance, resilience, and transition strategies serving to confer ecosystem resilience (Messier et al., 2019). Thus, reliance on multiple adaptation strategies that bridge or reflect more than one category within and across stands is emphasized to generate cross-scale functional linkages and dynamics that allow for rapid recovery and reassembly following disturbances or extreme climate events (Messier et al., 2019).

Resistance strategies focus on adaptation tactics designed to maintain the currently existing forest conditions on a site (Millar et al., 2007) and can be viewed as an expansion of silvicultural practices typically used to maintain and increase tree vigor and limit disturbance impacts (Chmura et al., 2011). A litmus test for a resistance strategy asks whether the forest is still maintaining development trends in structure and composition within observed ranges of variation after exposure to a given stressor relative to areas not experiencing these treatments. Many of these tactics focus therefore on reducing the impacts of stressors, e.g., extreme precipitation events, drought,

Table 13.2 Climate change adaptation strategies with associated goals, assumptions, and example management actions. Adapted from Millar et al. (2007) and Palik et al. (2020)

Strategy	Definition	Goal	Assumptions	Example actions
Passive	No actions specific to climate change are taken	Allow a response to climate change without direct intervention	High risk in the mid- to long term, low effort, good social acceptance (initially)	Harvest deferral on areas considered to have low vulnerability in the near term; reserve designation, particularly in areas expected to serve as climate refugia
Resistance	Improve the defense of a forest to change	Maintain relatively unchanged conditions over time	Low risk in the near term and moderate effort, high social acceptance	Density management and competition control to increase resource availability to crop trees; removal of nonnative species; reduction of fuel loading to minimize fire impacts; removal of low vigor and high-risk individuals through stand improvement treatments
Resilience	Accommodate some change but remain within the natural range of variability	Allow some change; encourage a return to a condition within the natural range of variability	Medium risk in the midterm and medium effort, good social acceptance	Regeneration methods that encourage and maintain multicohort and mixed-species forest conditions (selection, irregular shelterwood); deliberate retention and maintenance of diverse structural attributes and functional traits

(continued)

and disturbance agents, including fire, insects, and diseases, by manipulating tree- and stand-level structure and composition to reduce levels of risk (Swanston et al., 2016).

For example, treatments that increase the abundance of hardwood species in conifer-dominated boreal systems may be categorized as resistance approaches given that they decrease the risk and severity of wildfires (Johnstone et al., 2011) and

Table 13.2   (continued)

Strategy	Definition	Goal	Assumptions	Example actions
Transition	Accommodate change, allowing an adaptive response to new conditions	Actively facilitate the shift to a new condition to encourage adaptive responses	High risk in the near term and high effort, low social acceptance (initially)	Regeneration methods focused on encouraging genotypes and species expected to be adapted to future climate and disturbance regimes; generation of a wide range of environmental conditions in stands, ranging from high-resource, open areas to buffered reserve patches; can include enrichment planting as part of multi-aged systems or the establishment of novel plantations representing future-adapted individuals

reduce stand vulnerability to insect outbreaks, such as from spruce budworm, that target conifer components (Campbell et al., 2008). More generally, the application of thinning treatments to increase the available resources for residual trees and thus minimize drought and forest health impacts (Bottero et al., 2017; D'Amato et al. 2013) or fuel reduction treatments to reduce fire severity (Butler et al., 2013) represent resistance strategies broadly applicable to many forest systems. Regardless of the tactics employed, resistance strategies are generally viewed as limited to being near-term solutions but also represent low-risk approaches that are easily understood and implemented by foresters. Thus, they may be suitable for a stand close to the planned rotation age (Puettmann, 2011). Relying solely on resistance strategies is more problematic in the long term given the increasing difficulty and costs expected in maintaining current conditions as global change progresses (Elkin et al., 2015), particularly in boreal regions where the climate is and will be changing rapidly (Price et al., 2013).

As with resistance strategies, resilience strategies largely emphasize maintaining the characteristics of current forest systems; however, the latter differs somewhat by maintaining and enhancing ecosystem properties that support recovery. Therefore, these strategies allow for larger temporary deviations and thus a broader range of

Fig. 13.1 Gradient of adaptation strategies in a northern hardwood forest in New Hampshire, United States (*center column panels*) and red pine forests in northern Minnesota, United States (*right-hand column panels*) ranging from **a** passive, **b** resistance, **c** resilience, to **d** transition. *Left-hand column panels* represent kriged surfaces associated with tree (≥10 cm DBH) locations in a 1 ha portion of treatment units in the northern hardwood forests. The passive strategy represents a no action approach. Resistance strategies represent single-tree selection focused on maintaining low-risk individuals in northern hardwood forests (cf. Nolet et al., 2014) and thinning treatments in red pine forests to increase drought and pest resistance (D'Amato et al., 2013). For both examples, the resilience strategy comprises a single-tree and group selection with patch reserves—similar to variable-density thinning, cf. Donoso et al. (2020)—to increase spatial and compositional complexity (harvest gaps were planted in the red pine forests with future-adapted species found in the present ecosystem). The transition strategy represents continuous cover (northern hardwoods) or expanding gap (red pine) irregular shelterwoods with the planting of future-adapted species in harvest gaps (northern hardwood) or across the entire stand (red pine). Note that the photos in the *bottom row* are focused on the harvest gap portion or irregular shelterwoods. *Photo credits* Anthony W. D'Amato. Kriged surfaces created by Jess Wikle.

compositional and structural outcomes, often bounded by the range of natural variation for the ecosystem (Landres et al., 1999). A litmus test for a resilience strategy is to ask whether the forest conditions return to the ecosystem's existing range of conditions (or historical ranges) after stand response to a treatment and exposure to a given stressor. In contrast to resistance strategies, which try to minimize deviation from current or historic conditions and processes, resilience strategies aim to increase an ecosystem's ability to recover from disturbances or climate extremes in an attempt to return to pre-perturbation levels of different processes (e.g., aboveground productivity) and structural and compositional conditions (Gunderson, 2000).

Ecosystem attributes and conditions identified as conferring resilience include vegetation and physical structures surviving disturbance (i.e., biological legacies or ecological memory; Johnstone et al., 2016), as well as mixed species forest conditions in which there is a high degree of functional redundancy among constituent species (Bergeron et al., 1995; Biggs et al., 2020; Messier et al., 2019). Thus, most resilience strategies focus on creating two general stand conditions: mixed species and a heterogeneous structure. In the case of mixed-species conditions, resilience is conferred by including species having a range of functional responses, including different recovery mechanisms following climate extremes (e.g., drought tolerance; Ruehr et al., 2019) and reproductive strategies following disturbance events (e.g., sprouting or seed banking; Rowe, 1983). Approaches that encourage heterogeneous, multicohort structures can reduce vulnerabilities given that climate and disturbance impacts vary with tree size and age (Bergeron et al., 1995; Olson et al., 2018), and the presence of younger age classes provides a mechanism for the rapid replacement of overstory tree mortality via ingrowth (O'Hara and Ramage, 2013). Many of these approaches often build from and resemble ecological silviculture strategies developed to emulate outcomes of natural disturbance regimes for a given forest type (D'Amato & Palik, 2021).

Transition strategies represent the largest deviation from traditional silvicultural frameworks and are applied under the assumption that future climate conditions and prevailing disturbance regimes will become less suitable or even unsuitable for current species and existing forest structural conditions, such as the often high stocking levels used for timber management in many forest types (Rissman et al., 2018). A litmus test for a transition strategy asks whether the expected development of forest characteristics in response to the treatment will eventually fall outside the range of natural variation and accommodate novel conditions. These strategies focus therefore on transitioning forests to species and structural conditions that are predicted to be able to provide desired ecosystem services under future climate and disturbances (Millar et al., 2007). In many cases, transition strategies include the deliberate introduction of future-adapted genotypes or species (e.g., Muller et al., 2019), sometimes through assisted migration, thereby increasing the representation of species and functional attributes likely to be favored under future disturbance and climate regimes (e.g., Etterson et al., 2020). Correspondingly, transition approaches carry the most risk (Wilhelmi et al., 2017); they are often controversial (Neff & Larson, 2014), partly because of a lack of site-level guidance for determining the appropriate future species and provenances for a given region (Park & Talbot, 2018) and a general uncertainty surrounding how introduced species or genotypes may behave at a given site (Whittet et al., 2016; Wilhelmi et al., 2017).

Central to resilience and transition strategies is recognizing the functional responses associated with structural and compositional conditions created by a given set of silvicultural activities (Messier et al., 2015). This includes considering the response traits of species favored by a given practice, both in terms of their ability to persist in the face of changing climate regimes and their ability to respond and recover following future disturbances (Biggs et al., 2020; Elmqvist et al., 2003). Although an understanding of certain functional traits, namely shade tolerance, growth rate, and reproductive mechanisms, has always guided silvicultural activities (Dean, 2012), the novelty of global change impacts requires a broader integration of traits, such as migration potential, that emphasizes the mechanisms conferring adaptive potential within and across species (Aubin et al., 2016; Yachi & Loreau, 1999). Obtaining and summarizing the relevant trait values for many species remain critical challenges in many regions. However, the development of indices that rank species on the basis of suites of traits associated with key sensitivities and responses, such as regeneration modes (e.g., sprouting ability, seed banking) and drought and fire tolerance (e.g., Fig. 13.2; Boisvert-Marsh et al., 2020), may prove useful in guiding future species selection for a given ecosystem.

13.3 Examples and Outcomes of Adaptation in Temperate and Boreal Ecosystems

The fast pace of climate change is particularly challenging because of the long lag between the evaluation of an adaptation strategy through field observations and the ability to recommend and implement the strategy at a broad scale (Biggs et al., 2009, 2020). As a result, decisions surrounding the regional deployment of adaptation strategies are most often based on simulation studies of future landscape dynamics under different management regimes and climate conditions (Duveneck & Scheller, 2015; Dymond et al., 2014; Hof et al., 2017). Numerous studies applying landscape simulation and forest planning models (e.g., LANDIS-II) have demonstrated the potential for recommended strategies. For example, the broad-scale deployment of mixed-species plantings increased the resilience of biomass stocks and volume flows in temperate and boreal systems (Duveneck & Scheller, 2015; Dymond et al., 2014, 2020). Nevertheless, a key limitation of simulation modeling, as it relates to operationalizing any given practice, is the inability to fully capture uncertainties in the future social acceptance of an approach (Seidl & Lexer, 2013), including from forest managers (Hengst-Ehrhart, 2019; Sousa-Silva et al., 2018). Therefore, it remains critical to support these model outcomes with field-based applications that include managers and broader societal perspectives.

As an alternative to model simulations, numerous studies have used dendrochronological techniques to retrospectively evaluate the ability of adaptation strategies

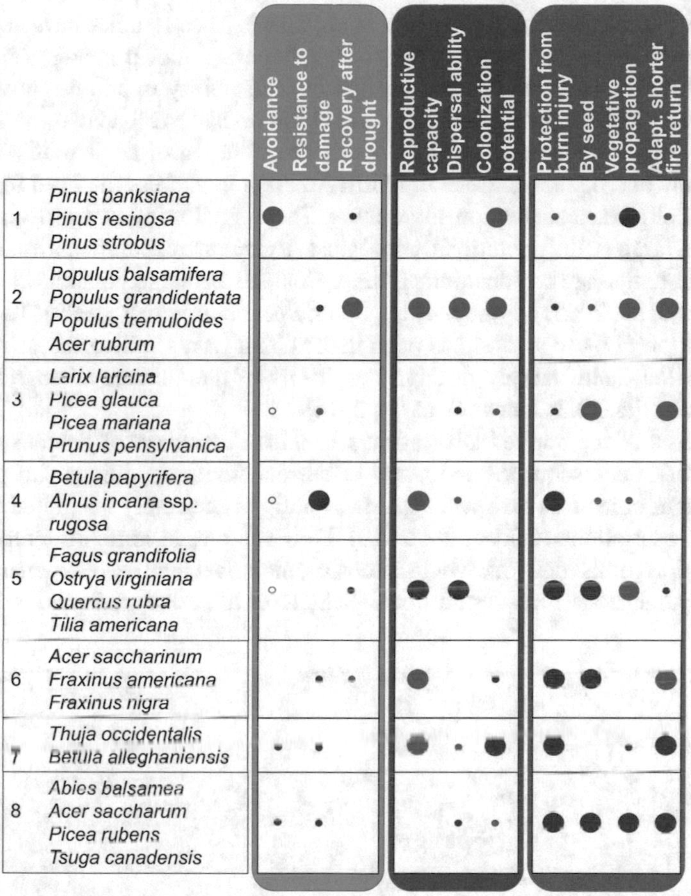

Fig. 13.2 Groupings of species from eastern Canada having similar sensitivities and responses to drought (*left blue column*), migration (*center green column*), and fire (*right brown column*) on the basis of functional traits. Average tolerance and sensitivity are denoted by *blue* (tolerance) and *red* (sensitivity) symbols; *larger symbols* indicate more extreme values. A *lack of a symbol* indicates intermediate values or the lack of a clear trend. Modified from Boisvert-Marsh et al. (2020), CC BY license

to confer resilience to stressors and climate extremes (e.g., severe drought). This research approach has confirmed the utility of commonly applied silvicultural treatments, such as thinning for density management (Bottero et al., 2017; D'Amato et al. 2013; Sohn et al., 2016) and mixed-species management (Bauhus et al., 2017; Drobyshev et al., 2013; Metz et al., 2016; Vitali et al., 2018) at promoting resistance and resilience to past drought events and insect outbreaks. Additionally, retrospective work examining the drought sensitivity of white spruce (*Picea glauca*) within common garden experiments in Québec, Canada, demonstrated the potential for deploying planting stock from drier locales to enhance the resilience to drought in boreal systems (Depardieu et al., 2020). These studies have collectively affirmed potential strategies suggested for addressing global change (Park et al., 2014). However, such studies are limited in their ability to address novel, future climate and socioecological conditions that have no historical analog.

Over the past decades, there has been a proliferation of adaptation silviculture experiments and demonstrations in North America to address the need for *forward thinking*, field-based adaptation silviculture. These studies follow from the legacy of numerous, large-scale, operational ecological silviculture experiments established in boreal and temperate regions during the 1990s and 2000s (e.g., Brais et al., 2004; Hyvärinen et al., 2005; Seymour et al., 2006; Spence & Volney, 1999). The greatest concentration of these studies has been in the Great Lakes and northeastern regions of the United States largely through the efforts of the Climate Change Response Framework (Fig. 13.3; Janowiak et al., 2014).

Syntheses of the applied adaptation strategies in a subset of demonstrations in this network underscore the influence of current forest conditions and prevailing management objectives on how climate adaptation is currently integrated into silvicultural prescriptions (Ontl et al., 2018). For example, in northern temperate and boreal regions of the network where intensive, historical land use has generated relatively homogeneous forest conditions (Schulte et al., 2007), adaptation strategies

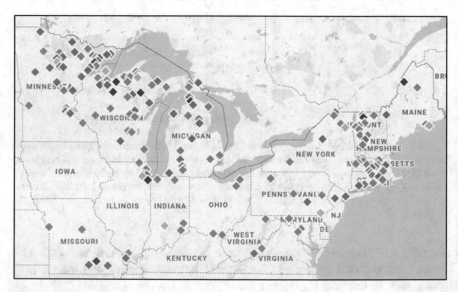

Fig. 13.3 Silvicultural experiments and demonstration areas evaluating various silvicultural adaptation strategies in the midwestern and northeastern United States as part of the Climate Change Response Framework (Janowiak et al., 2014). Since 2009, over 200 adaptation demonstrations have been established as part of this network, serving as early examples of how adaptation strategies can be operationalized across diverse forest conditions and ownership. Each area is designed with input from manager partners (i.e., *co-produced*) to ensure relevance to local ecological and operational contexts. Map obtained with permission from the Northern Institute of Applied Climate Science (NIACS)

have largely focused on increasing the diversity of canopy-tree species and the structural complexity of these forests (Ontl et al., 2018). In contrast, adaptation strategies in fire-adapted forests in the temperate region largely focus on the restoration of woodland structures and the introduction of prescribed fires (Ontl et al., 2018) to counter the long-standing outcomes of fire exclusion, e.g., higher tree densities and a greater abundance of mesophytic species (Hanberry et al., 2014). Overall, most adaptation strategies used by managers to date are best categorized as *resilience* approaches, highlighting a general reluctance to accept the initial risks and costs of more experimental transition strategies, a sentiment reflected in surveys of forest managers in other portions of the United States (Scheller & Parajuli, 2018) and Europe (Sousa-Silva et al., 2018).

The above summary highlights that many adaptation strategies will likely build off prevailing silvicultural approaches in a region, particularly in the near term. Some regions, such as boreal Canada, in which silvicultural systems rely heavily on artificial regeneration—either as part of plantation systems or as a supplement to natural regeneration—will have much greater capacity to implement resilience and transition strategies that rely on artificial regeneration than regions having historically relied solely on natural regeneration (Pedlar et al., 2012). Nonetheless for the boreal region and other regions, operationalizing novel transition strategies is not only hampered by a lack of experience but also by a limited nursery infrastructure and breeding programs. These programs would allow for species and genotypic selection to match projected climate and disturbance conditions for a given location (cf. O'Neill et al. 2017) and produce sufficient quantities to influence practices widely.

In many cases, the trigger for a more widespread application of novel adaptation strategies will likely be the realization that forest conditions are rapidly advancing toward undesirable thresholds because of changing climate, invasive species, and altered disturbance regimes. For instance, a fairly rapid shift toward applying transition strategies is underway in the Northern Lake States region in response to the threat to native black ash (*Fraxinus nigra*) wetlands from the introduced emerald ash borer (Rissman et al., 2018). The emerald ash borer is moving into the region in response to warming winters, and the habitat for native trees able to potentially replace black ash is rapidly declining because of climate change. In this example, novel enrichment plantings of climate-adapted, non-host species are being used as part of silvicultural treatments aimed at diversifying areas currently dominated by the host species and thus sustain post-invasion ecosystem functions (D'Amato et al., 2018).

13.4 Landscape and Regional Allocation of Adaptation Strategies

In addition to regional variation in the application of stand-scale adaptation strategies, within-region variation in ownership, management objectives, and the ability to absorb risks associated with experimental adaptation strategies may require landscape-level zonation into different intensities of adaptation (Park et al., 2014). The landscape is an important scale for adaptation planning because (1) major ecological processes such as metapopulation dynamics, species migration, and many natural disturbances occur at this scale; (2) forest habitat loss and fragmentation can only be addressed at large spatial scales; and (3) forest planning, including annual allowable harvest calculations, is multifaceted and depends on a variety of premises of current and targeted biophysical states as well as land ownership, policy, decision mandates, and governance mechanisms operating at the landscape scale. Correspondingly, the zonation of landscapes and regions into different silvicultural regimes has long been advocated as a strategy to achieve a diversity of objectives across ownerships (Seymour & Hunter, 1992; Tappeiner et al., 1986). In terms of application, zoning approaches are especially suitable to large areas under single ownership and characterized by low population densities (Sarr & Puettmann, 2008), such as for many boreal regions.

In most regions, including the boreal portions of Canada and Europe, zonation approaches have been motivated by potential incongruities between historical, commodity-focused objectives and those focused on broader nontimber objectives, including the maintenance of native biodiversity and cultural values (Côté et al., 2010; Messier et al., 2009; Naumov et al., 2018). Within the context of these often conflicting objectives, the TRIAD zonation model (Seymour & Hunter, 1992), has been popularized in parts of boreal Canada as a potential strategy for achieving diverse objectives over large landholdings. With this approach, landscapes are generally divided into intensive regions, characterized by high-input, production-focused silviculture (e.g., plantations), and extensive regions, where less-intensive approaches, such as ecological silviculture (sensu Palik et al., 2020), are used to attain nontimber objectives (e.g., biodiversity conservation, aesthetics) while also providing an opportunity for timber production (Fig. 13.4a). The third component of the TRIAD model—unmanaged, ecological reserves—are designated to protect unique ecological and cultural resources, enhance landscape connectivity, and serve as natural benchmarks to inform ecosystem management practices in extensively managed areas (Montigny & MacLean, 2005).

With its associated varying levels of silvicultural intensity and investment, the TRIAD zonation model is a useful construct for considering the opportunities and constraints to operationalizing adaptation strategies across large portions of the boreal forest (Park et al., 2014). For instance, high-input adaptation strategics, such as establishing future-adapted plantations, may be restricted to areas where intensive silvicultural regimes have predominated historically, such as lands proximate to mills. For instance, in western Canada, climate-informed reforestation strategies are most successful at minimizing drought-related reductions in timber volumes when resistant species and genotypes are planted proximate to mills and transportation routes, as opposed to more extensive planting approaches (Lochhead et al., 2019). In contrast, the financial and access constraints of extensively managed areas and the increasing risks of severe disturbance impacts (Boucher et al., 2017) argue for the use of a *portfolio* approach in these areas; this portfolio includes lower input resilience and

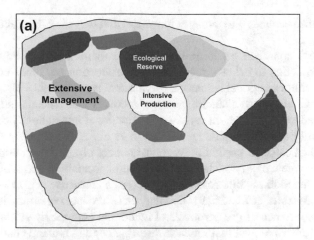

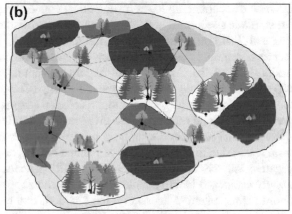

Fig. 13.4 (*top*) Forested landscape delineated according to TRIAD zonation (Seymour & Hunter, 1992), having zones of intensive production (*white polygons*), ecological reserves (*dark green polygons*), and extensive management in between. *Shades of green* within the extensive management zone indicate varying application levels of ecological silviculture (based on Palik et al., 2020). (*bottom*) Application of strategic, future-adapted planting across management intensities to generate functionally complex landscapes (sensu Messier et al., 2019); *tree size* depicts the level of deployed novel planting strategies, with intensive zones serving as central nodes of adaptation. *Solid lines* denote the functional connections between landscape elements, having similar response traits in planted species. *Dashed lines* represent long-term connections developed within unmanaged reserves, where no planting has occurred, because of the long-term colonization of the areas by future-adapted species planted in other portions of the landscape

higher input transition strategies that build from ecological silvicultural strategies, e.g., natural disturbance-based silvicultural systems, attributed to extensive zones under the TRIAD model (D'Amato & Palik, 2021). A key difference from the historical application of ecological silviculture is the integration of the targeted planting of future-adapted species—as enrichment plantings in actively managed stands or after natural disturbances—to increase functional diversity over time (Halofsky et al., 2020).

The TRIAD approach, as initially conceived, focused mainly on maximizing within-zone function to balance regional wood production and biodiversity conservation goals within a regional landscape (Seymour & Hunter, 1992). The emphasis of adaptation silviculture on enhancing potential recovery mechanisms and distributing risk has placed greater focus on cross-scale, functional interactions between zones when allocating adaptation strategies (Craven et al., 2016; Gömöry et al., 2020; Messier et al., 2019). In particular, a critical aspect of adaptation zonation is the strategic deployment of approaches, such as mixed-species plantations or enrichment plantings, to functionally link forest stands across a landscape (Fig. 13.4b; Aquilué et al., 2020; Messier et al., 2019). Guiding these recommendations is a recognition of the importance of greater levels of functional complexity at multiple scales to generate landscape-level resilience to disturbances and climate change (Messier et al., 2019). This includes enhancing levels of functional connectivity across landscape elements to facilitate species migration and recovery from disturbance (Millar et al., 2007; Nuñez et al., 2013) and designating central stands or *nodes* (sensu Craven et al., 2016) to serve as regional source populations for future-adapted species and key functional traits (Fig. 13.4b). Although still largely conceptual, future assessments of landscape-level functional connectivity and diversity (Craven et al., 2016) may be useful for prioritizing locations where more risk-laden adaptation strategies, such as novel species plantings, should occur in a given region (Aquilué et al., 2020). Note, however, that the risks associated with these strategies include not only financial and production losses due to maladaptation of planted species or genotypes but also potential negative impacts on forest-dependent wildlife species. Therefore, it becomes increasingly critical to identify strategies that maximize future adaptation potential while minimizing negative impacts on the functions and biota associated with ecological reserves and other portions of the landscape (cf. Tittler et al., 2015).

13.5 Conclusions

The application of silviculture has always assumed a level of uncertainty and risk in terms of ecological outcomes and socioeconomic feasibility and acceptability (Palik et al., 2020). Despite this uncertainty and risk, traditional silviculture approaches, after centuries of implementation, are well supported by long-term experience and research in many regions in the world. In contrast, in the context of rapid and novel global change, including climate, forest loss, disturbance, and invasive species, there is now an urgency to expand silviculture strategies to include high-risk *experimental* approaches, even if they are not well supported by long-term experience and research. Moreover, these approaches can still rely on the same framework for addressing uncertainty and risk that foresters have always used and understood (Palik et al., 2020). Given general aversions to risk, most field applications of adaptation strategies to date have built on past experiences and existing silvicultural practices; these include applying intermediate treatments to build resistance to change and ecological silvicultural practices to increase resilience. Although modeling exercises are useful for exploring responses to novel experimental strategies, like assisted migration, field experience with these approaches is currently limited, particularly at the operational scale. Given these challenges, we identify the following key needs to advance adaptation silviculture into widespread practice in forest landscapes:

- Integration of geospatial databases with disturbance and climate models to increase the spatial resolution of regional vulnerability assessments and allow a site-level determination of urgency and the appropriateness of adaptation strategies
- Improvement of existing modeling frameworks to better account for novel species interactions and potential feedbacks between future socioeconomic and ecological dynamics and adaptation practices over time
- Strategic investment in operational-scale adaptation experiments and demonstrations across regions, ecosystems, and site conditions, including high-risk strategies
- Coordination of the abovementioned experiments, trials, and demonstrations to allow for rapid information sharing among stakeholders and the adjustment of practices in response to observed outcomes and changing environmental dynamics
- Regional assessments of nursery capacity and novel stock availability in the context of adaptation plantings to prioritize investment in the propagation and wide distribution of desirable species and genotypes
- Continued development of trait-based indices to assist with operationalizing adaptation strategies focused on enhancing functional complexity across scales
- Consideration of relevant scales for the provision of ecosystem services to provide flexibility when applying adaptation strategies

Global change and its impacts appear to be greatly outpacing adaptation science, and investments in infrastructure must adapt. However, working to prioritize these scientific needs and investments, including deploying adaptation strategies in the near term that are compatible with current management frameworks, is critical to avoid crossing undesirable ecological thresholds. Seeing these rapidly approaching thresholds should serve as the primary motivating factor for moving forward with widespread adaptation to ensure the long-term sustainable production of goods and services.

References

Alotaibi, K., Ghumman, A. R., Haider, H., et al. (2018). Future predictions of rainfall and temperature using GCM and ANN for arid regions: A case study for the Qassim region. *Saudi Arabia Water, 10*, 1260. https://doi.org/10.3390/w10091260.
Aquilué, N., Filotas, É., Craven, D., et al. (2020). Evaluating forest resilience to global threats using functional response traits and network properties. *Ecological Applications, 30*, e02095. https://doi.org/10.1002/eap.2095.

Aubin, I., Boisvert-Marsh, L., Kebli, H., et al. (2018). Tree vulnerability to climate change: Improving exposure-based assessments using traits as indicators of sensitivity. *Ecosphere, 9*, e02108. https://doi.org/10.1002/ecs2.2108.

Aubin, I., Munson, A. D., Cardou, F., et al. (2016). Traits to stay, traits to move: A review of functional traits to assess sensitivity and adaptive capacity of temperate and boreal trees to climate change. *Environmental Reviews, 24*, 164–186. https://doi.org/10.1139/er-2015-0072.

Bauhus, J., Forrester, D. I., Gardiner, B., et al. (2017). Ecological stability of mixed-species forests. In H. Pretzsch, D. I. Forrester, & J. Bauhus (Eds.), *Mixed-species forests: Ecology and management* (pp. 337–382). Springer.

Bergeron, Y., Leduc, A., Joyal, C., et al. (1995). Balsam fir mortality following the last spruce budworm outbreak in northwestern Quebec. *Canadian Journal of Forest Research, 25*, 1375–1384. https://doi.org/10.1139/x95-150.

Biggs, C. R., Yeager, L. A., Bolser, D. G., et al. (2020). Does functional redundancy affect ecological stability and resilience? A Review and Meta-Analysis. *Ecosphere, 11*, e03184. https://doi.org/10.1002/ecs2.3184.

Biggs, R., Carpenter, S. R., & Brock, W. A. (2009). Turning back from the brink: Detecting an impending regime shift in time to avert it. *Proceedings of the National Academy of Sciences of the United States of America, 106*, 826–831. https://doi.org/10.1073/pnas.0811729106.

Boisvert-Marsh, L., Royer-Tardif, S., Nolet, P., et al. (2020). Using a trait-based approach to compare tree species sensitivity to climate change stressors in eastern Canada and inform adaptation practices. *Forests, 11*, 989. https://doi.org/10.3390/f11090989.

Bottero, A., D'Amato, A. W., Palik, B. J., et al. (2017). Density-dependent vulnerability of forest ecosystems to drought. *Journal of Applied Ecology, 54*, 1605–1614. https://doi.org/10.1111/1365-2664.12847.

Boucher, Y., Auger, I., Noël, J., et al. (2017). Fire is a stronger driver of forest composition than logging in the boreal forest of eastern Canada. *Journal of Vegetation Science, 28*, 57–68. https://doi.org/10.1111/jvs.12466.

Brais, S., Harvey, B. D., Bergeron, Y., et al. (2004). Testing forest ecosystem management in boreal mixedwoods of northwestern Quebec: Initial response of aspen stands to different levels of harvesting. *Canadian Journal of Forest Research, 34*, 431–446. https://doi.org/10.1139/x03-144.

Brokaw, N., & Busing, R. T. (2000). Niche versus chance and tree diversity in forest gaps. *Trends in Ecology and Evolution, 15*, 183–188. https://doi.org/10.1016/S0169-5347(00)01822-X.

Butler, B. W., Ottmar, R. D., Rupp, T. S., et al. (2013). Quantifying the effect of fuel reduction treatments on fire behavior in boreal forests. *Canadian Journal of Forest Research, 43*, 97–102. https://doi.org/10.1139/cjfr-2012-0234.

Camarero, J., Gazol, A., Sangüesa-Barreda, G., et al. (2018). Forest growth responses to drought at short- and long-term scales in Spain: Squeezing the stress memory from tree rings. *Frontiers in Ecology and Evolution, 6*, 9. https://doi.org/10.3389/fevo.2018.00009.

Campbell, E. M., MacLean, D. A., & Bergeron, Y. (2008). The severity of budworm-caused growth reductions in balsam fir/spruce stands varies with the hardwood content of surrounding forest landscapes. *Forestry Sciences, 54*, 195–205.

Chmura, D. J., Anderson, P. D., Howe, G. T., et al. (2011). Forest responses to climate change in the northwestern United States: Ecophysiological foundations for adaptive management. *Forest Ecology and Management, 261*(7), 1121–1142. https://doi.org/10.1016/j.foreco.2010.12.040.

Christensen, N. L., Bartuska, A. M., Brown, J. H., et al. (1996). The report of the Ecological Society of America Committee on the scientific basis for ecosystem management. *Ecological Applications, 6*, 665–691. https://doi.org/10.2307/2269460.

Côté, P., Tittler, R., Messier, C., et al. (2010). Comparing different forest zoning options for landscape-scale management of the boreal forest: Possible benefits of the TRIAD. *Forest Ecology and Management, 259*, 418–427. https://doi.org/10.1016/j.foreco.2009.10.038.

Craven, D., Filotas, E., Angers, V. A., et al. (2016). Evaluating resilience of tree communities in fragmented landscapes: Linking functional response diversity with landscape connectivity. *Diversity and Distributions, 22*, 505–518. https://doi.org/10.1111/ddi.12423.

D'Amato, A. W., Bradford, J. B., Fraver, S., et al. (2013). Effects of thinning on drought vulnerability and climate response in north temperate forest ecosystems. *Ecological Applications, 23*, 1735–1742. https://doi.org/10.1890/13-0677.1.

D'Amato, A. W., & Palik, B. J. (2021). Building on the last "new" thing: Exploring the compatibility of ecological and adaptation silviculture. *Canadian Journal of Forest Research, 51*, 172–180. https://doi.org/10.1139/cjfr-2020-0306.

D'Amato, A. W., Palik, B. J., Slesak, R. A., et al. (2018). Evaluating adaptive management options for black ash forests in the face of emerald ash borer invasion. *Forests, 9*, 348. https://doi.org/10. 3390/f9060348.

Daly, C., Conklin, D. R., & Unsworth, M. H. (2010). Local atmospheric decoupling in complex topography alters climate change impacts. *International Journal of Climatology, 30*, 1857–1864. https://doi.org/10.1002/joc.2007.

Dean, T. J. (2012). A simple case for the term "tolerance." *Journal of Forestry, 110*, 463–464. https://doi.org/10.5849/jof.12-039.

Depardieu, C., Girardin, M. P., Nadeau, S., et al. (2020). Adaptive genetic variation to drought in a widely distributed conifer suggests a potential for increasing forest resilience in a drying climate. *New Phytologist, 227*, 427–439. https://doi.org/10.1111/nph.16551.

Donoso, P. J., Puettmann, K. J., D'Amato, A. W., et al. (2020). Short-term effects of variable-density thinning on regeneration in hardwood-dominated temperate rainforests. *Forest Ecology and Management, 464*, 118058. https://doi.org/10.1016/j.foreco.2020.118058.

Drobyshev, I., Gewehr, S., Berninger, F., et al. (2013). Species specific growth responses of black spruce and trembling aspen may enhance resilience of boreal forest to climate change. *Journal of Ecology, 101*, 231–242. https://doi.org/10.1111/1365-2745.12007.

Duveneck, M. J., & Scheller, R. M. (2015). Climate-suitable planting as a strategy for maintaining forest productivity and functional diversity. *Ecological Applications, 25*, 1653–1668. https://doi. org/10.1890/14-0738.1.

Dymond, C. C., Giles-Hansen, K., & Asante, P. (2020). The forest mitigation-adaptation nexus: Economic benefits of novel planting regimes. *Forest Policy and Economics, 113*, 102124. https:// doi.org/10.1016/j.forpol.2020.102124.

Dymond, C. C., Tedder, S., Spittlehouse, D. L., et al. (2014). Diversifying managed forests to increase resilience. *Canadian Journal of Forest Research, 44*, 1196–1205. https://doi.org/10. 1139/cjfr-2014-0146.

Elkin, C., Giuggiola, A., Rigling, A., et al. (2015). Short- and long-term efficacy of forest thinning to mitigate drought impacts in mountain forests in the European Alps. *Ecological Applications, 25*, 1083–1098. https://doi.org/10.1890/14-0690.1.

Elmqvist, T., Folke, C., Nystrom, M., et al. (2003). Response diversity, ecosystem change, and resilience. *Frontiers in Ecology and the Environment, 1*, 488–494. https://doi.org/10.1890/1540-9295(2003)001[0488:RDECAR]2.0.CO;2.

Etterson, J. R., Cornett, M. W., White, M. A., et al. (2020). Assisted migration across fixed seed zones detects adaptation lags in two major North American tree species. *Ecological Applications, 30*, e02092. https://doi.org/10.1002/eap.2092.

Flint, C. G., McFarlane, B., & Müller, M. (2009). Human dimensions of forest disturbance by insects: An international synthesis. *Environmental Management, 43*, 1174–1186. https://doi.org/ 10.1007/s00267-008-9193-4.

Frey, S. J. K., Hadley, A. S., Johnson, S. L., et al. (2016). Spatial models reveal the microclimatic buffering capacity of old-growth forests. *Science Advances, 2*(4), e1501392. https://doi.org/10. 1126/sciadv.1501392.

Gauthier, S., Bernier, P., Burton, P. J., et al. (2014). Climate change vulnerability and adaptation in the managed Canadian boreal forest. *Environmental Reviews, 22*, 256–285. https://doi.org/10. 1139/er-2013-0064.

Gömöry, D., Krajmerová, D., Hrivnák, M., et al. (2020). Assisted migration vs. close-to-nature forestry: what are the prospects for tree populations under climate change? *Central European Forestry Journal, 66*(2), 63–70. https://doi.org/10.2478/forj-2020-0008.

Gunderson, L. H. (2000). Ecological resilience-in theory and application. *Annual Review of Ecology, Evolution, and Systematics, 31*, 425–439. https://doi.org/10.1146/annurev.ecolsys.31.1.425.

Halofsky, J. E., Peterson, D. L., & Harvey, B. J. (2020). Changing wildfire, changing forests: The effects of climate change on fire regimes and vegetation in the Pacific Northwest, USA. *Fire Ecology, 16*, 4. https://doi.org/10.1186/s42408-019-0062-8.

Hanberry, B. B., Kabrick, J. M., & He, H. S. (2014). Densification and state transition across the Missouri Ozarks landscape. *Ecosystems, 17*, 66–81. https://doi.org/10.1007/s10021-013-9707-7.

Hemery, G. (2008). Forest management and silvicultural responses to projected climate change impacts on European broadleaved trees and forests. *International Forestry Review, 10*, 591–607. https://doi.org/10.1505/ifor.10.4.591.

Hengst-Ehrhart, Y. (2019). Knowing is not enough: Exploring the missing link between climate change knowledge and action of German forest owners and managers. *Annals of Forest Science, 76*, 94. https://doi.org/10.1007/s13595-019-0878-z.

Hof, A. R., Dymond, C. C., & Mladenoff, D. J. (2017). Climate change mitigation through adaptation: The effectiveness of forest diversification by novel tree planting regimes. *Ecosphere, 8*, e01981. https://doi.org/10.1002/ecs2.1981.

Hyvärinen, E., Kouki, J., Martikainen, P., et al. (2005). Short-term effects of controlled burning and green-tree retention on beetle (Coleoptera) assemblages in managed boreal forests. *Forest Ecology and Management, 212*, 315–332. https://doi.org/10.1016/j.foreco.2005.03.029.

Janowiak, M. K., Swanston, C. W., Nagel, L. M., et al. (2014). A practical approach for translating climate change adaptation principles into forest management actions. *Journal of Forestry, 112*, 424–433. https://doi.org/10.5849/jof.13-094.

Johnstone, J. F., Allen, C. D., Franklin, J. F., et al. (2016). Changing disturbance regimes, ecological memory, and forest resilience. *Frontiers in Ecology and the Environment, 14*, 369–378. https://doi.org/10.1002/fee.1311.

Johnstone, J. F., Rupp, T. S., Olson, M., et al. (2011). Modeling impacts of fire severity on successional trajectories and future fire behavior in Alaskan boreal forests. *Landscape Ecology, 26*, 487–500. https://doi.org/10.1007/s10980-011-9574-6.

Kurz, W. A., Stinson, G., Rampley, G. J., et al. (2008). Risk of natural disturbances makes future contribution of Canada's forests to the global carbon cycle highly uncertain. *Proceedings of the National Academy of Sciences of the United States of America, 105*, 1551–1555. https://doi.org/10.1073/pnas.0708133105.

Landres, P. B., Morgan, P., & Swanson, F. J. (1999). Overview of the use of natural variability concepts in managing ecological systems. *Ecological Applications, 9*(4), 1179–1188. https://doi.org/10.1890/1051-0761(1999)009[1179:OOTUON]2.0.CO;2.

Lochhead, K., Ghafghazi, S., LeMay, V., et al. (2019). Examining the vulnerability of localized reforestation strategies to climate change at a macroscale. *Journal of Environmental Management, 252*, 109625. https://doi.org/10.1016/j.jenvman.2019.109625.

Messier, C., Bauhus, J., Doyon, F., et al. (2019). The functional complex network approach to foster forest resilience to global changes. *Forest Ecosystems, 6*, 21. https://doi.org/10.1186/s40663-019-0166-2.

Messier, C., Puettmann, K., Chazdon, R., et al. (2015). From management to stewardship: Viewing forests as complex adaptive systems in an uncertain world. *Conservation Letters, 8*, 368–377. https://doi.org/10.1111/conl.12156.

Messier, C., Tittler, R., Kneeshaw, D. D., et al. (2009). TRIAD zoning in Quebec: Experiences and results after 5 years. *The Forestry Chronicle, 85*, 885–896. https://doi.org/10.5558/tfc85885-6.

Metz, J., Annighöfer, P., Schall, P., et al. (2016). Site-adapted admixed tree species reduce drought susceptibility of mature European beech. *Global Change Biology, 22*, 903–920. https://doi.org/10.1111/gcb.13113.

Meyers, L. A., & Bull, J. J. (2002). Fighting change with change: Adaptive variation in an uncertain world. *Trends in Ecology and Evolution, 17*, 551–557. https://doi.org/10.1016/S0169-5347(02)02633-2.

Millar, C. I., Stephenson, N. L., & Stephens, S. L. (2007). Climate change and forests of the future: Managing in the face of uncertainty. *Ecological Applications, 17*, 2145–2151. https://doi.org/10.1890/06-1715.1.

Montigny, M. K., & MacLean, D. A. (2005). Using heterogeneity and representation of ecosite criteria to select forest reserves in an intensively managed industrial forest. *Biological Conservation, 125*, 237–248. https://doi.org/10.1016/j.biocon.2005.03.028.

Muller, J. J., Nagel, L. M., & Palik, B. J. (2019). Forest adaptation strategies aimed at climate change: Assessing the performance of future climate-adapted tree species in a northern Minnesota pine ecosystem. *Forest Ecology and Management, 451*, 117539. https://doi.org/10.1016/j.foreco.2019.117539.

Mumby, P. J., Chollett, I., Bozec, Y. M., et al. (2014). Ecological resilience, robustness and vulnerability: How do these concepts benefit ecosystem management? *Current Opinion in Environment Sustainability, 7*, 22–27. https://doi.org/10.1016/j.cosust.2013.11.021.

Nagel, L. M., Palik, B. J., Battaglia, M. A., et al. (2017). Adaptive silviculture for climate change: A national experiment in manager-scientist partnerships to apply an adaptation framework. *Journal of Forestry, 115*, 167–178. https://doi.org/10.5849/jof.16-039.

Naumov, V., Manton, M., Elbakidze, M., et al. (2018). How to reconcile wood production and biodiversity conservation? The Pan-European boreal forest history gradient as an "experiment." *Journal of Environmental Management, 218*, 1–13. https://doi.org/10.1016/j.jenvman.2018.03.095.

Navarro, L., Morin, H., Bergeron, Y., et al. (2018). Changes in spatiotemporal patterns of 20th century spruce budworm outbreaks in eastern Canadian boreal forests. *Frontiers in Plant Science, 9*, 1905. https://doi.org/10.3389/fpls.2018.01905.

Neff, M. W., & Larson, B. M. H. (2014). Scientists, managers, and assisted colonization: Four contrasting perspectives entangle science and policy. *Biological Conservation, 172*, 1–7. https://doi.org/10.1016/j.biocon.2014.02.001.

Nitschke, C. R., & Innes, J. L. (2008). Integrating climate change into forest management in South-Central British Columbia: An assessment of landscape vulnerability and development of a climate-smart framework. *Forest Ecology and Management, 256*, 313–327. https://doi.org/10.1016/j.foreco.2008.04.026.

Nolet, P., Doyon, F., & Messier, C. (2014). A new silvicultural approach to the management of uneven-aged Northern hardwoods: Frequent low-intensity harvesting. *Forestry, 87*, 39–48.

Nuñez, T. A., Lawler, J. J., McRae, B. H., et al. (2013). Connectivity planning to address climate change. *Conservation Biology, 27*, 407–416. https://doi.org/10.1111/cobi.12014.

O'Hara, K. L., & Ramage, B. S. (2013). Silviculture in an uncertain world: Utilizing multi-aged management systems to integrate disturbance. *Forestry, 86*, 401–410. https://doi.org/10.1093/forestry/cpt012.

O'Neill, G.A., Wang, T., Ukrainetz, N., et al. (2017). A proposed climate-based seed transfer system for British Columbia. Technical Report 099. Victoria: Province of British Columbia.

Olson, M. E., Soriano, D., Rosell, J. A., et al. (2018). Plant height and hydraulic vulnerability to drought and cold. *Proceeding of the National Academy of Sciences of the United States of America, 115*, 7551–7556. https://doi.org/10.1073/pnas.1721728115.

Ontl, T. A., Swanston, C., Brandt, L. A., et al. (2018). Adaptation pathways: Ecoregion and land ownership influences on climate adaptation decision-making in forest management. *Climatic Change, 146*, 75–88. https://doi.org/10.1007/s10584-017-1983-3.

Palik, B. J., D'Amato, A. W., Franklin, J. F., et al. (2020). *Ecological silviculture: Foundations and applications*. Waveland Press.

Park, A., Puettmann, K., Wilson, E., et al. (2014). Can boreal and temperate forest management be adapted to the uncertainties of 21st century climate change? *Critical Reviews in Plant Sciences, 33*, 251–285. https://doi.org/10.1080/07352689.2014.858956.

Park, A., & Talbot, C. (2018). Information underload: Ecological complexity, incomplete knowledge, and data deficits create challenges for the assisted migration of forest trees. *BioScience, 68*, 251–263. https://doi.org/10.1093/biosci/biy001.

Pedlar, J. H., McKenney, D. W., Aubin, I., et al. (2012). Placing forestry in the assisted migration debate. *BioScience, 62*, 835–842. https://doi.org/10.1525/bio.2012.62.9.10.

Price, D. T., Alfaro, R. I., Brown, K. J., et al. (2013). Anticipating the consequences of climate change for Canada's boreal forest ecosystems. *Environmental Reviews, 21*, 322–365. https://doi.org/10.1139/er-2013-0042.

Puettmann, K. J. (2011). Silvicultural challenges and options in the context of global change: Simple fixes and opportunities for new management approaches. *Journal of Forestry, 109*, 321–331.

Puettmann, K. J., Coates, K. D., & Messier, C. C. (2009). *A critique of silviculture: Managing for complexity*. Island Press.

Puettmann, K. J., & Messier, C. (2020). Simple guidelines to prepare forests for global change: The dog and the frisbee. *Northwest Science, 93*(3–4), 209–225. https://doi.org/10.3955/046.093.0305.

Rissman, A. R., Burke, K. D., Kramer, H. A. C., et al. (2018). Forest management for novelty, persistence, and restoration influenced by policy and society. *Frontiers in Ecology and the Environment, 16*, 454–462. https://doi.org/10.1002/fee.1818.

Rowe, J. S. (1983). Concepts of fire effects on plant individuals and species. In R. W. Wein & D. A. MacLean (Eds.), *The role of fire in northern circumpolar ecosystems* (pp. 135–153). John Wiley and Sons Ltd.

Ruehr, N. K., Grote, R., Mayr, S., et al. (2019). Beyond the extreme: Recovery of carbon and water relations in woody plants following heat and drought stress. *Tree Physiology, 39*, 1285–1299. https://doi.org/10.1093/treephys/tpz032.

Sarr, D. A., & Puettmann, K. J. (2008). Forest management, restoration, and designer ecosystems: Integrating strategies for a crowded planet. *Ecoscience, 15*, 17–26. https://doi.org/10.2980/1195-6860(2008)15[17:FMRADE]2.0.CO;2.

Scheller, R. M., & Parajuli, R. (2018). Forest management for climate change in New England and the Klamath ecoregions: Motivations, practices, and barriers. *Forests, 9*, 626. https://doi.org/10.3390/f9100626.

Schulte, L. A., Mladenoff, D. J., Crow, T. R., et al. (2007). Homogenization of northern US Great Lakes forests due to land use. *Landscape Ecology, 22*, 1089–1103. https://doi.org/10.1007/s10980-007-9095-5.

Seidl, R., & Lexer, M. J. (2013). Forest management under climatic and social uncertainty: Trade-offs between reducing climate change impacts and fostering adaptive capacity. *Journal of Environmental Management, 114*, 461–469. https://doi.org/10.1016/j.jenvman.2012.09.028.

Seidl, R., Schelhaas, M. J., Rammer, W., et al. (2014). Increasing forest disturbances in Europe and their impact on carbon storage. *Nature Climate Change, 4*, 806–810. https://doi.org/10.1038/nclimate2318.

Seidl, R., Thom, D., Kautz, M., et al. (2017). Forest disturbances under climate change. *Nature Climate Change, 7*(6), 395–402. https://doi.org/10.1038/nclimate3303.

Seymour, R. S., Guldin, J., Marshall, D., et al. (2006). Large-scale, long-term silvicultural experiments in the United States: Historical overview and contemporary examples. *Allgemeine Forst Jagdzeitung, 177*, 104–112.

Seymour, R. S., & Hunter, M. L. (1992). *New forestry in eastern spruce-fir forests: Principles and applications to Maine*. Orono: University of Maine.

Smith, D. M. (1962). *The practice of silviculture* (7th ed.). Wiley.

Sohn, J. A., Hartig, F., Kohler, M., et al. (2016). Heavy and frequent thinning promotes drought adaptation in *Pinus sylvestris* forests. *Ecological Applications, 26*, 2190–2205. https://doi.org/10.1002/eap.1373.

Sousa-Silva, R., Verbist, B., Lomba, Â., et al. (2018). Adapting forest management to climate change in Europe: Linking perceptions to adaptive responses. *Forest Policy and Economics, 90*, 22–30. https://doi.org/10.1016/j.forpol.2018.01.004.

Spence, J. R., & Volney, W. J. A. (1999). *EMEND: Ecosystem management emulating natural disturbance*. 1999–14 Edmonton: Sustainable Forest Management Network.

Swanston, C. W., Janowiak, M. K., & Brandt, L. A., et al. (2016). *Forest adaptation resources: climate change tools and approaches for land managers* (General Technical Report. NRS-GTR-87–2) (p. 161). Newtown Square: U.S. Department of Agriculture, Forest Service, Northern Research Station.

Tappeiner, J. C., Knapp, W. H., Wiermann, C. A., et al. (1986). Silviculture: The next 30 years, the past 30 years. Part II. The Pacific Coast. *Journal of Forestry, 84*, 37–46.

Thorne, J. H., Gogol-Prokurat, M., Hill, S., et al. (2020). Vegetation refugia can inform climate-adaptive land management under global warming. *Frontiers in Ecology and the Environment, 18*, 281–287. https://doi.org/10.1002/fee.2208.

Tittler, R., Filotas, É., Kroese, J., et al. (2015). Maximizing conservation and production with intensive forest management: It's all about location. *Environmental Management, 56*, 1104–1117. https://doi.org/10.1007/s00267-015-0556-3.

Vitali, V., Forrester, D. I., & Bauhus, J. (2018). Know your neighbours: Drought response of Norway spruce, silver fir and Douglas fir in mixed forests depends on species identity and diversity of tree neighbourhoods. *Ecosystems, 21*, 1215–1229. https://doi.org/10.1007/s10021-017-0214-0.

Whittet, R., Cavers, S., Cottrell, J., et al. (2016). Seed sourcing for woodland creation in an era of uncertainty: An analysis of the options for Great Britain. *Forestry, 90*, 163–173.

Wiens, J. A., Stralberg, D., Jongsomjit, D., et al. (2009). Niches, models, and climate change: Assessing the assumptions and uncertainties. *Proceeding of the National Academy of Sciences of the United States of America, 106*, 19729–19736. https://doi.org/10.1073/pnas.0901639106.

Wilhelmi, N. P., Shaw, D. C., Harrington, C. A., et al. (2017). Climate of seed source affects susceptibility of coastal Douglas-fir to foliage diseases. *Ecosphere, 8*(12), e02011. https://doi.org/10.1002/ecs2.2011.

Yachi, S., & Loreau, M. (1999). Biodiversity and ecosystem productivity in a fluctuating environment: The insurance hypothesis. *Proceeding of the National Academy of Sciences of the United States of America, 96*, 1463–1468. https://doi.org/10.1073/pnas.96.4.1463.

Chapter 14
Plantation Forestry, Tree Breeding, and Novel Tools to Support the Sustainable Management of Boreal Forests

Nelson Thiffault, Patrick R. N. Lenz, and Karin Hjelm

Abstract Successful stand regeneration is one of the keystone elements of sustainable forest management. It ensures that ecosystems submitted to stand-replacing disturbances return to a forested state so that they can maintain the provision of wood fiber, biodiversity, carbon sequestration, and other ecosystem services. This chapter describes how plantation forestry, including tree breeding, and novel tools, such as genomic selection, can support the sustainable management of boreal forests in the face of climate change by, among other benefits, reducing management pressure on natural forests and favoring ecosystem restoration.

N. Thiffault (✉)
Natural Resources Canada, Canadian Forest Service, Canadian Wood Fibre Centre, 1055 du PEPS, P.O. Box 10380, Stn. Sainte-Foy, Québec, QC G1V 4C7, Canada
e-mail: nelson.thiffault@nrcan-rncan.gc.ca

Centre for Forest Research, Université du Québec à Montréal, P.O. Box 8888, Stn. Centre-Ville, Montréal, QC H3C 3P8, Canada

IUFRO Task Force on Resilient Planted Forests Serving Society and Bioeconomy, Marxergasse 2, A-1030 Vienna, Austria

P. R. N. Lenz
Natural Resources Canada, Canadian Forest Service, Canadian Wood Fibre Centre, 1055 rue du PEPS, P.O. Box 10380, Stn. Sainte-Foy, Québec, QC G1V 4C7, Canada
e-mail: patrick.lenz@nrcan-rncan.gc.ca

K. Hjelm
Southern Swedish Forest Research Centre, Swedish University of Agricultural Sciences, Box 190, SE-234 22 Lomma, Sweden
e-mail: karin.hjelm@slu.se

M. M. Girona et al. (eds.), *Boreal Forests in the Face of Climate Change*,
Advances in Global Change Research 74,
https://doi.org/10.1007/978-3-031-15988-6_14

14.1   Introduction

Natural disturbances, such as wildfire, insect outbreaks, and windthrow, and anthropogenic disturbances, such as harvesting, are common in the boreal biome (Brandt et al., 2013; Shorohova et al., 2009). These events modify stand structure and affect the availability of environmental resources. Canopy removal increases light levels in the understory, modifying the microenvironment, plant community, and tree regeneration. These changes can have cascading effects on the capacity of forests to sustain their provision of ecosystem services. Vegetation can rapidly colonize the disturbed areas and prevent regeneration of the desired tree species or forest composition. For example, in some black spruce (*Picea mariana* (Mill.) BSP) stands of eastern Canada, the effects of harvesting on light levels and water table depth can trigger the growth of bryophyte communities; this shift favors paludification (Fenton & Bergeron, 2006), which in turn reduces forest productivity (Leroy et al., 2016). Similarly, site encroachment by ericaceous species such as *Kalmia angustifolia* L. or *Empetrum hermaphroditum* Hagerup can lead to a significant decline in soil fertility and conifer growth rates on some forest sites (Mallik, 2003).

Because successful stand regeneration can mitigate these effects, it is one of the keystone elements of sustainable forest management. In Canada, for example, regeneration success is used to monitor changes in conditions relevant to sustainable forest management under the Montreal Process (NRC 2020). It is also mandatory in Norway, Finland, and Sweden.

Successful regeneration, both from natural propagules and plantation practices, ensures that ecosystems submitted to stand-replacing disturbances return to a forested

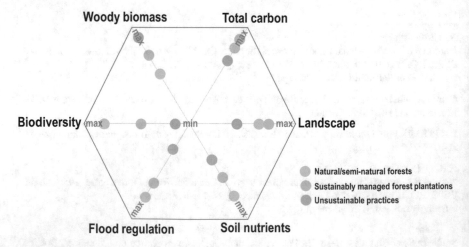

Fig. 14.1 Schematic representation of the relative effects of forest management intensity, including plantation silviculture, on the procurement of some ecosystem services. Concepts are detailed in Nijnik et al. (2014) and Freer-Smith et al. (2019)

state so that they can maintain wood fiber supply, biodiversity, carbon sequestration, wildlife habitat, spiritual values, social values, and other ecosystem services (Fig. 14.1).

Boreal ecosystems are dominated by conifer species having long-lived aerial seed banks, reproduction from vegetative growth (e.g., the formation of layers around trees), established dense understory seedling banks, or a combination of these characteristics (Thiffault et al., 2015). Depending on jurisdictions and years, natural regeneration can thus play a substantial role in forest renewal. For example, in Sweden, about 24% of the area harvested between 1999 and 2019 regenerated naturally, although showing a declining trend (SFA, 2020). In Canada, the average was about 44% between 2000 and 2020 (NRC, 2020). The remaining areas are regenerated with plantations, which can take various forms. They include intensively managed, even-aged forest areas planted with one or two species at a regular spacing. The primary objective of these regenerated areas is wood production (Fig. 14.2a, b), although there is also a consideration of environmental and societal values in most boreal jurisdictions. In contrast, less intensively managed plantations, resembling natural forests at stand maturity, have the main purpose of ecosystem restoration, the protection of soil and water values, and/or support of socioeconomic objectives (Fig. 14.2c). Globally, intensively managed and other planted forests cover about 291 million ha, an area that represents 7% of the world's forests (FAO, 2020).

The capability of forest plantations to fulfill their role relies on interdependent decisions and actions. In most forestry contexts in the boreal zone, this means selecting appropriate genotypes and seedling size, managing the soil and humus to create appropriate microsites, controlling competing vegetation, managing stand density, and, in some cases, increasing nutrient availability. Tree breeding, silviculture, and their interactions drive the production of ecosystem services from planted forests (e.g., Burdon et al., 2017). This chapter describes the actual and potential role that plantation silviculture, tree breeding, and novel tools such as genomic selection can play in supporting the sustainable management of boreal forests in the face of climate change. First, we summarize some of the fundamentals of plantation silviculture and show how various treatments support sustainable forest management objectives. Then, we explore the role of tree breeding and genomic tools in assisting forest management. Third, we provide examples illustrating the role plantation forestry plays in maintaining various ecosystem services from boreal stands in the context of global change. Finally, we identify some issues and challenges facing plantation forestry in the context of sustainable forest management.

14.2 Plantation Establishment and Silviculture

When a forest stand is harvested, the energy previously captured by the canopy now reaches the understory and the soil (Fig. 14.3); this exposure increases soil and air temperature and the evaporative demand of the air. For newly planted seedlings, a higher soil temperature can positively affect root growth and the uptake of water and

Fig. 14.2 Examples of various plantations, showing **a** an intensively managed white spruce (*Picea glauca* (Moench) Voss.) plantation in eastern Canada, **b** a mature Norway spruce (*Picea abies* (L.) Karst.) plantation in Sweden, **c** an extensively managed white spruce plantation in eastern Canada containing natural balsam fir (*Abies balsamea* (L.) Mill.) regeneration, and **d** a Norway spruce plantation in Sweden that was submitted to cleaning with the retention of naturally regenerated Scots pine (*Pinus sylvestris* L.). *Photo credits* **a, c** Nelson Thiffault, **b, d** Karin Hjelm

nutrients and reduce the risk of frosts. On the other hand, the increased evaporative demand and higher air temperatures can increase the risk of drought. Although soil water availability increases when trees are removed and more precipitation reaches the ground, colonizing vegetation competes for water in the rooting zone. Because the root functioning of newly planted seedlings is often poor (Grossnickle, 2005), this may further increase the impact of drought. Moreover, rapid changes in temperature between day and night can cause frost damage. Many nutrients are removed by harvesting the standing trees, although changes in the energy balance can heighten some nutrient cycle processes such as nitrification (Jerabkova et al., 2011).

In this context, plantation success in supporting sustainable forest management objectives relies on the proper selection and use of stock type and silviculture treatments (Rubilar et al., 2018). These decisions ensure that the planted seedlings have access to sufficient environmental resources from the time of planting until maturity

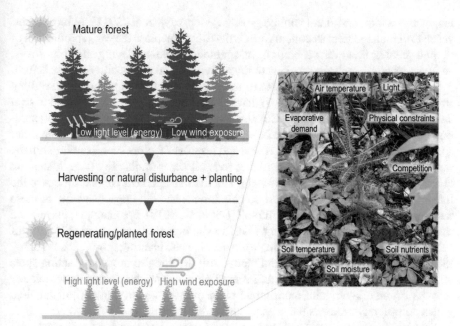

Fig. 14.3 Some environmental factors influenced by harvesting or natural disturbances that determine the establishment success of newly planted seedlings in the boreal forest. These factors can be manipulated using silvicultural treatments, such as mechanical site preparation, vegetation management, and fertilization, so that plantations can support sustainable forest management objectives. Tree-breeding programs can also select genotypes best adapted to sustain specific environmental conditions

so that survival is maximized and growth rates meet the silviculture and management objectives.

In general, seedling stock types vary in the size of the shoot/above ground biomass and the morphology of their root system. The choice of stock type size is generally based on the competing environment in which the seedlings are to be outplanted, as initial seedling size influences their inherent growth and capacity to compete for environmental resources (Jobidon et al., 2003). Larger seedlings are generally preferred on sites dominated, or that have the potential to be dominated, by fast-growing, light-demanding species. Smaller seedlings are ideal for sites where competition for light is low, as generally found at high latitudes where competing species consist mainly of shrubs, mosses, and lichens (Bell et al., 2011).

Most harvesting treatments alter vegetation dynamics; species well adapted to the new environmental conditions establish rapidly, occupy the site, and compete with planted seedlings for resources (Fig. 14.3). Vegetation management aims to direct the evolution of the forest succession to achieve a range of management objectives. The use of mechanical, chemical, biological treatments, or a combination thereof, applied during the various stages of early stand development can improve planted tree growth, vigor, resistance to damage from insects, survival, nutrient status, crown length and width, and stand volume growth (Wiensczyk et al., 2011). In most cases, vegetation management treatments are carried out to increase the wood production of desired species; however, vegetation management enables achieving other objectives, such as restoring declining species or diversifying stand composition and structure.

Mechanical site preparation is used to improve site factors and reduce seedling stress following planting (Fig. 14.3), leading to positive effects on seedling survival and growth (Sikström et al., 2020). In boreal contexts, mechanical site preparation increases the availability of site resources by reducing competition from other species colonizing the regeneration area and improves factors such as microclimate, nutrient mineralization, soil temperature, and soil water availability (e.g., Johansson et al., 2013). Site-specific characteristics and the management context influence the intensity of mechanical site preparation treatments and the impact severity of these treatments on the forest floor and the soil (Löf et al., 2012). For example, the applied treatment can consist of disturbing (locally) the organic layers through the use of motor manual equipment, mounding to create elevated planting spots, disk trenching to create linear rows of furrows and berms, soil inverting to produce planting spots with the mineral soil lying above an inverted humus layer, harrowing to completely mix the organic layers and incorporate them into the underlying soil, or blading, which completely removes the organic layer over large areas of soil.

Stand density (the number of stems growing per unit of space) influences productivity at the tree and stand levels. The size of individual trees is largest at low density because trees are exposed to low levels of intraspecific competition. At higher tree densities, volume production at the stand level is maximized because site occupancy is optimized (Groot & Cortini, 2016). Density management thus offers the opportunity to manipulate resource allocation to best fit the sustainable forest management objectives being pursued. In plantations, stand density is managed at the establishment phase by prescribing the planting distance between the seedlings. Thinning or cleaning treatments can later be used, either at the precommercial or commercial stage of stand development, to maintain or reduce stand density and select crop trees (Fig. 14.2d; Pelletier & Pitt, 2008). Thinning operations reduce competition between crop trees; hence, they improve the growth of the remaining stems. Although increased volume and radial growth rates generally lead to decreased wood density (Jaakkola et al., 2005), these effects can be nonsignificant (Franceschini et al., 2018; Vincent et al., 2011). The pruning of dead or living branches can also be used to increase wood quality and value (Mäkinen et al., 2014).

The availability of soil nitrogen is one of the major growth-limiting factors in boreal forests (Tamm, 1991). Fertilizers can be applied at planting, in the later stages of stand rotation, or at several points in time to promote plantation growth and achieve sustainable forest management objectives. Using fertilizers at planting can promote the rapid establishment and high initial growth of trees; for example, positive effects of amendments have been documented when used in combination with site preparation (Thiffault & Jobidon, 2006) or with nutrient irrigation (Johansson et al., 2012). The fertilization of mature stands, for its part, is seen as one of the most economically important measures to increase wood production. By adding nitrogen to middle-aged or older stands, usually a few years after the last thinning, tree growth can be increased significantly (Jacobson & Pettersson, 2010).

14.3   Tree Breeding and Genomic Selection

Millions of seedlings are planted each year in the boreal forest, with more than 450 million and 350 million seedlings planted annually in Canada and Sweden, respectively. In most northern countries, seedling material is improved for growth, or at least comes from known origins, to ensure its quality and adaptation to specific planting environments. This practice enhances plantation success and timely restocking so that ecosystem services can be fulfilled as quickly as possible after harvesting (Fig. 14.1). The use of improved planting material also protects investments and guarantees a future fiber supply of sufficient quantity and quality (Jansson et al., 2017).

Tree-breeding programs for boreal conifers have been established in many countries in the northern hemisphere to deliver improved seedlings for reforestation purposes (see Mullin et al., 2011 for an extensive review). The first tree breeding efforts comparing the growth of seed sources from different geographic origins go back to the early twentieth century. Structured tree-breeding programs for many commercial spruce (*Picea* spp.) and pine (*Pinus* spp.) species were initiated in the 1950 and 1960s by systematically sampling the genetic base (Fig. 14.4a). Hence, seeds and grafts from plus trees (particularly well growing trees in natural stands) were collected from the species' full distribution and planted in common garden experiments. These provenance studies determined the genetic variation within species to identify the best-growing seed sources and to study the genetic response to the environment (e.g., Li et al., 1997; Rehfeldt et al., 1999). These studies also established the foundation for crossing the best-performing individuals, leading to the beginning of a breeding population (Fig. 14.5).

Genetic trials of provenances or crosses follow distinct experimental designs that control within-site variation, determine genetic effects, and rank individual trees, families of crosses or their parents on the basis of their genetic merit (Fig. 14.4b). Measuring traits of interest in these experiments allows estimates of the different genetic parameters, e.g., their heritability, that determine the genetic gain expected through selection. For instance, height growth is typically between 20%–30% genetically controlled (e.g., Gamal El-Dien et al., 2015; Hong et al., 2014; Lenz et al., 2020a), whereas wood quality is under even stronger genetic control. In some cases, more than half of the observed variation in wood density and fiber dimension is attributed to genetics (e.g., Chen et al., 2014; Ivkovich et al., 2002). Adverse correlations between desired traits, such as growth and wood quality, require multitrait selection approaches to prevent wood quality degradation in planting stocks with enhanced growth (Hong et al., 2014; Lenz et al., 2020b).

Conventional tree breeding employs a recurrent cycle of evaluation, selection, and crossing of the best individuals, which are then re-evaluated (Fig. 14.5). Traditionally, selected individuals are multiplied and grafted into clone banks for next-generation

Fig. 14.4 a A 50-year-old mature white spruce (*Picea glauca*) provenance trial established at the Petawawa Research Forest in Ontario, Canada; and **b** a 16-year-old genetic trial of controlled crosses of Norway spruce (*P. abies*), an introduced species to eastern Canada for which several breeding programs are maintained. Trees in **a** and **b** were pruned to facilitate access for recurrent measurements of growth and for easier wood quality assessments. Pruning is also common practice in plantations for increasing wood quality. This treatment leads to fewer and smaller knots and hence stronger wood from the first log. *Photo credits* **a** Isabelle Duchesne, **b** Patrick Lenz

crosses and into seed orchards for seed production (White et al., 2007). For economic reasons, most seeds used for mass seedling production originate from open-pollinated seed orchards where only the maternal genetic value is well controlled. Other multiplication methods rely on sowing seeds and planting seedlings from controlled crosses or growing seedlings into hedges to produce rooted cuttings. Seedlings from cuttings or emblings obtained through somatic embryogenesis are significantly more expensive than standard material (Chamberland et al., 2020). Nevertheless, these clonal reproduction methods allow for the full control of the genetic makeup and, thus, maximize genetic gain (Park et al., 2016).

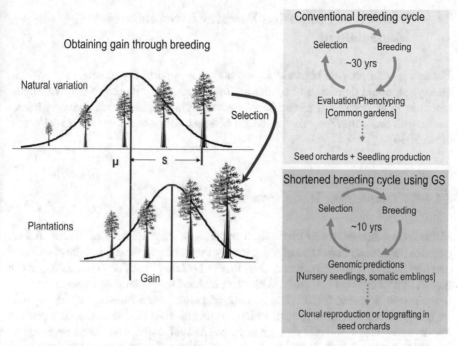

Fig. 14.5 Schematic illustration of the tree-breeding process, in which plus trees are identified to set the groundwork for crossing the best-performing individuals and, hence, establish a breeding population. Genomic selection (*GS*) can shorten the breeding cycle by decades. The genetic gain of a desired parameter (e.g., height) is thus dependent on the genetic control and the selection differential *s*, which is the difference between the general mean μ and the mean of the selected subpopulation

Over the last decade, genomic selection has been tested in forest tree breeding, with several proof-of-concept studies being published for boreal conifers (e.g., Beaulieu et al., 2014; Gamal El-Dien et al., 2015). Genomic selection relies on linking the genomic marker profiles of trees to their phenotypes. Once the models are calibrated, predictions are made only on the basis of marker profiles, which can already be obtained at the seedling stage or from embryonic tissue; this avoids the imperative establishment of field tests (Park et al., 2016). Hence, the evaluation time is reduced to a minimum, and completing a selection cycle lasts only 10–15 years until improved seedlings are available (Lenz et al., 2020a) (Fig. 14.5). Genetic gain is enhanced further when vegetative reproduction is used for valuable genotypes. Genomic selection also facilitates the screening for expensive phenotypes, such as resistance, and quality traits in breeding populations. Models can be calibrated for a representative subset of a breeding program, and predictions can be made for other genotyped trees in the same population.

14.4 Benefits of Plantation Forestry in Sustainable Management

Plantations can provide high yields and offer the opportunity to select for species, genotypes, stand density, and spatial arrangements. They therefore play an important role in augmenting, maintaining, and restoring forest productivity in boreal landscapes and improving the provision of other ecosystem services (Freer-Smith et al., 2019).

14.4.1 Increasing Wood Production

Plantations increase wood production per unit of area relative to natural forests because they make better use of the space by the desired species, and they are based on genetically improved material. Overall, fiber production of desired quality can be tripled in a plantation compared with that obtained from unmanaged natural stands (Paquette & Messier, 2010). The amount of gain in each breeding cycle depends on the genetic control of growth traits and on the selection intensity. Despite the low to moderate heritability of growth, substantial gains have been achieved in one to two selection cycles in northern conifer breeding programs. For example, Isaac-Renton et al. (2020) reported almost 30% volume gain in first-generation topcrosses of Douglas fir (*Pseudotsuga menziesii* (Mirb.) Franco) in western Canada. For Norway spruce in Sweden, Liziniewicz et al. (2018) reported volume gains of nearly 30% in realized gain trials in second-cycle seed orchards. Multivarietal forestry offers volume gains greater than 50% by multiplying top-performing clones (Weng et al., 2008). At the landscape level, and when integrated within a functional zoning approach, the enhanced productivity of plantations can reduce the management pressure on natural forests without affecting wood production within the forest management unit (Messier et al., 2003). Enhanced growth also leads to shorter rotations, thus reducing the duration that trees are exposed to biotic and abiotic risk factors, including those related to climate change.

14.4.2 Adapting Forests to Future Conditions with Trait Selection and Assisted Gene Flow

In addition to growth and stem form, tree breeding can screen for other traits adapted to novel climate or market conditions. More frequent and severe drought events and late and early frosts in the more northern regions will negatively affect tree regeneration at boreal latitudes (Boucher et al., 2020). Plantation forestry, using adapted planting stock, can help maintain forest productivity in these challenging conditions. Breeding and modern genomic tools make it possible to accelerate the selection for various breeding goals (Fig. 14.5), including improved resilience, adaptation, and resistance to climate extremes and more frequent biotic stresses, or particular wood attributes to respond to emerging markets. In recent years, several studies have coupled dendroecology with genetics and genomics to forecast the adaptive potential of populations on the basis of past growth responses extracted from tree-ring data (e.g., Montwé et al., 2018). Moreover, new classes of phenotypes have improved our understanding of the genetic underpinning of adaptation (Housset et al., 2018). Growth resilience and recovery after drought stress are under significant genetic control and can thus be used as breeding criteria for adapting seedling material to future climate conditions; thus, this selection of seedling material can support sustainable forest management (Depardieu et al., 2020). Breeding also provides opportunities to enhance the resistance of conifers to biotic stressors, such as insects (e.g., Lenz et al., 2020b).

14.4.3   Restoring and Maintaining Natural Species, Closed-Forest Landscapes, and Ecosystem Functions

Although forest plantations can be established with the objective of producing a maximum of wood fiber in the shortest period, forest plantations in the boreal zone are frequently used to compensate for deficient natural regeneration and maintain a closed-crown forest cover. For example, in northeastern Canada, black spruce seedlings are planted alone or in combination with eastern larch (*Larix laricina* (Du Roi) K. Koch) or other species after wildfires to limit the expansion of lichen woodlands (e.g., Thiffault & Hébert, 2017). Plantations are also used as a tool to address biodiversity issues under the paradigm of ecosystem-based management (Paquette & Messier, 2010). For instance, plantation forestry is used to regenerate species that reproduce through serotinous cones and that rarely reproduce without wildfire (Bouchard, 2008). Forest plantations are also established to restore or maintain certain declining species through enrichment planting (e.g., Neves Silva et al., 2019). White spruce is an example of a species native to boreal eastern Canada that is sensitive to environmental conditions and suffers from needle chlorosis and defoliation, a phenomenon that can potentially be accentuated by climate change in regions characterized by low base cation availability in the soil (Ouimet et al., 2013). Furthermore, natural regeneration of white spruce is impeded by harvesting effects on stand structure. In this context, enrichment planting (Fig. 14.2c) increases the proportion of white spruce in the landscape to restore the historical forest composition (Delmaire et al., 2020). Plantations can also serve to restore wildlife habitats; for example, ecosystems subjected to heavy browsing pressure from large ungulates may experience regeneration failure of palatable tree species (Beguin et al., 2016).

14.4.4 Capturing and Storing Carbon and Supporting the Bioeconomy

Because of their high productivity, plantations are frequently identified as a means of sequestering atmospheric carbon for mitigating climate change (Waring et al., 2020). This practice is of particular interest in afforestation contexts, i.e., establishing trees on areas previously deprived of a forest cover (e.g., Ouimet et al., 2007). Plantations can, however, show lower net primary production than naturally regenerated forests, resulting in a lower carbon stock (Liao et al., 2010). In the boreal biome, the benefits of using plantations for carbon sequestration can thus be realized only under specific conditions. For example, assuming that the albedo effect is taken into account (Bernier et al., 2011), a positive carbon sink can be observed after the reforestation of forest heaths, which results from cascades of natural or anthropogenic disturbance (Gaboury et al., 2009). Moreover, although site preparation and the planting of boreal stands prone to paludification can result in losses of soil carbon, plantations on such sites should be beneficial because of the increased carbon storage in tree biomass (Lavoie et al., 2005). For example, increasing site preparation intensity on mineral soils can significantly increase carbon stock in the forest ecosystem in the long-term (Mjöfors et al., 2017). The net effect of mechanical site preparation on carbon stock remains, however, dependent upon the initial humus content and the site-specific soil characteristics.

14.5 Risks and Challenges

Forests play a critical role in addressing many of the largest global challenges. These challenges include mitigating climate change, conserving biodiversity, and providing a variety of ecosystem services, including nutrient cycling, air and water purification, carbon sequestration and storage, and wildlife habitat. Forests also have social and spiritual benefits and are key to important cultural activities. Although plantations can support the delivery of these services, they are also associated with silvicultural regimes that have the greatest potential for the artificialization of natural forests (Barrette et al., 2014).

Silvicultural treatments necessary for establishing successful plantations can have undesired effects on ecosystems and result in unforeseen impacts on the silvicultural regimes themselves. For example, although site preparation improves the establishment of planted seedlings, the increased area of disturbed soil from this treatment favors the establishment of naturally regenerated seedlings (e.g., Johansson et al., 2013). Whereas natural regeneration can complement or replace planted seedlings if mortality occurs, this regeneration can increase the need for precommercial thinning and other silviculture investments. There are also concerns that mechanical site preparation can negatively affect long-term productivity by depleting soil nutrients through rapid decomposition and leaching. Although tree growth appears to persist many decades after treatment (e.g., Hjelm et al., 2019), long-term research is necessary to fully evaluate the legacy effects of mechanical site preparation on site productivity. Similarly, fertilization is one of the most questioned silvicultural measures. Whereas it is effective at increasing tree growth and stemwood production, fertilization may adversely affect the forest ecosystem through, for example, negative impacts on soil-solution chemistry. Fertilization usually modifies plant composition relative to natural succession because of the increased availability of nutrients and reduced light penetration through a denser tree canopy (Hedwall et al., 2010); these effects may remain after felling and regeneration (Strengbom & Nordin, 2008) and could potentially impact sites at a millennial time scale (Dupouey et al., 2002).

The breeding process involved in producing successful plant material for plantation forestry also raises issues and challenges. For example, the extended time frame that breeders must foresee is one of the greatest challenges for decision-making and the selection of the optimal traits. Genomic selection can provide part of the solution. Nonetheless, 10 years of breeding cycle added to 20 years until commercial thinning of improved plantations remains a long time horizon during which environmental or market conditions may likely change with potentially significant social and economic impacts and the risk that the selected genomes will not be adapted to unforeseen changes in biotic and abiotic disturbances. Currently, genomic selection models are not transferable among breeding populations, as they largely trace pedigree and, to a minor extent, marker-trait associations (Lenz et al., 2017). There is hence a continued need for developing appropriate genotyping and statistical methods.

The social acceptability of plantation forestry is ambiguous (e.g., Wyatt et al., 2011). Although plantation forestry is perceived positively in certain circumstances, it is often associated with industrial practices, monocultures, the use of chemicals, a deterioration of water quality, negative effects on biodiversity, fragmentation of the forest matrix, and other landscape-scale impacts (Paquette & Messier, 2010). The use of improved planting material through breeding programs and genomic selection (Fig. 14.5) may wrongly be associated with the use of genetically modified organisms (GMOs).

14.6   Conclusions and Perspectives

Overall, the use of plantations in the sustainable forest management of the boreal forest undoubtedly raises significant issues related to the scale, localization, and spatial arrangement of plantations, the key attributes and resilience of natural forests, social acceptability, and the productivity and profitability of plantations, particularly in the context of ecosystem-based management (Barrette et al., 2014) (Fig. 14.6). Whereas the role of plantations in supporting sustainable forest management in the boreal forest is undeniable, their use should thus consider the risks associated with their implementation. These risks are increasing as biotic (e.g., native or exotic pests) and abiotic (e.g., drought) hazards expand because of global change, while concurrent economic and social pressures evolve constantly.

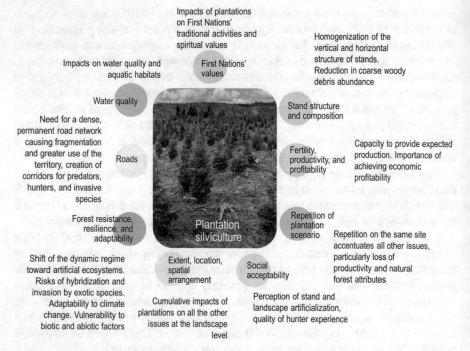

Fig. 14.6 Some issues related to plantation silviculture in the context of ecosystem-based management in the boreal forest. Adapted with permission from Barrette et al. (2014)

Plantation silviculture is compatible with and can support ecosystem-based management objectives (e.g., Barrette et al., 2019). When necessary, adaptive approaches can be applied at the stand level, such as establishing mixed- or multi-species plantations, maintaining biological legacies prior to establishing plantations, preserving patches of natural forest during site preparation, and favoring rare fruit-bearing tree species during cleaning treatments (Barrette et al., 2014). Maintaining or restoring the highest possible degree of naturalness within the forest matrix could address the complex issues associated with plantations at the landscape level (Tittler et al., 2012).

References

Barrette, M., Leblanc, M., Thiffault, N., et al. (2014). Issues and solutions for intensive plantation silviculture in a context of ecosystem management. *The Forestry Chronicle, 90*(6), 748–762. https://doi.org/10.5558/tfc2014-147.

Barrette, M., Thiffault, N., Tremblay, J. P., et al. (2019). Balsam fir stands of Northeastern North America are resilient to spruce plantation. *Forest Ecology and Management, 450*, 117504. https://doi.org/10.1016/j.foreco.2019.117504.

Beaulieu, J., Doerksen, T. K., MacKay, J., et al. (2014). Genomic selection accuracies within and between environments and small breeding groups in white spruce. *BMC Genomics, 15*(1), 1048. https://doi.org/10.1186/1471-2164-15-1048.

Beguin, J., Tremblay, J. P., Thiffault, N., et al. (2016). Management of forest regeneration in boreal and temperate deer–forest systems: Challenges, guidelines, and research gaps. *Ecosphere, 7*(10), e01488. https://doi.org/10.1002/ecs2.1488.

Bell, F. W., Thiffault, N., Szuba, K., et al. (2011). Synthesis of silviculture options, costs, and consequences of alternative vegetation management practices relevant to boreal and temperate conifer forests: Introduction. *The Forestry Chronicle, 87*(2), 155–160. https://doi.org/10.5558/tfc2011-005.

Bernier, P. Y., Desjardins, R. L., Karimi-Zindashty, Y., et al. (2011). Boreal lichen woodlands: A possible negative feedback to climate change in eastern North America. *Agricultural and Forest Meteorology, 151*(4), 521–528. https://doi.org/10.1016/j.agrformet.2010.12.013.

Bouchard, M. (2008). La sylviculture dans un contexte d'aménagement écosystémique en forêt boréale et en forêt mixte. In Gauthier S, Vaillancourt, M. A., Leduc, A., De Grandpré, L., Kneeshaw, D., Morin, H., Drapeau, P., & Bergeron, Y., (Eds.) *Aménagement écosystémique en forêt boréale*. Presses de l'Université du Québec (pp 335–359).

Boucher, D., Gauthier, S., Thiffault, N., et al. (2020). How climate change might affect tree regeneration following fire at northern latitudes: A review. *New Forests, 51*, 543–571. https://doi.org/10.1007/s11056-019-09745-6.

Brandt, J. P., Flannigan, M. D., Maynard, D. G., et al. (2013). An introduction to Canada's boreal zone: Ecosystem processes, health, sustainability, and environmental issues. *Environmental Reviews, 21*(4), 207–226. https://doi.org/10.1139/er-2013-0040.

Burdon, R. D., Li, Y., Suontama, M., et al. (2017). Genotype × site × silviculture interactions in radiata pine: Knowledge, working hypotheses and pointers for research. *New Zealand Journal of Forest Science, 47*, 6. https://doi.org/10.1186/s40490-017-0087-1.

Chamberland, V., Robichaud, F., Perron, M., et al. (2020). Conventional versus genomic selection for white spruce improvement: A comparison of costs and benefits of plantations on Quebec public lands. *Tree Genetics & Genomes, 16*(1), 17. https://doi.org/10.1007/s11295-019-1409-7.

Chen, Z. Q., Gil, M. R. G., Karlsson, B., et al. (2014). Inheritance of growth and solid wood quality traits in a large Norway spruce population tested at two locations in southern Sweden. *Tree Genetics & Genomes, 10*, 1291–1303. https://doi.org/10.1007/s11295-014-0761-x.

Delmaire, M., Thiffault, N., Thiffault, E., et al. (2020). White spruce enrichment planting in boreal mixedwoods as influenced by localized site preparation: 11-year update. *The Forestry Chronicle, 96*(1), 27–35. https://doi.org/10.5558/tfc2020-005.

Depardieu, C., Girardin, M. P., Nadeau, S., et al. (2020). Adaptive genetic variation to drought in a widely distributed conifer suggests a potential for increasing forest resilience in a drying climate. *New Phytologist, 227*(2), 427–439. https://doi.org/10.1111/nph.16551.

Dupouey, J. L., Dambrine, E., Laffite, J. D., et al. (2002). Irreversible impact of past land use on forest soils and biodiversity. *Ecology, 83*(11), 2978–2984. https://doi.org/10.1890/0012-9658(2002)083[2978:Iioplu]2.0.Co;2.

Food and Agriculture Organization of the United Nations (FAO). (2020). *The state of the world's forests. Forests, biodiversity and people*. Rome: Food and Agriculture Organization of the United Nations.

Fenton, N. J., & Bergeron, Y. (2006). Facilitative succession in a boreal bryophyte community driven by changes in available moisture and light. *Journal of Vegetation Science, 17*(1), 65–76. https://doi.org/10.1111/j.1654-1103.2006.tb02424.x.

Franceschini, T., Gauthray-Guyénet, V., Schneider, R., et al. (2018). Effect of thinning on the relationship between mean ring density and climate in black spruce (*Picea mariana* (Mill.) B.S.P.). *Forestry, 91*(3), 366–381. https://doi.org/10.1093/forestry/cpx040.

Freer-Smith, P., Muys, B., Bozzano, M., et al. (2019). *Plantation forests in Europe: challenges and opportunities. From Science to Policy 9*. (p. 52), Joensuu: European Forest Institute.

Gaboury, S., Boucher, J. F., Villeneuve, C., et al. (2009). Estimating the net carbon balance of boreal open woodland afforestation: A case-study in Québec's closed-crown boreal forest. *Forest Ecology and Management, 257*(2), 483–494. https://doi.org/10.1016/j.foreco.2008.09.037.

Gamal El-Dien, O., Ratcliffe, B., Klápště, J., et al. (2015). Prediction accuracies for growth and wood attributes of interior spruce in space using genotyping-by-sequencing. *BMC Genomics, 16*(1), 370. https://doi.org/10.1186/s12864-015-1597-y.

Groot, A., & Cortini, F. (2016). Effects of initial planting density on tree and stand development of planted black spruce up to age 30. *The Forestry Chronicle, 92*(2), 200–210. https://doi.org/10.5558/tfc2016-039.

Grossnickle, S. C. (2005). Importance of root growth in overcoming planting stress. *New Forests, 30*(2–3), 273–294. https://doi.org/10.1007/s11056-004-8303-2.

Hedwall, P. O., Nordin, A., Brunet, J., et al. (2010). Compositional changes of forest-floor vegetation in young stands of Norway spruce as an effect of repeated fertilisation. *Forest Ecology and Management, 259*(12), 2418–2425. https://doi.org/10.1016/j.foreco.2010.03.018.

Hjelm, K., Nilsson, U., Johansson, U., et al. (2019). Effects of mechanical site preparation and slash removal on long-term productivity of conifer plantations in Sweden. *Canadian Journal of Forest Research, 49*(10), 1311–1319. https://doi.org/10.1139/cjfr-2019-0081.

Hong, Z., Fries, A., & Wu, H. X. (2014). High negative genetic correlations between growth traits and wood properties suggest incorporating multiple traits selection including economic weights for the future Scots pine breeding programs. *Annals of Forest Science, 71*, 463–472. https://doi.org/10.1007/s13595-014-0359-3.

Housset, J. M., Nadeau, S., Isabel, N., et al. (2018). Tree rings provide a new class of phenotypes for genetic associations that foster insights into adaptation of conifers to climate change. *New Phytologist, 218*(2), 630–645. https://doi.org/10.1111/nph.14968.

Isaac-Renton, M., Stoehr, M., Bealle Statland, C., et al. (2020). Tree breeding and silviculture: Douglas-fir volume gains with minimal wood quality loss under variable planting densities. *Forest Ecology and Management, 465*, 118094. https://doi.org/10.1016/j.foreco.2020.118094.

Ivkovich, M., Namkoong, G., & Koshy, M. (2002). Genetic variation in wood properties of interior spruce. II. Tracheid characteristics. *Canadian Journal of Forest Research, 32*(12), 2128–2139. https://doi.org/10.1139/x02-139.

Jaakkola, T., Mäkinen, H., & Saranpää, P. (2005). Wood density in Norway spruce: Changes with thinning intensity and tree age. *Canadian Journal of Forest Research, 35*(7), 1767–1778. https://doi.org/10.1139/x05-118.

Jacobson, S., & Pettersson, F. (2010). An assessment of different fertilization regimes in three boreal coniferous stands. *Silva Fennica , 44*(5), 815–827. https://doi.org/10.14214/sf.123.

Jansson, G., Hansen, J. K., Haapanen, M., et al. (2017). The genetic and economic gains from forest tree breeding programmes in Scandinavia and Finland. *Scandinavian Journal of Forest Research, 32*(4), 273–286. https://doi.org/10.1080/02827581.2016.1242770.

Jerabkova, L., Prescott, C. E., Titus, B. D., et al. (2011). A meta-analysis of the effects of clearcut and variable-retention harvesting on soil nitrogen fluxes in boreal and temperate forests. *Canadian Journal of Forest Research, 41*(9), 1852–1870. https://doi.org/10.1139/x11-087.

Jobidon, R., Roy, V., & Cyr, G. (2003). Net effect of competing vegetation on selected environmental conditions and performance of four spruce seedling stock sizes after eight years in Québec (Canada). *Annals of Forest Science, 60*(7), 691–699. https://doi.org/10.1051/forest:2003063.

Johansson, K., Nilsson, U., & Örlander, G. (2013). A comparison of long-term effects of scarification methods on the establishment of Norway spruce. *Forestry, 86*(1), 91–98. https://doi.org/10.1093/forestry/cps062.

Johansson, K., Langvall, O., & Bergh, J. (2012). Optimization of environmental factors affecting initial growth of Norway spruce seedlings. *Silva Fennica, 46*(1), 64. https://doi.org/10.14214/sf.64.

Lavoie, M., Paré, D., & Bergeron, Y. (2005). Impact of global change and forest management on carbon sequestration in northern forested peatlands. *Environmental Reviews, 13*(4), 199–240. https://doi.org/10.1139/A05-014.

Lenz, P. R. N., Beaulieu, J., Mansfield, S. D., et al. (2017). Factors affecting the accuracy of genomic selection for growth and wood quality traits in an advanced-breeding population of black spruce (*Picea mariana*). *BMC Genomics, 18*(1), 335. https://doi.org/10.1186/s12864-017-3715-5.

Lenz, P. R. N., Nadeau, S., Mottet, M. J., et al. (2019). Multi-trait genomic selection for weevil resistance, growth, and wood quality in Norway spruce. *Evolutionary Applications, 13*(1), 76–94. https://doi.org/10.1111/eva.12823.

Lenz, P. R. N., Nadeau, S., Azaiez, A., et al. (2020a). Genomic prediction for hastening and improving efficiency of forward selection in conifer polycross mating designs: An example from white spruce. *Heredity, 124*(4), 562–578. https://doi.org/10.1038/s41437-019-0290-3.

Lenz, P. R. N., Nadeau, S., Mottet, M.-J., et al. (2020b). Multi-trait genomic selection for weevil resistance, growth, and wood quality in Norway spruce. *Evolutionary Applications, 13*(1), 76–94. https://doi.org/10.1111/eva.12823.

Leroy, C., Leduc, A., Thiffault, N., et al. (2016). Forest productivity after careful logging and fire in black spruce stands of the Canadian Clay Belt. *Canadian Journal of Forest Research, 46*(6), 783–793. https://doi.org/10.1139/cjfr-2015-0484.

Li, P., Beaulieu, J., & Bousquet, J. (1997). Genetic structure and patterns of genetic variation among populations in eastern white spruce (*Picea glauca*). *Canadian Journal of Forest Research, 27*(2), 189–198. https://doi.org/10.1139/x96-159.

Liao, C., Luo, Y., Fang, C., et al. (2010). Ecosystem carbon stock influenced by plantation practice: Implications for planting forests as a measure of climate change mitigation. *PLoS ONE, 5*, e10867. https://doi.org/10.1371/journal.pone.0010867.

Liziniewicz, M., Berlin, M., & Karlsson, B. (2018). Early assessments are reliable indicators for future volume production in Norway spruce (*Picea abies* L. Karst) genetic field trials. *Forest Ecology and Management, 411*, 75–81. https://doi.org/10.1016/j.foreco.2018.01.015.

Löf, M., Dey, D. C., Navarro, R. M., et al. (2012). Mechanical site preparation for forest restoration. *New Forests, 43*(5–6), 825–848. https://doi.org/10.1007/s11056-012-9332-x.

Mäkinen, H., Verkasalo, E., & Tuimala, A. (2014). Effects of pruning in Norway spruce on tree growth and grading of sawn boards in Finland. *Forestry, 87*(3), 417–424. https://doi.org/10.1093/forestry/cpt062.

Mallik, A. U. (2003). Conifer regeneration problems in boreal and temperate forests with ericaceous understory: Role of disturbance, seedbed limitation, and keystone species change. *Critical Reviews in Plant Sciences, 22*, 341–366. https://doi.org/10.1080/713610860.

Messier, C., Bigué, B., & Bernier, L. (2003). Using fast-growing plantations to promote forest ecosystem protection in Canada. *Unasylva, 54*(214/215), 59–63.

Mjöfors, K., Strömgren, M., Nohrstedt, H. Ö., et al. (2017). Indications that site preparation increases forest ecosystem carbon stocks in the long term. *Scandinavian Journal of Forest Research, 32*(8), 717–725. https://doi.org/10.1080/02827581.2017.1293152.

Montwé, D., Isaac-Renton, M., Hamann, A., et al. (2018). Cold adaptation recorded in tree rings highlights risks associated with climate change and assisted migration. *Nature Communications, 9*(1), 1574. https://doi.org/10.1038/s41467-018-04039-5.

Mullin, T. J., Andersson, B., Bastien, J., et al. (2011). Economic importance, breeding objectives and achievements. In C. Plomion, J. Bousquet, & C. Kole (Eds.), *Genetics, genomics and breeding of conifers* (pp. 40–127). Edenbridge Science Publishers & CRC Press.

Neves Silva, L., Freer-Smith, P., & Madsen, P. (2019). Production, restoration, mitigation: A new generation of plantations. *New Forests, 50*, 153–168. https://doi.org/10.1007/s11056-018-9644-6.

Nijnik, M., Slee, B., & Nijnik, A. (2014). Biomass production: Impacts on other ecosystem services. In Pelkonen, P., Mustonen, M., Asikainen, A., Egnell, G., Kant, P., Leduc, S., & Pettenella, D. (Eds.) *Forest bioenergy for Europe, what science can tell us* (Vol. 4, pp. 82–89). Joensuu: European Forest Institute.

Natural Resources Canada (NRC). (2020). *The state of Canada's forests. Annual report 2019.* (p. 80), Ottawa: Natural Resources Canada, Canadian Forest Service.

Ouimet, R., Tremblay, S., Périé, C., et al. (2007). Ecosystem carbon accumulation following fallow farmland afforestation with red pine in southern Quebec. *Canadian Journal of Forest Research, 37*(6), 1118–1133. https://doi.org/10.1139/x06-297.

Ouimet, R., Moore, J. D., Duchesne, L., et al. (2013). Etiology of a recent white spruce decline: Role of potassium deficiency, past disturbances, and climate change. *Canadian Journal of Forest Research, 43*(1), 66–77. https://doi.org/10.1139/cjfr-2012-0344.

Paquette, A., & Messier, C. (2010). The role of plantations in managing the world's forests in the Anthropocene. *Frontiers in Ecology and the Environment, 8*(1), 27–34. https://doi.org/10.1890/080116.

Park, Y. S., Beaulieu, J., & Bousquet, J. (2016). Multi-varietal forestry integrating genomic selection and somatic embryogenesis. In Park, Y. S., Bonga, J. M., & Moon, H. K. (Eds.) *Vegetative propagation of forest trees* (pp. 302–322). National Institute of Forest Science, Seoul.

Pelletier, G., & Pitt, D. G. (2008). Silvicultural responses of two spruce plantations to midrotation commercial thinning in New Brunswick. *Canadian Journal of Forest Research, 38*(4), 851–867. https://doi.org/10.1139/x07-173.

Rehfeldt, G. E., Ying, C. C., Spittlehouse, D. L., et al. (1999). Genetic responses to climate in *Pinus contorta*: Niche breadth, climate change, and reforestation. *Ecological Monographs, 69*(3), 375–407. https://doi.org/10.1890/0012-9615(1999)069[0375:GRTCIP]2.0.CO;2.

Rubilar, R. A., Lee Allen, H., Fox, T. R., et al. (2018). Advances in silviculture of intensively managed plantations. *Current Forestry Reports, 4*(1), 23–34. https://doi.org/10.1007/s40725-018-0072-9.

Shorohova, E., Kuuluvainen, T., Kangur, A., et al. (2009). Natural stand structures, disturbance regimes and successional dynamics in the Eurasian boreal forests: A review with special reference to Russian studies. *Annals of Forest Science, 66*(2), 201. https://doi.org/10.1051/forest/2008083.

Sikström, U., Hjelm, K., Holt Hanssen, K., et al. (2020). Influence of mechanical site preparation on regeneration success of planted conifers in clearcuts in Fennoscandia – a review. *Silva Fennica, 54*(2), 10172. https://doi.org/10.14214/sf.10172.

Strengbom, J., & Nordin, A. (2008). Commercial forest fertilization causes long-term residual effects in ground vegetation of boreal forests. *Forest Ecology and Management, 256*(12), 2175–2181. https://doi.org/10.1016/j.foreco.2008.08.009.

Swedish Forest Agency (SFA). (2020). *Quality of the regrowth.* Jönköping: Swedish Forest Agency.

Tamm, C. O. (1991). Nitrogen in terrestrial ecosystems. In *Questions of productivity, vegetational changes, and ecosystem stability.* New York: Springer.

Thiffault, N., & Hébert, F. (2017). Mechanical site preparation and nurse-plant facilitation for the restoration of subarctic forest ecosystems. *Canadian Journal of Forest Research, 47*(7), 926–934. https://doi.org/10.1139/cjfr-2016-0448.

Thiffault, N., & Jobidon, R. (2006). How to shift unproductive *Kalmia angustifolia - Rhododendron groenlandicum* heath to productive conifer plantation. *Canadian Journal of Forest Research, 36*(10), 2364–2376. https://doi.org/10.1139/x06-090.

Thiffault, N., Coll, L., & Jacobs, D. F. (2015). Natural regeneration after harvesting. In Peh, K. S. -H., R. T., Corlett, & Y. Bergeron (Eds.). *Routledge handbook of forest ecology,* (pp. 371–384). Routledge.

Tittler, R., Messier, C., & Fall, A. (2012). Concentrating anthropogenic disturbance to balance ecological and economic values: Applications to forest management. *Ecological Applications, 22*(4), 1268–1277. https://doi.org/10.1890/11-1680.1.

Vincent, M., Krause, C., & Koubaa, A. (2011). Variation in black spruce (*Picea mariana* (Mill.) BSP) wood quality after thinning. *Annals of Forest Science, 68*(6), 1115–1125. https://doi.org/10.1007/s13595-011-0127-6.

Waring, B., Neumann, M., Prentice, I. C., et al. (2020). Forests and decarbonization – roles of natural and planted forests. *Frontiers in Forests and Global Change, 3*, 58. https://doi.org/10.3389/ffgc.2020.00058.

Weng, Y., Park, Y., Krasowski, M., et al. (2008). Partitioning of genetic variance and selection efficiency for alternative vegetative deployment strategies for white spruce in Eastern Canada. *Tree Genetics & Genomes, 4*(4), 809–819. https://doi.org/10.1007/s11295-008-0154-0.

White, T. L., Neale, D. B., & Adams, W. T. (2007). *Forest genetics*. CABI Publishing.

Wiensczyk, A., Swift, K., Morneault, A. E., et al. (2011). An overview of the efficacy of vegetation management alternatives for conifer regeneration in boreal forests. *The Forestry Chronicle, 87*(2), 175–200. https://doi.org/10.5558/tfc2011-007.

Wyatt, S., Rousseau, M. H., Nadeau, S., et al. (2011). Social concerns, risk and the acceptability of forest vegetation management alternatives: Insights for managers. *The Forestry Chronicle, 87*(2), 274–289. https://doi.org/10.5558/tfc2011-014.

Chapter 15
Silviculture of Mixed-Species and Structurally Complex Boreal Stands

Patricia Raymond, Magnus Löf, Phil Comeau, Lars Rytter, Miguel Montoro Girona, and Klaus J. Puettmann

Abstract Understanding structurally complex boreal stands is crucial for designing ecosystem management strategies that promote forest resilience under global change. However, current management practices lead to the homogenization and simplification of forest structures in the boreal biome. In this chapter, we illustrate two options for managing productive and resilient forests: (1) the managing of two-aged mixed-species forests; and (2) the managing of multi-aged, structurally complex stands. Results demonstrate that multi-aged and mixed stand management are powerful silvicultural tools to promote the resilience of boreal forests under global change.

15.1 Introduction

Silvicultural practices have long been used to encourage the provision of desired ecosystem goods and services to landowners and society (Puettmann et al., 2009). The selection and implementation of specific practices are driven mainly by ownership objectives and logistical opportunities and constraints. Consequently, as management objectives have changed over the last few decades from a focus on timber production

P. Raymond (✉)
Direction de la recherche forestière, ministère des Forêts, de la Faune et des Parcs du Québec, 2700 rue Einstein, Québec G1P 3W8, Canada
e-mail: patricia.raymond@mffp.gouv.qc.ca

M. Löf
Southern Swedish Forest Research Centre, Swedish University of Agricultural Sciences (SLU), Box 190, 234 22 Lomma, Sweden
e-mail: magnus.lof@slu.se

P. Comeau
Department of Renewable Resources, University of Alberta, 751 General Services Bldg, Edmonton, AB T6G 2H1, Canada
e-mail: phil.comeau@ualberta.ca

L. Rytter
Forestry Research Institute of Sweden, Ekebo, 2250, SE-268 90 Svalöv, Sweden

© The Author(s) 2023
M. M. Girona et al. (eds.), *Boreal Forests in the Face of Climate Change*,
Advances in Global Change Research 74,
https://doi.org/10.1007/978-3-031-15988-6_15

to managing for a broader set of goals, e.g., biodiversity, recreation, and resilience, a more diverse suite of silvicultural practices had to be applied (Puettmann et al., 2009). On public lands, societal shifts have led to increased recognition of the importance of ecosystem services such as wildlife habitat, recreational opportunities, spiritual values, or biodiversity, in addition to or instead of timber production. Furthermore, recent concerns regarding biodiversity loss, reduced productivity (Chap 1; Table 1.1), and forest resilience in the face of global change (Chap. 1; Table 1.2) require applying a broader set of silvicultural practices than in the past to manage forests for a novel, uncertain future (Puettmann, 2011; Shvidenko & Apps, 2006).

The selection of silvicultural systems has traditionally been justified by understanding the dominant natural disturbance regimes (Bradshaw et al., 1994). In unmanaged boreal forests, natural regeneration is often initiated following disturbance by fire, insects, or windstorms (Kuuluvainen & Grenfell, 2012). The theory of natural disturbance emulation, holds that clear-cutting simulates large high-severity perturbations, e.g., fire, but this silvicultural approach leads to less standing and downed woody debris and different soil conditions than encountered following a fire (Bergeron et al., 2002; Kuuluvainen & Grenfell, 2012; Moussaoui et al., 2016a, 2020). Over the last few decades, ecosystem-based forest management has become a dominant management paradigm in many countries (Chap 1). Correspondingly, our understanding of natural disturbance regimes and their impacts on succession has expanded to underline the role and influence of spatial and temporal variability and environmental legacies (Bergeron & Harvey, 1997; Montoro Girona et al., 2018a). Thus, rather than having a narrow focus on variables such as the average fire return interval or fire size, silvicultural practices should reflect the full suite of disturbance frequencies and severities, especially small-scale disturbances (Kuuluvainen & Grenfell, 2012). Together with the shift in the abovementioned landowners' objectives, the recognition of the role of disturbances of wide-ranging severity and size has encouraged landowners to consider a more diverse range of silvicultural practices. As an example of the practical implications of this shift in thinking, variable retention has gained global attention (Gustafsson et al., 2012; Kuuluvainen & Grenfell, 2012; Moussaoui et al., 2016b).

M. M. Girona
Groupe de recherche en écologie de la MRC-Abitibi (GREMA), Forest Research Institute, Université du Québec en Abitibi-Témiscamingue, 341, rue Principale Nord, Amos Campus, Québec J9T 2L8, Canada
e-mail: miguel.montoro@uqat.ca

Department of Wildlife, Fish, and Environmental Studies, Swedish University of Agricultural Sciences (SLU), 901 83 Umeå, Sweden

Centre for Forest Research, Université du Québec à Montréal, Stn. Centre-Ville, P.O. Box 8888, Montréal, Québec 3C 3P8, Canada

K. J. Puettmann
Department of Forest Ecosystems and Society, Oregon State University, 321 Richardson Hall, Corvallis, OR 97331, USA
e-mail: Klaus.Puettmann@oregonstate.edu

Both the increased diversity of management objectives and an improved understanding of the variability created by natural disturbances present challenges, with their relative importance changing depending on ownership and the particular ecological and social context. Furthermore, addressing these factors will become even more complicated in response to social and ecological trends associated with global change (Puettmann, 2011). For example, although using the variability of natural disturbance patterns to manage for multiple ownership goals had received much attention in the past (Franklin et al., 2018; Kuuluvainen & Grenfell, 2012), practical suggestions to encourage the adaptive capacity, e.g., resilience, of forests to combat the negative impacts of global change are scarce (Puettmann & Messier, 2019). This adaptive capacity is of particular importance, as future conditions are expected to be increasingly influenced by human-caused rather than natural drivers; thus, managing for resilience and adaptive capacity will likely increase in importance (Puettmann, 2011).

An increased focus on a broader set of ecosystem services and the variability of natural disturbance regimes has led to an interest in managing forests within a wider envelope of structural and compositional conditions. This vision aligns with management approaches for resilience, as ecosystem adaptation mechanisms are based on maintaining or even enhancing functional diversity—species with different traits that, for example, respond differently to various disturbance agents—and cross-scale interactions, e.g., disturbances producing high structural and compositional variability within stands (Puettmann & Messier, 2019). In this context, this chapter highlights silvicultural practices aimed at encouraging heterogeneous species composition and stand structures in boreal forests, as quantified by tree species composition and vertical structure, respectively, to promote resilience to global change.

Compared with monocultures, mixed-species forests provide a more comprehensive suite of ecosystem services (Hector & Bagchi, 2007; Himes & Puettmann, 2020) and encourage a broader range of stand structures (Pretzsch et al., 2017). Stand structural variability is managed using a variety of approaches, from the classic uneven-aged management (*Plenterwald*) (O'Hara 2014) to variable-retention harvests (Gustafsson et al., 2012). In contrast to the classical *Plenterwald*, variable-retention harvests emphasize spatial variability and thus ensure that a variety of successional stages are present in stands, including early seral and older stages (Franklin et al., 2018). At the same time, the importance of ensuring a variety of ecosystem services, especially those related to biodiversity, leads to increased attention to other structural elements, such as understory vegetation, snags, and downed wood. Greater knowledge of species mixtures and heterogeneous stand structure supports practices that improve the resilience of forest stands, especially in a context of global change (Puettmann & Messier, 2019).

15.2   Silvicultural Systems and Complexity

The choice of a silvicultural system influences structural and compositional conditions and their evolution (Kuuluvainen et al., 2012; Puettmann et al., 2009; Raymond et al., 2009). Silvicultural systems influence structural diversity, which can range from simple single-canopy layer stands in even-aged systems to multiple canopy layers in uneven-aged stands. The spatiotemporal arrangement of management practices, e.g., gap creation or patch thinning, and the retention of structural attributes, e.g., choice of species and trees for retention at stand and landscape scales, can also maintain or increase complexity (e.g., Bauhus et al., 2009; Gustafsson et al., 2012). Furthermore, within-stand heterogeneity of topography, soil conditions, and available resources promote structural and species diversity, especially in late-successional forests (Moussaoui et al., 2019). In contrast to traditional efforts to homogenize forests for production efficiency (Puettmann et al., 2009), silvicultural systems that create diverse ecological niches (e.g., irregular shelterwood and hybrid selection-cutting systems) or that incorporate within-stand variability, such as canopy gaps or vertical structure in mixed-species stands are expected to facilitate species coexistence and diversity (Burton et al., 1999; Raymond & Bédard, 2017). Moreover, silvicultural systems that maintain continuous forest cover are more likely to sustain structural attributes, associated microhabitats, and, thus, biodiversity over time (Kim et al., 2021; Martin et al., 2020; Moussaoui et al., 2016b; Peura et al., 2018). The selection of a given silviculture option varies as a function of current stand and landscape conditions, ownership goals, and logistical opportunities and constraints. In the following sections, we illustrate two management examples to highlight options for managing productive, resilient boreal forests: (1) managing for two-aged mixed-species forests; and (2) managing for multi-aged, structurally complex forests.

15.3   Silviculture of Two-Aged Mixed Forests

Two-aged mixed stands, which combine fast-growing, early-successional, and light-demanding tree species (nurse trees) with late-successional and shade-tolerant tree species (target trees), is a management concept that has gained interest over the past two decades (Fig. 15.1; Paquette & Messier, 2010; Rytter et al., 2016). The faster-growing nurse trees provide shade to limit competing vegetation (Lieffers & Stadt, 1994) and protect smaller seedlings and saplings against late spring frost (Filipescu & Comeau, 2011). Nurse trees also facilitate the establishment of more slow-growing target trees and improve their stem form (Middleton & Munro, 2002; Paquette et al., 2006; Pommerening & Murphy, 2004). The risk of insect attack and the related impacts are reduced in mixed stands because the presence of multiple tree species reduces the impact of host-specific insects (Campbell et al., 2008; Lavoie et al., 2021; Taylor et al., 1996; Zhang et al., 2018). The risk of root disease is also reduced in mixed stands (Gerlach et al., 1997). Slow-growing crop tree species can

Fig. 15.1 Managing two-aged stands is more complex than managing monocultures; however, two-aged stands offer more adaptability to uncertain future conditions. **a** Silver birch (*Betula pendula* Roth)–Norway spruce in Sweden and **b** aspen–white spruce in Alberta, Canada are examples of boreal mixedwoods that can be managed as two-aged stands. *Photo credits* **a** Lars Rytter, **b** Phil Comeau

also be difficult to establish without protection from a nurse crop. In these conditions, facilitative interactions can be more prominent than competitive interactions, at least during the early stages of stand development (Pretzsch et al., 2017).

Under selected conditions, mixed species forests are often more productive than single-species forests (Pretzsch et al., 2017). This is particularly the case for two-aged stands where transgressive overyielding often occurs, i.e., the mixture is more productive than the monoculture of the most productive species in the mixture (Kweon & Comeau, 2019; Pretzsch et al., 2017). Two-aged management can also accelerate natural succession from shade-intolerant to mixedwood composition in second-growth forests (Prévost & DeBlois, 2014; Smith et al., 2016). Thus, with two-aged stands, greater biodiversity, resilience, and a more diversified portfolio of ecosystem services can be combined with increased stand growth and carbon sequestration (Felton et al., 2016; Pretzsch et al., 2017).

Several tree species combinations are relevant for this type of management, making it applicable to a range of site conditions. Such examples in Scandinavia are planted or naturally regenerated stands combining birch (*Betula* spp.) as nurse crops with Norway spruce (*Picea abies* L. Karst.) as the target tree species underneath (Mård, 1996). In Canada, similar stands with trembling aspen (*Populus tremuloides* Michx) and either planted white spruce (*Picea glauca* (Moench.) Voss.) (Kabzems et al., 2016; Lieffers et al., 2019; Pitt et al., 2015) or other natural mixtures of spruce and fir (Prévost & DeBlois, 2014; Smith et al., 2016) can be managed as two-aged stands. Such multispecies stands may be more productive than single-species stands, with a transgressive overyielding up to 20% (Kweon & Comeau, 2019). The use of a fast-growing nurse crop may be a cost-effective strategy for raising new forests because the nurse crop can be harvested during the early phase of stand development and provide earlier income for the manager (Löf et al., 2014). Nurse crops may benefit the establishment of the more shade-tolerant understory species on some sites experiencing global change.

Conceptually, the presence of more than one tree species may give managers greater flexibility in their future management through increased possibilities to adapt to changing societal objectives, especially if species and/or provenances are chosen to counter the potential impacts of global change (Puettmann, 2011; Puettmann & Messier, 2019). However, the management of such stands is more complicated than that for monocultures. The challenge occurs when facilitative interactions are overridden by competitive interactions, i.e., when the competition from the nurse crop decreases the growth of the understory tree species (Pretzsch et al., 2017). If thinning and harvesting of the nurse crop is not timed to the needs of the understory tree species, the latter may stagnate in growth, and mortality may increase. In most cases, the density management of the two (or more) tree species requires interventions at different times, resulting in multiple entries, each with smaller harvest yields, compared with even-aged monocultures. Despite the additional management costs, two-aged management can yield better economic results than monoculture stands (Valkonen & Valsta, 2001) and offset these higher management costs (Kabzems et al., 2016). For example, gains in volume in aspen–white spruce mixtures can yield up to 17% additional volume over that provided by a pure spruce stand (aspen plus spruce) when harvested at 90 years of age, and 41% more volume if aspen and spruce are harvested at 60 and 90 years of age, respectively (Kabzems et al., 2016).

Tending practices, including precommercial thinning, the removal of early-successional species within a prescribed radius of selected trees using herbicides, cutting or snapping treatments, and the application of herbicides in patches or strips, can be used to reduce the density of the early-successional species in the overstory and increase the growth of the subordinate species (Pitt et al., 2015; Prévost & Charette, 2017). Mixtures of faster-growing early-successional species with longer-lived late-successional species can also improve the self-pruning of the lower branches of dominant trees and the quality and value of stems because of the complementary use of vertical space and shading of lower boles by the conifers (Prévost & Charette, 2017; Puhlick et al., 2019). Precommercial thinning of shade-intolerant deciduous species, such as aspen and birch, taking care to protect advance conifer regeneration, can facilitate recruitment to upper classes and, in this way, accelerate natural succession and/or conversion of stands toward a more complex composition and structure (Prévost & Charette, 2017). Similarly, when trees reach commercial dimensions at later stages, partial cutting can promote advanced conifer regeneration growth—and limit suckering in aspen stands—before final overstory removal (Montoro Girona et al., 2018b; Prévost & DeBlois, 2014; Smith et al., 2016).

Managing two-aged stands is an appealing concept that merits further development, especially in boreal forests with their low taxonomic diversity but which contain species of contrasting growth habits. Additional gains in productivity and wood quality could, for example, be expected by combining this approach with genetically improved material, exotic tree species, e.g., *Poplar* spp. hybrids, and nitrogen-fixing

tree species. In addition, the nurse-crop system requires further development to identify appropriate regimes for the thinning of the nurse crops to support the successful
development of various target tree species. Improved knowledge of yield and those
factors influencing yield outcomes is needed to make and support economically
sound decisions. Despite the benefits, care must be exercised to avoid increasing the
risk of large catastrophic fires that may result from increased conifer abundance and
reduced broadleaf abundance and from greater aridity due to global change. Two-
aged stands also provide more structural diversity, habitats, and ecosystem services
than single-aged monocultures (Berger & Puettmann, 2000).

15.4 Silviculture of Structurally Complex Stands

Although stand-replacing fires are the main natural disturbance in boreal forests,
detailed investigations into the variability within and among fires have shown
that parts of these forests escape catastrophic fires and thus develop complex
multicohort, uneven-aged structures (Fig. 15.2; Boucher et al., 2003; Kuuluvainen & Grenfell, 2012). In the absence of stand-replacing disturbances, low- and
moderate-severity disturbances, caused by agents like wind, insects, and pathogens,
initiate regeneration processes (Kuuluvainen & Grenfell, 2012; Martin et al.,
2019, 2020; Pham et al., 2004). These findings suggest that silvicultural systems
other than clear-cutting could be applied to maintain or enhance forest structural
complexity (Bergeron & Harvey, 1997; Groot, 2002; Lieffers et al., 1996). Examples include traditional uneven-aged systems (e.g., selection cutting, Plenterwald)
that mimic small scale natural variability in boreal forests composed of long-lived
conifers, such as black spruce (Picea mariana; Groot, 2002; Ruel et al., 2013),
Norway spruce, and Scots pine (Pinus sylvestris) stands (Lähde et al., 2010; Pukkala
et al., 2010). In eastern Canada, operational selection-cutting systems maintain
complex stand structures, abundant coarse woody debris, and greater species diversity after the initial harvest in naturally uneven-aged black spruce forests (Ruel et al.,
2013). There is a lack of data on the long-term productivity of uneven-aged managed
boreal forests and, more broadly, for forests regenerated after partial cutting. Specific
concerns relate to post-harvest windthrow because of poor rooting conditions and the
slow growth rates observed under northern latitudes (Bose et al., 2014; Kuuluvainen
et al., 2012; Montoro Girona et al., 2019). However, the advantages of uneven-aged
managed forests in terms of maintaining wildlife habitat, species diversity, carbon
storage, and other ecosystem services can counterbalance the negative impacts of
partial cutting and justify management choices, especially when a variety of management goals are implicated (Ameray et al., 2021; Kuuluvainen et al., 2012; Montoro
Girona et al., 2016; Peura et al., 2018; Ruel et al., 2013).

Irregular shelterwood systems, originally called Femelschlag, can be more suitable to irregular uneven-aged stands—stands with heterogeneous spatial patterns,
stand structures, and species composition—than selection systems, especially when
these stands comprise species having a wide range of functional traits, e.g., life span

Fig. 15.2 In the absence of catastrophic stand-destroying disturbances, secondary disturbances enable the development of complex stand structures; **a** an old-growth unmanaged black spruce stand and **b** a balsam fir–yellow birch irregular stand managed by irregular shelterwood in Québec, Canada. *Photo credits* **a** Maxence Martin, **b** Patricia Raymond

and shade tolerance (Klopcic & Boncina, 2012; Lieffers et al., 1996; Raymond et al., 2009). The different variants and the potential range in resulting spatial and structural outcomes make irregular shelterwood systems highly adaptable and able to simultaneously address various management goals (Boncina, 2011; Raymond et al., 2009; Suffice et al., 2015). In eastern Canada, for example, continuous-cover irregular shelterwood can regenerate sub-boreal balsam fir (*Abies balsamea*)–yellow birch (*Betula alleghaniensis*) stands driven by cyclic moderate-severity disturbances, e.g., spruce budworm (*Choristoneura fumiferana*), while maintaining irregular stand structures and microhabitat diversity (Martin & Raymond, 2019; Raymond & Bédard, 2017). Expanding-gap irregular shelterwood systems have also proven useful for managing forests dominated by balsam fir and red spruce (*Picea rubens*) in North America (Saunders & Arseneault, 2013) and stands of silver fir (*Abies alba*) and Norway spruce in Europe (Heinrichs & Schmidt, 2009; Klopcic & Boncina, 2012). Several experiments and studies have documented the use of selection systems and irregular shelterwood systems to transform even-aged stands into uneven-aged stands. However, this process takes time and can be challenging, particularly for the establishment and development of regenerating cohorts (Heinrichs & Schmidt, 2009; Ligot et al., 2020).

Finally, partial-harvest operations, as an overarching concept that includes selection, shelterwood systems, and others, emphasize the importance of structural legacies (Franklin et al., 2018; McIntire et al., 2005) and provide a means of promoting structural and species diversity as an alternative to clear-cutting (Burton et al., 1999; Lieffers et al., 1996). Variable-retention cutting, a variant of clear-cut systems with the retention of overwood, can also increase structural and compositional diversity (Moussaoui et al., 2016a, 2016b). In a meta-analysis of retention harvests, species richness in retention patches was similar to that of primary boreal forests (Mori & Kitagawa, 2014), with mobile animals, such as birds and arthropods, doing well after retention cutting, whereas vascular plant diversity remained stable, and epiphyte diversity declined. This global analysis also indicated that responses did not differ between dispersed and aggregated retention. However, the highest variability of responses was found when both patterns were combined (Mori & Kitagawa, 2014), underscoring the benefit of flexibility in the layout of partial-harvest operations. Moreover, any silvicultural prescriptions designed in the context of sustainable forest management should include the retention of vital structural attributes, such as standing dead and large live trees, to prevent biodiversity loss (Burton et al., 1999; Puettmann & Messier, 2019).

15.5 Conclusions

The silviculture of boreal forests is dynamic because management objectives must constantly adjust to changing societal needs and ongoing global change but also maintain or enhance the adaptive capacity of forest ecosystems. The homogenization and the simplification of forest structures, caused by past harvesting and management practices, has induced a low resilience of boreal forests to global change (Felton et al., 2016). Consequently, productive boreal forests are being simplified, as areas are increasingly covered by even-aged stands of a limited number of conifer species and organized with little compositional and structural diversity (Felton et al., 2016). If simplification of the boreal forest ecosystems and biodiversity loss continues, forests will become less adaptable and resilient to global change (Puettmann & Messier, 2019). Relying on the principles of increasing within-stand compositional and structural variability, we encourage the use of multi-aged and mixed-species management approaches to increase resilience. However, it is essential to work at other scales by encouraging the diversification of forest structures, i.e., age classes and species, and by limiting fragmentation and biodiversity losses at the landscape scale. Silvicultural planning for sustainable management also requires accounting for global change, altered natural disturbance regimes and rapidly evolving socioeconomic needs. Consequently, it is necessary to work in the context of complex adaptive systems (nonlinearity, heterogeneity, and multiple scales), re-evaluate constantly forest management and silvicultural practices, and adopt resilience as main goal to ensure the long-term sustainability of boreal forests (Kuuluvainen et al., 2015; Montoro Girona et al., 2018b; Puettmann et al., 2009).

References

Ameray, A., Bergeron, Y., Valeria, O., et al. (2021). Forest carbon management: A review of silvicultural practices and management strategies across boreal, temperate and tropical forests. *Current Forestry Reports, 7*(4), 245–266. https://doi.org/10.1007/s40725-021-00151-w.

Bauhus, J., Puettmann, K., & Messier, C. (2009). Silviculture for old-growth attributes. *Forest Ecology and Management, 258*(4), 525–537. https://doi.org/10.1016/j.foreco.2009.01.053.

Berger, A., & Puettmann, K. J. (2000). Overstory composition and stand structure influence herbaceous plant diversity in the mixed aspen forest of northern Minnesota. *American Midland Naturalist, 143*(1), 111–125. https://doi.org/10.1674/0003-0031(2000)143[0111:OCASSI]2.0.CO;2.

Bergeron, Y., Leduc, A., & Harvey, B., et al. (2002). Natural fire regime: A guide for sustainable management of the Canadian boreal forest. *Silva Fennica, 36*(1), 553. https://doi.org/10.14214/sf.553.

Bergeron, Y., & Harvey, B. (1997). Basing silviculture on natural ecosystem dynamics: An approach applied to the southern boreal mixedwood forest of Quebec. *Forest Ecology and Management, 92*(1–3), 235–242. https://doi.org/10.1016/S0378-1127(96)03924-2.

Boncina, A. (2011). History, current status and future prospects of uneven-aged forest management in the Dinaric region: An overview. *Forestry, 84*(5), 467–478. https://doi.org/10.1093/forestry/cpr023.

Bose, A. K., Harvey, B. D., Brais, S., et al. (2014). Constraints to partial cutting in the boreal forest of Canada in the context of natural disturbance-based management: A review. *Forestry, 87*(1), 11–28. https://doi.org/10.1093/forestry/cpt047.

Boucher, D., De Grandpré, L., & Gauthier, S. (2003). Développement d'un outil de classification de la structure des peuplements et comparaison de deux territoires de la pessière à mousses du Québec. *The Forestry Chronicle, 79*(2), 318–328. https://doi.org/10.5558/tfc79318-2.

Bradshaw, R., Gemmel, P., & Björkman, L. (1994). Development of nature-based silvicultural models in southern Sweden: The scientific background. *Forest Landscape Research, 1*(2), 95–110.

Burton, P. J., Kneeshaw, D. D., & Coates, K. D. (1999). Managing forest harvesting to maintain old growth in boreal and sub-boreal forests. *The Forestry Chronicle, 75*(4), 623–631. https://doi.org/10.5558/tfc75623-4.

Campbell, E. M., Maclean, D. A., & Bergeron, Y. (2008). The severity of budworm-caused growth reductions in balsam fir/spruce stands varies with the hardwood content of surrounding forest landscapes. *Forest Science, 54*, 195–205.

Felton, A., Nilsson, U., & Sonesson, J. (2016). Replacing monocultures with mixed-species stands: Ecosystem service implications of two production forest alternatives in Sweden. *Ambio, 45*(S2), 124–139. https://doi.org/10.1007/s13280-015-0749-2.

Filipescu, C. N., & Comeau, P. G. (2011). Influence of *Populus tremuloides* density on air and soil temperature. *Scandinavian Journal of Forest Research, 26*(5), 421–428. https://doi.org/10.1080/02827581.2011.570784.

Franklin, J. F., Johnson, N. K., & Johnson, D. L. (2018). *Ecological forest management*. Waveland Press.

Gerlach, J. P., Reich, P. B., Puettmann, K., et al. (1997). Species, diversity, and density affect tree seedling mortality from *Armillaria* root rot. *Canadian Journal of Forest Research, 27*(9), 1509–1512. https://doi.org/10.1139/x97-098.

Groot, A. (2002). Is uneven-aged silviculture applicable to peatland black spruce (*Picea mariana*) in Ontario, Canada? *Forestry, 75*(4), 437–442. https://doi.org/10.1093/forestry/75.4.437.

Gustafsson, L., Baker, S. C., & Bauhus, J. (2012). Retention forestry to maintain multifunctional forests: A world perspective. *BioScience, 62*(7), 633–645. https://doi.org/10.1525/bio.2012. 62.7.6.

Hector, A., & Bagchi, R. (2007). Biodiversity and ecosystem multifunctionality. *Nature, 448*(7150), 188–190. https://doi.org/10.1038/nature05947.

Heinrichs, S., & Schmidt, W. (2009). Short-term effects of selection and clear cutting on the shrub and herb layer vegetation during the conversion of even-aged Norway spruce stands into mixed stands. *Forest Ecology and Management, 258*(5), 667–678. https://doi.org/10.1016/j.foreco.2009. 04.037.

Himes, A., & Puettmann, K. (2020). Tree species diversity and composition relationship to biomass, understory community, and crown architecture in intensively managed plantations of the coastal Pacific Northwest USA. *Canadian Journal of Forest Research, 50*, 1–12. https://doi.org/10.1139/ cjfr-2019-0236

Kabzems, R., Comeau, P. G., Filipescu, C. N., et al. (2016). Creating boreal mixedwoods by planting spruce under aspen: Successful establishment in uncertain future climates. *Canadian Journal of Forest Research, 46*(10), 1217–1223. https://doi.org/10.1139/cjfr-2015-0440.

Kim, S., Axelsson, E. P., Girona, M. M., et al. (2021). Continuous-cover forestry maintains soil fungal communities in Norway spruce dominated boreal forests. *Forest Ecology and Management, 480*, 118659. https://doi.org/10.1016/j.foreco.2020.118659.

Klopcic, M., & Boncina, A. (2012). Recruitment of tree species in mixed selection and irregular shelterwood forest stands. *Annals of Forest Science, 69*(8), 915–925. https://doi.org/10.1007/s13 595-012-0224-1

Kuuluvainen, T., & Grenfell, R. (2012). Natural disturbance emulation in boreal forest ecosystem management—theories, strategies, and a comparison with conventional even-aged management. *Canadian Journal of Forest Research, 42*(7), 1185–1203. https://doi.org/10.1139/x2012-064.

Kuuluvainen, T., Tahvonen, O., & Aakala, T. (2012). Even-aged and uneven-aged forest management in boreal Fennoscandia: A review. *Ambio, 41*(7), 720–737. https://doi.org/10.1007/s13280- 012-0289-y.

Kuuluvainen, T., Bergeron, Y., & Coates, K. D. (2015). Restoration and ecosystem-based management in the circumboreal forest: Background, challenges, and opportunities. In Stanturf, J. A. (Ed.), *Restoration of boreal and temperate forests* (pp. 251–270). Boca Raton: CRC Press.

Kweon, D., & Comeau, P. G. (2019). Factors influencing overyielding in young boreal mixedwood stands in western Canada. *Forest Ecology and Management, 432*, 546–557. https://doi.org/10. 1016/j.foreco.2018.09.053.

Lähde, E., Laiho, O., & Lin, C. J. (2010). Silvicultural alternatives in an uneven-sized forest dominated by *Picea abies*. *Journal of Forest Research, 15*(1), 14–20. https://doi.org/10.1007/s10310- 009-0154-4.

Lavoie, J., Montoro Girona, M., Grosbois, G., et al. (2021). Does the type of silvicultural practice influence spruce budworm defoliation of seedlings? *Ecosphere, 12*(4), 17. https://doi.org/10. 1002/ecs2.3506.

Lieffers, V. J., & Stadt, K. J. (1994). Growth of understory *Picea glauca, Calamagrostis canadensis* and *Epilobium angustifolium* in relation to overstory light. *Canadian Journal of Forest Research, 24*(6), 1193–1198. https://doi.org/10.1139/x94-157.

Lieffers, V., Stewart, J., Macmillan, R., et al. (1996). Semi-natural and intensive silvicultural systems for the boreal mixedwood forest. *The Forestry Chronicle, 72*(3), 286–292. https://doi.org/10.5558/ tfc72286-3.

Lieffers, V. J., Sidders, D., Keddy, T., et al. (2019). A partial deciduous canopy, coupled with site preparation, produces excellent growth of planted white spruce. *Canadian Journal of Forest Research, 49*(3), 270–280. https://doi.org/10.1139/cjfr-2018-0310.

Ligot, G., Balandier, P., Schmitz, S., et al. (2020). Transforming even-aged coniferous stands to multi-aged stands: An opportunity to increase tree species diversity? *Forestry, 93*(5), 616–629. https://doi.org/10.1093/forestry/cpaa004.

Löf, M., Bolte, A., Jacobs, D. F., et al. (2014). Nurse trees as a forest restoration tool for mixed plantations: Effects on competing vegetation and performance in target tree species. *Restoration Ecology, 22*(6), 758–765. https://doi.org/10.1111/rec.12136.

Mård, H. (1996). The influence of a birch shelter (*Betula* spp) on the growth of young stands of *Picea abies*. *Scandinavian Journal of Forest Research, 11*(1–4), 343–350. https://doi.org/10.1080/028 27589609382945.

Martin, M., & Raymond, P. (2019). Assessing tree-related microhabitat retention according to a harvest gradient using tree-defect surveys as proxies in Eastern Canadian mixedwood forests. *The Forestry Chronicle, 95*(3), 157–170. https://doi.org/10.5558/tfc2019-025.

Martin, M., Morin, H., & Fenton, N. J. (2019). Secondary disturbances of low and moderate severity drive the dynamics of eastern Canadian boreal old-growth forests. *Annals of Forest Science, 76*(4), 108. https://doi.org/10.1007/s13595-019-0891-2.

Martin, M., Boucher, Y., Fenton, N. J., et al. (2020). Forest management has reduced the structural diversity of residual boreal old-growth forest landscapes in Eastern Canada. *Forest Ecology and Management, 458*, 117765. https://doi.org/10.1016/j.foreco.2019.117765.

McIntire, E. J. B., Duchesneau, R., & Kimmins, J. P. (2005). Seed and bud legacies interact with varying fire regimes to drive long-term dynamics of boreal forest communities. *Canadian Journal of Forest Research, 35*(11), 2765–2773. https://doi.org/10.1139/x05-187.

Middleton, G. R., & Munro, B. D. (2002). *Wood density of Alberta white spruce—implications for silvicultural practices*. Vancouver: Forintek Canada Corporation.

Montoro Girona, M., Morin, H., Lussier, J. M., et al. (2016). Radial growth response of black spruce stands ten years after experimental shelterwoods and seed-tree cuttings in boreal forest. *Forests, 7*(10), 1–20. https://doi.org/10.3390/f7100240.

Montoro Girona, M., Navarro, L., & Morin, H. (2018a). A secret hidden in the sediments: Lepidoptera scales. *Frontiers in Ecology and Evolution, 6*, 2. https://doi.org/10.3389/fevo.2018. 00002.

Montoro Girona, M., Lussier, J. M., Morin, H., et al. (2018b). Conifer regeneration after experimental shelterwood and seed-tree treatments in boreal forests: Finding silvicultural alternatives. *Frontiers in Plant Science, 9*, 1145. https://doi.org/10.3389/fpls.2018.01145.

Montoro Girona, M., Morin, H., Lussier, J.-M., et al. (2019). Post-cutting mortality following experimental silvicultural treatments in unmanaged boreal forest stands. *Frontiers in Forests and Global Change, 2*, 4. https://doi.org/10.3389/ffgc.2019.00004.

Mori, A. S., & Kitagawa, R. (2014). Retention forestry as a major paradigm for safeguarding forest biodiversity in productive landscapes: A global meta-analysis. *Biological Conservation, 175*, 65–73. https://doi.org/10.1016/j.biocon.2014.04.016.

Moussaoui, L., Fenton, N. J., Leduc, A., et al. (2016a). Deadwood abundance in post-harvest and post-fire residual patches: An evaluation of patch temporal dynamics in black spruce boreal forest. *Forest Ecology and Management, 368*, 17–27. https://doi.org/10.1016/j.foreco.2016.03.012.

Moussaoui, L., Fenton, N. J., Leduc, A., et al. (2016b). Can retention harvest maintain natural structural complexity? A comparison of post-harvest and post-fire residual patches in boreal forest. *Forests, 7*(12), 243–260. https://doi.org/10.3390/f7100243.

Moussaoui, L., Leduc, A., Fenton, N. J., et al. (2019). Changes in forest structure along a chronosequence in the black spruce boreal forest: Identifying structures to be reproduced through silvicultural practices. *Ecological Indicators, 97*, 89–99. https://doi.org/10.1016/j.ecolind.2018. 09.059.

Moussaoui, L., Leduc, A., Montoro Girona, M., et al. (2020). Success factors for experimental partial harvesting in unmanaged boreal forest: 10-year stand yield results. *Forests, 11*, 1199. https://doi.org/10.3390/f11111199.

O'Hara, K. L. (2014). *Multiaged silviculture*. Oxford University Press.

Paquette, A., & Messier, C. (2010). The role of plantations in managing the world's forests in the Anthropocene. *Frontiers in Ecology and the Environment, 8*(1), 27–34. https://doi.org/10.1890/080116.

Paquette, A., Bouchard, A., & Cogliastro, A. (2006). Survival and growth of under-planted trees: A meta-analysis across four biomes. *Ecological Applications, 16*(4), 1575–1589. https://doi.org/10.1890/1051-0761(2006)016[1575:SAGOUT]2.0.CO;2.

Peura, M., Burgas, D., Eyvindson, K., et al. (2018). Continuous cover forestry is a cost-efficient tool to increase multifunctionality of boreal production forests in Fennoscandia. *Biological Conservation, 217*, 104–112. https://doi.org/10.1016/j.biocon.2017.10.018.

Pham, A. T., De Grandpré, L., Gauthier, S., et al. (2004). Gap dynamics and replacement patterns in gaps of the Northeastern boreal forest of Quebec. *Canadian Journal of Forest Research, 34*(2), 353–364. https://doi.org/10.1139/x03-265.

Pitt, D. G., Comeau, P. G., & Parker, W. C. (2015). Early vegetation control for regeneration of a single cohort, intimate mixture of white spruce and aspen on upland boreal sites – 10th year update. *The Forestry Chronicle, 91*(3), 238–251. https://doi.org/10.5558/tfc2015-045.

Pommerening, A., & Murphy, S. T. (2004). A review of the history, definitions and methods of continuous cover forestry with special attention to afforestation and restocking. *Forestry, 77*(1), 27–44. https://doi.org/10.1093/forestry/77.1.27.

Pretzsch, H., Forrester, D. I., & Bauhus, J. (Eds.). (2017). *Mixed-species forests* (p. 653). Ecology and management. Springer.

Prévost, M., & Charette, L. (2017). Precommercial thinning of overtopping aspen to release coniferous regeneration in a boreal mixedwood stand. *The Forestry Chronicle, 93*(3), 259–270. https://doi.org/10.5558/tfc2017-034.

Prévost, M., & DeBlois, J. (2014). Shelterwood cutting to release coniferous advance growth and limit aspen sucker development in a boreal mixedwood stand. *Forest Ecology and Management, 323*, 148–157. https://doi.org/10.1016/j.foreco.2014.03.015.

Puettmann, K. J. (2011). Silvicultural challenges and options in the context of global change: Simple fixes and opportunities for new management approaches. *Journal of Forestry, 109*, 321–331. https://doi.org/10.1093/jof/109.6.321

Puettmann, K. J., & Messier, C. (2019). Simple guidelines to prepare forests for global change: The dog and the frisbee. *Northwest Science, 93*(3–4), 209–225. https://doi.org/10.3955/046.093.0305.

Puettmann, K. J., Coates, K. D., & Messier, C. (2009). *A critique of silviculture: Managing for complexity*. Island Press.

Puhlick, J. J., Kuehne, C., & Kenefic, L. S. (2019). Crop tree growth response and quality after silvicultural rehabilitation of cutover stands. *Canadian Journal of Forest Research, 49*(6), 670–679. https://doi.org/10.1139/cjfr-2018-0248.

Pukkala, T., Lähde, E., & Laiho, O. (2010). Optimizing the structure and management of uneven-sized stands of Finland. *Forestry, 83*(2), 129–142. https://doi.org/10.1093/forestry/cpp037.

Raymond, P., & Bédard, S. (2017). The irregular shelterwood system as an alternative to clearcutting to achieve compositional and structural objectives in temperate mixedwood stands. *Forest Ecology and Management, 398*, 91–100. https://doi.org/10.1016/j.foreco.2017.04.042.

Raymond, P., Bédard, S., Roy, V., et al. (2009). The irregular shelterwood system: Review, classification, and potential application to forests affected by partial disturbances. *Journal of Forestry, 107*, 405–413. https://doi.org/10.1093/jof/107.8.405

Ruel, J. C., Fortin, D., & Pothier, D. (2013). Partial cutting in old-growth boreal stands: An integrated experiment. *The Forestry Chronicle, 89*(3), 360–369. https://doi.org/10.5558/tfc2013-066.

Rytter, L., Ingerslev, M., & Kilpeläinen, A. (2016). Increased forest biomass production in the Nordic and Baltic countries – a review on current and future opportunities. *Silva Fennica, 50*(5), 1660. https://doi.org/10.14214/sf.1660.

Saunders, M. R., & Arseneault, J. E. (2013). Potential yields and economic returns of natural disturbance-based silviculture: A case study from the Acadian Forest Ecosystem Research Program. *Journal of Forestry, 111*(3), 175–185. https://doi.org/10.5849/jof.12-059.

Shvidenko, A., & Apps, M. (2006). The International Boreal Forest Research Association: Understanding boreal forests and forestry in a changing world. *Mitigation and Adaptation Strategies for Global Change, 11*(1), 5–32. https://doi.org/10.1007/s11027-006-0986-6.

Smith, J., Harvey, B. D., Koubaa, A., et al. (2016). Sprucing up the mixedwoods: Growth response of white spruce (*Picea glauca*) to partial cutting in the Eastern Canadian boreal forest. *Canadian Journal of Forest Research, 46*(10), 1205–1215. https://doi.org/10.1139/cjfr-2015-0489.

Suffice, P., Joanisse, G., Imbeau, L., et al. (2015). Short-term effects of irregular shelterwood cutting on yellow birch regeneration and habitat use by snowshoe hare. *Forest Ecology and Management, 354*, 160–169. https://doi.org/10.1016/j.foreco.2015.06.025.

Taylor, S. P., Delong, C., Alfaro, R. I., et al. (1996). The effects of overstory shading on white pine weevil damage to white spruce and its effect on spruce growth rates. *Canadian Journal of Forest Research, 26*(2), 306–312. https://doi.org/10.1139/x26-034.

Valkonen, S., & Valsta, L. (2001). Productivity and economics of mixed two-storied spruce and birch stands in Southern Finland simulated with empirical models. *Forest Ecology and Management, 140*(2–3), 133–149. https://doi.org/10.1016/S0378-1127(00)00321-2.

Zhang, B., MacLean, D. A., Johns, R. C., et al. (2018). Effects of hardwood content on balsam fir defoliation during the building phase of a spruce budworm outbreak. *Forests, 9*(9), 530. https://doi.org/10.3390/f9090530.

Chapter 16
Innovative Silviculture to Achieve Sustainable Forest Management in Boreal Forests: Lessons from Two Large-Scale Experiments

Miguel Montoro Girona, Louiza Moussaoui, Hubert Morin, Nelson Thiffault, Alain Leduc, Patricia Raymond, Arun Bosé, Yves Bergeron, and Jean-Martin Lussier

Abstract Clear-cutting has been the dominant harvesting method used in boreal forest silviculture. Reducing the potential negative effects of intensive forestry activities on ecosystems, e.g., the simplification and homogenization of stand structure, requires diversifying silvicultural practices to promote forest resilience in the face of climate change. Priority therefore lies in developing, evaluating, and adapting partial cutting as a potential silvicultural option for ensuring the sustainable management of boreal forests. In this chapter, we summarize the findings of two large-scale experiments conducted in Canadian boreal forests that tested new silvicultural approaches and explore their implications for forest management. We discuss the effects of these treatments on tree growth, tree mortality, regeneration, and biodiversity, and we examine the challenges of existing silvicultural approaches in the context of climate change.

M. M. Girona (✉) · L. Moussaoui
Groupe de Recherche en Écologie de la MRC-Abitibi, Forest Research Institute, Université du Québec en Abitibi-Témiscamingue, Amos Campus, 341, rue Principale Nord, Amos, QC J9T 2L8, Canada
e-mail: miguel.montoro@uqat.ca

L. Moussaoui
e-mail: louiza.moussaoui@uqat.ca

M. M. Girona
Department of Wildlife, Fish, and Environmental Studies, Swedish University of Agricultural Sciences, 901 83 Umeå, Sweden

Y. Bergeron
Forest Research Institute, Université du Québec en Abitibi-Témiscamingue, 445 boul. de l'Université, Rouyn-Noranda, QC J9X 5E4, Canada
e-mail: yves.bergeron@uqat.ca

H. Morin
Département des Sciences fondamentales, Université du Québec à Chicoutimi, 555 boul. de l'université, Chicoutimi, QC G7H 2B1, Canada
e-mail: hubert_morin@uqac.ca

© The Author(s) 2023
M. M. Girona et al. (eds.), *Boreal Forests in the Face of Climate Change*,
Advances in Global Change Research 74,
https://doi.org/10.1007/978-3-031-15988-6_16

417

16.1 Context

Ecosystem-based management (EBM) is a vehicle for achieving sustainable forest management and aims to balance ecological, social, and economic objectives (Franklin et al., 2018; Gauthier et al., 2008; Palik & D'Amato, 2017). EBM emerged from the natural disturbance emulation paradigm (Bergeron et al., 2001), in which silvicultural treatments are used to mimic the main disturbances and the natural range of variation of the ecological attributes of a forest area (Angelstam, 1998; Kuuluvainen et al., 2012; Fig. 16.1). In the boreal forest, natural disturbances such as fire, insect outbreaks, and windthrow are the driving forces that generate significant ecosystem changes at various spatial and temporal scales, depending on their frequency and severity and the size of the affected area (De Grandpré et al., 2000). These disturbance patterns determine the dynamics, structure, and composition of forests. Thus, silvicultural practices can simulate the composition and structure of post-disturbance forests by modifying stand attributes and producing variability within forest landscapes (Lecomte & Bergeron, 2005; Puettmann et al., 2015).

Over the past two decades, timber harvesting has become the main disturbance in boreal forest ecosystems. Currently, clear-cutting remains the main silvicultural treatment within the boreal biome, used within 83% of the harvested area in Canadian forests (Fig. 16.2; CCFM, 2018). Clear-cut systems offer the advantage of low costs relative to the harvested volume (Rosenvald & Lõhmus, 2008). The regeneration is assured either by plantation or by protecting the natural advanced regeneration (Groot et al., 2005). It is also used to simulate stand-replacing disturbances such as wildfires, i.e., high severity events affecting extensive areas, although clear-cut systems cannot fully mimic all postfire characteristics (Buddle et al., 2006). Although fire is the most

N. Thiffault · J.-M. Lussier
Natural Resources Canada, Canadian Wood Fibre Centre, 1055 rue du PEPS, P.O. Box 10380, Stn Sainte Foy, Québec, QC G1V 4C7, Canada
e-mail: nelson.thiffault@nrcan-rncan.gc.ca

J.-M. Lussier
e-mail: jean-martin.lussier@nrcan-rncan.gc.ca

A. Leduc · Y. Bergeron
Département des Sciences Biologiques, Université du Québec à Montréal, P.O. Box 8888, Stn. Centre-Ville, Montréal, QC H3C 3P8, Canada
e-mail: leduc.alain@uqam.ca

P. Raymond
Direction de la recherche forestière, ministère des Forêts, de la Faune et des Parcs du Québec, 2700 rue Einstein, Québec, QC G1P 3W8, Canada
e-mail: patricia.raymond@mffp.gouv.qc.ca

A. Bosé
Swiss Federal Institute for Forest, Snow and Landscape Research WSL, Zürcherstrasse 111, 8903 Birmensdorf, Switzerland
e-mail: arun.bose@wsl.ch

Forestry and Wood Technology Discipline, Khulna University, Khulna 9208, Bangladesh

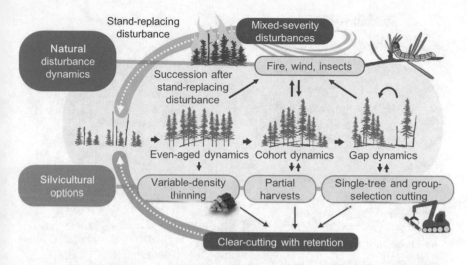

Fig. 16.1 This model represents silvicultural options for maintaining landscape-level forest structures and age distributions similar to those that would exist under a natural disturbance regime. Structural cohorts (*green bubbles*) correspond to the various postfire stand successional stages, and silvicultural options (*brown bubbles*) are presented along a gradient of harvest intensity. This illustration is inspired and adapted from the principle of the multicohort model and the ASIO model (Angelstam, 1998; Bergeron et al., 2002)

common disturbance in many boreal regions, this is not the case for all boreal forest landscapes. Consequently, forest management based entirely on clear-cut systems can alter structural and biodiversity characteristics at the stand and landscape scales (Bouchard & Pothier, 2011, Lindenmayer & Franklin, 2002). Consequently, even-aged management regimes having short forest rotations can produce habitat degradation, provoke the loss of productivity in some regions, and lead to structurally homogeneous stands (Fig. 16.2; Fischer & Lindenmayer, 2007; Nolet et al., 2018; Seedre et al., 2018). To address these concerns, forest management strategies in several boreal countries have prioritized the need to develop, diversify, and apply new silvicultural treatments within an EBM framework.

Partial-cutting treatments are a group of forestry practices included in existing boreal EBM strategies (Grenon et al., 2010). From an ecosystem management point of view, partial cuttings remove a portion of trees in a forest stand and maintain some characteristics of a closed forest cover (Fig. 16.2; Bose et al., 2014; Moussaoui et al., 2019). Partial cuttings that involve the removal of 30% to 50% of the stand basal area can therefore emulate natural disturbances of intermediate severity and extent, e.g., as observed following windstorms and insect outbreaks. Depending on management objectives, partial cutting is a generic term that can include commercial thinning (Nyland, 2016), selection cutting systems (Majcen, 1994), uniform and irregular shelterwood cutting systems (Raymond et al., 2009), HARP (harvesting with regeneration protection), and variable retention harvesting (Groot et al., 2005). Most of these treatments were initially developed in Europe and are being adapted

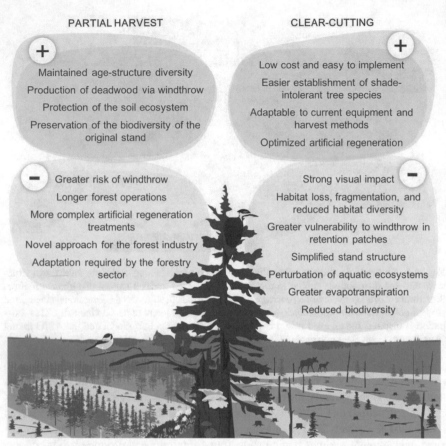

PARTIAL HARVEST

+
- Maintained age-structure diversity
- Production of deadwood via windthrow
- Protection of the soil ecosystem
- Preservation of the biodiversity of the original stand

–
- Greater risk of windthrow
- Longer forest operations
- More complex artificial regeneration treatments
- Novel approach for the forest industry
- Adaptation required by the forestry sector

CLEAR-CUTTING

+
- Low cost and easy to implement
- Easier establishment of shade-intolerant tree species
- Adaptable to current equipment and harvest methods
- Optimized artificial regeneration

–
- Strong visual impact
- Habitat loss, fragmentation, and reduced habitat diversity
- Greater vulnerability to windthrow in retention patches
- Simplified stand structure
- Perturbation of aquatic ecosystems
- Greater evapotranspiration
- Reduced biodiversity

Fig. 16.2 Some of the potential advantages (+) and disadvantages (−) of using partial and clear-cutting harvests in the boreal forest. Clear-cutting is represented as it is applied in boreal eastern Canada, where harvest trails are restricted to less than 25% of the area to protect soils and advance regeneration

to the context of the North American boreal forest, in particular adjustments related to mechanized operations. This adaptation to new boreal contexts requires an understanding of the effects of partial cutting on residual tree growth and mortality, natural regeneration, and biodiversity before this approach can be considered as a tool for ensuring the sustainable management of boreal forests.

In this chapter, we synthesize observations from two large-scale experiments undertaken in the Canadian boreal forest, which assessed the short-, medium-, and long-term effects of various experimental partial-cutting treatments on stand growth, mortality, regeneration, and biodiversity. We then provide a perspective on future research directions and the implementation of these harvesting approaches in the Canadian boreal forest.

16.2 Large-Scale Experiments and Innovative Silviculture: Two Case Studies in Black Spruce Forests

The need to evaluate the silvicultural potential of partial cutting in Canadian boreal forests led to the establishment of two large-scale experiments within three forest regions in Québec, Canada (Saguenay, Côte-Nord [North Shore]): *MISA* (Managing innovative silvicultural alternatives) and RECPA (Réseau expérimental de coupes partielles en Abitibi) *[Abitibi partial cutting network]*. Both experiments comprise multiple replicates and long-term monitoring plots to investigate partial-cutting modalities adapted to mechanized operations in the black spruce (*Picea mariana* (Mill.) BSP)–dominated forests of Québec.

The MISA experiment, established by the Canadian Forest Service of Natural Resources Canada in 2003, involved three novel shelterwood treatments adapted to mechanized harvesting (Meek, 2006), standard clear-cutting, a seed-tree method, and untreated controls (Fig. 16.3; Montoro Girona et al., 2017). The shelterwood system aims to promote natural regeneration in the understory before a final harvest through the gradual opening of the canopy (Larouche et al., 2013; Matthews, 1991; Raymond et al., 2013; Smith et al., 1997). This approach maintains part of the residual stand as a seed source and as a means of offering partial shade to protect seedlings during the regeneration period and preventing the establishment of competing early-successional shade intolerant species (Doucet et al., 1996; Raymond et al., 2000). Within an EBM context, shelterwood harvesting can be used to replicate the effect of a successional process occurring after low- to moderate-intensity secondary disturbances, such as insect outbreaks or windthrows, which promote the development of two-cohort stands (Drever et al., 2006; Kuuluvainen & Grenfell, 2012; Oliver & Larson, 1996; Smith et al., 1997). Incidentally, the shelterwood system allows residual trees to increase their volume before the final harvest and could be a promising silvicultural option for stands of black spruce—one of the most widely distributed species in North America (found from Québec to Alaska) and a shade-tolerant species that depends on exposed mineral soil for regeneration via seeds. Shelterwoods may provide an adequate solution for management strategies to maintain a high level of forest retention, particularly for the management of woodland caribou habitat (Courtois et al., 2004). This experiment was conducted in mature even-aged black spruce stands on upland sites, following a complete randomized block design with 36 experimental units of 3 ha each (Fig. 16.4).

The Abitibi partial cutting network (RECPA) was established in 1998 across northwestern Québec to test the operational feasibility of partial-cutting treatments in black spruce–feathermoss forests having an uneven-aged structure (Bescond et al., 2011; Fenton & Bergeron, 2007). RECPA included two experimental partial-cutting treatments: harvesting with advance regeneration protection (HARP) (Groot et al., 2005; Thorpe et al., 2007) and an experimental *conservation of canopy cover* (CCCC) treatment. This experiment also included clear-cut harvesting that removed

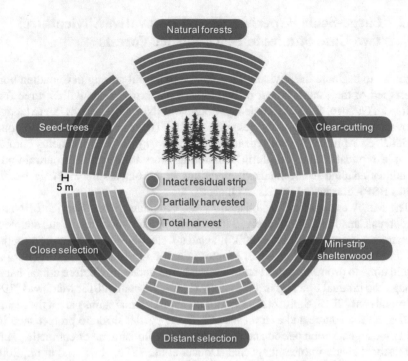

Fig. 16.3 Characteristics and spatial patterns of the three experimental shelterwood treatments and a seed-tree method applied in the MISA experiment

all merchantable stems (diameter at breast height; DBH > 9 cm). Both HARP and CCCC are applied to promote natural regeneration and stand growth. HARP is a partial-cutting treatment that involves removing stems with (generally) a DBH greater than 14 cm; this approach is used operationally in irregular boreal stands characterized by an abundance of saplings and small merchantable stems (Riopel et al., 2010). Moreover, HARP also promotes the development of both old-growth characteristics and the maintenance of high levels of biodiversity (Fenton et al., 2013; Opoku-Nyame et al., 2021). CCCC is a partial-cutting treatment in which stems from all diameter classes are harvested to maintain a similar proportion as that present before harvest (Arseneault et al., 2012). Although some partial-cutting treatments offer the potential of being effective at ensuring a regular input of deadwood and provide a compromise between conservation and harvesting in boreal forest stands (Fenton et al., 2013), some modalities, e.g., operational aspects, remain and affect the survival of residual stems. These modalities are insufficiently understood in the context of the Canadian boreal forest (Bose et al., 2014; Thorpe & Thomas, 2007). The RECPA experiment comprised six study sites, each site comprising three blocks (a partial cutting, clear-cutting, and untreated plot) of 50 ha each.

Fig. 16.4 **a** Naturally regenerating black spruce stand after clear-cutting, **b** partial cutting ten years post-treatment, **c** trail opening and canopy conditions after mini-strip shelterwood cuttings (50% removal). *Photo credits* Miguel Montoro Girona

16.3 Is Partial Cutting a Viable Alternative for Sustainable Forest Management?

16.3.1 Tree Growth

EBM aims to ensure both wood production and the maintenance of ecosystem functions. Thus, evaluating EBM performance must involve quantifying and understanding the effect of silvicultural treatments on tree growth. The effects of partial cutting on residual stand wood production in boreal forests are increasingly understood; several partial-cutting studies have been conducted involving various treatments and species in Scandinavia (Lähde et al., 2002; Pape, 1999; Peltola et al., 2002; Pukkala et al., 2009) and North America (Bourgeois et al., 2004; Goudiaby et al., 2012; Raulier et al., 2003; Schneider et al., 2008; Thorpe & Thomas, 2007). After partial cutting, residual stem growth generally increases because of decreased stand density and competition. For black spruce—the most harvested conifer in eastern Canadian forests owing to its excellent wood properties—residual stem growth depends on the partial-cutting intensity; the gain in tree growth is often marginal or insignificant for partial cuttings that involve about 30% of the basal area being removed, whereas marked residual stem growth is observed for cuts of 50% basal area (Goudiaby et al., 2012; Pamerleau-Couture et al., 2015; Soucy et al., 2012; Vincent et al., 2009). Normally, tree growth response is not consistent over time (i.e., the response is delayed by three to five years after treatment), across space (e.g., edge effect, site index and climate), and among stands (i.e., high individual tree variability related to ecological status, age, and genetics) (Montoro Girona et al., 2016, 2017).

In the MISA experiment, the novel shelterwood treatments enhanced the radial growth of black spruce stems, especially in younger stands (80–100 years old), and the growth response did not differ in relation to harvesting intensity or the type of silvicultural treatment applied among shelterwoods and seed trees. The radial growth response, 8 to 10 years postcutting, was 41% to 62% higher than that in untreated plots. The main factors affecting the growth response were stand structure, silvicultural treatment, tree position relative to skidding trails, growth before cutting, and time (Montoro Girona et al., 2016). Trees at the edge of the skidding trails showed twice the increase in growth compared with trees within residual strips, and this effect was greater in younger stands. Trail edges are characterized by less competition and a greater access to light and nutrients than within strips. On the other hand, trees located along trail edges may face greater exposure to wind and experience more frequent stem and root injuries caused by machinery during cutting and scarification operations than trees within strips (Cancino, 2005; Chen et al., 1993; Gardiner et al., 1997; Harper et al., 2016). In the MISA experiment, the positive response of black spruce along the trails suggests that the improved access to light and soil resources counterbalanced these potential trailside stresses (Fig. 16.5a). These observations confirm that silvicultural planning and stand selection in mature black spruce forests must consider both the spatial distribution of trails, to promote edge effect, and stand age, to maximize growth response.

In the RECPA experiment, residual stand volume showed net growth over the ten years that followed both the HARP and CCCC treatments in all studied sites (Moussaoui et al., 2020). Average tree-ring width after HARP in black spruce stands was double that of preharvest stands (Thorpe et al., 2007). In uneven-aged black spruce–dominated stands, greater tree radial growth after partial cutting can be limited by tree age and intertree competition (Pamerleau-Couture et al., 2015). In uneven-aged black spruce stands, although heavy partial cutting can re-attain the preharvest basal area 45 years after harvest (Groot, 2014), this return to the initial basal area can take 65 to 105 years on poorer quality sites corresponding to a more limited establishment of post-harvest growing stock (Thorpe et al., 2010). In Québec, HARP tends to reduce forest rotation. Results obtained 20 years post-treatment in the HARP experiment indicate that this reduction in forest rotation could be 40 to 50 years for stands having a median forest rotation of 80 to 90 years. This finding suggests, therefore, that in

Fig. 16.5 Response of black spruce stands after experimental silvicultural treatments in MISA; **a** tree rings show the strong radial growth response of edge trees during the first ten years after treatment; **b** windthrow damage observed ten years after the seed-tree method; **c** black spruce seedlings established ten years after the experimental shelterwood and scarification. *Photo credits* Miguel Montoro Girona

terms of residual tree growth, by considering site quality, partial-cutting treatments had a positive effect to promote radial growth after cutting.

16.3.2   Post-Harvest Mortality and Windthrow

A significant risk associated with partial cutting is post-harvest mortality due to windthrow disturbance (Fig. 16.5b). Partial cutting increases wind penetration into the residual stand, heightening the risk of windthrow (Gardiner, 1995; Riopel et al., 2010; Ruel, 1995). This effect is most evident during the first five years post-treatment (Jönsson et al., 2007; Macisaac & Krygier, 2009; Ruel, 2000; Thorpe et al., 2008) or where wind exposure is increased by large nearby openings or wind-favoring topography. Factors influencing windthrow include wind exposure (Ruel, 2000; Scott & Mitchell, 2005), edaphic conditions (Mitchell, 1995; Ruel, 1995; Stokes et al., 1995), stand composition (Burns & Honkala, 1990; Raymond et al., 2000; Riopel et al.,

2010), stand density (Cremer et al., 1982; Maccurrach, 1991), as well as stem mass, size, and height/diameter ratio (Riopel et al., 2010). In addition, tree injuries incurred during harvest operations may contribute to increase mortality for stems located next to skid trails (Bladon et al., 2008; Thorpe et al., 2008).

In the MISA experiment, 76% of the post-harvest mortality, ten years after treatment, could be explained by harvest treatment, machinery-caused injuries, and distance to adjacent cuts (Montoro Girona et al., 2019). Windthrow accounted for 80% of post-harvest mortality. Therefore, an expected increase in mortality with greater harvest intensity must be included in silvicultural guidelines for applying uniform shelterwood treatments and seed-tree harvesting (e.g., see Stathers et al., 1994). Retention levels should aim at 45% to 65% of the initial basal area to minimize losses, and stand selection should prioritize sites having conditions that favor the lowest probability of windthrow. Moreover, low retention levels increase the risk of tree mortality and produce high overturn rates, thereby compromising silvicultural objectives (Urgenson et al., 2013).

The success of partial cutting depends mainly on the survival of residual trees. Ten years after harvest, the uniform shelterwood treatments tested in the MISA experiment resulted in a mortality that was 15% to 20% higher than that observed in the control stands; however, this mortality was still within the range of that observed in natural stands in this area (De Grandpré et al., 2008). In the MISA experiment, seed-tree harvesting experienced the highest levels of mortality (45%–75% of the residual stems), primarily because of the higher exposure of residual trees to wind, relative to uniform shelterwood treatments, in which only trees along the trails and the edges close to clear-cut areas were highly exposed. Trees along smaller and more exposed residual strips are more vulnerable to wind damage (Jönsson et al., 2007) and experience higher rates of overturn in residual stands (Achim et al., 2005). Anyomi and Ruel (2015) and Urgenson et al. (2013) observed similar patterns, finding that high harvest intensity, such as that using seed trees and the removal of 75% of the basal area, produced 60% to 80% post-harvest mortality; these levels correspond to twice the amount of windthrow than that observed for intermediate intensity harvesting (40%–60% removal)—levels removed in the shelterwood system, for example.

Much of the research conducted in the RECPA experiment focused on the impacts of partial cutting on residual tree mortality over the short, medium, and long term (Lavoie et al., 2012). Moussaoui et al. (2020) demonstrated that ten years post-harvest, stem losses in black spruce forests depend largely on preharvest stand structures and site conditions, in agreement with previous results coniferous-dominated stands (Riopel et al., 2010). Moreover, the RECPA experiment showed that ten years after harvesting, depending on harvest treatment intensity, partial cutting could increase post-harvest tree recruitment and growth or reduce stand basal area because of a high rate of standing tree mortality (Moussaoui et al., 2020). For example, no cases of high mortality (basal area occupied by dead trees) were observed ten years after harvest when treatment intensity (HARP or CCCC), i.e., the percentage of harvested basal area, was ≤48%. In a study comparing mortality after dispersed and group-retention partial cuttings in Canadian boreal forests, Lavoie et al. (2012) observed that increased wind penetration into residual stands heightened post-harvest mortality; a combination of factors likely caused this increase, including fine-textured soils, flat topography, and the dominance of shade-intolerant species. Moussaoui et al. (2020) suggest that partial cutting in black spruce forests should be avoided in sites where the organic layer thickness approaches 17 cm or more to ensure an increase in the decennial stand yield after harvesting. From the results of both experiments, post-harvest tree mortality in boreal forests can be predicted using pre-existing stand conditions, even before considering the influence of the intensity and configuration of a partial-cutting treatment. Understanding the factors involved in this complex phenomenon is important for reducing post-harvest losses.

16.3.3   Regeneration

Successful natural regeneration is fundamental to sustainable forest management, as it enables the resilience of forest ecosystems. Natural regeneration is central to most management strategies in the boreal biome (Bose et al., 2014; Kuuluvainen, 1994; McDonald & Urban, 2004; Messier et al., 1999; Prévost, 1996; Prévost et al., 2010). Natural regeneration of boreal forests involves numerous processes, including seed production and dispersal, germination rates, seedling establishment, and early seedling and sapling growth and mortality (Blanco et al., 2009; Thiffault et al., 2015). Stand structure and silvicultural treatment determine ecological factors such as light availability, and substrate influences the quality of the environment for seedling establishment and growth. The availability and distribution of seedbeds composed of exposed mineral soil are crucial elements for the successful regeneration of boreal stands (Kolabinski, 1991; Martin et al., 2020; Raymond et al., 2000). Moreover, new openings in the forest cover caused by partial cuttings alter light availability and the physical conditions in the forest and its understory (Barik et al., 1992; Coates, 2000, 2002; Parent & Messier, 1995). Numerous studies have examined the role of increased light availability on the understory (Beaudet et al., 2011; Canham et al., 1990; Chazdon, 1988) and its effect on the growth of regenerating trees (Beaudet & Messier, 1998; Kobe et al., 1995). Studies have also quantified the influence of opening size on cohort biomass (Webster & Lorimer, 2002), variations in canopy openings after partial cutting (Beaudet & Messier, 2002; Domke et al., 2007), and gap formation rates (Raymond et al., 2006; Runkle, 2000; Van Der Meer & Bongers, 1996).

The MISA experiment evaluated how the creation of canopy openings from the uniform shelterwood treatments affected the density, stocking, and size of black spruce seedlings after partial cutting (Fig. 16.5c; Montoro Girona et al., 2018). The experiment demonstrated that uniform shelterwood and seed-tree treatments produced an abundant regeneration of black spruce seedlings and provided a more effective silvicultural option than clear-cutting in that regard; for example, experimental shelterwood-treated stands produced three times more regeneration outcomes than that observed in clear-cut stands (Montoro Girona et al., 2018). The shelterwood treatment involving a series of narrow cut strips (mini-strip shelterwood) was the most effective in terms of regeneration stocking after ten years. The MISA results indicated that regeneration outcomes depend more on substrate than light during the first ten years post-harvest. Following partial cutting, all sites were scarified using a 10-ton excavator equipped with a 1 m³ bucket in the skidding trails and along their edges where the residual spacing of trees allowed. Scarification promoted the stocking and density of black spruce regeneration by exposing the mineral soil, thereby emulating the effects of fire on the organic layer. Thus, partial cutting combined with patch scarification created the required substrate (mineral soil) and light conditions (lateral shadowing from the residual strip) to promote black spruce regeneration. No major competition with deciduous trees and shrubs was observed; however, future research must be undertaken to measure the changes over the longer term.

Piché (2017) described the effects of partial cutting on regeneration establishment ten years after treatments within the REPCA experiments. Stocking was similar between partial-cutting and clear-cutting treatments, whereas seedling growth remained low on paludified sites. The regional climate and physical characteristics of the soils found in the Clay Belt region of eastern Canada favor the accumulation of organic matter and the rise of the local water table (Bescond et al., 2011; Fenton et al., 2005, 2009; Payette & Rochefort, 2001). Moreover, the anchoring and intertwining of black spruce root systems is reduced; this leads to decreased natural establishment and productivity (Lafleur et al., 2010; Lecomte et al., 2009). Understanding the forest dynamics of this region and adapting partial-cutting modalities will require further studies, particularly in forest stands prone to paludification. Furthermore, the recent assessment of the effects of partial cutting ten years after treatment on stand development (recruitment, growth, and mortality) in RECPA stands revealed that tree recruitment increases significantly with greater residual sapling density. Moussaoui et al. (2020) found that a minimum density of 800 saplings/ha appears sufficient to promote the healthy recovery of black spruce stands and a high stand yield after partial cutting. Therefore, black spruce stands having a diversified diametrical structure with an abundance of saplings will respond positively to partial cutting over the short term.

16.3.4 Biodiversity

Silvicultural treatments modify the biotic (e.g., species composition, diversity, and community structure) and abiotic (e.g., light availability, soil temperature, and water availability) environment of forest stands (Kim et al., 2021). The effects of soil disturbance on vegetation colonization are site specific and depend largely on disturbance size and intensity, preharvest species composition, and species' functional traits. Seedbed conditions and existing seed banks also significantly influence shrub colonization following soil disturbance, as illustrated in multiple successional studies (Lafleur et al., 2010, 2015; Lecomte et al., 2006; Prévost, 1996).

Understory vegetation is a good indicator for understanding changes in forest dynamics caused by silviculture, as it is directly influenced by the dominant tree cover (Fraver et al., 2007; Hernández-Rodríguez et al., 2021; Macdonald & Fenniak, 2007). Previous studies in western Canada have focused on the response of plant communities after a clear-cutting of mixedwood forests in Manitoba (Kembel et al., 2008) and partial cutting in Alberta (Caners et al., 2013) and British Columbia (Man et al., 2010). In eastern Canada, forest succession after partial cutting has been studied in maple (*Acer saccharum* Marsh)-dominated stands (Archambault et al., 2003), mixed yellow birch (*Betula alleghaniensis* Britt.)–balsam fir (*Abies balsamea* (L.) Mill.) forests (Dubois et al., 2006), balsam fir–dominated stands (Raymond et al., 2000), and black spruce–dominated ecosystems (Fenton et al., 2013).

In the RECPA experiment, much of the research has focused on the impacts of partial cutting on biodiversity (Bescond et al., 2011; Fenton & Bergeron, 2007; Fenton et al., 2013; Paradis & Work, 2011). Partial cutting at a minimum of 40% to 60% retention maintained habitat attributes for various organisms (Bose et al., 2014; Fenton et al., 2013). Species-specific responses to partial cutting can be positive for both understory plants (Bescond et al., 2011) and small mammal populations (Cheveau et al., 2004), which are relatively resilient to low retention levels (<40%). Similar benefits were documented for vascular plants and mosses (Arseneault et al., 2012; Bescond et al., 2011; Fenton & Bergeron, 2007; Opoku-Nyame et al., 2021), epiphytic lichens (Boudreault et al., 2002), and birds (Lycke-Poulin, 2008) in the RECPA experiment. Maintaining arthropod assemblages similar to those of older, unmanaged forests requires, however, higher retention levels (>60%; Jacobs & Work, 2012; Paradis & Work, 2011). In the context of EBM, findings from the RECPA experiment for a variety of species groups indicate that a 50% retention level appears appropriate for maintaining biodiversity; however, long-term monitoring is required to inform adaptation strategies for sustainably managing forests in the context of a changing climate, as well as to include the larger body size species to improve our understanding of the biodiversity patterns at the landscape scale for different structural stands.

16.4 Research Perspectives

In this chapter, we have addressed some critical aspects of novel silvicultural treatments that aim to promote residual tree growth, minimize windthrow damage, favor regeneration, and maintain biodiversity in the boreal forest. Although we have focused on the MISA and RECPA experiments, other large-scale experiments, such as EMEND (Spence et al., 1999), SAFE (Brais et al., 2004), and EVO (Vanha-Majamaa et al., 2007), provide invaluable insights into alternative management systems applicable to the boreal forest. Nonetheless, a more complete assessment of sustainable forest management in the context of climate change requires additional research on a number of fronts:

Economic implications The selection and application of silvicultural treatments are highly dependent on financial and economic profitability. Future work should include analyses of the cost/benefit ratios related to implementing partial cutting in the boreal context and develop cost-effective means of planning, conducting, and monitoring partial-cutting treatments on an operational basis. Finally, the impact of partial cuttings on wood quality and value must be addressed to analyze final product values in the wood market.

Long-term monitoring of growth and yield Climate change will alter the growth dynamics of stands and species; new estimates of optimal rotations will probably be required. Large-scale experimental designs comprising permanent sampling plots offer the opportunity for a long-term monitoring of tree growth and stand dynamics (e.g., Achim et al., 2021; Pappas et al., 2022; Thiffault et al., 2021). Further investigations of the growth response in black spruce stands after partial cutting should examine the extent (distance) of the edge effect along residual strips; such research would help maximize post-treatment wood production. Another pressing question concerns the full estimate of postcutting growth response over time. Currently, existing dendrochronological series of black spruce from postcutting growth studies do not exceed 12 years (Montoro Girona et al., 2016, 2017; Pamerleau-Couture et al., 2015; Thorpe et al., 2007); longer-term assessments are required for planning the timing of the final cut and optimizing the effect of the treatment on radial growth.

Forest regeneration under climate change As climate change will lead to altered precipitation and temperature patterns and likely favor more frequent summer drought periods. Partial cutting could help reduce seedling vulnerability to drought; however, no studies have addressed this question to date. Climate change will create novel stand compositions, and the greater presence of hardwoods in boreal forest stands must be addressed (Brumelis & Carleton, 1988; Riopel et al., 2011; Solarik et al., 2020). The opening of the canopy, for example, stimulates the germination and survival of paper birch (*Betula papyrifera* Marsh.) (Perala & Alm, 1990). Even if hardwood species are generally found in relatively open areas receiving higher irradiance, some species can survive for a few years under conditions of 10% sunlight (Messier et al., 1999). Thus, early regeneration in unmanaged boreal forests characterized by infrequent fires is dominated by more shade-tolerant softwood species. Deciduous competition could affect the growth of black spruce seedlings and thus requires a careful analysis of regeneration after partial cutting.

Insect outbreaks Climate change is shifting spruce budworm (*Choristoneura fumiferana* (Clemens)) habitats to a more northern range and into areas currently dominated by black spruce (Navarro et al., 2018). Because of the potential significant impacts of spruce budworm outbreaks on residual stands and the postcutting regeneration, it is necessary to understand the spruce budworm–related effects in interaction with novel silvicultural practices. Between 1990 and 2016, harvesting affected 24 million ha in Canada. Consequently, a large portion of the North American boreal forest exists at an early development stage; it is thus important to understand the vulnerability of regeneration after cutting to insect outbreaks (Cotton-Gagnon et al., 2018; Lavoie et al., 2019). A recent study has demonstrated that partial cuttings can reduce the impact of insect outbreaks on regeneration (Lavoie et al., 2021); however, much more research is required to better understand these insect-regeneration interactions under future climate change.

Windthrow Climate change projections indicate increasing windthrow and wind damage in forests will significantly impact stand dynamics in the near future (Saad et al., 2017). Pursuing the research efforts conducted over the last 20 years (Achim et al., 2005; Gardiner et al., 2008; Solarik et al., 2012) is essential to understand better the factors driving forest vulnerability to wind damage, especially following partial cutting. These studies would contribute to minimizing the uncertainties associated with climate change in forest management strategies and help create decision support tools that consider those risks in planning. Finally, management measures that reduce windthrow risk must be tested further to provide effective tools for silviculturists in preventing losses due to wind damage.

Carbon sequestration Increasing the C sequestration capacity of a forest requires an understanding of the effects of management and climate together with predictions as to how these effects might change in forest ecosystems over both the short and long term (Hof et al., 2021). Silvicultural treatments and systems could create or maintain stands of suitable structure and composition to promote C sequestration and mitigate and adapt ecosystems to the effects of global change (Paradis et al., 2019). Partial cutting with cut-to-length or tree-length harvesting systems has been identified as a potential solution for increasing biomass and soil C content (Ameray et al., 2021); however, the long-term effects are not fully understood, particularly in terms of the modality and the intensity of partial cutting.

16.5   Conclusions

The large-scale experiments presented in this chapter are essential for quantifying the multiple ecological and economic outcomes of forest management alternatives. From the existing data, partial cutting offers viable silvicultural alternatives to clear-cutting when required by sustainable forest management objectives. The experimental treatments reviewed here promote residual tree growth, reduce windthrow-related losses, favor regeneration, and help maintain biodiversity. Nonetheless, the clear-cut system remains the main silvicultural regime within the boreal biome. Although it is an appropriate approach in many contexts, clear-cutting can create fragmented landscapes, promote young and even-aged stands to the detriment of multi-cohort stand structures, and benefit some commercial species rather than ensuring a more diverse composition. Moreover, current even-aged management tends to reduce forest structural variability. A more diverse silviculture that integrates partial cutting into the portfolio of available treatments could increase forests' adaptive capacity and resilience in the face of climate change, allowing to maintain a larger spectrum of forest composition and structures at different scales across the landscape.

References

Achim, A., Ruel, J. -C., Gardiner, B. A., et al. (2005). Modelling the vulnerability of balsam fir forests to wind damage. *Forest Ecology and Management, 204*, 37–52. https://doi.org/10.1016/j.foreco.2004.07.072.

Achim, A., Moreau, G., Coops, N. C., et al. (2021). The changing culture of silviculture. *Forestry, 95*(2), 143–152. https://doi.org/10.1093/forestry/cpab047.

Ameray, A., Bergeron, Y., Valeria, O., et al. (2021). Forest carbon management: A review of silvicultural practices and management strategies across boreal, temperate and tropical forests. *Current Forestry Reports, 7*(4), 245–266. https://doi.org/10.1007/s40725-021-00151-w.

Angelstam, P. K. (1998). Maintaining and restoring biodiversity in European boreal forests by developing natural disturbance regimes. *Journal of Vegetation Science, 9*, 593–602. https://doi.org/10.2307/3237275.

Anyomi, K. A., & Ruel, J. -C. (2015). A multiscale analysis of the effects of alternative silvicultural treatments on windthrow within balsam fir dominated stands. *Canadian Journal of Forest Research, 45*, 1739–1747. https://doi.org/10.1139/cjfr-2015-0221.

Arseneault, J., Fenton, N., & Bergeron, Y. (2012). Effects of variable canopy retention harvest on epixylic bryophytes in boreal black spruce—feathermoss forests. *Canadian Journal of Forest Research, 42*, 1467–1476. https://doi.org/10.1139/x2012-054.

Archambault, L., Bégin, J., Delisle, C., et al. (2003). Dynamique forestière après coupe partielle dans la Forêt expérimentale du Lac Édouard, Parc de la Mauricie, Québec. *The Forestry Chronicle, 79*, 672–684. https://doi.org/10.5558/tfc79672-3.

Barik, S. K., Pandey, H. N., Tripathi, R. S., et al. (1992). Microenvironmental variability and species diversity in treefall gaps in a sub-tropical broadleaved forest. *Vegetatio, 103*. https://doi.org/10.1007/BF00033414.

Beaudet, M., Harvey, B. D., Messier, C., et al. (2011). Managing understory light conditions in boreal mixedwoods through variation in the intensity and spatial pattern of harvest: A modelling approach. *Forest Ecology and Management, 261*, 84–94. https://doi.org/10.1016/j.foreco.2010.09.033.

Beaudet, M., & Messier, C. (1998). Growth and morphological responses of yellow birch, sugar maple, and beech seedlings growing under a natural light gradient. *Canadian Journal of Forest Research, 28*, 1007–1015. https://doi.org/10.1139/x98-077.

Beaudet, M., & Messier, C. (2002). Variation in canopy openness and light transmission following selection cutting in northern hardwood stands: An assessment based on hemispherical photographs. *Agricultural and Forest Meteorology, 110*, 217–228. https://doi.org/10.1016/S0168-1923(01)00289-1.

Bergeron, Y., Gauthier, S., Kafka, V., et al. (2001). Natural fire frequency for the eastern Canadian boreal forest: Consequences for sustainable forestry. *Canadian Journal of Forest Research, 31*, 384–391. https://doi.org/10.1139/cjfr-31-3-384.

Bergeron, Y., Leduc, A., Harvey, B., et al. (2002). Natural fire regime: A guide for sustainable management of the Canadian boreal forest. *Silva Fenncia, 36*(1), 553. https://doi.org/10.14214/sf.553.

Bescond, H., Fenton, N. J., & Bergeron, Y. (2011). Partial harvests in the boreal forest: Response of the understory vegetation five years after harvest. *The Forestry Chronicle, 87*, 86–98. https://doi.org/10.5558/tfc87086-1.

Bladon, K. D., Lieffers, V. J., Silins, U., et al. (2008). Elevated mortality of residual trees following structural retention harvesting in boreal mixedwoods. *The Forestry Chronicle, 84*, 70–75. https://doi.org/10.5558/tfc84070-1.

Blanco, J. A., Welham, C., Kimmins, J. P., et al. (2009). Guidelines for modeling natural regeneration in boreal forests. *The Forestry Chronicle, 85*, 427–439. https://doi.org/10.5558/tfc85427-3.

Bose, A. K., Harvey, B. D., Brais, S., et al. (2014). Constraints to partial cutting in the boreal forest of Canada in the context of natural disturbance-based management: A review. *Forestry: An International Journal of Forest Research, 87*(1), 11–28. https://doi.org/10.1093/forestry/cpt047.

Bouchard, M., & Pothier, D. (2011). Long-term influence of fire and harvesting on boreal forest age structure and forest composition in eastern Québec. *Forest Ecology and Management, 261*, 811–820. https://doi.org/10.1016/j.foreco.2010.11.020.

Boudreault, C., Bergeron, Y., Gauthier, S., et al. (2002). Bryophyte and lichen communities in mature to old-growth stands in eastern boreal forests of Canada. *Canadian Journal of Forest Research, 32*, 1080–1093. https://doi.org/10.1139/x02-027.

Bourgeois, L., Messier, C., & Brais, S. (2004). Mountain maple and balsam fir early response to partial and clear-cut harvesting under aspen stands of northern Quebec. *Canadian Journal of Forest Research, 34*, 2049–2059. https://doi.org/10.1139/X04-080.

Brais, S., Harvey, B. D., Bergeron, Y., et al. (2004). Testing forest ecosystem management in boreal mixedwoods of northwestern Quebec: Initial response of aspen stands to different levels of harvesting. *Canadian Journal of Forest Research, 34*(2), 431–446. https://doi.org/10.1139/x03-144.

Brumelis, G., & Carleton, T. J. (1988). The vegetation of postlogged black spruce lowlands in central Canada. I. Trees and tall shrubs. *Canadian Journal of Forest Research, 18*, 1470–1478. https://doi.org/10.1139/x88-226.

Buddle, C. M., Langor, D. W., Pohl, G. R., et al. (2006). Arthropod responses to harvesting and wildfire: Implications for emulation of natural disturbance in forest management. *Biological Conservation, 128*, 346–357. https://doi.org/10.1016/j.biocon.2005.10.002.

Burns, R. M., & Honkala, B. H. (1990). *Silvics of North America. Volume 2, Hardwoods*. U S Department of Agriculture (USDA), Forest Service, Agriculture Handbook 654.

Canadian Council of Forest Ministers (CCFM). (2018). Harvest: Forest area harvested on private and Crown lands in Canada. Ottawa: National Forestry Database, Natural Resources Canada. http://nfdp.ccfm.org/en/data/harvest.php.

Cancino, J. (2005). Modelling the edge effect in even-aged Monterey pine (*Pinus radiata* D. Don) stands. *Forest Ecology and Management, 210*, 159–172. https://doi.org/10.1016/j.foreco.2005.02.021.

Caners, R. T., Macdonald, S. E., & Belland, R. J. (2013). Bryophyte assemblage structure after partial harvesting in boreal mixedwood forest depends on residual canopy abundance and composition. *Forest Ecology and Management, 289*, 489–500. https://doi.org/10.1016/j.foreco.2012.09.044.

Canham, C. D., Denslow, J. S., Platt, W. J., et al. (1990). Light regimes beneath closed canopies and tree-fall gaps in temperate and tropical forests. *Canadian Journal of Forest Research, 20*, 620–631. https://doi.org/10.1139/x90-084.

Chazdon, R. L. (1988). Sunflecks and their importance to forest understorey plants. *Advances in Ecological Research, 18*, 1–63. https://doi.org/10.1016/S0065-2504(08)60179-8.

Chen, J., Franklin, J. F., & Spies, T. A. (1993). Contrasting microclimates among clearcut, edge, and interior of old-growth Douglas-fir forest. *Agricultural and Forest Meteorology, 63*, 219–237. https://doi.org/10.1016/0168-1923(93)90061-L.

Cheveau, M., Drapeau, P., Imbeau, L., et al. (2004). Owl winter irruptions as an indicator of small mammal population cycles in the boreal forest of eastern North America. *Oikos, 107*, 190–198. https://doi.org/10.1111/j.0030-1299.2004.13285.x.

Coates, K. D. (2000). Conifer seedling response to northern temperate forest gaps. *Forest Ecology and Management, 127*, 249–269. https://doi.org/10.1016/S0378-1127(99)00135-8.

Coates, K. D. (2002). Tree recruitment in gaps of various size, clearcuts and undisturbed mixed forest of interior British Columbia, Canada. *Forest Ecology and Management, 155*, 387–398. https://doi.org/10.1016/S0378-1127(01)00574-6.

Cotton-Gagnon, A., Simard, M., De Grandpré, L., et al. (2018). Salvage logging during spruce budworm outbreaks increases defoliation of black spruce regeneration. *Forest Ecology and Management, 430*, 421–430. https://doi.org/10.1016/j.foreco.2018.08.011.

Courtois, R., Quellet, J. P., Dussault, C., et al. (2004). Forest management guidelines for forest-dwelling caribou in Québec. *The Forestry Chronicle, 80*, 80. https://doi.org/10.5558/tfc80598-5.

Cremer, K. W., Borough, C. J., McKinnel, F. H., et al. (1982). Effects of stocking and thinning on wind damage in plantations. *New Zealand Journal of Forest Science, 12*, 245–268.

De Grandpré, L., Morissette, J., & Gauthier, S. (2000). Long-term post-fire changes in the north-eastern boreal forest of Quebec. *Journal of Vegetation Science, 11*, 791–800. https://doi.org/10.2307/3236549.

De Grandpré, L., Gauthier, S., & Allain, C. (2008). Vers un aménagement écosystémique de la forêt boréale de la Côte-Nord. In S. Gauthier, M. A. Vaillancourt, A. Leduc, L. De Grandpré, D. Kneeshaw, H. Morin, P. Drapeau, & Y. Bergeron (Eds.), *Aménagement écosystémique en forêt boréale* (pp. 241–268). Presses de l'Université du Québec.

Domke, G. M., Caspersen, J. P., & Jones, T. A. (2007). Light attenuation following selection harvesting in northern hardwood forests. *Forest Ecology and Management, 239*, 182–190. https://doi.org/10.1016/j.foreco.2006.12.006.

Doucet, R., Pineau, M., Ruel, J. -C., et al. (1996). Sylviculture appliquée. In L'Ordre des ingénieurs forestiers du Québec (Ed.) *Manuel de foresterie* (pp. 965–1004). Québec: Les Presses de l'Université Laval.

Drever, C. R., Peterson, G., Messier, C., et al. (2006). Can forest management based on natural disturbances maintain ecological resilience? *Canadian Journal of Forest Research, 36*, (9), 2285–2299. https://doi.org/10.1139/x06-132.

Dubois, J., Ruel, J. -C., Elie, J. G., et al. (2006). Dynamique et estimation du rendement des strates de retour après coupe totale dans la sapinière à bouleau jaune. *The Forestry Chronicle, 82*, 675–689. https://doi.org/10.5558/tfc82675-5.

Fenton, N. J., & Bergeron, Y. (2007). Sphagnum community change after partial harvest in black spruce boreal forests. *Forest Ecology and Management, 242*, 24–33. https://doi.org/10.1016/j.foreco.2007.01.028.

Fenton, N., Simard, M., & Bergeron, Y. (2009). Emulating natural disturbances: The role of silviculture in creating even-aged and complex structures in the black spruce boreal forest of eastern North America. *Journal of Forest Research, 14*, 258–267. https://doi.org/10.1007/s10310-009-0134-8.

Fenton, N. J., Lecomte, N., Legare, S., et al. (2005). Paludification in black spruce (*Picea mariana*) forests of eastern Canada: Potential factors and management implications. *Forest Ecology and Management, 213*, 151–159. https://doi.org/10.1016/j.foreco.2005.03.017.

Fenton, N. J., Imbeau, L., Work, T., et al. (2013). Lessons learned from 12 years of ecological research on partial cuts in black spruce forests of northwestern Québec. *The Forestry Chronicle, 89*, 350–359. https://doi.org/10.5558/tfc2013-065.

Fischer, J., & Lindenmayer, D. B. (2007). Landscape modification and habitat fragmentation: A synthesis. *Global Ecology and Biogeography, 16*(3), 265–280. https://doi.org/10.1111/j.1466-8238.2007.00287.x.

Franklin, J. F., Johnson, K. N., & Johnson, D. L. (2018). *Ecological forest management* (p. 646). Waveland Press.

Fraver, S., Seymour, R. S., Speer, J. H., et al. (2007). Dendrochronological reconstruction of spruce budworm outbreaks in northern Maine, USA. *Canadian Journal of Forest Research, 37*, 523–529. https://doi.org/10.1139/X06-251.

Gardiner, B. A. (1995). The interactions of wind and tree movement in forest canopies. In J. Grace & M. P. Coutts (Eds.), *Wind and Trees* (pp. 41–59). Cambridge University Press.

Gardiner, B. A., Stagey, G. R., Belcher, R. E., et al. (1997). Field and wind tunnel assessments of the implications of respacing and thinning for tree stability. *Forestry, 70*, 233–252. https://doi. org/10.1093/forestry/70.3.233.

Gardiner, B., Byrne, K., Hale, S., et al. (2008). A review of mechanistic modelling of wind damage risk to forests. *Forestry, 81*, 447–463. https://doi.org/10.1093/forestry/cpn022.

Goudiaby, V., Brais, S., Berninger, F., et al. (2012). Vertical patterns in specific volume increment along stems of dominant jack pine (*Pinus banksiana*) and black spruce (*Picea mariana*) after thinning. *Canadian Journal of Forest Research, 42*, 733–748. https://doi.org/10.1139/X2012-029.

Grenon, F., Jetté, J., & Leblanc, M. (2010) *Manuel de référence pour l'aménagement écosystémique des forêts au Québec–Module 1-Fondements et démarche de la mise en oeuvre*. Québec: CERFO Ministère des Ressources Naturelles et de la Faune.

Groot, A. (2014). Fifteen-year results of black spruce uneven-aged silviculture in Ontario, Canada. *Forestry, 87*, 99–107. https://doi.org/10.1093/forestry/cpt021.

Groot, A., Lussier, J. M., Mitchell, A. K., et al. (2005). A silvicultural systems perspective on changing Canadian forestry practices. *The Forestry Chronicle, 81*, 81. https://doi.org/10.5558/tfc 81050-1.

Harper, K. A., Drapeau, P., Lesieur, D., et al. (2016). Negligible structural development and edge influence on the understorey at 16–17-yr-old clear-cut edges in black spruce forest. *Applied Vegetation Science, 19*, 462–473. https://doi.org/10.1111/avsc.12226.

Hernández-Rodríguez, E., Escalera-Vázquez, L. H., García-ávila, D., et al. (2021). Reduced-impact logging maintain high moss diversity in temperate forests. *Forests, 12*(4), 383. https://doi.org/10. 3390/f12040383.

Hof, A. R., Montoro Girona, M., Fortin, M.-J., et al. (2021). Editorial: Using landscape simulation models to help balance conflicting goals in changing forests. *Frontiers in Ecology & Evolution*, 9.https://doi.org/10.3389/fevo.2021.795736.

Jacobs, J. M., & Work, T. T. (2012). Linking deadwood-associated beetles and fungi with wood decomposition rates in managed black spruce forests. *Canadian Journal of Forest Research, 42*, 1477–1490. https://doi.org/10.1139/X2012-073.

Jönsson, M. T., Fraver, S., Jonsson, B. G., et al. (2007). Eighteen years of tree mortality and structural change in an experimentally fragmented Norway spruce forest. *Forest Ecology and Management, 242*, 306–313. https://doi.org/10.1016/j.foreco.2007.01.048.

Kembel, S. W., Waters, I., & Shay, J. M. (2008). Short-term effects of cut-to-length versus full-tree harvesting on understorey plant communities and understorey-regeneration associations in Manitoba boreal forests. *Forest Ecology and Management, 255*, 1848–1858. https://doi.org/10. 1016/j.foreco.2007.12.006.

Kim, S., Axelsson, E. P., Girona, M. M., et al. (2021). Continuous-cover forestry maintains soil fungal communities in Norway spruce dominated boreal forests. *Forest Ecology and Management, 480*, 118659. https://doi.org/10.1016/j.foreco.2020.118659.

Kobe, R. K., Pacala, S. W., Silander, J. A., et al. (1995). Juvenile tree survivorship as a component of shade tolerance. *Ecological Applications, 5*, 517–532. https://doi.org/10.2307/1942040.

Kolabinski, V. S. (1991). *Effects of cutting method and seedbed treatment on black spruce regeneration in Manitoba*. Information Report NOR-X-316. Edmonton: Northern Forestry Centre.

Kuuluvainen, T. (1994). Gap disturbance, ground microtopography, and the regeneration dynamics of boreal coniferous forests in Finland: A review. *Annales Zoologici Fennici, 31*(1), 35–51.

Kuuluvainen, T., & Grenfell, R. (2012). Natural disturbance emulation in boreal forest ecosystem management—theories, strategies, and a comparison with conventional even-aged management. *Canadian Journal of Forest Research, 42*(7), 1185–1203. https://doi.org/10.1139/x2012-064.

Kuuluvainen, T., Tahvonen, O., & Aakala, T. (2012). Even-aged and uneven-aged forest management in boreal Fennoscandia: A review. *Ambio, 41*, 720–737. https://doi.org/10.1007/s13280-012-0289-y.

Lafleur, B., Fenton, N. J., Paré, D., et al. (2010). Contrasting effects of season and method of harvest on soil properties and the growth of black spruce regeneration in the boreal forested peatlands of Eastern Canada. *Silva Fennica, 44*. https://doi.org/10.14214/sf.122.

Lafleur, B., Cazal, A., Leduc, A., et al. (2015). Soil organic layer thickness influences the estab-lishment and growth of trembling aspen (*Populus tremuloides*) in boreal forests. *Forest Ecology and Management, 347*, 209–216. https://doi.org/10.1016/j.foreco.2015.03.031.

Lähde, E., Eskelinen, T., & Väänänen, A. (2002). Growth and diversity effects of silvicultural alternatives on an old-growth forest in Finland. *Forestry, 75*, 395–400. https://doi.org/10.1093/forestry/75.4.395.

Larouche, C., Guillemette, F., Raymond, P., et al. (Eds.). (2013). *Le guide sylvicole du Québec. Tome 2—Les concepts et l'application de la sylviculture.* Québec: Les Publications du Québec.

Lavoie, J., Montoro Girona, M., & Morin, H. (2019). Vulnerability of conifer regeneration to spruce budworm outbreaks in the eastern Canadian boreal forest. *Forests, 10*, 850. https://doi.org/10.3390/f10100850.

Lavoie, J., Montoro Girona, M., Grosbois, G., et al (2021). Does the type of silvicultural practice influence spruce budworm defoliation of seedlings? *Ecosphere, 12*. https://doi.org/10.1002/ecs2.3506.

Lavoie, S., Ruel, J. -C., Bergeron, Y., et al. (2012). Windthrow after group and dispersed tree retention in eastern Canada. *Forest Ecology and Management, 269*, 158–167. https://doi.org/10.1016/j.foreco.2011.12.018.

Lecomte, N., & Bergeron, Y. (2005). Successional pathways on different surficial deposits in the coniferous boreal forest of the Quebec Clay Belt. *Canadian Journal of Forest Research, 35*, 1984–1995. https://doi.org/10.1139/x05-114.

Lecomte, N., Simard, M., Fenton, N., et al. (2006). Fire severity and long-term ecosystem biomass dynamics in coniferous boreal forests of eastern Canada. *Ecosystems, 9*, 1215–1230. https://doi.org/10.1007/s10021-004-0168-x.

Lecomte, N., Simard, M., & Bergeron, Y. (2009). Effects of fire severity and initial tree composition on stand structural development in the coniferous boreal forest of northwestern Québec, Canada. *Ecoscience, 13*, 152–163. https://doi.org/10.2980/i1195-6860-13-2-152.1.

Lindenmayer, D. B., & Franklin, J. F. (2002). *Conserving forest biodiversity: A comprehensive multiscaled approach.* Island Press.

Lycke-Poulin, A. (2008). *Évaluation de l'impact de l'éclaircie commerciale sur le tétras du Canada (Falcipennis canadensis).* M.Sc. thesis, Université du Québec en Abitibi-Témiscamingue.

Maccurrach, R. (1991). Spacing: An option for reducing storm damage. *Scottish Forestry, 45*, 285–297.

Macdonald, S. E., & Fenniak, T. E. (2007). Understory plant communities of boreal mixedwood forests in western Canada: Natural patterns and response to variable-retention harvesting. *Forest Ecology and Management, 242*, 34–48. https://doi.org/10.1016/j.foreco.2007.01.029.

MacIsaac, D. A., & Krygier, R. (2009). Development and long-term evaluation of harvesting patterns to reduce windthrow risk of understorey spruce in aspen-white spruce mixedwood stands in Alberta, Canada. *Forestry, 82*, 323–342. https://doi.org/10.1093/forestry/cpp013.

Majcen, Z. (1994). History of selection cutting in forests with uneven aged trees in Quebec-Historique des coupes de jardinage dans les forêts inéquiennes au Québec. *Revue forestière française, 46*(4), 375–384. https://doi.org/10.4267/2042/26556.

Man, R., Rice, J. A., & MacDonald, G. B. (2010). Five-year light, vegetation, and regeneration dynamics of boreal mixedwoods following silvicultural treatments to establish productive aspen-spruce mixtures in northeastern Ontario. *Canadian Journal of Forest Research, 40*, 1529–1541. https://doi.org/10.1139/X10-088.

Martin, M., Girona, M. M., & Morin, H. (2020). Driving factors of conifer regeneration dynamics in eastern Canadian boreal old-growth forests. *PLoS ONE, 15*, e0230221. https://doi.org/10.1371/journal.pone.0230221.

Matthews, J. D. (1991). *Silvicultural systems.* Clarendon Press.

McDonald, R. I., & Urban, D. L. (2004). Forest edges and tree growth rates in the North Carolina Piedmont. *Ecology, 85*, 2258–2266. https://doi.org/10.1890/03-0313.

Meek, P. (2006). Trials of four trail layouts adapted to shelterwood cuts. Pointe-Claire: FPInnova-tions - Feric Division, *Advantage Report, 7*(8), 1–8.

Messier, C., Doucet, R., Ruel, J. -C., et al. (1999). Functional ecology of advance regeneration in relation to light in boreal forests. *Canadian Journal of Forest Research, 29*, 812–823. https://doi.org/10.1139/cjfr-29-6-812.

Mitchell, S. J. (1995). A synopsis of windthrow in British Columbia: Occurrence, implications, assessment and management. In M. P. Coutts & J. Grace (Eds.), *Wind and trees* (pp. 448–459). Cambridge University Press.

Montoro Girona, M., Morin, H., Lussier, J. M., et al. (2016). Radial growth response of black spruce stands ten years after experimental shelterwoods and seed-tree cuttings in boreal forest. *Forests, 7*, 240. https://doi.org/10.3390/f7100240.

Montoro Girona, M., Rossi, S., Lussier, J. M., et al. (2017). Understanding tree growth responses after partial cuttings: A new approach. *PLoS ONE, 12*, e0172653. https://doi.org/10.1371/journal.pone.0172653.

Montoro Girona, M., Lussier, J. M., Morin, H., et al. (2018). Conifer regeneration after experimental shelterwood and seed-tree treatments in boreal forests: Finding silvicultural alternatives. *Frontiers in Plant Science, 9*, 1145. https://doi.org/10.3389/fpls.2018.01145.

Montoro Girona, M., Morin, H., Lussier, J.-M., et al. (2019). Post-cutting mortality following experimental silvicultural treatments in unmanaged boreal forest stands. *Frontiers in Forests and Global Change, 2*, 4. https://doi.org/10.3389/ffgc.2019.00004.

Moussaoui, L., Leduc, A., Fenton, N. J., et al. (2019). Changes in forest structure along a chronosequence in the black spruce boreal forest: Identifying structures to be reproduced through silvicultural practices. *Ecological Indicators, 97*, 89–99. https://doi.org/10.1016/j.ecolind.2018.09.059.

Moussaoui, L., Leduc, A., & Montoro Girona, M. (2020). Success factors for experimental partial harvesting in unmanaged boreal forest: 10-year stand yield results. *Forests, 11*, 1199. https://doi.org/10.3390/f11111199.

Navarro, L., Morin, H., Bergeron, Y., et al. (2018). Changes in spatiotemporal patterns of 20th century spruce budworm outbreaks in eastern Canadian boreal forests. *Frontiers in Plant Science, 9*, 1905 https://doi.org/10.3389/fpls.2018.01905.

Nolet, P., Kneeshaw, D., Messier, C., et al. (2018). Comparing the effects of even- and uneven-aged silviculture on ecological diversity and processes: A review. *Ecology and Evolution, 8*(2), 1217–1226. https://doi.org/10.1002/ece3.3737.

Nyland, R. D. (2016). *Silviculture: Concepts and applications* (3rd ed., p. 680). Waveland Press.

Oliver, C. D., & Larson, B. C. (1996). *Forest stand dynamics* (Update). Yale School of the Environment Other Publications.

Opoku-Nyame, J., Leduc, A., & Fenton, N. J. (2021). Bryophyte conservation in managed boreal landscapes: Fourteen-year impacts of partial cuts on epixylic bryophytes. *Frontires in Forest and Global Change, 4*. https://doi.org/10.3389/ffgc.2021.674887.

Palik, B. J., & D'Amato, A. W. (2017). Ecological forestry: Much more than retention harvesting. *Journal of Forestry, 115*, 51–53. https://doi.org/10.5849/jof.16-057.

Pamerleau-Couture, É., Krause, C., Pothier, D., et al. (2015). Effect of three partial cutting practices on stand structure and growth of residual black spruce trees in north-eastern Quebec. *Forestry, 88*, 471–483. https://doi.org/10.1093/forestry/cpv017.

Pape, R. (1999). Influence of thinning and tree diameter class on the development of basic density and annual ring width in *Picea abies. Scandinavian Journal of Forest Research, 14*, 27–37. https://doi.org/10.1080/02827589908540806.

Pappas, C., Bélanger, N., Bergeron, Y., et al. (2022). Smartforests Canada: A network of monitoring plots for forest management under environmental change. In R. Tognetti, M. Smith, & P. Panzacchi (Eds.), *Climate-smart forestry in mountain regions* (pp. 521–543). Springer International Publishing.

Paradis, L., Thiffault, E., & Achim, A. (2019). Comparison of carbon balance and climate change mitigation potential of forest management strategies in the boreal forest of Quebec (Canada). *Forestry, 92*, 264–277. https://doi.org/10.1093/forestry/cpz004.

Paradis, S., & Work, T. T. (2011). Partial cutting does not maintain spider assemblages within the observed range of natural variability in Eastern Canadian black spruce forests. *Forest Ecology and Management, 262,* 2079–2093. https://doi.org/10.1016/j.foreco.2011.08.032.

Parent, S., & Messier, C. (1995). Effects of light gradient on height growth and crown architecture of a naturally regenerated Balsam fir. *Canadian Journal of Forest Research, 25,* 878–885.

Payette, S., & Rochefort, L. (2001). *Écologie des tourbières du Québec-Labrador.* (p. 644). Québec: Presses de l'Université Laval.

Peltola, H., Miina, J., Rouvinen, I., et al. (2002). Effect of early thinning on the diameter growth distribution along the stem of Scots pine. *Silva Fennica, 36*(4), 523. https://doi.org/10.14214/sf.523.

Perala, D. A., & Alm, A. A. (1990). Reproductive ecology of birch: A review. *Forest Ecology and Management, 32,* 1–38. https://doi.org/10.1016/0378-1127(90)90104-J.

Piché, R. (2017). *Abondance et croissance de l'épinette noire après coupe : une étude comparative entre la coupe adaptée avec maintien du couvert et la coupe avec protection de la régénération et des sols.* M.Sc. thesis, Université du Québec à Montréal.

Prévost, M. (1996). Effets du scarifiage sur les propriétés du sol et l'ensemencement naturel dans une pessière noire à mousses de la forêt boréale québécoise. *Canadian Journal of Forest Research, 26,* 72–86. https://doi.org/10.1139/x26-008.

Prévost, M., Raymond, P., & Lussier, J. M. (2010). Regeneration dynamics after patch cutting and scarification in yellow birch—Conifer stands. *Canadian Journal of Forest Research, 40,* 357–369. https://doi.org/10.1139/X09-192.

Puettmann, K. J., Wilson, S. M. G., Baker, S. C., et al. (2015). Silvicultural alternatives to conventional even-aged forest management—What limits global adoption? *Forest Ecosysytems, 2,* 8. https://doi.org/10.1186/s40663-015-0031-x.

Pukkala, T., Lähde, E., & Laiho, O. (2009). Growth and yield models for uneven-sized forest stands in Finland. *Forest Ecology and Management, 258,* 207–216. https://doi.org/10.1016/j.foreco.2009.03.052.

Raulier, F., Pothier, D., & Bernier, P. Y. (2003). Predicting the effect of thinning on growth of dense balsam fir stands using a process-based tree growth model. *Canadian Journal of Forest Research, 33*(3), 509–520. https://doi.org/10.1139/x03-009.

Raymond, P., Ruel, J. -C., & Pineau, M. (2000). Effet d'une coupe d'ensemencement et du milieu de germination sur la régénération des sapinières boréales riches de seconde venue du Québec. *The Forestry Chronicle, 76,* 643–652. https://doi.org/10.5558/tfc76643-4.

Raymond, P., Munson, A. D., Ruel, J. C., et al. (2006). Spatial patterns of soil microclimate, light, regeneration, and growth within silvicultural gaps of mixed tolerant hardwood—White pine stands. *Canadian Journal of Forest Research, 36,* 639–651. https://doi.org/10.1139/x05-269.

Raymond, P., Bédard, S., & Roy, V. (2009). The irregular shelterwood system: Review, classification, and potential application to forests affected by partial disturbances. *Journal of Forestry, 107,* 405–413 https://doi.org/10.1093/jof/107.8.405.

Raymond, P., Legault, I., Guay, L., et al. (2013). La coupe progressive régulière. In C. Larouche, F. Guillemette, P. Raymond, & J.-P. Saucier (Eds.), *Le guide sylvicole du Québec. Tome 2—Les concepts et l'application de la sylviculture* (pp. 410–453). Québec: Les Publications du Québec.

Riopel, M., Bégin, J., & Ruel, J. -C. (2010). Probabilités de pertes des tiges individuelles, cinq ans après des coupes avec protection des petites tiges marchandes, dans des forêts résineuses du Québec. *Canadian Journal of Forest Research, 40,* 1458–1472. https://doi.org/10.1139/X10-059.

Riopel, M., Bégin, J., & Ruel, J. -C. (2011). Coefficients de distribution de la régénération, cinq ans après des coupes avec protection des petites tiges marchandes appliquées dans des sapinières et des pessières noires du Québec. *The Forestry Chronicle, 87,* 669–683. https://doi.org/10.5558/tfc2011-073.

Rosenvald, R., & Lõhmus, A. (2008). For what, when, and where is green-tree retention better than clear-cutting? A review of the biodiversity aspects. *Forest Ecology and Management, 255,* 255. https://doi.org/10.1016/j.foreco.2007.09.016.

Ruel, J. -C. (1995). Understanding windthrow: Silvicultural implications. *The Forestry Chronicle, 71*, 434–445. https://doi.org/10.5558/tfc71434-4.

Ruel, J. -C. (2000). Factors influencing windthrow in balsam fir forests: From landscape studies to individual tree studies. *Forest Ecology and Management, 135*, 169–178. https://doi.org/10.1016/S0378-1127(00)00308-X.

Runkle, J. R. (2000). Canopy tree turnover in old-growth mesic forests of eastern North America. *Ecology, 81*, 554–567. https://doi.org/10.1890/0012-9658(2000)081[0554:CTTIOG]2.0.CO;2.

Saad, C., Boulanger, Y., Beaudet, M., et al. (2017). Potential impact of climate change on the risk of windthrow in eastern Canada's forests. *Climatic Change, 143*, 487–501. https://doi.org/10.1007/s10584-017-1995-z.

Schneider, R., Zhang, S. Y., Swift, D. E., et al. (2008). Predicting selected wood properties of jack pine following commercial thinning. *Canadian Journal of Forest Research, 38*, 2030–2043. https://doi.org/10.1139/X08-038.

Scott, R. E., & Mitchell, S. J. (2005). Empirical modelling of windthrow risk in partially harvested stands using tree, neighbourhood, and stand attributes. *Forest Ecology and Management, 218*, 193–209. https://doi.org/10.1016/j.foreco.2005.07.012.

Seedre, M., Felton, A., & Lindbladh, M. (2018). What is the impact of continuous cover forestry compared to clearcut forestry on stand-level biodiversity in boreal and temperate forests? A Systematic Review Protocol. *Environmental Evidence, 7*(1), 28. https://doi.org/10.1186/s13750-018-0138-y.

Smith, D., Larson, B., Kelty, M., et al. (1997). *The practice of silviculture: Applied forest ecology* (9th ed.). John Wiley and Sons Inc.

Solarik, K. A., Volney, W. J. A., Lieffers, V. J., et al. (2012). Factors affecting white spruce and aspen survival after partial harvest. *Journal of Applied Ecology, 49*(1), 145–154. https://doi.org/10.1111/j.1365-2664.2011.02089.x.

Solarik, K. A., Cazelles, K., Messier, C., et al. (2020). Priority effects will impede range shifts of temperate tree species into the boreal forest. *Journal of Ecology, 108*(3), 1155–1173. https://doi.org/10.1111/1365-2745.13311.

Soucy, M., Lussier, J. M., & Lavoie, L. (2012). Long-term effects of thinning on growth and yield of an upland black spruce stand. *Canadian Journal of Forest Research, 42*, 1669–1677. https://doi.org/10.1139/X2012-107.

Spence, J., Volney, W., Lieffers, V., Weber, M., Luchkow, S., & Vinge, T. (1999). The Alberta EMEND project: recipe and cooks' argument. In T. S. Veeman, D. W. Smith, B. G. Purdy, F. J. Salkie, & G. A. Larkin (Eds.), *Proceedings of the 1999 Sustainable Forest Management Network Conference, Science and Practice: Sustaining the Boreal Forest*. 14–17 February 1999. Edmonton: Sustainable Forest Management Network.

Stathers, R. J., Rollerson, T. P., & Mitchell, S. J. (1994). *Windthrow handbook for British Columbia forests*, Research Program Working Paper 9401. Victoria: Ministry of Forests Research Program.

Stokes, A., Fitter, A. H., & Courts, M. P. (1995). Responses of young trees to wind and shading: Effects on root architecture. *Journal of Experimental Botany, 46*, 1139–1146. https://doi.org/10.1093/jxb/46.9.1139.

Thiffault, N., Coll, L., & Jacobs, D. F. (2015). Natural regeneration after harvesting. In K. H.-S. Peh, R. Corlett, & Y. Bergeron (Eds.), *Routledge handbook of forest ecology* (pp. 371–384). London: Earthscan, Routledge.

Thiffault, N., Hoepting, M., Fera, J., et al. (2021). Managing plantation density through initial spacing and commercial thinning: Yield results from a 60-year-old red pine spacing trial experiment. *Canadian Journal of Forest Research, 51*, 181–189. https://doi.org/10.1139/cjfr-2020-0246.

Thorpe, H. C., & Thomas, S. C. (2007). Partial harvesting in the Canadian boreal: Success will depend on stand dynamic responses. *The Forestry Chronicle, 83*, 319–325. https://doi.org/10.5558/tfc83319-3.

Thorpe, H. C., Thomas, S. C., & Caspersen, J. P. (2007). Residual-tree growth responses to partial stand harvest in the black spruce (*Picea mariana*) boreal forest. *Canadian Journal of Forest Research, 37*(9), 1563–1571. https://doi.org/10.1139/X07-148.

Thorpe, H. C., Thomas, S. C., & Caspersen, J. P. (2008). Tree mortality following partial harvests is determined by skidding proximity. *Ecological Applications, 18*, 1652–1663. https://doi.org/10.1890/07-1697.1.

Thorpe, H. C., Vanderwel, M. C., Fuller, M. M., et al. (2010). Modelling stand development after partial harvests: An empirically based, spatially explicit analysis for lowland black spruce. *Ecological Modelling, 221*, 256–267. https://doi.org/10.1016/j.ecolmodel.2009.10.005.

Urgenson, L. S., Halpern, C. B., & Anderson, P. D. (2013). Level and pattern of overstory retention influence rates and forms of tree mortality in mature, coniferous forests of the Pacific Northwest, USA. *Forest Ecology and Management, 308*, 116–127. https://doi.org/10.1016/j.foreco.2013.07.021.

Van Der Meer, P. J., & Bongers, F. (1996). Formation and closure of canopy gaps in the rain forest at Nouragues, French Guiana. *Vegetatio, 126*, 167–179. https://doi.org/10.1007/bf00045602.

Vanha-Majamaa, I., Lilja, S., Ryömä, R., et al. (2007). Rehabilitating boreal forest structure and species composition in Finland through logging, dead wood creation and fire: The EVO experiment. *Forest Ecology and Management, 250*, 77–88. https://doi.org/10.1016/j.foreco.2007.03.012.

Vincent, M., Krause, C., & Zhang, S. Y. (2009). Radial growth response of black spruce roots and stems to commercial thinning in the boreal forest. *Forestry, 82*, 557–571. https://doi.org/10.1093/forestry/cpp025.

Webster, C. R., & Lorimer, C. G. (2002). Single-tree versus group selection in hemlock-hardwood forests: Are smaller openings less productive? *Canadian Journal of Forest Research, 32*, 591–604. https://doi.org/10.1139/x02-003.

Part VI
Ecological Restoration

Chapter 17
Strategies for the Ecological Restoration of the Boreal Forest Facing Climate Change

Timo Kuuluvainen and Petri Nummi

Abstract The large-scale simplification of boreal forest ecosystem structure, composition, and processes to boost timber production, combined with the increasing pressure of climate change, has created an urgent need to restore forest biodiversity and resilience. However, the issue of restoration is relatively new in boreal forests, and there are no established strategies to guide restoration planning and action. Here we provide an overview of suggested strategic concepts and approaches for boreal forest ecosystem restoration and discuss their applicability to various situations. The key strategic questions in restoration for attaining a favorable conservation status of native ecosystem types and their intrinsic dynamics in a given area are: what, how much, and when to restore? We conclude that adaptive capacity should serve as an overarching strategic framework in boreal forest restoration during times of rapid climate change.

17.1 Introduction

The boreal forest represents about one-third of the global forest, and it spreads across the boreal biome in Canada, Alaska, Russia, and Scandinavia (Kneeshaw et al., 2011). The boreal forest plays a crucial role in global climate regulation because it contains a large share of the global terrestrial carbon. This forest is vital for biodiversity, as it provides habitats for numerous species adapted to specific northern conditions (Bradshaw et al., 2009). Most boreal countries have long-standing and strong traditions in forestry education and related forest-dedicated institutions and timber-oriented forest management. They also produce a disproportionally large share of forest products for the global market (SNS, 2021).

T. Kuuluvainen (✉) · P. Nummi
Department of Forest Sciences, University of Helsinki, PO Box 27, 00014 Helsinki, Finland
e-mail: timo.kuuluvainen@helsinki.fi

P. Nummi
e-mail: petri.nummi@helsinki.fi

© The Author(s) 2023
M. M. Girona et al. (eds.), *Boreal Forests in the Face of Climate Change*,
Advances in Global Change Research 74,
https://doi.org/10.1007/978-3-031-15988-6_17

Although unmanaged forest still exists, especially in remote high-latitude areas of the boreal zone (Gauthier et al., 2015), the southern, more productive, and naturally species-rich forests are generally heavily exploited and currently under intensive management (Burton et al., 2010). Forest utilization has been most intensive and long-lasting in Fennoscandia, especially in Sweden and Finland where natural forest mostly remains only in remote high-latitude and high-altitude areas (Kuuluvainen et al., 2017). It is evident that to reach representativeness and favorable conservation status in these countries, there is a mounting need for forest protection and restoration, especially in the southern and middle boreal zones where natural biodiversity is high (Angelstam & Andersson, 2001; Angelstam et al., 2020; Berglund & Kuuluvainen, 2021; Vanha-Majamaa et al., 2007). The situation is also similar in Canada and Russia in that the southern boreal forests have been most intensively utilized, although extensive areas of natural forest remain in more northern boreal regions (Potapov et al., 2008).

Overall, the issue of forest restoration is relatively new in boreal forests (Stanturf & Madsen, 2005). The boreal countries have traditionally focused on timber manage-ment and timber-yield sustainability (Puettmann et al. 2009). This goal is met in many boreal countries using intensive even-aged management, where wood is harvested with short clear-cutting cycles relative to the natural longevity of stand develop-ment cycles. However, such agriculture-inspired crop management practices have turned vast areas of structurally diverse natural forests into production forests; the latter are structurally and compositionally simplified and lack vital structural legacy features, such as large old trees and abundant and diverse deadwood. The large-scale simplification of ecosystem structures to boost timber production has reduced biodi-versity, limited the ability of forests to deliver ecosystem services, and weakened forest resilience to perturbations (Angelstam et al., 2020; Berglund & Kuuluvainen, 2021). As climate change effects become increasingly evident, this simplification has created an urgent need for forest restoration and ecosystem management to increase resilience within large areas of the boreal forest (Burton & Macdonald, 2011; Kuuluvainen, 2009).

The extent and magnitude of change brought by intensive forestry is exemplified by the situation in Finland, where in a recent national assessment, 70% of forested habitat types on mineral soil were evaluated as threatened (constituting 49% of the country's forest area), mostly because of the low amounts of deadwood and simpli-fied structure of these forests (Kontula & Raunio, 2019). Such extensive degradation of forested ecosystems has taken place over most of boreal Fennoscandia within only the last 70 years (Keto-Tokoi & Kuuluvainen, 2014). Because of the inertia of forest ecosystems to environmental change, we have only seen part of the cumu-lative ecological effects of such large-scale alteration of northern forest ecosystem structures. Such delays in ecological responses are due, in part, to long successional sequences and long-lasting legacies from the more natural forest stages of the past, e.g., slowly decaying pools of fallen deadwood (Lilja & Kuuluvainen, 2005) and delayed population responses to habitat degradation and loss, a phenomenon known as extinction debt (Hanski, 2000).

Degradation of habitat quality and the anticipated rapid warming of climate at high latitudes threaten to accelerate biodiversity loss and boreal carbon pool depletion in the near future (Gauthier et al., 2015; Moen et al., 2014). At some point, the ecosystem may cross critical transition thresholds, resulting in large-scale ecosystem state shifts. Beyond this point, sustainable management may no longer be possible (Gauthier et al., 2015). Such ecosystem state shifts, from a closed forest to low productivity open woodland, are already evident in some parts of the boreal zone because of repeated high-severity fires (Girard et al., 2008; Jasinski & Payette, 2005).

As a response to ecosystem degradation, boreal forestry is confronted with increasing demands to restore structurally impoverished managed forests closer to their natural state of variability and complexity (Burton et al., 2010; Kuuluvainen, 2009; Messier et al., 2013). This pressure is challenging traditional forest manage-ment approaches, particularly the sustained yield paradigm of sustainability. The application of ecosystem management in boreal forestry calls for a paradigm change toward large-scale restoration and more diversified management approaches inspired by natural forest structure and dynamics (Berglund & Kuuluvainen, 2021; Burton & Macdonald, 2011).

In this chapter, we provide an overview of strategic concepts and approaches in boreal forest ecosystem restoration. We discuss under which circumstances the different approaches could be applicable and how to harness ecological knowledge in forest restoration and ecosystem management. The key strategic questions are: what, how much, and when to restore a forest to attain representativeness and continuity of native habitat types and ecological processes? However, the question of how to restore is an operational question that we do not address here. Finally, we discuss the importance of and prospects for forest restoration in the boreal zone in times of rapid climate change.

17.2 Development of Strategic Thinking in Conservation, Restoration, and Ecosystem Management

Restoration can be defined as a process aiding the recovery of degraded, damaged, or destroyed ecosystems toward their natural state. When we look into the past, it is possible to distinguish some broad developmental steps in conservation and restoration thinking, restoration concepts, and associated strategies reflecting the development of ecological science and an understanding of ecosystem structure and functioning (Fig. 17.1). The earliest strategic approach can be called *forest-as-museum*. This was founded on the Clementsian view of deterministic succession and its assumed natural and permanent static endpoint, the *climax* (Clements, 1916). According to this view, it was possible to protect or passively restore spectacular but often small remaining fragments of natural vegetation as examples of original local conditions.

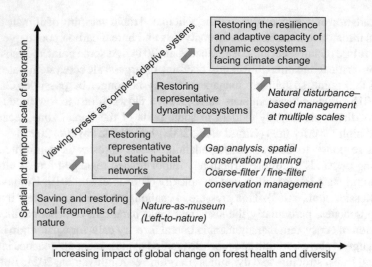

Fig. 17.1 A simplified illustration of the development of some key strategic approaches and concepts underlying forest conservation, restoration, and sustainable management. Modified with permission of Informa UK Ltd. through PLSclear from Kuuluvainen (2017)

However, because of the historical development of forest utilization and conservation, the remaining natural forest fragments are often too small and isolated to host their intrinsic dynamics and viable populations (Angelstam et al., 2020). Moreover, given the lack of proper conservation policies and planning, they are not representative of the original habitat distribution, as they are mostly forests located on marginal and remote sites of low productivity where biodiversity is naturally low (Angelstam et al., 2020; Kuuluvainen et al., 2017; Lilja & Kuuluvainen, 2005).

Such poorly connected and inadequately representative forest protection areas created a need to complement the conservation area network, where restoration of degraded forests and their connectivity played a central role (Halme et al., 2013; Ward et al., 2020); the ability of species to move between habitats was understood to be crucial for long-term viability of populations and must be taken into account in species conservation and restoration (Hanski, 2000; MacArthur & Wilson, 1967). This shift in thinking led to an emphasis on habitat quality and size, networks and connectivity, and, consequently, the spatial conservation planning necessary for ensuring a favorable conservation and viable populations within the habitats (Moilanen et al., 2011). Practical conservation measures in northern European forestry included the introduction of the landscape ecological planning framework based on the patch-corridor-matrix model (Forman 1995), the valuable key-habitat concept (Timonen et al., 2011), and retention-tree practices (Kuuluvainen et al., 2019; Simonsson et al., 2015).

With mounting concerns of the impacts of climate change on boreal forests, there has been an increased focus on dynamic properties of forest ecosystems as the basis of biodiversity conservation, ecosystem resilience, and adaptive capacity

(Bengtsson et al., 2003; Rist & Moen, 2013). This approach emphasizes the self-organized dynamic complexity of ecological systems (Levin, 1998) as a basis for ecosystem management and planning (Messier et al., 2013). Forested landscapes have been described as complex adaptive systems at multiple dynamically interacting scales (Gunderson & Holling, 2001) driven by local disturbance regimes affected by environmental fluctuations and global change (Messier et al., 2013). This view embraces the properties of complex ecological systems, such as emergent properties, self-organization, resilience, and adaptability (Filotas et al., 2014; Gunderson & Holling, 2001; Holling, 2001). This approach emphasizes the importance of the natural *adaptive cycle* in building *evolutionary resilience* (Sgrò et al., 2011) in restoration efforts and sustainable management. Interestingly, this echoes the early left-to-nature approach (Fig. 17.1) but now with a more robust ecological framework based on a novel understanding of the self-organization properties of ecosystems.

Although no overarching theory of restoration has emerged, several concepts are closely linked to and widely used in the context of ecological restoration. These include, in particular, the established concepts of (1) natural (historical) range of variation (NRV, Landres et al., 1999), (2) coarse- and fine-filter conservation management (Hunter, 1991, 1993), (3) natural disturbance emulation in forestry (Angelstam, 1998; Bergeron et al., 2002), (4) managing forests as complex adaptive systems (Messier et al., 2013), and (5) adaptive cycle and the concepts of *panarchy* (Gunderson & Holling, 2001). In the following section, we briefly explain these concepts and discuss them in relation to forest restoration with special reference to challenges brought by climate change.

17.3 Importance of Reference Conditions

Ecosystem restoration should ultimately be based on a thorough understanding of the structure and functioning of ecosystems and the habitat requirements and dynamics of species and communities. However, such knowledge of specific habitat requirements of the thousands of species living in every single forest stand is, and always will be, limited (Kuuluvainen & Siitonen, 2013). Moreover, we do not often know to which past or current conditions the species have adapted. This situation is complicated further by the possibility of rapid eco-evolutionary adaptation in some species populations (Rice & Emery, 2003; Sgrò et al., 2011). All this makes comprehensive species-by-species restoration challenging, if not impossible, and calls for more holistic approaches based on restoring habitats that emerge through the adaptive cycle.

For example, in northern Europe, boreal forests have, for the most part, been strongly transformed by a long history of intensive utilization and (more recently) by modern intensive forestry (Kuuluvainen, 2009; Linder & Östlund, 1998; Östlund et al., 1997). In this situation, knowledge of conditions characterizing boreal forest habitat types before the onset of intensive human usage is pivotal as a point of reference for conservation and restoration (Berglund & Kuuluvainen, 2021). This refers to conditions where—acknowledging that humans have to some degree probably been omnipresent in all boreal forests throughout history (Josefsson et al., 2010)—human influence has been negligible, and natural forest structure and dynamics have prevailed (Brūmelis et al., 2011).

It is worth noting that our understanding of reference conditions and their natural variation in boreal forests has been, to a large extent, revised by recent research (summarized in Berglund & Kuuluvainen, 2021). These new findings have important implications for answering the questions of what and how much to restore. For example, in boreal northern Europe, there has been a change from earlier perceptions of universal even-aged forest dynamics driven by stand-replacing disturbances toward current knowledge highlighting the role of non-stand-replacing disturbances and the resultant prevalence of old forests with complex structures and dynamics (Berglund & Kuuluvainen, 2021). A similar revision of reference forest conditions has taken place in North America (Bergeron & Fenton, 2012).

17.3.1   Natural Range of Variation and Coarse-Filter/Fine-Filter Management

The natural (historical) range of variation (NRV) concept of ecosystems provides knowledge of their past distribution, structure, and dynamics (Landres et al., 1999). This information can be used as a reference for guiding forest management, conservation, and restoration (Keane et al., 2009). The coarse- and fine-filter (CFF) approach builds upon a knowledge of NRV and separates two complementary strategies for sustaining ecosystems and their biodiversity. The coarse-filter strategy emphasizes the importance of maintaining natural ecosystem types and structures at large scales (Hunter, 1991, 1993). The assumption is that restoring or maintaining natural coarse-scale landscape conditions within a range to which the organisms are adapted will likely conserve most species and maintain sustainable ecosystems. The coarse-filter approach does not necessarily consider only reserves but rather recognizes ecological processes and the dynamic distribution of habitats across the entire landscape or region over time. The coarse-filter strategy has been recommended as a holistic approach, as it avoids the pitfalls of reductionist species-by-species planning (Table 17.1). The latter approach is also severely restricted by the lack of knowledge of ecological habitat requirements for most forest-dwelling species.

Complementary to this coarse-filter management, the fine-filter approach focuses on individual species or fine-scale elements of diversity, which are critical in conserving biodiversity and are not sufficiently accounted for by coarse-filter management. This fine-filter approach tries to safeguard, for example, those species that have very specialized habitat requirements or that do not tolerate any management actions and thus easily "fall through" the coarse filter. Thus, the fine-filter

Table 17.1 Comparison of different strategic approaches to forest conservation and restoration with indications of when a property is typical (*positive*) or not typical (*negative*) of a given restoration approach and when this relationship may be less strict (*parentheses*)

Approach	Static	Dynamic	Reductionist	Holistic	Multiple scales	Adaptive
Forest-as-museum	+	−	−	−	−	−
Coarse/fine filter	+	(−)	(+)	+	+	(+)
Disturbance-based	−	+	−	+	+	+
Adaptive cycle	−	+	−	+	+	+

approach consists of developing specific conservation strategies for specific species that are considered to be at particular risk under the coarse-filter approach. An example of a fine filter could be the provision of nesting boxes for cavity-nesting birds and mammals when cavity trees are not available, e.g., because of forest management actions.

The applicability of the coarse-filter/fine-filter approach depends on adequate knowledge of the past natural or historical conditions. Because of the long-term human impact, the protected forest fragments available as references may not be representative and large enough to harbor natural ecosystem structures, dynamics, and viable species assemblages (Lilja & Kuuluvainen, 2005; Nordén et al., 2013). In some cases, the system may have moved too far from its natural state to restore it to any past state (Hobbs et al., 2009; Jackson & Hobbs, 2009).

Another assumption of the coarse-filter/fine-filter approach is that the dynamics of future environments will be similar grosso modo to those of past environments. The approach is thus static and therefore does not provide a means of adapting to future changing environmental conditions. Accordingly, the main criticism of this approach is that because of rapid global change, knowledge of past ecosystem conditions cannot serve to guide forest restoration under future conditions, conditions that entail the development of novel ecosystems (Hobbs et al., 2011; Keane et al., 2009; McDowell et al., 2020). On the other hand, we may assume that native species are adapted to past ecological conditions, and knowledge of species habitat requirements can guide restoration even when future conditions differ from those of the past (Keane et al., 2009).

17.3.2 Natural Disturbance Emulation

Natural disturbance emulation (NDE) has become an important concept when aiming to implement the coarse-filter approach in forest ecosystem restoration and management (Angelstam, 1998; Bergeron et al., 2002; Berglund & Kuuluvainen, 2021; Kuuluvainen, 2009; Stockdale et al., 2016). Essentially, NDE is a strategy to implement the coarse filter over time. Thus, the coarse-filter concept emphasizes maintaining natural broad-scale habitat structures, whereas NDE focuses on the disturbance and successional processes and how to emulate them in forest restoration and management over time. NDE recognizes disturbance dynamics as a critical driver in forest dynamics and biodiversity maintenance (Kuuluvainen & Grenfell, 2012). According to this approach, management actions, especially timber harvesting, are planned to emulate the structural outcomes of natural disturbances typical of the forest landscape to be managed. Special attention is paid to the outcomes of such management at multiple scales, from deadwood microsites to landscape patterns, to provide natural habitat variability for various organisms.

Applying the NDE paradigm to forest restoration and management requires an adequate understanding of the past natural and potential future range of variation in forest structure and dynamics. Thus, it is possible to consider changes in forest disturbance dynamics likely to take place in the future (Cyr et al., 2009). In this manner, the ongoing rapid shifts in forest dynamics because of global stressors and climate warming, increasing disturbances, and land-use changes can be, in principle, incorporated into this approach. Moreover, natural disturbance can promote resilience by enhancing biodiversity and the ability of the ecosystems to resist or recover when hit by perturbations (Drever et al., 2006).

17.3.3 Complex Systems Framework

Traditionally, the boreal forest has been considered a simple system with few tree species and slow, predictable development. The complex systems approach challenges this view (Burton, 2013; Levin, 2005). Any single stand of boreal forest in Fennoscandia is estimated to contain some 2,500 to 5,000 species, with a large number of complex trophic and nontrophic interactions (Kuuluvainen & Siitonen, 2013). Thus, contrary to earlier perceptions, the boreal forest ecosystem holds a web of highly complex interactions (Burton, 2013), which in turn interact with the dynamic physical systems, such as climate and forest management. The boreal forest is, in essence, a complex system, and the goal should be to respect and restore this complexity (Drever et al., 2006; Filotas et al., 2014; Levin, 1998). This calls for a systems approach to the understanding, managing, and restoring of communities of trees and other plants, animals, and microorganisms that interact with their physical environment.

The ecological complexity perspective yields several implications for forest restoration and ecosystem management. First, a holistic approach is compulsory because of the complexity of the system. In landscape restoration, for example, it is necessary to pay attention to the cross-scale interactions of spatial and temporal heterogeneity. Second, the applied conceptual models, e.g., NRV and NDE, and the available reference ecosystems must be adequate to address ecologically important details and variations in the ecosystem (Berglund & Kuuluvainen, 2021). Overly simplistic conceptual approaches, such as conventional even-aged stand management, easily overlook critical details and interactions at spatial scales higher (landscape patterns) and lower (within-stand structures, microhabitats) than that of a tree stand (Kuuluvainen, 2009). Thus, management must address important structures and processes at multiple scales, from decaying logs to landscape patterns of habitats (Puettmann, 2014). Third, forest ecosystems are always undergoing change, and they must be managed by considering their long-term dynamics in a warming climate (Gauthier et al., 2015). Biodiversity and resilience are ultimately based on the adaptive cycle of disturbances and succession. Forests are indeed complex systems, and the challenge is to assimilate this complexity into forest ecosystem restoration and management (Messier et al., 2013).

17.4 When Are Different Strategies Applicable?

The applicability of specific restoration strategies and approaches is always sensitive to location and context (Fig. 17.2). Important issues are, first, how well the natural range of variability (NRV) and disturbance regime are known, and can we expect them to prevail as such in the future, or will they change (even radically), for example, because of climate change. Second, it is important to know the degree to which the restoration process is controllable. The lack of controllability may be because of a poor understanding of intrinsic system dynamics, e.g., in the case of novel ecosystems, or a shortage of managerial resources to carry out monitoring and control measures when needed.

In an ideal situation, NRV is well known and restoration is well controlled; then, it is possible to define specific targets and implement measures to obtain them (Fig. 17.2). Such a situation can be exemplified by developed countries like Sweden and Finland, which have both the research knowledge of reference systems and the resources to carry out and control the restoration process (Kuuluvainen & Siitonen, 2013).

If the controllability of restoration processes is high, but NRV is poorly understood, it becomes possible to apply a passive restoration based on ecosystem self-organization, or a coarse-filter/fine-filter approach based on a general knowledge of NRV (Clewell & McDonald, 2009). Such a situation can be typical in southern boreal forests, where reference systems are lacking because forests have been used for various purposes for hundreds (even thousands) of years. Finally, when the controllability of the restoration process is low, an adaptive management approach is preferable irrespective of the level of local knowledge of NRV. If the level is high, then NDE is applicable, and if the level is low, the robust coarse- and fine-filter approach can be used (Fig. 17.2). Restoration also needs to employ self-organization and be prepared for surprises and the emergence of novel species communities (Hobbs et al., 2011).

At large scales, the goal of restoration is typically to achieve a favorable conservation status for native ecosystem types—including the range of stages of forest succession—and their species assemblages over time. The measures to attain this goal depend not only on the properties of the area to be restored but also on the quality of the surrounding area (the forest matrix) (Fig. 17.3). This highlights that no restoration occurs in isolation and that a favorable conservation status can be

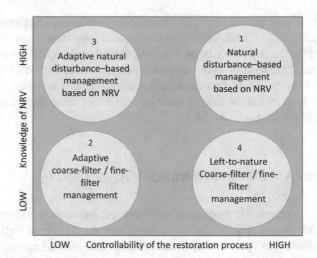

Fig. 17.2 Illustration of different forest restoration strategies in relation to the controllability of the restoration process and the level of understanding of the natural or historical range of variation (NRV). *1* When knowledge of NRV and controllability are both high, natural disturbance emulation is feasible. *2* When both the knowledge of NRV and controllability are low, it remains possible to practice adaptive coarse-filter/fine-filter management by applying a general understanding of NRV features. *3* Situations where knowledge of NRV is high but controllability is low favors use of an adaptive management approach to natural disturbance emulation. *4* A situation of low knowledge of NRV but high controllability allows applying a coarse-filter/fine-filter approach based on general understanding of NRV within similar systems. Modified from Allen et al. (2011) with permission from Elsevier

attained through different combinations of the managed matrix and core protected habitat, depending on their ecological quality. For example, if the forest matrix is under intensive plantation-type management and its habitat quality is low and does not provide habitat for native species, more pressure is put on restoring the full range of habitat and ecosystem types in the restored area (Fig. 17.3; Berglund & Kuuluvainen, 2021). On the other hand, if the matrix is managed on the basis of *ecological forestry* principles (Franklin et al., 2018), less intensive restoration may be required to obtain a favorable conservation status.

17.5 Adaptive Cycle and Adaptive Capacity: A Comprehensive Restoration Framework Under Climate Change Conditions

The adaptive cycle provides a universal metaphor and framework for describing and understanding long-term dynamics and change in complex socioecological systems (Fig. 17.4; Gunderson & Holling, 2001). The adaptive cycle is thought to be central

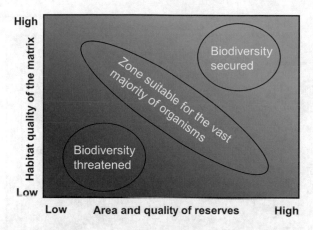

Fig. 17.3 Attaining the favorable conservation goals in restoration requires a landscape- or region-wide approach that includes both the area to be restored and the surrounding managed forest *matrix*. The level of biodiversity conservation can range from secured (*green*) to poorly secured or threatened (*red*). A reasonably favorable conservation status (*green–red transition*) can be achieved through various combinations of restored core area and ecosystem-based management of the surrounding managed landscape matrix. Modified by permission of Island Press, Washington, DC from Lindenmayer and Franklin (2002) *Conserving Forest Biodiversity*. Copyright © 2002 by the authors

to the endogenous processes of self-organization and evolution of complex systems through time (Levin, 1998, 2005). The widespread distribution of the adaptive cycle in ecosystem dynamics has been confirmed in many studies (Sundstrom & Allen, 2019). Thus, the adaptive cycle is useful as a conceptual model for management and restoration purposes, as it simplifies highly complex system behavior into four ubiquitous phases in ecosystem dynamics: (1) growth and development, (2) conservation, (3) release, and (4) reorganization. Most importantly, the adaptive cycle explains how ecosystems adapt to changing environmental conditions at multiple scales, a property necessary during periods of climate change. We suggest that the adaptive cycle and its extension, the nested adaptive cycle (*panarchy*), provide a comprehensive strategic ecological and evolutionary framework for forest ecosystem restoration (Fig. 17.4; Gunderson & Holling, 2001).

For forest restoration to be successful over the long term, a holistic approach is necessary where all four stages of the adaptive cycle and their dynamics are considered in relation to each other (Fig. 17.4; Box 17.1). This holistic approach stresses managing and restoring the naturally occurring processes and dynamics in the landscape. Forests are therefore allowed to complete full disturbance–succession cycles, from post-disturbance conditions to old-growth forest (Fig. 17.4). A landscape can also host multiple types of cycles depending on disturbance type. An example is provided by beaver pond dynamics, which can interact with and be embedded in the forest disturbance cycle (Box 17.2; Kivinen et al., 2020; Nummi & Kuuluvainen, 2013).

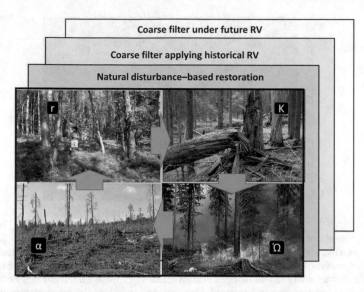

Fig. 17.4 Illustration of a typical adaptive cycle in the boreal forest with four stages: release, fire (Ω), reorganization (α), exploitation (r), and conservation (K). Such adaptive disturbance–succession cycles create the dynamic NRV that enables the adaptation of the ecosystem to changing environmental conditions. To become a useful strategic model in restoration and sustainable management, this conceptual model needs to be translated into a quantitative model based on knowledge of NRV, including the emulation of the natural disturbance regime. *Photo credits* Timo Kuuluvainen

---

Box 17.1: The Four Stages of the Adaptive Cycle (Gunderson & Holling, 2001)

Release, Ω Periodic disturbances are necessary to maintain habitat variation and biodiversity. Disturbances are also essential for the adaptive cycle, as they release growth resources and space and create critical post-disturbance structures, such as abundant deadwood and complex stand structures (Johnstone et al., 2016). The natural disturbance emulation approach is based on these notions and provides methods of implementing disturbance into forest restoration. The key NRV properties are the type, severity, size, and frequency of disturbances.

Reorganization, α This is a crucial stage of ecosystem dynamics promoting the adaptive capacity of the ecosystem to be restored. Here the critical process is the post-disturbance reassembly of the ecological community, as determined by the introduction or availability of new species and genotype assemblages, which through competition form novel communities potentially better adjusted to changing environmental conditions. This facilitates the system's adaptation to novel conditions, such as a warming and drying climate (Gauthier et al., 2015).

***Exploitation,* r** This part of ecosystem dynamics is perhaps most often addressed in restoring damaged or degraded ecosystems, as it represents the early-successional phase. Important natural processes in the adaptive cycle of ecosystems are the self-organization of species assemblages—through space filling and competitive self-thinning through multiple successional pathways—and the longevity of the successional processes at the landscape scale.

***Conservation,* K** This refers to the mature and late-successional phase where growth resources are tied to standing biomass, and changes in vegetation structure are slight and take place at a small scale. Here, the traditional example is old-growth forests and their restoration and conservation. Important NRV variables are the variation of old-growth types, their structure, and the areal proportions of old-growth forest in the past and present.

The adaptive cycle is closely related to and feeds on biodiversity (Fig. 17.5). The two main interrelated restoration goals are maintaining native biodiversity and keeping the adaptive cycle in operation to provide a continuity of habitats through disturbance and succession (Fig. 17.4). If successful, both goals will contribute to forest resilience and adaptive capacity. Resilience here means an ecosystem's resistance to disturbance (short-term) or its ability to recover when perturbed (medium-term). Adaptive capacity denotes the ecosystem's (long-term) evolutionary adaptive potential (Sgrò et al., 2011).

Fig. 17.5 Illustration of the relationships of key strategic concepts of ecological restoration; **a** the primary restoration goals maintain biodiversity by ensuring diverse habitats through the adaptive disturbance–succession cycle (see Fig. 17.4); **b** if successful, the goals improve forest resilience and adaptive capacity, properties that are crucial under conditions of rapid climate change

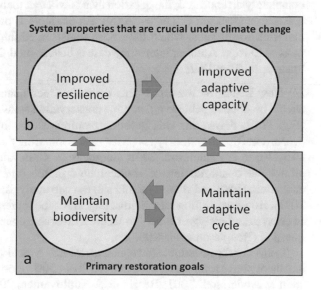

17.6 Strategic Questions: What, How Much, and When to Restore?

In practice, ecosystem types are often defined as comprehensive vegetation communities, or habitat types, which are used as proxies to represent natural ecosystems and their dynamics (Landres et al., 1999). The goal of restoration is the recovery of a favorable conservation status of all native ecosystem types and their natural dynamic stages in terms of the adaptive cycle (Fig. 17.4). At a large scale, this requires answers to three strategic questions: (1) What to restore? (2) How much to restore? (3) When to restore? The operational question, How to restore? must also be answered; however, most texts on restoration focus on this last question, and, from a policy and planning point of view, the three first questions are more important.

(1) What to Restore? This is a strategic question, as we can focus on restoring, for example, species, populations, communities, habitats, or full ecosystems at different scales. This chapter deals with forest ecosystem restoration, which requires an understanding of the ecosystem types and their intrinsic dynamics over time, as defined by the four stages of the adaptive ecosystem cycle (Fig. 17.4). This requires a classification of the ecosystem types to be restored (Fig. 17.6; Berglund & Kuuluvainen, 2021). Such a classification of forest ecosystem types could be based, for example, on soil fertility and moisture, tree species composition, disturbance origin, and successional stage, or a combination of these. The classification of ecosystem types should capture the essential features of NRV but also restrict the number of classes to allow their application in practical restoration (Fig. 17.6).

It should also be understood that some ecosystem types and dynamics may have completely disappeared; the question then is whether their restoration is possible and reasonable (Jackson & Hobbs, 2009). An example is provided by the most fertile herb-rich forest sites, which have been transformed into agricultural fields across most of Europe. Another interesting case is beaver-modified habitats in boreal forest landscapes (Box 17.2).

(2) How Much to Restore? This question urges managers to define how much should be done to achieve a favorable conservation status of native ecosystem types (Fig. 17.6). Answering this question requires estimating the natural or historical proportions of ecosystem types (NRV) and comparing them with their current extent (Angelstam & Andersson, 2001; Berglund & Kuuluvainen, 2021). This approach allows the quantification of shortcomings in the representativeness of different ecosystem types and their dynamic stages (gap analysis; Berglund & Kuuluvainen, 2021). However, setting quantitative targets for the representativeness and dynamics of ecosystem types over time—to secure a favorable conservation status—is a tricky question (Bengtsson et al., 2003).

To answer the question, How much to restore? typical values derived from ecological theory have referred to, for example, 20%–50% of original habitat cover (Angelstam & Andersson, 2001; Berglund & Kuuluvainen, 2021; Fahrig, 2001; Hanski, 2011; Wilson, 2016). Naturally, the issue is more complicated, including questions

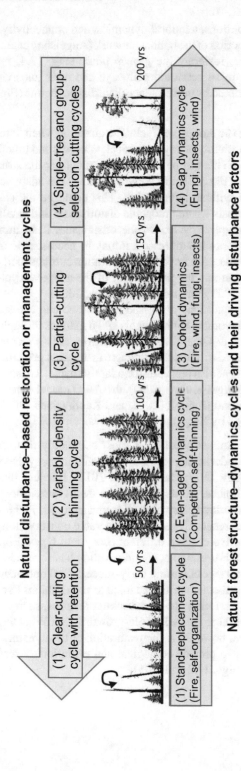

Fig. 17.6 The adaptive cycles in forest ecosystems are expressed as variable disturbance-driven forms that must be accounted for in restoration. An example of a strategic answer to the question, What to restore? is to divide the NRV of forest structure into four types of structure–dynamics cycles (*green arrow*) identified by their main driving disturbance factors. Embedded in the *1* stand-replacing fire-driven "large cycle" are three "small cycle" structure–dynamics cycles: *2* even-aged, *3* cohort, and *4* gap-dynamics cycles, with their respective restoration or management cutting cycles to be applied (*gray arrow*). The answer to, How much to restore? for each type of structure–dynamics cycle depends on their historical proportions, informed by the knowledge of the NRV of the associated disturbance regime. The answer to, When to restore? depends on the current state and proportions of habitats compared with the natural state and the disturbance return intervals

involving spatial and size distributions, temporal dynamics, and connectivity, all of which affect the conservation function of the habitat network (Angelstam et al., 2020; Ward et al., 2020). For example, restoring the *growth* phase (Fig. 17.4) requires information about the variation in successional pathways and their longevity and stand self-organization, as determined by the reference disturbance regime (Fig. 17.6; Kuuluvainen, 2009).

(3) When to Restore? Securing the Adaptive Cycle The question, when to restore? refers to the need to secure the continuity of key ecological processes and functioning of the adaptive cycle (Fig. 17.4). This can be realized by emulating the structure–dynamics cycles in management (Fig. 17.6). In some situations, the adaptive cycle can be restored and maintained without intervention. This could be the case, for example, in old humid spruce forests where autogenic disturbances and small-scale gap dynamics are driving the adaptive cycle. On the other hand, if an important natural disturbance driver such as fire is suppressed, it may be necessary to use fire as a restoration tool or at least emulate fire impacts in restoration cuttings to maintain the natural-like habitat distribution and the functioning of the adaptive cycle over time (Berglund & Kuuluvainen, 2021; Vanha-Majamaa et al., 2007).

To answer the three strategic questions discussed above, one can use strategic models to plan how to maintain favorable habitat status and dynamics through space and time (Fig. 17.6; Kuuluvainen & Grenfell, 2012). However, only a few specific strategic models have been proposed to transform strategic restoration principles into practical management solutions that also consider the adaptive cycle. These models include the multicohort model (Bergeron et al., 2002), the ASIO model (Angelstam, 1998) discussed in Kuuluvainen and Grenfell (2012) and Kuuluvainen (2017), and the revised-ASIO model proposed by Berglund and Kuuluvainen (2021).

(4) How to Restore? This operational question is most commonly addressed in restoration, which is a practical undertaking, and it is dealt with in many papers and textbooks (e.g., Allison & Murphy, 2017; Halme et al., 2013). Because the focus of this chapter is on strategic choices, we do not go into detail here. Some practical and tactical methods are described elsewhere (Kuuluvainen, 2017). However, the key issues are to strive for representativeness and favorable conservation status of the various phases of the adaptive cycle in different ecosystem types (Fig. 17.2; Berglund & Kuuluvainen, 2021). For example, if conservation has focused on old-growth forests, there may be a shortage of natural early-successional open-canopied forests that would secure ecological communities adapted to such habitats (Swanson et al., 2011). On the other hand, restoring old-forest habitat features, such as dead-wood structures, may be urgent if there is a lack of old-growth forest. What the adaptive cycle framework emphasizes, however, is the restoration of the dynamic continuity of all-natural habitat types to provide ecological and evolutionary resilience provided by the adaptive cycle (Figs. 17.2 and 17.5).

Box 17.2 Beaver: A Keystone Disturbance Agent of Boreal Landscapes

In the boreal forest, beavers (*Castor canadensis* and *C. fiber*) provide an example of a keystone ecosystem modifier (Johnston, 2017) extirpated from many parts of the boreal zone because of heavy exploitation for the animal's fur. Beavers are now returning to many parts of their original range in Eurasia and North America (Halley et al., 2021; Whitfield et al., 2015). This renewal provides a possibility of restoring former beaver habitats and their dynamics at landscape scales. Some beaver-affected areas may comprise up to 26% of the landscape (Naiman et al., 1988); in the boreal region, however, it is normally much less, e.g., 2.8% (Parker et al., 1999).

The power of ecological engineering by beavers is based on their ability to build dams (Johnston, 2017). The damming of creeks and ponds creates wetland habitats and successional pathways that otherwise would not exist in many boreal landscapes (Feldman et al., 2020). The beaver is unique in terms of its role in flooding riparian forests and transforming terrestrial habitats into aquatic ones. When beavers abandon a pond and the dam collapses, the terrestrial habitat returns. For both events, an early-successional stage is created.

Within a boreal forest matrix, beaver flooding can be viewed as a patch disturbance (Johnston & Naiman, 1990; Nummi & Kuuluvainen, 2013; Remillard et al., 1987). Whereas disturbances such as fire and windstorms mainly strike upland forests, beavers most pronouncedly affect lowland riparian stands. Fires and storms are very stochastic events, having local return times of tens to hundreds of years. In contrast, beaver disturbance in an active beaver landscape is more predictable, both spatially and temporally (Nummi & Kuuluvainen, 2013).

Kivinen et al. (2020) recently studied the effect of beavers on landscape heterogeneity in Finland using data of the yearly occupancies of beavers in a landscape over half a century. During this time, beavers colonized the landscape, and the number of beaver sites increased from 6 to 69. What is noteworthy in this boreal setting, however, is that at certain points, the amount of flooded land was much less than observed in more temperate areas (Naiman et al., 1988). Rather, the cumulative number of beaver-affected sites increased steadily. Along with this increase, different processes occurred, affecting the distribution and properties of the riparian habitats (Kivinen et al., 2020). First, the distance between beaver sites declined, and *hot spots* of spatially clustered beaver sites were then formed (Kivinen et al., 2020).

The biodiversity-enhancing impact of beavers relates to how they facilitate the establishment of numerous organisms (Stringer & Gaywood, 2016). Various species and species groups benefit from the different successional stages associated with beaver sites; these taxa benefiting from flooding phases include invertebrates (Bush et al., 2019; Nummi et al., 2021), frogs (Dalbeck et al.,

2007), and water birds (Nummi & Holopainen, 2014). In the terrestrial system, beaver-affected processes include the riparian forest becoming more dominated by deciduous trees after flooding (Hyvönen & Nummi, 2008). This shift in vegetation is notable because herbivores normally push forests toward a more coniferous direction because of the foraging preference of herbivores on deciduous trees. Greater amounts of deciduous trees benefit animals such as moose (Nummi et al., 2019) but also beaver itself when it reoccupies sites (Labrecque-Foy et al., 2020). Beaver flooding also creates deadwood (Thompson et al., 2016). Beaver sites may contain different-sized deadwood resulting from consecutive floods (Kivinen et al., 2020), and the presence of beavers in the landscape results in spatiotemporal deadwood continuity (Thompson et al., 2016).

17.7   Conclusions

The rapid loss of global forest area and degradation of forest ecological conditions because of global change (Gauthier et al., 2015; McDowell et al., 2020) have evoked an urgent need to develop and apply ecologically sound restoration and sustainable management approaches that can be applied to various situations and scaled over large areas (Angelstam et al., 2020; Berglund & Kuuluvainen, 2021; Kuuluvainen, 2017). Over time, different approaches to conservation and restoration have been proposed, reflecting the increasing understanding of forest ecosystems and their intrinsic structure and dynamics (Fig. 17.1).

Forest restoration strategies and tactics should be based on an ecological and evolutionary understanding of long-term ecosystem structure and dynamics (Rice & Emery, 2003; Sgrò et al., 2011), as the species, communities, and ecosystems being restored have emerged via evolutionary processes operating under past environmental conditions. The assembly of species and communities in ecosystems is by and large regulated by properties of the prevailing disturbance–succession cycles and the resulting habitat mosaic structure at multiple scales (Johnstone et al., 2016). Therefore, a holistic understanding of ecosystem structure and function at multiple scales is necessary to set tangible restoration targets and effective actions.

In a time of climate change, the adaptive capacity of ecosystems should become the fundamental strategic priority and guiding principle of ecological restoration. This requires viewing and understanding forests as complex adaptive systems where the most important long-term goal is to restore and maintain the favorable conservation status of habitat types and the native adaptive cycles driven by the cyclic dynamics of forest disturbance and succession (Fig. 17.4).

References

Allen, C. R., Fontaine, J. J., Pope, K. L., et al. (2011). Adaptive management for a turbulent future. *Journal of Environmental Management, 92*, 1339–1345. https://doi.org/10.1016/j.jenvman.2010.11.019.

Allison, S. K., & Murphy, S. D. (Eds.). (2017). *Routledge handbook of ecological restoration* (p. 620). Routledge.

Angelstam, P. (1998). Maintaining and restoring biodiversity in European boreal forests by developing natural disturbance regimes. *Journal of Vegetation Science, 9*, 593–602. https://doi.org/10.2307/3237275.

Angelstam, P., & Andersson, L. (2001). Estimates of the needs for forest reserves in Sweden. *Scandinavian Journal of Forest Research, 16*, 38–51. https://doi.org/10.1080/028275801300090582.

Angelstam, P., Manton, M., Green, M., et al. (2020). Sweden does not meet agreed national and international forest biodiversity targets: A call for adaptive landscape planning. *Landscape and Urban Planning, 202*, 103838. https://doi.org/10.1016/j.landurbplan.2020.103838.

Bengtsson, J., Angelstam, P., Elmqvist, T., et al. (2003). Reserves, resilience and dynamic landscapes. *Ambio, 32*, 389–396. https://doi.org/10.1579/0044-7447-32.6.389.

Bergeron, Y., & Fenton, N. (2012). Boreal forests of eastern Canada revisited: Old growth, nonfire disturbances, forest succession and biodiversity. *Botany, 90*, 509–523. https://doi.org/10.1139/b2012-034.

Bergeron, Y., Leduc, A., Harvey, B. D., et al. (2002). Natural fire regime: A guide for sustainable management of the Canadian boreal forest. *Silva Fennica, 36*, 81–95. https://doi.org/10.14214/sf.553.

Berglund, H., & Kuuluvainen, T. (2021). Representative boreal forest habitats in northern Europe, and a revised model for ecosystem management and biodiversity conservation. *Ambio, 50*, 1003–1017. https://doi.org/10.1007/s13280-020-01444-3.

Bradshaw, C. J. A., Warkentin, I. G., & Sodhi, N. S. (2009). Urgent preservation of boreal carbon stocks and biodiversity. *Trends in Ecology & Evolution, 24*, 541–548. https://doi.org/10.1016/j.tree.2009.03.019.

Brūmelis, G., Jonsson, B. G., Kouki, J., et al. (2011). Forest naturalness in northern Europe: Perspectives on processes, structures and species diversity. *Silva Fennica, 45*, 807–821. https://doi.org/10.14214/sf.446.

Burton, P. J. (2013). Exploring complexity in boreal forests. In C. Messier, K. J. Puettmann, & K. D. Coates (Eds.), *Managing forests as complex adaptive systems: Building resilience to the challenge of global change* (pp. 79–109). Routledge.

Burton, P. J., & Macdonald, S. E. (2011). The restorative imperative: Challenges, objectives and approaches in restoring naturalness in forests. *Silva Fennica, 45*(5), 843–863. https://doi.org/10.14214/sf.74.

Burton, P. J., Bergeron, Y., Bogdanski, B. E. C., et al. (2010). Sustainability of boreal forests and forestry in a changing environment. In G. Mery, P. Katila, G. Galloway, R.I. Alfaro, M. Kanninen, M. Lobovikov, & J . Varjo (Eds.), *Forests and society—responding to global drivers of change.* (pp. 249–282). International Union of Forest Research Organizations.

Bush, B. M., Stenert, C., Maltchik, L., et al. (2019). Beaver-created successional gradients increase β-diversity of invertebrates by turnover in stream-wetland complexes. *Freshwater Biology, 64*(7), 1265–1274. https://doi.org/10.1111/fwb.13302.

Clements, F. E. (1916). Plant succession: an analysis of the development of vegetation (No. 242). Washington: Carnegie Institution of Washington Publication.

Clewell, A., & McDonald, T. (2009). Relevance of natural recovery to ecological restoration. *Ecological Restoration, 27*(2), 122–124. https://doi.org/10.3368/er.27.2.122.

Cyr, D. S., Gauthier, S., Bergeron, Y., et al. (2009). Forest management is driving the eastern North American boreal forest outside its natural range of variability. *Frontiers in Ecology and the Environment, 7*(10), 519–524. https://doi.org/10.1890/080088.

Dalbeck, L., Lüscher, B., & Ohlhoff, D. (2007). Beaver ponds as habitats of amphibian communities in a central European highland. *Amphibia-Reptilia, 28*(4), 493–501. https://doi.org/10.1163/156 853807782152561.

Drever, C. R., Peterson, G., Messier, C., et al. (2006). Can forest management based on natural disturbances maintain ecological resilience? *Canadian Journal of Forest Research, 36,* 2285–2299. https://doi.org/10.1139/x06-132.

Fahrig, L. (2001). How much habitat is enough? *Biological Conservation, 100,* 65–74. https://doi.org/10.1016/S0006-3207(00)00208-1.

Feldman, M. J., Girona, M. M., Grosbois, G., et al. (2020). Why do beavers leave home? Lodge abandonment in an invasive population in Patagonia. *Forests, 11*(11), 1161. https://doi.org/10.3390/f11111161.

Filotas, E., Parrott, L., Burton, P. J., et al. (2014). Viewing forests through the lens of complex systems science. *Ecosphere, 5*(1):art1 https://doi.org/10.1890/ES13-00182.1.

Franklin, J. F., Johnson, K. N., & Johnson, D. L. (2018). *Ecological forest management.* Waveland Press.

Gauthier, S., Bernier, P., Kuuluvainen, T., et al. (2015). Boreal forest health and global change. *Science, 349,* 819–822. https://doi.org/10.1126/science.aaa9092.

Girard, F., Payette, S., & Gagnon, R. (2008). Rapid expansion of lichen woodlands within the closed-crown boreal forest zone over the last 50 years caused by stand disturbances in eastern Canada. *Journal of Biogeography, 35,* 529–537. https://doi.org/10.1111/j.1365-2699.2007.018 16.x.

Gunderson, L. H., & Holling, C. S. (Eds.). (2001). *Panarchy: Understanding transformations in human and natural systems* (p. 450). Island Press.

Halley, D. J., Saveljev, A. P., & Rosell, F. (2021). Population and distribution of beavers *Castor fiber* and *Castor canadensis* in Eurasia. *Mammal Review, 51*(1), 1–24. https://doi.org/10.1111/mam.12216.

Halme, P., Allen, K. A., Auniņš, A., et al. (2013). Challenges of ecological restoration: Lessons from forests in northern Europe. *Biological Conservation, 167,* 248–256. https://doi.org/10.1016/j.biocon.2013.08.029.

Hanski, I. (2000). Extinction debt and species credit in boreal forests: Modelling the consequences of different approaches to biodiversity conservation. *Annales Zoologici Fennici, 37,* 271–280.

Hanski, I. (2011). Habitat loss, the dynamics of biodiversity, and a perspective on conservation. *Ambio, 40,* 248–255. https://doi.org/10.1007/s13280-011-0147-3.

Hobbs, R. J., Higgs, E., & Harris, J. A. (2009). Novel ecosystems: Implications for conservation and restoration. *Trends in Ecology & Evolution, 24*(11), 599–605. https://doi.org/10.1016/j.tree.2009.05.012.

Hobbs, R. J., Hallett, L. M., Ehrlich, P. R., et al. (2011). Intervention ecology: Applying ecological science in the twenty-first century. *BioScience, 61*(6), 442–450. https://doi.org/10.1525/bio.2011.61.6.6.

Holling, C. S. (2001). Understanding the complexity of economic, ecological, and social systems. *Ecosystems, 4,* 390–405. https://doi.org/10.1007/s10021-001-0101-5.

Hunter, M. L., Jr. (1991). Coping with ignorance: The coarse filter strategy for maintaining biodiversity. In K. A. Kohm (Ed.), *Balancing on the brink of extinction* (pp. 266–281). Island Press.

Hunter, M. L., Jr. (1993). Natural fire regimes as spatial models for managing boreal forests. *Biological Conservation, 65,* 115–120. https://doi.org/10.1016/0006-3207(93)90440-C.

Hyvönen, T., & Nummi, P. (2008). Habitat dynamics of beaver *Castor canadensis* at two spatial scales. *Wildlife Biology, 14,* 302–308. https://doi.org/10.2981/0909-6396(2008)14[302:HDO BCC]2.0.CO;2.

Jackson, S. T., & Hobbs, R. J. (2009). Ecological restoration in the light of ecological history. *Science, 325,* 567–569. https://doi.org/10.1126/science.1172977.

Jasinski, J. P. P., & Payette, S. (2005). The creation of alternative stable states in the southern boreal forest, Québec, Canada. *Ecological Monographs, 75,* 561–583. https://doi.org/10.1890/04-1621.

Johnston, C. A. (2017). *Beavers: Boreal ecosystem engineers*. Springer International Publishing.

Johnston, C. A., & Naiman, R. J. (1990). Aquatic patch creation in relation to beaver population trends. *Ecology, 71*, 1617–1621. https://doi.org/10.2307/1938297.

Johnstone, J. F., Allen, C. D., Franklin, J. F., et al. (2016). Changing disturbance regimes, ecological memory, and forest resilience. *Frontiers in Ecology and the Environment, 14*(7), 369–378. https://doi.org/10.1002/fee.1311.

Josefsson, T., Olsson, J., & Östlund, L. (2010). Linking forest history and conservation efforts: Long-term impact of low-intensity timber harvest on forest structure and wood-inhabiting fungi in northern Sweden. *Biological Conservation, 143*(7), 1803–1811. https://doi.org/10.1016/j.biocon.2010.04.020.

Keane, R. E., Hessburg, P. F., Landres, P. B., et al. (2009). The use of historical range and variability (HRV) in landscape management. *Forest Ecology and Management, 258*(7), 1025–1037. https://doi.org/10.1016/j.foreco.2009.05.035.

Keto-Tokoi, P., & Kuuluvainen, T. (2014) *Primeval forests of Finland, cultural history, ecology and conservation* (p. 302). Helsinki: Maahenki.

Kivinen, S., Nummi, P., & Kumpula, T. (2020). Beaver-induced spatiotemporal patch dynamics affect landscape-level environmental heterogeneity. *Environmental Research Letters, 15*, 094065. https://doi.org/10.1088/1748-9326/ab9924.

Kneeshaw, D. D., Bergeron, Y., & Kuuluvainen, T. (2011). Forest ecosystem structure and disturbance dynamics across the circumboreal forest. In A. C. Millington, M. Blumler, & U. Schickhoff (Eds.), *The SAGE handbook of biogeography* (pp. 263–280). SAGE.

Kontula, T., & Raunio, A. (Eds.). (2019). *Threatened habitat types in Finland 2018—Red list of habitats results and basis for assessment* (p. 258). Finnish Environment Institute and Ministry of the Environment.

Kuuluvainen, T. (2009). Forest management and biodiversity conservation based on natural ecosystem dynamics in northern Europe: The complexity challenge. *Ambio, 38*, 309–315. https://doi.org/10.1579/08-A-490.1.

Kuuluvainen, T. (2017). Restoration and ecosystem management in the boreal forest. From ecological principles to tactical solutions. In S. Allison & S. Murphy (Eds.), *Routledge handbook of ecological and environmental restoration* (pp. 93–112). Routledge.

Kuuluvainen, T., & Grenfell, R. (2012). Natural disturbance emulation in boreal forest ecosystem management: Theories, strategies and a comparison with conventional even-aged management. *Canadian Journal of Forest Research, 42*, 1185–1203. https://doi.org/10.1139/x2012-064.

Kuuluvainen, T., & Siitonen, J. (2013). Fennoscandian boreal forests as complex adaptive systems. Properties, management challenges and opportunities. In C. Messier, K.J. Puettman, & K. D. Coates (Eds.), *Managing forests as complex adaptive systems. Building resilience to the challenge of global change* (pp. 244–268). London: The Earthscan forest library, Routledge.

Kuuluvainen, T., Hofgaard, A., Aakala, T., et al. (2017). North Fennoscandian mountain forests: History, composition, disturbance dynamics and the unpredictable future. *Forest Ecology and Management, 385*, 140–149. https://doi.org/10.1016/j.foreco.2016.11.031.

Kuuluvainen, T., Lindberg, H., Vanha-Majamaa, I., et al. (2019). Low level retention forestry, certification, and biodiversity: Case Finland. *Ecological Processes, 8*(1), 47. https://doi.org/10.1186/s13717-019-0198-0.

Labrecque-Foy, J.-P., Morin, H., & Girona, M. M. (2020). Dynamics of territorial occupation by North American beavers in canadian boreal forests: A novel dendroecological approach. *Forests, 11*(2), 221. https://doi.org/10.3390/f11020221.

Landres, P. B., Morgan, P., & Swanson, F. J. (1999). Overview of the use of natural variability concepts in managing ecological systems. *Ecological Applications, 9*, 1179–1188.

Levin, S. A. (1998). Ecosystems and the biosphere as complex adaptive systems. *Ecosystems, 1*, 431–436. https://doi.org/10.1007/s100219900037.

Levin, S. A. (2005). Self-organization and the emergence of complexity in ecological systems. *BioScience, 55*, 1075–1079. https://doi.org/10.1641/0006-3568(2005)055[1075:SATEOC]2.0.CO;2.

Lilja, S., & Kuuluvainen, T. (2005). Structure of old *Pinus sylvestris* dominated foreststands along a geographic and human impact gradient in mid-boreal Fennoscandia. *Silva Fennica, 39*, 407–428. https://doi.org/10.14214/sf.377.

Lindenmayer, D. B., & Franklin, J. F. (2002). *Conserving forest biodiversity—A comprehensive multiscaled approach* (p. 351). Island Press.

Linder, P., & Östlund, L. (1998). Structural changes in three mid-boreal Swedish forest landscapes, 1885–1996. *Biological Conservation, 85*, 9–19. https://doi.org/10.1016/S0006-3207(97)001 68-7.

MacArthur, R. H., & Wilson, E. O. (1967). *The theory of island biogeography.* Princeton University Press.

McDowell, N. G., Allen, C. D., Anderson-& Teixeira, K., et al. (2020). Pervasive shifts in forest dynamics in a changing world. *Science, 368*(6494), eaaz9463 https://doi.org/10.1126/science.aaz 9463.

Messier, C., Puettmann, K. J., & Coates, K. D. (Eds.). (2013). *Managing forests as complex adaptive systems: Building resilience to the challenge of global change* (p. 368). Routledge.

Moen, J., Rist, L., Bishop, K., et al. (2014). Eye on the taiga: Removing global policy impediments to safeguard the boreal forest. *Conservation Letters, 7*(4), 408–418. https://doi.org/10.1111/conl. 12098.

Moilanen, A., Leathwick, J. R., & Quinn, J. M. (2011). Spatial prioritization of conservation management. *Conservation Letters, 4*, 383–393. https://doi.org/10.1111/j.1755-263X.2011.001 90.x.

Naiman, R. J., Johnston, C. A., & Kelley, J. C. (1988). Alteration of North American streams by beaver: The structure and dynamics of streams are changing as beaver recolonize their historic habitat. *BioScience, 38*(11), 753–762. https://doi.org/10.2307/1310784.

Nordén, J., Penttilä, R., Siitonen, J., et al. (2013). Specialist species of wood-inhabiting fungi struggle while generalists thrive in fragmented boreal forests. *Journal of Ecology, 101*(3), 701–712. https://doi.org/10.1111/1365-2745.12085.

Nordic Forest Research (SNS). (2021). *Nordic countries' share of world's forest resources.* Alnarp: Nordic Forest Research.

Nummi, P., & Holopainen, S. (2014). Whole-community facilitation beaver: Ecosystem engineer increases waterbird diversity. *Aquatic Conservation, 24*, 623–633. https://doi.org/10.1002/aqc. 2437.

Nummi, P., & Kuuluvainen, T. (2013). Forest disturbance by an ecosystem engineer: Beaver in boreal forest landscapes. *Boreal Environment Research, 18*, 13–24.

Nummi, P., Liao, W., Huet, O., et al. (2019). The beaver facilitates species richness and abundance of terrestrial and semi-aquatic mammals. *Global Ecology and Conservation, 20*, e00701. https://doi.org/10.1016/j.gecco.2019.e00701.

Nummi, P., Liao, W., van der Schoor, J., et al. (2021). Beaver creates early successional hotspots for water beetles. *Biodiversity and Conservation, 30*, 2655–2670. https://doi.org/10.1007/s10531-021-02213-8.

Östlund, L., Zackrisson, O., & Axelsson, A. L. (1997). The history and transformation of a Scandinavian boreal forest landscape since the 19th century. *Canadian Journal of Forest Research, 27*, 1198–1206. https://doi.org/10.1139/x97-070.

Parker, H., Haugen, A., Kristensen, Ø., et al (1999). Landscape use and economic value of Eurasian beaver (*Castor fiber*) on large forest in southeast Norway. 1st Euro-American Beaver Congress. Kazan.

Potapov, P., Yaroshenko, A., Turubanova, S., et al. (2008). Mapping the world's intact forest landscapes by remote sensing. *Ecology and Society, 13*(2), 51. https://doi.org/10.5751/ES-02670-130251.

Puettmann, K. J. (2014). Restoring the adaptive capacity of forest ecosystems. *Journal of Sustainable Forestry, 33*, 15–27. https://doi.org/10.1080/10549811.2014.884000.

Puettmann, K. J., Coates, K. D., & Messier, C. (2009). *A critique of silviculture: Managing for complexity.* Island Press.

Remillard, M. M., Gruendling, G. K., & Bogucki, D. J. (1987). Disturbance by beaver (*Castor canadensis* Kuhl) and increased landscape heterogeneity. In M. G. Turner (Ed.), *Landscape heterogeneity and disturbance ecological studies*. (pp. 103–122). Ecological Studies vol. 64, New York: Springer.

Rice, K. S., & Emery, N. C. (2003). Managing microevolution: Restoration in the face of global change. *Frontiers in Ecology and the Environment, 1*(9), 469–478. https://doi.org/10.1890/1540-9295(2003)001[0469:MMRITF]2.0.CO;2.

Rist, L., & Moen, J. (2013). Sustainability in forest management and a new role for resilience thinking. *Forest Ecology and Management, 310*, 416–427. https://doi.org/10.1016/j.foreco.2013.08.033.

Sgrò, C. M., Lowe, A. J., & Hoffmann, A. A. (2011). Building evolutionary resilience for conserving biodiversity under climate change. *Evolutionary Applications, 4*, 326–337. https://doi.org/10.1111/j.1752-4571.2010.00157.x.

Simonsson, P., Gustafsson, L., & Östlund, L. (2015). Retention forestry in Sweden: Driving forces, debate and implementation 1968–2003. *Scandinavian Journal of Forest Research, 30*(2), 154–173. https://doi.org/10.1080/02827581.2014.968201.

Stanturf, J. A., & Madsen, P. (Eds.). (2005). *Restoration of boreal and temperate forests*. CRC Press.

Stockdale, C., Flannigan, M., & Macdonald, S. E. (2016). Is the END (emulation of natural disturbance) a new beginning? A critical analysis of the use of fire regimes as the basis of forest ecosystem management with examples from the Canadian western Cordillera. *Environmental Reviews, 24*, 233–243. https://doi.org/10.1139/er-2016-0002.

Stringer, A., & Gaywood, M. (2016). The impacts of beavers *Castor* spp. on biodiversity and the ecological basis for their reintroduction to Scotland, UK. *Mammal Review, 46*(4), 270–283. https://doi.org/10.1111/mam.12068.

Sundstrom, S. M., & Allen, C. R. (2019). The adaptive cycle: More than a metaphor. *Ecological Complexity, 39*, 100767. https://doi.org/10.1016/j.ecocom.2019.100767.

Swanson, M. E., Franklin, J. F., Beschta, R. L., et al. (2011). The forgotten stage of forest succession: Early successional ecosystems on forest sites. *Frontiers in Ecology and the Environment, 9*(2), 117–125. https://doi.org/10.1890/090157.

Thompson, S., Vehkaoja, M., & Nummi, P. (2016). Beaver-created deadwood dynamics in the boreal forest. *Forest Ecology and Management, 360*, 1–8. https://doi.org/10.1016/j.foreco.2015.10.019.

Timonen, J., Gustafsson, L., Kotiaho, J. S., et al. (2011). Hotspots in a cold climate: Conservation value of woodland key habitats in boreal forests. *Biological Conservation, 144*, 2061–2067. https://doi.org/10.1016/j.biocon.2011.02.016.

Vanha-Majamaa, I., Lilja, S., Ryömä, R., et al. (2007). Rehabilitating boreal forest structure and species composition in Finland through logging, dead wood creation and fire: The EVO experiment. *Forest Ecology and Management, 250*, 77–88. https://doi.org/10.1016/j.foreco.2007.03.012.

Ward, M., Saura, S., Williams, B., et al. (2020). Just ten percent of the global terrestrial protected area network is structurally connected via intact land. *Nature Communications, 11*(1), 4563. https://doi.org/10.1038/s41467-020-18457-x.

Whitfield, C. J., Baulch, H. M., Chun, K. P., et al. (2015). Beaver-mediated methane emission: The effects of population growth in Eurasia and the Americas. *Ambio, 44*(1), 7–15. https://doi.org/10.1007/s13280-014-0575-y.

Wilson, E. O. (2016). *Half-Earth: Our planet's fight for life*. Liveright Publishing Corporation.

Chapter 18
Ecological Restoration of the Boreal Forest in Fennoscandia

Joakim Hjältén, Jari Kouki, Anne Tolvanen, Jörgen Sjögren, and Martijn Versluijs

Abstract Mixed-severity disturbances have historically shaped boreal forests, creating a dynamic mosaic landscape. In Fennoscandia, however, intensive even-aged forest management has simplified the forest landscape, threatening biodiversity. To safeguard this biodiversity, we therefore need to restore structural complexity in hitherto managed forests. Knowledge generated from relevant case studies on natural disturbance emulation–based ecological restoration suggests that prescribed burning positively affects many early-successional organisms. Gap cutting benefits some insects and wood fungi but has a limited effect on birds, bryophytes, and vascular plants. Restoration of deciduous forests appears to benefit light- and deciduous tree–associated insect species and some forest birds.

J. Hjältén (✉) · J. Sjögren
Department of Wildlife, Fish, and Environmental Studies, Swedish University of Agricultural Sciences, 901 83 Umeå, Sweden
e-mail: joakim.hjalten@slu.se

J. Sjögren
e-mail: jorgen.sjogren@slu.se

J. Kouki
School of Forest Sciences, University of Eastern Finland, P.O. Box 111, 80101 Joensuu, Finland
e-mail: jari.kouki@uef.fi

A. Tolvanen
Natural Resources Institute Finland (Luke), Paavo Havaksen tie 3, 90570 Oulu, Finland
e-mail: anne.tolvanen@luke.fi

M. Versluijs
The Helsinki Lab of Ornithology, Finnish Museum of Natural History, University of Helsinki, 00014 Helsinki, Finland
e-mail: martijnversluijs@hotmail.com

467
M. M. Girona et al. (eds.), *Boreal Forests in the Face of Climate Change*,
Advances in Global Change Research 74,
https://doi.org/10.1007/978-3-031-15988-6_18

18.1   Background

18.1.1   Natural Disturbance

Both large-scale and small-scale disturbances have shaped boreal forests. Large-scale disturbances include, for example, fire, windstorms, and insect outbreaks, all believed to be important forces structuring the boreal forest (Attiwill, 1994; Bonan & Shugart, 1989; Kuuluvainen & Aakala, 2011). Small-scale disturbances, such as gap dynamics, local flooding events, smaller windthrow events, and localized insect and fungi damage, contribute to creating a dynamic mosaic boreal landscape with many ecological niches (Berglund & Kuuluvainen, 2021). This spatiotemporal variability has structured boreal communities and maintains the typical biodiversity of these ecosystems (see Chap. 19 for details).

18.1.2   Forestry

Current forestry practices in boreal Fennoscandia are highly mechanized and dominated by even-aged forest management where the typical management unit, the forest stand, is most often a few hectares in size (Fig. 18.1). Active management that promotes conifers and actively removes deciduous trees during thinning has created homogeneous stands with reduced tree species diversity and has led to the loss of ancient trees. Changes in forest structure and dynamics can be seen as transforming formerly complex forest ecosystems characterized by considerable variations in habitat type, including vertical structure, tree species composition, age distribution, and deadwood dynamics, into simplified forest habitats (Esseen et al., 1997; Kuuluvainen, 2009). Commercially managed forests are also denser, have less variation in tree height, and are less permeable to sunlight than natural forests. For example, stand-level timber volumes in Sweden have increased 40%–80% since the 1950s (SLU, 2012), and this increase has led to an impoverished flora and fauna of species associated with sun-exposed conditions and deciduous broadleaf trees (Berg et al., 1994; Bernes, 2011).

Fire was the predominant large-scale disturbance in boreal forests; however, as observed in most areas of Fennoscandia, fire frequency has dropped dramatically during the past century because of effective fire-suppression measures (Zackrisson, 1977). For example, less than 0.02% of the forest area burns each year in Sweden compared with approximately 1% before CE 1900 (Granström, 2001; Zackrisson, 1977). Many boreal species are strongly favored by fire or prefer charred substrates (Granström & Schimmel, 1993), and some fire-associated species reproduce almost exclusively in burned forest, including many invertebrate species and fungi (Heikkala et al., 2017; Kouki & Salo, 2020).

The reduced habitat diversity is considered a key factor behind the species' decline in managed boreal forest ecosystems (Buddle et al., 2006; Hjältén et al., 2012;

Fig. 18.1   Forest stand subjected to clear-felling and stump harvest. *Photo credit* Jon Andersson

Jonsson et al., 2005; Kouki & Salo, 2020; Kuuluvainen, 2009; Paillet et al., 2010; Siitonen, 2001; Stenbacka et al., 2010). Forests are the most important habitat for red-listed and threatened species in Sweden and Finland. In Finland, 32% (2,133 species) of red-listed species are forest dwelling (Hyvärinen et al., 2019). Similarly, 43% (2,041 species) of the red-listed species in Sweden are forest dwelling (Artdata-banken, 2020). The species most negatively affected by silviculture are old-growth specialists dependent on a long forest continuity and old trees and species associated with deciduous trees (Artdatabanken, 2020; Bernes, 2011; Hyvärinen et al., 2019). Efforts to mitigate these adverse effects on biodiversity have been introduced to limit the harmful effects of the prevailing forestry practices on species and habitats.

18.1.3   Mitigation Strategies

Over the last three decades, Fennoscandia has experienced an increased interest in a forest management approach that aims to mitigate the negative effects of forestry on biodiversity. This change has come about through a combination of updated legislation, e.g., the Finnish Forest Act updated in 2014 and the Swedish Forestry Act updated in 1993, revised management recommendations in forestry, and higher consumer awareness that demands products from environmentally certified forestry, such as the Forest Stewardship Council (FSC) and the Programme for the Endorsement of Forest Certification (PEFC). Currently, the forest industry is required to apply a variety of conservation measures to improve conditions for biodiversity to fulfill certification demands and legal requirements (Johansson et al., 2013). In boreal regions, these measures include setting forest stands aside from ordinary forestry, leaving buffer zones of trees alongside wetlands and water bodies, leaving snags and logs on clear-cuts, and also actively creating deadwood in connection to final harvesting (Gustafsson et al., 2012; Johansson et al., 2013); the latter measure often occurs in the form of artificially created high stumps of trees. The prescribed burning of clear-cuts and, to some extent, standing forests are also included in the Swedish FSC standard (FSC, 2020). Although these efforts may increase the availability of vital forest habitat structures, they are likely insufficient to sustain viable populations of all forest-dwelling species (Johansson et al., 2013). Moreover, formally protected forests have increased in area in both Sweden and Finland (Hohti et al., 2019), albeit very slowly. Despite these efforts, the Swedish environmental objectives of "a rich diversity of plant and animal life" and "a living forest" are not being fulfilled, as shown by, for example, the high number of threatened species in the forest landscape (Naturvårdsverket, 2019). One reason for the lack of progress in biodiversity conservation may be that a large part of the currently protected forest areas was managed before becoming established as reserves. Consequently, they do not contain forest habitats or forest legacies that would prevail in a corresponding truly natural forest. For example, in Finland, about half of the protected forests in the southern part of the country are young, often less than 100 years old, and were intensively managed—including clear-cutting in many cases—before being established as reserves.

18.1.3.1 Why Restoration?

Since very few unmanaged forest habitats remain globally, including in Fennoscandia, conserving biodiversity can no longer rely on passive conservation measures, i.e., setting aside conservation areas under a free-development philosophy to reach conservation goals (Aronson & Alexander, 2013; Millennium Ecosystem Assessment, 2005). To achieve conservation goals, we require methods for restoring hitherto managed forest and applying an active management of forest reserves.

18.1.3.2 How to Restore?

It has been argued that reintroducing natural disturbances, referred to as *natural disturbance emulation* (NDE), is an ideal management approach when restoring natural systems (Attiwill, 1994; Kuuluvainen, 2002; Lindenmayer et al., 2006). In Fennoscandian boreal forests, appropriate NDE restoration efforts should include both large- and small-scale disturbances, e.g., introducing fire through prescribed burning and emulating gap dynamics by creating gaps in the canopy and generating coarse woody debris (CWD; Kuuluvainen, 2002). Both restoration methods accelerate the production and structural variability of CWD (Hekkala et al., 2016;

Kuuluvainen, 2002; Laarmann et al., 2013) and create more diversified forest habitats. The conceptual and practical aspects of NDE in the boreal forest are elaborated further in Chap 19.

A major challenge in forest restoration is that identifying and emulating natural disturbances is not always straightforward. These co-occur at different spatial and temporal scales, and it appears evident that natural disturbances per se can experience shifts in disturbance regimes, e.g., because of climate effects, or can present context-specific patterns related to soil or topographical factors. If specific natural disturbance processes have almost completely disappeared from managed forests, then restoring any of such features should be beneficial. Fire is an excellent example in this context for situations where fires have been completely suppressed from managed forests. Although it may be challenging to fully restore fire disturbances or fire regimes over large landscapes and at different time scales, the reintroduction of fire, even within small areas or in young forests (Hägglund et al., 2015; Hekkala et al., 2014a; Hjältén et al., 2017), can have rapid and beneficial effects on species. However, it is also clear that the benefits differ depending on the regional and local conditions (Kouki et al., 2012).

Additionally, as NDE is often introduced into landscapes containing both managed and protected areas, the actual restoration method may need to be adjusted accordingly. In situations where land-sharing prevails, a gradient of restoration methods can be implemented so that full NDE is likely only in the protected areas, whereas more nuanced measures may be more applicable to the managed parts of the landscape where timber production may continue to be the dominant land use. Overall, the difficulty of having a realistic NDE model (but see Chap. 19) and incorporating any existing limitations associated with prevailing land-use patterns and land-use history is that this quickly leads to applying a low- to high-intensity NDE gradient among the various landscapes. It is clear, however, that there are no general ecological principles or practical guidelines on how to achieve the optimal combination of different NDE methods in such a landscape mosaic. Achieving this requires a better understanding of specific case studies that can highlight how restoration can occur in an ecologically effective manner.

18.2   NDE of Large-Scale Disturbances: Prescribed Burning

Wildfires are major natural disturbances across the boreal region (Bonan & Shugart, 1989; Kouki et al., 2012; Kuuluvainen & Aakala, 2011). Because fires have been suppressed in many intensively managed landscapes, the reintroduction of fire is a promising method for NDE, and prescribed burning is required by the Swedish and Finnish FSC certification standards (FSC, 2020). Relative to many other restoration methods, prescribed burning is generally technically more challenging to apply. For example, prescribed restoration burns require large numbers of skilled fire managers and for the fires to be set during specific weather conditions. Furthermore, prescribed fires always involve a safety risk, and it is also not exactly certain how prescribed burns should be conducted to mimic natural disturbance conditions.

Prescribed burning of clear-cut areas is a traditional management method in forestry in Fennoscandia (Fig. 18.2). Its primary purpose is to modify soil properties and promote the establishment of a new tree cohort; however, the method was abandoned because of pest- and pathogen-related damage and high labor costs. Therefore, this technique was replaced by mechanical site preparation methods (Löf et al., 2015). Methods of prescribed burning for ecological restoration vary and involve different levels of tree retention. A few recent experiments have explored the effects of prescribed burning on biodiversity patterns. Most of these studies have included various types or intensities of tree harvests combined with prescribed burns; however, some studies also included comparisons with other restoration methods. The consequences of prescribed burns have been monitored, at least, for birds, beetles and other invertebrates, wood-associated and other macrofungi, vascular plants, bryophytes and lichens, and tree seedlings. The treated forest stands typically cover 2 to 25 ha and can be regarded as large-scale experiments in a Fennoscandian context.

Fig. 18.2 Prescribed burn of a forest stand as part of an ecological restoration experiment in Sweden (see e.g., Hägglund et al., 2020; Hjältén et al., 2017; Versluijs et al., 2017) *Photo credit* Joakim Hjältén

18.2.1   *Response of Insects and Fungi*

The effect of fire on biodiversity patterns and forest dynamics is generally always strong and immediate. For example, beetle assemblages are altered dramatically when a forest is burned. This change is evident regardless of the level of harvesting (Hyvärinen et al., 2005) or the amount of fuelwood created (Hekkala et al., 2014a). Notably, the use of fire appears to favor rare and threatened coleopteran species. For example, Hyvärinen et al. (2006) found that a forest stand burned in Finland immediately harbored about four times more rare or threatened beetle species than comparable unburned forest stands. Thus, fire has a significant biodiversity conservation effect, as these species are usually the rarest of all threatened species and are in the most urgent need of conservation actions. Hekkala et al. (2014a) observed, however, that the initial and rapid increase in the richness of saproxylic and fire-dependent beetle species declined to pretreatment levels only a few years after a prescribed burn. Thus, they suggest that fire should be introduced into neighboring areas at five-year intervals to maintain populations of the most fire-dependent pyrophilous species (Hekkala et al., 2014a).

The prescribed burning of spruce-dominated forests in Sweden also revealed a strong short-term effect on saproxylic assemblages and an increase in species richness and abundance of several functional groups of beetles and flat bugs (Hägglund et al., 2015, 2020; Hjältén et al., 2017). Fire-favoring and fire-dependent beetles and flat bugs benefit in particular from prescribed burns (Hägglund et al., 2015, 2020). Contrary to the results from Finland, however, no strong short-term effect on red-listed species has been detected. A possible explanation for these differing outcomes is that the effects of prescribed burns depend on landscape quality. In landscapes with a long history of intensive management and fire suppression, the insect community may be impoverished, making it more difficult for threatened fire-dependent species to find and colonize burned sites (Johansson et al., 2013; Kouki et al., 2012). This underlines the importance of considering both temporal aspects and landscape in restoration planning.

Prescribed burns of coniferous stands affect tree mortality and thus modify the dynamics of resources available for species (Hämäläinen et al., 2016; Heikkala et al., 2014). A decadal follow-up study of the same sites of Hyvärinen et al. (2006) showed that beetle assemblages remained more diverse on burned sites (Heikkala et al., 2016). However, several obligate fire-associated species were very ephemeral in their occurrence within the burned stands. Flat bugs are good examples of this phenomenon. They were observed to efficiently colonize the burned forests (Hägglund et al., 2015; Heikkala et al., 2017), but they also disappeared only a few years after the fire (Fig. 18.3; Heikkala et al., 2017).

Macrofungi also presented several similar fire-associated species taxa that colonized quickly after a fire but then also disappeared rapidly from the assemblages, and several soil fungi were noted during the initial years after a wildfire or prescribed burn (Salo & Kouki, 2018; Salo et al., 2019). Contrary to soil fungi, wood-associated fungi responded to fire over a much more extended period and typically required a

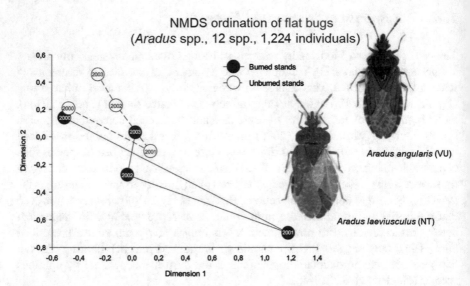

Fig. 18.3 Assemblage dynamics of fire-associated flat bugs (*Aradus* spp.) after forest harvests with burning (*black circles*) and after harvests without burning (*white circles*). Analysis is based on nonmetric multidimensional scaling (NMDS); assemblages sharing a similar composition are clustered together. *Circles* also include survey years: pretreatment (*2000*) of harvests and burning) and post-treatment survey years (*2001–2003*). Fire-associated species quickly colonized the burned areas in 2001; however, these species also disappeared quickly in 2002–2003, and the assemblage became similar to unburned sites. Modified with permission from John Wiley and Sons from Heikkala et al. (2017). *Photo credits* Petri Martikainen

decade or longer to become established (Junninen et al., 2008; Salo & Kouki, 2018; Suominen et al., 2015). The effect of fire was nevertheless very evident for these species. For example, even the stumps of harvested trees maintained a higher species richness for wood-associated fungi when the stumps were burned (Suominen et al., 2018).

18.2.2 Response of Vegetation and Pollinators

Boreal forest vegetation, dominated by coniferous trees (*Picea abies* (L.) H. Karst. and *Pinus sylvestris* L.) and dwarf shrubs such as *Calluna* spp., *Vaccinium vitis-idaea* L., and *V. myrtillus* L., is highly resilient and adapted to recurrent natural disturbances (Rydgren et al., 2004; Zackrisson, 1977). Depending on their intensity and magnitude, disturbances can affect the composition of the vegetation so that it resembles earlier stages along the successional path. Fire is the most intense natural disturbance, which may remove late-successional dwarf shrubs, mosses, lichens, and trees and replace them with seed-dispersing birch and other pioneer species (Schimmel & Granström, 1996). Restoration experiments of varying fire intensity confirm this pattern and show that severe prescribed burning is most effective at initiating natural vegetation succession, whereas tree felling varies in its impact on the vegetation composition depending on the number of felled trees (Espinosa del Alba et al., 2021; Hekkala et al., 2014b; Johnson et al., 2014; Tatsumi et al., 2020). The effect of fire appears very organism dependent, and the time for recovery after disturbance can vary accordingly. Espinosa del Alba et al. (2021) show that ground-living bryophytes are severely adversely affected by prescribed burning and that the bryophyte community had yet to recover eight years after a fire. In contrast, after an initial decrease, the species richness of vascular plants was greater eight years postfire than pretreatment richness. However, epiphytic lichens also appear to be very sensitive to a fire's direct heat and burn effects (Hämäläinen et al., 2016). If such species groups occur in an area planned for fire restoration, special attention must be paid to the design and execution of this intervention to avoid risks to rare and threatened species. Unlike invertebrates, few plant species depend on fire in Fennoscandian boreal forests, although given the absence of natural fires, these species are increasingly threatened. Examples include the annual herbs *Geranium bohemicum* L. and *G. lanuginosum* Lam., which require high temperatures for their seeds to germinate.

Pollinators are expected to respond to vegetational changes due to fire or other disturbances (Rodríguez & Kouki, 2017). In a Finnish study, parasitoids—potential regulators of eruptive species—and pollinators of major forest dwarf shrubs bilberry (*Vaccinium myrtillus*) and lingonberry (*V. vitis-idaea*) were more diverse after a prescribed burn (Rodríguez et al., 2019). Prescribed burns expose mineral soils that provide sites for pollinators' nests. In addition, leaving dead trees or producing deadwood during prescribed burns provides nesting sites for pollinators. Prescribed burning can therefore be important for maintaining forest ecosystem functioning and providing ecosystem services continuously.

18.2.3 Response of Birds

The presence and distribution of forest birds are largely associated with stand-scale habitat structures, such as tree species diversity, the quantity of deciduous trees, the quantity of deadwood, and understory density (Hurlbert, 2004). Forest fires create various vital structures because there is a postfire shift toward an early-successional stage of the vegetation (Schimmel & Granström, 1996), the creation of deadwood, and an enhanced regeneration of deciduous trees, particularly aspen (*Populus tremula* L.) (Hekkala et al., 2014b). Nevertheless, changes in bird assemblages after prescribed burns have rarely been studied, possibly because burned areas tend to be small relative to the general habitat requirements of birds and other vertebrates. One of these rare prescribed burn–bird studies investigated bird assemblage changes after a prescribed burn in northern Sweden (Versluijs et al., 2017). Prescribed burning created habitat for long-distance migrants, ground breeders, and species preferring early-successional habitats. Moreover, Versluijs et al. (2020) showed that prescribed burns represent an effective means of fostering a rapid and long-lasting enrichment of important forest structures for woodpeckers. This benefit to woodpeckers is caused mainly by the large numbers of killed and weakened trees, which facilitates their colonization by saproxylic insect populations (Kärvemo et al., 2017; Morissette et al., 2002). Phloem sap from fire-damaged Scots pine has also been shown to provide instant foraging opportunities for Three-toed Woodpeckers (*Picoides tridactylus*) (Pakkala et al., 2017). In the short-term, fire decreases the abundance of healthy trees and reduces understory density. These stands normally constitute important breeding and feeding habitats for birds preferring early-successional habitat; thus, off-ground breeders and species closely connected with mature forest occurred in lower numbers (Versluijs et al., 2017).

Most other studies on this topic have only explored the responses of birds to wild-fire. A study from northern Sweden found that wildfire positively affected ground-feeding insectivorous species (Edenius, 2011). Similarly, several studies from other forest systems showed that fire clearly benefits numerous bird species (Clavero et al., 2011; Hutto, 1995; Lowe et al., 2012). Although prescribed burns should mimic wildfire, it is unknown whether prescribed fires provoke the same response in bird assemblages as natural fires. The main difference is that, in most cases, mixed-severity wildfires produce a mosaic of variably burned areas (Salo & Kouki, 2018), whereas prescribed burns often result in low-intensity fires. Several studies have shown that fire intensity is also an important variable affecting bird responses to fire (Hutto & Patterson, 2016; Lindenmayer et al., 2014).

18.2.4   Management Considerations

Fire severity is important to consider when targeting expected biodiversity responses. Fire in the NDE of managed landscapes is often applied in a spectrum of severity so that the amount of timber left on the burned areas varies across a given site. In nature conservation areas where timber is not typically removed, the abundance of cut trees can be altered to modify the quantity of burning load and, subsequently, the amount of burned wood (Hekkala et al., 2014a, 2016). Although the use of fire represents an effective tool for restoring lost properties of a boreal stand, it is also evident that all aspects of wildfires are hard to emulate in a controlled fashion. Above all, if too few trees are left, there is unlikely a local continuity to the structures created by fire (Hämäläinen et al., 2016; Heikkala et al., 2014; Hyvärinen et al., 2005). Second, wildfires vary in severity, and the local variation in fire severity has significant consequences on biodiversity (Salo & Kouki, 2018; Salo et al., 2019;). In prescribed burns, fire severity often remains or is actively kept at a low level because of the risks associated with high-severity fires, which are typically canopy-destroying or stand-replacing fires. Despite these shortcomings, prescribed burns effectively restore lost forest properties and enhance biodiversity across landscapes that include both managed and protected sites.

18.3 NDE of Small-Scale Disturbances: Gap Cuttings and Deadwood Creation

18.3.1 Response of Insect and Fungi

Gap cuttings have been implemented to reduce the adverse effects of clear-felling on biodiversity and ecological processes by reducing clear-cut size. However, they also serve as a direct ecological restoration method that mimics gap dynamics and small-scale disturbances, e.g., windthrow and localized insect outbreaks (Fig. 18.4). Pasanen et al. (2016) found that gaps and the deadwood in gaps diversified wood-associated fungi five years after gap creation. On the other hand, Hägglund et al. (2020) found that ecological restoration involving the creation of small gaps (20 m in diameter) and deadwood had no significant effect on the overall stand-level species richness of beetles; however, gap-cut stands had a higher species richness for cambivores and known fire-favoring species than observed within reference stands. Moreover, coleopteran species composition differed significantly between stand types. A marginal increase of flat bugs has also been observed after gap cutting (Hägglund et al., 2015). Joelsson et al. (2018) confirmed the importance of stand heterogeneity for insect diversity by showing that harvest trails supported a different beetle assemblage than the surrounding intact forest. A likely explanation for these patterns is that the degree of sun exposure on deadwood has a strong effect on the saproxylic assemblage colonizing the deadwood (Hjältén et al., 2012; Lindhe et al., 2005; Seibold et al., 2016). This effect is potentially mediated by changes in the fungal community, as sun exposure also strongly determines the fungal composition and fungal growth rate in deadwood (Bouget & Duelli, 2004). Consistent with these observations, deadwood created in gaps favors numerous saproxylic beetles, including some fire-favored and fire-dependent species (Hägglund & Hjältén, 2018), suggesting these gaps attract species associated with more open-forest habitats. However, Hägglund and Hjältén (2018) also found significant differences in beetle assemblages in deadwood because of tree species and stature (standing or downed logs), consistent with earlier findings (Hjältén et al., 2012; Seibold et al., 2016). Pasanen et al. (2014) also observed that although wood fungi diversity was enhanced by gaps, red-listed wood fungi did not occur in gaps, most likely because of the lack of qualitatively suitable deadwood in the gaps during the five years of the study. These observations suggest that a high diversity of deadwood forms and quality must be available within the landscape to maintain saproxylic biodiversity (Penttilä et al., 2004; Similä et al., 2003).

18.3.2 Response of Birds

A bird study in the same stands in Sweden did not support the prediction that gaps attract species found in more open-forest habitats (Versluijs et al., 2017). They found that gap cutting did not affect bird assemblages; this pattern likely relates to the

Fig. 18.4 Gap cutting includes creating deadwood as part of an ecological restoration experiment in Sweden (see e.g., Hägglund et al., 2020; Hjältén et al., 2017; Versluijs et al., 2017) *Photo credit* Joakim Hjältén

combined effect of too-small gaps to attract open-area or edge specialists and a lack of response in the understory vegetation. Forsman et al. (2013), studying larger gaps, also did not find any general effect of gap disturbance on the overall abundance and richness of boreal-forest bird species. This could suggest that organisms such as birds that have larger home ranges, a larger spatial scale must be considered for restoration efforts and subsequent assessment (Hof & Hjältén, 2018).

18.3.3   Response of Vegetation

The documented effects of gap cutting on vascular plants provide ample theoretical support for this intervention being beneficial for species diversity per the intermediate disturbance hypothesis (Connell, 1978), which states that disturbances of intermediate frequency and severity maintain higher levels of diversity. Gap cutting could, in this sense, be viewed as an intermediate severity disturbance. However, the scientific literature is rather scarce for empirical studies on this topic (Eckerter et al., 2019). When gaps are formed in the canopy following tree felling or natural disturbances, light penetration is increased on the forest floor. In North American studies, thinning or partial cutting increases the total cover of vegetation and understory species diversity (Burke et al., 2008; Thomas et al., 1999). One of the few studies from Fennoscandia demonstrated that felling 20%–40% of the initial stand volume does not affect the understory vegetation up to seven years after treatment (Hekkala et al.,

2014b). However, the uprooting of trees—to simulate storm felling—increased the species richness of vascular plants (Hekkala et al., 2014b). The authors conclude that the exposure of soil from the uprooting increases microsite heterogeneity and, there-fore, greater habitat availability for pioneer seeds. The effects on vegetation have also been studied in the abovementioned Swedish experiment (Hägglund et al., 2020). However, Espinosa del Alba et al. (2021) found that 20 m diameter gaps had no signif-icant impact on species richness or the composition of vascular plant assemblages or ground-living bryophyte assemblages up to eight years post-treatment.

In principle, canopy gaps also provide sites for tree-seedling regeneration and, thus, maintain continuous cover forests that appear as typical natural boreal land-scapes. However, canopy gaps alone may be insufficient to facilitate regeneration unless the soils are also disturbed (Pasanen et al., 2016). Seed germination and seedling establishment in boreal forests generally require exposing the mineral soil to alleviate competition with the dense understory vegetation (Eriksson & Fröborg, 1996; Hautala et al., 2001, 2008). Moreover, restoration studies indicate that simu-lated storms that expose the soil through tree uprooting increase species diversity and the number of tree seedlings more than restoration by only cutting trees (Hekkala et al., 2014b). Pasanen et al. (2016) reported a low overall establishment of pine trees in canopy gaps despite a good initial regeneration rate. The lack of long-term success may have been caused by intensified root competition even though the soil was slightly modified in this experiment. Therefore, the use of fire in combination with small gap creation may enhance recruitment (Pasanen et al., 2015).

18.4   NDE: Restoration of Deciduous Forest Stands

In Sweden, stand-level timber volumes have increased 40%–80% since the 1950s (SLU 2012) owing to an increased production of conifers at the expense of broadleaf trees that are disfavored by modern forestry, e.g., during thinning. Commercial forests are therefore denser and less permeable to sunlight. Broadleaf trees are also disad-vantaged when natural disturbance regimes, such as recurrent wildfires in upland forests and seasonal floods in riparian environments, are suppressed or altered (Hell-berg, 2004; Johansson & Nilsson, 2002; Linder et al., 1997). These changes have led to an impoverished fauna of species associated with sun-exposed conditions and broadleaf trees (Bernes, 2011). Additionally, the abundance of large deciduous trees may also decline in protected areas, probably because these areas in Fennoscandia are often too small to sustain natural disturbance regimes (Hardenbol et al., 2020). Restoring broadleaf stands is therefore instrumental for biodiversity conservation.

During the last decades, large areas (much greater than 10,000 ha) have been restored in Sweden to benefit the White-backed Woodpecker (*Dendrocopos leucotos*), a critically endangered species with a population consisting of only a handful of breeding pairs. This species was once widespread throughout most of Sweden but declined rapidly during the past century because of intensified forest management (Aulén, 1988; Stighäll et al., 2011). To restore habitats for the White-backed Woodpecker, forest managers in Sweden and Finland have created deadwood from broadleaf trees and selectively harvested spruce trees to open up forests and make deciduous trees more competitive (Blicharska et al., 2014; Hämäläinen et al., 2020). The White-backed Woodpecker has not yet recovered, but other less area-demanding and fast-responding species having similar habitat requirements have benefited from these restoration actions. Bell et al. (2015) found that the species richness of saproxylic beetles associated with deciduous deadwood and greater sun exposure was higher in the restored stands than unrestored ones, as were red-listed saproxylic beetle species. In addition, the availability of suitable insect food for White-backed Woodpeckers increased in restored areas, suggesting that when a sufficient area has been restored, the area-demanding White-backed Woodpecker can recover (Hof & Hjältén, 2018); nonetheless, the response at lower trophic levels are stronger indicators of ecosystem recovery (Fig. 18.5).

For a wide range of bird species, the occurrence of large-diameter deciduous trees is a critical habitat component. Although there is not much known about how restoring broadleaf stands influences bird assemblages, habitat specialists such as the White-backed Woodpecker are favored by an increased availability of deciduous trees. Aspen (*Populus tremula* L.) is particularly preferred as a nesting tree (Angel-stam & Mikusiński, 1994) and is frequently used for foraging by the White-backed Woodpecker (Stenberg & Hogstad, 2004). Additionally, the presence of this wood-pecker indicates a high species richness for forest birds, red-listed cryptogams, and

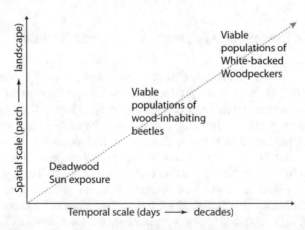

Fig. 18.5 Hypothetical example of the spatiotemporal-scaled response of organism groups, differing in their spatial requirements and reproduction rates, to the restoration of deciduous stands. Many species are resource and process limited, albeit at different spatial and temporal scales. Local restoration will not necessarily fulfill the habitat requirements of top predators, such as the White-backed Woodpecker; however, less area-demanding species, e.g., many saproxylic beetles, respond more rapidly. Forest restoration likely produces a bottom-up effect on top predators within saprox-ylic food webs. Under such circumstances, the recovery of umbrella species could testify to a full ecosystem recovery. The recovery of species at lower levels in the food chain could provide robust indicators of the onset of ecosystem recovery. Modified from Bell et al. (2015), CC BY 3.0 license

saproxylic beetles (Bell et al., 2015; Mikusiński et al., 2001; Roberge et al., 2008). Different species of deciduous trees also contribute to a high variability in saproxylic beetles, as beetle composition differs between aspen and birch stands. High variation in deciduous tree species, age, and deadwood at different decay stages will positively influence other bird species, especially bark-feeding and secondary cavity nesters. Eggers and Low (2014) observed that 83% of Willow Tits (*Poecile montanus*) excavate cavities mainly in birch and that the diameter of the nesting tree at nest height has a positive relationship with nest survival. Restoring the diversity and abundance of deciduous trees in boreal forests is thus likely crucial for the conservation of boreal-forest birds, in particular as deciduous trees improve foraging and breeding opportunities.

18.5   Risks Associated with Ecological Restoration

Prescribed burns in spruce-dominated forests in southern Finland have increased attacks by bark beetles (*Ips typographus*, *Pityogenes chalcographus*), although the harmful effects on tree survival in neighboring forests have generally been low (Eriksson et al., 2006). This observation suggests that restored areas do not provide significant refugia for the bark beetle populations unless restoration actions are repeated over consecutive years within a small area, allowing for bark beetle populations to build over time (Toivanen et al., 2009). Prescribed burns in Swedish spruce forests produced similar results, with a marked but short-lived increase in bark beetle abundance. Bark beetle densities had already decreased dramatically in the second year postfire, and five years after burning, the bark beetle densities were lower than those in the control areas, although the abundance of the natural predators of bark beetles was greater than in the controls (Hekkala et al., 2021; Kärvemo et al., 2017).

Tomicus spp., pine shoot beetles, are potentially harmful pests that may reduce the growth of Scots pine, although they usually do not kill healthy trees. In small gaps within pine forests, restoring deadwood increased *Tomicus* bark beetle numbers; however, these Coleoptera did not spread into adjacent forests, showing less than a few tens of meters of incursion into these neighboring sites. Thus, the effect was highly localized to the immediate neighborhood of restored sites. Additionally, the eruptive phase of *Tomicus* and the effects on adjacent trees typically last only a couple of years (Komonen & Kouki, 2008; Komonen et al., 2009; Martikainen et al., 2006). These observations suggest that restoring deadwood in pine forests does not increase the risk of bark beetle–related damage.

Besides the potential risk of pest outbreaks, there is also a risk of adverse effects on nontarget species. There is ample evidence that prescribed burns can harm species associated with old-growth forests and long forest continuity. For some species groups, such as epiphytic lichens and bryophytes, it is clear that they are susceptible to the direct heat and burn effects of fire (Hämäläinen et al., 2014). However, species from many other groups are disfavored by postfire conditions, and the adverse effects on some beetle groups may be transitional (Hyvärinen et al., 2009). Particular attention should be paid to the design and execution of prescribed burns so that rare and threatened species are not disfavored.

18.6   Conclusions

Ample evidence exists that ecological restoration within a NDE framework benefits biodiversity. However, there remain considerable gaps in our knowledge regarding the effect of different restoration methods on specific taxa and the duration of restoration benefits. Furthermore, most studies assessing the effects of NDE incorporating large-scale disturbances such as fire have been conducted at the plot or stand scale, whereas our knowledge of landscape-scale effects remains very limited. Thus, there is an urgent need to study the landscape-scale effect of ecological restoration (but see Kouki et al., 2012). Most studies have investigated the effects of large-scale disturbances associated with NDE—generally how prescribed burning affects biodiversity—whereas our knowledge of the impact of NDE on small-scale disturbances and the restoration of deciduous forest stands is more limited. Prescribed burns benefit many fire-adapted species; however, the restoration outcome depends on fire severity and landscape properties, including management history. The more limited number of assessments of deciduous forest stands suggests that restoration benefits light-demanding species associated with deciduous trees. However, more studies assessing this type of restoration and the response of different taxa are needed. The effect of gap cutting on biodiversity appears weak, and outcomes vary among studies and taxa. Moreover, the number of studies that have evaluated the impacts of gap cutting remains low, and the applied restoration methods differ among these studies, highlighting the need for additional and more comparative studies. Overall, the active restoration of critical habitats and substrates appears to be the only feasible way of alleviating and reducing the ongoing and projected biodiversity loss in degraded forest landscapes. Relying on passive restoration, i.e., waiting for natural structures to reappear through natural successional processes, is a painfully slow means of mitigating the rapidly advancing threat to forest biodiversity.

References

Angelstam, P., & Mikusiński, G. (1994). Woodpecker assemblages in natural and managed boreal and hemiboreal forest—a review. *Annales Zoologici Fennici, 31*, 157–172.

Aronson, J., & Alexander, S. (2013). Ecosystem restoration is now a global priority: Time to roll up our sleeves. *Restoration Ecology, 21*(3), 293–296. https://doi.org/10.1111/rec.12011.

Artdatabanken. (2020). *The Swedish redlist*. Rodlistade arter i Sverige. Uppsala: ArtDatabanken SLU

Attiwill, P. M. (1994). The disturbance of forest ecosystems: The ecological basis for conservative management. *Forest Ecology and Management, 63*, 247–300. https://doi.org/10.1016/0378-112 7(94)90114-7.

Bell, D., Hjalten, J., Nilsson, C., et al. (2015). Forest restoration to attract a putative umbrella species, the white-backed woodpecker, benefited saproxylic beetles. *Ecosphere, 6*(12), 278. https://doi.org/10.1890/os14 00551.1.

Berg, Å., Ehnström, B., Gustafsson, L., et al. (1994). Threatened plant, animal, and fungus species in Swedish forests: Distribution and habitat associations. *Conservation Biology, 8*(3), 718–731. https://doi.org/10.1046/j.1523-1739.1994.08030718.x.

Berglund, H., & Kuuluvainen, T. (2021). Representative boreal forest habitats in northern Europe, and a revised model for ecosystem management and biodiversity conservation. *Ambio, 50*, 1003–1017. https://doi.org/10.1007/s13280-020-01444-3.

Bernes, C. (2011). *Biodiversity in Sweden. Monitor 22*. Stockholm: Swedish Environmental Protection Agency.

Blicharska, M., Baxter, P., & Mikusiński, G. (2014). Practical implementation of species' recovery plans—lessons from the White-backed Woodpecker Action Plan in Sweden. *Ornis Fennica, 91*(2), 108–128.

Bonan, G. B., & Shugart, H. H. (1989). Environmental-factors and ecological processes in boreal forests. *Annual Review of Ecology and Systematics, 20*, 1–28. https://doi.org/10.1146/annurev. es.20.110189.000245.

Bouget, C., & Duelli, P. (2004). The effects of windthrow on forest insect communities: A literature review. *Biological Conservation, 118*, 281–299. https://doi.org/10.1016/j.biocon.2003.09.009.

Buddle, C. M., Langor, D. W., Pohl, G. R., et al. (2006). Arthropod responses to harvesting and wildfire: Implications for emulation of natural disturbance in forest management. *Biological Conservation, 128*(3), 346–357. https://doi.org/10.1016/j.biocon.2005.10.002.

Burke, D. A., Elliott, K. A., Holmes, S. B., et al. (2008). The effects of partial harvest on understory vegetation of southern Ontario woodlands. *Forest Ecology and Management, 255*(7), 2204–2212. https://doi.org/10.1016/j.foreco.2007.12.032.

Clavero, M., Brotons, L., & Herrando, S. (2011). Bird community specialization, bird conservation and disturbance: The role of wildfires. *Journal of Animal Ecology, 80*(1), 128–136. https://doi.org/10.1111/j.1365-2656.2010.01748.x.

Connell, J. H. (1978). Diversity in tropical rain forests and coral reefs. *Science, 199*, 1302–1310. https://doi.org/10.1126/science.199.4335.1302.

Eckerter, T., Buse, J., Forschler, M., et al. (2019). Additive positive effects of canopy openness on European bilberry (*Vaccinium myrtillus*) fruit quantity and quality. *Forest Ecology and Management, 433*, 122–130. https://doi.org/10.1016/j.foreco.2018.10.059.

Edenius, L. (2011). Short-term effects of wildfire on bird assemblages in old pine- and spruce-dominated forests in northern Sweden. *Ornis Fennica, 88*, 71–79.

Eggers, S., & Low, M. (2014). Differential demographic responses of sympatric Parids to vegetation management in boreal forest. *Forest Ecology and Management, 319*, 169–175. https://doi.org/10.1016/j.foreco.2014.02.019.

Eriksson, M., Lilja, S., & Roininen, H. (2006). Dead wood creation and restoration burning: Implications for bark beetles and beetle induced tree deaths. *Forest Ecology and Management, 231*(1–3), 205–213. https://doi.org/10.1016/j.foreco.2006.05.050.

Eriksson, O., & Fröborg, H. (1996). "Windows of opportunity" for recruitment in long-lived clonal plants: Experimental studies of seedling establishment in *Vaccinium* shrubs. *Canadian Journal of Botany, 74*(9), 1369–1374. https://doi.org/10.1139/b96-166.

Espinosa del Alba, C., Hjältén, J., & Sjögren, J. (2021). Restoration strategies in boreal forests: Differing field and ground layer response to ecological restoration by burning and gap cutting. *Forest Ecology and Management, 494*, 119357. https://doi.org/10.1016/j.foreco.2021.119357.

Esseen, P.-A., Ehnström, B., Ericson, L., et al. (1997). Boreal forests. *Ecological Bulletins, 46*, 16–47.

Forest Stewardship Council. (2020). *The FSC National Forest Stewardship Standard of Sweden FSC-STD-SWE-03–2019*. Bonn: Forest Stewardship Council.

Forsman, J. T., Reunanen, P., Jokimäki, J., et al. (2013). Effects of canopy gap disturbance on forest birds in boreal forests. *Annales Zoologici Fennici, 50*(5), 316–326. https://doi.org/10.5735/085.050.0506.

Granström, A. (2001). Fire management for biodiversity in the European boreal forest. *Scandinavian Journal of Forest Research, 16*(sup003), 62–69. https://doi.org/10.1080/028275801300090627.

Granström, A., & Schimmel, J. (1993). Heat effects on seeds and rhizomes of a selection of boreal forest plants and potential reaction to fire. *Oecologia, 94*(3), 307–313. https://doi.org/10.1007/BF00317103.

Gustafsson, L., Baker, S. C., Bauhus, J., et al. (2012). Retention forestry to maintain multifunctional forests: A world perspective. *BioScience, 62*(7), 633–645. https://doi.org/10.1525/bio.2012.62.7.6.

Hägglund, R., & Hjältén, J. (2018). Substrate specific restoration promotes saproxylic beetle diversity in boreal forest set-asides. *Forest Ecology and Management, 425*, 45–58. https://doi.org/10.1016/j.foreco.2018.05.019.

Hägglund, R., Hekkala, A. M., Hjältén, J., et al. (2015). Positive effects of ecological restoration on rare and threatened flat bugs (Heteroptera: Aradidae). *Journal of Insect Conservation, 19*(6), 1089–1099. https://doi.org/10.1007/s10841-015-9824-z.

Hägglund, R., Dynesius, M., Löfroth, T., et al. (2020). Restoration measures emulating natural disturbances alter beetle assemblages in boreal forest. *Forest Ecology and Management, 462*, 117934. https://doi.org/10.1016/j.foreco.2020.117934.

Hämäläinen, A., Kouki, J., & Löhmus, P. (2014). The value of retained Scots pines and their dead wood legacies for lichen diversity in clear-cut forests: The effects of retention level and prescribed burning. *Forest Ecology and Management, 324*, 89–100. https://doi.org/10.1016/j.foreco.2014.04.016.

Hämäläinen, A., Hujo, M., Heikkala, O., et al. (2016). Retention tree characteristics have major influence on the post-harvest tree mortality and availability of coarse woody debris in clear-cut areas. *Forest Ecology and Management, 369*, 66–73. https://doi.org/10.1016/j.foreco.2016.03.037.

Hämäläinen, K., Junninen, K., Halme, P., et al. (2020). Managing conservation values of protected sites: How to maintain deciduous trees in white-backed woodpecker territories. *Forest Ecology and Management, 461*, 117946. https://doi.org/10.1016/j.foreco.2020.117946.

Hardenbol, A. A., Junninen, K., & Kouki, J. (2020). A key tree species for forest biodiversity, European aspen (*Populus tremula*), is rapidly declining in boreal old-growth forest reserves. *Forest Ecology and Management, 462*, 118009. https://doi.org/10.1016/j.foreco.2020.118009.

Hautala, H., Tolvanen, A., & Nuortila, C. (2001). Regeneration strategies of dominant boreal forest dwarf shrubs in response to selective removal of understorey layers. *Journal of Vegetation Science, 12*(4), 503–510. https://doi.org/10.2307/3237002.

Hautala, H., Tolvanen, A., & Nuortila, C. (2008). Recovery of pristine boreal forest floor community after selective removal of understorey, ground and humus layers. *Plant Ecology, 194*(2), 273–282. https://doi.org/10.1007/s11258-007-9290-0.

Heikkala, O., Suominen, M., Junninen, K., et al. (2014). Effects of retention level and fire on retention tree dynamics in boreal forests. *Forest Ecology and Management, 328*, 193–201. https://doi.org/10.1016/j.foreco.2014.05.022.

Heikkala, O., Martikainen, P., & Kouki, J. (2016). Decadal effects of emulating natural disturbances in forest management on saproxylic beetle assemblages. *Biological Conservation, 194*, 39–47. https://doi.org/10.1016/j.biocon.2015.12.002.

Heikkala, O., Martikainen, P., & Kouki, J. (2017). Prescribed burning is an effective and quick method to conserve rare pyrophilous forest-dwelling flat bugs. *Insect Conservation Diversity, 10*(1), 32–41. https://doi.org/10.1111/icad.12195.

Hekkala, A. M., Paatalo, M. L., Tarvainen, O., et al. (2014a). Restoration of young forests in eastern Finland: Benefits for saproxylic beetles (Coleoptera). *Restoration Ecology, 22*(2), 151–159. https://doi.org/10.1111/rec.12050.

Hekkala, A. M., Tarvainen, O., & Tolvanen, A. (2014b). Dynamics of understory vegetation after restoration of natural characteristics in the boreal forests in Finland. *Forest Ecology and Management, 330*, 55–66. https://doi.org/10.1016/j.foreco.2014.07.001.

Hekkala, A. M., Ahtikoski, A., Paatalo, M. L., et al. (2016). Restoring volume, diversity and continuity of deadwood in boreal forests. *Biodiversity and Conservation, 25*(6), 1107–1132. https://doi.org/10.1007/s10531-016-1112-z.

Hekkala, A. M., Kärvemo, S., Versluijs, M., et al. (2021). Ecological restoration for biodiversity conservation triggers response of bark beetle pests and their natural predators. *Forestry, 94*(1), 115–126. https://doi.org/10.1093/forestry/cpaa016.

Hellberg, E. (2004). *Historical variability of deciduous trees and deciduous forests in northern Sweden. Effects of forest fires, land-use and climate.* Ph.D. thesis, Swedish University of Agricultural Sciences.

Hjältén, J., Stenbacka, F., Pettersson, R. B., et al. (2012). Micro and macro-habitat associations in saproxylic beetles: Implications for biodiversity management. *PLoS ONE, 7*(7), e41100. https://doi.org/10.1371/journal.pone.0041100.

Hjältén, J., Hägglund, R., Löfroth, T., et al. (2017). Forest restoration by burning and gap cutting of voluntary set-asides yield distinct immediate effects on saproxylic beetles. *Biodiversity and Conservation, 26*(7), 1623–1640. https://doi.org/10.1007/s10531-017-1321-0.

Hof, A. R., & Hjältén, J. (2018). Are we restoring enough? Simulating impacts of restoration efforts on the suitability of forest landscapes for a locally critically endangered umbrella species. *Restoration Ecology, 26*(4), 740–750. https://doi.org/10.1111/rec.12628.

Hohti, J., Halme, P., & Hjelt, M., et al. (2019). *Ten years of METSO—An interim review of the first decade of the Forest Biodiversity Programme for Southern Finland.* Helsinki: Publications of the Ministry of Environment

Hurlbert, A. H. (2004). Species–energy relationships and habitat complexity in bird communities. *Ecology Letters, 7*(8), 714–720. https://doi.org/10.1111/j.1461-0248.2004.00630.x.

Hutto, R. L. (1995). Composition of bird communities following stand-replacement fires in northern Rocky Mountain (U.S.A.) conifer forests. *Conservation Biology, 9*(5), 1041–1058. https://doi.org/10.1046/j.1523-1739.1995.9051033.x-i1.

Hutto, R. L., & Patterson, D. A. (2016). Positive effects of fire on birds may appear only under narrow combinations of fire severity and time-since-fire. *International Journal of Wildland Fire, 25*(10), 1074–1085. https://doi.org/10.1071/WF15228.

Hyvärinen, E., Kouki, J., & Martikainen, P. (2006). Fire and green-tree retention in conservation of red-listed and rare deadwood-dependent beetles in Finnish boreal forests. *Conservation Biology, 20*(6), 1711–1719. https://doi.org/10.1111/j.1523-1739.2006.00511.x.

Hyvärinen, E., Kouki, J., Martikainen, P., et al. (2005). Short-term effects of controlled burning and green-tree retention on beetle (Coleoptera) assemblages in managed boreal forests. *Forest Ecology and Management, 212*(1–3), 315–332. https://doi.org/10.1016/j.foreco.2005.03.029.

Hyvärinen, E., Kouki, J., & Martikainen, P. (2009). Prescribed fires and retention trees help to conserve beetle diversity in managed boreal forests despite their transient negative effects on some beetle groups. *Insect Conservation and Diversity, 2*(2), 93–105. https://doi.org/10.1111/j.1752-4598.2009.00048.x.

Hyvärinen, E., Juslén, A., & Kemppainen, E., et al. (2019). *Suomen lajien uhanalaisuus— Punainen kirja 2019/The 2019 Red List of Finnish species.* Ympäristöministeriö and Suomen ympäristökeskus/Ministry of the Environment and Finnish Environment Institute.

Joelsson, K., Hjältén, J., & Work, T. (2018). Uneven-aged silviculture can enhance within stand heterogeneity and beetle diversity. *Journal of Environmental Management, 205,* 1–8. https://doi.org/10.1016/j.jenvman.2017.09.054.

Johansson, M. E., & Nilsson, C. (2002). Responses of riparian plants to flooding in free-flowing and regulated boreal rivers: An experimental study. *Journal of Applied Ecology, 39*(6), 971–986. https://doi.org/10.1046/j.1365-2664.2002.00770.x.

Johansson, T., Hjältén, J., de Jong, J., et al. (2013). Environmental considerations from legislation and certification in managed forest stands: A review of their importance for biodiversity. *Forest Ecology and Management, 303,* 98–112. https://doi.org/10.1016/j.foreco.2013.04.012.

Johnson, S., Strengbom, J., & Kouki, J. (2014). Low levels of tree retention do not mitigate the effects of clearcutting on ground vegetation dynamics. *Forest Ecology and Management, 330,* 67–74. https://doi.org/10.1016/j.foreco.2014.06.031.

Jonsson, B. G., Kruys, N., & Ranius, T. (2005). Ecology of species living on dead wood—lessons for dead wood management. *Silva Fennica, 39*(2), 289–309. https://doi.org/10.14214/sf.390.

Junninen, K., Kouki, J., & Renvall, P. (2008). Restoration of natural legacies of fire in European boreal forests: An experimental approach to the effects on wood-decaying fungi. *Canadian Journal of Forest Research, 38*(2), 202–215. https://doi.org/10.1139/X07-145.

Kärvemo, S., Björkman, C., Johansson, T., et al. (2017). Forest restoration as a double-edged sword: The conflict between biodiversity conservation and pest control. *Journal of Applied Ecology, 54*(6), 1658–1668. https://doi.org/10.1111/1365-2664.12905.

Komonen, A., & Kouki, J. (2008). Do restoration fellings in protected forests increase the risk of bark beetle damages in adjacent forests? A case study from Fennoscandian boreal forest. *Forest Ecology and Management, 255*(11), 3736–3743. https://doi.org/10.1016/j.foreco.2008.03.029.

Komonen, A., Laatikainen, A., & Similä, M., et al. (2009). *Ytimennävertäjien kasvainsyönti trombin kaataman suojelumännikön ympäristössä Höytiäisen saaressa Pohjois-Karjalassa.* Metsätieteen Aikakauskirja.

Kouki, J., Hyvarinen, E., Lappalainen, H., et al. (2012). Landscape context affects the success of habitat restoration: Large-scale colonization patterns of saproxylic and fire-associated species in boreal forests. *Diversity and Distributions, 18*(4), 348–355. https://doi.org/10.1111/j.1472-4642.2011.00839.x.

Kouki, J., & Salo, K. (2020). Forest disturbances affect functional groups of macrofungi in young successional forests—harvests and fire lead to different fungal assemblages. *Forest Ecology and Management, 463,* 118039. https://doi.org/10.1016/j.foreco.2020.118039.

Kuuluvainen, T. (2002). Natural variability of forests as a reference for restoring and managing biological diversity in boreal Fennoscandia. *Silva Fennica, 26*(1), 97–125. https://doi.org/10.14214/sf.552.

Kuuluvainen, T. (2009). Forest management and biodiversity conservation based on natural ecosystem dynamics in northern Europe: The complexity challenge. *Ambio, 38*(6), 309–315. https://doi.org/10.1579/08-A-490.1.

Kuuluvainen, T., & Aakala, T. (2011). Natural forest dynamics in boreal Fennoscandia: A review and classification. *Silva Fennica, 45*(5), 823–841. https://doi.org/10.14214/sf.73.

Laarmann, D., Korjus, H., Sims, A., et al. (2013). Initial effects of restoring natural forest structures in Estonia. *Forest Ecology and Management, 304,* 303–311. https://doi.org/10.1016/j.foreco.2013.05.022.

Lindenmayer, D. B., Franklin, J. F., & Fischer, J. (2006). General management principles and a checklist of strategies to guide forest biodiversity conservation. *Biological Conservation, 131*(3), 433–445. https://doi.org/10.1016/j.biocon.2006.02.019.

Lindenmayer, D. B., Blanchard, W., McBurney, L., et al. (2014). Complex responses of birds to landscape-level fire extent, fire severity and environmental drivers. *Diversity and Distributions, 20*(4), 467–477. https://doi.org/10.1111/ddi.12172.

Linder, P., Elfving, B., & Zackrisson, O. (1997). Stand structure and successional trends in virgin boreal forest reserves in Sweden. *Forest Ecology and Management, 98*(1), 17–33. https://doi.org/10.1016/s0378-1127(97)00076-5.

Lindhe, A., Lindelow, A., & Asenblad, N. (2005). Saproxylic beetles in standing dead wood density in relation to substrate sun-exposure and diameter. *Biodiversity and Conservation, 14*(12), 3033–3053. https://doi.org/10.1007/s10531-004-0314-y.

Löf, M., Erson, B., & Hjältén, J., et al. (2015). Site preparation techniques for forest restoration. In J. A. Stanturf (Ed.) *Restoration of boreal and temperate forests.* 2nd edition, (pp. 85–103). Boca Raton: CRC Press.

Lowe, J., Pothier, D., Rompré, G., et al. (2012). Long-term changes in bird community in the unmanaged post-fire eastern Québec boreal forest. *Journal of Ornithology, 153*(4), 1113–1125. https://doi.org/10.1007/s10336-012-0841-3.

Martikainen, P., Kouki, J., & Heikkala, O., et al. (2006). Effects of green tree retention and prescribed burning on the crown damage caused by the pine shoot beetles (*Tomicus* spp.) in pine-dominated timber harvest areas. *Journal of Applied Entomology, 130*(1), 37–44. https://doi.org/10.1111/j.1439-0418.2005.01015.x.

Mikusiński, G., Gromadzki, M., & Chylarecki, P. (2001). Woodpeckers as indicators of forest bird diversity. *Conservation Biology, 15*(1), 208–217. https://doi.org/10.1111/j.1523-1739.2001.99236.x.

Millennium Ecosystem Assessment. (2005). *Ecosystems and human well-being: Biodiversity synthesis.* World Resources Institute.

Morissette, J. L., Cobb, T. P., Brigham, R. M., et al. (2002). The response of boreal forest songbird communities to fire and post-fire harvesting. *Canadian Journal of Forest Research, 32*(12), 2169–2183. https://doi.org/10.1139/x02-134.

Naturvårdsverket. (2019). *Fördjupad utvärdering av miljömålen 2019.* Stockholm: Naturvårdsverket.

Paillet, Y., Bergès, L., Hjältén, J., et al. (2010). Biodiversity differences between managed and unmanaged forests: Meta analysis of species richness in Europe. *Conservation Biology, 24*(1), 101–112. https://doi.org/10.1111/j.1523-1739.2009.01399.x.

Pakkala, T., Kouki, J., & Piha, M., et al. (2017). Phloem sap in fire-damaged Scots pine trees provides instant foraging opportunities for Three toed Woodpeckers *Picoides tridactylus. Ornis Svecica, 27*(2–4), 144–149. https://doi.org/10.34080/os.v27.19568.

Pasanen, H., Junninen, K., & Kouki, J. (2014). Restoring dead wood in forests diversifies wood-decaying fungal assemblages but does not quickly benefit red-listed species. *Forest Ecology and Management, 312*, 92–100. https://doi.org/10.1016/j.foreco.2013.10.018.

Pasanen, H., Rehu, V., Junninen, K., et al. (2015). Prescribed burning of canopy gaps facilitates tree seedling establishment in restoration of pine-dominated boreal forests. *Canadian Journal of Forest Research, 45*(9), 1225–1231. https://doi.org/10.1139/cjfr-2014-0460.

Pasanen, H., Rouvinen, S., & Kouki, J. (2016). Artificial canopy gaps in the restoration of boreal conservation areas: Long-term effects on tree seedling establishment in pine-dominated forests. *European Journal of Forest Research, 135*(4), 697–706. https://doi.org/10.1007/s10342-016-0965-8.

Penttilä, R., Siitonen, J., & Kuusinen, M. (2004). Polypore diversity in managed and old-growth boreal *Picea abies* forests in southern Finland. *Biological Conservation, 117*(3), 271–283. https://doi.org/10.1016/j.biocon.2003.12.007.

Roberge, J. M., Angelstam, P., & Villard, M. A. (2008). Specialised woodpeckers and natural-ness in hemiboreal forests—Deriving quantitative targets for conservation planning. *Biological Conservation, 141*(4), 997–1012. https://doi.org/10.1016/j.biocon.2008.01.010.

Rodríguez, A., & Kouki, J. (2017). Disturbance-mediated heterogeneity drives pollinator diversity in boreal managed forest ecosystems. *Ecological Applications, 27*(2), 589–602. https://doi.org/10.1002/eap.1468.

Rodríguez, A., Pohjoismaki, J. L. O., & Kouki, J. (2019). Diversity of forest management promotes parasitoid functional diversity in boreal forests. *Biological Conservation, 238*, 108205. https://doi.org/10.1016/j.biocon.2019.108205.

Rydgren, K., Okland, R. H., & Hestmark, G. (2004). Disturbance severity and community resilience in a boreal forest. *Ecology, 85*(7), 1906–1915. https://doi.org/10.1890/03-0276.

Salo, K., & Kouki, J. (2018). Severity of forest wildfire had a major influence on early successional ectomycorrhizal macrofungi assemblages, including edible mushrooms. *Forest Ecology and Management, 415–416*, 70–84. https://doi.org/10.1016/j.foreco.2017.12.044.

Salo, K., Domisch, T., & Kouki, J. (2019). Forest wildfire and 12 years of post-disturbance succession of saprotrophic macrofungi (Basidiomycota, Ascomycota). *Forest Ecology and Management, 451*, 117454. https://doi.org/10.1016/j.foreco.2019.117454.

Schimmel, J., & Granström, A. (1996). Fire severity and vegetation response in the boreal Swedish forest. *Ecology, 77*(5), 1436–1450. https://doi.org/10.2307/2265541.

Seibold, S., Bassler, C., Brandl, R., et al. (2016). Microclimate and habitat heterogeneity as the major drivers of beetle diversity in dead wood. *Journal of Applied Ecology, 53*(3), 934–943. https://doi.org/10.1111/1365-2664.12607.

Siitonen, J. (2001). Forest management, coarse woody debris and saproxylic organisms: Fennoscandian boreal forest as an example. *Ecological Bulletins, 49*, 11–41.

Similä, M., Kouki, J., & Martikainen, P. (2003). Saproxylic beetles in managed and seminatural Scots pine forests: Quality of dead wood matters. *Forest Ecology and Management, 174*(1–3), 365–381. https://doi.org/10.1016/S0378-1127(02)00061-0.

Stenbacka, F., Hjältén, J., Hilszczański, J., et al. (2010). Saproxylic and non-saproxylic beetle assemblages in boreal spruce forests of different age and forestry intensity. *Ecological Applications, 20*(8), 2310–2321. https://doi.org/10.1890/09-0815.1.

Stenberg, I., & Hogstad, O. (2004). Sexual dimorphism in relation to winter foraging in the white-backed woodpecker (*Dendrocopos leucotos*). *Journal of Ornithology, 145*(4), 321–326. https://doi.org/10.1007/s10336-004-0045-6.

Stighäll, K., Roberge, J.-M., Andersson, K., et al. (2011). Usefulness of biophysical proxy data for modelling habitat of an endangered forest species: The white-backed woodpecker *Dendrocopos leucotos*. *Scandinavian Journal of Forest Research, 26*(6), 576–585. https://doi.org/10.1080/02827581.2011.599813.

Suominen, M., Junninen, K., Heikkala, O., et al. (2015). Combined effects of retention forestry and prescribed burning on polypore fungi. *Journal of Applied Ecology, 52*(4), 1001–1008. https://doi.org/10.1111/1365-2664.12447.

Suominen, M., Junninen, K., Heikkala, O., et al. (2018). Burning harvested sites enhances polypore diversity on stumps and slash. *Forest Ecology and Management, 414*, 47–53. https://doi.org/10.1016/j.foreco.2018.02.007.

Swedish University of Agricultural Sciences (SLU). (2012). *Forest statistics 2012. Official statistics of Sweden.* Umeå: Swedish University of Agricultural Sciences.

Tatsumi, S., Strengbom, J., Čugunovs, M., et al. (2020). Partitioning the colonization and extinction components of beta diversity across disturbance gradients. *Ecology, 101*(12), e03183. https://doi.org/10.1002/ecy.3183.

Thomas, S. C., Halpern, C. B., Falk, D. A., et al. (1999). Plant diversity in managed forests: Understory responses to thinning and fertilization. *Ecological Applications, 9*(3), 864–879. https://doi.org/10.1890/1051-0761(1999)009[0864:pdimfu]2.0.co;2.

Toivanen, T., Liikanen, V., & Kotiaho, J. S. (2009). Effects of forest restoration treatments on the abundance of bark beetles in Norway spruce forests of southern Finland. *Forest Ecology and Management, 257*(1), 117–125. https://doi.org/10.1016/j.foreco.2008.08.025.

Versluijs, M., Eggers, S., Hjältén, J., et al. (2017). Ecological restoration in boreal forest modifies the structure of bird assemblages. *Forest Ecology and Management, 401*, 75–88. https://doi.org/10.1016/j.foreco.2017.06.055.

Versluijs, M., Eggers, S., & Mikusiński, G., et al. (2020). Foraging behavior of the Eurasian Three-toed Woodpecker (*Picoides tridactylus*) and its implications for ecological restoration and sustainable boreal forest management. *Avian Conservation and Ecology, 15*(1):art6. https://doi.org/10.5751/ACE-01477-150106.

Zackrisson, O. (1977). Influence of forest fires on the North Swedish boreal forest. *Oikos, 29*, 22–32. https://doi.org/10.2307/3543289.

Chapter 19
Boreal Forest Landscape Restoration in the Face of Extensive Forest Fragmentation and Loss

Johan Svensson, Grzegorz Mikusiński, Jakub W. Bubnicki, Jon Andersson, and Bengt Gunnar Jonsson

Abstract Historical conditions that provide a natural legacy for defining restoration targets are not applicable without adjusting these targets to expected future conditions. Prestoration approaches, defined as restoration that simultaneously considers past, present, and future conditions with a changing climate, are necessary to advance the protection of biodiversity and the provisioning of ecosystem services. Large areas of boreal forest landscapes are transformed and degraded by industrial forestry practices. With largely fragmented and too-small areas of remaining high conservation value forests, protection and preservation are insufficient and must be complemented by active restoration in the managed forest matrix. Successful forest landscape restoration incorporates varied spatiotemporal scales and resolutions to compose restoration routes that best reflect the expected future sustainability challenges as well as planning and governance frameworks.

J. Svensson (✉) · B. G. Jonsson
Department of Wildlife, Fish, and Environmental Studies, Swedish University of Agricultural Sciences, SE-90183 Umeå, Sweden
e-mail: johan.svensson@slu.sc

B. G. Jonsson
e-mail: bengt-gunnar.jonsson@miun.se

G. Mikusiński
School for Forest Management, Swedish University of Agricultural Sciences, 739 21 Skinnskatteberg, Sweden
e-mail: grzegorz.mikusinski@slu.se

J. W. Bubnicki
Mammal Research Institute, Polish Academy of Sciences, 17-230 Białowieża, Poland
e-mail: kbubnicki@ibs.bialowieza.pl

J. Andersson
Borcasts, Kungsgatan 85A, 903 30 Umeå, Sweden
e-mail: jon.pm.andersson@outlook.com

B. G. Jonsson
Department of Natural Sciences, Mid Sweden University, SE-851 70 Sundsvall, Sweden

© The Author(s) 2023
M. M. Girona et al. (eds.), *Boreal Forests in the Face of Climate Change*,
Advances in Global Change Research 74,
https://doi.org/10.1007/978-3-031-15988-6_19

19.1 Introduction

In the face of climate change, the challenges for sustainable forest and landscape management become even more pronounced (Hlásny et al., 2017; Kremen & Merenlender, 2018). Forest landscapes characterized by the effects of long-term and intensive forest logging dominate vast areas in northern boreal Europe but are also increasingly common in all boreal regions (Curtis et al., 2018). In addition to extensive forest harvesting and other land-use impacts, a changing climate puts into place often unknown or difficult-to-predict trajectories of ecosystem response to disturbance (Kuuluvainen et al., 2017; Lindner et al., 2010; Scheffer et al., 2012). Forest fragmentation and loss are integrated with and respond to climate change through multiple unforeseeable feedback effects on forest conditions (e.g., Wang et al., 2020). Thus, the circumstances for biodiversity conservation, ecosystem service provisioning, as well as forestry and other land uses may differ markedly in the near future from the present and past circumstances (Frelich et al., 2020). Consequently, current and future landscape analysis and integrated planning oriented toward stand and landscape restoration are critical for maintaining viable and resilient boreal landscapes (Arts et al., 2017; Svensson et al., 2019a). Thus, climate adaptation and mitigation approaches must be integrated into green infrastructure planning, defined as a spatiotemporally functional planning framework for maintaining biodiversity and ecosystem services in landscapes affected by climate change and land use (Mikusiński et al., 2021; Stanturf, 2015).

There is much evidence for the loss of natural, near-natural, and intact forest landscapes and the associated negative consequences for biodiversity, ecosystem services, and other benefits to people (e.g., Potapov et al., 2017; Zanotti & Knowles, 2020). In Europe, most forest types have little to no remaining natural forests (Sabatini et al., 2020). Consequently, and recognized for example in the UN Decade on Ecosystem Restoration 2021–2030 (FAO, 2020) and the European Union Biodiversity Strategy for 2030 (EC, 2020), the current levels of protection, combined with often limited conservation functionality in the existing protected areas (Halme et al., 2013; Watson et al., 2014), are insufficient. Here we define limited functionality as areas that are too small and too fragmented to develop or maintain a favorable conservation status. Additionally, it is increasingly recognized that effective conservation of protected areas depends not only on the intrinsic values within these areas but also on the quality of the landscape matrix (Orlikowska et al., 2020; Ward et al., 2020). Thus, landscape restoration has a central role in green infrastructure planning.

New and innovative avenues need to be explored locally, nationally, and globally to preserve functional ecosystems for future generations. In addition to more and larger protected areas and greater consideration of nature conservation in standard forestry practices, active measures must include restoring forest patches and forest landscapes within sustainable management and governance strategies and plans (Mansourian, 2017; Stanturf et al., 2014). The preservation of forest ecosystem functions, biodiversity, and the naturally rich pools of ecosystem services and nature's contribution to people requires more active and progressive restoration approaches (IPBES, 2018).

Moreover, as land-use pressure is high and increasing from multiple, varying, and sometimes conflicting interests (Knoot et al., 2010; Svensson et al., 2020b), restoration must be oriented not only toward nature conservation values but also toward sociocultural and economic values associated with a broadening and diversifying of the forest landscape value chains (Jonsson et al., 2019; Stanturf, 2015). That is, restoration should aim at supporting a multifunctional forest use rather than a single-use orientation of a service or good, such as wood biomass for timber, pulpwood, or energy production.

In this chapter, we explore various aspects and routes forward for forest landscape restoration in the context of climate change. We benefit from recent research on Sweden's boreal and subalpine regions, which exemplifies a geographically broad case that harbors both generic and specific boreal characteristics. The study region encompasses around 27 million ha, of which 19 million ha is forest (Fig. 19.1; Mikusiński et al., 2021). Distinct gradients in historical and current land use provide representative examples of forest landscapes characterized by different biogeographical contexts and intensities of human exploitation. The loss of intact forest landscapes caused by the dominant systematic forest clear-cutting system has largely transformed forest landscapes across vast areas. The current Swedish Red List (Artdatabanken, 2020) encompasses 1,400 species listed as a direct and indirect consequence of this forestry approach. About 1,100 species of these listed species are found in northern Sweden. Only a narrow hinterland belt in the mountainous area, the *Scandinavian Mountains Green Belt*, can be considered intact (Fig. 19.2; Svensson et al., 2020a). The loss of natural forests, the geographically imbalanced conditions of the remaining intact forest landscapes in northern Sweden, and areas where landscape restoration is critically needed are illustrated in Fig. 19.1.

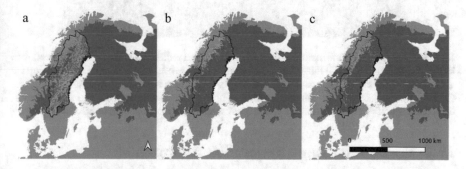

Fig. 19.1 Northern Sweden (*black delimiting line*) with the surrounding terrestrial areas (*gray shading*) and boreal biome (*dark green shading*) delimited; the illustrations show the structural connectivity of **a** all forest land, **b** protected forestland, and **c** remaining forestland not subjected to clear-cutting since the introduction of systematic forest clear-cutting in Sweden in the middle of the twentieth century (i.e., proxy continuity forests; see Svensson et al., 2019a). Connectivity was calculated using circuit theory (McRae et al., 2008), where structural connectivity implies that all forests are treated as a single entity, i.e., without separating the area into ecologically different forest types. Figure modified from Mikusiński et al. (2021), CC BY license

Fig. 19.2 Large areas of the mountain foothill forests are part of the Scandinavian Mountains Green Belt (Svensson et al., 2020a) intact forest landscape; (*top*) Laxbäcken, Vilhelmina, overlooking the gradual change from coniferous-dominated forests to the broadleaf alpine tree line woodlands; (*bottom*) the landscape-scale mixture of forests, open mires and grasslands, and water bodies, toward the Marsfjället nature reserve. *Photo credits top* Jon Andersson, *bottom* Mikael Strömberg

19.2   Forest Landscape Restoration Approaches

Strategic planning at the landscape scale is critical for effectively securing representative aspects of biodiversity and forest ecosystem services (Mansourian et al., 2017). Restoration must simultaneously target different spatial scales, from individual trees to stands to landscapes. However, a landscape cannot be constrained by a single definition, as it is inherently context-dependent. For example, the term is used generically for defining a geographical area, for describing a spatial extent between *local* and *regional*, as an ecological term representing the spatiotemporal gradient in energy flow, nutrient cycling, and species interactions, and as a socioecological system in which different actors perceive and influence the spatial composition and functioning of various landscape elements. The term landscape may also refer to an older delineation of administrative units. Thus, any landscape approach is defined by the specific questions, species, habitats, and contexts being addressed.

Similar to the definition of landscape, its scale, i.e., the spatiotemporal extent of a landscape, is also conditioned by the habitat and species context. For forest areas, the extent should be sufficiently large to include an adequate range of different naturally occurring forest types and connected landscape elements that represent a relevant and practical scale for actors such as forest management planners or administrative authorities working with green infrastructure planning. In the context of boreal Europe, this normally translates into areas of a few tens of thousands of hectares. At a global scale, analyses of intact forest landscapes tend to address significantly larger areas and may include several hundreds of thousands of hectares (e.g., Potapov et al., 2017), i.e., the size perceived to encompass large-scale natural dynamics linked to disturbance regimes.

Forest landscape restoration encompasses a range of measures at various scales for numerous specific purposes (Chazdon et al., 2016) and with various specific measures and activities, such as restoration fire, the production of deadwood, and green tree retention. Below, we detail some of the more central terms, approaches and measures for boreal forests and forest landscape restoration (drawing from Mansourian, 2018), where, for the purpose of this chapter, we have clustered similar and related terms. In addition to various active measures that aid the development of forest habitats to improve biodiversity and resilience, passive strategies, allowing natural processes and dynamics to act, are optional or preferred in many situations.

Forest landscape restoration/ecological restoration: Traditionally, this approach relies on an understanding of historical landscape composition as a model for moving landscape structures closer to a historical baseline, often referring to a natural range of variability in terms of the extent of forest types and disturbance processes (Kuuluvainen et al., 2015; Pennanen, 2002). Thus, landscape restoration is a planning process rather than direct actions within individual stands, which include, for example, applying relevant data and the active participation of various landowners and decision-makers and planning according to given regulations and policies.

Prestoration: This approach is defined as restoration that simultaneously relies on past and present states that impact the present and future stages as expected by climate change while using as a starting point the species' need for suitable habitats (Butterfield et al., 2017; Mansourian, 2018). Prestoration aims to support biodiversity and ecosystem services given the anticipated effects, i.e., restoration with a target into an expected future given current knowledge and projections. Therefore, a central question is which tree species or genotypes should be planted or promoted for restoration to match the climatic conditions in 100 years or more (Halme et al., 2013; Kuuluvainen et al., 2017). Prestoration can be applied at the landscape scale and at the scale of specific stands and habitats; it should be explicitly sensitive to temporal dimensions, particularly for ecosystems that recover slowly such as the boreal forest.

Therefore, specific restoration actions can be performed in recently planted forest and during precommercial thinning and thinning stages and in the form of translocating biodiversity attributes such as snags and logs, i.e., ecological compensation approaches. The planting of tree species beyond their current distribution also forms part of this approach.

Habitat restoration/habitat reconstruction/rehabilitation: These measures include promoting structures and processes that have been lost through forestry or other land-use transformations of natural landscapes, normally within currently existing forest areas or landscapes dominated by forests. For boreal forests, measures include creating multilayered forest canopies, increasing volumes of deadwood, veteranization of living trees, reintroducing forest fires, and applying other stand- and tree-level measures. The veteranization of trees collectively includes measures that damage or affect living trees in ways that advance aging qualities, such as bark damage to create sap flow or cavity development. Measures also include the mitigation of degraded habitats by restoring soils through revitalizing or translocating soil biota, restoring hydrology through the blocking of ditches or restoring streams modified by timber floating, and establishing or replacing existing vegetation cover in forest edges and other transition zones. Habitat restoration, reconstruction, and rehabilitation can also be achieved through natural development, with or without minor active interventions, if conservation attributes and ecological processes have been maintained.

Reclamation/reconciliation/reallocation/reforestation/afforestation: These measures encompass the artificial planting or seeding of trees and the promotion of natural tree regeneration in areas that historically have been transformed from forests to other land cover types for longer or shorter time periods. The planting of a selected tree species can extend forest habitat areas and provide new habitat patches for associated species. This transformation of previously open areas to forest usually leads to decreases in values associated with open land cover, e.g., grassland biodiversity, landscape vistas, or farmland for food production. Thereby, explicit concern must be accounted for, e.g., natural or cultural values, and any potential trade-offs must be managed.

19.3   Dimensions in Forest Landscape Restoration

Except for historical slash-and-burn cultivation and wood for iron mining, which resulted in localized long and intense forest use that left extensive degraded areas (Angelstam et al., 2013), the transformation of boreal forests in northern Europe is relatively recent. Most transformation has occurred during the last two centuries with increasing intensity during the twentieth century. In particular, the systematic clear-cut rotation system—fully implemented after the mid-twentieth century— represented a shift from a continuous forest cover with multi-aged, multispecies stands to even-aged monocultural forests (Kuuluvainen et al., 2012). This shift has produced a severely fragmented landscape structure across vast areas of the boreal region. Only fragmented and small remnants of old and natural forests of high natural value are preserved (Fig. 19.3). With such a landscape configuration as a starting point for forest landscape restoration—with high conservation value forest only occupying a low share of the remaining forestlands—combined with the low growth rates of boreal trees and their limited dispersal capacities, restoration takes time and requires long-term planning.

A fully restored landscape should deliver the attributes of a naturally dynamic landscape, including living space for all native species, a full representation of different habitat types, and the presence of all-natural processes essential for ecosystem functioning. This restoration must also maintain natural disturbances to the extent possible, given societal risks with wildfires, for example. In boreal landscapes characterized by high levels of spatiotemporal randomness for the main natural disturbance agent, i.e., fire, extensive areas should be restored or protected to secure a continuous availability of all naturally occurring habitats. For example, Andrew et al. (2014) proposed that such a *minimum dynamic area* for the Canadian boreal forest should be at least 20,000 km^2. If wildfires are absent or occur too rarely or at a too-low intensity, as in Sweden because of effective long-term fire suppression, natural succession with broadleaf dominance is very rare and leads to a generically low abundance of broadleaf trees in the boreal tree species mixture (Bengtsson et al., 2000; Mikusiński et al., 2003). As a remedy, a forest management system based on mimicking natural disturbance regimes has been promoted for many years (e.g., Angelstam, 1998; Bergeron et al., 2002). However, the situation has not changed much despite such early promotion, as broadleaf species are not a central resource for the Swedish forest industry. Thus, aging broadleaf trees and stands remain critically rare in Swedish boreal landscapes (Mikusiński et al., 2021).

Old-growth forests are focal biodiversity nodes within boreal forest landscapes and have long been protected; in Sweden, however, their spatial distribution is highly skewed toward the northwestern mountain areas (Angelstam et al., 2020), i.e., the Scandinavian Mountains Green Belt. Protected forests are much less extensive in the other parts of the country. Achieving old-growth conditions in boreal forests after clear-cutting forestry or a major natural disturbance may take centuries (e.g., Hedwall & Mikusiński, 2015; Lilja et al., 2006).

Conservation planning tools that extend from the remaining ecological mainlands, i.e., geographically large nodes of intact forests and forest landscapes, must be used to embrace the temporal and spatial complexity of restoration in the boreal forest. This is particularly true from a green infrastructure perspective (Snäll et al., 2016) that supports the spread and migration of species into the surrounding landscape matrix (Mikusiński et al., 2007). Enhancing the functionality of the few remaining old-growth, primary or natural forest patches outside such mainlands and building the future green infrastructure pool, requires new protected areas having robust existing conservation values and also enhancing restoration efforts when the temporal transition to strong conservation conditions can be foreseen. Thus, in everyday forest-production landscapes where the transformation of natural conditions is substantial,

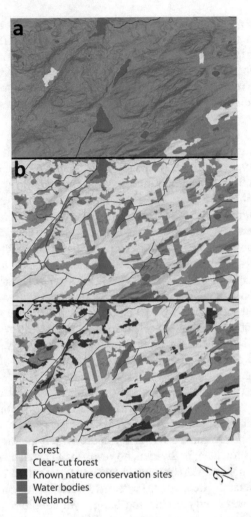

Forest
Clear-cut forest
Known nature conservation sites
Water bodies
Wetlands

Fig. 19.3 Clear-cutting forestry was introduced at a large scale in northern Sweden in the mid-twentieth century; most forested areas have since been clear-cut. **a** Map of forest and clear-cut areas in 1958 and **b** 2016; **c** the locations of all known nature conservation areas, protected and not protected, are superimposed on the map to illustrate both the overlap and the remaining share of non-high-conservation value forests, as determined through inventories. The study area is situated 60 km west of the city of Umeå, east coast of northern Sweden, and covers about 3,000 ha. The area was previously 90% forested, whereas 72% was clear-cut by 2016. (see Svensson et al., 2019b)

restoration must become a natural part of landscape planning and include a broader spectrum of approaches and measures.

Boreal landscapes typically include land cover types other than forest, such as water bodies, open mires and grasslands, and subalpine environments. This heterogeneity represents a natural level of forest fragmentation in an intact dynamic landscape to which, in a broad sense, the associated forest species adapt. Thus, land cover types other than forests contribute significantly to landscape-level biodiversity and ecosystem services. The transition zones to forests, i.e., forest edges, are by themselves essential habitats for biodiversity and ecosystem services but also function as bridging elements (Harper et al., 2015). Consequently, effective restoration requires a holistic approach with integrated planning and policies across land cover types (Chazdon et al., 2017).

The current landscape configuration represents a natural, seminatural, or artificial land cover distribution that may be stable for a particular duration. However, a landscape may have had another configuration historically, where the land cover and the modifying agent that generated the configuration have left both natural and anthropogenic legacies. These legacies have relevance for the present state and premises for restoration. For example, northern Sweden is currently experiencing a loss of open habitats in rural areas and thus a loss in the biodiversity, cultural values, and ecosystem services associated with open habitats. The recent red list (Artdatabanken, 2020) includes around 1,400 species as direct and indirect consequences of the loss of open and semi-open landscapes being transformed into forests. Habitats with a certain value, for example providing rich winter grazing resources for ungulate species such as reindeer (*Rangifer tarandus*), may be replaced by dense, fast-growing forest stands (cf. Sandström et al., 2016). Many of these open habitats were naturally open in the distant past because of poor site conditions or were created by the active removal of forests to increase farmland. Over the last century, these areas transformed back to forest, either naturally or through silvicultural reforestation measures. Forests not currently being used for forestry, including other woodlands, sites having a low tree growth capacity and limited natural values, and single trees and tree groups in other land cover types, can connect spatially disrupted old-growth forest patches and decrease adverse effects from fragmentation. Thus, in landscape restoration, the land cover composition represents the first dimension that must be considered (Fig. 19.4).

The tree-age distribution of the forest represents the second dimension in landscape restoration. In managed forest landscapes, much of the old forest and forests composed of mixed-age assemblages have been transformed into young and middle-aged, fast-growing, and dense forests with management oriented toward wood biomass production. In northern Sweden, currently only 15% of all forests are 140 years or older, including forests that are of no interest to production forestry, i.e., mean annual wood biomass growth ≤ 1 m^3 · ha^{-1} over the rotation period. It should be noted, however, that 140-year-old forests often have not (yet) developed old-growth boreal characteristics; thus, *old* refers more to the forestry rotation cycle, i.e., stands at the final logging stage, than to a biologically significant status. With a focus on forest age, restoration activities can be directed toward a diversified and broader tree- and stand-age distribution on proportionally larger forest areas than at present. However, because boreal species are adapted to landscape compositions of low predictability and structure because of the stochasticity of the main large-scale disturbances, adequate trajectories of landscape and forest restoration must be promoted to attain a more varied forest-age distribution (Berglund & Kuuluvainen,

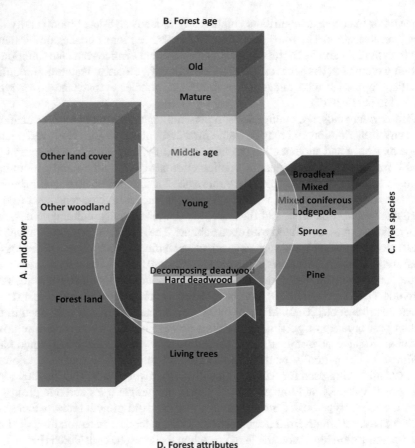

Fig. 19.4 Forest landscape restoration that considers four dimensions: land cover type, forest-age distribution, dominating tree species, and forest attributes; the illustration is derived from data of the Swedish National Forest Inventory for northern Sweden, roughly representing the boreal biome distribution (SLU 2020). **a** The distribution of forests (56%), other woodland areas (sparse and low growth), other land cover and land-use types (mainly alpine and open mires); **b** stand age based on data from all forests with young ($\leq$20 years), middle-aged (21 to $\leq$ 0 years), mature (81 to $\leq$ 140 years), and old ($\geq$140 years) forest; **c** stand-scale dominant ($\geq$65%) tree species from data on forestry lands (productive, not formally protected) of Scots pine, Norway spruce, lodgepole pine, mixed coniferous, mixed, and broadleaf forest; **d** volume of living trees (m^2 basal area), hard (m^3) and decomposing deadwood (m^3) from all forests. The layer thickness in each bar is proportional to the abundance of the illustrated component. Dimension **a** is illustrated based on land surface area (27 million ha), dimensions **b** and **d** on forest land area (19 million ha), and dimension **c** on productive forest land (15 million ha). Here we apply data commonly recorded in national forest inventories and, hence, similar assessments can be made for other boreal regions

2021). From this perspective, it can be noted that regenerating young forests dominated by broadleaf trees can play a role as natural fire barriers and, accordingly, form
part of the forest-age distribution at the landscape scale.

Dominant tree species represent a third landscape restoration dimension. Systematic clear-cutting forestry with a regular harvesting rotation has resulted in forest-
stand monocultures. In northern Sweden, only 7% of forests are truly mixed, and
only 5% are dominated by broadleaf species. Under natural conditions, succession,
dynamics, and the various natural disturbances, ranging from small-scale treefalls
to wind-felled stands to extensive burned regions, create tree species conglomerates
that vary across time and space. From a site's disturbance dynamics and soil/bedrock
conditions, different tree species naturally occur in a mosaic of stands having a single,
few, or multiple species where the configuration can only be predicted at a very large
scale (Pennanen, 2002). Clearly, restoration aiming for a more balanced and mixed
tree species composition results in niche separation that supports broader pools of
forest biodiversity and ecosystem services. Moreover, it provides prerequisites for
diversified management strategies and innovative value chains.

The fourth dimension is exemplified by deadwood, representing a key biodiversity attribute, and other attributes typical of old-growth characteristics, i.e., multiple
forest layers, old trees, horizontal heterogeneity, and broad substrate diversity. Deadwood is lacking in northern Swedish boreal forests but is slowly increasing, averaging presently around 8 $m^3 \cdot ha^{-1}$ (SLU, 2020). This quantity of deadwood is
very low compared with natural conditions where deadwood volumes can be 50–
80 $m^3 \cdot ha^{-1}$ for comparable forest types (Siitonen, 2001). Note, however, that an
overall increase and general improvement of the ecosystem attributes that intrinsically support biodiversity and ecosystem services, e.g., deadwood as a colonizing
substrate, the functionality of a given substrate in a specific site, are determined not
only by site-intrinsic characteristics but also by characteristics in the surrounding
habitats and landscapes.

19.4   Forest Landscape Prestoration to Mitigate Clear-Cutting Debt

The dominance of the rotation clear-cutting system has led the Swedish boreal forest
landscape to lose most of its historical configuration. Outside the intact forest landscapes of the Scandinavian Mountains Green Belt, only fragments remain of forests
that have never been subjected to clear-cutting. At the landscape scale, the dominance of young to middle-aged planted forests has led to a connectivity loss between
the remaining old-growth patches. This resulting lack of connectedness between
non-clear-cut forests represents a significant challenge, given that intact forest landscapes were historically dominated by older forests (Berglund & Kuuluvainen, 2021;
Pennanen, 2002). As a parallel to the *extinction debt* related to species loss through
habitat destruction (Hanski, 2000; Tilman et al., 1994), we may consider clear-cutting

as the cause of a broader debt in terms of deteriorated or lost natural processes, structures, and other, not-yet-fully-known ecosystem changes. From this perspective, landscape restoration is required at much broader scales and higher rates than at present to manage and mitigate this debt.

However, for practical, economic, and climatic reasons, the target of landscape restoration cannot be to return to a pristine historical situation. Instead, what is needed is a careful consideration of those restoration measures able to provide components of natural forests that are sufficiently robust and resilient for any particular landscape given its natural settings, legacies of land use, and current socioeconomic situation. These considerations must then be placed and evaluated against climate change scenarios, i.e., restoration in the sense of prestoration. Thus, any restoration planning must consider climate change–driven biogeographical translocations and, therefore, include climate models as input data. The different available targets (Fig. 19.5) provide a gradient in segregating and integrating conservation goals (Bollmann et al., 2020), where some targets are relevant in areas primarily managed for biodiversity, i.e., protected areas, and other targets more suitable for mitigating adverse effects on natural values in more-or-less intensively managed forests. Regardless of the conservation goals, climate change will affect all forest types and their adaptative potential. Although historical knowledge of past natural conditions provides a critical reference state for species and biodiversity, we must now also address future conditions. Any prestoration targets for the future must include forms of *secondary natural forests* and novel, designed managed forests that ensure the full range of ecosystem services from the forest landscape (Bollmann et al., 2020).

Their remaining natural forests in boreal Sweden are, in most cases, restricted to the lower end of the site productivity gradient, i.e., mainly occurring on marginal lands (e.g., Andrew et al., 2014; Angelstam et al., 2020). Yet, these forests still provide crucial elements. Here, prestoration may complement nonintervention management and include promoting tree species that may be important biodiversity structures under future conditions and, if introduced, may prepare the ground for future range shifts of associated species. Prestoration in natural forests may also include the translocation of species to habitats outside their current distribution ranges.

For forests having been subject to a relatively limited impact from recent forestry, possible measures include habitat restoration through the veteranization of trees, prescribed fires, retention measures, and increased volumes of deadwood. When these measures are carefully applied and well placed at the landscape scale, they will also support landscape restoration and prestoration. The range of options for forests having a recent harvest history is likely to be greater, although the positive effects on biodiversity are delivered in a more distant future. A careful choice of tree species for regeneration and active measures to create structural and functional diversity in forests across broader spatial scales exemplifies different possible reforestation, rehabilitation, and habitat restoration measures.

Forested areas not used for commercial forestry can play a crucial role. These sites include woodlands in remote places, technically challenging sites such as steep slopes, and less fertile sites having a poor tree growth capacity. Such low-production forests often occur as islands or belts within productive forest landscapes. With

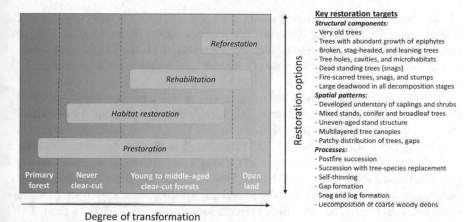

Fig. 19.5 A generic forest landscape composed of four categories of forests representing degrees of transformation: (1) primary forests with no or very limited human impact; (2) forests that have never been clear-cut; (3) young and middle-aged forests regenerated after clear-cutting; and (4) open lands that potentially could become forested through natural or silvicultural measures. The relative area of each category broadly reflects the current situation in Fennoscandian (Norway, Sweden, Finland) boreal forests. The figure exemplifies the most relevant type of restoration for each category that collectively represents landscape restoration opportunities if carefully planned at the landscape scale. The list (*right*) includes factors known to be essential for boreal forest biodiversity (after Esseen et al., 1997) and hence represent targets for restoration activities. Various types and groups of targets can be implemented at different degrees of transformation and at a range of spatiotemporal scales

careful consideration of the structures and habitats produced by these woodlands, it is possible to identify stepping-stone and corridor functions to improve landscape connectivity. Directed habitat restoration measures can enhance their functionality in cases where historical land use caused a loss of certain structures. From a prestoration perspective, it is also possible to increase structures beyond natural levels to compensate for the intensively managed forests in the surrounding areas.

19.5 Forest Landscape Restoration to Meet Global Sustainability and Conservation Targets

Sustainability, ecosystem services, and biodiversity are widely recognized on global agendas. The UN Sustainable Development Goals (SDG; FAO, 2020) and the Convention on Biological Diversity Aichi targets (CBD, 2010) have been paramount in setting this global policy agenda. The linkages between biodiversity and the fulfillment of the SDGs are apparent at multiple levels (Blicharska et al., 2019). Both the 2030 European Union Biodiversity Strategy (EC, 2020) and the new CBD framework further highlight the importance and challenges that humanity must consider moving from a net loss of natural values to a net gain. From this perspective, forest restoration and prestoration represent major opportunities given the high level of potential multifunctionality through these approaches, the inherent effects on biodiversity, and the generic applicability of measures and targets to local conditions and circumstances.

Intensive forest management has caused a loss of boreal biodiversity and reduced the provision of ecosystem services. Structurally and compositionally simplified forests and landscapes can only deliver some of the services essential for human well-being (e.g., Gamfeldt et al., 2013; Jonsson et al., 2020). Successful restoration of boreal forests and landscapes will, directly and indirectly, generate positive progress toward achieving several SDGs (Table 19.1). Whereas the positive impacts of restoration on biodiversity (SDGs 14 and 15) are obvious, the delivery of many other services, such as the securing of diverse food resources (SDG 2), health (SDG 3), clean water and energy (SDGs 6 and 7), and climate actions (SDG 13) are all, in some manner, linked to the successful restoration of vital forests (Table 19.1). Because achieving the full palette of services from forest environments that support multiple SDGs can be assumed to be impossible at the local level, diversification of management regimes at the broader landscape level has been advised (e.g., Felton et al., 2020; Triviňo et al., 2017). Unlike clear-cutting forestry, continuous cover forestry has a particular role in restoring multiple services within boreal landscapes (Eyvindson et al., 2021).

Restoration of boreal forests and landscapes clearly affects the ability to achieve Aichi Target 7 of sustainable forest management and Target 11 of setting aside 17% of all ecosystems for biodiversity conservation (CBD, 2010). The future management and conservation of forests in Sweden are currently at a crossroad between intensified wood production and multiple-use forests (Felton et al., 2020; Jonsson et al., 2019). Restoration aiming to improve a greater expanse of available habitat and securing their functional connectivity—along with safeguarding the long-term provision of these features within multiple-use forest landscapes—is a viable and successful means for achieving the Aichi targets.

19.6   Conclusions

Like the two-faced Roman god Janus, restoration must also look simultaneously in different directions. This reality means building on the historical understanding of species' habitat- and landscape-level requirements and considering climate change and future conditions, which we assume will differ substantially from the past and present. Thus, a relevant temporal resolution is necessary to reflect a slow ecosystem response where the net effects of restoration may lie far into the future. Spatial scaling is also necessary to reflect species' niches and behaviors in terms of movement,

Table 19.1 Examples of the benefits of forest landscape restoration in relation to 11 of the 17 UN Sustainable Development Goals, separated into the dimensions of the biosphere, society, and economy

SDG #	Biosphere	Society	Economy
1. No poverty	Increased pools of ecosystem services	Revitalize degenerated land for labor opportunities	Subsistence economy and multiple value chains
2. Zero hunger	Edible plants and hunting opportunities	Prevent erosion of agroforestry land	Subsistence economy and multiple value chains
3. Good health and well-being	Increased pools of ecosystem services	Forest medicinal plants, forests as de-stressing space	Ecotourism opportunities
4. Clean water and sanitation	Natural hydrological filtering of freshwater	Local freshwater accessibility	Local freshwater accessibility
5. Affordable and clean energy	Sustained growth of local bioenergy	Fossil-free neutral bioenergy production	Bio-economy options
6. Decent work and economic growth	Increased pools of ecosystem services	Rural development based on natural forest resources	Rural economy based on sustainable forestry and forest product processing
7. Sustainable cities and communities	Increasing green structures in urban and peri-urban areas	Accessibility to forest-based provisioning and cultural ecosystem services	Bio-based construction material of high quality
8. Responsible consumption and production	Sustainable forestry practices	Access to forest products based on sustainable use	Added market value from sustainably used forests
9. Climate action	Increased carbon storage in growing forests	Climate mitigation and offsetting	Payment for net carbon storage, fossil fuel substitution
10. Life below water	Ecological integrity of riparian and shoreline forestlands	Access to fresh and marine waters free from eutrophication	Revitalized aquatic systems with fish and other ecosystem services
11. Life on land	Improved habitats, biodiversity, ecosystem services	Provision of a full range of forest ecosystem services	Ecotourism opportunities, integrity of ecosystem functions and values

migration, and seasonal distribution patterns. Operating at a landscape scale is necessary, adding factors such as land cover types, landowners, policies, and decision-making. Restoration must target both natural forests and managed forests as important parts of the landscape, covering transformed and degraded landscapes. Boreal biodiversity and ecosystem services cannot be preserved solely through protecting the remaining high conservation value forests. An active restoration that mitigates fragmentation and the loss of intact forest landscapes and natural forest habitat values has a core role in integrated, green infrastructure–oriented landscape planning. Prestoration approaches, which acknowledge forest restoration across multiple spatiotemporal scales on the basis of past legacies and expected future situations, should be promoted and included within the governance and management of forests and forest landscapes.

References

Andrew, M. E., Wulder, M. A., & Cardille, J. A. (2014). Protected areas in boreal Canada: A baseline and considerations for the continued development of a representative and effective reserve network. *Environmental Reviews, 22*, 135–160. https://doi.org/10.1139/er-2013-0056.

Angelstam, P. K. (1998). Maintaining and restoring biodiversity in European boreal forests by developing natural disturbance regimes. *Journal of Vegetation Science, 9*(4), 593–602. https://doi.org/10.2307/3237275.

Angelstam, P., Andersson, K., Isacson, M., et al. (2013). Learning about the history of landscape use for the future: Consequences for ecological and social systems in Swedish Bergslagen. *Ambio, 42*(2), 146–159. https://doi.org/10.1007/s13280-012-0369-z.

Angelstam, P., Manton, M., Green, M., et al. (2020). Sweden does not meet agreed national and international forest biodiversity targets: A call for adaptive landscape planning. *Landscape and Urban Planning, 202*, 103838. https://doi.org/10.1016/j.landurbplan.2020.103838.

Artdatabanken. (2020). *The Swedish redlist. Rödlistade arter i Sverige.* Uppsala: Swedish University of Agricultural Sciences (SLU).

Arts, B., Buizer, M., Horlings, L., et al. (2017). Landscape approaches: A state-of-the-art review. *Annual Review of Environment and Resources, 42*, 439–463. https://doi.org/10.1146/annurev-env iron-102016-060932.

Bengtsson, J., Nilsson, S. G., Franc, A., et al. (2000). Biodiversity, disturbances, ecosystem function and management of European forests. *Forest Ecology and Management, 132*(1), 39–50. https://doi.org/10.1016/S0378-1127(00)00378-9.

Bergeron, Y., Leduc, A., Harvey, B. D., et al. (2002). Natural fire regime: A guide for sustainable management of the Canadian boreal forest. *Silva Fennica, 36*(1), 81–95. https://doi.org/10.14214/sf.553.

Berglund, H., & Kuuluvainen, T. (2021). Representative boreal forest habitats in northern Europe, and a revised model for ecosystem management and biodiversity conservation. *Ambio, 50*, 1003–1017. https://doi.org/10.1007/s13280-020-01444-3.

Blicharska, M., Smithers, R. J., Mikusiński, G., et al. (2019). Biodiversity's contributions to sustainable development. *Nature Sustainability, 2*(12), 1083–1093. https://doi.org/10.1038/s41893-019-0417-9.

Bollmann, K., Kraus, D., & Paillet, Y. (2020). A unifying framework for the conservation of biodiversity in multi-functional forest landscapes. In F. Krumm, A. Schuck, & A. Rigling (Eds.), *How to balance forestry and biodiversity conservation—A view across Europe* (pp. 27–45). Birmensdorf: European Forest Institute. Swiss Federal Institute for Forest, Snow and Landscape Research.

Butterfield, B. J., Copeland, S. M., Munson, S. M., et al. (2017). Prestoration: Using species in restoration that will persist now and into the future. *Restoration Ecology, 25*, S155–S163. https://doi.org/10.1111/rec.12381.

Chazdon, R. L., Brancalion, P. H. S., Laestadius, L., et al. (2016). When is a forest a forest? Forest concepts and definitions in the era of forest and landscape restoration. *Ambio, 45*(5), 538–550. https://doi.org/10.1007/s13280-016-0772-y.

Chazdon, R. L., Brancalion, P. H. S., Lamb, D., et al. (2017). A policy-driven knowledge agenda for global forest and landscape restoration. *Conservation Letters, 10*(1), 125–132. https://doi.org/10.1111/conl.12220.

Convention on Biological Diversity (CBD). (2010). *Convention on biological diversity. Strategic plan for biodiversity 2011–2020, including Aichi Biodiversity Targets.* Montréal: Secretariat of the Convention on Biological Diversity.

Curtis, P. G., Slay, C. M., Harris, N. L., et al. (2018). Classifying drivers of global forest loss. *Science, 361*(6407), 1108–1111. https://doi.org/10.1126/science.aau3445.

Esseen, P. A., Ehnström, B., Ericson, L., et al. (1997). Boreal forests. *Ecological Bulletins, 46,* 16–47.

European Commission (EC). (2020). *Bringing nature back into our lives.* Communication from the commission to the European Parliament, the Council, the European Economic and Social Committee and the Committee of the Regions—EU Biodiversity Strategy for 2030. Brussels: European Commission.

Eyvindson, K., Duflot, R., Triviňo, M., et al. (2021). High boreal forest multifunctionality requires continuous cover forestry as a dominant management. *Land Use Policy, 100,* 104918. https://doi.org/10.1016/j.landusepol.2020.104918.

Felton, A., Löfroth, T., Angelstam, P., et al. (2020). Keeping pace with forestry: Multi-scale conservation in a changing production forest matrix. *Ambio, 49*(5), 1050–1064. https://doi.org/10.1007/s13280-019-01248-0.

Food and Agriculture Organization of the United Nations (FAO). (2020, February). *The UN Decade on ecosystem restoration 2021–2030* (p. 5). Factsheet. Rome: UNEP/FAO.

Frelich, L. E., Jogiste, K., Stanturf, J., et al. (2020). Are secondary forests ready for climate change? It depends on magnitude of climate change, landscape diversity and ecosystem legacies. *Forests, 11*(9), 965. https://doi.org/10.3390/f11090965.

Gamfeldt, L., Snäll, T., Bagchi, R., et al. (2013). Higher levels of multiple ecosystem services are found in forests with more tree species. *Nature Communications, 4,* 1340. https://doi.org/10.1038/ncomms2328.

Halme, P., Allen, K. A., Auniņš, A., et al. (2013). Challenges of ecological restoration: Lessons from forests in northern Europe. *Biological Conservation, 167,* 248–256. https://doi.org/10.1016/j.biocon.2013.08.029

Hanski, I. (2000). Extinction debt and species credits in boreal forests: Modelling the consequences of different approaches to biodiversity conservation. *Annales Zoologici Fennici, 37,* 241–280.

Harper, K. A., Macdonald, S. E., Mayerhofer, M. S., et al. (2015). Edge influence on vegetation at natural and anthropogenic edges of boreal forests in Canada and Fennoscandia. *Journal of Ecology, 103,* 550–562. https://doi.org/10.1111/1365-2745.12398.

Hedwall, P. O., & Mikusiński, G. (2015). Structural changes in protected forests in Sweden: Implications for conservation functionality. *Canadian Journal of Forest Research, 45,* 1215–1224. https://doi.org/10.1139/cjfr-2014-0470.

Hlásny, T., Barka, I., Kulla, L., et al. (2017). Sustainable forest management in a mountain region in the Central Western Carpathians, northeastern Slovakia: The role of climate change. *Regional Environmental Change, 17*(1), 65–77. https://doi.org/10.1007/s10113-015-0894-y.

Intergovernmental Science-Policy Platform on Biodiversity and Ecosystem Services (IPBES). (2018). *Summary for policymakers of the regional assessment report on biodiversity and ecosystem services for Europe and Central Asia of the Intergovernmental Science-Policy Platform on Biodiversity and Ecosystem Services.* In M. Fischer, M. Rounsevell, A. T.-M. Rando, A. Mader, A. Church, M. Elbakidze, V. Elias, T. Hahn, P. A. Harrison, J. Hauck, B. Martín-López, I. Ring, C. Sandström, I. S. Pinto, P. Visconti, N. E. Zimmermann, & M. Christie (Eds.) (p. 48). Bonn: IPBES Secretariat.

Jonsson, B. G., Svensson, J., Mikusiński, G., et al. (2019). European Union's last intact forest landscapes are at a value chain crossroad between multiple use and intensified wood production. *Forests, 10*(7), 564. https://doi.org/10.3390/f10070564.

Jonsson, M., Bengtsson, J., & Moen, J., et al. (2020). Stand age and climate influence forest ecosystem service delivery and multifunctionality. *Environmental Research Letters, 15*(9), 0940a0948. https://doi.org/10.1088/1748-9326/abaf1c.

Knoot, T. G., Schulte, L. A., & Rickenbach, M. (2010). Oak conservation and restoration on private forestlands: Negotiating a social-ecological landscape. *Environmental Management, 45*, 155–164. https://doi.org/10.1007/s00267-009-9404-7.

Kremen, C., & Merenlender, A. M. (2018). Landscapes that work for biodiversity and people. *Science, 362*(6412), eaau6020. https://doi.org/10.1126/science.aau6020.

Kuuluvainen, T., Tahvonen, O., & Aakala, T. (2012). Even-aged and uneven-aged forest management in boreal Fennoscandia: A review. *Ambio, 41*(7), 720–737. https://doi.org/10.1007/s13280-012-0289-y.

Kuuluvainen, T., Bergeron, Y., & Coates, K. D. (2015). Restoration and ecosystem-based management in the circumboreal forest: Background, challenges, and opportunities. In J. A. Stanturf (Ed.), *Restoration of boreal and temperate forests* (2nd edition, pp. 251–270). Boca Raton: CRC Press.

Kuuluvainen, T., Hofgaard, A., Aakala, T., et al. (2017). North Fennoscandian mountain forests: History, composition, disturbance dynamics and the unpredictable future. *Forest Ecology and Management, 385*, 140–149. https://doi.org/10.1016/j.foreco.2016.11.031.

Lilja, S., Wallenius, T., & Kuuluvainen, T. (2006). Structure and development of old *Picea abies* forests in northern boreal Fennoscandia. *Ecoscience, 13*(2), 181–192. https://doi.org/10.2980/i1195-6860-13-2-181.1.

Lindner, M., Maroschek, M., Netherer, S., et al. (2010). Climate change impacts, adaptive capacity, and vulnerability of European forest ecosystems. *Forest Ecology and Management, 259*(4), 698–709. https://doi.org/10.1016/j.foreco.2009.09.023.

Mansourian, S. (2017). Governance and forest landscape restoration: A framework to support decision-making. *Journal for Nature Conservation, 37*, 21–30. https://doi.org/10.1016/j.jnc.2017.02.010.

Mansourian, S. (2018). In the eye of the beholder: Reconciling interpretations of forest landscape restoration. *Land Degradation and Development, 29*, 2888–2898. https://doi.org/10.1002/ldr.3014.

Mansourian, S., Stanturf, J. A., Derkyi, M. A. A., et al. (2017). Forest landscape restoration: Increasing the positive impacts of forest restoration or simply the area under tree cover? *Restoration Ecology, 25*(2), 178–183. https://doi.org/10.1111/rec.12489.

McRae, B. H., Dickson, B. G., Keitt, T. H., et al. (2008). Using circuit theory to model connectivity in ecology, evolution, and conservation. *Ecology, 89*, 2712–2724. https://doi.org/10.1890/07-1861.1.

Mikusiński, G., Angelstam, P., & Sporrong, U. (2003). Distribution of deciduous stands in villages located in coniferous forest landscapes in Sweden. *Ambio, 32*, 520–526. https://doi.org/10.1579/0044-7447-32.8.520.

Mikusiński, G., Pressey, R. L., Edenius, L., et al. (2007). Conservation planning in forest landscapes of Fennoscandia and an approach to the challenge of countdown 2010. *Conservation Biology, 21*, 1445–1454. https://doi.org/10.1111/j.1523-1739.2007.00833.x.

Mikusiński, G., Orlikowska, E. H., Bubnicki, J. W., et al. (2021). Strengthening the network of high conservation value forests in boreal landscapes. *Frontiers in Ecology and Evolution, 8*, 595730. https://doi.org/10.3389/fevo.2020.595730.

Orlikowska, E. H., Svensson, J., Roberge, J.-M., et al. (2020). Hit or miss? Evaluating the effectiveness of Natura 2000 for conservation of forest bird habitat in Sweden. *Global Ecology and Conservation, 22*, e00939. https://doi.org/10.1016/j.gecco.2020.e00939.

Pennanen, J. (2002). Forest age distribution under mixed-severity fire regimes—A simulation-based analysis for middle boreal Fennoscandia. *Silva Fennica, 36*, 213–231. https://doi.org/10.14214/sf.559.

Potapov, P., Hansen, M. C., Laestadius, L., et al. (2017). The last frontiers of wilderness: Tracking loss of intact forest landscapes from 2000 to 2013. *Science Advances, 3*(1), e1600821. https://doi.org/10.1126/sciadv.1600821.

Sabatini, F. M., Keeton, W. S., Lindner, M., et al. (2020). Protection gaps and restoration opportunities for primary forests in Europe. *Diversity and Distributions, 26*(12), 1646–1662. https://doi.org/10.1111/ddi.13158.

Sandström, P., Cory, N., Svensson, J., et al. (2016). On the decline of ground lichen forests in the Swedish boreal landscape: Implications for reindeer husbandry and sustainable forest management. *Ambio, 45*(4), 415–429. https://doi.org/10.1007/s13280-015-0759-0.

Scheffer, M., Hirota, M., Holmgren, M., et al. (2012). Thresholds for boreal biome transitions. *Proceedings of the National Academy of Sciences of the United States of America, 109*, 21384–21389. https://doi.org/10.1073/pnas.1219844110.

Siitonen, J. (2001). Forest management, coarse woody debris and saproxylic organisms: Fennoscandian boreal forests as an example. *Ecological Bulletins, 49*, 11–41.

Snäll, T., Lehtomäki, J., Arponen, A., et al. (2016). Green infrastructure design based on spatial conservation prioritization and modeling of biodiversity features and ecosystem services. *Environmental Management, 57*(2), 251–256. https://doi.org/10.1007/s00267-015-0613-y.

Stanturf, J. A. (2015). Future landscapes: Opportunities and challenges. *New Forests, 46*, 615–644. https://doi.org/10.1007/s11056-015-9500-x.

Stanturf, J. A., Palik, B. J., & Dumroese, R. K. (2014). Contemporary forest restoration: A review emphasizing function. *Forest Ecology and Management, 331*, 292–323. https://doi.org/10.1016/j.foreco.2014.07.029.

Svensson, J., Andersson, J., Sandström, P., et al. (2019a). Landscape trajectory of natural boreal forest loss as an impediment to green infrastructure. *Conservation Biology, 33*(1), 152–163. https://doi.org/10.1111/cobi.13148.

Svensson, J., Mikusiński, G., & Jonsson, B. G. (2019b). *Green infrastructure in the boreal forest landscape* [in Swedish] (p. 76). Stockholm: Naturvårdsverket, Swedish Environmental Protection Agency.

Svensson, J., Bubnicki, J., Jonsson, B. G., et al. (2020a). Conservation significance of intact forest landscapes in the Scandinavian Mountains Green Belt. *Landscape Ecology, 35*(9), 2113–2131. https://doi.org/10.1007/s10980-020-01088-4

Svensson, J., Neumann, W., Bjärstig, T., et al. (2020b). Landscape approaches to sustainability—Aspects of conflict, integration and synergy in national public land-use interests. *Sustainability, 12*(12), 5113. https://doi.org/10.3390/su12125113.

Swedish University of Agricultural Sciences (SLU). (2020). *Forest statistics 2020 Official statistics of Sweden*. Umeå: Swedish University of Agricultural Sciences.

Tilman, D., May, R. M., Lehman, C. L., et al. (1994). Habitat destruction and the extinction debt. *Nature, 371*(6492), 65–66. https://doi.org/10.1038/371065a0.

Triviño, M., Pohjanmies, T., Mazziotta, A., et al. (2017). Optimizing management to enhance multifunctionality in a boreal forest landscape. *Journal of Applied Ecology, 54*(1), 61–70. https://doi.org/10.1111/1365-2664.12790.

Wang, J. A., Sulla-Menashe, D., Woodcock, C. E., et al. (2020). Extensive land cover change across Arctic-Boreal Northwestern North America from disturbance and climate forcing. *Global Change Biology, 26*(2), 807–822. https://doi.org/10.1111/gcb.14804.

Ward, M., Saura, S., Williams, B., et al. (2020). Just ten percent of the global terrestrial protected area network is structurally connected via intact land. *Nature Communications, 11*(1), 4563. https://doi.org/10.1038/s41467-020-18457-x.

Watson, J. E. M., Dudley, N., Segan, D. B., et al. (2014). The performance and potential of protected areas. *Nature, 515*(7525), 67–73. https://doi.org/10.1038/nature13947.

Zanotti, L., & Knowles, N. (2020). Large intact forest landscapes and inclusive conservation: A political ecological perspective. *Journal of Political Ecology, 27*, 539–557. https://doi.org/10.2458/v27i1.23165.

Chapter 20
Governance in the Boreal Forest: What Role for Local and Indigenous Communities?

Sara Teitelbaum, Hugo Asselin, Jean-François Bissonnette, and Denis Blouin

Abstract This chapter describes key trends in boreal forest governance in the twenty-first century and implications for the engagement of local and Indigenous communities. By focusing on three global trends—internationalization, marketization, and decentralization—we highlight the evolving role of local and Indigenous communities in increasingly hybrid and multiscale governance arrangements. We present two case studies, community forests in Canada and Sami–industry collaborative planning in Sweden, to analyze the qualities of local governance initiatives and how they seek to transform conventional approaches to economic development and land-use practices according to the values and priorities of local and Indigenous communities.

S. Teitelbaum (✉)
Département de Sociologie, Université de Montréal, Pavillon Lionel-Groulx, P.O. Box 6128, Stn. Centre-Ville, Montréal, QC H3C 3J7, Canada
e-mail: sara.teitelbaum@umontreal.ca

H. Asselin
School of Indigenous Studies, Université du Québec en Abitibi-Témiscamingue, 445, boulevard de l'Université, Rouyn-Noranda, QC J9X 5E4, Canada
e-mail: hugo.asselin@uqat.ca

J.-F. Bissonnette · D. Blouin
Département de Géographie, Faculté de Foresterie, de Géographie et de Géomatique, Université Laval, 2405 Rue de la Terrasse, Québec, QC G1V 0A6, Canada
e-mail: jean-francois.bissonnette@ggr.ulaval.ca

D. Blouin
e-mail: deblo27@ulaval.ca

© The Author(s) 2023
M. M. Girona et al. (eds.), *Boreal Forests in the Face of Climate Change*,
Advances in Global Change Research 74,
https://doi.org/10.1007/978-3-031-15988-6_20

20.1   Introduction

The boreal forest has taken center stage in environmental politics because of its status as one of the world's largest "intact" forest landscapes, its unique wildlife, and its role in the fight against climate change (Watson et al., 2018). While boreal forest conservation has become a key global priority, the boreal forest is also an inhabited landscape, which includes many culturally diverse communities with long-standing ties to the forest for cultural and subsistence purposes and other communities more actively engaged in the industrial development of natural resources (Nitoslawski et al., 2019). There are many Indigenous communities in the boreal forest whose identities, cultures, and livelihoods are closely connected to the land.

Conciliating environmental conservation and socioeconomic well-being in the boreal forest is a major challenge, especially in the context of global climate change (Gauthier et al., 2015). It requires coordinated efforts among a diversity of actors working at multiple scales. The term *environmental governance* is often invoked to describe the myriad processes through which decisions regarding the management and stewardship of the boreal forest are taken. According to Larson and Petkova (2011), "Governance refers to who makes decisions and how decisions are made, from national to local scale, including formal and informal institutions and rules, power relations and practices of decision-making" (p. 87).

This chapter looks at the evolution of boreal forest governance, with a specific focus on the role and influence of local and Indigenous communities. While the conflicts between environmentalists and industrialists over boreal forest protection are widely publicized, it is more difficult to characterize engagement on the part of local communities (Jensen, 2000; Patriquin et al., 2007; Willow, 2012). This may be related to the cultural diversity of communities, which includes Indigenous and non-Indigenous peoples. It may also be a consequence of variable histories of engagement in resource development and the presence of diverse and sometimes divergent sets of social values. However, what is clear is that since the introduction of the sustainability paradigm in the late 1980s, the notion that local people should be included in decision-making processes has been increasingly regarded as a priority. Local participation is lauded for several reasons: its purported ability to enhance accountability by bringing decisions closer to affected people; the improved integration of time- and place-specific knowledge—thereby enhancing environmental benefits; and its reduction of potential conflict by enhancing local buy-in (Lemos & Agrawal, 2006). Thus, across the board, new governance approaches, both corporate-driven, e.g., forestry certification, and government-driven, e.g., policies, regulations, emphasize advancing community participation.

The turn toward community participation in the forest sector has become the subject of a wide-ranging academic literature, which adopts a variety of lenses. The literature describes a range of governance approaches, from the more unidirectional processes associated with the public review of plans to the more institutionalized power-sharing arrangements, such as co-management boards and community forests

(Kittredge, 2003; Teitelbaum, 2016). However, despite important structural differences in objectives and design, research reflects the common observation that many arrangements do not meet local communities' expectations (Fuss et al., 2019). The notion of power sharing or devolution of authority is key to successful governance in many instances (Berkes, 2010). Indeed, given the long history of industrialized resource development in many boreal regions, it raises the question, central to this chapter, *To what extent has boreal forest governance evolved to include community-based approaches, and what do these look like?* Our analysis is based on an examination of the recent research literature in two major boreal forest countries (Canada and Sweden), including more than 95 articles, book chapters, and reports produced by academics and policymakers.

This chapter begins with a description of some of the historical experiences of local and Indigenous communities in each country. We then set the broader context for participatory governance in the boreal forest through a description of some key trends in forest governance since the 1980s and the implications for the participation of local and Indigenous communities. We finally describe local governance initiatives in each jurisdiction and focus on how they are seeking to transform local economies and predominant forms of land use, despite what are considerable obstacles.

20.2   A Portrait of Forest-Dependent Communities in the Boreal Regions of Canada and Sweden

In Canada, the vast majority (94%) of forests are under public ownership (Natural Resources Canada, 2020). Most fall under the jurisdiction of provincial governments and are allocated to forestry companies under long-term licenses. Historically, many industry-based communities prospered under the patronage of forestry companies, supported by governmental investment in wood processing and manufacturing under an *even-flow* policy regime. However, since the 1980s, the forestry industry has undergone significant structural changes because of the growing influence of global market forces and the introduction of neoliberal policies. While this has led to mill closures and consolidations, it has also resulted in a less *hands-on* approach on the part of forestry companies. This has revealed some of the underlying weaknesses within forest-dependent communities, including insufficient economic diversification, a lack of skilled labor, and limited community capacity (Patriquin et al., 2007). Communities are also increasingly facing risks associated with climate change, e.g., increased incidence of forest fire and insect outbreaks and changes in species composition, creating problems related to wood supply (Davidson et al., 2003; Podur et al., 2002).

Indigenous communities in the boreal region have much longer relationships to forest lands and a different relationship to the forest sector. The traditional territories of Indigenous boreal peoples in Canada cover vast forest areas and continue to support livelihoods and cultures (Saint-Arnaud et al., 2009; Smith, 2015). In most boreal forest regions in Canada, Indigenous communities are covered by historical treaties, which set out certain limited hunting, fishing, trapping, and gathering rights (RCAP, 1996). However, Indigenous people face legal barriers to having their treaty rights respected, as the courts tend to lean on the side of extractive industries (McCrossan, 2018). Other nations have entered into modern-day agreements, such as the Innu in Labrador, the Cree and Naskapi in Québec, the Tłįchǫ in the Northwest Territories, and several First Nations in the Yukon Territory (see Samson, 2016). Finally, some Indigenous peoples—mostly in Québec and British Columbia—have yet to sign any form of land-claim agreement with the government.

Historically, Indigenous communities were excluded from the benefits of resource development and suffered many negative impacts from resource development (Teitelbaum, 2015). This continues to be the case, as evidenced by the ongoing campaigns of Indigenous groups, including in the courts, to block resource development or to have their grievances addressed. Many Indigenous communities face high unemployment and see little direct economic benefit from resource development (Proulx et al., 2020). However, in recent decades, Indigenous peoples have strengthened their political actions in pursuit of the recognition of Indigenous rights, the settlement of outstanding land claims, and the redistribution of resources (Lawler & Bullock, 2017; Pinkerton, 2019; Wyatt et al., 2019). More recently, Indigenous participation in the forest sector has increased in some provinces, in part through the allocation of forest tenures. Indigenous-held forest tenures increased from 7 to 19 million $m^3 \cdot yr^{-1}$ between 2002 and 2017, i.e., from 4 to 10.5% of the Canadian total of forest tenures (NAFA, 2003, 2018). Some Indigenous communities are developing alternative avenues to forestry development, for example through offering tourism and recreation activities or the development of nontimber forest products. Indeed, there are increasing calls for community-centered approaches to land use, stewardship, and local development to foster reconciliation and more sustainable patterns of land use (Baldwin, 2003; Patriquin et al., 2007).

In Sweden, ownership patterns and the history of forestry development differ markedly from that of the Canadian context. Roughly 50% of Swedish boreal forests are owned by small and family enterprises, whereas the other half is split more or less equally between large companies and the state (Skogsstyrelsen, 2015; Stjernström et al., 2017). Small-scale forest farms are thus an important economic model, which often combines forestry with farming activities. Since the early twentieth century, forest owners have collectivized their activities by creating forest cooperatives or associations that use management techniques akin to those of large forestry companies. There are three forest-owner associations in Sweden, which collectively represent approximately 112,000 members who own and manage 6.2 million ha (Lantbrukarnas Riksförbund, 2014). The goal of the owner associations is to ensure better market access, offer forest management services, and play an advocacy role in defending the rights of forest owners. Some have also invested in mills and installations for the energy sector (Skogsstyrelsen, 2015). Forest cooperatives are represented by the National Federation of Family Forest Owners, which has a national and international presence with the European Union. Indeed, forests in Sweden are greatly valued for the recreational opportunities they provide to all citizens (known as *allemansrätten,* or right of public access; Stjernström et al., 2017). In Sweden, usufructuary rights to the forest are granted to all citizens. This allows them to access land (whether public or private) to pick berries, gather mushrooms, camp, or pursue outdoor activities. Hunting is also very popular, especially moose hunting. This ethic of public access is highly developed in Sweden and has been likened to a type of collective responsibility. "The idea of everyman's right forms the basis for a culture of stewardship. It defines a framework for community access to public forest lands, and indeed to the landscape as a whole" (Bullock & Hanna, 2012, p. 149). Reindeer husbandry is under the exclusive, constitutionally protected rights of the Sami Indigenous people (Moen & Keskitalo, 2010). Reindeer-herding areas cover approximately 55% of the Swedish land base, i.e., 23 million ha (Skogsstyrelsen, 2015). The territory used by reindeer herders is divided into 51 reindeer-herding communities, many of which overlap with commercial forestry lands. Reindeer husbandry relies on large grazing grounds, as only natural low productive vegetation is used for forage. Forestry operations can affect reindeer husbandry through forest fragmentation, forest age structure changes, and increased infrastructure, such as roads (Berg et al., 2008; Kivinen et al., 2012). Thus, the Swedish Forest Agency (*Skogsstyrelsen*) and certification systems have helped implement a consultation regime between Sami and forestry companies.

Although both Canada and Sweden have seen an increase in consultative requirements, Indigenous peoples and local communities continue to express ongoing concerns in regard to their real level of influence in forest-related decisions (Reed, 2010; Sandström & Widmark, 2007) and in relation to ecological degradation from resource development and the lack of tangible benefits from this resource exploitation. Many communities are seeking avenues to assert greater influence over forestry governance processes and build economic development strategies that are in line with community aspirations.

20.3 Global Governance Trends: Internationalization, Marketization, Decentralization/Devolution

Since the 1980s, the overarching political and economic context for forestry governance in the boreal forest has shifted considerably. This has created new opportunities for community participation and institutional innovation, but it has also created new challenges related to what is an increasingly globally competitive and technologically intensive environment. There is also increasing pressure for governance initiatives to demonstrate their adherence to sustainable development objectives, including biodiversity conservation, climate change mitigation, and social justice. This procedural shift toward a sustainability paradigm is reflected at different scales, from the local to the global, and has had impacts on the forms of governance being promoted and experimented with by government, industry, and civil society actors.

In the following section, we describe three trends in forest governance in the twenty-first century: (1) *internationalization*, (2) *marketization*, and (3) *decentralization* (Fuss et al., 2019; Lemos & Agrawal, 2006); we also reflect on how these trends are influencing opportunities for local and Indigenous communities.

20.3.1 Internationalization

The conservation and protection of forests have been part of the global environmental agenda since the 1980s. However, building a consensus around an international forest policy agenda has proved challenging. So far, efforts to convene a legally binding international agreement for forests have been unsuccessful. Indeed, international forest policy has been described as a "fragmented regime with a conflictive rather than cooperative architecture" (Howlett et al., 2010, p. 93).

Instead, international forest policy reflects a multipronged strategy that combines a number of instruments, including sectoral agreements, multilateral policies, and programs, many based on voluntary or soft policy approaches. These have been classified in multiple ways. For example, according to Humphreys in McDermott et al. (2010), the international forest regime covers:

- a growing body of soft international law focused on forests such as Chapter 11 of Agenda 21 and the United Nations Strategic Plan for Forests adopted in 2017 (see Sotirov et al., 2020 for further examples)
- hard international legal instruments with a forest-related mandate (e.g., the Convention on Biological Diversity; United Nations Framework Convention on Climate Change)
- voluntary private sector regulation, such as the Forest Stewardship Council principles for forest management

All these international instruments encompass commitments aimed at protecting the rights of local and Indigenous communities. Broad goals, such as the preservation of traditional knowledge, the promotion of equitable sharing of benefits, poverty eradication, and support for forest-based development and the rights to enhanced participation in forest governance—including the right to free, prior, and informed consent—are part of many international policy initiatives (Arts & Babili, 2013). One example is the Convention on Biological Diversity, signed in 1992 at the Earth Summit in Rio de Janeiro (United Nations, 1992). There are a number of social goals inscribed within the agreement, including Article 8j, which seeks to preserve and maintain traditional knowledge for the conservation and sustainable use of biodiversity, and Article 10c, which seeks to promote traditional cultural practices that meet conservation or sustainable use requirements. International agreements on climate change have also paid increasing attention to the role of forests and forest-dependent communities in the fight against climate change (Rayner et al., 2010).

However, international forest policy has been criticized for focusing predominantly on tropical forests and underrepresenting the importance of boreal forests both from a climate change and biodiversity perspective (Moen et al., 2014; Warkentin & Bradshaw, 2012). According to Moen et al. (2014), the escalating impacts of climate change in boreal forests, e.g., increased severity and frequency of forest fires, insect outbreaks, combined with accelerated harvesting justifies rapid international policy action to offset ecological risks and capitalize on existing management approaches and institutions in boreal countries. Warketin and Bradshaw (2012) argue that this requires more extensive forest reserve systems to foster carbon sequestration, the incorporation of climate and predictions about shifts in ecosystem dynamics into management, and a stronger focus on reforestation, especially in Russia where deforestation and fragmentation are most pronounced. International policy and instruments are recognized as a potentially important source of financing to support action on climate change and the preservation of biodiversity (Hoogeveen & Verkooijen, 2010). Climate commitments should also create new business opportunities for local communities, for example through the development of wood as a source of renewable bioenergy (Fuss et al., 2019).

There is also a growing network of nongovernmental organizations collaborating on issues related to boreal forest conservation and operating across national boundaries, such as the Nature Conservancy, the World Wildlife Fund, and Greenpeace. A notable example was the establishment of the Canadian Boreal Forest Agreement (CBFA), described as "the world's largest conservation agreement which incorporates both environmental and economic values" (CPAWS-Saskatchewan, 2021). Funded, in part, by foundations from the United States, this was a voluntary agreement involving forestry companies and environmental NGOs aimed at protecting habitat for woodland caribou through restrictions on forest harvesting in sensitive habitats in exchange for the suspension of environmental NGO campaigns against industry. However, the CBFA suffered from a loss of credibility related, in part, to the exclusion of Indigenous peoples from negotiations, and this agreement was ultimately unsuccessful (Fuss et al., 2019). There have also been international efforts to build research capacity and to help actors measure and track progress toward implementing sustainable forest management (Linser et al., 2018). Organizations such as the United Nations Intergovernmental Forum on Forests (IFF) and the Food and Agriculture Organization (FAO) have supported processes to develop criteria and indicators of sustainable forest management at regional levels, including their integration into Agenda 21 (McDermott et al., 2010).

20.3.2   Marketization

Another dominant strategy in the pursuit of sustainable forest management in the boreal forest is the use of market-based approaches. Rather than being founded in traditional legal or regulatory approaches, these are driven by the private sector and civil society actors and focus on enhancing corporate responsibility. Most are based on voluntary corporate action via a commitment to a sustainability-centered norm or policy through an incentive-based system. Market-based approaches are described as part of the neoliberal turn within environmental governance because they minimize/displace the role of government as the central source of decision-making authority (Krott et al., 2014). They are often promoted as "win–win" opportunities that marry economic efficiency—market-based approaches are described as innovative and cost-effective—and environmental protection. There is increasing evidence of government involvement and support of market-based approaches, leading some to describe regulatory regimes as *hybrid* or *intersecting* (Bostrom, 2003; Schneiberg & Bartley, 2008). Examples include corporate codes of conduct and forest certification. Market-based approaches have spurred important debates within civil society and academic circles regarding the degree to which they facilitate systemic change in practices (Klooster, 2010; McCarthy, 2006).

In boreal regions, including Canada, Sweden, and Russia, the leading example of market-based instruments are forestry certification standards (Chap. 21), which have made huge gains in recent decades. Forestry certification is based on corporate conformance to a forest management standard, covering social, environmental, and economic aspects. Performance is most often verified by third-party auditors, and the successful adherence to the standard is rewarded through the opportunity to use the logo, which in theory confers a certain market advantage (Rametsteiner & Simula, 2003). Several certification systems compete for space internationally, including Programme for the Endorsement of Forest Certification (PEFC) standards and the Forest Stewardship Council (FSC). The FSC standard was created in 1994 by environmental and civil society groups along with industry partners in response to the failure of governments to develop a binding international forest agreement. PEFC, an industry-based system, was created a few years later; it is based on the endorsement of existing certification standards that comply with PEFC's international and regional criteria.

Both FSC and PEFC include commitments to local and Indigenous communities. FSC, widely considered the most stringent in this area, integrates the principle of free, prior, and informed consent (FPIC) into its international and national standards (Mahanty & McDermott, 2013; Teitelbaum et al., 2021), whereas PEFC recognizes Indigenous rights through written policies, communications, and the protection of cultural sites. However, Indigenous people have expressed discontent with certification, questioning its ability to adequately protect their rights and denouncing power asymmetries in favor of industrial stakeholders (Johansson, 2014; Tikina et al., 2010). Furthermore, although certification systems require forestry companies to engage in public participation and include measures for the protection of local forest-based activities, Indigenous people have raised concerns over the insufficient evaluation of their use of land and resources (Teitelbaum & Wyatt, 2013), leading to inadequate protection and rehabilitation of biocultural landscapes (Meadows et al., 2019). From a governance perspective, certification has helped reconfigure relationships and adds a new level of oversight and transparency to forest management (Johansson, 2014; Sandström & Widmark, 2007; Tikina et al., 2010). For example, the FSC's Permanent Indigenous Peoples Committee allows Indigenous people to be involved in standard development and review (Meadows et al., 2019).

20.3.3 Decentralization

Decentralization can be defined as the transfer of powers from the central government to lower-level actors and institutions (Agrawal & Ribot, 1999). Others use the terms *devolution* or *community-based management* to refer to initiatives that provide enhanced decision-making authority to local communities (Ambus & Hoberg, 2011). What unites these different approaches is the dispersion of points of decision-making to new actors and institutions, usually toward the local or regional level (Bissonnette et al., 2020). Since the 1990s, decentralization of natural resource governance has become a popular approach with international organizations, aid agencies, and state-based agencies, especially in the global South (World Bank, 1999). Disappointed with the shortcomings of centralized and top down resource governance, decentralization was seen as an avenue with the potential to enhance participation and equity in resource management (Larson & Petkova, 2011). Whether through local government agencies or community-based institutions, these organizations were seen as being closer to affected populations and thus better able to include their views, reflect their concerns, and capitalize on local knowledge and priorities when designing appropriate development strategies. However, decentralization is often administrative (from central governments to local branches of central governments) rather than political (from central governments to local communities; Ribot et al., 2006). This has created difficulties for local governments and community organizations who often find themselves charged with operational responsibilities, whereas the more strategic aspects remain in the hands of central governments. Some researchers observe that the rise of decentralization is synonymous with a neoliberal shift within policymaking, which has resulted in the imposition of administrative responsibilities on lower institutional levels without the corresponding authority, political power, or financial resources to manage forests effectively (McCarthy, 2006).

Both in Sweden and Canada, comparisons of various management scenarios and forest simulation studies have shown that taking into account the needs and views of Indigenous people only marginally reduces profits from logging, while increasing social acceptability and maintaining cultural and biological diversity (Asselin et al., 2015; Dhital et al., 2013; Horstkotte et al., 2016; Korosuo et al., 2014). With that in mind, decentralization could theoretically allow for greater autonomy and self-government for Indigenous communities. In Québec, different types of delegation agreements are defined in the forest management regime, many of which have been used by band councils, including those of the Atikamekw (Fortier & Wyatt, 2019) and the Mi'kmaq (Blouin et al., 2020). The Cree Nation of Québec, working with provincial authorities, has been crafting culturally sensitive forestry arrangements on its ancestral lands, which are under a modern treaty (Jacqmain et al., 2012). In northern Saskatchewan, forestry co-management has been in place through Mistik Management, which is based on a participatory approach led by industry. However, in most cases, forest-related arrangements between Indigenous communities and provincial authorities remain small-scale and with a limited scope in terms of land control and governance (Blouin et al., 2020). Moreover, administrative procedures required to set up and pursue even limited agreements involve costs that are often prohibitive (Lawler & Bullock, 2019). In addition, there are concerns that upon signing delegation agreements, Indigenous peoples in Canada are forced to accept institutional parameters of the state "whose strategy consists essentially in consolidating its colonial (and racist) sway over Indigenous peoples" (Salée & Lévesque, 2010, p. 101). Nevertheless, it can also be argued that Indigenous peoples have the capacity "to advance their cause and navigate efficiently and creatively past the state's roadblocks on the path to political autonomy" (Salée & Lévesque, 2010, p. 101).

20.4 Further Examination of Decentralization in Boreal Regions

20.4.1 Case Study of Community Forests in Canada

Community forestry is a broadly accepted if somewhat mythologized term in Canada. Both rural and Indigenous communities across Canada have manifested their discontent with the industrial–corporate model of forestry stemming from the perception that insufficient benefits are being retained in local communities and because of concerns that forestry is causing long-term damage to ecosystems, including water quality (Teitelbaum, 2016). Thus, community forestry is synonymous with an alternative form of development, which is seen to increase local decision-making over forest resource use and management by developing forestry practices that reflect community objectives and values while improving cultural, ecological, and economic sustainability (Bullock & Hanna, 2017; McIlveen & Rhodes, 2016). A variety of practices and institutional arrangements fall under the umbrella term of community forestry (Teitelbaum et al., 2006). However, most arrangements take place between local communities, usually represented by an organization, i.e., municipality, NGO, Indigenous band council, provincial public land management authorities, and, in some cases, private forestry companies. As a result, community forest initiatives mainly rest on complex arrangements that often require the devolution of power by provincial authorities to local organizations (Fuss et al., 2019).

Progress toward implementing community forestry in Canada has mainly occurred in provinces that have made legal reforms in that direction, often in response to conflicts surrounding forest use and community dissatisfaction with the extent of participation in decision-making (Bullock & Hanna, 2012; Lawler & Bullock, 2017). In the case of Indigenous-run forests, some arrangements have come about as the result of political negotiations regarding land rights. One well-known initiative in Canada is the *British Columbia Community Forest Agreement*. Initially started as a pilot program in 1998, it was eventually made an official tenure, allowing the provincial government to grant community forest tenures to organizations such as local governments, Indigenous communities, and community groups through 25-year renewable leases. There are now more than 50 community forests of this type in

British Columbia (Government of British Columbia, 2020a) of which approximately one-quarter are held by Indigenous communities. Many of these initiatives have achieved their goals, such as increasing local benefits from forestry and providing jobs in small timber-dependent communities. The Burns Lake Community Forest, located in the north-central interior of British Columbia, is often cited as an exemplary case of a successful community-based forestry operator (McIlveen & Bradshaw, 2009). However, pressure on the forest sector exerted by the mountain pine beetle epidemic and forest fires has endangered the economic stability of the Burns Lake Community Forest, revealing some of the vulnerabilities of operating on a smaller scale in a context dominated by large and highly industrialized firms (McIlveen & Rhodes, 2016). Indeed, despite its success, the design of the BC community tenure has been criticized, as it is seen as replicating pre-existing provincial industrial land-based forest tenures, which provide limited flexibility and authority to tenure holders (Ambus & Hoberg, 2011). British Columbia has also created a tenure for Indigenous communities—the *First Nations Woodland Licences*—of which there are 19, covering an area of 3,795,000 ha (Government of British Columbia, 2020b). Nevertheless, the extent of governmental devolution is also criticized here. As with the Burns Lake Community Forest, "the emphasis remains on timber production with all final decisions regarding forest management continuing to be held by the Ministry of Forests and Range" (Trosper & Tindall, 2013, p. 313).

Ontario and Québec have also made reforms in the direction of community forestry. The 2009 Ontario Forest Tenure Modernization Act, although yielding mixed results, exemplifies some of the measures deployed by provincial governments to reform forest tenure and grant more power to resource-dependent communities (Palmer et al., 2016). For example, the province created Local Forest Management Corporations (LFMCs), Crown agencies responsible for vast forest territories, which include community and Indigenous representatives on the board of directors. It is difficult to compare these LFMCs with community forests elsewhere in Canada, as they remain very much in line with the large-scale and industrialized approach to forest management.

In Québec, the 2010 Forest Regime includes a provision on community forests through the concept of *Local Forest* (usually referred to as *forêt de proximité*), which involves the extensive delegation of responsibilities. Although this possibility has elicited much enthusiasm among forest stakeholders, its large-scale implementation has been delayed numerous times (Bissonnette et al., 2020). Nevertheless, in Ontario, as in Québec, community forest initiatives have been implemented on public lands, often through community-based or municipal management corporations that established mutually beneficial partnerships with logging companies, outside formal arrangements provided by existing legal frameworks, i.e., Maria-Chapdelaine in Saguenay-Lac-Saint-Jean, Québec, and the Enhanced Sustainable Forest License, in northern Ontario (Fournier, 2013; Lachance, 2017). One of the greatest barriers to the development of community forestry in the boreal forest is the organization of the forestry sector around large-scale industrial logging, which has constrained the capacity for innovation in tenure. In southern Québec and Ontario, municipal forests exhibit innovative forms of community-based governance processes that depend on

local citizen participation (Bissonnette et al., 2020). A primary concern of municipal, Indigenous, and conservation authorities in Ontario and Québec, for example,
is the protection and enhancement of ecosystem services (Teitelbaum & Bullock,
2012; Uprety et al., 2017). In these cases, clear tenure rights and the absence of pre-
existing area-based agreements with logging companies provide local stakeholders
with more power to implement community forest practices and allow communities
to set up alternatives to the productivist forest regime present in Canada since the
beginning of the industrial era (Blais & Boucher, 2013).

20.4.2 Case Study of Collaborative Planning Between the Forest Industry and the Sami in Sweden

The Sami Indigenous people have a usufructuary right to practice reindeer husbandry,
which takes place in about 75% of the forest area in northern Sweden, including
both public and private lands (Johansson, 2014). However, frequent conflicts arise
with forest companies, making it difficult for the Sami to assert their rights, despite
compulsory consultation procedures having been introduced by the Swedish government in the 1970s in year-round grazing areas (Widmark, 2006) and recently extended
to all grazing areas by the FSC certification standard (FSC Sweden, 2010). On the
one hand, timber harvesting removes not only trees but also lichen, the reindeer's
preferred winter food. On the other hand, preserving older, lichen-rich forests exclusively for reindeer grazing leads to lost timber revenues (Bostedt et al., 2003). Joint
management could simultaneously benefit the forest industry and the Sami reindeer
herders by using selective cuts instead of clear-cuts (Berg et al., 2008; Korosuo et al.,
2014); however, there are currently no joint management initiatives in Sweden, except
for a few experiments (e.g., Stjernström et al., 2020). Moreover, selective cuts are
not allowed by the Swedish Forestry Act (*Skogsvårdslagen*) because they allegedly
do not allow for meeting forest regeneration objectives. In one joint management
experiment, Sandström et al. (2006) used a collaborative learning technique to bring
together five forestry representatives and five Sami representatives to evaluate seven
scenarios describing alternative future relationships. They identified six overarching
needs that should be addressed to improve relationships: (1) agree on a common definition of what consultation is; (2) adopt a long-term perspective; (3) consult earlier
in the planning process; (4) improve consultation tools, e.g., maps, by using both
scientific and Indigenous knowledge; (5) value different activities on the land; and (6)
elaborate a conflict resolution strategy. The importance of adopting co-management
is increasingly evident, as climate change affects both the forest industry and reindeer
husbandry, both of which would benefit from working together toward the adaptation of the entire socioecological system (Moen & Keskitalo, 2010; Pape & Löffler,
2012).

In 2017, the Swedish government proposed a bill on the obligation to consult Sami people (Larsen & Raitio, 2019). The proposal was severely criticized on both industrial and Sami fronts. First, the industry feared the bill would increase uncertainty over resource access and threaten economic interests. Second, the Sami parliament denounced the first draft of the bill, arguing that it did not allow meaningful influence on decision-making and failed to comply with international standards for protecting Indigenous cultures and rights. When this chapter was written, the bill had yet to be adopted. Meanwhile, the FSC National Forest Stewardship Standard of Sweden was revised in 2020 and now requires large forest owners to engage in a participatory planning process with reindeer-herding communities, which "can choose to give consent to the proposed management activity, together with the considerations and any adaptations that are agreed upon in the participatory planning process, or choose not to give consent to the activity" (FSC, 2020, p. 25). However, the new standard goes on to specify that, in case of dispute, if the parties cannot reach an agreement despite all the conflict resolution and mediation measures in place, "it is up to [the company] to either: (a) raise the management activity for participatory planning again once the forest grazing conditions have changed or; (b) carry out the activity without the consent of the [reindeer-herding community]" if the company can show that the Sami demands would substantially affect long-term forest management or that the Sami did not provide a sufficiently clear account of how the activity would disturb reindeer herding (FSC, 2020, p. 26). Hence, while timid advances are being made toward increased Sami participation in decision making, the search continues for an effective collaborative planning process. To this end, a pilot project of innovative land-use planning was undertaken in the municipality of Vilhelmina (Bjärstig et al., 2019). The project revealed the importance of (1) personally contacting participants and making sure all interest groups are represented; (2) jointly establishing a timeline; (3) agreeing on responsibilities; (4) setting clear objectives; (5) building capacity and involving the locals in drafting the plan (rather than merely being consulted on it); (6) providing participatory mapping tools; (7) relying on a neutral external moderator to facilitate the meetings; and (8) providing multiple occasions for participants to react on and validate the plan, both individually and during group meetings.

20.5   Conclusions

In this chapter, we explored the role played by local and Indigenous communities in boreal forest governance, focusing on the Canadian and Swedish contexts. The impending transformation of the boreal forest because of global environmental change will require making difficult management decisions to ensure boreal forests continue to play their key ecological functions, e.g., contribution to biodiversity and carbon sequestration. It is increasingly recognized that local and Indigenous communities must be involved in forest governance, in a bottom-up manner, for management decisions to be in phase with the local context and garner social acceptability. Moreover, it is now widely recognized that local and Indigenous ecological knowledge can significantly contribute to improving forest ecosystem management and reduce the impacts associated with large-scale industrial logging (Angelstam et al., 2011; Asselin, 2015). We emphasized three key trends that influence the level of involvement of local and Indigenous communities in boreal forest governance: internationalization, marketization, and decentralization. These trends reveal the growing importance of nonstate actors in boreal forest governance and hence the complex interactions among environmental NGOs, public authorities, Indigenous communities, and forest industries. This analysis revealed that governance in boreal forests is fragmented and is characterized by a diverse set of national and global policy instruments, including voluntary approaches. The limited reach of international regulation frameworks in boreal forest management, i.e., the Convention on Biological Diversity's Aichi Targets, highlights the need for public participation to elaborate management guidelines to ensure the resilience of the socioecological system. The role of national governmental authorities in boreal forest management, whether planning forestry activities or devolving this responsibility to industry, has raised concerns over the possibility of ensuring adequate participation of local and Indigenous communities. The alleged insufficiency of national regulatory frameworks and a lack of international hard law on boreal forest governance have partly been filled in Canada and Sweden by market-driven initiatives such as certification standards. Although coupled with the pursuit of forest exploitation, the most stringent standards, e.g., FSC, can, in some cases, be more rigorous and demanding than national forest laws, allowing for greater protection of biological and cultural diversity. Indigenous peoples have found through certification a forum to not only express their views and needs but to directly influence policymaking. However, there remains an important gap between the aspirations of Indigenous peoples with regard to land stewardship and the progressive changes brought about through certification (Johansson, 2014; Teitelbaum & Wyatt, 2013). More importantly, certification standards purportedly deepen market-based relations and reinforce a neoliberal logic that is considered contrary to values defended by many local and Indigenous communities (Klooster, 2010).

A growing community of researchers and advocacy groups is calling for the implementation of community forest initiatives that support local and Indigenous visions, recognize the value of community involvement in forest management, and support the diversification of forest uses to enhance social and ecological resilience. However, decentralization still too often equates with the mere transfer of power from the central government to its regional constituents instead of a real devolution to local and Indigenous communities (Ribot et al., 2006). The large-scale industrial forest exploitation model is embedded in production-based boreal forest tenure systems, which drastically constrains local communities' involvement and the diversification of forestry practices. As we have demonstrated, the case studies examined here face a number of challenges related to the scope of decision-making authority, regulatory flexibility, and economies of scale. However, these initiatives nonetheless represent clear examples of institutional innovations, which are forging a new path in regard to the conciliation of timber-related objectives with other community priorities related to the integration of sociocultural values and the protection of ecosystems.

References

Agrawal, A., & Ribot, J. C. (1999). Accountability in decentralization: A framework with South Asian and West African cases. *Journal of Developing Areas, 33*(4), 473–502.

Ambus, L., & Hoberg, G. (2011). The evolution of devolution: A critical analysis of the Community Forest Agreement in British Columbia. *Society & Natural Resources, 24*(9), 933–950. https://doi.org/10.1080/08941920.2010.520078.

Angelstam, P., Axelsson, R., Elbakidze, M., et al. (2011). Knowledge production and learning for sustainable forest management on the ground: Pan-European landscapes as a time machine. *Forestry, 84*(5), 581–596. https://doi.org/10.1093/forestry/cpr048.

Arts, B., & Babili, I. (2013). Global forest governance: Multiple practices of policy performance. In B. Arts, J. Behagel, S. van Bommel, J. de Koning, & E. Turnhout (Eds.), *Forest and nature governance. A practice based approach* (pp. 111–32). Dordrecht: Springer.

Asselin, H. (2015). Indigenous forest knowledge. In K. S.-H. Peh, R. Corlett, & Y. Bergeron (Eds.), *Routledge handbook of forest ecology* (pp. 586–596). London: Earthscan, Routledge.

Asselin, H., Larouche, M., & Kneeshaw, D. (2015). Assessing forest management scenarios on an Aboriginal territory through simulation modeling. *The Forestry Chronicle, 91*(4), 426–435. https://doi.org/10.5558/tfc2015-072.

Baldwin, A. (2003). The nature of the boreal forest: Governmentality and forest-nature. *Space and Culture, 6*(4), 415–428. https://doi.org/10.1177/1206331203253189.

Berg, A., Östlund, L., Moen, J., et al. (2008). A century of logging and forestry in a reindeer herding area in northern Sweden. *Forest Ecology and Management, 256*(5), 1009–1020. https://doi.org/10.1016/j.foreco.2008.06.003.

Berkes, F. (2010). Devolution of environment and resources governance: Trends and future. *Environmental Conservation, 37*(4), 489–500. https://doi.org/10.1017/S037689291000072X.

Bissonnette, J.-F., Blouin, D., Bouthillier, L., et al. (2020). Vers des forêts de proximité en terres publiques? Le « bricolage » institutionnel comme vecteur d'innovation en foresterie communautaire au Québec, Canada. *Revue Gouvernance, 17*(2), 52–77. https://doi.org/10.7202/1073111ar.

Bjärstig, T., Thellbro, C., Zachrisson, A., et al. (2019). Implementing collaborative planning in the Swedish mountains—The case of Vilhelmina. *WIT Transactions on Ecology and the Environment, 217*, 781–795.

Blais, R., & Boucher, J. L. (2013). Les temps des régimes forestiers au Québec. In G. Chiasson & E. Leclerc (Eds.), *La gouvernance locale des forêts publiques québécoises: Une avenue de développement des régions périphériques?* (pp. 53–84). Québec: Presses de l'Université du Québec.

Blouin, D., Bissonnette, J.-F., & Bouthillier, L. (2020). Vers l'émergence d'une gouvernance territoriale régionale autochtone ? Parcours des Mi'gmaq de Gespeg pour transformer la gestion des forêts publiques de leur territoire ancestral au Québec, Canada. *Revue Gouvernance, 17*(2), 78–104. https://doi.org/10.7202/1073112ar.

Bostedt, G., Parks, P. J., & Boman, M. (2003). Integrated natural resource management in northern Sweden: An application to forestry and reindeer husbandry. *Land Economics, 79*(2), 149–159. https://doi.org/10.2307/3146864.

Bostrom, M. (2003). How state-dependent is a non-state-driven rule-making project? The case of forest certification in Sweden. *Journal of Environmental Policy & Planning, 5*(2), 165–180. https://doi.org/10.1080/1523908032000121184.

Bullock, R. C. L., & Hanna, K. S. (2012). *Community forestry: Local values, conflict and forest governance.* Cambridge: Cambridge University Press.

Bullock, R. C. L., & Hanna, K. S. (2017). Community forestry: Mitigating or creating conflict in British Columbia? *Society & Natural Resources, 21*(1), 77–85. https://doi.org/10.1080/08941920701561007.

Canadian Parks and Wilderness Society—Saskatchewan Chapter (CPAWS-Saskatchewan). (2021). *Canadian Boreal Forest Agreement: 11 years later*. Saskatoon: CPAWS Saskatchewan. Retrieved July16, 2022, from https://cpaws-sask.org/canadian-boreal-forest-agreement-11-years-later/.

Davidson, D. J., Williamson, T., & Parkins, J. R. (2003). Understanding climate change risk and vulnerability in northern forest-based communities. *Canadian Journal of Forest Research, 33*(11), 2252–2261. https://doi.org/10.1139/x03-138.

Dhital, N., Raulier, F., Asselin, H., et al. (2013). Emulating boreal forest disturbance dynamics: Can we maintain timber supply, aboriginal land use, and woodland caribou habitat? *The Forestry Chronicle, 89*(1), 54–65. https://doi.org/10.5558/tfc2013-011.

Forest Stewardship Council (FSC). (2020). *The FSC national forest stewardship standard of Sweden FSC-STD-SWE-03-2019 EN* (p. 102). Bonn: Forest Stewardship Council.

Forest Stewardship Council (FSC) Sweden. (2010). *Swedish FSC standard for forest certification including SLIMF indicators, FSC-STD-SWE-02-04-2010 Sweden Natural, Plantations and SLIMF EN* (p. 99). Uppsala: Forest Stewardship Council.

Fortier, J. F., & Wyatt, S. (2019). Émergence et évolution de la collaboration dans la planification forestière du Nitaskinan. *Canadian Journal of Forest Research, 49*(4), 350–360. https://doi.org/10.1139/cjfr-2018-0290.

Fournier, J. (2013). *Facteurs de succès et contraintes à la foresterie communautaire: Étude de cas et évaluation de deux initiatives*. M.Sc. thesis, Université du Québec à Montréal.

Fuss, G. E., Steenberg, J. W. N., Weber, M. L., et al. (2019). Governance as a driver of change in the Canadian boreal zone. *Environmental Reviews, 27*(3), 318–332. https://doi.org/10.1139/er-2018-0057.

Gauthier, S., Bernier, P., Kuuluvainen, T., et al. (2015). Boreal forest health and global change. *Science, 349*(6250), 819–822. https://doi.org/10.1126/science.aaa9092.

Government of British Columbia. (2020a, April 24). Community Forest Agreements. Issued and invited community forests.

Government of British Columbia. (2020b, April 24). Community Forest Agreements. Issued First Nations woodland licences.

Hoogeveen, H., Verkooijen, P. V. J. D. (2010). *Transforming sustainable development diplomacy: Lessons learned from global forest governance*. Internal Ph.D., Wageningen University.

Horstkotte, T., Lind, T., & Moen, J. (2016). Quantifying the implications of different land users' priorities in the management of boreal multiple-use forests. *Environmental Management, 57*(4), 770–783. https://doi.org/10.1007/s00267-015-0643-5.

Howlett, M., Rayner, J., & Goehler, D., et al. (2010). Overcoming the challenges to integration: Embracing complexity in forest policy design through multi-level governance. In J. Rayner, A. Buck, & P. Katila (Eds.), *Embracing complexity: Meeting the challenges of international forest governance. A global assessment report* (pp. 93–107). Vienna: IUFRO.

Jacqmain, H., Bélanger, L., Courtois, R., et al. (2012). Aboriginal forestry: Development of a socioecologically relevant moose habitat management process using local Cree and scientific knowledge in Eeyou Istchee. *Canadian Journal of Forest Research, 42*(4), 631–641. https://doi.org/10.1139/x2012-020.

Jensen, E. L. (2000). Negotiating forests: The concept of biodiversity in Swedish forestry debate. In A. Hornborg, G. Pálsson (Eds.), *Negotiating nature: Culture, power, and environmental argument*. Lund: Lund University Press.

Johansson, J. (2014). Towards democratic and effective forest governance? The discursive legitimation of forest certification in northern Sweden. *Local Environment, 19*(7), 803–819. https://doi.org/10.1080/13549839.2013.792050.

Kittredge, D. B. (2003). Private forestland owners in Sweden: Large-scale cooperation in action. *Journal of Forestry, 101*(2), 41–46.

Kivinen, S., Berg, A., Moen, J., et al. (2012). Forest fragmentation and landscape transformation in a reindeer husbandry area in Sweden. *Environmental Management, 49*(2), 295–304. https://doi.org/10.1007/s00267-011-9788-z.

Klooster, D. (2010). Standardizing sustainable development? The Forest Stewardship Council's plantation policy review process as neoliberal environmental governance. *Geoforum, 41*(1), 117–129. https://doi.org/10.1016/j.geoforum.2009.02.006.

Korosuo, A., Sandström, P., Öhman, K., et al. (2014). Impacts of different forest management scenarios on forestry and reindeer husbandry. *Scandinavian Journal of Forest Research, 29*(Supp 1), 234–251. https://doi.org/10.1080/02827581.2013.865782.

Krott, M., Bader, A., Schusser, C., et al. (2014). Actor-centred power: The driving force in decentralised community based forest governance. *Forest Policy and Economics, 49*, 34–42. https://doi.org/10.1016/j.forpol.2013.04.012.

Lachance, C. (2017). Northeast Superior Regional Chiefs' Forum (NSRCF): A community forestry framework development process. In R. Bullock, G. Broad, L. Palmer, & P. Smith (Eds.), *Growing community forests: Practice, research and advocacy in Canada* (pp. 118–125). Winnipeg: University of Manitoba Press.

Lantbrukarnas Riksförbund. (2014). *Swedish family forestry*. Stockholm: Lantbrukarnas Riksförbund.

Larsen, R. K., & Raitio, K. (2019). Implementing the state duty to consult in land and resource decisions: Perspectives from Sami communities and Swedish state officials. *Arctic Review on Law and Politics, 10*, 4–23. https://doi.org/10.23865/arctic.v10.1323.

Larson, A. M., & Petkova, E. (2011). An introduction to forest governance, people and REDD+ in Latin America: Obstacles and opportunities. *Forests, 2*(1), 86–111. https://doi.org/10.3390/f20 10086.

Lawler, J. H., & Bullock, R. C. (2019). Indigenous control and benefits through small-scale forestry: A multi-case analysis of outcomes. *Canadian Journal of Forest Research, 49*(4), 404–413. https://doi.org/10.1139/cjfr-2018-0279.

Lawler, J. H., & Bullock, R. C. L. (2017). A case for Indigenous community forestry. *Journal of Forestry, 115*(2), 117–125. https://doi.org/10.5849/jof.16-038.

Lemos, M. C., & Agrawal, A. (2006). Environmental governance. *Annual Review of Environment and Resources, 31*, 297–325. https://doi.org/10.1146/annurev.energy.31.042605.135621.

Linser, S., Wolfslehner, B., Bridge, S. R. J., et al. (2018). 25 years of criteria and indicators for sustainable forest management: How intergovernmental C&I processes have made a difference. *Forests, 9*(9), 578. https://doi.org/10.3390/f9090578.

Mahanty, S., & McDermott, C. L. (2013). How does 'Free, Prior and Informed Consent' (FPIC) impact social equity? Lessons from mining and forestry and their implications for REDD+. *Land Use Policy, 35*, 406–416. https://doi.org/10.1016/j.landusepol.2013.06.014.

McCarthy, J. (2006). Neoliberalism and the politics of alternatives: Community forestry in British Columbia and the United States. *Annals of the Association of American Geographers, 96*(1), 84–104. https://doi.org/10.1111/j.1467-8306.2006.00500.x.

McCrossan, M. (2018). Eviscerating historic treaties: Judicial reasoning, settler colonialism, and 'legal' exercises of exclusion. *Journal of Law and Society, 45*(4), 589–616. https://doi.org/10.1111/jols.12131.

McDermott, C., Humphreys, D., Wildburger, C., et al. (2010). Mapping the core actors and issues defining international forest governance. In J. Rayner, A. Buck, & P. Katila (Eds.), *Embracing complexity: Meeting the Challenges of International Forest Governance* (pp. 19–36). Vienna: IUFRO.

McIlveen, K., & Bradshaw, B. (2009). Community forestry in British Columbia, Canada: The role of community support and local participation. *Local Environment, 14*(2), 193–205. https://doi.org/10.1080/13549830802522087.

McIlveen, K., & Rhodes, M. (2016). Community forestry in an age of crisis: Structural change, the mountain pine beetle, and the evolution of the Burns Lake community forest. In S. Teitelbaum (Ed.), *Community forestry: Lessons from policy and practice* (pp. 179–207). Vancouver: UBC Press.

Meadows, J., Annandale, M., & Ota, L. (2019). Indigenous Peoples' participation in sustainability standards for extractives. *Land Use Policy, 88*, 104118. https://doi.org/10.1016/j.landusepol.2019. 104118.

Moen, J., & Keskitalo, E. C. H. (2010). Interlocking panarchies in multi-use boreal forests in Sweden. *Ecology & Society, 15*, 3. https://doi.org/10.5751/ES-03444-150317.

Moen, J., Rist, L., Bishop, K., et al. (2014). Eye on the Taiga: Removing global policy impediments to safeguard the boreal forest. *Conservation Letters, 7*(4), 408–418. https://doi.org/10.1111/conl. 12098.

National Aboriginal Forestry Association (NAFA). (2003). *Aboriginal-held forest tenures in Canada 2002–2003*. Ottawa: National Aboriginal Forestry Association.

National Aboriginal Forestry Association (NAFA). (2018). *Fourth report on Indigenous-held forest tenures in Canada*. Ottawa: National Aboriginal Forestry Association.

Natural Resources Canada. (2020). *Forest land ownership*. Ottawa: Natural Resources Canada.

Nitoslawski, S. A., Chin, A. T. M., Chan, A., et al. (2019). Demographics and social values as drivers of change in the Canadian boreal zone. *Environmental Reviews, 27*(3), 377–392. https://doi.org/10.1139/er-2018-0063.

Palmer, L., Smith, M. A., & Chander, S. (2016). Community forestry on Crown land in Northern Ontario: Emerging paradigm or localized anomaly? In S. Teitelbaum (Ed.), *Community forestry in Canada: Lessons from policy and practice* (pp. 94–135). Vancouver: UBC Press.

Pape, R., & Löffler, J. (2012). Climate change, land use conflicts, predation and ecological degradation as challenges for reindeer husbandry in northern Europe: What do we really know after half a century of research? *Ambio, 41*(5), 421–434. https://doi.org/10.1007/s13280-012-0257-6.

Patriquin, M. N., Parkins, J. R., & Stedman, R. C. (2007). Socio-economic status of boreal communities in Canada. *Forestry, 80*(3), 279–291. https://doi.org/10.1093/forestry/cpm014.

Pinkerton, E. (2019). Benefits of collaboration between Indigenous and non-Indigenous communities through community forests in British Columbia. *Canadian Journal of Forest Research, 49*(4), 387–394. https://doi.org/10.1139/cjfr-2018-0154.

Podur, J., Martell, D. L., & Knight, K. (2002). Statistical quality control analysis of forest fire activity in Canada. *Canadian Journal of Forest Research, 32*(2), 195–205. https://doi.org/10.1139/x01-183.

Proulx, G., Beaudoin, J. M., Asselin, H., et al. (2020). Untapped potential? Attitudes and behaviours of forestry employers toward the Indigenous workforce in Quebec, Canada. *Canadian Journal of Forest Research, 50*(4), 413–421. https://doi.org/10.1139/cjfr-2019-0230.

Rametsteiner, E., & Simula, M. (2003). Forest certification—An instrument to promote sustainable forest management? *Journal of Environmental Management, 67*(1), 87–98. https://doi.org/10.1016/S0301-4797(02)00191-3.

Rayner, J., Buck, A., & Katila, P. (Eds.). (2010). *Embracing complexity: Meeting the challenges of international forest governance* (p. 172). Vienna: IUFRO.

Reed, M. G. (2010). Guess who's (not) coming for dinner: Expanding the terms of public involvement in sustainable forest management. *Scandinavian Journal of Forest Research, 25*, 45–54. https://doi.org/10.1080/02827581.2010.506785.

Ribot, J. C., Agrawal, A., & Larson, A. M. (2006). Recentralizing while decentralizing: How national governments reappropriate forest resources. *World Development, 34*(11), 1864–1886. https://doi.org/10.1016/j.worlddev.2005.11.020.

Royal Commission on Aboriginal Peoples (RCAP). (1996). *Report of the Royal Commission on Aboriginal Peoples*. Final Report. Ottawa: Government of Canada.

Saint-Arnaud, M., Asselin, H., Dubé, C., et al. (2009). Developing criteria and indicators for Aboriginal forestry: Mutual learning through collaborative research. In M. G. Stevenson & D. C. Natcher (Eds.), *Changing the culture of forestry in Canada: Building effective institutions for Aboriginal engagement in sustainable forest management* (pp. 85–105). Edmonton: Canadian Circumpolar Institute Press.

Salée, D., & Lévesque, C. (2010). Representing aboriginal self-government and First Nations/State relations: Political agency and the management of the boreal forest in Eeyou Istchee. *International Journal of Canadian Studies, 41*, 99–135. https://doi.org/10.7202/044164ar.

Samson, C. (2016). Canada's strategy of dispossession: Aboriginal land and rights cessions in comprehensive land claims. *Canadian Journal of Law and Society, 31*(1), 87–110. https://doi.org/10.1017/cls.2016.2.

Sandström, C., & Widmark, C. (2007). Stakeholders' perceptions of consultations as tools for co-management—A case study of the forestry and reindeer herding sectors in northern Sweden. *Forest Policy and Economics, 10*(1–2), 25–35. https://doi.org/10.1016/j.forpol.2007.02.001.

Sandström, C., Moen, J., Widmark, C., et al. (2006). Progressing toward co-management through collaborative learning: Forestry and reindeer husbandry in dialogue. *International Journal of Biodiversity Science & Management, 2*(4), 326–333. https://doi.org/10.1080/17451590609618153.

Schneiberg, M., & Bartley, T. (2008). Organizations, regulation and economic behaviour: Regulatory dynamics and forms from the nineteenth to twenty-first century. *Annual Review of Law and Social Science, 4*, 31–61. https://doi.org/10.1146/annurev.lawsocsci.4.110707.172338.

Skogsstyrelsen (Swedish Forest Agency). (2015). *Forests and forestry in Sweden* (p. 24). Stockholm: The Royal Swedish Academy of Agriculture and Forestry.

Smith, M. A. (2015). A reflection on First Nations in their boreal homelands in Ontario: Between a rock and a caribou. *Conservation & Society, 13*(1), 23–38. https://doi.org/10.4103/0972-4923.161214.

Sotirov, M., Pokorny, B., Kleinschmit, D., et al. (2020). International forest governance and policy: Institutional architecture and pathways of influence on global sustainability. *Sustainability, 12*, 7010. https://doi.org/10.3390/su12177010.

Stjernström, O., Ahas, R., Bergstén, S., et al. (2017). Multi-level planning and conflicting interests in the forest landscape. In C. H. Keskitalo (Ed.), *Globalisation and change in forest ownership and forest use* (pp. 225–259). London: Palgrave Macmillan.

Stjernström, O., Pashkevich, A., & Avango, D. (2020). Contrasting views on co-management of indigenous natural and cultural heritage—Case of Laponia World Heritage site, Sweden. *Polar Record, 56*, E4. https://doi.org/10.1017/S0032247420000121.

Teitelbaum, S. (2015). Le respect des droits des peuples autochtones dans le régime forestier québécois: Quelle évolution (1960–2014)? *Recherches Sociographiques, 56*(2–3), 299–323. https://doi.org/10.7202/1034209ar.

Teitelbaum, S. (Ed.). (2016). *Community forestry in Canada: Lessons from policy and practice* (p. 416). Vancouver: UBC Press.

Teitelbaum, S., Beckley, T., & Nadeau, S. (2006). A national portrait of community forestry on public land in Canada. *The Forestry Chronicle, 82*(3), 416–428. https://doi.org/10.5558/tfc82416-3.

Teitelbaum, S., & Bullock, R. (2012). Are community forestry principles at work in Ontario's County, Municipal, and Conservation Authority forests? *The Forestry Chronicle, 88*(6), 697–707. https://doi.org/10.5558/tfc2012-136.

Teitelbaum, S., & Wyatt, S. (2013). Is forest certification delivering on First Nation issues? The effectiveness of the FSC standard in advancing First Nations' rights in the boreal forests of Ontario and Quebec, Canada. *Forest Policy and Economics, 27*, 23–33. https://doi.org/10.1016/j.forpol.2012.09.014.

Teitelbaum, S., Tysiachniouk, M., McDermott, C., et al. (2021). Articulating FPIC through transnational sustainability standards: A comparative analysis of Forest Stewardship Council's standard development processes in Canada, Russia and Sweden. *Land Use Policy, 109*, 105631. https://doi.org/10.1016/j.landusepol.2021.105631.

Tikina, A. V., Innes, J. L., & Trosper, R. L., et al. (2010). Aboriginal peoples and forest certification: A review of the Canadian situation. *Ecology & Society, 15*(3), art33. https://doi.org/10.5751/ES-03553-150333.

Trosper, R. L., & Tindall, D. B. (2013). Consultation and accommodation: Making losses visible. In D. B. Tindall, R. L. Trosper, & P. Perreault (Eds.), *Aboriginal peoples and forest lands in Canada* (pp. 313–325). Vancouver: UBC Press.

United Nations. (1992). *Convention on biological diversity*. New York: United Nations.

Uprety, Y., Asselin, H., & Bergeron, Y. (2017). Preserving ecosystem services on Indigenous territory through restoration and management of a cultural keystone species. *Forests, 8*(6), 194. https://doi.org/10.3390/f8060194.

Warkentin, I. G., & Bradshaw, C. J. A. (2012). A tropical perspective on conserving the boreal 'lung of the planet.' *Biological Conservation, 151*(1), 50–52. https://doi.org/10.1016/j.biocon.2011.10.025.

Watson, J. E. M., Evans, T., Venter, O., et al. (2018). The exceptional value of intact forest ecosystems. *Nature Ecology and Evolution, 2*(4), 599–610. https://doi.org/10.1038/s41559-018-0490-x.

Widmark, C. (2006). Forestry and reindeer husbandry in northern Sweden—The development of a land use conflict. *Rangifer, 26*(2), 43–54. https://doi.org/10.7557/2.26.2.187.

Willow, A. J. (2012). Re(con)figuring alliances: Place membership, environmental justice, and the remaking of Indigenous-environmentalist relationships in Canada's boreal forest. *Human Organization, 71*(4), 371–382. https://doi.org/10.17730/humo.71.4.x267775756735078.

World Bank. (1999). *Entering the 21st century, world development report 1999/2000: Entering the 21st century*. New York: Oxford University Press.

Wyatt, S., Hébert, M., Fortier, J. F., et al. (2019). Strategic approaches to Indigenous engagement in natural resource management: Use of collaboration and conflict to expand negotiating space by three Indigenous Nations in Quebec, Canada. *Canadian Journal of Forest Research, 49*(4), 375–386. https://doi.org/10.1139/cjfr-2018-0253.

Chapter 21
Forest Certification in Boreal Forests: Current Developments and Future Directions

Constance L. McDermott, Marine Elbakidze, Sara Teitelbaum, and Maria Tysiachniouk

Abstract Forest certification has expanded rapidly in boreal forests as a means to verify responsible management. It was spearheaded in the early 1990s by civil society organizations concerned about the negative impacts of industrial forestry on biodiversity and the rights of Indigenous and local communities. Certification standards are agreed by multistakeholder groups and outline a set of environmental and social requirements. Forest companies that meet those standards can put a green label on their wood products, thus gaining market recognition for good forest practice. This chapter reviews the particular challenges facing certification in the boreal region and the ongoing debates about how best to address those challenges. It examines differences between certification schemes and variations in requirements across world regions on key issues, such as protecting the rights of Indigenous and local communities and management of woodland caribou. It finds, for example, that the recognition and protection of Indigenous rights are more comprehensive in Canada than in Russia. This highlights the political and dynamic nature of certification as it evolves and adapts to changing social and environmental contexts.

C. L. McDermott (✉)
Environmental Change Institute, School of Geography and the Environment, University of Oxford, OUCE 5 South Parks Road, Oxford OX1 3QY, UK
e-mail: constance.mcdermott@ouce.ox.ac.uk

M. Elbakidze
Faculty of Forest Sciences, Swedish University of Agricultural Science, Herrgårdsvägen 8, 739 31 Skinnskatteberg, Sweden
e-mail: marine.elbakidze@slu.se

Faculty of Geography, Ivan Franko National University of Lviv, Doroshenko str., 41, Lviv 29000, Ukraine

S. Teitelbaum
Département de Sociologie, Université de Montréal, Pavillon Lionel-Groulx, P.O. Box 6128, Stn. Centre-Ville, Montréal, QC H3C 3J7, Canada
e-mail: sara.teitelbaum@umontreal.ca

© The Author(s) 2023
M. M. Girona et al. (eds.), *Boreal Forests in the Face of Climate Change*,
Advances in Global Change Research 74,
https://doi.org/10.1007/978-3-031-15988-6_21

21.1   Introduction

Forest certification is a system for labeling forest products produced in accordance with environmental and social standards of responsible forestry. Forest certification first emerged in the late 1980s and early 1990s in response to rising concerns over the negative impacts of industrial wood production, particularly in tropical and temperate old-growth forests. Tropical deforestation was accelerating at this time, as were conflicts over the logging of old-growth stands on the Pacific Coast of North America (Cashore et al., 2010). This era was also pivotal in the struggle over Indigenous rights to land and territory, with the adoption by the International Labour Organization's (ILO) of the Indigenous and Tribal Peoples Convention in 1989. The strong presence of Indigenous peoples and other local communities in forest areas around the world led to the inclusion of Indigenous rights and community well-being in forest certification standards from the outset of the forest certification movement.

More recently, global attention has expanded to encompass boreal forests. These forests contain nearly the same percentage of intact forest landscapes as the tropics and hence are considered important biodiversity "hot spots" (Potapov et al., 2008). They are also home to large numbers of Indigenous and forest-dependent communities. As industrial harvesting in the boreal region intensifies, large boreal forest companies are under increasing pressure to become certified to demonstrate responsible practice that protects biodiversity and does not harm the rights and livelihoods of local communities. This pressure is reflected by the countries with the largest boreal forest areas, such as Canada and Russia, leading the world in area of certified forest (FSC, 2019, PEFC, 2020).

Forest certification is frequently referred to as *nonstate market-driven* (NSMD) forest governance (Cashore et al., 2004) because it is spearheaded by nongovernmental actors and is focused on market incentives. The oldest global scheme is the *Forest Stewardship Council* (FSC), which was founded in 1993 by a consortium of environmental and social nongovernmental organizations (NGOs) and concerned members of the wood products and retail sectors. These stakeholders were frustrated by the failure of governments to agree on a global forest convention that would protect the world's forests and by the limited effectiveness of boycotts and other negative pressure campaigns to arrest forest loss (Auld et al., 2008). They were also alarmed by the growing number of private labels and claims being made about the sustainability of wood products and the lack of transparency about what was behind these claims (Elliott, 2000).

Hence, the FSC was designed as a global multistakeholder institution that sets environmental and social standards of responsible forest practice, e.g., see Pattberg

M. Tysiachniouk
University of Eastern Finland, Yliopistokatu 2, P.O. Box 111, FI-80101 Joensuu, Finland
e-mail: maria.tysiachniouk@uef.fi

Nelson Institute, University of Wisconsin-Madison, 550 N. Park Street, Madison,
WI 53706-1491, USA

(2005). It is a membership organization divided into three chambers—environmental, social, and economic—each with equal voting power. Voting power is likewise evenly split between the global North and South. At the international level, the FSC has created ten principles and criteria (*FSC P&C*) for good responsible forest management (FSC, 2015a). They are supplemented with national indicators developed by national working groups to guide the interpretation of the FSC P&C in particular country contexts.

All FSC standards are subject to revisions every five years to improve and update their relevance to contemporary forest challenges (FSC, 2008). As discussed in more detail in Sect. 21.3, the FSC has recently introduced a set of *international generic indicators* (IGIs) to harmonize standards across countries. This follows a general trend among certification standards toward increasing detail and prescription to ensure consistent interpretation (Judge-Lord et al., 2020). Debates over the correct level of harmonization between national standards and the correct level of prescription or flexibility in certification standards have generated considerable conflict and dynamism in certification rule-making over time.

In addition to its multistakeholder standards, the FSC's claim to legitimacy is also based on a system for accrediting and monitoring third-party auditors to assess the compliance of forest companies to its standards. It likewise oversees the *chain of custody* (CoC) of wood products leaving certified forests and entering the marketplace, requiring formal monitoring and verification of product claims involving the FSC label.

Despite the FSC's efforts to serve as the single go-to label for responsible forestry, its lack of government authority and reliance on market support leave it open to competition from other schemes. In particular, the *Programme for the Endorsement of Forest Certification Schemes* (PEFC) has gained widespread industry support as the FSC's main competitor. The PEFC is a global organization that endorses national forest certification schemes which meet its rules and guidelines. Over time, the FSC and PEFC have competed with each other for market dominance, engaging in claims and counterclaims about the relative stringency or appropriateness of their respective standards and procedures (Judge-Lord et al., 2020). As of April 2020, the FSC had certified roughly 211 million ha in 82 countries compared with PEFC having certified about 325 million ha in over 70 countries (FSC, 2020c).

Whatever the differences between the FSC and PEFC, both schemes share several core challenges. First, the distribution of forest certification worldwide is highly uneven, with most certified areas located in developed countries in the global North and involving large, high-capacity producers able to (1) meet the extensive requirements for formal documentation of forestry management planning and forestry impacts and (2) absorb the high costs of annual auditing. As the relatively lucrative wood product markets in developed countries are increasingly demanding certification, this can exclude many small-scale and community-based producers from these markets, even if these producers practice responsible forest management. Indeed, these kinds of inequalities are common to sustainability certification across a range of sectors beyond the wood products industry (McDermott, 2013).

In part because of these inequalities, the presence of large areas of forest not managed primarily for timber production, as well as factors such as low demand for certified wood products and the low industrial capacity in the global South (Ebeling & Yasué, 2009), growth in certified forest area worldwide has slowed (FSC, 2020c). This fuels concern that, even if certification succeeds in promoting good practice within certified forest areas, it could displace rather than eradicate bad practices beyond its borders. It also feeds debates over the difficulty and stringency of certification standards—standards that are very stringent and expensive to implement may have limited market uptake, whereas standards that are very flexible may do little to change status quo forest practice (Cashore et al., 2007a). Yet regardless of these ongoing challenges and debates, some of certification's greatest impacts may be on forest governance as a whole through the creation of new norms for stakeholder participation and the protection of a wide range of forest values (Auld et al., 2008).

All these issues serve as a backdrop to the particular case of certification in boreal forests. The next section outlines some of the key forest management challenges relevant to boreal forests, including respecting and protecting the rights of Indigenous peoples and local communities and protecting the remaining large, intact boreal forest landscapes. We then delve into the FSC's recent introduction of IGIs and discuss the pros and cons of harmonizing standards between countries and how harmonizing efforts have played out differently across the boreal forest countries of Russia, Canada, and Sweden. This is followed by a discussion of other key trends, including the expansion of forest certification to encompass additional environmental priorities, e.g., climate change and ecosystem services, the role of new monitoring technologies to improve credibility and lower costs, and the efforts to increase access to certification for smallholders and low-intensity forest producers. We then conclude with some general reflections on the dynamic and evolving nature of forest certification in boreal forests and beyond.

21.2 Key Challenges in the Certification of Boreal Forests

21.2.1 Respecting the Rights of Indigenous Peoples

Both major forestry certification systems in boreal regions, the FSC- and PEFC-endorsed national certification schemes, address the rights of Indigenous peoples in their forest management standards. However, the conciliation of industrial forestry with the livelihood practices of Indigenous peoples represents a significant challenge and land-use conflicts are frequent (Huseman & Short, 2012; Johnson & Miyanishi, 2012; Tulaeva & Tysiachniouk, 2017). In some countries such as Canada and Sweden, legal systems provide the foundations for arbitrating relationships between Indigenous peoples and resource development; however, there is also a role played by market-based initiatives such as forest certification.

The FSC has been described as leader in the area of Indigenous rights because of its governance structure and standard design (Mahanty & McDermott, 2013; Meadows et al., 2019). Within governance structures at international and national levels, Indigenous peoples are usually represented within the FSC's social chamber. In Canada, however, a fourth Aboriginal chamber was created. In 2013, the FSC International Board created a *Permanent Indigenous Peoples Committee* to advise the board on issues affecting Indigenous rights. In regard to standards, the international FSC P&C include "Principle 3: Indigenous peoples' rights" along with multiple associated criteria, such as requirements to uphold the legal and customary rights of Indigenous peoples through *Free, Prior, and Informed Consent* (FPIC), adherence to United Nations Declaration on the Rights of Indigenous Peoples (UNDRIP) and the International Labor Organizations' (ILO) Convention 169, and the protection of special sites and traditional ecological knowledge (FSC, 2015a).

The 2018 international PEFC benchmark standard also calls for compliance with ILO 169, UNDRIP, and FPIC under Sect. 6.3 on "compliance requirements" (PEFC, 2018a). The PEFC's endorsement of national certification schemes requires demonstrating compliance with the PEFC benchmarks, but there is flexibility in translating these benchmarks into national standards (Judge-Lord et al., 2020). In Canada, consultation requirements are framed around stipulations for developing Indigenous policies, conferring with Indigenous peoples, and responding to inquiries and concerns (Smith & Perreault, 2017). In Sweden, the PEFC standard addresses Sami rights through provisions requiring large forest owners to obtain agreements through consultation with Sami peoples before establishing exotic species on sites of special importance to reindeer herding and requiring compliance with provisions from the Swedish Forest Act (PEFC Sweden, 2016).

Only a few studies have looked at the impacts of certification on Indigenous peoples in boreal regions, and some only address Indigenous rights as part of a larger suite of issues. Most of these studies have focused on the Forest Stewardship Council rather than the PEFC.

Research addressing FPIC in Canada includes a study by Mahanty and McDermott (2013) that compared the FSC FPIC standards and implementation in Canada and Brazil. These authors found that contextual factors, such as the strength of government laws and policies, play a key role in either supporting or undermining FPIC requirements. Similarly, Teitelbaum et al. (2019) and Wyatt and Teitelbaum (2018) provided examples of the politicization of FSC certification resulting from a "regulatory gap" between the FSC's Indigenous consent requirements and governmental practices of consultation. In one case, this culminated in a high-profile dispute between a well-known forestry company and an Indigenous nation (the James Bay Cree), which, while instigated by an FSC-certification decision, was only resolved through high-level negotiations at a governmental level (Teitelbaum et al., 2019).

In another Canadian study, Masters et al. (2010) found that the FSC's Indigenous requirements are some of the most challenging to achieve, observing that Principle 3 accrued the second-largest number of mandatory corrective action requests compared with the nine other FSC principles. A review of audit reports and an in-depth qualitative study of one audit by Teitelbaum and Wyatt (2013) showed a tendency for auditors to issue minor nonconformances and to accept evidence of "work in progress" rather than outright compliance. Several studies have also indicated that Indigenous-owned forestry companies, many of which are small-scale, face barriers to certification because of the high financial costs and administrative burdens associated with certification (Collier et al., 2002; Mahanty & McDermott, 2013).

In regard to PEFC in Canada, a study of the PEFC-endorsed Canadian Standards Association (CSA) forest certification scheme found evidence that Aboriginal organizations were not satisfied with CSA standards for Indigenous consultation. As a result, the National Aboriginal Forestry Association withdrew from participating in the CSA review process because of the lack of a distinct Aboriginal criterion (Smith, 2004; Tikina et al., 2010). For both the PEFC and FSC, Indigenous communities face challenges related to insufficient knowledge and information concerning forest certification (Johansson, 2014; Kant & Brubacher, 2008).

In Russia, a recent study examined the effects of FSC certification on the Evenk community in Tokma, Siberia. In this remote community, local consultations conducted as part of efforts to meet the FSC Russia standard led an FSC-certified company to construct a winter road for local residents, contribute to renovations of the post office and airport, provide essential medical equipment, and respond to community requests for lumber (Tysiachniouk & Henry, 2019).

Several studies in Sweden focus on the perceptions and experiences of Sami reindeer-herding organizations having FSC certification. Overall, the research findings reveal mixed reactions. On the one hand, there is a recognition among Sami interview respondents that the FSC has improved the consultation processes. For example, the introduction of FSC requirements increased the geographic area included under forestry industry consultations to include those forests used during the winter— forests previously excluded from the consultations required under the Swedish Forest Act (Sandström & Widmark, 2007; Keskitalo et al., 2009). Another study, focused on a single Swedish county, found that Sami respondents felt forestry companies had become more aware of issues faced by reindeer herders (Johansson, 2014). On the other hand, several studies report dissatisfaction among Sami respondents with the consultation processes owing to a lack of influence (Keskitalo et al., 2009). Sami respondents in Johansson's study (2014) reported that current forest management practices were resulting in the progressive degradation of key grazing habitat because of the lack of real integration of Sami concerns, creating "very pessimistic views on the long-term effects of FM [forest management] in this county" (Johansson, 2014, p. 184). Similarly, in a study by Sandström and Widmark (2007) covering territories under both government consultation and FSC regimes, respondents from reindeer-herding communities described consultations as a form of "information sharing" or "dialogue" with little real influence over decisions.

21.2.2   Local Forest-Dependent Communities

The certification standards of both the FSC and PEFC cover a range of issues relating to local communities, including the protection of local people's livelihoods, workers' rights, and the protection and use of nontimber forest products. However, the ways in which local communities value and use local resources vary significantly among countries and regions.

In Russia, approximately 20% of the population lives in forested areas. Many of these communities depend directly on forest resources (e.g., mushrooms, berries, and bushmeat) for their basic subsistence. Life in these rural areas is affected by general institutional turbulence at the national level, the restructuring of Russian state agencies with constantly changing jurisdictions, and the domination of large international companies in the forest sector (Kotilainen et al., 2008). Local people in the more marginalized and remote areas suffer from poor infrastructure, poor development of local small and medium businesses, and severe unemployment. Logging rights to Russia's state-owned forests are generally allocated to large-scale timber concessions and generate minimal local employment (Tysiachniouk & McDermott, 2016). Although local workers may be employed for low-skill, low-wage work in harvesting and wood processing, skilled workers, such as those required to run harvesters and forwarders, are typically hired from outside the local communities (Tysiachniouk, 2012).

Despite forest certification standards calling on forest companies to consult with local communities about logging impacts, research on standards' implementation suggests such consultation is often minimal. As standard practice, companies may make formal announcements in the newspapers to invite local residents to consultations; however, attendance at such meetings is low. Those who do attend may focus on grievances such as road damage and dust from logging activities, poor village infrastructure, high prices for sawed wood, and the lack of firewood, but many of these complaints are likely to go unresolved (Tysiachniouk, 2012).

Some FSC certificate holders employ social experts in community organizing to better comply with certification requirements, and there is some evidence this has led to significant improvements in community outreach (Maletz, 2013; Maletz & Tysiachniouk, 2009; Tysiachniouk, 2012). These improvements have been achieved through the extensive and proactive engagement of community members, informing them of the FSC standards and what rights they have within the FSC system (Tysiachniouk & Henry, 2015; Meidinger & Tysiachniouk, 2006). Such cases are, however, more the exception than the rule, and most communities lack the institutional capacity to engage effectively with companies in the absence of external support (Keskitalo et al., 2009; Tysiachniouk, 2012).

Whereas there is limited evidence that the FSC's generalized requirements for community consultation have had much effect on forest practices in Russia, FSC Principle 9 requirements for designating socially valuable *high conservation value* (HCV) forests have shown more promise. In some cases, villagers have participated actively in the HCV process to allocate places on company leaseholds for the special protection of sites where communities gather mushrooms and berries or where there are historically valuable territories, such as battlegrounds, cemeteries, and places of religious significance (Maletz & Tysiachniouk, 2009; Tysiachniouk, 2012; Tysiachniouk & Henry, 2015). There has been less success, however, in allocating hunting grounds as HCV. Hunters are often reluctant to disclose their hunting sites, which may be spread across many localities and include sheds that are considered illegal in Russian legislation. Similar to the findings on Principle 3 and Indigenous rights detailed above, these observations testify to the importance of legal recognition of customary rights in shaping the implementation of FSC standards (Shmatkov et al., 2014).

Apart from Russian-based studies, there is very little research looking at the certification impacts on communities outside of Indigenous communities, a gap acknowledged in the literature (Sténs et al., 2016). One Swedish case study revealed a concern among some local stakeholders that attention to Indigenous rights and forest protection would adversely impact forestry activities, in turn having negative consequences for local economies. Another study from Sweden found that private forest owners were favorable to certification, in part because of the perception of offering enhanced protection of social values such as recreation (Bjärstig & Kvastegård, 2016). In Canada, a study from Québec looked at the role of stakeholders in the implementation phase of certification processes—including the FSC, CSA, and the United States–based Sustainable Forestry Initiative (SFI) systems—through a province-wide survey. Respondents reported that although certification created opportunities for participation, this was at a consultative level; the respondents did not perceive that they had significantly influenced decisions (Roberge et al., 2011).

21.2.3   Intact Forest Landscapes

As discussed in the introduction to this chapter, the conservation of highly valued habitats, such as old-growth forests, has been a driving concern of forest certification since its inception. This concern has been addressed under FSC Principle 9, which initially designated "large" and relatively undisturbed "landscape-level forests" as another key type of HCV forest (FSC, 2002). More specific requirements for the protection of *intact forest landscapes* (IFLs) have since been introduced into the 2015 FSC P&C and associated guidance documents (FSC, 2015a, 2020a).

The evolution from the protection of *large landscape-level forests* (FSC, 2002) to the more precise concept of IFLs (FSC, 2015a) can be traced to the work of Greenpeace Russia in defining IFLs within Russian boreal forests (Yaroshenko et al., 2001). IFLs are defined as a natural environment having no signs of significant human impacts or habitat fragmentation. IFLs are also of sufficient size to contain, support, and maintain a viable complex of native biodiversity, including sufficient populations of a wide range of genera and species (Potapov et al., 2008). An operational definition of IFL has been developed that defines IFL as a territory having an area of at least 500 km^2 (50,000 ha) and a minimal width of 10 km, located within today's global extent of forest cover, and containing forest and nonforest ecosystems that have been minimally influenced by human economic activity (Yaroshenko et al., 2001). IFLs, as the last remaining large unfragmented forest areas on Earth, have been identified as critical for biodiversity conservation, climate mitigation, the maintenance of ecological processes, and the supply of ecosystem services at multiple scales (Watson et al., 2018).

IFLs are estimated to cover 23% of forest ecosystems (13.1 million km^2). Two biomes hold almost all these IFLs: dense tropical and subtropical forests (45%) and boreal forests (44%). Three countries—Canada, Russia, and Brazil—contain 64% of the total IFL area (Potapov et al., 2008). Approximately 19% of the global IFL area is under some form of legal protection; however, about 80% of IFLs are open for any human activities, including mining, oil and gas extraction, and commercial forestry. Currently, powerful short-term economic interests, intensified forest management, natural resource extraction, globalization, and other drivers create multiple challenges for the maintenance of IFLs (IPBES 2018). According to Potapov et al. (2017), industrial timber extraction, resulting in forest landscape alteration and fragmentation, was the primary cause of the global decline of IFL area. From 2000 to 2013, the global IFL area decreased by 7.2%, a reduction of 919,000 km^2. Three countries are responsible for 52% of the total loss of IFLs: Russia (179,000 km^2 lost), Brazil (157,000 km^2), and Canada (142,000 km^2) (Potapov et al., 2017). Environmental NGOs have played a vital role in using forest certification schemes to reduce logging in the remaining IFLs.

During the last decade, the FSC certification system has been widely criticized for failures to protect IFLs, and several prominent environmental NGOs (e.g., Greenpeace International, Greenpeace Russia) have left the FSC processes in protest of the inability of the FSC to stop logging in IFLs. In 2014, the FSC approved Policy Motion 65 to strengthen the protection of IFLs within their forestry standards. In 2017, the preliminary directives came into effect, instructing that forest management cannot reduce an IFL below 50,000 ha or impact more than 20% of IFLs within a forest management unit. World Wildlife Fund (WWF)-Russia with the FSC Standards Development Group invented another approach to IFL protection, called *80-50-30*. This approach requires forest managers to set aside 80% of the area of an IFL within their forest management unit when a rigorous IFL zoning process with relevant stakeholders is not conducted. If the manager is committed to reaching an agreement with stakeholders and conducting such a process—the process should identify priority areas for conservation and adapted methods for timber harvesting in the remaining areas—then the threshold of full protection can be brought down to 50%. If the forest manager is also willing to jointly lobby with stakeholders to have the IFL "core area" set aside as an officially protected area, and this is successful, then the threshold can go as low as 30% (WWF, 2018).

However, at least four challenging issues have provoked conflicting debates among relevant stakeholders. The first relates to the agreed threshold of 50,000 ha. Some stakeholders, including academics and environmental NGOs, claim that this threshold is inadequate to meet the very broad objectives of protecting all biodiversity and ecological processes, particularly in boreal forests (Bernier et al., 2017; Venier et al., 2018). For example, the scale at which the most extensive natural processes, e.g., fire and insects, occur and the size of habitat required by some species, e.g., woodland caribou, is likely greater than 50,000 ha (e.g., Venier et al., 2018). Thus, the rigid IFL requirements are useful for global tracking of IFLs but may be inadequate for biodiversity conservation at the regional level.

The second challenge concerns contrasting opinions on acceptable measures related to the conservation and protection of IFLs. Some stakeholders, mainly forest companies, complain about the prescriptiveness of the FSC resolution in Motion 65 that could protect the vast majority of IFLs, which might negatively affect their economic viability without using some portion of the IFL on their territory. Other stakeholders, e.g., Greenpeace Russia, argue to the contrary that the FSC should demand a stop to all logging of IFLs.

The third challenging issue highlights the difficulty in translating the global-scale conceptual idea of IFLs to a practical operational definition at a regional scale. This is particularly relevant for countries where the large majority of forests are publicly owned, such as in Russia or Canada. The challenge is that forest operators do not have the authority to prevent logging of IFLs located outside of their forest management units or to stop resource extraction or the creation of roads from other industries or governments within their leased areas.

Finally, the fourth challenging issue is integrating the protection of IFLs with traditional land uses of boreal forests by Indigenous communities. For example, many IFLs are used by Indigenous communities for their traditional activities, such as hunting, fishing, and wild food/medicine gathering. In Canada, for example, a critical element of the IFL debate has become the concept of *Indigenous Cultural Landscapes* (ICL) developed by representatives of First Nations communities. The ICL concept seeks to ensure that Indigenous communities' rights, interests, and values, including economic development, are considered when decisions are made about land use in FSC-certified forests.

21.3 To Harmonize or Not to Harmonize?

21.3.1 Debates Over Consistency Versus Diversity

21.3.1.1 FSC Versus PEFC and Differences Within These Schemes

Two obvious conclusions can be drawn from the above debates: (1) the environmental and social context of a particular country or company matters in shaping what standards are appropriate or achievable, and (2) there is considerable variation in stakeholder perspectives on the best way to address key challenges for boreal forest certification. Yet global certification labels such as the FSC and PEFC were designed to communicate a consistent, global message of good forestry performance, wherever their point of origin. This creates tension between both schemes and countries.

FSC and PEFC compete for the reputation of having high standards and attracting companies to their schemes through affordable prices and achievable requirements. At the same time, both FSC and PEFC face pressure to justify to consumers and producers any variation in standards between countries. The FSC addresses this balance between global consistency and local context by supplementing its international P&C with national indicators. The PEFC system, which is more decentralized, endorses national schemes based on their consistency with PEFC guidelines but does not require that all countries adopt these guidelines verbatim.

21.3.2 How is the Consistency/Diversity Tension Playing Out in Russia, Canada, and Sweden?

The following case studies draw on research within the FSC system to compare and contrast how this tension between ensuring global consistency, keeping costs down, and accommodating diversity in the local context have played out in regard to key challenges in the boreal forest countries of Canada, Russia, and Sweden. We take as our starting point the most recent FSC standards revision processes and the obligation for national standards to integrate the FSC's new IGIs as a means to strengthen and harmonize national standards. It was a FSC requirement that each country either adopt the IGIs verbatim or justify why they should be adapted, dropped, or have new indicators added (FSC, 2016). The differing responses of stakeholders in these countries and the resulting differences in their revised standards speak to the diversity of contexts in which boreal forest certification takes place. At this chapter's writing, Russia and Sweden had yet to implement their new standards, whereas Canada had just started transitioning to its new standards on January 1, 2020. It remains to be seen how and to what degree differences in standards requirements result in differences in on-the-ground performance.

> **Box 21.1 Comparing the Treatment of FPIC in FSC in Russia, Canada, and Sweden**
>
> A study by Teitelbaum et al. (2021) reveals some differences between the treatment of *Free, Prior, and Informed Consent* (FPIC) within Forest Stewardship Council (FSC) certification standards in three boreal countries: Canada, Russia, and Sweden. The study looks specifically at the process of developing the latest national FSC standards in these three countries, which also have the highest proportion of FSC-certified forests in the world. These national standards, which are elaborated by a chamber-balanced group of FSC members, are based on the new version of FSC's Principles & Criteria (P&C) through the addition of context-specific national indicators. The process was guided by FSC's international generic indicators (IGIs). The new P&C include a

strengthened commitment to FPIC for both Indigenous and local communities through adherence to the United Nations Declaration on the Rights of Indigenous Peoples (UNDRIP) and through the development of a process covering information sharing, impact assessment, and explicit consent for management operations.

The researchers conducted interviews with participants in standard development processes ($n = 49$) in all three countries and compared the written standards approved by each nation. Teitelbaum et al. (2021) found a different dynamic within each of the standard development groups (SDG). In Canada, much emphasis was placed on building consensus around a "relational" approach to FPIC, meaning a process that emphasizes building meaningful relationships between Indigenous peoples and forestry companies through ongoing engagement. The resulting national standard stays close to the wording of the IGI, although at times adopts stronger language, e.g., terms like "dialogue" rather than "informing" Indigenous communities. FSC Canada also pushed for more flexible timelines associated with FPIC processes to accommodate differences in time, capacity, and priority among Indigenous communities.

In Russia, negotiations around FPIC were more conflictual. Members of FSC Russia's economic chamber resisted the integration of FPIC for both Indigenous and local communities on the basis that FPIC could contravene Russian law and result in Indigenous or local communities placing a veto on forestry operations. The resulting national standard in Russia is also more restrictive in its application of FPIC. It limits the applicability of FPIC to customary rights that are *not governed by law*. It also outlines several circumstances where FPIC need not apply, such as when FPIC obligations conflict with other requirements of the standard (e.g., causing significant job losses) or when obtaining FPIC will lead to a conflict between the forest company and rights-holders or between different groups of rights-holders.

In Sweden, where Sami reindeer herding overlaps with forestry operations, a subgroup of the SDG was instrumental in developing appropriate wording for the indicators. The approach taken in Sweden was much more prescriptive, designed to integrate FPIC into an existing process of participatory planning that is applied uniformly across all Sami reindeer-herding territories. The national standard sets out a more operational approach to engagement around FPIC, including specifying which activities should be included under participatory planning, what the timelines should be, and what conflict resolution processes are to be followed in cases where FPIC has not been achieved.

Box 21.2 Comparing the Treatment of Caribou and Reindeer Habitat in FSC Canada and FSC Sweden

A paper by Elbakidze et al. (2022) analyses why, and to what degree, current FSC standards harmonization efforts at the global level have changed because of national contextual factors. Among the debated issues during the negotiation processes in Canada and Sweden was how to improve forest practices to maintain habitats of *Rangifer tarandus*, known as boreal woodland caribou in Canada and reindeer in Sweden. In both countries, *R. tarandus* is recognized as an important species because of its ecological and social significance, its status as a hallmark species, and its presence serving as an indicator of forest ecosystem integrity. However, conservation and maintenance of this species are addressed differently in the new national FSC standards in Canada and Sweden, partly owing to sociolegal differences between these two countries. In Canada, most indicators related to *R. tarandus* are included in Principle 6 "Environmental values and impacts," whereas in Sweden, they are included in Principle 3 "Indigenous people's rights."

In Canada, the Committee on the Status of Endangered Wildlife designated woodland caribou as a threatened species, and the species was included in the Federal Species at Risk Act in 2012. This act triggered the development of the *Recovery Strategy for the Woodland Caribou* (2012). However, despite these new government policies, the protection of caribou habitats remains an issue of significant debate among diverse stakeholders in Canada. During the latest FSC-standard development process, the main discussions among forestry-related stakeholders were on maintaining the intactness of boreal forests needed for caribou while maintaining timber production and socioeconomic benefits for local and Indigenous communities. From interview data, Elbakidze et al. (2022) identify two main factors that helped lead to agreement on the maintenance and protection of caribou in Canada's FSC standard: the Federal Recovery Strategy for the Woodland Caribou and the availability of scientific evidence.

The Federal Recovery Strategy for Woodland Caribou formed the basis for three main management options for caribou under the new FSC Canada standards. These options are outlined under Indicator 6.4.5, which is devoted entirely to the management of habitat for boreal woodland caribou. The first management option requires that caribou habitat be managed according to a Species at Risk Act (SARA)–compliant range plan that is consistent with the content, measures, and objectives in the Range Plan Guidance for Woodland Caribou (ECCC, 2016). The second option might be applied in a case when a SARA-compliant range plan does not yet exist and sets out requirements based on a management template put forward in the Federal Recovery Strategy for the boreal population of woodland caribou (Environment Canada, 2012). Finally, the third option is to use an engagement process to develop other approaches that are consistent with the Range Plan Guidance requirements. Agreement on these caribou habitat requirements was further reinforced by cutting-edge scientific knowledge used by experts involved in the standard development process.

In Sweden, reindeers are semi-domesticated animals that belong to the Sami. The protection of reindeer habitats is an integrated part of the criteria and indicators (C&I) related to the protection of Sami rights as Indigenous People in Sweden (Principle 3). Sami reindeer herding, including the management and protection of reindeer habitats, is implemented through the participatory planning process as a part of FPIC (see Box 21.1 about the planning process). The participatory planning process is conducted using a landscape perspective, allowing the forest management activities to be analyzed in a larger context.

21.4 Other Key Trends

Sections 21.2 and 21.3 examined forest certification's evolving response to three issues of long-standing concern: Indigenous rights, the welfare of local communities, and the protection of large and relatively undisturbed forest landscapes. A review of both the FSC (https://fsc.org) and PEFC (https://pefc.org) websites and strategic plans (FSC, 2015b; PEFC, 2018b) reveals several other recent developments that illustrate the dynamic and evolving nature of forest certification. The following sections divide these developments into three general categories: (1) the expansion of certification focus from timber to a broader suite of forest-related values; (2) the use of new technologies; and (3) innovations to enhance the reach and accessibility of certification schemes.

21.4.1 Changing Climate, Changing Values: New Standards for Ecosystem Services

Environmental concerns, social values, and economies change, and, likewise, certification schemes must adapt. Forest certification has initially focused on timber and wood products as a means to promote sustainable forest management. Although these schemes intend to recognize diverse forest values, this initial focus on timber reflects relatively long-standing societal concerns about the role of wood products in deforestation and forest degradation. Hence, timber producers presumably have market incentives to become certified to enhance their reputation in ways that those managing forests for nontimber forest products, e.g., mushrooms, berries, and game, conservation, and recreation, for example, may not.

Over time, however, forest certification schemes have been criticized for focusing too heavily on timber. In particular, rising concerns about climate change and biodiversity loss have driven the development of new markets for forest carbon and other *ecosystem services* that forests provide. These ecosystem service payment schemes, which like forest certification are generally voluntary, face their own credibility challenges and need to distinguish themselves in the marketplace. In response, both the FSC and PEFC have been developing new standards and processes that move beyond their traditional focus on timber production. The FSC has launched processes for certifying nontimber forest products and the ecosystem services of biodiversity conservation, carbon storage and sequestration, soil conservation, and recreation services (FSC, 2018a). Likewise, the PEFC has launched task forces to address ecosystem services and trees outside of forests.

21.4.2 New Technologies—Enhancing Efficiency or Reliability? Experimenting with Remote Sensing, DNA Testing, Blockchain, etc.

Another key development for certification schemes stems from the increasing use of advanced technologies to improve credibility and potentially lower the costs of certification. This use of technology includes experimentation with remote sensing to monitor forest cover change (Lopatin et al., 2016), the testing of wood samples for DNA as a means to verify claims regarding the origin of wood products and track the *chain of custody* of certified wood products back to their point of origin, and the use of blockchain to increase the efficiency of financial transactions and/or guard against fraud (FSC, 2020d). All of these advances coincide with the expansion of certification into more remote regions, including large expanses of remote boreal forests where traditional methods of on-the-ground monitoring and sampling may be cost prohibitive.

21.4.3 Expanding Certification Access: New Approaches for Smallholders

As discussed in Sect. 20.1, forest certification and other sustainable certification schemes can create disproportionate barriers to entry for small-scale, low-intensity, and community-based forest operators because of heavy reporting requirements, economies of scale, and other factors. The FSC and PEFC approach this problem in different ways. The more decentralized PEFC system has supported the use of simplified standards and highly reduced certification requirements for small-scale operators or family forest associations in some countries. Examples of these approaches include the American Tree Farm Association, which focuses on small private forests in the

United States, or the Finnish national standard, which allows simultaneous certification at the level of forestry associations encompassing thousands of individual forest ownerships (Cashore et al., 2007b).

The FSC has taken a somewhat different approach to improving smallholder access. This includes incorporating the concept of *scale, intensity, and risk* into FSC standards, whereby requirements are adjusted on the basis of the risk of the proposed forestry activities. This enables a lessening of certain requirements for smaller landholdings if forest management activities on those landholdings are considered to pose a lower risk. Other important strategies include *group certification* and *resource manager certification*, whereby organized groups of forest owners, or forest managers who manage multiple properties, apply for certification on behalf of all of the properties who opt for certification. More recently, the FSC has launched its "New Approaches" project to experiment with more radical innovations. These include pilot tests to simplify the content and language used in the standards, improve procedures for certifying groups, and divide responsibilities across forest owners, group entities, and forestry contractors (FSC, 2020b).

21.5   Closing Reflections

Forest certification has become an increasingly influential tool to address boreal forest challenges. Whereas global interest in forest certification may have initially been sparked by concerns over tropical and temperate old-growth forests, certification has since expanded at an exceptionally rapid rate in boreal forests. As a *nonstate market-driven* form of governance, certification has been promoted by civil society as a means to pressure companies to prove that the forest products they produce do not contribute to the loss or degradation of boreal forests, or violate the rights of the many thousands of Indigenous and local communities dependent on these forests. However, precisely how certification should provide that assurance and what constitutes genuinely "sustainable" boreal forest management remains a subject of ongoing debate.

One overarching source of such debate is the degree to which certification standards should be prescriptive or flexible, harmonized or locally adapted in relation to key issues such as IFLs and the rights of Indigenous and local communities. These debates relate, in turn, to ongoing concerns over rising costs and other barriers of access to certification, especially for small and low-intensity forest producers. Meanwhile, shifting societal values and priorities and rapidly changing technologies are pushing forest certification schemes to expand their focus beyond timber to other ecosystem services and develop new verification systems. All these pressures contribute to the dynamism and change in certification standards and procedures.

In general, forest certification requirements have become more complex and prescriptive over time. However, there are signs this trend could change. For example, the FSC, which has historically supported more prescriptive forest standards, is currently transitioning toward a *risk-based approach*, which could help simplify the standards across some criteria in countries where the likelihood and impact of noncompliance are deemed to be low (FSC, 2018b). Hence, just as political disputes push and pull on the nature and degree of governmental forest regulation, forest certification faces its own political tensions. This dynamic highlights the need to continually monitor and adapt forest certification to ensure positive impacts on boreal forests and the people who depend on them.

References

Auld, G., Gulbrandsen, L. H., & McDermott, C. (2008). Certification schemes and the impact on forests and forestry. *Annual Review of Environment and Resources, 33*(1), 187–211. https://doi.org/10.1146/annurev.environ.33.013007.103754.

Bernier, P. Y., Paré, D., Stinson, G., et al. (2017). Moving beyond the concept of "primary forest" as a metric of forest environment quality. *Ecological Applications, 27*(2), 349–354. https://doi.org/10.1002/eap.1477.

Bjärstig, T., & Kvastegård, E. (2016). Forest social values in a Swedish rural context: The private forest owners' perspective. *Forest Policy and Economics, 65*, 17–24. https://doi.org/10.1016/j.forpol.2016.01.007.

Cashore, B., Auld, G., Bernstein, S., et al. (2007a). Can non-state governance 'ratchet up' global environmental standards? Lessons from the forest sector. *Review of European Community & International Environmental Law, 16*(2), 158–172. https://doi.org/10.1111/j.1467-9388.2007.00560.x.

Cashore, B. W., Egan, E., Auld, G., et al. (2007b). Revising theories of nonstate market-driven (NSMD) governance: Lessons from the Finnish forest certification experience. *Global Environmental Politics, 7*(1), 1–44. https://doi.org/10.1162/glep.2007.7.1.1.

Cashore, B., Auld, G., & Newsom, D. (2004). *Governing through markets: Forest certification and the emergence of non-state authority* (p. 345). New Haven: Yale University Press.

Cashore, B., Galloway, G., Cubbage, F., et al. (2010). Ability of institutions to address new challenges. In G. Mery, P. Katila, G. Galloway, R. I. Alfaro, M. Kanninen, M. Lobovikov, & J. Varjo (Eds.), *Forests and society–responding to global drivers of change* (pp. 441–486). Vienna: IUFRO-WFSE.

Collier, R., Parfitt, B., & Woollard, D. (2002). *A voice on the land: An Indigenous peoples' guide to forest certification in Canada* (p. 107). Vancouver: National Aboriginal Forestry Association and Ecotrust Canada.

Ebeling, J., & Yasué, M. (2009). The effectiveness of market-based conservation in the tropics: Forest certification in Ecuador and Bolivia. *Journal of Environmental Management, 90*, 1145–1153. https://doi.org/10.1016/j.jenvman.2008.05.003.

Elbakidze, M., Dawson, L., McDermott, C., et al. (2022). Biodiversity conservation through forest certification: Key factors shaping national FSC standard development processes in Canada, Sweden and Russia. *Ecology & Society, 27*(1), 9. https://doi.org/10.5751/ES-12778-270109.

Elliott, C. (2000). *Forest certification: A policy perspective* (p. 329). Jakarta: Center for International Forestry Research (CIFOR).

Environment and Climate Change Canada (ECCC). (2016). *Range plan guidance for woodland caribou, boreal population* (p. 26). Species at Risk Act, Policies and guidelines series. Ottawa: Environment and Climate Change Canada.

Environment Canada. (2012). Recovery strategy for the Woodland Caribou (*Rangifer tarandus caribou*), boreal population in Canada (p. 138). Species at Risk Act Recovery Strategy Series. Ottawa: Environment Canada.

Forest Stewardship Council (FSC). (2002). *FSC principles and criteria for forest stewardship.* FSC-STD-01-001 V4-0 EN. Bonn: The Forest Stewardship Council.

Forest Stewardship Council (FSC). (2008). *Advice note: Expiry of national/regional FSC-endorsed Forest Stewardship standards.* Bonn: The Forest Stewardship Council.

Forest Stewardship Council (FSC). (2015a). *FSC principles and criteria for forest stewardship* (p. 32). FSC-STD-01-001 V5-2 EN. Bonn: The Forest Stewardship Council.

Forest Stewardship Council (FSC). (2015b). *FSC Global strategic plan 2015–2020: Delivering forests for all forever.* Bonn: The Forest Stewardship Council.

Forest Stewardship Council (FSC). (2016). *International generic indicators.* FSC-STD-60-004 V1-0 EN. Bonn: The Forest Stewardship Council.

Forest Stewardship Council (FSC). (2018a). *Guidance for demonstrating ecosystem services impacts* (p. 62). FSC-GUI-30-006 V1-0 EN. Bonn: The Forest Stewardship Council.

Forest Stewardship Council (FSC). (2018b). *Guideline for standard developers on incorporating a risk-based approach in national forest stewardship standards* (p. 35). FSC-GUI-60-010 V1-0 EN. Bonn: The Forest Stewardship Council.

Forest Stewardship Council (FSC). (2019). *FSC facts and figures, December 4, 2019* (p. 12). Bonn: The Forest Stewardship Council.

Forest Stewardship Council (FSC). (2020a). *Guidance for standard developers to develop a national threshold for the core area of intact forest landscapes (IFL) within the management unit.* FSC-GUI-60–004 V1–0 EN. Bonn: The Forest Stewardship Council.

Forest Stewardship Council (FSC). (2020b). *New approaches for smallholders and communities certification project 2019 annual report.* Bonn: The Forest Stewardship Council.

Forest Stewardship Council (FSC). (2020c). *Facts & figures.* Bonn: Forest Stewardship Council International. Retrieved May 29, 2020, from https://fsc.org/en/facts-figures.

Forest Stewardship Council (FSC). (2020d). *Innovation.* Bonn: Forest Stewardship Council International. Retrieved May 29, 2020, from https://fsc.org/en/innovation.

Huseman, J., & Short, D. (2012). 'A slow industrial genocide': Tar sands and the indigenous peoples of northern Alberta. *The International Journal of Human Rights, 16*(1), 216–237. https://doi.org/10.1080/13642987.2011.649593.

Johansson, J. (2014). Towards democratic and effective forest governance? The discursive legitimation of forest certification in northern Sweden. *Local Environment, 19*(7), 803–819. https://doi.org/10.1080/13549839.2013.792050.

Johnson, E. A., & Miyanishi, K. (2012). The boreal forest as a cultural landscape. *Annals of the New York Academy of Sciences, 1249,* 151–165. https://doi.org/10.1111/j.1749-6632.2011.06312.x.

Judge-Lord, D., McDermott, C. L., & Cashore, B. (2020). Do private regulations ratchet up? How to distinguish types of regulatory stringency and patterns of change. *Organization & Environment, 33*(1), 96–125. https://doi.org/10.1177/1086026619858874.

Kant, S., & Brubacher, D. (2008). Aboriginal expectations and perceived effectiveness of forest management practices and forest certification in Ontario. *The Forestry Chronicle, 84,* 378–391. https://doi.org/10.5558/tfc84378-3.

Keskitalo, E., Sandström, C., Tysiachniouk, M., et al. (2009). Local consequences of applying international norms: Differences in the application of forest certification in northern Sweden, northern Finland, and northwest Russia. *Ecology & Society, 14*(2), art 1. https://doi.org/10.5751/ES-02893-140201.

Kotilainen, J., Tysiachniouk, M., Kuliasova, A., et al. (2008). The potential for ecological modernisation in Russia: Scenarios from the forest industry. *Environmental Politics, 17*(1), 58–77. https://doi.org/10.1080/09644010701811665.

Lopatin, E., Trishkin, M., & Gavrilova, O. (2016). Assessment of compliance with PEFC forest certification indicators with remote sensing. *Forests, 7*(12), 85. https://doi.org/10.3390/f7040085.

Mahanty, S., & McDermott, C. L. (2013). How does 'free, prior and informed consent' (FPIC) impact social equity? Lessons from mining and forestry and their implications for REDD+. *Land Use Policy, 35,* 406–416. https://doi.org/10.1016/j.landusepol.2013.06.014.

Maletz, O. (2013). The translation of transnational voluntary standards into practices: Civil society and the forest stewardship council in Russia. *Journal of Civil Society, 9*(3), 300–324. https://doi.org/10.1080/17448689.2013.816538.

Maletz, O., & Tysiachniouk, M. (2009). The effect of expertise on the quality of forest standards implementation: The case of FSC forest certification in Russia. *Forest Policy and Economics, 11*(5–6), 422–428. https://doi.org/10.1016/j.forpol.2009.03.002.

Masters, M., Tikina, A., & Larson, B. (2010). Forest certification audit results as potential changes in forest management in Canada. *The Forestry Chronicle, 86*, 455–460. https://doi.org/10.5558/tfc86455-4.

McDermott, C. L. (2013). Certification and equity: Applying an "equity framework" to compare certification schemes across product sectors and scales. *Environmental Science & Policy, 33*, 428–437. https://doi.org/10.1016/j.envsci.2012.06.008.

Meadows, J., Annandale, M., & Ota, L. (2019). Indigenous Peoples' participation in sustainability standards for extractives. *Land Use Policy, 88*, 104118. https://doi.org/10.1016/j.landusepol.2019.104118.

Meidinger, E., & Tysiachniouk, M. S. (2006). *Using forest certification to strengthen rural communities: Cases from northwest Russia.* Buffalo legal studies research paper 2006–2011. Rochester: Social Science Research Network.

Pattberg, P. (2005). What role for private rule-making in global environmental governance? Analysing the Forest Stewardship Council (FSC). *International Environmental Agreements: Politics, Law and Economics, 5*(2), 175–189 . https://doi.org/10.1007/s10784-005-0951-y.

Potapov, P., Yaroshenko, A., Turubanova, S., et al. (2008). Mapping the world's intact forest landscapes by remote sensing. *Ecology & Society, 13*(2), art 51. https://doi.org/10.5751/ES-02670-130251.

Potapov, P., Hansen, M. C., Laestadius, L., et al. (2017). The last frontiers of wilderness: Tracking loss of intact forest landscapes from 2000 to 2013. *Science Advances, 3*(1), e1600821. https://doi.org/10.1126/sciadv.1600821.

Programme for the Endorsement of Forest Certification (PEFC). (2018a). *Sustainable forest management–requirements.* PEFC benchmark standard. ST 1003:2018. Geneva: Programme for the Endorsement of Forest Certification.

Programme for the Endorsement of Forest Certification (PEFC). (2018b). PEFC strategy 2018–2022. Geneva: Programme for the Endorsement of Forest Certification.

Programme for the Endorsement of Forest Certification (PEFC). (2020). PEFC global statistics: Data: March 2020. Geneva: Programme for the Endorsement of Forest Certification.

Programme for the Endorsement of Forest Certification Sweden (PEFC Sweden). (2016). PEFC Sweden forest standard. Geneva: Programme for the Endorsement of Forest Certification.

Roberge, A., Bouthillier, L., & Mercier, J. (2011). The gap between theory and reality of governance: The case of forest certification in Quebec (Canada). *Society & Natural Resources, 24*(7), 656–671. https://doi.org/10.1080/08941920.2010.483244.

Sandström, C., & Widmark, C. (2007). Stakeholders' perceptions of consultations as tools for co-management–a case study of the forestry and reindeer herding sectors in northern Sweden. *Forest Policy and Economics, 10*(1–2), 25–35. https://doi.org/10.1016/j.forpol.2007.02.001.

Shmatkov, N. M., Kulyasova, A. A., & Korchagov, S. A. E. (2014). Regulatory framework and development perspectives of the mechanism of public participation in the management of Russia's forests. *Ekonomicheskie i Sotsialnye Peremeny: Facts Trends Forecast, 1*(31), 78–86 . https://doi.org/10.15838/esc/2014.1.31.9.

Smith, M. A., & Perreault, P. (2017). *Are all forest certification systems equal? An opinion on Indigenous engagement in the Forest Stewardship Council and the Sustainable Forestry Initiative.* Ottawa: National Aboriginal Forestry Association.

Smith, P. (2004). Inclusion before streamlining: The status of data collection on Aboriginal issues for sustainable forest management in Canada. In J. L. Innes, G. M. Hickey, & B. Wilson (Eds.),

International perspectives on streamlining local-level information for sustainable forest management, a selection of papers from a conference held in Vancouver, Canada, August 28 and 29, 2000. Victoria: Pacific Forestry Centre, Canadian Forest Service, Natural Resources Canada.

Sténs, A., Bjärstig, T., Nordström, E. M., et al. (2016). In the eye of the stakeholder: The challenges of governing social forest values. *Ambio, 45*, 87–99. https://doi.org/10.1007/s13280-015-0745-6.

Teitelbaum, S., & Wyatt, S. (2013). Is forest certification delivering on First Nation issues? The effectiveness of the FSC standard in advancing First Nations' rights in the boreal forests of Ontario and Quebec, Canada. *Forest Policy & Economics, 27*, 23–33. https://doi.org/10.1016/j.forpol.2012.09.014.

Teitelbaum, S., Wyatt, S., Saint-Arnaud, M., et al. (2019). Regulatory intersections and Indigenous rights: Lessons from Forest Stewardship Council certification in Quebec, Canada. *Canadian Journal of Forest Research, 49*, 414–422. https://doi.org/10.1139/cjfr-2018-0240.

Teitelbaum, S., Tysiachniouk, M., McDermott, C., et al. (2021). Articulating FPIC through transnational sustainability standards: A comparative analysis of Forest Stewardship Council's standard development processes in Canada, Russia and Sweden. *Land Use Policy, 109*, 105631. https://doi.org/10.1016/j.landusepol.2021.105631.

Tikina, A., Innes, J., Trosper, R., et al. (2010). Aboriginal peoples and forest certification: A review of the Canadian situation. *Ecology & Society, 15*(3), art 33. https://doi.org/10.5751/ES-03553-150333.

Tulaeva, S., & Tysiachniouk, M. (2017). Benefit-sharing arrangements between oil companies and indigenous people in Russian northern regions. *Sustainability, 9*, 1326. https://doi.org/10.3390/su9081326.

Tysiachniouk, M. S. (2012). *Transnational governance through private authority: The case of Forest Stewardship Council certification in Russia.* Ph.D. thesis, Wageningen University

Tysiachniouk, M., & Henry, L. A. (2015). Managed citizenship: Global forest governance and democracy in Russian communities. *International Journal of Sustainable Development & World Ecology, 22*(6), 476–489. https://doi.org/10.1080/13504509.2015.1065520.

Tysiachniouk, M., & Henry, L. A. (2019). Benefit-sharing in Russia's resource extractive industries: When global standards meet local communities. *Finnish Business Law Journal (Liikejuridiikka), 2*, 137–167.

Tysiachniouk, M., & McDermott, C. L. (2016). Certification with Russian characteristics: Implications for social and environmental equity. *Forest Policy and Economics, 62*, 43–53. https://doi.org/10.1016/j.forpol.2015.07.002.

Venier, L. A., Walton, R., Thompson, I. D., et al. (2018). A review of the intact forest landscape concept in the Canadian boreal forest: Its history, value, and measurement. *Environmental Reviews, 26*(4), 369–377. https://doi.org/10.1139/er-2018-0041.

Watson, J. E. M., Evans, T., Venter, O., et al. (2018). The exceptional value of intact forest ecosystems. *Nature Ecology & Evolution, 2*(4), 599–610. https://doi.org/10.1038/s41559-018-0490-x.

World Wildlife Fund (WWF). (2018). *Comparative analysis of land use options within intact forest landscapes: How can FSC make a difference?* (p. 23). World Wildlife Fund Forest practice white paper.

Wyatt, S., & Teitelbaum, S. (2018). Certifying a state forestry agency in Quebec: Complementarity and conflict around government responsibilities, Indigenous rights, and certification of the state as forest manager: Certifying state forestry in Quebec. *Regulation & Governance, 14*(3), 551–567. https://doi.org/10.1111/rego.12229.

Yaroshenko, A., Potapov, P., & Turubanova, S. (2001). *The last intact forest landscapes of Northern European Russia* (p. 75). Moscow: Greenpeace Russia.

Chapter 22
Gender and the Imaginary of Forestry in Boreal Ecosystems

Maureen G. Reed and Gun Lidestav

Abstract In this chapter, we examine forestry work in two boreal regions—Canada and Sweden—where gender mainstreaming has long been established in government policy. Despite having policies that support gender equality in both countries, the roles, opportunities, remuneration, and expectations of women and men engaged in forestry work are highly differentiated by gender. We explain this discrepancy by considering the way in which forestry work has been and continues to be imagined. The narrow interpretation of forestry as "tree cutting" has reduced the visibility of women and continues to narrow the range of activities deemed valuable to the forestry sector. By asking questions about how forestry has been imagined, we seek to catalyze fresh thinking about the nature of forestry work and the capacity of the forest industry in both countries to adapt to climate change.

22.1 Introduction

For many years, scholars from the global North have described the industrial forest sector as characterized by a highly gendered division of labor, which has contributed to (and even valorized) a masculine identity built on dangerous, physically demanding, "dirty" work associated with timber harvesting or "piling up the timber" (e.g., Ager, 2014; Lidestav et al., 2019; Reed, 2003a). Men working in forestry in boreal regions in the early twentieth century were described as "robust, hardy and able to bear up against natural forces like rain, snow, storms and frost"

M. G. Reed (✉)
UNESCO Co-Chair in Biocultural Diversity, Sustainability, Reconciliation and Renewal, School of Environment and Sustainability, University of Saskatchewan, 117 Science Place, Saskatoon, SK S7N 5C8, Canada
e-mail: mgr774@mail.usask.ca

G. Lidestav
Swedish University for Agricultural Sciences, SE-901 83 Umeå, Sweden
e-mail: gun.lidestav@slu.se

© The Author(s) 2023
M. M. Girona et al. (eds.), *Boreal Forests in the Face of Climate Change*,
Advances in Global Change Research 74,
https://doi.org/10.1007/978-3-031-15988-6_22

(Brandth & Haugen, 2005, pp. 16–17). Although they too have long worked in forests, women and the work they undertook have largely remained invisible or, at best, have been characterized as helpmates to male workers (Johansson, 1994; Östlund et al., 2020).

In the second half of the twentieth century onward, the work of forestry was no longer the job of individual, hardy men. Restructuring of the forest industry, the introduction of mechanization and new technologies, economic globalization and the outsourcing of operations, and the enhanced regulation of the industry to demonstrate its environmental sustainability led to other competencies and skill sets being identified and required in forestry (Ager, 2014; Hayter, 2000; Ross, 1997). Social reorganization of employment and family structures also created greater demand for employment opportunities for diverse groups previously excluded from the forest economy, including women and Indigenous Peoples. By the late twentieth century, new employment avenues opened up in information science, planning, monitoring, regulation, management, and policymaking for professionals—experts and supervisors with appropriate academic degrees—who could, more or less, perform their job from offices in urban environments rather than in the field and the rural environments of traditional forestry (Brandth & Haugen, 2005; Reed, 2003a). Governments took a greater interest in forest management and planning and promoted their position through new policy statements about sustainable forest management (Canadian Council of Forest Ministers, 2003; Sveriges Riksdag, 2008). While the location of key decisions about forests has moved to company boardrooms, government offices, and computer labs, the iconic image of forestry remains the rough-and-ready, male logger. Indeed, forestry is an industry with a highly gendered division of labor in which men dominate across a range of key activities, including harvesting, production, silviculture, and regulation and management (Häggström et al., 2013; Johansson, 2020; Reed, 2003a; Wyatt et al., 2021). Unless this pattern changes, this male dominance will also have implications for who determines climate-related adaptation measures in forestry and may affect the capacity to engage in innovative and effective adaptation strategies.

In this chapter, we explain the emergence and persistence of gendered roles, relationships, and identities in the forest sector in two boreal regions—Canada and Sweden. We have selected these two countries because of the shared importance of boreal forests to the respective national economies and the shared significance of forests more generally to their cultural identity. Additionally, federal governments of both countries have made public commitments to gender equality in all sectors. We use the term *imaginary* to help explain how forestry work is imagined. This idea follows other political and sociological theorists such as Steger and James (2013, p. 23) who explain imaginaries as "patterned convocations of the social whole. These deep-seated modes of understanding provide largely pre-reflexive parameters within which people imagine their social existence." In our case, the imaginary of forestry establishes expectations of what kinds of work qualify as forestry, who undertakes that work, and how it is accomplished.

In both countries, a narrow set of masculine and feminine norms have long undergirded the industrial model of forestry. Although forest ownership and management practices differ in each country, the industrial model of forestry continues to shape the perception that forestry is man's work—and not just any man's work, but a man who exhibits particular characteristics of masculinity. The values and assumptions associated with these characteristics, then, help define who is considered a legitimate worker in the forest and helps form a masculine imaginary of forestry. By reviewing the history and persistence of masculinity in forestry, we reveal underlying assumptions and explore both opportunities and constraints to establishing a forest sector that is both environmentally and socially sustainable.

We have organized our chapter in the following way. First, we characterize the sector in both countries, demonstrating the presence of imbalances between women and men in key occupational categories in each country. Next, we explore how forestry has shaped gender roles, relationships, and identities, revealing a distinctive form of masculinity associated with forestry work. We call this the *forestry imaginary*. We then argue for the need to consider fundamental research questions to better understand how the forestry imaginary has restricted the discussion of forestry and gender to a rudimentary counting of women and men. We pose questions of our own to help explain dominant perceptions of gender and forestry. We consider how these perceptions also affect key socioenvironmental issues, such as the need for the industry in both countries to adapt to climatic change. Finally, we invite our readers to pose their own questions and begin questioning the fundamental assumptions that have shaped the contemporary forestry identity in boreal regions. By offering alternative framings of forestry, we seek to catalyze fresh thinking about the nature of forestry work and the capacity of the forest industry in both countries to adapt to climate change.

22.2   A Tale of Two Countries: Characterizing the Gender Balance in the Forest Sectors in Canada and Sweden

22.2.1   Canada

Approximately 28% or 307 million ha of the world's boreal forest is located in Canada (NRC, 2020). In fact, the boreal forest makes up 75% of Canada's forest lands, encompassing all but three of Canada's provinces. The vast majority of these forests are located on public lands where provincial governments grant licenses to, and regulate the activities of, large-scale, often multinational, forest companies.[1] Although employment in the forestry sector has declined in the twenty-first century, the 2016 census revealed that about 205,890 workers are considered part of Canada's forestry workforce. Additionally, although Indigenous people make up about 4.9% of Canada's population, approximately 70% of Indigenous communities are located in

---

[1] For the country as a whole, 94% of commercial forest land is publicly owned and managed. The largest proportion of private forest land ownership is in the Atlantic provinces. In British Columbia, 2% of the forest land has been dedicated to community forest licenses.

Canada's forested regions. Forests are important for Indigenous Peoples and communities for cultural, spiritual, and economic reasons (Sherry et al., 2005). Indeed, census data revealed that in 2016, 7% of forest sector employees were Indigenous compared with 4% for the total Canadian workforce (Wyatt et al., 2021).

Despite changes in the structure of the industry, forestry has been remarkably male dominated. It is not possible to separate out jobs data in the boreal region from the country as a whole; however, there is no reason to believe that the structure of the industry is different for commercial boreal forests than for other commercial forests in Canada. Data from the 2016 census show that women make up only 17% of forestry jobs in Canada. These data have not changed much since 1996, when 14% of employees in all forest industries were female. Women working in all parts of the forest industry have, on average, higher levels of formal education than men. Yet, they continue to be overrepresented in clerical and administrative occupations and underrepresented in operations, scientific, and management categories. For example, according to Statistics Canada, women represent 91% of accounting and related clerks and 92% of executive assistants in the forest sector. Men, on the other hand, represent 98% of logging-machine operators, 93% of sawmill-machine operators, and 92% of supervisors in logging and forestry. Women are also underrepresented in professional and managerial roles in both the private and public sectors. Within the total cohort of women in the sector, about 20% across the country are registered professional foresters,[2] and many leave the profession over the course of their working lives. For example, data collected by the Association of BC Forest Professionals for 2021 indicate that in British Columbia, whereas just over 40% of professional foresters under the age of 30 are women, approximately 12% are women over the age of 50 (Christine Gelowitz, personal communication, February 2021).

As a consequence of this division of labor and a myriad of other social factors, women across all job categories have always earned significantly less than men in the forestry workforce (Baruah, 2018). For example, in a survey of 500 women working in the forest sector and arboriculture across Canada and the United States, 60% of female respondents reported earning less than their male counterparts (Bardekjian et al., 2018). Calculations of wages in natural resource sectors as a whole (including mining and forestry) indicate that the average weekly wage for women has increased over time from $666 in 2000 to $938 in 2015, whereas men's weekly wages have increased from $1,342 to $1,608 over the same period (Baruah, 2018). There is also evidence that women progress through the pay hierarchy more slowly than men. Data from surveys conducted by the Association of BC Forest Professionals show that among professional foresters in that province, men have higher salaries than women who graduated at the same time. Compounding the fact that women take home less money during good times, past recessions have revealed that women are also more likely than men to lose their jobs or take pay cuts during economic restructuring

[2] Professional foresters are regulated by the provinces. Some provinces do not report by gender, but it appears that the proportions of women working as professional foresters vary between 15% in Québec and 21% in British Columbia.

(Barnes et al., 1999; Commission on Resources and Environment, 1994; Hayter, 2000).

Inequities also persist for Indigenous Peoples who have sought forestry employment as a means for economic well-being. While Indigenous Peoples appear to be employed in forestry in higher numbers than their population might suggest, Indigenous workers continue to face job segregation, with their jobs typically concentrated in forest activities (forest management, logging) and wood product manufacturing, which are often lower paid and less secure than other occupations. Indigenous women are doubly disadvantaged. More than twice as many Indigenous women in the forest sector hold university degrees, trade school certificates, or college diplomas as Indigenous men, and yet they are typically hired in more precarious positions, such as working in nurseries or gathering always closed up (except non-analog, non-native) timber forest products, e.g., mushrooms. They have even been excluded from typically female-dominated occupations, such as clerical and secretarial services, which are mostly dominated by white women (Mills, 2006). Indigenous men and women also continue to face challenges as a consequence of stereotyping, discrimination, and a lack of accessible training (Proulx et al., 2020).

22.2.2 Sweden

Similar to Canada, the boreal forest is a main feature of the Swedish landscape and represents an important resource for processing industries and export income. Of the 28 million ha covered by different types of forests (corresponding to 69% of the Swedish land surface), 24 million ha is considered productive forest. Dissimilar to Canada, less than a quarter of the forestland is publicly owned, and another quarter is owned by large-scale private companies, leaving 52% to some 330,000 small-scale private forest landowners (SLU, 2019). Yet the large-scale industrial forestry model has influenced and been "incorporated" by this land and ownership.

In terms of Indigenous communities, the 51 reindeer-herding communities in northern Sweden have grazing rights on all forest land within their reindeer-herding districts.[3] These communities are organized into about 1,000 enterprises with some 4,600 reindeer owners. Of these owners, 2,500 depend on incomes generated from reindeer husbandry (Sametinget, 2020). This means that about 10% of the Samí people (who have the exclusive right to reindeer husbandry) are associated with forest land use through reindeer husbandry. In comparison, about 3% of the entire Swedish population[4] (of whom some are Samí) are forest owners, and their combined

---

[3] Reindeer husbandry can be carried out on 22.6 million ha of mountain and forest land equal to 55% of the Swedish land base, and more than 50% of the productive forest land. As reindeer husbandry is always carried out in conjunction with other land uses, forestry has a major impact on the conditions for reindeer husbandry and for maintaining a reindeer husbandry–based Sami culture (Buchanan et.al. 2016; Sandström 2015).

[4] Sweden does not categorize citizens by ethnicity.

work in their own forest is estimated at 6,345 days of full-time work, which corresponds to 38% of the total day labor in forest operations. However, most work, both in terms of area and volume, is done by contractors and their employees (8,762 or 55%). In addition, there are 1,249 employees (7%) in large-scale forestry (Skogsstyrelsen, 2020). When including the timber processing industries, transportation, and the other logistic and required services, there are nearly 60,000 people directly employed in the forestry sector. With subcontractors, there are about 200,000 employees who make up 4% of the Swedish workforce. All in all, the primary production of trees, the secondary production of timber (harvesting and transportation), and tertiary production of forestry-derived products makes Sweden the world's third-largest exporter of pulp, paper, and sawed wood products (Skogssverige, 2020).

While societal influences on gender equality have brought about an increase in the number of women forest owners from 20% in 1976 (Lidestav & Ekström, 2000) to 38% four decades later (Follo et al., 2017), women's participation in the forest workforce remains low and focuses on particular segments of the sector (Johansson et al., 2020). In self-employed forestry work, for example, harvesting is much more likely to be done by men, whereas in planting and desk work, the involvement of women and men is more equal (Lidestav & Nordfjell, 2005).

Of the total workforce in forest contracting firms working in silviculture (planting and cleaning), only 4% are women (Häggström et al., 2013; Wide & Nordin, 2019). In large-scale forestry companies and forest-owner associations, 15% of the staff are women, who, to a large extent, have an academic education in forestry or a similar program. Indeed, in academia, women have made up approximately and 20% and 33% of students enrolled in bachelor and master programs in forestry, respectively, over the last two decades (SLU, 2015). Despite a growing number of women having training in forestry, patterns of gender segregation in employment, work tasks, and roles remain. For example, women are more involved in training/consultancy, administration, and forest preservation, whereas men numerically dominate work that is more closely associated with production-oriented forestry, e.g., harvesting and wood processing (Lidestav et al., 2011).

It is difficult to directly compare education and employment prospects between Sweden and Canada because the structures of the industry differ between the two countries, and official sources collect different types of data. Nevertheless, it is evident that the industrial model of forestry has created a strong division of labor whereby timber harvesting defines forest management and remains "men's work," whereas administrative work, which appears less distinctive to forestry (involving secretarial, accounting, or human resource–related tasks), remains "women's work." How this division of labor has come about and established the forestry imaginary has been theorized through a series of explanations, as described below.

22.3   Explanations for the Gendered Aspects of Forestry

These observable differences in the opportunities and experiences of women and men in the forest sector have been interpreted through different theories and concepts. For example, *labor-market segmentation theory* was first developed outside of forestry to explain employment and income disparities by distinguishing between primary and secondary sectors, with the primary segment characterized by "high wages, good working conditions, employment stability, chances of advancement, equity, and due process in the administration of work rules" and the secondary segment having "low wages and fringe benefits, poor conditions, high labor turnover, little chance of advancement, and often arbitrary and capricious supervision" (Doeringer & Piore, 1971, p. 165). This theory has been applied to forestry to explain employment and income opportunities for women and men in forestry. Jobs in primary resource extraction and processing have typically been classified as primary, and they have been characterized by trade unions that have secured high wages, a seniority system, and relative job security for men. Jobs in administrative and service segments of the industry have been considered secondary. They have typically not been unionized and confer lower wages and more precarious employment opportunities for women (Reed, 2003a, 2003b; 2008).

Additionally, the concepts of *gender order* and *workplace culture* were advanced by feminist scholars to explain women's disadvantage in "nontraditional work settings," explaining that organizational rules and values are responsible for creating and perpetuating perceptions of *maleness* and *femaleness*—perceptions that can reinforce barriers to the inclusion of women as equal partners in the workforce (Gherardi & Poggio, 2001, p. 246; Johansson, 2020). In Canada's forest sector, this "traditional" division of labor of the male breadwinner and the female homemaker has remained remarkably persistent (both empirically and discursively), particularly in rural areas where many of the "primary jobs" are located. This scenario is true even where women have been engaged in paid work, as they continue to carry a disproportionate share of childcare and other domestic duties (Martz et al., 2006; Preston et al., 2000). Both labor-market and gender-order theories have been used to explain the masculinized work culture, systemic discrimination and harassment, and barriers to advancement and training in forestry and in the cognate resource sector of mining (Cox & Mills, 2015; Mills et al., 2013; Parmenter, 2011; Reed, 2003b). But these findings are not unique to Canada. An international survey of gender in forestry conducted by the "Team of Specialists on Gender and Forestry for the United Nations Economic Commission for Europe" also revealed that "a gendered organizational logic [was] at work, which not only reproduces a structure of gender division but also, paradoxically, and simultaneously, makes gender invisible" (FAO, 2006, p. 1).

Masculinity theories have also been used to understand how the organization of forestry work over time has shaped different ideals of masculinity. For example, Nordin (2006) identified four modes of masculinity that have emerged as forestry work has been restructured: the combat pilot (machine operator linked to technology and performance), the man of the forest (manual laborer working close to nature and freedom), the business executive (organizational and management expertise), and the contractor in crisis (an entrepreneur with very limited opportunities to control his own, and his employees' work situations). Similarly, Brandth and Haugen (2000) argued that the dominant representations of masculinity in Norwegian forestry have changed over time from the logger, who is a nature-mastering man with a body marked by hard work, to the machine operator mastering chain saws and tractors, to the organizational man with his business management skills. These studies suggest that as forestry work has changed, the significance of gendered stereotypes has not diminished. Rather, the number of masculine norms has multiplied, drawing on different forms of knowledge of forestry: practical/manual, technological, and theoretical/administrative. To some extent, these theoretical concepts have a longitudinal dimension in the sense that manual labor implies *traditional* forestry and theoretical/ administrative work implies *modern* forestry. However, these concepts are best understood as overlapping, as technological developments have not eliminated the ideals and assumptions regarding the "real" work of forestry.

Associated with these depictions are the ways in which skill sets have been gendered. For example, men have been ascribed technical job skills and are assumed to be competent in them. By contrast, women are assumed not to have such skills. Hence, they still have to prove that they are capable more frequently than their male colleagues (Lu & Sexton, 2010; Navarro-Astor et al., 2017; Smith, 2013). Research about women in "nontraditional" employment sectors has demonstrated that women are perceived as having stronger emotional and supporting skills; hence, they have been viewed as having a positive effect on men's behavior, which in turn is likely to have an effect on the overall productivity of the company (Eveline & Booth, 2002). Using policy analyses, Mayes and Pini (2014) argued that the "business case" for gender equality used in the mining industry in Australia describes women as bringing something different from men, such as other types of communication and decision-making. Similar conditions in Swedish forestry work organizations have been reported by Johansson and Ringblom (2017). These kinds of findings suggest that women are viewed as having the potential to "civilize the workforce and the workplace" (Mayes & Pini, 2014, p. 538). These depictions ultimately do not challenge gendered values, skills, and division of labor; instead, they potentially burden women professionals with the requirement and responsibility to change dominant discourses.

Lastly, the concept of *intersectionality* now encourages researchers to examine labor-market inequalities in resource sectors by considering how a range of social identity factors, as well as institutions, structures, norms, and power dynamics at different scales, operate to create advantages or disadvantages for different social groups (Cox & Mills, 2015; Hankivsky, 2014; Manning, 2014; Mills et al., 2013; Parmenter, 2011; Ringblom & Johansson, 2020). Consideration of intersecting factors in forestry is revealing. Mills (2007), for example, used census data to compare the employment profiles of Indigenous and non-Indigenous men and women within the forest industry in Saskatchewan, Canada. She found evidence that gender, class, and racialized identity work together to the general disadvantage of women and to the greater detriment of Indigenous women and men. Such a disadvantage is demonstrated in the employment opportunities and job security offered to them and in the wage differentials they experience.

While these theories and concepts help us to understand possible root causes and impacts of gender inequality in the forest sector, we also need to recognize that official statistics, and the ways in which these data are used, reflect, and influence societal values (Waring, 1988). For example, for years national forestry associations have collected data without accounting for gender, reflecting and reinforcing the view that gender is not important and/or that gender bias in the workforce does not exist. In Canada, we have a very sketchy picture of the diversity of social groups employed in the forest industry, and national data have not historically been collated by resource sector, job classification, location, and gender. Reliable, commensurable data can provide more detailed information about who is working in the forest sector, where jobs and workers are, and what training or retention strategies might be needed. Uncritical use of data has contributed to *gender-neutral* policies and programs that have typically favored men and maintained an ongoing cycle of marginalization of certain groups (Reed, 2008; Walker et al., 2019).

22.4 The *Making* of Men and Women in Industrial Forestry

Empirical research has employed these theories to understand how forestry work is gendered. Such research moves well beyond documenting the numbers of women and men employed in the forestry sector. Rather, it helps us consider how forestry is *imagined*, how this imaginary "makes" male and female forestry workers, and how it ascribes value to the work they undertake.

In Sweden, forestry has been represented as a modernizing force that lifted the country, particularly the northern part, from poverty by the end of the nineteenth century to prosperity 50 years later (Kardell, 2004). This large-scale activity, although geographically scattered across state, company, and private lands, became a way for up to 200,000 men, both locals and migrant male workers, to support themselves through seasonal work in harvesting operations during winter and log-driving during spring and summer (Johansson, 1994). In contrast to the traditional farmer society, forestry work was organized in such a way that it offered freedom from paternalistic relationships between landed and landless men and gave the latter access to an independent and equal social and economic status. The piece-rate system, i.e., payment by performance in terms of logs or "piece" processed per day's work, increased the predictability of income and recognized meritorious conduct. The lumberjacks who worked hardest and were considered most skillful were able to pile up the largest stack of timber and thereby received the most earnings. The "contracts for harvesting," on the one hand between the forest company and the contractor (log driver) and, on the other hand, between the contractor (log driver) and the lumberjacks, were the cornerstone of forestry work and economic organization. The piece-rate-system also decisively influenced the social organization in the forest camp (Ager, 2014). According to Johansson (1994), in these seemingly all-male settings out in the forest, a new type of masculinity was constructed on the basis of the individual work performance rather than on class and property. Thus, men in forestry defined the *modern man*[5] and rural masculinity as being closely associated with the male body and the capacity to master harsh working conditions. Such a man not only tolerated but even glorified the crowded and unhealthy living conditions in the forest hut.[6] As a large part of the male population in northern Sweden was involved in winter logging operations, this critical mass induced a material and mental change in the perception of man and manhood, i.e., an individual, who by his own ability and performance, mastered the environment and thereby contributed to the co-construction and imaginary of gender in forestry.

Major changes from the 1950s to today, i.e., mechanization and digitalization, mean that some tasks can now be organized and executed from the office instead of the field (Ager, 2014). These changes have, to some extent, modified the perception of the *forest man* and his performance. Yet, the volume of timber produced remains a central feature of "performance in forestry" (Hugosson, 1999). Furthermore, the "management masculinity" described by Brandth and Haugen (2000) and the four modes of contractor masculinity identified by Nordin (2006) continuously emphasize physical capacity, technical skills, and practical experience of physical forest work associated with logging. These features continue to be central aspects of forestry work, and they persist in bringing legitimacy to the carrier of those attributes in contemporary forestry work organizations (Brandth & Haugen, 2000, 2005; Nordin, 2006). Similar attributes and conceptions are found within small-scale family forestry. Although self-employment in forestry operations is declining, the image of the active forest owner as male still represents the norm, and the division of work between women and men forest owners is significant (Lidestav, 2001; Lidestav & Nordfjell, 2005; Westin et al., 2017).

Therefore, if forestry is about men's work, male collectivism, and male interactions, how have women been represented in the core activities in forestry? According to Ella Johansson (1994), a lack of visibility and recognition[7] of women in forestry work should not be understood as their nonexistence. Rather, it should be interpreted as "women in forestry work appear to have crossed a boundary in at least the male classification system," and the idea of them as women becomes preposterous and therefore something to conceal (Johansson, 1994, p. 135). She argued that logging represents the kind of hard work that requires periods of rest, e.g., evenings and Sundays, for men. In the then rural society, however, a "proper woman" was never

---

[5] Here, we consider a *modern man* as an individual who explores and transforms nature in accordance with the idea that planning, calculation, and rational decisions will lead to progress, a better life, and a better society.

[6] A forest hut was a simple log house for 5 to 15 men who had to share beds and cooking facilities, with a fireplace serving as the only source of heat.

[7] In the nearly 200 narratives about forestry that constitute the main source of Ella Johansson's thesis research (1994), women and girls are mentioned in only a handful of occasions. Yet, she refers to other records that describe women hauling timber, barking, and even fulfilling the husband's logging contract if he became ill. There is also photographic evidence of women and girls doing afforestation.

supposed to rest, and therefore a women logger became an anomaly. Accordingly, if a woman rested, she would act unwomanly and above all violate her own self-respect. In contrast, if she worked longer days in the forest than men, the men would seem unmanly. Consequently, we can assume that there was a mutual interest of women and men *not* to mention or pay tribute to women who worked in the forest.

Other research focused on more recent times have also found evidence that both men and women in forestry share a common interest in gender invisibility (Lidestav & Sjölander, 2007). In practice, this means that women active in forestry should try to look like men and effectively uphold the male imaginary. Such efforts have an important effect on where we find women and men in the forestry workforce. Johansson (2020, p. 4) found that "when the ideal image of the forestry worker or forestry professional is based on the male body, women are not assumed to possess the right kind of skills or experiences, are expected to need additional help and thereby are not understood as carriers of knowledge." Consequently, the spaces that are accessible for women in forestry are constrained; "women are more often found to work in areas related to forest preservation, communication, or administration and in public organisations such as Swedish Forestry Agency while they are less likely to work in harvesting, processing, or as managers"(Johansson, 2020, p. 4). Although these findings are from Sweden, Canadian census data reveal a similar division of labor.

In Canada, while land ownership and corporate structures are quite different, there remain important similarities to how *the modern man* was created in northern Sweden and in the consequences for the division of labor in the sector. In keeping with other natural resource sectors in Canada, Quam-Wickham (1999) argued that the acquisition and practice of skills in the lumber, mining, and oil industries have been the pivotal means by which male workers construct their masculinity. For example, during the 1930s and the Great Depression, the government of British Columbia (a province where forestry has long dominated the economy) promoted its forestry work programs on the grounds that "this forestry programme offers them [young men] useful work under conditions that must benefit them physically and mentally, leaving them more self-reliant and with a saner outlook towards the future." (cited by Ekers, 2009, p. 309). Similarly, its report of the newly established Young Men's Forestry Training program involved "all outdoor work, well calculated to improve young men mentally and physically and to develop initiative and self-reliance" (Department of Lands 1936; cited by Ekers, 2009, p. 309). Ekers' review of these programs (2009, p. 309) revealed that

> it was not simply being in nature that engendered this construction of identity, but rather, it was getting men out of cities and having them work in nature that was deemed expedient. It was working in jobs that were traditionally—and continue to be masculinized—that conferred the central features of masculinity onto the subjects who laboured.

Furthermore, according to Ekers (2009, p. 309) "'nature' was assumed to have essential (rather than socially constructed) characteristics that would aid the men in finding their 'true' masculine selves."

Histories of more recent events have built on these ideas by describing how forestry cultures have established monolithic ideals of masculinity—intersecting with those of ethnicity and gender—that delineate clear lines of inclusion and exclusion (Dunk, 1991; Mills, 2006). For example, Coen et al., (2013, p. 98) described masculinity in a forestry town in northern British Columbia, wherein they argued, "in many ways the idealized Prince George man and the cold, rugged landscape merged and mirrored each other: the iconic male was strong, indefatigable, impenetrable; he commanded nature, extracting natural resources and turning them into consumer products." Perhaps surprisingly, Coen et al. (2013) found that these old stereotypes continued to be reproduced by men and women living in forestry towns and contributed to both depression and the unwillingness of men to seek outside help to address their mental health. Their research reinforces that the dominant framing of masculinity is not a "natural" creation nor merely a historical anecdote but one forged from human ideals and normative structures that persist well into the twenty-first century. Another consequence has been the erasure, marginalization, and discrimination of women and nonconforming men (including Indigenous Peoples) from the dominant narrative of forestry work.

22.5 More Fundamental Questions: Seeing the Forest and not just the Trees

These examples illustrate that despite economic and technological change in forestry, an enduring imaginary of masculinity pervades the industry. This imaginary then influences where and how women and men are employed in the forest sector. We pose questions about the aims and activities of forestry that attempt to challenge dominant masculine norms and consider the implications of a male-dominated workforce for the forestry sector to adapt to climate change:

> Why is forestry in boreal ecosystems still defined by cutting trees and not by planting trees?
> How does this *imaginary* impact on how women and men identify themselves with forestry?

These questions are highly relevant when we consider that reforestation has been a core activity in Swedish forestry for more than a century. A review of the Swedish situation reveals a somewhat different trajectory around tree planting than in North America. Since the advent of the first modern forestry act in 1903, it has been compulsory to reforest land that has been clear-cut, and reforestation by planting has been the dominant method since the 1950s (Enander, 2007). Prior to that period, great efforts were made by public and private organizations to reforest large areas that had been deforested because of grazing and firewood chopping (Enander, 2007). Thus, from a legal and policy point of view, and also from a business perspective, tree planting could be regarded as both a profitable investment (assuming revenues from future harvesting) and a prerequisite for timber harvesting, i.e., an integrated and indispensable activity of forestry. Yet, the status of tree planting and tree planters is inferior to that of harvesting and transporting timber. The forestry imaginary, therefore, appears to be colored by a principle that the closer you get in time and space to a full-grown tree (timber), and thus the harvest, the more valuable the associated work. This, in turn, can be understood by the dominant forest economy approach, in which a log represents income in the near future, whereas a seedling represents a cost that hopefully will pay off in the long run.

The organization of tree planting in Sweden presents both similarities and differences with planting and harvesting. Major similarities are the employment of contractors and the piece-rate system, whereby payment is provided according to performance (trees planted). A key difference is that whereas planting is still carried out manually, harvesting is almost all mechanized, completed by operators with harvesters and forwarders (Ager, 2014). Hence, planting remains hard, physical work. Until two decades ago, local young women and men were recruited for this summer seasonal work, but recently they have been replaced by migrant workers from Eastern Europe who are considered to be used to hard physical labor (Ager, 2016). In this regard, there are apparent similarities with the working conditions of the logging camps a century ago, and it would be very interesting to study (and compare with Johansson, 1994) how masculinity is constructed in this context.

In Sweden, while there are stories of *logger heroes* that connect logging with stereotypical male characteristics, the same cannot be said for *planting heroes*. Given the higher proportion of women in tree planting, does the lower status of tree planting reflect the perceived qualifications for undertaking the work? For example, no specific education or previous training is required for individuals to be hired for forest planting, whereas there have always been educational requirements for machine operators. To work as a harvest operator in Sweden is considered a highly qualified task because of the range and number of decisions that the operator must make, and according to Nordin (2006), some operators make the comparison with that of a combat pilot. This reference not only reflects the notion of the interacting stressors of the work but also expresses the construction of masculinity and the gendering of forestry work. According to our knowledge, there are no similar metaphoric references related to the forest planter, and very little research has been carried out regarding regenerating work (force) as compared with harvesting work (force). Without negating the difficulties of logging work, we also note that perceptions of the necessary qualifications for and intensity of particular work tasks are also gendered.

In Canada, tree planting also has gender dimensions, although these dimensions play out somewhat differently. As in Sweden, tree planting is undertaken through independent contractors and employs a large proportion of women, although workers are not necessarily local or rural. Young people, typically from middle-class or upper-class socioeconomic backgrounds, take up the physical and precarious work during summer months, often while pursuing higher education (Ekers & Farnan, 2010; Sweeney, 2009a). The volume of research about tree planting is a minute fraction of that about timber harvesting (Sweeney, 2009b), leading Sweeney (2009b, p. 47) to assert that "researchers have all but ignored the workers who plant the trees." Research that touches on gender in tree planting explains that the proportion of women planters is higher than for the forest sector as a whole; however, specific numbers for the sector are not provided. Current estimates by promotional outlets suggest that at least 30% of tree planters are women, and the proportion is growing (Silviculture Canada, 2022). Nevertheless, researchers also suggest that tree planting activities reinforce gendered identities of the larger forestry culture. For example, researchers have explained how women have negotiated their roles within a masculine-dominated industry, aimed to work like men, and addressed explicit forms of gender-based discrimination and sexual objectification and even violence (e.g., Long, 2021; Main, 2009). While the findings of Main's research may appear dated, current news reports and research related to gender-based violence in tree planting worksites suggest that objectification of women tree planters remains significant today (Long, 2021; Trumpener, 2020).

The overall lack of workforce research on tree planting in boreal forestry (including gender-disaggregated data) in both Sweden and Canada means that questions relating to how the status of tree planting impacts how women and men identify themselves with forestry cannot be answered. However, the theories introduced above can guide us to make assumptions, set up hypotheses, specify research questions, search for "hidden" data, and interpret our findings.

These reflections might also shape the readiness of the forest sector to adapt to climate change. *What are the implications of the forestry imaginary for the sector's ability to adapt to climate change?* Despite legislation and policy commitments to gender equality in Canada and Sweden, the forest sectors in both countries continue to limit the engagement of women in forestry work and management decisions. For example, a recent Canadian report used census data to explain that "in 2016 women occupied 17% of the jobs in Canada's forest sector. This is an improvement over 1996 but maintaining the same rate of change suggests that it would take another 200 years to reach parity" (Wyatt et al., 2021, p. 1). The proportion of women working in decision-making positions is even lower. The minimal representation of women (and of Indigenous and other racialized peoples) is coupled with how forest management is viewed as a highly technical exercise whereby forest companies are granted licenses to meet annual harvesting targets. As detailed above when describing *management masculinity*, such technical expertise and narrow focus have been demonstrated in both Canada and Scandinavia to be characteristics of a masculine enterprise (Brandth & Haugen, 2005; Reed, 2003b), effectively restricting who is deemed to have appropriate expertise to contribute to management decisions. This narrow problem formulation and assessment of appropriate expertise may limit the capacity of the sector to adapt to rapidly changing circumstances. For example, the demands to assess the vulnerabilities associated with climate change and to determine adaptation or mitigation strategies are urgent. Yet, criticisms have been leveled at the pace of adaptation in both countries. In Sweden, for example, following a study involving 15 forest organizations across Sweden's forest sector, Andersson and Keskitalo (2018) remained pessimistic about the capacity of the sector to adapt, suggesting that "business-as-usual remains the logical choice in Swedish forestry" (p. 75) for the near future. Similarly, Andersson et al. (2017) provide a long list of social and institutional barriers to taking up bold climate adaptation measures (particularly those that might not conform to the present economic logic), including a lack of relevant expertise and alternative management practices for forest owners. In Canada, Johnston and Hesseln (2012) also found several institutional and financial barriers related to tenure and regulations and investment that limited the capacity of forest managers to adapt to climate change. A more recent review by Williamson et al. (2019) is more optimistic but indicates that adaptation measures related to policy, practices, and approaches remain in their early stages. They argue that to enable adaptation would require (1) engaging with Indigenous Peoples through collaboration; (2) revising institutions such as regulations, tenure structures, and definitions of sustainable forest management; (3) improving communication between scientists and decision-makers; (4) raising awareness within and beyond forest companies; (5) and providing resources and leadership for local innovation and experimentation.

Interestingly, in quite a separate literature, research has found that a more diverse workforce is more likely to embrace innovation and change. For example, a recent study by McKinsey & Company (2020, p. ii) argued that "diverse teams have been shown to be more likely to radically innovate and anticipate consumer needs and consumption patterns—helping their companies to gain a competitive edge." In Canada, at least, there is a change in the proportions of women seeking forestry training, at least within the professional job categories. For example, the proportion of women graduating from professional programs in forestry was 48% in 2020 (Wyatt et al., 2021). If coupled with strategies that encourage the retention of women and a rethinking of the managerial models of forestry, improvements in equity, diversity, and inclusion around decision-making tables may also enhance the capacity of the forest sector to adapt to climate change.

22.6 Conclusions

In this chapter, we have reviewed research from Canada and Scandinavia that has demonstrated that although the nature of forestry work has changed through economic and technological restructuring, the significance of gender has not diminished. Rather than reducing the salience of *masculinity*, restructuring has resulted in the multiplication of masculine norms and ideals, with different forms of knowledge (practical, technological, administrative) being reimagined through a masculine lens. New forms of masculinity have emerged, and yet an overall *imaginary* remains that continues to valorize timber production over a broader suite of possible forestry activities. Furthermore, despite multiple opportunities to engage and employ women, the industry remains remarkably male dominated and women continue to be subjected to discrimination, sexualization, and harassment (Johansson et al., 2018, 2019; Long, 2021; Trumpener, 2020). These findings are true in Canada and Sweden, two countries where boreal forests dominate and where governments pride themselves on supporting gender equality.

There is hope that public demands to address climate change and embrace a broader agenda for forest management are starting to be realized through certification requirements and policy changes. Nevertheless, if we continue to promote *suitable men* in the roles of producers, decision-makers, and managers, women will remain grossly underrepresented in these positions, and the forest industry's capacity to adapt and innovate may be stifled.

This is no time for complacency. Feminist theories and empirical analyses to date have unpacked the assumptions of this imaginary and have documented material consequences for the sector and the workers who labor in it. While feminist scholars have discovered an extensive literature filled with male *heroes*, they have not found a corresponding literature about women who work in the sector. While the value of representing individuals as heroes and heroines may be disputed, the task of documenting the critical roles played by women through the phases and levels of industrial practice is just beginning. Perhaps even more necessary is the need to open the sector to women in roles that can help shape the industry's capacity to adapt to urgent priorities such as climate change. By combining feminist theories with stories of women who work in different aspects of forestry, we can render women's expertise and contributions to the forest sector more visible, while also addressing some of the fundamental questions related to what forests produce and how forest companies can adapt to climate change. We hope that readers of this chapter also begin unpacking assumptions that have yet to be questioned. We invite you to use our reflections to generate more fundamental questions of your own and pursue a deeper agenda with respect to gender and forestry across boreal landscapes.

References

Ager, B. (2014). *The humanisation and rationalisation of forestry work, from 1900 and onwards* [in Swedish]. Ph.D. thesis, Luleå University of Technology.

Ager, B. (2016). *Structural and organisational renewal of forestry operations-history and future.* Department of Forest Biomaterials and Technology, Swedish University of Agricultural Sciences.

Andersson, E., & Keskitalo, E. C. H. (2018). Adaptation to climate change? Why business-as-usual remains the logical choice in Swedish forestry. *Global Environmental Change, 48*, 76–85. https://doi.org/10.1016/j.gloenvcha.2017.11.004.

Andersson, E., Keskitalo, E. C. H., & Lawrence, A. (2017). Adaptation to climate change in forestry: A perspective on forest ownership and adaptation responses. *Forests, 8*(12), 493. https://doi.org/10.3390/f8120493.

Bardekjian, A., Nesbitt, L., Konijnendijk, C., et al. (2018). Girls talk trees: Examining barriers to women in arboriculture and urban forestry across Canada and the United States. In *World forum on urban forestry: Changing the nature of cities.* Mantua

Barnes, T. J., Hayter, R., & Hay, E. (1999). "Too young to retire, too bloody old to work": Forest industry restructuring and community response in Port Alberni, British Columbia. *The Forestry Chronicle, 75*(5), 781–787. https://doi.org/10.5558/tfc75781-5.

Baruah, B. (2018). *Barriers and opportunities for women's employment in natural resources industries in Canada.* Ottawa: Natural Resources Canada.

Brandth, B., & Haugen, M. S. (2000). From lumberjack to business manager: Masculinity in the Norwegian forestry press. *Journal of Rural Studies, 16*(3), 343–355. https://doi.org/10.1016/s0743-0167(00)00002-4.

Brandth, B., & Haugen, M. (2005). Text, body and tools: Changing mediations of rural masculinity. *Men and Masculinities, 8*, 148–163. https://doi.org/10.1177/1097184X05277716.

Buchanan, A., Reed, M. G., & Lidestav, G. (2016). What's counted as a reindeer herder? Gender and the adaptive capacity of Sami reindeer herding communities in Sweden. *Ambio, 45*, 352–362. https://doi.org/10.1007/s13280-016-0834-1.

Canadian Council of Forest Ministers (CCFM). (2003). *Defining sustainable forest management in Canada: Criteria and indicators.* Ottawa: Canadian Council of Forest Ministers, Natural Resources Canada.

Coen, S. E., Oliffe, J. L., Johnson, J. L., et al. (2013). Looking for Mr. PG: Masculinities and men's depression in a northern resource-based Canadian community. *Health & Place, 21*, 94–101. https://doi.org/10.1016/j.healthplace.2013.01.011.

Commission on Resources and Environment. (1994). Vancouver Island land use plan. Victoria: *BC commission on resources and environment.*

Cox, D., & Mills, S. (2015). Gendering environmental assessment: Women's participation and employment outcomes at Voisey's Bay. *Arctic, 68*(2), 246–260. https://doi.org/10.14430/arctic 4478.

Doeringer, P. B., & Piore, M. J. (1971). *Internal labor markets and manpower analysis* (p. 214). Lexington: DC Heath & Company.

Dunk, T. W. (1991). *It's a working man's town. Male working-class culture in northwestern Ontario.* Montréal: McGill-Queen's University Press.

Ekers, M. (2009). The political ecology of hegemony in depression-era British Columbia, Canada: Masculinities, work and the production of the forestscape. *Geoforum, 40*(3), 303–315. https:// doi.org/10.1016/j.geoforum.2008.09.011.

Ekers, M., & Farnan, M. (2010). Planting the nation: Tree planting art and the endurance of Canadian nationalism. *Space and Culture, 13*(1), 95–120. https://doi.org/10.1177/1206331209358348.

Enander, K. G. (2007). *Skogsbruk på samhällets villkor. Skogsskötsel och skogspolitik under 150 år Umeå* [in Swedish]. Umeå: Department of Forest Ecology and Management, Swedish University of Agricultural Sciences.

Eveline, J., & Booth, M. (2002). Gender and sexuality in discourses of managerial control: The case of women miners. *Gender, Work and Organization, 9*(5), 556–578. https://doi.org/10.1111/ 1468-0432.00175.

Follo, G., Lidestav, G., Ludvig, A., et al. (2017). Gender in European forest ownership and management: Reflections on women as "new forest owners". *Scandinavian Journal of Forest Research, 32*(2), 174–184. https://doi.org/10.1080/02827581.2016.1195866.

Food and Agriculture Organization of the United Nations (FAO). (2006). *UNECE FAO team of specialists on gender and forestry: Time for action changing the gender situation in forestry* (p. 186). Rome: Food and Agriculture Organization of the United Nations.

Gherardi, S., & Poggio, B. (2001). Creating and recreating gender order in organizations. *Journal of World Business, 36*(3), 245–259. https://doi.org/10.1016/s1090-9516(01)00054-2.

Häggström, C., Kawasaki, A., & Lidestav, G. (2013). Profiles of forestry contractors and development of the forestry-contracting sector in Sweden. *Scandinavian Journal of Forest Research, 28*(4), 395–404. https://doi.org/10.1080/02827581.2012.738826.

Hankivsky, O. (2014). *Intersectionality 101.* The Institute for Intersectionality Research & Policy, Burnaby: Simon Fraser University.

Hayter, R. (2000). *Flexible crossroads: The restructuring of British Columbia's forest economy* (p. 448). Vancouver: UBC Press.

Hugosson, M. (1999). *Constructing cultural patterns from actors' views on industrial forestry in Sweden: An interpretive study based on assessments of conceptualizations and definitions in organizational culture theory.* Ph.D. thesis, Acta Universitatis Agriculturae Sueciae Silvestria 113 (Swedish University of Agricultural Sciences).

Johansson, E. (1994). *The free sons of the forests: Masculinity and modernity in forest work of Norrland* [in Swedish]. Stockholm: Nordiska Museet.

Johansson, K., Andersson, E., Johansson, M., et al. (2019). The discursive resistance of men to gender-equality interventions: Negotiating "unjustness" and "unnecessity" in Swedish forestry. *Men and Masculinities, 22*(2), 177–196. https://doi.org/10.1177/1097184x17706400.

Johansson, K., Andersson, E., Johansson, M., et al. (2020). Conditioned openings and restraints: The meaning-making of women professionals breaking into the male-dominated sector of forestry. *Gender, Work and Organization, 27*(6), 927–943. https://doi.org/10.1111/gwao.12403.

Johansson, M. (2020). *Business as usual?: Doing gender equality in Swedish forestry work organisations.* Ph.D. thesis, Luleå University of Technology.

Johansson, M., & Ringblom, L. (2017). The business case of gender equality in Swedish forestry and mining-restricting or enabling organizational change. *Gender, Work and Organization, 24*(6), 628–642. https://doi.org/10.1111/gwao.12187.

Johansson, M., Johansson, K., & Andersson, E. (2018). #Metoo in the Swedish forest sector: Testimonies from harassed women on sexualised forms of male control. *Scandinavian Journal of Forest Research, 33*(5), 419–425. https://doi.org/10.1080/02827581.2018.1474248.

Johnston, M., & Hesseln, H. (2012). Climate change adaptive capacity of the Canadian forest sector. *Forest Policy and Economics, 24*, 29–34. https://doi.org/10.1016/j.forpol.2012.06.001.

Kardell, L. (2004). *Svenskarna och skogen* [in Swedish] (p. 303). Jönköping: Skogsstyrelsens förl.

Lidestav, G. (2001). Does the forest demand men, or will a woman do? [in Swedish]. In B. Liljewall, K. Niskanen, & M. Sjöberg (Eds.), *Kvinnor och jord. Arbete och ägande från medeltid till nutid. Skrifter om skogs-och lantbrukshistoria nr. 15* (pp. 159–173). Lund: Nordiska museets förlag.

Lidestav, G., & Ekström, M. (2000). Introducing gender in studies on management behaviour among non-industrial private forest owners. *Scandinavian Journal of Forest Research, 15*(3), 378–386. https://doi.org/10.1080/028275800448011.

Lidestav, G., & Nordfjell, T. (2005). A conceptual model for understanding social practices in family forestry. *Small-scale Forest Economics, Management and Policy, 4*(4), 391–408. https://doi.org/10.1007/s11842-005-0024-7.

Lidestav, G., & Sjölander, A. (2007). Gender and forestry: A critical discourse analysis of forestry professions in Sweden. *Scandinavian Journal of Forest Research, 22*(4), 351–362. https://doi.org/10.1080/02827580701504928.

Lidestav, G., Andersson, E., Lejon, S. B., et al. (2011). *Jämställt arbetsliv i skogssektorn* [in Swedish]. Umeå: Swedish University of Agricultural Sciences.

Lidestav, G., Johansson, M., & Huff, E. S. (2019). Gender perspectives on forest services in the rise of a bioeconomy discourse. In T. Hujala, A. J. Toppinen, & B. Butler (Eds.), *Services in family forestry* (pp. 307–325). Cham: Springer International Publishing.

Long, J. (2021). *Not so clear cut: Gender-based violence in BC's tree planting industry.* M.A. thesis, University of Victoria.

Lu, S. L., & Sexton, M. (2010). Career journeys and turning points of senior female managers in small construction firms. *Construction Management and Economics, 28*(2), 125–139. https://doi.org/10.1080/01446190903280450.

Main, C. (2009). *The experiences of women tree planters in northern Ontario.* M.A. thesis, Lakehead University.

Manning, S. (2014). *A FemNorthNet fact sheet. Feminist intersectional policy analysis: Resource development and extraction framework* (p. 4). Ottawa: Canadian Research Institute for the Advancement of Women.

Martz, D., Reed, M. G., Brueckner, I., et al. (2006). *Hidden actors, muted voices: The employment of rural women in Canadian forestry and agri-food industries.* Ottawa: Policy Research Fund.

Mayes, R., & Pini, B. (2014). The Australian mining industry and the ideal mining woman: Mobilizing a public business case for gender equality. *Journal of Industrial Relations, 56*(4), 527–546. https://doi.org/10.1177/0022185613514206.

McKinsey & Company. (2020). *Diversity wins: How inclusion matters* (p. 52). London: McKinsey & Company.

Mills, S., Dowsley, M., & Cameron, E. (2013). *Gender in research on northern resource development.* Gap analysis report #14. Whitehorse: Resources and Sustainable Development in the Arctic (ReSDA).

Mills, S. E. (2006). Segregation of women and aboriginal people within Canada's forest sector by industry and occupation. *The Canadian Journal of Native Studies, 26*, 147–171.

Mills, S. E. (2007). *Women's experiences and representations of diversity management and organizational restructuring in a multinational forest company.* Ph.D. thesis, University of Saskatchewan.

Natural Resources Canada (NRC). (2020). 8 facts about Canada's boreal forest. Ottawa: Natural Resources Canada. https://www.nrcan.gc.ca/our-natural-resources/forests-forestry/sustainable-forest-management/boreal-forest/8-facts-about-canadas-boreal-forest/17394.

Navarro-Astor, E., Román-Onsalo, M., & Infante-Perea, M. (2017). Women's career development in the construction industry across 15 years: Main barriers. *Journal of Engineering, Design and Technology, 15*(2), 199–221. https://doi.org/10.1108/jedt-07-2016-0046.

Nordin, M. H. (2006). *'They use to compare it to a fighter pilot': Notions of work environments and risks in forestry machine work* [in Swedish]. Ph.D. thesis, Umeå University.

Östlund, L., Öbom, A., Löfdahl, A., et al. (2020). Women in forestry in the early twentieth century–new opportunities for young women to work and gain their freedom in a traditional agrarian society. *Scandinavian Journal of Forest Research, 35*(7), 403–416. https://doi.org/10.1080/028 27581.2020.1808054.

Parmenter, J. (2011). Experiences of Indigenous women in the Australian mining industry. In K. Lahiri-Dutt (Ed.), *Gendering the field: Towards sustainable livelihoods for mining communities* (pp. 67–86). ANU Press.

Preston, V., Rose, D., Norcliffe, G., et al. (2000). Shift work, childcare and domestic work: Divisions of labour in Canadian paper mill communities. *Gender, Place and Culture, 7*(1), 5–29. https://doi.org/10.1080/09663690024843.

Proulx, G., Beaudoin, J.-M., Asselin, H., et al. (2020). Untapped potential? Attitudes and behaviours of forestry employers toward the Indigenous workforce in Quebec, Canada. *Canadian Journal of Forest Research, 50*(4), 413–421. https://doi.org/10.1139/cjfr-2019-0230.

Quam-Wickham, N. (1999). Rereading man's conquest of nature. *Men and Masculinities, 2*(2), 135–151. https://doi.org/10.1177/1097184x99002002002.

Reed, M. G. (2003a). Marginality and gender at work in forestry communities of British Columbia, Canada. *Journal of Rural Studies, 19*(3), 373–389. https://doi.org/10.1016/S0743-0167(03)000 21-4.

Reed, M. G. (2003b). *Taking stands.* Vancouver: UBC Press.

Reed, M. G. (2008). Reproducing the gender order in Canadian forestry: The role of statistical representation. *Scandinavian Journal of Forest Research, 23*(1), 78–91. https://doi.org/10.1080/02827580701745778.

Richardson, K. (2008). *A gendered perspective of learning and representation on forest management advisory committees in Canada.* Master of Natural Resources Management, University of Manitoba.

Ringblom, L., & Johansson, M. (2020). Who needs to be 'more equal' and why? Doing gender equality in male-dominated industries. *Equality, Diversity and Inclusion: An International Journal, 39*, 337–353. https://doi.org/10.1108/EDI-01-2019-0042.

Ross. M. (1997). *A history of forest legislation in Canada 1867–1996.* Occasional Paper No. 2. Calgary: Canadian Institute of Resources Law.

Sametinget (Sami Parliament). (2020). *Sametinget.* Giron/Kiruna: Sametinget.

Sandström, P. (2015). *A toolbox for co-production of knowledge and improved land use dialogues–the perspective of reindeer husbandry.* Ph.D. thesis, Swedish University of Agricultural Sciences.

Sherry, E., Halseth, R., Fondahl, G., et al. (2005). Local-level criteria and indicators: An Aboriginal perspective on sustainable forest management. *Forestry, 78*(5), 513–539. https://doi.org/10.1093/forestry/cpi048.

Silviculture Canada. (2022). Tree planting: It's a tough job but somebody has to do it. Silviculture Canada. Retrieved March 30, 2022, from, http://www.silviculturecanada.ca/treeplanting.html#:~:text=What%20does%20a%20typical%20Canadian,becoming%20more%20balanced%20every%20year.

Skogsstyrelsen. (2020). *Sysselsättning I Skogsbruket.* Jönköping: Skogsstyrelsen.

Skogssverige. (2020). *Skogen & ekonomin.* Stockholm: Skogssverige.

Smith, L. (2013). Working hard with gender: Gendered labour for women in male dominated occupations of manual trades and information technology (IT). *Equality, Diversity and Inclusion: An International Journal, 32*(6), 592–603. https://doi.org/10.1108/edi-12-2012-0116.

Steger, M. B., & James, P. (2013). Levels of subjective globalization: Ideologies, imaginaries, ontologies. *Perspectives on Global Development and Technology, 12*(1–2), 17–40. https://doi.org/10.1163/15691497-12341240.

Sveriges Riksdag. (2008). *Regeringens proposition 2007/08:108*. En skogspolitik i takt med tiden Prop., Stockholm.

Swedish University of Agricultural Sciences (SLU). (2015). *Plan för lika villkor 2015–2017– Fakulteten för skogsvetenskap [Equality plan for the Faculty of forest science 2015 – 2017]. Steering document.*Uppsala: Swedish University of Agricultural Sciences.

Swedish University of Agricultural Sciences (SLU). (2019). *Forest statistics 2019 official statistics of Sweden*. Umeå: Swedish University of Agricultural Sciences.

Sweeney, B. (2009a). Sixty years on the margin: The evolution of Ontario's tree planting industry and labour force: 1945–2007. *Labour/Le Travail, 63*, 47–78.

Sweeney, B. (2009b). Producing liminal space: Gender, age and class in northern Ontario's tree planting industry. *Gender, Place and Culture, 16*(5), 569–586. https://doi.org/10.1080/09663690903148432.

Trumpener, B. (2020). Accounts of sex assaults in B.C. tree planter camps 'deeply disturbing' CBC News, January 31, 2020, Prince George.

Walker, H. M., Culham, A., Fletcher, A. J., et al. (2019). Social dimensions of climate hazards in rural communities of the global north: An intersectionality framework. *Journal of Rural Studies, 72*, 1–10. https://doi.org/10.1016/j.jrurstud.2019.09.012.

Waring, M. (1988). *If women counted*. San Francisco: Harper & Row.

Westin, K., Eriksson, L., Lidestav, G., et al. (2017). Individual forest owners in context. In E. Keskitalo (Ed.), *Globalisation and change in forest ownership and forest use: Natural resource management in transition* (pp. 57–95). London: Palgrave Macmillan.

Wide, R., & Nordin, M. H. (2019). *Jämställd skogssektor*. Jönköping: Skogsstyrelsen.

Williamson, T. B., Johnston, M. H., Nelson, H. W., et al. (2019). Adapting to climate change in Canadian forest management: Past, present and future. *The Forestry Chronicle, 95*(2), 76–90. https://doi.org/10.5558/tfc2019-015.

Wyatt, S., Reed, M., Feng, X., et al. (2021). *Evidence on diversity in Canada's forest sector* (p. 43). Ottawa: Forest Products Association of Canada and Canadian Institute of Forestry.

Chapter 23
Public Participation at a Crossroads: Manipulation or Meaningful Engagement in the Boreal Region

John R. Parkins and A. John Sinclair

Abstract Advances in public participation are stimulated by multiple drivers, including public concern for environmental degradation, conflict between forest users, Indigenous rights, and international agreements. Yet, with many notable advances, innovation has stagnated, and the quality of participatory processes in forest management is highly variable. The body of evidence to date demonstrates weaknesses in the design and implementation of participatory processes. With examples from Europe and North America, in this chapter we note that public engagement is often mostly about legitimating predefined plans and policies, narrow technical discussions that malign the inherently political nature of forest management, and participants that are not representative of the general public. To move beyond these challenges, we propose several changes, including technological innovations such as web-based and emerging social media platforms and institutional innovations such as episodic and punctuated modes of engagement that are part of an overall participation plan.

23.1 Introduction

Public participation in forestry is ubiquitous throughout the boreal region with initiatives that engage citizens in countries across Europe and Canada. These initiatives take many forms, ranging from short-term workshops and planning sessions (Pappila & Pölönen, 2012) to complex institutional arrangements and enhanced Indigenous engagement (Klenk et al., 2013). In all countries, we observe advances in

J. R. Parkins (✉)
Department of Resource Economics and Environmental Sociology, University of Alberta, 515 General Services Building, Edmonton, AB T6G 2H1, Canada
e-mail: jparkins@ualberta.ca

A. J. Sinclair
Natural Resources Institute, University of Manitoba, Winnipeg, MB R3T 2M6, Canada
e-mail: john.sinclair@umanitoba.ca

© The Author(s) 2023
M. M. Girona et al. (eds.), *Boreal Forests in the Face of Climate Change*,
Advances in Global Change Research 74,
https://doi.org/10.1007/978-3-031-15988-6_23

public participation that are stimulated by multiple drivers, including public concern for environmental degradation, the desire to be involved in forestry decisions, conflict between forest users, scientific complexity and uncertainty, Indigenous rights, and international agreements that codify public participation as a key dimension of sustainable forest management.

In some respects, advances in participatory processes are noteworthy and reflect a deepening of democracy in boreal regions; thousands of citizens have at least some opportunities to learn and influence the direction of forest management in their region. Improvements in public participation have the potential to enhance transparency, improve accountability, reduce conflict, and improve overall decision-making (Hanna, 2015). This is the promise and potential of public participation in forest management.

Yet, as discussed in this chapter, although there is an increase in venues for public engagement in forest management, the quality of these engagements is highly variable, and the body of evidence to date demonstrates weaknesses in design and implementation. These weaknesses are largely consistent across the boreal region in Europe and Canada. Borrowing a term from the "eight rungs of the ladder of citizen participation" (Arnstein, 1969), what passes as public participation in forestry is often little more than "manipulation" or "therapy." It is a type of engagement that is synonymous with tokenism. Within this chapter, we use the term *manipulation* intentionally to represent forms of participation that meet minimal standards for citizen engagement and procedures but are designed to control the outcome of engagement processes. Given the achievements in public participation across the boreal region, our criticisms may sound harsh or overreaching. Yet, we make this claim in part because we observe little innovation or improvement to engagement processes in recent decades. In spite of more than 20 years of critical analysis and repeated recommendations for improvements, we observe almost no change (Lindgren et al., 2019).

This chapter considers the status of public participation in forest management by highlighting several policy developments and catalysts from the onset of environmentalism in the 1970s to the present day. We note several key achievements and also the challenges that lie ahead for democratizing forest management. Although we review the literature on this topic drawing from examples in the boreal regions in Europe (highlighting commonalities), this chapter focuses mainly on the Canadian context with attention to current challenges and opportunities for advancing public engagement. With topics on Indigenous forestry (Chap. 20) and gender aspects (Chap. 22) covered in other chapters in this volume, this chapter pays particular attention to issues of representation, meaningful participation, and possibilities for institutional innovation.

23.2   Public Participation in the Boreal Forest: International Perspectives

The FAO Joint Committee (2000, p. 7) defines public participation as:

> Various forms of direct public involvement where people, individually or through organized groups, can exchange information, express opinions, and articulate interests, and have the potential to influence decisions or the outcome of specific forestry issues.

This definition gets at the heart of the concept emphasizing two-way flows of information as a distinct alternative to one-way flows of information, i.e., communications. The other key part of this definition involves the potential to influence decisions. This definition of public participation relates to a range of consultation and engagement procedures in forestry or other regulatory settings, such as impact assessment and public hearings. Here we distinguish between public participation and more direct forms of delegated authority, direct democracy, or citizen control. The basic thrust behind public participation involves a sense that citizen engagement is meaningful. This idea of meaningfulness is derived from mutual learning and the possibility of linking directly with policy or management decisions (Sinclair et al., 2017).

Considering public participation from a more theoretical perspective, Pappila and Pölönen (2012) refer to the environmental, integrative, and democratic functions of participation. In particular, they note that public participation consists of "(1) access to information, (2) participation in decision-making and (3) access to justice in keeping with the terms of the Aarhus Convention, the most important international agreement on public participation" (p. 178). This convention (UNECE, 1998) provides a framework for environmental management within European countries that involves access to information, public participation in decision-making, and access to justice. The agreement mobilizes forest policy development in boreal regions such as Finland, Norway, and Sweden (Lindstad & Solberg, 2012).

The rationale for public participation in European countries is partly a function of democratic impulses at the national level as well as international agreements, but there are also more pragmatic reasons for implementing participatory processes in forest management. These reasons are summarized by Kleinschmit et al. (2018) to include empowerment, influence, legitimacy, representation, transparency, accountability, and effectiveness. Added to this list is the role that participatory processes can play in achieving sustainability. "It is argued that gathering, integrating and taking into account society's perspectives in relation to a forest-related objective or problem makes the resulting policy decisions and implementation more sustainable" (Kleinschmit et al., 2018, p. 7). Toward this end, the aspirational aspects of public participation lead some Finnish researchers to declare promising directions toward sustainability in the form of an improved implementation of regulations, enhanced conflict resolution, identification of shared interests, and possibilities for the acceptance of decisions as they emerge more transparently from participatory processes (Pappila & Pölönen, 2012).

Researchers have also recognized the importance of individual, social, and mutual learning to meaningful participation and effective resource management, including forest management in both Europe and Canada (e.g., Romina, 2014; Van der Wal et al., 2014). As established by Woodhil and Röling (2000, p. 54), such learning can "help improve the quality and wisdom of the decisions we take when faced with complexity, uncertainty, conflict, and paradox." Important social learning outcomes through involvement in forest management indicate that learning can result in collective action outcomes, such as protecting cultural heritage, acquiring new knowledge about forests and forestry, and building relationships (e.g., Assuah & Sinclair, 2019).

Notwithstanding the promise of public participation, researchers have identified a plethora of ongoing challenges. These issues are clearly evident across the boreal region, and we summarize some of the most salient issues here. First, consistent with problems of tokenism, we note that public engagement is often little more than a process of *legitimation*. It serves to legitimize the dominant discourses of elite interests rather than allowing for the consideration of alternatives to the status quo (Parkins & Sinclair, 2014). Studies suggest "that uneven power relations, unclear mandates and vague forms of accountability favor the state, forest owners and forest industry" (Lindahl et al., 2017, p. 54), while at the same time discrediting local knowledge, local users, and local systems of forest governance.

Second, participation often involves *depoliticization*. Working with interested citizens is an inherently political process involving careful consideration of contending and legitimate values, ideas, and supporting evidence. Although participation is inherently political in this way, we observe efforts to depoliticize processes of participation through highly technical or science-based decisions. This includes efforts to get past the messiness of politics with big assumptions about how to implement unbiased and clear-cut options presented within scientific data. This technical approach often maligns the complex and contested nature of forest management. In this context, authors such as Klenk et al. (2013) conclude that limitations on participation "effectively curtailed the advocacy of participants' political interests" (p. 172).

Third, moving to more functional and pragmatic challenges, many scholars note a clear lack of *representation* in participatory processes. Numerous studies in the Canadian context, for example, highlight a limited range of public values within participatory processes (e.g., McFarlane & Boxall, 2000), as well as procedural aspects of engagement processes that curtail the capacity of specific individuals to participate effectively in group processes (Parkins & Sinclair, 2014).

Lastly, there are *functional challenges* related to implementing participatory processes in forestry. Some of these challenges involve the timing of engagement processes and misalignment with specific points of decision-making. Much of the literature noted above establishes that participation is often relegated to operational decisions at best, i.e., what trees to cut and when. For many, this operational discussion comes too late in the decision process and represents the thinking behind the initial phases of participation in forest management that leaned toward trying to protect cultural values on the land, i.e., cut around them. Furthermore, especially in Europe, with fragmented ownership structures, the opportunity for meaningful public engagement is difficult to implement on operations occurring on small tracts of land (Pappila & Pölönen, 2012).

23.3 Public Participation in Canadian Forestry

When we focus more specifically on the Canadian context, it is helpful first to understand how we have arrived at this point in time and identify the challenges ahead. Public participation in Canada enjoyed initial momentum from many of the same international movements that propelled civic engagement on environmental issues. This engagement started with the environmental movement in the 1970s, reflecting broad-based public concern for environmental degradation and demands for regulatory oversight of industrial activity. Synonymous with this movement was the establishment of departments of the environment in many jurisdictions throughout North America (McKenzie, 2002). Similarly, in the 1980s, the push to clarify what it means to undertake sustainable development (Brundtland Commission, 1987) propelled a number of international initiatives, such as the Earth Summit in Rio de Janeiro in 1992. One legacy of Rio was the Montréal (2015), which formalized criteria and indicators for conservation and sustainable forest management in boreal forests. These efforts were instrumental in propelling specifically Canadian responses to these international initiatives.

In Table 23.1, we summarize key incentives that stem from these initial developments, with an attempt to focus (mostly) on national events or initiatives having a national impact. In this section, readers will observe an emphasis on events in the 1990s. This decade was a particularly challenging and innovative time for Canadian forestry, with several institutional and regulatory developments taking place throughout the boreal region and beyond.

To provide a backdrop to these developments, we make particular note of the Clayoquot Sound blockades in 1993 (Hayter, 2003). Although the conflict reached a peak in the early 1990s, Clayoquot Sound is emblematic of a decade-long conflict between environmentalists and the forest industry in British Columbia, spilling out into other parts of the country. Often dubbed the "war in the woods," this persistent and high-profile conflict was a catalyst for changes to forest policy, including a series of initiatives to integrate citizens and key stakeholders into forest management in a more meaningful way.

There are three key initiatives through the 1990s that warrant specific attention here. First, in terms of forest policy, Canada is perhaps best known worldwide for its leadership in establishing *Canada's Model Forest Program* (LaPierre, 2003). Initiated in 1991 with financial support from the Canadian Forest Service, the program had ten sites across the country. It also established an international presence and propagated the idea of model forest institutions in many other countries. The idea behind the program was to develop and showcase a new institutional model to foster

Table 23.1 Key incentives for public participation in Canadian forestry

Year	Event/Topic	Description	References
1970s	Departments of environment	Second-wave environmentalism established environment departments and laws around broad-based public concern for environmental degradation	McKenzie (2002)
1987	Sustainable development	Global movement to define sustainability and establish measures of progress toward this goal	Brundtland (1987)
1990s	Public advisory committees	Advisory committees associated with tenure holders became a policy requirement, culminating in more than 100 committees nationally by the year 2000	Parkins et al. (2006)
1991	Model forest program	Diverse representation on boards and local stakeholder engagement	LaPierre (2003)
1993	Clayoquot Sound blockades	Decades of conflict characterized by a "war in the woods," which culminated in blockades, mass civil unrest, and the resolve to improve public engagement in forest management	Hayter (2003)
1995	Sustainable forest management	Inspired by the Montréal Process, detailed criteria and indicators for sustainable forest management, including fair, effective, and informed decision-making	Bridge et al. (2005)
1996	Forest certification	Canadian Standards Association CAN/CSA Z809 established standards for public participation in forest management	Clark and Kozar (2011)
1998	Community forests	Community forest licenses started in several provinces leading to some decentralized decision-making	Teitelbaum et al. (2006)
2000s	Indigenous forest tenure	Indigenous communities establish direct management control of local forest resources	O'Flaherty et al. (2008)

local innovation and influence the management of public forest lands with closer collaboration between key forest stakeholders (Sinclair & Lobe, 2005; Sinclair & Smith, 1999).

Second, because of ongoing contraction in the forest sector and a sense of urgency to reconnect forestry with community development, by the late 1990s we observe the flourishing of *community forests* (see Chap. 21 for more details). By one estimate, there were 100 community forests across the country, the majority located in the provinces of British Columbia, Ontario, and Québec (Teitelbaum et al., 2006). Whereas model forests connected multiple stakeholders and land managers in partnership (including public and private sectors), community forests allowed municipalities and Indigenous communities to establish long-term lease agreements with provincial governments to manage local forest landscapes for multiple values. Local control and local benefits define the nature of community forests and, in many ways, bring rural municipal leadership into the governance model of forestry in Canada. A number of Indigenous communities are involved in community forests, and this model of Indigenous forest tenure has evolved over the last decade to include direct "nation-to-nation" agreements (O'Flaherty et al., 2008).

Third, although the abovementioned initiatives enjoy a higher profile, the last initiative we describe here is arguably the most ubiquitous in terms of public participation in the forest sector. Established in the early 1990s, the *public advisory committee* has become the default mode of public engagement across the country. Relying on a national survey of advisory committees (Parkins et al., 2006), we identified more than 100 committees tied directly to the industrial forest land base. These committees have similar mandates to support two-way flows of information and facilitate public influence over decision-making. In some jurisdictions such as Ontario, *local citizen committees* are organized and managed by the provincial government, and these citizen bodies are intended to contribute to area-based forest management plans. By contrast, in Alberta, advisory committees are sponsored by private firms, which have the responsibility for planning and managing area-based tenures in the province. As a process for local engagement, advisory committees include stakeholders, such as municipal leaders, recreation groups, environmental organizations, and sometimes educational and religious leaders.

Evolving from forest governance post-Rio and the Montréal Process, public advisory committees received further support from several national initiatives in the forest sector. First, in 1995, the Canadian Council of Forest Ministers established an influential set of criteria and indicators of sustainable forest management (Bridge et al., 2005). These indicators included the goal of fair, effective, and informed decision-making as a key dimension of sustainability. At about the same time, the Canadian Standards Association established a formal standard for forest certification (CAN/CSA Z809), and this standard included a key role for public engagement in defining and monitoring sustainable forest management. Taken together, these public and private sector initiatives further cemented the public advisory committee as a key component of forest governance in Canada.

582 J. R. Parkins and A. J. Sinclair

23.4 Public Participation at a Crossroads

This brief review reflects a burst of innovation in forest governance during the 1990s. Advances during this time were remarkable, partly because of the wide-ranging initiatives that promised to strengthen citizen engagement across a range of institutions. Over the past 20 years, however, the evolution of forest governance has faltered. On the one hand, endless critiques and invitations for improvement are accumulating in the published literature (e.g., Ambus & Hoberg, 2011; McGurk et al., 2006). On the other hand, little has changed, perhaps with the exception of Indigenous forestry initiatives (Wyatt et al., 2019), covered in Chap. 20.

In a recent review of public advisory committees in Canada, authors from multiple regions of the country have become more strident in their critiques. Lindgren et al. (2019) state that little has changed "since 2004 in terms of representativeness, insufficient public outreach and transparency, and indeterminate effectiveness in influencing forest management" (p. 37). The authors also claim that public advisory committees,

> are not likely to deliver on many of the complex issues facing forest managers such as consideration of the impacts of and adaptation to climate change, reconciliation with Indigenous people, and meaningful consideration of gender and other diversity factors in decision-making. (p. 37).

Other researchers echoed these sentiments when assessing the efficacy of participatory processes in the forest sector. For example, Miller and Nadeau (2017, p. 19) examined 15 years of participatory processes in the boreal forest of Nova Scotia and New Brunswick, identifying: (1) the importance of understanding the historic power imbalances that continue to shape dialogue and spaces for participation; (2) repeated attempts to enhance engagement followed by,

> disappointments in implementation that have led to feelings of meaningless involvement, a closed system, and mistrust in the government and industry; (3) a system of privileged access that runs counter to the ideals of deliberative democracy and an equitable decision-making process. (p. 19).

These recent critiques represent just a small sampling of concern from many researchers about the state of public engagement in Canadian forestry. These concerns are echoed across parts of Europe, with Lindahl et al. (2017) and others identifying enduring struggles for meaningful local engagement. With the foregrounding of these concerns, most analysts would agree we are at a crossroads: one path leads to further manipulation, legitimation, and degrees of tokenism that erode forest governance, whereas a second path involves a meaningful response to the challenges ahead. For brevity, we address two of these challenges here as a signpost for the work that lies ahead.

23.4.1   Representation—Broader Community Involvement

Much of the literature on participation in forest management relates to "marginal-ized" groups and the need for group diversity. Concerns are often associated with women, Indigenous people, youth, and sometimes local environmental organizations that reject local engagement processes (Nenko et al., 2019; Reed, 2010; Reed & Varghese, 2007). Research shows that even when these voices are at the table, they often have trouble gaining voice (Parkins & Sinclair, 2014). Studies also indicate that participants within such groups and roundtables are often "representatives" of other constituents; however, there is no, or little, capacity to actually help these people communicate with their constituents, share information, and collect feedback (Lindgren et al., 2019; McGurk et al., 2006).

In addition to sociodemographic diversity and the representation of marginalized groups, representation is also associated with the diversity of values that are repre-sented by specific participants. This aspect of representation can be more challenging to characterize, but researchers often identify values on a spectrum from biocentric to anthropocentric and can characterize these values within an advisory committee in comparison with values within the general population (McFarlane and Boxall 2000). This approach to understanding representation is less common in practice but is no less important in bringing diverse interests together to discuss forest management issues.

The involvement of Indigenous people has also been vexing. Indigenous voices are often those noted as missing by participants in forest management (Nenko et al., 2019; Parkins et al., 2006). One cannot overstate the importance of international agree-ments, such as the UN Declaration on the Rights of Indigenous Peoples (UNDRIP) and associated imperatives regarding Free Prior and Informed Consent (FPIC) in terms of their influence on the current and future engagement of Indigenous people. Coupled with these international commitments, Indigenous peoples within Canada retain rights and privileges within the Canadian Constitution that hold implications for forest management. We view the future in regard to respecting these constitutional rights and the UN declaration in two ways: (1) the continued development of robust government-to-government agreements on the management of forest lands (e.g., the Whitefeather Forest Initiative described by O'Flaherty et al., 2008); (2) the partic-ipation of Indigenous people in forest management in other settings. For example, Alberta has Guidelines on Consultation with First Nations on Land and Natural Resource Management that forest product companies must follow (Government of Alberta, 2014). Manitoba has the Community Timber Agreements with Indigenous and northern communities that show potential for meaningful engagement (Lawler & Bullock, 2019).

Finally, the importance of local environmental organizations is noteworthy. From Clayoquot Sound conflicts in the early 1990s to the present, local environmental organizations are often at the forefront of changing environmental practices in forest management. Because of a perceived lack of political efficacy, some of these local organizations have withdrawn from public advisory committees to the detriment of diverse local representation.

23.4.2 Meaningful Dialogue—Engagement on Issues of Public Concern

Bringing people to the table, i.e., representation, is one part of the process; what they do when they are there is also an important consideration (Romina, 2014). As indicated in the FAO Joint Committee (2000) definition stated earlier in this chapter, the purpose of public participation is to "exchange information, express opinions, and articulate interests." This purpose cannot be achieved if participation is limited to one-way forms of communication and is dominated by "information out" from those leading committees or round tables.

To foster meaningful engagement, we suggest a set of overarching principles that should guide participatory processes in forest management. These include:

- Adequate and appropriate notice of engagement opportunities is provided
- Participation begins early in the decision process and builds public confidence
- Public input can influence or change the outcome/decision being considered
- Processes are fair and transparent and allow for the local acceptance of final decisions
- Opportunities for public comment are open to all interested parties, are varied and flexible, include openings for face-to-face discussions, and involve the public in the actual design of an appropriate participation program
- Formal processes of engagement, such as forums of dispute resolution, are available
- Participant assistance and capacity building are available for informed dialogue and discussion
- Participation programs are oriented toward learning for all participants, including governments, proponents, and participants.
- Information about the decision in question is available and in local languages.

23.5 Moving Forward

As a final word, it is often much easier to envision a meaningful process of public participation than it is to implement the process. The busyness of life, the competition for people's time, and the histories of mistrust and conflict, particularly in the forest sector, often result in suboptimal outcomes, even with the best of intentions. The work of public participation is not easy. Well-trained practitioners and professionals coupled with well-resourced engagement processes can help to some extent, as do the examples of practice within the boreal forest (e.g., Kleinschmit et al., 2018; Pappila & Pölönen, 2012), but we may also need to envision entirely new ways of engaging citizens on forest management issues. One solution might involve technological innovation. Social media platforms, such as Instagram, along with virtual conferencing and webinar platforms, such as Zoom, have proliferated in recent years, allowing for many new ways of linking people with forest landscapes and decision-making (e.g., Sherren et al., 2017; Sinclair et al., 2017). These technological innovations hold much promise but remain underutilized in the forest sector.

A second solution might involve new institutional designs. With a new generation of emerging professionals and leaders, the culture of engagement may need to shift. It may no longer be sufficient to assume participation within established and long-term processes, such as public advisory committees or working groups. Less permanent and more episodic modes of engagement that are part of an overall participant plan may be needed; for example, rather than hosting monthly information sessions, one could build a brief program of engagement that is focused on a specific point of decision-making, where it makes sense to link the engagement process directly to the point of decision-making. It also seems critical to ask at least some of the people whom you want to engage what types of participatory processes they are most likely to want to be part of, not just assume what is best.

Moving public participation in forestry beyond the crossroads and down the more enlightened path will, in many cases, require a complete rethink of participatory processes from government, industry, nongovernmental agencies, and the public. We believe that the principles of meaningful participation outlined above provide a framework for action and would help ensure that the promise of participation is met. Such action will necessarily include rethinking approaches to engagement to address functional challenges, paying attention to representation, producing meaningful dialogue, and establishing new institutional designs. There is now more than ample experience through training with professional organizations such as the International Association for Public Participation (IAP2), not to mention guidebooks and handbooks, to help frame new engagement designs (Heierbacher, 2010). We only have to be willing to think beyond open houses aimed at placating the public toward a more civic approach to engagement (Sinclair & Diduck, 2017). This change will require academics to re-engage in scholarship regarding approaches to participation in forest management because little has been written on innovative approaches in the last 20 years.

References

Ambus, L., & Hoberg, G. (2011). The evolution of devolution: A critical analysis of the community forest agreement in British Columbia. *Society and Natural Resources, 24*(9), 933–950. https://doi.org/10.1080/08941920.2010.520078.

Arnstein, S. R. (1969). A ladder of citizen participation. *Journal of the American Institute of Planners, 35*(4), 216–224. https://doi.org/10.1080/01944366908977225.

Assuah, A., & Sinclair, A. J. (2019). Unraveling the relationship between collective action and social learning: Evidence from community forest management in Canada. *Forests, 10*(6), 494. https://doi.org/10.3390/f10060494.

Bridge, S. R., Cooligan, D., Dye, D., et al. (2005). Reviewing Canada's national framework of criteria and indicators for sustainable forest management. *The Forestry Chronicle, 81*(1), 73–80. https://doi.org/10.5558/tfc81073-1.

Brundtland Commission. (1987). *Our common future* (p. 400). Oxford: Oxford University Press.

Clark, M. R., & Kozar, J. S. (2011). Comparing sustainable forest management certifications standards: A meta-analysis. *Ecology & Society, 16*(1):art3. https://doi.org/10.5751/ES-03736-160103.

FAO Joint Committee. (2000). *Public participation in forestry in Europe and North America*. Report of the FAO/ECE/ILO Joint committee on forest technology management and training, team of specialists on participation in forestry. Geneva: Sectoral Activities Department, Food & Agriculture Organization.

Government of Alberta. (2014). *The Government of Alberta's Guidelines on consultation with First Nations on land and natural resource management* (p. 29). Edmonton: Government of Alberta.

Hanna, K. S. (2015). The enduring importance of Canada's forest sector. In B. Mitchell (Ed.), *Resource and environmental management in Canada: Addressing conflict and uncertainty*. 5th ed. Don Mills: Oxford University Press.

Hayter, R. (2003). "The war in the woods": Post-Fordist restructuring, globalization, and the contested remapping of British Columbia's forest economy. *Annals of the Association of American Geographers, 93*(3), 706–729. https://doi.org/10.1111/1467-8306.9303010.

Heierbacher, S. (2010). *Resource guide on public engagement* (p. 17). Boiling Springs: National Coalition for Dialogue & Deliberation.

Kleinschmit, D., Pülzl, H., Secco, L., et al. (2018). Orchestration in political processes: Involvement of experts, citizens, and participatory professionals in forest policy making. *Forest Policy Economics, 89*, 4–15. https://doi.org/10.1016/j.forpol.2017.12.011.

Klenk, N. L., Reed, M. G., Lidestav, G., et al. (2013). Models of representation and participation in Model Forests: Dilemmas and implications for networked forms of environmental governance involving indigenous people. *Environmental Policy and Governance, 23*(3), 161–176. https://doi.org/10.1002/eet.1611.

LaPierre, L. (2003). Canada's model forest program. *The Forestry Chronicle, 79*(4), 794–798. https://doi.org/10.5558/tfc79794-4.

Lawler, J. H., & Bullock, R. C. (2019). Indigenous control and benefits through small-scale forestry: A multi-case analysis of outcomes. *Canadian Journal of Forest Research, 49*(4), 404–413. https://doi.org/10.1139/cjfr-2018-0279.

Lindahl, K. B., Sténs, A., Sandström, C., et al. (2017). The Swedish forestry model: More of everything? *Forest Policy Economics, 77*, 44–55. https://doi.org/10.1016/j.forpol.2015.10.012.

Lindgren, A., Robson, J. P., Reed, M. G., et al. (2019). *Engaging the public in sustainable forest management in Canada: Results from a national survey of advisory committees*. Information report LAU-X-142E (p. 79). Québec: Canadian Forest Service, Natural Resources Canada.

Lindstad, B. H., & Solberg, B. (2012). Influences of international forest policy processes on national forest policies in Finland, Norway and Sweden. *Scandinavian Journal of Forest Research, 27*(2), 210–220. https://doi.org/10.1080/02827581.2011.635079.

McFarlane, B. L., & Boxall, P. C. (2000). Forest values and attitudes of the public, environmentalists, professional foresters, and members of public advisory groups in Alberta. *The Government of Alberta's Guidelines on consultation with First Nations on land and natural resource management*. Northern Forestry Centre Information Report NOR-X-374 (p. 17). Edmonton: Canadian Forest Service, Natural Resources Canada.

McGurk, B., Sinclair, J. A., & Diduck, A. (2006). An assessment of stakeholder advisory committees in forest management: Case studies from Manitoba, Canada. *Society & Natural Resources, 19*(9), 809–826. https://doi.org/10.1080/08941920600835569.

McKenzie, J. I. (2002). *Environmental Politics in Canada: Managing the commons into the twenty-first century*. Oxford University Press.

Miller, L. F., & Nadeau, S. (2017). Participatory processes for public lands: Do provinces practice what they preach? *Ecology & Society, 22*(2), 19. https://doi.org/10.5751/ES-09142-220219.

Nenko, A., Parkins, J. R., & Reed, M. G. (2019). Indigenous experiences with public advisory committees in Canadian forest management. *Canadian Journal of Forest Research, 49*(4), 331–338. https://doi.org/10.1139/cjfr-2018-0235.

O'Flaherty, R. M., Davidson-Hunt, I. J., & Manseau, M. (2008). Indigenous knowledge and values in planning for sustainable forestry: Pikangikum First Nation and the Whitefeather Forest Initiative. *Ecology & Society, 13*(1), 6. https://doi.org/10.5751/ES-02284-130106.

Pappila, M., & Pölönen, I. (2012). Reconsidering the role of public participation in the Finnish forest planning system. *Scandinavian Journal of Forest Research, 27*(2), 177–185. https://doi.org/10.1080/02827581.2011.635084.

Parkins, J. R., Hunt, L., & Nadeau, S., et al. (2006). *Public participation in forest management: Results from a national survey of advisory committees.* Northern Forestry Centre Information Report NOR-X-409E (p. 74). Edmonton: Canadian Forest Service, Natural Resources Canada.

Parkins, J. R., & Sinclair, A. J. (2014). Patterns of elitism within participatory environmental governance. *Environment and Planning C: Politics and Space, 32*(4), 746–761. https://doi.org/10.1068/c1293.

Reed, M. G. (2010). Guess who's (not) coming for dinner: Expanding the terms of public involvement in sustainable forest management. *Scandinavian Journal of Forest Research, 25*, 45–54. https://doi.org/10.1080/02827581.2010.506785.

Reed, M. G., & Varghese, J. (2007). Gender representation on Canadian forest sector advisory committees. *The Forestry Chronicle, 83*(4), 515–525. https://doi.org/10.5558/tfc83515-4.

Romina, R. (2014). Social learning, natural resource management, and participatory activities: A reflection on construct development and testing. *NJAS: Wageningen Journal of Life Sciences, 69*(1), 15–22. https://doi.org/10.1016/j.njas.2014.03.004.

Sherren, K., Parkins, J. R., Smit, M., et al. (2017). Digital archives, big data and image-based culturomics for social impact assessment: Opportunities and challenges. *Environmental Impact Assessment Review, 67*, 23–30. https://doi.org/10.1016/j.eiar.2017.08.002.

Sinclair, A. J., & Diduck, A. P. (2017). Reconceptualizing public participation in environmental assessment as EA civics. *Environmental Impact Assessment Review, 62*(1), 174–182. https://doi.org/10.1016/j.eiar.2016.03.009.

Sinclair, A. J., & Lobe, K. (2005). Canada's Model Forests: Public involvement through partnership. *Environments, 33*(2), 35–56.

Sinclair, A. J., Peirson-Smith, T. J., & Boerchers, M. (2017). Environmental assessment in the internet age: The role of e-governance and social media in creating platforms for meaningful participation. *Impact Assessment and Project Appraisal, 35*(2), 148–157. https://doi.org/10.1080/14615517.2016.1251697.

Sinclair, A. J., & Smith, D. L. (1999). The model forest program in Canada: Building consensus on sustainable forest management? *Society & Natural Resources, 12*(2), 121–138. https://doi.org/10.1080/089419299279795.

Teitelbaum, S., Beckley, T., & Nadeau, S. (2006). A national portrait of community forestry on public land in Canada. *The Forestry Chronicle, 82*(3), 416–428. https://doi.org/10.5558/tfc824 16-3.

The Montréal Process. (2015). *Criteria and indicators for the conservation and sustainable management of temperate and boreal forests.* Rotorua: 5th Montréal Process Liaison Office.

United Nations Economic Commission for Europe (UNECE). (1998). *Aarhus Convention on access to information, public participation in decision-making and access to justice in environmental matters.* Brussels: Directorate-General for Environment, European Commission.

van Der Wal, M., De Kraker, J., Offermans, A., et al. (2014). Measuring social learning in participatory approaches to natural resource Management. *Environmental Policy and Governance, 24*(1), 1–15. https://doi.org/10.1002/eet.1627.

Woodhil, J., & Röling, N. G. (2000). The second wing of the eagle: Human dimension in learning our way to sustainable futures. In N. G. Röling & M. A. E. Wagemakers (Eds.), *Facilitating sustainable agriculture: Participatory learning and adaptive management in times of environmental uncertainty* (pp. 46–69). Cambridge University Press.

Wyatt, S., Hébert, M., Fortier, J. F., et al. (2019). Strategic approaches to Indigenous engagement in natural resource management: Use of collaboration and conflict to expand negotiating space by three Indigenous nations in Quebec. *Canada. Canadian Journal of Forest Research, 49*(4), 375–386. https://doi.org/10.1139/cjfr-2018-0253.

Part VIII
New Tools for Monitoring Climate Change Effects

Chapter 24
Modeling Natural Disturbances in Boreal Forests

Rupert Seidl, Marie-Josée Fortin, Juha Honkaniemi, and Melissa Lucash

Abstract Natural disturbances such as wildfires, insect outbreaks, and windthrow are important processes shaping the structure and functioning of boreal forests. Disturbances are expected to intensify in the future, and this change will have profound consequences on the supply of ecosystem services to society. Consequently, models are needed to project future disturbance trajectories and quantify disturbance impacts on boreal forests. Here, we summarize key concepts of modeling natural disturbances in boreal forests. We focus specifically on disturbances from wildfire, wind and snow, and herbivores and discuss the different approaches used to capture their dynamics in models.

R. Seidl (✉)
School of Life Sciences, Technical University of Munich, Hans-Carl-von-Carlowitz-Platz 2, 85354 Freising, Germany
e-mail: rupert.seidl@tum.de

Berchtesgaden National Park, Doktorberg 6, 83471 Berchtesgaden, Germany

M.-J. Fortin
Department of Ecology and Evolutionary Biology, University of Toronto, 25 Willcocks Street, Toronto, ON M5S 3B2, Canada
e-mail: mariejosee.fortin@utoronto.ca

J. Honkaniemi
Natural Resources Institute Finland (Luke), Latokartanonkaari 9, 00790 Helsinki, Finland
e-mail: juha.honkaniemi@luke.fi

M. Lucash
Department of Geography, University of Oregon, 1251 University of Oregon, 97403-1251, Eugene, OR, USA
e-mail: mlucash@uoregon.edu

© The Author(s) 2023
M. M. Girona et al. (eds.), *Boreal Forests in the Face of Climate Change*,
Advances in Global Change Research 74,
https://doi.org/10.1007/978-3-031-15988-6_24

24.1   Introduction

In boreal forests, stand composition and structure are influenced directly by fire events, insect outbreaks, windstorms, and industrial harvesting (James et al., 2011). Hence, the current state of the forest is affected by the spatial heterogeneity (patchiness and mosaics) of past forest disturbances; in turn, this spatial heterogeneity forms the template for future disturbance dynamics. Forest landscapes have a long memory of disturbances (Peterson, 2002), and their patterns can persist for several decades or even centuries (James et al., 2007). Consequently, disturbances are a crucial driver of forest landscape dynamics.

Forest disturbance regimes around the globe are changing rapidly because of climate change (Seidl et al., 2017). In boreal forest ecosystems, the coincidence of warmer and drier than average conditions consistently leads to increased disturbance activity (Seidl et al., 2020). As the climate system will likely continue to change in the coming decades and boreal regions warm more rapidly than other parts of the world (IPCC, 2013), forest disturbances in the boreal zone could increase in the future (Boulanger et al., 2014; Flannigan et al., 2009; Gauthier et al., 2015). Understanding and projecting future forest disturbance regimes is of paramount importance, as disturbances shape ecosystem structure and function and influence the ability of forests to provide important ecosystem services to society (Thom & Seidl, 2016). The main tools for making inferences on potential future disturbance trajectories and their impacts are models. This chapter reviews different approaches to modeling the most important natural disturbance agents in boreal forests: wildfire, wind and snow, and herbivory from pathogens, insects, and mammals. Given that the main platforms for modeling boreal forest disturbances are forest landscape models, we precede our discussion of modeling individual disturbance agents with a short introduction to forest landscape modeling.

24.2   Forest Landscape Modeling

Forest landscape models simulate forest dynamics beyond the stand scale in a spatially explicit manner and consider landscape-scale processes, such as the dispersal of seeds and the spread of fire across the landscape (Shifley et al., 2017). Because forest structures vary over time and space and landscape patterns create feedbacks affecting the frequency and severity of subsequent disturbances (James et al., 2007), simulating the spatiotemporal interactions between vegetation and disturbances has become a central purpose of forest landscape models. Several spatially explicit landscape models have been developed over the last decades, ranging from individual-based (SORTIE-NT, Beaudet et al., 2011; iLand, Seidl et al., 2012) to cohort-based models (LANDIS, He & Mladenoff, 1999). Many of these models aim to investigate the synergistic effects of the apparent stochasticity of natural disturbances (fire events, insect outbreaks) and scheduled human activities (forest management). Here, we refer the reader to excellent reviews that synthesize the wide range of landscape and disturbance models available to date (Keane et al., 2015; Perera et al., 2015; Seidl et al., 2011; Shifley et al., 2017, among others).

Most forest landscape models build on a common structure and add modules to incorporate more processes or features. For example, LANDIS (He & Mladenoff, 1999; He et al., 1999) models species by age cohort on a lattice where several processes must occur (seed dispersal, succession), while additional disturbances are optional, including fire, insect outbreaks, and harvesting). The PnET module of LANDIS-II (de Bruijn et al., 2014; Scheller et al., 2007) improved on the original LANDIS approach by adding ecophysiology and successional models to model biomass per age class and tree species. Similarly, LANDIS PRO (Wang et al., 2014; Xiao et al., 2017), which is derived from LANDIS, models biomass on the basis of tree density and size per cell; this makes it possible to interface the model directly with forest inventory data. Other forest landscape models have been developed using dedicated modeling languages, such as SELES (Spatially Explicit Landscape Event Simulator, Fall & Fall, 2001). Another example is the Vermillion Landscape Model (VLM; James et al., 2007, 2011), which simulates the effects of fire events, insect outbreaks, and harvesting. Other models have used a state-and-transition modeling approach in which vegetation dynamics are simulated as transitions between discrete vegetation states (e.g., ST-SIM). For instance, Daniel et al. (2017) incorporated both deterministic state transitions from forest management plans and stochastic effects of fire events in their simulation of a study area in the boreal forest of Ontario. They showed the importance of including stochasticity caused by disturbances within forest management planning and highlighted the potential of disturbances to create shortfalls in timber harvest. We also note that in the context of forest management, many stand-level models are applied (e.g., Díaz-Yáñez et al., 2019; Valinger & Fridman, 2011).

Regardless of the selected modeling approach, all models must be parameterized, and an understanding of the direct and indirect relationships among the modeled processes is a prerequisite. Models developed for a specific region usually cannot be applied to another region without first calibrating the model with new data or evaluating it against independent data obtained from the new study area. A key difference exists between empirical models, which are fit to available data, and process-based models, which are built from a quantitative understanding of the processes underlying forest dynamics. Whereas empirical models use the available data for model building—and are thus often more precise in their projections—process-based models, which are more general and can also robustly capture the effects of future environmental conditions not represented in past data, require data from the study area to evaluate whether the model can reproduce observed patterns (Grimm et al., 2005).

24.3 Fire

A tough challenge in projecting future changes in boreal forests is the inclusion
of disturbance processes, such as wildland fire, that are driven by extreme events,
exhibit nonlinear dynamics, and involve spatial relationships. Simulating fire involves
emulating dynamic and sometimes stochastic processes of fire ignition, spread, and
extinguishment controlled by a host of climatic, geoenvironmental, and societal
factors, such as wind speed, slope, aspect, and proximity to human development.
These factors vary widely in their relative importance depending on the type of simulated fire, i.e., natural wildfires (lightning-caused), human-caused accidental fires,
and prescription burning. Lightning, for example, is the cause of most fires in Alaska
(Kasischke et al., 2010); however, the majority of fires in Siberia are human-caused
(Achard et al., 2008). This causes differences in the spatial and temporal pattern of
ignitions and affects the rate of spread and the potential for fire suppression. Across
all types, fire is sensitive to vegetation composition and structure (Johnson, 1996),
but fire also has a significant effect on the rate and successional sequence of vegetation and carbon cycling (Agee, 1996; DeBano et al., 1998). Creating models that
simulate the timing, pattern, and severity of different fire types, while allowing for
nonlinear changes in vegetation responses to climate change is not a simple task
(Fig. 24.1).

One of the most effective tools for simulating wildfire is the landscape fire succession model, which runs the gamut from simple models of successional pathways and
stochastic wildfire (e.g., SIMPLEE, Chew et al., 2004) to complex models simulating individual trees, biogeochemistry, and climate (e.g., Fire-BGC, Keane et al.,
1996). These models vary in their ability to simulate different ignition types, i.e.,
natural or lightning-caused, suppression activities, the degree to which they rely on

Fig. 24.1 The complex patterns created by wildfires in boreal forests (*left*) and their impacts on
vegetation (*right*) as evidenced by the Hess Creek Fire, which burned 76,634 ha in central Alaska
in 2019. *Photo credits* Melissa Lucash

first principles, the level of stochasticity, and the appropriate scale to which they should be applied (Keane et al., 2004; McKenzie & Perera, 2015). Despite the wide range of approaches and applications, all landscape fire models share four essential components: (1) fire ignition, (2) fire spread, (3) vegetation, and (4) fire effects.

The ignition component of a fire model simulates the initiation of a fire event, which has both spatial and temporal aspects owing to variations in climate, vegetation, and topography, which affect the probability of a successful ignition. Wildfire ignitions are often simulated stochastically by applying a user-defined number of ignitions combined with a probability distribution function (e.g., Weibull, zero-inflated Poisson, and Pareto). These functions are then calibrated to match fire data from a fire-history database or perimeter atlas for the study region. However, this simulates the pattern but not necessarily the underlying mechanisms. Models differ in the factors that influence burn probability; some include weather (e.g., BFOLDS, Perera et al., 2002, 2008; FlamMap, Finney, 2006), a flammability coefficient or stand age (ALFRESCO, Rupp et al., 2000b), fire return interval (FireBGCv2, Keane et al., 2011), or fuel moisture and type (BFOLDS, Perera et al., 2002, 2008). Most models do not account for the different spatial and temporal ignition patterns between human and natural fires. Human-caused fires often occur in areas of high accessibility and on holidays and weekends (Beale & Jones, 2011; Maingi & Henry, 2007), whereas natural fires are driven more by fuel type, fuel moisture, and climatic conditions that favor lightning. High-quality data are needed to parameterize or calibrate approaches that explicitly account for factors controlling ignitions (Prestemon et al., 2013), and recent efforts have compiled large databases for public use in the United States and Canada (e.g., NRC, 2020; Short, 2017). These databases are not available for all circumboreal forests, notably Siberia, and have limitations; for example, accidental fires may be reported within minutes of ignition, but lightning fires may not always be detected because they can smolder for days before growing to a detectable size often in remote locations. A decision-tree analysis is often employed for simulating human-prescribed burning in boreal ecosystems, whereby a maximum allowable number of ignitions is user-prescribed, and weather-conditional statements are applied to determine whether the fire ignites (e.g., SCRPPLE, Scheller et al., 2019). Including all physical processes that affect wildfire ignitions for the various causes of fire is a complicated task (Prestemon et al., 2013) and has yet to be fully integrated into forest simulation models.

Once a fire ignites, the spread components of the model determine the shape and extent of the fire, applying either a lattice approach or a vector strategy (Gardner et al., 1999). The lattice approach simulates fire spread from one raster pixel to another using cellular automata (EMBYR, Hargrove et al., 2000) or bond percolation (SpaDES, Marchal et al., 2020). These models allow the stochastic spread of fires between raster pixels on the basis of (1) probability distributions (e.g., Base Fire in LANDIS-II, He & Mladenoff, 1999; WMFire, Kennedy & McKenzie, 2010, McKenzie & Kennedy, 2012), (2) stochasticity combined with empirical relationships derived from laboratory experiments or field data (e.g., iLand, Seidl et al., 2012; SCRPPLE in LANDIS-II, ALFRESCO, Rupp et al., 2000b), or (3) physics-based combustion and spread models (e.g., the WFDS, Mell et al., 2007; Coupled Atmo-spheric Weather-Fire Experiment, Coen et al., 2013). Vector strategies use raster maps of ignitions, but they allow fire to spread using two-dimensional vertices that increase in number as the fire grows (Finney, 1998). The spread is driven stochas-tically, empirically with generalized linear modeling, or via algorithms of physical processes (e.g., FARSITE, Finney, 1998; FARSITE in Fire-BGC, Keane et al., 1996). Spread in both cellular automata and vector approaches is influenced by vegetation succession, which can be simulated using (1) a state-and-transition model of user-defined community types and pathways, (2) a cohort model of species and age, or (3) an individual plant model that simulates each tree or plant on the landscape. State-and-transition models, like ALFRESCO, have been widely used in boreal forests to char-acterize changes in vegetation type (Johnstone et al., 2011; Rupp et al., 2000a), tree-line expansion (Hewitt et al., 2016), and vegetation-climate feedbacks (Euskirchen et al., 2016) in response to climate change. Cohort models, e.g., LANDIS-II, have seldom been used in boreal forests; an exception is their application to character-izing the importance of timber harvesting in driving long-term succession in Siberia (Gustafson et al., 2011). Although studies of postfire boreal succession have, to date, relied primarily on simpler, more deterministic models, future studies will focus on modeling fire spread and vegetation development to capture the emerging nonlinear dynamics stemming from the increased fire frequency in these systems (Johnstone et al., 2010; Kasischke et al., 2010).

Fire effects are often simulated very simplistically using either rule-based method-ology (e.g., SIMMPLE, Chew et al., 2004; LANDSUM, Keane et al., 2006, 2008; TELSA, Klenner et al., 2000, Kurz et al., 2000) or mechanistic mortality proba-bilities (Fire-BGC, Keane et al., 1996, SCRAPPLE in LANDIS-II, Scheller et al., 2019). In some individual models, all trees die if a fire burns in a cell (e.g., Base Fire in LANDIS-II, He & Mladenoff, 1999; He et al., 1999), whereas state-and-transition models use rules to determine the fate of a vegetation type (i.e., transi-tion to a different state, Rupp et al., 2001). A more mechanistic approach relies on empirically derived logistic regression probabilities to model species or age-specific mortality (e.g., SCRAPPLE in LANDIS-II, Scheller et al., 2019); however, this has not been used in boreal ecosystems to date.

Future attempts to project how boreal forests will be affected by wildland fire and climate change could be improved by (1) capturing the different mechanisms and spatial patterns between human-caused fires and wildfires, (2) establishing direct linkages to smoke models to estimate the impacts of smoke on human health, (3) creating models that couple processes of fire, vegetation, permafrost, and hydrology, and (4) ensuring the models capture nonlinear, emergent fire and vegetation behavior under a changing climate. Improved projections of wildland fire and smoke in boreal forests will help identify urban and rural communities at risk and determine the most effective strategies for developing future land-use plans.

24.4 Wind and Snow

Wind is a major disturbance agent in coastal forests around the globe. The risk of wind disturbance generally decreases with distance from the coast. High gust speeds are the primary trigger of wind disturbances, with individual trees falling when gusts exceed approximately 30 m·s^{-1}, and marked wind impacts occur when gusts exceed 40 m·s^{-1} (Gardiner et al., 2010). The main causes of strong winds are (1) cyclonal storms resulting from large-scale pressure differences in the atmosphere; these storms are generally responsible for the most extensive wind disturbances in forest ecosystems; (2) thunderstorms, often with very high wind speeds but only local impacts; (3) katabatic winds resulting from cold air pooling over ice masses; and (4) winds resulting from weather differences between the windward and leeward sides of mountain ranges (e.g., foehn, Chinook). Strong winds can generally cause a wide variety of disturbance patterns in forest ecosystems, ranging from small-scale canopy openings via the replacement of individual trees to large-scale, high-severity disturbance patches (Fig. 24.2). Wind impact is strongly modulated by forest structure, with tree height and species identity being the most prominent predictors (Díaz-Yáñez et al., 2017; Valinger & Fridman, 2011). The main impact of wind on trees is stem breakage and uprooting. As this fundamental impact is the same for snow disturbances, the two agents are often modeled similarly and are addressed jointly here. Snow-related disturbances require the presence of snow, which limits them to areas having frequent snowfall or long periods of snow cover, such as the boreal zone. However, critical for the occurrence of snow-related disturbances are individual heavy snowfall events or rain-on-snow events, which cause heavy snow loads in tree canopies. The risk of snow-related disturbance is generally considered high when the cumulative snow load exceeds 20 kg m^2 (Kilpeläinen et al., 2010).

Three crucial processes must be addressed to capture the dynamics of wind and snow disturbances in models: (1) the occurrence of strong winds or heavy snow loads, (2) the susceptibility of forests to wind and snow disturbance, and (3) the

Fig. 24.2 The impacts of wind and snow on forest ecosystems range from individual tree death or damage (*left*) to the killing of trees at the stand to landscape scale (*right*). *Photo credits* Rupert Seidl

impacts of wind and snow on vegetation. Disturbances by wind and snow are triggered by climatic extremes, such as high winds and extreme snow loads. The occurrence probability of such events can be derived from statistical analyses of climate data, e.g., using extreme value theory (Bengtsson & Nilsson, 2007). However, good climate observations—a prerequisite for such analyses—are often not available for remote forested areas. The occurrence of extreme wind and snow conditions has thus frequently been modeled as dependent on topographic variables (Ruel et al., 1997; Suárez et al., 1999), describing the varying exposure to such disturbances in a landscape. Most existing dynamic forest landscape models trigger wind and snow disturbances stochastically. Increasingly, however, detailed local airflow models are used to derive critical windspeeds for forest landscapes. Approaches such as WAsP (Zeng et al., 2006) and MS-Micro/3 (Talkkari et al., 2000) have been applied to model the wind development over forest canopies, accounting for the effects of topography and forest structure at the landscape scale. Such models can also be applied to downscale projections from regional climate models to obtain detailed wind projections for forests (WINDA, Blennow & Olofsson, 2008).

Forest structure and composition strongly determine how vegetation responds to strong winds and high snow loads. In general, the susceptibility of forests to wind increases with tree height (Valinger & Fridman, 2011). Crown shape, stem taper, and species-specific wood properties also influence the sensitivity of forests to wind and snow. A common approach to modeling these susceptibility differences is fitting regression models to observational data (e.g., Díaz-Yáñez et al., 2017; Jalkanen & Mattila, 2000). Such models can then be implemented in simulation models that dynamically project forest structure and composition. Dose–response models are a more mechanistic approach to modeling vegetation susceptibility to wind and snow. These models typically quantify tree and stand attributes related to resisting the physical forces of wind and snow, such as tree height, modulus of rupture, and rooting strength. They subsequently determine the critical loads required for breaking or overturning a tree (GALES, Gardiner et al., 2000, 2008; HWIND, Peltola et al., 1999a; see also Fig. 24.3). Tree-pulling experiments provide an important empirical database for the parameterization of these models (Nicoll et al., 2006). Because frozen soil can considerably improve the anchoring of trees, soil frost has also been considered in modeling wind disturbances (Peltola et al., 1999b; Seidl et al., 2014). Furthermore, the spatial context of a stand is an important factor determining its wind risk, e.g., whether there is a large upwind gap or not, a situation considered explicitly in some simulation frameworks (HWIND, Zeng et al., 2009; iLand, Seidl et al., 2014).

The impacts of wind and snow on forests can be manifold, ranging from broken branches and roots to stem breakage and the uprooting of trees. Furthermore, the frequent exposure to wind and snow can result in acclimation processes within a tree, e.g., increased allocation of carbohydrates to roots, changed canopy structure. To date, these processes have rarely been explicitly considered in disturbance models. Most models of wind and snow disturbance impacts consider only stem breakage and uprooting and determine whether a tree survives a given event or not. Some models additionally consider that trees can be killed by falling neighboring trees

INPUTS   DERIVED VALUES   MODEL   OUTPUTS
                     AND   CALCULATION
                RELATIONSHIPS

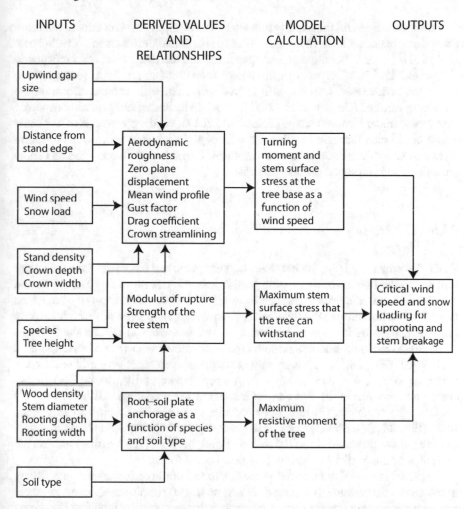

Fig. 24.3 The conceptual design of the HWIND dose–response model applied to simulate wind and snow disturbance in Finland. Modified with permission from Canadian Science Publishing, permission conveyed through Copyright Clearance Center, Inc., from Peltola et al. (1999a)

in a windthrow (ForGEM-W, Schelhaas et al., 2007). In cohort-based approaches, wind disturbances reset the forest development of a cohort (LANDIS, He et al., 1999), whereas in structurally simple *big leaf* ecosystem models, wind impacts are simulated by removing biomass from the respective pools (BiomeBGC, Lindroth et al., 2009). In contrast to fire, the spatial extent of wind and snow disturbances is usually not determined through an active spread process. However, by updating the vegetation structure *during* a wind event and accounting for newly exposed trees, the landscape patterns created by windthrow can be mimicked closely in simulations (iLand, Seidl et al., 2014).

Climate change will profoundly influence the occurrence and severity of wind and snow disturbances. Changes in peak wind speeds remain difficult to project (Shaw et al., 2016); nonetheless, warmer temperatures and increased levels of atmospheric CO_2 could lead to taller trees in the boreal forest, which are more susceptible to windthrow. Moreover, warming will reduce soil frost, with adverse effects on the anchoring of trees (Gregow et al., 2011). Snow disturbances are generally expected to decrease under climate change (Seidl et al., 2017), yet the prevalence of wet snow events could also increase locally with warming. To improve projections of future wind and snow disturbances, we need better information on future wind and snow conditions and improved process models.

24.5 Herbivory

Biotic disturbance agents, such as insects, pathogens, and mammals, consume plant biomass in the boreal forest, causing growth loss and tree mortality. The extent and severity of biotic disturbances range from small, low-severity, gap-type dynamics, typical in Fennoscandia (Kuuluvainen & Aakala, 2011), to large-scale high-severity outbreaks, such as the spruce budworm (*Choristoneura* sp.) in Canada (Navarro et al., 2018) or the Siberian silkmoth (*Dendrolimus superans*) in Siberia (Kharuk et al., 2007) (Fig. 24.4). Although a tiny bark beetle may seem very different from a gigantic moose (*Alces alces*), they—from a perspective of disturbance modeling—share common processes that can be harnessed for modeling. All herbivores (1) disperse, (2) establish, (3) reproduce and die, and (4) affect their host in various ways (Fig. 24.5). The details of each process vary between agents and systems; however, these processes form the basic building blocks of models that simulate the dynamics of biotic disturbance agents in the boreal forest.

Dispersal is one of the relevant processes for all biotic disturbance agents. Many agents move autonomously (using their feet or wings) for dispersal, whereas others rely on external aid, such as wind or water. Dispersal can be modeled simply by the distance an individual moves in each time step. Many landscape models use probability density functions to describe the probability of an individual moving from point A to point B in time t (e.g., Pukkala et al., 2014). However, these models neglect the direct effects of landscape structure, the size of the individual agent, and the prevailing weather conditions. Consequently, more detailed approaches have been developed to consider these factors (e.g., Norros et al., 2014; Sturtevant et al., 2013).

Once the agent has moved into a new area, whether it can establish itself in that location is constrained by two factors: (1) habitat quality, i.e., host availability and climatic suitability, and (2) population density, i.e., the number of individuals required to maintain a viable population. These are most often included in models as Boolean filters to indicate the success or failure of establishment or are represented in indices of varying levels of complexity (e.g., Lustig et al., 2017; Sturtevant et al., 2004).

Fig. 24.4 Disturbances caused by biotic agents vary by extent and severity; (*left*) a low-severity, small-scale disturbance caused by the European spruce bark beetle (*Ips typographus*) in Finland, and (*right*) a high-severity, large-scale disturbance caused by eastern spruce budworm (*Choristoneura fumiferana*) in Quebec, Canada. *Photo credits* Juha Honkaniemi (*left*), Miguel Montoro Girona and Janie Lavoie (*right*)

Population growth over time is crucial for an established population to thrive, and reproduction and mortality are central processes determining population growth. One of the simplest ways to model population dynamics is through logistic growth equations, where the population growth rate is defined by the birth and death rates of the agent, and the population size is constrained by external factors setting the carrying capacity of an area. However, population dynamics of well-studied biotic disturbance agents can be modeled more explicitly, e.g., using detailed phenological models (Baier et al., 2007; Bentz et al., 1991).

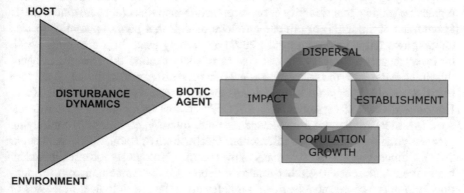

Fig. 24.5 The disturbance dynamics involving biotic agents are the result of close interactions between the agent, its host, and the surrounding environment. The common processes shared by biotic agents, together with the interactive variables of host vegetation and environment, form the basic building blocks for model development

By definition, biotic disturbance agents affect their host species; these impacts can vary from decaying the root system to consuming foliage. Depending on the intensity of the biomass consumption and the compartment that is affected, these impacts can lead to growth loss or tree mortality. Modeling tree mortality probabilistically at different scales is one of the most common approaches to simulate biotic disturbance impacts (e.g., iLand, Seidl & Rammer, 2017; LANDIS BDA, Sturtevant et al., 2004). Some models incorporate more detailed approaches, such as the herbivory of foliage (Régnière & You, 1991) or the decay of root systems (WINDROT, Honkaniemi et al., 2017), and are thus capable of quantifying more detailed effects and various impact pathways in models.

The changing climate affects biotic disturbance agents in various ways. The ongoing changes in environmental conditions are causing the poleward migration of many species, introducing pests to new environments (Bebber et al., 2013; Økland et al., 2019). Many insect species also respond to increased temperatures with accelerated reproduction and increased winter survival (Bale et al., 2002). In general, insects and pathogens will adapt faster to a changing climate than their hosts, increasing the risk for large-scale, high-severity outbreaks in the future. Mammalian herbivores, such as moose and voles, do not generally respond to climate change by rapid range shifts. They may, however, change their behavior with a warming climate, which may affect where and when these species cause disturbance (Korpela et al., 2013; Melin et al., 2014). In addition to climate change, global trade has accelerated the introduction of nonnative pests and pathogens to new ecosystems, creating novel threats to forests globally (Chapman et al., 2017; Santini et al., 2013; Seebens et al., 2017). Both climate change and invasive alien species are serious challenges for boreal forests, for which simulation models can make a significant contribution.

The structure and type of models used for simulating biotic disturbances depend on the questions that must be answered. Some models are aimed to simulate the potential distribution, occurrence, and population dynamics of a pest, whereas others focus on quantifying pest impacts on forest ecosystems. Statistical models, such as logistic regression models that quantify short-term disturbance risks (e.g., Jalkanen, 2001; Magnussen et al., 2004) or climate-envelope models that predict potential species distributions (e.g., Vanhanen et al., 2007) are widely used. In a changing world, however, statistical models are not able to robustly capture the changes in agent dynamics. Moreover, the interaction effects between different disturbance agents are often impossible to capture with statistical models because of data limitations (e.g., Honkaniemi et al., 2018; James et al., 2011). Process-based models have been developed to simulate potential future scenarios more robustly, capturing the underlying processes and their cross-scale interactions (Malmström & Raffa, 2000; Seidl et al., 2011). Climate conditions drive many of the common biological processes shared by biotic agents, most often weather-related variables such as temperature or precipitation. In process-based models, these variables drive the physiological processes of agent dynamics, also making these models applicable for simulations under future climate conditions.

The processes of biotic disturbances are often complex and must be simplified in models. The specific processes needing to be simplified and to what degree they can be simplified is best decided on the basis of the model's aims. Following the principles of pattern-oriented modeling (Grimm et al., 2005), a suitable trade-off between model complexity and payoff must be determined. For most biotic disturbance agents, however, we lack sufficient information on the species biology and ecology to freely choose a level of complexity in model development. Model complexity is dictated by the limited information available for parametrizing key processes foremost disturbance agents.

Agent-based modeling, in which the behavior of individuals or groups in an environment is simulated (Grimm & Railsback, 2006), is one of the most common types of process-based models of biotic disturbance agents. In particular, well-studied biotic disturbance agents, such as the European spruce bark beetle in Europe and the mountain pine beetle in North America, are simulated in highly detailed agent-based models. These models can answer a broad range of questions from agent dynamics to disturbance impacts on ecosystem services (e.g., Bone & Altaweel, 2014; Jönsson et al., 2012; Powell & Bentz, 2014; Seidl & Rammer, 2017).

However, the focus of model development on a small number of well-known agents is problematic, as it can overstate the susceptibility of certain host species over other hosts having less-known biotic disturbance agents. Therefore, the goal in developing process-based models for biotic disturbance agents in the future should be an inclusive and broad consideration of a wide variety of disturbance agents (see Honkaniemi et al., 2021; Lustig et al., 2017; Sturtevant et al., 2004 as examples of such models). General approaches to simulate the dynamics of different biotic disturbance agents will also help estimate the potential impacts of invasive alien pests in novel environments.

Projections of the future disturbance regimes in boreal forests suggest a marked increase of pests and pathogens (e.g., Seidl et al., 2017; Weed et al., 2013). The harsh climate of the boreal region is becoming more favorable to many species, and the poleward migration of species will increase pest introductions and their successful establishment (Hof & Svahlin, 2016; Vanhanen et al., 2007). As boreal forests usually have low tree species diversity, they are particularly vulnerable to changing biotic disturbances. Considering a broader spectrum of biotic disturbance agents and their potential interactions is essential for future improvements of disturbance modeling of boreal forests.

24.6 Outlook

An important challenge for modeling boreal disturbance regimes is capturing the interactions between disturbances. Frequently, disturbance agents do not act in isolation but are influenced by other disturbances (Buma, 2015). Many of these relationships are amplifying interactions, but dampening interactions can also occur, e.g., when disturbance agents compete for the same resource. As disturbances increase under climate change, interactions between disturbances are also likely to increase (Seidl et al., 2017). It is thus essential to address disturbance interactions in modeling.

Dynamic landscape simulation models, such as LANDIS and iLand, provide a robust framework for addressing interactive disturbances because they can simulate multiple disturbance agents and their impacts on vegetation as an emergent property at the landscape scale (e.g., Lucash et al., 2018; Seidl & Rammer, 2017).

A second significant challenge for disturbance modeling is scaling. As the impact of forest disturbances for the provisioning of ecosystem services is increasingly recognized (Thom & Seidl, 2016), disturbance effects must be considered in assessments at policy-relevant scales (i.e., national to global scales). Disturbance modules are, for instance, currently being developed for many Dynamic Global Vegetation models (e.g., Huang et al., 2020; Kautz et al., 2018), which are used inter alia to inform climate policy in regard to the strength of the global vegetation carbon sink. Nonetheless, given the complex interplay between vegetation, climate, and disturbance processes, the scaling of disturbance dynamics is not trivial. An important tool for scaling could be metamodeling (Urban, 2005), i.e., deriving (more broadly applicable) models from existing models. Utilizing emerging machine-learning techniques can further contribute to scalable vegetation and disturbance models (Rammer & Seidl, 2019).

Finally, robust and powerful models are highly dependent on the data and information available for modeling. To improve process-based models of forest disturbance regimes, we need a better understanding of critical processes, such as the dispersal of biotic disturbances and the interactions between disturbance agents. The elimination of such knowledge gaps and the improvement of process-based modeling of forest disturbance regimes requires more experimental research. Notwithstanding limitations in process understanding, the development of forest models is progressing at an accelerating pace, fueled by an increasing computational capacity and the growing availability of data. Remote sensing is increasingly important, as forest disturbances across the globe can now be continuously detected and measured from space (Hansen et al., 2013; Senf & Seidl, 2021). Overall, the modeling of natural disturbances can make an important contribution to an improved understanding of boreal forest dynamics in a changing world and can inform decision-makers in forest management and forest policy regarding the potential consequences of and responses to changing forest disturbance regimes.

References

Achard, F., Eva, H. D., Mollicone, D., et al. (2008). The effect of climate anomalies and human ignition factor on wildfires in Russian boreal forests. *Philosophical Transactions of the Royal Society of London. Series B, Biological Sciences, 363*, 2331–2339. https://doi.org/10.1098/rstb.2007.2203.

Agee, J. K. (1996). *Fire ecology of Pacific Northwest forests*. Washington: Island Press.

Baier, P., Pennerstorfer, J., & Schopf, A. (2007). PHENIPS—A comprehensive phenology model of *Ips typographus* (L.) (Col., Scolytinae) as a tool for hazard rating of bark beetle infestation. *Forest Ecology and Management, 249*, 171–186. https://doi.org/10.1016/j.foreco.2007.05.020.

Bale, J. S., Masters, G. J., Hodkinson, I. D., et al. (2002). Herbivory in global climate change research: Direct effects of rising temperature on insect herbivores. *Global Change Biology, 8,* 1–16. https://doi.org/10.1046/j.1365-2486.2002.00451.x.

Beale, J., & Jones, W. (2011). Preventing and reducing bushfire arson in Australia: A Review of what is known. *Fire Technology, 47,* 507–518. https://doi.org/10.1007/s10694-010-0179-4.

Beaudet, M., Harvey, B. D., Messier, C., et al. (2011). Managing understory light conditions in boreal mixedwoods through variation in the intensity and spatial pattern of harvest: A modelling approach. *Forest Ecology and Management, 261,* 84–94. https://doi.org/10.1016/j.foreco.2010.09.033.

Bebber, D. P., Ramotowski, M. A. T., & Gurr, S. J. (2013). Crop pests and pathogens move polewards in a warming world. *Nature Climate Change, 3,* 985–988. https://doi.org/10.1038/nclimate1990.

Bengtsson, A., & Nilsson, C. (2007). Extreme value modelling of storm damage in Swedish forests. *Natural Hazards and Earth System Sciences, 7,* 515–521. https://doi.org/10.5194/nhess-7-515-2007.

Bentz, B. J., Logan, J. A., & Amman, G. D. (1991). Temperature-dependent development of the mountain pine beetle (Coleoptera: Scolytidae) and simulation of its phenology. *The Canadian Entomologist, 123,* 1083–1094. https://doi.org/10.4039/Ent1231083-5.

Blennow, K., & Olofsson, E. (2008). The probability of wind damage in forestry under a changed wind climate. *Climatic Change, 87,* 347–360. https://doi.org/10.1007/s10584-007-9290-z.

Bone, C., & Altaweel, M. (2014). Modeling micro-scale ecological processes and emergent patterns of mountain pine beetle epidemics. *Ecological Modelling, 289,* 45–58. https://doi.org/10.1016/j.ecolmodel.2014.06.018.

Boulanger, Y., Gauthier, S., & Burton, P. J. (2014). A refinement of models projecting future Canadian fire regimes using homogeneous fire regime zones. *Canadian Journal of Forest Research, 44,* 365–376. https://doi.org/10.1139/cjfr-2013-0372.

Buma, B. (2015). Disturbance interactions: Characterization, prediction, and the potential for cascading effects. *Ecosphere, 6:*art70. https://doi.org/10.1890/ES15-00058.1.

Chapman, D., Purse, B. V., Roy, H. E., et al. (2017). Global trade networks determine the distribution of invasive non-native species. *Global Ecology and Biogeography, 26,* 907–917. https://doi.org/10.1111/geb.12599.

Chew, J. D., Stalling, C., & Moeller, K. (2004). Integrating knowledge for simulating vegetation change at landscape scales. *Western Journal of Applied Forestry, 19,* 102–108. https://doi.org/10.1093/wjaf/19.2.102.

Coen, J. L., Cameron, M., Michalakes, J., et al. (2013). WRF-Fire: Coupled weather–wildland fire modeling with the weather research and forecasting model. *Journal of Applied Meteorology and Climatology, 52,* 16–38. https://doi.org/10.1175/JAMC-D-12-023.1.

Daniel, C. J., Ter-Mikaelian, M. T., Wotton, B. M., et al. (2017). Incorporating uncertainty into forest management planning: Timber harvest, wildfire and climate change in the boreal forest. *Forest Ecology and Management, 400,* 542–554. https://doi.org/10.1016/j.foreco.2017.06.039.

de Bruijn, A., Gustafson, E. J., Sturtevant, B. R., et al. (2014). Toward more robust projections of forest landscape dynamics under novel environmental conditions: Embedding PnET within LANDIS-II. *Ecological Modelling, 287,* 44–57. https://doi.org/10.1016/j.ecolmodel.2014.05.004.

DeBano, L. F., Neary, D. G., & Ffolliott, P. F. (1998). *Fire effects on ecosystems.* John Wiley & Sons.

Díaz-Yáñez, O., Mola-Yudego, B., González-Olabarria, J. R., et al. (2017). How does forest composition and structure affect the stability against wind and snow? *Forest Ecology and Management, 401,* 215–222. https://doi.org/10.1016/j.foreco.2017.06.054.

Díaz-Yáñez, O., Mola-Yudego, B., & González-Olabarria, J. R. (2019). Modelling damage occurrence by snow and wind in forest ecosystems. *Ecological Modelling, 408,* 108741. https://doi.org/10.1016/j.ecolmodel.2019.108741.

Euskirchen, E. S., Bennett, A. P., Breen, A. L., et al. (2016). Consequences of changes in vegetation and snow cover for climate feedbacks in Alaska and northwest Canada. *Environmental Research Letters, 11*, 105003. https://doi.org/10.1088/1748-9326/11/10/105003.

Fall, A., & Fall, J. (2001). A domain-specific language for models of landscape dynamics. *Ecological Modelling, 141*, 1–18. https://doi.org/10.1016/S0304-3800(01)00334-9.

Finney, M. A. (1998). *FARSITE: Fire area simulator-model development and evaluation*. Ogden: U.S. Department of Agriculture, Forest Service, Rocky Mountain Research Station, (p. 47).

Finney, M. A. (2006). An overview of FlamMap fire modeling capabilities. In P. L. Andrews, & B. W. Butler (Eds.), *Fuels management-How to measure success: Conference proceedings*, Proceedings RMRS-P-41 (pp. 213–220). Portland: U.S. Department of Agriculture, Forest Service, Rocky Mountain Research Station.

Flannigan, M., Stocks, B., Turetsky, M., et al. (2009). Impacts of climate change on fire activity and fire management in the circumboreal forest. *Global Change Biology, 15*, 549–560. https://doi.org/10.1111/j.1365-2486.2008.01660.x.

Gardiner, B., Peltola, H., Kellomäki, S. (2000). Comparison of two models for predicting the critical wind speeds required to damage coniferous trees. *Ecological Modelling, 129*, 1–23. https://doi.org/10.1016/S0304-3800(00)00220-9.

Gardiner, B., Byrne, K., Hale, S., et al. (2008). A review of mechanistic modelling of wind damage risk to forests. *Forestry, 81*(3), 447–463. https://doi.org/10.1093/forestry/cpn022.

Gardiner, B., Blennow, K., Carnus, J. M., et al. (2010). *Destructive storms in European forests: Past and forthcoming impacts*. Final report to European Commission DG Environment (p. 138). EFIATLANTIC. Joensuu: European Forest Institute.

Gardner, R. H., Romme, W. H., & Turner, M. G. (1999). *Predicting forest fire effects at landscape scales*. Spatial modeling of forest landscape change: approaches and applications. Cambridge: Cambridge University Press.

Gauthier, S., Bernier, P., Kuuluvainen, T., et al. (2015). Boreal forest health and global change. *Science, 349*, 819–822. https://doi.org/10.1126/science.aaa9092.

Gregow, H., Peltola, H., Laapas, M., et al. (2011). Combined occurrence of wind, snow loading and soil frost with implications for risks to forestry in Finland under the current and changing climatic conditions. *Silva Fennica, 45*, 35–54. https://doi.org/10.14214/sf.30.

Grimm, V., & Railsback, S. F. (2006). Agent-based models in ecology: Patterns and alternative theories of adaptive behaviour. In F. C. Billari, T. Fent, A. Prskawetz, & J. Scheffran (Eds.), *Agent-based computational modelling: Applications in demography, social, economic and environmental sciences* (pp. 139–152). Physica-Verlag HD.

Grimm, V., Revilla, E., Berger, U., et al. (2005). Pattern-oriented modeling of agent-based complex systems: Lessons from ecology. *Science, 310*, 987–991. https://doi.org/10.1126/science.1116681.

Gustafson, E. J., Shvidenko, A. Z., & Scheller, R. M. (2011). Effectiveness of forest management strategies to mitigate effects of global change in south-central Siberia. *Canadian Journal of Forest Research, 41*, 1405–1421. https://doi.org/10.1139/x11-065.

Hansen, M. C., Potapov, P. V., Moore, R., et al. (2013). High-resolution global maps of 21st-century forest cover change. *Science, 342*, 850–853. https://doi.org/10.1126/science.1244693.

Hargrove, W. W., Gardner, R., Turner, M., et al. (2000). Simulating fire patterns in heterogeneous landscapes. *Ecological Modelling, 135*, 243–263. https://doi.org/10.1016/S0304-3800(00)00368-9.

He, H. S., & Mladenoff, D. J. (1999). Spatially explicit and stochastic simulation of forest-landscape fire disturbance and succession. *Ecology, 80*, 81–99. https://doi.org/10.1890/0012-9658(1999)080[0081:SEASSO]2.0.CO;2.

He, H. S., Mladenoff, D. J., & Boeder, J. (1999). An object-oriented forest landscape model and its representation of tree species. *Ecological Modelling, 119*, 1–19. https://doi.org/10.1016/S0304-3800(99)00041-1.

Hewitt, R. E., Bennett, A. P., Breen, A. L., et al. (2016). Getting to the root of the matter: Landscape implications of plant-fungal interactions for tree migration in Alaska. *Landscape Ecology, 31*, 895–911. https://doi.org/10.1007/s10980-015-0306-1.

Hof, A. R., & Svahlin, A. (2016). The potential effect of climate change on the geographical distribution of insect pest species in the Swedish boreal forest. *Scandinavian Journal of Forest Research, 31*(1), 29–39. https://doi.org/10.1080/02827581.2015.1052751.

Honkaniemi, J., Lehtonen, M., Väisänen, H., et al. (2017). Effects of wood decay by *Heterobasidion annosum* on vulnerability of Norway spruce stands to wind damage: A mechanistic modelling approach. *Canadian Journal of Forest Research, 47*, 777–787. https://doi.org/10.1139/cjfr-2016-0505.

Honkaniemi, J., Ojansuu, R., Kasanen, R., et al. (2018). Interaction of disturbance agents on Norway spruce: A mechanistic model of bark beetle dynamics integrated in simulation framework WINDROT. *Ecological Modelling, 388*, 45–60. https://doi.org/10.1016/j.ecolmodel.2018.09.014.

Honkaniemi, J., Rammer, W., & Seidl, R. (2021). From mycelia to mastodons—A general approach for simulating biotic disturbances in forest ecosystems. *Environmental Modelling & Software, 138*, 104977. https://doi.org/10.1016/j.envsoft.2021.104977.

Huang, J., Kautz, M., Trowbridge, A. M., et al. (2020). Tree defence and bark beetles in a drying world: Carbon partitioning, functioning and modelling. *New Phytologist, 225*, 26–36. https://doi.org/10.1111/nph.16173.

Intergovernmental Panel on Climate Change (IPCC) (Ed.). (2013). *Climate change 2013: The physical science basis. Contribution of Working Group I to the fifth assessment report of the Intergovernmental Panel on Climate Change* (p. 1535). Cambridge and New York: Cambridge University Press.

Jalkanen, A. (2001). The probability of moose damage at the stand level in southern Finland. *Silva Fennica, 35*(2), 593. https://doi.org/10.14214/sf.593.

Jalkanen, A., & Mattila, U. (2000). Logistic regression models for wind and snow damage in northern Finland based on the National Forest Inventory data. *Forest Ecology and Management, 135*, 315–330. https://doi.org/10.1016/S0378-1127(00)00289-9.

James, P. M., Fortin, M. J., Fall, A., et al. (2007). The effects of spatial legacies following shifting management practices and fire on boreal forest age structure. *Ecosystems, 10*, 1261–1277. https://doi.org/10.1007/s10021-007-9095-y.

James, P. M. A., Fortin, M. J., Sturtevant, B. R., et al. (2011). Modelling spatial interactions among fire, spruce budworm, and logging in the boreal forest. *Ecosystems, 14*, 60–75. https://doi.org/10.1007/s10021-010-9395-5.

Johnson, E. A. (1996). *Fire and vegetation dynamics: Studies from the North American boreal forest*. Cambridge University Press.

Johnstone, J. F., Hollingsworth, T. N., Chapin III, F. S., et al. (2010). Changes in fire regime break the legacy lock on successional trajectories in Alaskan boreal forest. *Global Change Biology, 16*, 1281–1295. https://doi.org/10.1111/j.1365-2486.2009.02051.x.

Johnstone, J. F., Rupp, T. S., Olson, M., et al. (2011). Modeling impacts of fire severity on successional trajectories and future fire behavior in Alaskan boreal forests. *Landscape Ecology, 26*, 487–500. https://doi.org/10.1007/s10980-011-9574-6.

Jönsson, A. M., Schroeder, L. M., Lagergren, F., et al. (2012). Guess the impact of *Ips typographus*— An ecosystem modelling approach for simulating spruce bark beetle outbreaks. *Agricultural and Forest Meteorology, 166–167*, 188–200. https://doi.org/10.1016/j.agrformet.2012.07.012.

Kasischke, E., Verbyla, D. L., Rupp, T. S., et al. (2010). Alaska's changing fire regime—implications for the vulnerability of its boreal forests. *Canadian Journal of Forest Research, 40*, 1313–1324. https://doi.org/10.1139/X10-098.

Kautz, M., Anthoni, P., Meddens, A. J. H., et al. (2018). Simulating the recent impacts of multiple biotic disturbances on forest carbon cycling across the United States. *Global Change Biology, 24*, 2079–2092. https://doi.org/10.1111/gcb.13974.

Keane, R. E., Morgan, P., & Running, S. W. (1996). *Fire-BGC: A mechanistic ecological process model for simulating fire succession on coniferous forest landscapes of the northern Rocky Mountains*. PB-96–158357/XAB; FSRP/INT-484, TRN: 61211352. Ogden: US Forest Service, Intermountain Research Station.

Keane, R. E., Cary, G. J., Davies, I. D., et al. (2004). A classification of landscape fire succession models: Spatial simulations of fire and vegetation dynamics. *Ecological Modelling, 179*, 3–27. https://doi.org/10.1016/j.ecolmodel.2004.03.015.

Keane, R. E., Holsinger, L. M., & Pratt, S. D. (2006). *Simulating historical landscape dynamics using the landscape fire succession model LANDSUM version 4.0.* General Technical Report RMRS-GTR-171 (p. 73). Fort Collins: U.S. Department of Agriculture, Forest Service, Rocky Mountain Research Station.

Keane, R. E., Holsinger, L. M., Parsons, R. A., et al. (2008). Climate change effects on historical range and variability of two large landscapes in western Montana, USA. *Forest Ecology and Management, 254*, 375–389. https://doi.org/10.1016/j.foreco.2007.08.013.

Keane, R. E., Loehman, R. A., Holsinger, L. M. (2011). *The FireBGCv2 landscape fire and succession model: a research simulation platform for exploring fire and vegetation dynamics.* General Technical Report RMRS-GTR-255 (p. 137). Fort Collins: U.S. Department of Agriculture, Forest Service, Rocky Mountain Research Station.

Keane, R. E., McKenzie, D., Falk, D. A., et al. (2015). Representing climate, disturbance, and vegetation interactions in landscape models. *Ecological Modelling, 309–310*, 33–47. https://doi.org/10.1016/j.ecolmodel.2015.04.009.

Kennedy, M. C., & McKenzie, D. (2010). Using a stochastic model and cross-scale analysis to evaluate controls on historical low-severity fire regimes. *Landscape Ecology, 25*, 1561–1573. https://doi.org/10.1007/s10980-010-9527-5.

Kharuk, V. I., Ranson, K. J., & Fedotova, E. V. (2007). Spatial pattern of Siberian silkmoth outbreak and taiga mortality. *Scandinavian Journal of Forest Research, 22*, 531–536. https://doi.org/10.1080/02827580701763656.

Kilpeläinen, A., Gregow, H., Strandman, H., et al. (2010). Impacts of climate change on the risk of snow-induced forest damage in Finland. *Climatic Change, 99*, 193–209. https://doi.org/10.1007/s10584-009-9655-6.

Klenner, W., Kurz, W., & Beukema, S. (2000). Habitat patterns in forested landscapes: Management practices and the uncertainty associated with natural disturbances. *Computers and Electronics in Agriculture, 27*, 243–262. https://doi.org/10.1016/S0168-1699(00)00110-1.

Korpela, K., Delgado, M., Henttonen, H., et al. (2013). Nonlinear effects of climate on boreal rodent dynamics: Mild winters do not negate high-amplitude cycles. *Global Change Biology, 19*, 697–710. https://doi.org/10.1111/gcb.12099.

Kurz, W. A., Beukema, S. J., Klenner, W., et al. (2000). TELSA: The tool for exploratory landscape scenario analyses. *Computers and Electronics in Agriculture, 27*, 227–242. https://doi.org/10.1016/S0168-1699(00)00109-5.

Kuuluvainen, T., & Aakala, T. (2011). Natural forest dynamics in boreal Fennoscandia: A review and classification. *Silva Fennica, 45*, 823–841. https://doi.org/10.14214/sf.73.

Lindroth, A., Lagergren, F., Grelle, A., et al. (2009). Storms can cause Europe-wide reduction in forest carbon sink. *Global Change Biology, 15*, 346–355. https://doi.org/10.1111/j.1365-2486.2008.01719.x.

Lucash, M., Scheller, R. M., Sturtevant, B. R., et al. (2018). More than the sum of its parts: How disturbance interactions shape forest dynamics under climate change. *Ecosphere, 9*, e02293. https://doi.org/10.1002/ecs2.2293.

Lustig, A., Worner, S. P., Pitt, J. P. W., et al. (2017). A modeling framework for the establishment and spread of invasive species in heterogeneous environments. *Ecology and Evolution, 7*, 8338–8348. https://doi.org/10.1002/ece3.2915.

Magnussen, S., Boudewyn, P. A., & Alfaro, R. I. (2004). Spatial prediction of the onset of spruce budworm defoliation. *The Forestry Chronicle, 80*(4), 485–494. https://doi.org/10.5558/tfc80485-4.

Maingi, J. K., & Henry, M. C. (2007). Factors influencing wildfire occurrence and distribution in eastern Kentucky, USA. *International Journal of Wildland Fire, 16*, 23–33. https://doi.org/10.1071/WF06007.

Malmström, C. M., & Raffa, K. F. (2000). Biotic disturbance agents in the boreal forest: Consider-ations for vegetation change models. *Global Change Biology, 6*, 35–48. https://doi.org/10.1046/j.1365-2486.2000.06012.x.

Marchal, J., Cumming, S. G., & McIntire, E. J. B. (2020). Turning down the heat: Vegetation feedbacks limit fire regime responses to global warming. *Ecosystems, 23*, 204–216. https://doi.org/10.1007/s10021-019-00398-2.

McKenzie, D., & Kennedy, M. C. (2012). Power laws reveal phase transitions in landscape controls of fire regimes. *Nature Communications, 3*, 726. https://doi.org/10.1038/ncomms1731.

McKenzie, D., & Perera, A. H. (2015). Modeling wildfire regimes in forest landscapes: Abstracting a complex reality. In A. H. Perera, B. R. Sturtevant, & L. J. Buse (Eds.), *Simulation modeling of forest landscape disturbances* (pp. 73–92). Springer International Publishing.

Melin, M., Matala, J., Mehtätalo, L., et al. (2014). Moose (*Alces alces*) reacts to high summer temperatures by utilizing thermal shelters in boreal forests—an analysis based on airborne laser scanning of the canopy structure at moose locations. *Global Change Biology, 20*, 1115–1125. https://doi.org/10.1111/gcb.12405.

Mell, W., Jenkins, M. A., Gould, J., et al. (2007). A physics-based approach to modelling grassland fires. *International Journal of Wildland Fire, 16*, 1–22. https://doi.org/10.1071/WF06002.

Natural Resources Canada. (2020). Canadian national fire database (CNFDB). https://cwfis.cfs.nrcan.gc.ca/ha/nfdb. Accessed July 19, 2020.

Navarro, L., Morin, H., Bergeron, Y., et al. (2018). Changes in spatiotemporal patterns of 20th century spruce budworm outbreaks in eastern Canadian boreal forests. *Frontiers in Plant Science, 9*, 1905. https://doi.org/10.3389/fpls.2018.01905.

Nicoll, B., Gardiner, B., Rayner, B., et al. (2006). Anchorage of coniferous trees in relation to species, soil type, and rooting depth. *Canadian Journal of Forest Research, 36*, 1871–1883. https://doi.org/10.1139/x06-072.

Norros, V., Rannik, U., Hussein, T., et al. (2014). Do small spores disperse further than large spores? *Ecology, 95*, 1612–1621. https://doi.org/10.1890/13-0877.1.

Økland, B., Flø, D., Schroeder, M., et al. (2019). Range expansion of the small spruce bark beetle *Ips amitinus*: A newcomer in northern Europe. *Agricultural and Forest Entomology, 21*(3), 286–298. https://doi.org/10.1111/afe.12331.

Peltola, H., Kellomäki, S., Väisänen, H., et al. (1999a). A mechanistic model for assessing the risk of wind and snow damage to single trees and stands of scots pine, Norway spruce, and birch. *Canadian Journal of Forest Research, 29*, 647–661. https://doi.org/10.1139/x99-029.

Peltola, H., Kellomäki, S., & Väisänen, H. (1999b). Model computations of the impact of climatic change on the windthrow risk of trees. *Climatic Change, 41*, 17–36. https://doi.org/10.1023/A:1005399822319.

Perera, A., Ouellette, M., Cui, W., et al. (2008). *BFOLDS 1.0: a spatial simulation model for exploring large scale fire regimes and succession in boreal forest landscapes.* Forest Research Report 152 (p. 50). Sault Ste Marie: Ontario Ministry of Natural Resources, Ontario Forest Research Institute.

Perera, A., Yemshanov, D., Schnekenburger, F., et al. (2002). *Boreal FOrest Landscape Dynamics Simulator (BFOLDS): a grid-based spatially stochastic model for predicting crown fire regime and forest cover transition.* Forest Research Information Paper 155. Sault Ste Marie: Ontario Ministry of Natural Resources.

Perera, A. H., Sturtevant, B. R., & Buse, L. J. (Eds.). (2015). *Simulation modelling forest landscape disturbances.* Springer International Publishing.

Peterson, G. D. (2002). Contagious disturbance, ecological memory, and the emergence of landscape pattern. *Ecosystems, 5*, 329–338. https://doi.org/10.1007/s10021-001-0077-1.

Powell, J. A., & Bentz, B. J. (2014). Phenology and density-dependent dispersal predict patterns of mountain pine beetle (*Dendroctonus ponderosae*) impact. *Ecological Modelling, 273*, 173–185. https://doi.org/10.1016/j.ecolmodel.2013.10.034.

Prestemon, J. P., Hawbaker, T. J., Bowden, M., et al. (2013). *Wildfire ignitions: A review of the science and recommendations for empirical modeling*. General Technical Report SRS-GTR-171 (p. 20). Asheville: U.S. Department of Agriculture, Forest Service, Southern Research Station.

Pukkala, T., Möykkynen, T., & Robinet, C. (2014). Comparison of the potential spread of pinewood nematode (*Bursaphelenchus xylophilus*) in Finland and Iberia simulated with a cellular automaton model. *Forest Pathology, 44*, 341–352. https://doi.org/10.1111/efp.12105.

Rammer, W., & Seidl, R. (2019). Harnessing deep learning in ecology: An example predicting bark beetle outbreaks. *Frontiers in Plant Science, 10*, 1327. https://doi.org/10.3389/fpls.2019.01327.

Régnière, J., & You, M. (1991). A simulation model of spruce budworm (Lepidoptera: Tortricidae) feeding on balsam fir and white spruce. *Ecological Modelling, 54*, 277–297. https://doi.org/10.1016/0304-3800(91)90080-K.

Ruel, J. C., Pin, D., Spacek, L., et al. (1997). The estimation of wind exposure for windthrow hazard rating: Comparison between Strongblow, MC2, Topex and a wind tunnel study. *Forestry, 70*, 253–266. https://doi.org/10.1093/forestry/70.3.253.

Rupp, T. S., Chapin, F. S., & Starfield, A. M. (2000a). Response of subarctic vegetation to transient climatic change on the Seward Peninsula in north-west Alaska. *Global Change Biology, 6*, 541–555. https://doi.org/10.1046/j.1365-2486.2000.00337.x.

Rupp, T. S., Starfield, A. M., & Chapin III, F. S. (2000b). A frame-based spatially explicit model of subarctic vegetation response to climatic change: Comparison with a point model. *Landscape Ecology, 15*, 383–400. https://doi.org/10.1023/A:1008168418778.

Rupp, T. S., Chapin, F. S., & Starfield, A. M. (2001). Modeling the influence of topographic barriers on treeline advance at the forest-tundra ecotone in northwestern Alaska. *Climatic Change, 48*, 399–416. https://doi.org/10.1023/A:1010738502596.

Santini, A., Ghelardini, L., De Pace, C., et al. (2013). Biogeographical patterns and determinants of invasion by forest pathogens in Europe. *New Phytologist, 197*(1), 238–250. https://doi.org/10.1111/j.1469-8137.2012.04364.x.

Schelhaas, M. J., Kramer, K., Peltola, H., et al. (2007). Introducing tree interactions in wind damage simulation. *Ecological Modelling, 207*, 197–209. https://doi.org/10.1016/j.ecolmodel.2007.04.025.

Scheller, R. M., Domingo, J. B., Sturtevant, B. R., et al. (2007). Design, development, and application of LANDIS-II, a spatial landscape simulation model with flexible temporal and spatial resolution. *Ecological Modelling, 201*, 409–419. https://doi.org/10.1016/j.ecolmodel.2006.10.009.

Scheller, R., Kretchun, A., Hawbaker, T. J., et al. (2019). A landscape model of variable social-ecological fire regimes. *Ecological Modelling, 401*, 85–93. https://doi.org/10.1016/j.ecolmodel.2019.03.022.

Seebens, H., Blackburn, T. M., Dyer, E. E., et al. (2017). No saturation in the accumulation of alien species worldwide. *Nature Communications, 8*(1), 14435. https://doi.org/10.1038/ncomms14435.

Seidl, R., & Rammer, W. (2017). Climate change amplifies the interactions between wind and bark beetle disturbances in forest landscapes. *Landscape Ecology, 32*, 1485–1498. https://doi.org/10.1007/s10980-016-0396-4.

Seidl, R., Fernandes, P. M., Fonseca, T. F., et al. (2011). Modelling natural disturbances in forest ecosystems: A review. *Ecological Modelling, 222*, 903–924. https://doi.org/10.1016/j.ecolmodel.2010.09.040.

Seidl, R., Rammer, W., Scheller, R. M., et al. (2012). An individual-based process model to simulate landscape-scale forest ecosystem dynamics. *Ecological Modelling, 231*, 87–100. https://doi.org/10.1016/j.ecolmodel.2012.02.015.

Seidl, R., Rammer, W., & Blennow, K. (2014). Simulating wind disturbance impacts on forest landscapes: Tree-level heterogeneity matters. *Environmental Modelling & Software, 51*, 1–11. https://doi.org/10.1016/j.envsoft.2013.09.018.

Seidl, R., Thom, D., Kautz, M., et al. (2017). Forest disturbances under climate change. *Nature Climate Change, 7*, 395–402. https://doi.org/10.1038/nclimate3303.

Seidl, R., Honkaniemi, J., Aakala, T., et al. (2020). Globally consistent climate sensitivity of natural disturbances across boreal and temperate forest ecosystems. *Ecography, 43*(7), 967–978. https://doi.org/10.1111/ecog.04995.

Senf, C., & Seidl, R. (2021). Mapping the forest disturbance regimes of Europe. *Nature Sustainability, 4*(1), 63–70. https://doi.org/10.1038/s41893-020-00609-y.

Shaw, T. A., Baldwin, M., Barnes, E. A., et al. (2016). Storm track processes and the opposing influences of climate change. *Nature Geoscience, 9*, 656–664. https://doi.org/10.1038/ngeo2783.

Shifley, S. R., He, H. H., Lischke, H., et al. (2017). The past and future of modeling forest dynamics: From growth and yield curves to forest landscape models. *Landscape Ecology, 32*, 1307–1325. https://doi.org/10.1007/s10980-017-0540-9.

Short, K. C. (2017). *Spatial wildfire occurrence data for the United States, 1992–2015 [FPA_FOD_20170508]* (4th ed.). Fort Collins: U.S. Department of Agriculture, Forest Service, Rocky Mountain Research Station. https://doi.org/10.2737/RDS-2013-0009.4.

Sturtevant, B. R., Gustafson, E. J., Li, W., et al. (2004). Modeling biological disturbances in LANDIS: A module description and demonstration using spruce budworm. *Ecological Modelling, 180*, 153–174. https://doi.org/10.1016/j.ecolmodel.2004.01.021.

Sturtevant, B. R., Achtemeier, G. L., Charney, J. J., et al. (2013). Long-distance dispersal of spruce budworm (*Choristoneura fumiferana* Clemens) in Minnesota (USA) and Ontario (Canada) via the atmospheric pathway. *Agricultural and Forest Meteorology, 168*, 186–200. https://doi.org/10.1016/j.agrformet.2012.09.008.

Suárez, J., Gardiner, B. A., & Quine, C. P. (1999). A comparison of three methods for predicting wind speeds in complex forested terrain. *Meteorological Applications, 6*, 329–342. https://doi.org/10.1017/S1350482799001267.

Talkkari, A., Peltola, H., Kellomaki, S., et al. (2000). Integration of component models from the tree, stand and regional levels to assess the risk of wind damage at forest margins. *Forest Ecology and Management, 135*, 303–313. https://doi.org/10.1016/S0378-1127(00)00288-7.

Thom, D., & Seidl, R. (2016). Natural disturbance impacts on ecosystem services and biodiversity in temperate and boreal forests. *Biological Reviews, 91*, 760 781. https://doi.org/10.1111/brv.12193.

Urban, D. L. (2005). Modeling ecological processes across scales. *Ecology, 86*, 1996–2006. https://doi.org/10.1890/04-0918.

Valinger, E., & Fridman, J. (2011). Factors affecting the probability of windthrow at stand level as a result of Gudrun winter storm in southern Sweden. *Forest Ecology and Management, 262*, 398–403. https://doi.org/10.1016/j.foreco.2011.04.004.

Vanhanen, H., Veteli, T., Päivinen, S., et al. (2007). Climate change and range shifts in two insect defoliators: Gypsy moth and nun moth—a model study. *Silva Fennica, 41*(4), 469. https://doi.org/10.14214/sf.469.

Wang, W. J., He, H. S., Fraser, J. S., et al. (2014). LANDIS PRO: A landscape model that predicts forest composition and structure changes at regional scales. *Ecography, 37*, 225–229. https://doi.org/10.1111/j.1600-0587.2013.00495.x.

Weed, A. S., Ayres, M. P., & Hicke, J. A. (2013). Consequences of climate change for biotic disturbances in North American forests. *Ecological Monographs, 83*(4), 441–470. https://doi.org/10.1890/13-0160.1.

Xiao, J., Liang, Y., He, H. S., et al. (2017). The formulations of site-scale processes affect landscape-scale forest change predictions: A comparison between LANDIS PRO and LANDIS-II forest landscape models. *Landscape Ecology, 32*, 1347–1363. https://doi.org/10.1007/s10980-016-0442-2.

Zeng, H., Peltola, H., Talkkari, A., et al. (2006). Simulations of the influence of clear-cutting on the risk of wind damage on a regional scale over a 20-year period. *Canadian Journal of Forest Research, 36*, 2247–2258. https://doi.org/10.1139/x06-123.

Zeng, H., Peltola, H., Väisänen, H., et al. (2009). The effects of fragmentation on the susceptibility of a boreal forest ecosystem to wind damage. *Forest Ecology and Management, 257*, 1165–1173. https://doi.org/10.1016/j.foreco.2008.12.003.

Chapter 25
Modeling the Impacts of Climate Change on Ecosystem Services in Boreal Forests

Anouschka R. Hof, Johanna Lundström, and Matthew J. Duveneck

Abstract With the increasing effects of climate change, a rapid development of effective approaches and tools are needed to maintain forest biodiversity and ecosystem functions. The response, or lack thereof, of forest managers to climate change and its impacts on ecosystem services will have broad ramifications. Here we give an overview of approaches used to predict impacts of climate change and management scenarios for a range of ecosystem services provided by the boreal forest, including timber supply, carbon sequestration, bioenergy provision, and habitat for wildlife and biodiversity. We provide examples of research in the field and summarize the outstanding challenges.

A. R. Hof (✉)
Wildlife Ecology and Conservation Group, Wageningen University, Droevendaalsesteeg 3, 6708 PB Wageningen, The Netherlands
e-mail: Anouschka.Hof@wur.nl

Department of Wildlife, Fish, and Environmental Studies, Swedish University of Agricultural Sciences, 901 83 Umeå, Sweden

J. Lundström
Department of Forest Resource Management, Swedish University of Agricultural Sciences, 901 83 Umeå, Sweden
e-mail: johanna.lundstrom@slu.se

M. J. Duveneck
Harvard Forest, Harvard University, 324 North Main Street, Petersham, MA 01366-9504, USA
e-mail: mduveneck@gmail.com

New England Conservatory, 290 Huntington Ave, Boston, MA 02115, USA

© The Author(s) 2023
M. M. Girona et al. (eds.), *Boreal Forests in the Face of Climate Change*,
Advances in Global Change Research 74,
https://doi.org/10.1007/978-3-031-15988-6_25

25.1   Introduction

Climate change and intensive forestry are important drivers of altered forest dynamics and related changes in the provision of ecosystem services in boreal forests. As defined by the Millennium Ecosystem Assessment (2005),

> Ecosystem services are the benefits people obtain from ecosystems. These include provisioning services such as food and water; regulating services such as regulation of floods, drought, land degradation, and disease; supporting services such as soil formation and nutrient cycling; and cultural services such as recreational, spiritual, religious, and other nonmaterial benefits.

Boreal forests provide a large variety of ecosystem services. These include timber, food, bioenergy, carbon sequestration, habitat for wildlife, water regulation, as well as recreational, spiritual, and religious experiences (Fig. 25.1; Shvidenko et al., 2005). As discussed in the earlier chapters of this book, climate change will not only affect the distribution of tree species but will likely affect disturbances, such as the frequency of forest fires, the importance of windthrow, and the severity of insect infestations (Chaps. 3 and 4). These effects will, in turn, disrupt ecosystem services provided by the forest. Much evidence exists that the boreal forest has already responded to climatic changes. Soja et al. (2007) reviewed observed shifts in tree line in Siberia, decreased growth of white spruce (*Picea glauca*), increases in extreme fire years in Siberia, Alaska, and Canada, and multiyear outbreaks of the spruce beetle (*Dendroctonus rufipennis*) in Alaska. Since this review, much more evidence has emerged (see Brecka et al., 2018). Outbreaks of mountain pine beetle (*Dendroctonus ponderosae*) have occurred in large parts of Canada since the 1990s, affecting more than 18 million ha of forest (NRC, 2020). Extreme fire events and severe outbreaks of insect pests that alter entire ecosystems are expected to become even more common in the future (Safranyik et al., 2010; Stocks et al., 1998; Wolken et al., 2011). Trees will therefore experience increased stress levels (Rebetez & Dobbertin, 2004; Schlyter et al., 2006), likely enhancing their sensitivity to damage (Schlyter et al., 2006). Such events can have devastating impacts on forest ecosystem services through tree mortality and subsequent economic losses, reduced wildlife habitat, and decreased carbon storage capacity (Chan-McLeod, 2006; Kurz et al., 2008; Nealis & Peter, 2008).

Although the risk of natural disturbances in forests may increase, management practices will alter the extent of the damage (Schlyter et al., 2006). The effect of these changes is likely to have significant consequences for, among others, the forestry sector. There is, therefore, an increasing awareness of the necessity to adapt forest management practices to mitigate the adverse effects of climate change (Keenan, 2015) through increasing the uptake of carbon by vegetation (Lindenmayer et al., 2012) and reducing storm- and insect-related tree damage (Felton et al., 2016; Imai et al., 2009). However, the particular choice of management strategy to use in forest ecosystems will have marked consequences on the responses of forest ecosystems and, therefore, the range of ecosystem services provided by forests (Imai et al., 2009; Schlyter et al., 2006). A solid understanding of how climate change will affect forest

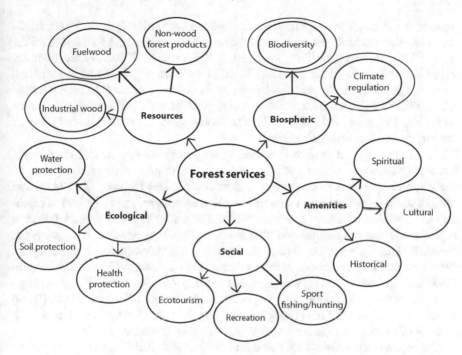

Fig. 25.1 Major classes of forestry services as defined by Shvidenko et al. (2005), including the ecosystem services discussed in this chapter (*red circles*). Modified from Fig. 21.6 in Shvidenko et al. (2005). *Chapter 21 Forest and woodland systems. Ecosystems and Human Well-Being. Current States & Trends* by the Millennium Ecosystem Assessment. Copyright © 2005 Millennium Ecosystem Assessment. Reproduced by permission of Island Press, Washington, DC

dynamics is therefore required if we want to safeguard the ecosystem services offered by boreal forests.

Forest landscape models and decision-support systems can assess the effects of future climate change, shifts in disturbances and management practices, and the establishment of new floral and faunal species within boreal forest ecosystems (Borges et al., 2014). The extensive range of relevant economic, ecological, and social aspects incorporated within long-term forest management planning can be overwhelming for decision-makers; therefore, multiple forest landscape models and forest decision-support systems have been developed globally over the last decades (Borges et al., 2014; Xi et al., 2009). A forest landscape model simulates the survival, growth, and mortality of trees (or stands of trees) over time at relatively large spatial scales (He, 2008). There are many different forest landscape models. In the 1970s, forest gap models were developed to simulate the within-site survival, growth, and mortality of individual trees. An example of such a model is JABOWA (Botkin et al., 1972). A couple of decades later, gap models were developed to assess the (long-term) impacts of climate change on forests; examples of such models include LINKAGES (Post & Pastor, 1996) and ZELIG (Miller & Urban, 1999). Since the early 1990s, many forest landscape models have been developed to simulate forest ecosystems at larger scales. These models include LANDIS (He et al., 1999), which then served as the basis for other, more recent models, such as LANDIS-II (Scheller et al., 2007) and FIN-LANDIS (Pennanen & Kuuluvainen, 2002). Moreover, many extensions able to be coupled to these models have been built to simulate additional processes, such as the impacts on carbon pools (Forest Carbon Succession Extension v.2.0 ForCS, Dymond et al., 2016) and impacts of ungulate browsing (Browsing extension, De Jager et al., 2017). Xi et al. (2009) provide an overview of many of the earlier forest landscape models.

A decision-support system is a model-based software system that combines a knowledge system consisting of forest data, models (e.g., a forest landscape model that simulates tree survival, growth, and mortality), and methods (e.g., statistical computations or optimization solvers) with a problem-processing system to calculate the outcome of management scenarios. The entire decision process becomes reproducible and rational through this decision-support system. An example is the Swedish Heureka program, which can handle simulations of forest management practices, timber production, carbon sequestration, bioenergy, biodiversity, recreation, and economic values (Wikström et al., 2011). Orazio et al. (2017) provide an overview of many decision-support systems used in Europe. Nordström et al. (2019) reviewed the capacity of nine European forest decision-support systems to cope with impacts of climate change for a variety of ecosystem services.

The required information for forest landscape models and forest decision-support systems generally includes detailed information on the initial conditions of the study site (e.g., biomass and age composition of the dominant tree species) as well as such parameters as tree species' longevity, seed dispersal, shade and fire tolerance, vegetative reproduction probability, minimum sprout age, and growth rates. Furthermore, detailed information regarding the environmental conditions, such as precipitation, temperature regimes, and soil properties, are often required. Additional information may be sought, depending on the study aims. Such information generally requires the availability of detailed forest survey data or extensive fieldwork, and a lack of this vital information can therefore hamper the reliability of outcomes.

Multiple studies have relied on forest landscape models or decision-support systems to evaluate the consequences of forestry practices on ecosystem services, as it is generally not possible to examine the various effects of different management strategies on the forest using field studies. Such studies are generally focused on stemwood production rather than other ecosystem services, such as bioenergy harvesting or the provision of wildlife habitat (Biber et al., 2015; Eyvindson et al., 2018; Garcia-Gonzalo et al., 2015). These studies do, however, provide valuable information on, for example, possibilities for increased carbon sequestration (Lucash et al., 2017), optimal restoration practices for forests having sustained damage (Xi et al., 2008), and wildlife preservation (Hof & Hjältén, 2018). They can thus be used to identify *best practice* management strategies to safeguard high levels of several ecosystem services. In this chapter, we provide an overview of the current status and examples of studies that use forest landscape models or decision-support systems to assess the impacts of climate change and management scenarios on several of the most commonly studied ecosystem services provided by the boreal forest: timber supply,

Fig. 25.2 Timber transport on the Northern Dvina River in Arkhangelsk, Russian Federation, June 2010. *Photo credit* Anouschka Hof

carbon sequestration, bioenergy provision, and wildlife habitat. We then discuss some current challenges (Fig. 25.2).

25.2 Timber Production

The past few decades have shown that climate change will likely have a large impact on timber production in boreal forests. These impacts will vary according to geographic location, the dominant tree species, insect and disease outbreaks, and management strategies. The growth rates of boreal forests are limited mainly by short growing seasons. Assuming adequate water supplies, increased temperatures (and carbon dioxide fertilization) may enhance growth and timber volume. However, a more prolonged and enhanced growing season alone does not explain all the uncertainty in projections of future boreal forests. Indeed, changing temperature and precipitation regimes may have both positive and negative effects on future tree growth; for example, summer temperatures may heighten tree respiration that will result in reduced growth. Alternatively, growth may be enhanced if summer respiration demands are offset by a longer growing season (because of an earlier spring, longer autumn, or both). The effects on timber production are therefore uncertain.

Goldblum and Rigg (2005) found, for instance, that whereas commercially valuable sugar maple (*Acer saccharum*) was predicted to experience an increase in its growth rate in the deciduous–boreal forest study site in Ontario, Canada, balsam fir (*Abies balsamea*), another commercially valuable species, was likely to experience decreased growth rates. Uncertainties related to how climate will change compound this challenge of projecting the future growth of boreal taxa.

Ultimately, the most significant impacts of climate change on future timber supply in the boreal forest may be linked to indirect effects, such as insect outbreaks (Safranyik et al., 2010). For example, invasive insects (either currently known or unknown from the boreal region) may increase drastically under a warmer climate to cause the widespread mortality of a commercially valuable species. This event could then lead to extensive salvage harvesting of dead and dying individuals of these tree species and overwhelm timber markets. The hemlock woolly adelgid (*Adelges tsugae*), currently kept in check by cold winters, offers an example of an insect pest found south of the boreal forest in North America that is moving northward because of climate warming and causing the large-scale mortality of eastern hemlock (*Tsuga canadensis*). As another possible scenario, land-use changes driven by economics related to global climate change may cause landowners in the boreal forest to abandon current silvicultural systems for other land uses or ecosystem services, such as land development, agriculture, carbon sequestration, and water protection.

The need to understand the interacting effects of climate change with insects, timber markets, land use, and other disturbances on the timber supply of boreal forests makes forest decision-support systems and landscape models able to incorporate multiple interacting drivers well poised to explore multiple scenarios and tease apart these drivers. In addition to the general tree species' parameters and the environmental conditions needed to run such models, parameters related to, for example, merchantable age, market prices, and management strategies are commonly required. Over the next decade, we expect much research in this area to help explore the uncertainty in future boreal forest timber supply. Multiple interacting effects currently lead to much uncertainty in regard to the outcomes, and current modeling efforts are just beginning to address these numerous interactions (Duveneck & Scheller, 2016; Dymond et al., 2014; Hof et al., 2021; Orazio et al., 2017).

Meanwhile, the call to use alternative management strategies is increasing. Specific adaptive management strategies have been proposed, including those of Spittlehouse and Stewart (2003) and Millar et al. (2007). These strategies include (1) shorter rotation times to decrease the period of stand vulnerability or facilitate a shift to more climate-suitable species; (2) assisted migration of tree species or provenances in anticipation of future losses of productivity with existing species/varieties; (3) tree species diversification strategies aimed at increasing forest resilience; and (4) conservation of corridors to facilitate species migration. However, similar to the direct effects of climate change, such strategies may have large impacts on timber production and other ecosystems services (Felton et al., 2016; Lindenmayer et al., 2012; Noss, 2001). Robust predictions for a range of scenarios are therefore required. Multiple examples of such studies exist, particularly from Canada and Finland, whereby researchers have attempted to predict forest landscape response to climate change and various management strategies, including those proposed above, using various forest decision-support systems and landscape models.

Brecka et al. (2018) reviewed the impacts of climate change on ecological processes in established boreal forest stands and the effects on timber supply and forestry. They found that climate change has led to a reduced rate of volume accumulation and, thus, less timber available for harvest (Fig. 25.3). Their review suggests that climate change, although spatially variable, has already produced significant adverse effects on the timber supply in the boreal forest. Although not necessarily expected, Brecka et al. (2018) found that climate change favored pioneer species, such as pine (*Pinus* spp.) and poplar (*Populus* spp.), over late-successional species, such as spruce (*Picea* spp.) and fir (*Abies* spp.). Climate-suitable species may have been correlated with shade tolerance, thus affecting successional dynamics at the stand level. Ultimately, the boreal forest industry may need to adapt silviculture systems to incorporate and find markets for climate-adapted species. Incorporating alternative tree species more suited to future climate regimes has been proposed (Millar et al., 2007) and simulated as an alternative climate change adaptation strategy by several studies focused on Siberia (e.g., Nadezda et al., 2006) and North America (e.g., Duveneck & Scheller, 2016; Hof et al., 2017).

In Fennoscandia, almost all forested land is managed, and the boreal forest is dominated mainly by Norway spruce (*Picea abies*), Scots pine (*Pinus sylvestris*), and to a smaller extent by birch species (*Betula* spp.). All these species are heavily exploited for producing sawtimber and wood pulp (Esseen et al., 1992). With changing climatic conditions, tree species composition is expected to change under baseline forest management strategies, with birch and Scots pine increasing at the expense of Norway spruce. At the same time, future timber production is expected to increase significantly because of the longer growing seasons, thereby increasing growing stock

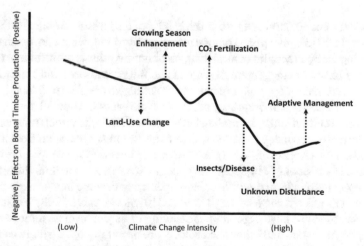

Fig. 25.3 Theoretical interacting effects of climate change and disturbances on boreal timber supply. Indirect effects of climate change (*dotted lines*) may have positive or negative effects on timber supply. These effects will vary depending on climate change intensity

(Peltola et al., 2010). Garcia-Gonzalo et al. (2007) assessed the impacts of climate change and management practices on the timber yield of a 1,450 ha forest management unit in Finland dominated by Scots pine, Norway spruce, and silver birch (*Betula pendula*); they estimated an enhanced growth of 22% to 26% resulting in a 12% to 13% increase in timber yield. The greatest yields were obtained when a thinning regime with high stocking was used with a 100-year rotation period. Subramanian (2016) showed that changing the thinning regime, shortening rotation periods, and planting hybrid tree species such as hybrid aspen (*Populus tremula* × *P. tremuloides*) and hybrid larch (*Larix* × *Eurolepis*) could increase future timber yields. In northern Minnesota, at the boreal-temperate forest ecotone, Duveneck and Scheller (2016) used LANDIS-II coupled to the Biomass-Succession extension and found that climate change had a negative effect on simulated aboveground and harvested biomass. They explored multiple alternative silviculture systems, including expanding reserve areas and changing rotation lengths. However, the climate-suitable planting of broadleaf species currently found immediately south of the studied landscape resulted in the greatest increase in harvested and aboveground biomass. In a Finnish study using the simulation–optimization software Monsu (Pukkala, 2004), Díaz-Yáñez et al. (2020) simulated five separate management scenarios under climate change and found that silviculture systems that used thinning from above were profitable and provided more additional ecosystem services. However, as outlined here, future modeling work will continue to address multiple interacting effects of climate change on boreal forest timber supply and other ecosystem services (Fig. 25.4; Hof et al., 2021).

25.3 Carbon Sequestration

Boreal forests, encompassing approximately 30% of the forested area across the globe (Brandt et al., 2013), may play a pivotal role in halting or slowing climate change by sequestering carbon (Melillo et al., 1993). As about two-thirds of the boreal forest is under some form of management, its current and future resilience and ability to store carbon depend mainly on which management strategies are chosen (Gauthier et al., 2015) and the severity and frequency of natural disturbances. Although forest ecosystems are often carbon sinks, natural disturbances such as large insect outbreaks, e.g., that are currently occurring in Canada, can change a forest ecosystem from a carbon sink into a source (Kurz et al., 2008). Altered fire regimes may also have large impacts on boreal carbon stocks (Miquelajauregui et al., 2019a). As potential disturbances, such as insect outbreaks and wildfires, are expected to become increasingly frequent with future climate change (Safranyik et al., 2010; Stocks et al., 1998; Wolken et al., 2011), it is crucial to have a good understanding of carbon cycles in forest ecosystems. Forest managers and other stakeholders are becoming increasingly aware that alternative management practices may need to be applied in forest ecosystems to increase carbon uptake by vegetation (Ameray et al., 2021; Fares et al., 2015) or reduce emissions caused by natural disturbances (Noss, 2001). However, different

Fig. 25.4 Harvested trees, northern Minnesota, USA. The availability of future timber supply from boreal forests will depend on many interacting factors. *Photo credit* Matthew Duveneck

management practices can have different consequences for the ability of forests to sequester carbon; forest management can increase or reduce sinks and emissions (Kurz et al., 2008). Furthermore, as mentioned in the previous section, climate change effects on growth rates remain uncertain and may increase or decrease the capacity of forests to store carbon.

As it is difficult to assess the impacts of all these environmental changes on the capacity of boreal forests to sequester carbon in field studies at large scales, several studies have aimed to predict such impacts using forest landscape models or decision-support systems. In addition to the general parameters mentioned in Sect. 25.1, such models generally require parameters on deadwood matter and tree species decay rates under various conditions.

Several modeling studies have investigated boreal carbon storage under future scenarios for European and Asian forests. Ito (2005) developed a coupled carbon cycle and fire regime model for the larch (*Larix gmelinii* and *Larix cajanderi*)-dominated boreal forest of eastern Siberia and found that fire events, which are expected to become more frequent in the future (Stocks et al., 1998; Wolken et al., 2011), released approximately 12% of the carbon fixed by the vegetation and lead to an accelerated carbon cycling in the forest. Gustafson et al. (2011) used LANDIS-II coupled with Biomass-Succession to predict the effects of climate change in another region in Siberia. They assessed climate impacts on timber harvesting and insect outbreaks on a Scots pine–dominated boreal forest landscape in the Chuno-Angarsky region. They predicted direct effects of climate change on the forest ecosystem and modeled that changes to the forest's ability to hold carbon would be relatively minor compared with the effects of (novel) forest management practices and the increased risk of insect pest outbreaks. Jiang et al. (2002) used the ecosystem process model CENTURY 4.0 to assess the impact of various harvest disturbance regimes on the carbon stocks and fluxes of a boreal forest landscape in China. They concluded that carbon stocks in their landscape could markedly increase under harvests at 100- to 200-year rotations. The FINNFOR model when applied to implemented thinning regimes in Finland, which allow a higher stocking of trees relative to current practices in the managed boreal forest, indicated that this approach would increase the amount of carbon in the forest ecosystem by up to 11% and the timber yield by up to 14%, depending on the climate trajectory (Garcia-Gonzalo et al., 2007). Therefore, this modeling suggests that carbon sequestration in boreal forest ecosystems may be enhanced using certain management regimes without loss of timber production, even with future climate change. The land ecosystem model JSBACH and the stand growth model PREBAS assessed the impacts of current management practices in Finland and future climate change on ecosystem services provided by the boreal forest; the models indicated a potential increase in the annual carbon sink of approximately 40% at the end of the twenty-first century because of increased annual forest growth (Holmberg et al., 2019).

Examples of modeling boreal forest response to climate change in North America include that of Lucash et al. (2017). They used LANDIS-II coupled with the Century Succession extension to simulate the capacity of various forest management strategies to maintain or increase forest landscape resilience along the broadleaf–boreal transition zone in north-central Minnesota over the next century. They found that climate change would lower forest resilience and that a scenario aimed at maximizing carbon storage by harvesting 30% less land and increasing rotation length did not perform better than their business-as-usual scenario. Forest resilience did increase through use of adapted management strategies and the planting of species adapted to expected future conditions. In the same forest landscape, Lucash et al. (2018) then simulated the effects of fire, insect, wind, and forest management disturbances under changing climatic conditions on this forest. They found that whereas changing climate was the most important driver of soil carbon—leading to smaller future stocks because of increased heterotrophic respiration—simulated future disturbance regimes affected aboveground carbon stocks to a greater extent. In Canada, Miquela-jauregui et al. (2019b) simulated the response of carbon stocks to climate change in a black spruce (*Picea mariana*)–dominated landscape in northern Québec. They used an R software–based simulation model having three interacting modules: patch, fire, and carbon dynamics. Their simulations showed that climate change reduces carbon storage by 10% by the end of 2100.

Results across the boreal forest biome suggest that whereas the capacity of boreal forests to store carbon may increase as a result of greater annual forest growth related to a changing climate, increases in natural disturbances, i.e., fire and pest outbreaks, and the potential augmentation in harvest rates may have significant effects on boreal forest carbon stocks; thus, much of the boreal forest may instead act as a carbon source. The increased uncertainty in regard to the direction, severity, and frequency of a multitude of natural and anthropogenic disturbances and their effect on the capacity of forests to sequester carbon under future warming both heighten the complexity of decision-making for forest managers when selecting and implementing management strategies. Several studies from across the boreal region (as well as other regions) suggest that management strategies aimed at increasing species diversity and resilience may effectively reduce the risks of increased greenhouse gas emissions (Fig. 25.5; Duveneck & Scheller, 2016; Dymond et al., 2014; Hof et al., 2017).

Management strategies aimed at increasing species diversity and resilience may not be sufficient, however, to fully offset the impacts of climate change and natural disturbances and should be tailored to the individual ecosystems to be most effective. Ontl et al. (2020) developed a Forest Carbon Management Menu to help managers identify forest adaptation strategies beneficial for storing forest carbon by reducing climate change–related losses of carbon, sustaining forest health, and enhancing productivity. In addition to simulations of how landscapes may respond to climate change, such tools can guide decision-makers and help mitigate climate change by increasing carbon stocks in the boreal forest through the appropriate selection of forest management strategies.

Fig. 25.5 An increase in species diversity may help boreal forests respond to climate change. Photo of the boreal–deciduous transition zone, northern Minnesota. *Photo credit* Matthew Duveneck

25.4   Bioenergy

From the moment humans discovered fire, our species began using biomass for energy; however, bioenergy has gained increasing attention as a more climate-neutral energy source relative to fossil fuels in recent years. The share of bioenergy, defined as renewable energy from biological sources, is expected to increase and contribute to mitigating climate change (IPCC 2014; Creutzig et al., 2015). However, there are uncertainties, and several factors affect whether bioenergy use helps or hinders climate change mitigation.

The assumption that bioenergy is carbon neutral is based on biogenic CO_2 emissions from the use of harvested biomass eventually being absorbed by biomass regrowth through photosynthesis (Ragauskas et al., 2006). If proper management is adopted, there is a possibility to ensure that the harvesting and use of bioenergy are carbon negative (Lehmann, 2007). Nonetheless, carbon loss from forests used for bioenergy is distinctly possible, contradicting its effectiveness in mitigating climate change. The mitigation role of forest bioenergy thus depends on case-specific factors, such as the biophysical features of the biomass production system and the greenhouse gas intensity of the energy source that bioenergy replaces (Cowie et al., 2019). Moreover, time is an important aspect. Harvesting biomass for bioenergy decreases the forest carbon stock—compared with not harvesting for bioenergy. This decrease balances out over time by lowering greenhouse gas production by replacing fossil fuels. Until this point in time, however, more CO_2 is released from the bioenergy-based system than the fossil fuel system. The length of this carbon payback time varies and depends on many factors, including forest characteristics, the type of biomass used, the fossil fuel being replaced, and alternative land use; for the boreal forest, this payback is likely over several decades (Agostini et al., 2014).

Despite the uncertainties related to the effect of bioenergy on climate, harvesting for bioenergy is expected to increase in the future. As the current outtake is far less than its potential because of low demand, increasing prices for biomass feedstock will substantially increase the outtake. Woody biomass used for energy can be from primary sources available after harvesting operations, such as branches and treetops as well as stumps and stems, all produced from operations focused on harvesting biomass during early thinning or stems that are downgraded from other assortments (Fig. 25.6). Secondary sources include industrial by-products, such as bark, sawdust, and shavings, that accumulate during processing operations. Finally, a third source, including end-of-life wood products, such as construction and demolition wood, are also usable for energy. Extraction of woody biomass has consisted mainly of logging residues; however, the share of stumps and roundwood is most likely to increase with a greater demand (Díaz-Yáñez et al., 2013).

Since forestry has such large impacts on other ecosystem services, adding residue extraction has only a minimal additional effect (de Jong & Dahlberg, 2017). Deadwood-dependent species are adversely affected, whereas the effect on other species is not as uniform; some are impacted positively, others negatively. Although most ecosystem services are adversely affected by residue extraction (e.g., soil

Fig. 25.6 Conventional forwarder loaded with small-diameter trees from a bioenergy thinning operation in Bräcke, northwestern Sweden, October 2019. *Photo credit* Raul Fernandez Lacruz

quality, productivity, and water quality), for recreation and pest–fungi control the relationship can be the opposite (Ranius et al., 2018).

Most studies investigating the consequences of bioenergy harvesting have focused on stand-scale and short-term impacts (Ranius et al., 2018). The boreal forest management system is slow, with rotation times of up to almost 100 years. Evaluating the long-term consequences of harvesting for bioenergy on other ecosystem services and any future potential for biomass harvest relies on simulations with forest landscape models or forest decision-support systems (Borges et al., 2014). Additional parameters beyond the general tree-species-specific parameters and environmental conditions include biomass pools of all parts of the tree, e.g., branches, foliage, and bark. Examples of using such a model or system to evaluate consequences from bioenergy extraction include that of Repo et al. (2020). They applied the SIMO modeling framework, complemented with Yasso07 for soil carbon modeling, to simulate forest development over 100 years in Finland, both with and without residue extraction. They concluded that biodiversity, especially deadwood-associated species, is threatened by residue harvest. They also showed that the forest carbon balance is affected. The emission savings from bioenergy is reduced because of lower carbon stocks when harvesting residues, especially during the first decade post-extraction, supporting the carbon payback-time assumption. Furthermore, they conclude that stands having a high biodiversity potential also often have a high potential for producing bioenergy, which complicates the trade-off between management strategies.

In a study in central Sweden, Hof et al. (2018) used LANDIS-II coupled with the Forest Carbon Succession Extension to simulate the effects of various bioenergy extraction scenarios on deadwood availability and the subsequent habitat suitability for saproxylic species. They found that—in their landscape already largely depleted of deadwood—even a scenario aimed at species conservation only led to about 10 m^3 deadwood per ha, a low value relative to deadwood volumes in many other parts of the boreal region. In a study in northern Sweden, Eggers et al. (2020) evaluated an intensive bioenergy harvest strategy, also using small-diameter trees in early thinning, and compared this approach with the prevailing strategy where only residues (tops and branches) are harvested. The simulations, using Heureka and covering 100 years, found a considerable potential to increase bioenergy harvest with the more intensive strategy, without substantial adverse effects on biodiversity and carbon storage. There was also room for a simultaneous increase in harvest residue extraction, improved conditions for biodiversity, and increased carbon stocks relative to current levels; however, this scenario requires effective forest management planning that considers all critical aspects.

25.5 Wildlife Habitat and Biodiversity

In addition to the direct impacts of climate change on species inhabiting forest ecosystems, forest management adaptations may affect the wildlife inhabiting boreal forests. These effects may be negative (Lindenmayer et al., 2012; Noss, 2001) or positive (Imai et al., 2009), and under current forest management practices, national environmental objectives to conserve wildlife are not being attained (e.g., Swedish Environmental Protection Agency, 2022). Thus far, scenario-based assessments of the impacts of climate change adaptation strategies on forests have mainly targeted ecosystem services limited to carbon uptake and forest productivity, ignoring wildlife. A likely reason for this exclusion is that forest landscape models and decision-support systems used for scenario-based assessments were initially developed to investigate the impacts on timber and pulp production (Borges et al., 2014; Xi et al., 2009). Studies investigating the effects of climate change on forest wildlife frequently ignore the indirect impacts that climate change may have on the various ecological processes within forest landscapes (Keenan, 2015). Instead, modeling efforts often rely on species distribution models and decision trees or focus on the population viability of targeted species on the basis of their known suitability to various environmental conditions. Therefore, such studies ignore the impacts of climate change and forest management strategies on important ecological processes affecting forest dynamics.

The modeling community has recently started to integrate wildlife models into forest landscape models and decision-support systems to conduct scenario-based assessments of wildlife response to climate change adaptation strategies in forests (e.g., He, 2009; Kolström & Lumatjärvi, 1999). The required parameters for such an exercise, in addition to the general tree-species-specific parameters and environmental conditions, are heavily dependent on those wildlife species on which studies have focused. Deadwood-related parameters are needed when studying the effects of climate change and forest management on saproxylic species, and data on understory vegetation is required for modeling climate impacts involving browsers. Several efforts have used landscape models to infer habitat suitability for wildlife in the boreal forest. For instance, Tremblay et al. (2018) projected the cumulative impacts of climate change and forest management strategies in the boreal forest of eastern Canada on the Black-backed Woodpecker (*Picoides arcticus*). They simulated forest attributes relevant to this woodpecker using LANDIS-II and PICUS to infer future landscape suitability for the species under various climate change scenarios. Tremblay et al. (2018) found that such cumulative impacts produced significant adverse effects on the woodpecker and on the biodiversity associated with deadwood and old-growth boreal forests. To help mitigate these negative impacts, they suggested adaptations to current management practices, including reduced harvesting levels and strategies to promote coniferous species. Pearman-Gillman et al. (2020) used the land change model Dinamica EGO and several future forest scenarios—developed using LANDIS-II in combination with species distribution models—to assess how species distributions for nine mammal and one bird species changed under five trajectories of modified landscapes in New England in the northeastern United States. They predicted that seven of these species would experience regional declines irrespective of the landscape change trajectory. A similar approach was used by Hof and Hjältén (2018) in Sweden. They also used LANDIS-II coupled with Biomass-Succession to simulate the effects of different levels of restoration on a boreal forest landscape in central Sweden and inferred the landscape suitability for the White-backed Woodpecker (*Dendrocopos leucotos*); an umbrella species in the boreal forest of Fennoscandia; its protection may serve to preserve a range of species that favor high amounts of deadwood and old-growth forest (Fig. 25.7). This study, however, did not incorporate the possible impacts of future climate change. De Jager et al. (2020) also used LANDIS-II coupled with Biomass-Succession and the PnET-II ecophysiology model to simulate how climate change and different wolf (*Canis lupus*) management intensities would affect moose (*Alces alces*) densities and the subsequent impacts on the forests of Isle Royale National Park, Michigan, United States. They found that irrespective of predation pressure, browsing by moose under projected changes in climate leads to strong declines in total forest biomass. Lagergren and Jönsson (2017) used the biogeochemical ecosystem model LPJ-GUESS to study the impact of climate change and alternative management strategies on timber production, carbon storage, and biodiversity in one nemoral and two boreal forest landscapes in Sweden. Their simulations, using the fraction of broadleaf forest, the proportion of old trees, the proportion of old broadleaf trees, and stem litter as proxies for biodiversity, demonstrated that increasing the proportion of broadleaf trees, associated with increasing levels of biodiversity, can promote the storm resistance of a forest landscape.

Fig. 25.7 Forest restoration in the boreal forest landscape in Sweden simulated by Hof and Hjältén (2018)—eliminating coniferous trees and creating deadwood to benefit the White-backed Woodpecker in Sweden. *Photo credit* Anouschka Hof

Several forest management practices have thus far been applied with the aim of mitigating climate change and simultaneously increasing biodiversity in forests. Whereas the study by Tremblay et al. (2018) in Canada advocated strategies to promote coniferous species to benefit the Black-backed Woodpecker, strategies to promote broadleaf species to benefit the White-backed Woodpecker (Hof & Hjältén, 2018) and biodiversity in general (Lagergren & Jönsson, 2017) were suggested for Sweden. Both strategies would, however, lead to a more diverse forest in their respective settings, and the diversification of forests is commonly cited as a climate change mitigation strategy and a means to generate the highest possible levels of various ecosystem services, including habitat provision for wildlife (Lagergren & Jönsson, 2017; van der Plas et al., 2016).

Other climate change mitigation strategies that could benefit biodiversity include thinning practices. Thinning practices can promote high carbon sequestration rates and enhance the structural and compositional complexity in forests (D'Amato et al., 2011), both of which may be good indicators of high forest biodiversity (Lindenmayer et al., 2000). Uneven instead of even-aged management and prescribed fire regimes have also been proposed as mitigation measures to benefit biodiversity (Millar et al., 2007). Moreover, tree retention practices are frequently used to alleviate the adverse effects of felling on species (Gustafsson et al., 2010). Such measures are commonly introduced to promote structural complexity, forest continuity, and the availability of deadwood and old-growth forest patches, which are all generally related to high levels of species diversity (Paillet et al., 2010). A literature review by Felton et al. (2016) assessed the implications for biodiversity of several climate change adaptation and mitigation strategies implemented in Swedish production forests. They concluded that forest managers will be obliged to accept trade-offs to implement climate change adaptation strategies and meet the biodiversity goals set by the Swedish government. This scenario is likely to hold for other boreal countries as well.

25.6 Outlook and Challenges

Predictions of climate change impacts on the provision of boreal ecosystem services face several major challenges. These include uncertainties surrounding the potential distribution and productivity of future boreal forests related to uncertainties in the projections of future climate (Hof et al., 2021; Keenan, 2015; Prestele et al., 2016). Such uncertainties complicate decisions and developments regarding (novel) adaptation and mitigation strategies for managing the forest. Trade-off analyses and multiobjective optimization techniques can evaluate the consequences of conflicting management strategies on multiple ecosystem services (Chen et al., 2016). Tools such as decision-support systems may play a role in performing optimization across multiple objectives; however, to our knowledge, no system currently exists that incorporates the multitude of ecosystem services provided by the boreal forest. Furthermore, the numerous existing models and decision-support systems are very data hungry. It is questionable whether high-quality data are available throughout the boreal forest biome for all required parameters. Much time and effort are likely needed to collect the essential data to set up reliable models and support systems. However, once set up, they should be able to guide decision-makers in selecting appropriate management strategies. Data obtained via remote sensing, such as through LiDAR, MODIS, and Landsat, may play prominent roles in the future in regions where field data are not readily available.

Screening the published literature, we find that most studies focus on boreal forest landscapes in Europe and northern North America. As more than half of all boreal forests occur in the Russian Federation, producing about 20% of the world's timber resources (Krankina et al., 1997), it is paramount that we have a good understanding of how climate change may affect boreal forests and the associated ecosystem services in this region. However, studies related to the boreal forest in the Russian Federation and published in the peer-reviewed literature in English are severely lacking. Furthermore, a quick search in Web of Science illustrates that the primary focus of the research community (to the present) in regard to studies of climate change impacts on ecosystem services provided by boreal forests is mainly on carbon sequestration. We found 53% of the hits addressed carbon, 20% timber, and 5% wildlife. Few studies focused on bioenergy provisions in the context of boreal forests facing climate change (3%, Fig. 25.8). Harvesting for bioenergy appears less developed in Siberia

Fig. 25.8 Number of hits in May 2020 for the search string "Topic: boreal forest AND climate change AND ecosystem services" refined by "carbon," "timber," "wildlife," and "bioenergy" in Web of Science

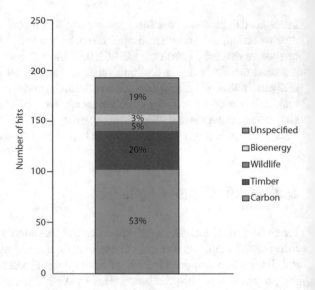

and North America than in Fennoscandia, as inferred by the number of studies in the published English-language literature from Sweden and Finland in regard to this particular ecosystem service.

Here we have overviewed the complexities and uncertainties of the boreal forest under climate change. It is clear that interacting, indirect effects of climate change on the boreal forest will be significant, which has a large impact on simulation outcomes (Lucash et al., 2018). Future modeling studies will undoubtedly need to address these compounding effects. Massive challenges lie ahead for forest managers to safeguard boreal ecosystem services while also maintaining ecosystem resilience. Fortunately, multiple tools exist to aid their decision-making. Furthermore, frameworks to incorporate uncertainty within forest management facing the additional challenge of climate change have been developed (Daniel et al., 2017) and provide additional guidance to forest managers.

References

Agostini, A., Giuntoli, J., & Boulamanti, A. (Eds.). (2014). *Carbon accounting of forest bioenergy: Conclusions and recommendations from a critical literature review*. Publications Office of the European Union.

Ameray, A., Bergeron, Y., Valeria, O., et al. (2021). Forest carbon management: A review of silvicultural practices and management strategies across boreal, temperate and tropical forests. *Current Forestry Reports, 7*(4), 245–266. https://doi.org/10.1007/s40725-021-00151-w.

Biber, P., Borges, J. G., Moshammer, R., et al. (2015). How sensitive are ecosystem services in European forest landscapes to silvicultural treatment? *Forests, 6*(5), 1666–1695. https://doi.org/10.3390/f6051666.

Borges, J. G., Nordstrom, E. M., Garcia-Gonzalo, J., et al. (2014). *Computer-based tools for supporting forest management* (p. 503). Umeå: Department of Forest Resource Management, Swedish University of Agricultural Sciences.

Botkin, D. B., Janak, J. F., & Wallis, J. R. (1972). Some ecological consequences of a computer model of forest growth. *Journal of Ecology, 60*, 849–872. https://doi.org/10.2307/2258570.

Brandt, J. P., Flannigan, M. D., Maynard, D. G., et al. (2013). An introduction to Canada's boreal zone: Ecosystem processes, health, sustainability, and environmental issues. *Environmental Reviews, 21*, 207–226. https://doi.org/10.1139/er-2013-0040.

Brecka, A. F., Shahi, C., & Chen, H. Y. (2018). Climate change impacts on boreal forest timber supply. *Forestry Policy and Economics, 92*, 11–21. https://doi.org/10.1016/j.forpol.2018.03.010.

Chan-McLeod, A. C. A. (2006). A review and synthesis of the effects of unsalvaged mountain-pine-beetle-attacked stands on wildlife and implications for forest management. *BC Journal Ecosystem Management, 7*, 119–132.

Chen, S., Shahi, C., & Chen, H. Y. (2016). Economic and ecological trade-off analysis of forest ecosystems: Options for boreal forests. *Environmental Reviews, 24*, 348–361. https://doi.org/10.1139/er-2015-0090.

Cowie, A. L., Brandão, M., & Soimakallio, S. (2019). Quantifying the climate effects of forest-based bioenergy. In T. M. Letcher (Ed.), *Managing Global Warming* (pp. 399–418). Academic Press.

Creutzig, F., Ravindranath, N. H., Berndes, G., et al. (2015). Bioenergy and climate change mitigation: An assessment. *GCB Bioenergy, 7*(5), 916–944. https://doi.org/10.1111/gcbb.12205.

D'Amato, A. W., Bradford, J. B., Fraver, S., et al. (2011). Forest management for mitigation and adaptation to climate change: Insights from long-term silviculture experiments. *Forest Ecology and Management, 262*, 803–816. https://doi.org/10.1016/j.foreco.2011.05.014.

Daniel, C. J., Ter-Mikaelian, M. T., Wotton, B. M., et al. (2017). Incorporating uncertainty into forest management planning: Timber harvest, wildfire and climate change in the boreal forest. *Forest Ecology and Management, 400*, 542–554. https://doi.org/10.1016/j.foreco.2017.06.039.

De Jager, N. R., Drohan, P. J., Miranda, B. M., et al. (2017). Simulating ungulate herbivory across forest landscapes: A browsing extension for LANDIS-II. *Ecological Modelling, 350*, 11–29. https://doi.org/10.1016/j.ecolmodel.2017.01.014.

De Jager, N. R., Rohweder, J. J., & Duveneck, M. J. (2020). Climate change is likely to alter future wolf-moose-forest interactions at Isle Royale National Park, United States. *Frontiers in Ecology and Evolution, 8*, 290. https://doi.org/10.3389/fevo.2020.543915.

de Jong, J., & Dahlberg, A. (2017). Impact on species of conservation interest of forest harvesting for bioenergy purposes. *Forest Ecology and Management, 383*, 37–48. https://doi.org/10.1016/j.foreco.2016.09.016.

Díaz-Yáñez, O., Mola-Yudego, B., Anttila, P., et al. (2013). Forest chips for energy in Europe: Current procurement methods and potentials. *Renewable and Sustainable Energy Reviews, 21*, 562–571. https://doi.org/10.1016/j.rser.2012.12.016.

Díaz-Yáñez, O., Pukkala, T., Packalen, P., et al. (2020). Multifunctional comparison of different management strategies in boreal forests. *Forestry, 93*, 84 95. https://doi.org/10.1093/forestry/cpz053.

Duveneck, M. J., & Scheller, R. M. (2016). Measuring and managing resistance and resilience under climate change in northern Great Lake forests (USA). *Landscape Ecology, 31*, 669–686. https://doi.org/10.1007/s10980-015-0273-6.

Dymond, C. C., Tedder, S., Spittlehouse, D. L., et al. (2014). Diversifying managed forests to increase resilience. *Canadian Journal of Forest Research, 44*, 1196–1205. https://doi.org/10.1139/cjfr-2014-0146.

Dymond, C. C., Beukema, S., Nitschke, C. R., et al. (2016). Carbon sequestration in managed temperate coniferous forests under climate change. *Biogeosciences, 13*, 1933–1947. https://doi.org/10.5194/bg-13-1933-2016.

Eggers, J., Melin, Y., Lundström, J., et al. (2020). Management strategies for wood fuel harvesting—Trade-offs with biodiversity and forest ecosystem services. *Sustainability, 12*, 4089. https://doi.org/10.3390/su12104089.

Esseen, P. A., Ehnström, B., & Ericson, L. (1992). Boreal forests—the focal habitats of Fennoscandia. In L. Hansson (Ed.), *Ecological principles of nature conservation. Conservation ecology series: Principles, practices and management* (pp. 252–325). Boston: Springer.

Eyvindson, K., Repo, A., & Mönkkönen, M. (2018). Mitigating forest biodiversity and ecosystem service losses in the era of bio-based economy. *Forest Policy and Economics, 92*, 119–127. https://doi.org/10.1016/j.forpol.2018.04.009.

Fares, S., Mugnozza, G. S., Corona, P., et al. (2015). Sustainability: Five steps for managing Europe's forests. *Nature, 519*, 407–409. https://doi.org/10.1038/519407a.

Felton, A., Gustafsson, L., Roberge, J. M., et al. (2016). How climate change adaptation and mitigation strategies can threaten or enhance the biodiversity of production forests: Insights from Sweden. *Biological Conservation, 194*, 11–20. https://doi.org/10.1016/j.biocon.2015.11.030.

Garcia-Gonzalo, J., Peltola, H., Zubizarreta Gerendiain, A., et al. (2007). Impacts of forest landscape structure and management on timber production and carbon stocks in the boreal forest ecosystem under changing climate. *Forest Ecology and Management, 241*(1), 243–257. https://doi.org/10.1016/j.foreco.2007.01.008.

Garcia-Gonzalo, J., Bushenkov, V., McDill, M. E., et al. (2015). A decision support system for assessing trade-offs between ecosystem management goals: An application in Portugal. *Forests, 6*, 65–87. https://doi.org/10.3390/f6010065.

Gauthier, S., Bernier, P., Kuuluvainen, T., et al. (2015). Boreal forest health and global change. *Science, 349*, 819–822. https://doi.org/10.1126/science.aaa9092.

Goldblum, D., & Rigg, L. S. (2005). Tree growth response to climate change at the deciduous boreal forest ecotone, Ontario, Canada. *Canadian Journal of Forest Research, 35*, 2709–2718. https://doi.org/10.1139/x05-185.

Gustafson, E. J., Shvidenko, A. Z., & Scheller, R. M. (2011). Effectiveness of forest management strategies to mitigate effects of global change in south-central Siberia. *Canadian Journal of Forest Research, 41*, 1405–1421. https://doi.org/10.1139/x11-065.

Gustafsson, L., Kouki, J., & Sverdrup-Thygeson, A. (2010). Tree retention as a conservation measure in clear-cut forests of northern Europe: A review of ecological consequences. *Scandinavian Journal of Forest Research, 25*, 295–308. https://doi.org/10.1080/02827581.2010.497495.

He, H. S. (2008). Forest landscape models: Definitions, characterization, and classification. *Forest Ecology and Management, 254*, 484–498. https://doi.org/10.1016/j.foreco.2007.08.022.

He, H. S. (2009). A review of LANDIS and other forest landscape models for integration with wildlife models. In J. J. Millspaugh & F. R. Thompson (Eds.), *Models for planning wildlife conservation in large landscapes* (pp. 321–338). Burlington: Academic Press.

He, H. S., Mladenoff, D. J., & Crow, T. R. (1999). Linking an ecosystem model and a landscape model to study forest species response to climate warming. *Ecological Modelling, 114*, 213–233. https://doi.org/10.1016/S0304-3800(98)00147-1.

Hof, A. R., & Hjältén, J. (2018). Are we restoring enough? Simulating impacts of restoration efforts on the suitability of forest landscapes for a locally critically endangered umbrella species. *Restoration Ecology, 26*, 740–750. https://doi.org/10.1111/rec.12628.

Hof, A. R., Dymond, C. C., & Mladenoff, D. J. (2017). Climate change mitigation through adaptation: The effectiveness of forest diversification by novel tree planting regimes. *Ecosphere, 8*, e01981. https://doi.org/10.1002/ecs2.1981.

Hof, A. R., Löfroth, T., Rudolphi, J., et al. (2018). Simulating long-term effects of bioenergy extraction on dead wood availability at a landscape scale in Sweden. *Forests, 9*, 457. https://doi.org/10.3390/f9080457.

Hof, A. R., Montoro Girona, M., Fortin, M.-J., et al. (2021). Editorial: Using landscape simulation models to help balance conflicting goals in changing forests. *Frontiers in Ecology and Evolution*. https://doi.org/10.3389/fevo.2021.795736.

Holmberg, M., Aalto, T., Akujärvi, A., et al. (2019). Ecosystem services related to carbon cycling–modeling present and future impacts in boreal forests. *Frontiers in Plant Science, 10*, 343. https://doi.org/10.3389/fpls.2019.00343.

Imai, N., Samejima, H., Langner, A., et al. (2009). Co-benefits of sustainable forest management in biodiversity conservation and carbon sequestration. *PLoS ONE, 4*, e8267. https://doi.org/10.1371/journal.pone.0008267.

Intergovernmental Panel on Climate Change (IPCC). (2014). Summary for policymakers. In O. Edenhofer, R. Pichs-Madruga, Y. Sokona, E. Farahani, S. Kadner, K. Seyboth, A. Adler, I. Baum, S. Brunner, P. Eickemeier, B. Kriemann, J. Savolainen, S. Schlömer, C. von Stechow, T. Zwickel, & J. C. Minx (Eds.), *Climate change 2014: Mitigation of climate change. Contribution of Working Group III to the fifth assessment report of the Intergovernmental Panel on Climate Change.* Cambridge and New York: Cambridge University Press.

Ito, A. (2005). Modelling of carbon cycle and fire regime in an east Siberian larch forest. *Ecological Modelling, 187*, 121–139. https://doi.org/10.1016/j.ecolmodel.2005.01.037.

Jiang, H., Apps, M. J., Peng, C., et al. (2002). Modelling the influence of harvesting on Chinese boreal forest carbon dynamics. *Forest Ecology and Management, 169*, 65–82. https://doi.org/10.1016/S0378-1127(02)00299-2.

Keenan, R. J. (2015). Climate change impacts and adaptation in forest management: A review. *Annals of Forest Science, 72*, 145–167. https://doi.org/10.1007/s13595-014-0446-5.

Kolström, M., & Lumatjärvi, J. (1999). Decision support system for studying effect of forest management on species richness in boreal forests. *Ecological Modelling, 119*, 43–55. https://doi.org/10.1016/S0304-3800(99)00060-5.

Krankina, O. N., Dixon, R. K., Kirilenko, A. P., et al. (1997). Global climate change adaptation: Examples from Russian boreal forests. *Climatic change, 36*, 197–215. https://doi.org/10.1023/A:1005348614843.

Kurz, W. A., Dymond, C. C., Stinson, G., et al. (2008). Mountain pine beetle and forest carbon feedback to climate change. *Nature, 452*, 987–990. https://doi.org/10.1038/nature06777.

Lagergren, F., & Jönsson, A. M. (2017). Ecosystem model analysis of multi-use forestry in a changing climate. *Ecosystem Services, 26*, 209–224. https://doi.org/10.1016/j.ecoser.2017.06.007.

Lehmann, J. (2007). A handful of carbon. *Nature, 447*, 143–144. https://doi.org/10.1038/447143a.

Lindenmayer, D. B., Margules, C. R., & Botkin, D. B. (2000). Indicators of biodiversity for ecologically sustainable forest management. *Conservation Biology, 14*, 941–950. https://doi.org/10.1046/j.1523-1739.2000.98533.x.

Lindenmayer, D. B., Franklin, J. F., Lõhmus, A., et al. (2012). A major shift to the retention approach for forestry can help resolve some global forest sustainability issues. *Conservation Letters, 5*, 421–431. https://doi.org/10.1111/j.1755-263X.2012.00257.x.

Lucash, M. S., Scheller, R. M., Gustafson, E. J., et al. (2017). Spatial resilience of forested landscapes under climate change and management. *Landscape Ecology, 32*, 953–969. https://doi.org/10.1007/s10980-017-0501-3.

Lucash, M. S., Scheller, R. M., Sturtevant, B. R., et al. (2018). More than the sum of its parts: How disturbance interactions shape forest dynamics under climate change. *Ecosphere, 9*, e02293. https://doi.org/10.1002/ecs2.2293.

Melillo, J. M., McGuire, A. D., Kicklighter, D. W., et al. (1993). Global climate change and terrestrial net primary production. *Nature, 363*, 234–240. https://doi.org/10.1038/363234a0.

Millar, C. I., Stephenson, N. L., & Stephens, S. L. (2007). Climate change and forests of the future: Managing in the face of uncertainty. *Ecological Applications, 17*, 2145–2151. https://doi.org/10.1890/06-1715.1.

Millennium Ecosystem Assessment. (2005). *Ecosystems and human well-being: Scenarios.* Washington: Island Press.

Miller, C., & Urban, D. L. (1999). A model of surface fire, climate and forest pattern in the Sierra Nevada, California. *Ecological Modelling, 114*, 113–135. https://doi.org/10.1016/S0304-3800(98)00119-7.

Miquelajauregui, Y., Cumming, S. G., & Gauthier, S. (2019a). Sensitivity of boreal carbon stocks to fire return interval, fire severity and fire season: A simulation study of black spruce forests. *Ecosystems, 22*, 544–562. https://doi.org/10.1007/s10021-018-0287-4.

Miquelajauregui, Y., Cumming, S. G., & Gauthier, S. (2019b). Short-term responses of boreal carbon stocks to climate change: A simulation study of black spruce forests. *Ecological Modelling, 409*, 108754. https://doi.org/10.1016/j.ecolmodel.2019.108754.

Nadezda, M. T., Gerald, E. R., & Elena, I. P. (2006). Impacts of climate change on the distribution of *Larix* spp. and *Pinus sylvestris* and their climatypes in Siberia. *Mitigation and Adaptation Strategies for Global Change, 11*, 861–882. https://doi.org/10.1007/s11027-005-9019-0.

Natural Resources Canada (NRC). (2020). *Mountain pine beetle.* Ottawa: Natural Resources Canada. https://www.nrcan.gc.ca/our-natural-resources/forests/wildland-fires-insects-disturban ces/top-forest-insects-and-diseases-canada/mountain-pine-beetle/13381. Accessed May 29, 2020.

Nealis, V. G., & Peter, B. (2008). *Risk assessment of the threat of mountain pine beetle to Canada's boreal and eastern pine resources* (p. 38). Victoria: Canadian Forest Service, Pacific Forestry Centre.

Nordström, E.-M., Nieuwenhuis, M., Başkent, E. Z., et al. (2019). Forest decision support systems for the analysis of ecosystem services provisioning at the landscape scale under global climate and market change scenarios. *European Journal of Forest Research, 138*(4), 561–581. https:// doi.org/10.1007/s10342-019-01189-z.

Noss, R. F. (2001). Beyond Kyoto: Forest management in a time of rapid climate change. *Conservation Biology, 15*, 578–590. https://doi.org/10.1046/j.1523-1739.2001.015003578.x.

Ontl, T. A., Janowiak, M. K., Swanston, C. W., et al. (2020). Forest management for carbon seques-tration and climate adaptation. *Journal of Forestry, 118*, 86–101. https://doi.org/10.1093/jofore/ fvz062.

Orazio, C., Cordero Montoya, R., Régolini, M., et al. (2017). Decision support tools and strategies to simulate forest landscape evolutions integrating forest owner behaviour: A review from the case studies of the European project INTEGRAL. *Sustainability, 9*(4), 599.

Paillet, Y., Bergès, L., Hjältén, J., et al. (2010). Biodiversity differences between managed and unmanaged forests: Meta-analysis of species richness in Europe. *Conservation Biology, 24*, 101–112. https://doi.org/10.1111/j.1523-1739.2009.01399.x.

Pearman-Gillman, S. B., Duveneck, M. J., Murdoch, J. D., et al. (2020). Drivers and consequences of alternative landscape futures on wildlife distributions in New England, USA. *Frontiers in Ecology and Evolution, 8*, 164. https://doi.org/10.3389/fevo.2020.00164.

Peltola, H., Ikonen, V. P., Gregow, H., et al. (2010). Impacts of climate change on timber production and regional risks of wind-induced damage to forests in Finland. *Forest Ecology and Management, 260*, 833–845. https://doi.org/10.1016/j.foreco.2010.06.001.

Pennanen, J., & Kuuluvainen, T. (2002). A spatial simulation approach to natural forest landscape dynamics in boreal Fennoscandia. *Forest Ecology and Management, 164*, 157–175. https://doi. org/10.1016/S0378-1127(01)00608-9.

Post, W. M., & Pastor, J. (1996). Linkages—An individual-based forest ecosystem model. *Climatic Change, 34*, 253–261. https://doi.org/10.1007/BF00224636.

Prestele, R., Alexander, P., Rounsevell, M. D. A., et al. (2016). Hotspots of uncertainty in land-use and land-cover change projections: A global-scale model comparison. *Global Change Biology, 22*(12), 3967–3983. https://doi.org/10.1111/gcb.13337.

Pukkala, T. (2004). Dealing with ecological objectives in the Monsu planning system. *Silva Lusitana, 12*, 1–15.

Ragauskas, A. J., Williams, C. K., Davison, B. H., et al. (2006). The path forward for biofuels and biomaterials. *Science, 311*, 484–489. https://doi.org/10.1126/science.1114736.

Ranius, T., Hämäläinen, A., Egnell, G., et al. (2018). The effects of logging residue extrac-tion for energy on ecosystem services and biodiversity: A synthesis. *Journal of Environmental Management, 209*, 409–425. https://doi.org/10.1016/j.jenvman.2017.12.048.

Rebetez, M., & Dobbertin, M. (2004). Climate change may already threaten Scots pine stands in the Swiss Alps. *Theoretical and Applied Climatology, 79*, 1–9. https://doi.org/10.1007/s00704-004-0058-3.

Repo, A., Eyvindson, K. J., Halme, P., et al. (2020). Forest bioenergy harvesting changes carbon balance and risks biodiversity in boreal forest landscapes. *Canadian Journal of Forest Research, 50*, 1184–1193. https://doi.org/10.1139/cjfr-2019-0284.

Safranyik, L., Carroll, A. L., Régnière, J., et al. (2010). Potential for range expansion of mountain pine beetle into the boreal forest of North America. *The Canadian Entomologist, 142*, 415–442. https://doi.org/10.4039/n08-CPA01.

Scheller, R. M., Domingo, J. B., Sturtevant, B. R., et al. (2007). Design, development, and application of LANDIS-II, a spatial landscape simulation model with flexible temporal and spatial resolution. *Ecological Modelling, 201*, 409–419. https://doi.org/10.1016/j.ecolmodel.2006.10.009.

Schlyter, P., Stjernquist, I., Bärring, L., et al. (2006). Assessment of the impacts of climate change and weather extremes on boreal forests in northern Europe, focusing on Norway spruce. *Climate Research, 31*, 75–84. https://doi.org/10.3354/cr031075.

Shvidenko, A., Barber, C. V., Persson, R., et al. (2005). Forest and woodland systems. In *Millennium ecosystem assessment Volume 1. Current state & trends assessment* (pp. 585–621). Washington: Island Press.

Soja, A. J., Tchebakova, N. M., French, N. H., et al. (2007). Climate-induced boreal forest change: Predictions versus current observations. *Global and Planetary Change, 56*, 274–296. https://doi.org/10.1016/j.gloplacha.2006.07.028.

Spittlehouse, D. L., & Stewart, R. B. (2003). Adaptation to climate change in forest management. *BC Journal Ecosystem Management, 4*, 1–11.

Stocks, B. J., Fosberg, M. A., Lynham, T. J., et al. (1998). Climate change and forest fire potential in Russian and Canadian boreal forests. *Climatic Change, 38*, 1–13. https://doi.org/10.1023/A:1005306001055.

Subramanian, N. (2016). *Impacts of climate change on forest management and implications for Swedish forestry: An analysis based on growth and yield models.* Ph.D. Thesis, Swedish University of Agricultural Sciences.

Swedish Environmental Protection Agency. (2022). *The environmental objectives system.* Stockholm: Swedish Environmental Protection Agency. https://www.sverigesmiljomal.se/environmental-objectives/. Accessed March 6, 2022.

Tremblay, J. A., Boulanger, Y., Cyr, D., et al. (2018). Harvesting interacts with climate change to affect future habitat quality of a focal species in eastern Canada's boreal forest. *PLoS ONE, 13*, e0191645. https://doi.org/10.1371/journal.pone.0191645.

van der Plas, F., Manning, P., Allan, E., et al. (2016). Jack-of-all-trades effects drive biodiversity–ecosystem multifunctionality relationships in European forests. *Nature Communications, 7*(1), 11109. https://doi.org/10.1038/ncomms11109.

Wikström, P., Edenius, L., Elfving, B., et al. (2011). The Heureka forestry decision support system: An overview. *Mathematical and Computational Forestry & Natural-Resource Sciences, 3*(2), 87–95.

Wolken, J. M., Hollingsworth, T. N., Rupp, T. S., et al. (2011). Evidence and implications of recent and projected climate change in Alaska's forest ecosystems. *Ecosphere, 2*(11):art124. https://doi.org/10.1890/ES11-00288.1.

Xi, W., Coulson, R. N., Waldron, J. D., et al. (2008). Landscape modeling for forest restoration planning and assessment: Lessons from the southern Appalachian mountains. *Journal of Forestry, 106*, 191–197.

Xi, W., Coulson, R. N., Birt, A. G., et al. (2009). Review of forest landscape models: Types, methods, development and applications. *Acta Ecologica Sinica, 29*, 69–78. https://doi.org/10.1016/j.chnaes.2009.01.00.

Chapter 26
Remote Sensing Tools for Monitoring Forests and Tracking Their Dynamics

Richard Massey, Logan T. Berner, Adrianna C. Foster, Scott J. Goetz, and Udayalakshmi Vepakomma

Abstract Remote sensing augments field data and facilitates foresight required for forest management by providing spatial and temporal observations of forest characteristics at landscape and regional scales. Statistical and machine-learning models derived from plot-level field observations can be extrapolated to larger areas using remote sensing data. For example, instruments such as light detection and ranging (LiDAR) and hyperspectral sensors are frequently used to quantify forest characteristics at the stand to landscape level. Moreover, multispectral imagery and synthetic aperture radar (SAR) data sets derived from satellite platforms can be used to extrapolate forest resource models to large regions. The combination of novel remote sensing technologies, expanding computing capabilities, and emerging geospatial methods ensures a data-rich environment for effective strategic, tactical, and operational planning and monitoring in forest resource management.

R. Massey · L. T. Berner · S. J. Goetz (✉)
School of Informatics, Computing, and Cyber Systems, Northern Arizona University, 1295 S Knoles Drive, Flagstaff, AZ 86011, USA
e-mail: scott.goetz@nau.edu

R. Massey
e-mail: rm885@nau.edu

L. T. Berner
e-mail: logan.berner@nau.edu

A. C. Foster
Climate and Global Dynamics Laboratory, National Center for Atmospheric Research, 1850 Table Mesa Dr, Boulder, CO 80305, USA
e-mail: afoster@ucar.edu

U. Vepakomma
FPInnovations, 570 Boulevard St-Jean, Pointe-Claire, QC H9R 3J9, Canada
e-mail: udayalakshmi.vepakomma@fpinnovations.ca

© The Author(s) 2023
M. M. Girona et al. (eds.), *Boreal Forests in the Face of Climate Change*,
Advances in Global Change Research 74,
https://doi.org/10.1007/978-3-031-15988-6_26

26.1   Introduction

Forests play a primary role for life on Earth. Measuring and quantifying the state of forests at the landscape scale is critical from both a strategic and tactical perspective. Management-oriented forest monitoring efforts include complex and evolving objectives, such as timber production, environmental protection, biodiversity preservation, forest fire prevention, wilderness and open spaces, and adaptation to a changing climate. Forest monitoring approaches have continuously improved over the last few decades with innovations in remote sensing and computing methods. Although field surveys and inventories remain invaluable sources of information, the use of in situ methods to monitor critical forest metrics is limited at larger scales. However, with spaceborne and airborne remote sensing technology, forest monitoring efforts have advanced rapidly in terms of capacity, scale, and detail. For example, large swaths of land are imaged every day by Earth observation (EO) satellites, enabling the constant monitoring of global forest conditions (Mitchell et al., 2017). Such sizable remote sensing data sets provide opportunities to extrapolate the results of models derived from spatially limited field data to the landscape level and permit the observation of large-scale changes.

26.2   Remote Sensing of Forests

Earth observation satellites offer great opportunities to quantify landscape and regional land cover, composition, and change. Some of the commonly used satellite imagery include that from the National Aeronautics and Space Administration (NASA), National Oceanic and Atmospheric Administration (NOAA), European Space Agency (ESA), Indian Space Research Organization (ISRO), Canadian Space Agency (CSA), China National Space Administration (CNSA), and Japan Aerospace Exploration Agency (JAXA) (Table 26.1). Additionally, several commercial satellite imagery providers offer cutting-edge satellite data with higher spatial and temporal resolution and, in many cases, customized monitoring solutions. Some prominent commercial satellite imagery providers include DigitalGlobe from Maxar, Planet Labs, and Airbus. Commercial and openly available EO data are used in a wide variety of Earth science, forestry, agriculture, and geological applications by research, government, and commercial entities.

Some of the most common types of EO data include multispectral and synthetic aperture radar (SAR) systems. Examples of multispectral satellites include Sentinel-1 and 2, Landsat, the Moderate Resolution Imaging Spectroradiometer (MODIS), and the Advanced Very High-Resolution Radiometer (AVHRR). Of these, the higher spatial resolution satellites (e.g., Landsat and Sentinel) are generally more useful from a forest management perspective.

Table 26.1 List of satellite and airborne remote sensing instruments frequently used for forestry and land-cover applications; *RGB*, red, green, blue; *NIR*, near infrared; *SWIR*, shortwave infrared; *TIR*, thermal infrared; *MIR*, middle infrared; *Pan*, panchromatic; *SAR*, synthetic aperture radar; *LiDAR*, light detection and ranging; *NASA*, National Aeronautics and Space Administration; *NOAA*, National Oceanic and Atmospheric Administration; *ESA*, European Space Agency; *JAXA*, Japan Aerospace Exploration Agency; *USGS*, United States Geological Survey; *ISRO*, Indian Space Research Organization; *CSA*, Canadian Space Agency

Platform	Sensor	Agency	Imaging modes	Bands	Revisit time (days)	Spatial resolution (m)	Further information
Landsat-5	Thematic mapper (TM)	NASA/USGS	Multispectral	RGB, NIR, SWIR, TIR	16	30	https://www.usgs.gov/land-resources/nli/landsat/landsat-5
Landsat-7	Enhanced thematic mapper Plus (ETM+)	NASA/USGS	Multispectral	RGB, NIR, SWIR, TIR	16	30	https://www.usgs.gov/land-resources/nli/landsat/landsat-7
Landsat-8	Operational land imager (OLI)	NASA/USGS	Multispectral	RGB, NIR, SWIR, TIR, Pan, Aerosol	16	30	https://www.usgs.gov/land-resources/nli/landsat/landsat-8
Terra, Aqua	Moderate resolution spectroradiometer (MODIS)	NASA	Multispectral	RGB, NIR, SWIR, TIR, Vapor, Clouds, Aerosol, Snow, Ice, Ozone	1–2	250, 500, 1000	https://modis.gsfc.nasa.gov

(continued)

Table 26.1 (continued)

Platform	Sensor	Agency	Imaging modes	Bands	Revisit time (days)	Spatial resolution (m)	Further information
NOAA-satellites 6–19	Advanced very high-resolution radiometer (AVHRR)	NOAA	Multispectral	R, NIR, MIR, TIR	<1	1,000	https://www.avl class.noaa.gov/rel ease/data_available/ avhrr/index.htm
ALOS-2	Phased array type L-band synthetic aperture radar (PALSAR)	JAXA	SAR	L-band	14	>7	https://www.eorc. jaxa.jp/ALOS/en/ about/palsar.htm
Sentinel-1a and 1b	C-band synthetic aperture radar (C-SAR)	ESA	SAR	C-band	6	>5	https://sentinel.esa. int/web/sentinel/ missions/sentinel-1
Sentinel-2a and 2b	Multispectral instrument (MSI)	ESA	Multispectral	RGB, NIR, Red-edge, Aerosol, SWIR	5	10, 20, 60	https://sentinel.esa. int/web/sentinel/ missions/sentinel-2
Airborne	Airborne visible/infrared imaging spectrometer (AVIRIS)	NASA	Hyperspectral	224 bands	–	20	https://aviris.jpl.nas a.gov
Airborne	Land, vegetation, and ice sensor (LVIS)	NASA	LiDAR altimeter	–	–	5	https://lvis.gsfc.nas a.gov

(continued)

Table 26.1 (continued)

Platform	Sensor	Agency	Imaging modes	Bands	Revisit time (days)	Spatial resolution (m)	Further information
International Space Station	Global ecosystem dynamics investigation (GEDI)	NASA	LiDAR	Full waveform	–	25	https://gedi.umd.edu
ICESat-2	Advanced topographic laser altimeter system (ATLAS)	NASA	LiDAR	–	–	13	https://icesat-2.gsfc.nasa.gov
Cartosat-3	Multispectral VNIR (MX), Panchromatic instrument (PAN)	ISRO	Multispectral, Panchromatic	Pan, RGB, NIR, MIR	4	0.25, 1.2, 6, 12	https://www.isro.gov.in/Spacecraft/cartosat-3
RADARSAT-2	C-band synthetic aperture radar	CSA	SAR	C-band	24	1–100	www.asc-csa.gc.ca/eng/satellites/radarsat2
Dove, SuperDove, CubeSats	Planetscope imagery product	Planet Labs Inc	Multispectral	RGB, NIR	<1	3	https://www.planet.com/products/planet-imagery/
SkySat	SkySat	Planet Labs Inc	Multispectral, Video	RGB, NIR, Pan	4–5	0.5	
RapidEye	RapidEye	Planet Labs Inc		RGB, NIR, Red-edge	5–6	5–6.5	
Worldview-2	Worldview-2	Maxar Technologies	Multispectral	Pan, RGB, Red-edge, Coastal, NIR		0.5, 2	https://www.digitalglobe.com/products/satellite-imagery

(continued)

Table 26.1 (continued)

Platform	Sensor	Agency	Imaging modes	Bands	Revisit time (days)	Spatial resolution (m)	Further information
Worldview-3	Worldview-3	Maxar Technologies	Multispectral	Pan, RGB, Red-edge, Coastal, NIR, SWIR, Cloud, Aerosol, Vapor, Ice, snow		0.5, 2	
Worldview-4	Worldview-4	Maxar Technologies	Multispectral	Pan, RGB, NIR		0.31, 1.24	

The Sentinel satellites are part of ESA's Copernicus Program, one of the most recent and ambitious EO programs. Currently, there are two series of Sentinel satellites that provide data to users around the globe: Sentinel-1 and Sentinel-2. The former consists of a constellation of two satellites, 1A and 1B, carrying C-band SAR with a lower spatial resolution limit at 5 m. Spaceborne SAR is an active radar system that can image the Earth's surface with or without cloud cover and through smoke and other aerosols. Depending on the wavelength, microwaves from a SAR system can even penetrate the top layers of soil and vegetation and provide useful information regarding the soil's physical properties, such as soil moisture. The Sentinel-1 constellation can revisit the same location about every six days. Sentinel-2 consists of two multispectral satellites, 2A and 2B, having a spatial resolution of 10 m and a revisit time of five days. Multispectral satellite sensors measure how sunlight is reflected by the Earth's surface across a range of wavelengths and are passive in nature, i.e., without an active source of electromagnetic radiation. Figure 26.1 shows an example of SAR and multispectral images showing variations in backscatter and reflectance, respectively.

The Landsat satellite program offers the richest and longest-running historical archive of satellite data with observations since the 1970s (Wulder et al., 2019). This archive provides unique opportunities to study the mechanisms and extent of past and present forest dynamics. The moderately fine (30 m) spatial resolution of the Landsat Thematic Mapper sensors and their revisit time of 16 days make them uniquely suited for longer-term forest monitoring and management applications from space (Hansen & Loveland, 2012). The Landsat data archive became publicly available in 2008, which, combined with ready access via Google's Earth Engine

0   5
L_____I KM

Fig. 26.1 Forest clear-cuts in western Oregon, United States, shown using (*left*) a Sentinel-1 SAR image composite with spring, late spring, and summer images as three bands in VV polarization and (*right*) a true-color image from a Sentinel-2 composite image using red, green, and blue bands from the summer 2018. SAR data are sensitive to topography, biomass, and water content. Recent timber harvest units have a higher color intensity in the SAR and lose seasonal variation as the forest regrows

platform (Gorelick et al., 2017), promoted the widespread use of these data for many research and commercial applications. Coarser spatial resolution satellites, such as MODIS and AVHRR, have historically been used to map and classify land cover at spatial resolution scales ranging between 250 m and 10 km. These coarse spatial resolution satellites have a high temporal resolution, with near-daily imagery, but their coarse resolution makes it challenging to derive reliable map-based estimates of forest characteristics and change (Chen et al., 2018). Although they have limited utility at local scales, MODIS and AVHRR satellites provide frequent remote sensing data that are useful for disaster monitoring systems and as inputs and validation for ecosystem models evaluating land-cover changes over large areas.

In addition to multispectral imagery, newer remote sensing technologies, such as light detection and ranging (LiDAR), provide emerging opportunities to assess boreal forest characteristics and can be used to quantify changes in forests over time (Dubayah & Drake, 2000). LiDAR can be used to estimate a variety of forest structural attributes across large areas, including canopy height, cover, volume, and biomass. Unlike multispectral sensors, LiDAR instruments actively emit photons via infrared lasers and then measure the amount of time required for the photons to strike a target and return to the sensor. Photon return time indicates the distance between the sensor and the target. LiDAR can be used to assess forest structure, including forest aboveground biomass (AGB), and reproduce subcanopy surface topography. LiDAR instruments can be airborne, spaceborne, or land-based (terrestrial) and can be used at different levels of detail to provide forestry-relevant management and inventory information (Magnussen et al., 2018; Shendryk et al., 2014; Zhao et al., 2018). Spaceborne LiDAR is increasingly being used to assess forest structure and biomass around the world. It has become progressively more feasible with photon-counting technology onboard ICESat-1/GLAS (2003–2009) and ICESat-2 (2017–present) (Popescu et al., 2018), and with the full waveform capability of the GEDI instrument (2019–present) (Dubayah et al., 2020). These new technologies are typically used to infer forest attributes at field sampling locations, which are further extended via remote sensing imagery across a larger area of interest on the basis of empirical relationships (Margolis et al., 2015; Neigh et al., 2013). Such methods can enable the rapid, robust, and cost-effective characterization of forest attributes across large areas.

In combination with the ever-increasing geospatial data being made available by multiple remote sensing and non–remote sensing sources, geographic information systems (GIS) are used as visualization, data manipulation, and processing tools for a wide range of data sets. The coevolution of GIS and remote sensing technologies has augmented field and inventory data with satellite imagery for map production, spatial visualization and query, and decision support (Sonti, 2015). Linking field inventory, aerial surveys, and remote sensing with GIS tools has helped foresters and ecologists develop more accurate records of forest cover, composition, and configuration for strategic (long-term) and tactical (short-term) planning.

26.3   Forest Biodiversity

Forest biodiversity is an essential consideration for sustainable forest management. Assessments of spatial heterogeneity in biodiversity commonly use satellite and aerial remote sensing data. In boreal environments, which have a moderate to low tree species diversity, such assessments may involve supervised and unsupervised classifications of high spatial resolution multispectral and hyperspectral data (Baldeck & Asner, 2013). Recent advances in data fusion techniques have enabled the use of high spatial and temporal resolution data combined with LiDAR (Fig. 26.2) from aerial platforms to measure and relate plot-level variations of species composition to environmental and physical factors (Powers et al., 2013; Rocchini et al., 2015). In addition to the analysis of tree species, remote sensing indicators have been used to model and map animal species diversity across large landscapes (Davies & Asner, 2014); for example, Coops et al. (2009) predicted bird species richness in Ontario, Canada, using productivity, topography, and land cover derived from remote sensing. Similarly, Kerr et al. (2001) modeled butterfly species richness on the basis of remote sensing–derived land cover and climate data. These studies also indicate that although biodiversity assessments can incorporate remote sensing approaches, it is seldom trivial to select remotely sensed indicators of biodiversity, and this approach requires a combination of traditional ecological knowledge and mathematical modeling.

0   100
└──────────────────┘ m

Fig. 26.2 (*left*) A LiDAR three-dimensional (3D) point cloud, color-coded by height from a baseline, of a small area in a Douglas fir–dominated forest. Data obtained from the United States Geological Survey's 3D elevation program (3DEP). (*right*) Airborne false-color imagery of the same area at a 1 m spatial resolution. Data from the United States Department of Agriculture's National Agricultural Imagery Program (NAIP) for 2018. In addition to the spatial location of individual trees, airborne LiDAR can capture the 3D structure of the forest

26.4 Forest Disturbances

Forest disturbances, such as wildfire, windthrow, and insect outbreaks, are integral and natural components of forest ecosystem dynamics. They impact forest species composition, structure, above- and belowground carbon storage (Alexander & Mack, 2016), forest regeneration and successional dynamics (Johnstone et al., 2010), as well as water and energy cycling (Goetz et al., 2012). Remote sensing methods can be used to detect and monitor forest disturbance across large areas and thus inform sustainable forest management policies and practices (Guindon et al., 2018; Hall et al., 2016). Ecosystem responses to disturbance events can be assessed using data from multiple satellite missions with field and airborne campaigns to monitor changes in connectivity, complexity, and heterogeneity across a region (Skidmore et al., 2015). The fusion of multilevel and multiresolution data can inform tactical and strategic management efforts. Such data collections can also be used to model ecosystem response to climate and help the strategic planning of resources in relation to future climate scenarios (Whitman et al., 2019).

26.4.1 Fire Detection and Risk

Fire occurrence and severity can be detected using multispectral satellite imagery by observing the difference in pre- and postfire indices, such as the normalized burn ratio (NBR) and infrared bands (Key & Benson, 2005). For fire management and detection, satellites such as MODIS. PlanetScope, and SkySat provide a daily updated stream of satellite images, which, when combined with aerial imagery, can be used to monitor fire progression (Giglio et al., 2016). During tactical planning stages, fire risk can be evaluated by assessing species composition, forest density, forest structure, and fuel conditions using a combination of airborne and spaceborne remote sensing data. The information storage and analysis capabilities of GIS tools are particularly useful for decision-making in tactical situations and emergencies where fire management and prevention, prescribed burning, and postfire recovery actions are planned by integrating GIS and remote sensing data to, for example, prepare maps of burn severity (Wulder & Franklin, 2006).

26.4.2 Monitoring Forest Health

Nonstand replacing disturbances, such as windthrow, insect outbreaks, and disease, often disproportionately impact certain tree species or sizes, leading to shifts in species composition, stand structure, and productivity (Goetz et al., 2012). Insect disturbances are usually observed indirectly in satellite imagery using specialized methods for each insect type (Senf et al., 2017; White et al., 2007). For example, once

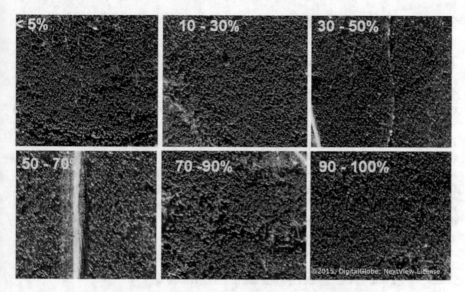

Fig. 26.3 Coniferous stands affected by the gradient of percent cumulative insect defoliation in a Canadian boreal forest as seen using 10 cm high spatial resolution false-color image and pan-sharpened shortwave infrared (SWIR) bands from the WorldView-3 satellite

infested by bark beetles, such as the mountain pine beetle (*Dendroctonus ponderosae*) or the spruce beetle (*Dendroctonus rufipennis*), the tree moisture status is often impacted through stomatal closure and secondary infection by fungal pathogens. Needle color changes from green to red (*red-attack* stage) or gray (*gray-attack* stage) depending on the tree species (Hall et al., 2016). The change in needle color, especially at the red-attack stage, is detectable in high-resolution multispectral imagery and can indicate insect infestation (Coops et al., 2006). The calibration and validation of insect disturbance mapping efforts are often achieved through comparisons with field data (Senf et al., 2017). For the long-term monitoring and detection of infestations, detection programs can employ annual aerial detection surveys as starting points to digitize validation polygons from the photointerpretation of high spatial resolution imagery (Meddens et al., 2012). Because of the inherent multi-scale nature of insect outbreaks, infestations occur at the individual tree scale but can quickly spread across landscapes (Raffa et al., 2008). Given that insect outbreaks often progress over several years, multidate time-series observations are usually required to detect and observe the complete response of a forest to an outbreak (Senf et al., 2017). Additionally, many infestation cases warrant the use of higher resolution remote sensing imagery (Fig. 26.3) from multispectral satellites, e.g., WorldView, hyperspectral imaging satellites, e.g., Hyperion, and aerial-based remote sensing instruments, e.g., airborne visible infrared imaging spectrometer, AVIRIS; Senf et al. (2017) and Makoto et al. (2013).

26.4.3   Invasive Species

The expansion of invasive species decreases the diversity of native plants and thus presents a threat to overall ecosystem resilience (Harrod & Reichard, 2001). Multi-date observations of changes in vegetation indices, such as the normalized difference vegetation index (NDVI) or enhanced vegetation index (EVI), can indicate an increased dominance of invasive species, especially where the invasive species are spectrally distinct from the native population. In cases where the invasive species are not spectrally distinct within multispectral imagery, hyperspectral imagery may be required to develop suitable models for detecting and mapping the encroachment of these invasive species (Huang & Asner, 2009). When there are structural or height differences between the native and invasive species, multispectral or hyperspectral data can be augmented using LiDAR or SAR data.

26.5   Forest Characteristics and Productivity

Forest management objectives are often achieved by monitoring and controlling forest characteristics in a stand to influence growth and yield. As large parts of the boreal forest are managed for wood production (Gauthier et al., 2015), remote sensing technologies provide effective tools to monitor stands, particularly when combined with field surveys and forest inventory data.

26.5.1   Assessing Forest Productivity with Remote Sensing

At landscape scales, remote sensing assessments of boreal forest productivity often rely on repeat measurements of coarse- or moderate-resolution multispectral data and vegetation indices (e.g., NDVI, EVI), SAR data, and airborne LiDAR. At the stand scale, however, tree species distributions derived from high-resolution multispectral data are often used to describe and project forest growth and yield (Modzelewska et al., 2020). Although field surveys traditionally determine species composition within stands, remote sensing tools can expand the field-derived models to larger scales. Such maps derived from remote sensing data also provide forest managers with the spatial distribution of products likely to be produced from the forest and also the vulnerability of stands to disturbance on the basis of tree species.

When it comes to measuring harvest potential and products that can be derived from forest stands, terrestrial and airborne LiDAR instruments are of primary importance. LiDAR data form an important part of growth and yield modeling simulations. Furthermore, tree-level data inform on both timber assortments and biodiversity. In addition to species distribution, stand density plays a major role in forest yield assessments and the monitoring of growth. In recent decades, airborne laser scanning (ALS) has emerged as a promising technology for estimating stand density within forested areas. ALS is a LiDAR approach that uses an airborne platform to transmit and measure returns from tree canopies in the near-infrared range. These returns can be used to accurately estimate the number of trees and stand density using spatial relationships between LiDAR points and "point clouds" (Næsset, 2004).

ALS-derived LiDAR point clouds can also be used to derive aboveground biomass and estimate stand age. As trees age, they typically grow in height up to a (usually species-specific) point, after which their vertical growth slows even as their carbon accumulation rate may continue to increase (Stephenson et al., 2014). Species distribution maps combined with stand-age data can be used to identify the site index, which is a measure of projected height at an index age (typically 25, 50, or 100 years). The site index is typically used as an indirect measure of site quality and its ability to produce specific wood products. Site quality is an essential parameter for forest managers as it can help determine the quantification of merchantable timber and is an essential input for the strategic planning of forest resources. Stand-age and site-index maps derived using remote sensing and GIS can also be used to identify harvest locations in the forest by identifying optimal mean annual increments (MAI) to maximize sustained volume productivity.

26.5.2   Mapping Forest Aboveground Biomass Using Remote Sensing

Forest aboveground biomass (AGB) describes the total dry weight of live trees per unit area and is related to structural metrics such as tree density, diameter, height, and composition. Forest AGB is useful for forest managers to consider because it provides additional information for volume estimates for timber production purposes, such as stand carbon sequestration and storage. Forest AGB maps may also serve as tools to identify areas of high conservation priority or with high intraspecific competition having a potential need of management treatments. Forest AGB can be mapped over an area of interest by linking plot-level forest inventories with remote sensing measurements related to forest canopy cover, structure, and composition (Berner et al., 2012; Puliti et al., 2020). Various remote sensing instruments are used to measure and map boreal forest AGB, including LiDAR, SAR, and multispectral sensors, often in combination with one another.

In addition to LiDAR, SAR data are used to map boreal forest AGB. Live-tree growing stock volume (GSV) is an important parameter for predicting forest AGB and can be mapped across large areas using SAR data (Santoro et al., 2015). Forest AGB can then be predicted by combining GSV with information related to land cover, land cover–specific wood density, and biomass allocation (Fig. 26.4; Thurner et al., 2014). Multispectral satellite imagery is an inexpensive means of extending AGB estimates from plot-level measurements through the use of airborne and terrestrial LiDAR,

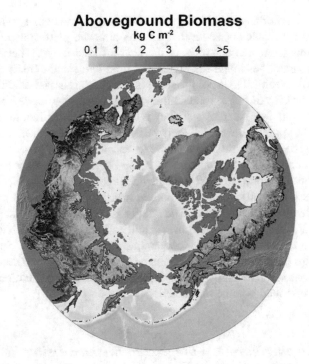

Fig. 26.4 Aboveground biomass mapped across the boreal forest biome using the synthetic aperture radar (SAR) and ancillary information (Santoro et al., 2015; Thurner et al., 2014). The boreal biome extent is based on the boreal ecoregions mapped by the World Wildlife Fund (Olson et al., 2001)

very high-resolution imagery, and field inventories. This multisource, multiscale approach can also be used to monitor changes in forest AGB over time. LiDAR data are particularly useful for augmenting multispectral imagery, as forest canopy closure obscures forest structure (Wulder et al., 2020). Ancillary geospatial information can also improve model predictions of boreal forest AGB (Puliti et al., 2020).

The integration of multispectral imagery with repeated LiDAR and SAR also provides emerging opportunities to assess boreal forest productivity by quantifying net changes in boreal forest AGB over time (ΔAGB), typically using either a *direct* or *indirect* approach (Karila et al., 2019; McRoberts et al., 2015). The direct approach involves predicting ΔAGB on the basis of differences in forest canopy structure between successive remote sensing measurements. The indirect approach involves predicting forest AGB at two points in time using remote sensing measurements and then computing ΔAGB by differencing the two predictions. The direct approach requires measurements from the same ground location during each survey, although prediction errors are easier to estimate. From an inventory standpoint, both methods can increase the precision of ΔAGB estimates relative to relying exclusively on field inventory measurements (McRoberts et al., 2015). Remote sensing efforts to quantify ΔAGB in the boreal forest have primarily relied on repeat airborne LiDAR surveys over small landscapes (Hopkinson et al., 2016; McRoberts et al., 2015). Recent efforts have also demonstrated the utility of spaceborne SAR to quantify ΔAGB, which allows for larger-scale mapping (Askne et al., 2018; Karila et al., 2019). Spaceborne LiDAR, e.g., ICESat-2, could also enable large-scale, sample-based estimates of ΔAGB in the boreal forest. The combination of satellite and airborne remote sensing provides a suite of tools for assessing boreal forest productivity at both the local and landscape scales.

26.6   Novel Technologies in Remote Sensing

Satellite programs such as Landsat, Copernicus, and MODIS provide a high degree of homogeneity of the data sets over time by ensuring fixed observation conditions, regular sensor and data calibration, and minimal geolocation errors. Such repeat observations are invaluable for evaluating and monitoring forest conditions at landscape scales over longer periods. Moreover, commercial petabyte-scale satellite archives of daily high-resolution images from providers such as Planetscope and SkySat from Planet Labs provide a constantly updated satellite data stream of the entire planet, ensuring global monitoring and data continuity for both tactical and strategic planning. These microsatellite constellations allow obtaining multiple measures of the same area of interest throughout a single season, which enables the study of phenological metrics in forest plots over several years.

However, large archives of satellite data present a major challenge in regard to processing and handling the collected imagery. Although large-capacity computing solutions can be built, such tools require significant time and resources and are only generally available within large institutions. Recent rapid advances in cloud-computing technology have increased the availability of on-demand computing capabilities for research and commercial users alike. Recently evolved cloud-computing technologies from Google Cloud Platform, Amazon Web Services, and DigitalOcean, to name a few commercially available platforms, are enabling researchers to push the boundaries of science by providing pay-per-use computing infrastructure. Cloud-computing services provide managed computational tools and platforms that can process large amounts of data without the need to install local computing infrastructure. Cloud-computing platforms, e.g., Google Earth Engine (GEE) (Gorelick et al., 2017), can help resolve challenges associated with the large amounts of computing required for working with and analyzing petabyte-scale satellite imagery data without interacting with it on a local computer. Many openly available satellite data collections, including Landsat, MODIS, Sentinel-1 and 2, and many derived regional and global products, are now available on GEE for user-defined processing and computation. Moreover, tools such as GEE are highly scalable and process satellite imagery in parallel, thereby markedly reducing time for many workflows. The scalable nature of GEE permits machine-learning workflows for classifying images.

The ability of forest managers to respond to the effects of changing climate on forests depends on effective data collection, processing, and derivation of actionable insights. Because it is not feasible to frequently or even infrequently census an entire forest using field surveys, it becomes necessary to monitor large tracts of forests for changes through remote sensing platforms and instruments. Models developed using a combination of field and remote sensing data can provide avenues for keeping forest managers informed of changes in biodiversity, biomass, vulnerability, stand density, and other forest characteristics. Future advances in remote sensing technologies, computing platforms, and geospatial software will further advance monitoring and mapping capabilities toward more sustainable planning and management of boreal forest resources and better equip forest managers for mitigating the consequences of ongoing climatic change.

References

Alexander, H. D., & Mack, M. C. (2016). A canopy shift in interior Alaskan boreal forests: Consequences for above-and belowground carbon and nitrogen pools during post-fire succession. *Ecosystems, 19*(1), 98–114. https://doi.org/10.1007/s10021-015-9920-7.

Askne, J. I., Persson, H. J., & Ulander, L. M. (2018). Biomass growth from multi-temporal TanDEM-X interferometric synthetic aperture radar observations of a boreal forest site. *Remote Sensing, 10*(4), 603. https://doi.org/10.3390/rs10040603.

Baldeck, C. A., & Asner, G. P. (2013). Estimating vegetation beta diversity from airborne imaging spectroscopy and unsupervised clustering. *Remote Sensing, 5*(5), 2057–2071. https://doi.org/10.3390/rs5052057.

Berner, L. T., Beck, P. S. A., Loranty, M. M., et al. (2012). Cajander larch (*Larix cajanderi*) biomass distribution, fire regime and post-fire recovery in northeastern Siberia. *Biogeosciences, 9*(10), 3943–3959. https://doi.org/10.5194/bg-9-3943-2012.

Chen, X., Wang, D., Chen, J., et al. (2018). The mixed pixel effect in land surface phenology: A simulation study. *Remote Sensing of Environment, 211*, 338–344. https://doi.org/10.1016/j.rse.2018.04.030.

Coops, N. C., Johnson, M., Wulder, M. A., et al. (2006). Assessment of QuickBird high spatial resolution imagery to detect red attack damage due to mountain pine beetle infestation. *Remote Sensing of Environment, 103*(1), 67–80. https://doi.org/10.1016/j.rse.2006.03.012.

Coops, N. C., Wulder, M. A., & Iwanicka, D. (2009). Exploring the relative importance of satellite-derived descriptors of production, topography and land cover for predicting breeding bird species richness over Ontario, Canada. *Remote Sensing of Environment, 113*(3), 668–679. https://doi.org/10.1016/j.rse.2008.11.012.

Davies, A. B., & Asner, G. P. (2014). Advances in animal ecology from 3D-LiDAR ecosystem mapping. *Trends in Ecology & Evolution, 29*(12), 681–691. https://doi.org/10.1016/j.tree.2014.10.005.

Dubayah, R. O., & Drake, J. B. (2000). Lidar remote sensing for forestry. *Journal of Forestry, 98*(6), 44–46.https://doi.org/10.1093/jof/98.6.44.

Dubayah, R., Blair, J. B., Goetz, S., et al. (2020). The global ecosystem dynamics investigation: High-resolution laser ranging of the Earth's forests and topography. *Science of Remote Sensing, 1*, 100002. https://doi.org/10.1016/j.srs.2020.100002.

Gauthier, S., Bernier, P., Kuuluvainen, T., et al. (2015). Boreal forest health and global change. *Science, 349*(6250), 819–822. https://doi.org/10.1126/science.aaa9092.

Giglio, L., Schroeder, W., & Justice, C. O. (2016). The collection 6 MODIS active fire detection algorithm and fire products. *Remote Sensing of Environment, 178*, 31–41. https://doi.org/10.1016/j.rse.2016.02.054.

Goetz, S. J., Bond-Lamberty, B., Law, B. E., et al. (2012). Observations and assessment of forest carbon dynamics following disturbance in North America. *Journal of Geophysical Research: Biogeosciences, 117*(G02022), 1–17. https://doi.org/10.1029/2011JG001733.

Gorelick, N., Hancher, M., Dixon, M., et al. (2017). Google Earth Engine: Planetary-scale geospatial analysis for everyone. *Remote Sensing of Environment, 202*, 18–27. https://doi.org/10.1016/j.rse.2017.06.031.

Guindon, L., Bernier, P., Gauthier, S., et al. (2018). Missing forest cover gains in boreal forests explained. *Ecosphere, 9*(1), e02094. https://doi.org/10.1002/ecs2.2094.

Hall, R. J., Castilla, G., White, J. C., et al. (2016). Remote sensing of forest pest damage: A review and lessons learned from a Canadian perspective. *The Canadian Entomologist, 148*, S296–S356. https://doi.org/10.4039/tce.2016.11.

Hansen, M. C., & Loveland, T. R. (2012). A review of large area monitoring of land cover change using Landsat data. *Remote Sensing of Environment, 122*, 66–74. https://doi.org/10.1016/j.rse.2011.08.024.

Harrod, R. J., & Reichard, S. (2001). Fire and invasive species within the temperate and boreal coniferous forests of western North America. In K. E. M. Galley, & T. P. Wilson (Eds.), *Proceedings of the Invasive Species Workshop: The Role of Fire in the Control and Spread of Invasive Species. Fire Conference 2000: The First National Congress on Fire Ecology, Prevention, and Management*, Miscellaneous Publication No. 11 (pp. 95–101). Tallahassee: Tall Timbers Research Station.

Hopkinson, C., Chasmer, L., Gynan, C., et al. (2016). Multisensor and multispectral LiDAR characterization and classification of a forest environment. *Canadian Journal of Remote Sensing, 42*(5), 501–520. https://doi.org/10.1080/07038992.2016.1196584.

Huang, C. Y., & Asner, G. P. (2009). Applications of remote sensing to alien invasive plant studies. *Sensors, 9*(6), 4869–4889. https://doi.org/10.3390/s90604869.

Johnstone, J. F., Hollingsworth, T. N., Chapin, F. S., et al. (2010). Changes in fire regime break the legacy lock on successional trajectories in Alaskan boreal forest. *Global Change Biology, 16*(4), 1281–1295. https://doi.org/10.1111/j.1365-2486.2009.02051.x.

Karila, K., Matikainen, L., Litkey, P., et al. (2019). The effect of seasonal variation on automated land cover mapping from multispectral airborne laser scanning data. *International Journal of Remote Sensing, 40*(9), 3289–3307. https://doi.org/10.1080/01431161.2018.1528023.

Kerr, J. T., Southwood, T. R. E., & Cihlar, J. (2001). Remotely sensed habitat diversity predicts butterfly species richness and community similarity in Canada. *Proceedings of the National Academy of Sciences of the United States of America, 98*(20), 11365–11370. https://doi.org/10.1073/pnas.201398398.

Key, C. H., & Benson, N. C. (Eds.). (2005). *Landscape assessment: Remote sensing of severity, the normalized burn ratio* (p. LA1–LA51). Ogden: USDA Forest Service, Rocky Mountain Research Station.

Magnussen, S., Nord-Larsen, T., & Riis-Nielsen, T. (2018). Lidar supported estimators of wood volume and aboveground biomass from the Danish national forest inventory (2012–2016). *Remote Sensing of Environment, 211*, 146–153. https://doi.org/10.1016/j.rse.2018.04.015.

Makoto, K., Tani, H., & Kamata, N. (2013). High-resolution multispectral satellite image and a postfire ground survey reveal prefire beetle damage on snags in Southern Alaska. *Scandinavian Journal of Forest Research, 28*(6), 581–585. https://doi.org/10.1080/02827581.2013.793387.

Margolis, H. A., Nelson, R. F., Montesano, P. M., et al. (2015). Combining satellite lidar, airborne lidar, and ground plots to estimate the amount and distribution of aboveground biomass in the boreal forest of North America. *Canadian Journal of Forest Research, 45*(7), 838–855. https://doi.org/10.1139/cjfr-2015-0006.

McRoberts, R. E., Næsset, E., Gobakken, T., et al. (2015). Indirect and direct estimation of forest biomass change using forest inventory and airborne laser scanning data. *Remote Sensing of Environment, 164*, 36–42. https://doi.org/10.1016/j.rse.2015.02.018.

Meddens, A. J., Hicke, J. A., & Ferguson, C. A. (2012). Spatiotemporal patterns of observed bark beetle-caused tree mortality in British Columbia and the western United States. *Ecological Applications, 22*(7), 1876–1891. https://doi.org/10.1890/11-1785.1.

Mitchell, A. L., Rosenqvist, A., & Mora, B. (2017). Current remote sensing approaches to monitoring forest degradation in support of countries measurement, reporting and verification (MRV) systems for REDD. *Carbon Balance and Management, 12*(1), 9. https://doi.org/10.1186/s13021-017-0078-9.

Modzelewska, A., Fassnacht, F. E., & Stereńczak, K. (2020). Tree species identification within an extensive forest area with diverse management regimes using airborne hyperspectral data. *ITC Journal, 84*, 101960. https://doi.org/10.1016/j.jag.2019.101960.

Næsset, E. (2004). Practical large-scale forest stand inventory using a small-footprint airborne scanning laser. *Scandinavian Journal of Forest Research, 19*(2), 164–179. https://doi.org/10.1080/02827580310019257.

Neigh, C. S., Nelson, R. F., Ranson, K. J., et al. (2013). Taking stock of circumboreal forest carbon with ground measurements, airborne and spaceborne LiDAR. *Remote Sensing of Environment, 137*, 274–287. https://doi.org/10.1016/j.rse.2013.06.019.

Olson, D. M., Dinerstein, E., Wikramanayake, E.D., et al. (2001). Terrestrial ecoregions of the world: A new map of life on Earth. A new global map of terrestrial ecoregions provides an innovative tool for conserving biodiversity. *Bioscience, 51*(11), 933–938. https://doi.org/10.1641/0006-3568(2001)051[0933:TEOTWA]2.0.CO;2.

Popescu, S. C., Zhou, T., Nelson, R., et al. (2018). Photon counting LiDAR: An adaptive ground and canopy height retrieval algorithm for ICESat-2 data. *Remote Sensing of Environment, 208*, 154–170. https://doi.org/10.1016/j.rse.2018.02.019.

Powers, R. P., Coops, N. C., Morgan, J. L., et al. (2013). A remote sensing approach to biodiversity assessment and regionalization of the Canadian boreal forest. *Progress in Physical Geography, 37*(1), 36–62. https://doi.org/10.1177/0309133312457405.

Puliti, S., Hauglin, M., Breidenbach, J., et al. (2020). Modelling above-ground biomass stock over Norway using national forest inventory data with ArcticDEM and Sentinel-2 data. *Remote Sensing of Environment, 236*, 111501. https://doi.org/10.1016/j.rse.2019.111501.

Raffa, K. F., Aukema, B. H., Bentz, B. J., et al. (2008). Cross-scale drivers of natural disturbances prone to anthropogenic amplification: The dynamics of bark beetle eruptions. *BioScience, 58*(6), 501–517. https://doi.org/10.1641/B580607.

Rocchini, D., Hernández-Stefanoni, J. L., & He, K. S. (2015). Advancing species diversity estimate by remotely sensed proxies: A conceptual review. *Ecological Informatics, 25*, 22–28. https://doi.org/10.1016/j.ecoinf.2014.10.006.

Santoro, M., Beaudoin, A., Beer, C., et al. (2015). Forest growing stock volume of the northern hemisphere: Spatially explicit estimates for 2010 derived from Envisat ASAR. *Remote Sensing of Environment, 168*, 316–334. https://doi.org/10.1016/j.rse.2015.07.005.

Senf, C., Seidl, R., & Hostert, P. (2017). Remote sensing of forest insect disturbances: Current state and future directions. *International Journal of Applied Earth Observation and Geoinformation, 60*, 49–60. https://doi.org/10.1016/j.jag.2017.04.004.

Shendryk, I., Hellström, M., Klemedtsson, L., et al. (2014). Low-density LiDAR and optical imagery for biomass estimation over boreal forest in Sweden. *Forests, 5*(5), 992–1010. https://doi.org/10.3390/f5050992.

Skidmore, A. K., Pettorelli, N., Coops, N. C., et al. (2015). Environmental science: Agree on biodiversity metrics to track from space. *Nature, 523*(7561), 403–405. https://doi.org/10.1038/523403a.

Sonti, S. H. (2015). Application of geographic information system (GIS) in forest management. *Journal of Geography & Natural Disasters, 5*(3), 1000145. https://doi.org/10.4172/2167-0587.1000145.

Stephenson, N. L., Das, A. J., Condit, R., et al. (2014). Rate of tree carbon accumulation increases continuously with tree size. *Nature, 507*(7490), 90–93. https://doi.org/10.1038/nature12914.

Thurner, M., Beer, C., Santoro, M., et al. (2014). Carbon stock and density of northern boreal and temperate forests. *Global Ecology and Biogeography, 23*(3), 297–310. https://doi.org/10.1111/geb.12125.

White, J. C., Coops, N. C., Hilker, T., et al. (2007). Detecting mountain pine beetle red attack damage with EO-1 Hyperion moisture indices. *International Journal of Remote Sensing, 28*(10), 2111–2121. https://doi.org/10.1080/01431160600944028.

Whitman, E., Parisien, M. A., Thompson, D. K., et al. (2019). Short-interval wildfire and drought overwhelm boreal forest resilience. *Scientific Reports, 9*(1), 18796. https://doi.org/10.1038/s41598-019-55036-7.

Wulder, M. A., & Franklin, S. E. (Eds.). (2006). *Understanding forest disturbance and spatial pattern: Remote sensing and GIS approaches* (p. 268). Boca Raton: CRC Press.

Wulder, M. A., Loveland, T. R., Roy, D. P., et al. (2019). Current status of Landsat program, science, and applications. *Remote Sensing of Environment, 225*, 127–147. https://doi.org/10.1016/j.rse.2019.02.015.

Wulder, M. A., Hermosilla, T., White, J. C., et al. (2020). Biomass status and dynamics over Canada's forests: Disentangling disturbed area from associated aboveground biomass conse-quences. *Environmental Research Letters, 15*(9), 094093. https://doi.org/10.1088/1748-9326/ab8b11.

Zhao, K., Suarez, J. C., Garcia, M., et al. (2018). Utility of multitemporal lidar for forest and carbon monitoring: Tree growth, biomass dynamics, and carbon flux. *Remote Sensing of Environment, 204*, 883–897. https://doi.org/10.1016/j.rse.2017.09.007.

Chapter 27
Remote Sensing at Local Scales for Operational Forestry

Udayalakshmi Vepakomma, Denis Cormier, Linnea Hansson, and Bruce Talbot

Abstract The success of current and future forest management, particularly when dealing with triggered changes stemming from extreme climate change–induced events, will require prompt, timely, and reliable information obtained at local scales. Remote sensing platforms and sensors have been evolving, emerging, and converging with enabling technologies that can potentially have an enormous impact in providing reliable decision support and making forest operations more coherent with climate change mitigation and adaptation objectives.

27.1 Introduction

Forest operations are fundamental to the management needs specifically designed to respond to a trigger. These triggers are a planned sequence of events along the developmental stages of the stand that are set by the forest management plan during tactical or operational planning. Forest operations can also be a response

U. Vepakomma (✉) · D. Cormier
FPInnovations, 570 Boulevard St-Jean, Pointe-Claire, QC H9R 3J9, Canada
e-mail: udayalakshmi.vepakomma@fpinnovations.ca

D. Cormier
e-mail: denis.cormier@fpinnovations.ca

L. Hansson
Skogforsk, The Forestry Research Institute of Sweden, Uppsala Science Park, SE-751 83 Uppsala, Sweden
e-mail: linnea.hansson@skogforsk.se

B. Talbot
Faculty of AgriSciences, Department of Forest and Wood Science, Stellenbosch University, Private Bag X1, Matieland, Stellenbosch 7602, South Africa
e-mail: bruce@sun.ac.za

Division of Forest and Forest Resources, Norwegian Institute of Bioeconomy Research, Høgskoleveien 8, 1433 Ås, Norway

© The Author(s) 2023
M. M. Girona et al. (eds.), *Boreal Forests in the Face of Climate Change*,
Advances in Global Change Research 74,
https://doi.org/10.1007/978-3-031-15988-6_27

657

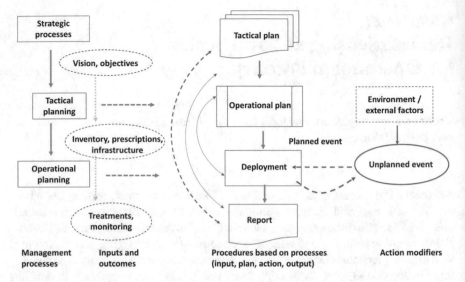

Fig. 27.1 Schematic of the general information flow (*gray arrows*), feedback loops (*blue arrows*), and feedback loops induced by a modifier (*dashed red arrows*)

to an unplanned change (unplanned trigger) that could alter the decision process and operation deployment, generating a feedback loop to the execution of the plan (Fig. 27.1).

Forest operations include timber harvests, fiber recovery, site preparation for suitable establishment (natural regeneration, seeding, or planting), thinning, pruning, timber stand improvement, competitive vegetation control, sanitization, and salvage (Fig. 27.2). They are designed to meet management needs (Fig. 27.1) on the basis of the targeted ecological response, technical applicability, and economic feasibility within compliance standards (Rummer, 2002). For example, harvesting within ecosystem-based management often prescribes the retention of legacy trees and the use of suitable techniques to avoid any damage to these trees. Whereas operations are a response to a planned trigger, they can also cause significant expected changes to the environment within a very short time; these changes also require tracking. For example, harvesting a matured stand will reset (change) the developmental process to its early-successional stages.

The effective implementation of sustainable forest management depends largely on carrying out sustainable forest operations (Marchi et al., 2018), which can prove to be more challenging in the context of climate change. The intensity and frequency of extreme climate events and severe insect outbreaks are predicted consequences of climate change and will alter the natural dynamics of the forests and drastically alter the local environment (Spittlehouse, 2005). For instance, operational deployment could be impeded by sudden flooding, early thawing, catastrophic tree damage, etc. The feedback loop to tactical planning in such situations happens rapidly and more frequently (Fig. 27.1).

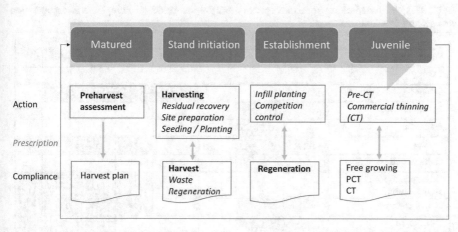

Fig. 27.2 Schematic showing various forest operations along the stand development stages (*dark gray boxes*), including the critical operations (*bold*)

The success and efficient deployment and the completion of any response or action depend on a prompt, timely, and reliable information feed at the planning, deployment, and operational stages. The status of vegetation (e.g., tree species or stem quality) and terrain (e.g., slope or ground-bearing capacity) features are critical information needs (Table 27.1). Their level of detail, intensity, and periodicity is defined by the complexity of the type of operation or the environmental conditions in which the operation must be completed along the stand developmental stages (Table 27.1). For example, harvesting a sustainably managed mixedwood stand growing mainly on complex terrain conditions requires safe access to the site and detailed information on the targeted species, e.g., stem quality. It is critical to properly identify the seed trees and create microsites that favor natural regeneration during operations. Hence, information needs tend to relate to planning, i.e., a priori, and during the actual operation, i.e., real time. The recentness of the acquired data is also important. Moreover, detail intensity increases from a homogeneous plantation to a heterogeneous natural stand. The level of detail for planning a harvest operation may be at the tree to stand level for vegetation, whereas accessibility (surface, slope, skid trails, landings, and wood catchment zones) is generally required at the block level (Table 27.2). However, during the harvest operation itself, the required details are instantaneous, repetitive, and intense within the operator's line of sight.

Traditionally, data used for planning purposes has been based mainly on a priori ground surveys (e.g., walk-throughs, cruising—a method to determine value of a specific area—or inventory of plot installations) or coarse interpreted images. Treatment execution is completed using visual assessments and compliance reporting with independent surveys. Recent innovations in remote sensing technology for rapidly gathering, processing, and accessing information have significantly modernized how forest operations are planned and conducted. This chapter documents current remote

Table 27.1 Required information for vegetation and the post-harvest to complete a forest operation along stand development stages

	Matured	Stand initiation			Establishment		Juvenile	
Vegetation features	PHS	Harv	FibRec	SitePrep	Sd/Plnt	Compet	PCT	CT
Canopy								
Competing vegetation				▦	▦	▦	▦	
Crown balance	▓	≡						≡
Retention		▦						
Species	▓	▦			▦	▦		▦
Stand structure	▓	≡						
Stem spacing/occupancy/voids					▦	▦		
Stem location	▓	‖			▦	▦	▦	▦
Tree height	▓	‖				▦	▦	▦
Vigor	▓					▦	▦	▦
Stem								
Stem quality	▓	▦						
Tree damage						▓	▓	
Tree form		‖						‖
Wood catchment/volume								
Post-harvest								
Bucking/log sort		‖						‖
Log scaling		▦	▦					▦
Residue distribution			▦	▦				
Residue geometry			▦					
Stump				▦				

Legend: ▓ Action / Report / Both; ≡ A priori; ‖ Real time; ▦ Both

PHS, preharvest survey; *Harv*, harvesting; *FibRec*, fiber recovery, i.e., the process of calculating the recovery rate, removing residual fiber, secondary use of fiber, piling, burning; *SitePrep*, site preparation; *Sd/Plnt*, seeding/planting; *Compet*, competition, i.e., weed control; *PCT*, precommercial thinning; *CT*, commercial thinning

sensing technologies suitable for understanding, monitoring, and mapping forest conditions at local scales to plan, perform, and report forest operations successfully.

Table 27.2   Required information for the terrain and derived features to complete a forest operation along stand development stages

	Matured	Stand Init			Establishment		Juvenile	
Terrain features	PHS	Harv	FibRec	SitePrep	Sd/Plnt	Compet	PCT	CT
Roughness	■	■						■
Ground bearing	■	■						■
Obstacle		■					■	■
Soil disturbance		■		■	■	■		
Slope	■	■						■
Skid trails		■	■	■	■	■		
Drainage	■	■				■		■
Derived features								
Accessibility		■	■	■	■	■	■	■
Safety		■	■	■	■	■	■	■
Cutblock boundary		■						
Hot spot				■				
Trafficability		■						■
Protected zones	■	■				■		
Microsite availability				■	■			
Legend								

Legend: Action   Report   Both
A priori |||||| Real time   ++++++ Both

PHS, preharvest survey; *Harv*, harvesting; *FibRec*, fiber recovery, i.e., the process of calculating the recovery rate, removing residual fiber, secondary use of fiber, piling, burning; *SitePrep*, site preparation; *Sd/Plnt*, seeding/planting; *Compet*, competition, i.e., weed control; *PCT*, precommercial thinning; *CT*, commercial thinning

27.2   Remote Sensing Platforms for Operational Forestry

Remote sensing is a platform-sensor combination (PSC) used to gather information about an object without being in physical contact with the object. PSC has the advantage of providing quick, synoptic, and repeated information over large and multiple spaces. The level of detail (coverage, resolution, timing, and frequency) varies with combinations of these various parameters (Table 26.2). Sensors are either passive (e.g., imaging/reflectance and thermal/radiation) or active (e.g., LiDAR or laser scanners and RaDAR or microwave scanners). The periodicity of the satellite data is fixed on the basis of constellations (daily to a few days; see Table 26.2), whereas all other acquisitions are programmed as per need (Tables 27.3 and 27.4).

Perception sensors help describe surface objects or perceive the environment. Positional or pose sensors (e.g., Global Navigation Satellite System–Global Positioning System, GNSS-GPS; Realtime Kinematics, RTK; Inertial Navigation System, INS; Inertial Measurement Unit, IMU; gyros and wheel encoders) determine the location and pose of the platform. A sensor platform refers to its carrier; these include

Table 27.3 Remote sensing platforms, their properties, and best uses for forestry applications

	Platform	Distance to object	Spatial resolution	Optimal temporal scale	Forest spatial scale under management	Sample applications
Mobile	Satellite (S)	700–35,000 km	Low to medium	15 days–twice daily	Landscape	Forest extent, biodiversity, drought, fragmentation, ecoclimatic zones, land-use class, disturbance regimes
	Aircraft/helicopter (A)	250 m to 12 km	Medium to high	>Daily	Stand–landscape	Species groups, roads/trails, drainage, terrain: elevation, slope, reconnaissance, disease, fire
	Drone (D)	<150 m	High to ultra-high	Near-real time	Tree–stand	Individual trees, tree species, biophysical estimates, hot spots, tree damage, subcanopy

(continued)

Table 27.3 (continued)

	Platform	Distance to object	Spatial resolution	Optimal temporal scale	Forest spatial scale under management	Sample applications
Horizontal range	Machine (M)	<30 m	Ultra-high	Real time	Tree–subtree	Branching, tree form, occlusions, obstacles, regeneration, soil bearing capacity, soil disturbance, machine pose
	Human (H)	<30 m	Ultra-high	Real time	Tree–subtree	Branching, tree form, occlusions, obstacles, regeneration, soil bearing capacity, soil disturbance, bi-pied pose
	Stationary (T)	<30 m	Ultra-high	Real time	Tree–subtree	Branching, tree form, occlusions, obstacles, regeneration, soil bearing capacity

Table 27.4 Sensors and their platforms, vegetation analyzed with the sensors, and examples of use for vegetation and terrain analyses. Abbreviations for available platforms are given in Table 27.3

Suitable for forest operations–feature extraction

		Sensors	Available platform	Vegetation	Examples	Terrain	Examples
Perception	Imaging	RGB	All	Species/species groups, individual stem and location,	Puliti et al. (2015), Natesan et al. (2020)	Slope, drainage, skid trails	
		MSS	All	stem spacing, occupancy, voids, tree height, vigor, retention, competing vegetation, residues:	Coops et al. (2007), Pouliot et al. (2019), Vepakomma et al. (2021)		
		Hyper	S, A, D, T	location, competing vegetation, residues:	Modzelewska et al. (2020), Fassnacht et al. (2014)		
		RGB-D	D, M, H, T	location, geometry, stump	Chandail and Vepakomma (2020), Li and Vepakomma (2020)		Murphy et al. (2008), Ågren et al. (2014)
	Radiation	Thermal Infrared	All	Hot spot: pile burn, disease, stem deterioration	Lagouarde et al. (2000)		
	Ranging	LiDAR	All	Stem quality, tree damage, tree form, volume, stand structure, stem and location, species, log scaling	Næsset (2007), White et al. (2013), Sibona et al. (2017), Holmgren et al. (2019), Vepakomma and Cormier (2017)	Roughness, obstacle, ground bearing, soil disturbance,	Ring et al. (2020), Rönnqvist et al. (2020), Talbot & Rahif, (2017), Chhatkuli et al. (2012)
		RaDAR	S, A	Species group, stand structure	Andersen et al. (2008), Tighe et al. (2009)	Trafficability, bearing capacity, water depth	

Fig. 27.3 Comparison and coregistration of LiDAR point clouds of a coniferous stand as captured from various platforms. *ALS*, aerial LiDAR; *ULS*, UAV-based LiDAR; *TLS*, terrestrial LiDAR)

mobile platforms, such as satellites, airborne platforms (aircraft and drones, also known as *unmanned aerial vehicles* or UAV), manned or unmanned ground vehicles, and human or stationary platforms, such as towers and tripods. Sensors are also sometimes distinguished by the platform they carry; for example, Liang et al. (2015) classified laser scanners as being either airborne (ALS), terrestrial (TLS), mobile (MLS), or personal (PLS). Platforms above the forest canopy can provide a synoptic view over large contiguous areas to provide a top-to-bottom description. In contrast, platforms in proximity, below the canopy, or closer to the ground provide vertical stem information and a detailed terrain description that is not feasible or possible from above canopy platform systems (Fig. 27.3). For example, Kankare et al. (2014) demonstrated that TLS produces preharvest tree- and stand-level bucking details at a greater degree of accuracy than conventional means. Such data can help estimate the stumpage value of a stand or more suitable wood assortments.

Optical sensors capture the reflectance from materials within, e.g., standard digital RGB camera, and beyond the visible spectrum, e.g., infrared, whereas thermal sensors capture radiation from materials. Multispectral sensors (MSS) capture reflectance in limited or broad spectral regions (bands), and hyperspectral sensors have narrower but multiple bands. The spatial resolution of the image represents the ground sampling distance (GSD), which varies on the basis of the focal length of the sensor, the altitude at which the sensor is placed, and the speed with which the platform moves (Table 26.2). For instance, depending on the sensor platform, GSD may vary from a subcentimeter (e.g., drone), to submeter (e.g., WorldView series) to kilometer (e.g., AVHRR) scale. GSD is important in determining the spatial resolvability (*mappability*) of the feature on the image. Typically, assuming a reasonable contrast of the target feature from its background, more than 3 cm GSD is recommended for manually discerning trees as small as 0.4 m in height with a 35 cm crown diameter or form on an image (Pitt et al., 1997).

LiDAR (light detection and ranging) and RaDAR (radio detection and ranging) are active ranging sensors. RaDAR transmits microwave radio signals, whereas LiDAR transmits infrared energy. Both emit pulses that can penetrate through smoke, cloud, and small openings in tree canopies to reach the forest floor as well as measure the reflected backscatter. The range is converted to distance to provide precise locations (x, y, z) of the point of interaction with an object in space, and these sensors are best suited to describing the structure of an object, e.g., crown shape or tree height. Available radar systems provide a spatial resolution larger than 1 m and are better suited for large-scale mapping relevant to strategic or tactical forest management.

In terms of the capabilities of data recording, LiDAR systems can be fullwave (complete distribution of intercepted and returned laser pulse along the pathway) or discrete return (few observations are recorded from a laser pulse that is intercepted and reflected from targets). As they record the entire pathway along with the additional attributes of amplitude and intensity, fullwave systems are better suited for detailed above- and below-canopy characterization. Fullwave recording requires large-scale data management and algorithms, and this approach still remains at the experimental stage; however, discrete LiDAR is currently in operational use (Crespo-Peremarch et al., 2020). LiDAR systems are also differentiated by laser footprint size. A small footprint (less than one meter) on the ground provides a good link between the LiDAR beam and the structural vegetation attributes that are subtle among or within individual trees. By segregating the returns, e.g., vegetation versus ground, the points can be interpolated to describe continuous object (*digital surface model, DSM*) and terrain (*digital terrain model, DTM*) surfaces. Their arithmetic differences represent the aboveground surfaces, e.g., *canopy height model (CHM)*. Point clouds, as well as surface models, are used to extract features. Point density and the power of the laser signal to penetrate through the canopy define feature resolvability and the estimated dimensions. Because imaging sensors receive the resulting light reaction from a particular surface, they tend to be best suited for understanding floristic compositional/structural characteristics related to the object, e.g., species, vigor, canopy cover, and density. Imagery is a 2D raster, and LiDAR is 3D point data or vertical profile; however, when images are gathered either as a stereo or overlapping sequence, they can provide photogrammetric 3D data useful for describing the structure of objects, such as canopy structure. Table 26.2 highlights the estimable direct/indirect features relevant to forestry on commonly available platforms.

The selection of PSC for a forest operation depends on the spatial extent and patterns of the area of interest, the timing, the recentness of the acquired information, and the repetitiveness between triggers for the required monitoring/reporting, specifically for vegetation status. Preharvest surveys should be within a year of the operation, whereas site preparation for competition control is conducted within a month, and regeneration surveys are two to five years after stand initiation (Table 26.1). The availability of certain RS platforms, such as satellite or aerial platforms, may be limited. Similarly, the phenology of target vegetation is an important consideration, as coniferous crops remain distinctly visible during the early spring or late fall, whereas deciduous vegetation is transparent during these periods.

Measurements with static terrestrial platforms provide a single or a small number of viewpoints and hence are limited to the validation or calibration of models built on higher platforms (Liang et al., 2018). On the other hand, mobile systems have the potential to provide near-real-time data and detailed below-canopy data relevant for operational decisions (Holmgren et al., 2019).

27.2.1 Positioning and Tracking Systems

Precise positioning in space and navigation is essential for safe and effective target action through localization (e.g., fire, salvage, or herbicide sprays) or tracking activity data (e.g., machine movements or harvest operations) for assessing efficiency, quality, and productivity (Keefe et al., 2019). The most commonly used positioning tools are *global positioning systems* (GPS) operating via a constellation of satellites, 24 as in GNSS. A GPS receiver can provide latitude, longitude, elevation, and the vector heading to monitor one's location on topographic/thematic maps or imagery (Picchio et al., 2019). This information is generally integrated with a geographic information system (GIS) and can be visualized by the operator. Differential GPS or RTK systems can improve locational accuracy. Given the poor precision in forested environments because of canopies blocking satellite signals, additional sensors like INS can be used to estimate relative position and orientation of a mobile vehicle, e.g., an operating forestry machine. The heart of INS is the IMU—a combination of gyroscopes, accelerometers, and magnetic sensors used for determining translational and rotational velocity to provide a navigational solution. More recently, simultaneous localization and mapping (SLAM), a technique popular with autonomous systems, has also been tested for use in forestry (Chandail & Vepakomma, 2020; Tang et al., 2015). SLAM involves creating a map for an unknown environment while simultaneously determining the agent's location using a laser or RGB-D camera to estimate depth in combination with other location sensors, such as GPS and IMU.

27.3 Remote Sensing–Based Feature Extraction for Forest Operations

In the context of a forest operation deployment sequence, we can essentially discuss remote sensing technologies as those (1) providing information on the forest environment for operational planning, monitoring, or assessing the effectiveness of an operation and/or reporting compliance; (2) gathering environmental information during the operations; and (3) relating to the operations themselves. Essentially, planning is a priori information that has a recentness from the day to a few months previous and helps determine the selection and use of machine systems. Information used for compliance or when monitoring effectiveness following a treatment must also be recent, whereas data needs are real time to near–real time for the deployment of actual operations. Above-canopy platforms are more suitable for planning and monitoring, especially in contiguous spaces, whereas close-range, terrestrial, and mobile platforms are most suitable for real-time operations. The following subsections are organized to understand how remote sensing, especially using platforms closer to the canopies, can be used for information feed, particularly in relation to vegetation and the underlying terrain, as highlighted in Table 27.1, along the sequence of a forest operation deployment. We provide, where possible, examples of different applications.

27.3.1   Vegetation Features

During the stand development stages, many prescriptions call for vegetation changes to the established stands, e.g., harvesting of crop trees, regeneration cuttings in shelterwood or group selection systems, thinning, sanitation removals of diseased or infested trees, or the spraying of herbicides on competing shrubs that affect crop tree growth. These prescriptions require accurate data at the subtree, tree, or, at the least, microstand level for efficient and effective management. This relevant data includes assessing tree height, form, quality, vigor, and species, as well as the tree's surrounding environment, e.g., stocking, growing space, species mix, and competition.

27.3.1.1   Pretreatment Assessment

Given its ability to reconstruct 3D forest structures and reliably estimate several biophysical parameters describing within- and below-canopy structure and function, LiDAR has become an essential component of operational forest inventories in numerous countries (Maltamo et al., 2021; Næsset, 2007; White et al., 2016). Two main approaches for LiDAR have been developed: an *area-based* and an *individual-tree* approach. The former is aimed at large-scale assessments that have a coarse point density effective for producing a stand portrait. As the name suggests, the individual-tree approach relies on identifying and delineating trees, including species identification, direct estimation of height and crown parameters, modeled diameter at breast height, basal area, and volume. The area-based approach (ABA) is a model-based estimate in which canopy descriptors or metrics are predicted on the basis of regression or discriminant analysis using accurate in situ plot data and height distribution (quantiles, percentiles, etc.) of LiDAR beam reflection (White et al., 2013). This method has demonstrated an accuracy of 4–8% for stem height, 6–12% for mean stem diameter, 9–12% for basal area, 17–22% for stem density, and 11–14% for volume estimates in boreal forest studies attempting to capture within-stand variability (Holmgren, 2004; Maltamo et al., 2010; Næsset, 2007; Sibona et al., 2017; White et al., 2013). In the absence of tree-level information, this stand or microstand level of characterization has been applied in eastern Canada to aid silvicultural prescriptions, such as commercial thinning or salvaging (Lussier & Meek, 2014; Meek & Lussier, 2008). Integrating vigor information with LiDAR canopy stratification helped machine operators improve productivity by 4% (Fig. 27.4; Gaudreau & Lirette, 2020). Area-based estimates using digital aerial photogrammetry collected across a range of boreal forest types is comparable with that obtained via aerial LiDAR (Goodbody et al., 2019). McRoberts et al. (2018) and Fekety et al. (2015) note, however, some challenges of using ABA models in relation to their shelf life and temporal transferability.

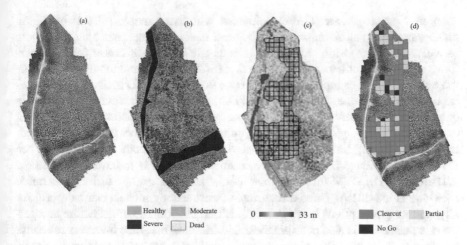

Fig. 27.4 LiDAR-based stratification combined with image-based vigor for silvicultural prescription and operator assistance **a** ortho image, **b** vigor class, **c** canopy height model, and **d** logging map

The extended history of aerial and satellite platforms carrying optical sensors, more recently combined with LiDAR, has produced a large body of work demonstrating the successful implementation of remote sensing to studies of canopy vegetation (Cerrejón et al., 2021). Multiple approaches exist for quantifying and estimating the structural and compositional parameters of interest and spatially mapping these parameters at various spatial scales. Generally, very-high-resolution imagery in 2D, stereoscopic, or overlapping imagery in 3D is visually interpreted based on the calibration of a series of field plots combined with guidelines related to the vegetation in terms of foliage color, texture, crown shape, and branching structure (Corbane et al., 2015). Semi- or fully automated workflows can be summarized as segmenting the image into homogeneous objects (a tree, a collection of trees, or a stand) and then (1) estimating directly the structural or compositional parameters of interest or (2) estimating these parameters indirectly through proxy variables. Segmentation, in particular individual tree crowns (ITC), is 2D raster-based (either multispectral images, grayscale images, or CHM) and 3D point clouds (photogrammetric or LiDAR).

Separating vegetation from its background and assuming the brightest pixel to the highest point of the foliage on high-resolution 2D images (similar to raster-based CHM models), rule-based semi- or fully automated approaches can then extract tree crowns. Accuracy varies with GSD and by partitioning images into homogeneous forest stands; for instance, an accuracy of 60% (70 cm resolution) to 89% (31 cm) has been estimated for open coniferous to more complex mixedwood boreal forests, respectively (Katoh & Gougeon, 2012; Leckie et al., 2005). The number of trees per species depends on ITC accuracy, which improves when understory species are eliminated. Two-dimensional models may help with segmentation when estimating species density, although the structural assessment of canopies also requires determining canopy height.

Digital aerial photos (DAPs) combined with stereoscopic (visual) or digital photogrammetry can reconstruct a 3D forest canopy. Image matching and, more recently, computer-vision techniques such as SIFT (*scale-invariant feature transform*) combined with *structure from motion* (SfM) are very commonly used to estimate the 3D forest canopy from sequences of overlapping 2D images, e.g., images captured from a drone. If an accurate DTM is derivable, which can be difficult in complex, mature stands, DAPs can estimate the structural variables of the uppermost canopy, e.g., height, basal area, volume, quite accurately, comparable with the accuracy obtained using aerial LiDAR (Baltsavias, 1999; Goodbody et al., 2019). When DTM from an image is not derivable, a simple solution is to have a coregistered LiDAR or SRTM DTM for DAP point normalization or canopy surface generation (St-Onge et al., 2015). Three-dimensional forest canopy models can be useful for silvicultural prescriptions when data acquisition is optimally timed before an operation is planned. It is also possible that rapid and near-real-time inventory measurements, e.g., canopy cover, based on ocular estimates are made with an improved precision using nadir—the sensor looking vertically downward—images from a drone. UAV-SfM estimates of several inventory variables are comparable to those of LiDAR in terms of root mean square error for dominant height (3.5%), Lorey's height (13.3–14.4%), stem density (38.6%), basal area (15.4–23.9%), and timber volume (14.9–26.1%) (Puliti et al., 2015; Tuominen et al., 2015).

Although raster-based ITC approaches can segment most of the top canopy, potential segmentation within the multilayered vertical structure of the canopy to capture subcanopy elements—especially using LiDAR echoes from above-canopy platforms—is possible through point-based clustering. Hamraz et al. (2016) obtained >94% detection rate for dominant and codominant trees in complex stands. Because of the ultra-high-density data in current LiDAR systems, there is also a greater possibility of extending techniques to direct and nondestructive estimates of a suite of stem-quality determinants with a high level of accuracy, including estimates of crown base, clear stem, stem taper, stem straightness, and branchiness (Vepakomma & Cormier, 2017, 2019). This offers great potential in the more refined selection of trees on the basis of target mill product specifications and automated bucking, where each tree can be analyzed at the stump to optimize its market value (Fig. 27.5).

Distinct tree architecture and branching patterns can be observed from high density LiDAR (Fig. 27.5). A 77.8% accuracy has been achieved in distinguishing predominantly boreal tree species by correlating estimated LiDAR features to vertical and horizontal foliage patterns (Li & Hu, 2012). Use of the textural or spectral intensity of multiwavelength LiDAR improved the accuracy (Budei et al., 2018). However, given the easy availability of optical images, spectral-based species discrimination is the most suitable, rapid, and pragmatic means of mapping large forest spaces.

Accuracy in tree species classification has improved through a priori crown extraction (Dalponte et al., 2014), although Heinzel and Koch (2011) found pixel-based classification improved the undersegmentation of crowns. While very-high spatial and super-spectral-resolution images such as the Worldview series are promising, the automated discrimination of more than ten tree species has been achieved with 82% accuracy using high-spatial-resolution MSS data (Immitzer et al., 2019). Accuracy

Fig. 27.5 Estimating wood-quality determinants using ultra-high-density LiDAR. Modified from Vepakomma and Cormier (2019), CC BY 4.0 license

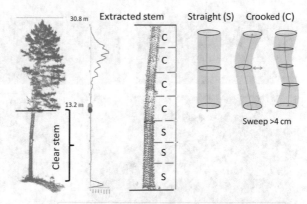

	Along the clear stem	Along the crown		
	Consecutive 2 m straight Sections*	Number of 2.44 m logs		
TREEID	S, straight; C, crooked, N, no data available	HIGH	MED	LOW
1667	C,S,S,S,N,N,S,C,S	4	0	0
1958	S,C,C,C,C,C,C,C,C,S,C,C,S	2	0	0
297	S,C,N	3	1	0
906	C,C,S,S,N	4	2	0
2603	S,N,S,S,S,C,C,S,N	4	0	0
1804	C,S,S,N,S,S,S,S,C,C,S,C,N	1	0	0
1742	S,S,S,S,S,S,S,C,C,C,S	3	1	0
817	C,S,S,S,S,S,S,C,C,N	3	0	1
1162	S,C,S,C,S,C,S,N	2	0	0
1141	C,C,S,N	3	0	0
1590	C,C,C,C,N,N,N,N,N,N,N,N,N	1	0	0

*Example: In tree 1742, a consecutive S,S,S,S,S,S of 2 m logs results in a 12 m straight section of the stem

improved greatly when models were adapted to narrowband hyperspectral sensors (Fassnacht et al., 2014; Modzelewska et al., 2020). Hyperspectral data, nevertheless, is data- and process-intensive and is restricted to being the most successful when collected in bright light conditions. Some researchers have found that sensor-fusion approaches, such as MSS or hyperspectral data with LiDAR, have improved species discriminability in boreal regions (Dalponte et al., 2014; Trier et al., 2018). These models identified as many as 19 species at 87% accuracy. Because temporal variability is a critical factor for species discrimination and there is an existing insufficiency of training samples, drone-based solutions can serve to map at local scales and develop a reference database (Fassnacht et al., 2014; Natesan et al., 2020). After iteratively building tree libraries from drone-based simple RGB images acquired in variable light-season-year conditions, Natesan et al. (2020) discriminated five conifer species at 73–91% accuracy (Fig. 27.6) and adaptively improved this library to identify six more deciduous taxa at over 79% accuracy in boreal regions.

An indicator of forest health is a forest's resistance and resilience to disturbances and its ability to adapt to climate change over the long term. Altered structure, functioning, or taxonomy because of the physiological stress of resource limitation, disease, or disturbances occur at all spatial (vertical and horizontal) and temporal

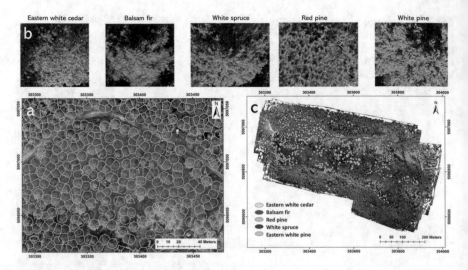

Fig. 27.6 Tree species recognition using an extensive image library from UAV-based RGB images; **a** automated crown delineation, **b** extracted crowns for training, **c** softwood species classification. Modified with permission from Canadian Science Publishing, permission conveyed through Copyright Clearance Center, Inc., from Natesan et al. (2020)

scales; nevertheless, capturing early signs can help minimize disturbance-related damage. Using the concepts of spectral traits and their variability (direct or proxy variables of forest health), Lausch et al. (2013) conducted an extensive review of the best PSC and available techniques for quantifying or qualifying short- to long-term monitoring of vigor. Close-range sensing improves precision or calibrates spectral responses of stress or disturbance in airborne remote sensing (Fassnacht et al., 2014). Slight declines in chlorophyll or moisture levels (identifiable through hyperspectral sensing) have helped provide early warnings of bark beetle (Fassnacht et al., 2014; Safonova et al., 2019) and herbivorous insect (Cardil et al., 2017; Meng et al., 2018; Vepakomma et al., 2021) infestations. There has been some success in identifying isolated impacted trees to group mortality (Fassnacht et al., 2014; Sylvain et al., 2019) and distinguishing the effects of multiple disturbances, e.g., pine blister rust and mountain pine beetle (Coops et al., 2003; Hatala et al., 2010).

27.3.1.2 During Treatment

Planning and executing forest operations is as much about following best practices as it is avoiding changing or damaging cultural remnants and special biotopes or transgressing property borders. LiDAR has been used to detect cultural heritage sites (Risbøl et al., 2014) and map habitat characteristics with the possibility of earmarking areas that must be avoided (Evju & Sverdrup-Thygeson, 2016). Proximal scanning also shows a strong potential in providing decision support during operations. For example, the rapid detection and estimates of stand density, tree position, and stem diameter (Holmgren et al., 2019) has helped with thinning tree selection and allowed data to be collected on individual tree selection by harvester operators through modeling of which tree the operator might select a priori (Brunner & Gizachew, 2014) or during operations (Gaudreau & Lirette, 2020).

27.3.1.3 Post-operation Monitoring

In most jurisdictions, standard practice involves compliance of contractual or regulatory frameworks and a post-operation follow-up; for example, these can entail assessment of post-harvest renewal or establishment monitoring to ensure sustainable production. The desired management objectives are typically to control stocking, species composition, survival, and growth. The distinct conical shape of conifer seedlings allows their easy detection for both planted trees and natural irregularly spaced stems. Vepakomma et al. (2015) distinguished conifer seedlings at least 0.3 m in height and estimated their size with a low average bias of 0.02 m through simple RGB images obtained from a drone. The data formed a basis for evaluating stocking, growing space, and regeneration gaps. By distinguishing competitive species, the models were further extended to qualify free-growing trees and assess regeneration compliance (Fig. 27.7). Pouliot et al. (2002) found that although the automated detection of six-year-old planted conifers was significantly high (at 91%), crown size extraction was sensitive to pixel resolution. In their case, they noted an 18% error compared with field assessments.

27.3.2 Terrain Features

The cost, efficiency, and potential environmental impact of forest operations all depend greatly on terrain features, e.g., surface roughness, slope, obstacles, and hydrographic data (flow channels, slope, drainage, and wet areas). These features can be described at macro-, meso-, and microlevels. DTM at corresponding resolutions are derivable using RaDAR interferometry, e.g., inSAR from satellite, available SRTM (Shuttle RaDAR Topographic Mission) data, LiDAR, or images using photogrammetric techniques where the ground is visible (Talbot & Rahif, 2017). Given the current technologies, LiDAR has proven to be the best available and most accurate tool for terrain assessments of mature stands. However, ALS with coarse data density can still provide a resolution greater than what planning methods can actually use (Talbot & Rahif, 2017).

Knowledge of terrain surface roughness and the number of potential hazards, especially under a dense canopy or on steep slopes, is critical for operational safety. Full-waveform LiDAR has a higher chance of returns from dense terrain and enables the successful detection of hazards, such as protruding rocks over 2 m wide (Chhatkuli

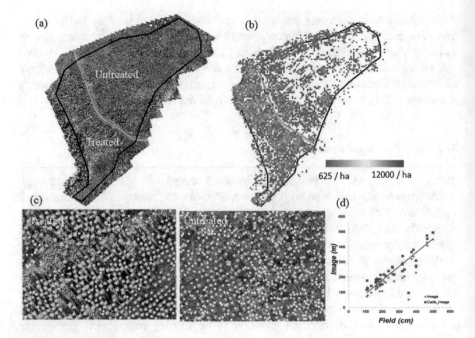

Fig. 27.7 Automated coniferous regeneration assessment for UAV-based RGB images for reporting compliance; **a** orthorectified image, **b** species group map, **c** species-wise detected individual stems, **d** stem height calibration model

et al., 2012). Slope and hydrographic features are directly derivable from a DTM, and model-based indices are used to identify wet areas and surface roughness (Ågren et al., 2014; Murphy et al., 2008). Identifying surface features that could be potential hazards helps ensure the safe driving of machinery at a harvest site (Fig. 27.8; Li and Vepakomma, 2020). Wet-area maps, which are characterized by indices such as cartographic depth-to-water (DTW) or the topographic wetness index (TWI) help to assess soil, vegetation, and drainage type and are used by the machine operators during forest operations to avoid or mitigate site damage (Ring et al., 2020). Such maps constitute a considerable improvement in recent forest management data and the planning of forest operations (Talbot & Astrup, 2021).

27.3.2.1 Pretreatment Assessment

Harvesting planning. Forest operations alter the environment, which, most often, is desired and intended. Undesirable impacts occur in particular when moving material or equipment into the forest (Rummer, 2002). In practice, machine operators in Europe and the Americas use tree cover and ground information as a canvas to plan harvesting or the moving of equipment. Providing an automated feature extraction

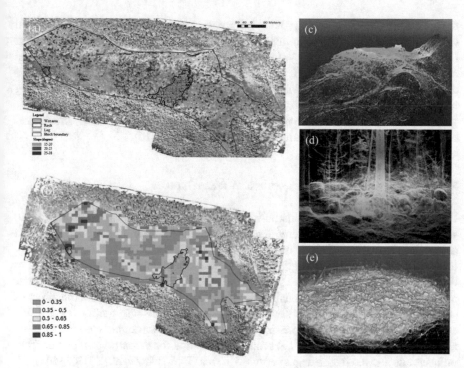

Fig. 27.8 Hazard detection and model-based drivability map using ultra-high-density LiDAR data from a UAV; **a** detected hazards overlying an orthorectified RGB image, **b** modeled drivability index showing a gradient of drivable (*green*) to no-go areas (*red*), **c** digital elevation model, **d** rock outcrops captured by mobile LiDAR, **e** extracted stumps. Modified with permission from Li and Vepakomma (2020)

from remote sensing as part of harvest planning can significantly minimize detrimental factors. These features can then be fed into algorithms to identify potential landings adjacent to roads or aid route optimization (Flisberg et al., 2021). In ground-based harvesting, the skid trail layout should be adapted to both the topography and the soil bearing capacity. Rönnqvist et al. (2020) combined digital elevation models, depth-to-water maps, and LiDAR-based tree volumes to spatially optimize extraction routes. In steep terrain, identifying suitable load paths for cable yarding and maximizing the use of each yarder setup is essential for optimizing economic performance. Detection of suitable end trees (tail spars) and intermediate support trees and discerning actual terrain form between contour lines—previously carried out by manual profile surveys—can now be easily replaced by LiDAR assessments (Dupire et al., 2015; Søvde, 2015).

Roads and transport. Monitoring forest road conditions includes gathering information on road geometries, surface conditions, condition of the drainage system, the presence of vegetation, and seasonal damage (Talbot & Rahif, 2017). Regular geometric shapes such as roads are easily discernable on images and high-resolution

LiDAR, which can be used to provide information on widths, curve geometries, and slope (White et al., 2010). Similar to the in-field driving applications, Waga et al. (2020) used LiDAR-derived TWI models to predict road quality. They obtained an accuracy of up to 70 and 86% when the models were combined with other variables, such as surface quality index and soil type. Surface quality factors, including roughness, gradient, and camber, can also be recorded from a vehicle using a profilograph and then entered into the model to determine the effect of these factors on fuel consumption during timber hauling (Svenson & Fjeld, 2016).

27.3.2.2 Post-treatment Assessment of Disturbances

Harvest compliance in many jurisdictions includes minimizing rutting and damage to soils (Talbot & Rahif, 2017). Remote sensing can help locate and characterize ruts and assess the level of soil disturbance caused by an operation (Pierzchała et al., 2016). Haas et al. (2016) used photogrammetry to quantify variations in rutting related to tires of differing dimensions and the use of steel bands on forwarders. In a similar analysis, Marra et al. (2018) considered differences in tire pressure and the effect of several forwarder passes on rut development. Although this information is helpful at an individual rut level, remote sensing can also help locate and characterize ruts at the site level. Nevalainen et al. (2017) proposed a method for measuring rut depths from point clouds derived from images captured from a UAV. Talbot et al. (2018) used UAV-based orthomosaics to determine the extent and severity of rutting at a stand level and developed a method to reduce the need for field sampling in assessing site impacts. However, photogrammetry-based solutions have their limitations; for example, light and weather conditions can affect accuracy. Moreover, although surface models can be generated, occlusion greatly limits information related to site conditions for sites under a partial canopy or under brush mats on the ground.

27.4 Remote Sensing–Enabling Autonomy

The automation of the remote sensing information feed for active decision support and adaptive forest management is very close to reality. Embedded sensors that were used to remotely monitor hazardous or inaccessible environments (e.g., nuclear reactors or rail tracks) are being applied to the proximal monitoring of machine movements and perception of surrounding forest environments (Holmgren et al., 2019). SLAM, through onboard sensors such as 2D LiDAR scanners and stereocameras, has demonstrated its potential in estimating machine pose with reference to its complex unstructured forest surroundings either in combination with GPS (Pierzchała et al., 2018; Tang et al., 2015) or using only visual odometry in GPS-denied environments (Chandail & Vepakomma, 2020).

The last decade has seen a gradual paradigm shift toward developing "intelligent" machines converging with sensing systems, thereby moving from automation to autonomously navigating and negotiating different entities. In the aerial sector, miniaturized vision systems and artificial intelligence (AI) combined with remotely piloted systems or fully autonomous UAV swarms can now monitor and provide real-time situational awareness. For example, Hummingbird drones mounted with infrared-sensing instruments and AI are now used for fire monitoring (www.hum mingbirddrones.ca). There is a movement away from man-heavy to man-light operations in manufacturing, agriculture, and mining sectors focusing on improving productivity or safety under challenging conditions. Although the forestry sector mandates environmentally friendly systems, the harsh diverse forest environment and obstacle-ridden forest floor may tax the limits and the reliability of all types of instruments (Billingsley et al., 2008). Although challenging to implement, the automation and autonomizing of future forestry is the focus of considerable research through programs, such as Forestry 4.0 (Canada, https://web.fpinnovations.ca/for est-operations-solutions-to-help-the-canadian-forest-industry/forestry-4-0/, https://www.youtube.com/watch?v=r4vhLQ8OEP0) or Auto2 (Sweden, Gelin et al., 2021), and the application of such programs to forestry issues (e.g., forest fire management, Sahal et al., 2021).

The success of any current or future forest management, particularly when dealing with triggered changes from extreme climate change–induced events, will require a prompt, timely, and reliable information feed. Remote sensing has been evolving, emerging, and converging with enabling technologies and offers reliable decision support and can ensure safer forest operations.

References

Ågren, A. M., Lidberg, W., Strömgren, M., et al. (2014). Evaluating digital terrain indices for soil wetness mapping—A Swedish case study. *Hydrology and Earth System Sciences, 18,* 3623–3634. https://doi.org/10.5194/hess-18-3623-2014.

Andersen, H. E., McGaughey, R. J., & Reutebuch, S. E. (2008). Assessing the influence of flight parameters, interferometric processing, slope and canopy density on the accuracy of X-band IFSAR-derived forest canopy height models. *International Journal of Remote Sensing, 29,* 1495–1510. https://doi.org/10.1080/01431160701736430.

Baltsavias, E. P. (1999). A comparison between photogrammetry and laser scanning. *ISPRS Journal of Photogrammetry and Remote Sensing, 54,* 83–94. https://doi.org/10.1016/S0924-2716(99)000 14-3.

Billingsley, J., Visala, A., & Dunn, M. (2008). Robotics in agriculture and forestry. In B. Siciliano & O. Khatib (Eds.), *Springer handbook of robotics* (pp. 1065–1077). Berlin, Heidelberg: Springer.

Brunner, A., & Gizachew, B. (2014). Rapid detection of stand density, tree positions, and tree diameter with a 2D terrestrial laser scanner. *European Journal of Forest Research, 133*(5), 819–831. https://doi.org/10.1007/s10342-014-0799-1.

Budei, B. C., St-Onge, B., Hopkinson, C., et al. (2018). Identifying the genus or species of individual trees using a three-wavelength airborne LiDAR system. *Remote Sensing of Environment, 204,* 632–647. https://doi.org/10.1016/j.rse.2017.09.037.

Cardil, A., Vepakomma, U., & Brotons, L. (2017). Assessing pine processionary moth defoliation using unmanned aerial systems. *Forests, 8*(10), 402. https://doi.org/10.3390/f8100402.

Cerrejón, C., Valeria, O., Marchand, P., et al. (2021). No place to hide: Rare plant detection through remote sensing. *Diversity and Distributions, 27*, 948–961. https://doi.org/10.1111/ddi.13244.

Chandail, R., & Vepakomma, U. (2020). *Enhanced COGNIMO: an integrated system for real-time information update and navigation, working towards robust localisation* (p. 36). Technical Report TR 2020, No. 65. Pointe-Claire: FPInnovations.

Chhatkuli, S., Mano, K., & Kogure, T., et al. (2012). Full waveform LiDAR and its evaluation in the mixed forest hilly region. *ISPRS International Archives of the Photogrammetry, Remote Sensing and Spatial Information Sciences, XXXIX*, Part B7, 505–509.

Coops, N., Stanford, M., Old, K., et al. (2003). Assessment of dothistroma needle blight of *Pinus radiata* using airborne hyperspectral imagery. *Phytopathology, 93*, 1524–1532. https://doi.org/10.1094/PHYTO.2003.93.12.1524.

Coops, N. C., Hilker, T., Wulder, M. A., et al. (2007). Estimating canopy structure of Douglas-fir forest stands from discrete-return LiDAR. *Trees, 21*(3), 295–310. https://doi.org/10.1007/s00468-006-0119-6.

Corbane, C., Lang, S., Pipkins, K., et al. (2015). Remote sensing for mapping natural habitats and their conservation status—New opportunities and challenges. *ITC Journal, 37*, 7–16. https://doi.org/10.1016/j.jag.2014.11.005.

Crespo-Peremarch, P., Fournier, R. A., Nguyen, V.-T., et al. (2020). A comparative assessment of the vertical distribution of forest components using full-waveform airborne, discrete airborne and discrete terrestrial laser scanning data. *Forest Ecology and Management, 473*, 118268. https://doi.org/10.1016/j.foreco.2020.118268.

Dalponte, M., Ørka, H. O., Ene, L. T., et al. (2014). Tree crown delineation and tree species classification in boreal forests using hyperspectral and ALS data. *Remote Sensing of Environment, 140*, 306–317. https://doi.org/10.1016/j.rse.2013.09.006.

Dupire, S., Bourrier, F., & Berger, F. (2015). Predicting load path and tensile forces during cable yarding operations on steep terrain. *Journal of Forest Research, 21*, 1–14. https://doi.org/10.1007/s10310-015-0503-4.

Evju, M., & Sverdrup-Thygeson, A. (2016). Spatial configuration matters: A test of the habitat amount hypothesis for plants in calcareous grasslands. *Landscape Ecology, 31*, 1891–1902. https://doi.org/10.1007/s10980-016-0405-7.

Fassnacht, F. E., Latifi, H., Ghosh, A., et al. (2014). Assessing the potential of hyperspectral imagery to map bark beetle-induced tree mortality. *Remote Sensing of Environment, 140*, 533–548. https://doi.org/10.1016/j.rse.2013.09.014.

Fekety, P. A., Falkowski, M. J., Hudak, A. T. (2015). Temporal transferability of LiDAR-based imputation of forest inventory attributes. *Canadian Journal of Forest Research, 45*, 422–435. https://doi.org/10.1139/cjfr-2014-0405.

Flisberg, P., Rönnqvist, M., Willén, E., et al. (2021). Optimized locations of landings in forest operations. *Canadian Journal of Forest Research, 52*(1), 59–69. https://doi.org/10.1139/cjfr-2021-0032.

Gaudreau, J. P., & Lirette, J. (2020). *Utilisation de la cartographie de la vigueur des tiges pour guider les operations de récolte* (p. 4). InfoNote. Pointe-Claire: FPInnovations.

Gelin, O., Rossander, M., Semberg, T., et al. (2021). *Automation for autonomous terrain mobility (AUTO2)*. Final Report–Stage 2. Arbetsrapport 1077-2021. Uppsala: Skogforsk.

Goodbody, T. R. H., Coops, N. C., & White, J. C. (2019). Digital aerial photogrammetry for updating area-based forest inventories: A review of opportunities, challenges, and future directions. *Current Forestry Reports, 5*, 55–75. https://doi.org/10.1007/s40725-019-00087-2.

Haas, J., Hagge Ellhöft, K., Schack-Kirchner, H., et al. (2016). Using photogrammetry to assess rutting caused by a forwarder—A comparison of different tires and bogie tracks. *Soil and Tillage Research, 163*, 14–20. https://doi.org/10.1016/j.still.2016.04.008.

Hamraz, H., Contreras, M. A., & Zhang, J. (2016). A robust approach for tree segmentation in deciduous forests using small-footprint airborne LiDAR data. *ITC Journal, 52*, 532–541. https://doi.org/10.1016/j.jag.2016.07.006.

Hatala, J. A., Crabtree, R. L., Halligan, K. Q., et al. (2010). Landscape-scale patterns of forest pest and pathogen damage in the Greater Yellowstone Ecosystem. *Remote Sensing of Environment, 114*, 375–384. https://doi.org/10.1016/j.rse.2009.09.008.

Heinzel, J., & Koch, B. (2011). Exploring full-waveform LiDAR parameters for tree species classification. *ITC Journal, 13*, 152–160. https://doi.org/10.1016/j.jag.2010.09.010.

Holmgren, J. (2004). Prediction of tree height, basal area and stem volume in forest stands using airborne laser scanning. *Scandinavian Journal of Forest Research, 19*, 543–553. https://doi.org/10.1080/02827580410019472.

Holmgren, J., Tulldahl, M., Nordlöf, J., et al. (2019). Mobile laser scanning for estimating tree stem diameter using segmentation and tree spine calibration. *Remote Sensing, 11*(23), 2781. https://doi.org/10.3390/rs11232781.

Immitzer, M., Neuwirth, M., Böck, S., et al. (2019). Optimal input features for tree species classification in central Europe based on multi-temporal Sentinel-2 data. *Remote Sensing, 11*(22), 2599. https://doi.org/10.3390/rs11222599.

Kankare, V., Vauhkonen, J., Tanhuanpaa, T., et al. (2014). Accuracy in estimation of timber assortments and stem distribution—A comparison of airborne and terrestrial laser scanning techniques. *ISPRS Journal of Photogrammetry and Remote Sensing, 97*, 89–97. https://doi.org/10.1016/j.isprsjprs.2014.08.008.

Katoh, M., & Gougeon, F. A. (2012). Improving the precision of tree counting by combining tree detection with crown delineation and classification on homogeneity guided smoothed high resolution (50 cm) multispectral airborne digital data. *Remote Sensing, 4*, 1411–1424. https://doi.org/10.3390/rs4051411.

Keefe, R., Wempe, A., Becker, R., et al. (2019). Positioning methods and the use of location and activity data in forests. *Forests, 10*, 458. https://doi.org/10.3390/f10050458.

Lagouarde, J.-P., Ballans, H., Moreau, P., et al. (2000). Experimental study of brightness surface temperature angular variations of maritime pine (*Pinus pinaster*) stands. *Remote Sensing of Environment, 72*(1), 17–34. https://doi.org/10.1016/S0034-4257(99)00085-1.

Lausch, A., Pause, M., Merbach, I., et al. (2013). A new multiscale approach for monitoring vegetation using remote sensing-based indicators in laboratory, field, and landscape. *Environmental Monitoring and Assessment, 185*, 1215–1235. https://doi.org/10.1007/s10661-012-2627-8.

Leckie, D. G., Gougeon, F. A., Tinis, S., et al. (2005). Automated tree recognition in old growth conifer stands with high resolution digital imagery. *Remote Sensing of Environment, 94*(3), 311–326. https://doi.org/10.1016/j.rse.2004.10.011.

Li, J., & Hu, B. (2012). Exploring high-density airborne light detection and ranging data for classification of mature coniferous and deciduous trees in complex Canadian forests. *Journal of Applied Remote Sensing, 6*(1), 063536. https://doi.org/10.1117/1.JRS.6.063536.

Li, J., & Vepakomma, U. (2020). *Automatic terrain feature detection and drivability assessment in forest* (p. 18). Technical Report TR2020, No. 65. Pointe-Claire: FPInnovations.

Liang, X., Wang, Y., Jaakkola, A., et al. (2015). Forest data collection using terrestrial image-based point clouds from a handheld camera compared to terrestrial and personal laser scanning. *IEEE Transactions on Geoscience and Remote Sensing, 53*(9), 5117–5132. https://doi.org/10.1109/TGRS.2015.2417316.

Liang, X., Hyyppä, J., Kaartinen, H., et al. (2018). International benchmarking of terrestrial laser scanning approaches for forest inventories. *ISPRS Journal of Photogrammetry and Remote Sensing, 144*, 137–179. https://doi.org/10.1016/j.isprsjprs.2018.06.021.

Lussier, J. M., & Meek, P. (2014). Managing heterogeneous stands using a multiple-treatment irregular shelterwood method. *Journal of Forestry, 112*, 287–295. https://doi.org/10.5849/jof.13-041.

Maltamo, M., Bollandsås, O. M., Næsset, E., et al. (2010). Different plot selection strategies for field training data in ALS-assisted forest inventory. *Forestry, 84*(1), 23–31. https://doi.org/10.1093/forestry/cpq039.

Maltamo, M., Packalen, P., & Kangas, A. (2021). From comprehensive field inventories to remotely sensed wall-to-wall stand attribute data—A brief history of management inventories in the Nordic countries. *Canadian Journal of Forest Research, 51*(2), 257–266. https://doi.org/10.1139/cjfr-2020-0322.

Marchi, E., Chung, W., Visser, R., et al. (2018). Sustainable forest operations (SFO): A new paradigm in a changing world and climate. *Science of the Total Environment, 634*, 1385–1397. https://doi.org/10.1016/j.scitotenv.2018.04.084.

Marra, E., Cambi, M., Fernandez-Lacruz, R., et al. (2018). Photogrammetric estimation of wheel rut dimensions and soil compaction after increasing numbers of forwarder passes. *Scandinavian Journal of Forest Research, 33*(6), 613–620. https://doi.org/10.1080/02827581.2018.1427789.

McRoberts, R. E., Chen, Q., Gormanson, D. D., et al. (2018). The shelf-life of airborne laser scanning data for enhancing forest inventory inferences. *Remote Sensing of Environment, 206*, 254–259. https://doi.org/10.1016/j.rse.2017.12.017.

Meek, P., & Lussier, J. M. (2008). Trials of partial cuts in heterogeneous forests using the multiple-treatment approach. Pointe-Claire: FPInnovations - Feric Division, *Advantage Report, 10*(2), 1–16.

Meng, R., Dennison, P. E., Zhao, F., et al. (2018). Mapping canopy defoliation by herbivorous insects at the individual tree level using bi-temporal airborne imaging spectroscopy and LiDAR measurements. *Remote Sensing of Environment, 215*, 170–183. https://doi.org/10.1016/j.rse.2018.06.008.

Modzelewska, A., Fassnacht, F. E., & Stereńczak, K. (2020). Tree species identification within an extensive forest area with diverse management regimes using airborne hyperspectral data. *ITC Journal, 84*, 101960. https://doi.org/10.1016/j.jag.2019.101960.

Murphy, P. N. C., Ogilvie, J., Castonguay, M., et al. (2008). Improving forest operations planning through high-resolution flow-channel and wet-areas mapping. *The Forestry Chronicle, 84*, 568–574. https://doi.org/10.5558/tfc84568-4.

Næsset, E. (2007). Airborne laser scanning as a method in operational forest inventory: Status of accuracy assessments accomplished in Scandinavia. *Scandinavian Journal of Forest Research, 22*, 433–442. https://doi.org/10.1080/02827580701672147.

Natesan, S. A., Armenakis, C., & Vepakomma, U. (2020). Individual tree species identification using Dense Convolutional Network (DenseNet) on multitemporal RGB images from UAV. *Journal of Unmanned Vehicle Systems, 8*, 310–333. https://doi.org/10.1139/juvs-2020-0014.

Nevalainen, P., Salmivaara, A., Ala-Ilomäki, J., et al. (2017). Estimating the rut depth by UAV photogrammetry. *Remote Sensing, 9*(12), 1279. https://doi.org/10.3390/rs9121279.

Picchio, R., Proto, A. R., Civitarese, V., et al. (2019). Recent contributions of some fields of the electronics in development of forest operations technologies. *Electronics, 8*, 1465. https://doi.org/10.3390/electronics8121465.

Pierzchała, M., Talbot, B., & Astrup, R. (2016). Measuring wheel ruts with close-range photogrammetry. *Forestry, 89*(4), 383–391. https://doi.org/10.1093/forestry/cpw009.

Pierzchała, M., Giguère, P., & Astrup, R. (2018). Mapping forests using an unmanned ground vehicle with 3D LiDAR and graph-SLAM. *Computers and Electronics in Agriculture, 145*, 217–225. https://doi.org/10.1016/j.compag.2017.12.034.

Pitt, D. G., Wagner, R. G., Hall, R. J., et al. (1997). Use of remote sensing for forest vegetation management: A problem analysis. *The Forestry Chronicle, 73*(4), 459–477. https://doi.org/10.5558/tfc73459-4.

Pouliot, D. A., King, D. J., Bell, F. W., et al. (2002). Automated tree crown detection and delineation in high-resolution digital camera imagery of coniferous forest regeneration. *Remote Sensing of Environment, 82*, 322–334. https://doi.org/10.1016/S0034-4257(02)00050-0.

Pouliot, D., Latifovic, R., Pasher, J., et al. (2019). Assessment of convolution neural networks for wetland mapping with Landsat in the central Canadian boreal forest region. *Remote Sensing, 11*(7), 772. https://doi.org/10.3390/rs11070772.

Puliti, S., Ørka, H. O., Gobakken, T., et al. (2015). Inventory of small forest areas using an unmanned aerial system. *Remote Sensing, 7*, 9632–9654. https://doi.org/10.3390/rs70809632.

Ring, E., Ågren, A., Bergkvist, I., et al. (2020). *A guide to using wet area maps in forestry.* Arbetsrapport 1051-2020. Uppsala: Skogforsk.

Risbøl, O., Briese, C., Doneus, M., et al. (2014). Monitoring cultural heritage by comparing DEMs derived from historical aerial photographs and airborne laser scanning. *Journal of Cultural Heritage, 16*(2), 202–209. https://doi.org/10.1016/j.culher.2014.04.002.

Rönnqvist, M., Flisberg, P., & Willén, E. (2020). Spatial optimization of ground based primary extraction routes using the BestWay decision support system. *Canadian Journal of Forest Research, 51*(5), 675–691. https://doi.org/10.1139/cjfr-2020-0238.

Rummer, B. (2002). Forest operations technology. In D. N. Wear, & J. G. Greis (Eds.), *Southern Forest Resource Assessment* (p. 635). General Technical Report SRS-53. Asheville: U.S. Department of Agriculture, Forest Service, Southern Research Station.

Safonova, A., Tabik, S., Alcaraz-Segura, D., et al. (2019). Detection of fir trees (*Abies sibirica*) damaged by the bark beetle in unmanned aerial vehicle images with deep learning. *Remote Sensing, 11*(6), 643. https://doi.org/10.3390/rs11060643.

Sahal, R., Alsamhi, S. H., Breslin, J. G., et al. (2021). Industry 4.0 towards Forestry 4.0: Fire detection use case. *Sensors, 21*, 694. https://doi.org/10.3390/s21030694.

Sibona, E., Vitali, A., Meloni, F., et al. (2017). Direct measurement of tree height provides different results on the assessment of LiDAR accuracy. *Forests, 8*, 7. https://doi.org/10.3390/f8010007.

Søvde, N. E. (2015). Algorithms for estimating the suitability of potential landing sites. *Mathematical and Computational Forestry & Natural-Resource Science, 7*(1), 1–8.

Spittlehouse, D. L. (2005). Integrating climate change adaptation into forest management. *The Forestry Chronicle, 81*(5), 691–695. https://doi.org/10.5558/tfc81691-5.

St-Onge, B., Audet, F. A., & Bégin, J. (2015). Characterizing the height structure and composition of a boreal forest using an individual tree crown approach applied to photogrammetric point clouds. *Forests, 6*, 3899–3922. https://doi.org/10.3390/f6113899.

Svenson, G., & Fjeld, D. (2016). The impact of road geometry and surface roughness on fuel consumption of logging trucks. *Scandinavian Journal of Forest Research, 31*(5), 526–536. https://doi.org/10.1080/02827581.2015.1092574.

Sylvain, J. D., Drolet, G., & Brown, N. (2019). Mapping dead forest cover using a deep convolutional neural network and digital aerial photography. *ISPRS Journal of Photogrammetry and Remote Sensing, 156*, 14–26. https://doi.org/10.1016/j.isprsjprs.2019.07.010.

Talbot, B., & Astrup, R. (2021). A review of sensors, sensor-platforms and methods used in 3D modelling of soil displacement after timber harvesting. *Croatian Journal of Forest Engineering, 42*(1), 149–164. https://doi.org/10.5552/crojfe.2021.837.

Talbot, B., & Rahif, J. (2017). Applications of remote and proximal sensing for improved precision in forest operations. *Croatian Journal of Forest Engineering, 38*, 327–336.

Talbot, B., Pierzchała, M., & Astrup, R. (2018). An operational UAV-based approach for stand-level assessment of soil disturbance after forest harvesting. *Scandinavian Journal of Forest Research, 33*, 387–396. https://doi.org/10.1080/02827581.2017.1418421.

Tang, J., Chen, Y., Kukko, A., et al. (2015). SLAM-aided stem mapping for forest inventory with small-footprint mobile LiDAR. *Forests, 6*, 4588–4606. https://doi.org/10.3390/f6124390.

Tighe, M. L., Balzter, H., & McNairn, H. (2009). Comparison of X/C-HH InSAR and L-PolInSAR for canopy height estimation in a lodgepole pine forest. In H. Lacoste, & L. Ouwehand (Eds.), *Proceedings of the 4th International Workshop on Science and Applications of SAR Polarimetry and Polarimetric Interferometry (PolInSAR 2009)*. Paris: European Space Agency.

Trier, O. D., Salberg, A. B., Kermit, M., et al. (2018). Tree species classification in Norway from airborne hyperspectral and airborne laser scanning data. *European Journal of Remote Sensing, 51*, 336–351. https://doi.org/10.1080/22797254.2018.1434424.

Tuominen, S., Balazs, A., Saari, H., et al. (2015). Unmanned aerial system imagery and photogrammetric canopy height data in area-based estimation of forest variables. *Silva Fennica, 49*(5),1348. https://doi.org/10.14214/sf.1348.

Vepakomma, U., & Cormier, D. (2017). Potential of multi-temporal UAV-borne LiDAR in assessing effectiveness of silvicultural treatments. *International Archives of the Photogrammetry, Remote Sensing and Spatial Information Sciences, XLII-2*, 393–397. https://doi.org/10.5194/isprs-archives-XLII-2-W6-393-2017.

Vepakomma, U., & Cormier, D. (2019). Valuing forest stand at a glance with UAV-based LIDAR. *ISPRS International Archives of the Photogrammetry, Remote Sensing and Spatial Information Sciences, XLII-2/W13*, 643. https://doi.org/10.5194/isprs-archives-XLII-2-W13-643-2019.

Vepakomma, U., Cormier, D., & Thiffault, N. (2015). Potential of UAV based convergent photogrammetry in monitoring regeneration standards. *International Archives of the Photogrammetry, Remote Sensing and Spatial Information Sciences, XL-1*, 281–285. https://doi.org/10.5194/isprsarchives-XL-1-W4-281-2015.

Vepakomma, U., Chandail, R., Evans, C., et al. (2021). *SBW AIDD: Developing an adaptive and intelligent defoliator—Predicting tree level SBW annual defoliation at a UAV platform* (p. 39). Technical report TR2021 No 74. Pointe-Claire: FPInnovations.

Waga, K., Malinen, J., & Tokola, T. (2020). A topographic wetness index for forest road quality assessment: An application in the lakeland region of Finland. *Forests, 11*, 1165. https://doi.org/10.3390/f11111165.

White, R. A., Dietterick, B. C., Mastin, T., et al. (2010). Forest roads mapped using LiDAR in steep forested terrain. *Remote Sensing, 2*, 1120–1141. https://doi.org/10.3390/rs2041120.

White, J. C., Wulder, M. A., Varhola, A., et al. (2013). *A best practices guide for generating forest inventory attributes from airborne laserscanning data using the area-based approach* (p. 50). Information Report FI-X-10. Victoria: Natural Resources Canada, Canadian Forest Service, Canadian Wood Fibre Centre, Pacific Forestry Centre.

White, J. C., Coops, N. C., Wulder, M. A., et al. (2016). Remote sensing technologies for enhancing forest inventories: A review. *Canadian Journal of Remote Sensing, 42*, 619–641. https://doi.org/10.1080/07038992.2016.1207484.

Part IX
Trends and Challenges

Chapter 28
Network Framework for Forest Ecology and Management

Élise Filotas, Isabelle Witté, Núria Aquilué, Chris Brimacombe, Pierre Drapeau, William S. Keeton, Daniel Kneeshaw, Christian Messier, and Marie-Josée Fortin

Abstract Applications of network science to forest ecology and management are rapidly being adopted as important conceptualization and quantitative tools. This chapter highlights the potential of network analysis to help forest managers develop strategies that foster forest resilience in our changing environment. We describe how networks have been used to represent different types of associations within forest ecosystems by providing examples of species interaction networks, spatial and spatiotemporal networks, and social and social-ecological networks. We then review basic measures used to describe their topology and explain their relevance to different management situations. We conclude by presenting the challenges and potential opportunities for an effective integration of network analysis with forest ecology and management.

É. Filotas (✉) · I. Witté · N. Aquilué · D. Kneeshaw · C. Messier
Centre for Forest Research, Université du Québec à Montréal, P.O. Box 8888, Stn. Centre-Ville, Montréal, QC H3C 3P8, Canada
e-mail: efilotas@teluq.ca

I. Witté
e-mail: isabelle.witte@mnhn.fr

N. Aquilué
e-mail: nuria.aquilue@ctfc.cat

D. Kneeshaw
e-mail: kneeshaw.daniel@uqam.ca

C. Messier
e-mail: messier.christian@uqam.ca; christian.messier@uqo.ca

É. Filotas
Département Science et Technologie, Université TÉLUQ, 5800 St-Denis, Montréal, QC H2S 3L4, Canada

I. Witté
PatriNat, OFB-CNRS-MNHN, 36 Rue Geoffroy St Hilaire, CEDEX 05, CP 41, 75231 Paris, France

© The Author(s) 2023
M. M. Girona et al. (eds.), *Boreal Forests in the Face of Climate Change*,
Advances in Global Change Research 74,
https://doi.org/10.1007/978-3-031-15988-6_28

685

28.1 Introduction

Understanding how human activities modify the structure and function of forest ecosystems is a central challenge for achieving sustainable forest management. To this end, in recent decades, forest scientists have started applying network theory to ecosystem management (Dale & Fortin, 2010, 2021; Fall et al., 2007; Hamilton et al., 2019; Martin & Eadie, 1999; Rayfield et al., 2011). Network theory provides a novel framework for designing effective strategies intended to maintain forest functions while conserving biodiversity (Aquilué et al., 2020; D'Aloia et al., 2019; Messier et al., 2019; Ruppert et al., 2016).

Forest ecosystems are composed of highly heterogeneous elements—organisms to forest stands—that interact through ecological processes over a wide range of temporal, spatial, and organizational scales (Filotas et al., 2014). Specifically, network theory can be used to model forest ecosystems as ensembles of connected elements (Aquilué et al., 2020; Mina et al., 2021; Ruppert et al., 2016). Examples include food webs linking species across several trophic levels (Eveleigh et al., 2007), nest webs linking species across microhabitat structures such as tree cavities (Martin et al., 2004), isolated forest fragments connected by wind or animal dispersed seeds (Aquilué et al., 2020), and social organizations engaged in a common management effort (Fischer & Jasny, 2017). Network analysis focuses on describing the topology of interactions linking elements together and can establish a relationship between this network topology and forest functions for management purposes (Ruppert et al., 2016). In particular, network analysis can be used to quantify the alteration of forest

C. Brimacombe · M.-J. Fortin
Department of Ecology and Evolutionary Biology, University of Toronto, 25 Willcocks Street, Toronto, ON M5S 3B2, Canada
e-mail: chris.brimacombe@mail.utoronto.ca

M.-J. Fortin
e-mail: mariejosee.fortin@utoronto.ca

P. Drapeau
Centre for Forest Research and UQAT-UQAM Chair in Sustainable Forest Management, Université du Québec à Montréal, P.O. Box 8888, Stn. Centre-Ville, Montréal, QC H3C 3P8, Canada
e-mail: drapeau.pierre@uqam.ca

W. S. Keeton
Rubenstein School of Environment and Natural Resources and Gund Institute for Environment, University of Vermont, 81 Carrigan Drive, Burlington, VT 05405, USA
e-mail: william.keeton@uvm.edu

C. Messier
Département des Sciences Naturelles, Institut des Sciences de La Forêt Tempérée, Université du Québec en Outaouais, 58, Rue Principale, Ripon, QC J0V 1V0, Canada

M.-J. Fortin
Département des Sciences Naturelles, Institut des Sciences de La Forêt Tempérée, 58 Rue Principale, Ripon, QC J0V 1V0, Canada

functions resulting from human-mediated and natural disturbances that directly or indirectly modify the ecological components of forest ecosystems, including their interactions and spatial setting (Aquilué et al., 2020).

A network is a simplified representation of a system based on connections—links—among its component elements—nodes. A food web, for example, is a network representing the trophic interactions among an ecosystem's constituent species (Pimm et al., 1991). Each element in a network is represented by a node, also called a vertex, which may be connected to other nodes by links, also called edges, representing potential or realized interactions between two elements. Nodes are defined by one or more attributes and their connections to other nodes. Links may be unidirectional or bidirectional and may be weighted to express the strength of an interaction. In a food web, for instance, nodes represent species, and links represent predator–prey interactions among species (Ings et al., 2009). A unidirectional link would represent a predator species feeding on a prey, whereas a bidirectional link could represent a mutual interaction or dependency between two species. Moreover, a node could be characterized by its species' abundance, and a link could be weighted to represent a predator's relative preference for a given prey.

Network science originates from graph theory, a fundamental topic in the field of discrete mathematics that can be traced to the work of Euler in the eighteenth century (Newman, 2003). Nowadays, the study of networks is pervasive across all fields of science, including molecular biology, neuroscience, linguistics, and epidemiology (Newman, 2003; Strogatz, 2001; Turnbull et al., 2018). The World Wide Web, social media networks, and global plane travel networks are only a few of many examples of networks present in our everyday life.

Network science continues to develop tools that characterize the topology of networks, a concept referring to the architecture of nodes and links. Moreover, it studies the possible relationships between a network topology and the ability of the corresponding system to function and adapt to disturbances. Generally, the strength of network science is the universality of tools available for studying disparate systems, varying widely in their nature and scale (Albert & Barabási, 2002). For example, the structure of a network can provide information about its vulnerability or adaptability to the loss or addition of nodes and links or the efficiency with which resources and information are propagated within the network (Fig. 28.1; Barabási & Albert, 1999; Watts & Strogatz, 1998). Will a food web collapse following the extinction of a given species? Is an epidemic more likely to spread within a population if a given demographic group is infected? Can consensus within a community divided over an environmental issue be improved by creating new communication channels? These and other important basic and applied science questions can be answered using the methods from network science.

The application of network theory to ecology and evolutionary biology has seen a remarkable development over the past 20 years (Dale & Fortin, 2010; Kool et al., 2013; Proulx et al., 2005). Well-studied ecological networks include protein and gene networks (Jeong et al., 2001; Vidal et al., 2011), pollination networks (Bascompte et al., 2003; Memmott et al., 2004), food webs (Dunne et al., 2002a), nest webs (Martin et al., 2004), and habitat conservation networks (Urban & Keitt, 2001).

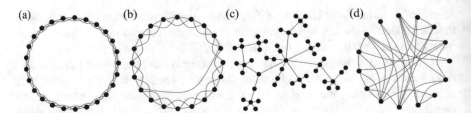

Fig. 28.1 Different topologies of undirected networks. **a** Regular network in which all nodes have the same number of connections; **b** a *small-world* network constructed by rewiring a few nodes of a regular network, thereby reducing its diameter and making each node easily accessible from any other nodes of the network (Watts & Strogatz, 1998); **c** a *scale-free* network created by adding connections to nodes with a probability that increases with their number of connections such that well-connected nodes become even more connected. Such networks are more vulnerable to disturbances that target hubs (Barabási & Albert, 1999). **d** Random networks in which the number of connections is randomly assigned to each node

Specific applications of network science to forest ecology and management are more recent, but this approach is rapidly gaining adoption as an important conceptualization and quantitative tool. For example, networks are used to understand how locally interacting entities drive forest ecosystem functions and inform management strategies that more directly integrate cross-scale interactions (Messier et al., 2019).

This chapter highlights the potential of network thinking to address key issues of cross-scale interactions in forest ecology and management. First, we describe how networks have been used to represent different types of associations within forest ecosystems by reviewing examples of species interaction networks, spatial and spatiotemporal networks, and social and social-ecological networks. We explain how nodes and links can be defined and synthesize the particular features that characterize each network type. Then, we review basic measures used to quantify the structure of networks and explain their relevance to different management situations. We conclude by presenting the challenges and potential opportunities for an effective integration of network analysis with forest ecology and management. The network framework may prove invaluable in helping forest managers to better anticipate and adapt to global change.

28.2 Representing Forests with Networks

As with any network, the identification of nodes and links varies with the questions of interest and with respect to how the system under study can be decomposed into sets of distinct and interacting components (Table 28.1). Here, we describe different network categories employed in forest ecology and management. These categories differ in the nature of nodes, including individual species, forest stands, and governance institutions. Consequently, the type and scale of interaction among nodes also vary between these categories.

Table 28.1 Different categories of networks employed in forest ecology and management. For each category, examples of nodes, links, and relevant references are provided

Network category	Nodes	Links	Examples
Species interaction networks			
Plant–herbivore or plant–frugivore	Tree or fruiting plant species and herbivore or frugivore species	Feeding interactions	• Variations in the level of specialization of plant–frugivore networks across a fragmented rainforest landscape (Chama et al., 2013) • Effect of species life-history traits on the structure of plant–leafminer networks in dry open woodlands (Cagnolo et al., 2011)
Pollination	Flower plant species and pollinating species	Pollination visits	• Variations in the structure of bumblebee pollination networks across an agricultural landscape of fragmented north-temperate mixed forests (Gómez-Martínez et al., 2020) • Temporal changes in the networks of pollen-carrying moth and pine species in a boreal forest (Devoto et al., 2011)

(continued)

Table 28.1 (continued)

Network category	Nodes	Links	Examples
Host–parasitoids	Host species and parasitoid species; sometimes including the lower (plant–herbivore) and/or higher (parasitoid–hyperparasitoid) trophic levels	Parasitoid attacks	• Changes in the structure of plant-host–parasitoid networks in a rainforest following the introduction of an alien caterpillar as an agricultural biological agent (Henneman & Memmott, 2001) • Changes in the structure of the balsam fir-host–parasitoid-hyperparasitoid network during a spruce budworm outbreak in a temperate forest (Eveleigh et al., 2007)
Nest webs	Cavity-bearing tree species and vertebrate cavity users: excavators and nonexcavators	Vertebrate–tree cavities interactions (excavation, breeding, roosting)	• Robustness of nest webs to selective logging in a subtropical forest (Ruggera et al., 2016) • Changes in nest webs structure during an outbreak of mountain pine beetle in a temperate forest (Cockle & Martin, 2015)
Fungal network	All tree genotypes within a sampling plot	A common fungal genotype shared between two different tree genotypes	• Change in the mycorrhizal networks formed between *Rhizopogon* and Douglas fir 25 years after selective logging (Van Dorp et al., 2020)
	Plant and lichen species and their associated fungal species	Symbiotic plant/lichen–fungi interactions	• Chagnon et al. (2012) used a network approach to demonstrate that a community of plants linked by arbuscular mycorrhizal fungi in a hemiboreal forest was highly nested and modular

(continued)

Table 28.1 (continued)

Network category	Nodes	Links	Examples
Spatial graphs			
Habitat-patch networks	Patches of habitat for a given wildlife species or group of species	Potential direct animal movement between habitat patches	• Trade-offs between timber harvesting and maintaining habitat connectivity for woodland caribou in a boreal forest (Ruppert et al., 2016) • Identification of key nodes and links for providing the connectivity of a network of protected forest areas fragmented by highways (Gurrutxaga et al., 2011)
Forest insect pests	Forest patches	Potential insect dispersal between forest patches	• de la Fuente et al. (2018) used a network-based model to predict the areas and speed of the natural spread of the pine wood nematode and suggested which forest stands should be harvested to stop the spread of the nematode by breaking the connectivity between the remaining forest stands • Wildemeersch et al. (2019) used a network-of-networks approach to simulate the outbreak dynamics of the geometrid moth in birch forests and the European spruce bark beetle

(continued)

Table 28.1 (continued)

Network category	Nodes	Links	Examples
Forest-stand networks	Forest patches	Potential seed dispersal between forest fragments	• Identification of key forest patches in which to focus conservation or restoration efforts to maintain a high functional response diversity across a fragmented agricultural landscape (Craven et al., 2016) • Trade-offs between different forest landscape management scenarios for maintaining the connectivity of a fragmented forest network subjected to potential disturbances, i.e., drought, pest outbreak, timber harvesting (Aquilué et al., 2020)
Spatiotemporal networks	Habitat patches at different moments in time	Potential animal movement between habitat patches at different moments that can be direct or indirect (through patches that were gained or lost)	• Martensen et al. (2017) proposed a new spatiotemporal connectivity algorithm that accounts for both spatial and temporal search windows to estimate how dynamic landscapes of forested patches can maintain animal movements • Huang et al. (2020) used the spatiotemporal connectivity algorithm to determine how future dynamic landscapes of forested patches combined with climate changes will maintain predators' range shifts while also accounting for their preys' range shifts under the same conditions

(continued)

Table 28.1 (continued)

Network category	Nodes	Links	Examples
Social networks			
Organizational networks	Actors or group of actors engaged in a common environmental or management issue	Associations between organizations, e.g., sharing information, working with each other	• Adaptation capacity of a network of organizations (local government agencies, nonprofit groups, federal agencies, universities, etc.) that share concerns about increasing wildfire risk (Fischer & Jasny, 2017) • Knoot and Rickenbach (2014) used a social network to examine the collaborative capacity of public- and private-sector foresters to help facilitate a landscape-scale multifunctional management of privately owned forests
Social-ecological networks			
Forest governance networks	Social nodes: organizations involved in the management of forests Ecological nodes: bounded forest lands managed by one or more organizations	Ecological links: ecological processes linking ecological nodes, e.g., seed dispersal, forest fire Social links: interorganizational interactions	• Likelihood of social organizations coordinating their management of wildfire risks on the basis of the spatial configuration of their risk interdependence (Hamilton et al., 2019) • Governance of small forests in an agricultural landscape among actors having varying forest uses and kinship relations (Bodin & Tengö, 2012)

28.2.1   Species Interaction Networks

In networks of species interactions, a single species sometimes provides a natural unit for denoting a node. This is the case, for example, in pollination networks (Devoto et al., 2011; Vázquez et al., 2009), host–parasitoid networks (Memmott et al., 1994), and nest webs (Martin & Eadie, 1999; Martin et al., 2004). However, other systems may highlight the need for different aggregation units, such as species playing a common function or a guild of species with a similar trophic position (Dunne et al., 2002a). Links between nodes denote potential or realized interspecific interactions that may or may not involve biomass transfer, including antagonistic (e.g., plant–herbivore and host–parasitoid networks), mutualistic (e.g., pollination networks), symbiotic (e.g., mycorrhizal network), and commensal associations (e.g., nest webs) (Delmas et al., 2019).

Networks of interspecific interactions may be unipartite, meaning that any two nodes may interact, or they may be ordered over multiple hierarchical levels where only nodes in different levels can interact (Fig. 28.2a; Delmas et al., 2019). In nest webs, which represent the relationships among tree species and cavity-nesting verte-brates, links connect tree species to one or more *nidic* levels (Martin & Eadie, 1999; Martin et al., 2004; Ruggera et al., 2016). These levels consist of cavities that originate either from tree decay or from animal excavators, and also include obligate cavity users, which cannot excavate a cavity and thus depend entirely on existing cavities for nesting (Cockle et al., 2019; Martin et al., 2004). Host–parasitoid networks may also encompass lower (plant–herbivore) and higher (parasitoid–hyperparasitoid) trophic levels (Eveleigh et al., 2007). On the other hand, some networks focus on repre-senting the associations between two levels only, such as pollinator–plant (Devoto et al., 2011; Gómez-Martínez et al., 2020), plant–herbivore (Cagnolo et al., 2011), and plant–frugivore networks (Chama et al., 2013). Such networks, termed bipartite networks (Fig. 28.2b), can also be used to represent nest webs (Cockle & Martin, 2015; Ruggera et al., 2016) and host–parasitoid networks over narrower scales of interspecific organization (Tylianakis et al., 2007; Van Veen et al., 2008).

Ecological networks can also be used to represent mycorrhizal associations between plant roots and fungi, or relationships among algae, fungi, and sometimes bacteria within lichen (Southworth et al., 2005). Two different approaches may be

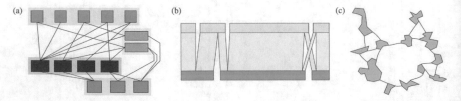

(a)   (b)   (c)

Fig. 28.2 Different categories of network in forest ecology. **a** Network extending over multiple hierarchical levels, e.g., food webs and nest webs; **b** a bipartite network where nodes are separated into two levels, e.g., pollinators (*yellow*) and plants (*green*); **c** an undirected spatial network where links denote potential least-cost movement between patches of habitat (*green polygons*)

adopted (Table 28.1). The first studies the bipartite network formed by the symbiotic interactions between the hosts (plant or algae species) and their associated (endo-phytic or endolichenic) fungi (Chagnon et al., 2012; Toju et al., 2015). The second adopts a phytocentric perspective where tree boles in a sampling plot correspond to nodes and links. This represents the pairwise connection of trees through the same fungal genet (Beiler et al., 2010, 2015; Simard, 2009; Van Dorp et al., 2020).

28.2.2 Spatial and Spatiotemporal Networks of Forest Ecosystems

In spatial and spatiotemporal networks, nodes are conceptualized as spatially local-ized units of contiguous area, such as forest stands that, when aggregated, compose forested landscapes (Table 28.1; Bunn et al., 2000; Fall et al., 2007; Pelletier et al., 2017; Urban & Keitt, 2001). We can distinguish between habitat-patch networks and forest-stand networks. The former stresses the relationship between habitat patches for wildlife connectivity—usually for the conservation of a specific species or group of species of concern (Gurrutxaga et al., 2011; James et al., 2005; Ruppert et al., 2016) or to predict the spread of undesirable species (de la Fuente et al., 2018; Ferrari et al., 2014; Wildemeersch et al., 2019)—whereas the latter focuses on the connectivity of tree communities (Aquilué et al., 2020; Craven et al., 2016; Saura et al., 2011). Nodes are defined either by the GPS locations of organisms, bird nests (Melles et al., 2012), and territories/home ranges or by delineated forested patches according to specific criteria, e.g., stand age, structure, and species composition (Aquilué et al., 2020). Nodes can be characterized by spatial, e.g., area, shape, edge/area ratio, and nonspatial attributes, e.g., species diversity, habitat quality.

In spatial networks, links between nodes denote the movement of animals or plant seeds, either as a potential or a relative measure (Bunn et al., 2000; Fall et al., 2007; Urban & Keitt, 2001). Links can be determined according to species' dispersal abilities and behavioral responses to the intervening landscape that facilitates or impedes organism movement (i.e., functional connectivity; Rayfield et al., 2010). Thus, links can be represented by the Euclidean distances between patches or as a function of movement cost. In this case, the distance between patches is weighted by the additional difficulty for a given species to disperse through the given matrix cover types (James et al., 2005). Consequently, spatial networks provide a framework to evaluate the functional connectivity of a landscape for a particular species or tree community, transcending simpler structural connectivity assessments.

Unlike species interaction networks where links are mostly directed, thereby expressing relationships between consumers and their resource, spatial networks can have both directed and nondirected links and do not form a hierarchical structure (Fig. 28.2c). In habitat networks, links are nondirected because an animal's ability to move between two habitat patches can, theoretically, be assumed to be the same in both directions (Ruppert et al., 2016). On the other hand, in forest-stand networks, a node contains a community of tree species that differ in their seed dispersal ability (Tamme et al., 2014). Thus, the flux of seeds dispersing from one stand to another is not equivalent in both directions, leading to directed links between nodes (Aquilué et al., 2020).

Box 28.1 Spatiotemporal networks

In spatiotemporal networks, habitat patches or forest stands are dynamic, where: **a** the weights of both nodes and links change through time but not the network topology, **b** as in **a** although the topology changes through time, and **c** the nodes and links are given by organisms' movements.

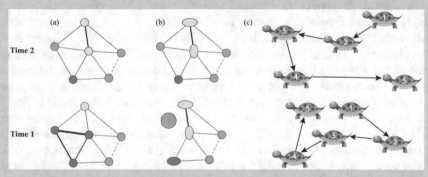

To determine the degree of functional connectivity of a habitat network and how it changes through time, one can quantify connectivity at specific times as a series of static snapshots. However, the degree of connectivity can be affected by the temporal dimension of the forest dynamics relative to the species' longevity (Zeigler & Fagan, 2014). For this reason, one cannot treat habitat networks at different times as independent static snapshots. To address the effects of such transient dynamics of habitat patches, Martensen et al. (2017) proposed a novel spatiotemporal connectivity algorithm to quantify the sequential spatial overlaps of habitat patches that are available to account for a temporal window matching species life history. Martensen et al. (2017) showed—by considering explicitly in their algorithm the spatiotemporal dimension of habitat patches and species dispersal abilities—that the transient use of habitat patches can favor a higher degree of connectivity compared with static spatial connectivity values.

Spatiotemporal networks express relationships within and between spatial networks at different times (Huang et al., 2020; Martensen et al., 2017). They add the temporal dimension to spatial networks by integrating the dynamic nature of forest stands and habitat patches (Box 28.1). They capture the fact that ecological processes and disturbances affect the persistence and attributes of spatial nodes. For example, tree communities within forest stands and habitat patches undergo succes-sional changes and are modified by natural disturbances (e.g., fire, insect outbreak, drought, and windthrow) and human activities (e.g., harvesting and land-use change). If network nodes change too quickly or are destroyed, organisms may not have time to reach other suitable nodes. Therefore, nodes and links that were present in a static spatial network could be absent in a spatiotemporal network. Moreover, this frame-work allows for the representation of indirect links between patches to indicate that an organism has moved through an intermediate stepping-stone patch that has been gained or lost during the two different time observations.

28.2.3   Social and Social-Ecological Networks

Nodes in social networks represent any social entity, from single individuals, e.g., a forest owner or user, to collectives of individuals, e.g., forest management organiza-tions, forest-based communities, or groups of stakeholders sharing similar interests or belonging to the same governance sectors (Guerrero et al., 2020). Links between these social entities can correspond to both formal and informal relationships and represent (1) flows (e.g., information, resources, and money), (2) social relations (e.g., employee of, neighbor of) and interactions (e.g., work with, share information to), and (3) similarities (e.g., same location, same attitude) (Borgatti et al., 2009; Guer-rero et al., 2020). Nodes may be characterized by demographic and social/cultural attributes (e.g., age and occupation), attitudes and behaviors toward a management or conservation issue, and features of the corresponding organization, e.g., size, mission, and governance level. Links can be weighted according to the strength of the relationship or frequency of the interaction (Guerrero et al., 2020).

Depending on the social system under study and the types of relationships consid-ered, links in social networks can be directed, e.g., sharing information to, or undi-rected, e.g., same conservation goal as another entity, and form different hierarchical structures ranging from one to multiple levels of governance that include several jurisdictions and geographic areas (Fischer, 2018; Guerrero et al., 2020). Moreover, social networks are shaped by processes specific to human and social interactions, such as *homophily*, *intentionality*, and *reciprocity* (Fischer & Jasny, 2017; Guerrero et al., 2020; Knoot & Rickenbach, 2014). Homophily refers to the tendency to be connected to people having similar values and goals, whereas intentionality refers to the conscious choice to associate (or not) with someone else, and reciprocity is the tendency for mutual interactions.

Social-ecological systems can also be represented by networks (Folke, 2006; Kleindl et al., 2018) and aim to capture the interplay and possible feedbacks between human decisions and actions in managing an ecosystem and the structure and function of that ecosystem (Bodin, 2017; Bodin & Tengö, 2012; Fischer, 2018; Janssen et al., 2006). Generally, social-ecological networks are used in the context of governance challenges emerging from (1) a scale mismatch between the ecological and the social processes operating in the system, (2) competition for access, use, or management of a shared ecological resource, and (3) sensitivity to the order with which management activities are realized, e.g., steps to take to reduce risk (Bodin et al., 2019; Hamilton et al., 2019). Therefore, social-ecological networks comprise both ecological and social nodes and focus on the interdependencies between these various kinds of nodes. For example, a social-ecological link could represent timber harvesting by a forest owner (social node) in their forest stand (ecological node). Ecological nodes usually consist of groups of plants or animals, or have a spatial dimension, such as specific forest patches. However, more aggregated biophysical forms, e.g., ecosystem services (Dee et al., 2017), may be a more appropriate node representation when social-ecological interactions are associated with specific ecological functions that are produced by multiple ecological entities (Bodin et al., 2019).

Social-ecological networks may develop via human activities that create interactions between ecological elements (Janssen et al., 2006). For example, firewood movement between localities is associated with the wide dispersal of emerald ash borer (*Agrilus planipennis*) across North American forests (Siegert et al., 2015), and the construction of forest roads has been associated with increased gray wolf (*Canis lupus*) movement across managed forest stands (Courbin et al., 2014). Social interactions may also emerge from ecological connections. For instance, when two organizations managing distinct forest lands decide to collaborate on a wildfire risk mitigation strategy following a forest fire that has burned across both lands (Hamilton et al., 2019). Sayles et al. (2019) distinguished between different kinds of social-ecological networks depending on how nodes and links are defined: (1) *multiplex networks* in which all nodes can be connected by social and ecological links; (2) *multi-level networks* in which social and ecological nodes are viewed as being on different layers and only one interaction between any two nodes is considered; and (3) *multi-dimensional networks* in which nodes are represented as in multilevel networks, but multiple interactions between nodes are possible (Fig. 28.3).

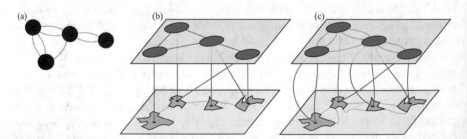

Fig. 28.3 Different frameworks to represent social-ecological networks, as suggested by Sayles et al. (2019). **a** Multiplex network where nodes (*black circles*) can be either social or ecological and connected by both social and ecological links; **b** and **c** multilevel networks with social (*blue circles*) and ecological (*green polygons*) nodes on different layers. Nodes are connected by intralayer links (*blue* or *green*) and/or interlayer links (*orange*). In **b,** only one link exists between pairs of nodes, and in **c** multiple interactions between nodes are possible, as in multiplex networks

28.3   Network Analysis

Multiple statistical measures can be extracted from networks to describe their architecture and determine the degree to which a system is connected, how interactions
are distributed among nodes, and whether specific nodes occupy important positions.
These measures are then used as indicators to better understand the system's function
and its robustness or capacity to adapt to changing conditions and disturbances. Table
28.2 presents a few key fundamental network metrics; more in-depth discussions
can be found in the literature on networks (Newman, 2003; Strogatz, 2001) and their
application to community ecology (Bersier et al., 2002; Blüthgen et al., 2006; Delmas
et al., 2019; Proulx et al., 2005), conservation biology (Dale & Fortin, 2010; D'Aloia
et al., 2019; Galpern et al., 2011; Rayfield et al., 2011) and social (Bodin et al., 2006)
and social-ecological systems (Janssen et al., 2006). Most measures described in
Table 28.2 are general and apply to all types of networks, emphasizing the universality of many network metrics. However, a small number are specific to certain types
of networks. For example, specialization is a measure used in bipartite networks,
whereas connectivity is used in spatial networks. Network measures are termed
qualitative when they apply to binary networks, i.e., networks with unweighted links
that only report the presence or absence of interactions, or quantitative when they
apply to weighted networks in which links represent the strength or frequency of
interactions. Table 28.2 largely focuses on qualitative measures but includes some
quantitative measures, e.g., specialization.

The most general measures used to describe a network are its *order*, meaning
the number of nodes in the network, and its *size*, which is the number of links.
These descriptors already provide an idea of the extent and possible complexity
of the network. The average number of links per node measures the *density* of the
network. In social webs, a high density is often associated with a better exchange
of information among actors. This can facilitate the development of new ideas and
also improve collective actions in natural resource governance (Bodin et al., 2006).
Conversely, an extremely dense network of actors can homogenize information and
impede the development of new knowledge. It can also be associated with a reduced
diversity of management practices that could lead to lock-in and limit the capacity
of actors and organizations to come up with novel strategies to adapt to changing
conditions (Bodin et al., 2006; Janssen et al., 2006).

A measure similar to linkage density is *connectance*, the proportion of potential
interactions that occur. Connectance is the term used to determine species interactions
within ecological networks. It can be a good indicator of the sensitivity of ecological
communities to disturbances resulting in the loss of species (Dunne et al., 2002b;
Montoya et al., 2006). Connectance is also associated with community dynamics
and may be used to understand variations in population density or infer potential
indirect interactions (Van Veen et al., 2008). A spatial analog of connectance is
functional *connectivity*, which applies to species-habitat and forest-stand networks.
Functional connectivity is a species-specific measurement, as species perceive forest
fragmentation differently depending on their movement ability. Multiple indices
of connectivity exist, all with the general purpose of determining the availability

Table 28.2 Measures to quantify network structures and the corresponding illustrations. **a** Two networks having the same order, but the smaller-sized network (*right*) has a smaller linkage density and connectance but a longer diameter (depicted by the number of links separating the two blue nodes); **b** two spatial networks of the same order but one (*left*) has a larger size and a higher connectivity. **c** The yellow node has a lower degree than the red node. The vulnerability (*number of blue links*) of the red node is identical to its generality (*number of purple links*). **d** Pink nodes are generalists, whereas blue nodes are specialists. **e** The red node (*right*) has a low clustering coefficient, whereas the red node (*left*) has a high clustering coefficient; thus, it forms a clique with its neighbors. **f** The red node has the highest betweenness centrality in this network. **g** The degree distribution is homogeneous when all nodes have similar degrees (*left*) and is heterogeneous when degrees vary among nodes. **h** The bipartite network (*left*) has a nested structure contrary to that on the right. **i** A high modularity network (*left*) contains six modules, whereas the other network (*right*) lacks a modular structure. This table constitutes a nonexhaustive list of measures. Interested readers should consult references cited in the main text for a deeper exploration of network measures

General Measures	
Order Total number of nodes	**(a)**
Size Total number of links	
Linkage density Average number of links per node	
Diameter Longest of the shortest paths between all pairs of nodes in the network	
Connectance Number of links over the total possible number of links	
Connectivity The degree to which spatial nodes are reachable through internode movement	**(b)**
Node-Level Measures	
Degree Number of links for a specific node	**(c)**
Specialization Diversity of partners for a given species	**(d)**

(continued)

Table 28.2   (continued)

Clustering coefficient Degree to which neighbors of a node are connected	**(e)**
Betweenness centrality Number of times a node sits in a path between all pairs of nodes in the network	**(f)**
Network-Level Measures	
Degree distribution Frequency distribution of degrees in the network	**(g)**
Nestedness Degree to which specialist species interact with a subset of the group of species with which generalists interact	**(h)**
Modularity How closely connected nodes are divided into modules	**(i)**

of habitat for a given species (Rayfield et al., 2011). Therefore, these indices are modulated not only by the number of patches and their connections but also by their area. For example, the probability of connectivity index is a quantitative measure that corresponds to the probability that two individuals randomly placed in habitat patches across the landscape can reach each other (Saura & Pascual-Hortal, 2007).

Functional connectivity measures are useful for conservation planning, such as designing reserve networks (D'Aloia et al. 2019; James et al., 2005; Saura & Pascual-Hortal, 2007) or evaluating changes in forest connectivity over time (Saura et al., 2011). They may also be used when planning harvesting operations. Ruppert et al. (2016) developed a heuristic procedure to schedule timber harvesting on the basis of a trade-off between wood volume and habitat connectivity for the woodland caribou (*Rangifer tarandus caribou*). Tittler et al. (2015) compared the habitat connectivity of various wildlife species across management strategies that differed in their distribution and aggregation of forest cuts. Functional connectivity is also considered a critical component of forest resilience (Box 28.2; Aquilué et al., 2020; Craven et al., 2016; Mina et al., 2021). High connectivity implies that source–sink dynamics may be possible in a fragmented forest landscape whereby disturbed forest patches can regenerate by receiving seeds from unaltered patches (Craven et al., 2016).

Box 28.2 Effect of Landscape Management and Disturbances on a Forest Patch Network

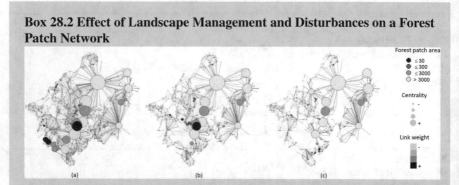

The variations in patch centrality in a spatial network of forest patches across increasing levels of timber harvesting; **a** no harvesting, harvesting at **b** 5%, and **c** 10% tree cover. Nodes are colored according to the size of their corresponding forest patch. The diameter of each node is proportional to its betweenness centrality. Links between patches are directed and weighted according to the tree species composition within each node as well as the seed dispersal capacity of each tree species

Spatial networks can guide landscape-scale forest management. Aquilué et al. (2020) used a network approach to model fragmented forest patches in a rural landscape in central Québec, southeastern Canada, and explore how connectivity among patches varied according to different management strategies—functional enrichment of current forest patches, plantations in newly created forest patches—and under different disturbance scenarios— timber harvesting, drought-induced mortality, and pest outbreak. Interested readers should read Chap. 31 for a discussion of the effect of functional enrichment on the resilience of fragmented landscapes.

The above figure illustrates how tree harvesting affects the betweenness centrality of forest patches. Indeed, cutting trees has the effect of removing small patches and reducing the flux of seeds that can travel among patches, thereby affecting the entire functional connectivity of the landscape. As a result, the importance of each patch in maintaining connectivity is altered by disturbance. This figure also illustrates that small patches (*dark color*) can have a high centrality value (*large diameter*), emphasizing that the contribution of a forest patch to the connectivity of the landscape is not simply based on its surface area.

The *diameter* of a network is the maximum number of links between any two network nodes and thus measures the extent to which nodes are accessible to each other (Janssen et al., 2006). The small diameter of a habitat network can indicate its susceptibility to the rapid spread of an invasive plant (Minor et al., 2009). In organizational networks, a small diameter implies the existence of efficient channels to diffuse information (Bodin et al., 2006). Although a short diameter may be correlated with network density, this is not always the case. Linkage density does not account for how links are distributed among nodes. Therefore, it is possible to have a dense network characterized by a large diameter whenever nodes are distributed in a few well-connected clusters that are isolated from each other (Janssen et al., 2006). In social networks characterized by such a topology, shared information will tend to remain within clusters.

Network analysis allows for the identification of nodes that play a key role in structuring the system. The *degree* is a node-level measure that, in an undirected network, corresponds to the number of links that connect a node. In directed networks, the degree can be decomposed into in-degree and out-degree. In food webs, the former is an indication of the vulnerability of a species, i.e., the number of predators, whereas the latter relates to its generality, i.e., the number of resources (Delmas et al., 2019). In a bipartite network, such as pollination, frugivore, and host–parasitoid networks, a similar concept is that of *specialization*, which corresponds to the diversity of interacting partners of a species (Blüthgen et al., 2006; Chama et al., 2013; Gómez-Martínez et al., 2020). Gómez-Martínez et al. (2020) found that the level of specialization in bumblebee pollination networks decreased with the increased fragmentation of the surrounding forest landscape. Another related descriptor used in nest web studies is the *species importance index*. When measured for a tree species, this index corresponds to the proportion of bird species that use the particular tree species' cavities relative to the number of other tree species used by the same bird species. Identifying keystone tree species that cavity users and excavators routinely use is essential to define more specific conservation guidelines (Ruggera et al., 2016).

The degree of a node is, therefore, a measure of its influence on other nodes and is one of multiple measures assessing the *centrality* of a node. For instance, in studying the mycorrhizal networks of interior Douglas fir (*Pseudotsuga menziesii*), Beiler et al. (2015) found that large trees had a higher *degree centrality* in xeric plots compared with mesic plots. This analysis suggests that the role of large trees—in facilitating the survival and productivity of newly established seedlings through shared myccorhizal fungi—is more important under water-deficit conditions.

In social webs, the organization having the highest degree can play a determining role in coordinating a group of organizations with diverging opinions on the best risk mitigation strategy, e.g., forest fire, toward a consensus (Bodin et al., 2006; Hamilton et al., 2019). Yet, a node with a low degree can also exert a central importance within the network if, for example, its position connects clusters of nodes that would otherwise be isolated. Such nodes are said to have a high *betweenness centrality*. Actors or organizations that occupy these bridging positions in social webs are essential for developing trust among parties holding conflicting views. In species-habitat and forest-stand networks, determining patches of high betweenness centrality helps identify patches that are not necessarily large but that still have a high conservation value because they enable wildlife species to move across the landscape from one region of well-connected patches to another (Aquilué et al., 2020; Gurrutxaga et al., 2011). A spatial network is more vulnerable to the destruction of nodes having a high betweenness centrality because their loss can cause the fragmentation of the landscape into unconnected components (Box 28.2; Aquilué et al., 2020). This destruction could result, for example, from harvesting, pest infestation, or forest fire.

Different measurements can provide information about the possible asymmetric distribution of interactions within networks. The simplest approach is to derive the frequency distribution of degrees within a network, i.e., its *degree distribution*, which describes the level of degree heterogeneity. For example, scale-free networks (Fig. 28.1c) are characterized by a few highly connected nodes and a large number of poorly connected nodes. In one example, the degree distribution of the mycorrhizal networks of Douglas-fir trees followed a scale-free distribution (Beiler et al., 2010). The large mature trees in a plot had the most connections, suggesting that such networks are robust to the random loss of trees but fragile to the loss of large trees with consequences for the regeneration of the entire community of connected trees (Beiler et al., 2010).

In weighted networks, one can measure the interaction diversity—a Shannon diversity of links—to quantify how degrees are distributed among nodes. For example, Cockle and Martin (2015) found that the interaction diversity of a nest web increased during a mountain pine beetle outbreak because the greater availability of cavity trees allowed for a wider variety of excavators and new opportunities of interactions with secondary cavity nesters.

The *clustering coefficient* of a node measures the extent to which neighbors of that node are closely connected. In social webs, interconnected nodes with high clustering coefficients are said to form a *clique*. The formation of cliques, or clusters, results from the tendency of social partners to interact, a property of social interactions called *transitivity*. The presence of clusters may help maintain a heterogeneity of knowledge and experiences across the network. This may prove essential for innovation and adaptation to novel environmental conditions (Bodin et al., 2006).

Modularity measures the extent to which a network is divided into modules of well-connected nodes (also called compartments). Modularity is thus a concept similar to clustering. But while clustering applies to neighboring nodes, modularity is measured at the scale of the entire network (Delmas et al., 2019; Guimerà & Amaral, 2005). The modularity of a nest web, for example, can indicate whether a conservation strategy for a particular tree species will have a positive influence on an entire bird community of cavity excavators and nesters (if no modules are present) or whether strategies focusing on tree species in other modules are needed (Ruggera et al., 2016). For example, in analyzing the nest web of an Argentinian tropical forest, Ruggera et al. (2016) found that woodpeckers and nonexcavator birds formed distinct modules because the former interacts with both living and standing dead trees. In contrast, the nonexcavator birds use only decay-formed cavities in living trees. Consequently, they suggested that conservation efforts for cavity-nesting birds should focus on standing dead trees as much as on certain alive tree species. In species interaction networks and in spatial networks, a certain degree of modularity is beneficial to the system's stability or resilience because it impedes the negative cascading effects of species' extinction or prevents disturbances from rapidly propagating across the network of forest patches (Messier et al., 2019; Stouffer & Bascompte, 2011).

Nestedness is a characteristic of bipartite networks in which specialist species interact with a subset of the group of species with which generalists interact (Almeida-Neto et al., 2008; Delmas et al., 2019). Devoto et al. (2011) found strong nestedness in the moth pollination network of a boreal pine forest, which was associated with the dominance of a small core of generalist species that also interacted with the more specialized species. This asymmetric pattern made this hub of species, as well as the pollination service they provided, vulnerable to poor weather conditions.

Networks can also be analyzed by measuring the frequency with which different *motifs* appear in their architecture (Delmas et al., 2019). Motifs are smaller subsets of interacting nodes that are viewed as the building blocks of networks (Table 28.3; Milo et al., 2002). By simulating random networks that conserve some key properties of the observed network, e.g., order, size, and connectance, one can determine whether a particular motif occurs more frequently in the network than what would be expected by chance (Bodin & Tengö, 2012; Robins et al., 2007). In species interaction networks, motifs can be used to derive the different roles that individual species play in a network from their position in motifs (Stouffer et al., 2012). For example, Baker et al. (2015) found that despite variability in species composition in a host–parasitoid community with time and along a gradient of forest fragmentation, the role of species remained largely stable. In social webs, the analysis of motifs can be used to understand the relationships between structuring interactions and the ability of the system to adapt their management of natural resources. For instance, Fischer and Jasny (2017) found that homophily was a strong structuring pattern in the network formed by organizations concerned about increased wildfire risk. In this example, homophily may insulate organizations from being exposed to a diversity of ideas, thereby impeding their capacity to develop novel management strategies. In a social-ecological network that combined fire transmission (ecological links, see Table 28.1) and coordination of fire risk mitigation (social links), Hamilton et al. (2019) found

Table 28.3 Examples of possible motifs for different categories of network. Species interaction networks: **a** apparent competition of two consumers for a single resource; **b** a linear three-level food chain; and **c** omnivory in a three-level food chain. Social networks: **d** reciprocal interactions between three actors; **e** an actor acts as an intermediate between two actors that are otherwise disconnected; and **f** two disconnected actors report to a third that exerts a leadership role. Motifs of natural resource access in social-ecological networks (from Bodin & Tengö, 2012): **g** each actor manages their own resource independently even if their resources are ecologically linked, e.g., a spreading disturbance; **h** one actor depends on the other actor for access to the resource; **i** both actors compete for access to a single resource and are not engaged in any dialogue for co-management

Network category	Examples of possible motif
Species interaction networks	
Social networks	
Social-ecological networks	

that actors favored interactions with their immediate geographic neighbors, which constitutes an important challenge for the large-scale governance of wildfire risk.

28.4 Discussion

28.4.1 Challenges

The use of networks in forest ecology and management presents multiple challenges, the most important being the difficulty in identifying appropriate nodes and links. Creating species interaction networks requires intensive sampling to obtain high-resolution data. For instance, nest webs are constructed by identifying cavity-bearing trees and inferring excavator species by relying on the correlation between their body size and the diameter of cavity entrances. Interaction between tree and bird species is then determined by routinely inspecting in the field cavities using a camera system to observe signs of breeding or roosting (eggs, feathers, nestlings, etc.), which is labor intensive (Cockle & Martin, 2015; Ouellet-Lapointe et al., 2012; Ruggera et al., 2016). Investigating mycorrhizal networks requires sampling needles and cambium tissue from study trees, as well as an intensive sampling of the forest floor to collect tuberculate mycorrhizae (Beiler et al., 2010). Moreover, the identification of polli-nator–plant interactions requires field observations of flower visits (Gómez-Martínez et al., 2020; Memmott, 1999) or the capture of pollinator organisms to identify pollen on their body and quantification of the interaction by counting pollen grains (Devoto et al., 2011). Similarly, host–parasitoid interactions are determined by collecting host organisms in the field and then rearing parasitoids in the lab (Cagnolo et al., 2011; Van Veen et al., 2008). Accurate identification of species may rely on DNA barcoding, especially in species-rich systems where morphologically similar species abound (Smith et al., 2011). Obviously, reconstructing species interaction networks is sensitive to sampling efforts such that abundant species may receive more attention than rare ones (Cagnolo et al., 2011; Van Veen et al., 2008).

In spatial networks, nodes are identified from raster images, such as remote-sensing data, by aggregating adjacent cells that satisfy environmental criteria to be considered as forest or habitat patches, e.g., forest cover type and age, tree density (Bunn et al., 2000). Patches may be easily identified in landscapes presenting a dichotomous vegetation cover, such as fragmented forests in agricultural landscapes or urban settings. However, this task is more difficult in heterogeneous and contin-uous forest landscapes and for wildlife species whose habitat includes a diversity of cover types with varying preferences, e.g., the woodland caribou (*Rangifer tarandus caribou*; Galpern et al., 2011; O'Brien et al., 2006). In these cases, edge detection methods can be used to delineate patches from the matrix (Fortin, 1994). Moreover, field observations and expert opinions may be necessary to make certain assump-tions regarding how organisms interact with their environment, such that cover types and patches can be ranked according to their quality or relative use by the species of interest (O'Brien et al., 2006; Pascual-Hortal & Saura, 2007; Saura & Pascual-Hortal, 2007).

Links in spatial networks are generally identified by least-cost paths between nodes. This approach assumes that organisms travel between nodes using the most risk-free and efficient route, which may not always be the case for organisms charac-terized by anisotropic or passive dispersal. Moreover, in networks where links denote the movement of wind-dispersed seeds or certain bird and insect species, least-cost paths can be estimated using Euclidean distances between patches. However, for most wildlife species, links will correspond to nonlinear paths that consider the envi-ronmental heterogeneity of the matrix and the biological traits that influence their dispersal ability (Fall et al., 2007). Therefore, the determination of least-cost paths can be sensitive to the values and resolution of the resistance surface (Etherington, 2016;

Rayfield et al., 2010), which, in turn, requires intensive parameterization efforts; thus, the results may be prone to bias (Etherington, 2016).

Defining and identifying nodes and links is also a pervasive challenge when translating social and social-ecological systems into networks. Not unlike ecological data, collecting social data to identify entities and their interconnections involves substantial investment and is prone to errors. For example, a common approach to identify nodes in social networks is snowball sampling, an approach based on multiple steps (Doreian & Woodard, 1992; Fischer & Jasny, 2017; Hamilton et al., 2019; Knoot & Rickenbach, 2014). In the first step, a subset of key actors (single individuals or organizations) is selected and interviewed to obtain the names of other actors with whom they interact. In the second step, these additional actors are then interviewed to obtain yet again other names. The process continues until no new actors are identified (Fischer & Jasny, 2017). Depending on the number of sampling waves or the depth of the interviews, this approach can be time-consuming and subject to selection bias, e.g., well-connected individuals being identified more easily, and bias in reporting (or not) certain conflicting relationships, e.g., between opposing individuals or organizations (Doreian & Woodard, 1992).

Additionally, the construction of social-ecological networks often requires a certain level of aggregation of the ecological or social units determined by the system and the question under study. For example, studying coordination between forest management organizations may require that ecological nodes be scaled up to represent forests within the jurisdictional boundaries over which these organizations interact, thereby losing the local environmental specificity and limiting the utility of the network approach for managers working at local scales (Hamilton et al., 2019; Sayles et al., 2019). Bodin et al. (2019) proposed that a starting point in conceptualizing nodes and links is defining the social-ecological interdependencies central to the investigated management issue. Focusing on these connections will facilitate identifying the most relevant nodes and choosing the appropriate level of aggregation. Likewise, creating a network necessarily requires bounding the system under study. This bounding imposes an artificial frontier with the implicit assumption that connections beyond its limit have negligible impacts on the system's structure and function (Sayles et al., 2019). Given that any network analyses are sensitive to the number of nodes and links, which are themselves the result of the bounding choice, Sayles et al. (2019) suggested that studying the effects of different bounding approaches is needed to advance the field of social-ecological networks.

Common natural resource governance challenges occur in a variety of ecosystems, contexts, and scales (Ostrom, 2009). One goal of social-ecological network research is to understand the causal pathways between network structures and how these challenges emerge or are solved (Bodin & Tengö, 2012; Groce et al., 2019; Guerrero et al., 2020; Janssen et al., 2006). However, because of the numerous methodological choices involved in translating a social-ecological system into a network, the resulting network analysis tends to be specific to the studied system, limiting the ability to compare studies (Bodin et al., 2019; Young et al., 2006). Recently, Bodin et al. (2019) emphasized the need to develop a set of research design guidelines, applicable across contexts and scales, to facilitate synthesis and gain insights from diverse studies. These authors also suggested that advancing social-ecological network research requires a systematic classification of different basic causal relationships between simple patterns of network structure and environmental outcomes to help researchers make clearer assumptions about causality when more complex pathways are operating in their system (Bodin et al., 2019; Groce et al., 2019).

28.4.2   Benefits and Potential of the Network Approach in Forest Management

Managing for forest resilience has become imperative in a changing environment (Gauthier et al., 2015; Trumbore et al., 2015). Many symptoms of climate effects, invasive insects and diseases, and expanding land use are already evident within forest ecosystems (Hansen et al., 2013; Prăvălie, 2018). Moreover, due to global change, boundary conditions are shifting for many ecological processes, including disturbance regimes, species ranges, phenology, and carbon flux dynamics (Ramsfield et al., 2016; Seidl et al., 2017; Vose et al., 2019). Transition zones and locations where species exist at the limits of their current ecological tolerances, such as portions of the hemiboreal ecotone of eastern North America, may be particularly sensitive to these shifts (Thom et al., 2019). Management decisions that we make today must account for the uncertainty in future environmental threats, and they must anticipate uncertainty related to the rapidly changing economic and social context affecting demand for forest services and products. Network theory could help forest managers identify sensitivities and vulnerabilities linked with these changes and mitigate their effects accordingly, for instance through adaptive forest management (Gauthier et al., 2008; Millar et al., 2007). Moreover, the application of network theory could also likely be a key for monitoring biodiversity and projecting the future state of biodiversity in managed forests (Mina et al., 2021).

Recent applications of network theory have been proposed for evaluating and managing the resilience of large tracts of forests to global change stressors (Box 28.2; Aquilué et al., 2020, 2021; Mina et al., 2021). In these approaches, resilience is viewed as a multidimensional concept combining biodiversity and network topology measures likely to positively influence the capacity of spatial forest networks to cope with future disturbances (Messier et al., 2019). More precisely, resilience accounts for functional redundancy, the functional response diversity of forest metacommunities (Mori et al., 2013), their network connectivity, mean centrality, and modularity (Gonzalès & Parrott, 2012). Management strategies that modify one or more of these resilience-based properties can then be tested against scenarios of climate change and disturbance, e.g., drought, insect outbreak (Aquilué et al., 2020, 2021; Mina et al., 2021). For example, this approach can determine whether establishing plantations of functionally rare species or enriching forest stands to increase the variety of response traits, at locations that also improve forest network connectivity,

provide forest ecosystems the ability to resist or adapt to future environmental conditions. One can then use this approach to identify the management strategy that best
conserves the forest landscape under a range of possible but uncertain disturbances.

Network theory is useful for managing ecological recovery from natural disturbances, especially in the context of climate change where many disturbances are
expected to increase in occurrence, severity, and size (Prăvălie, 2018). For example,
following large forest fires, seeds, fungal spores, and organisms often disperse from
natural fire refugia (nodes in spatial forest networks) and then interact demograph-
ically as the landscapes recover through succession (Keeton & Franklin, 2004;
Krawchuk et al., 2020). In landscapes where fire management is used to mitigate
fire risk, network theory can assist in designing strategies, e.g., location and size of
prescribed burns, that preserve habitat connectivity for wildlife species (Sitters & Di
Stefano, 2020).

Spatiotemporal and spatial networks employed together with simulation models
could help predict shifting conditions in forest ecosystems to adapt management practices accordingly. For example, Huang et al. (2020) used spatiotemporal networks
together with species distribution models to determine how future climates will
affect habitat availability for terrestrial mammals experiencing range shifts in North
America. Future applications of networks are, therefore, expected to be used in
conjunction with other models to better integrate changing environmental conditions and ecological processes occurring at different spatial or temporal scales. For
example, Mina et al. (2021) coupled a spatial network approach with a spatially
explicit simulation model of forest dynamics (LANDIS-II, Mladenoff, 2004) to determine how climate-induced changes in forest cover influence landscape connectivity.
Wildemeersch et al. (2019) used a network-of-networks approach to simulate forest
pest outbreaks. Their model included a landscape-scale network of forest patches as
well as a stand-scale network within each patch. The small scale captured the local
pest pressure, whereas the large scale captured the influence of landscape connectivity
on the spreading behavior of the pest.

Networks are promising tools for multifunctional forest management because
they effectively integrate the interactions between social and ecological elements.
Spatial networks can help assess trade-offs between conflicting management goals.
For example, they can be used to determine management strategies that account for
ecological connectivity to satisfy conservation and economic targets (Ruppert et al.,
2016) or optimize the provision of multiple ecosystem services (Vogdrup-Schmidt
et al., 2019). Moreover, social-ecological networks can be used to identify linkages
that would foster coordinated efforts in the management of natural disturbance risks
(Hamilton et al., 2019), such as reducing fire risk hazards within the wildland–urban
interface (Keeton et al., 2007; Vilà-Vilardell et al., 2020).

To summarize, this chapter has demonstrated the richness and flexibility of the
network framework for forest management. Further applications of network theory
to forest management will necessitate an adaptive approach, accounting for shifting

dynamics and interactions among nodes, be they ecological or social. Network analysis is a powerful tool for identifying sensitivities and vulnerabilities within networks. It may prove invaluable in helping forest managers to better anticipate and adapt to global change.

References

Albert, R., & Barabási, A. L. (2002). Statistical mechanics of complex networks. *Reviews of Modern Physics, 74*, 47–97. https://doi.org/10.1103/RevModPhys.74.47.

Almeida-Neto, M., Guimarães, P., Guimarães, P. R., et al. (2008). A consistent metric for nestedness analysis in ecological systems: Reconciling concept and measurement. *Oikos, 117*, 1227–1239. https://doi.org/10.1111/j.0030-1299.2008.16644.x.

Aquilué, N., Filotas, É., Craven, D., et al. (2020). Evaluating forest resilience to global threats using functional response traits and network properties. *Ecological Applications, 30*, e02095. https://doi.org/10.1002/eap.2095.

Aquilué, N., Messier, C., Martins, K. T., et al. (2021). A simple-to-use management approach to boost adaptive capacity of forests to global uncertainty. *Forest Ecology and Management, 481*, 118692. https://doi.org/10.1016/j.foreco.2020.118692.

Baker, N. J., Kaartinen, R., Roslin, T., et al. (2015). Species' roles in food webs show fidelity across a highly variable oak forest. *Ecography, 38*, 130–139. https://doi.org/10.1111/ecog.00913.

Barabási, A. L., & Albert, R. (1999). Emergence of scaling in random networks. *Science, 286*, 509–512. https://doi.org/10.1126/science.286.5439.509.

Bascompte, J., Jordano, P., Melián, C. J., et al. (2003). The nested assembly of plant-animal mutualistic networks. *Proceedings of the National Academy of Sciences of the United States of America, 100*, 9383–9387. https://doi.org/10.1073/pnas.1633576100.

Beiler, K. J., Durall, D. M., Simard, S. W., et al. (2010). Architecture of the wood-wide web: *Rhizopogon* spp. genets link multiple Douglas-fir cohorts. *New Phytologist, 185*, 543–553. https://doi.org/10.1111/j.1469-8137.2009.03069.x.

Beiler, K. J., Simard, S. W., & Durall, D. M. (2015). Topology of tree–mycorrhizal fungus interaction networks in xeric and mesic Douglas-fir forests. *Journal of Ecology, 103*, 616–628. https://doi.org/10.1111/1365-2745.12387.

Bersier, L. F., Banašek-Richter, C., & Cattin, M. F. (2002). Quantitative descriptors of food-web matrices. *Ecology, 83*, 2394–2407. https://doi.org/10.1890/0012-9658(2002)083[2394:QDOFWM]2.0.CO;2.

Blüthgen, N., Menzel, F., & Blüthgen, N. (2006). Measuring specialization in species interaction networks. *BMC Ecology, 6*, 9. https://doi.org/10.1186/1472-6785-6-9.

Bodin, Ö. (2017). Collaborative environmental governance: Achieving collective action in social-ecological systems. *Science, 357*, 659. https://doi.org/10.1126/science.aan1114.

Bodin, Ö., & Tengö, M. (2012). Disentangling intangible social–Ecological systems. *Global Environmental Change, 22*, 430–439. https://doi.org/10.1016/j.gloenvcha.2012.01.005.

Bodin, O., Crona, B., & Ernstson, H. (2006). Social networks in natural resource management: What is there to learn from a structural perspective? *Ecology & Society, 11*, r2. https://doi.org/10.5751/ES-01808-1102r02.

Bodin, Ö., Alexander, S. M., Baggio, J., et al. (2019). Improving network approaches to the study of complex social–ecological interdependencies. *Nature Sustainability, 2*, 551–559. https://doi.org/10.1038/s41893-019-0308-0.

Borgatti, S. P., Mehra, A., Brass, D. J., et al. (2009). Network analysis in the social sciences. *Science, 323*, 892–895. https://doi.org/10.1126/science.1165821.

Bunn, A. G., Urban, D. L., & Keitt, T. H. (2000). Landscape connectivity: A conservation application of graph theory. *Journal of Environmental Management, 59*, 265–278. https://doi.org/10.1006/jema.2000.0373.

Cagnolo, L., Salvo, A., & Valladares, G. (2011). Network topology: Patterns and mechanisms in plant-herbivore and host-parasitoid food webs. *Journal of Animal Ecology, 80*, 342–351. https://doi.org/10.1111/j.1365-2656.2010.01778.x.

Chagnon, P. L., Bradley, R. L., & Klironomos, J. N. (2012). Using ecological network theory to evaluate the causes and consequences of arbuscular mycorrhizal community structure. *New Phytologist, 194*, 307–312. https://doi.org/10.1111/j.1469-8137.2011.04044.x.

Chama, L., Berens, D. G., Downs, C. T., et al. (2013). Habitat characteristics of forest fragments determine specialisation of plant-frugivore networks in a mosaic forest landscape. *PLoS ONE, 8*, e54956. https://doi.org/10.1371/journal.pone.0054956.

Cockle, K. L., & Martin, K. (2015). Temporal dynamics of a commensal network of cavity-nesting vertebrates: Increased diversity during an insect outbreak. *Ecology, 96*, 1093–1104. https://doi.org/10.1890/14-1256.1.

Cockle, K. L., Ibarra, J. T., Altamirano, T. A., et al. (2019). Interspecific networks of cavity-nesting vertebrates reveal a critical role of broadleaf trees in endangered Araucaria mixed forests of South America. *Biodiversity and Conservation, 28*, 3371–3386. https://doi.org/10.1007/s10531-019-01826-4.

Courbin, N., Fortin, D., Dussault, C., et al. (2014). Logging-induced changes in habitat network connectivity shape behavioral interactions in the wolf-caribou-moose system. *Ecological Monographs, 84*, 265–285. https://doi.org/10.1890/12-2118.1.

Craven, D., Filotas, E., Angers, V. A., et al. (2016). Evaluating resilience of tree communities in fragmented landscapes: Linking functional response diversity with landscape connectivity. *Diversity and Distributions, 22*, 505–518. https://doi.org/10.1111/ddi.12423.

D'Aloia, C. C., Naujokaitis-Lewis, I., Blackford, C., et al. (2019). Coupled networks of permanent protected areas and dynamic conservation areas for biodiversity conservation under climate change. *Frontiers in Ecology and Evolution, 7*, 7. https://doi.org/10.3389/fevo.2019.00027.

Dale, M. R. T., & Fortin, M. J. (2010). From graphs to spatial graphs. *Annual Review of Ecology, Evolution, and Systematics, 41*, 21–38. https://doi.org/10.1146/annurev-ecolsys-102209-144718.

Dale, M. R. T., & Fortin, M. J. (2021). *Quantitative analysis of ecological networks*. Cambridge: Cambridge University Press. https://doi.org/10.1017/9781108649018.

de la Fuente, B., Saura, S., & Beck, P. S. A. (2018). Predicting the spread of an invasive tree pest: The pine wood nematode in Southern Europe. *Journal of Applied Ecology, 55*, 2374–2385. https://doi.org/10.1111/1365-2664.13177.

Dee, L. E., Allesina, S., Bonn, A., et al. (2017). Operationalizing network theory for ecosystem service assessments. *Trends in Ecology & Evolution, 32*, 118–130. https://doi.org/10.1016/j.tree.2016.10.011.

Delmas, E., Besson, M., Brice, M. H., et al. (2019). Analysing ecological networks of species interactions. *Biological Reviews , 94*, 16–36. https://doi.org/10.1111/brv.12433.

Devoto, M., Bailey, S., & Memmott, J. (2011). The 'night shift': Nocturnal pollen-transport networks in a boreal pine forest. *Ecological Entomology, 36*, 25–35. https://doi.org/10.1111/j.1365-2311.2010.01247.x.

Doreian, P., & Woodard, K. L. (1992). Fixed list versus snowball selection of social networks. *Social Science Research, 21*, 216–233. https://doi.org/10.1016/0049-089X(92)90016-A.

Dunne, J. A., Williams, R. J., & Martinez, N. D. (2002a). Food-web structure and network theory: The role of connectance and size. *Proceedings of the National Academy of Sciences of the United States of America, 99*, 12917–12922. https://doi.org/10.1073/pnas.192407699.

Dunne, J. A., Williams, R. J., & Martinez, N. D. (2002b). Network structure and biodiversity loss in food webs: Robustness increases with connectance. *Ecology Letters, 5*, 558–567. https://doi.org/10.1046/j.1461-0248.2002.00354.x.

Etherington, T. R. (2016). Least-cost modelling and landscape ecology: Concepts, applications, and opportunities. *Current Landscape Ecology Reports, 1*, 40–53. https://doi.org/10.1007/s40 823-016-0006-9.

Eveleigh, E. S., McCann, K. S., McCarthy, P. C., et al. (2007). Fluctuations in density of an outbreak species drive diversity cascades in food webs. *Proceedings of the National Academy of Sciences of the United States of America, 104*, 16976–16981. https://doi.org/10.1073/pnas.0704301104.

Fall, A., Fortin, M. J., Manseau, M., et al. (2007). Spatial graphs: Principles and applications for habitat connectivity. *Ecosystems, 10*, 448–461. https://doi.org/10.1007/s10021-007-9038-7.

Ferrari, J. R., Preisser, E. L., & Fitzpatrick, M. C. (2014). Modeling the spread of invasive species using dynamic network models. *Biological Invasions, 16*, 949–960. https://doi.org/10.1007/s10 530-013-0552-6.

Filotas, E., Parrott, L., Burton, P. J., et al. (2014). Viewing forests through the lens of complex systems science. *Ecosphere 5*, art 1. https://doi.org/10.1890/ES13-00182.1.

Fischer, A. P. (2018). Forest landscapes as social-ecological systems and implications for management. *Landscape and Urban Planning, 177*, 138–147. https://doi.org/10.1016/j.landurbplan. 2018.05.001.

Fischer, A. P., & Jasny, L. (2017). Capacity to adapt to environmental change: Evidence from a network of organizations concerned with increasing wildfire risk. *Ecology & Society, 22*(1), 23. https://doi.org/10.5751/ES-08867-220123.

Folke, C. (2006). Resilience: The emergence of a perspective for social–ecological systems analyses. *Global Environmental Change, 16*, 253–267. https://doi.org/10.1016/j.gloenvcha.2006.04.002.

Fortin, M. J. (1994). Edge detection algorithms for two-dimensional ecological data. *Ecology, 75*, 956–965. https://doi.org/10.2307/1939419.

Galpern, P., Manseau, M., & Fall, A. (2011). Patch-based graphs of landscape connectivity: A guide to construction, analysis and application for conservation. *Biological Conservation, 144*, 44–55. https://doi.org/10.1016/j.biocon.2010.09.002.

Gauthier, S., Bernier, P., Kuuluvainen, T., et al. (2015). Boreal forest health and global change. *Science, 349*, 819–822. https://doi.org/10.1126/science.aaa9092.

Gauthier, S., Vaillancourt, M. A., Leduc, A., et al. (Eds.). (2008). *Aménagement écosystémique en forêt boréale*. Québec: Presses de l'Université du Québec.

Gómez-Martínez, C., Aase, A. L. T. O., Totland, Ø., et al. (2020). Forest fragmentation modifies the composition of bumblebee communities and modulates their trophic and competitive interactions for pollination. *Scientific Reports, 10*, 10872. https://doi.org/10.1038/s41598-020-67447-y.

Gonzalès, R., & Parrott, L. (2012). Network theory in the assessment of the sustainability of social-ecological systems. *Geography Compass, 6*, 76–88. https://doi.org/10.1111/j.1749-8198.2011. 00470.x.

Groce, J. E., Farrelly, M. A., Jorgensen, B. S., et al. (2019). Using social-network research to improve outcomes in natural resource management. *Conservation Biology, 33*, 53–65. https:// doi.org/10.1111/cobi.13127.

Guerrero, A. M., Barnes, M., Bodin, Ö., et al. (2020). Key considerations and challenges in the application of social-network research for environmental decision making. *Conservation Biology, 34*, 733–742. https://doi.org/10.1111/cobi.13461.

Guimerà, R., & Amaral L. A. N. (2005). Cartography of complex networks: Modules and universal roles. *Journal of Statistical Mechanics: Theory and Experiment, 02*, P02001. https://doi.org/10. 1088/1742-5468/2005/02/p02001.

Gurrutxaga, M., Rubio, L., & Saura, S. (2011). Key connectors in protected forest area networks and the impact of highways: A transnational case study from the Cantabrian Range to the Western Alps (SW Europe). *Landscape and Urban Planning, 101*, 310–320. https://doi.org/10.1016/j.lan durbplan.2011.02.036.

Hamilton, M., Fischer, A. P., & Ager, A. (2019). A social-ecological network approach for understanding wildfire risk governance. *Global Environmental Change, 54*, 113–123. https://doi.org/ 10.1016/j.gloenvcha.2018.11.007.

Henneman, M. L., & Memmott, J. (2001). Infiltration of a Hawaiian community by introduced biological control agents. *Science, 293*(5533), 1314–1316. https://doi.org/10.1126/science.106 0788.

Hansen, M. C., Potapov, P. V., Moore, R., et al. (2013). High-resolution global maps of 21st-century forest cover change. *Science, 342*, 850–853. https://doi.org/10.1126/science.1244693.

Huang, J. L., Andrello, M., Martensen, A. C., et al. (2020). Importance of spatio-temporal connectivity to maintain species experiencing range shifts. *Ecography, 43*, 591–603. https://doi.org/10. 1111/ecog.04716.

Ings, T. C., Montoya, J. M., Bascompte, J., et al. (2009). Ecological networks—Beyond food webs. *Journal of Animal Ecology, 78*, 253–269. https://doi.org/10.1111/j.1365-2656.2008.01460.x.

James, P., Rayfield, B., Fall, A., et al. (2005). Reserve network design combining spatial graph theory and species' spatial requirements. *Geomatica, 59*, 121–129.

Janssen, M. A., Bodin, Ö., Anderies, J. M., et al. (2006). Toward a network perspective of the study of resilience in social-ecological systems. *Ecology & Society, 11*(1), 15. https://doi.org/10.5751/ ES-01462-110115.

Jeong, H., Mason, S. P., Barabási, A. L., et al. (2001). Lethality and centrality in protein networks. *Nature, 411*, 41–42. https://doi.org/10.1038/35075138.

Keeton, W. S., & Franklin, J. F. (2004). Fire-related landform associations of remnant old-growth trees in the southern Washington Cascade range. *Canadian Journal of Forest Research, 34*, 2371–2381. https://doi.org/10.1139/x04-111.

Keeton, W. S., Mote, P. W., & Franklin, J. F. (2007). Climate variability, climate change, and western wildfire with implications for the urban–wildland interface. In T. Austin & G. K. Roger (Eds.), *Living on the Edge* (pp. 225–253). Somerville: Emerald Group Publishing Limited.

Kleindl, W. J., Stoy, P. C., Binford, M. W., et al. (2018). Toward a social-ecological theory of forest macrosystems for improved ecosystem management. *Forests, 9*, 200. https://doi.org/10.3390/f90 40200.

Knoot, T. G., & Rickenbach, M. (2014). Forester networks: The intersection of private lands policy and collaborative capacity. *Land Use Policy, 38*, 388–396. https://doi.org/10.1016/j.landusepol. 2013.11.025.

Kool, J. T., Moilanen, A., & Treml, E. A. (2013). Population connectivity: Recent advances and new perspectives. *Landscape Ecology, 28*, 165–185. https://doi.org/10.1007/s10980-012-9819-z.

Krawchuk, M. A., Meigs, G. W., Cartwright, J. M., et al. (2020). Disturbance refugia within mosaics of forest fire, drought, and insect outbreaks. *Frontiers in Ecology and the Environment, 18*, 235–244. https://doi.org/10.1002/fee.2190.

Martensen, A. C., Saura, S., & Fortin, M. J. (2017). Spatio-temporal connectivity: Assessing the amount of reachable habitat in dynamic landscapes. *Methods in Ecology and Evolution, 8*, 1253–1264. https://doi.org/10.1111/2041-210X.12799.

Martin, K., Aitken, K. E. H., & Wiebe, K. L. (2004). Nest sites and nest webs for cavity-nesting communities in interior British Columbia, Canada: Nest characteristics and niche partitioning. *The Condor, 106*, 5–19. https://doi.org/10.1093/condor/106.1.5.

Martin, K., & Eadie, J. M. (1999). Nest webs: A community-wide approach to the management and conservation of cavity-nesting forest birds. *Forest Ecology and Management, 115*, 243–257. https://doi.org/10.1016/S0378-1127(98)00403-4.

Melles, S., Fortin, M. J., Badzinski, D., et al. (2012). Relative importance of nesting habitat and measures of connectivity in predicting the occurrence of a forest songbird in fragmented landscapes. *Avian Conservation & Ecology, 7*(2), 3. https://doi.org/10.5751/ACE-00530-070203.

Memmott, J. (1999). The structure of plant-pollinator food web. *Ecology Letters, 2*, 276–280. https:// doi.org/10.1046/j.1461-0248.1999.00087.x.

Memmott, J., Godfray, H. C. J., & Gauld, I. D. (1994). The structure of a tropical host-parasitoid community. *Journal of Animal Ecology, 63*, 521–540. https://doi.org/10.2307/5219.

Memmott, J., Waser, N. M., & Price, M. V. (2004). Tolerance of pollination networks to species extinctions. *Proceedings of the Royal Society of London. Series B, Biological Sciences, 271*, 2605–2611. https://doi.org/10.1098/rspb.2004.2909.

Messier, C., Bauhus, J., Doyon, F., et al. (2019). The functional complex network approach to foster forest resilience to global changes. *Forest Ecosystems, 6*, 21. https://doi.org/10.1186/s40663-019-0166-2.

Millar, C. I., Stephenson, N. L., & Stephens, S. L. (2007). Climate change and forests of the future: Managing in the face of uncertainty. *Ecological Applications, 17*, 2145–2151. https://doi.org/10.1890/06-1715.1.

Milo, R., Shen-Orr, S., Itzkovitz, S., et al. (2002). Network motifs: Simple building blocks of complex networks. *Science, 298*, 824–827. https://doi.org/10.1126/science.298.5594.824.

Mina, M., Messier, C., Duveneck, M., et al. (2021). Network analysis can guide resilience-based management in forest landscapes under global change. *Ecological Applications, 31*(1), e2221. https://doi.org/10.1002/eap.2221.

Minor, E. S., Tessel, S. M., Engelhardt, K. A. M., et al. (2009). The role of landscape connectivity in assembling exotic plant communities: A network analysis. *Ecology, 90*, 1802–1809. https://doi.org/10.1890/08-1015.1.

Mladenoff, D. J. (2004). LANDIS and forest landscape models. *Ecological Modelling, 180*, 7–19. https://doi.org/10.1016/j.ecolmodel.2004.03.016.

Montoya, J. M., Pimm, S. L., & Solé, R. V. (2006). Ecological networks and their fragility. *Nature, 442*, 259–264. https://doi.org/10.1038/nature04927.

Mori, A. S., Furukawa, T., & Sasaki, T. (2013). Response diversity determines the resilience of ecosystems to environmental change. *Biological Reviews, 88*, 349–364. https://doi.org/10.1111/brv.12004.

Newman, M. E. J. (2003). The structure and function of complex networks. *SIAM Review, 45*, 167–256. https://doi.org/10.1137/S003614450342480.

O'Brien, D., Manseau, M., Fall, A., et al. (2006). Testing the importance of spatial configuration of winter habitat for woodland caribou: An application of graph theory. *Biological Conservation, 130*, 70–83. https://doi.org/10.1016/j.biocon.2005.12.014.

Ostrom, E. (2009). A general framework for analyzing sustainability of social-ecological systems. *Science, 325*, 419–422. https://doi.org/10.1126/science.1172133.

Ouellet-Lapointe, U., Drapeau, P., Cadieux, P., et al. (2012). Woodpecker excavations suitability for and occupancy by cavity users in the boreal mixedwood forest of eastern Canada. *Ecoscience, 19*, 391–397. https://doi.org/10.2980/19-4-3582.

Pascual-Hortal, L., & Saura, S. (2007). Impact of spatial scale on the identification of critical habitat patches for the maintenance of landscape connectivity. *Landscape and Urban Planning, 83*, 176–186. https://doi.org/10.1016/j.landurbplan.2007.04.003.

Pelletier, D., Lapointe, M. É., Wulder, M. A., et al. (2017). Forest connectivity regions of Canada using circuit theory and image analysis. *PLoS ONE, 12*, e0169428. https://doi.org/10.1371/journal.pone.0169428.

Pimm, S. L., Lawton, J. H., & Cohen, J. E. (1991). Food web patterns and their consequences. *Nature, 350*, 669–674. https://doi.org/10.1038/350669a0.

Prăvălie, R. (2018). Major perturbations in the Earth's forest ecosystems. Possible implications for global warming. *Earth-Science Reviews, 185*, 544–571. https://doi.org/10.1016/j.earscirev.2018.06.010.

Proulx, S. R., Promislow, D. E. L., & Phillips, P. C. (2005). Network thinking in ecology and evolution. *Trends in Ecology & Evolution, 20*, 345–353. https://doi.org/10.1016/j.tree.2005.04.004.

Ramsfield, T. D., Bentz, B. J., Faccoli, M., et al. (2016). Forest health in a changing world: Effects of globalization and climate change on forest insect and pathogen impacts. *Forestry, 89*, 245–252. https://doi.org/10.1093/forestry/cpw018.

Rayfield, B., Fortin, M. J., & Fall, A. (2010). The sensitivity of least-cost habitat graphs to relative cost surface values. *Landscape Ecology, 25*, 519–532. https://doi.org/10.1007/s10980-009-9436-7.

Rayfield, B., Fortin, M. J., & Fall, A. (2011). Connectivity for conservation: A framework to classify network measures. *Ecology, 92*, 847–858. https://doi.org/10.1890/09-2190.1.

Robins, G., Pattison, P., Kalish, Y., et al. (2007). An introduction to exponential random graph (p*) models for social networks. *Social Networks, 29*, 173–191. https://doi.org/10.1016/j.socnet.2006. 08.002.

Ruggera, R. A., Schaaf, A. A., Vivanco, C. G., et al. (2016). Exploring nest webs in more detail to improve forest management. *Forest Ecology and Management, 372*, 93–100. https://doi.org/10. 1016/j.foreco.2016.04.010.

Ruppert, J. L. W., Fortin, M. J., Gunn, E. A., et al. (2016). Conserving woodland caribou habitat while maintaining timber yield: A graph theory approach. *Canadian Journal of Forest Research, 46*, 914–923. https://doi.org/10.1139/cjfr-2015-0431.

Saura, S., Estreguil, C., Mouton, C., et al. (2011). Network analysis to assess landscape connectivity trends: Application to European forests (1990–2000). *Ecological Indicators, 11*, 407–416. https:// doi.org/10.1016/j.ecolind.2010.06.011.

Saura, S., & Pascual-Hortal, L. (2007). A new habitat availability index to integrate connectivity in landscape conservation planning: Comparison with existing indices and application to a case study. *Landscape and Urban Planning, 83*, 91–103. https://doi.org/10.1016/j.landurbplan.2007. 03.005.

Sayles, J. S., Mancilla Garcia, M., Hamilton, M., et al. (2019). Social-ecological network analysis for sustainability sciences: A systematic review and innovative research agenda for the future. *Environmental Research Letters, 14*, 093003. https://doi.org/10.1088/1748-9326/ab2619.

Seidl, R., Thom, D., Kautz, M., et al. (2017). Forest disturbances under climate change. *Nature Climate Change, 7*, 395–402. https://doi.org/10.1038/nclimate3303.

Siegert, N. W., Mercader, R. J., & McCullough, D. G. (2015). Spread and dispersal of emerald ash borer (Coleoptera: Buprestidae): Estimating the spatial dynamics of a difficult-to-detect invasive forest pest. *The Canadian Entomologist, 147*, 338–348. https://doi.org/10.4039/tce.2015.11.

Simard, S. W. (2009). The foundational role of mycorrhizal networks in self-organization of interior Douglas-fir forests. *Forest Ecology and Management, 258*, S95–S107. https://doi.org/10.1016/j. foreco.2009.05.001.

Sitters, H., & Di Stefano, J. (2020). Integrating functional connectivity and fire management for better conservation outcomes. *Conservation Biology, 34*, 550–560. https://doi.org/10.1111/cobi. 13446.

Smith, M. A., Eveleigh, E. S., McCann, K. S., et al. (2011). Barcoding a quantified food web: Crypsis, concepts, ecology and hypotheses. *PLoS ONE, 6*, e14424. https://doi.org/10.1371/jou rnal.pone.0014424.

Southworth, D., He, X. H., Swenson, W., et al. (2005). Application of network theory to potential mycorrhizal networks. *Mycorrhiza, 15*, 589–595. https://doi.org/10.1007/s00572-005-0368-z.

Stouffer, D. B., & Bascompte, J. (2011). Compartmentalization increases food-web persistence. *Proceedings of the National Academy of Sciences of the United States of America, 108*, 3648–3652. https://doi.org/10.1073/pnas.1014353108.

Stouffer, D. B., Sales-Pardo, M., Sirer, M. I., et al. (2012). Evolutionary conservation of species' roles in food webs. *Science, 335*, 1489–1492. https://doi.org/10.1126/science.1216556.

Strogatz, S. H. (2001). Exploring complex networks. *Nature, 410*, 268–276. https://doi.org/10.1038/ 35065725.

Tamme, R., Götzenberger, L., Zobel, M., et al. (2014). Predicting species' maximum dispersal distances from simple plant traits. *Ecology, 95*, 505–513. https://doi.org/10.1890/13-1000.1.

Thom, D., Golivets, M., Edling, L., et al. (2019). The climate sensitivity of carbon, timber, and species richness covaries with forest age in boreal-temperate North America. *Global Change Biology, 25*, 2446–2458. https://doi.org/10.1111/gcb.14656.

Tittler, R., Filotas, É., Kroese, J., et al. (2015). Maximizing conservation and production with intensive forest management: It's all about location. *Environmental Management, 56*, 1104–1117. https://doi.org/10.1007/s00267-015-0556-3.

Toju, H., Guimarães, P. R., Olesen, J. M., et al. (2015). Below-ground plant-fungus network topology is not congruent with above-ground plant-animal network topology. *Science Advances, 1*, e1500291. https://doi.org/10.1126/sciadv.1500291.

Trumbore, S., Brando, P., & Hartmann, H. (2015). Forest health and global change. *Science, 349*, 814–818. https://doi.org/10.1126/science.aac6759.

Turnbull, L., Hütt, M. T., Ioannides, A. A., et al. (2018). Connectivity and complex systems: Learning from a multi-disciplinary perspective. *Applied Network Science, 3*, 11. https://doi.org/10.1007/s41109-018-0067-2.

Tylianakis, J. M., Tscharntke, T., & Lewis, O. T. (2007). Habitat modification alters the structure of tropical host-parasitoid food webs. *Nature, 445*, 202–205. https://doi.org/10.1038/nature05429.

Urban, D., & Keitt, T. (2001). Landscape connectivity: A graph-theoretic perspective. *Ecology, 82*, 1205–1218. https://doi.org/10.1890/0012-9658(2001)082[1205:LCAGTP]2.0.CO;2.

Van Dorp, C. H., Simard, S. W., & Durall, D. M. (2020). Resilience of *Rhizopogon*-Douglas-fir mycorrhizal networks 25 years after selective logging. *Mycorrhiza, 30*, 467–474. https://doi.org/10.1007/s00572-020-00968-6.

Van Veen, F. J. F., Müller, C. B., Pell, J. K., et al. (2008). Food web structure of three guilds of natural enemies: Predators, parasitoids and pathogens of aphids. *Journal of Animal Ecology, 77*, 191–200. https://doi.org/10.1111/j.1365-2656.2007.01325.x.

Vázquez, D. P., Blüthgen, N., Cagnolo, L., et al. (2009). Uniting pattern and process in plant-animal mutualistic networks: A review. *Annals of Botany, 103*, 1445–1457. https://doi.org/10.1093/aob/mcp057.

Vidal, M., Cusick, M. E., & Barabási, A. L. (2011). Interactome networks and human disease. *Cell, 144*, 986–998. https://doi.org/10.1016/j.cell.2011.02.016.

Vilà-Vilardell, L., Keeton, W. S., Thom, D., et al. (2020). Climate change effects on wildfire hazards in the wildland-urban-interface—Blue pine forests of Bhutan. *Forest Ecology and Management, 461*, 117927. https://doi.org/10.1016/j.foreco.2020.117927.

Vogdrup-Schmidt, M., Olsen, S. B., Dubgaard, A., et al. (2019). Using spatial multi-criteria decision analysis to develop new and sustainable directions for the future use of agricultural land in Denmark. *Ecological Indicators, 103*, 34–42. https://doi.org/10.1016/j.ecolind.2019.03.056.

Vose, J., Peterson, D., Domke, G., et al. (2019). Forests. In D. R. Reidmiller, C. W. Avery, D. R. Easterling, et al. (Eds.), *Impacts, risks, and adaptation in the United States: Fourth national climate assessment* (pp. 232–267). Washington: Global Change Research Program.

Watts, D. J., & Strogatz, S. H. (1998). Collective dynamics of 'small-world' networks. *Nature, 393*, 440–442. https://doi.org/10.1038/30918.

Wildemeersch, M., Franklin, O., Seidl, R., et al. (2019). Modelling the multi-scaled nature of pest outbreaks. *Ecological Modelling, 409*, 108745. https://doi.org/10.1016/j.ecolmodel.2019.108745.

Young, O. R., Lambin, E. F., Alcock, F., et al. (2006). A portfolio approach to analyzing complex human-environment interactions: Institutions and land change. *Ecology & Society, 11*(2), art 31. https://doi.org/10.5751/ES-01799-110231.

Zeigler, S. L., & Fagan, W. F. (2014). Transient windows for connectivity in a changing world. *Movement Ecology, 2*(1), 1. https://doi.org/10.1186/2051-3933-2-1.

Chapter 29
Land and Freshwater Complex Interactions in Boreal Forests: A Neglected Topic in Forest Management

Guillaume Grosbois, Danny Chun Pong Lau, Martin Berggren, Miguel Montoro Girona, Willem Goedkoop, Christian Messier, Joakim Hjältén, and Paul del Giorgio

Abstract Aquatic and terrestrial habitats are interdependent components of the boreal forest landscape involving multiple dynamic interactions; these are manifested particularly in riparian areas, which are key components in the forest landscape. However, this interdependence between aquatic and terrestrial habitats is not adequately accounted for in the current management of forest ecosystems. Here we review the impacts of land disturbances on the optical and physicochemical properties of water bodies, aquatic food web health, and the ecological functioning of these freshwaters. We also describe how freshwaters influence the adjacent terrestrial ecosystems. A better understanding of these dynamic biotic and abiotic interactions between land and freshwater of the boreal forest is a first step toward including these freshwaters in the sustainable management of the boreal forest.

G. Grosbois (✉) · M. M. Girona
Groupe de Recherche en Écologie de la MRC-Abitibi, Institut de recherche sur les forêts, Université du Québec en Abitibi-Témiscamingue, Amos Campus, 341, rue Principale Nord, Amos, QC J9T 2L8, Canada
e-mail: guillaume.grosbois@uqat.ca

M. M. Girona
e-mail: miguel.montoro@uqat.ca

G. Grosbois · D. C. P. Lau · W. Goedkoop
Department of Aquatic Sciences and Assessment, Swedish University of Agricultural Sciences, P.O. Box 7050, SE-75007 Uppsala, Sweden
e-mail: danny.lau@slu.se

W. Goedkoop
e-mail: willem.goedkoop@slu.se

D. C. P. Lau
Department of Ecology and Environmental Science, Umeå University, Linnaeus väg 6, 90187 Umeå, Sweden

© The Author(s) 2023   719
M. M. Girona et al. (eds.), *Boreal Forests in the Face of Climate Change*,
Advances in Global Change Research 74,
https://doi.org/10.1007/978-3-031-15988-6_29

29.1   Introduction

The boreal forest, the world's largest land biome, is characterized by a high density and diversity of freshwater environments. These water bodies form a complex aquatic network that interacts dynamically with the surrounding terrestrial environment. However, scientists have traditionally kept to their respective areas of interest, and research in the respective fields of terrestrial and aquatic ecology has remained separate. Consequently, there is a knowledge gap in our understanding of the interactions between aquatic and terrestrial habitats as well as those processes specific to the aquatic–terrestrial ecotones (Hjältén et al., 2016). Nevertheless, environments such as the littoral zone in lakes and the shoreline area in forests are habitats rich in biological diversity and are sites where essential processes occur, e.g., the primary production of macrophytes and benthic algae or tree species associated with wet areas. Land–water interactions, comprising energy and matter fluxes, occur mainly in this ecotone (Fig. 29.1). The largest fluxes from land toward freshwater are the dissolved and particulate organic and inorganic matter carried by surface runoff, groundwater, and wind (Vander Zanden & Gratton, 2011). The freshwater to land fluxes are smaller in volume but greater in energy and are higher in nutritional quality. These fluxes follow animal movements, such as insect emergence or terrestrial predation on aquatic prey.

Both types of fluxes are intimately entangled, connecting the land and water environments in the boreal forest (Baxter et al., 2005); however, these habitats are commonly viewed as separate ecosystems and are thus managed by distinct environmental institutions or agencies. The aquatic habitats are therefore usually not directly included in sustainable forest management. This view may pose risks of underestimating the complexity of the structure and functioning of the forest and

M. Berggren
Department of Physical Geography and Ecosystem Science, Lund University, Sölvegatan 12, S-223 62 Lund, Sweden
e-mail: martin.berggren@nateko.lu.se

M. M. Girona · J. Hjältén
Restoration Ecology Group, Department of Wildlife, Fish, and Environmental Studies, Swedish University of Agricultural Sciences, 901 83 Umeå, Sweden
e-mail: Joakim.Hjalten@slu.se

M. M. Girona · C. Messier
Centre d'étude de la forêt/Centre for Forest Research, Université du Québec à Montréal, P.O. Box 8888, Stn. Centre-Ville, Montréal, QC H3C 3P8, Canada
e-mail: christian.messier@uqo.ca

C. Messier
Département des Sciences Naturelles, Institut des Sciences de la Forêt Tempérée, Université du Québec en Outaouais, 58 Rue Principale, Ripon, QC J0V 1V0, Canada

P. del Giorgio
Department of Biological Sciences, Université du Québec à Montréal, P.O. Box 8888, Stn. Centre-Ville, Montréal, QC H3C 3P8, Canada
e-mail: del_giorgio.paul@uqam.ca

Fig. 29.1 a Connectivity between the aquatic and terrestrial habitats of the boreal forest; **b** forest harvesting in a lake watershed. *Photo credits* Miguel Montoro Girona and Janie Lavoie

hinder successful management (Gauthier et al., 2009; Messier et al., 2013). For an improved representation of reality, we highlight in this chapter that terrestrial and freshwater habitats are integral parts of the same boreal forest ecosystem, and, as such, we will no longer refer to terrestrial and aquatic ecosystems as separate entities.

One way to consider land–water interactions is by implementing riparian buffer strips, which are commonly included in forest management. Although the direct and indirect effects of these buffer strips on aquatic habitats have been studied extensively (Lidman et al., 2017a, b; Peterjohn & Correll, 1984), there remains an absence of information of the site-specific role of riparian forests, e.g., on different types of forest soil, with different slopes or different stand compositions (Kuglerová et al., 2014); this knowledge gap impedes the advancement of best practices in boreal forest management (Kuglerová et al., 2017). Furthermore, the effectiveness of riparian forests in moderating the impacts of harvesting on the terrestrial part of the boreal forest has rarely been tested. The construction of roads related to forestry practices and the type and timing of harvesting may affect how forestry machinery modifies stream habitats, both directly and indirectly, by increasing turbidity (Reiter et al., 2009). Extensive cutting may raise the water table and increase the amount of organic matter exported to aquatic habitats (Laudon et al., 2009; Sun et al., 2000). In addition, increases in the number of water pools in the tracks of the machines may increase mercury methylation and the related contamination of the landscape (de Wit et al., 2014; Sørensen et al., 2009). Fertilization, liming, and fire control may also affect the leaching of nutrients and other constituents into the water (Bisson et al., 1992; Degerman et al., 1995). This chapter therefore aims to describe the existing natural and anthropogenic interactions between the aquatic and terrestrial habitats of the boreal forest and highlight the importance of integrating the aquatic environments into the sustainable management of the forest. We demonstrate that revising the existing forest management paradigm, which currently considers land and freshwater habitats as isolated ecosystems, is essential to acknowledge the interdependence between these components and enhance forest management success.

29.2   The Browning of Boreal Freshwaters

A portion of the brown color of boreal lakes and rivers is caused by dissolved organic matter (DOM) leaching from the soil and litter of forests (Roulet & Moore, 2006). The DOM concentration, including the dissolved organic carbon (DOC) component, has increased in freshwaters during the last decades to shift the water color of boreal water bodies toward brown, a phenomenon called *browning* or *brownification* (Fig. 29.2; Lennon et al., 2013). The browning of freshwater is a serious concern for many ecosystem services, including drinking water supply (Haaland et al., 2010), as treatment of brown water rich in DOM is costly and may produce chlorinated carcinogenic compounds (Richardson et al., 2007). Whereas the extent and intensity of browning remain poorly studied at the global scale, the most dramatic changes in DOC concentrations (>0.15 mg·L^{-1}·yr^{-1}) have been reported in temperate and boreal regions (de Wit et al., 2016; Monteith et al., 2007).

Browning is associated with natural and anthropogenic changes in land cover as well as other factors that modify the amount and nature of organic matter exported from the watershed into freshwater habitats (Kritzberg, 2017). In boreal forests, increasing DOC concentrations in lakes is a direct consequence of forest harvesting. Harvesting raises the water table, thereby increasing the hydrological connectivity between shallow DOC-rich soils and the recipient freshwaters (Glaz et al., 2015). Organic matter export may also increase because of greater terrestrial primary production in the forest (Larsen et al., 2011), enhanced by increased atmospheric carbon dioxide levels (Campbell et al., 2017). Other factors include increased nitrogen deposition (Rowe et al., 2014), higher precipitation (Hongve et al.,

Fig. 29.2 Lake water with low (*left*) and high (*right*) concentrations of dissolved organic matter. *Photo credit* Guillaume Grosbois

2004), and greater export of iron, which can form stable, highly colored complexes with DOM (Weyhenmeyer et al., 2014). Several recent studies have attributed the DOC increase to recovery from atmospheric acid deposition, which enhances organic matter solubility and, therefore, the mobility of DOC from forest to aquatic environments (Meyer-Jacob et al., 2019). These researchers questioned whether the current browning of freshwaters reflected a re-browning to reach past levels of water color or whether this represented a new trend toward unprecedented browning (Meyer-Jacob et al., 2020). Both trends have been observed, and whereas re-browning is associated with regions having received moderate to high levels of acid deposition, higher than preindustrial DOC concentrations have been recorded in regions characterized by low levels of acid deposition (Meyer-Jacob et al., 2019). Thus, recovery from acidification is insufficient to explain browning in all lakes, and the trends in water color may be more related to changes in land use, e.g., from agriculture to modern forestry (Kritzberg, 2017).

An increase in DOM in freshwaters can drastically change the structure and functioning of inland waters having a darker water color by increasing the surface water temperature and modifying thermal stratification (Williamson et al., 2015). Although higher temperatures can potentially enhance algal production, browner waters also attenuate the penetration of photosynthetically active radiation and therefore diminish a water body's suitability for aquatic primary production (Creed et al., 2018). Increasing DOM in freshwater strongly affects microbial processes by shifting basal pelagic production from phytoplankton to bacteria (Jansson et al., 2000). Although the biodegradation coefficient per unit of DOM is reduced in brown lakes and rivers (Berggren & Al-Kharusi, 2020; Berggren et al., 2020), the absolute amount of organic carbon that is assimilated and metabolized by microorganisms increases with greater DOM concentrations (Lennon, 2004). This has consequences for higher trophic levels, affecting zooplankton production and fish survival and growth (Hedström et al., 2017; Taipale et al., 2018). Browning therefore radically alters the functioning and subsidies of aquatic food webs (Hayden et al., 2019).

In the near future, green (eutrophication) and brown (browning) lakes are expected to replace blue lakes in the boreal forest (Leech et al., 2018). The increase in DOM export in some regions is clear, particularly in northern ecosystems where higher temperatures and shorter winters will allow a greater water movement through litter and soil and thus increase DOM export to freshwaters (Laudon et al., 2012). These major changes in DOC could modify the overall C budget of the boreal forest, and future research must focus on this issue to better understand how these new conditions will influence the future C sinks and sources.

29.3   Nutrients in Freshwaters: Eutrophication and Oligotrophication

Lake and river environments depend strongly on nutrient export from their associated watersheds (Hynes, 1975). In boreal forests, nutrients are fixed via atmospheric deposition, N-fixation by cyanobacteria, or weathering, and a portion is exported to aquatic ecosystems through runoff (Hall, 2003). Nutrients are not only transported by freshwaters but are also physically, chemically, and biologically processed. One way of estimating these biotic contributions is to experimentally manipulate the watershed, for example by preventing acid rain (Hultberg & Skeffington, 1998) or harvesting the trees (Likens et al., 1970). During the Hubbard Brook Experiment, the cutting of all watershed vegetation demonstrated the key role of terrestrial vegetation in retaining nutrients within the watershed (Likens et al., 1970). Moreover, nutrient export to streams and lakes drastically increased after tree harvesting; this highlighted the terrestrial origin of most nutrients. Rivers and lakes are, therefore, highly dependent on watershed sources of nutrients. The Global Nutrient Export from Watersheds (NEWS) model estimated that, at the global scale, the export of dissolved inorganic nitrogen (N) and phosphorus (P) is dominated by anthropogenic sources, whereas 80% of the dissolved organic N and P is dominated by natural sources (Seitzinger et al., 2005). Bernhardt et al. (2003) also demonstrated that rivers react to major forest disturbances, reducing nitrogen concentrations downstream of the disturbance. Thus, aquatic environments are strongly dependent on the export of forest nutrients and are biogeochemically reactive habitats where algae and grazers/predators have a high capacity for assimilating and storing nutrients.

Increased exploitation of natural resources, e.g., forestry and mining, and associated infrastructure development in boreal regions has increased the leakage of carbon, nitrogen, and phosphorus from catchments into watercourses (Payette et al., 2001). This increase of biologically reactive dissolved inorganic nitrogen has been measured over large areas globally and has been found to affect nutrient ratios in freshwaters and lead to lake eutrophication (Hessen, 2013). Such an increase in nutrient inputs may shift an initial N-limitation to P-limitation for primary producers and consumers; this alteration causes major changes to the base of food webs in lakes (Elser et al., 2009, 2010) and rivers (Chen et al., 2013). Additionally, terrestrial organic matter is associated with DOM, nitrogen, and phosphorus (Aitkenhead-Peterson et al., 2003), and their respective inputs increase with browning (see Sect. 29.2). Many boreal freshwater ecosystems will experience greater nutrient inputs in the future; however, the effect on the functioning of food webs remains relatively unknown. Prepas et al. (2001) demonstrated that harvesting around the headwater lakes of Alberta's boreal plain increased total phosphorus concentration in the lakes, triggering an increase in cyanobacteria and cyanotoxins. Consequently, zooplankton biomass and abundance decreased in these boreal lakes impacted by forest harvesting. Given that N and P export to freshwaters is predicted to increase over the next decades, terrestrial–aquatic interactions must be accounted for in the future management of the boreal forest.

In already-nutrient-poor systems in many boreal regions of the world, e.g., Sweden, Canada, and Finland, lakes have experienced a substantial, long-term decline in total phosphorus concentrations ($2.1\%\cdot yr^{-1}$ since the 1980s), reaching values close to the detection limit; this process is referred to as lake oligotrophication (Arvola et al., 2011; Eimers et al., 2009; Huser et al., 2018). These declines are linked to the recovery from acidification, the increased climate change–induced trapping of nutrients by plant roots (Elmendorf et al., 2012), changes in watershed processes related to soil properties (Gustafsson et al., 2012), or a combination thereof. Cyanobacterial blooms are often triggered by an excess of nutrients in lakes; however, they can also be enhanced by N-limitation, as cyanobacteria can fix environmental N and are therefore more competitive than other algae (Berman-Frank et al., 2007). Diehl et al. (2018) report that cyanobacteria make up more than 50% of the biomass in shallow, epilithic biofilms in northern Swedish lakes. Cyanobacteria-dominated communities in oligotrophic waters negatively impact lake food webs through a lower trophic transfer efficiency from primary producers to consumers (Brett & Muller-Navarra, 1997). Declining total phosphorus concentrations could therefore act in synchrony with increasing DOM and suppress aquatic productivity.

When harvesting a significant area of a watershed as part of a management strategy in boreal forests, it is essential to consider the effect on the already fragile balance of nutrients resulting from browning, eutrophication, or oligotrophication.

29.4   Metabolism in Freshwaters: Heterotrophy Versus Autotrophy

Inland waters process, store, and outgas most of the carbon (C) exported from land, thereby playing a significant role in the global C cycle (Battin et al., 2009; Raymond et al., 2013). These external carbon inputs can vary from being a primary energy source in freshwater metabolism (*heterotrophy*) to being less important relative to aquatic photosynthetic production (*autotrophy*). This reliance depends on many factors that determine the interactions between the aquatic and forest components of the boreal landscape, such as ecosystem size. For example, small streams connected to the landscape mainly emit terrestrially derived carbon dioxide (CO_2), whereas larger rivers mostly re-emit CO_2 previously fixed by aquatic primary producers (Hotchkiss et al., 2015). Inland waters are classified as net heterotrophic when metabolism based on external watershed inputs causes the CO_2 emissions from aquatic organisms, i.e., respiration R, to be greater than aquatic gross primary production (GPP). In contrast, waters are classified as net autotrophic when the metabolism is predominantly based on aquatic production and, therefore, R is less than GPP (Jansson et al., 2000). New inputs of organic matter from the boreal forest are also a source of nutrients to the water and may increase GPP even though they may not change the net ecosystem production (NEP, i.e., GPP–R), as it may also increase R (Cole et al., 2000). New inputs from forests, such as those following forest harvesting, thus affect the basal production of aquatic ecosystems, potentially increasing heterotrophic bacterial metabolism as well as photosynthetic production. Clapcott and Barmuta (2010) demonstrated that forest harvesting increases metabolism and organic matter processes in small headwater streams; however, it remains unclear whether this is the case in larger ecosystems (Klaus et al., 2018). The impact of forest harvesting on aquatic metabolism is, therefore, very difficult to predict because organic matter inputs from the watershed are susceptible to switching from an equilibrated ecosystem, i.e., $R = GPP$, toward net heterotrophy when bacterial metabolism is favored but also toward net autotrophy when the associated nutrients favor photosynthetic algal production.

Inland waters are classified in terms of their trophic state as eutrophic (very productive), mesotrophic (moderately productive), or oligotrophic (unproductive) environments. The impact of forest harvesting is expected to differ according to the trophic state of lakes and rivers. Extra watershed inputs of organic matter to eutrophic lakes may increase an already high GPP but still trigger a switch of an ecosystem from a sink to a source of greenhouse gases because of increased production and diffusion of methane (Grasset et al., 2020). Greater nutrient and organic matter inputs from the catchment can unbalance autotrophy versus heterotrophy processes and lead to an accumulation of organic matter in the sediments. The degradation of this extra t-OM under anoxic conditions in the sediments is likely to produce methane (Donis et al., 2017). Mesotrophic and oligotrophic lakes can alternate from autotrophic to heterotrophic states, in different seasons, from one year to another, or after meteorological events such as storms that uncouple R and GPP (Richardson et al., 2017; Vachon & del Giorgio, 2014; Vachon et al., 2017). Given that even within the same ecosystem, C from the forest and C synthesized by algae can be used differently by bacteria—forest C for biomass and algal C for respiration (Guillemette et al., 2016)—the impact of forest harvesting on aquatic bacterial communities is difficult to predict.

Limnologists often consider terrestrial DOM (t-DOM) to be homogeneous; in reality, however, t-DOM represents a mix of material of very different origins and differing states of degradation (Berggren et al., 2010b). It includes low molecular weight compounds, such as carboxylic acids, amino acids, and carbohydrates, which bacteria easily utilize and eventually transfer to higher trophic levels (Berggren et al., 2010a). Molecules are referred to as labile, semi-labile, or recalcitrant according to how easily they can be degraded by bacteria (Kragh & Sondergaard, 2004). The terrestrial share of DOM is composed of compounds that originate from terrestrial plant tissues, and soil microorganism communities often modify these compounds before entering inland waters (Solomon et al., 2015). Generally, t-DOM is composed of humic and fulvic acids that contain aromatic hydrocarbons, including phenols, carboxylic acids, quinones, and catechol, and a nonhumic fraction characterized by lipids, carbohydrates, polysaccharides, amino acids, proteins, waxes, and resins (McDonald et al., 2004). Among these compounds, low molecular weight carboxylic acids, amino acids, and carbohydrates can potentially support all bacterial production in boreal ecosystems (Berggren et al., 2010a). The remaining incoming terrestrial organic matter is highly concentrated in tannins and represents organic compounds not degraded by microbial fauna in the soil (Daniel, 2005). Degradation processes preferentially remove oxidized, aromatic compounds, whereas reduced, aliphatic, and N-containing compounds are either resistant to degradation or tightly cycled; they therefore persist in aquatic systems (Kellerman et al., 2015). The role of allochthonous carbon in aquatic ecosystems is closely related to bacterial abundance, biomass, and production (Azam et al., 1983; Roiha et al., 2012), as most DOM decomposition is undertaken by planktonic bacteria (Daniel, 2005; Wetzel, 1975). Depending on the quality of the organic carbon, bacterial productivity might change, whereas the quantity of the organic carbon appears to define community composition (Roiha et al., 2012). Glaz et al. (2015) demonstrated that, in contrast to DOM quantity, the nature of t-DOM in lakes affected by forest harvesting did not change. Therefore, forest harvesting triggers extra inputs of DOM into aquatic environments that can directly influence and foster bacterial communities and modify the entire-lake metabolism.

29.5   Allochthony in Boreal Aquatic Consumers

In aquatic habitats within forests, organic matter (OM) inputs from the adjoining terrestrial counterparts are *allochthonous*, whereas aquatic primary production is referred to as *autochthonous*. Terrestrial OM eventually enters the aquatic food webs, and its use by aquatic organisms for biomass production is referred to as *allochthony* The significance of allochthony in supporting aquatic food webs has been shown recently for bacteria (Berggren et al., 2010a; Guillemette et al., 2016), zooplankton (Berggren et al., 2014; Grosbois et al., 2017a), invertebrates (Hayden et al., 2016), and fish (Glaz et al., 2012; Tanentzap et al., 2014).

Although all aquatic organisms may play a role in processing terrestrial OM directly or indirectly, zooplankton occupy a strategic position in aquatic food webs. Zooplankton can consume both autochthonous and allochthonous OM for biomass production and are key organisms responsible for transferring OM to higher trophic levels (Grosbois et al., 2020). Whereas earlier studies found significant trophic transfer of terrestrial particulate OM to zooplankton (Cole et al., 2006), recent studies demonstrate that this direct use of terrestrial particulate OM by pelagic food webs is rather limited (Mehner et al., 2016; Wenzel et al., 2012). However, the assimilation of terrestrial OM in aquatic food webs can follow different pathways and may begin at the lower tropic levels, i.e., through the microbial degradation of terrestrial OM. The terrestrial OM is then transferred to the trophic levels comprising ciliates, flagellates, or rotifers through their consumption of the microbes (Jansson et al., 2007; Masclaux et al., 2013). The allochthonous OM assimilated by these species is then consumed by larger zooplankton taxa that are, in turn, available for zooplanktivores, including numerous fish species, e.g., *Perca flavescens*, and invertebrates, e.g., *Chaoborus obscuripes* and *Leptodora kindtii* (Tanentzap et al., 2014). Thus, zooplankton can serve as indicators of allochthony in aquatic food webs and have been the focus of multiple studies of allochthony (Cole et al., 2011; Lee et al., 2013; Perga et al.,

2006). Allochthony in zooplankton varies widely, from less than 5% (Francis et al., 2011) to 100% (Rautio et al., 2011). Although there remains some debate about the significance of the terrestrial OM contribution for zooplankton in different lake types, e.g., large clear-water lakes versus small humic lakes, it is now increasingly accepted that allochthony can often be very significant for many zooplankton taxa, especially in lakes that receive large terrestrial OM inputs, which limit light availability and aquatic primary production (Cole et al., 2011; Emery et al., 2015; Wilkinson et al., 2013).

The allochthony of benthic invertebrates in rivers and lakes has often been well defined because of the remarkable feeding adaptations of these taxa (grazers, filterers, shredders, and predators). Therefore, we can observe an allochthony gradient in these organisms, from the least allochthonous grazers to the most allochthonous shredders (Rasmussen, 2010). However, the diet of each feeding group can include a mix of food sources with organisms feeding mainly on autochthonous material and also assimilating allochthonous OM. For example, grazers can consume and assimilate terrestrial OM deposited on benthic algal mats, and filter-feeders can obtain suspended terrestrial OM. Inversely, animals feeding on allochthonous material may ingest autochthonous material, e.g., shredders can consume plant litter together with periphytic algae growing on the litter surface. The use of allochthonous or autochthonous OM by benthic invertebrates also depends on the riparian vegetation and the aquatic habitat size. The river continuum concept (RCC) (Vannote et al., 1980) describes a decreasing allochthony gradient from small forested headwater streams to large autochthonous rivers having minimal canopy cover. However, recent studies have also shown strong autochthony rather than allochthony in headwater streams (Lau et al., 2009a, b; Torres-Ruiz et al., 2007). Erdozain et al. (2019) demonstrated, contrary to RCC predictions, that allochthony in aquatic food webs is low in forest headwaters and increases with greater harvesting intensity and delivery of terrestrial OM. It is, therefore, very likely that silvicultural practices can strongly impact autochthony and allochthony in aquatic food webs.

Changes in the degree of allochthony in zooplankton, benthic invertebrate communities, or both will be reflected in their predators, e.g., fish. Moreover, detritivorous fish in reservoir ecosystems, such as gizzard shad (*Dorosoma cepedianum*), can ingest terrestrial detritus directly, accounting for about 35% of their biomass (Babler et al., 2011). Pelagic fish in a temperate lake had an estimated minimum allochthony of 44% in young bluegill (*Lepomis macrochirus*) and 43% in young yellow perch (*Perca flavescens*), whereas in older individuals, allochthony was 53% (Weidel et al., 2008). The quantification of fish allochthony can be complicated by the diet contribution from terrestrial insects, as fish can feed directly on terrestrial prey items that fall into the water. Neglecting to include the terrestrial prey will likely underestimate the terrestrial trophic support for fish. Allan et al. (2003) estimated that terrestrial and aquatic prey contributed equally to the diet of juvenile coho (*Oncorhynchus kisutch*) in Alaskan streams, and the fish ingested an average total of 12 mg of insect dry mass per day. Forest harvesting affects terrestrial invertebrate communities, such as red-listed beetles (Franc & Götmark, 2008), and, therefore, can strongly disturb the direct subsidy of terrestrial insects for fish growth.

29.6   Health of the Aquatic–Terrestrial Food Web

The health of food webs is largely determined by the nutritional quality of the available food resources. Terrestrial OM inputs comprise mainly biochemically recalcitrant lignocellulose and lack biomolecules such as polyunsaturated fatty acids (PUFA), which are essential for animal growth and reproduction (Schneider et al., 2016, 2017; Taipale et al., 2015). Because essential PUFA are mainly synthesized de novo by algae, their acquisition in aquatic food webs is negatively related to the consumers' degree of allochthony (Jardine et al., 2015; Lau et al., 2014). However, other studies have demonstrated that most aquatic biomass has a terrestrial origin in temperate and boreal lakes, illustrating the strong physical and hydrological connections to OM-rich catchments (Cole et al., 2011; Wilkinson et al., 2013). The effect of this substantial diet contribution of terrestrial OM on the health of aquatic consumers in the natural environment remains unknown; however, laboratory-based feeding experiments have shown lower survival, growth, and reproduction for benthic and pelagic invertebrates when they are fed only with terrestrial plant litter (Brett et al., 2009; Lau et al., 2013). Moreover, the survival of many zooplankton species in boreal lakes depends on the opportunity to accumulate PUFA from algae in autumn and under the lake ice at the beginning of winter (Grosbois et al., 2017b). Furthermore, algae sustain the production of benthic consumers and supply them the necessary PUFA in headwater streams despite the often dense riparian canopy cover over these streams (Lau et al., 2009a, b). Although many species of fungi degrade the detrital material of plant litter, few fungal species exist in aquatic environments to consume the lignified OM and make it accessible to consumers at higher trophic levels (however, see Masclaux et al. (2013) for pollen degradation by aquatic chytrids to access essential fatty acids). In forests, plant litter is degraded initially by microbial communities and then transported to aquatic habitats by runoff. This runoff carries only the "leftovers" of microbial degradation and hence the most recalcitrant molecules (Brett et al., 2017). Moreover, a large portion of terrestrial material can be deposited onto the anoxic bottom of lakes, greatly reducing or stopping its degradation, and lake metabolism becomes directed mainly toward methanogenesis (Schink, 1997).

Inputs of terrestrial DOM to freshwaters considerably affect the productivity of the entire ecosystem, thereby affecting the growth and reproduction of organisms. First, these inputs can diminish benthic and pelagic algal primary production, with the increases in t-DOM, i.e., the browning of waters, limiting light penetration through the water column (Fig. 29.3; Karlsson et al., 2009). Lower primary production in lakes and rivers will reduce the availability of essential molecules, e.g., PUFA, synthesized by algae and the transfer of these molecules to higher trophic levels (Strandberg et al., 2015). Kelly et al. (2014) showed that t-DOM concentrations are negatively correlated to zooplankton production. Karlsson et al. (2015) confirmed this observation, finding lower fish production in small boreal lakes with high t-DOM concentrations. Moreover, Grosbois et al. (2020) demonstrated that zooplankton production based on t-DOM (defined as the allochtrophy) in a boreal lake was lower, on average, throughout the year. Still, they remained in the same range as zooplankton production based on algae. Estimating secondary production in aquatic habitats is challenging, as traditional methods require identifying, counting, and measuring all individuals of a community during a long period and sampling at a high frequency (Runge & Roff, 2000). New methods to measure secondary production, such as the use of the chitobiase enzyme (Yebra et al., 2017), will help quantify the effects of terrestrial OM inputs on aquatic food webs and consumer production.

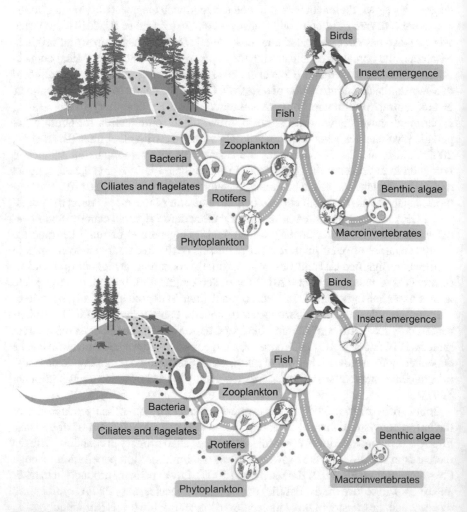

Fig. 29.3 Aquatic–terrestrial food web in an undisturbed (*top*) and a harvested watershed (*bottom*). Dissolved organic matter (*brown dots*) and nutrient inputs (*red dots*) to lakes increase with forest harvesting, leading to the browning of waters, increased bacterial metabolism, decreased light penetration into the water, and suppressed benthic primary and secondary production. Health consequences for terrestrial and aquatic organisms are expected with the altered transfer of essential fatty acids (*yellow dots*) within the aquatic–terrestrial food web

The concept of *one health* highlights that the health of one component of a system depends on that of the other components (Destoumieux-Garzón et al., 2018). This is particularly true for habitats interacting dynamically within the same landscape, such as observed in the boreal forest. Many terrestrial organisms depend on the aquatic environment to access important resources such as food and water for their survival, with some species, e.g., amphibians and insects, having life stages in water. Although subsidies from terrestrial to aquatic habitats are much higher in quantity, the aquatic subsidies to land are of higher nutritional quality (Martin-Creuzburg et al., 2017), energy density, and nutrient concentration; the result is a similar subsidy to animal carbon in both directions (Bartels et al., 2012). Most aquatic subsidies must go against gravity and therefore often rely on animal movement. One of the most observed examples is the emergence of insects from lakes or rivers; these insects eventually feed terrestrial insectivores (Fig. 29.3; Muehlbauer et al., 2014; Paetzold et al., 2005). Emerging aquatic insects are essential prey, as they have a high content of long-chain PUFA. Preliminary estimates of long-chain PUFA export to terrestrial ecosystems range from 0.1 to 672.2 mg dry weight·m^{-2}·yr^{-1} (Gladyshev et al., 2009). Dreyer et al. (2015) also estimated that the whole-lake emergence of aquatic insects in an Islandic lake (3.1–76.0 Mg·yr^{-1}) had deposited 100 kg ha^{-1} yr^{-1} of insect biomass within 50 m of the lake shoreline, corresponding to 10 kg N·ha^{-1} yr^{-1} and 1 kg of P ha^{-1} yr^{-1}. Fluxes from aquatic to terrestrial habitats are therefore potentially significant for many terrestrial animals. More studies are required to quantify the aquatic biomass export to terrestrial animals feeding on fish, e.g., bears, bald eagles, and herons, or on macrophytes, e.g., moose, although this type of export is more difficult to estimate.

Aquatic habitats also influence nutrient cycling in the forest landscape, with lotic waters transporting nutrients that are deposited outside of the river bed during flooding (Jacobson et al., 2000) and are made accessible to plant roots in the subsurface water, i.e., hyporheic zone (Pinay et al., 2009). Trees can therefore use aquatic nutrients for growth. In a boreal watershed in Alaska, about 25% of the foliar nitrogen of trees and shrubs was derived from salmon near the fish spawning sites (Helfield & Naiman, 2002). A greater understanding of the land–water interactions and the subsidies from aquatic to terrestrial habitats is essential for the sustainable management and maintaining the health of the forest ecosystem (Schindler & Smits, 2017).

Despite the general high quality of aquatic subsidies to the terrestrial food webs, anoxic zones may form in particular aquatic environments, e.g., ponds, wetlands, deep lake water, where bioavailable methylmercury is produced and bioaccumulates in aquatic food webs (Downs et al., 1998). Contaminants such as methylmercury can therefore be exported to terrestrial food webs with aquatic-feeding species (Cristol et al., 2008). However, this contamination of terrestrial food webs depends more on the complex trophic structure and interactions than on the aquatic prey contribution, such as through insect emergence or predation on fish (Schindler & Smits, 2017). For example, Bartrons et al. (2015) demonstrated that spiders feeding on contaminated aquatic midges had lower methylmercury concentrations than spiders feeding on terrestrial prey because the aquatic prey had a lower position in the food web than their terrestrial counterparts. The trophic position and food-chain length are stronger determinants of methylmercury contamination than trophic reliance on aquatic prey, as longer food chains enhance biomagnification. Furthermore, methylmercury production and accumulation in aquatic food webs can increase with more frequent and severe forest disturbances and climate change. For example, forest harvesting and fires increase the OM inputs in lakes and, therefore, the methylmercury contamination in zooplankton (Garcia & Carignan, 1999) and fish (Garcia & Carignan, 2000). Therefore, the production, assimilation, and export of contaminants is a complex process that depends on both aquatic and terrestrial components. This dependence highlights the need for considering the aquatic and terrestrial components as integral parts of the boreal forest ecosystem.

29.7   The Role and Impacts of Forest Disturbances on Aquatic Ecosystem Services

Aquatic environments provide an essential resource for humanity: freshwater. Freshwater is used for drinking water, watering crops, fisheries, and many important human activities; however, aquatic environments are highly sensitive to disturbances and management occurring on the adjacent forested land. The most well-known example of lake reaction is the temporal increase of lake turbidity after forest harvesting (Glaz et al., 2015) or the more permanent increase in cyanobacterial blooms caused by heightened nutrient inputs from residential areas or agricultural fields (Paerl & Otten, 2013). Thus, a greater consideration of interactions between terrestrial and aquatic environments is required to manage the forest ecosystem. In this respect, it is crucial to account for the role of both natural and anthropogenic forest disturbances and how changes to either of these disturbance regimes will affect aquatic environments. Natural disturbances that shape the structure and ecological functioning of the terrestrial component of the boreal forest have been well covered by Gauthier et al. (2009) and are already included in the sustainable management of boreal forests. Nonetheless, their effects on aquatic habitats have not been fully discussed, partly because of knowledge gaps, and it is crucial to consider possible effects in future management frameworks.

The effects of natural and anthropogenic disturbances, such as fire and forest harvesting, on freshwaters in the boreal forests have been studied previously (Carignan et al., 2000; Garcia & Carignan, 1999; Patoine et al., 2000; Pinel-Alloul et al., 1998). This research has shown that fires strongly affect the water biogeochemistry (Lamontagne et al., 2000; Olefeldt et al., 2013) through, for example, the increase of additional inputs such as phosphorus, which can be directly related to the proportion of burned areas (Carignan et al., 2000). Recently burned watersheds and their associated organic inputs into the surrounding freshwaters can increase lake metabolism and, therefore, the liberation of carbon through aquatic respiration (Marchand et al., 2009). Carignan et al. (2000) proposed a simple model to estimate the impact of fires on lakes on the basis of the burned watershed area divided by the lake's area or volume. Although the effects of fire on lake biogeochemistry have been addressed in these studies, research into the impact on food web functioning and ecology is almost nonexistent. Planas et al. (2000), however, demonstrated that fires modify the biomass and composition of algal communities. Moreover, Patoine et al. (2000) showed that zooplankton biomass is affected after a forest fire, albeit for only a few years, before returning to a normal level; this pattern suggests a high resilience capacity for lakes.

Both forest harvesting and fires can physically disturb aquatic ecosystems, as the removal of forest biomass by burning or harvesting increases a water body's exposure to wind (Montoro Girona et al., 2019; Scully et al., 2000). However, the impacts of forest harvesting on aquatic habitats often differ from that of fires, causing water tables and streamflow to fluctuate more if harvesting is carried out on unfrozen soil (Veny, 1986). Forest harvesting also affects water physicochemical properties by increasing dissolved organic carbon (DOC), algal biomass, and total nitrogen (Steedman, 2000). The concentration of DOC increases significantly one year after logging without changing its characteristics and usually decreases and returns to normal levels one to two years later (Glaz et al., 2015). The additional nutrients and DOC inputs may positively impact juvenile fish growth via an increased primary production in the lake and reduced prey visibility in browner waters (Leclerc et al., 2011). However, these effects depend on the relative increase of nutrients versus DOC because the growth rate and abundance of fish are negatively affected by DOC (Benoît et al., 2016). An important aspect to consider in the sustainable management of forests is that forest harvesting increases methylmercury bioaccumulation in zooplankton and fish via the mercury and methylmercury loadings associated with t-DOM inputs from land to aquatic systems (Garcia & Carignan, 2000; Wu et al., 2018). The effect can be mitigated by an associated higher nutrient input that alters food web structure and productivity (de Wit et al., 2014). To our knowledge, the impact of disturbances other than fire and harvesting, such as insect outbreaks or windthrows, on freshwater ecology has yet to be appropriately studied in the boreal forest, highlighting the need for more research to better understand the possible effects of disturbance on aquatic ecosystems.

It is of utmost importance to consider the potential impacts of forestry on aquatic habitats in future sustainable forest management guidelines because of the role of aquatic habitats in the boreal region in providing important ecosystem services. The consequences of not taking potential impacts into account include browner and more nutrient-rich water caused by anthropogenic, e.g., forest harvesting, or natural disturbances, e.g., fires, affecting drinking water supplies, commercial fishing, and recreational activities (Kritzberg et al., 2020). The browning of waters is associated with a higher concentration of aromatic molecules, which are more difficult to degrade by aquatic microbes (Koehler et al., 2012). Brown waters containing greater amounts of dissolved organic molecules require more treatment with chemical precipitation using $FeCl_3$, $AlCl_3$, or $Al_2(SO_4)_3$, which increases water treatment costs and the risks to human health (Kritzberg et al., 2020). Fisheries are susceptible to forestry practices because logging alters water properties, diminishes fish growth, and increases methylmercury in aquatic food webs. Sportfishing is an important activity in the economy and culture of boreal countries, generating in Canada, in 2018 alone, $3.5 billion CDN in income and supporting 58,000 jobs (Conference Board of Canada, 2019). Finally, many water-associated recreational activities, such as boating, kayaking, and swimming, are potentially impacted by logging. Brown lake water having a high DOC concentration is not considered a high-quality environment for water sports, and the increased nutrients in water may cause cyanobacterial blooms and liberate toxins harmful to humans. It is therefore essential to include considerations of freshwater ecology in the management of the boreal forest to achieve sustainable management goals.

Over the last decades, several tools and approaches have been developed in various parts of the boreal biome to minimize the impact of forestry practices on aquatic habitats, including riparian buffer strips, partial cutting (Fig. 29.4), and continuous cover forestry. Riparian buffer strips have served as the main silvicultural tool to preserve aquatic habitats from harvesting activities. The riparian strip is a physical barrier of vegetation and trees between uplands and rivers or lakes. This vegetation prevents or diminishes the input of organic and inorganic materials into the adjacent freshwater by reducing erosion and runoff (Kuglerová et al., 2014). Riparian strips also represent a refuge habitat for numerous groups of species and are used as a corridor to improve the connectivity across boreal landscapes (Barton et al., 1985; Machtans et al., 1996). Nonetheless, riparian strips are very vulnerable to wind exposure; therefore, their width and configuration must be modified in accordance with the topographic and forest stand conditions to guarantee their viability over the long term (Ruel et al., 2001). Partial cutting, where between 30 and 70% of the forest cover is logged, is also used as an intermediate disturbance to reduce the negative effects of logging on adjacent freshwaters (Montoro Girona et al., 2017, 2018; Moussaoui et al., 2020). One aspect that has been little studied is tree species composition and tree diversity around freshwaters and their role and impact on the

Fig. 29.4 **a** Partial cutting on the North Shore as part of the MISA experiment, Québec, Canada; **b** Forest buffers in a landscape in Abitibi-Témiscamingue, Québec, Canada. *Photo credits* **a** Miguel Montoro Girona, **b** Guillaume Grosbois

aquatic ecology. However, recent studies have suggested that species composition and diversity could play an important role (Kärnä et al., 2019; López-Rojo et al., 2019).

29.8   Conclusions

As boreal stands lie within a watershed, all natural and anthropogenic disturbances on the land within the watershed also influence the associated aquatic environments. Nonetheless, aquatic environments have been overlooked when assessing current silvicultural practices in boreal forests. In the sustainable management of the boreal forest, forestry operations emulate natural disturbances to reduce harvesting effects on forests. This approach also reduces the impact on freshwaters, as aquatic environments also experience the effects of natural disturbances on land, despite being largely untouched directly by anthropogenic disturbances. However, there remains a lack of information about the consequences of both natural and anthropogenic disturbances on aquatic environments. In this chapter, we have described the main effects of terrestrial changes to the watershed on aquatic habitats and how these changes in turn affect the forest. The complex interconnectivity between aquatic and terrestrial habitats should ultimately be included in the sustainable management of the boreal forest to preserve the health of the boreal biome. A healthy boreal biome will be crucial in mitigating climate change and managing the increased intensity and frequency of natural disturbances likely to heighten the vulnerability of freshwaters.

References

Aitkenhead-Peterson, J. A., McDowell, W. H., & Neff, J. C. (2003). Sources, production, and regulation of allochthonous dissolved organic matter inputs to surface waters. In S. E. G. Findlay & R. L. Sinsabaugh (Eds.), *Aquatic ecosystems* (pp. 25–70). Academic Press.

Allan, J. D., Wipfli, M. S., Caouette, J. P., et al. (2003). Influence of streamside vegetation on inputs of terrestrial invertebrates to salmonid food webs. *Canadian Journal of Fisheries and Aquatic Sciences, 60*, 309–320. https://doi.org/10.1139/f03-019.

Arvola, L., Jarvinen, M., & Tulonen, T. (2011). Long-term trends and regional differences of phytoplankton in large Finnish lakes. *Hydrobiologia, 660*, 125–134. https://doi.org/10.1007/s10 750-010-0410-9.

Azam, F., Fenchel, T., Field, J., et al. (1983). The ecological role of water-column microbes in the sea. *Marine Ecology Progress Series, 10*, 257–263. https://doi.org/10.3354/meps010257.

Babler, A. L., Pilati, A., & Vanni, M. J. (2011). Terrestrial support of detritivorous fish populations decreases with watershed size. *Ecosphere, 2*(7), 1–23. https://doi.org/10.1890/ES11-00043.1.

Bartels, P., Cucherousset, J., Steger, K., et al. (2012). Reciprocal subsidies between freshwater and terrestrial ecosystems structure consumer resource dynamics. *Ecology, 93*, 1173–1182. https://doi.org/10.1890/11-1210.1.

Barton, D. R., Taylor, W. D., & Biette, R. M. (1985). Dimensions of riparian buffer strips required to maintain trout habitat in southern Ontario streams. *North American Journal of Fisheries Management, 5*, 364–378. https://doi.org/10.1577/1548-8659(1985)5<364:DORBSR>2.0.CO;2.

736   G. Grosbois et al.

Bartrons, M., Gratton, C., Spiesman, B. J., et al. (2015). Taking the trophic bypass: Aquatic-terrestrial linkage reduces methylmercury in a terrestrial food web. *Ecological Applications, 25*, 151–159. https://doi.org/10.1890/14-0038.1.

Battin, T. J., Luyssaert, S., Kaplan, L. A., et al. (2009). The boundless carbon cycle. *Nature Geoscience 2*, 598–600. https://doi.org/10.1038/ngeo618.

Baxter, C. V., Fausch, K. D., & Carl Saunders, W. (2005). Tangled webs: Reciprocal flows of invertebrate prey link streams and riparian zones. *Freshwater Biology, 50*, 201–220. https://doi.org/10.1111/j.1365-2427.2004.01328.x.

Benoît, P. O., Beisner, B. E., & Solomon, C. T. (2016). Growth rate and abundance of common fishes is negatively related to dissolved organic carbon concentration in lakes. *Canadian Journal of Fisheries and Aquatic Sciences, 73*, 1230–1236. https://doi.org/10.1139/cjfas-2015-0340.

Berggren, M., & Al-Kharusi, E. (2020). Decreasing organic carbon bioreactivity in European rivers. *Freshwater Biology, 65*, 1128–1138. https://doi.org/10.1111/fwb.13498.

Berggren, M., Laudon, H., Haei, M., et al. (2010a). Efficient aquatic bacterial metabolism of dissolved low-molecular-weight compounds from terrestrial sources. *ISME Journal, 4*, 408–416. https://doi.org/10.1038/ismej.2009.120.

Berggren, M., Strom, L., Laudon, H., et al. (2010b). Lake secondary production fueled by rapid transfer of low molecular weight organic carbon from terrestrial sources to aquatic consumers. *Ecology Letters, 13*, 870–880. https://doi.org/10.1111/j.1461-0248.2010.01483.x.

Berggren, M., Ziegler, S. E., St-Gelais, N. F., et al. (2014). Contrasting patterns of allochthony among three major groups of crustacean zooplankton in boreal and temperate lakes. *Ecology, 95*, 1947–1959. https://doi.org/10.1890/13-0615.1.

Berggren, M., Gudasz, C., Guillemette, F., et al. (2020). Systematic microbial production of optically active dissolved organic matter in subarctic lake water. *Limnology and Oceanography, 65*, 951–961. https://doi.org/10.1002/lno.11362.

Berman-Frank, I., Quigg, A., Finkel, Z. V., et al. (2007). Nitrogen-fixation strategies and Fe requirements in cyanobacteria. *Limnology and Oceanography, 52*, 2260–2269. https://doi.org/10.4319/lo.2007.52.5.2260.

Bernhardt, E. S., Likens, G. E., Buso, D. C., et al. (2003). In-stream uptake dampens effects of major forest disturbance on watershed nitrogen export. *Proceedings of the National Academy of Sciences of the United States of America, 100*, 10304–10308. https://doi.org/10.1073/pnas.1233676100.

Bisson, P. A., Ice, G. G., Perrin, C. J., et al. (1992). Effects of forest fertilization on water quality and aquatic resources in the Douglas-fir region. In H. N. Chappell, G. F. Weetman GF & R. E. Miller (Eds.), *Forest fertilization: sustaining and improving nutrition and growth of western forests*. Seattle: University of Washington.

Brett, M. T., & Muller-Navarra, D. C. (1997). The role of highly unsaturated fatty acids in aquatic food web processes. *Freshwater Biology, 38*, 483–499. https://doi.org/10.1046/j.1365-2427.1997.00220.x.

Brett, M. T., Kainz, M. J., Taipale, S. J., et al. (2009). Phytoplankton, not allochthonous carbon, sustains herbivorous zooplankton production. *Proceedings of the National Academy of Sciences of the United States of America, 106*, 21197–21201. https://doi.org/10.1073/pnas.0904129106.

Brett, M. T., Bunn, S. E., Chandra, S., et al. (2017). How important are terrestrial organic carbon inputs for secondary production in freshwater ecosystems? *Freshwater Biology, 62*, 833–853. https://doi.org/10.1111/fwb.12909.

Campbell, J. E., Berry, J. A., Seibt, U., et al. (2017). Large historical growth in global terrestrial gross primary production. *Nature, 544*, 84–87. https://doi.org/10.1038/nature22030.

Carignan, R., D'Arcy, P., & Lamontagne, S. (2000). Comparative impacts of fire and forest harvesting on water quality in Boreal Shield lakes. *Canadian Journal of Fisheries and Aquatic Sciences, 57*, 105–117. https://doi.org/10.1139/f00-125.

Chen, N., Peng, B., Hong, H., et al. (2013). Nutrient enrichment and N: P ratio decline in a coastal bay-river system in southeast China: The need for a dual nutrient (N and P) management strategy. *Ocean and Coastal Management, 81*, 7–13. https://doi.org/10.1016/j.ocecoaman.2012.07.013.

Clapcott, J. E., & Barmuta, L. A. (2010). Forest clearance increases metabolism and organic matter processes in small headwater streams. *Journal of the North American Benthological Society, 29*, 546–561. https://doi.org/10.1899/09-040.1.

Cole, J. J., Pace, M. L., Carpenter, S. R., et al. (2000). Persistence of net heterotrophy in lakes during nutrient addition and food web manipulations. *Limnology and Oceanography, 45*, 1718–1730. https://doi.org/10.4319/lo.2000.45.8.1718.

Cole, J. J., Carpenter, S. R., Pace, M. L., et al. (2006). Differential support of lake food webs by three types of terrestrial organic carbon. *Ecology Letters, 9*, 558–568. https://doi.org/10.1111/j.1461-0248.2006.00898.x.

Cole, J. J., Carpenter, S. R., Kitchell, J., et al. (2011). Strong evidence for terrestrial support of zooplankton in small lakes based on stable isotopes of carbon, nitrogen, and hydrogen. *Journal of the North American Benthological Society, 108*, 1975–1980. https://doi.org/10.1073/pnas.101 2807108.

Conference Board of Canada. (2019). *The economic footprint of angling, hunting, trapping and sport shooting in Canada* (p. 40). Ottawa: The Conference Board of Canada.

Creed, I. F., Bergstrom, A. K., Trick, C. G., et al. (2018). Global change-driven effects on dissolved organic matter composition: Implications for food webs of northern lakes. *Global Change Biology, 24*, 3692–3714. https://doi.org/10.1111/gcb.14129.

Cristol, D. A., Brasso, R. L., Condon, A. M., et al. (2008). The movement of aquatic mercury through terrestrial food webs. *Science, 320*, 335–335. https://doi.org/10.1126/science.1154082.

Daniel, C. B. (2005). *The importance of terrestrial carbon in plankton food webs*. Ph.D. thesis, Lund University.

de Wit, H. A., Granhus, A., Lindholm, M., et al. (2014). Forest harvest effects on mercury in streams and biota in Norwegian boreal catchments. *Forest Ecology and Management, 324*, 52–63. https://doi.org/10.1016/j.foreco.2014.03.044.

de Wit, H. A., Valinia, S., Weyhenmeyer, G. A., et al. (2016). Current browning of surface waters will be further promoted by wetter climate. *Environmental Science & Technology Letters, 3*, 430–435. https://doi.org/10.1021/acs.estlett.6b00396.

Degerman, E., Henrikson, L., Herrmann, J., et al. (1995). The effects of liming on aquatic fauna in L. Henrikson & Y. W. Brodin (Eds.), *Liming of acidified surface waters: A Swedish synthesis* (pp. 221–282). Springer.

Destoumieux-Garzón, D., Mavingui, P., Boetsch, G., et al. (2018). The one health concept: 10 years old and a long road ahead. *Frontiers in Veterinary Science, 5*, 14. https://doi.org/10.3389/fvets.2018.00014.

Diehl, S., Thomsson, G., Kahlert, M., et al. (2018). Inverse relationship of epilithic algae and pelagic phosphorus in unproductive lakes: Roles of N_2 fixers and light. *Freshwater Biology, 63*, 662–675. https://doi.org/10.1111/fwb.13103.

Donis, D., Flury, S., Stockli, A., et al. (2017). Full-scale evaluation of methane production under oxic conditions in a mesotrophic lake. *Nature Communications, 8*, 1661. https://doi.org/10.1038/s41467-017-01648-4.

Downs, S., MacLeod, C., & Lester, J. (1998). Mercury in precipitation and its relation to bioaccumulation in fish: A literature review. *Water, Air, and Soil Pollution, 108*, 149–187. https://doi.org/10.1023/A:1005023916816.

Dreyer, J., Townsend, P. A., Hook, J. C., et al. (2015). Quantifying aquatic insect deposition from lake to land. *Ecology, 96*, 499–509. https://doi.org/10.1890/14-0704.1.

Eimers, M. C., Watmough, S. A., Paterson, A. M., et al. (2009). Long-term declines in phosphorus export from forested catchments in south-central Ontario. *Canadian Journal of Fisheries and Aquatic Sciences, 66*, 1682–1692. https://doi.org/10.1139/F09-101.

Elmendorf, S. C., Henry, G. H., Hollister, R. D., et al. (2012). Plot-scale evidence of tundra vegetation change and links to recent summer warming. *Nature Climate Change, 2*, 453–457. https://doi.org/10.1038/nclimate1465.

Elser, J. J., Andersen, T., Baron, J. S., et al. (2009). Shifts in lake N: P stoichiometry and nutrient limitation driven by atmospheric nitrogen deposition. *Science, 326*, 835–837. https://doi.org/10.1126/science.1176199.

Elser, J. J., Peace, A. L., Kyle, M., et al. (2010). Atmospheric nitrogen deposition is associated with elevated phosphorus limitation of lake zooplankton. *Ecology Letters, 13*, 1256–1261. https://doi.org/10.1111/j.1461-0248.2010.01519.x.

Emery, K. A., Wilkinson, G. M., Ballard, F. G., et al. (2015). Use of allochthonous resources by zooplankton in reservoirs. *Hydrobiologia, 758*, 257–269. https://doi.org/10.1007/s10750-015-2338-6.

Erdozain, M., Kidd, K., Kreutzweiser, D., et al. (2019). Increased reliance of stream macroinvertebrates on terrestrial food sources linked to forest management intensity. *Ecological Applications, 29*, e01889. https://doi.org/10.1002/eap.1889.

Franc, N., & Götmark, F. (2008). Openness in management: Hands-off vs partial cutting in conservation forests, and the response of beetles. *Biological Conservation, 141*, 2310–2321. https://doi.org/10.1016/j.biocon.2008.06.023.

Francis, T. B., Schindler, D. E., Holtgrieve, G. W., et al. (2011). Habitat structure determines resource use by zooplankton in temperate lakes. *Ecology Letters, 14*, 364–372. https://doi.org/10.1111/j.1461-0248.2011.01597.x.

Garcia, E., & Carignan, R. (1999). Impact of wildfire and clear-cutting in the boreal forest on methyl mercury in zooplankton. *Canadian Journal of Fisheries and Aquatic Sciences, 56*, 339–345. https://doi.org/10.1139/f98-164.

Garcia, E., & Carignan, R. (2000). Mercury concentrations in northern pike (*Esox lucius*) from boreal lakes with logged, burned, or undisturbed catchments. *Canadian Journal of Fisheries and Aquatic Sciences, 57*, 129–135. https://doi.org/10.1139/f00-126.

Gauthier, S., Vaillancourt, M. A., Leduc, A., et al. (Eds.). (2009). *Ecosystem management in the boreal forest* (p. 572). Presses de l'Université du Québec.

Gladyshev, M., Arts, M., & Sushchik, N. I. (2009). Preliminary estimates of the export of omega-3 highly unsaturated fatty acids (EPA+ DHA) from aquatic to terrestrial ecosystems. In M. T. Arts, M. T. Brett, & M. Kainz (Eds.), *Lipids in aquatic ecosystems* (pp. 179–210). Springer.

Glaz, P., Sirois, P., & Nozais, C. (2012). Determination of food sources for benthic invertebrates and brook trout *Salvelinus fontinalis* in Canadian Boreal Shield lakes using stable isotope analysis. *Aquatic Biology, 17*, 107–117. https://doi.org/10.3354/ab00465.

Glaz, P., Gagne, J. P., Archambault, P., et al. (2015). Impact of forest harvesting on water quality and fluorescence characteristics of dissolved organic matter in eastern Canadian Boreal Shield lakes in summer. *Biogeosciences, 12*, 6999–7011. https://doi.org/10.5194/bg-12-6999-2015.

Grasset, C., Sobek, S., Scharnweber, K., et al. (2020). The CO_2-equivalent balance of freshwater ecosystems is non-linearly related to productivity. *Global Change Biology, 26*, 5705–5715. https://doi.org/10.1111/gcb.15284.

Grosbois, G., del Giorgio, P. A., & Rautio, M. (2017a). Zooplankton allochthony is spatially heterogeneous in a boreal lake. *Freshwater Biology, 62*, 474–490. https://doi.org/10.1111/fwb.12879.

Grosbois, G., Mariash, H., Schneider, T., et al. (2017b). Under-ice availability of phytoplankton lipids is key to freshwater zooplankton winter survival. *Scientific Reports, 7*, 11543. https://doi.org/10.1038/s41598-017-10956-0.

Grosbois, G., Vachon, D., del Giorgio, P. A., et al. (2020). Efficiency of crustacean zooplankton in transferring allochthonous carbon in a boreal lake. *Ecology, 101*, e03013. https://doi.org/10.1002/ecy.3013.

Guillemette, F., Leigh McCallister, S., & del Giorgio, P. A. (2016). Selective consumption and metabolic allocation of terrestrial and algal carbon determine allochthony in lake bacteria. *ISME Journal, 10*, 1373–1382. https://doi.org/10.1038/ismej.2015.215.

Gustafsson, J. P., Mwamila, L. B., & Kergoat, K. (2012). The pH dependence of phosphate sorption and desorption in Swedish agricultural soils. *Geoderma, 189–190*, 304–311. https://doi.org/10.1016/j.geoderma.2012.05.014.

Haaland, S., Hongve, D., Laudon, H., et al. (2010). Quantifying the drivers of the increasing colored organic matter in boreal surface waters. *Environmental Science and Technology, 44*, 2975–2980. https://doi.org/10.1021/es903179j.

Hall, R. O. (2003). A stream's role in watershed nutrient export. *Proceedings of the National Academy of Sciences of the United States of America, 100*, 10137–10138. https://doi.org/10.1073/pnas.1934477100.

Hayden, B., McWilliam-Hughes, S. M., & Cunjak, R. A. (2016). Evidence for limited trophic transfer of allochthonous energy in temperate river food webs. *Freshwater Science, 35*, 544–558. https://doi.org/10.1086/686001.

Hayden, B., Harrod, C., Thomas, S. M., et al. (2019). From clear lakes to murky waters - tracing the functional response of high-latitude lake communities to concurrent 'greening' and 'browning.' *Ecology Letters, 22*, 807–816. https://doi.org/10.1111/ele.13238.

Hedström, P., Bystedt, D., Karlsson, J., et al. (2017). Brownification increases winter mortality in fish. *Oecologia, 183*, 587–595. https://doi.org/10.1007/s00442-016-3779-y.

Helfield, J. M., & Naiman, R. J. (2002). Salmon and alder as nitrogen sources to riparian forests in a boreal Alaskan watershed. *Oecologia, 133*, 573–582. https://doi.org/10.1007/s00442-002-1070-x.

Hessen, D. O. (2013). Inorganic nitrogen deposition and its impacts on N:P-ratios and lake productivity. *Water, 5*, 327–341. https://doi.org/10.3390/w5020327.

Hjältén, J., Nilsson, C., Jorgensen, D., et al. (2016). Forest-stream links, anthropogenic stressors, and climate change: Implications for restoration planning. *BioScience, 66*, 646–654. https://doi.org/10.1093/biosci/biw072.

Hongve, D., Riise, G., & Kristiansen, J. F. (2004). Increased colour and organic acid concentrations in Norwegian forest lakes and drinking water—a result of increased precipitation? *Aquatic Sciences, 66*, 231–238. https://doi.org/10.1007/s00027-004-0708-7.

Hotchkiss, E. R., Hall, R. O., Sponseller, R. A., et al. (2015). Sources of and processes controlling CO$_2$ emissions change with the size of streams and rivers. *Nature Geoscience, 8*, 696–699. https://doi.org/10.1038/ngeo2507.

Hultberg, H., & Skeffington, R. (Eds.). (1998). *Experimental reversal of acid rain effects: The Gardsjon Roof Project* (p. 484). John Wiley and Sons Inc.

Huser, B. J., Futter, M. N., Wang, R., et al. (2018). Persistent and widespread long-term phosphorus declines in boreal lakes in Sweden. *Science of the Total Environment, 613–614*, 240–249. https://doi.org/10.1016/j.scitotenv.2017.09.067.

Hynes, H. (1975). The stream and its valley. *Verhandlungen der Internationalen Vereinigung für Theoretische und Angewandte Limnologie, 19*, 1–15.

Jacobson, P., Jacobson, K., Angermeier, P., et al. (2000). Hydrologic influences on soil properties along ephemeral rivers in the Namib Desert. Journal of Arid Environments. *Journal of Arid Environments, 45*, 21–34. https://doi.org/10.1006/jare.1999.0619.

Jansson, M., Bergstrom, A. K., Blomqvist, P., et al. (2000). Allochthonous organic carbon and phytoplankton/bacterioplankton production relationships in lakes. *Ecology, 81*, 3250–3255. https://doi.org/10.1890/0012-9658(2000)081[3250:AOCAPB]2.0.CO;2.

Jansson, M., Persson, L., De Roos, A. M., et al. (2007). Terrestrial carbon and intraspecific size-variation shape lake ecosystems. *Trends in Ecology & Evolution, 22*, 316–322. https://doi.org/10.1016/j.tree.2007.02.015.

Jardine, T. D., Woods, R., Marshall, J., et al. (2015). Reconciling the role of organic matter pathways in aquatic food webs by measuring multiple tracers in individuals. *Ecology, 96*, 3257–3269. https://doi.org/10.1890/14-2153.1.

Karlsson, J., Bystrom, P., Ask, J., et al. (2009). Light limitation of nutrient-poor lake ecosystems. *Nature, 460*, 506–509. https://doi.org/10.1038/nature08179.

Karlsson, J., Bergstrom, A. K., Bystrom, P., et al. (2015). Terrestrial organic matter input suppresses biomass production in lake ecosystems. *Ecology, 96*, 2870–2876. https://doi.org/10.1890/15-0515.1.

Kärnä, O.-M., Heino, J., Laamanen, T., et al. (2019). Does catchment geodiversity foster stream biodiversity? *Landscape Ecology, 34*, 2469–2485. https://doi.org/10.1007/s10980-019-00901-z.

Kellerman, A. M., Kothawala, D. N., Dittmar, T., et al. (2015). Persistence of dissolved organic matter in lakes related to its molecular characteristics. *Nature Geoscience, 8*, 454–457. https://doi.org/10.1038/ngeo2440.

Kelly, P. T., Solomon, C. T., Weidel, B. C., et al. (2014). Terrestrial carbon is a resource, but not a subsidy, for lake zooplankton. *Ecology, 95*, 1236–1242. https://doi.org/10.1890/13-1586.1.

Klaus, M., Geibrink, E., Jonsson, A., et al. (2018). Greenhouse gas emissions from boreal inland waters unchanged after forest harvesting. *Biogeosciences, 15*, 5575–5594. https://doi.org/10.5194/bg-15-5575-2018.

Koehler, B., von Wachenfeldt, E., Kothawala, D., et al. (2012). Reactivity continuum of dissolved organic carbon decomposition in lake water. *Journal of Geophysical Research. Biogeosciences, 117*, G01024. https://doi.org/10.1029/2011JG001793.

Kragh, T., & Sondergaard, M. (2004). Production and bioavailability of autochthonous dissolved organic carbon: Effects of mesozooplankton. *Aquatic Microbial Ecology, 36*, 61–72. https://doi.org/10.3354/ame036061.

Kritzberg, E. S. (2017). Centennial-long trends of lake browning show major effect of afforestation. *Limnology Oceanography Letters, 2*, 105–112. https://doi.org/10.1002/lol2.10041.

Kritzberg, E. S., Hasselquist, E. M., Skerlep, M., et al. (2020). Browning of freshwaters: Consequences to ecosystem services, underlying drivers, and potential mitigation measures. *Ambio, 49*, 375–390. https://doi.org/10.1007/s13280-019-01227-5.

Kuglerová, L., Agren, A., Jansson, R., et al. (2014). Towards optimizing riparian buffer zones: Ecological and biogeochemical implications for forest management. *Forest Ecology and Management, 334*, 74–84. https://doi.org/10.1016/j.foreco.2014.08.033.

Kuglerová, L., Hasselquist, E. M., Richardson, J. S., et al. (2017). Management perspectives on *Aqua incognita*: Connectivity and cumulative effects of small natural and artificial streams in boreal forests. *Hydrological Processes, 31*, 4238–4244. https://doi.org/10.1002/hyp.11281.

Lamontagne, S., Carignan, R., D'Arcy, P., et al. (2000). Element export in runoff from eastern Canadian Boreal Shield drainage basins following forest harvesting and wildfires. *Canadian Journal of Fisheries and Aquatic Sciences, 57*, 118–128. https://doi.org/10.1139/f00-108.

Larsen, S., Andersen, T. O. M., & Hessen, D. O. (2011). Climate change predicted to cause severe increase of organic carbon in lakes. *Global Change Biology, 17*, 1186–1192. https://doi.org/10.1111/j.1365-2486.2010.02257.x.

Lau, D. C. P., Leung, K. M. Y., & Dudgeon, D. (2009a). Are autochthonous foods more important than allochthonous resources to benthic consumers in tropical headwater streams? *Journal of the North American Benthological Society, 28*, 426–439. https://doi.org/10.1899/07-079.1.

Lau, D. C. P., Leung, K. M. Y., & Dudgeon, D. (2009b). What does stable isotope analysis reveal about trophic relationships and the relative importance of allochthonous and autochthonous resources in tropical streams? A synthetic study from Hong Kong. *Freshwater Biology, 54*, 127–141. https://doi.org/10.1111/j.1365-2427.2008.02099.x.

Lau, D. C. P., Goedkoop, W., & Vrede, T. (2013). Cross-ecosystem differences in lipid composition and growth limitation of a benthic generalist consumer. *Limnology and Oceanography, 58*, 1149–1164. https://doi.org/10.4319/lo.2013.58.4.1149.

Lau, D. C. P., Sundh, I., Vrede, T., et al. (2014). Autochthonous resources are the main driver of consumer production in dystrophic boreal lakes. *Ecology, 95*, 1506–1519. https://doi.org/10.1890/13-1141.1.

Laudon, H., Hedtjarn, J., Schelker, J., et al. (2009). Response of dissolved organic carbon following forest harvesting in a boreal forest. *Ambio, 38*, 381–386. https://doi.org/10.1579/0044-7447-38.7.381.

Laudon, H., Buttle, J., Carey, S. K., et al. (2012). Cross-regional prediction of long-term trajectory of stream water DOC response to climate change. *Geophysical Research Letters, 39*, L18404. https://doi.org/10.1029/2012GL053033.

Leclerc, V., Sirois, P., & Bérubé, P. (2011). Impact of forest harvesting on larval and juvenile growth of yellow perch (*Perca flavescens*) in boreal lakes. *Boreal Environment Research, 16*, 417–429.

Lee, J., Lee, Y., Jang, C., et al. (2013). Stable carbon isotope signatures of zooplankton in some reservoirs in Korea. *Journal of Ecology Environment, 36*, 183–191. https://doi.org/10.5141/eco env.2013.183.

Leech, D. M., Pollard, A. I., Labou, S. G., et al. (2018). Fewer blue lakes and more murky lakes across the continental U.S.: Implications for planktonic food webs. *Limnology and Oceanography, 63*, 2661–2680. https://doi.org/10.1002/lno.10967.

Lennon, J. T. (2004). Experimental evidence that terrestrial carbon subsidies increase CO_2 flux from lake ecosystems. *Oecologia, 138*, 584–591. https://doi.org/10.1007/s00442-003-1459-1.

Lennon, J. T., Hamilton, S. K., Muscarella, M. E., et al. (2013). A source of terrestrial organic carbon to investigate the browning of aquatic ecosystems. *PLoS ONE, 8*, e75771. https://doi.org/10.1371/journal.pone.0075771.

Lidman, F., Boily, Å., Laudon, H., et al. (2017a). From soil water to surface water-how the riparian zone controls element transport from a boreal forest to a stream. *Biogeosciences, 14*, 3001–3014. https://doi.org/10.5194/bg-14-3001-2017.

Lidman, J., Jonsson, M., Burrows, R. M., et al. (2017b). Composition of riparian litter input regulates organic matter decomposition: Implications for headwater stream functioning in a managed forest landscape. *Ecology and Evolution, 7*, 1068–1077. https://doi.org/10.1002/ece3.2726.

Likens, G. E., Bormann, F. H., Johnson, N. M., et al. (1970). Effects of forest cutting and herbicide treatment on nutrient budgets in the Hubbard Brook watershed-ecosystem. *Ecological Monographs, 40*, 23–47. https://doi.org/10.2307/1942440.

López-Rojo, N., Pozo, J., Pérez, J., et al. (2019). Plant diversity loss affects stream ecosystem multifunctionality. *Ecology, 100*, e02847. https://doi.org/10.1002/ecy.2847.

Machtans, C. S., Villard, M. A., & Hannon, S. J. (1996). Use of riparian buffer strips as movement corridors by forest birds. *Conservation Biology, 10*, 1366–1379. https://doi.org/10.1046/j.1523-1739.1996.10051366.x.

Marchand, D., Prairie, Y. T., & del Giorgio, P. A. (2009). Linking forest fires to lake metabolism and carbon dioxide emissions in the boreal region of Northern Quebec. *Global Change Biology, 15*, 2861–2873. https://doi.org/10.1111/j.1365-2486.2009.01979.x.

Martin-Creuzburg, D., Kowarik, C., & Straile, D. (2017). Cross-ecosystem fluxes: Export of polyunsaturated fatty acids from aquatic to terrestrial ecosystems via emerging insects. *Science of the Total Environment, 577*, 174–182. https://doi.org/10.1016/j.scitotenv.2016.10.156.

Masclaux, H., Perga, M. E., Kagami, M., et al. (2013). How pollen organic matter enters freshwater food webs. *Limnology and Oceanography, 58*, 1185–1195. https://doi.org/10.4319/lo.2013.58.4.1185.

McDonald, S., Bishop, A. G., Prenzler, P. D., et al. (2004). Analytical chemistry of freshwater humic substances. *Analytica Chimica Acta, 527*, 105–124. https://doi.org/10.1016/j.aca.2004.10.011.

Mehner, T., Attermeyer, K., Brauns, M., et al. (2016). Weak response of animal allochthony and production to enhanced supply of terrestrial leaf litter in nutrient-rich lakes. *Ecosystems, 19*, 311–325. https://doi.org/10.1007/s10021-015 9933-2.

Messier, C., Puettmann, K. J., & Coates, K. D. (2013). *Managing forests as complex adaptive systems: Building resilience to the challenge of global change*. Routledge.

Meyer-Jacob, C., Michelutti, N., Paterson, A. M., et al. (2019). The browning and re-browning of lakes: Divergent lake-water organic carbon trends linked to acid deposition and climate change. *Scientific Reports, 9*, 16676. https://doi.org/10.1038/s41598-019-52912-0.

Meyer-Jacob, C., Labaj, A. L., Paterson, A. M., et al. (2020). Re-browning of Sudbury (Ontario, Canada) lakes now approaches pre-acid deposition lake-water dissolved organic carbon levels. *Science of the Total Environment, 725*, 138347. https://doi.org/10.1016/j.scitotenv.2020.138347.

Monteith, D. T., Stoddard, J. L., Evans, C. D., et al. (2007). Dissolved organic carbon trends resulting from changes in atmospheric deposition chemistry. *Nature, 450*, 537–540. https://doi.org/10.1038/nature06316.

Montoro Girona, M., Rossi, S., Lussier, J. M., et al. (2017). Understanding tree growth responses after partial cuttings: A new approach. *PLoS ONE, 12*, e0172653. https://doi.org/10.1371/journal.pone.0172653.

Montoro Girona, M., Lussier, J. M., Morin, H., et al. (2018). Conifer regeneration after experimental shelterwood and seed-tree treatments in boreal forests: Finding silvicultural alternatives. *Frontiers in Plant Science, 9*, 1145. https://doi.org/10.3389/fpls.2018.01145.

Montoro Girona, M., Morin, H., Lussier, J.-M., et al. (2019). Post-cutting mortality following experimental silvicultural treatments in unmanaged boreal forest stands. *Frontiers in Forests and Global Change, 2*, 4. https://doi.org/10.3389/ffgc.2019.00004.

Moussaoui, L., Leduc, A., Montoro Girona, M., et al. (2020). Success factors for experimental partial harvesting in unmanaged boreal forest: 10-year stand yield results. *Forests, 11*, 1199. https://doi.org/10.3390/f11111199.

Muehlbauer, J. D., Collins, S. F., Doyle, M. W., et al. (2014). How wide is a stream? Spatial extent of the potential "stream signature" in terrestrial food webs using meta-analysis. *Ecology, 95*, 44–55. https://doi.org/10.1890/12-1628.1.

Olefeldt, D., Devito, K. J., & Turetsky, M. R. (2013). Sources and fate of terrestrial dissolved organic carbon in lakes of a Boreal Plains region recently affected by wildfire. *Biogeosciences, 10*, 6247–6265. https://doi.org/10.5194/bg-10-6247-2013.

Paerl, H. W., & Otten, T. G. (2013). Harmful cyanobacterial blooms: Causes, consequences, and controls. *Microbial Ecology, 65*, 995–1010. https://doi.org/10.1007/s00248-012-0159-y.

Paetzold, A., Schubert, C., & Tockner, K. (2005). Aquatic terrestrial linkages along a braided-river: Riparian arthropods feeding on aquatic insects. *Ecosystems, 8*, 748–759. https://doi.org/10.1007/s10021-005-0004-y.

Patoine, A., Pinel-Alloul, B., Prepas, E. E., et al. (2000). Do logging and forest fires influence zooplankton biomass in Canadian Boreal Shield lakes? *Canadian Journal of Fisheries and Aquatic Sciences, 57*, 155–164. https://doi.org/10.1139/f00-105.

Payette, S., Fortin, M. J., & Gamache, I. (2001). The subarctic forest-tundra: The structure of a biome in a changing climate: The shifting of local subarctic tree lines throughout the forest-tundra biome, which is linked to ecological processes at different spatiotemporal scales, will reflect future global changes in climate. *BioScience, 51*, 709–718. https://doi.org/10.1641/0006-3568(2001)051[0709:TSFTTS]2.0.CO;2.

Perga, M. E., Kainz, M., Matthews, B., et al. (2006). Carbon pathways to zooplankton: Insights from the combined use of stable isotope and fatty acid biomarkers. *Freshwater Biology, 51*, 2041–2051. https://doi.org/10.1111/j.1365-2427.2006.01634.x.

Peterjohn, W. T., & Correll, D. L. (1984). Nutrient dynamics in an agricultural watershed: Observations on the role of a riparian forest. *Ecology, 65*, 1466–1475. https://doi.org/10.2307/1939127.

Pinay, G., O'Keefe, T., Edwards, R., et al. (2009). Nitrate removal in the hyporheic zone of a salmon river in Alaska. *River Research and Applications, 25*, 367–375. https://doi.org/10.1002/rra.1164.

Pinel-Alloul, B., Patoine, A., Carignan, R., et al. (1998). Responses of lake zooplankton to natural fire and forest harvesting in the boreal ecozone in Quebec: Preliminary study. *Annales de Limnologie, 34*, 401–412.

Planas, D., Desrosiers, M., Groulx, S. R., et al. (2000). Pelagic and benthic algal responses in eastern Canadian Boreal Shield lakes following harvesting and wildfires. *Canadian Journal of Fisheries and Aquatic Sciences, 57*, 136–145. https://doi.org/10.1139/f00-130.

Prepas, E. E., Pinel-Alloul, B., Planas, D., et al. (2001). Forest harvest impacts on water quality and aquatic biota on the Boreal Plain: Introduction to the TROLS lake program. *Canadian Journal of Fisheries and Aquatic Sciences, 58*, 421–436. https://doi.org/10.1139/f00-259.

Rasmussen, J. B. (2010). Estimating terrestrial contribution to stream invertebrates and periphyton using a gradient-based mixing model for delta ^{13}C. *Journal of Animal Ecology, 79*, 393–402. https://doi.org/10.1111/j.1365-2656.2009.01648.x.

Rautio, M., Mariash, H., & Forsstrom, L. (2011). Seasonal shifts between autochthonous and allochthonous carbon contributions to zooplankton diets in a subarctic lake. *Limnology and Oceanography, 56*, 1513–1524. https://doi.org/10.4319/lo.2011.56.4.1513.

Raymond, P. A., Hartmann, J., Lauerwald, R., et al. (2013). Global carbon dioxide emissions from inland waters. *Nature, 503*, 355–359. https://doi.org/10.1038/nature12760.

Reiter, M., Heffner, J. T., Beech, S., et al. (2009). Temporal and spatial turbidity patterns over 30 years in a managed forest of western Washington. *Journal of the American Water Resources Association, 45*, 793–808. https://doi.org/10.1111/j.1752-1688.2009.00323.x.

Richardson, S. D., Plewa, M. J., Wagner, E. D., et al. (2007). Occurrence, genotoxicity, and carcinogenicity of regulated and emerging disinfection by-products in drinking water: A review and roadmap for research. *Mutation Research, 636*, 178–242. https://doi.org/10.1016/j.mrrev.2007.09.001.

Richardson, D. C., Carey, C. C., Bruesewitz, D. A., et al. (2017). Intra-and inter-annual variability in metabolism in an oligotrophic lake. *Aquatic Sciences, 79*, 319–333. https://doi.org/10.1007/s00027-016-0499-7.

Roiha, T., Tiirola, M., Cazzanelli, M., et al. (2012). Carbon quantity defines productivity while its quality defines community composition of bacterioplankton in subarctic ponds. *Aquatic Sciences, 74*(3), 513–525.

Roulet, N., & Moore, T. R. (2006). Environmental chemistry: Browning the waters. *Nature, 444*, 283–284. https://doi.org/10.1038/444283a.

Rowe, E. C., Tipping, E., Posch, M., et al. (2014). Predicting nitrogen and acidity effects on long-term dynamics of dissolved organic matter. *Environmental Pollution, 184*, 271–282. https://doi.org/10.1016/j.envpol.2013.08.023.

Ruel, J. C., Pin, D., & Cooper, K. (2001). Windthrow in riparian buffer strips: Effect of wind exposure, thinning and strip width. *Forest Ecology and Management, 143*, 105–113. https://doi.org/10.1016/S0378-1127(00)00510-7.

Runge, J., & Roff, J. (2000). The measurement of growth and reproductive rates. In R. Harris, P. Wiebe, J. Lenz, H. R. Skjoldal, & M. Huntley (Eds.), *ICES Zooplankton methodology manual* (pp. 401–454). Academic Press.

Schindler, D. E., & Smits, A. P. (2017). Subsidies of aquatic resources in terrestrial ecosystems. *Ecosystems, 20*, 78–93. https://doi.org/10.1007/s10021-016-0050-7.

Schink, B. (1997). Energetics of syntrophic cooperation in methanogenic degradation. *Microbiology and Molecular Biology Reviews, 61*, 262–280. https://doi.org/10.1128/.61.2.262-280.1997.

Schneider, T., Grosbois, G., Vincent, W. F., et al. (2016). Carotenoid accumulation in copepods is related to lipid metabolism and reproduction rather than to UV-protection. *Limnology and Oceanography, 61*, 1201–1213. https://doi.org/10.1002/lno.10283.

Schneider, T., Grosbois, G., Vincent, W. F., et al. (2017). Saving for the future: Pre-winter uptake of algal lipids supports copepod egg production in spring. *Freshwater Biology, 62*, 1063–1072. https://doi.org/10.1111/fwb.12925.

Scully, N., Leavitt, P., & Carpenter, S. (2000). Century-long effects of forest harvest on the physical structure and autotrophic community of a small temperate lake. *Canadian Journal of Fisheries and Aquatic Sciences, 57*(S2), 50–59. https://doi.org/10.1139/f00-115.

Seitzinger, S. P., Harrison, J. A., Dumont, E., et al. (2005). Sources and delivery of carbon, nitrogen, and phosphorus to the coastal zone: An overview of Global Nutrient Export from Watersheds (NEWS) models and their application. *Global Biogeochemical Cycles, 19*, GB4S01 https://doi.org/10.1029/2005GB002606.

Solomon, C. T., Jones, S. E., Weidel, B. C., et al. (2015). Ecosystem consequences of changing inputs of terrestrial dissolved organic matter to lakes: Current knowledge and future challenges. *Ecosystems, 18*, 376–389. https://doi.org/10.1007/s10021-015-9848-y.

Sørensen R, Meili M, Lambertsson L, et al (2009) The effects of forest harvest operations on mercury and methylmercury in two boreal streams: Relatively small changes in the first two years prior to site preparation. *Ambio, 38*(7), 364–372 https://doi.org/10.1579/0044-7447-38.7.364.

Steedman, R. J. (2000). Effects of experimental clearcut logging on water quality in three small boreal forest lake trout (*Salvelinus namaycush*) lakes. *Canadian Journal of Fisheries and Aquatic Sciences, 57*, 92–96. https://doi.org/10.1139/f00-119.

Strandberg, U., Taipale, S. J., Hiltunen, M., et al. (2015). Inferring phytoplankton community composition with a fatty acid mixing model. *Ecosphere, 6*(1), 16. https://doi.org/10.1890/ES14-00382.1.

Sun, G., Riekerk, H., & Kornhak, L. V. (2000). Ground-water-table rise after forest harvesting on cypress-pine flatwoods in Florida. *Wetlands, 20*, 101–112. https://doi.org/10.1672/0277-5212(2000)020[0101:GWTRAF]2.0.CO;2.

Taipale, S. J., Kainz, M. J., & Brett, M. T. (2015). A low ω-3:ω-6 ratio in *Daphnia* indicates terrestrial resource utilization and poor nutritional condition. *Journal of Plankton Research, 37*, 596–610. https://doi.org/10.1093/plankt/fbv015.

Taipale, S. J., Kahilainen, K. K., Holtgrieve, G. W., et al. (2018). Simulated eutrophication and browning alters zooplankton nutritional quality and determines juvenile fish growth and survival. *Ecology and Evolution, 8*, 2671–2687. https://doi.org/10.1002/ece3.3832.

Tanentzap, A. J., Szkokan-Emilson, E. J., Kielstra, B. W., et al. (2014). Forests fuel fish growth in freshwater deltas. *Nature Communications, 5*, 4077. https://doi.org/10.1038/ncomms5077.

Torres-Ruiz, M., Wehr, J. D., & Perrone, A. A. (2007). Trophic relations in a stream food web: Importance of fatty acids for macroinvertebrate consumers. *Journal of North American Benthological Society, 26*, 509–522. https://doi.org/10.1899/06-070.1.

Vachon, D., & del Giorgio, P. A. (2014). Whole-lake CO_2 dynamics in response to storm events in two morphologically different lakes. *Ecosystems, 17*, 1338–1353. https://doi.org/10.1007/s10021-014-9799-8.

Vachon, D., Solomon, C. T., & del Giorgio, P. A. (2017). Reconstructing the seasonal dynamics and relative contribution of the major processes sustaining CO_2 emissions in northern lakes. *Limnology and Oceanography, 62*, 706–722. https://doi.org/10.1002/lno.10454.

Vander Zanden, M. J., & Gratton, C. (2011). Blowin' in the wind: Reciprocal airborne carbon fluxes between lakes and land. *Canadian Journal of Fisheries and Aquatic Sciences, 68*, 170–182. https://doi.org/10.1139/F10-157.

Vannote, R. L., Minshall, G. W., Cummins, K. W., et al. (1980). River continuum concept. *Canadian Journal of Fisheries and Aquatic Sciences, 37*, 130–137. https://doi.org/10.1139/f80-017.

Veny, E. S. (1986). Forest harvesting and water: The lake states experience. *Journal of the American Water Resources Association, 22*, 1039–1047. https://doi.org/10.1111/j.1752-1688.1986.tb00775.x.

Weidel, B., Carpenter, S., Cole, J., et al. (2008). Carbon sources supporting fish growth in a north temperate lake. *Aquatic Sciences, 70*, 446–458. https://doi.org/10.1007/s00027-008-8113-2.

Wenzel, A., Bergstrom, A. K., Jansson, M., et al. (2012). Poor direct exploitation of terrestrial particulate organic material from peat layers by *Daphnia galeata*. *Canadian Journal of Fisheries and Aquatic Sciences, 69*, 1870–1880. https://doi.org/10.1139/f2012-110.

Wetzel, R. G. (1975). *Limnology*. WB Saunders Company.

Weyhenmeyer, G. A., Prairie, Y. T., & Tranvik, L. J. (2014). Browning of boreal freshwaters coupled to carbon-iron interactions along the aquatic continuum. *PLoS ONE, 9*, e88104. https://doi.org/10.1371/journal.pone.0088104.

Wilkinson, G. M., Carpenter, S. R., Cole, J. J., et al. (2013). Terrestrial support of pelagic consumers: Patterns and variability revealed by a multilake study. *Freshwater Biology, 58*, 2037–2049. https://doi.org/10.1111/fwb.12189.

Williamson, C. E., Overholt, E. P., Pilla, R. M., et al. (2015). Ecological consequences of long-term browning in lakes. *Scientific Reports, 5*, 18666. https://doi.org/10.1038/srep18666.

Wu, P., Bishop, K., von Brömssen, C., et al. (2018). Does forest harvest increase the mercury concentrations in fish? Evidence from Swedish lakes. *Science of the Total Environment, 622–623*, 1353–1362. https://doi.org/10.1016/j.scitotenv.2017.12.075.

Yebra, L., Kobari, T., Sastri, A. R., et al. (2017). Advances in biochemical indices of zooplankton production. *Advances in Marine Biology, 76*, 157–240. https://doi.org/10.1016/bs.amb.2016. 09.001.

Chapter 30
Current Symptoms of Climate Change in Boreal Forest Trees and Wildlife

Loïc D'Orangeville, Martin-Hugues St-Laurent, Laura Boisvert-Marsh, Xianliang Zhang, Guillaume Bastille-Rousseau, and Malcolm Itter

Abstract Measuring climate change impacts on forest ecosystems can be challenging, as many of these changes are imperceptible within the typical time scale of short-term (e.g., 3–4 years) funding of research projects. Boreal trees are notoriously imperturbable, given their tolerance to harsh conditions and their adaptability. However, the buildup of decades of warming should now translate into measurable alterations of boreal ecosystem processes. The boreal forest is host to numerous northern animals; therefore, any change in boreal forest dynamics should affect wildlife. In this chapter, we aim to provide a nonexhaustive synthesis of documented impacts of climate change on selected key processes driving boreal forest ecosystem dynamics. We focus on the themes of plant and wildlife range shifts and stand growth and death, as they are keystone parameters of boreal forest ecosystem health that are symptomatic of climate change impacts on the boreal biota. For each theme, we introduce the general concepts and processes, convey some of the limitations of current assessments, and suggest future pressing challenges.

L. D'Orangeville (✉)
Faculty of Forestry and Environmental Management, University of New Brunswick, 28 Dineen Drive, P.O. Box 4400, Fredericton, NB E3B 5A3, Canada
e-mail: loic.dorangeville@unb.ca

M.-H. St-Laurent
Département de biologie, chimie et géographie, Université du Québec à Rimouski, Stn. Rimouski, Rimouski, QC G5L 3A1, Canada
e-mail: martin-hugues_st-laurent@uqar.ca

Centre for Forest Research, Université du Québec à Montréal, P.O. Box 8888, Stn. Centre-Ville, Montréal, QC H3C 3P8, Canada

L. Boisvert-Marsh
Great Lakes Forestry Centre, Canadian Forest Service, Natural Resources Canada, 1219 Queen Street East, Sault Ste. Marie, ON P6A 2E5, Canada
e-mail: laura.boisvert-marsh@nrcan-rncan.gc.ca

X. Zhang
College of Forestry, Hebei Agricultural University, 289 Linguysi Street, Baoding, Hebei, China
e-mail: zhxianliang@126.com

© The Author(s) 2023   747
M. M. Girona et al. (eds.), *Boreal Forests in the Face of Climate Change*,
Advances in Global Change Research 74,
https://doi.org/10.1007/978-3-031-15988-6_30

30.1   Theme 1: Plants and Wildlife Range Shifts, Expansions, and Contractions

30.1.1   General Concepts and Processes at Play

For decades, wildlife range shifts have fascinated biologists, biogeographers, hunters, and conservationists given the forces at play in nature that dictate the rules of wildlife–habitat relationships. Fundamentally dynamic (Laliberte & Ripple, 2004), species ranges respond to a variety of factors, including geological forcing, climate change, and the forces driving the recent and projected biodiversity decline, namely anthropogenic habitat loss, overexploitation, and invasive species (Purvis et al., 2000). Moreover, many of these drivers interact; for example, an ecosystem can show a greater vulnerability or an enhanced resistance to invasive species because of climate change (Walther et al., 2009). Understanding the factors contributing to species range shifts, contractions, or expansions is crucial not only to develop rigorous predictions of future changes in species distributions but also to implement conservation and management strategies that can slow these changes or at least mitigate the added influence of humans on flora and fauna (Laliberte & Ripple, 2004). This is especially relevant now that many authors consider human-related activities as the ultimate driver explaining worldwide range shifts, contractions, and expansions (Channell & Lomolino, 2000; Lawler et al., 2009).

In this section, we focus on range shifts driven by climate change–including the synergistic effects of other drivers of change in local biodiversity–by briefly surveying the relevant methods for studying, modeling, and predicting future shifts, expansions, and contractions in species distributions. We also discuss the limitations of these approaches and document three contrasting case studies to illustrate different scenarios. To facilitate an understanding of the mechanisms at play, we propose using the following definitions of concepts central to the study of range shifts. First, we use the term *effect* to refer to a change in the environment that results from a disturbance and the term *impact* to represent the consequences of this change for wildlife populations (Wärnbäck & Hilding-Rydevik, 2009). We also differentiate between *climate* and *weather*. According to Watson, (1963), climate refers to the interplay between solar energy, temperature, air movement, rain and snow, atmospheric humidity, and mist and fog. The seasonal pattern of change in these variables and the similar interactions from year to year constitute the climate per se in its

G. Bastille-Rousseau
School of Biological Sciences, Southern Illinois University, 1263 Lincoln Drive, Carbondale, IL 62901, USA
e-mail: gbr@siu.edu

M. Itter
Department of Environmental Conservation, University of Massachusetts Amherst, 160 Holdsworth Way, Amherst, MA 01003, USA
e-mail: mitter@umass.edu

geographical meaning. The short-term (daily, weekly) deviations from this pattern are called weather. Finally, we refer to several common concepts in ecology, e.g., physiological constraints, competition, predation, diseases, parasites, as potential mechanisms linking drivers of change, e.g., climate change, to the species of interest via their interplay with other species and components of the ecosystems.

Box 30.1 Observed Climate Change in Boreal Forest Ecosystems

North American boreal forests have experienced an average 2 °C warming since the 1950s, with the greatest warming observed in northern Canada (Zhang et al., 2015). Eurasian boreal forests are also warming rapidly, with 1.35 and 2.00 °C increases in summer and winter temperatures since 1881, respectively, and the warming rate is accelerating (Groisman & Soja, 2009). Warming patterns vary seasonally; winter temperatures have been increasing faster than summer temperatures across boreal forests, with over 4 °C warming in some areas of North America since the late 1940s. Concurrently, annual precipitation has remained constant in Eurasia, whereas precipitation increases of 5–30% have been recorded in North American boreal forests, albeit with wide spatial variations and great uncertainty (Zhang et al., 2015). Extreme temperatures have also shifted toward more extreme-warm days and fewer extreme-cold days. Such warming has lengthened the growing season by approximately two weeks over the last 30 years, concurrent with a reduction in spring snow cover. Such warming, combined with limited changes in precipitation, has led to increased heat-induced drought and a rapid degradation of the permafrost.

30.1.2   Brief Overview of the Methods (and Limits)

Multiple approaches are used to assess species' vulnerability to environmental change, thereby paving the way to understand range shifts, contractions, and expansions induced by climate change or other drivers. Pacifici et al. (2015) categorized these approaches into four types: correlative, mechanistic, trait-based, and a combination of different approaches.

Correlative approaches, often referred to as species distribution models (SDMs) or ecological niche models (ENMs), are frequently used to assess the impacts of disturbances on species distribution across a geographical range within the limits of the species' realized niche (Guisan & Thuiller, 2005). These models are supported by correlations between current species distributions and current environmental covariates; the models are then run with the predicted changes in environmental covariates to extrapolate future species distributions for a variety of taxa (Pacifici et al., 2015) and across spatial scales (Harrison et al., 2006). Although correlative models have the advantages of being spatially explicit, generally user-friendly, and adaptable to various types of data, they are also limited by their inherent correlative nature (Sinclair et al., 2010), debatable underlying assumptions, e.g., the current distribution of a species reflects an equilibrium with its environment, and the inability to capture the complexity of the biological processes driving shifts in species distributions over time.

Mechanistic models require more parameters than correlative approaches to document the behavior or mechanisms developed by organisms to cope with changing environmental conditions (Huey et al., 2012). They rely on empirical relationships linking climate parameters and demographic rates or physiological tolerances and generally focus on a single species (Deutsch et al., 2008). Mechanistic species distribution models are viewed as being more robust than correlative models for predicting species' responses to climate change (Evans et al., 2015). These models provide insights into the fundamental niche of the climatic space that an organism can occupy rather than the realized niche of a species—the latter more commonly obtained through a correlative approach (Morin & Thuiller, 2009). Nonetheless, mechanistic niche models are limited by the need for large amounts of data, being often species-specific, and not fully accounting for the dispersal ability or biotic interactions of the modeled species.

Trait-based approaches rely on the representative biological characteristics of a species, which translate into the sensitivity and adaptability of a species to future change (Aubin et al., 2016; Moyle et al., 2013; Rowland et al., 2011). Contrary to the two abovementioned approaches, trait-based models are simpler, easier to use, and apply to multiple species. These advantages could account for the popularity of trait-based approaches among practitioners in conservation and management agencies. However, the accuracy of trait-based approaches is limited by the arbitrary selection of vulnerability thresholds for the various traits under analysis, as thresholds are based on expert opinion or observations for which environmental variations are poorly understood. This arbitrary selection can add uncertainty to predictions (Foden et al., 2013), lead to inconsistent results within species (Lankford et al., 2014), and produce incoherent comparisons between taxonomic groups when different traits are selected. Trait-based approaches are often difficult to validate, have low explanatory power, and are of limited utility for conservation and management (Angert et al., 2011).

Given these limitations, there is growing consensus that combining different types of models and data is the most suitable approach (Pacifici et al., 2015), including criteria-based approaches (Thomas et al., 2011), correlative trait–based approaches (Barbet-Massin et al., 2012), and mechanistic–correlative approaches (Dullinger et al., 2012). The latter include the very effective dynamic-range models (Lurgi et al., 2015), which are supported by spatially explicit demographic individual-based models (McLane et al., 2011).

30.1.3   Case Study 1.1 The Northern Biodiversity Paradox

Although we are currently observing a worldwide loss of biodiversity, some northern regions are experiencing (or will experience) an intriguing phenomenon: an increase in local species richness. This phenomenon, called the *northern biodiversity paradox*, involves climate change–induced increases in local biodiversity in the northern latitudes. Given that the ranges of several species are currently limited by low temperatures, e.g., ectotherms (Araújo et al., 2006), this concept holds that climate warming will lead to a northern range shift of many species (Parmesan & Yohe, 2003).

Berteaux et al., (2018) applied climate–niche modeling, using 1961–1990 data, to assess the potential impacts of climate change on the probability of occurrence for 529 species within 1,749 protected areas spread over approximately 600,000 km² in Québec (Canada). This extensive study area encompassed the northern limit of the distribution of several species of animals (birds and amphibians) and trees and other vascular plants. The regional climate is currently characterized by cold winters and short summers, limiting several species that are poorly adapted to the harsh winter conditions and short growing seasons. Berteaux et al.'s modeling suggested that a major species turnover is very likely within a 50- to 80-year horizon (CE 2071–2100), assuming all studied species can track their suitable climatic conditions. Depending on the specific protected area, their model projects either a relative gain in species diversity (12−530%) or relative loss (7−55%). The greatest gains are predicted for the northernmost parts of Québec's protected area network (approximately 50°−52°N), and losses will occur mainly in the southern areas. Overall, average species richness is predicted to increase in this northeastern region of North America because of climate warming, illustrating well the northern biodiversity paradox.

Berteaux et al.'s results nonetheless suggest that the arrival of several new colonizing species in northern areas could alter the structure and functioning of northern ecosystems, as observed by others (Elmhagen et al., 2015; Foxcroft et al., 2017; Gallant et al., 2020), compromising further several *at-risk* species in northern environments (Alda et al., 2013). Many studies confirm that wildlife species are expanding their distribution range at their high-latitude or high-elevation (cool) margins, whereas these species' low-latitude and low-elevation (warm) margins are retracting to higher latitudes and elevations. For example, Gallant et al., (2020) demonstrated that the impressive >1,700 km poleward range shift of the red fox (*Vulpes vulpes*) into the Arctic over the last century occurred both during cooling and warming climate phases; however, they showed that the highest migration rate of the red fox occurred during warmer winters. More globally, the meta-analysis conducted by Chen et al., (2011) determined that the range limits of 764 plant and animal species have moved, on average, 16.9 km northward per decade owing to climate warming. Wilson et al., (2005) highlighted range contractions along the warmer margins of distribution ranges for 16 butterflies species. The altitude rise of the butterflies' lower elevation limit (averaging 212 m over 30 years) caused a 33% loss in suitable habitat for these taxa. Serious issues arise when the southern (warmer) edge of a range contracts faster than the northern (cooler) margin can extend (Jackson & Sax, 2010), as shown by Wiens (2016) for many taxa that were part of a survey of 976 animal and plant species. Over the long term, such an asymmetrical shift in range boundaries results in species becoming *trapped* by the displacement of suitable habitat conditions. This situation can counterbalance or even reverse the current regional increase in species richness observed in northern latitudes (Berteaux et al., 2018).

30.1.4 Case Study 1.2 Compositional Shifts in Tree Regeneration

Poleward migration in response to warming is an expected response of tree species as suitable climate conditions shift northward in the boreal forest (Périé & de Blois, 2016); however, trees and plant species require multiple generations of dispersal and a successful establishment at previously unoccupied sites before sustained range shifts can be detected. An approach for detecting range shifts is the percentile method, which links changes in species presence to latitude. This method provides evidence of range shifts when combined with broad spatial assessments of plot occupancy (gain, loss, or unchanged; Boisvert-Marsh et al., 2014). Analyzing latitudinal tree shifts across broad geographic areas requires consistent survey methods through time (Woodall et al., 2009), the precise recording of survey locations (Tingley & Beissinger, 2009), and extensive data coverage (Shoo et al., 2006). In 1970, the Québec Ministry of Forests, Wildlife and Parks established an extensive network of inventory plots south of 53°N to characterize forest resources for commercial purposes in the province. At present, four inventories have been completed across more than 6,200 permanent plots (approximately 761,000 km²), of which over 70% are located in the boreal forest. Recent studies in forest ecology have used this exceptional data set, demonstrating that tree regeneration patterns and overall community dynamics are shifting mainly at the transition between the northern temperate forest and the southern boreal forest (Boisvert-Marsh, 2020; Boisvert-Marsh et al., 2014; Duchesne & Ouimet, 2008).

Recruitment patterns of juvenile stages, e.g., saplings, can provide early evidence of shifting regeneration patterns and migration trends, revealing the biotic or abiotic factors that facilitate or hinder range shifts. Using this approach, Boisvert-Marsh (2020) detected southward latitudinal shifts between 1970 and 2015 for black spruce (*Picea mariana*), white spruce (*Picea glauca*), and balsam fir (*Abies balsamea*). These shifts were driven by occupancy gains into mixed temperate balsam fir–yellow birch (*Betula alleghaniensis*) domains and southern boreal balsam fir–white birch (*B. papyrifera*) domains. White birch and trembling aspen (*Populus tremuloides*) showed evidence of a northward latitudinal shift combined with occupancy gains toward the northern edge of the inventory area. Red maple (*Acer rubrum*) also showed a northward shift, increasing its presence within the southern edge of the boreal forest. Climate is not the only factor, however, that can drive such large-scale migrations in regeneration. Notably, changes to stand dynamics precipitated by disturbances elicit species turnover and can break the inertia that inhibits more southerly species from moving northward. Although there is some evidence that sugar maple (*Acer saccharum*) could colonize sites with conditions typical of the boreal forest (Kellman, 2004), edaphic and climatic factors interact to mitigate the extent of its northward migration, at least for now (Boisvert-Marsh & de Blois, 2021; Boisvert-Marsh et al. 2019; Collin et al., 2018). Moderate to major distur-bances are accelerating species turnover toward stands dominated by red maple, white birch, and trembling aspen (Brice et al., 2019), particularly in the southern boreal where balsam fir is common. Harvesting is linked to red maple regeneration in this area, and the expansion of red maple is aided further by its ability to recruit into plots with white birch. As expected, white birch and trembling aspen recruitment in the boreal forest has occurred into plots where black spruce or balsam fir were formerly present but that have been removed through harvesting (Brice et al., 2019). As fires, insect outbreaks, and windthrow are expected to become more frequent and intense with climate change—coupled with the northward expansion of harvesting activities—disturbances could create conditions for other temperate species to follow suit. Such changes to plant and tree communities influence the spatial arrangement and availability of wildlife habitat, e.g., resources and shelter, such that range shifts in one species can trigger shifts in others.

30.1.5   Case Study 1.3. Under Pressure: The Case of Boreal Populations of Woodland Caribou

Although several studies have documented recent shifts in species' ranges, range expansion or contraction, and projections of future range shifts related to climate change, many other species have experienced displacement by a combination of the five main drivers of biodiversity loss (listed above), in particular anthropogenic habitat loss. Across the Northern Hemisphere, we have witnessed a global decline of caribou (in North America) and reindeer (in Eurasia) subspecies, mainly driven by human-induced disturbances (Vors & Boyce, 2009). In North America, the boreal populations of woodland caribou, an ecotype of the *Rangifer tarandus caribou* subspecies, has been historically associated with the pan-Canadian belt of boreal and temperate forests, extending its range as far south as New England (Fig. 30.1; reviewed in Bergerud & Mercer, 1989). The current distribution range of boreal caribou is considerably reduced relative to its historical range, as this ecotype has been extirpated from much of the Maritimes, the northeastern United States, and south of the St. Lawrence River (Fig. 30.1; COSEWIC, 2014). Similar range contractions have been recorded in southwestern Canada and the northwestern United States where local populations of caribou have disappeared (Grant et al., 2019; Seip & Cichowski, 1996).

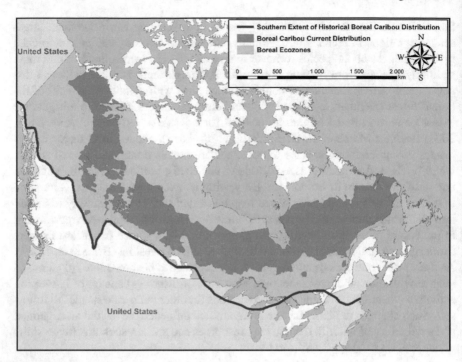

Fig. 30.1 Historical and current distribution (*orange shading*) of boreal caribou across the boreal ecozones (*green shading*) of Canada. Note that because of the lack of information on the historical distribution of boreal caribou in British Columbia, the historical southern extent in that province tracks the boreal ecozone boundary (*red line*). Adapted from Environment Canada (2011), permission courtesy of Environment and Climate Change Canada (ECCC)

Multiple studies have highlighted the major role of industrial activities in causing this range contraction (e.g., Rudolph et al., 2017; Schaefer, 2003; Vors & Boyce, 2009)—related to an increase in the number and efficacy of predators (e.g., Seip, 1992; Whittington et al., 2011) and maladaptive caribou behavior in human-disturbed landscapes (e.g., Lafontaine et al., 2019; Leclerc et al., 2014; Losier et al., 2015). In contrast, few studies have identified climate change as a driver of this caribou range contraction (however, see Yannic et al., (2014) for the last 21,000 years). Untangling the respective roles of past climate change and anthropogenic-related habitat loss as drivers of boreal caribou range contraction is not an easy task, as both have occurred and expanded their influence simultaneously during the last decades. Land use–related impacts, because of their significance within the boundaries of the caribou range, have received most of the research effort. Schaefer (2003), for example, related the 1880–1990 southern range contraction (34,800 km^2 per decade) of caribou in Ontario to the northern advance of timber harvesting, reporting a northward range recession of 34 km per decade for the caribou; however, climate change was not linked to this shift. Such conclusions are not unique, as most recent short-term studies have struggled to isolate the effects of a changing climate as a potential explanation of caribou decline. This is not surprising considering that climate change impacts are more complex to study over the long term than human-induced disturbances. Nevertheless, a smaller (but growing) number of researchers have recently focused their efforts on distinguishing the potential impacts of climate change from the effect of anthropogenic habitat loss along the boundaries of current and future caribou distribution ranges.

Using multiple environmental suitability models, Murray et al. (2015) suggested that the distribution range of woodland caribou in the boreal forest will decrease by approximately 29–52% by 2080 under various climate change scenarios. Barber et al. (2018) applied the analytical framework of Whitman et al. (2017) to model changes to the extent of future caribou habitat in Alberta. They point out that the boreal caribou range in Alberta will experience a severe contraction under various climate change scenarios. This decrease is triggered by a marked increase in grassland vegetation by the 2080s that results in the contraction of mixedwood and coniferous forests, which are suitable habitats for caribou (following Schneider et al., 2009). This projected shift in vegetation favors an increase in white-tailed deer (*Odocoileus virginianus*) and its predators (gray wolf, *Canis lupus*; see Latham et al., 2011) and also an increased prevalence of deer-related disease, as highlighted previously (Pickles et al., 2013).

These indirect effects are all expected to compromise the long-term persistence of caribou in the boreal forest landscapes of this Canadian province (Barber et al., 2018). These examples predict coarse changes in caribou distribution through correlative or mechanistic models; however, we still lack sufficient knowledge of the fine-scale mechanisms linking the behavior and demography of caribou, predators, and their alternate prey under variable weather (e.g., Bastille-Rousseau et al., 2018; Leclerc et al., 2019). These relationships should be linked with changing climate, especially in regard to a synergy with intensive land use within the caribou range, e.g., timber harvesting, mining, and oil and gas extraction (Festa-Bianchet et al., 2011), to assist in orienting conservation efforts and ensure the persistence of this iconic species.

30.1.6 Future Challenges

Understanding species range shifts, expansions, and contractions has broad implications for scientists but also for politicians, industries, NGOs, and many other stakeholders. These implications range from predictions of the functioning of ecosystems, sources of food provisioning, conservation efforts for protecting currently endangered species, and the management of potentially depredating species. Researchers have provided decision-makers with various modeling tools that can support predictions of future range displacement; despite all these efforts, however, uncertainty remains and requires greater attention. Although climate change could facilitate species' range expansion into regions where they are currently unable to survive and reproduce (Walther et al., 2009), other facets of change could limit a species' ability to track suitable climate and habitat. Schloss et al. (2012) estimate that across the Americas, approximately 9.2% of mammals will likely be unable to keep pace with projected climate change. Among the 87% of mammalian species that will suffer range contraction, 20% will likely be limited by their dispersal capacities. The authors conclude that mammalian vulnerability to climate change may be more extensive than previously anticipated; therefore, dispersal capacities must be included in range-shift models to improve our projections of species distribution and vulnerabilities as well as our conservation efforts, thus joining conclusions obtained by other research teams (e.g., Barbet-Massin et al., 2012; Sinclair et al., 2010).

Similarly, integrating the level of plasticity and local adaptations to our understanding of current distributions can improve future predictions (Peterson et al., 2019; Valladares et al., 2014). As our world is profoundly impacted by the human footprint—characterized by impressive levels of loss and fragmentation of natural habitats (Fahrig, 2003)—a better understanding of the interactions between these key pressures on biodiversity and climate change is urgently needed (Hof et al., 2011; Howard et al., 2020; Opdam & Wascher, 2004).

30.2 Theme 2: The Life and Death of Warmer Boreal Forests

30.2.1 General Concepts and Processes at Play

Shifts in demographic indices, e.g., growth and mortality rates, are major indicators of changes in stand health (Berdanier & Clark, 2016), whereas the relationship between these indices and environmental drivers yields information on the capacity of species to cope with climate change (Buechling et al., 2017; Foster et al., 2016). In this sense, changes in demographic performance indices may be easier to detect than actual range shifts, which require the local extirpation of all individuals or a marked migratory movement (Vanderwel & Purves, 2014).

All trees found in the boreal forest ecosystem must confront the challenge of surviving long, cold winters and then reacting quickly to the ticking clock during the brief warm summer when species must complete their life cycle. Therefore, a reasonable expectation is that boreal plants should thrive under a warmer climate. Warming has been shown to release some constraints, such as low soil fertility and the short growing season, imposed by harsh climates (D'Orangeville et al., 2014; Myneni et al., 1997; Peñuelas et al., 2009). However, these benefits may be outweighed by the increased metabolic cost to the plant under warmer, drier conditions; for example, an earlier spring increases summer drought stress (see Buermann et al., 2018) and frost damage (Marquis et al., 2020). Species may also struggle to maintain their competitive fitness under a warmer climate (Clark et al., 2014) and intensified disturbance regime (Gauthier et al., 2015).

Under controlled conditions, some physiological processes of boreal tree species can tolerate warming as long as the tree has access to sufficient resources to sustain the co-occurring increase in metabolic cost. For example, a mechanistic model calibrated with physiological data demonstrated that the growth of black spruce is optimal at temperatures found at its southern range limit; however, at these southern latitudes, the tree requires much greater amounts of resources (Bonan & Sirois, 1992). Nonetheless, other life cycle processes of boreal tree species, such as bud break or seed production, may be poorly adapted to warming. As competition for water, nutrients, light, and space resources is already the main driver of closed-crown forest mortality and growth (Franklin et al., 1987; Oliver & Larson, 1996), climate-related shifts in resource availability, e.g., water, have the potential to radically transform a species' competitive fitness, with the specific consequences dependent on stand composition, structure, and density (Clark et al., 2016). Notably, warming-induced drought stress can halt photosynthesis and deplete carbohydrate reserves in trees, thus reducing carbon allocation to growth or defensive compounds (Anderegg et al., 2015; Waring, 1987). In turn, these weakened defense mechanisms, coupled with warming-induced shifts in the range limits of certain pests and pathogens, can heighten a tree species' susceptibility to secondary stressors and damage from insect outbreaks (Anderegg et al., 2015; Kurz et al., 2008; Navarro et al., 2018). Drought-induced tree mortality often takes years or decades to occur, and this slow drought-imposed trajectory to tree death has been referred to as the *death spiral* (Franklin et al., 1987; Manion, 1991).

30.2.2 Brief Overview of Methods (and Limits)

Three data streams are commonly used to monitor how boreal forest ecosystems adjust to ongoing climate change: tree-ring records, remote sensing information, and permanent sample-plot data (Marchand et al., 2018). By matching annual tree-ring width to its year of formation, we can turn back time and reconstruct growth trends by relating ring width to climate (Girardin et al., 2016). Nonetheless, the usual lack of concurrent information on past changes in stand structure—the *fading record* problem—can introduce considerable bias when linking long-term, i.e., several decades, growth changes to climate when such growth changes can also be due to variations in stand density and stand development (Swetnam et al., 1999). Although remote sensing can capture information over quite large areas, several trade-offs for this broad coverage remain, including a coarse spatial and temporal resolution and the challenge of relating productivity indices, such as the Normalized Difference Vegetation Index, to specific ecosystem processes (National Academies of Sciences, Engineering, and Medicine, 2019). Permanent sample plots (PSP) provide a coarser time resolution than tree-ring records; nonetheless, they offer an exhaustive record of all changes in growth and mortality within a given plot over time. Another approach is to use snapshot data, i.e., a single measurement in time, to measure impacts among a diversity of forest stands following climatic anomalies (Michaelian et al., 2011).

Impacts from extreme climatic events are easier to detect than long-term, gradual climate change–driven shifts in stand demographics. Stand demographics vary naturally with stand development following stand-replacing disturbances such as fire or forest management (Lutz & Halpern, 2006). These processes may further be affected by secondary stressors, e.g., pests/pathogens, which are often difficult to detect. Researchers are faced with the challenging task of assessing the interactions between climate and stand processes or controlling for all these driving factors, sometimes leading to the exclusion of more than 95% of initial study plots; these issues raise the question of the representativeness of the obtained conclusions (Ma et al., 2012; Peng et al., 2011).

30.2.3   Case Study 2.1 Mortality in the Boreal Forest of North America

Here, we review the recent mortality trends and pulses for boreal forest trees of North America and the role of climate change in explaining these patterns. The 2001–2002 drought that affected boreal aspen stands in western Canada offers a striking glimpse of the possible impact of future climate anomalies on northern forest health. Precipitation was halved that year, and extensive mortality—up to 80% in some stands—quickly followed in this water-limited boreal ecosystem (Hogg et al., 2008; Michaelian et al., 2011). Climate warming has also been related to the unprecedented severity of recent insect outbreaks. The most recent mountain pine beetle (*Dendroctonus ponderosae*) outbreak in western Canada was ten times larger than all previous recorded outbreaks (Kurz et al., 2008), related to a combination of warming, which enabled the survival of the insect outside its typical range, and recent drought, which weakened the host trees (Taylor et al., 2006). By 2013, 53% of all merchantable pine in British Columbia had been killed by the insect (Walton, 2013). This deadly combination of warmer temperatures and drought similarly favored the expansion of the spruce beetle (*D. rufipennis*) into colder areas of Alaska's boreal forest (Berg et al., 2006). The punctual nature of insect outbreaks makes it difficult to identify warming-associated trends over time; however, the robust correlations between temperature and outbreak events provide a compelling case for a climate influence on both spruce (Berg et al., 2006) and mountain pine beetle (Logan & Powell, 2001). Similar conclusions can be reached regarding forest fires. Temperature is one of the best predictors of long-term trends of area burned (Flannigan et al., 2005); thus, unsurprisingly, annual burned forest areas have increased markedly to 2.5 million ha·yr^{-1} since the 1970s, closely tracking regional human-induced warming (Gillett et al., 2004). There is, however, marked regional variability in some regions of Canada, particularly in eastern Canada, which shows a decrease in annual area burned (Hanes et al., 2019).

Whereas linkages between climate change and disturbance-induced mortality are unanimously supported by the peer-reviewed literature, the exact role of climate change on mortality within undisturbed boreal stands remains more tenuous. The monitoring of permanent sample plots has revealed a threefold increase in mortality rates across western boreal North America since the 1950s (e.g., Hember et al., 2017; Luo & Chen, 2013, 2015; Peng et al., 2011; Thorpe & Daniels, 2012; Zhang et al., 2015). In comparison, eastern Canada shows no strong evidence of increasing mortality rates, from a weak 0.2% increase in annual mortality (Peng et al., 2011) to no change at all (Ma et al., 2012). Is climate change linked to this recent increase in mortality? In the cool and wet foothills of west-central Alberta, Thorpe and Daniels, (2012) could not detect any relationship between climate and increasing mortality rates. Rather, stand development processes, mainly tree size and basal area, appear to drive these mortality increases. In drier boreal forests, the water deficit displays only a weak covariation with long-term mortality trends (Luo & Chen, 2015; Peng et al., 2011; Zhang et al., 2015). Given the low temperatures observed in boreal forests, if a water deficit is not predominant, warming effects could be affecting multiple stand processes, such as growth or competition, thereby affecting our ability to establish clear causal relationships. For instance, analysis of large-scale tree-ring collections and exceptionally old trees has established that higher growth rates reduce tree longevity (Black et al., 2008; Di Filippo et al., 2015). Such concomitant increases in growth and mortality rates in western Canada have been observed by some researchers (Chen & Luo, 2015; Luo & Chen, 2015; Searle & Chen, 2018) but not all (Ma et al., 2012; Zhang et al., 2015). If true, this would support the hypothesis of a temperature-driven acceleration of stand developmental processes, potentially related to improved water-use efficiency with increasing CO_2 (Giguère-Croteau et al., 2019), a longer growing season (D'Orangeville et al , 2016, 2018), or increased microbial activity in soils releasing more nutrients (D'Orangeville et al., 2014). Similarly, joint increases in competition and mortality over time have also been reported, despite limited evidence for their interaction with climate (Luo & Chen, 2013; Zhang et al., 2015). Given the control exerted by competition over tree growth response to climate (Clark et al., 2011; Ford et al., 2017), our poor understanding of the mechanisms behind this increase in competition over time is astonishing (Price et al., 2015). Uncovering these mechanisms will require the acquisition of species-specific demographic response curves to determine the interactive effects of warming and drying.

30.2.4   Case Study 2.2 Recent Growth Trends in Asian Boreal Forests

Whereas Scots pine (*Pinus sylvestris*) is associated with the southern boreal forests of Asia, Dahurian larch (*Larix gmelinii*) dominates the northern forests, where its growth is restricted by the presence of permafrost. The high productivity reported for larch trees growing on upland sites is contrasted by the extremely low productivity of this species on permafrost plains and wetlands (Gauthier et al., 2015).

Regional climate warming, via the thawing of permafrost to greater depths, is triggering complex shifts in tree growth. Larch has been experiencing large increases in growth on the plains (Fig. 30.2; Zhang et al., 2019a) and decreased growth in wetlands (Juřička et al., 2020). The positive growth response observed on permafrost plains could be transitory, however, as climate warming is likely to convert some areas of the permafrost plains into waterlogged wetlands. These results stress the important role of microtopography and permafrost type for predicting future larch tree growth in the region.

In the southern boreal forests of Asia, Scots pine also displays similarly contrasted growth responses to warming. Although previous studies observed a negative effect of temperature on tree growth across most Scots-pine populations (e.g., Reich & Oleksyn, 2008), increased growth has been reported for the northern part of boreal Asian Scots-pine forests (Zhang et al., 2019b). In this region, the recent rapid warming is advancing the growth onset sufficiently to overlap temporally with the snowmelt period; this overlap allows Scots pine to benefit from the warmer climate and have new access to an additional water resource. Hence, these Scots-pine forests may be in a unique position of withstanding or even benefiting, at least temporarily, from the current rise in temperatures.

Birch (*Betula pendula*), Siberian fir (*Abies sibirica*), and Norway spruce (*Picea abies*) are important pioneer and accompanying species in southern Asian boreal forests. Warming-induced extensions of the growing season have heightened the growth of birch forests of western Siberia. In contrast, the same species has experienced a decline in the drier regions of the Trans-Baikal forest–steppe ecotone because

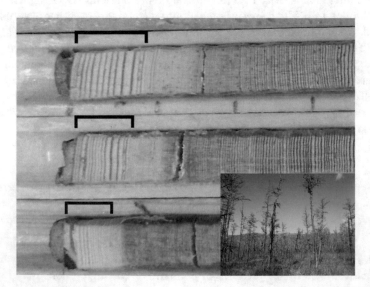

Fig. 30.2 Annual growth rings from three Asian larch trees (51°36′N, 121°25′E), continuous permafrost region of northeastern China (*inset picture*) showing a marked growth increase in recent years (*brackets*) associated with warming. *Photo credits* Xianliang Zhang

of warming-induced water stress (Kharuk et al., 2014). Using a network of 34 tree-ring chronologies for Siberian larch (*Larix sibirica*) and Siberian pine (*Pinus sibirica*) in the Altai mountains of central Asia, Kang et al. (2021) suggested that tree radial growth in the region may decline with future projected climate change. Similarly, drought stress appears to drive a negative growth response for fir (Kharuk et al., 2017) and Norway spruce (Kharuk et al., 2015) in other arid parts of the Asian boreal forest; this pattern contrasts with the trend of increased growth found in the colder northern regions (Schaphoff et al., 2016). The current decline in arid Asian boreal forest species is likely to turn the Eurasian carbon sink into a source by 2100 (Kicklighter et al., 2014).

30.2.5   Future Challenges

The effective management of boreal forest ecosystems to minimize tree growth decline and mortality losses under rapidly warming global temperatures and increased biotic disturbances—including pests and pathogens—is a key challenge for the future. Our ability to develop effective management strategies is limited by a lack of formal understanding of the potentially strong interactive effects of climate change, insects, and disease (Allen et al., 2015; Anderegg et al., 2015). If we are to maintain the health, vigor, and ecosystem services provided by the boreal forest, we must improve our understanding of forest demographic responses to novel climate and disturbance regimes. Improving this understanding will require a combination of large-scale studies of sufficient spatial and temporal scale to allow for a meaningful inference of the key drivers of forest demographics and improved forest models able to approximate the hypothesized complex interactions (Anderegg et al., 2015). Traditional forestry models, including growth and yield models that have been parameterized using historical experimental data, may not be capable of representing such complex interactions or may not be applicable under novel *non-analog* future conditions. These model limitations lead to significant uncertainty and variability in forecasts of future boreal forest dynamics (Purves & Pacala, 2008). One potentially fruitful avenue for improving our understanding of boreal forest response to future climate and disturbance is to apply recent methodological advances in ecological forecasting (Dietze et al., 2018). These approaches allow large-scale historical and experimental data to be synthesized or fused with existing forest models to improve the model-based representation of complex forest responses to changing climate and disturbance regimes. Thus, predictions of future forest conditions are more informed and accurate. Development of these types of model frameworks and their application to new, broad-scale boreal forest data sets and experimental data on species-specific demographic parameters (reproduction, growth, seed production) will be key to sustainably managing boreal forest ecosystems well into the future.

30.3   Conclusions

Climate change is not the only driver of human-related change in boreal ecosystems. Nearly two-thirds of boreal forests are under some form of management, e.g., timber harvesting, plantations, fire suppression, or insect control (Gauthier et al., 2015). Despite the negative impacts of historical management regimes on critical aspects of boreal ecosystems, e.g., species diversity and structure conservation, management is perhaps our best ally to help forests adapt to ongoing changes. Silvicultural interventions such as assisted migration could help implement better warm-adapted genotypes of indigenous tree species to maintain continuous closed-crown forest habitats under climate change. With nearly 600 million trees planted each year in Canada following harvesting, the infrastructure to grow and transport these trees is already in place. Yet, our current knowledge gap in terms of species' abilities to cope with ongoing changes and the feasibility of assisted migration are urgent matters that remain to be addressed with adequate resources. From a wildlife management perspective, various timber-harvesting strategies that reflect the variability of forest attributes resulting from natural disturbances (Gauthier et al., 2009) should be used to generate a range of landscapes and stand structures that reflect the variability of forest attributes resulting from natural disturbances and that are likely to be ecologically sustainable for wildlife (Drapeau et al., 2016), especially for threatened and endangered species (e.g., Nadeau Fortin et al., 2016). Such an increase in forest structural complexity could also provide additional resilience against global change (Messier et al., 2013).

References

Alda, F., González, M. A., Olea, P. P., et al. (2013). Genetic diversity, structure and conservation of the endangered Cantabrian Capercaillie in a unique peripheral habitat. *European Journal of Wildlife Research, 59*, 719–728. https://doi.org/10.1007/s10344-013-0727-6.

Allen, C. D., Breshears, D. D., & McDowell, N. G. (2015). On underestimation of global vulnerability to tree mortality and forest die-off from hotter drought in the Anthropocene. *Ecosphere, 6*(8), 129. https://doi.org/10.1890/ES15-00203.1.

Anderegg, W. R. L., Hicke, J. A., Fisher, R. A., et al. (2015). Tree mortality from drought, insects, and their interactions in a changing climate. *New Phytologist, 208*, 674–683. https://doi.org/10.1111/nph.13477.

Angert, A. L., Crozier, L. G., Rissler, L. J., et al. (2011). Do species' traits predict recent shifts at expanding range edges? *Ecology Letters, 14*, 677–689. https://doi.org/10.1111/j.1461-0248.2011.01620.x.

Araújo, M. B., Thuiller, W., & Pearson, R. G. (2006). Climate warming and the decline of amphibians and reptiles in Europe. *Journal of Biogeography, 33*, 1712–1728. https://doi.org/10.1111/j.1365-2699.2006.01482.x.

Aubin, I., Munson, A. D., Cardou, F., et al. (2016). Traits to stay, traits to move: A review of functional traits to assess sensitivity and adaptive capacity of temperate and boreal trees to climate change. *Environmental Reviews, 24*, 164–186. https://doi.org/10.1139/er-2015-0072.

Barber, Q. E., Parisien, M. A., Whitman, E., et al. (2018). Potential impacts of climate change on the habitat of boreal woodland caribou. *Ecosphere, 9*, e02472. https://doi.org/10.1002/ecs2.2472.

Barbet-Massin, M., Thuiller, W., & Jiguet, F. (2012). The fate of European breeding birds under climate, land-use and dispersal scenarios. *Global Change Biology, 18*, 881–890. https://doi.org/10.1111/j.1365-2486.2011.02552.x.

Bastille-Rousseau, G., Schaefer, J. A., Peers, M. J. L., et al. (2018). Climate change can alter predator-prey dynamics and population viability of prey. *Oecologia, 186*, 141–150. https://doi.org/10.1007/s00442-017-4017-y.

Berdanier, A. B., & Clark, J. S. (2016). Multiyear drought-induced morbidity preceding tree death in southeastern U.S. forests. *Ecological Applications, 26*, 17–23. https://doi.org/10.1890/15-0274.

Berg, E. E., David Henry, J., Fastie, C. L., et al. (2006). Spruce beetle outbreaks on the Kenai Peninsula, Alaska, and Kluane National Park and Reserve, Yukon territory: Relationship to summer temperatures and regional differences in disturbance regimes. *Forest Ecology and Management, 227*, 219–232. https://doi.org/10.1016/j.foreco.2006.02.038.

Bergerud, A., & Mercer, W. (1989). Caribou introductions in eastern North America. *Wildlife Society Bulletin, 17*, 111–120,

Berteaux, D., Ricard, M., St-Laurent, M. H., et al. (2018). Northern protected areas will become important refuges for biodiversity tracking suitable climates. *Scientific Reports, 8*, 4623. https://doi.org/10.1038/s41598-018-23050-w.

Black, B. A., Colbert, J. J., & Pederson, N. (2008). Relationships between radial growth rates and lifespan within North American tree species. *Ecoscience, 15*, 349–357. https://doi.org/10.2980/15-3-3149.

Boisvert-Marsh, L. (2020). *Patterns and processes affecting northward migration of tree species in a changing climate*. Ph.D. thesis, McGill University.

Boisvert-Marsh, L., & de Blois, S. (2021). Unravelling potential northward migration pathways for tree species under climate change. *Journal of Biogeography, 48*, 1088–1100. https://doi.org/10.1111/jbi.14060.

Boisvert-Marsh, L., Périé, C., & de Blois, S. (2014). Shifting with climate? evidence for recent changes in tree species distribution at high latitudes. *Ecosphere, 5*, art83. https://doi.org/10.1890/ES14-00111.1.

Boisvert-Marsh, L., Périé, C., & de Blois, S. (2019). Divergent responses to climate change and disturbance drive recruitment patterns underlying latitudinal shifts of tree species. *Journal of Ecology, 107*, 1956–1969. https://doi.org/10.1111/1365-2745.13149.

Bonan, G. B., & Sirois, L. (1992). Air temperature, tree growth, and the northern and southern range limits to Picea mariana. *Journal of Vegetation Science, 3*, 495–506. https://doi.org/10.2307/3235806.

Brice, M. H., Cazelles, K., Legendre, P., et al. (2019). Disturbances amplify tree community responses to climate change in the temperate–boreal ecotone. *Global Ecology and Biogeography, 28*, 1668–1681. https://doi.org/10.1111/geb.12971.

Buechling, A., Martin, P. H., & Canham, C. D. (2017). Climate and competition effects on tree growth in Rocky Mountain forests. *Journal of Ecology, 105*, 1636–1647. https://doi.org/10.1111/1365-2745.12782.

Buermann, W., Forkel, M., O'Sullivan, M., et al. (2018). Widespread seasonal compensation effects of spring warming on northern plant productivity. *Nature, 562*, 110–114. https://doi.org/10.1038/s41586-018-0555-7.

Channell, R., & Lomolino, M. V. (2000). Trajectories to extinction: Spatial dynamics of the contraction of geographical ranges. *Journal of Biogeography, 27*, 169–179. https://doi.org/10.1046/j.1365-2699.2000.00382.x.

Chen, H. Y. H., & Luo, Y. (2015). Net aboveground biomass declines of four major forest types with forest ageing and climate change in western Canada's boreal forests. *Global Change Biology, 21*, 3675–3684. https://doi.org/10.1111/gcb.12994.

Chen, I. C., Hill, J. K., Ohlemüller, R., et al. (2011). Rapid range shifts of species associated with high levels of climate warming. *Science, 333*, 1024–1026. https://doi.org/10.1126/science.1206432.

Clark, J. S., Bell, D. M., Hersh, M. H., et al. (2011). Climate change vulnerability of forest biodiversity: Climate and competition tracking of demographic rates. *Global Change Biology, 17,* 1834–1849. https://doi.org/10.1111/j.1365-2486.2010.02380.x.

Clark, J. S., Bell, D. M., Kwit, M. C., et al. (2014). Competition-interaction landscapes for the joint response of forests to climate change. *Global Change Biology, 20,* 1979–1991. https://doi.org/10.1111/gcb.12425.

Clark, J. S., Iverson, L., Woodall, C. W., et al. (2016). The impacts of increasing drought on forest dynamics, structure, and biodiversity in the United States. *Global Change Biology, 22,* 2329–2352. https://doi.org/10.1111/gcb.13160.

Collin, A., Messier, C., Kembel, S. W., et al. (2018). Can sugar maple establish into the boreal forest? Insights from seedlings under various canopies in southern Quebec. *Ecosphere, 9,* e02022. https://doi.org/10.1002/ecs2.2022.

COSEWIC. (2014). *COSEWIC assessment and status report on the Caribou* Rangifer tarandus, *Newfoundland population, Atlantic-Gaspésie population and Boreal population, in Canada.* Ottawa: Committee on the Status of Endangered Wildlife in Canada.

D'Orangeville, L., Houle, D., Côté, B., et al. (2014). Soil response to a 3-year increase in temperature and nitrogen deposition measured in a mature boreal forest using ion-exchange membranes. *Environmental Monitoring and Assessment, 186,* 8191–8202. https://doi.org/10.1007/s10661-014-3997-x.

D'Orangeville, L., Duchesne, L., Houle, D., et al. (2016). Northeastern North America as a potential refugium for boreal forests in a warming climate. *Science, 352,* 1452–1455. https://doi.org/10.1126/science.aaf4951.

D'Orangeville, L., Houle, D., Duchesne, L., et al. (2018). Beneficial effects of climate warming on boreal tree growth may be transitory. *Nature Communications, 9,* 3213. https://doi.org/10.1038/s41467-018-05705-4.

Deutsch, C. A., Tewksbury, J. J., Huey, R. B., et al. (2008). Impacts of climate warming on terrestrial ectotherms across latitude. *Proceedings of the National Academy of Sciences of the United States of America, 105,* 6668–6672. https://doi.org/10.1073/pnas.0709472105.

Di Filippo, A., Pederson, N., Baliva, M., et al. (2015). The longevity of broadleaf deciduous trees in Northern Hemisphere temperate forests: Insights from tree-ring series. *Frontiers in Ecology and Evolution, 3,* 46. https://doi.org/10.3389/fevo.2015.00046.

Dietze, M. C., Fox, A., Beck-Johnson, L. M., et al. (2018). Iterative near-term ecological forecasting: Needs, opportunities, and challenges. *Proceedings of the National Academy of Sciences of the United States of America, 115,* 1424–1432. https://doi.org/10.1073/pnas.1710231115.

Drapeau, P., Villard, M. A., Leduc, A., et al. (2016). Natural disturbance regimes as templates for the response of bird species assemblages to contemporary forest management. *Diversity and Distributions, 22,* 385–399. https://doi.org/10.1111/ddi.12407.

Duchesne, L., & Ouimet, R. (2008). Population dynamics of tree species in southern Quebec, Canada: 1970–2005. *Forest Ecology and Management, 255,* 3001–3012. https://doi.org/10.1016/j.foreco.2008.02.008.

Dullinger, S., Gattringer, A., Thuiller, W., et al. (2012). Extinction debt of high-mountain plants under twenty-first-century climate change. *Climate Change, 2,* 619–622. https://doi.org/10.1038/nclimate1514.

Elmhagen, B., Kindberg, J., Hellström, P., et al. (2015). A boreal invasion in response to climate change? Range shifts and community effects in the borderland between forest and tundra. *Ambio, 44,* S39–S50. https://doi.org/10.1007/s13280-014-0606-8.

Environment Canada. (2011). *Scientific assessment to inform the identification of critical habitat for woodland caribou* (Rangifer tarandus caribou)*, boreal population, in Canada: 2011 update* (p. 102). Environment Canada.

Evans, T. G., Diamond, S. E., & Kelly, M. W. (2015). Mechanistic species distribution modelling as a link between physiology and conservation. *Conservation Physiology, 3,* cov056 https://doi.org/10.1093/conphys/cov056.

Fahrig, L. (2003). Effects of habitat fragmentation on biodiversity. *Annual Review of Ecology, Evolution, and Systematics, 34*, 487–515. https://doi.org/10.1146/annurev.ecolsys.34.011802. 132419.

Festa-Bianchet, M., Ray, J. C., Boutin, S., et al. (2011). Conservation of caribou (Rangifer tarandus) in Canada: An uncertain future. *Canadian Journal of Zoology, 89*, 419–434. https://doi.org/10. 1139/z11-025.

Flannigan, M. D., Logan, K. A., Amiro, B. D., et al. (2005). Future area burned in Canada. *Climate Change, 72*, 1–16. https://doi.org/10.1007/s10584-005-5935-y.

Foden, W. B., Butchart, S. H. M., Stuart, S. N., et al. (2013). Identifying the world's most climate change vulnerable species: A systematic trait-based assessment of all birds, amphibians and corals. *PLoS ONE, 8*, e65427. https://doi.org/10.1371/journal.pone.0065427.

Ford, K. R., Breckheimer, I. K., Franklin, J. F., et al. (2017). Competition alters tree growth responses to climate at individual and stand scales. *Canadian Journal of Forest Research, 47*, 53–62. https:// doi.org/10.1139/cjfr-2016-0188.

Foster, J. R., Finley, A. O., D'Amato, A. W., et al. (2016). Predicting tree biomass growth in the temperate-boreal ecotone: Is tree size, age, competition, or climate response most important? *Global Change Biology, 22*, 2138–2151. https://doi.org/10.1111/gcb.13208.

Foxcroft, L. C., Pyšek, P., Richardson, D. M., et al. (2017). Plant invasion science in protected areas: Progress and priorities. *Biological Invasions, 19*, 1353–1378. https://doi.org/10.1007/s10 530-016-1367-z.

Franklin, J. F., Shugart, H. H., & Harmon, M. E. (1987). Tree death as an ecological process. *BioScience, 37*, 550–556. https://doi.org/10.2307/1310665.

Gallant, D., Lecomte, N., & Berteaux, D. (2020). Disentangling the relative influences of global drivers of change in biodiversity: A study of the twentieth-century red fox expansion into the Canadian Arctic. *Journal of Animal Ecology, 89*, 565–576. https://doi.org/10.1111/1365-2656. 13090.

Gauthier, S., Vaillancourt, M. A., Leduc, A., et al. (Eds.). (2009). *Ecosystem management in the boreal forest* (p. 572). Presses de l'Université du Québec.

Gauthier, S., Bernier, P., Kuuluvainen, T., et al. (2015). Boreal forest health and global change. *Science, 349*, 819–822. https://doi.org/10.1126/science.aaa9092.

Giguère-Croteau, C., Boucher, É., Bergeron, Y., et al. (2019). North America's oldest boreal trees are more efficient water users due to increased [CO_2], but do not grow faster. *Proceedings of the National Academy of Sciences of the United States of America, 116*, 2749–2754. https://doi.org/ 10.1073/pnas.1816686116.

Gillett, N. P., Weaver, A. J., Zwiers, F. W., et al. (2004). Detecting the effect of climate change on Canadian forest fires. *Geophysical Research Letters, 31*, L18211. https://doi.org/10.1029/200 4GL020876.

Girardin, M. P., Bouriaud, O., Hogg, E. H., et al. (2016). No growth stimulation of Canada's boreal forest under half-century of combined warming and CO_2 fertilization. *Proceedings of the National Academy of Sciences of the United States of America, 113*(52), E8406–E8414. https://doi.org/10. 1073/pnas.1610156113.

Grant, L., Johnson, C., & Thiessen, C. (2019). Evaluating the efficacy of translocation: Maintaining habitat key to long-term success for an imperiled population of an at-risk species. *Biodiversity and Conservation, 28*, 2727–2743. https://doi.org/10.1007/s10531-019-01789-6.

Groisman, P., & Soja, A. J. (2009). Ongoing climatic change in Northern Eurasia: Justification for expedient research. *Environmental Research Letters, 4*(4), 045002. https://doi.org/10.1088/1748-9326/4/4/045002.

Guisan, A., & Thuiller, W. (2005). Predicting species distribution: Offering more than simple habitat models. *Ecology Letters, 8*, 993–1009. https://doi.org/10.1111/j.1461-0248.2005.00792.x.

Hanes, C. C., Wang, X., Jain, P., et al. (2019). Fire-regime changes in Canada over the last half century. *Canadian Journal of Forest Research, 49*(3), 256–269. https://doi.org/10.1139/cjfr-2018-0293.

Harrison, P. A., Berry, P. M., Butt, N., et al. (2006). Modelling climate change impacts on species' distributions at the European scale: Implications for conservation policy. *Environmental Science & Policy, 9*, 116–128. https://doi.org/10.1016/j.envsci.2005.11.003.

Hember, R. A., Kurz, W. A., & Coops, N. C. (2017). Relationships between individual-tree mortality and water-balance variables indicate positive trends in water stress-induced tree mortality across North America. *Global Change Biology, 23*, 1691–1710. https://doi.org/10.1111/gcb.13428.

Hof, C., Levinsky, I., Araújo, M. B., et al. (2011). Rethinking species' ability to cope with rapid climate change. *Global Change Biology, 17*, 2987–2990. https://doi.org/10.1111/j.1365-2486.2011.02418.x.

Hogg, E. H., Brandt, J. P., & Michaelian, M. (2008). Impacts of a regional drought on the productivity, dieback, and biomass of western Canadian aspen forests. *Canadian Journal of Forest Research, 38*, 1373–1384. https://doi.org/10.1139/X08-001.

Howard, C., Stephens, P. A., Pearce-Higgins, J. W., et al. (2020). Disentangling the relative roles of climate and land cover change in driving the long-term population trends of European migratory birds. *Diversity and Distributions, 26*, 1442–1455. https://doi.org/10.1111/ddi.13144.

Huey, R. B., Kearney, M. R., Krockenberger, A., et al. (2012). Predicting organismal vulnerability to climate warming: Roles of behaviour, physiology and adaptation. *Philosophical Transactions of the Royal Society of London. Series B, Biological Sciences, 367*, 1665–1679. https://doi.org/10.1098/rstb.2012.0005.

Jackson, S. T., & Sax, D. F. (2010). Balancing biodiversity in a changing environment: Extinction debt, immigration credit and species turnover. *Trends in Ecology & Evolution, 25*, 153–160. https://doi.org/10.1016/j.tree.2009.10.001.

Juřička, D., Novotná, J., Houška, J., et al. (2020). Large-scale permafrost degradation as a primary factor in Larix sibirica forest dieback in the Khentii massif, northern Mongolia. *Journal of Forest Research, 31*, 197–208. https://doi.org/10.1007/s11676-018-0866-4.

Kang, J., Jiang, S., Tardif, J. C., et al. (2021). Radial growth responses of two dominant conifers to climate in the Altai Mountains Central Asia. *Agricultural and Forest Meteorology, 298–299*, 108297. https://doi.org/10.1016/j.agrformet.2020.108297.

Kellman, M. (2004). Sugar maple (*Acer saccharum* Marsh.) establishment in boreal forest: Results of a transplantation experiment. *Journal of Biogeography, 31*, 1515–1522. https://doi.org/10.1111/j.1365-2699.2004.01128.x.

Kharuk, V. I., Kuzmichev, V. V., Im, S. T., et al. (2014). Birch stands growth increase in Western Siberia. *Scandinavian Journal of Forest Research, 29*, 421–426. https://doi.org/10.1080/02827581.2014.912345.

Kharuk, V. I., Im, S. T., Dvinskaya, M. L., et al. (2015). Climate-induced mortality of spruce stands in Belarus. *Environmental Research Letters, 10*, 125006. https://doi.org/10.1088/1748-9326/10/12/125006.

Kharuk, V. I., Im, S. T., Petrov, I. A., et al. (2017). Fir decline and mortality in the southern Siberian Mountains. *Regional Environmental Change, 17*, 803–812. https://doi.org/10.1007/s10113-016-1073-5.

Kicklighter, D. W., Cai, Y., Zhuang, Q., et al. (2014). Potential influence of climate-induced vegetation shifts on future land use and associated land carbon fluxes in Northern Eurasia. *Environmental Research Letters, 9*, 035004. https://doi.org/10.1088/1748-9326/9/3/035004.

Kurz, W. A., Dymond, C. C., Stinson, G., et al. (2008). Mountain pine beetle and forest carbon feedback to climate change. *Nature, 452*, 987–990. https://doi.org/10.1038/nature06777.

Lafontaine, A., Drapeau, P., Fortin, D., et al. (2019). Exposure to historical burn rates shapes the response of boreal caribou to timber harvesting. *Ecosphere, 10*, e02739. https://doi.org/10.1002/ecs2.2739.

Laliberte, A. S., & Ripple, W. J. (2004). Range contractions of North American carnivores and ungulates. *BioScience, 54*, 123–138. https://doi.org/10.1641/0006-3568(2004)054[0123:RCONAC]2.0.CO;2.

Lankford, A. J., Svancara, L. K., Lawler, J. J., et al. (2014). Comparison of climate change vulnerability assessments for wildlife. *Wildlife Society Bulletin, 38*, 386–394. https://doi.org/10.1002/wsb.399.

Latham, A. D. M., Latham, M. C., McCutchen, N. A., et al. (2011). Invading white-tailed deer change wolf–caribou dynamics in northeastern Alberta. *Journal of Wildlife Management, 75*, 204–212. https://doi.org/10.1002/jwmg.28.

Lawler, J. J., Shafer, S. L., White, D., et al. (2009). Projected climate-induced faunal change in the Western Hemisphere. *Ecology, 90*, 588–597. https://doi.org/10.1890/08-0823.1.

Leclerc, M., Dussault, C., & St-Laurent, M. H. (2014). Behavioural strategies towards human disturbances explain individual performance in woodland caribou. *Oecologia, 176*, 297–306. https://doi.org/10.1007/s00442-014-3012-9.

Leclerc, M., Tarroux, A., Fauchald, P., et al. (2019). Effects of human-induced disturbances and weather on herbivore movement. *Journal of Mammalogy, 100*, 1490–1500. https://doi.org/10.1093/jmammal/gyz101.

Logan, J. A., & Powell, J. A. (2001). Ghost forests, global warming, and the mountain pine beetle (Coleoptera: Scolytidae). *American Entomologist, 47*, 160–173. https://doi.org/10.1093/ae/47.3.160.

Losier, C. L., Couturier, S., St-Laurent, M. H., et al. (2015). Adjustments in habitat selection to changing availability induce fitness costs for a threatened ungulate. *Journal of Applied Ecology, 52*, 496–504. https://doi.org/10.1111/1365-2664.12400.

Luo, Y., & Chen, H. Y. H. (2013). Observations from old forests underestimate climate change effects on tree mortality. *Nature Communications, 4*, 1655. https://doi.org/10.1038/ncomms2681.

Luo, Y., & Chen, H. Y. H. (2015). Climate change-associated tree mortality increases without decreasing water availability. *Ecology Letters, 18*, 1207–1215. https://doi.org/10.1111/ele.12500.

Lurgi, M., Brook, B. W., Saltré, F., et al. (2015). Modelling range dynamics under global change: Which framework and why? *Methods in Ecology and Evolution, 6*, 247–256. https://doi.org/10.1111/2041-210X.12315.

Lutz, J. A., & Halpern, C. B. (2006). Tree mortality during early forest development: A long-term study of rates, causes, and consequences. *Ecological Monographs, 76*, 257–275. https://doi.org/10.1890/0012-9615(2006)076[0257:TMDEFD]2.0.CO;2.

Ma, Z., Peng, C., Zhu, Q., et al. (2012). Regional drought-induced reduction in the biomass carbon sink of Canada's boreal forests. *Proceedings of the National Academy of Sciences of the United States of America, 109*, 2423–2427. https://doi.org/10.1073/pnas.1111576109.

Manion, P. D. (1991). *Tree disease concepts.* Prentice-Hall.

Marchand, W., Girardin, M. P., Gauthier, S., et al. (2018). Untangling methodological and scale considerations in growth and productivity trend estimates of Canada's forests. *Environmental Research Letters, 13*(9), 093001. https://doi.org/10.1088/1748-9326/aad82a.

Marquis, B., Bergeron, Y., Simard, M., et al. (2020). Growing-season frost is a better predictor of tree growth than mean annual temperature in boreal mixedwood forest plantations. *Global Change Biology, 26*(11), 6537–6554. https://doi.org/10.1111/gcb.15327.

McLane, A. J., Semeniuk, C., McDermid, G. J., et al. (2011). The role of agent-based models in wildlife ecology and management. *Ecological Modelling, 222*, 1544–1556. https://doi.org/10.1016/j.ecolmodel.2011.01.020.

Messier, C., Puettmann, K., & Coates, K. D. (2013). *Managing forests as complex adaptive systems: Building resilience to the challenge of global change* (p. 368). Routledge.

Michaelian, M., Hogg, E. H., Hall, R. J., et al. (2011). Massive mortality of aspen following severe drought along the southern edge of the Canadian boreal forest. *Global Change Biology, 17*, 2084–2094. https://doi.org/10.1111/j.1365-2486.2010.02357.x.

Morin, X., & Thuiller, W. (2009). Comparing niche- and process-based models to reduce prediction uncertainty in species range shifts under climate change. *Ecology, 90*, 1301–1313. https://doi.org/10.1890/08-0134.1.

Moyle, P. B., Kiernan, J. D., Crain, P. K., et al. (2013). Climate change vulnerability of native and alien freshwater fishes of California: A systematic assessment approach. *PLoS ONE, 8*, e63883. https://doi.org/10.1371/journal.pone.0063883.

Murray, D. L., Majchrzak, Y. N., Peers, M. J. L., et al. (2015). Potential pitfalls of private initiatives in conservation planning: A case study from Canada's boreal forest. *Biological Conservation, 192*, 174–180. https://doi.org/10.1016/j.biocon.2015.09.017.

Myneni, R. B., Keeling, C. D., Tucker, C. J., et al. (1997). Increased plant growth in the northern high latitudes from 1981 to 1991. *Nature, 386*, 698–702. https://doi.org/10.1038/386698a0.

Nadeau Fortin, M. A., Sirois, L., & St-Laurent, M. H. (2016). Extensive forest management contributes to maintain suitable habitat characteristics for the endangered Atlantic-Gaspésie caribou. *Canadian Journal of Forest Research, 46*, 933–942. https://doi.org/10.1139/cjfr-2016-0038.

National Academies of Sciences, Engineering, and Medicine. (2019). *Understanding northern latitude vegetation greening and browning: Proceedings of a workshop.* Washington: The National Academies Press.

Navarro, L., Morin, H., Bergeron, Y., et al. (2018). Changes in spatiotemporal patterns of 20th century spruce budworm outbreaks in eastern Canadian boreal forests. *Frontiers in Plant Science, 9*, 1905. https://doi.org/10.3389/fpls.2018.01905.

Oliver, C. D., & Larson, B. C. (1996). *Forest stand dynamics* (Update). Yale School of the Environment Other Publications.

Opdam, P., & Wascher, D. (2004). Climate change meets habitat fragmentation: Linking landscape and biogeographical scale levels in research and conservation. *Biological Conservation, 117*, 285–297. https://doi.org/10.1016/j.biocon.2003.12.008.

Pacifici, M., Foden, W. B., Visconti, P., et al. (2015). Assessing species vulnerability to climate change. *Nature Climate Change, 5*, 215–224. https://doi.org/10.1038/nclimate2448.

Parmesan, C., & Yohe, G. (2003). A globally coherent fingerprint of climate change impacts across natural systems. *Nature, 421*, 37–42. https://doi.org/10.1038/nature01286.

Peng, C., Ma, Z., Lei, X., et al. (2011). A drought-induced pervasive increase in tree mortality across Canada's boreal forests. *Nature Climate Change, 1*, 467–471. https://doi.org/10.1038/nclimate1293.

Peñuelas, J., Rutishauser, T., & Filella, I. (2009). Phenology feedbacks on climate change. *Science, 324*, 887–888. https://doi.org/10.1126/science.1173004.

Périé, C., & de Blois, S. (2016). Dominant forest tree species are potentially vulnerable to climate change over large portions of their range even at high latitudes. *PeerJ, 4*, e2218. https://doi.org/10.7717/peerj.2218.

Peterson, M. L., Doak, D. F., & Morris, W. F. (2019). Incorporating local adaptation into forecasts of species' distribution and abundance under climate change. *Global Change Biology, 25*, 775–793. https://doi.org/10.1111/gcb.14562.

Pickles, R. S. A., Thornton, D., Feldman, R., et al. (2013). Predicting shifts in parasite distribution with climate change: A multitrophic level approach. *Global Change Biology, 19*, 2645–2654. https://doi.org/10.1111/gcb.12255.

Price, D. T., Cooke, B. J., Metsaranta, J. M., et al. (2015). If forest dynamics in Canada's west are driven mainly by competition, why did they change? Half-century evidence says: Climate change. *Proceedings of the National Academy of Sciences of the United States of America, 112*, E4340–E4340. https://doi.org/10.1073/pnas.1508245112.

Purves, D., & Pacala, S. (2008). Predictive models of forest dynamics. *Science, 320*, 1452–1453. https://doi.org/10.1126/science.1155359.

Purvis, A., Jones, K. E., & Mace, G. M. (2000). Extinction. *Bioessays, 22*, 1123–1133. https://doi.org/10.1002/1521-1878(200012)22:12%3c1123::AID-BIES10%3e3.0.CO;2-C.

Reich, P. B., & Oleksyn, J. (2008). Climate warming will reduce growth and survival of Scots pine except in the far north. *Ecology Letters, 11*, 588–597. https://doi.org/10.1111/j.1461-0248.2008.01172.x.

Rowland, E. L., Davison, J. E., & Graumlich, L. J. (2011). Approaches to evaluating climate change impacts on species: A guide to initiating the adaptation planning process. *Environmental Management, 47*, 322–337. https://doi.org/10.1007/s00267-010-9608-x.

Rudolph, T. D., Drapeau, P., Imbeau, L., et al. (2017). Demographic responses of boreal caribou to cumulative disturbances highlight elasticity of range-specific tolerance thresholds. *Biodiversity and Conservation, 26*, 1179–1198. https://doi.org/10.1007/s10531-017-1292-1.

Schaefer, J. A. (2003). Long-term range recession and the persistence of caribou in the taiga. *Conservation Biology, 17*, 1435–1439. https://doi.org/10.1046/j.1523-1739.2003.02288.x.

Schaphoff, S., Reyer, C. P. O., Schepaschenko, D., et al. (2016). Tamm review: Observed and projected climate change impacts on Russia's forests and its carbon balance. *Forest Ecology and Management, 361*, 432–444. https://doi.org/10.1016/j.foreco.2015.11.043.

Schloss, C. A., Nuñez, T. A., & Lawler, J. J. (2012). Dispersal will limit ability of mammals to track climate change in the Western Hemisphere. *Proceedings of the National Academy of Sciences of the United States of America, 109*, 8606–8611. https://doi.org/10.1073/pnas.1116791109.

Schneider, R. R., Hamann, A., Farr, D., et al. (2009). Potential effects of climate change on ecosystem distribution in Alberta. *Canadian Journal of Forest Research, 39*, 1001–1010. https://doi.org/10.1139/X09-033.

Searle, E. B., & Chen, H. Y. H. (2018). Temporal declines in tree longevity associated with faster lifetime growth rates in boreal forests. *Environmental Research Letters, 13*, 125003. https://doi.org/10.1088/1748-9326/aaea9e.

Seip, D. R. (1992). Factors limiting woodland caribou populations and their interrelationships with wolves and moose in southeastern British Columbia. *Canadian Journal of Zoology, 70*, 1494–1503. https://doi.org/10.1139/z92-206.

Seip, D. R., & Cichowski, D. B. (1996). Population ecology of caribou in British Columbia. *Rangifer, 16*, 73–80. https://doi.org/10.7557/2.16.4.1223.

Shoo, L. P., Williams, S. E., & Hero, J. M. (2006). Detecting climate change induced range shifts: Where and how should we be looking? *Austral Ecology, 31*, 22–29. https://doi.org/10.1111/j.1442-9993.2006.01539.x.

Sinclair, S. J., White, M. D., & Newell, G. R. (2010). How useful are species distribution models for managing biodiversity under future climates? *Ecology & Society, 15*(1), 8. https://doi.org/10.5751/ES-03089-150108.

Swetnam, T. W., Allen, C. D., & Betancourt, J. L. (1999). Applied historical ecology: Using the past to manage for the future. *Ecological Applications, 9*, 1189–1206. https://doi.org/10.1890/1051-0761(1999)009[1189:AHEUTP]2.0.CO;2.

Taylor, S. W., Carroll, A. L., Alfaro, R. I., et al. (2006). *The mountain pine beetle: A synthesis of biology, management and impacts in lodgepole pine.* Victoria: Canadian Forest Service, Natural Resources Canada.

Thomas, C. D., Hill, J. K., Anderson, B. J., et al. (2011). A framework for assessing threats and benefits to species responding to climate change. *Methods in Ecology and Evolution, 2*, 125–142. https://doi.org/10.1111/j.2041-210X.2010.00065.x.

Thorpe, H. C., & Daniels, L. D. (2012). Long-term trends in tree mortality rates in the Alberta foothills are driven by stand development. *Canadian Journal of Forest Research, 42*(9), 1687–1696. https://doi.org/10.1139/x2012-104.

Tingley, M. W., & Beissinger, S. R. (2009). Detecting range shifts from historical species occurrences: New perspectives on old data. *Trends in Ecology & Evolution, 24*, 625–633. https://doi.org/10.1016/j.tree.2009.05.009.

Valladares, F., Matesanz, S., Guilhaumon, F., et al. (2014). The effects of phenotypic plasticity and local adaptation on forecasts of species range shifts under climate change. *Ecology Letters, 17*, 1351–1364. https://doi.org/10.1111/ele.12348.

Vanderwel, M. C., & Purves, D. W. (2014). How do disturbances and environmental heterogeneity affect the pace of forest distribution shifts under climate change? *Ecography, 37*, 10–20. https://doi.org/10.1111/j.1600-0587.2013.00345.x.

Vors, L. S., & Boyce, M. S. (2009). Global declines of caribou and reindeer. *Global Change Biology, 15*, 2626–2633. https://doi.org/10.1111/j.1365-2486.2009.01974.x.

Walther, G. R., Roques, A., Hulme, P. E., et al. (2009). Alien species in a warmer world: Risks and opportunities. *Trends in Ecology & Evolution, 24*, 686–693. https://doi.org/10.1016/j.tree.2009.06.008.

Walton, A. (2013). *Provincial-level projection of the current mountain pine beetle outbreak.* Victoria: British Columbia Forest Service.

Waring, R. H. (1987). Characteristics of trees predisposed to die. *BioScience, 37*, 569–574. https://doi.org/10.2307/1310667.

Wärnbäck, A., & Hilding-Rydevik, T. (2009). Cumulative effects in Swedish EIA practice–Difficulties and obstacles. *Environmental Impact Assessment Review, 29*, 107–115. https://doi.org/10.1016/j.eiar.2008.05.001.

Watson, D. J. (1963). Climate, weather, and plant yield. In L. T. Evans (Ed.), *Environmental control of plant growth* (pp. 337–350). Academic Press.

Whitman, E., Parisien, M. A., Price, D. T., et al. (2017). A framework for modeling habitat quality in disturbance-prone areas demonstrated with woodland caribou and wildfire. *Ecosphere, 8*, e01787. https://doi.org/10.1002/ecs2.1787.

Whittington, J., Hebblewhite, M., DeCesare, N. J., et al. (2011). Caribou encounters with wolves increase near roads and trails: A time-to-event approach. *Journal of Applied Ecology, 48*, 1535–1542. https://doi.org/10.1111/j.1365-2664.2011.02043.x.

Wiens, J. J. (2016). Climate-related local extinctions are already widespread among plant and animal species. *PLoS Biology, 14*, e2001104. https://doi.org/10.1371/journal.pbio.2001104.

Wilson, R. J., Gutiérrez, D., Gutiérrez, J., et al. (2005). Changes to the elevational limits and extent of species ranges associated with climate change. *Ecology Letters, 8*, 1138–1146. https://doi.org/10.1111/j.1461-0248.2005.00824.x.

Woodall, C. W., Oswalt, C. M., Westfall, J. A., et al. (2009). Tree migration detection through comparisons of historic and current forest inventories. In W. McWilliams, G. Moisen, R. Czaplewski (Eds.), *Forest Inventory and Analysis (FIA) Symposium 2008* (p. 9). Fort Collins: U.S. Department of Agriculture, Forest Service, Rocky Mountain Research Station.

Yannic, G., Pellissier, L., Ortego, J., et al. (2014). Genetic diversity in caribou linked to past and future climate change. *Nature Climate Change, 4*, 132–137. https://doi.org/10.1038/nclimate2074.

Zhang, J., Huang, S., & He, F. (2015). Half-century evidence from western Canada shows forest dynamics are primarily driven by competition followed by climate. *Proceedings of the National Academy of Sciences of the United States of America, 112*, 4009–4014. https://doi.org/10.1073/pnas.1420844112.

Zhang, X., Bai, X., Hou, M., et al. (2019a). Warmer winter ground temperatures trigger rapid growth of dahurian larch in the permafrost forests of northeast China. *Journal of Geophysical Research. Biogeosciences, 124*, 1088–1097. https://doi.org/10.1029/2018JG004882.

Zhang, X., Manzanedo, R. D., D'Orangeville, L., et al. (2019b). Snowmelt and early to mid-growing season water availability augment tree growth during rapid warming in southern Asian boreal forests. *Global Change Biology, 25*, 3462–3471. https://doi.org/10.1111/gcb.14749.

Chapter 31
Challenges for the Sustainable Management of the Boreal Forest Under Climate Change

Miguel Montoro Girona, Tuomas Aakala, Núria Aquilué,
Annie-Claude Bélisle, Emeline Chaste, Victor Danneyrolles,
Olalla Díaz-Yáñez, Loïc D'Orangeville, Guillaume Grosbois, Alison Hester,
Sanghyun Kim, Niko Kulha, Maxence Martin, Louiza Moussaoui,
Christoforos Pappas, Jeanne Portier, Sara Teitelbaum,
Jean-Pierre Tremblay, Johan Svensson, Martijn Versluijs, Märtha Wallgren,
Jicjie Wang, and Sylvie Gauthier

Abstract The increasing effects of climate and global change oblige ecosystem-based management to adapt forestry practices to deal with uncertainties. Here we provide an overview to identify the challenges facing the boreal forest under projected future change, including altered natural disturbance regimes, biodiversity loss, increased forest fragmentation, the rapid loss of old-growth forests, and the need to develop novel silvicultural approaches. We specifically address subjects previously lacking from the ecosystem-based management framework, e.g., Indigenous communities, social concerns, ecological restoration, and impacts on aquatic ecosystems. We conclude by providing recommendations for ensuring the successful long-term management of the boreal biome facing climate change.

M. M. Girona (✉) · V. Danneyrolles · G. Grosbois · S. Kim · L. Moussaoui
Groupe de Recherche en Écologie de la MRC-Abitibi, Forest Research Institute, Université du Québec en Abitibi-Témiscamingue, Amos Campus, 341, rue Principale Nord, Amos, QC J9T 2L8, Canada
e-mail: miguel.montoro@uqat.ca

V. Danneyrolles
e-mail: victor.danneyrolles@uqat.ca

G. Grosbois
e-mail: guillaume.grosbois@uqat.ca

S. Kim
e-mail: sanghyun.kim@uqat.ca

L. Moussaoui
e-mail: louiza.moussaoui@uqat.ca

© The Author(s) 2023   773
M. M. Girona et al. (eds.), *Boreal Forests in the Face of Climate Change*,
Advances in Global Change Research 74,
https://doi.org/10.1007/978-3-031-15988-6_31

M. M. Girona · J. Svensson · M. Wallgren
Department of Wildlife, Fish, and Environmental Studies, Swedish University of Agricultural
Sciences, 901 83 Umeå, Sweden
e-mail: johan.svensson@slu.se

M. Wallgren
e-mail: martha.wallgren@slu.se

M. M. Girona · N. Aquilué · A.-C. Bélisle · V. Danneyrolles · S. Kim · M. Martin · C. Pappas
Centre for Forest Research, Université du Québec à Montréal, P.O. Box 8888, Stn. Centre-Ville,
Montréal, QC H3C 3P8, Canada
e-mail: nuria.aquilue@ctfc.cat

A.-C. Bélisle
e-mail: annieclaude.belisle@uqat.ca

M. Martin
e-mail: maxence.martin@uqat.ca

C. Pappas
e-mail: cpappas@upatras.gr

T. Aakala
School of Forest Sciences, University of Eastern Finland, P.O. Box 111, FI-80101 Joensuu,
Finland
e-mail: tuomas.aakala@uef.fi

A.-C. Bélisle · M. Martin
Forest Research Institute, Université du Québec en Abitibi-Témiscamingue, 445 boul. de
l'Université, Rouyn-Noranda, QC J9X 5E4, Canada

E. Chaste
Université de Lorraine, AgroParisTech, INRAE, SILVA, 14 Rue Girardet, 54000 Nancy, France
e-mail: emelinechaste6@hotmail.com

O. Díaz-Yáñez
Department of Environmental Systems Science, ETH Zürich, Forest Ecology, Universitätstrasse
16, 8092 Zürich, Switzerland
e-mail: olalla.diaz@usys.ethz.ch

L. D'Orangeville · J. Wang
Faculty of Forestry and Environmental Management, University of New Brunswick, 28 Dineen
Drive, P.O. Box 4400, Fredericton, NB E3B 5A3, Canada
e-mail: loic.dorangeville@unb.ca

J. Wang
e-mail: jiejie.wang@unb.ca

G. Grosbois
Department of Aquatic Sciences and Assessment, Swedish University of Agricultural Sciences,
P.O. Box 7050, 75007 Uppsala, Sweden

31.1 How Did We Get Here? A Perspective on Boreal Forest Management

Ecosystem degradation has intensified because of increased human pressure on natural systems worldwide (Foley et al., 2005; Rands et al., 2010). During the twentieth century, the world's population increased from 1.6 to 7.7 billion people (Lutz et al., 2004), resulting in a greater demand for natural resources to meet the needs

A. Hester
James-Hutton Institute, Craigiebuckler, Aberdeen AB15 8QH, Scotland, UK
e-mail: alison.hester@hutton.ac.uk

N. Kulha
Natural Resources Institute Finland (Luke), Latokartanonkaari 9, FI-00790 Helsinki, Finland
e-mail: niko.kulha@luke.fi

Zoological Museum, Biodiversity Unit, University of Turku, Natura, FI-20014 Turku, Finland

M. Martin
Département des Sciences fondamentales, Université du Québec à Chicoutimi, 555 boul. de l'Université, Chicoutimi, QC G7H 2B1, Canada

C. Pappas
Department of Civil Engineering, University of Patras, 26504 Rio Patras, Greece

J. Portier
Swiss Federal Institute for Forest, Snow and Landscape Research WSL, Zürcherstrasse 111, 8903 Birmensdorf, Switzerland
e-mail: jeanne.portier@wsl.ch

S. Teitelbaum
Département de Sociologie, Université de Montréal, P.O. Box 6128, Stn. Centre-Ville, Pavillon Lionel-Groulx, Montréal, QC H3C 3J7, Canada
e-mail: sara.teitelbaum@umontreal.ca

J.-P. Tremblay
Département de biologie, Faculté des sciences et de génie, Université Laval, 1045 avenue de la Médecine, Québec, QC G1V 0A6, Canada
e-mail: jean-pierre.tremblay@bio.ulaval.ca

Centre d'étude de la forêt, Centre d'études nordiques, Université Laval, 2405 Rue de la Terrasse, Québec, QC G1V 0A6, Canada

M. Versluijs
The Helsinki Lab of Ornithology, Finnish Museum of Natural History, University of Helsinki, FI-00014 Helsinki, Finland
e-mail: martijnversluijs@hotmail.com

J. Wang
Natural Resources Canada, Canadian Forest Service—Atlantic Forestry Centre, 1350 Regent Street, PO Box 4000, Fredericton, NB E3B 5P7, Canada

S. Gauthier
Natural Resources Canada, Canadian Forest Service, Laurentian Forestry Centre, 1055 rue du PEPS, P.O. Box 10380, Stn. Sainte-Foy, Québec, QC G1V 4C7, Canada
e-mail: Sylvie.gauthier@nrcan-rncan.gc.ca

of this expanding population. Technological advances have increased our efficiency in exploiting ecosystems; humans now alter the environment faster and at a greater scale than ever before (Boserup, 1981; Puettmann et al., 2009). Nonetheless, societies depend on finite natural resources and ecosystem services (Perrow & Davy, 2002). The main activities causing ecosystem impacts are agriculture, industry, forestry, and urbanization, and this economic development has therefore led to the alteration of the original ecosystems across a large portion of the planet. This loss of an ecological–economic equilibrium has led to a need to further develop and apply the concept of sustainable development as a means of balancing resource exploitation, biological conservation, and social conditions for future generations (Quarrie, 1992; Rockström et al., 2009b; Steffen et al., 2015).

Forests account for 31% of the world's land area (FAO, 2016), and forest resources are vital to the development of human societies (FAO, 2014). At least 18% of the world's population uses wood to build their homes, 2.4 billion people cook by burning woody materials, and 90 million people in Europe and the United States use wood as an energy source for domestic heating. The boreal forest is the second-largest terrestrial biome in the world (Teodoru et al., 2009), covering 14 million km^2, distributed in a circumpolar forest belt (Burton et al., 2003), and representing about 25% of the world's forest (Dunn et al., 2007).

Currently, two-thirds of this biome is managed mainly for timber production (Gauthier et al., 2015b). Boreal forests are critical for the global wood supply, producing 37% of the world's wood (Gauthier et al., 2015b). During the last century, forest management practices had timber production as their main goal. Logging activities to meet the demand for timber significantly affect this biome (Halme et al., 2013; Kuuluvainen & Siitonen, 2013; Messier et al., 2013; Puettmann et al., 2009). From 1990 to 2000, Canada recorded the most intense period of logging operations in the world, with forests harvested at more than two hectares per minute (Perrow & Davy, 2002).

The current global demand for wood is 1.5 billion m^3, whereas it is expected to increase to between 2.3 and 3.5 billion m^3 by 2050 (Smeets & Faaij, 2007). Thus, logging and related activities will likely have an ever-greater impact on the boreal forest in the near future, continuing the twentieth-century trend of expanded exploitation (Park & Wilson, 2007). For example, in Québec (Canada), the total volume of wood harvested over the past century increased steadily until 2005. In 1924, it was 13.9 million m^3, rising to 21.9 million m^3 in 2011. It should be noted that between 1997 and 2005, the volume harvested was more than 40 million m^3, reaching its peak in 2005 at 45.64 million m^3 (Duchesne & Ouimet, 2007; National Research Council, 2016). In the intensively managed boreal forests of Finland and Sweden, annual growth and harvesting have been increasing in the past 100 years. In the recent past (averaged over 2013–2017), an average of 68.3 and 82.8 million m^3 were harvested, corresponding to 75% and 78% of the annual growth in these two countries (Korhonen et al., 2021; SLU, 2020). In Sweden, 20 of 28 million ha of forest is accessible for intensive forestry; therefore, these forested areas are currently on a transformation trajectory away from natural and resilient ecosystem conditions having multiple value chains (Angelstam et al., 2020).

The reduction, modification, and loss of forests are not recent phenomena; these human-related alterations to forest ecosystems trace the evolution and migration of human populations. Human activity is one of the key processes in the history of forest land transformation (Williams, 2003). The net loss in the global forest area between 2000 and 2010 was 5.2 million ha/yr (roughly the size of Costa Rica). This loss was 8.3 million ha/yr between 1990 and 2000 (FAO, 2011). However, the State of Canada's Forests report (Natural Resources Canada, 2020) maintains that changes in forest area caused by deforestation are not significant in Canada, as it would take 40 years for Canada to lose 1% of its forest area under the most pessimistic forest harvesting scenario (Guindon et al., 2018). In Fennoscandian forests, deforestation due to forestry is prevented by legislation, i.e., clear-cut harvesting must be followed by forest regeneration; the main sustainability issues are related more to loss of biodiversity, ecosystem services, multiple value chains, and Indigenous and local cultures and less to sustained yield or deforestation.

Regardless of the region, climate change intensifies threats to forest health (Trumbore et al., 2015). The intensity and frequency of forest fires have increased in both Canada and the United States, exacerbated by prolonged drought episodes (attributed to climate change) and fire-suppression policies that have increased the amount of available fuel loads (FAO, 2009). From the projected cumulative impacts of fire, drought, and insects on timber volumes across North American boreal forest, the current level of harvesting could thus be difficult to maintain without implementing of adaptative measures (Boucher et al., 2018). In the Fennoscandian forests, the past decade has seen several years of exceptional forest fires, storms, and insect outbreaks, particularly in Sweden (Hlásny et al., 2021; Krikken et al., 2021; Valinger & Fridman, 2011). These events raise questions in regard to the vulnerability of a ubiquitous simplified forest management system—and also from a sustainable timber yield viewpoint—and advocate for the application of more diverse, ecosystem- and disturbance-based management perspectives (Berglund & Kuuluvainen, 2021). Droughts, insect outbreaks, and windstorms are particularly problematic for Norway spruce, which is favored as a commercial tree species in both countries, partly due to the extremely high ungulate browsing pressure on deciduous trees and pines in some parts of the region. The uncertainty associated with disturbances and their potential trajectories in future climates requires a profound reflection on the challenges faced by the boreal biome to achieve sustainable forest management in terms of wood material supply, biodiversity conservation, maintenance, and enhancement of forest carbon sinks, and the cultural values of forests.

Forestry activities in the recent decades have contributed to a decline in habitat diversity and productivity of forest ecosystems around the world, a phenomenon accelerated by climate change (Fischer & Lindenmayer, 2007; Lindenmayer & Fischer, 2007; Schütz, 1997). With increased social concerns about protecting biodiversity (Franklin et al., 2002), boreal forestry has begun to address goods and services other than timber production (Dobson et al., 1997; Gauthier et al., 2009; Halme et al., 2013; Kuuluvainen, 2002; Puettmann et al., 2009). The preservation of biodiversity and the modification of forestry practices to reduce their impact on ecosystems have emerged as two key issues in forest management (FAO, 2009; Lindenmayer & Franklin, 2002; Myers et al., 2000). These concerns confront traditional forest management, which focuses on a deterministic planning of harvesting and exploitation without considering changes, natural disturbances, social issues, uncertainty, and nonlinearity. Forest ecosystem-based management (FEM), in contrast, aims to bridge the gap between natural and managed forests to maintain the ecological integrity and biodiversity of ecosystems. FEM was specifically defined as

> a management approach that aims to maintain healthy and resilient ecosystems by reducing the gaps between natural and managed landscapes to ensure, in the long term, the maintenance of multiple ecosystem functions and, consequently, to maintain the social and economic benefits derived from them (Gauthier et al., 2009).

This approach stems from the reflections on sustainable forest development, which emerged from the 1992 Earth Summit in Rio de Janeiro, and FEM has become increasingly applied within the boreal biome, especially in North America (Burton et al., 2003; Mitchell & Beese, 2002). FEM applies an ecosystem model to reconcile timber harvesting with the long-term maintenance of the structure, functioning, and ecological processes responsible for maintaining ecosystem services. This approach manages the forest through a holistic view (Kimmins, 1997) to ensure its integrity, biodiversity, and sustainability (Gauthier et al., 2009). However, the question arises: Is FEM a useful framework to deal with climate change and the associated impacts?
Our new definition considers FEM as

> an adaptative management approach that aims to promote healthy and resilient forests under climate change to ensure the long-term maintenance of ecosystem functions and thereby retain the social and economic benefits they provide to society.

Thus, FEM is a promising solution for achieving sustainable forest management within a context of climate change, an approach able to include responses and solutions for all the challenges facing the boreal biome (Grenon et al., 2010). In this critical moment, scientific cooperation is essential to adapt forest management practices for the future. FEM within the boreal forest provides one of the last remaining global opportunities to proactively plan forest management for sustainable ecosystem and economic development. In this final chapter, we present the most important challenges facing the future boreal forest (Fig. 31.1). Our goal is to provide helpful recommendations and tools to reduce uncertainty and to justify how FEM can address future challenges within the second-largest terrestrial ecosystem in the world, the boreal biome.

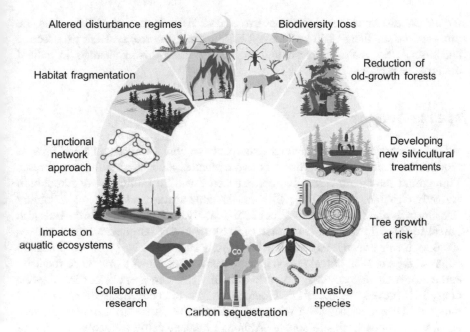

Fig. 31.1 The challenges facing boreal forests under global climate change

31.2 The Challenges of Sustainable Management in Boreal Forests Facing Climate Change

The need to adapt FEM to the future and develop measures to achieve sustainable forest management in the face of climate change led to a discussion among 147 researchers in forest sciences to build this book and produce a list outlining the challenges facing the boreal biome in terms of natural disturbances, silvicultural practices, biodiversity, landscape ecology, economy, and society (Fig. 31.1). In this section, we outline and assess the current state as a starting point to establish future research directions and applications.

31.2.1 Natural Disturbance Regime Change

Boreal forests are affected by various natural disturbances, including wildfires, weather-related disturbances (heat, drought, snow, and wind), insect outbreaks, and disease (Gauthier et al., 2015b; Price et al., 2013). These disturbances operate over a wide range of spatial and temporal scales and are among the core factors driving landscape dynamics and the structure, composition, and biodiversity in these forests (Berglund & Kuuluvainen, 2021; Price et al., 2013; Shorohova et al., 2011). Climate change can impact these forests by modifying the timing, extent, and severity of the natural disturbance regimes (Navarro et al., 2018a; Seidl et al., 2017). Thus, the most serious challenge will be to adapt FEM to this new reality. Here we describe and discuss the main observed and expected climate-driven changes in natural disturbance regimes.

31.2.1.1 Wildfires

Wildfires constitute a major natural disturbance in boreal forests, a natural process required to maintain the biodiversity and dynamics associated with these forests. Throughout the circumboreal biome, between 9 and 20 million ha of forest burn annually (Robinne et al., 2018) with considerable spatial and temporal variability (De Groot et al., 2013a; Gauthier et al., 2015a; Ryan, 2002). The North American boreal forest, for example, is characterized by relatively infrequent, high-intensity, stand-replacing crown fires that often completely burn extensive patches of forest. In contrast, the Eurasian boreal region experiences repeated low- to moderate-intensity surface fires characterized by low tree mortality (De Groot et al., 2013a; Robinne et al., 2018; Rogers et al., 2015). Human-caused fires occur mostly in areas where they are likely to be detected; this permits a rapid response from fire-suppression agencies, to often limit fire spread. Although humans cause a large proportion of fires in the boreal forest, lightning-caused fires account for most of the area burned in a given season (Robinne et al., 2018; Stocks et al., 2003).

Natural fire ignition and propagation depend on a combination of factors related to, on one hand, climate and weather (Eden et al., 2020; Seidl et al., 2017) and, on the other hand, fuel type and availability (De Groot et al., 2013a; Rogers et al., 2015). Consequently, ongoing climate change is expected to markedly alter future fire regimes. Fire weather could become more severe in the coming years (De Groot et al., 2013b; Flannigan et al., 2016), leading to drier fuels and favoring easier fire ignition and propagation (Flannigan et al., 2016). Despite uncertainties related to differences in climate change scenarios, fire weather is expected to be more severe in western Canada than in central Russia (De Groot et al., 2013b). Nevertheless, Russia is also expected to experience a marked increase in fire activity (De Groot et al., 2013b) and possibly a substantial increase in stand-replacing fires (Gauthier et al., 2015b). Although the fuel consumption rate will be higher in boreal stands in Canada, total carbon emissions could be higher from Russian boreal fires owing to a larger annually burned area (De Groot et al., 2013b). Longer and more active fire seasons will significantly affect boreal forests. Forest composition would shift toward an increased proportion of fire-tolerant and fire-resistant species (De Groot et al., 2013b). The amount of old-growth forest, associated with high biodiversity, would be greatly reduced to give way to landscapes dominated by young forests (Kuuluvainen & Gauthier, 2018). This phenomenon—resulting in fewer mature trees across the landscape because of the repeated occurrence of fires—increases the risks of regeneration failure, thereby leading to a gradual opening of forests (Jasinski & Payette, 2005; Kuuluvainen & Gauthier, 2018). For example, Splawinski et al. (2019) projected a progressive increase in the area affected by natural regeneration failure under climate change for northern Québec, culminating with a 65.8% loss under the worst-case scenario.

The global carbon cycle would also be affected as larger, more frequent, and more severe wildfires release higher levels of carbon into the atmosphere. Boreal forests therefore risk shifting from being carbon sinks to carbon sources (Walker et al., 2019), thereby amplifying this positive climate feedback.

31.2.1.2   Weather- and Climate-Related Disturbances: Heat, Drought, Wind, Floods, and Snow

Boreal forests are expected to experience large increases in temperature over the twenty-first century, accompanied by modest increases in precipitation in some regions (IPCC, 2014). These changes will lead to higher frequencies and intensities of extreme heat and drought events (Price et al., 2013). Heat- and drought-induced tree mortality has already increased over the last two decades (Allen et al., 2010). This phenomenon will likely be further exacerbated in the twenty-first century in the boreal biome (Gauthier et al., 2015b; Liu et al., 2013). Forest sensitivity to heat and drought events depends on such factors as the intensity and frequency of these events and the tolerance of tree species to heat/drought. Drought-intolerant aspen-dominated forests in western Canada, for example, have experienced very severe drought-induced diebacks at levels similar to postfire mortality (Michaelian et al., 2011). Overall, the driest regions of the boreal biome have been shown to be more sensitive to weather induced diebacks (Gauthier et al., 2015b). Western Canadian boreal forests associated with a drier climate are already experiencing increased mortality rates (Boucher et al., 2018; Peng et al., 2011), whereas, at least for now, the moister forests of eastern Canada have been less affected (D'Orangeville et al., 2018).

Wind and snow, common disturbances in boreal forests, cause the uprooting and breakage of trees (Lavoie et al., 2019; Mitchell, 2013; Montoro Girona et al., 2019; Saad et al., 2017; Valinger & Fridman, 2011). These damages can rapidly alter forest structure, species composition, and the spatial and temporal availability of resources, in turn disrupting forest management and planning. Boreal forests have been recurrently affected by severe storms in the past, such as the Gudrun storm in 2005 in northern Europe and the Great Ice Storm of 1998 in eastern Canada. It remains unclear how storm regimes will be affected by climate change (Feser et al., 2015; Mölter et al., 2016); however, it is expected that increasing temperatures will favor an increased frequency and intensity of winds and greater snow loads (Gregow et al., 2011). Warmer winters will lead to shorter periods with frozen soil and greater loads of heavy humid snow; this combination heightens the likelihood of trees being uprooted and suffering stem breakage (Nykänen et al., 1997; Peltola et al., 1999). Furthermore, expected changes in tropical cyclone regimes could also increase windthrow impacts on boreal forests, as observed during the Sandy and Ophelia storm events reaching, respectively, Canadian forests in 2012 and Norwegian forests in 2017. Riparian forests may also be affected by an increase in flooding as unprecedented low and high spring discharge in recent decades—relative to the historical natural variability of the last 250 years—also suggests that the increase in flood frequency and magnitude originates from climate change (Nolin et al., 2021).

31.2.1.3 Insect Outbreaks and Diseases

Biotic agents, such as native or non-native insects and pathogens, constitute major disturbances in boreal forests. The most damaging insects for boreal tree species are defoliators, which eat leaves or needles, and bark beetles, which feed on phloem and cambium (MacLean, 2016). On the other hand, pathogens cause significant damage to all tree parts, i.e., foliage, stem, and roots, leading to reduced photosynthetic activity and water/nutrient uptake and producing structural problems (Malmström & Raffa, 2000; Natural Resources Canada, 2020). Climate and weather conditions affect the distributions and ecological dynamics of insects and pathogens and those of their hosts (Dukes et al., 2009; Malmström & Raffa, 2000). Although the life cycle of insects and pathogens responds mainly to temperature, pathogens are also sensitive to precipitation and humidity (Price et al., 2013). Pathogen-induced diseases reduce growth and productivity (Price et al., 2013) and cause widespread forest decline and mortality when co-occurring with other disturbance agents (Dukes et al., 2009). Insect outbreaks markedly impact forest productivity and dynamics by affecting tree growth, seed production, tree regeneration, and successional processes. Outbreaks are cyclic and often synchronous over large geographic areas; this leads to a region-wide mortality of host trees in a relatively short period. Spruce budworm (*Choristoneura fumiferana*) is the main defoliator of spruce and fir forests in North America, affecting extensive areas and causing important losses of timber supplies (Montoro Girona et al., 2018b; Régnière et al., 2012). In Eurasia, there are no comparable records of large-scale outbreaks of defoliators as that of the spruce budworm in North America. Although insect outbreaks generally have less severe punctual impacts on forest productivity and dynamics than fires, they often affect larger areas. For example, the Canadian Forest Service calculated that insects affected 15.6 million ha of Canadian forest in 2017, whereas 3.4 million ha of forests burned that same year (Natural Resources Canada, 2020).

Defoliating insect outbreaks have shown an increase in severity, extent, and duration over the recent decades, and their frequency and severity are expected to increase further (Navarro et al., 2018c; Zhang et al., 2014). Climate change will likely modify the geographic distribution of host trees within boreal forests and may alter both the range and outbreak potential of their associated insects and pathogens (Malmström & Raffa, 2000). Models project that future warmer winters and longer growing seasons will favor the northward expansion of the northern range limits of many insect pests (Dukes et al., 2009; Pureswaran et al., 2015; Régnière et al., 2012). Spruce budworm in eastern Canadian boreal forests experienced such a northward shift during the twentieth century (Navarro et al., 2018c). Similarly, an expansion in climatically suitable habitats at the beginning of the twenty-first century for the mountain pine beetle has facilitated the northward and higher-elevation expansion of outbreaks in western Canadian boreal forests (Kurz et al., 2008). Projections of climate change impacts on pathogen populations remain uncertain, as pathogen outbreaks are less predictable than insect outbreaks given the links of the former to precipitation levels (Dukes et al., 2009; Pautasso et al., 2015; Price et al., 2013). Researchers do agree, however, that pathogen activity in circumboreal forests will likely increase (Price et al., 2013).

31.2.1.4 Interactions Between Natural Disturbances

Interactions between natural disturbances are common across the boreal biome. The most frequently reported interaction is the increased flammability of forests induced by drought events, which enhances the frequency and severity of fires (Flannigan et al., 2016). Interactions between biotic, i.e., insect outbreaks and pathogens, and abiotic disturbances are also very frequent and are critical to the dynamics of biotic disturbance agents (Canelles et al., 2021; Nolin et al., 2021; Seidl et al., 2017). For example, the large number of dead trees resulting from drought, fire, or windfall can trigger a strong increase in insect populations and amplify the spread and intensity of outbreaks (Marini et al., 2013; Price et al., 2013). On the other hand, stands affected by insect outbreaks significantly increase the amount of flammable fuel and, therefore, a fire's potential spread and severity (James et al., 2017; Perrakis et al., 2014). Most interactions between disturbances tend to amplify the mutual effects of disturbance agents (Seidl et al., 2017). Outbreak severity may vary with forest composition at the landscape level (Lavoie et al., 2021). The long fire cycle may favor increasing outbreak severity stemming from the abundance of late-successional host trees (Bergeron & Leduc, 1998; Navarro et al., 2018c). On the contrary, a short fire cycle may increase the abundance of broadleaf stands and favor a better control by natural enemies (Cappuccino et al., 1998). Shifts between short and long fire cycles explain the dynamics of defoliators through the Holocene (see Chap. 2). The observations are of critical concern, as climate-induced increases in natural disturbances may be further intensified by such interactions, heightening the risk of exceeding ecological thresholds and tipping points.

31.2.2 Biodiversity Loss

Boreal forest landscapes have transformed rapidly over the last decades (Mori et al., 2021). In Fennoscandia, extensive changes have occurred since the 1950s when forestry methods became more mechanized and efficient, and clear-cutting was introduced, i.e., even-aged management (Esseen et al., 1997). Moreover, forestry and fire suppression have led to the disappearance of natural disturbances, replaced instead by anthropogenic disturbances, including thinning, clear-cutting, soil scarification, and the planting of conifers mostly in monospecific regimes (Esseen et al., 1997; Wallenius, 2011). Unlike forests structured by natural dynamics, managed forests often consist of even-aged monocultures that lack structural complexity, including an absence of coarse woody debris, snags, and old trees (Bengtsson et al., 2000; Löfman & Kouki, 2001). In North American boreal forest regions, forest composition and age structure are predicted to change over the twenty-first century because of climate change and increased anthropogenic pressure (Boulanger et al., 2016). These changes in forest composition and structures will have/have had serious consequences for the many species that rely on deciduous trees, deadwood, large-diameter trees, and complex horizontal and vertical structures of tree vegetation (Kuuluvainen, 2009; Ram et al., 2017; Regos et al., 2018; Virkkala, 2016).

In general, the structural complexity of habitats is strongly correlated with species richness for most taxonomic groups (Honnay et al., 2003; Lassau et al., 2005). In Fennoscandia, forest conditions remain determined by the transformation to even-aged, single-species conifer forests. This simplistic forest management has had severe consequences for forest biodiversity. In Finland, for example, the forest is the primary habitat for 31% of threatened species. For almost five-sixths of these threatened species, the primary driver of the population decrease is a change in forest habitat (Hyvärinen et al., 2019). This biodiversity loss continues to increase (Fig. 31.2). In Sweden, more than 50% of red-listed species are connected to forest habitats, 43% are dependent on forests, and 1,400 species are directly threatened by forest clear-cutting (Artdatabanken, 2020). In North American boreal forests, for a variety of species, it is predicted that populations will decrease, and their ranges, in terms of size and distribution, will shift through the loss of climate suitability and greater anthropogenic influence, i.e., habitat degradation and fragmentation (Cadieux et al., 2020; Woo-Durand et al., 2020).

Biodiversity plays a vital role in the functioning of forest ecosystems and is closely related to the health status of an ecosystem. Highly diverse systems are expected to be less prone to perturbations such as pest outbreaks. A healthier ecosystem provides

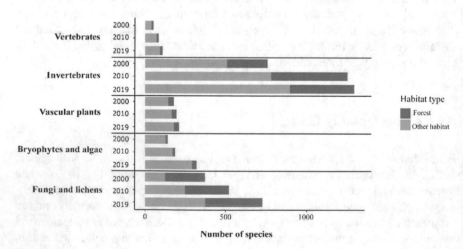

Fig. 31.2 The number of threatened species (*bars*) in Finland by phylum over time in forests (*red*) and other habitats (*blue*). Data obtained from the Natural Resources Institute Finland (2021)

higher-quality ecosystem services, and robust forest health correlates with forest productivity, thereby maximizing resource exploration in well-maintained forests (Bohn & Huth, 2017; Zhang et al., 2012). Ecosystems that deteriorate to an unsustainable level result in problems that are often very expensive, economically speaking, to reverse. Hence, the integration of biodiversity management as a target into current forestry practices becomes a priority. Over the last decades, forestry has drastically decreased legacies crucial for biodiversity. The challenge lies in bringing these critical components back into the boreal forest ecosystem without jeopardizing commercial forestry practices. Forests harbor multiple species, each with its own environmental and ecological requirements. This is particularly true for habitat specialists, which commonly require a precise range of environmental conditions or a specific diet. Forest structures and environmental conditions, e.g., moisture and light conditions, are important factors severely affected by forestry but are, nonetheless, critical for the occurrence of species from different taxonomic groups. The main challenge is to improve the overall habitat and landscape conditions that favor greater biodiversity. For example, deadwood quantity and quality, e.g., of varying decay stages, including standing and fallen trees, recently dead or decomposing stems, influence the occurrence of a variety of beetle species, polypores, and bird species. In Fennoscandia, managed forests harbor on average only 4 to 5 m^3 of deadwood per hectare, whereas in natural forests, deadwood can exceed 100 m^3 per hectare. Additionally, vertical structures, e.g., understory and uneven-aged forests, are important for food availability, nesting opportunities, and hiding spots for many boreal forest birds (Brokaw & Lent, 1999; Culbert et al., 2013; Eggers & Low, 2014).

The landscape structure plays an important role, as lichen and polypores are sensitive to habitat fragmentation and the lack of old-growth forests; therefore, large conservation areas benefit these species (Junninen & Komonen, 2011). For managed boreal forest landscapes to play a vital role in conserving biodiversity, these must include a mixture of habitats of varied successional stages, e.g., containing both early- and late-successional forest stages. Clear-cutting has long been held as an appropriate method for emulating natural disturbances (Mielikäinen & Hynynen, 2003). In naturally dynamic boreal forest systems, disturbances such as fire, insect outbreaks, and windfall contribute to high structural complexity; however, the ecological effects of clear-cutting differ from those of wildfire (Heikkala et al., 2016). Furthermore, forest dynamics in the boreal forest are also driven by various other natural, small-scale disturbance agents, like wind, pathogens, and insects, that have maintained a semicontinuous forest cover containing small gaps. The main challenge for forestry lies in reintroducing natural dynamics and restoring the natural systems while also minimizing any damage to forest production.

31.2.3   Loss of Old-Growth Forests

Old-growth forests are generally defined as stands at the end of forest succession, where post-disturbance cohorts are beginning to be replaced by new trees, human impacts are negligible, and low-severity disturbances are the primary drivers (Knee-shaw & Gauthier, 2003; Wirth et al., 2009). Specific structural attributes often distin-guish these forests from younger stands, including a higher deadwood volume or a more complex structure (Kulha et al., 2019; Martin et al., 2018; Paillet et al., 2015; Wirth et al., 2009). Even within a given landscape, the concept of old-growth forest actually refers to a wide diversity of structures and composition that vary over time, depending on environmental conditions and the local disturbance history (Kulha et al., 2020; Martin et al., 2020a, d; Meigs et al., 2017; Portier et al., 2018). Old-growth forests hence typically consist of complex mosaics of uneven-aged stands. The long continuity of the forested state in these ecosystems is also vital for many disturbance-sensitive and low-dispersal species. For these reasons, old-growth forests provide a wide range of habitats, increase biodiversity, and provide numerous ecosystem services, such as carbon sequestration, water filtration, and cultural and aesthetic values (see Chap. 7; Keeton, 2018; Warren et al., 2018; Watson et al., 2018). Hence, old-growth forests are key elements of natural landscapes (Fig. 31.3). However, the climatic and fertility constraints in boreal landscapes may inhibit the development of old-growth attributes common to other biomes, such as very large trees or a complex vertical structure (Bergeron & Harper, 2009; Martin et al., 2020b, 2021b).

The decreased area, diversity, connectivity, and functionality of boreal old-growth forests in managed landscapes represent major issues facing the boreal biome. Because of their remoteness and low productivity, boreal forests have long remained undisturbed by logging activities, particularly in the northern and eastern parts of Eurasia and in northern North America (Potapov et al., 2017; Venier et al., 2018; Wells et al., 2020). The development of industrial-scale forest management has nevertheless led to increased exploitation of these territories, especially since the mid-twentieth century (Boucher et al., 2017; Dupuis et al., 2020; Ostlund et al., 1997). Logging activities have therefore led to a loss of old-growth forest coverage (Cyr et al., 2009; Grondin et al., 2018; Ostlund et al., 1997), changes in tree-species composition (Boucher & Grondin, 2012; Boucher et al., 2015; Kuuluvainen et al., 2017), landscape homogenization and fragmentation (Haeussler & Kneeshaw, 2003; Löfman & Kouki, 2001; Schmiegelow & Mönkkönen, 2002), and decreased dead-wood availability (Jonsson & Siitonen, 2012; Moussaoui et al., 2016). In certain regions, such as Fennoscandia, old-growth forests have almost completely disap-peared and now represent a minimal part of the total forest cover (Forest Europe, 2015; Kuuluvainen et al., 2017; O'Brien et al., 2021; Potapov et al., 2017). Conse-quently, many species that depend on old-growth forests or associated elements, such as deadwood, are now threatened in the European boreal forests (Esseen et al., 1992; Jonsson & Siitonen, 2012; Tikkanen et al., 2006). In Canada and Russia, old-growth forests remain relatively abundant, but their areas are rapidly decreasing, and this is already causing major biodiversity issues (Aksenov et al., 1999; Bergeron et al., 2017; Cyr et al., 2009). For example, the level of fragmentation and degradation of old-growth forests in Canada has caused a collapse of woodland caribou (*Rangifer tarandus caribou*) populations (Venier et al., 2014). In Russia, saproxylic species are now facing a similar threat as they have experienced in Europe because of the development of forestry (Wallenius et al., 2010). Moreover, concerns have recently

◄**Fig. 31.3** Old-growth boreal forests in eastern Canada are dominated by black spruce *(Picea mariana* (Mill.) BSP) and, to a lesser extent, balsam fir *(Abies balsamea* (L.) Mill.). These forests are defined by a strong heterogeneity of structures and microhabitats, shaped by specific natural disturbance histories and abiotic characteristics. The low diversity of tree species and their limited size may nevertheless make these stands appear—erroneously—as homogeneous. The difficulty in correctly identifying their heterogeneity may eventually lead to the disappearance of old-growth forests defined by distinctive functions and habitats not found elsewhere. *Photo credits* **1–9** Maxence Martin, **10** Frédéric L. Tremblay

been raised about the characteristics of remnant old-growth forests in managed land-scapes; these remnants are often defined by lower productivity or different structures than those observed in natural landscapes (Martin et al., 2020c, 2021a; Price et al., 2020). This implies that some habitats or ecosystem services specific to old-growth forests with higher economic value may be particularly at risk. A conservation approach focusing only on the area of old-growth forest to be conserved, without considering its quality, becomes insufficient. The challenges related to the protection of boreal old-growth forests therefore concern not only their size but also their diversity, connectivity, and functionality.

The expected effects of climate change on old-growth forests remain hard to project, as they can often be contradictory (Fig. 31.4). Late-successional boreal species may benefit or suffer under these future conditions (D'Orangeville et al., 2016, 2018; Thom et al., 2019). The warmer temperatures and longer growing season may enable a northward range expansion for southern boreal or hemibo-real species, such as European beech *(Fagus sylvatica* L.) or sugar maple *(Acer saccharum* Marsh) (Bouchard et al., 2019; D'Orangeville et al., 2018; Kramer et al., 2010). The replacement of shade-tolerant boreal species with new late-successional species, however, depends on the migration capacity of the latter, which remains uncertain (Bouchard et al., 2019). If the late-successional species cannot be replaced, developing old-growth forests dominated by pioneer species could be possible. Accordingly, Cumming et al. (2000) have observed forests dominated by trembling aspen *(Populus tremuloides* Michx.), driven by low- and moderate-severity secondary disturbances. Currently, this type of forest remains rare (Bergeron & Harper, 2009).

Climate change is also expected to increase the frequency and severity of natural disturbances in the coming decades (Bergeron et al., 2017; Kuuluvainen & Gauthier, 2018). This change in disturbance regime may potentially reduce remnant old-growth areas or increase the abundance of forests degraded by recurrent secondary distur-bances, thereby eventually overwhelming stand resistance (Bergeron et al., 2017; Martin et al., 2019). Nevertheless, Bergeron et al. (2017) highlighted that in boreal landscapes, industrial-scale forest management based on short-rotation clear-cuts, i.e., rotation periods well below those of regional fire cycles, will remain the prin-cipal agent of the loss of old-growth forest, more than the projected increase in fire frequency and other changes in disturbance regimes (Fig. 31.4). Therefore, forest management will certainly have a much greater, immediate, and predictable impact than climate change on boreal old-growth forests. Although we require a

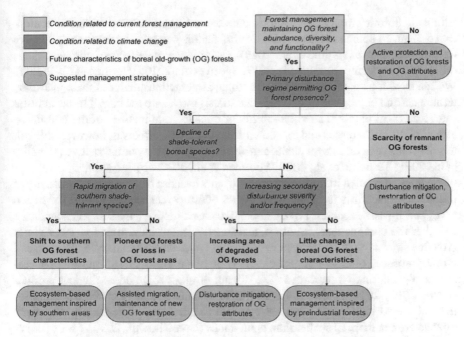

Fig. 31.4 Conceptual and simplified flowchart of possible changes in the characteristics of boreal old-growth (*OG*) forests in the context of future climate change, including processes (*italics*), old-growth forest characteristics (**bold**), and possible management solutions (*rounded boxes*)

better understanding of how climate change affects old-growth forests, the more immediate and pressing need is the proposal of management strategies that efficaciously protect these ecosystems. Combining forest monitoring networks (e.g., SmartForests Canada; Pappas et al., 2022) and mechanistic modeling (e.g., Fatichi et al., 2019) could enhance our process understanding and facilitate the development of sustainable forest management under environmental change.

31.2.4 Biotic Stress Factors as Underlying Drivers of Ecological Change

The ecosystem-based management of boreal forests emphasizes emulating natural disturbances to maintain the ecological composition, structure, functioning, and services provided by the boreal forest relative to historical baselines (Berglund & Kuuluvainen, 2021; Jackson et al., 2001; Kuuluvainen & Grenfell, 2012). Taking climate change into consideration when developing and implementing ecosystem-based management strategies involves integrating the effects of biotic stress factors that may slowly build up or involve sudden, and likely cumulative, extreme events.

Similar to abiotic stress factors, biotic ones can act as underlying drivers of ecological change. In some situations, biotic stress factors can even override the effects of natural disturbances (Nuttle et al., 2013) or counteract the effects of climate change on forest ecosystems (Seidl et al., 2017; Speed et al., 2010; Vuorinen et al., 2020a). We can expect a spatiotemporal lag in trophic interactions under climate warming, as animals and pathogens from temperate forests move into the boreal forest at faster rates than most plants. The vulnerability of trees to combinations of the bottom-up effects of abiotic stresses and top-down effects of biotic effects is, however, difficult to predict, as such effects on plants can be positive, negative, or interactive (Canelles et al., 2021; Teshome et al., 2020; Vuorinen et al., 2020a, b).

Ecological change in the boreal biome occurs because of an increased intensity of biotic stressors, shifts in the climatic niches of temperate species toward the boreal zone, and invasive species. For example, increased moose (*Alces alces*) abundance can alter the composition, structure, and functioning of balsam fir forests in North America (McLaren & Peterson, 1994; Pastor et al., 1998) and mixed Scots pine and Norway spruce forests in Fennoscandia (Lorentzen et al., 2018), pushing succession along alternative pathways (De Vriendt et al., 2021) and ecological regimes (Gosse et al., 2011; Smith et al., 2010). In Poland, moose altered their behavior in response to higher temperatures by more frequent use of dense forests, which provide greater thermal shelter than open stands (Borowik et al., 2020). Moose populations are also appearing to shift to higher latitudes and altitudes in North America and China in response to warmer late-spring temperatures (Dou et al., 2013; Tape et al., 2016). In Canada, the distribution range of white-tailed deer (*Odocoileus virginianus*) has expanded northward since the mid-twentieth century, and climate data better explain this range shift than land-use metrics (Dawe & Boutin, 2016). Selective browsing by an introduced population of white-tailed deer north of this species' historical range and in an area lacking predators has led to the conversion of balsam fir forest to spruce-dominated forest and parkland (Barrette et al., 2014, 2017). Climate warming will likely reduce forage limitations for ungulates in the boreal zone. Although a corresponding increase in predation pressure may counteract the population growth of these herbivores, full compensation is unlikely, as it would require apex predator populations to attain levels above historical numbers, an improbable scenario (Pasanen-Mortensen et al., 2017). Climate change, combined with globalization in the transportation of people and goods, has increased global incidents of invasions by alien species. These introductions now constitute one of the major threats to global biodiversity and planetary ecosystems (Bellard et al., 2013; Vitousek et al., 1996). Seidl et al. (2017) predicted that boreal forests will experience the most pronounced future changes in disturbances of all forest types. In boreal forests, exotic species of defoliator insects, earthworms, slugs, and pathogens, known drivers of major ecological change in forest dynamics, are increasingly observed (Sanderson et al., 2012).

31.2.5 Future Boreal Tree Growth at Risk

Wood formation is highly sensitive to temperature, and dendroecologists routinely use the growth rings of trees to reconstruct climate prior to the instrumental period (Fritts, 2001). Nonetheless, predicting the future productivity of boreal forest stands under projected increases in temperature and aridity remains extremely challenging. Changes in forest productivity will likely vary because of differences in stand composition and structure, site characteristics, variations in disturbance regimes, and regional and local climate anomalies.

Currently, forest productivity appears stimulated with warming in many cold-limited boreal regions of eastern Canada, Finland, Russia, and Asia (D'Orangeville et al., 2016, 2018; Kauppi et al., 2014; Lapenis et al., 2005; Loehle & Solarik, 2019; Myneni et al., 1997; Zhang et al., 2019a, b) despite important variations in climate sensitivity according to species, competition or stand development (Marchand et al., 2019); however, indicators of reduced growth have been observed in warmer or drier boreal regions (Barber et al., 2000; Beck et al., 2011). Potential growth increases may not necessarily translate into carbon storage gains. Indeed, a higher forest carbon pool hinges on higher mean tree longevity (Körner, 2017); however, higher growth rates have been observed in all forest ecosystems, including boreal forests, to reduce the longevity of trees as trees complete their natural life-span faster, e.g., higher susceptibility to windthrow with increasing size, or get harvested earlier (Brienen et al., 2020; Körner, 2017).

In addition to site factors, species-specific characteristics in terms of climate tolerance, adaptation, and migration capacity can also confound tree growth projections. In eastern Canada, drought-adapted species, such as jack pine (*Pinus banksiana*), are projected to cope well with projected future climate change (Aubin et al., 2018; Marchand et al., 2021). However, such predictions are based on historical data, whereas all boreal regions are on a warming trajectory that goes beyond that of the observed or reconstructed climate space. In addition, species' vulnerabilities could depend on the future trajectory of disturbance regimes. The increased frequency and severity of climate anomalies, such as the 2004 drought in central boreal Canada that killed up to 80% of trees in some areas (Michaelian et al., 2011), wildfires (De Groot et al., 2013b) and the expansion of native or exotic forest pests, as seen with the recent mountain pine beetle outbreak in western North America (Cullingham et al., 2011; Robertson et al., 2009), could cancel out any gains in growth within the boreal forest.

31.2.6 Carbon Sequestration

Carbon sequestration, i.e., the storage of atmospheric CO_2 in forested ecosystems (e.g., as tree biomass) at climate-relevant time scales (often referred to as a carbon sink), is one of the many services that forests offer to humanity. The terrestrial carbon sink presents a prominent natural climate solution, contributing to climate change mitigation by absorbing part of anthropogenic CO_2 emissions (e.g., Cook-Patton et al., 2020). The circumboreal region includes more than 30% of the Earth's forested area and represents one of the largest carbon storage pools (Pan et al., 2013). However, the strength of the boreal forest as a C sink varies markedly among regions, and the C sink response to global change remains uncertain. Extrapolating ecophysiological understanding of tree growth (e.g., Hilty et al., 2021) to the landscape and regional scales remains challenging; hence, predicting the fate of the terrestrial carbon sink under global change remains uncertain (Hof et al., 2021). Tree growth provides a carbon sink, yet the resulting ecosystem-level carbon sink stems from numerous interacting processes, including growth (and forest productivity, e.g., Hilty et al., 2021) and respiration (autotrophic and heterotrophic). Natural and anthropogenic disturbances, e.g., fires, insect outbreaks, could abruptly alter the ecosystem-level carbon balance, releasing part of the sequestered carbon back to the atmosphere (Ameray et al., 2021). Quantifying the carbon sequestration potential in the boreal region thus requires an accurate description of the residence times of the carbon stored in different pools (e.g., above vs. belowground; Friend et al., 2014; Pappas et al., 2020). Disentangling the C balance pools could facilitate a robust and quantitative description of the C sink strength and its fate under global change (see Chap. 10).

31.2.7   New Silvicultural Practices

Over the last century, anthropogenic disturbances have had a stronger impact than climate change on boreal forest/stand compositional changes because of the extensive use of even-aged approaches and the application of clear-cutting as the main silvicultural treatment e.g., in northeastern Canadian forests (Danneyrolles et al., 2019). The urgent need to diversify harvest treatments to reduce the homogenization of forest landscapes has become a priority. The difficulty stems from the past approaches to forest management, which failed to consider cumulative landscape-scale effects in forest planning and heavily favored the use of even-aged approaches and short rotation in boreal stands during the last half-century (see Chaps. 15, 16). The consequences of these decisions reduced the resilience of forest ecosystems facing climate change due to the homogenization and simplification of forest structure and composition across the forest landscapes (Franklin et al., 1997; Puettmann, 2011). This homogenization—promoting the dominance of even-aged stands—and simplification of forest structures affected many species that depend on deadwood, large-diameter trees, and complex horizontal and vertical structures of tree vegetation (Kuuluvainen, 2009). Climate change will alter tree growth and gradually replace existing tree species with more climatically suitable vegetation. Accordingly, these modifications will significantly impact post-disturbance recovery potential (Splawinski et al., 2019) and could affect post-harvest forest resilience and dynamics. Under climate change, the sustainability of forest management in North America is at risk, especially in regions currently characterized by a short fire cycle and low productivity

(Gauthier et al., 2015a; Johnstone et al., 2016). Hence, new silvicultural tools and approaches will be required to maintain forest resilience and increase the adaptability of forest ecosystems to novel future conditions (Montoro Girona, 2017; Puettmann, 2011; Spies et al., 2012; Stephens et al., 2013). For Fennoscandian boreal forests, continuous-cover forestry is increasingly promoted to mitigate the loss of biodiversity, ecosystem services, and multiple value chains and ensure a long-term sustained timber yield (Peura et al., 2018; Pukkala, 2016). This shift calls for alternatives to the dominating, intensive rotation forestry. Berglund and Kuuluvainen (2021) proposed a shared (each one-third) distribution between clear-cutting, partial cutting and gap cutting combined with selective thinning on the basis of natural disturbance dynamics.

Partial harvests and variable retention forestry, in which only part of the stand is harvested, represent promising silvicultural approaches that can ensure a more diverse structure in managed forests by maintaining specific ecosystem attributes in the boreal biome, such as large living, dying, and dead trees, to favor greater biodiversity (Gustafsson et al., 2020; Kuuluvainen & Grenfell, 2012; Shorohova et al., 2019). Partial harvests and variable retention forestry include a broad range of treatments, which include commercial thinning (Gagné et al., 2012), selection cutting (Majcen, 1994), and shelterwood cuttings (Montoro Girona et al., 2017, 2018a, 2019; Prévost & DeBlois, 2014; Raymond & Bédard, 2017). Over the last decades, these treatments have been applied in North American and European boreal forests as an alternative to conventional clear-cutting to maintain structural attributes and ensure biodiversity and a continued timber supply (Hernández Rodríguez et al., 2021, Kim et al., 2021; Moussaoui et al., 2020). However, these silvicultural treatments were initially developed in Europe, and there remains many questions in regard to their potential adaptability to North American boreal forests because of the few experiments and limited long-term monitoring of partial harvests in Canada and the United States (Bose et al., 2014; Montoro Girona et al., 2017).

Over the last 20 years, studies in both Canadian and European boreal forests have attempted to understand the effects of partial cutting on biodiversity and stand yields and have reported several sustainable benefits (Bescond et al., 2011; Brais et al., 2004; Webb et al., 2008). Partial harvesting preserves more favorable habitat attributes for various organisms by maintaining some residual stand structures within cutblocks (Fenton et al., 2013; Kim et al., 2021; Kuuluvainen & Grenfell, 2012; Moussaoui et al., 2016; Ruel et al., 2013). Partial harvesting also promotes increased residual tree growth in boreal forests by light thinning (Montoro Girona et al., 2016, 2017; Pothier et al., 2003; Thorpe et al., 2007). Despite these benefits, the main challenge in implementing partial harvesting in a context of climate change will be adapting this silvicultural treatment to the future conditions of North American boreal forests (species, stands, growth ratio) to develop its potential as a tool for ensuring carbon sequestration, biodiversity, and timber production while being implemented in an economic and financially cost-effective manner. Although additional research is required, these forest practices appear as means of providing an increased forest resistance and resilience to change and facilitating the boreal ecosystem's ability to adapt to future conditions, e.g., drought and insect outbreaks.

31.2.8   Including Freshwater Systems Within Forest Management

Water-covered lands represent about 30% of the world boreal forest area (Benoy et al., 2007). In North America, for example, of the 6.3 million km^2 of boreal forest, 850,000 km^2 is covered by fresh surface waters, and about 1.27 million km^2 is covered by peatlands, which, when combined, is equal in size to the country of Indonesia (Gingras et al., 2018). Aquatic and forest environments of the boreal landscape are highly connected because of a high number of contact zones that form a long and complex ecotone in which most aquatic–terrestrial interactions take place. Most organic matter and energy fluxes take their sources in forests and are transported toward aquatic environments by precipitation, freshets, and wind. Once the forest-sourced organic carbon reaches the aquatic habitat, one fraction is processed by the aquatic food web (Grosbois et al., 2020)—either assimilated into biomass or respired—one fraction is stored in the sediments of lakes, rivers, or wetlands, and another fraction is transferred to the atmosphere (Cole et al., 2007). Feedback fluxes from aquatic to forest environments are lower in magnitude because they rely on a faunal transfer of biomass, e.g., aquatic predation by terrestrial consumers and insect emergence from lakes, rivers, and ponds. These fluxes are of higher nutritional quality, however, than the organic matter transported from the forest to aquatic habitats; the latter is in an advanced state of decomposition. Because of these aquatic–terrestrial links and their major implications for forest ecosystem functions, the ecosystem-based management of boreal forests must include freshwater environments. Despite the aim of original ecosystem-based frameworks to manage the forest as a whole, these forest–aquatic system interactions have been largely neglected (see Chap. 29).

Given this strong connectivity between the aquatic and forest environments in the boreal biome, all disturbances affecting the boreal forest also impact aquatic environments within the same watershed. For example, wildfires increase the export of nitrogen and phosphorus to lakes (Lamontagne et al., 2000). Although this quantity represents a negligible amount of nutrients for forest ecosystems, the export of these new wildfire-released nutrients is an important complementary input to freshwaters. Wildfires also influence lake metabolism (Marchand et al., 2009), phytoplankton (Planas et al., 2000), zooplankton (Patoine et al., 2000), and, most likely, the entire aquatic food web. Nonetheless, very little information is available in the literature. This lack of data stems from the logistical challenge of studying wildfires and the traditional separation in the study of land-based disturbances and aquatic habitats. Forestry activities represent an anthropogenic disturbance that, in addition to causing an increased export of nutrients to lakes, produces an additional input of dissolved organic carbon (DOC), which affects the physiochemistry of a water body. The removal of wood drastically diminishes the forest's capacity to retain precipitation and organic matter. The result is an enhanced release and transport into aquatic environments of terrestrial dissolved molecules and particles of inorganic and organic carbon. The new inputs of forest-derived dissolved organic molecules alter the water color (*browning*), and the added particles also increase turbidity in the water column. Both changes affect light penetration in the water column and therefore influence primary productivity, e.g., algal abundance (Steedman, 2000). Higher DOC concentrations can impact the entire food web and diminish fish growth (Benoît et al., 2016). Also, forest harvesting increases the mobility of methylmercury and increases its assimilation into the aquatic biomass of plankton and fish (Garcia & Carignan, 2000; Wu et al., 2018). The impact of other disturbances, such as insect outbreak or windthrow, on aquatic environments, has yet to be studied. Their influence on freshwater physicochemical properties, freshwater metabolism, and food webs therefore remains unknown. It is thus essential for future research to investigate how freshwater systems and their associated food webs react to land disturbances and adjust forestry operations to ensure the sustainable management of the boreal forest, especially in the context of future climate change.

31.2.9 Connectivity and Fragmentation

Connectivity and fragmentation are natural components in the continuum of habitat types and their transition zones, which comprise ecosystem configurations at multiple spatial and temporal scales. Both connectivity and fragmentation vary and change in response to natural disturbances and dynamics as well as anthropogenic influences. Intact forest landscapes and primary forests provide an ecological legacy that harbors intrinsic ecosystem resilience and adaptive capacity to withstand degradation (Lindner et al., 2010; Potapov et al., 2017; Sabatini et al., 2020; Venier et al., 2018) and avoid a sledgehammer effect where ecosystems risk entering new and potentially irreversible ecological states (Barnosky et al., 2012); this is particularly important in forests facing future climate change. With greater than 70% of the Earth's land surface (Barnosky et al., 2012) and more than 80% of the remaining forests (Watson et al., 2018) modified by land use, however, the Anthropocene human footprint critically influences key ecological functions (Tucker et al., 2018). In addition to relocating natural forest frontiers (Potapov et al., 2017) and decreasing remaining intact forest landscape area (Svensson et al., 2020), extensive land use has also generically affected landscape matrix functionality and the remaining protected areas (Heino et al., 2015; Jones et al., 2018).

The consequences of anthropogenic-related forest fragmentation are increasingly debated (Ward et al., 2020). In many regions of the boreal biome, the natural configuration of forest landscapes has become seriously marked by systematic clear-cutting and monoculture–rotation forestry systems (Boucher et al., 2009; Peura et al., 2018) at rates beyond those of sustainability and biodiversity policies and environmental targets (Chazdon, 2018; Jonsson et al., 2019; Selva et al., 2020). The consequent forest landscape fragmentation has had consequences beyond the actual loss of primary forest area and the separation of a few larger areas into many smaller retained and set-aside patches. Edge effects penetrating the remaining patches also generate a proportionally larger loss of functional core areas (Harper et al., 2015;

Pfeifer et al., 2017). A study of boreal Sweden, for example, revealed that systematic forest clear-cutting since the middle of the twentieth century has left behind only 6% primary forest core area for this biome (Svensson et al., 2019). In addition, artificial forest edges are created; these edges do not harbor the natural ecological attributes associated with natural edges (Haddad et al., 2015), further affecting biodiversity and ecosystem resilience. Hence, the re-creation of both structural and functional connectivity in landscapes typified by extensive clear-cutting forestry is challenging (Chazdon, 2018; Ward et al., 2020).

31.2.10   Collaborative Research and Indigenous Peoples

The relationship with the land is a foundation of Indigenous peoples' identity, culture, and livelihood in boreal regions. Boreal landscapes are places of hunting, trapping, fishing, cultural and language learning, and healing. These practices provide access to a multitude of ecosystem services, both tangible and intangible, and contribute to the well-being of Indigenous people (Davidson-Hunt & Berkes, 2003; Saint-Arnaud et al., 2009). However, climate change and forest management drive major transformations of boreal forests and affect Indigenous people's relationship with the land (Fuentes et al., 2020; Turner & Clifton, 2009). Although Indigenous institutions play an increasing role in forest governance (Wyatt et al., 2019), the consideration of Indigenous values and perspectives remains the exception rather than the rule.

Collaborative research contributes to bridging the different perspectives on boreal landscapes and facilitates forest co-management through various means (Blackstock et al., 2007):

- Collaborative research calls for the complementarity between Indigenous knowledge and science-based knowledge (e.g., Asselin, 2015; Suffice et al., 2017). Their combination extends the spatial and temporal scales of observations and provides a wider and comprehensive understanding of boreal environments (Bartlett et al., 2012; Lyver et al., 2018).
- The bridging of Indigenous and scientific knowledge is based on the premise that every knowledge is situated within a knowledge system and is partial, and that there is no hierarchy between knowledge systems (Ericksen & Woodley, 2005). Following this principle, collaborative research legitimizes the knowledge creation process and the associated land management decisions.
- Indigenous people are underrepresented in scientific research institutions, as are Indigenous concepts, methodologies, and ethics (Littlechild et al., 2021; McGregor, 2018). Collaborative research contributes to increasing the research capacity within Indigenous institutions and the Indigenous representation within scientific institutions.

Collaborative research, however, faces challenges. First, Indigenous and science-based knowledge belong to different knowledge systems, with their ontologies, epistemologies, and methodologies (Bartlett et al., 2012). Knowledge co-production requires sustained work at the boundary between knowledge systems to formulate research questions, make explicit the different perspectives of a phenomenon, and develop appropriate methodologies (Dam Lam et al., 2019; Robinson & Wallington, 2012). Second, research partnerships need to overcome existing mistrusts. On the one hand, Indigenous people have experienced negative relationships with scientists in the past, and confidence often needs to be rebuilt and be based on stronger research ethics (Canadian Institutes of Health Research, Natural Sciences and Engineering Research Council of Canada and Social Sciences and Humanities Research Council, 2018). On the other hand, scientists often seek to validate Indigenous knowledge with the same methods as for validating experimental ecological data (Gilchrist et al., 2005). Such an exercise is problematic because a piece of knowledge is taken out of the knowledge system that defines its meaning, sense, and scope and is imported into a different knowledge system without the contextual information required to appreciate its value and validity (Castleden et al., 2017). Extended collaborative work is thus needed to co-produce better-informed, legitimate, and valid knowledge, following both knowledge systems.

31.2.11   Resilient Forest Landscapes

The long-standing forest management approaches targeting only highly productive monocultures have homogenized and simplified forest ecosystems worldwide (Puettmann et al., 2009). Given that climate and global change entail multiple environmental and socioeconomic uncertainties, we urgently need novel forest management paradigms that (1) acknowledge this future uncertainty; (2) promote forest ecosystem resilience to altered disturbance regimes and climate change mainly through the functional diversification of tree communities; and (3) scale up the objectives and impacts of these novel silvicultural interventions from the stand to the landscape scale (Messier et al., 2019). The challenge lies in adopting a trait-based approach to rethink and redesign sustainable forest management plans (Cadotte et al., 2011). New management paradigms should foster functional diversity and redundancy within tree communities to actively turn boreal forests into ecosystems that are more resistant to known and unknown disturbances and have a higher capacity for adapting to novel environmental conditions.

Plant functional traits determine, on one level, the fitness of each individual via their effects on growth, reproduction, and survival and thus their influence on ecosystem functioning, including productivity, competition for resources, and nutrient balancing (Lavorel & Garnier, 2002). On another level, traits also determine the strategies and responses of the species to changing environmental conditions and determine how tree populations respond to different environmental factors and recover from disturbances (Lavorel & Garnier, 2002). When explicitly considering effect traits, the higher the functional diversity of an ecosystem, the higher its overall productivity (Tilman et al., 1997); but when accounting for response traits, functional diversity indicates the variety of forms possessed by a community to resist environmental changes, recover from disturbances, and adapt to novel environmental conditions (Elmqvist et al., 2003). Functional redundancy acts as insurance for the ecosystem because the loss of a redundant species, i.e., a species that performs similar roles within the ecosystem and responds similarly to environmental stressors and disturbances, will not compromise ecosystem functioning and resistance. Therefore, high-functional diversity and redundancy translate into a more resilient ecosystem (Mori et al., 2013).

Specifying the management guidelines for a particular forest region requires not only quantifying the functional diversity and redundancy of current tree communities and the entire landscape (Aquilué et al., 2020) and determining other suitable species that could establish and grow in the region either naturally or through assisted migration. Some community-level functional dissimilarity measures can be partitioned in a way to quantify species-level contributions to overall functional diversity (Pavoine & Ricotta, 2019). Another way to approach this task is to group species not using a taxonomic-based method but rather via a trait-based method by clustering species into functionally homogeneous groups (Fig. 31.5). Species of the same functional group share similar functioning at the individual level, similar responses to environmental variations, and similar effects on the ecosystem (Cornelissen et al., 2003). Once tree species are clustered into functional groups, it is possible to identify the surplus functions and functional groups, as well as those less represented, and later target species having a greater potential to maximize functional diversity at both the community and landscape scales (Aquilué et al., 2021).

31.3 How to Face the Challenges Confronting the Boreal Biome? Looking Toward the Future: Implications for Ecosystem-Based Management

The future holds many challenges and much uncertainty. In this section, we outline the main directions and perspectives required to confront these challenges and the implications for ecosystem-based forest management in the future boreal biome.

31.3.1 Alteration of Natural Disturbance Regimes

As climate change is expected to trigger important changes in future disturbance regimes, forest management will face notable economic and ecological challenges. Setting management objectives is becoming increasingly complex, as forests are expected to provide various ecosystem services other than timber, including carbon emissions mitigation, biodiversity, protection, and recreative roles (Ameray et al., 2021; Thom & Seidl, 2016). With the increased risk of boreal forests losing large amounts of trees to natural disturbances, the main concern for the forestry sector is the possibility of significant timber shortfalls in the near future (Boulanger et al.,

Species common characteristics	Representative species
Conifers, shade tolerant, drought and flood intolerant, wind dispersed e.g. Black spruce, Balsam fir, Hemlock, White pine	
Conifers, shade intolerant to mid-shade tolerant, drought tolerant, wind dispersed e.g. Jack pine, Red pine, Tamarack	
Deciduous, shade tolerant, drought and flood intolerant, wind dispersed e.g. Red maple, Sugar maple, White ash, Yellow birch	
Deciduous, shade intolerant, drought intolerant, wind dispersed e.g. Paper birch, Balsam poplar, Bigtooth aspen, Trembling aspen	
Deciduous, shade tolerant, mid-drought tolerant, flood intolerant, animal dispersed e.g. American beech, Black walnut, Black cherry	
Deciduous, mid-shade tolerant, drought tolerant, animal dispersed e.g. Red oak, white oak, Canadian hawthorn	

Tree community	# species	# Functional groups	Functional diversity	Functional redundancy
a	4	1	LOW	HIGH
b	3	2	LOW	LOW
c	4	3	MOD.	LOW
d	6	4	MOD.	MOD.
e	7	3	MOD.	HIGH
f	6	5	HIGH	LOW

◄**Fig. 31.5** Functional group classification of tree species commonly found in Canada's Boreal Shield and Mixedwood Plains ecozones. The eight functional traits used to classify the species are drought tolerance, shade tolerance, waterlogging tolerance, main seed dispersal vector, seed mass, wood density, leaf mass area, and taxonomic division. Of these eight traits, drought, shade, and waterlogging tolerance reflect a species susceptibility to environmental conditions (Niinemets & Valladares, 2006), whereas the other five relate to the capacity and mechanisms of a species to respond to natural disturbances. The bottom diagram presents the number of species, the number of functional groups, the functional diversity, and the functional redundancy of six typical tree communities. Vertical lines between the tree illustrations separate the functional groups. **a** This community consists of four species of the same functional group; it therefore has a very low functional diversity, although the functional redundancy is very strong because if a species disappears, the main functional traits remain in the stand. **b** This three-species community is relatively functionally poor, and the functional redundancy is very low because the loss of a species may compromise community functioning. **c** Functional diversity is moderate as up to three functional groups are present; however, as the loss of a species will certainly entail the loss of important traits, functional redundancy is very low. **d** This community is taxonomically and functionally richer than that of **c** and because half of the functional groups are represented by more than one species, redundancy is moderate. **e** In the most species-rich community, functional redundancy is very high because the three present functional groups are represented by more than one species. **d** This tree community is the most functionally diverse; however, six species are not sufficient to maintain all the diversity when a species is lost. Consequently, functional redundancy is very low

2019; Daniel et al., 2017; De Grandpré et al., 2018). Although large uncertainties remain, evaluating future risks at meaningful spatial and temporal scales is a crucial first step (Daniel et al., 2017; De Grandpré et al., 2018).

Ecosystem-based forest management aims to preserve natural forest attributes and processes by setting forestry strategies and targets on the basis of the variability of past disturbance regimes (Landres et al., 1999); for example, given the dominant role of wildfires in driving the dynamics of Canadian boreal forests, forest managers in Canada rely heavily on presettlement wildfire regimes to develop ecosystem-based management strategies (Thom & Seidl, 2016). In practice, this implies, for example, that the total annually burned or harvested area should not exceed that having burned under past fire regimes. Simultaneously integrating the risk of all-natural disturbances into ecosystem-based management strategies will be essential to ensure the applicability of this management approach under future conditions (Thom & Seidl, 2016).

Several management actions can be undertaken to handle the risks associated with future natural disturbance regimes in boreal forests. In areas facing high fire risks, governments should invest in fuel management to reduce the potential of fire occurrence and spread (De Groot et al., 2013b). Raising public awareness and producing prevention campaigns that explain the effects of forest fires can limit the occurrence of human-induced fires, likely to increase in frequency in drier conditions in the future. Other mitigation measures, such as favoring tree species that are less sensitive to insect outbreak, disease, drought, or fire, can be applied in these high-risk areas. Protecting foliage, reducing forest homogeneity, strategically removing fallen and weakened trees, thinning, debarking, and applying biological controls could all help reduce the risk of insect outbreak and disease (Ivantsova et al., 2019; Sturte-vant et al., 2015). The rescheduling of harvests must also be considered to maximize timber production during a budworm outbreak (Sturtevant et al., 2015). Changes in the timing and intensity of management actions can also heighten the resistance of boreal forests to snow- and wind-related damage. Specifically, this resistance could be achieved by changing forest landscape-level structures, such as the distance to stand edges, decreasing stand height differences, or shortening harvest rotations (Díaz-Yáñez et al., 2019; Zeng et al., 2007; Zubizarreta-Gerendiain et al., 2017).

31.3.2 Biodiversity Loss and Forest Attributes

Ecosystem-based management could mediate between preserving ecological processes, economic goals, and social values. We may need to increase uneven-aged and continuous-cover forest management approaches to maintain desired levels of late-successional trees in forest landscapes. Moreover, uneven-aged managed stands tend to be less prone to windthrow than even-aged stands, and landscapes with larger old-forest patches are less vulnerable to fire (Leduc et al., 2015; Nevalainen, 2017; Pukkala, 2016). Measures to rehabilitate important forest structures and improve habitat quality for biodiversity conservation involve ecological restoration through simulating natural disturbances, e.g., prescribed burning and gap cutting (Hägglund et al., 2015; Hekkala et al., 2014; Hjältén et al., 2017; Versluijs et al., 2017), and the use of longer rotation cycles to ensure sufficient and adequate habitats for forest specialists (Roberge et al., 2018).

From a biodiversity perspective, the top priority is to ensure a sufficient proportion of old-growth and uneven-aged forests in the landscape (Kuuluvainen & Gauthier, 2018). To this end, an obvious measure would be to reduce harvesting levels (Daniel et al., 2017; Kuuluvainen & Gauthier, 2018). Management practices aiming to conserve post-disturbance legacy structures and old-forest attributes should also be promoted (Boulanger et al., 2019; De Grandpré et al., 2018; Kuuluvainen & Gauthier, 2018). At the stand scale, this can be achieved in part by promoting partial cutting over clear-cuts (Bose et al., 2014). In addition, salvage logging, i.e., the harvesting of disturbed forests, is a relatively common and increasingly used post-disturbance management strategy (Leverkus et al., 2018; Sturtevant et al., 2015). The use of salvage logging should, however, be carefully prescribed depending upon management objectives; for instance, as windthrow creates large amounts of dead-wood, salvage logging should be avoided when priority is given to habitats that favor biodiversity (Nappi et al., 2011; Thorn et al., 2020). Nonetheless, salvage logging could offer a preferable treatment in recreational areas, for example, where large numbers of trees have been weakened by insects or disease and threaten to produce tree fall–related accidents (Ivantsova et al., 2019). All these measures could potentially generate economic losses; these losses could be compensated, to some degree, by valuing timber quality rather than quantity. Finally, a better understanding of fire–carbon feedbacks and deadwood dynamics resulting from insect outbreaks, disease, or weather-related disturbances would be necessary to better monitor carbon storage and release in boreal forests. This data would help develop improved, adapted management strategies that can limit carbon emissions.

31.3.3 Ecosystem-Based Management of Boreal Old-Growth Forests

Even-aged management based on short-rotation periods and clear-cutting has been, and will continue to be, the main cause of the loss of old-growth forest areas (Bergeron et al., 2017; Kuuluvainen & Gauthier, 2018; Martin et al., 2020c). The implementation of ecosystem-based management in boreal landscapes therefore requires a profound change in forestry practices. A combination of continuous-cover forestry, salvage logging, and clear-cutting with longer rotation periods, associated with the proactive mitigation of severe natural disturbances, has the potential to attain a balance between sustainable wood provision and environmental objectives in the context of climate change (Bergeron et al., 2006; Kuuluvainen, 2009; Leduc et al., 2015). Continuous-cover forestry can also be used to restore old-growth attributes in areas where these forests are currently absent, while limiting the anthropogenic impact on remnant old-growth forests (Eyvindson et al., 2021; Fenton et al., 2013; Montoro Girona et al., 2017). The remaining intact forests have an invaluable role as a natural reference from which to learn (Watson et al., 2018), and the conservation of the last large tracts of boreal old-growth forest must be a priority. In addition to the retention and continuous-cover forestry, natural disturbance–based management may provide a framework to increase the biodiversity, resiliency, and adaptive capacity of boreal forests (Kuuluvainen et al., 2021). Development of these management approaches based on scientific knowledge will heighten the flexibility to forest management, allowing it to maintain the structural diversity observed in old-growth forests and to better adapt to the new constraints caused by climate change (Fig. 31.4; Kuuluvainen et al., 2019). For example, in the event of conifer decline, it may be possible to carry out the assisted migration of shade-intolerant species following partial cutting.

Successful ecosystem-based management and the application of appropriate conservation and restoration activities needed to sustain old-growth forests require a detailed ecological understanding of the occurrence and dynamics of these key ecosystems. For example, the accurate identification and mapping of boreal old-growth forests, including their structural diversity, are necessary steps to define relevant protected old-growth areas that are representative of the preindustrial landscapes and to establish effective restoration targets. However, the efficacy of current survey methods, e.g., aerial photographic surveys, has been questioned (Martin et al., 2020b). Therefore, it is necessary to develop new tools, e.g., using LiDAR or UAV, to better identify the diversity and dynamics of old-growth forests in boreal landscapes. Furthermore, old-growth definitions can greatly vary between jurisdictions, including within the same country (see Chap. 7). This barrier severely limits the effectiveness of any strategy for old-growth conservation, for instance by reducing the coherence of measures taken by different actors and their degree of complexity. In the future, it will be necessary to continue the scientific effort initiated during the last decades to broaden our knowledge of old-growth boreal forests, a key for their preservation and sustainable management.

31.3.4 Tree Growth and Productivity

Effective management strategies require large-scale studies to better estimate key drivers of forest demographics under global change; these drivers include warmer temperatures, potential frost damage, and more severe water deficits. In turn, these drivers should be integrated into the next generation of forest growth models to simulate the interactive effects of climate change, insects, and disease on tree growth and mortality (Anderegg et al., 2015; Fatichi et al., 2019; McDowell et al., 2020). Until these data are available, risk reduction through portfolio diversification should drive management strategies for maintaining growth under global change. This strategy can include diversifying the composition of the boreal forest—a biome in which a few tree species dominate large tracts of landscapes—toward drought-adapted species and genotypes, thereby increasing the structural complexity of the forest (Messier et al., 2019). Another strategy is to increase the abundance of warm-adapted boreal broadleaf species, such as aspen and birch, to reduce fire risks while also increasing surface albedo (Astrup et al., 2018).

Maintaining stand productivity can also be ensured through stand-level interventions. Across a range of forest types in the United States, thinning has been shown to significantly increase resistance to greater water deficits (Bottero et al., 2017; D'Amato et al., 2013; Grant et al., 2013). In the boreal region of Canada, thinned balsam fir stands experiencing drought have shown increased resistance relative to natural stands, although important interactions with tree position and size, where larger or suppressed trees displayed lower resistance to these climate anomalies (Fig. 31.6). Revisiting long-term silvicultural studies, improving forest models, and moving away from business-as-usual forest management could yield the critical solutions for increasing growth resilience in the boreal forest in the context of global change.

31.3.5 Biotic Stressors

New challenges arising from shifting environmental conditions, species invasions, and human impacts will likely modify trophic interactions (Filbee-Dexter et al., 2017). Accounting for biotic stressors (in combination with abiotic ones) is essential when implementing an ecosystem-based forest management strategy, despite the

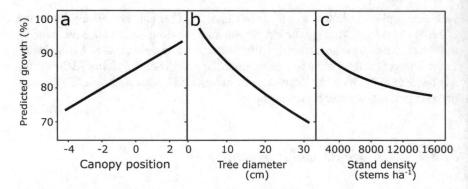

Fig. 31.6 Variation in the drought response of balsam fir tree growth (1999–2017) in terms of **a** tree canopy position (ratio of tree height to average stand height; positive values indicate dominant trees), **b** tree size (diameter at breast height), and **c** stand density. The model was fitted to the standardized annual growth of 247 balsam fir trees sampled across 22 stands within the Montmorency Forest (Québec, Canada) between 1999 and 2017. A value of 100% predicted growth corresponds to the standardized mean annual diameter growth

risk of complicating the achievement of management targets (De Vriendt et al., 2021). These various ecological drivers and processes operate at different and often specific scales. The linking of management actions relative to wildlife and forests at appropriate spatial and temporal scales, as illustrated by Beguin et al. (2016), is essential for achieving ecosystem-based management goals (Fig. 31.7). This will require transdisciplinary collaboration to develop a consensus regarding those factors driving ecosystem succession and the identified management challenges, objectives, targets, and indicators (Gunderson, 2015). At the same time, we must develop flexible and adaptable processes able to deal with ecological surprises (Filbee-Dexter et al., 2017), alternative successional pathways (Hidding et al., 2013), and regime shifts (Folke et al., 2004; Gauthier et al., 2015b).

31.3.6 Forest Plantations as Silvicultural Tools for Adapting to New Conditions

Forest plantations, i.e., cultivated forest ecosystems established by planting, seeding, or both in the process of afforestation and reforestation, are often put forward in many parts of the world as a sustainable silvicultural tool for reconciling wood production and preserving the original natural forest (Paquette & Messier, 2010). Whereas forest plantations comprise less than 5% of forested lands, these areas provide 15% of the world's timber production (Carnus et al., 2006). Nonetheless, biodiversity–ecosystem functioning must be considered when implementing plantations to ensure diverse forests, rather than large-scale monocultures, to favor a greater resistance to insect

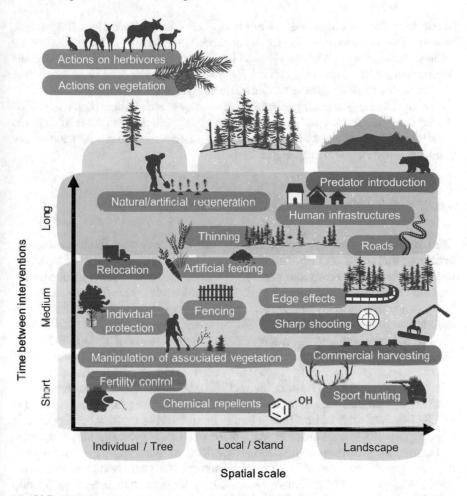

Fig. 31.7 Wildlife and forest management actions at multiple temporal and spatial scales. Modified from Beguin et al. (2016), CC BY 3.0 license

pests and diseases (Carnus et al., 2006; Paquette & Messier, 2010). Including conservation and various objectives into tree plantation planning will be a major future challenge for diversifying plantations. In addition to reducing pressure on natural forests, the use of forest plantations can restore some lost ecological forest services (Sedjo & Botkin, 1997), such as carbon sequestration (Ameray et al., 2021). Preventive planning is needed to deal with both the beneficial and detrimental effects of climate change. Managing plantations in an ecologically sustainable manner and planning for predicted future climate conditions will promote the maintenance of biodiversity at all scales (stand, landscape) and for all components (genes, species, communities). For example, by controlling invasive species, forest plantations can reduce non-natural competition and thus protect natural habitats, refuge networks, and tree populations that are either isolated or at margins of their distributions (Paquette & Messier, 2010). Moreover, we can accommodate the new climatic conditions by reforesting with local or locally adapted seed sources suitable for future conditions (Puettmann, 2011). However, natural disturbances regimes must to be taken into consideration, because if the fire regimes are shorter in the future, plantations could have a low chance of reaching commercial maturity in areas where burn rate will be shorter than the harvest age. The main challenge of forest plantations is to maintain the desired levels of stands in an ecologically sustainable manner by anticipatory planning and scientific management.

31.3.7   Silviculture in a Changing Context

Climate change heightens the variability and uncertainty of future ecological, economic, and social contexts. Maintaining and restoring biodiversity and the structural complexity of forests can promote forest resilience to human-induced pressures and impacts from climate change (Thompson et al., 2009). The structural and functional attributes of boreal forests are being altered quickly and significantly in response to a changing climate. FEM, however, promotes the diversification of silvicultural practices and the maintenance of forest stands of various structures. Given that these structures are in constant evolution in a changing context, FEM offers a means of prioritizing ecological processes that ensure the functioning of forest ecosystems. In this context, we must also consider factors other than timber production, including recreation, wildlife, and biodiversity conservation. The range of EBM-derived silvicultural practices can be a valuable tool for helping forests respond to a changing climate because these treatments generally aim to manage and sustain the growth, structure, and composition of forest vegetation for multiple objectives, including wildlife habitat, timber production, water resources, and recreation (Millar et al., 2007; Rist & Moen, 2013). There is thus a need to develop new silvicultural practices to increase forest resilience and promote the conservation of the ecosystem goods (wood and nonwood products) and services provided by forests, such as climate regulation, carbon sequestration, soil stabilization, nutrient cycling, and recreational values (Ameray et al., 2021; Hassan et al., 2005). To ensure that silviculture becomes a useful tool in rapidly changing realities, this discipline must build a new framework with a focus in three axes (Achim et al., 2021): (1) observational (monitor forest, detect changes, update data, integrate knowledge); (2) anticipative (integrate climate projections, predict future scenarios); and (3) adaptive (flexible implementation, risk acceptance, consideration of social acceptance). This new framework could be helpful to adapt silviculture to respond to new trends, needs, and preoccupations at the ecological, social, and economic scales, including a holistic conception of forest ecosystems under climate change.

31.3.8 Accounting for Freshwaters in the Sustainable Management of Forests

Forestry practices result in ecological consequences not only for the forest from which the trees are harvested but also for freshwaters within the harvested watershed. Including freshwater systems into the sustainable management of boreal forests is therefore vital in future management decisions. Maintaining healthy boreal freshwater systems benefits terrestrial food webs, which depend on the high nutritional quality feedback fluxes from the aquatic to land environments. Aquatic environments are key contributors to the ecological functioning of terrestrial food webs and are critical to many key boreal species; for example, moose and beaver feed largely on aquatic macrophytes (Bergman et al., 2018; Feldman et al., 2020; Fraser et al., 1984; Labrecque-Foy et al., 2020). Many terrestrial bird species also depend on the emergence of aquatic insects (Murakami & Nakano, 2002). Ensuring a healthy aquatic environment positively affects the health of terrestrial environments and, thus, of the entire boreal biome.

Currently, riparian buffer strips represent the best means of protecting aquatic habitats from terrestrial human disturbances such as forest harvesting. These buffer strips are applied universally to all forest types and ecosystems; however, they do not respond to all specific protection needs. Riparian buffer strips must be adapted to the physical and biological landscape. For example, strip width and composition should be adapted to the slope, soil type, and forest stand to reduce windthrow and high tree mortality. It is now crucial to understand the impact of forest disturbances on aquatic environments, as changing land use and climate are expected to increase the terrestrial influence on freshwaters, e.g., browning. Determining the resiliency of aquatic habitats and food webs will be especially important to define the maximum supported human pressure on boreal forests. Accounting for freshwaters within the forestry plans will therefore be essential for attaining the sustainable management of the boreal forest and ensuring the health of the entire boreal biome.

31.3.9 Green Infrastructure

Green infrastructure is a strategically planned management of natural and anthropic lands that aims to secure biodiversity, habitat resilience, and ecosystem services at multiple spatial scales (EC, 2013; Liquete et al., 2015; Wang & Banzhaf, 2018). Thus, green infrastructure promotes landscape-scale and holistic planning approaches that are based on remaining biodiversity and ecosystem-service hot spots and their functional connectivity within a landscape matrix subjected to ongoing climate and land-use changes. In forest landscapes having a legacy of extractive forestry, applying green infrastructure implies that forest patches are restored, matrix quality is improved, and connectedness is strengthened (Dondina et al., 2017; Heller & Zavaleta, 2009). In addition to preserving existing intact forest landscapes and primary forest as connectivity nodes, networks can be re-created within degraded or extensively transformed forest landscapes to secure functional habitats for species to move and spread. Forest edges, for example, are important transitional biotopes but also hold specific conservation values (Harper et al., 2015) and provide a functional green infrastructure in natural as well as need-to-be-restored anthropic forest landscapes in which clear-cut edges dominate (Esseen et al., 2016). Approaches are required that range from local species occurrence and microsites to habitats, landscapes, and entire regions to attain functional connectivity, i.e., a connectivity that supports the representative traits of species composition, habitat structures, and ecological processes (Heller & Zavaleta, 2009) and that goes beyond the protection of remaining biodiversity key habitats to forest landscape restoration and prestoration (Mansourian, 2018). Nonetheless, pan-national policies and policy implementation instruments and routines are needed to ensure that fragmentation (as a consequence) and connectivity (as a necessity) are accounted for in boreal sustainable forest management and governance.

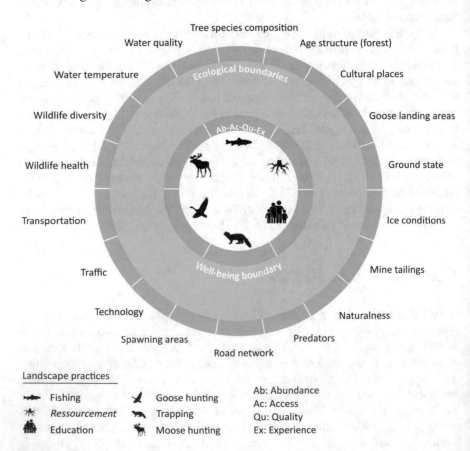

Fig. 31.8 Boreal landscape sustainability according to Raworth's (2017) doughnut economics and land-use experts from the Abitibiwinni and Ouje-Bougoumou First Nations

31.3.10 A Sustainability Framework Emerging from Collaborative Research

An example of collaborative research is the partnership (2015–2021) between university researchers and the Abitibiwinni (Anishnaabeg) and Ouje-Bougoumou (Cree) First Nations, all located in boreal Québec, Canada, that evaluated the effects of environmental change on Indigenous landscapes. The main results can be synthesized in a sustainability framework inspired by Kate Raworth's doughnut-shaped sustainability economics diagram (Raworth, 2017) (Fig. 31.8). The inner boundary of the doughnut represents the limits of human well-being (Nussbaum, 2011; Sen, 1980) and is defined as the capacity of the practices that are important for an Indigenous community, e.g., moose hunting, trapping, and education. The well-being boundary can be delineated with indicators of the resources abundance of, access to the land, quality of resources, and lived experience while on the land (Bélisle et al., 2021). The outer boundary represents the ecological limits of boreal landscapes (O'Neill et al., 2018; Rockström et al., 2009a). It is delineated by a set of influential factors identified by Indigenous land-use experts, e.g., forest road density, forest composition, and water quality (Bélisle & Asselin, 2021). In between these two zones, the sustainability zone represents the zone to be targeted and monitored for ecosystem management purposes.

31.3.11 The Functional Network Approach

From an ecosystem-based management perspective, forest resilience to global threats and uncertainty can be achieved through the functional diversification of tree communities. This diversification requires, critically, the identification of both functionally redundant and functionally rare species to prevent their likely decrease in abundance and promote their regeneration either naturally or by planting (Aquilué et al., 2021). Network theory can then be applied to target locations to optimize silvicultural interventions and attain greater levels of landscape resilience.

In the functional network approach, a forest landscape is represented as a network of forest stands. Tree species dispersal capacity dictates whether forest stands are connected; thus, the relative dispersal capacity within and between adjacent stands influences the dispersion of the associated functional traits for each tree species. Three network/landscape-level properties contribute to forest ecosystem resilience: functional connectivity, modularity, and node-level centrality (Fig. 31.9; Aquilué et al., 2020; Gonzalès & Parrott, 2012). A higher functional connectivity between forest stands facilitates the exchange of functionally distinct species and genes better adapted to novel environmental conditions and ensures a rapid tree recolonization of disturbed stands by seeds coming from the surrounding intact stands; these conditions contribute to an efficient reorganization of the ecosystem. Modular systems are organized in clusters of highly interconnected nodes that are loosely connected to nodes of other clusters. Modularity acts as an effective defense against the spread of pathogen outbreaks because modules buffer the spread of a perturbation and minimize the risk of ecosystem collapse (Gilarranz et al., 2017). Therefore, hypothetical modular forest landscapes will be functionally structured in clusters and less vulnerable to rapid, synchronized insect outbreaks. Finally, node centrality accounts for the different roles an element plays in the flow of energy, nutrients, organisms, and genes across the landscape (Bodin & Saura, 2010). Generally, central nodes concentrate most connections and/or bridge two subsets of nodes that would otherwise be disconnected. In a forest landscape, central stands (regardless of size) act as both source and sink for a large proportion of species and functional traits within the landscape.

Practitioners can evaluate current management plans and silvicultural practices at the landscape scale through the analyses of these five indicators—functional diversity, functional redundancy, functional connectivity, modularity, and centrality (Fig. 31.9). They can also assess novel, untested alternative management regimes, in particular where network analysis is combined with a modeling framework to evaluate future projections and determine the compounding impacts of natural disturbances, climate change, and silvicultural interventions on forest ecosystems (Hof et al., 2021; Mina et al., 2021). If tree planting is envisaged, it is advisable to consider species vulnerability to these natural disturbances that regularly or will likely impact a region in the near future (Mina et al., 2021). This extra challenge makes it even more important to merge environmental model predictions with expert knowledge. In short, foresters, climatologists, entomologists, modelers, economists, and other relevant experts must collaborate within a transdisciplinary framework to design ecosystem-based management plans that foster forest resilience to global change.

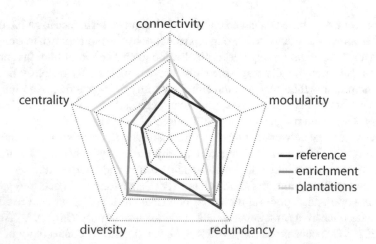

Fig. 31.9 Characterization of a fragmented forest landscape relying on two functional indicators (*diversity, redundancy*) and three properties of the spatial network (*connectivity, centrality, modularity*). The *reference* is a forested landscape in an agroforest mosaic in which forest stands are separated by agricultural fields, roads, and small villages. The *enrichment* strategy promotes natural regeneration of functionally rare species within the existing forest stands, and the *plantations* strategy aims to convert bare or agricultural land into woodlands using multispecies plantations to enhance the functional diversification of the landscape

31.3.12   Restoration Under Climate Change

Restoration theory is generally based on the emulation of natural disturbances (Kuuluvainen, 2002). However, climate change will most likely impact disturbance regimes, such as by modifying the frequency and severity of forest fires, windthrow, floods, and insect infestations. Thus, future restoration efforts also must confront the challenge of climate change; we require a solid understanding of how global change may affect forest dynamics if we wish to safeguard ecosystem services and the biodiversity hosted by forests (Aerts & Honnay, 2011; Harris et al., 2006). The question is whether we should adapt ecological restoration to fit this new natural disturbance regime. One obstacle in this respect is that climate change effects are likely to be abrupt and unpredictable, where sudden and significant changes are inevitable in the next 20 years (Harris et al., 2006). As climate changes, the patterns of natural disturbances are also shifting, increasing uncertainty in terms of how the dynamics of forest ecosystems will evolve in the future (Johnstone et al., 2016). Paleoecological studies have demonstrated that the boreal forest has a substantial resilience to natural disturbances (Aakala et al., 2018; Carcaillet et al., 2001; Navarro et al., 2018a). However, alterations in disturbance regimes can also increase the vulnerability of forest ecosystems to degradation (Seidl et al., 2016). For example, a hurricane damaged 75 million m^3 of Swedish spruce forests in 2005, resulting in an economic loss of $2 billion US (Valinger & Fridman, 2011). Consequently, ecological restoration must also consider the effects of the altered natural disturbances on ecosystem services. Disturbances operate at several and nested spatiotemporal scales, causing the heterogeneity of stands (age structure) and landscape patches (Kuuluvainen & Gauthier, 2018). Thus, to effectively restore forest ecosystems subjected to climate change, we must understand the factors involved in determining when, how, and where fire, insect outbreaks, and windthrow regimes have altered and will alter forest dynamics to restore for future scenarios rather than previous reference states.

31.3.13   Governance Policies

Governance policies have been essential for the implementation of FEM and in attempts to minimize future impacts of climate change. Moving forward, international and national legislative frameworks must better integrate climate change scenarios to enhance the protection of habitat types, species, and carbon sequestration. Local communities are also vulnerable to climate change through, for example, impacts on livelihoods and forest-based activities. These must also be considered in land-use planning at different scales. To the best of our knowledge, some topics have yet to be addressed within any country, such as ecological restoration within climate change strategies or functional landscape planning strategies. Moreover, some direct or partial conflicts remain between climate change adaptation and mitigation strategies and forest biodiversity goals, e.g., keeping deadwood versus removing logging residues. This is reflected in differences in policy direction among boreal countries. For example, the European Union considers climate change adaptation, natural resource sustainability, and restoration as central priorities, thereby becoming foci of environmental policy. In Canada, general plans and actions around climate change adaptations are occurring at multiple scales, including specific actions in the area of forest management adaptation, however there is little evidence of coordination or an overarching framework. In the case of restoration policies in Canada, the main focus is on specific species, rather than on risks to ecosystems, i.e., static conservation goals (Harris et al., 2006; Parks Canada, 2008). In the Russian Federation, ecological restoration continues to be missing from governance and climate policies, and commitments vary from minimal efforts to no action and are not at all consistent with the Paris Agreement. Given this situation, we suggest creating an active strategy for boreal countries to coordinate their sustainable forest management actions and ecological restoration policies under the umbrella of climate change.

31.4 Conclusions

Ongoing intensive forestry practices have significantly altered boreal forest ecosystems and will continue to do so, illustrated in part by the increased landscape fragmentation and homogenization of forest stands and a reduced species diversity. FEM is a promising approach for maintaining all ecosystem services, habitats, and functions provided by forests. However, in addition to intensive forestry, climate change is altering forest ecosystems around the world. Boreal forests face multiple challenges for which unpredictability and uncertainty will be the new baseline conditions, further complicating sustainable forest management. Thus, existing paradigms and the implementation of FEM must be sufficiently flexible and adaptable given that we will experience novel, previously unknown or unobserved scenarios. Solutions for safeguarding ecosystem services and biodiversity require a solid understanding of how global change will affect forest dynamics and processes (structure and function) (Table 31.1).

Table 31.1 Challenges, directions, and solutions facing boreal forests

Challenges	How to confront the challenges facing the boreal biome. Emerging directions in forest management	References
Natural disturbance regime change	• Evaluate future risks and integrate risk factors into FEM planning • Increase fuel management efforts limiting human fire right • Raise public awareness and produce awareness campaigns to limit the occurrence of human-induced fires • Favor tree species that are less sensitive to natural disturbances • Protect foliage, reduce forest homogeneity, strategically remove fallen and weakened trees, apply thinning, debarking, and biological controls, as well as preventive silviculture • Consider the impact of natural disturbances on rotation and final product and reschedule harvests after disturbances • Improve our understanding of disturbance interactions at multiple temporal and spatial scales	Daniel et al. (2017), De Grandpré et al. (2018), De Groot et al. (2013b), Díaz-Yáñez et al. (2019), Hof et al. (2021), Ivantsova et al. (2019), Montoro Girona et al. (2018b), Navarro et al. (2018b, c), Sturtevant et al. (2015), Zeng et al. (2007), Zubizarreta-Gerendiain et al. (2017)
Biodiversity loss	• Apply forest management approaches that favor increased uneven-aged and continuous-cover stands • Rehabilitate important forest structures and improve habitat quality, e.g., prescribed burning and gap cutting • Use longer rotation cycles to ensure sufficient and adequate habitats for forest specialists • Reduce harvesting levels and carefully prescribe salvage logging • Promote management practices aimed at conserving post-disturbance legacy structures and old-forest attributes • Promote partial cutting over clear-cutting • Value timber quality rather than quantity	Hägglund et al. (2015), Hekkala et al. (2014), Hjältén et al. (2017), Versluijs et al. (2017), Bosc et al. (2014), Daniel et al. (2017), Kuuluvainen and Gauthier (2018), Montoro Girona et al. (2016), Roberge et al. (2018)
Loss of old-growth forest	• Limit human impacts and increase the conservation areas of remaining old-growth forest • Use continuous-cover forestry to restore old-growth attributes • Harmonize old-growth forest definitions and conservation strategies between jurisdictions • Accurately identify old-growth forests and their structural diversity in boreal landscapes to ensure the sustainability of management and conservation strategies • Develop new tools, e.g., LiDAR or UAV, to better identify old-growth stand diversity and dynamics • Increase ecological knowledge of boreal old-growth forests	Bergeron et al. (2006), Fenton et al. (2009, 2013), Kuuluvainen (2009), Kuuluvainen et al. (2021), Leduc et al. (2015), Martin et al. (2018, 2021a), Montoro Girona et al. (2017)

(continued)

Table 31.1 (continued)

Challenges	How to confront the challenges facing the boreal biome. Emerging directions in forest management	References
Biotic stress factors as underlying drivers of ecological change	• Account for biotic stresses in combination with abiotic ones • Link management actions related to wildlife and forests at appropriate spatial and temporal scales • Implement transdisciplinary collaborations to develop a consensus regarding factors driving ecosystem succession and identify management challenges, objectives, targets, and indicators • Develop flexible and adaptable processes able to deal with ecological surprises, alternative successional pathways, regime shifts	Beguin et al. (2016), De Vriendt et al. (2021), Filbee-Dexter et al. (2017), Folke et al. (2004), Gauthier et al. (2015b), Gunderson (2015), Hidding et al. (2013)
Reduced boreal tree growth	• Require large-scale studies to better estimate key drivers of forest demographics under warmer temperatures and integrate these into forest management • Reduce risks through portfolio diversification • Diversify the composition of the boreal forest • Increase the abundance of warm-adapted boreal broadleaf taxa, such as aspen and birch, to reduce fire risks while also increasing surface albedo • Continuous and campaign-based observation networks of boreal forest function with high-precision monitoring • Partitioned estimates of boreal forest carbon sources and sinks • Undertake thinning operations to increase resistance to heightened water deficits	Anderegg et al. (2015), Astrup et al. (2018), Bottero et al. (2017), D'Amato et al. (2013), Grant et al. (2013), Messier et al. (2019), Pappas et al. (2020, 2022), D'Orangeville et al. (2016, 2018), Marchand et al. (2019, 2021)
New silvicultural practices	• Develop new silvicultural practices to increase forest resilience to ensure the ecosystem goods and services • Implement FEM paradigms to promote the diversification of silvicultural practices • Consider factors other than timber production, including recreation, wildlife, and biodiversity conservation • Adopt nature-based silvicultural solutions • Develop the potential of partial harvesting and plantations for carbon sequestration • Revisit long-term silvicultural studies and improve forest models • Understand the interactions between natural and anthropic disturbances • Create a new silvicultural framework based on the three foci: observe (monitor, detect change, update data), anticipate (to integrate climate projections and to predict future scenarios) and adapt (accept risk, social acceptance)	Achim et al. (2021), Thiffault and Pinno (2021), Gustafsson et al. (2020), Kim et al. (2021), Berglund and Kuuluvainen (2021), Montoro Girona et al. (2018a, 2019), Paradis et al. (2019), O'Hara, (2016), Pukkala, (2016), D'Amato et al. (2011), Rist and Moen, (2013), Nagel et al. (2017), Senez-Gagnon et al. (2018)

(continued)

Table 31.1 (continued)

Challenges	How to confront the challenges facing the boreal biome. Emerging directions in forest management	References
Including freshwater systems within forest management	• Adapt riparian buffer strips to the physical and biological landscape: buffer width and composition to the slope, soil type, and forest stand to reduce windthrow • Understand the impact of forest disturbances on aquatic environments • Determine the resilience of aquatic habitats and food webs to forest disturbances • Account for freshwater ecosystems as a new paradigm within the forestry framework	Carignan and Steedman (2000), Glaz et al. (2014, 2015), Klaus et al. (2018), Kritzberg et al. (2020), Pinel-Alloul et al. (2002), Ruel et al. (2001), Wang and Banzhaf (2018)
Connectivity and fragmentation	• Maintain existing intact forest landscapes and primary forest as connectivity nodes • Account for fragmentation as a consequence and connectivity as a necessity in sustainable forest management	Mikusiński et al. (2021), Svensson et al. (2019, 2020)
Collaborative research and Indigenous peoples	• Provide a framework to increase the biodiversity, resilience, and adaptive capacity in boreal forests • Establish and maintain long-term collaborations between scientific research and Indigenous institutions • Develop and monitor sustainability indicators in collaboration with Indigenous institutions • Consider Indigenous values and perspectives in the definition of ecosystem management and sustainability	Bélisle et al. (2021), Bélisle and Asselin (2021), Saint-Arnaud et al. (2009)
Make forest landscapes resilient: the functional network approach	• Apply the functional network approach to improve adaptations to new environmental conditions • Agree on a list of key functional traits and obtain trait values from the field or from freely available trait databases • Adopt an existing functional group classification on the basis of relevant traits • Consider species vulnerability to natural disturbances, e.g., wildfires, ice storms, drought, and browsing • Consider forests as complex systems	Aquilué et al. (2021), Messier et al. (2013), Mina et al. (2021)

Altered climatic conditions will affect the distribution of forest species and impact disturbances, modify tree growth, and favor the arrival of invasive species; these consequences with therefore have significant economic and ecological implications for human societies. It is thus critical to (1) establish long-term monitoring networks documenting forest functioning across different boreal forest landscapes and environmental conditions; (2) rapidly develop reliable indicators of climate change to identify shifts within forest ecosystems; (3) develop new decision-support tools that can predict future scenarios in forest ecosystems; (4) promote international collaboration and cooperation among boreal countries; and (5) build governance policies (at international, national, and regional levels) that provide a legislative framework to ensure the application of sustainable forest management in the context of the greatest challenge currently facing humanity: climate change.

References

Aakala, T., Pasanen, L., Helama, S., et al. (2018). Multiscale variation in drought controlled historical forest fire activity in the boreal forests of eastern Fennoscandia. *Ecological Monographs, 88*(1), 74–91. https://doi.org/10.1002/ecm.1276.

Achim, A., Moreau, G., Coops, N. C., et al. (2021). The changing culture of silviculture. *Forestry, 95*(2), 143–152. https://doi.org/10.1093/forestry/cpab047.

Aerts, R., & Honnay, O. (2011). Forest restoration, biodiversity and ecosystem functioning. *BMC Ecology, 11*(1), 29. https://doi.org/10.1186/1472-6785-11-29.

Aksenov, D., Karpachevskiy, M., Lloyd, S., et al. (1999). *The last of the last: the old-growth forests of boreal Europe*. Helsinki: Taiga Rescue Network.

Allen, C. D., Macalady, A. K., Chenchouni, H., et al. (2010). A global overview of drought and heat-induced tree mortality reveals emerging climate change risks for forests. *Forest Ecology and Management, 259*(4), 660–684. https://doi.org/10.1016/j.foreco.2009.09.001.

Ameray, A., Bergeron, Y., Valeria, O., et al. (2021). Forest carbon management: A review of silvicultural practices and management strategies across boreal, temperate and tropical forests. *Current Forestry Reports, 7*(4), 245–266. https://doi.org/10.1007/s40725-021-00151-w.

Anderegg, W. R. L., Hicke, J. A., Fisher, R. A., et al. (2015). Tree mortality from drought, insects, and their interactions in a changing climate. *New Phytologist, 208*(3), 674–683. https://doi.org/10.1111/nph.13477.

Angelstam, P., Manton, M., Green, M., et al. (2020). Sweden does not meet agreed national and international forest biodiversity targets: A call for adaptive landscape planning. *Landscape and Urban Planning, 202*, 103838. https://doi.org/10.1016/j.landurbplan.2020.103838.

Aquilué, N., Filotas, É., Craven, D., et al. (2020). Evaluating forest resilience to global threats using functional response traits and network properties. *Ecological Applications, 30*(5), e02095. https://doi.org/10.1002/eap.2095.

Aquilué, N., Messier, C., Martins, K. T., et al. (2021). A simple-to-use management approach to boost adaptive capacity of forests to global uncertainty. *Forest Ecology and Management, 481*, 118692. https://doi.org/10.1016/j.foreco.2020.118692.

Artdatabanken. (2020). *The Swedish redlist. Rodlistade arter i Sverige*. ArtDatabanken SLU, Uppsala.

Asselin, H. (2015). Indigenous forest knowledge. In K. H.-S., Peh, R. Corlett, & Y. Bergeron (Eds.), *Routledge handbook of forest ecology* (pp. 586–596). London: Earthscan, Routledge.

Astrup, R., Bernier, P. Y., Genet, H., et al. (2018). A sensible climate solution for the boreal forest. *Nature Climate Change, 8*, 11–12. https://doi.org/10.1038/s41558-017-0043-3.

Aubin, I., Boisvert-Marsh, L., Kebli, H., et al. (2018). Tree vulnerability to climate change: Improving exposure-based assessments using traits as indicators of sensitivity. *Ecosphere, 9*(2), e02108. https://doi.org/10.1002/ecs2.2108.

Barber, V. A., Juday, G. P., & Finney, B. P. (2000). Reduced growth of Alaskan white spruce in the twentieth century from temperature-induced drought stress. *Nature, 405*(6787), 668–673. https://doi.org/10.1038/35015049.

Barnosky, A. D., Hadly, E. A., Bascompte, J., et al. (2012). Approaching a state shift in Earth's biosphere. *Nature, 486*(7401), 52–58. https://doi.org/10.1038/nature11018.

Barrette, M., Bélanger, L., De Grandpré, L., et al. (2014). Cumulative effects of chronic deer browsing and clear-cutting on regeneration processes in second-growth white spruce stands. *Forest Ecology and Management, 329*, 69–78. https://doi.org/10.1016/j.foreco.2014.06.020.

Barrette, M., Bélanger, L., De Grandpré, L., et al. (2017). Demographic disequilibrium caused by canopy gap expansion and recruitment failure triggers forest cover loss. *Forest Ecology and Management, 401*, 117–124. https://doi.org/10.1016/j.foreco.2017.07.012.

Bartlett, C., Marshall, M., & Marshall, A. (2012). Two-Eyed Seeing and other lessons learned within a co-learning journey of bringing together indigenous and mainstream knowledges and ways of knowing. *Journal of Environmental Studies and Sciences, 2*(4), 331–340. https://doi.org/10.1007/s13412-012-0086-8.

Beck, P. S. A., Juday, G. P., Alix, C., et al. (2011). Changes in forest productivity across Alaska consistent with biome shift. *Ecology Letters, 14*(4), 373–379. https://doi.org/10.1111/j.1461-0248.2011.01598.x.

Beguin, J., Tremblay, J. P., Thiffault, N., et al. (2016). Management of forest regeneration in boreal and temperate deer-forest systems: Challenges, guidelines, and research gaps. *Ecosphere, 7*(10), e01488. https://doi.org/10.1002/ecs2.1488.

Bélisle, A. C., & Asselin, H. (2021). A collaborative typology of boreal Indigenous landscapes. *Canadian Journal of Forest Research, 51*(9), 1253–1262. https://doi.org/10.1139/cjfr-2020-0369.

Bélisle, A. C., Wapachee, A., & Asselin, H. (2021). From landscape practices to ecosystem services: Landscape valuation in Indigenous contexts. *Ecological Economics, 179*, 106858. https://doi.org/10.1016/j.ecolecon.2020.106858.

Bellard, C., Thuiller, W., Leroy, B., et al. (2013). Will climate change promote future invasions? *Global Change Biology, 19*(12), 3740–3748. https://doi.org/10.1111/gcb.12344.

Bengtsson, J., Nilsson, S. G., Franc, A., et al. (2000). Biodiversity, disturbances, ecosystem function and management of European forests. *Forest Ecology and Management, 132*(1), 39–50. https://doi.org/10.1016/S0378-1127(00)00378-9.

Benoît, P. O., Beisner, B. E., & Solomon, C. T. (2016). Growth rate and abundance of common fishes is negatively related to dissolved organic carbon concentration in lakes. *Canadian Journal of Fisheries and Aquatic Sciences, 73*(8), 1230–1236. https://doi.org/10.1139/cjfas-2015-0340.

Benoy, G., Cash, K., McCauley, E., et al. (2007). Carbon dynamics in lakes of the boreal forest under a changing climate. *Environmental Reviews, 15*, 175–189. https://doi.org/10.1139/a07-006.

Bergeron, Y., & Harper, K. A. (2009). Old-growth forests in the Canadian boreal: The exception rather than the rule? In C. Wirth, G. Gleixner, & M. Heimann (Eds.), *Old-growth forests. Ecological studies (analysis and synthesis)*. Berlin, Heidelberg: Springer.

Bergeron, Y., & Leduc, A. (1998). Relationships between change in fire frequency and mortality due to spruce budworm outbreak in the southeastern Canadian boreal forest. *Journal of Vegetation Science, 9*(4), 492–500. https://doi.org/10.2307/3237264.

Bergeron, Y., Cyr, D., Drever, C. R., et al. (2006). Past, current, and future fire frequencies in Quebec's commercial forests: Implications for the cumulative effects of harvesting and fire on age-class structure and natural disturbance-based management. *Canadian Journal of Forest Research, 36*(11), 2737–2744. https://doi.org/10.1139/x06-177.

Bergeron, Y., Vijayakumar, D. B. I. P., Ouzennou, H., et al. (2017). Projections of future forest age class structure under the influence of fire and harvesting: Implications for forest management in the boreal forest of eastern Canada. *Forestry, 90*(4), 485–495. https://doi.org/10.1093/forestry/cpx022.

Berglund, H., & Kuuluvainen, T. (2021). Representative boreal forest habitats in northern Europe, and a revised model for ecosystem management and biodiversity conservation. *Ambio, 50*, 1003–1017. https://doi.org/10.1007/s13280-020-01444-3.

Bergman, B. G., Bump, J. K., & Romanski, M. C. (2018). Revisiting the role of aquatic plants in beaver habitat selection. *American Midland Naturalist, 179*(2), 222–246. https://doi.org/10.1674/0003-0031-179.2.222.

Bescond, H., Fenton, N. J., & Bergeron, Y. (2011). Partial harvests in the boreal forest: Response of the understory vegetation five years after harvest. *The Forestry Chronicle, 87*(1), 86–98. https://doi.org/10.5558/tfc87086-1.

Blackstock, K. L., Kelly, G. J., & Horsey, B. L. (2007). Developing and applying a framework to evaluate participatory research for sustainability. *Ecological Economics, 60*(4), 726–742. https://doi.org/10.1016/j.ecolecon.2006.05.014.

Bodin, Ö., & Saura, S. (2010). Ranking individual habitat patches as connectivity providers: Integrating network analysis and patch removal experiments. *Ecological Modelling, 221*(19), 2393–2405. https://doi.org/10.1016/j.ecolmodel.2010.06.017.

Bohn, F. J., & Huth, A. (2017). The importance of forest structure to biodiversity-productivity relationships. *Royal Society Open Science, 4*(1), 160521. https://doi.org/10.1098/rsos.160521.

Borowik, T., Ratkiewicz, M., Maślanko, W., et al. (2020). Too hot to handle: Summer space use shift in a cold-adapted ungulate at the edge of its range. *Landscape Ecology, 35*(6), 1341–1351. https://doi.org/10.1007/s10980-020-01018-4.

Bose, A. K., Harvey, B. D., Brais, S., et al. (2014). Constraints to partial cutting in the boreal forest of Canada in the context of natural disturbance-based management: A review. *Forestry, 87*(1), 11–28. https://doi.org/10.1093/forestry/cpt047.

Boserup, E. (1981). *Population and technological change: A study of long-term trends.* University of Chicago Press.

Bottero, A., D'Amato, A. W., Palik, B. J., et al. (2017). Density-dependent vulnerability of forest ecosystems to drought. *Journal of Applied Ecology, 54*(6), 1605–1614. https://doi.org/10.1111/1365-2664.12847.

Bouchard, M., Aquilué, N., Périé, C., et al. (2019). Tree species persistence under warming conditions: A key driver of forest response to climate change. *Forest Ecology and Management, 442*, 96–104. https://doi.org/10.1016/j.foreco.2019.03.040.

Boucher, D., De Grandpré, L., Kneeshaw, D. D., et al. (2015). Effects of 80 years of forest management on landscape structure and pattern in the eastern Canadian boreal forest. *Landscape Ecology, 30*, 1913–1929. https://doi.org/10.1007/s10980-015-0220-6.

Boucher, D., Boulanger, Y., Aubin, I., et al. (2018). Current and projected cumulative impacts of fire, drought, and insects on timber volumes across Canada. *Ecological Applications, 28*(5), 1245–1259. https://doi.org/10.1002/eap.1724.

Boucher, Y., & Grondin, P. (2012). Impact of logging and natural stand-replacing disturbances on high-elevation boreal landscape dynamics (1950–2005) in eastern Canada. *Forest Ecology and Management, 263*, 229–239. https://doi.org/10.1016/j.foreco.2011.09.012.

Boucher, Y., Arseneault, D., Sirois, L., et al. (2009). Logging pattern and landscape changes over the last century at the boreal and deciduous forest transition in Eastern Canada. *Landscape Ecology, 24*(2), 171–184. https://doi.org/10.1007/s10980-008-9294-8.

Boucher, Y., Perrault-Hébert, M., Fournier, R., et al. (2017). Cumulative patterns of logging and fire (1940–2009): Consequences on the structure of the eastern Canadian boreal forest. *Landscape Ecology, 32*(2), 361–375. https://doi.org/10.1007/s10980-016-0448-9.

Boulanger, Y., Taylor, A. R., Price, D. T., et al. (2016). Climate change impacts on forest landscapes along the Canadian southern boreal forest transition zone. *Landscape Ecology, 32*(7), 1415–1431. https://doi.org/10.1007/s10980-016-0421-7.

Boulanger, Y., Arseneault, D., Boucher, Y., et al. (2019). Climate change will affect the ability of forest management to reduce gaps between current and presettlement forest composition in southeastern Canada. *Landscape Ecology, 34*(1), 159–174. https://doi.org/10.1007/s10980-018-0761-6.

Brais, S., Harvey, B. D., Bergeron, Y., et al. (2004). Testing forest ecosystem management in boreal mixedwoods of northwestern Quebec: Initial response of aspen stands to different levels of harvesting. *Canadian Journal of Forest Research, 34*(2), 431–446. https://doi.org/10.1139/x03-144.

Brienen, R. J. W., Caldwell, L., Duchesne, L., et al. (2020). Forest carbon sink neutralized by pervasive growth-lifespan trade-offs. *Nature Communications, 11*(1), 4241. https://doi.org/10.1038/s41467-020-17966-z.

Brokaw, N. V. L., & Lent, R. A. (1999). Vertical structure. In M. Hunter (Ed.), *Maintaining biodiversity in forest ecosystems* (pp. 373–399). Cambridge University Press.

Burton, P. J., Messier, C., Weetman, G., et al. (2003). The current state of boreal forestry and the drive for change. In P. J. Burton, C. Messier, D. W. Smith, & W. L. Adamowicz (Eds.), *Towards sustainable management of the boreal forest* (pp. 1–40). NRC Research Press.

Cadieux, P., Boulanger, Y., Cyr, D., et al. (2020). Projected effects of climate change on boreal bird community accentuated by anthropogenic disturbances in western boreal forest, Canada. *Diversity and Distributions, 26*(6), 668–682. https://doi.org/10.1111/ddi.13057.

Cadotte, M. W., Carscadden, K., & Mirotchnick, N. (2011). Beyond species: Functional diversity and the maintenance of ecological processes and services. *Journal of Applied Ecology, 48*(5), 1079–1087. https://doi.org/10.1111/j.1365-2664.2011.02048.x.

Canadian Institutes of Health Research, Natural Sciences and Engineering Council of Canada, Social Sciences and Humanities Research Council. (2018). *Tri-council policy statement ethical conduct for research involving humans Secretariat on Responsible Conduct of Research*, Ottawa, p. 231.

Canelles, Q., Aquilué, N., James, P. M. A., et al. (2021). Global review on interactions between insect pests and other forest disturbances. *Landscape Ecology, 36*(4), 945–972. https://doi.org/10.1007/s10980-021-01209-7.

Cappuccino, N., Lavertu, D., Bergeron, Y., et al. (1998). Spruce budworm impact, abundance and parasitism rate in a patchy landscape. *Oecologia, 114*(2), 236–242. https://doi.org/10.1007/s004420050441.

Carcaillet, C., Bergeron, Y., Richard, P. J. H., et al. (2001). Change of fire frequency in the eastern Canadian boreal forests during the Holocene: Does vegetation composition or climate trigger the fire regime? *Journal of Ecology, 89*(6), 930–946. https://doi.org/10.1046/j.0022-0477.2001.00614.x https://doi.org/10.1111/j.1365-2745.2001.00614.x.

Carignan, R., & Steedman, R. J. (2000). Impacts of major watershed perturbations on aquatic ecosystems. *Canadian Journal of Fisheries and Aquatic Sciences, 57*(S2), 1–4. https://doi.org/10.1139/f00-172.

Carnus, J. M., Parrotta, J., Brockerhoff, E., et al. (2006). Planted Forests and Biodiversity. *Journal of Forestry, 104*(2), 65–77. https://doi.org/10.1093/jof/104.2.65.

Castleden, H. E., Hart, C., Harper, S., et al. (2017). Implementing indigenous and western knowledge systems in water research and management (Part 1): A systematic realist review to inform water policy and governance in Canada. *International Indigenous Policy Journal, 8*(4).1–33 https://doi.org/10.18584/iipj.2017.8.4.6.

Chazdon, R. L. (2018). Protecting intact forests requires holistic approaches. *Nature Ecology and Evolution, 2*(6), 915–915. https://doi.org/10.1038/s41559-018-0546-y.

Cole, J. J., Prairie, Y. T., Caraco, N. F., et al. (2007). Plumbing the global carbon cycle: Integrating inland waters into the terrestrial carbon budget. *Ecosystems, 10*(1), 172–185. https://doi.org/10.1007/s10021-006-9013-8.

Cook-Patton, S. C., Leavitt, S. M., Gibbs, D., et al. (2020). Mapping carbon accumulation potential from global natural forest regrowth. *Nature, 585*(7826), 545–550. https://doi.org/10.1038/s41586-020-2686-x.

Cornelissen, J. H. C., Lavorel, S., Garnier, E., et al. (2003). A handbook of protocols for standardised and easy measurement of plant functional traits worldwide. *Australian Journal of Botany, 51*(4), 335–380. https://doi.org/10.1071/BT02124.

Culbert, P. D., Radeloff, V. C., Flather, C. H., et al. (2013). The influence of vertical and horizontal habit at structure on nationwide patterns of avian biodiversity. *The Auk, 130*(4), 656–665. https://doi.org/10.1525/auk.2013.13007.

Cullingham, C. I., Cooke, J. E. K., Dang, S., et al. (2011). Mountain pine beetle host-range expansion threatens the boreal forest. *Molecular Ecology, 20*(10), 2157–2171. https://doi.org/10.1111/j.1365-294X.2011.05086.x.

Cumming, S. G., Schmiegelow, F. K. A., & Burton, P. J. (2000). Gap dynamics in boreal aspen stands: Is the forest older than we think? *Ecological Applications, 10*(3), 744–759. https://doi.org/10.1890/1051-0761(2000)010[0744:GDIBAS]2.0.CO;2.

Cyr, D., Gauthier, S., Bergeron, Y., et al. (2009). Forest management is driving the eastern North American boreal forest outside its natural range of variability. *Frontiers in Ecology and the Environment, 7*(10), 519–524. https://doi.org/10.1890/080088.

D'Amato, A. W., Bradford, J. B., Fraver, S., et al. (2011). Forest management for mitigation and adaptation to climate change: Insights from long-term silviculture experiments. *Forest Ecology and Management, 262*, 803–816. https://doi.org/10.1016/j.foreco.2011.05.014.

D'Amato, A. W., Bradford, J. B., Fraver, S., et al. (2013). Effects of thinning on drought vulnerability and climate response in north temperate forest ecosystems. *Ecological Applications, 23*(8), 1735–1742. https://doi.org/10.1890/13-0677.1.

D'Orangeville, L., Duchesne, L., Houle, D., et al. (2016). Northeastern North America as a potential refugium for boreal forests in a warming climate. *Science, 352*(6292), 1452–1455. https://doi.org/10.1126/science.aaf4951.

D'Orangeville, L., Houle, D., Duchesne, L., et al. (2018). Beneficial effects of climate warming on boreal tree growth may be transitory. *Nature Communications, 9*(1), 3213. https://doi.org/10.1038/s41467-018-05705-4.

Dam Lam, R., Gasparatos, A., Chakraborty, S., et al. (2019). Multiple values and knowledge integration in indigenous coastal and marine social-ecological systems research: A systematic review. *Ecosystem Services, 37*, 100910. https://doi.org/10.1016/j.ecoser.2019.100910.

Daniel, C. J., Ter-Mikaelian, M. T., Wotton, B. M., et al. (2017). Incorporating uncertainty into forest management planning: Timber harvest, wildfire and climate change in the boreal forest. *Forest Ecology and Management, 400*, 542–554. https://doi.org/10.1016/j.foreco.2017.06.039.

Danneyrolles, V., Dupuis, S., Fortin, G., et al. (2019). Stronger influence of anthropogenic disturbance than climate change on century-scale compositional changes in northern forests. *Nature Communications, 10*(1), 1265. https://doi.org/10.1038/s41467-019-09265-z.

Davidson-Hunt, I., & Berkes, F. (2003). Learning as you journey: Anishinaabe perception of social-ecological environments and adaptive learning. *Conservation Ecology, 8*(1):art5. https://doi.org/10.5751/es-00587-080105.

Dawe, K. L., & Boutin, S. (2016). Climate change is the primary driver of white-tailed deer (*Odocoileus virginianus*) range expansion at the northern extent of its range; land use is secondary. *Ecology and Evolution, 6*(18), 6435–6451. https://doi.org/10.1002/ece3.2316.

De Grandpré, L., Waldron, K., Bouchard, M., et al. (2018). Incorporating insect and wind disturbances in a natural disturbance-based management framework for the boreal forest. *Forests, 9*(8), 471. https://doi.org/10.3390/f9080471.

De Groot, W. J., Cantin, A. S., Flannigan, M. D., et al. (2013a). A comparison of Canadian and Russian boreal forest fire regimes. *Forest Ecology and Management, 294*, 23–34. https://doi.org/10.1016/j.foreco.2012.07.033.

De Groot, W. J., Flannigan, M. D., & Cantin, A. S. (2013b). Climate change impacts on future boreal fire regimes. *Forest Ecology and Management, 294*, 35–44. https://doi.org/10.1016/j.foreco.2012.09.027.

De Vriendt, L., Lavoie, S., Barrette, M., et al. (2021). From delayed succession to alternative successional trajectory: How different moose browsing pressures contribute to forest dynamics following clear-cutting. *Journal of Vegetation Science, 32*(1), e12945. https://doi.org/10.1111/jvs.12945.

Díaz-Yáñez, O., Arias-Rodil, M., Mola-Yudego, B., et al. (2019). Simulating the effects of wind and snow damage on the optimal management of Norwegian spruce forests. *Forestry, 92*(4), 406–416. https://doi.org/10.1093/forestry/cpz031.

Dobson, A. P., Bradshaw, A. D., & Baker, A. J. M. (1997). Hopes for the future: Restoration ecology and conservation biology. *Science, 277*(5325), 515–522. https://doi.org/10.1126/science.277.5325.515.

Dondina, O., Orioli, V., D'Occhio, P., et al. (2017). How does forest species specialization affect the application of the island biogeography theory in fragmented landscapes? *Journal of Biogeography, 44*(5), 1041–1052. https://doi.org/10.1111/jbi.12827.

Dou, H., Jiang, G., Stott, P., et al. (2013). Climate change impacts population dynamics and distribution shift of moose (*Alces alces*) in Heilongjiang Province of China. *Ecological Research, 28*(4), 625–632. https://doi.org/10.1007/s11284-013-1054-9.

Duchesne, L., & Ouimet, R. (2007). *Une histoire de perturbations! Les changements de composition dans la forêt du Québec méridional au cours des 30 dernières années.* Québec: Direction de la recherche forestière, ministère des Forêts, de la Faune et des Parcs.

Dukes, J. S., Pontius, J., Orwig, D., et al. (2009). Responses of insect pests, pathogens, and invasive plant species to climate change in the forests of northeastern North America: What can we predict? *Canadian Journal of Forest Research, 39*, 231–248. https://doi.org/10.1139/X08-171.

Dunn, A. L., Barford, C. C., Wofsy, S. C., et al. (2007). A long-term record of carbon exchange in a boreal black spruce forest: Means, responses to interannual variability, and decadal trends. *Global Change Biology, 13*(3), 577–590. https://doi.org/10.1111/j.1365-2486.2006.01221.x.

Dupuis, S., Danneyrolles, V., Laflamme, J., et al. (2020). Forest transformation following European settlement in the Saguenay-Lac-St-Jean valley in eastern Québec, Canada. *Frontiers in Ecology and Evolution, 8*(August), 1–13. https://doi.org/10.3389/fevo.2020.00257.

Eden, J. M., Krikken, F., & Drobyshev, I. (2020). An empirical prediction approach for seasonal fire risk in the boreal forests. *International Journal of Climatology, 40*(5), 2732–2744. https://doi.org/10.1002/joc.6363.

Eggers, S., & Low, M. (2014). Differential demographic responses of sympatric Parids to vegetation management in boreal forest. *Forest Ecology and Management, 319*, 169–175. https://doi.org/10.1016/j.foreco.2014.02.019.

Elmqvist, T., Folke, C., Nystrom, M., et al. (2003). Response diversity, ecosystem change, and resilience. *Frontiers in Ecology and the Environment, 1*, 488–494. https://doi.org/10.1890/1540-9295(2003)001[0488:RDECAR]2.0.CO;2.

Ericksen, P., & Woodley, E. (2005). Using multiple knowledge systems: Benefits and challenges. In A. M. E. Assessment (Ed.), *Ecosystems and human well-being* (pp. 85–117). World Resources Institute.

Esseen, P. A., Ehnström, B., Ericson, L., et al. (1992). Boreal forests—the focal habitats of Fennoscandia. In L. Hansson (Ed.), *Ecological principles of nature conservation* (pp. 252–325). Elsevier Applied Science.

Esseen, P. A., Ehnström, B., Ericson, L., et al. (1997). Boreal forests. *Ecological Bulletins, 46*, 16–47.

Esseen, P. A., Hedström Ringvall, A., Harper, K. A., et al. (2016). Factors driving structure of natural and anthropogenic forest edges from temperate to boreal ecosystems. *Journal of Vegetation Science, 27*(3), 482–492. https://doi.org/10.1111/jvs.12387.

European Commission (EC). (2013). *Communication from the Commission to the European Parliament, the Council, The European Economic and Social Committee and the Committee of the Regions: Green infrastructure (GI)—Enhancing Europe's natural capital: COM/2013/0249.* Brussels: European Commission.

Eyvindson, K., Duflot, R., Triviño, M., et al. (2021). High boreal forest multifunctionality requires continuous cover forestry as a dominant management. *Land Use Policy, 100*, 104918. https://doi.org/10.1016/j.landusepol.2020.104918.

Fatichi, S., Pappas, C., Zscheischler, J., et al. (2019). Modelling carbon sources and sinks in terrestrial vegetation. *New Phytologist, 221*(2), 652–668. https://doi.org/10.1111/nph.15451.

Feldman, M. J., Girona, M. M., Grosbois, G., et al. (2020). Why do beavers leave home? Lodge abandonment in an invasive population in Patagonia. *Forests, 11*(11), 1161. https://doi.org/10.3390/f11111161.

Fenton, N., Simard, M., & Bergeron, Y. (2009). Emulating natural disturbances: The role of silviculture in creating even-aged and complex structures in the black spruce boreal forest of eastern North America. *Journal of Forest Research, 14*, 258–267. https://doi.org/10.1007/s10310-009-0134-8.

Fenton, N. J., Imbeau, L., Work, T., et al. (2013). Lessons learned from 12 years of ecological research on partial cuts in black spruce forests of northwestern Québec. *The Forestry Chronicle, 89*(03), 350–359. https://doi.org/10.5558/tfc2013-065.

Feser, F., Barcikowska, M., Krueger, O., et al. (2015). Storminess over the North Atlantic and northwestern Europe—A review. *Quarterly Journal Royal Meteorological Society, 141*(687), 350–382. https://doi.org/10.1002/qj.2364.

Filbee-Dexter, K., Pittman, J., Haig, H. A., et al. (2017). Ecological surprise: Concept, synthesis, and social dimensions. *Ecosphere, 8*(12), e02005. https://doi.org/10.1002/ecs2.2005.

Fischer, J., & Lindenmayer, D. B. (2007). Landscape modification and habitat fragmentation: A synthesis. *Global Ecology and Biogeography, 16*(3), 265–280. https://doi.org/10.1111/j.1466-8238.2007.00287.x.

Flannigan, M. D., Wotton, B. M., Marshall, G. A., et al. (2016). Fuel moisture sensitivity to temperature and precipitation: Climate change implications. *Climatic Change, 134*(1–2), 59–71. https://doi.org/10.1007/s10584-015-1521-0.

Foley, J. A., DeFries, R., Asner, G. P., et al. (2005). Global consequences of land use. *Science, 309*(5734), 570–574. https://doi.org/10.1126/science.1111772.

Folke, C., Carpenter, S., Walker, B., et al. (2004). Regime shifts, resilience, and biodiversity in ecosystem management. *Annual Review of Ecology, Evolution, and Systematics, 35*(1), 557–581. https://doi.org/10.1146/annurev.ecolsys.35.021103.105711.

Food and Agriculture Organization of the United Nations (FAO). (2009). *State of the world's forests 2009* (p. 168). Electronic Publishing Policy and Support Branch, Communication Division.

Food and Agriculture Organization of the United Nations (FAO). (2011). *State of the world's forest 2011*. Electronic Publishing Policy and Support Branch.

Food and Agriculture Organization of the United Nations (FAO). (2014). *State of the world's forest: Enhancing the socioeconomic benefits from forests*. Electronic Publishing Policy and Support Branch.

Food and Agriculture Organization of the United Nations (FAO). (2016). *State of the world's forest 2016*. Electronic Publishing Policy and Support Branch.

Forest Europe. (2015). State of Europe's forests 2015. In *Ministerial Conference on the Protection of Forests in Europe*, Madrid: Forest Europe, Liaison Unit Madrid.

Franklin, J. F., Berg, D. F., Thornburg, D., et al. (1997). Alternative silvicultural approaches to timber harvesting: Variable retention harvest systems. In K. A. Kohm & J. F. Franklin (Eds.), *Creating a forestry for the 21st century: The science of ecosystem management* (pp. 111–140). Island Press.

Franklin, J. F., Spies, T. A., Pelt, R. V., et al. (2002). Disturbances and structural development of natural forest ecosystems with silvicultural implications, using Douglas-fir forests as an example. *Forest Ecology and Management, 155*(1–3), 399–423. https://doi.org/10.1016/S0378-1127(01)00575-8.

Fraser, D., Chavez, E. R., & Paloheimo, J. E. (1984). Aquatic feeding by moose: Selection of plant species and feeding areas in relation to plant chemical composition and characteristics of lakes. *Canadian Journal of Zoology, 62*(1), 80–87. https://doi.org/10.1139/z84-014.

Friend, A. D., Lucht, W., Rademacher, T. T., et al. (2014). Carbon residence time dominates uncertainty in terrestrial vegetation responses to future climate and atmospheric CO_2. *Proceedings of the National Academy of Sciences of the United States of America, 111*(9), 3280. https://doi.org/10.1073/pnas.1222477110.

Fritts, H. C. (2001). *Tree rings and climate* (p. 584). Caldwell: Blackburn Press.

Fuentes, L., Asselin, H., Bélisle, A. C., et al. (2020). Impacts of environmental changes on well-being in Indigenous communities in Eastern Canada. *International Journal of Environmental Research and Public Health, 17*(2), 637. https://doi.org/10.3390/ijerph17020637.

Gagné, L., Lavoie, L., & Binot, J. M. (2012). Growth and mechanical properties of wood after commercial thinning in a 32-year-old white spruce (*Picea glauca*) plantation. *Canadian Journal of Forest Research, 42*(2), 291–302. https://doi.org/10.1139/X11-181.

Garcia, E., & Carignan, R. (2000). Mercury concentrations in northern pike (*Esox lucius*) from boreal lakes with logged, burned, or undisturbed catchments. *Canadian Journal of Fisheries and Aquatic Sciences, 57*, 129–135. https://doi.org/10.1139/f00-126.

Gauthier, S., Villancourt, M. A., Leduc, A., et al. (2009). *Ecosystem management in the boreal forest*. Presses de l'Université de Québec.

Gauthier, S., Bernier, P. Y., Boulanger, Y., et al. (2015a). Vulnerability of timber supply to projected changes in fire regime in Canada's managed forests. *Canadian Journal of Forest Research, 45*, 1439–1447. https://doi.org/10.1139/cjfr-2015-0079.

Gauthier, S., Bernier, P., Kuuluvainen, T., et al. (2015b). Boreal forest health and global change. *Science, 349*, 819–822. https://doi.org/10.1126/science.aaa9092.

Gilarranz, L. J., Rayfield, B., Liñán-Cembrano, G., et al. (2017). Effects of network modularity on the spread of perturbation impact in experimental metapopulations. *Science, 357*(6347), 199–201. https://doi.org/10.1126/science.aal4122.

Gilchrist, G., Mallory, M., & Merkel, F. (2005). Can local ecological knowledge contribute to wildlife management? Case studies of migratory birds. *Ecology and Society, 10*(1), 12. https://doi.org/10.5751/ES-01275-100120.

Gingras, B., Slattery, S., Smith, K., et al. (2018). Boreal wetlands of Canada and the United States of America. In C. M. Finlayson, G. R. Milton, R. C. Prentice, & N. C. Davidson (Eds.), *The wetland book: II: Distribution, description, and conservation* (pp. 521–542). Springer.

Glaz, P., Sirois, P., Archambault, P., et al. (2014). Impact of forest harvesting on trophic structure of eastern Canadian Boreal Shield lakes: Insights from stable isotope analyses. *PLoS ONE, 9*(4), e96143. https://doi.org/10.1371/journal.pone.0096143.

Glaz, P., Gagné, J. P., Archambault, P., et al. (2015). Impact of forest harvesting on water quality and fluorescence characteristics of dissolved organic matter in eastern Canadian Boreal Shield lakes in summer. *Biogeosciences, 12*(23), 6999–7011. https://doi.org/10.5194/bg-12-6999-2015.

Gonzalès, R., & Parrott, L. (2012). Network theory in the assessment of the sustainability of social ecological systems. *Geography Compass, 6*(2), 76–88. https://doi.org/10.1111/j.1749-8198.2011.00470.x.

Gosse, J., Hermanutz, L., McLaren, B., et al. (2011). Degradation of boreal forests by nonnative herbivores in Newfoundland's national parks: Recommendations for ecosystem restoration. *Natural Areas Journal, 31*(4), 331–339. https://doi.org/10.3375/043.031.0403.

Grant, G. E., Tague, C. L., & Allen, C. D. (2013). Watering the forest for the trees: An emerging priority for managing water in forest landscapes. *Frontiers in Ecology and the Environment, 11*(6), 314–321. https://doi.org/10.1890/120209.

Gregow, H., Peltola, H., Laapas, M., et al. (2011). Combined occurrence of wind, snow loading and soil frost with implications for risks to forestry in Finland under the current and changing climatic conditions. *Silva Fennica, 45*(1), 30. https://doi.org/10.14214/sf.30.

Grenon, F., Jetté, J., & Leblanc, M. (2010). *Manuel de référence pour l'aménagement écosystémique des forêts au Québec–Module 1-Fondements et démarche de la mise en oeuvre*. Québec: CERFO, ministère des Ressources naturelles et de la Faune.

Grondin, P., Gauthier, S., Poirier, V., et al. (2018). Have some landscapes in the eastern Canadian boreal forest moved beyond their natural range of variability? *Forest Ecosystems, 5*(1), 30. https://doi.org/10.1186/s40663-018-0148-9.

Grosbois, G., Vachon, D., Del Giorgio, P. A., et al. (2020). Efficiency of crustacean zooplankton in transferring allochthonous carbon in a boreal lake. *Ecology, 101*(6), e03013. https://doi.org/10.1002/ecy.3013.

Guindon, L., Bernier, P., Gauthier, S., et al. (2018). Missing forest cover gains in boreal forests explained. *Ecosphere, 9*(1), e02094. https://doi.org/10.1002/ecs2.2094.

Gunderson, L. (2015). Lessons from adaptive management: Obstacles and outcomes. In C. R. Allen & A. S. Garmestani (Eds.), *Adaptive management of social-ecological systems* (pp. 27–38). Springer.

Gustafsson, L., Bauhus, J., Asbeck, T., et al. (2020). Retention as an integrated biodiversity conservation approach for continuous-cover forestry in Europe. *Ambio, 49*(1), 85–97. https://doi.org/10.1007/s13280-019-01190-1.

Haddad, N. M., Brudvig, L. A., Clobert, J., et al. (2015). Habitat fragmentation and its lasting impact on Earth's ecosystems. *Science Advances, 1*(2), e1500052. https://doi.org/10.1126/sciadv.1500052.

Haeussler, S., & Kneeshaw, D. (2003). Comparing forest management to natural processes. In P. J. Burton, C. Messier, D. W. Smith, & W. L. Adamowicz (Eds.), *Towards sustainable management of the boreal forest* (pp. 307–368). NRC Research Press.

Hägglund, R., Hekkala, A. M., Hjältén, J., et al. (2015). Positive effects of ecological restoration on rare and threatened flat bugs (Heteroptera: Aradidae). *Journal of Insect Conservation, 19*(6), 1089–1099. https://doi.org/10.1007/s10841-015-9824-z.

Halme, P., Allen, K. A., Auniņš, A., et al. (2013). Challenges of ecological restoration: Lessons from forests in northern Europe. *Biological Conservation, 167*, 248–256. https://doi.org/10.1016/j.biocon.2013.08.029.

Harper, K. A., Macdonald, S. E., Mayerhofer, M. S., et al. (2015). Edge influence on vegetation at natural and anthropogenic edges of boreal forests in Canada and Fennoscandia. *Journal of Ecology, 103*, 550–562. https://doi.org/10.1111/1365-2745.12398.

Harris, J. A., Hobbs, R. J., Higgs, E., et al. (2006). Ecological restoration and global climate change. *Restoration Ecology, 14*(2), 170–176. https://doi.org/10.1111/j.1526-100X.2006.00136.x.

Hassan, R. M., Scholes, R. J., Ash, N., et al. (2005). *Ecosystems and human well-being : Current state and trends : Findings of the Condition and Trends Working Group of the Millennium Ecosystem Assessment.* Island Press.

Heikkala, O., Seibold, S., Koivula, M., et al. (2016). Retention forestry and prescribed burning result in functionally different saproxylic beetle assemblages than clear-cutting. *Forest Ecology and Management, 359*, 51–58. https://doi.org/10.1016/j.foreco.2015.09.043.

Heino, M., Kummu, M., Makkonen, M., et al. (2015). Forest loss in protected areas and intact forest landscapes: A global analysis. *PLoS ONE, 10*(10), e0138918. https://doi.org/10.1371/journal.pone.0138918.

Hekkala, A. M., Tarvainen, O., & Tolvanen, A. (2014). Dynamics of understory vegetation after restoration of natural characteristics in the boreal forests in Finland. *Forest Ecology and Management, 330*, 55–66. https://doi.org/10.1016/j.foreco.2014.07.001.

Heller, N. E., & Zavaleta, E. S. (2009). Biodiversity management in the face of climate change: A review of 22 years of recommendations. *Biological Conservation, 142*(1), 14–32. https://doi.org/10.1016/j.biocon.2008.10.006.

Hernández-Rodríguez, E., Escalera-Vázquez, L. H., García-Ávila, D., et al. (2021). Reduced-impact logging maintain high moss diversity in temperate forests. *Forests, 12*(4), 383. https://doi.org/10.3390/f12040383.

Hidding, B., Tremblay, J. P., & Côté, S. D. (2013). A large herbivore triggers alternative successional trajectories in the boreal forest. *Ecology, 94*(12), 2852–2860. https://doi.org/10.1890/12-2015.1.

Hilty, J., Muller, B., Pantin, F., et al. (2021). Plant growth: The what, the how, and the why. *New Phytologist, 232*(1), 25–41. https://doi.org/10.1111/nph.17610.

Hjältén, J., Hägglund, R., Löfroth, T., et al. (2017). Forest restoration by burning and gap cutting of voluntary set-asides yield distinct immediate effects on saproxylic beetles. *Biodiversity and Conservation, 26*(7), 1623–1640. https://doi.org/10.1007/s10531-017-1321-0.

Hlásny, T., König, L., Krokene, P., et al. (2021). Bark beetle outbreaks in Europe: State of knowledge and ways forward for management. *Current Forestry Reports, 7*(3), 138–165. https://doi.org/10.1007/s40725-021-00142-x.

Hof, A. R., Montoro Girona, M., Fortin, M.-J., et al. (2021). Editorial: Using landscape simulation models to help balance conflicting goals in changing forests. *Frontiers in Ecology and Evolution.* https://doi.org/10.3389/fevo.2021.795736.

Honnay, O., Piessens, K., Van Landuyt, W., et al. (2003). Satellite based land use and landscape complexity indices as predictors for regional plant species diversity. *Landscape and Urban Planning, 63*(4), 241–250. https://doi.org/10.1016/S0169-2046(02)00194-9.

Hyvärinen, E., Juslén, A., Kemppainen, E., et al. (2019). *Suomen lajien uhanalaisuus—Punainen kirja 2019/The 2019 Red List of Finnish Species.* Ympäristöministeriö and Suomen ympäristökeskus/Ministry of the Environment and Finnish Environment Institute.

Intergovernmental Panel on Climate Change (IPCC). (2014). *Climate Change 2014: Synthesis report. Contribution of Working Groups I, II and III to the fifth assessment report of the Intergovernmental Panel on Climate Change* (p. 151). Geneva: IPCC.

Ivantsova, E. D., Pyzhev, A. I., & Zander, E. V. (2019). Economic consequences of insect pests outbreaks in boreal forests: A literature review. *Journal of Siberian Federal University. Humanities & Social Sciences, 12*(4), 627–642. https://doi.org/10.17516/1997-1370-0417.

Jackson, J. B. C., Kirby, M. X., Berger, W. H., et al. (2001). Historical overfishing and the recent collapse of coastal ecosystems. *Science, 293*(5530), 629–637. https://doi.org/10.1126/science.1059199.

James, P. M. A., Robert, L. E., Wotton, B. M., et al. (2017). Lagged cumulative spruce budworm defoliation affects the risk of fire ignition in Ontario, Canada. *Ecological Applications, 27*(2), 532–544. https://doi.org/10.1002/eap.1463.

Jasinski, J. P. P., & Payette, S. (2005). The creation of alternative stable states in the southern boreal forest, Québec, Canada. *Ecological Monographs, 75*(4), 561–583. https://doi.org/10.1890/04-1621.

Johnstone, J. F., Allen, C. D., Franklin, J. F., et al. (2016). Changing disturbance regimes, ecological memory, and forest resilience. *Frontiers in Ecology and the Environment, 14*(7), 369–378. https://doi.org/10.1002/fee.1311.

Jones, K. R., Venter, O., Fuller, R. A., et al. (2018). One-third of global protected land is under intense human pressure. *Science, 360*(6390), 788–791. https://doi.org/10.1126/science.aap9565.

Jonsson, B. G., & Siitonen, J. (2012). Dead wood and sustainable forest management. In B. G. Jonsson, J. N. Stokland, & J. Siitonen (Eds.), *Biodiversity in dead wood* (pp. 302–337). Cambridge: Cambridge University Press.

Jonsson, B. G., Svensson, J., Mikusiński, G., et al. (2019). European Union's last intact forest landscapes are at a value chain crossroad between multiple use and intensified wood production. *Forests, 10*(7), 564. https://doi.org/10.3390/f10070564.

Junninen, K., & Komonen, A. (2011). Conservation ecology of boreal polypores: A review. *Biological Conservation, 144*(1), 11–20. https://doi.org/10.1016/j.biocon.2010.07.010.

Kauppi, P. E., Posch, M., & Pirinen, P. (2014). Large impacts of climatic warming on growth of boreal forests since 1960. *PLoS ONE, 9*(11), e111340. https://doi.org/10.1371/journal.pone.0111340.

Keeton, W. S. (2018). Source or sink? Carbon dynamics in eastern old-growth forests and their role in climate change mitigation. In A. M. Barton & W. S. Keeton (Eds.), *Ecology and recovery of eastern old-growth forests* (pp. 267–288). Island Press/Center for Resource Economics.

Kim, S., Axelsson, E. P., Girona, M. M., et al. (2021). Continuous-cover forestry maintains soil fungal communities in Norway spruce dominated boreal forests. *Forest Ecology and Management, 480*, 118659. https://doi.org/10.1016/j.foreco.2020.118659.

Kimmins, J. P. (1997). *Forest ecology—a foundation for sustainable forest management and environmental ethics in forestry.* Prentice Hall.

Klaus, M., Geibrink, E., Jonsson, A., et al. (2018). Greenhouse gas emissions from boreal inland waters unchanged after forest harvesting. *Biogeosciences, 15*(18), 5575–5594. https://doi.org/10.5194/bg-15-5575-2018.

Kneeshaw, D. D., & Gauthier, S. (2003). Old growth in the boreal forest: A dynamic perspective at the stand and landscape level. *Environmental Reviews, 11*, S99–S114. https://doi.org/10.1139/a03-010.

Korhonen, K., Ahola, A., Heikkinen, J., et al. (2021). Forests of Finland 2014–2018 and their development 1921–2018. *Silva Fennica, 55*(5), 10662. https://doi.org/10.14214/sf.10662.

Körner, C. (2017). A matter of tree longevity. *Science, 355*(6321), 130–131. https://doi.org/10.1126/science.aal2449.

Kramer, K., Degen, B., Buschbom, J., et al. (2010). Modelling exploration of the future of European beech (*Fagus sylvatica* L.) under climate change-Range, abundance, genetic diversity and adaptive response. *Forest Ecology and Management, 259*(11), 2213–2222. https://doi.org/10.1016/j.foreco.2009.12.023.

Krikken, F., Lehner, F., Haustein, K., et al. (2021). Attribution of the role of climate change in the forest fires in Sweden 2018. *Natural Hazards and Earth System Sciences, 21*(7), 2169–2179. https://doi.org/10.5194/nhess-21-2169-2021.

Kritzberg, E. S., Hasselquist, E. M., Škerlep, M., et al. (2020). Browning of freshwaters: Consequences to ecosystem services, underlying drivers, and potential mitigation measures. *Ambio, 49*(2), 375–390. https://doi.org/10.1007/s13280-019-01227-5.

Kulha, N., Pasanen, L., Holmström, L., et al. (2019). At what scales and why does forest structure vary in naturally dynamic boreal forests? An analysis of forest landscapes on two continents. *Ecosystems, 22*(4), 709–724. https://doi.org/10.1007/s10021-018-0297-2.

Kulha, N., Pasanen, L., Holmström, L., et al. (2020). The structure of boreal old-growth forests changes at multiple spatial scales over decades. *Landscape Ecology, 35*(4), 843–858. https://doi.org/10.1007/s10980-020-00979-w.

Kurz, W. A., Stinson, G., Rampley, G. J., et al. (2008). Risk of natural disturbances makes future contribution of Canada's forests to the global carbon cycle highly uncertain. *Proceedings of National Academy of Sciences of the United States of America, 105*(5), 1551–1555. https://doi.org/10.1073/pnas.0708133105.

Kuuluvainen, T. (2002). Natural variability of forests as a reference for restoring and managing biological diversity in boreal Fennoscandia. *Silva Fennica, 36*(1), 552. https://doi.org/10.14214/sf.552.

Kuuluvainen, T. (2009). Forest management and biodiversity conservation based on natural ecosystem dynamics in northern Europe: The complexity challenge. *Ambio, 38*(6), 309–315. https://doi.org/10.1579/08-A-490.1.

Kuuluvainen, T., & Grenfell, R. (2012). Natural disturbance emulation in boreal forest ecosystem management—theories, strategies, and a comparison with conventional even-aged management. *Canadian Journal of Forest Research, 42*(7), 1185–1203. https://doi.org/10.1139/x2012-064.

Kuuluvainen, T., & Siitonen, J. (2013). Fennoscandian boreal forests as complex adaptive systems. Properties, management challenges and opportunities. In C. Messier, K. J. Puettman, & K. D. Coates (Eds.), *Managing forests as complex adaptive systems. Building resilience to the challenge of global change* (pp. 244–268). London: The Earthscan Forest Library, Routledge.

Kuuluvainen, T., & Gauthier, S. (2018). Young and old forest in the boreal: Critical stages of ecosystem dynamics and management under global change. *Forestry Ecosystems, 5*(1), 26. https://doi.org/10.1186/s40663-018-0142-2.

Kuuluvainen, T., Hofgaard, A., Aakala, T., et al. (2017). North Fennoscandian mountain forests: History, composition, disturbance dynamics and the unpredictable future. *Forest Ecology and Management, 385*, 140–149. https://doi.org/10.1016/j.foreco.2016.11.031.

Kuuluvainen, T., Lindberg, H., Vanha-Majamaa, I., et al. (2019). Low-level retention forestry, certification, and biodiversity: Case Finland. *Ecological Processes, 8*(1), 47. https://doi.org/10.1186/s13717-019-0198-0.

Kuuluvainen, T., Angelstam, P., Frelich, L., et al. (2021). Natural disturbance-based forest management: Moving beyond retention and continuous-cover forestry. *Frontiers in Forests and Global Change, 4*(24), 629020. https://doi.org/10.3389/ffgc.2021.629020.

Labrecque-Foy, J.-P., Morin, H., & Girona, M. M. (2020). Dynamics of territorial occupation by North American beavers in Canadian boreal forests: A novel dendroecological approach. *Forests, 11*(2), 221. https://doi.org/10.3390/f11020221.

Lamontagne, S., Carignan, R., D'Arcy, P., et al. (2000). Element export in runoff from eastern Canadian Boreal Shield drainage basins following forest harvesting and wildfires. *Canadian Journal of Fisheries and Aquatic Sciences, 57*(S2), 118–128. https://doi.org/10.1139/f00-108.

Landres, P. B., Morgan, P., & Swanson, F. J. (1999). Overview of the use of natural variability concepts in managing ecological systems. *Ecological Applications, 9*(4), 1179–1188. https://doi.org/10.1890/1051-0761(1999)009[1179:OOTUON]2.0.CO;2.

Lapenis, A., Shvidenko, A., Shepaschenko, D., et al. (2005). Acclimation of Russian forests to recent changes in climate. *Global Change Biology, 11*(12), 2090–2102. https://doi.org/10.1111/j.1365-2486.2005.001069.x.

Lassau, S. A., Hochuli, D. F., Cassis, G., et al. (2005). Effects of habitat complexity on forest beetle diversity: Do functional groups respond consistently? *Diversity and Distributions, 11*(1), 73–82. https://doi.org/10.1111/j.1366-9516.2005.00124.x.

Lavoie, J., Girona, M. M., & Morin, H. (2019). Vulnerability of conifer regeneration to spruce budworm outbreaks in the Eastern Canadian boreal forest. *Forests, 10*(10), 1–14. https://doi.org/10.3390/f10100850.

Lavoie, J., Montoro Girona, M., Grosbois, G., et al. (2021). Does the type of silvicultural practice influence spruce budworm defoliation of seedlings? *Ecosphere, 12*(4), 17. https://doi.org/10.1002/ecs2.3506.

Lavorel, S., & Garnier, E. (2002). Predicting changes in community composition and ecosystem functioning from plant traits: Revisiting the Holy Grail. *Functional Ecology, 16*(5), 545–556. https://doi.org/10.1046/j.1365-2435.2002.00664.x.

Leduc, A., Bernier, P. Y., Mansuy, N., et al. (2015). Using salvage logging and tolerance to risk to reduce the impact of forest fires on timber supply calculations. *Canadian Journal of Forest Research, 45*(4), 480–486. https://doi.org/10.1139/cjfr-2014-0434.

Leverkus, A. B., Lindenmayer, D. B., Thorn, S., et al. (2018). Salvage logging in the world's forests: Interactions between natural disturbance and logging need recognition. *Global Ecology and Biogeography, 27*(10), 1140–1154. https://doi.org/10.1111/geb.12772.

Lindenmayer, D. B., & Franklin, J. F. (2002). *Conserving forest biodiversity: A comprehensive multiscaled approach.* Island Press.

Lindenmayer, D. B., & Fischer, J. (2007). Tackling the habitat fragmentation panchreston. *Trends in Ecology & Evolution, 22*(3), 127–132. https://doi.org/10.1016/j.tree.2006.11.006.

Lindner, M., Maroschek, M., Netherer, S., et al. (2010). Climate change impacts, adaptive capacity, and vulnerability of European forest ecosystems. *Forest Ecology and Management, 259*(4), 698–709. https://doi.org/10.1016/j.foreco.2009.09.023.

Liquete, C., Kleeschulte, S., Dige, G., et al. (2015). Mapping green infrastructure based on ecosystem services and ecological networks: A Pan-European case study. *Environmental Science & Policy, 54*, 268–280. https://doi.org/10.1016/j.envsci.2015.07.009.

Littlechild, D. B., Finegan, C., & McGregor, D. (2021). "Reconciliation" in undergraduate education in Canada: The application of Indigenous knowledge in conservation. *Facets, 6*(1), 665–685. https://doi.org/10.1139/facets-2020-0076.

Liu, H., Park Williams, A., Allen, C. D., et al. (2013). Rapid warming accelerates tree growth decline in semi-arid forests of Inner Asia. *Global Change Biology, 19*(8), 2500–2510. https://doi.org/10.1111/gcb.12217.

Loehle, C., & Solarik, K. A. (2019). Forest growth trends in Canada. *The Forestry Chronicle, 95*(03), 183–195. https://doi.org/10.5558/tfc2019-027.

Löfman, S., & Kouki, J. (2001). Fifty years of landscape transformation in managed forests of southern Finland. *Scandinavian Journal of Forest Research, 16*(1), 44–53. https://doi.org/10.1080/028275801300004406.

Lorentzen Kolstad, A., Austrheim, G., Solberg, E. J., et al. (2018). Pervasive moose browsing in boreal forests alters successional trajectories by severely suppressing keystone species. *Ecosphere, 9*(10), e02458. https://doi.org/10.1002/ecs2.2458.

Lutz, W., Sanderson, W. C., & Scherbov, S. (2004). *The end of world population growth in the 21st century: New challenges for human capital formation and sustainable development.* Routledge.

Lyver, P. O. B., Richardson, S. J., Gormley, A. M., et al. (2018). Complementarity of indigenous and western scientific approaches for monitoring forest state. *Ecological Applications, 28*(7), 1909–1923. https://doi.org/10.1002/eap.1787.

MacLean, D. A. (2016). Impacts of insect outbreaks on tree mortality, productivity, and stand development. *The Canadian Entomologist, 148*, S138–S159. https://doi.org/10.4039/tce.2015.24.

Majcen, Z. (1994). History of selection cutting in uneven-aged forests in Quebec (Historique des coupes de jardinage dans les forêts inéquiennes au Québec). *Revue Forestière Française, 45*(4), 375–384. https://doi.org/10.4267/2042/26556.

Malmström, C. M., & Raffa, K. F. (2000). Biotic disturbance agents in the boreal forest: Considerations for vegetation change models. *Global Change Biology, 6*, 35–48. https://doi.org/10.1046/j.1365-2486.2000.06012.x.

Mansourian, S. (2018). In the eye of the beholder: Reconciling interpretations of forest landscape restoration. *Land Degradation and Development, 29*(9), 2888–2898. https://doi.org/10.1002/ldr.3014.

Marchand, D., Prairie, Y. T., & del Giorgio, P. A. (2009). Linking forest fires to lake metabolism and carbon dioxide emissions in the boreal region of Northern Québec. *Global Change Biology, 15*(12), 2861–2873. https://doi.org/10.1111/j.1365-2486.2009.01979.x.

Marchand, W., Girardin, M. P., Hartmann, H., et al. (2019). Taxonomy, together with ontogeny and growing conditions, drives needleleaf species' sensitivity to climate in boreal North America. *Global Change Biology, 25*(8), 2793–2809. https://doi.org/10.1111/gcb.14665.

Marchand, W., Girardin, M. P., Hartmann, H., et al. (2021). Contrasting life-history traits of black spruce and jack pine influence their physiological response to drought and growth recovery in northeastern boreal Canada. *Science of the Total Environment, 794*, 148514. https://doi.org/10.1016/j.scitotenv.2021.148514.

Marini, L., Lindelöw, Å., Jönsson, A. M., et al. (2013). Population dynamics of the spruce bark beetle: A long-term study. *Oikos, 122*(12), 1768–1776. https://doi.org/10.1111/j.1600-0706.2013.00431.x.

Martin, M., Fenton, N., & Morin, H. (2018). Structural diversity and dynamics of boreal old-growth forests case study in Eastern Canada. *Forest Ecology and Management, 422*, 125–136. https://doi.org/10.1016/j.foreco.2018.04.007.

Martin, M., Morin, H., & Fenton, N. J. (2019). Secondary disturbances of low and moderate severity drive the dynamics of eastern Canadian boreal old-growth forests. *Annals of Forest Science, 76*(4), 108. https://doi.org/10.1007/s13595-019-0891-2.

Martin, M., Boucher, Y., Fenton, N. J., et al. (2020a). Forest management has reduced the structural diversity of residual boreal old-growth forest landscapes in Eastern Canada. *Forest Ecology and Management, 458*, 117765. https://doi.org/10.1016/j.foreco.2019.117765.

Martin, M., Fenton, N. J., & Morin, H. (2020b). Boreal old-growth forest structural diversity challenges aerial photographic survey accuracy. *Canadian Journal of Forest Research, 50*(2), 155–169. https://doi.org/10.1139/cjfr-2019-0177.

Martin, M., Girona, M. M., & Morin, H. (2020c). Driving factors of conifer regeneration dynamics in eastern Canadian boreal old-growth forests. *PLoS ONE, 15*(7), e0230221. https://doi.org/10.1371/journal.pone.0230221.

Martin, M., Krause, C., Fenton, N. J., et al. (2020d). Unveiling the diversity of tree growth patterns in boreal old-growth forests reveals the richness of their dynamics. *Forests, 11*(3), 252. https://doi.org/10.3390/f11030252.

Martin, M., Fenton, N. J., & Morin, H. (2021a). Tree-related microhabitats and deadwood dynamics form a diverse and constantly changing mosaic of habitats in boreal old-growth forests. *Ecological Indicators, 128*, 107813. https://doi.org/10.1016/j.ecolind.2021.107813.

Martin, M., Grondin, P., Lambert, M.-C., et al. (2021b). Compared to wildfire, management practices reduced old-growth forest diversity and functionality in primary boreal landscapes of eastern Canada. *Frontiers in Forests and Global Change, 4*, 1–16. https://doi.org/10.3389/ffgc.2021.639397.

McDowell, N. G., Allen, C. D., Anderson-Teixeira, K., et al. (2020). Pervasive shifts in forest dynamics in a changing world. *Science, 368*(6494), eaaz9463. https://doi.org/10.1126/science.aaz9463.

McGregor, D. (2018). From "decolonized" to reconciliation research in Canada: Drawing from indigenous research paradigms. *Acme, 17*(3), 810–831.

McLaren, B. E., & Peterson, R. O. (1994). Wolves, moose, and tree rings on Isle Royale. *Science, 266*(5190), 1555–1558. https://doi.org/10.1126/science.266.5190.1555.

Meigs, G. W., Morrissey, R. C., Bače, R., et al. (2017). More ways than one: Mixed-severity disturbance regimes foster structural complexity via multiple developmental pathways. *Forest Ecology and Management, 406*, 410–426. https://doi.org/10.1016/j.foreco.2017.07.051.

Messier, C., Puettmann, K. J., & Coates, K. D. (Eds.). (2013). *Managing forests as complex adaptive systems: Building resilience to the challenge of global change* (p. 368). Routledge.

Messier, C., Bauhus, J., Doyon, F., et al. (2019). The functional complex network approach to foster forest resilience to global changes. *Forest Ecosystems, 6*(1), 21. https://doi.org/10.1186/s40663-019-0166-2.

Michaelian, M., Hogg, E. H., Hall, R. J., et al. (2011). Massive mortality of aspen following severe drought along the southern edge of the Canadian boreal forest. *Global Change Biology, 17*(6), 2084–2094. https://doi.org/10.1111/j.1365-2486.2010.02357.x.

Mielikäinen, K., & Hynynen, J. (2003). Silvicultural management in maintaining biodiversity and resistance of forests in Europe–boreal zone: Case Finland. *Journal of Environmental Management, 67*(1), 47–54. https://doi.org/10.1016/S0301-4797(02)00187-1.

Mikusiński, G., Orlikowska, E. H., Bubnicki, J. W., et al. (2021). Strengthening the network of high conservation value forests in boreal landscapes. *Frontiers in Ecology and Evolution, 8*, 595730. https://doi.org/10.3389/fevo.2020.595730.

Millar, C. I., Stephenson, N. L., & Stephens, S. L. (2007). Climate change and forests of the future: Managing in the face of uncertainty. *Ecological Applications, 17*(8), 2145–2151. https://doi.org/10.1890/06-1715.1.

Mina, M., Messier, C., Duveneck, M., et al. (2021). Network analysis can guide resilience-based management in forest landscapes under global change. *Ecological Applications, 31*(1), e2221. https://doi.org/10.1002/eap.2221.

Mitchell, S. J. (2013). Wind as a natural disturbance agent in forests: A synthesis. *Forestry, 86*(2), 147–157. https://doi.org/10.1093/forestry/cps058.

Mitchell, S. J., & Beese, W. J. (2002). The retention system: Reconciling variable retention with the principles of silvicultural systems. *The Forestry Chronicle, 78*(3), 397–403. https://doi.org/10.5558/tfc78397-3.

Mölter, T., Schindler, D., Albrecht, A. T., et al. (2016). Review on the projections of future storminess over the North Atlantic European region. *Atmosphere, 7*(4), 60. https://doi.org/10.3390/atmos7040060.

Montoro Girona, M. (2017). *À la recherche de l'aménagement durable en forêt boréale: croissance, mortalité et régénération des pessières noires soumises à différents systèmes sylvicoles*. Ph.D. thesis, Université du Québec à Chicoutimi.

Montoro Girona, M., Morin, H., Lussier, J. M., et al. (2016). Radial growth response of black spruce stands ten years after experimental shelterwoods and seed-tree cuttings in boreal forest. *Forests, 7*(10), 1–20. https://doi.org/10.3390/f7100240.

Montoro Girona, M., Rossi, S., Lussier, J. M., et al. (2017). Understanding tree growth responses after partial cuttings: A new approach. *PLoS ONE, 12*(2), e0172653. https://doi.org/10.1371/journal.pone.0172653.

Montoro Girona, M., Lussier, J. M., Morin, H., et al. (2018a). Conifer regeneration after experimental shelterwood and seed-tree treatments in boreal forests: Finding silvicultural alternatives. *Frontiers in Plant Science, 9*(August), 1145. https://doi.org/10.3389/fpls.2018.01145.

Montoro Girona, M., Navarro, L., & Morin, H. (2018b). A secret hidden in the sediments: Lepidoptera scales. *Frontiers in Ecology and Evolution, 6*, 2. https://doi.org/10.3389/fevo.2018.00002.

Montoro Girona, M., Morin, H., Lussier, J.-M., et al. (2019). Post-cutting mortality following experimental silvicultural treatments in unmanaged boreal forest stands. *Frontiers in Forests and Global Change, 2*, 4. https://doi.org/10.3389/ffgc.2019.00004.

Mori, A. S., Furukawa, T., & Sasaki, T. (2013). Response diversity determines the resilience of ecosystems to environmental change. *Biological Reviews, 88*(2), 349–364. https://doi.org/10.1111/brv.12004.

Mori, A. S., Dee, L. E., Gonzalez, A., et al. (2021). Biodiversity–productivity relationships are key to nature-based climate solutions. *Nature Climate Change, 11*(6), 543–550. https://doi.org/10.1038/s41558-021-01062-1.

Moussaoui, L., Fenton, N. J., Leduc, A., et al. (2016). Can retention harvest maintain natural structural complexity? A comparison of post-harvest and post-fire residual patches in boreal forest. *Forests, 7*(10), 243. https://doi.org/10.3390/f7100243.

Moussaoui, L., Leduc, A., Girona, M. M., et al. (2020). Success factors for experimental partial harvesting in unmanaged boreal forest: 10-year stand yield results. *Forests, 11*(11), 1199. https://doi.org/10.3390/f11111199.

Murakami, M., & Nakano, S. (2002). Indirect effect of aquatic insect emergence on a terrestrial insect population through by birds predation. *Ecology Letters, 5*(3), 333–337. https://doi.org/10.1046/j.1461-0248.2002.00321.x.

Myers, N., Mittermeier, R. A., Mittermeier, C. G., et al. (2000). Biodiversity hotspots for conservation priorities. *Nature, 403*(6772), 853–858. https://doi.org/10.1038/35002501.

Myneni, R. B., Keeling, C. D., Tucker, C. J., et al. (1997). Increased plant growth in the northern high latitudes from 1981 to 1991. *Nature, 386*(6626), 698–702. https://doi.org/10.1038/386698a0.

Nagel, L. M., Palik, B. J., Battaglia, M. A., et al. (2017). Adaptive silviculture for climate change: A national experiment in manager-scientist partnerships to apply an adaptation framework. *Journal of Forestry, 115*(3), 167–178. https://doi.org/10.5849/jof.16-039.

Nappi, A., Dery, S., Bujold, F., et al. (2011). *Harvesting in burned forests—Issues and orientations for ecosystem-based management*. Québec: Ministère des Ressources naturelles et de la Faune, Direction de l'Environnement et de la Protection des Forêts.

National Research Council (NRC). (2016). Données statistiques des ressources forestières. https://scf.rncan.gc.ca/profilstats.

Natural Resources Canada. (2020). *The state of Canada's forests. Annual report 2019* (p. 80). Ottawa: Natural Resources Canada, Canadian Forest Service.

Natural Resources Institute Finland. (2021). *Statistics database: Number of threatened species*. Helsinki: Natural Resources Institute Finland.

Navarro, L., Harvey, A. É., & Morin, H. (2018a). Lepidoptera wing scales: A new paleoecological indicator for reconstructing spruce budworm abundance. *Canadian Journal of Forest Research, 48*(3), 302–308. https://doi.org/10.1139/cjfr-2017-0009.

Navarro, L., Harvey, A. É., Ali, A., et al. (2018b). A Holocene landscape dynamic multiproxy reconstruction: How do interactions between fire and insect outbreaks shape an ecosystem over long time scales? *PLoS ONE, 13*(10), e0204316. https://doi.org/10.1371/journal.pone.0204316.

Navarro, L., Morin, H., Bergeron, Y., et al. (2018c). Changes in spatiotemporal patterns of 20th century spruce budworm outbreaks in eastern Canadian boreal forests. *Frontiers in Plant Science 9*, 1905. https://doi.org/10.3389/fpls.2018c.01905.

Niinemets, Ü., & Valladares, F. (2006). Tolerance to shade, drought, and waterlogging of temperate northern hemisphere trees and shrubs. *Ecological Monographs, 76*(4), 521–547. https://doi.org/10.1890/0012-9615(2006)076[0521:TTSDAW]2.0.CO;2.

Nolin, A. F., Tardif, J. C., Conciatori, F., et al. (2021). Spatial coherency of the spring flood signal among major river basins of eastern boreal Canada inferred from flood rings. *Journal of Hydrology, 596*, 126084. https://doi.org/10.1016/j.jhydrol.2021.126084.

Nussbaum, M. C. (2011). *Creating capabilities: The human development approach*. Harvard University Press.

Nuttle, T., Royo, A. A., Adams, M. B., et al. (2013). Historic disturbance regimes promote tree diversity only under low browsing regimes in eastern deciduous forest. *Ecological Monographs, 83*(1), 3–17. https://doi.org/10.1890/11-2263.1.

Nykänen M.-L., Broadgate, M., Kellomäki, S., et al. (1997). Factors affecting snow damage of trees with particular reference to European conditions. *Silva Fennica, 31*(2), 5618. https://doi.org/10.14214/sf.a8519.

O'Hara, K. L. (2016). What is close-to-nature silviculture in a changing world? *Forestry, 89*(1), 1–6. https://doi.org/10.1093/forestry/cpv043.

O'Brien, L., Schuck, A., Fraccaroli, C., et al. (2021). *Protecting old-growth forests in Europe—a review of scientific evidence to inform policy implementation*. Final report. Joensuu: European Forest Institute.

O'Neill, D. W., Fanning, A. L., Lamb, W. F., et al. (2018). A good life for all within planetary boundaries. *Nature Sustainability, 1*(2), 88–95. https://doi.org/10.1038/s41893-018-0021-4.

Ostlund, L., Zackrisson, O., & Axelsson, A. L. (1997). The history and transformation of a Scandinavian boreal forest landscape since the 19th century. *Canadian Journal of Forest Research, 27*(8), 1198–1206. https://doi.org/10.1139/x97-070.

Paillet, Y., Pernot, C., Boulanger, V., et al. (2015). Quantifying the recovery of old-growth attributes in forest reserves: A first reference for France. *Forest Ecology and Management, 346*, 51–64. https://doi.org/10.1016/j.foreco.2015.02.037.

Pan, Y., Birdsey, R. A., Phillips, O. L., et al. (2013). The structure, distribution, and biomass of the world's forests. *Annual Review of Ecology, Evolution, and Systematics, 44*(1), 593–622. https://doi.org/10.1146/annurev-ecolsys-110512-135914.

Pappas, C., Maillet, J., Rakowski, S., et al. (2020). Aboveground tree growth is a minor and decoupled fraction of boreal forest carbon input. *Agricultural and Forest Meteorology, 290*, 108030. https://doi.org/10.1016/j.agrformet.2020.108030.

Pappas, C., Bélanger, N., Bergeron, Y., et al. (2022). Smartforests Canada: A network of monitoring plots for forest management under environmental change. In R. Tognetti, M. Smith, & P. Panzacchi (Eds.), *Climate-smart forestry in mountain regions* (pp. 521–543). Springer International Publishing.

Paquette, A., & Messier, C. (2010). The role of plantations in managing the world's forests in the Anthropocene. *Frontiers in Ecology and the Environment, 8*(1), 27–34. https://doi.org/10.1890/080116.

Paradis, L., Thiffault, E., & Achim, A. (2019). Comparison of carbon balance and climate change mitigation potential of forest management strategies in the boreal forest of Quebec (Canada). *Forestry, 92*(3), 264–277. https://doi.org/10.1093/forestry/cpz004.

Park, A., & Wilson, E. R. (2007). Beautiful plantations: Can intensive silviculture help Canada to fulfill ecological and timber production objectives? *The Forestry Chronicle, 83*(6), 825–839. https://doi.org/10.5558/tfc83825-6.

Parks Canada. (2008). *Principles and guidelines for ecological restoration in Canada's protected natural areas.* Gatineau: National Parks Directorate, Parks Canada.

Pasanen-Mortensen, M., Elmhagen, B., Lindén, H., et al. (2017). The changing contribution of top-down and bottom-up limitation of mesopredators during 220 years of land use and climate change. *Journal of Animal Ecology, 86*(3), 566–576. https://doi.org/10.1111/1365-2656.12633.

Pastor, J., Dewey, B., Moen, R., et al. (1998). Spatial patterns in the moose-forest-soil ecosystem on Isle Royale, Michigan, USA. *Ecological Applications, 8*(2), 411–424. https://doi.org/10.1890/1051-0761(1998)008[0411:SPITMF]2.0.CO;2.

Patoine, A., Pinel-Alloul, B., Prepas, E. E., et al. (2000). Do logging and forest fires influence zooplankton biomass in Canadian Boreal Shield lakes? *Canadian Journal of Fisheries and Aquatic Sciences, 57*(S2), 155–164. https://doi.org/10.1139/f00-105.

Pautasso, M., Schlegel, M., & Holdenrieder, O. (2015). Forest health in a changing world. *Microbial Ecology, 69*(4), 826–842. https://doi.org/10.1007/s00248-014-0545-8.

Pavoine, S., & Ricotta, C. (2019). Measuring functional dissimilarity among plots: Adapting old methods to new questions. *Ecological Indicators, 97*, 67–72. https://doi.org/10.1016/j.ecolind.2018.09.048.

Peltola, H., Kellomäki, S., Väisänen, H., et al. (1999). A mechanistic model for assessing the risk of wind and snow damage to single trees and stands of Scots pine, Norway spruce, and birch. *Canadian Journal of Forest Research, 29*(6), 647–661. https://doi.org/10.1139/x99-029.

Peng, C., Ma, Z., Lei, X., et al. (2011). A drought-induced pervasive increase in tree mortality across Canada's boreal forests. *Nature Climate Change, 1*(9), 467–471. https://doi.org/10.1038/nclimate1293.

Perrakis, D. D. B., Lanoville, R. A., Taylor, S. W., et al. (2014). Modeling wildfire spread in mountain pine beetle-affected forest stands, British Columbia, Canada. *Fire Ecology, 10*(2), 10–35. https://doi.org/10.4996/fireecology.1002010.

Perrow, M. R., & Davy, A. J. (Eds.), (2002). *Handbook of ecological restoration* (p. 444). Cambridge University Press.

Peura, M., Burgas, D., Eyvindson, K., et al. (2018). Continuous cover forestry is a cost-efficient tool to increase multifunctionality of boreal production forests in Fennoscandia. *Biological Conservation, 217*, 104–112. https://doi.org/10.1016/j.biocon.2017.10.018.

Pfeifer, M., Lefebvre, V., Peres, C. A., et al. (2017). Creation of forest edges has a global impact on forest vertebrates. *Nature, 551*(7679), 187–191. https://doi.org/10.1038/nature24457.

Pinel-Alloul, B., Prepas, E., Planas, D., et al. (2002). Watershed impacts of logging and wildfire: Case studies in Canada. *Lake and Reservoir Management, 18*(4), 307–318. https://doi.org/10.1080/07438140209353937.

Planas, D., Desrosiers, M., Groulx, S. R., et al. (2000). Pelagic and benthic algal responses in eastern canadian Boreal Shield Lakes following harvesting and wildfires. *Canadian Journal of Fisheries and Aquatic Sciences, 57,* 136–145. https://doi.org/10.1139/cjfas-57-s2-136 https://doi.org/10.1139/f00-130.

Portier, J., Gauthier, S., Cyr, G., et al. (2018). Does time since fire drive live aboveground biomass and stand structure in low fire activity boreal forests? Impacts on their management. *Journal of Environmental Management, 225,* 346–355. https://doi.org/10.1016/j.jenvman.2018.07.100.

Potapov, P., Hansen, M. C., Laestadius, L., et al. (2017). The last frontiers of wilderness: Tracking loss of intact forest landscapes from 2000 to 2013. *Science Advances, 3*(1), e1600821. https://doi.org/10.1126/sciadv.1600821.

Pothier, D., Prévost, M., & Auger, I. (2003). Using the shelterwood method to mitigate water table rise after forest harvesting. *Forest Ecology and Management, 179*(1–3), 573–583. https://doi.org/10.1016/S0378-1127(02)00530-3.

Prévost, M., & DeBlois, J. (2014). Shelterwood cutting to release coniferous advance growth and limit aspen sucker development in a boreal mixedwood stand. *Forest Ecology and Management, 323,* 148–157. https://doi.org/10.1016/j.foreco.2014.03.015.

Price, D. T., Alfaro, R. I., Brown, K. J., et al. (2013). Anticipating the consequences of climate change for Canada's boreal forest ecosystems. *Environmental Reviews, 21*(4), 322–365. https://doi.org/10.1139/er-2013-0042.

Price, K., Holt, R., & Daust, D. (2020). *BC's old growth forest : A last stand for biodiversity.* Victoria: Sierra Club BC.

Puettmann, K. J. (2011). Silvicultural challenges and options in the context of global change: "simple" fixes and opportunities for new management approaches. *Journal of Forestry, 109*(6), 321–331. https://doi.org/10.1093/jof/109.6.321.

Puettmann, K. J., Coates, K. D. D., & Messier, C. C. (2009). *A critique of silviculture: Managing for complexity.* Island Press.

Pukkala, T. (2016). Which type of forest management provides most ecosystem services? *Forestry Ecosystems, 3*(1), 9. https://doi.org/10.1186/s40663-016-0068-5.

Pureswaran, D. S., De Grandpré, L., Paré, D., et al. (2015). Climate-induced changes in host tree–insect phenology may drive ecological state-shift in boreal forests. *Ecology, 96*(6), 1480–1491. https://doi.org/10.1890/13-2366.1.

Quarrie, J. (Ed.). (1992). *Earth Summit '92.* The United Nations Conference on Environment and Development, Rio de Janeiro 1992. Rome: Food and Agriculture Organization of the United Nations.

Ram, D., Axelsson, A. L., Green, M., et al. (2017). What drives current population trends in forest birds—forest quantity, quality or climate? A large-scale analysis from northern Europe. *Forest Ecology and Management, 385,* 177–188. https://doi.org/10.1016/j.foreco.2016.11.013.

Rands, M. R. W., Adams, W. M., Bennun, L., et al. (2010). Biodiversity conservation: Challenges beyond 2010. *Science, 329*(5997), 1298–1303. https://doi.org/10.1126/science.1189138.

Raworth, K. (2017). *Doughnut economics: Seven ways to think like a 21st century economist.* Chelsea Green Publishing.

Raymond, P., & Bédard, S. (2017). The irregular shelterwood system as an alternative to clearcutting to achieve compositional and structural objectives in temperate mixedwood stands. *Forest Ecology and Management, 398,* 91–100. https://doi.org/10.1016/j.foreco.2017.04.042.

Régnière, J., St-Amant, R., & Duval, P. (2012). Predicting insect distributions under climate change from physiological responses: Spruce budworm as an example. *Biological Invasions, 14*(8), 1571–1586. https://doi.org/10.1007/s10530-010-9918-1.

Regos, A., Imbeau, L., Desrochers, M., et al. (2018). Hindcasting the impacts of land-use changes on bird communities with species distribution models of Bird Atlas data. *Ecological Applications, 28*(7), 1867–1883. https://doi.org/10.1002/eap.1784.

Rist, L., & Moen, J. (2013). Sustainability in forest management and a new role for resilience thinking. *Forest Ecology and Management, 310,* 416–427. https://doi.org/10.1016/j.foreco.2013. 08.033.

Roberge, J. M., Öhman, K., Lämås, T., et al. (2018). Modified forest rotation lengths: Long-term effects on landscape-scale habitat availability for specialized species. *Journal of Environmental Management, 210,* 1–9. https://doi.org/10.1016/j.jenvman.2017.12.022.

Robertson, C., Nelson, T. A., Jelinski, D. E., et al. (2009). Spatial-temporal analysis of species range expansion: The case of the mountain pine beetle, *Dendroctonus ponderosae. Journal of Biogeography, 36*(8), 1446–1458.https://doi.org/10.1111/j.1365-2699.2009.02100.x.

Robinne, F. N., Burns, J., Kant, P., et al. (2018). *Global fire challenges in a warming world.* IUFRO.

Robinson, C. J., & Wallington, T. J, (2012). Boundary work: Engaging knowledge systems in co-management of feral animals on Indigenous lands. *Ecology and Society, 17*(2), 16. https://doi. org/10.5751/ES-04836-170216.

Rockström, J., Steffen, W., Noone, K., et al. (2009a). Planetary boundaries: Exploring the safe operating space for humanity. *Ecology and Society, 14*(2), 32. https://doi.org/10.5751/ES-03180-140232.

Rockström, J., Steffen, W., Noone, K., et al. (2009b). A safe operating space for humanity. *Nature, 461*(7263), 472–475. https://doi.org/10.1038/461472a.

Rogers, B. M., Soja, A. J., Goulden, M. L., et al. (2015). Influence of tree species on continental differences in boreal fires and climate feedbacks. *Nature Geoscience, 8*(3), 228–234. https://doi. org/10.1038/ngeo2352.

Ruel, J. C., Pin, D., & Cooper, K. (2001). Windthrow in riparian buffer strips: Effect of wind exposure, thinning and strip width. *Forest Ecology and Management, 143,* 105–113. https://doi. org/10.1016/S0378-1127(00)00510-7.

Ruel, J. C., Fortin, D., & Pothier, D. (2013). Partial cutting in old-growth boreal stands: An integrated experiment. *The Forestry Chronicle, 89*(3), 360–369. https://doi.org/10.5558/tfc2013-066.

Ryan, K. C. (2002). Dynamic interactions between forest structure and fire behavior in boreal ecosystems. *Silva Fennica, 36*(1), 548. https://doi.org/10.14214/sf.548.

Saad, C., Boulanger, Y., Beaudet, M., et al. (2017). Potential impact of climate change on the risk of windthrow in eastern Canada's forests. *Climatic Change, 143,* 487–501. https://doi.org/10.1007/ s10584-017-1995-z.

Sabatini, F. M., Keeton, W. S., Lindner, M., et al. (2020). Protection gaps and restoration opportunities for primary forests in Europe. *Diversity and Distributions, 26*(12), 1646–1662. https://doi. org/10.1111/ddi.13158.

Saint-Arnaud, M., Asselin, H., Dubé, C., et al. (2009). Developing criteria and indicators for aboriginal forestry: Mutual learning through collaborative research. In M. Stevenson & D. C. Natcher (Eds.), *Changing the culture of forestry in Canada: Building effective institutions for Aboriginal engagement in sustainable forest management* (pp. 85–105). Canadian Circumpolar Institute Press.

Sanderson, L. A., Mclaughlin, J. A., & Antunes, P. M. (2012). The last great forest: A review of the status of invasive species in the North American boreal forest. *Forestry, 85*(3), 329–340. https:// doi.org/10.1093/forestry/cps033.

Schmiegelow, F. K. A., & Mönkkönen, M. (2002). Habitat loss and fragmentation in dynamic landscapes: Avian perspectives from the boreal forest. *Ecological Applications, 12*(2), 375–389. https://doi.org/10.1890/1051-0761(2002)012[0375:HLAFID]2.0.CO;2.

Schütz, J. P. (1997). *Sylviculture 2: La gestion des forets irregulieres et melangees.* Presses Polytechniques et Universitaires Romandes.

Sedjo, R. A., & Botkin, D. (1997). Using forest plantations to spare natural forests. *Environment, 39*(10), 14–30. https://doi.org/10.1080/00139159709604776.

Seidl, R., Spies, T. A., Peterson, D. L., et al. (2016). Searching for resilience: Addressing the impacts of changing disturbance regimes on forest ecosystem services. *Journal of Applied Ecology, 53*(1), 120–129. https://doi.org/10.1111/1365-2664.12511.

Seidl, R., Thom, D., Kautz, M., et al. (2017). Forest disturbances under climate change. *Nature Climate Change, 7*(6), 395–402. https://doi.org/10.1038/nclimate3303.

Selva, N., Chylarecki, P., Jonsson, B.-G., et al. (2020). Misguided forest action in EU Biodiversity Strategy. *Science, 368*(6498), 1438–1439. https://doi.org/10.1126/science.abc9892.

Sen, A. K. (1980). Equality of what? In S. McMurrin (Ed.), *Tanner Lectures on Human Values* (pp. 197–220). Cambridge University Press.

Senez-Gagnon, F., Thiffault, E., Paré, D., et al. (2018). Dynamics of detrital carbon pools following harvesting of a humid eastern Canadian balsam fir boreal forest. *Forest Ecology and Management, 430*, 33–42. https://doi.org/10.1016/j.foreco.2018.07.044.

Shorohova, E., Kneeshaw, D., Kuuluvainen, T., et al. (2011). Variability and dynamics of old-growth forests in the circumboreal zone: Implications for conservation, restoration and management. *Silva Fennica, 45*(5), 72. https://doi.org/10.14214/sf.72.

Shorohova, E., Sinkevich, S., Kryshen, A., et al. (2019). Variable retention forestry in European boreal forests in Russia. *Ecological Processes, 8*(1), 34. https://doi.org/10.1186/s13717-019-0183-7.

Smeets, E. M. W., & Faaij, A. P. C. (2007). Bioenergy potentials from forestry in 2050. *Climatic Change, 81*(3–4), 353–390. https://doi.org/10.1007/s10584-006-9163-x.

Smith, C., Beazley, K. F., & Duinker, P., et al (2010). The impact of moose (*Alces alces andersoni*) on forest regeneration following a severe spruce budworm outbreak in the Cape Breton highlands, Nova Scotia, Canada. *Alces, 46*, 135–150.

Speed, J. D. M., Austrheim, G., Hester, A. J., et al. (2010). Experimental evidence for herbivore limitation of the treeline. *Ecology, 91*(11), 3414–3420. https://doi.org/10.1890/09-2300.1.

Spies, T. A., Lindenmayer, D. B., Gill, A. M., et al. (2012). Challenges and a checklist for biodiversity conservation in fire-prone forests: Perspectives from the Pacific Northwest of USA and southeastern Australia. *Biological Conservation, 145*(1), 5–14. https://doi.org/10.1016/j.biocon.2011.09.008.

Splawinski, T. B., Cyr, D., Gauthier, S., et al. (2019). Analyzing risk of regeneration failure in the managed boreal forest of northwestern Quebec. *Canadian Journal of Forest Research, 49*(6), 680–691. https://doi.org/10.1139/cjfr-2018-0278.

Steedman, R. J. (2000). Effects of experimental clearcut logging on water quality in three small boreal forest lake trout (*Salvelinus namaycush*) lakes. *Canadian Journal of Fisheries and Aquatic Sciences, 57*(S2), 92–96. https://doi.org/10.1139/f00-119.

Steffen, W., Richardson, K., Rockström, J., et al. (2015). Sustainability. Planetary boundaries: Guiding human development on a changing planet. *Science, 347*(6223), 1259855. https://doi.org/10.1126/science.1259855.

Stephens, S. L., Agee, J. K., Fulé, P. Z., et al. (2013). Managing forests and fire in changing climates. *Science, 342*(6154), 41–42. https://doi.org/10.1126/science.1240294.

Stocks, B. J., Mason, J. A., Todd, J. B., et al. (2003). Large forest fires in Canada, 1959–1997. *Journal of Geophysical Research, 108*(1), 8149. https://doi.org/10.1029/2001jd000484.

Sturtevant, B. R., Cooke, B. J., Kneeshaw, D. D., et al. (2015). Modeling insect disturbance across forested landscapes: Insights from the spruce budworm. In A. H. Perera, B. R. Sturtevant & L. J. Buse (Eds.), *Simulation modeling of forest landscape disturbances* (pp. 93–134). Springer International Publishing.

Suffice, P., Asselin, H., Imbeau, L., et al. (2017). More fishers and fewer martens due to cumulative effects of forest management and climate change as evidenced from local knowledge. *Journal of Ethnobiology and Ethnomedicine, 13*(1), 51. https://doi.org/10.1186/s13002-017-0180-9.

Svensson, J., Andersson, J., Sandström, P., et al. (2019). Landscape trajectory of natural boreal forest loss as an impediment to green infrastructure. *Conservation Biology, 33*(1), 152–163. https://doi.org/10.1111/cobi.13148.

Svensson, J., Bubnicki, J. W., Jonsson, B. G., et al. (2020). Conservation significance of intact forest landscapes in the Scandinavian Mountains Green Belt. *Landscape Ecology, 35*(9), 2113–2131. https://doi.org/10.1007/s10980-020-01088-4.

Swedish University of Agricultural Sciences (SLU). (2020). *Forest statistics 2020 Official statistics of Sweden*. Umeå: Swedish University of Agricultural Sciences.

Tape, K. D., Gustine, D. D., Ruess, R. W., et al. (2016). Range expansion of moose in arctic Alaska linked to warming and increased shrub habitat. *PLoS ONE, 11*(4), e0152636. https://doi.org/10.1371/journal.pone.0152636.

Teodoru, C. R., del Giorgio, P. A., Prairie, Y. T., et al. (2009). Patterns in pCO$_2$ in boreal streams and rivers of northern Quebec, Canada. *Global Biogeochemical Cycles, 23*(2):n/a. https://doi.org/10.1029/2008GB003404.

Teshome, D. T., Zharare, G. E., & Naidoo, S. (2020). The threat of the combined effect of biotic and abiotic stress factors in forestry under a changing climate. *Frontiers in Plant Science, 11*, 601009. https://doi.org/10.3389/fpls.2020.601009.

Thiffault, N., & Pinno, B. D. (2021). Enhancing forest productivity, value, and health through silviculture in a changing world. *Forests, 12*(11), 1550. https://doi.org/10.3390/f12111550.

Thom, D., & Seidl, R. (2016). Natural disturbance impacts on ecosystem services and biodiversity in temperate and boreal forests. *Biological Reviews, 91*(3), 760–781. https://doi.org/10.1111/brv.12193.

Thom, D., Golivets, M., Edling, L., et al. (2019). The climate sensitivity of carbon, timber, and species richness covaries with forest age in boreal-temperate North America. *Global Change Biology, 25*, 2446–2458. https://doi.org/10.1111/gcb.14656.

Thompson, I., Mackey, B., McNulty, S., et al. (2009). *Forest resilience, biodiversity, and climate change: a synthesis of the biodiversity/resilience/stability relationship in forest ecosystems* (pp. 1–67). Montreal: Secretariat of the Convention on Biological Diversity.

Thorn, S., Chao, A., Georgiev, K. B., et al. (2020). Estimating retention benchmarks for salvage logging to protect biodiversity. *Nature Communications, 11*(1), 4762. https://doi.org/10.1038/s41467-020-18612-4.

Thorpe, H. C., Thomas, S. C., & Caspersen, J. P. (2007). Residual tree growth responses to partial stand harvest in the black spruce (*Picea mariana*) boreal forest. *Canadian Journal of Forest Research, 37*(9), 1563–1571. https://doi.org/10.1139/X07-148.

Tikkanen, O. P., Martikainen, P., Hyvärinen, E., et al. (2006). Red-listed boreal forest species of Finland: Associations with forest structure, tree species, and decaying wood. *Annales Zoologici Fennici, 43*(4), 373–383.

Tilman, D., Knops, J., Wedin, D., et al. (1997). The influence of functional diversity and composition on ecosystem processes. *Science, 277*, 1300–1302. https://doi.org/10.1126/science.277.5330.1300.

Trumbore, S., Brando, P., & Hartmann, H. (2015). Forest health and global change. *Science, 349*(6250), 814–818. https://doi.org/10.1126/science.aac6759.

Tucker, M. A., Böhning-Gaese, K., Fagan, W. F., et al. (2018). Moving in the Anthropocene: Global reductions in terrestrial mammalian movements. *Science, 359*(6374), 466–469. https://doi.org/10.1126/science.aam9712.

Turner, N. J., & Clifton, H. (2009). "It's so different today": Climate change and indigenous lifeways in British Columbia, Canada. *Global Environmental Change, 19*(2), 180–190. https://doi.org/10.1016/j.gloenvcha.2009.01.005.

Valinger, E., & Fridman, J. (2011). Factors affecting the probability of windthrow at stand level as a result of Gudrun winter storm in southern Sweden. *Forest Ecology and Management, 262*(3), 398–403. https://doi.org/10.1016/j.foreco.2011.04.004.

Venier, L. A., Thompson, I. D., Fleming, R., et al. (2014). Effects of natural resource development on the terrestrial biodiversity of Canadian boreal forests. *Environmental Reviews, 22*(4), 457–490. https://doi.org/10.1139/er-2013-0075.

Venier, L. A., Walton, R., Thompson, I. D., et al. (2018). A review of the intact forest landscape concept in the Canadian boreal forest: Its history, value, and measurement. *Environmental Reviews, 26*(4), 369–377. https://doi.org/10.1139/er-2018-0041.

Versluijs, M., Eggers, S., Hjältén, J., et al. (2017). Ecological restoration in boreal forest modifies the structure of bird assemblages. *Forest Ecology and Management, 401*, 75–88. https://doi.org/ 10.1016/j.foreco.2017.06.055.

Virkkala, R. (2016). Long-term decline of southern boreal forest birds: Consequence of habitat alteration or climate change? *Biodiversity and Conservation, 25*(1), 151–167. https://doi.org/10. 1007/s10531-015-1043-0.

Vitousek, P. M., D'Antonio, C. M., Loope, L. L., et al. (1996). Biological invasions as global environmental change. *American Scientist, 84*(5), 468–478.

Vuorinen, K. E. M., Kolstad, A. L., De Vriendt, L., et al. (2020a). Cool as a moose: How can browsing counteract climate warming effects across boreal forest ecosystems? *Ecology, 101*(11), e03159. https://doi.org/10.1002/ecy.3159.

Vuorinen, K. E. M., Rao, S. J., Hester, A. J., et al. (2020b). Herbivory and climate as drivers of woody plant growth: Do deer decrease the impacts of warming? *Ecological Applications, 30*(6), e02119. https://doi.org/10.1002/eap.2119.

Walker, X. J., Baltzer, J. L., Cumming, S. G., et al. (2019). Increasing wildfires threaten historic carbon sink of boreal forest soils. *Nature, 572*(7770), 520–523. https://doi.org/10.1038/s41586-019-1474-y.

Wallenius, T. (2011). Major decline in fires in coniferous forests—reconstructing the phenomenon and seeking for the cause. *Silva Fennica, 45*(1), 36. https://doi.org/10.14214/sf.36.

Wallenius, T., Niskanen, L., Virtanen, T., et al. (2010). Loss of habitats, naturalness and species diversity in Eurasian forest landscapes. *Ecological Indicators, 10*(6), 1093–1101. https://doi.org/ 10.1016/j.ecolind.2010.03.006.

Wang, J., & Banzhaf, E. (2018). Towards a better understanding of Green Infrastructure: A critical review. *Ecological Indicators, 85*, 758–772. https://doi.org/10.1016/j.ecolind.2017.09.018.

Ward, M., Saura, S., Williams, B., et al. (2020). Just ten percent of the global terrestrial protected area network is structurally connected via intact land. *Nature Communications, 11*(1), 4563. https://doi.org/10.1038/s41467-020-18457-x.

Warren, D. R., Keeton, W. S., Bechtold, H. A., et al. (2018). Forest-stream interactions in eastern old-growth forests. In A. M. Barton & W. S. Keeton (Eds.), *Ecology and recovery of eastern old-growth forests* (pp. 159–178). Island Press/Center for Resource Economics.

Watson, J. E. M., Evans, T., Venter, O., et al. (2018). The exceptional value of intact forest ecosystems. *Nature Ecology and Evolution, 2*(4), 599–610. https://doi.org/10.1038/s41559-018-0490-x.

Webb, A., Buddle, C. M., Drapeau, P., et al. (2008). Use of remnant boreal forest habitats by saprox-ylic beetle assemblages in even-aged managed landscapes. *Biological Conservation, 141*(3), 815–826. https://doi.org/10.1016/j.biocon.2008.01.004.

Wells, J. V., Dawson, N., Culver, N., et al. (2020). The state of conservation in North America's boreal forest: Issues and opportunities. *Frontiers in Forests and Global Change, 3*, 90. https:// doi.org/10.3389/ffgc.2020.00090.

Williams, M. (2003). *Deforesting the earth: From prehistory to global crisis.* University of Chicago Press.

Wirth, C., Messier, C., Bergeron, Y., et al. (2009). Old-growth forest definitions: A pragmatic view. In C. Wirth, G. Gleixner, & M. Heimann (Eds.), *Old-growth forests: Function, fate and value* (pp. 11–33). Springer.

Woo-Durand, C., Matte, J.-M., Cuddihy, G., et al. (2020). Increasing importance of climate change and other threats to at-risk species in Canada. *Environmental Reviews, 28*(4), 449–456. https:// doi.org/10.1139/er-2020-0032.

Wu, P., Bishop, K., von Brömssen, C., et al. (2018). Does forest harvest increase the mercury concentrations in fish? Evidence from Swedish lakes. *Science of the Total Environment, 622–623*, 1353–1362. https://doi.org/10.1016/j.scitotenv.2017.12.075.

Wyatt, S., Hebert, M., Fortier, J. F., et al. (2019). Strategic approaches to Indigenous engagement in natural resource management: Use of collaboration and conflict to expand negotiating space by three Indigenous nations in Quebec, Canada. *Canadian Journal of Forest Research, 49*(4), 375–387. https://doi.org/10.1139/cjfr-2018-0253.

Zeng, H., Pukkala, T., & Peltola, H. (2007). The use of heuristic optimization in risk management of wind damage in forest planning. *Forest Ecology and Management, 241*(1–3), 189–199. https://doi.org/10.1016/j.foreco.2007.01.016.

Zhang, X., Lei, Y., Ma, Z., et al. (2014). Insect-induced tree mortality of boreal forests in eastern Canada under a changing climate. *Ecology and Evolution, 4*(12), 2384–2394. https://doi.org/10.1002/ece3.988.

Zhang, X., Bai, X., Hou, M., et al. (2019a). Warmer winter ground temperatures trigger rapid growth of dahurian larch in the permafrost forests of northeast China. *Journal of Geophysical Research. Biogeosciences, 124*(5), 1088–1097. https://doi.org/10.1029/2018JG004882.

Zhang, X., Manzanedo, R. D., D'Orangeville, L., et al. (2019b). Snowmelt and early to mid-growing season water availability augment tree growth during rapid warming in southern Asian boreal forests. *Global Change Biology, 25*(10), 3462–3471. https://doi.org/10.1111/gcb.14749.

Zhang, Y., Chen, H. Y. H., & Reich, P. B. (2012). Forest productivity increases with evenness, species richness and trait variation: A global meta-analysis. *Journal of Ecology, 100*(3), 742–749. https://doi.org/10.1111/j.1365-2745.2011.01944.x.

Zubizarreta-Gerendiain, A., Pukkala, T., & Peltola, H. (2017). Effects of wind damage on the optimal management of boreal forests under current and changing climatic conditions. *Canadian Journal of Forest Research, 47*(2), 246–256. https://doi.org/10.1139/cjfr-2016-0226.

Printed in the United States
by Baker & Taylor Publisher Services